P9-DEI-271

APPLIED COLLEGE ALGEBRA

a graphing approach

APPLIED COLLEGE ALGEBRA

a graphing approach

GARETH WILLIAMS

Stetson University

Saunders College Publishing

A DIVISION OF HARCOURT COLLEGE PUBLISHERS

Fort Worth Philadelphia San Diego New York Orlando Austin
San Antonio Toronto Montreal London Sydney Tokyo

To Eirwen, Glyn, and Nia

Publisher: Emily Barrosse
Executive Editor: Angus McDonald
Senior Acquisitions Editor: Liz Covello
Senior Marketing Strategist: Julia Downs
Senior Associate Editor: Alexa Epstein
Project Editor: Theodore Lewis
Production Manager: Alicia Jackson
Art Director and Cover Designer: Lisa Adamitis
Text Designers: Merry Obrecht Sawdy and Lisa Adamitis

Cover credits: Cliff Palace, Mesa Verde National Park, Colorado. Front cover photo: 1998, ©Terrence Moore. Back cover photo: 1907, ©Mesa Verde National Park/Nusbaum print 9523, no. 8.

Applied College Algebra: A Graphing Approach
ISBN: 0-03-026026-4
Library of Congress Card Number: 99-067054

Address for domestic orders:
Saunders College Publishing, 6277 Sea Harbor Drive, Orlando, FL 32887-6777
1-800-782-4479
e-mail collegesales@harcourt.com

Address for international orders:
International Customer Service, Harcourt, Inc.
6277 Sea Harbor Drive, Orlando, FL 32887-6777
Phone (407) 345-3800 · Fax (407) 345-4060
e-mail hbintl@harcourt.com

Address for editorial correspondence:
Saunders College Publishing,
Public Ledger Building, Suite 1250, 150 S. Independence Mall West,
Philadelphia, PA 19106-3412

Web Site address: http://www.harcourtcollege.com

Printed in the United States of America

9012345678 032 10987654321

CONTENTS

2 Solving Equations and Inequalities 162

3 Further Development of Functions 217

4 Further Theory of Polynomials 323

7 Matrices 476

8 Sequences and Series 535

9 Permutations, Combinations, and Probability 582

PREFACE

This text on algebra is designed for a course at the college level. The book has been written with the following points in mind.

AUDIENCE

Level of Rigor The course is taken by many students who are not strong in mathematics. An understanding of concepts is important; however, complicated explanations are to be avoided. There is a fine line between appropriate explanation and too much rigor. Many explanations are given but, as far as possible, great rigor has been avoided. One must tread gently in places.

Backgrounds of the Students The course is taken by students with a variety of backgrounds in mathematics. The early material will give the weaker student a solid foundation in algebra. Students with a good background in intermediate algebra can use the early material to reinforce their knowledge. The real applications should make this material refreshing and interesting for everyone. The book takes the student from an introductory level to being fluent in understanding and using algebra, with the pace gradually increasing. By the time the student reaches Chapter 3 for an in-depth discussion of functions, he or she should be able to handle some of the more sophisticated concepts, such as shifts, function composition, and function inverse. The student should be ready for the exponential function and logarithmic function (always a difficult topic) in Chapter 5.

Amount of Material There is much material to cover in this course. I have selected topics and examples carefully, being conscious that time is a factor. Much space is often devoted in college algebra books to integer exponents, radicals, rational exponents, and rational expressions. I think that these topics are often covered in far too much depth, with overly complicated expressions. Students can be unnecessarily overwhelmed early in the course by these topics. This book moves quickly on topics such as these.

Preparation Time The course is taught not only by full-time professors but by adjuncts and graduate students who have limited preparation time. The book is straightforward for the teacher to use.

APPLICATIONS

Applying Mathematics Interesting applications are included in the text. Students should be taught when and how to apply mathematics. To give the student confidence in applying mathematics, I revisit certain applications, such as motion of an object under gravity, when a new concept is first introduced in an application. The student then sees how to interpret the idea in a familiar setting before using it in a new setting.

Use of Applications for Motivation I use applications to motivate some mathematical concepts. An application is used to motivate the idea of the inverse of a function, for example. Not all mathematical concepts lend themselves to this approach. Instructors may want to glance ahead when a certain mathematical concept is introduced, to give the students an idea of applications that involve that concept. I favor this approach.

Variety of Applications The course is a wonderful opportunity to introduce the student to the role of mathematics in many fields. I have included real applications from such disciplines as bacteriology, demography, economics, relativity, and pharmacology. These have come from consultations with experts in these fields. Many first-year undergraduates have little idea of where they want to go in their careers. There are interesting fields of which they have never heard. The course could guide a student toward becoming a pharmacologist, for example.

The applications are diverse. There should be something here to interest every reader. No student who has taken this course should have grounds for asking "Why am I learning this mathematics; what use is it?" The instructor can select applications that are of interest to the class. Many applications lend themselves to reading assignments with follow-up exercises. Instructors should train students to read and understand mathematics themselves. What better place than in the new, interesting "applied algebra" environment. Other applications can be suggested for general reading to help the student appreciate the key role mathematics plays in many fields. My hope is that students will find this material interesting enough to want to browse through the book themselves. An Index of Applications follows the Walkthrough. This index can be used to locate specific applications, to obtain an overview of the variety of applications, and to plan the course.

EMPHASIS ON FUNCTIONS

Library of Functions The student is encouraged to build a library of functions that can be quickly graphed. This collection includes linear functions, quadratic functions, general polynomial functions, the square root function, the absolute value function, exponential functions, and logarithmic functions.

Construction of Functions I believe that it is important for students to see how functions that describe real situations are constructed. Sometimes, the functions are constructed from classes of functions in our library that are known to have the desired characteristics; at other times, they are constructed from data using regression. I have demonstrated how regression is used to arrive at functions that describe, for example, the dissipation of a drug in the body, the distribution of velocity for a liquid flowing through a pipe, and the number of people in the United States with four years of college education. Some instructors may want to arrive at these functions using regression on a calculator. Others may only indicate how the functions are constructed. These approaches are more satisfying than producing a function out of thin air. Some data have been taken from the Web. In these cases, URLs are given on the companion Web site: http://www.saunderscollege.com/math/williams (but be aware that URLs can change).

USE OF TECHNOLOGY

Graphing Calculator A graphing calculator (or computer) can enhance a college algebra course. It can be used to:

- teach the student to look at a mathematical idea from both algebraic and geometrical viewpoints. Algebra and geometry so often go side by side.
- perform computations that are tedious or cannot be accomplished by hand.
- teach students when and how to implement the mathematics; many students will be using this mathematics later in their careers on calculators or on computers.
- free students to focus on the art of translating a problem from words to mathematics. The calculator can be used to do the dirty work (or work that is impossible to do by hand at this level because it may involve calculus).
- motivate students. Students enjoy the hands-on, exploratory aspect of working with a graphing calculator.

Although the calculator can give marvelous geometrical insight into algebraic concepts, it should not relieve the student of the responsibility of mastering basic algebra. The role of the calculator in the course should be natural, not forced; it should be introduced when useful. In this day and age, students should be taught when and how to use technology effectively.

At times, the instructor may be forced to decide between two or more calculator techniques. It should always be remembered that the aim is to teach mathematics. For example, one can choose between (1) tracing and zooming in on a point on a graph a number of times until a required degree of accuracy is reached or (2) using a max/min finder or zero finder. Iterating in on a point until one has the desired degree of accuracy is an important mathematical process to learn. Teachers should include this discussion if at all possible. However, once the method is seen and appreciated, the max/min or zero finder feature is probably the best way of actually arriving at max/mins and zeros from that time on.

The calculator arithmetic in Sections R.1 and R.2 may not be new to students. I have included it because in my experience, students who do not master this basic material at this time encounter calculator problems throughout the course. They are often worrying about elementary errors in calculator entries and are missing the learning opportunity. Useful calculator reference material is included in Appendix B.

There are numerous graphing calculators on the market. Even if a specific calculator is designated the "official calculator" for the course, a variety of calculators appear; many students enter the class already owning a graphing calculator. This text makes use of calculator concepts (e.g., viewing window, graph, zoom, and trace) that are common to many graphing calculators. Screens (customized TI-82/TI-83 screens) are displayed in a format that can be easily interpreted by users of the different calculators.

Technology Tips Helpful graphing calculator (or computer) information, marked by ▨, appears throughout the text. For example, a square setting can be used to give a true geometrical picture. A max/min finder can be used to find the maxima and minima of a graph. Technology can be used to find a function that fits data quickly. Students should determine whether the suggested ideas can be implemented using their technology tools.

Explorations Mathematics can be learned and students can be encouraged to mature mathematically in many ways. Explorations give students the opportunity to discover (with guidance) mathematical concepts for themselves. They discover, for example, how the sign of the coefficient of x^2 affects the graph of a parabola and how to shift graphs vertically and horizontally.

SPECIAL FEATURES

Chapter Introductions Chapter introductions provide an informal overview of the chapter so that the student will have some idea of where he or she is heading. Each introduction contains an interesting historical tidbit. Mathematics has been developed by all kinds of people living in many parts of the world at different ages. There is an important human face to the field. Occasional footnotes continue this historical flavor in the sections that follow. The work of such giants as Gauss and Einstein are discussed. The introductions also contain a summary of an interesting application that occurs in the chapter.

Examples Careful examples have been constructed to illustrate important techniques. Often, more can be learned through a good example than a lengthy theoretical discussion.

Self-Check Self-check exercises follow selected examples to reinforce the ideas presented. The exercises provide practice at the concept introduced. They can be completed by students in or out of class. I have found in-class use of some of the self-checks to be an extremely effective way of actively involving students in the class. For ease of use, I have included the answers to the exercises at the ends of the sections in which they appear.

Further Ideas Further ideas are exercises that take the students a little further than the simple self-check of a concept. They are opportunities for students to better understand the ideas underlying selected examples using the exploration ideas mentioned. After looking at an example of a quadratic equation with repeated roots, the student is given the chance to discover how this fact is revealed by the graph of the equation. An example on prediction using an exponential function is followed by the opportunity to explore the danger of using mathematics to predict too far into the future.

Exercises Good exercises are essential. The exercises range from routine drill based strictly on examples given in the sections, to straightforward exercises for which no examples are given, to more demanding exercises. Exercises help students mature. The more difficult exercises can be used to challenge the better students. Answers to odd-numbered exercises are included at the back of the text.

Chapter Highlights Highlights with references are given at the end of each chapter. All the important concepts and results are included. The highlights can be used to reinforce the important ideas and to pinpoint gaps in knowledge. They can also be used to obtain a global perspective on the material in the chapter. One can appreciate the sequence in which the material develops and see how it all fits together.

Review Exercises and Tests These are included at the end of each chapter.

Cumulative Tests Cumulative Tests are included after Chapters 2, 5, and 9.

Caution Comments Caution Comments are included to alert the student to possible sources of error.

Full-color artwork Graphs and pictures play an important role in describing and understanding college algebra. To help strengthen students' graphical intuition, learn graphing skills and interpretation, and understand the relevance of mathematics, this text includes hundreds of full-color graphs, pictures, and photographs. All graphing calculator screens are based on the TI-82/TI-83 graphing calculator, but are easy to understand regardless of which technology tool is available.

Layout A textbook at this level must be clear and easy to read. The layout, paragraphs, equations, and so on affect the readability of the book. Much effort has gone into producing a clean, uncomplicated text.

COMMUNICATION AND GROUP WORK

Writing Skills Students often find it difficult to express mathematical ideas in writing. This course should train the student to analyze and think clearly. Being able to explain lines of reasoning and results by means of words and mathematical symbols is important. I have included a number of writing exercises. These are of a brief nature, taking into account the backgrounds of the students. I often lead into such exercises with phrases like "Discuss . . . ", "Explain . . . ", "Describe . . . ", or "Write a brief essay on . . . ". Sometimes these writing opportunities are given in the Section Projects and Chapter Projects. I have also given the student the chance to write brief essays on such topics as π and the Nobel Prize.

Section Projects and Chapter Projects Projects have been included at the ends of certain sections and at the end of each chapter. Some of these projects are theoretical in nature; some involve applications. They are often open-ended and give the student the opportunity to delve and to do original work. They can be assigned on an individual basis or to groups.

Group Discussions Discussion opportunities are presented after many sections. The discussions give the students the chance to "talk mathematics" among themselves. They can be used for small groups or for the class as a whole. Some involve groups arriving at conclusions and then interacting with other groups.

Web Projects Chapters 1 through 9 each have an innovative Web-based project extending a key idea of the chapter. The project, related problems, and URLs necessary to complete the project are located on the companion Web site: http://www.saunderscollege.com/math/williams.

CONTENTS AND ORGANIZATION

Sequence of Topics The sequence in which the topics are introduced in a course is important. Much thought has gone into arriving at the order of topics in this book. One is confronted with the necessity of covering background algebra as quickly as possible, getting to graphing with a calculator as soon as possible, and introducing functions as early as possible. It is advantageous, for example, to have a knowledge of equations when functions are introduced; however, as soon as one starts to discuss the values of y for various values of x in an equation, one wants to talk in terms of functions. Furthermore, as soon as the table facility of a calculator is introduced, one wants to use the language of functions. After trying many arrangements of topics, I feel that the pieces have fallen into place. The review chapter presents the algebraic aspects of equations followed by that of functions so that these tools can be developed and used in a natural way thereafter, side by side.

A major aim of the book is the simultaneous presentation of algebraic and geometrical ways of looking at things. With this goal in mind, parabolas are introduced in Chapter 1. They are then available to illustrate algebraic results, such as solutions of quadratic equations, in Chapter 2. It can be readily appreciated why quadratic equations have two, one, or no solutions when these solutions are viewed as the x-intercepts of a parabola.

Overview of the Contents Chapter R is a review chapter that introduces functions and reinforces such topics as rational exponents, factoring, and linear equations that the student has encountered before. This chapter lays the groundwork for the numerical use of the calculator and the use of tables in the course. The instructor should spend as much time as is needed on this material. Chapter 1 introduces a geometrical flavor to the mathematics. Linear and quadratic equations are discussed from algebraic and geometrical viewpoints. The equation concept is extended naturally to functions, and the method of using least-squares for constructing functions to fit data is introduced. Chapter 2 focuses on algebraic and geometrical methods of solving nonlinear equations and inequalities. Chapter 3 may be the most important chapter in the book. This chapter gives a comprehensive presentation of functions. Although the student will have been working with functions for much of the course, it is in this chapter that one develops a mature understanding of functions. Such concepts as shifts, composition, iterative processes, and inverses of functions are discussed. Chapter 4 extends the knowledge of polynomials and introduces the student to fractal geometry. Section 4.2, on complex numbers, is valuable for students planning to enter the sciences or mathematics. Other sections are optional. Chapter 5 completes the discussion of functions with an introduction to exponential and logarithmic functions. Chapters 6 and 7 introduce the student to elementary linear algebra. Systems of linear equations are solved, and matrix techniques are introduced. Chapters 8 and 9 complete the introduction to algebra at the college level. Sequences, series, permutations, combinations, and probability are presented. Appendix A provides a brief introduction to conic sections. All self-check exercises, further ideas, projects, group discussions, and applications are optional.

My hope is that students and instructors will find this material readable, interesting, and useful. Applying mathematics can be fun!

ACKNOWLEDGMENTS

It is a pleasure to acknowledge the support of many of my colleagues in preparing this book. One advantage of teaching at a small university like Stetson is that one gets to know experts in various fields. I thank my colleagues in mathematics for interesting insights in integrating technology and applications into mathematics courses. I am grateful to the following friends from various walks of life who have guided me to interesting applications: Terence Farrell and Peter May (biology); Duane Cochran and Richard Kindred (psychology); John Schorr (sociology); Ken Everett (chemistry); Gary Maris (political science); Donna Williams (computer science); Michael Branton, Lisa Coulter, Harold Danzburger, Erich

Friedman, Margie Hale, and Dennis Kletzing (mathematics); Alan Wise (teacher); Meira Edwards (housewife with an interest in mathematics); Nick Colden (Eckerd Pharmacy); and Jeff Williams (actuary).

My thanks are extended to the reviewers of this book for their many constructive suggestions:

David Anderson, *South Suburban College*
Judy Barclay, *Cuesta College*
Daniel C. Biles, *Western Kentucky University*
Diane W. Burleson, *Central Piedmont Community College*
Dick J. Clark, *Portland Community College*
Brenda Diesslin, *Iowa State University*
Marcia Drost, *Texas A&M University*
Patricia Dueck, *Arizona State University*
Larry Friesen, *Butler County Community College*
William Grimes, *Central Missouri State University*
Sue Haller, *St. Cloud State University*
Celeste Hernandez, *Richland College*
James Kenneth Johnston, *Hinds Community College*
John Martin, *Santa Rosa Junior College*
Michael Montano, *Riverside Community College*
Ferne Mizell, *Austin Community College,* Rio Grande
Kathy C. Nickell, *College of DuPage*
Nancy Olson, *Johnson County Community College*
Joan Raines, *Middle Tennessee State University*
Henry M. Smith, *Southeastern Louisiana University*
Dottie H. Vaughn, *Louisiana State University*

I thank Barbara Fogelman, Ann Huffer, and Jeff Williams for working out the answers to many of the exercises. I thank Bruce Kessler, Mike Montano, and Larry Friesen for checking the accuracy of all the examples, the odd-numbered exercise answers, and the end-of-chapter Review Exercises, Chapter Test problems, and Cumulative Test exercises. Their attention to detail throughout the manuscript, galley, and page proof stages have helped ensure that this text is as free of errors as possible.

I also want to express my gratitude to the folks at Saunders College Publishing: Angus McDonald, for inviting me to write for the company; Stephanie Klein and Ted Lewis, who patiently guided the production of the book; Lisa Adamitis, for the careful art direction; and Julia Downs and Andrea Garrett, for the marketing. I am grateful to Alexa Epstein, who helped so much in many ways: editing; making mathematical, layout, and art suggestions; encouragement; and supervision of the whole project. It has been a pleasure working with you all.

A special thanks to my wife Donna for her encouragement, her input into the mathematics and the way technology can be used to enhance the understanding of mathematics, and her help in proofreading.

Gareth Williams
September, 1999

SUPPLEMENTS

SUPPLEMENTS FOR INSTRUCTORS

Instructors who adopt this text may receive, free of charge, the following items.

Instructor's Resource Manual Written by the author, this manual contains detailed solutions to all the Self-check exercises, Further Ideas exercises, regular exercises and end-of-chapter Review Exercises, Chapter Test problems, and Cumulative Review exercises to assist the instructor in the classroom and in grading assignments. Each manual also includes the **Digital Video Applications CD-ROM,** the related problems and detailed solutions, as well as solutions to the Web projects.

Digital Video Applications CD-ROM This innovative ancillary is designed to show students how and where college algebra concepts arise in real life. More than 20 engaging interviews, conducted by Lori Palmer of Utah Valley State College, with professionals from such fields as aviation, restaurant management, banking, and environmental science motivate the key concepts from the text. Each vignette is accompanied by two problems, written by Carolyn Hamilton of Utah Valley State College, to test students' understanding of the underlying mathematical ideas and skills. Answers are provided on disk; detailed solutions are in the Instructor's Resource Manual.

Test Bank Written by Henry Smith of Southeastern Louisiana University, this manual provides almost 1700 multiple choice and open-ended questions arranged in six forms per chapter, each form containing approximately 25 questions. Master answer sheets and a complete answer section are included. All answers were independently checked for accuracy by Daniel Biles of Western Kentucky University.

ESATEST 2000 Computerized Test Bank CD-ROM Available for both Macintosh and Windows formats, the **ESATEST 2000 Computerized Test Bank CD-ROM** contains all the test bank questions and allows instructors to prepare quizzes and examinations quickly and easily. Approximately 40% of the questions are also algorithmic, allowing greater testing flexibility. Instructors may add questions or modify existing ones. ESATEST 2000 has gradebook capabilities for recording and tracking students' grades. Instructors have the opportunity to post and administer a test over a network or on the Web. ESATEST 2000's user-friendly printing capability accommodates all printing platforms.

Video Series The videotape package consists of 10 VHS videotapes, one for each chapter in the book. Each one-hour tape offers numerous worked examples to reinforce the key concepts of the chapter. The tapes are an ideal way for students to review for examinations or make up for missed classes. On-location footage is used to introduce an extended application at the beginning of each tape. This application and a related problem are explored fully at the end of the tape. (See also the **Core Concepts Videotape.**)

Web site The companion Web site offers additional resources to students and instructors who adopt the text. An on-line glossary allows students to review and confirm their understanding of important terms and concepts in the text. The necessary URLs for all Web-related text exercises are given, as well as the innovative Web projects created by William Grimes of Central Missouri State University. These applied projects extend a key concept from each chapter and pose a series of related problems, which students can complete only by surfing the given URLs for relevant data. Detailed solutions to these Web projects are in the Instructor's Resource Manual.

SUPPLEMENTS FOR STUDENTS

An Electronic Companion to College Algebra CD-ROM This interactive and comprehensive CD-ROM accompanies each copy of the text. Through multiple representations of the key concepts (graphically, symbolically, numerically, and verbally), students strengthen their algebra skills and their conceptual understanding. The CD-ROM is divided into two parts: Review Topics, which present the main ideas of college algebra interactively; and Test Yourself, which offers problems and questions about the material in the Review Topics section. The CD-ROM is ideal for preparing for class and for reviewing for tests.

Students using *Applied College Algebra: A Graphing Approach* may purchase the following additional supplements through their bookstore or the general Web site: http://www.harcourtcollege.com/store.

Student Resource Manual This manual comprises two distinct parts. Detailed solutions to the odd-numbered exercises and end-of-chapter Review Exercises, Chapter Test questions, and Cumulative Test exercises are included in the first part. The manual also contains the **Digital Video Applications CD-ROM** (described above) and accompanying problems with sufficient space to show student work.

TI-83 Graphing Calculator Manual Written by Johanna Halsey of Dutchess Community College, this manual gives complete instructions for using the TI-83 to explore and solve college algebra problems. It helps solidify the students' understanding of the concepts in the text while strengthening their confidence using the graphing calculator.

ESATUTOR 2000 CD-ROM This computer software package contains algorithmically generated multiple-choice and open-ended questions referenced to every section of the text. Students can complete a pretest to evaluate their level of understanding of the concepts in each chapter and identify weak areas. When they are ready, students can take posttests to ensure that they have grasped the primary learning objectives. In addition, the software comes with a built-in graphing calculator. Students who are interested in purchasing this software package should refer to the insert at the back of this text.

Core Concepts Videotape This two-hour videotape contains the most important topics covered in the full video available to adopters and instructors. The take-home tutorial can be used to preview the classroom discussion, as an aid in completing homework assignments, to review for a test, or to make up for missed classes. On-location footage is used to introduce an extended application at the beginning of each tape. This application and a related problem are explored fully at the end of the tape.

Saunders College Publishing, a division of Harcourt College Publishers, may provide complimentary instructional aids and supplements or supplement packages to those adopters qualified under our adoption policy. Please contact your sales representative for more information. If as an adopter or potential user you receive supplements you do not need, please return them to your sales representative or send them to

Attn: Returns Department
Troy Warehouse
465 South Lincoln Drive
Troy, MO 63379

WALKTHROUGH

CHAPTER OPENERS

Chapter Introductions offer an informal overview, including some historical background of the chapter's concepts. Students learn that algebra is a cumulative achievement spanning many peoples, regions, and periods.

Further Development of Functions

Leonhard Euler (1707 – 1783)
(©1994 Northwind Picture Archives)

The concept of a function permeates much of mathematics and is an important tool in applying mathematics. We have seen that a function can be used to describe how one variable varies with respect to another. The function notation $f(x)$ (we say "f is a function of x") suggests this dependency on x. This notation was first used by the Swiss mathematician Leonhard Euler. Euler was born in Basel, Switzerland, in 1707. His father, who was a minister, gave him his first instruction in mathematics and later sent him to the university in Basel. Euler wrote his dissertation at the age of 19 on the masting of ships. He served as professor in St. Petersburg, Russia, and later as head of the Prussian Academy in Berlin. His amazing productivity was not in the least impaired when he had the misfortune to become totally blind.

In this chapter, we look into algebraic and geometrical techniques for working with functions. We will see how different functions are related and how the graphs of functions can be shifted vertically and horizontally, reflected in the x- and y-axes, and stretched. Through such operations as addition, multiplication, and composition, functions will be combined to produce new functions that are useful in applications.

We will see how functions can be used to help a cosmetics company decide how often to advertise its product on television, and by scientists to learn more about animals.

Often we need to know where functions are increasing in value, where they are decreasing, and when they reach their maximum or minimum values. For example, the spread of an epidemic may be described by a function, allowing authorities to estimate its duration and intensity. Functions are also used to tell doctors how rapidly a drug dissipates in the body. These applications and the following analysis of the path of an as- ... are discussed in this chapter.

217

APPLICATION	Sizes of Skeletons

The brachiosaurus is believed to be the largest land animal ever to exist. (*©The Field Museum, GN88141.13C, John Weinstein*)

The world we live in is governed by the laws of chemistry and physics. Because animals, among other creatures, live within the bounds of these laws, biologists attempt to discover and understand these laws. For example, they study the relationships between sizes of animals and their various biologic components. Scientists have found that the skeleton/body weight ratio of a land animal of weight x pounds is given by the expression

$$\frac{0.061x^{1.09}}{x}$$

A table of values of this expression shows how the skeleton weight becomes an increasingly larger fraction of the total weight of an animal. The table shows that at 243 pounds the skeleton weight becomes one tenth of the total body weight of a human being.

A brachiosaurus was 70 feet long and weighed approximately 170,000 pounds. Scientists found the skeleton/body weight ratio of a brachiosaurus to be 0.1803. They believe that this number represents the largest such ratio for a land animal. Because land animals cannot sustain a skeleton that is a larger fraction of body weight, no larger land animal can exist. *(This application is discussed in Section R.7, Example 8, on pages 72–73.)*

2

Chapter Applications are also presented at the beginning of each chapter. These applications motivate the concepts to be covered and are continued in detail later in the chapter.

USING TECHNOLOGY

Graphing Calculators, integrated throughout the book, are an important tool for exploring and motivating algebra concepts. Students learn to balance algebraic and geometrical viewpoints and where and when to use technology effectively.

Chapter 2 Solving Equations and Inequalities

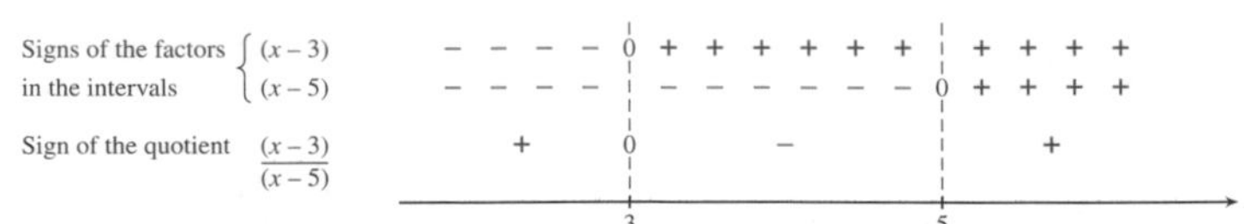

Figure 2.48

It can be seen that $(x-3)/(x-5) < 0$ for values of x in the interval (3, 5). The set of solutions is thus (3, 5).

We check this result using geometry. Graph $y = 2x/(x-5)$ (Figure 2.49). Observe that the graph is below the line $y = -3$ over the interval (3, 5). Thus, $2x/(x-5) < -3$ over (3, 5). The set of solutions is (3, 5).

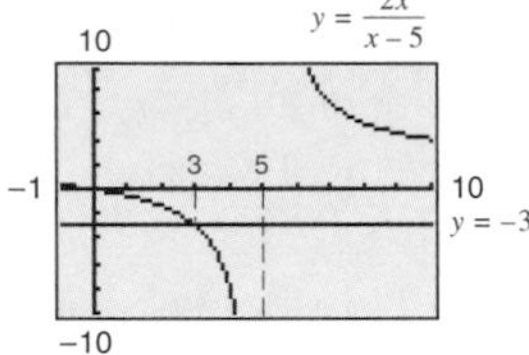

Figure 2.49 $\frac{2x}{x-5} < -3$ when $3 < x < 5$.

VERTEX WITH A CALCULATOR

The vertex of a parabola is called a **maximum** or a **minimum** depending on whether the parabola opens down or up. (The vertex is the highest or lowest point on the graph.) Graphing calculators have a **max/min finder** that can be used to find the vertex. Let us graph and find the vertex of the parabola $y = -2.1x^2 + 9.7x + 5$ to four decimal places (Figure 1.49). The vertex is (2.3095, 16.2012).

Trace and *zoom* can also be used together to find the vertex but is not as efficient as using a max/min finder. We show the results of tracing and zooming three times in Figure 1.49. We still have a way to go before obtaining the vertex to four decimal places.

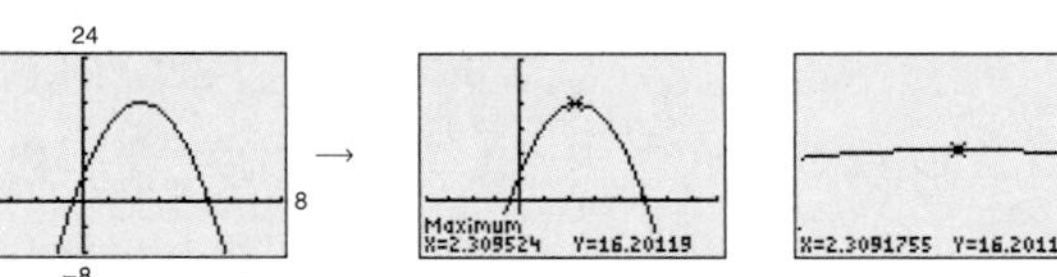

Vertex using max/min finder

Vertex using trace/zoom three times

Figure 1.49 Using a calculator to determine the vertex of a parabola.

Technology Tips offer students helpful suggestions about using the graphing calculator effectively.

Explorations guide students to discover mathematical ideas on their own.

Exploration

Consider the graph of

$$f(x) = k(x-a)(x-b)(x-c)$$

Prove that the x-intercepts are a, b, and c.

Use a calculator to explore the graph of

$$f(x) = k(x-1)(x-2)(x-5)$$

for various values of k. How does the value of k affect the graph?

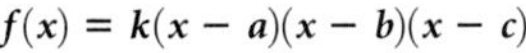

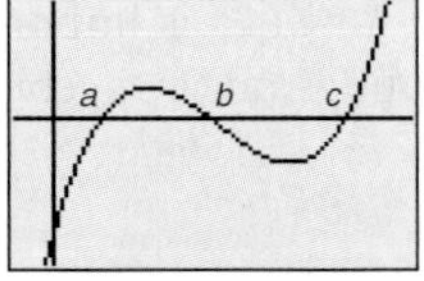

x-Intercepts are a, b, and c.

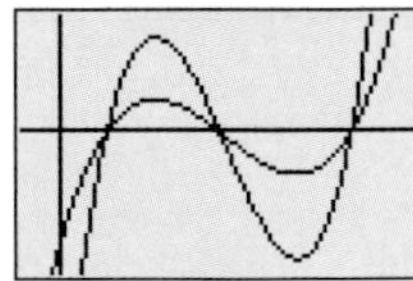

k is a stretching/compressing factor.

APPLICATIONS AND ARTWORK

Applications from a variety of disciplines motivate the mathematical concepts and make them relevant to students. Some are revisited as students develop new skills. A complete index of these innovative applications, with page references, follows this Walkthrough.

76. Advertising Expenditures The Aloma Advertising Company arrived at the following statistics relating the amount of money spent on advertising the Flex Exercise Machine to the realized sale of that product. **(a)** Find the least-squares line. **(b)** Use this line to predict the sales when $4000 is spent on advertising.

Dollars spent	1000	1500	2000	2500	3000	3500
Units sold	525	540	582	600	610	634

77. GPA Analysis Winchester College compiled the following data on the verbal SAT scores of ten entering students and their graduating GPA scores 4 years later. **(a)** Determine the least-squares line. **(b)** Use it to predict the graduating GPA of a student who entered with an SAT score of 500. **(c)** Consider a student who graduates with a GPA of 3.7 but who did not take the SAT examination. What would the student probably have scored on the examination? **(d)** If you took the SAT, what would be your predicted GPA according to these statistics?

SAT	490	450	640	510	680	610	470	450	600	650
GPA	3.2	2.8	3.8	2.9	4.0	3.1	2.2	2.3	3.5	3.8

78. Mortality Analysis The Acropolis Insurance Company computed the following percentages of people dying at various ages from 20 to 65. Use linear regression to estimate the percentage of deaths at 43 years of age.

Age (years)	20	25	30	35	40	45	50	55	60	65
Deaths (%)	0.18	0.19	0.21	0.32	0.41	0.64	0.78	0.81	0.79	0.83

79. Production Costs The production numbers and costs of the Summit Company for the last 6-month period were as follows. **(a)** Use linear regression to find the best linear function for these data (called the cost–volume equation). **(b)** Use this function to predict total costs for the next month (to the nearest $10) if production was planned to rise sharply to 165 units.

	July	Aug	Sept	Oct	Nov	Dec
Units of production	120	140	155	135	110	105
Total cost ($)	6300	6500	6670	6450	6100	5950

80. Enrollment Projections The following were projections of college enrollment (in units of 1 million) for men and women for the years 1994 to 2004 in the United States (*The Chronicle of Higher Education Almanac,* September 1, 1997). **(a)** Use linear regression to find the best linear functions for the data, one for men and one for women. **(b)** Observe that there were 6.831 million men enrolled in colleges in 1994, whereas there were 8.174 million women. There were more women than men enrolled. Is this trend likely to continue, or is the trend toward more men than women? (Use the variable x, time measured from 1994, as the independent variable.)

Full-color Artwork is important in describing and understanding college algebra. Hundreds of graphing calculator screens, graphs, pictures, and photographs strengthen students' graphical intuition and understanding of the relevance of mathematics.

Example 8 | **Modeling a Population of Trout**

Let $f(x) = 2.5x(1 - x)$ describe the behavior of the trout population of a mountain lake. If the population at a certain time is 0.72, let us predict the populations of the next three generations using a calculator.

We start with the initial value of x, namely 0.72, and compute $f(0.72)$. This answer then becomes the new x, and the new value of f is computed and so on. To carry out this process on a calculator, we write the equation $f(x) = 2.5x(1 - x)$ in the form

$$\text{New result} = 2.5(\text{Previous answer})[1 - (\text{Previous answer})]$$

The iterative process can then be implemented as shown in Figure 3.68. We see that the population fluctuates. The first four generations (rounded to two decimal places) are 0.72, 0.50, 0.62, and 0.59.

EXAMPLES AND PRACTICE

Examples with detailed solutions have been carefully chosen to illustrate important techniques and ideas. Numerous full-color graphs, data tables, and graphing calculator screens are included.

Example 3 | Scuba Diving

Nan Kuznicki goes scuba diving as often as she can. The air tank cost her \$257.78, and it costs \$22.50 to fill the tank each time she goes out. Construct a function that describes these expenses. Represent the information graphically.

SOLUTION Let Nan go scuba diving x times. Let the total expenses then be $f(x)$. Then

$$f(x) = \text{Cost of } x \text{ fill-ups} + \text{Cost of the tank}$$
$$= 22.50x + 257.78$$

Nan's expenses are described by the linear function $f(x) = 22.50x + 257.78$, with graph shown in Figure 1.43.

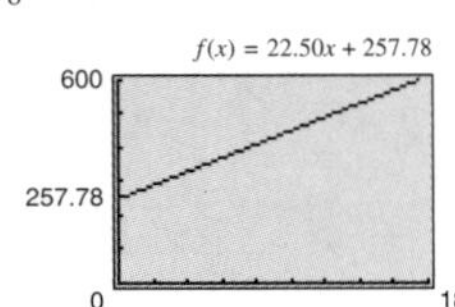

Figure 1.43 y-intercept = cost of tank. Slope = cost of an air fill-up.

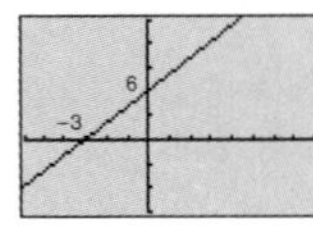

Figure 1.13

Example 5 | Description of a Viewing Window

Consider the graph of the line shown in Figure 1.13. The x-intercept and the y-intercept are displayed. Use this information to describe the viewing window (identify X_{min}, X_{max}, X_{scl}, Y_{min}, Y_{max}, and Y_{scl}).

SOLUTION Three tick marks on the x-axis lead to $x = -3$. Thus, $X_{scl} = 1$. Two tick marks on the y-axis give $y = 6$; thus, $Y_{scl} = 3$. Counting tick marks, we get $X_{min} = -6$, $X_{max} = 8$, $Y_{min} = -9$, $Y_{max} = 15$.

Figure 1.14

Self-Check 2 Consider the graph of the line shown in Figure 1.14. Use information from the display to describe the viewing window.

As we proceed in the course and develop more algebraic ideas, we will be able to use some of these new ideas to select suitable windows for graphs. We now introduce a method that is particularly appropriate for selecting windows for lines.

Self-Check exercises follow selected examples to reinforce the ideas presented. They are ideal for practice problems or as an in-class activity. Answers appear at the end of the section.

Further Idea exercises extend the concepts of the examples and encourage student exploration. Answers for these exercises are also at the end of the section.

Further Ideas 2

(a) Draw a table to predict Nan's costs for the first eight times she goes diving (Example 3).

(b) If Nan's friend Jane Robinson manages to buy the same air tank at a discount price of \$225, discuss how Jane's graph would differ from Nan's graph.

Example 4 | Reservoir Monitoring

Water flows into the Squaw Ridge Reservoir at the rate of 30 gallons per minute. It can be pumped out at the rate of 55 gallons per minute. The pumping process is started when there are 50,000 gallons of water in the reservoir. Construct a function that gives the number of gallons of water in the reservoir at time t minutes after the pumping starts. Use the function to determine the amount of water in the reservoir after 5 hours of pumping. What are the practical interpretations of the y-intercept and the slope of the graph of the function?

COMMUNICATION AND GROUP WORK

Section Projects and **Chapter Projects,** included in selected sections and at the end of every chapter, cover a mixture of theoretical and applied topics. These open-ended activities encourage exploration and are ideal for individual or group work.

SECTION PROJECT
Half-Timber Designs

Houses and churches in many towns and villages in Europe are decorated with *fachwerk* (Germany) or *half-timber* (England). Fine examples are to be found in the university town of Tübingen in Germany and in the cathedral city of Chester in England. Figure 3.52 shows popular fachwerk designs. Find the equations that give these designs and reproduce them on your calculator using the graphing window $X_{min} = 0$, $X_{max} = 5$, $X_{scl} = 2$, $Y_{min} = 0$, $Y_{max} = 5$, and

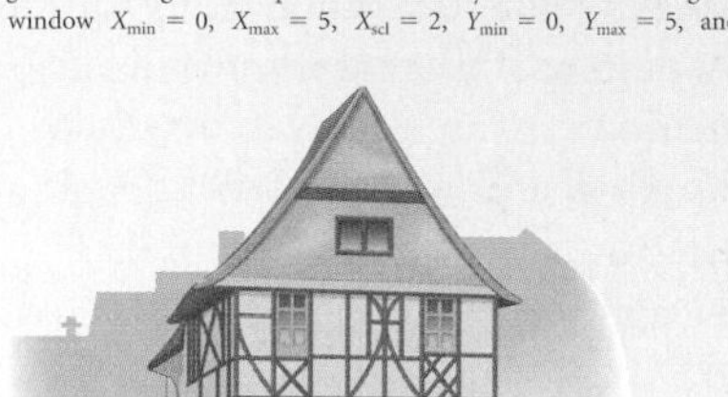

154 Chapter 1 Interplay Between Algebra and Geometry

CHAPTER 1 PROJECT | Real and Subjective Times

Scientists have studied the sense of time in humans and have concluded that there is a relationship between real time and a person's estimate of time (called subjective time). In most cases, this relationship is thought to be approximately linear. Let t be real time, s be subjective time, and $t = ms + b$ be the linear equation that describes the connection between the two times. The constants m and b vary from person to person.

The author determined his subjective time line as follows. Starting at a given time (while watching television), he estimated intervals of 10 minutes for a total of 60 minutes. A coworker recorded the actual times. The results were as follows:

Subjective time, s (minutes)	10	20	30	40	50	60
Real time, t (minutes)	8.5	22	32	38	50	63

These data were plotted and found to lie approximately on a line. The graph of the data points is called a **scatter diagram.** The least-squares line was then computed (Figure 1.75).

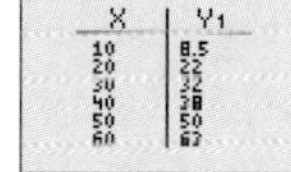

Enter data.

Plot data on a calculator or by hand. Points lie approximately on a line.

Compute least-squares line. $y = 1.04x + 0.67$ (to two decimal places).

Plot points and line. Line is a good fit.

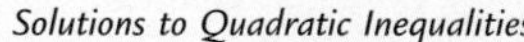

Figure 1.75 Relationship between real time and subjective time.

GROUP DISCUSSION | *Solutions to Quadratic Inequalities*

All the quadratic inequalities that we discussed have had many solutions. On the other hand, we have seen that quadratic equations can have a single solution or no solutions. This prompts us to ask whether quadratic inequalities that have a single solution or no solution can exist. Investigate this. We suggest that you be guided by geometry. If you are convinced that such inequalities can exist, use geometry to construct examples. Can questions about the motion of an object fired vertically from ground level be expressed in terms of inequalities that turn out to have a single solution or no solutions? Discuss.

Group Discussion problems train students to analyze, think, and communicate clearly. They can be used by small groups or individually.

WEB PROJECT 2

To explore the concepts of this chapter further, try the interesting WebProject. Each activity relates the ideas and skills you have learned in this chapter to a real-world applied situation. Visit the Williams Web site at

www.saunderscollege.com/math/williams

to find the complete WebProject, related questions, and live links to the Internet.

Web Projects, located on the companion Web site, extend a key chapter concept and include related problems and the necessary URLs to solve them.

EXERCISES

Exercises, in every section and at the end of every chapter, range from routine to challenging. Their number and variety, including straightforward, conceptual, applied, and writing, will appeal to students of every need and ability. Odd-numbered exercise answers follow the text.

202 **Chapter 2** Solving Equations and Inequalities

In Exercises 19–24, solve the double inequalities. Write the solutions in interval notation. Support your answers using graphs.

19. $2 < -2x + 4 \le 12$ **20.** $7 > x - 3 \ge 4$

21. $13 \ge -3x + 1 \ge 4$ **22.** $-3 \le 4t - 1 < 7$

23. $1 \le 2x + 3 \le 5$ **24.** $-4 < -x - 3 < 1$

In Exercises 25–28, use the displays to estimate the values of x for which the inequalities are true. Confirm your answers using algebra.

25. $-4 < x + 2 < 8$ **26.** $-5 \le 2x - 1 \le 11$

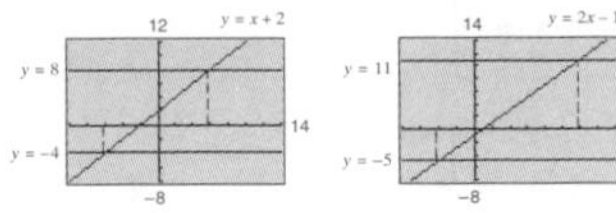

27. $2 < -x + 4 < 8$ **28.** $-5 \le -2x + 3 < 13$

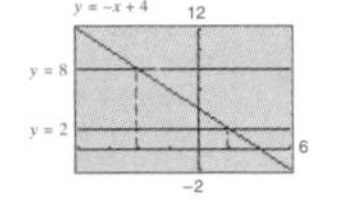

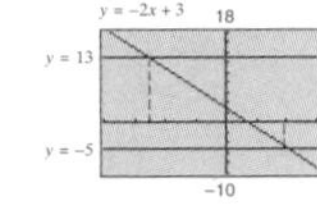

43. $-x^2 + 8x - 12 < 0$ **44.** $-x^2 - 2x + 48 \ge 0$

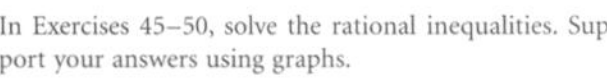

In Exercises 45–50, solve the rational inequalities. Support your answers using graphs.

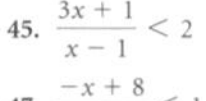

45. $\frac{3x + 1}{x - 1} < 2$ **46.** $\frac{2x - 1}{x - 3} \ge 1$

47. $\frac{-x + 8}{x - 2} < 1$ **48.** $\frac{5x + 2}{x + 2} \le 3$

49. $\frac{6x}{x - 4} > -6$ **50.** $\frac{-4x + 2}{x + 3} \le -2$

In Exercises 51–54, use the given displays to solve the inequalities. The x-intercepts of the graphs are given.

51. $x^3 - 2x^2 - 5x + 6 > 0$

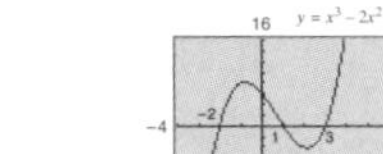

52. $-x^3 + 15x^2 - 68x + 96 \le 0$

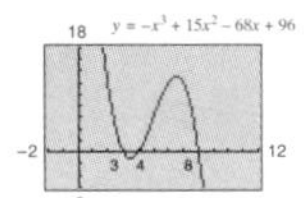

ic inequalities. Sup-

$4x + 3 \ge 0$

$6x + 8 \le 0$

$+ 5x - 4 \ge 0$

ic inequalities. Sup-

$4x - 3 \le 2$

$+ x + 1 > 6x - 2$

to estimate the val-
true. Confirm your

$x^2 + 2x - 15 \ge 0$

254 **Chapter 3** Further Development of Functions

76. Sizes of Eggs Data on the weights of the eggs of 800 species of birds were collected and analyzed. It was found that egg weight was the following function of body weight (all in grams):

$$M_{egg} = 0.277(M_{body})^{0.77}$$

(a) A hummingbird weighs 2.5 grams. What is its egg weight, rounded to two decimal places?

(b) The largest bird that ever lived was the elephant bird from Madagascar. It weighed about 500 kilograms. What was the approximate weight of its egg both in grams and pounds

(1 kilogram = 2.2046 pounds)?

(c) An ostrich egg weighs 1 kilogram. Use a graph to determine the weight of the ostrich in kilograms, to two decimal places. (Research on these topics was carried out by H. Rahn and his associates and is reported in "Relation of Avian Egg Weight to Body Weight," *Auk* **92:** 750–765, 1975).

77. Metabolic Rates Physiologists have given much study to the metabolic rates of mammals and birds. The metabolic rate R (in appropriate units) has been found to be the following function of body mass M (in kilograms):

$$R = 73.3(M)^{0.74}$$

(a) Find the metabolic rates of a 20-gram mouse and a 300-kilogram cow, rounded to two decimal places.

(b) The metabolic rate per unit mass is called the specific metabolic rate (Royal Society Symbols Committee, 1975). Find the function that describes specific metabolic rate as a function of body weight. Plot the graph of this function. Discuss why the graph tells you that the metabolic rate per unit mass decreases with body size. What does this imply in terms of the relative restlessness of mice, sheep, horses, and elephants?

units of 10...

(a) Estimate the number of new cars produced annually from 1989 to 1995 and the total number of new cars manufactured during this period.

(b) If current trends continue, when will the number of new cars produced annually exceed 1,200,000?

62. Ideal Body Weight

(a) The function $M(x) = 5.5x - 220$ describes the ideal body weight of men, where x is height in inches and $M(x)$ is the corresponding weight in pounds. This equation is usually valid for heights of 60 to 84 inches. Compute M(72). What does the result mean? Give the domain and range of the function M in this application.

(b) The function $W(x) = 5.4x - 231$ describes the ideal body weight of women, where x is height in inches and $W(x)$ is the corresponding weight in pounds. This equation is usually valid for heights of 54 to 78 inches. Compute $W(66)$. What does the result mean? Give the domain and range of the function W.

63. Cable Television The function $P(t) = 0.0057t^3 - 0.302t^2 + 6.087t + 22.6$ closely de-

(c) Compute $P(35)$. How do you interpret this number? Discuss the problem of using the function $P(t)$ to make long-term predictions.

64. Area of a Rectangle A rectangle with one side of length x units has a perimeter of length 72 units.

(a) Express the area A as a function of x.

(b) Determine $A(6)$, $A(8)$, and $A(30)$.

(c) Why do you expect $A(6)$ to be equal to $A(30)$?

(d) What is the domain of A?

65. Volume of a Box The surface area of four sides and the base of an open box having a square base of side x inches is 64 inches.

(a) Express the volume V as a function of x.

(b) Compute $V(0)$, $V(1)$, $V(2)$, ..., $V(8)$.

(c) What is the domain of V in this application?

(d) For which integer value of x is the volume largest? What is that volume?

66. Housing Costs A builder constructs houses on lots that cost $25,000.

(a) If the houses cost $32.50 per square foot, determine the cost of houses as a function of area.

(b) Models are available with areas of 1800, 1900, 2000, 2100, 2200, 2300, and 2400 square feet. If you plan to spend $100,000 for a house, how large a house can you afford?

END-OF-CHAPTER REVIEW

Chapter Highlights summarize the important concepts and results with references to the chapter. Students should use the highlights to reinforce and identify gaps in their knowledge.

Review Exercises and **Tests,** with problems similar to those in the examples and section exercises, are included for each chapter. Answers to review exercises and chapter test problems follow the text.

CHAPTER 3 HIGHLIGHTS

Section 3.1

Vertical Line Test: A set of points in the plane is the graph of a function if and only if every vertical line intersects the graph in at most one point.
Linear Function: $f(x) = ax + b$, where a and b are real numbers. The graph is a line.
Quadratic Function: $f(x) = ax^2 + bx + c$, where a, b, and c are real numbers with $a \neq 0$. The graph is a parabola.
Polynomial Function of Degree n: $f(x) = a_nx^n + a_{n-1}x^{n-1} + \cdots + a_1x + a_0$, with $a_n \neq 0$, and n is a nonnegative integer.
Intercepts: The y-intercept of the graph of $y = f(x)$ is $f(0)$. The x-intercepts of the graph of $y = f(x)$ are the roots of $f(x) = 0$. The x-intercepts are also called zeros of f.

Section 3.2

Increasing (Decreasing): f is increasing (decreasing) over an interval if graph is rising (descending) in going left to right over the interval.
Local Maximum (Minimum): $f(a)$ is greater (smaller) than the value of f at neighboring points.
Max/Min of a Polynomial Function of Degree n: The graph has at most $n - 1$ local maxima and minima.

Square Root Function: $f(x) = \sqrt{x}$. The domain and range are both $[0, \infty)$.
Symmetry: Function is even if $f(-x) = f(x)$ for every x in the domain. An even function is symmetric with respect to the y-axis; e.g. $f(x) = x^2$. Function is odd if $f(-x) = -f(x)$ for every x in the domain. An odd function is symmetric with respect to the origin; e.g., $f(x) = x^3$.
Absolute Value Function: $f(x) = |x|$. The domain is $(-\infty, \infty)$, and the range is $[0, \infty)$.
Piecewise-Defined Function: Function is defined by different equations over

from
$- c$,
nits.
ed from
zontal
$f(x)$ by a

CHAPTER 3 REVIEW EXERCISES

1. Determine the x- and y-intercepts of the graphs of the following functions, and sketch the graphs using algebraic techniques. Support your answers by duplicating the graphs on your calculator.
   (a) $f(x) = (x + 1)(x - 2)(x - 5)$
   (b) $f(x) = x^2(x - 3)(x + 2)$
   (c) $g(x) = x^3 + 3x^2 - 4x$
2. Construct a polynomial function of degree 3 with a graph having the following properties. Three x-intercepts at $x = -3$, 2, and 4, y-intercept $y = 48$, with f being negative on $(-\infty, -3)$ and $(2, 4)$ and positive on $(-3, 2)$ and $(4, \infty)$. Check your answers with a calculator display.

CHAPTER 3 TEST

1. Determine the x- and y-intercepts of
$$f(x) = (x + 2)(x - 4)(x - 6)$$
and sketch the graph using algebraic techniques. Support your answer by duplicating the graph on a calculator.
2. Sketch the graph of the function
$$f(x) = x^2 + 6x + 1$$
with domain $[-9, 1]$. Give the range of f. Support your results using a graphing calculator. Use the domain and the range as the x and y intervals of your window.
3. Use algebra to show that the function
$$f(x) = x^4 - 3x^2$$
is an even function. Sketch the graph of f with the aid of a graphing calculator, using the window $X_{min} = -4$, $X_{max} = 4$, $X_{scl} = 1$, $Y_{min} = -4$, $Y_{max} = 4$, and $Y_{scl} = 1$. Indicate any symmetry in the graph that supports your conclusion that f is even.

the

graph of g can be obtained from the graph of f using translations.
5. Consider the following tables of $f(x) = x^2$ and $g(x)$. What is the function $g(x)$? Check your answer by duplicating the table on your calculator.

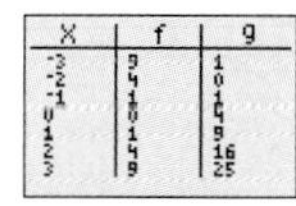

6. Give the equations of any vertical or horizontal asymptotes, identify the domain, and sketch the graph of the function
$$f(x) = \frac{2}{x^2 - 3x - 4}$$
Support your answer by duplicating the graph on your calculator.
7. Graph $f(x) = 2x^3 + 3x^2 - 5x + 4$ on your graphing calculator. Find the intervals over which f is increasing, where it is decreasing, the values of the local maxima and minima, and the zeros, rounded to two decimal places.

CUMULATIVE TEST* CHAPTERS R, 1, AND 2

1. Compute the following expression, rounded to four decimal places, using a calculator.
   (a) $(\sqrt{8.3} + 0.57^{-1}) \div 2.91$ (b) $\sqrt[5]{27.38^5}$
2. Compute the following by hand. Support your answers using a calculator.
   (a) $\left(\sqrt[4]{\frac{16}{81}}\right)^3$ (b) $(-125)^{2/3}$
3. Simplify the following algebraic expression. Express the result in a form that involves each variable once, to a positive exponent.
$$\frac{(x^{-2}y^3)^4}{(x^{-2}y^5)^{-2}}$$
4. Multiply $(2x - 3)(3x + 5)$.
5. Factor. (a) $12x^2 - 5x - 2$ (b) $9x^2 - 25y^2$
6. Find the domains and ranges of the following functions.
   (a) $f(x) = -2x^2 + 4$ (b) $f(x) = \sqrt{x - 1} + 3$
7. Find the equation of the line through the point $(3, -1)$ having slope 2. Graph the equation on your calculator, and use the display to support your answer.
8. Find the eq

calculator and use the display to support your answers.
9. Find the x-intercept of the line through the point $(-1, 5)$ parallel to the line $10x - 2y + 4 = 0$.
10. Find the vertex, y-intercept, and axis, and sketch the graph of $y = x^2 - 8x - 4$ using algebra. Verify your graph using a graphing calculator.
11. Use the distance formula to compute the distance between the points $(1, 5)$ and $(-2, 8)$. Round the answer to four decimal places.
12. Find the centers and radii of the circles having the following equations. Graph the equations on your calculator, and use the displays to support your answers.
   (a) $(x - 3)^2 + (y + 7)^2 = 25$
   (b) $x^2 + y^2 + 2x - 6y + 1 = 0$
13. Solve the equation $2x^2 + 5x - 12 = 0$ by factoring. Use a graph to support your answer.
14. Solve (if possible) the following quadratic equations using the quadratic formula. Round answers to four decimal places. Use graphs to support your answers.
   (a) $x^2 + 4x - 7 = 0$ (b) $x^2 - 3x + 8 = 0$
15. Solve x^4 ... using an appropriate

Cumulative Test exercises, included after Chapters 2, 5, and 9, offer another opportunity for students to check skills and understanding.

Index of Applications

ARCHAEOLOGY

ARCHITECTURE AND CONSTRUCTION

ASTRONOMY

BIOLOGICAL AND LIFE SCIENCES

BUSINESS AND INDUSTRY

COMPUTERS AND COMPUTER SCIENCE

CONSUMER APPLICATIONS

EARTH SCIENCE

ECONOMICS

EDUCATION

ELECTRONICS AND ENGINEERING

GEOMETRY

HEALTH AND MEDICINE

MATHEMATICS

MISCELLANEOUS

PHYSICAL SCIENCES

PSYCHOLOGY

SOCIETY

SPORTS AND RECREATION

TRAVEL

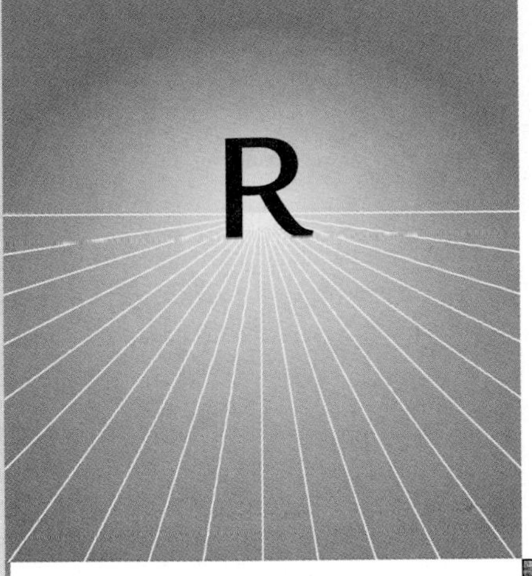

Review of Basic Algebra

Pythagoras (circa 572 BCE – circa 500 BCE) *(©1994 Northwind Picture Archives)*

Algebra has a fascinating history that can be traced back to ancient times. Many diverse cultures and innumerable personalities have contributed to the modern field that we now call algebra. The Chinese, Persians, Hindus, and Babylonians used algebra thousands of years ago. Algebra appears in the writings of Ahmes, an Egyptian mathematician who lived around 1700 BCE. Later the Greeks contributed to the field. The Greek Pythagoras was a mystic who believed that the key to understanding the universe lies in our ability to see how things are built up from and resolve back to numbers. Later, in the third century, the Greek mathematician Diophantus used quadratic equations and, consequently, has been called the father of algebra. The Arabs contributed to algebra in the 7th to 15th centuries. The word **algebra** is derived from the Arabic *al'jabr,* which means "reduction" in the sense of solving an equation. Additionally, the Persian astronomer Omar Khayyam wrote a book on algebra in the 12th century.

The 16th century marked the beginning of the use of symbols in algebra. Christoff Rudolff of Germany introduced the radical sign "$\sqrt{}$" in 1525, Richard Stifel of Germany introduced the signs "+" and "−" in 1544, and Robert Recorde of England introduced the modern equal sign "=" in 1557. René Descartes of France was the first to use the last letters of the alphabet to represent unknown numbers in the late 1500s.

The practice of algebra has changed considerably since these early times. We can now use graphing calculators to generate numbers that satisfy the Pythagorean theorem (and we do so in this chapter). Algebra also is used widely. We have found applications for it in such fields as business, medicine, the natural sciences, engineering, and industry. For example, businesspeople need to know how revenue and profit depend on the cost of an item. These quantities are represented by letters, and algebra is used to determine the relationships among them. Algebra is also applied in medicine. For example, after a drug such as sodium phenobarbital, which is used to treat cirrhosis of the liver, is administered, the concentration of the drug in the blood gradually decreases with time. This behavior can be described using algebra, leading to a knowledge of when the drug should be administered. Algebra is used to study the growth of bacteria such as *Escherichia coli.* This is the bacterium that causes milk to sour and that has lately been contaminating hamburgers in certain parts of the country. Additionally, molecular biologists use *E. coli* in experiments involving genetic manipulation. The popular typesetting software T_EX uses algebra to produce figures for overhead slides, manuscripts and textbooks. In this course, you will learn algebra and you will also learn how to apply algebra by looking at interesting applications such as these.

In this chapter, we will review many of the basic ideas of algebra, such as rules for exponents, multiplying and adding polynomials, and factoring, equations, and variation. The last section of this chapter introduces the concept of the function, which enables us to study how one variable affects another. The graphing calculator is introduced throughout the chapter when appropriate. We use the calculator to evaluate exponents, compute function values, and construct tables of functions. We strongly encourage the student to master the basic ideas of algebra and the techniques for using a calculator in this chapter, because the course builds on this material.

APPLICATION | Sizes of Skeletons

The brachiosaurus is believed to be the largest land animal ever to exist. *(©The Field Museum, GN88141.13C, John Weinstein)*

The world we live in is governed by the laws of chemistry and physics. Because animals, among other creatures, live within the bounds of these laws, biologists attempt to discover and understand these laws. For example, they study the relationships between sizes of animals and their various biologic components. Scientists have found that the skeleton/body weight ratio of a land animal of weight x pounds is given by the expression

$$\frac{0.061x^{1.09}}{x}$$

A table of values of this expression shows how the skeleton weight becomes an increasingly larger fraction of the total weight of an animal. The table shows that at 243 pounds the skeleton weight becomes one tenth of the total body weight of a human being.

A brachiosaurus was 70 feet long and weighed approximately 170,000 pounds. Scientists found the skeleton/body weight ratio of a brachiosaurus to be 0.1803. They believe that this number represents the largest such ratio for a land animal. Because land animals cannot sustain a skeleton that is a larger fraction of body weight, no larger land animal can exist. *(This application is discussed in Section R.7, Example 8, on pages 72–73.)*

R.1 Real Numbers and Their Properties

• Real Numbers • Inequalities • Intervals • Calculators and Computers • Rounding Numbers • A Snapshot of π • Absolute Value • Algebraic Properties of Addition and Multiplication • Powers of a Number • Hierarchy of Operations • Recognizing Mathematical Patterns

In this section, we look at real numbers from algebraic and geometric viewpoints and discuss their properties. We often use the complete set of real numbers in our discussions; however, at times, we only use a subset. The following are the most important subsets of the set of real numbers.

Natural numbers are the numbers 1, 2, 3, 4, . . . ; they are used for counting.

Integers are the numbers . . . , $-3, -2, -1, 0, 1, 2, 3, \ldots$. These are the natural numbers together with their negatives and zero.

Rational Numbers are numbers of the form $\frac{a}{b}$, where a and b are integers with b nonzero. For example, $-\frac{3}{4}, \frac{1}{2}, \frac{21}{5}$ are rational numbers. Every natural number and every integer is a rational number because any such number a can be written $\frac{a}{1}$. Thus, the set of natural numbers and the set of integers are subsets of the set of rational numbers.

Irrational numbers are real numbers that cannot be expressed in the form $\frac{a}{b}$. The difference between rational and irrational numbers can be seen by using their decimal forms. A rational number has a decimal form that either terminates or involves a pattern that repeats itself. Conversely, an irrational number has a decimal form that does not terminate and does not involve a repeating pattern.

The following examples of rational and irrational numbers illustrate these concepts.

Rational Numbers		**Irrational Numbers**
$\frac{1}{4} = 0.25$	terminates	$\sqrt{2} = 1.41421356\ldots$
$\frac{1}{3} = 0.33333\ldots$	3 repeated	$\pi = 3.14159265\ldots$
$\frac{8}{11} = 0.727272\ldots$	72 repeated	These numbers do not terminate and
$\frac{79}{108} = 0.73148148\ldots$	148 repeated	have no repeating pattern

The first eight decimal places of $\sqrt{2}$ and π are displayed. These decimal forms do not terminate, and there would be no repeated pattern to observe no matter how many decimal places were displayed.

The appearance of some irrational numbers in our everyday lives is thought provoking. We cannot give the exact length in decimal form of the diagonal of the right triangle whose other two sides are 1. The length of the diagonal is $\sqrt{2}$ (Figure R.1). Neither can we give the exact area in decimal form of a circle whose radius is 2. Using the formula $A = \pi r^2$, we find that the area of this circle is 4π (Figure R.2). We can express both of these numbers to any number of decimal places that we like, but we cannot say, "This is the decimal number!"

The rational numbers together with the irrational numbers make up the set of **real numbers.** We illustrate the relationships among the various subsets of real numbers in Figure R.3.

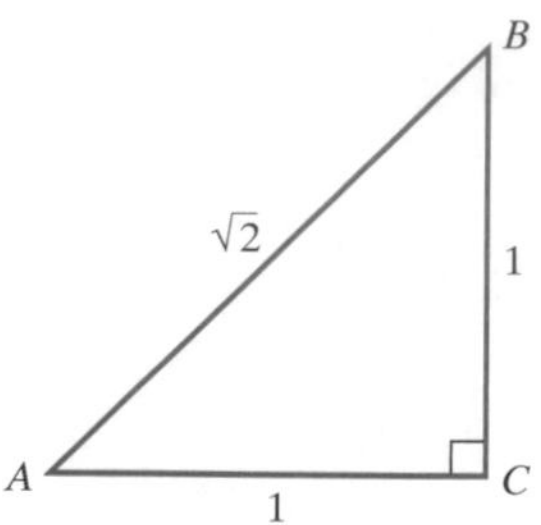

By the Pythagorean Theorem
$AB = \sqrt{(AC)^2 + (BC)^2} = \sqrt{1^2 + 1^2} = \sqrt{2}$,
an irrational number.

Figure R.1

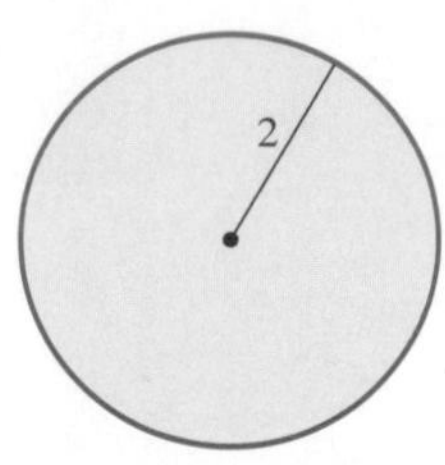

Area of circle radius 2
$= \pi r^2 = \pi(2)^2 = 4\pi$,
an irrational number.

Figure R.2

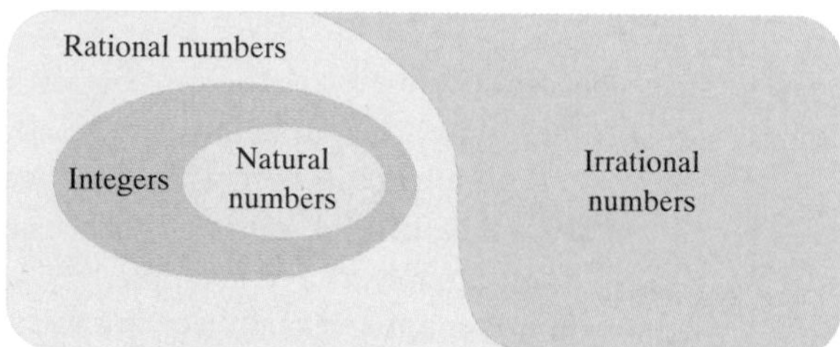

Figure R.3 The real numbers.

Example 1 **Recognizing Various Types of Real Numbers**

Consider the set

$$A = \{-7, -\tfrac{11}{3}, -\tfrac{1}{2}, 0, 1, \tfrac{5}{4}, 4\}$$

List all the elements of A that are (a) natural numbers, (b) integers, (c) integers that are not natural numbers, (d) rational numbers that are not integers.

SOLUTION

(a) The natural numbers are 1, 4.
(b) The integers are -7, 0, 1, 4.
(c) The integers that are not natural numbers are -7, 0.
(d) The rational numbers that are not integers are $-\frac{11}{3}, -\frac{1}{2}, \frac{5}{4}$.

The Real Number Line

The real numbers can be conveniently represented by points on a line. Select a point on the line to correspond to the integer 0. Call this point the **origin.** Place the positive numbers to the right of the origin and the negative numbers to the left [see Figure R.4]. We show the locations of the integers. Other real numbers are represented by points on the line that lie between these points. Note the approximate locations of $\sqrt{2}$ and π.

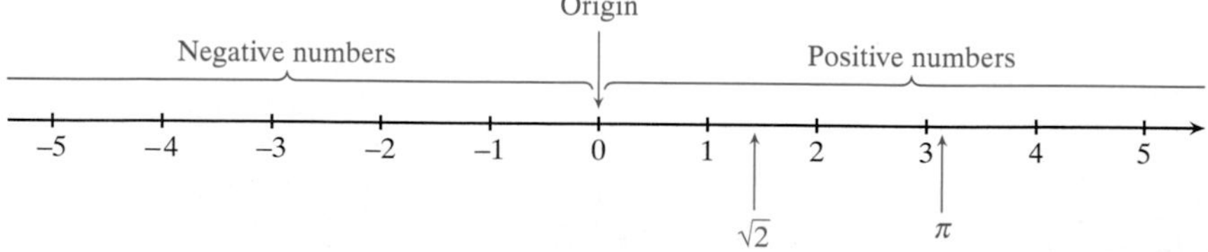

Figure R.4 The real number line.

INEQUALITIES

Consider the real number line in Figure R.4. This geometric representation of the set of real numbers leads to the following order relations between real numbers, called *inequalities.*

Inequality	Meaning	Example
$a < b$, a is less than b	a lies to the left of b	$2 < 5$
$a \le b$, a is less than or equal to b	a coincides with or lies to the left of b	$-3 \le 0$
$a > b$, a is greater than b	a lies to the right of b	$17 > 10$
$a \ge b$, a is greater than or equal to b	a coincides with or lies to the right of b	$8 \ge 8$

Because the numbers to the left of zero are negative and those to the right of zero are positive, we can relate negative and positive numbers to inequalities as follows:

> ***Negative and Positive Numbers*** $a < 0$ means that a is **negative,** and $a > 0$ means that a is **positive.**

INTERVALS

An **interval** is a set of numbers on the real line that forms a line segment, a half line, or an entire line. The numbers that define the ends of an interval are called the **end points** of the interval. There are three types of intervals; namely, **closed intervals, open intervals,** and **half-open intervals**, depending on whether the end points are included in the interval. A square bracket is used if an end point is included, whereas a parenthesis is used if it is not.

For example, the interval $[-2, 6)$ consists of all the real numbers that are greater than or equal to -2 and less than 6. The number -2 is in the interval, but 6 is not. It is a half-open interval with end points -2 and 6. The **graph** of $[-2, 6)$ is

The symbols ∞ and $-\infty$, called "infinity" and "negative infinity," are used to describe half lines and the real line. These are not real numbers, but they indicate that the interval extends without bound. For example, the interval $(12, \infty)$ consists of all the real numbers that are greater than 12. It is an open interval with end point 12. The graph is

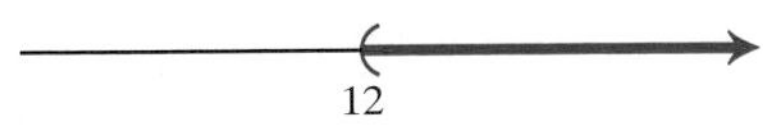

We illustrate various types of intervals that can arise for two real numbers a and b in Figure R.5.

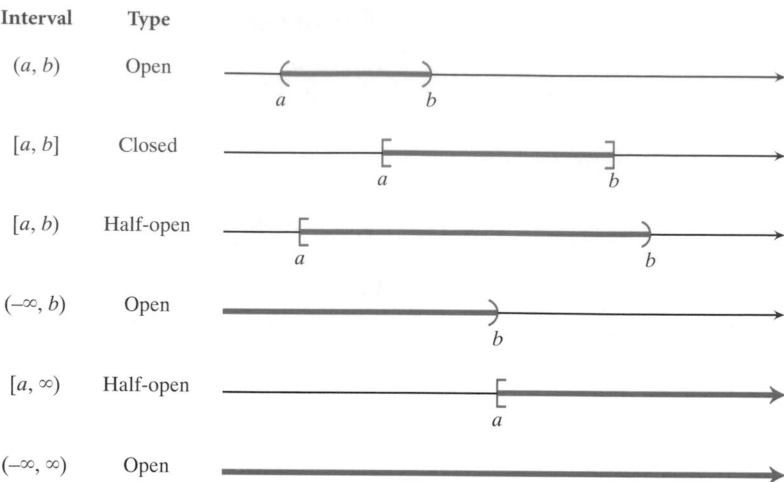

Figure R.5 Intervals on the real line.

CALCULATORS AND COMPUTERS

Calculations involving real numbers are usually performed using calculators or computers. A graphing calculator (or a computer*) can be used in this course. Important graphing concepts (e.g., the window for displaying a graph, tracing, and zooming) common to all graphing calculators are discussed. The methods of implementing these concepts vary from calculator to calculator. We use straightforward displays that students can easily interpret for their different calculators. *You should determine how to implement the ideas on your own calculator.*

Calculators can only compute and display a finite number of digits. Numbers having many decimal places are "rounded" and answers are often only very good approximations. To get an understanding of how calculators round numbers, let us enter and display the numbers, $\frac{1}{3}$ and $\frac{2}{3}$ on a calculator.

Rounding Down $\frac{1}{3}$ is a rational number where the 3 in 0.3333 ... should be repeated indefinitely (Figure R.6). The number has been **rounded down.** The numbers that have been discarded do not warrant increasing the last digit displayed to 4. Rounding down takes place if the first digit discarded is less than 5.

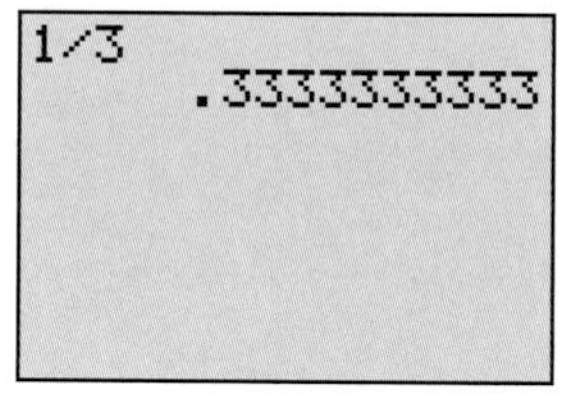

$1 \div 3$

Figure R.6 Rounding down.

Rounding Up $\frac{2}{3}$ is a rational number where the 6 in 0.6666 ... should be repeated indefinitely (Figure R.7). The last digit is **rounded up** to 7 to make up for the numbers that were discarded. Rounding up takes place if the first digit discarded is greater than or equal to 5.

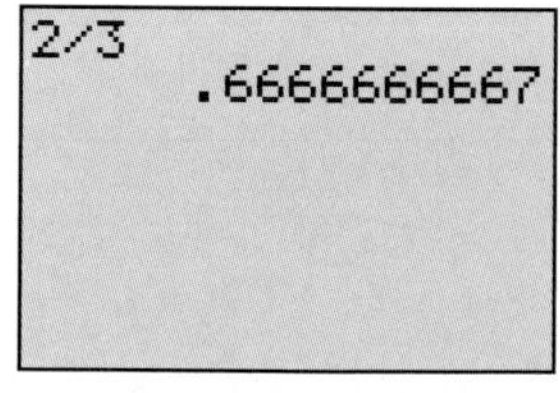

$2 \div 3$

Figure R.7 Rounding up.

*We will refer to graphing calculators in this text, but the graphing ideas can also be implemented on a computer.

The concept of rounding applies whether one works with calculators or not if one wants to express a number to a certain number of digits. The rules for rounding numbers follow.

Rules for Rounding

1. Determine the number of digits to be displayed.
2. If the first digit to be discarded is less than 5, round down by leaving the last digit unchanged.
3. If the first digit to be discarded is greater than or equal to 5, round up by increasing the last digit by 1.

Example 2 | Rounding Numbers

Round 7.834854231683 **(a)** to eight decimal places **(b)** to ten decimal places.

SOLUTION

(a) Count eight decimal places.

$$7.\underbrace{83485423}_{\text{Eight decimal places}}|\underset{\uparrow \text{ First discarded digit}}{1}683$$

The first discarded digit is 1. Because $1 < 5$, leave the last digit 3 unchanged.

$$7.834854231683 = 7.83485423, \text{ to eight decimal places}$$

(b) Count ten decimal places.

$$7.\underbrace{834852316}_{\text{Ten decimal places}}|\underset{\uparrow \text{ First discarded digit}}{8}3$$

The first discarded digit is 8. Because $8 \geq 5$, increase the last digit 6 to 7.

$$7.834854231683 = 7.8348542317, \text{ to ten decimal places.}$$

Further Ideas 1* Find the square root key on your calculator and determine how to use it.

(a) Find the approximation to $\sqrt{2}$ given by your calculator. To how many decimal places is the approximation given?

(b) Use your calculator to compute $\sqrt{\sqrt{2}}$ and $\sqrt{\sqrt{\sqrt{2}}}$.

A Snapshot of π

The history of π is fascinating. The early mathematicians thought that π was a rational number. Not until 1767 did they discover that π is an irrational number. The following landmarks provide a glimpse of the interest it has aroused over the years.

2000 BCE: Egyptian mathematicians, $\pi = \frac{256}{81}$.
250 BCE: Archimedes, π lies between $\frac{223}{71}$ and $\frac{22}{7}$.
150 CE: Ptolemy, $\pi = \frac{377}{120}$.
480 CE: Tsu Chung-Chih computed π to six decimal places.

*"Further Ideas" indicates an exercise that immediately follows an example to reinforce and extend slightly the ideas presented in the example. Answers are given at the end of the section.

1596: Ludolph van Ceulen, a Dutchman, computed π to 35 decimal places.
1706: William Jones, a Welshman, broke the 100-decimal-place barrier for π.
1767: Johann Heinrich Lambert, first showed that π was an irrational number.
1844: Johann Dase, a 16-year-old German boy, computed π to 200 decimal places.
1949: the first electronic calculation of π with a computer, to 2,037 decimal places.
1989: David and Gregory Chudnovsky computed π to 1,011,196,691 decimal places.
1994: The Chudnovsky brothers (Columbia University) computed π to 2,260,000,000 decimal places.
1995: Yasumasa Kanada (University of Tokyo) computed π on a Hitachi supercomputer to more than six billion decimal places—6,442,450,938 places to be exact.

```
22/7
      3.142857143
π
      3.141592654
```

Figure R.8 Comparison of values of π.

Many people use the approximation $\frac{22}{7}$ for π, going back to the work of Archimedes. Let us determine the accuracy of this value of π using a calculator (Figure R.8.) We see that $\frac{22}{7}$ is accurate to two decimal places. We leave it to you to determine the accuracies of the Egyptian and Ptolemaic values of π.

Use of π to Test Hardware and Software

"The Quest for Pi," a fascinating article in the *Mathematical Intelligencier,** discusses not only the history of π, but also why one would be interested in computing π to so many places. Such calculations are, for example, excellent tests of the integrity of computer hardware and software. If two independent computations of digits of π on two computers agree, then most likely both computers performed billions or even trillions of operations flawlessly. Conversely, if they disagree, the computers and the software need closer examination. For example, in 1986 a π-calculating program detected some obscure hardware problems in one of the original Cray-2 supercomputers.

*Exploration***

Getting the Most out of Your Calculator

Calculators and computers are limited in the number of digits they can display. Nevertheless, numbers often are stored to a larger number of digits in the memory of the machine than are displayed. For example, the author's calculator displays π to nine decimal places, showing $\pi = 3.141592654$. This calculator, however, "knows" the irrational number π to more digits than this. Find out how to display π on your calculator and determine whether there is a way to extract another decimal place for π out of your calculator! (*Hint:* You already know the first digit in π. It is occupying unnecessary display space. Displaying $\pi - 3$ may reveal more digits. See exercise 79 at the end of this section to display even more digits of π.)

*D. H. Bailey et al., "The Quest for Pi," *Mathematical Intelligencier,* Winter 1997, volume 19, number 1, page 50.

**Exploration opportunities are presented to give you the chance to discover concepts for yourself.

ABSOLUTE VALUE

The absolute value of a real number a, denoted $|a|$, is the **size** or **magnitude** of the number. It gives a measure of how large that number is. For example, the magnitude of 5 is 5, thus $|5| = 5$. The magnitude of -3 is 3 giving $|-3| = 3$.

These examples illustrate that the absolute value of a positive number is that number, whereas the absolute value of a negative number is the negative of that number.

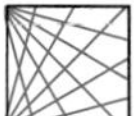

*ABSOLUTE VALUE WITH A CALCULATOR

An absolute value can be displayed on a calculator. We support the result

$|9| = 9$ and $|-9| = 9$

with a calculator.

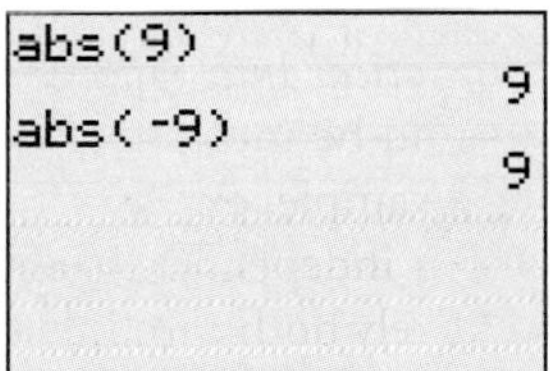

Figure R.9 Absolute value with a calculator.

Absolute Value If a is a real number,

$$|a| = \begin{cases} a \text{ if } a \geq 0 \\ -a \text{ if } a < 0 \end{cases}$$

Absolute value has the following properties.

$|a| \geq 0$, because magnitude cannot be negative.

$|a| = |-a|$, because the magnitudes of a and $-a$ are the same.

Geometric Interpretation of Absolute Value

The absolute value of the real number a can be thought of as the distance of a from the origin on the real line. For example, consider the two real numbers -3 and 5 as points on the real line (Figure R.10). Observe that

$$|-3| = 3, \text{ the distance of } -3 \text{ from the origin and}$$
$$|5| = 5, \text{ the distance of 5 from the origin.}$$

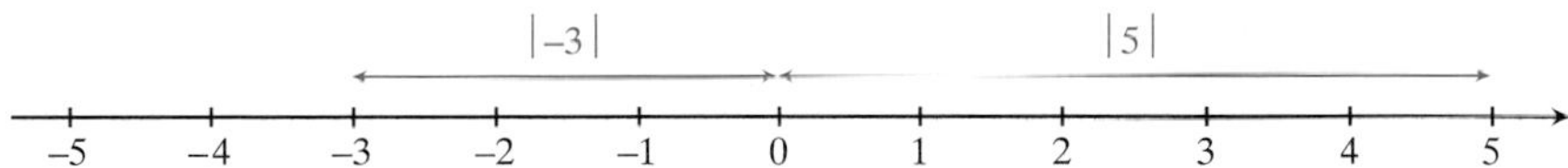

Figure R.10 $|a|$ = distance of a from the origin.

ALGEBRAIC PROPERTIES OF ADDITION AND MULTIPLICATION

Let a, b, and c denote real numbers. Let $a + b$ denote the sum of a and b, and ab (or $a \cdot b$) denote the product of a and b. These operations have the following properties. Parentheses are used to indicate the order in which operations are to be performed, with operations inside parentheses to be performed first.

* [calculator icon] indicates useful calculator information. Determine how to implement these concepts on your calculator.

Properties of Addition and Multiplication

	Addition	Multiplication
Commutative properties	$a + b = b + a$	$ab = ba$
Associative properties	$a + (b + c) = (a + b) + c$	$a(bc) = (ab)c$
Identity properties	$a + 0 = a$ and $0 + a = a$	$a \cdot 1 = a$ and $1 \cdot a = a$
Inverse properties	$a + (-a) = 0$ and $(-a) + a = 0$	$a\left(\frac{1}{a}\right) = 1$ and $\left(\frac{1}{a}\right)a = 1$, when $a \neq 0$
Distributive properties	$a(b + c) = ab + ac$, $(a + b)c = ac + bc$	

Example 3 **Recognizing Algebraic Properties**

Indicate the algebraic property that is illustrated by each of the following statements.

(a) $2 \cdot (4 + 3) = (2 \cdot 4) + (2 \cdot 3)$
(b) $-1 + (2 + 3) = (-1 + 2) + 3$
(c) $4 \cdot 6 = 6 \cdot 4$
(d) $(x + y)\,z = xz + yz$, where x, y, and z are real numbers

SOLUTION

(a) Distributive property
(b) Associative property of addition
(c) Commutative property of multiplication
(d) Distributive property

Self-Check 1* Arithmetic operations usually are performed on calculators or computers. Familiarize yourself with your calculator by performing the following computations.

(a) $5.379 + 4.218$ **(b)** 4.29×5.38 **(c)** $11.18 \div 4.3$
(d) $2.73(4.21 - 1.37)$ **(e)** $(9.743 - 3.7269) \times (4.23 + 1.97)$

(Calculators have parentheses keys "("and")".)

Powers of a Number

Consider the arithmetic expression

$$3 \cdot 3 \cdot 3 \cdot 3$$

*"Self-Check" indicates an exercise that immediately follows an example to reinforce the ideas presented in the example. Answers are given at the end of the section.

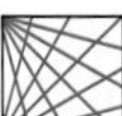

Powers with a Calculator

Powers can be computed with a calculator. We use a calculator to compute $(3.72)^4$.

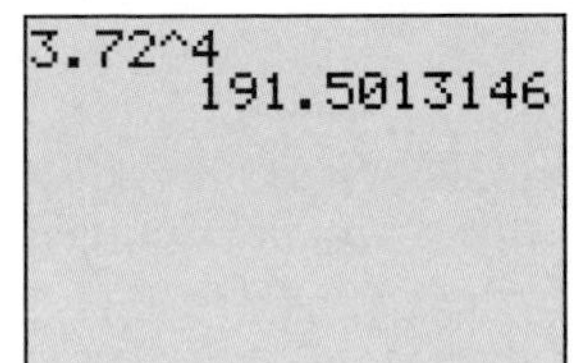

Figure R.11 Computing $(3.72)^4$.

This expression involves multiplying 3 four times. The value of the expression is 81. A convenient notation has been developed for expressing such a repeated product. We write

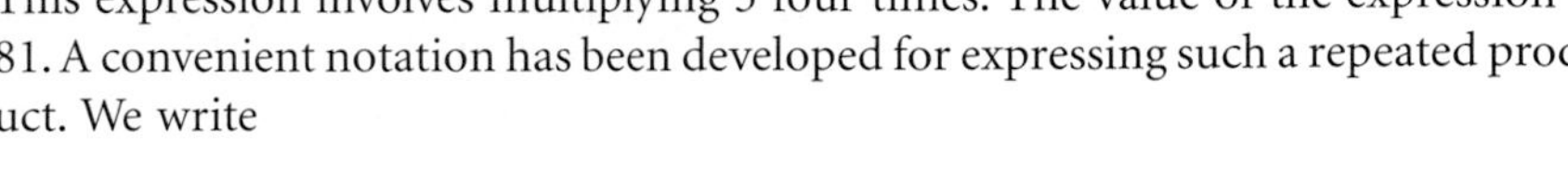

$$3^4 \quad \text{for} \quad 3 \cdot 3 \cdot 3 \cdot 3$$

Notation for Powers

Let a be a real number and n be a positive integer. Then

$$a^n = \underbrace{a \cdot a \cdot a \cdot \ldots \cdot a}_{n \text{ times}}$$

a^n is the ***n*th power** of a. The number a is called the **base,** and the integer n is called the **exponent.**

Example 4 **Computing Powers of Numbers**

Write each of the following expressions as a repeated product and evaluate. Support your answer on a calculator.

(a) 2^5 (b) $(-3)^4$ (c) -3^4

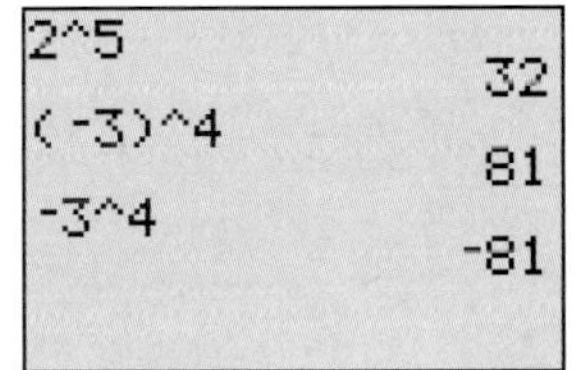

Figure R.12 Computing powers.

SOLUTION

(a) $2^5 = 2 \cdot 2 \cdot 2 \cdot 2 \cdot 2 = 32$; the base is 2 and the exponent is 5.

(b) $(-3)^4 = (-3) \cdot (-3) \cdot (-3) \cdot (-3) = 81$; the base is -3 and the exponent is 4.

(c) $-3^4 = -(3 \cdot 3 \cdot 3 \cdot 3) = -81$; the base is 3 and the exponent is 4.

A calculator supports these answers (Figure R.12).

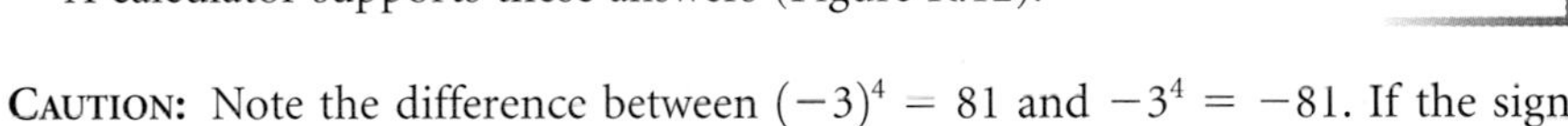

Caution: Note the difference between $(-3)^4 = 81$ and $-3^4 = -81$. If the sign is to be included in the base, as in the case of say $(-3)^4$, parentheses must be used.

Self-Check 2 Evaluate $-4.23^2\,(2.7^3)$.

Hierarchy of Operations

When an arithmetic expression is entered into a calculator on a line, the keying sequence always reads from left to right. There is a standard order (hierarchy) in which exponentiations, multiplications, divisions, additions, and subtractions are evaluated.

Hierarchy of Operations

All exponentiations are performed first; then multiplications and divisions are carried out, and finally additions and subtractions are done.

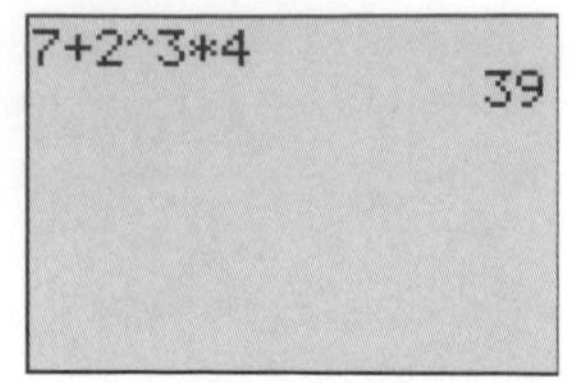

Figure R.13 Hierarchy of operations.

For example, if we key in the sequence 7 + 2 ^ 3 * 4, we get the answer 39 (Figure R.13). To arrive at this result, the calculator scans the line from left to right and performs the exponentiation first; then it multiplies and finally adds as follows.

$$7 + \underbrace{2 \wedge 3}_{\substack{\text{exponentiation}\\\text{first}}} * 4 \rightarrow 7 + \underbrace{8 * 4}_{\substack{\text{multiplication}\\\text{second}}} \rightarrow \underbrace{7 + 32}_{\substack{\text{addition}\\\text{third}}} \rightarrow 39$$

CAUTION: If an expression involves a computation that does not follow this hierarchy of operations, use parentheses to override the hierarchy. For example, suppose we want to compute

$$\frac{19.6 + 8.4}{4}$$

The addition in the numerator has to be completed before the division is performed. We indicate this by means of parentheses in Figure R.14(a). If we omit the parentheses we get the incorrect answer, as shown in Figure R.14(b).

(a)

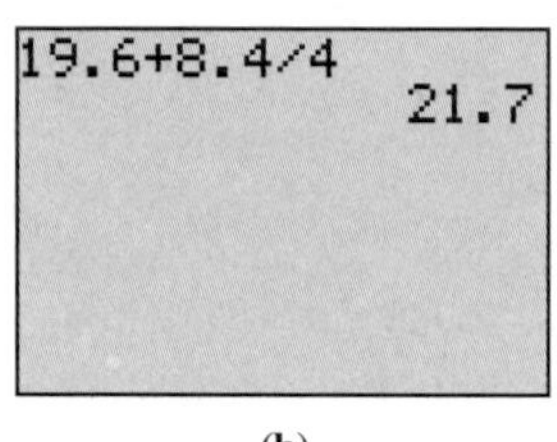

(b)

Figure R.14 Calculation of $\frac{19.6 + 8.4}{4}$.

(a) Correct sequence of operations
(19.6 + 8.4)/4 → 28/4 → 7

(b) Incorrect sequence
19.6 + 8.4/4 → 19.6 + 2.1 → 21.7

Using the hierarchy of operations enables us to enter expressions compactly on a calculator; however, much care must be taken when parentheses are omitted. When in doubt, use parentheses to enclose quantities that are to be evaluated first.

Self-Check 3

(a) Write the calculator expression 16 ÷ 2 ^ 3 + 5 in standard algebraic form and compute the result. Check your answer by performing the sequence of keystrokes on your calculator.

(b) Compute the value of the expression for $\dfrac{-a + 2b}{\sqrt{a^2 + 4}}$ for $a = 2.1$, $b = 9.2$, expressed to four decimal places.

Patterns can occur when working with real numbers. It is important to look for patterns instinctively and to recognize them when they occur. When a pattern occurs in the mathematics being used to describe a real situation, it usually reveals something about that situation. We will meet such applications in this course. The following example introduces you to the art of recognizing patterns in sequences of numbers.

Example 5 Recognizing Mathematical Patterns

Determine the relationships among consecutive terms in the following sequence. Use your calculator to predict the sixth term to four decimal places.

$$256, 16, 4, 2, \ldots$$

SOLUTION Observe that each term is the square root of the previous term. Thus we get

$$256, 16, 4, 2, \sqrt{2}, \sqrt{\sqrt{2}}, \ldots$$

Recognizing this pattern enables us to produce as many terms as we like of this sequence. The sixth term $\sqrt{\sqrt{2}} = 1.1892$, expressed to four decimal places.

GROUP DISCUSSION

Order of Operations

Determine the order of operations that the calculator uses in computing an expression of the form a^{b^c}. For example, if we enter 4 ^ 3 ^ 2, do we get $(4^3)^2$ or $4^{(3^2)}$? If you were designing the calculator, which order would you have selected and why?

Answers to Self-Check Exercises

1. **(a)** 9.597, **(b)** 23.082, **(c)** 2.6, **(d)** 7.7532, **(e)** 37.29982.
2. $-4.23^2\,(2.7^3) = -352.1859507$. See screen.

```
-4.23^2(2.7^3)
          -352.1859507
```

3. **(a)** 16 ÷ 2 ^ 3 + 5 is equivalent to $(16/2^3) + 5 = 7$. **(b)** 5.6207.

Answers to Further Ideas Exercises

1. **(a)** $\sqrt{2} = 1.414213562$ (nine decimal places). **(b)** $\sqrt{\sqrt{2}} = 1.189207115$, $\sqrt{\sqrt{\sqrt{2}}} = 1.090507733$. See screen.

```
√(2)
           1.414213562
√(Ans)
           1.189207115
           1.090507733
```

EXERCISES R.1

1. Let $A = \{-9, -\frac{7}{3}, -2, 0, 2, \frac{8}{3}, 9\}$. List all the elements of A that are **(a)** natural numbers, **(b)** integers, **(c)** rational numbers.
2. Let $C = \{-7, -\sqrt{2}, -\frac{1}{3}, 1, \pi, 4\}$. List all the elements of C that are **(a)** rational numbers, **(b)** irrational numbers, **(c)** integers or irrational numbers.
3. Let $B = \{-60, -21, -\frac{15}{3}, 0, \frac{4}{2}, 5, 9, \frac{17}{6}\}$. List all the elements of B that are **(a)** natural numbers, **(b)** integers, and **(c)** rational numbers but not natural numbers.
4. Let $X = \{-4, -\frac{3}{2}, -\sqrt{2}, 1, \frac{5}{3}, \pi, \sqrt{2}, 12\}$. List all the elements of X that are **(a)** not natural numbers, **(b)** not integers, **(c)** not rational numbers, **(d)** not irrational numbers.

In Exercises 5–10, graph each of the intervals.

5. $[2, 9)$ 6. $(-\infty, -3]$ 7. $(-5, \infty)$
8. $(-4, 3]$ 9. $(-\infty, 8)$ 10. $[2, \infty)$

In Exercises 11–16, round the numbers to the given number of decimal places.

11. 23.45687, to four decimal places.
12. 4.5372923, to three decimal places.

13. 29.5555555, to six decimal places.

14. 32.373737, to four decimal places.

15. 539.649, to one decimal place.

16. 758.6748, to two decimal places.

In Exercises 17–20, perform the divisions on a calculator. Give your answers to four decimal places.

17. $12.64 \div 9.23$ **18.** $3.725 \div 10.978$

19. $53.16 \div -4.37$ **20.** $57.2374 \div 21.3598$

In Exercises 21–28, indicate which algebraic property of real numbers is used to justify each statement.

21. $3 \cdot 4 = 4 \cdot 3$

22. $5 + 6 = 6 + 5$

23. $2 \cdot (2 + 9) = (2 \cdot 2) + (2 \cdot 9)$

24. $7 + (3 + 1) = (7 + 3) + 1$

25. $x + (y + z) = (x + y) + z$

26. $(a + b)c = ac + bc$

27. $(a + b) + 4 = a + (b + 4)$

28. $xy = yx$

In Exercises 29–32, compute the expressions involving parentheses using a calculator.

29. $2.3 \times (4.15 + 7.93)$ **30.** $3.96 \times (9.21 - 7.45)$

31. $1.23 \times (-4.56 + 7.91 - 2.3576)$

32. $(3.42 + 5.79) \times (-1.78 + 9.23)$

In Exercises 33–40, compute the expressions on a calculator. Round the answers to the indicated number of decimal places.

33. $\sqrt{3}$, to four decimal places.

34. $\sqrt{37}$, to six decimal places.

35. $(\sqrt{37} + 3.2) \div 1.67$, to two decimal places.

36. $-\sqrt{21} + 2.7) \div 1.3$, to four decimal places.

37. $\frac{1}{2.35}$, to two decimal places.

38. $\frac{1}{0.07836}$, to six decimal places.

39. $|2.67 - 9.86|$, to two decimal places.

40. $|3.45 - 1.23| \div (|4.56 - 7.83| + 1.54)$, to four decimal places.

In Exercises 41–46, write each of the exponential forms as a repeated product and evaluate. Support your answer on a calculator.

41. 2^6 **42.** 3^5 **43.** -4^3

44. $-(2)^5$ **45.** $(-3)^4$ **46.** $(\frac{1}{4})^3$

In Exercises 47–52, write the given calculator expressions in standard algebraic form and compute the result. Check your answers by performing the sequences of keystrokes on a calculator.

47. $8 + 9 \div 3 - 1$ **48.** $3 \times 6 + 2 \div 10$

49. $5 + 9 \div 3 \times 4$ **50.** $5 \times 2 + 4 \times 3$

51. $2 + 7 - 3 \times 5$ **52.** $24 \div 4 \div 2$

In Exercises 53–58, write the given calculator expressions in standard algebraic form and compute the result. Check your answers by performing the sequences of keystrokes on a calculator.

53. 7 + 2 ^ 3 **54.** 4 × 3 ^ 2 ÷ 2

55. 1 + 4 × 5 ^ 2 **56.** 3 ^ 2 ÷ 2 ^ 4

57. 2 × 7 − 2 ^ 5 + 1 **58.** 6 + 8 − 6 ^ 2 ÷ 3

In Exercise 59–62, compute the values of the expressions for the given values of a, b, and c to four decimal places using a calculator.

59. $\dfrac{1}{\sqrt{a}} - |b - c|$, $a = 7$, $b = -3$, $c = 5$

60. $\dfrac{b - c}{\sqrt{a + 3.29}}$, $a = 1.25$, $b = -3.47$, $c = 9.81$

61. $\sqrt{|a - 3.2|} + bc$, $a = -5.8$, $b = 2.4$, $c = -8.3$

62. $\sqrt{|abc|}$, $a = -3.1$, $b = 7.8$, $c = 3$

In Exercises 63–66, consider the right triangle ABC where a, b, and c are the lengths of the sides opposite the angles A, B, and C, as shown in Figure R.15. You are given two sides of the triangle. Use the Pythagorean Theorem ($a^2 = b^2 + c^2$) to compute the third side. Use your calculator and express the answer to four decimal places.

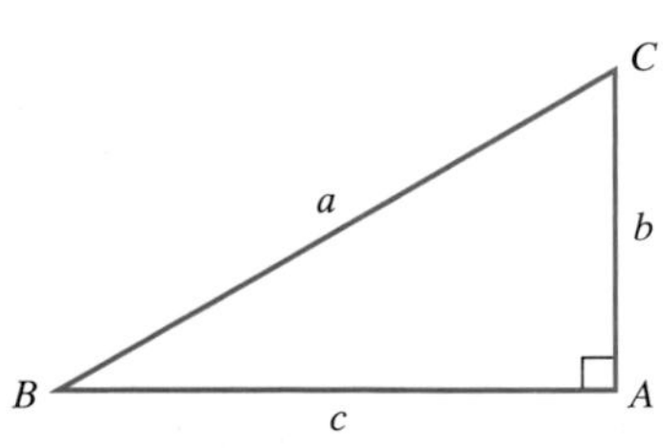

Figure R.15

63. If $b = 3.2$ and $c = 4.8$, determine a.

64. If $a = 9.76$ and $b = 4.35$, determine c.

65. If $b = 4.365$ and $c = 9.871$, determine a.

66. If $a = 7.2345$ and $c = 5.7032$, determine b.

In Exercises 67–70, use a calculator to compute $\frac{\sqrt{a^2 + 2bc}}{a + 4b}$ for the given values of a, b, and c and express the answers to four decimal places. (After you have entered this algebraic expression for the numbers of Exercise 67 and computed the result, determine whether you can duplicate the expression on the next line in a form that can be edited and then used for Exercises 68 to 70. (You can do this with the ENTRY key on some calculators.)

67. $a = 4, b = 1.12, c = 3.6.$

68. $a = 8, b = -2.7, c = 1.5$

69. $a = 6, b = -1.3, c = 8.7$

70. $a = 7, b = 1.51, c = -12$

71. There will be a key for the number π on your calculator. Find the π key and determine how to use it.

 (a) π is an irrational number. Determine the approximation to π given by your calculator. To how many decimal places is the approximation given?

 (b) Give the approximation of π rounded to five decimal places.

 (c) Give the approximation of π to seven decimal places.

 (d) Compute $(\sqrt{2} - 9\pi)^2 \div 3$ to four decimal places.

72. The largest planet in the solar system is Jupiter. Its equatorial diameter is approximately 143,650 kilometers. The smallest planet is Pluto, with a diameter of approximately 6000 kilometers.

 (a) Compute, to the nearest 10 kilometers, the equatorial circumference of each planet.

 (b) Find out how to convert kilometers to miles. What is each circumference in miles, to the nearest 10 miles?

 (c) Use your library or other information resource to determine the equatorial circumference of Earth for comparison.

73. Compute the following expressions (if possible) on a calculator to four decimal places. Explain the outputs. Do not perform any preliminary algebra such as replacing $|-3|$ by 3. The purpose of this exercise is to understand how to use your calculator.

 (a) $\sqrt{3}$ **(b)** $\sqrt{-3}$ **(c)** $\sqrt{|-3|}$

74. Use a calculator to compute $\sqrt{3}$, $\sqrt{\sqrt{3}}$, and $\sqrt{\sqrt{\sqrt{3}}}$. Predict the pattern if one continues indefinitely to take the square root. Check your answer with a calculator.

75. Determine the relationship between consecutive terms of the following sequence. Use a calculator to predict the sixth term.

$$2, \quad 4, \quad 16, \quad 256, \quad \ldots$$

76. Determine the relationship between consecutive terms of the following sequence. Use a calculator to predict the next term.

$$1.2, \quad 1.728, \quad 5.159780352, \quad 137.370552, \quad \ldots$$

77. **(a)** The French mathematician Fermat (1601–1665) believed that the numbers $2^{2^n} + 1$, where n is a natural number, are primes. (A number is prime if it is divisible only by 1 and itself. For example, the numbers 1, 2, 3, 5, 7, and 11 are prime numbers, but 6 is not prime because it is divisible by 2 and by 3.) Compute $2^{2^n} + 1$ for $n = 1, 2, 3, 4$, and 5.

(b) The Swiss mathematician Euler (1707–1783) later verified that $2^{2^n} + 1$ is prime for $n = 1, 2, 3$, and 4, but not for $n = 5$. Verify that $2^{2^5} + 1$ is not prime by showing that it is divisible by 641.

78. Consider the following rational numbers. Use a calculator to find a decimal form of each number. To how many places is the result given? Use the display to determine each number to the indicated number of decimal places and explain your reasoning. (*Hint*: Decimal forms of rational numbers involve a pattern that repeats itself.)

(a) $\frac{1}{41}$, to 18 decimal places.
(b) $\frac{1}{21}$, to 24 decimal places.
(c) $\frac{3}{13}$, to 24 decimal places.
(d) $\frac{57}{63}$, to 34 decimal places.

79. Calculators and computers are limited in the number of digits they can display; however, numbers are often stored to a larger number of digits in memory than are displayed. For example, our calculator displays $\pi = 3.141592654$. It "knows" π to more decimal places than this. Extract the best possible value of π you can from your calculator and explain your method. [*Hint:* Compute and explain the following sequence of numbers.

$$\pi - 3, \quad 10(\pi - 3.1), \quad 100(\pi - 3.14), \quad 1000(\pi - 3.141) \ldots$$

They may reveal more digits of π.]

80. Use the method of exercise 79 to compute the value of $\sqrt{2}$ to as many decimal places as you can using your calculator.

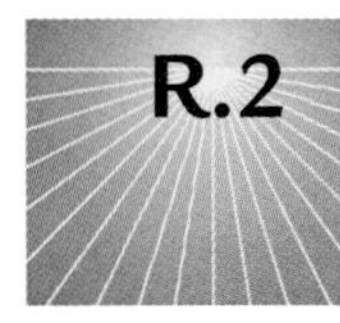

R.2 Rational Exponents, Scientific Notation, and Doses of Medicine

• Radicals • Rational Exponents • Rules for Exponents • Scientific Notation • Distance, Speed and Time • Compound Interest • Doses of Medicine • Economic Ordering Quantity

In the previous section, we looked at the notation for working with integer exponents. If a is a real number and n is a positive integer, then a^n represents the product of the number a, n times.

$$a^n = \underbrace{a \cdot a \cdot a \cdot \ldots \cdot a}_{n \text{ times}}$$

We now embark on two related themes, namely the development of rules for working with exponents and the extension of exponential notation to include rational values for n.

Product Rule

For all positive integers m and n and every real number a,

$$a^m \cdot a^n = a^{m+n}$$

This result is derived as follows:

$$a^m \cdot a^n = \underbrace{(a \cdot a \cdot \ldots \cdot a)}_{m \text{ times}} \cdot \underbrace{(a \cdot a \cdot \ldots \cdot a)}_{n \text{ times}} = \underbrace{(a \cdot a \cdot \ldots \cdot a)}_{m+n \text{ times}} = a^{m+n}$$

For example, $2^3 \cdot 2^4 = 2^7$.

```
2^3*2^4
                128
2^7
                128
```

Is $a^m \cdot a^n \cdot a^s = a^{m+n+s}$? Investigate.

Power Rule

For all positive integers m and n and every real number a:

$$(a^m)^n = a^{mn}$$

This result is obtained as follows:

$$(a^m)^n = \underbrace{(a^m) \ldots (a^m)}_{n \text{ times}} = \underbrace{(a \cdot \ldots \cdot a)}_{mn \text{ times}} = a^{mn}$$

For example, $(5^2)^3 = 5^6$.

```
(5^2)^3
              15625
5^6
              15625
```

Is $[(a^m)^n]^s = a^{mns}$? Investigate.

Zero Exponents

For any nonzero real number a,

$$a^0 = 1$$

For the product rule to apply to zero exponents, we require that

$$a^n \cdot a^0 = a^{n+0} = a^n$$

This will be true if $a^0 = 1$ for $a \neq 0$.

Caution: Note that 0^0 is not defined.

For example, $6^0 = 1$.

What does your calculator give for 0^0?

NEGATIVE EXPONENTS

For any nonzero real number a and any positive integer n,

$$a^{-n} = \frac{1}{a^n}$$

For the product rule to work for negative integer exponents we need

$$a^n \cdot a^{-n} = a^{n-n} = a^0 = 1$$

This would hold if $a^{-n} = \dfrac{1}{a^n}$ for $a \neq 0$.

CAUTION: Note that 0^{-n} is not defined.

Example: $4^{-2} = \dfrac{1}{4^2}$.

```
4^-2
                .0625
1/4^2
                .0625
```

What would you expect $(a^{-1})^{-1}$ to be? Investigate.

QUOTIENT RULE

For all integers m and n and every nonzero number a,

$$\frac{a^m}{a^n} = a^{m-n}$$

This rule follows from the definition of negative exponent and the product rule.

$$\frac{a^m}{a^n} = a^m \cdot a^{-n} = a^{m-n}$$

For example, $\dfrac{2^7}{2^4} = 2^{7-4}$.

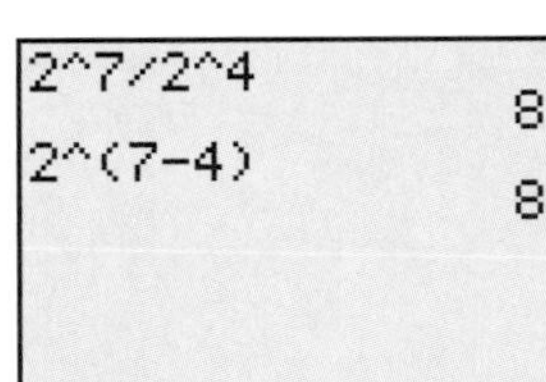

We now illustrate how these rules and definitions are used.

Example 1 Working with Exponents

Evaluate the following expressions by hand using these rules. Support the results using a calculator.

(a) 2^{-3} (b) $4^3 \cdot 4^{-1}$ (c) $\dfrac{1}{5^{-4}}$ (d) $6^0 \cdot 2^5 \cdot \dfrac{1}{2^3}$

SOLUTION We get

(a) $2^{-3} = \frac{1}{2^3} = \frac{1}{8} = 0.125$

(b) $4^3 \cdot 4^{-1} = 4^{3-1} = 4^2 = 16$

(c) $\frac{1}{5^{-4}} = 5^{-(-4)} = 5^4 = 625$

(In general, $\frac{1}{a^{-n}} = a^n$.)

(d) $6^0 \cdot 2^5 \cdot \frac{1}{2^3} = 1 \cdot 2^5 \cdot 2^{-3} = 2^{5-3} = 2^2 = 4$

A calculator supports these answers.

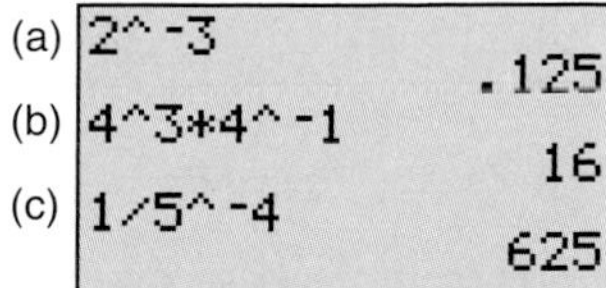

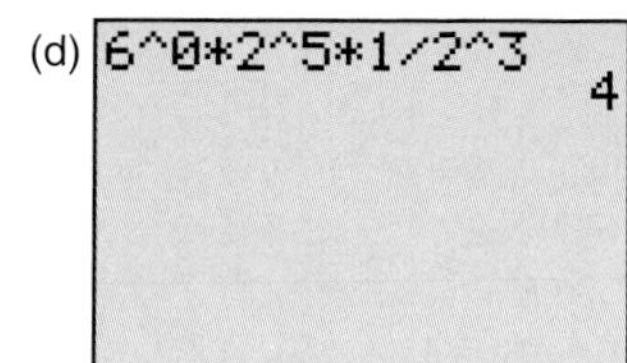

Figure R.16

Note how we have used the hierarchy of arithmetic operations in evaluating these expressions on the calculator.

Self-Check 1 Evaluate each of the following by hand. Support your answers using a calculator. (a) $5^{-2} \cdot 3^4$ and (b) $\frac{5^3}{5^2} \cdot 2^{-4}$.

Radicals

The student is familiar with the concept of a square root of a number. For example, 2 is a square root of 4 because we can write $2^2 = 4$, and 5 is a square root of 25 because $5^2 = 25$. We say that 3 is a **cube root** of 27 because $3^3 = 27$, and we say that 2 is a **fourth root** of 16 because $2^4 = 16$ and so on.

> ***The nth Root*** For real numbers a and b and positive integer n, b is an ***n*th root** of a if $b^n = a$.

A given real number may have a single real root, two real roots, or no real roots. For example,

27 has a single cube root because 3 is the only real number that satisfies $b^3 = 27$;

4 has two real square roots, namely 2 and -2, because both numbers satisfy $b^2 = 4$;

-5 has no square root because there is no real number b that satisfies $b^2 = -5$.

If a number has two nth roots, one is always the negative of the other, as in the example for the two square roots of 4. The positive root is then called the **principal root.** We use the radical sign for the root of a real number (principal root if there are two roots). Thus, we write $\sqrt[3]{27} = 3$ and $\sqrt{4} = 2$.

The Radical Sign

The nth root of a is written $\sqrt[n]{a}$ (the principal root if there are two roots).

Example 2 **Evaluating Roots**

Evaluate each of the following roots when possible.

(a) $\sqrt[3]{125}$ **(b)** $\sqrt{16}$ **(c)** $\sqrt[3]{-8}$ **(d)** $\sqrt{-12}$

SOLUTION

(a) $\sqrt[3]{125} = 5$ because $5^3 = 125$.
(b) $\sqrt{16} = 4$, because $4^2 = 16$ and $4 > 0$. (4 is the principal root of 16.)
(c) $\sqrt[3]{-8} = -2$ because $(-2)^3 = -8$.
(d) $\sqrt{-12}$ is not a real number. There is no real number b such that $b^2 = -12$.

The Radical $\sqrt[n]{a^n}$

The radical $\sqrt[n]{a^n}$ is worthy of special mention. Consider $\sqrt[4]{(-2)^4}$, where $a = -2$ and $n = 4$. This becomes

$$\sqrt[4]{(-2)^4} = \sqrt[4]{16} = 2, \qquad \text{which we can write as } |-2|.$$

An analysis of various possibilities that can arise for radicals of the form $\sqrt[n]{a^n}$ leads to the following result:

The Radical $\sqrt[n]{a^n}$

Let a be a real number and n be a positive integer; Then

$$\sqrt[n]{a^n} = \begin{cases} a \text{ if } n \text{ is odd} \\ |a| \text{ if } n \text{ is even} \end{cases}$$

Self-Check 2 Evaluate each of the following roots using the preceding result.

(a) $\sqrt{(5)^2}$ **(b)** $\sqrt{(-5)^2}$ **(c)** $\sqrt[3]{(7)^3}$ **(d)** $\sqrt[3]{(-7)^3}$

Rational Exponents

We now extend our definition of exponents to include rational numbers. (One reason for using rational exponents is that they allow us to compute radicals on a calculator.) We first define $a^{1/n}$. The power rule then enables us to write $a^{m/n}$ as $(a^{1/n})^m$.

Rational Exponents

Let a be a real number, m be an integer, and n be a positive integer. Then

$$a^{1/n} = \sqrt[n]{a}$$

$$a^{m/n} = (a^{1/n})^m = (\sqrt[n]{a})^m \qquad (\text{if } \sqrt[n]{a} \text{ exists})$$

Note that the power rule also enables us to write also $a^{m/n} = (a^m)^{1/n} = (\sqrt[n]{a^m})$.

Example 3 Working with Rational Exponents

Evaluate each of the following by hand when possible. Support the answers using a calculator.

(a) $25^{1/2}$ (b) $(-8)^{1/3}$ (c) $(-36)^{1/2}$

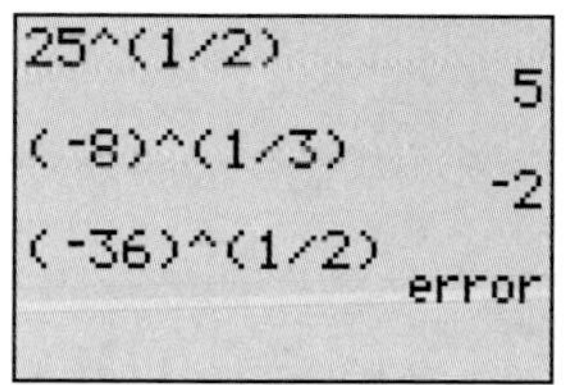

Figure R.17 Working with rational exponents

SOLUTION Using these definitions, we get

(a) $25^{1/2} = \sqrt{25} = 5$

(b) $(-8)^{1/3} = \sqrt[3]{-8} = -2$

(c) $(-36)^{1/2} = \sqrt{-36}$. This is not a real number.

See Figure R.17.

Example 4 Working with Rational Exponents

Evaluate each of the following by hand when possible. Support the answers using a calculator.

(a) $4^{5/2}$ (b) $(-27)^{1/3}$ (c) $(-16)^{5/4}$

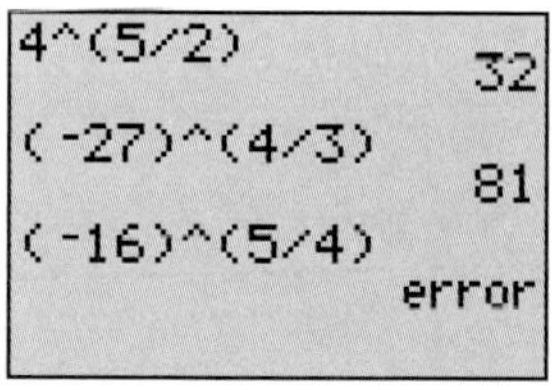

4^(5/2)
32
(-27)^(4/3)
81
(-16)^(5/4)
error

Figure R.18 Working with rational exponents

SOLUTION Using definitions we get

(a) $4^{5/2} = (4^{1/2})^5 = 2^5 = 32$

(b) $(-27)^{4/3} = [(-27)^{1/3}]^4 = (-3)^4 = 81$

(c) $(-16)^{5/4} = [(-16)^{1/4}]^5 = (\sqrt[4]{-16})^5$, which is not a real number. (Figure R.18).

Self-Check 3 Evaluate each of the following by hand. Support the answers using a calculator.

(a) $16^{1/4}$ (d) $49^{-1/2}$

(b) $(-125)^{1/3}$ (e) $81^{5/4}$

(c) $\left(\frac{1}{8}\right)^{2/3}$ (f) $-8^{7/3}$

We now summarize the set of general rules that govern exponents.

Rules for Exponents

Let a and b be real numbers and m and n be rational numbers. Then, provided all expressions are defined,

	Rule	Example
1. Product	$a^m \cdot a^n = a^{m+n}$	$4^{1/2} \cdot 4^{5/2} = 4^{1/2+5/2} = 4^{6/2} = 4^3$
2. Quotient	$\dfrac{a^m}{a^n} = a^{m-n}, a \neq 0$	$\dfrac{3^{7/3}}{3^{2/3}} = 3^{7/3-2/3} = 3^{5/3}$
3. Power	i) $(ab)^m = a^m \cdot b^m$	$(4 \cdot 2)^3 = 4^3 \cdot 2^3$
	ii) $\left(\dfrac{a}{b}\right)^m = \dfrac{a^m}{b^m}, b \neq 0$	$\left(\dfrac{2}{5}\right)^3 = \dfrac{2^3}{5^3}$
	iii) $(a^m)^n = a^{mn}$	$(3^4)^{1/2} = 3^{4 \cdot 1/2} = 3^2$

Example 5 **Using the Rules for Exponents**

Evaluate by hand (a) $7^{1/3} \cdot 7^{5/3}$ (b) $\left(\dfrac{2}{3}\right)^4 \cdot 3^6$.

SOLUTION

(a) $7^{1/3} \cdot 7^{5/3} = 7^{1/3+5/3} = 7^{6/3} = 7^2 = 49$

(b) $\left(\dfrac{2}{3}\right)^4 \cdot 3^6 = \dfrac{2^4}{3^4} \cdot 3^6 = 2^4 \cdot 3^{6-4} = 2^4 \cdot 3^2 = 16 \cdot 9 = 144$

Support these answers using a calculator.

Self-Check 4 Evaluate $\dfrac{81^{1/4} \cdot (-32)^{2/5}}{4^{-3/2}}$ by hand. Support the answer using a calculator.

Let us now look at expressions that involve variables. We remind the student that we often omit the symbol $\cdot$ for multiplication when working with variables and merely write one letter after the other. For example, $a^{-2} \cdot b$ will be written $a^{-2}b$.

Example 6 **Simplifying Expressions That Involve Exponents**

Use the rules for exponents to simplify the following algebraic expressions. The final forms should involve only positive exponents. Show all the steps you have taken.

(a) $(x^2)^{-3}$ **(b)** $\left(\dfrac{3x^4y^{-1}}{x^{-2}y^3}\right)^2$ **(c)** $\dfrac{x(x^{1/2} - 4x^{-5})}{x^{-3}}$

SOLUTION

(a) $(x^2)^{-3} = x^{-6} = \dfrac{1}{x^6}$

(b) Using the rules for exponents, write all the x terms in the numerator so that they may be combined into one term having a positive exponent. Write all the y terms in the denominator because this is the way to get a positive y exponent.

$$\left(\frac{3x^4y^{-1}}{x^{-2}y^3}\right)^2 = \left(\frac{3x^4x^2}{y^3y}\right)^2 = \left(\frac{3x^6}{y^4}\right)^2 = \frac{9x^{12}}{y^8}$$

(c) $$\frac{x(x^{1/2} - 4x^{-5})}{x^{-3}} = xx^3(x^{1/2} - 4x^{-5}) = x^4(x^{1/2} - 4x^{-5})$$

$$= x^{9/2} - 4x^{-1} = x^{9/2} - \frac{4}{x}$$

Self-Check 5 Write $\left(\frac{x^{-2}y^3}{2^{-1}x^{-3}y}\right)^2$ in a form that involves each variable once, to a positive exponent.

Example 7 Removing Terms from Within Radicals

Simplify the following expressions by removing as many factors as possible from under the radical.

(a) $\sqrt[3]{x^3y^4z^7}$ **(b)** $\sqrt[4]{x^6y^5}$

SOLUTION

(a) Because the root is a cube, we factor $x^3y^4z^7$ as a product containing cubes:

$$\sqrt[3]{x^3y^4z^7} = \sqrt[3]{x^3y^3yz^6z} = \sqrt[3]{x^3y^3y(z^2)^3z} = xyz^2\sqrt[3]{yz}$$

(b) Because the root is a fourth, we factor x^6y^5 as a product containing fourth powers:

$$\sqrt[4]{x^6y^5} = \sqrt[4]{x^4x^2y^4y}$$

At this time, recall that $\sqrt[n]{a^n} = |a|$ if n is even. Thus we can write

$$\sqrt[4]{x^6y^5} = |x|\,|y|\sqrt[4]{x^2y} = |xy|\sqrt[4]{x^2y}$$

Scientific Notation

Scientists often work with very large and very small numbers. For example, the distance from the sun to the planet Uranus is 1,785,000,000 miles. The radius of an electron is approximately 0.0000000002 centimeters. These numbers are written in a convenient compact form involving exponents, called scientific notation, as 1.785×10^9 and 2×10^{-10}.

Scientific Notation The **scientific notation** form of a number is

$$\pm a \times 10^n$$

where $1 \le a < 10$ and n is an integer. $\pm a$ is called the **mantissa** of the number, and n is the **characteristic**.

Example 8

(a) **Intergalactic Distances** Show that the distance 1,785,000,000 miles between the sun and Uranus can be written in scientific notation as 1.785×10^9.

(b) **Sizes of Electrons** Show that the approximate radius of an electron, 0.0000000002 centimeters, can be written in scientific notation as 2×10^{-10}.

SOLUTION

(a) View 1,785,000,000 as

$$\underbrace{1,785,000,000}_{9 \text{ digits}}.$$

To write this number in scientific notation, position the decimal point nine places to the left and multiply by 10^9 to compensate for this shift. Thus $1,785,000,000 = 1.785 \times 10^9$.

(b) View 0.0000000002 as

$$0.\underbrace{0000000002}_{10 \text{ digits}}.$$

To write in scientific notation, position the decimal point ten digits to the right and multiply by 10^{-10} to make up for this shift. Thus $0.0000000002 = 2 \times 10^{-10}$.

Self-Check 6 Express 9816.542 in scientific notation.

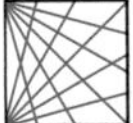

DISPLAYING LARGE AND SMALL NUMBERS ON A CALCULATOR

Calculators use scientific notation to display numbers that are too large or too small to fit in the calculator window.

For example, if we compute

$$3,465,481 \times 23,791,456$$

on our calculator, we get the result shown in Figure R.19. The result is to be interpreted as $8.244883873 \times 10^{13}$.

3465481*23791456
8.244883873E13

Figure R.19 Scientific notation

Self-Check 7 Large and small numbers are displayed in scientific notation in various ways on calculators. Determine the convention used on your calculator. Compute (a) $435,213 \times 546,179$ (b) $0.000006713 \times 0.00004532$.

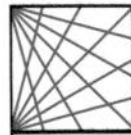

EXPONENT KEY

Calculators have a special key, usually labeled EE or EXP, for entering numbers in scientific notation. Find the exponent key on your calculator and determine how to use it to enter 5.2369×10^5. Use the exponent key on your calculator to verify the following result:

$$\frac{(5.2369 \times 10^5)}{(23.46 \times 1862.78)} = 11.9835, \text{ to four decimal places.}$$

Applications . . .

Example 9 Distance, Speed, and Time

Speed is a rate of change, a quantity that tells us how fast something is changing. We often use it in the context of how fast distance is changing. The following equation relates distance, speed, and time:

$$\text{Distance} = \text{Speed} \times \text{Time}$$

Thus, if one traveled at a speed of 55 miles per hour for 3 hours, the distance traveled would be

$$\text{Distance} = 55 \times 3 = 165 \text{ miles}$$

In applying the equation, we will be given two of the variables—distance, speed, and time—and want to find the third. The equation is rewritten in the appropriate form, expressing the unknown variable in terms of the known variables. For example, let us use the equation to determine the time it takes light to reach us from the sun. The average distance from the sun to the Earth is 9.295×10^7 miles. Light travels at a speed of 186,282 miles per second. Rewriting the equation to express time in terms of distance and speed, we get

$$\text{Time} = \frac{\text{Distance}}{\text{Speed}} = \frac{9.295 \times 10^7}{186282} = 498.9746728 \text{ seconds}$$

Divide this number by 60 to get the time in minutes. The time it takes light to reach us from the sun is 8.32 minutes, to two decimal places.

Example 10 Compound Interest

If a sum of money is deposited in a bank, the bank will pay **interest.** The interest is, in effect, payment made by the bank for the use of this money. Let a sum P be deposited in an account that pays interest at an annual rate r. Let the interest be computed annually and left in the account. If interest is paid on both the principal P and the accumulated interest each time, it is called **compound interest.** We now derive a formula for the amount that will be in the account at the end of t years.

The amount in the account after 1 year is

$$\underset{\text{principal}}{P} + \underset{\text{interest}}{Pr} = P(1 + r).$$

The amount in the account after 2 years is

$$\underset{\text{new principal}}{P(1 + r)} + \underset{\text{new interest}}{P(1 + r)r} = P(1 + r)^2.$$

Following this pattern, the amount in the account after t years is $P(1 + r)^t$.

Compound Interest Formula

A sum P deposited for t years at $r\%$ compounded annually will grow to an amount

$$A = P(1 + r)^t$$

Let's apply this formula. If \$2500 is deposited in a savings account that pays an interest rate of 6% compounded annually, determine the amount in the account after 7 years. We know that $P = 2500$, $r = 6\% = \frac{6}{100} = 0.06$, and $t = 7$. Thus,

$$A = P(1 + r)^t = 2500(1 + 0.06)^7 = 2500(1.06)^7 = 3759.075647$$

The amount in the account after 7 years is \$3759.08, to the nearest cent.

Further Ideas 1 Although it is impossible to make exact predictions concerning many future financial situations, investment and insurance companies use the following type of mathematical model to get estimates. Consider a 22-year-old graduate of the University of Wisconsin who starts her career earning \$30,000 per annum in the year 2000. If she gets constant raises of 4% per annum, what will her final annual salary be if she retires at the age of 65 in the year 2043?

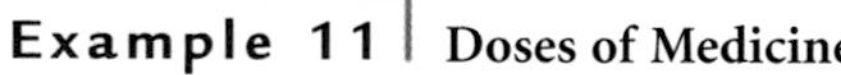

Example 11 **Doses of Medicine**

Children require and tolerate relatively larger medical doses per pound than adults. Body surface area (SA) has been found to be a better correlate of dosing requirements in children than body weight, possibly because it is a better correlate of cardiac output. A child's maintenance dose is calculated using the formula

$$\text{Child's dose} = \left(\frac{SA_{\text{child}}}{SA_{\text{average adult}}}\right) \times \text{Adult dose}$$

Body surface area (SA) is calculated using the following height–weight formula:

$$SA = \left(\frac{w}{2.2}\right)^{0.5378} \times \left(\frac{h}{0.4}\right)^{0.3964} \times 0.2402235$$

where w is the weight of the individual in pounds and h is the height in inches. The resulting units of body surface are square feet. We now apply these formulas.

Let's compute the surface area of an average adult who weighs 148 pounds and is 66 inches tall. We get

$$SA_{\text{average adult}} = \left(\frac{148}{2.2}\right)^{0.5378} \times \left(\frac{66}{0.4}\right)^{0.3964} \times 0.2402235 = 17.48 \text{ square feet}$$

Let us now compute the surface area of a 3-month-old child who weighs 14 pounds and is 24 inches long. We get

$$SA_{\text{child}} = \left(\frac{14}{2.2}\right)^{0.5378} \times \left(\frac{24}{0.4}\right)^{0.3964} \times 0.2402235 = 3.29 \text{ square feet}$$

Now we can calculate the child's maintenance dose:

$$\text{Child's dose} = \left(\frac{SA_{\text{child}}}{SA_{\text{average adult}}}\right) \times \text{Adult dose} = \frac{3.29}{17.48} \times \text{Adult dose}$$
$$= 0.2 \text{ Adult dose}$$

The child's dose of medicine would be 0.2 of that recommended for the average adult. (Note that the ratio of child to adult weight is $14/148 = 0.1$. If weight rather than surface area had been used the child's dose would have been 0.1 of the adult's dose. The child's dose is, per pound, twice as large as the adult's.)

Self-Check 8 Compute your own surface area. (The author's surface area is 17.92 square feet!)

Example 12 Economic Ordering Quantity

Businesses are concerned about storage costs and ordering costs. If too many items are ordered at once, they can remain in storage a long time, and storage costs become high. Conversely, if too few items are ordered, there can be frequent reordering, and ordering costs become high. An optimal shipping size that leads to a balance beween storage costs and odering costs must be found. Let a store sell N units of a product per year. Suppose it costs $\$p$ to store one unit for a year. There are two parts to the ordering cost, a fixed part that is independent of the number of items ordered and a variable part that depends on the number ordered. The optimal shipping size depends only on the fixed cost. Let the fixed cost for ordering one shipment be $\$q$. Companies use the formula $\sqrt{2qN/p}$ units for the optimal shipment size. This is called the **economic ordering quantity.**

For example, suppose an appliance store plans to sell 1560 refrigerators in a year. The storage cost of each refrigerator is \$35 per year, and the fixed cost for ordering a shipment is \$175. Therefore, $N = 1560$, $p = 35$, and $q = 175$. We get

$$\sqrt{\frac{2qN}{p}} = \sqrt{\frac{2(175)(1560)}{35}} = 124.90$$

This number is rounded to 125. The economic ordering quantity is 125 refrigerators. The store should plan to order 125 refrigerators at a time when inventory is getting low.

Answers to Self-Check Exercises

1. (a) $\frac{81}{25} = 3.24$ (b) $\frac{5}{16} = 0.3125$
2. (a) 5 (b) 5 (c) 7 (d) -7
3. (a) 2 (b) -5 (c) $\frac{1}{4}$ (d) $\frac{1}{7}$ (e) 243 (f) -128
4. $81^{1/4} \cdot (-32)^{2/5} \cdot 4^{3/2} = 3 \cdot (-2)^2 \cdot (2)^3 = 3 \cdot 4 \cdot 8 = 96$
5. $(2x^{-2}x^3y^3y^{-1})^2 = (2xy^2)^2 = 4x^2y^4$
6. 9.816542×10^3
7. (a) $2.377042011 \times 10^{11}$ (b) $3.0423316 \times 10^{-10}$

Answer to Further Ideas Exercise

1. $30{,}000(1 + 0.04)^{43} = 162{,}014.858.$ $162,014.86 per annum.

EXERCISES R.2

In Exercises 1–6, evaluate each expression by hand using the rules for exponents. Show all the steps, as in Example 1. Support your answers using a calculator.

1. 2^{-3} **2.** $3^2 \cdot 3^{-4}$ **3.** $3^{-2} \cdot 3^0 \cdot 3^5$

4. $\dfrac{1}{2^{-4}} \cdot 2^3$ **5.** $5^{-3} \cdot 5 \cdot 25^2$ **6.** $\dfrac{2^4}{2^{-2}} \cdot 2^{-3}$

In Exercises 7–16, evaluate each expression by hand (if it is a real number).

7. $\sqrt{25}$ **8.** $\sqrt[3]{-27}$ **9.** $\sqrt[3]{-1000}$ **10.** $\sqrt{4^3}$

11. $(\sqrt[3]{27})^5$ **12.** $\sqrt{3^2}$ **13.** $\sqrt[4]{-16}$ **14.** $\sqrt[5]{-32}$

15. $(\sqrt[5]{(3)})^5$ **16.** $\sqrt[4]{(-3)^4}$

In Exercises 17–22, compute each root by hand (if it is a real number). Support your answers using a calculator.

17. $16^{1/2}$ **18.** $8^{1/3}$ **19.** $(-27)^{1/3}$

20. $(-9)^{1/2}$ **21.** $81^{1/4}$ **22.** $(-32)^{1/5}$

In Exercises 23–30, evaluate each expression (if it is a real number) using a calculator. Support your answers by hand showing all the steps.

23. $16^{3/2}$ **24.** $8^{4/3}$ **25.** $27^{-2/3}$ **26.** $(-81)^{3/4}$

27. $-9^{5/2}$ **28.** $(\frac{1}{25})^{3/2}$ **29.** $(-16)^{-5/4}$ **30.** $(-32)^{4/5}$

In Exercises 31–36, evaluate each expression by hand using the rules for exponents. Show all the steps and support your answers using a calculator.

31. $3^2 \cdot 3^{-3}$ **32.** $(\frac{3}{2})^2 \cdot 2^3$ **33.** $4^3 \cdot (\frac{3}{4})^2$

34. $(\frac{5}{4})^7 \cdot (\frac{5}{4})^{-6}$ **35.** $(7 \cdot 2)^3 \cdot (\frac{2}{7})^4$ **36.** $(2 \cdot 4)^3 \cdot 4^{-5}$

In Exercises 37–42, evaluate each expression by hand using the rules for exponents. Show all the steps and support your answers using a calculator.

37. $3^{1/2} \cdot 3^{3/2}$ **38.** $7^{5/2} \cdot 7^{-3/2}$ **39.** $6^{1/4} \cdot 6^{1/2} \cdot 6^{-3/4}$

40. $\dfrac{2^{7/2}}{2^{3/2}}$ **41.** $\dfrac{27^{4/3}}{27^{2/3}}$ **42.** $(-27)^{2/3} \cdot 3^{-1}$

In Exercises 43–48, use the rules for exponents to write each of the algebraic expressions in a form that involves each variable once to a positive exponent.

43. x^3x^{-1} **44.** $x^3x^2x^{-2}$ **45.** $x^2y^3x^{-7}$

46. $\dfrac{x^5x^{-1}}{x^2}$ **47.** $\dfrac{x^4x^{-2}}{x^{-3}}$ **48.** $\dfrac{4^{-2}x^3y^2}{2^{-3}x^{-1}y^{-3}}$

In Exercises 49–54, simplify the algebraic expressions. Express each result in a form that involves each variable once to a positive exponent.

49. $x^{1/3}x^{5/6}$ **50.** $x^{1/4}x^{-1/2}$ **51.** $x^2x^{-5/6}$

52. $\dfrac{x^{3/2}}{x^{1/4}}$ **53.** $\dfrac{(x^2y^{1/3})^2}{y^{2/3}}$ **54.** $(x^4x^{-2/3}y^{5/3})^3$

In Exercises 55–60, simplify the expressions by removing as many factors as possible from under the radical.

55. $\sqrt[3]{8x^5}$ **56.** $\sqrt{4x^3y^5}$ **57.** $\sqrt{x^4y^7}$

58. $\sqrt[3]{x^4y^6}$ **59.** $\sqrt[4]{2x^3y^5z^{13}}$ **60.** $\sqrt[3]{16x^4y^6z^7}$

In Exercises 61–64, assume that all the terms exist.

61. Show that $\dfrac{1}{x^{-n}} = x^n$.

62. Show that $\dfrac{1}{x^n} = \left(\dfrac{1}{x}\right)^n$.

63. Show that $\dfrac{1}{\left(\dfrac{1}{x}\right)} = x$.

64. Show that $\left(\dfrac{x}{y}\right)^{-n} = \left(\dfrac{y}{x}\right)^n$.

In Exercises 65–72, use your calculator to perform the computations, rounding the results to four decimal places.

65. 4.1^5 **66.** $3.742^{2.8}$ **67.** $5.18^{4.7}$

68. $0.0781^{-3.2}$ **69.** $4.932^{2/5}$ **70.** $5.932^{7/3}$

71. $3.5426^{-5/7}$ **72.** $0.09876^{-5/9}$

In Exercises 73–78, use your calculator to perform the computations, rounding the results to four decimal places.

73. $\sqrt{5.62}$ **74.** $\sqrt[3]{12.7891}$ **75.** $\sqrt[4]{16.4532}$

76. $\sqrt[4]{12.46^3}$ **77.** $\sqrt[5]{(-19.81^3)}$ **78.** $\sqrt[5]{\sqrt[3]{(-13.79)}}$

In Exercises 79–84, express each of the numbers in scientific notation.

79. 25,000,000 **80.** 236 **81.** 0.00043

82. 5,932,600 **83.** 63,400,000,000 **84.** 0.000000026

In Exercises 85–90, express each of the numbers in decimal notation.

85. 2×10^4 **86.** 5.36×10^5 **87.** -4.7×10^3

88. 6.324×10^2 **89.** 9.3471×10^{-4} **90.** 3×10^{-6}

In Exercises 91–94, evaluate each of the expressions, giving your answers in scientific notation.

91. $(4 \times 10^2) \times (3 \times 10^3)$

92. $(3 \times 45 \times 10^4) \times (9.81 \times 10^5)$

93. $(5 \times 36 \times 10^4) \times (2.81 \times 10^{-5})$

94. $(-6.34 \times 10^5) \times (9.37 \times 10^{-2})$

95. Intergalactic Distances The distances between stars and galaxies are vast. Light years are used to describe such distances. A light year is the distance light travels in one year.

(**a**) The speed of light is 186,262 miles per second. How many miles are in a light year?

(**b**) The star cluster Pleiades in the constellation Taurus is 410 light years from Earth. How many miles from earth is Pleiades?

96. Intergalactic Travel The bright star Capella is 45 light years away from Earth. How long would it take a space ship traveling at 10^{-1} times the speed of light to get to the star?

97. Protons What is the total mass of 5.26×10^{21} protons, each of which has a mass of 1.673×10^{-24} grams?

98. Compound Interest A sum of \$3750 is deposited in a savings account that pays an interest rate of 6.5% compounded annually. Determine the amount in the account (**a**) after 5 years and (**b**) for each of the first 5 years.

99. Compound Interest If a sum P is deposited in an account that pays interest at an annual rate r, compounded n times per year, the amount in the account after t years is

$$A = P\left(1 + \frac{r}{n}\right)^{nt}.$$

(**a**) Derive this result.

(**b**) A sum of \$8750 is deposited in a savings account that pays an interest rate of 8% compounded quarterly. What is the amount in the account after 5 years?

(**c**) What is the amount if compounding takes place monthly?

100. Compound Interest A bank has high-interest accounts of 7% compounded annually or 6.75% compounded monthly for 10 years. You want to invest \$20,000 in one of these accounts. Which would you take?

101. Volume of a Cylinder The volume of a cylinder is $\pi r^2 h$, where r is the radius of the base and h is the height. Your calculator will have a π key for computing with π. Determine the volume of a cylinder with a radius of 5 inches and a height of 18 inches. Express the result to two decimal places.

102. Dosage of Medicine Compute the surface area of a 6-year-old child (Turner Swann of DeLand,

Florida, in 1999) with a weight of 42 pounds and height of 47 inches using the formula of example 8. Determine the medicine dosage relative to the adult dosage that the child should receive. Express the result to two decimal places.

103. Economic Ordering A store plans to sell 2500 television sets a year. The storage cost of each television is $15 per year, and the fixed cost for ordering a shipment is $225. Determine the number of television sets that should be ordered at any one time (the economic ordering quantity) to optimize the cost of storage and ordering.

104. Economic Ordering A store plans to sell 600 Astra II computers a year. The storage cost of each computer is $32 per year and the fixed cost for ordering a shipment is $150. Determine the number of computers that should be ordered at any one time (the economic ordering quantity) to optimize the cost of storage and ordering.

SECTION PROJECT

Motion of a Pendulum

The period of a pendulum is the time it takes to perform one complete oscillation. Consider a pendulum of length L (see Figure R.20). Physicists have shown that the period T in seconds, when the length is in feet, is given by

$$T = 2\pi\sqrt{\frac{L}{32}}$$

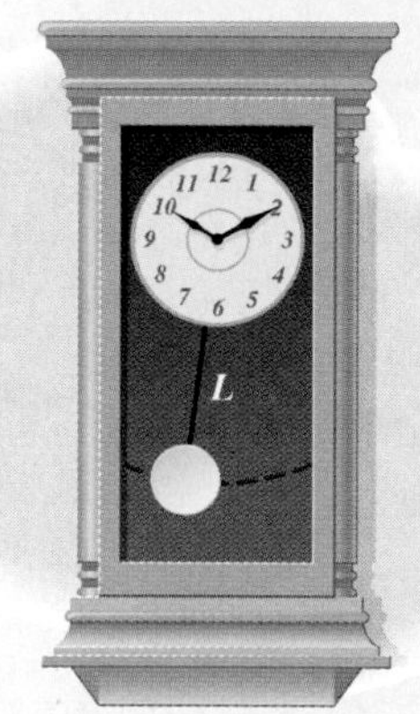

Figure R.20

Using string and a suitable object, such as a pen, construct a pendulum whose length can be varied. Time the periods for various lengths. Draw up a table of lengths and corresponding periods. Confirm that the data fits the equation.

R.3 Polynomials

• Polynomials • Addition and Subtraction of Polynomials • Multiplication of Polynomials • The FOIL Method • Special Products of Polynomials • Long Division

Polynomials are the most basic form of algebraic expression. We introduce them in this section and use them throughout the course. The following algebraic expressions are all examples of polynomials:

$$4x^2 - 7x - 1, \qquad 6x^3 + 2x^2 + x + 3, \qquad 7x^8 - 3x^4 + 2$$

Observe that all the exponents are integers and that none of these exponents is negative. The following algebraic expressions are not polynomials:

$$\sqrt{x^2 + 1}, \qquad x^2 + x + x^{-1}, \qquad \frac{x}{x^2 + x + 3}$$

Polynomial A **polynomial of degree *n* in the variable *x*** is an algebraic expression of the form

$$a_n x^n + a_{n-1}x^{n-1} + \cdots + a_1 x + a_0$$

where $a_n, \ldots, a_0$ are real numbers, with $a_n \neq 0$, and where n is a nonnegative integer.

The **degree** (or **order**) of a polynomial is the highest power of the variable in the polynomial. The polynomial $a_n x^n + a_{n-1}x^{n-1} + \cdots + a_1 x + a_0$ with $a_n \neq 0$ is of degree n. The numbers $a_n, \ldots, a_1$ are the **coefficients** of the polynomial. $a_n x^n$, $a_{n-1}x^{n-1}, \ldots, a_0$ are **terms** of the polynomial. a_0 is the **constant term.**

In addition to all these types of polynomials, there is a special polynomial that has all zero coefficients. This is the **zero polynomial** and is denoted by 0. There is no degree assigned to this polynomial.

Example 1 Describing a Polynomial

The algebraic expression $2x^5 - 7x^3 + 2x + 3$ is a polynomial of degree 5. The coefficient of x^5 is 2, the coefficient of x^3 is -7, and the coefficient of x is 2. The first term is $2x^5$, the second term is $-7x^3$, the third term is $2x$, and the constant term is 3.

The expression $7x^4 + 2x^{-3} - 7x + 6$ is not a polynomial because the exponent -3 is negative. $2x^3 + 4x^{1/3} + 5$ is not a polynomial because of the exponent $\frac{1}{3}$.

We now introduce rules for adding, subtracting, and multiplying polynomials. These rules are based on the properties of real numbers. Because the coefficients of a polynomial are real numbers and the variable represents a real number, the properties of real numbers are valid for polynomials. To define the rules, we use the concept of **like terms.** Two terms are said to be like terms if they contain the same variable with the same exponent. For example, consider the two polynomials

$$P = 4x^3 + 6x^2 + 3x + 5 \qquad \text{and} \qquad Q = 7x^3 - 8x^2 + 4x + 1;$$

the terms $6x^2$ of P and $-8x^2$ of Q are like terms because both contain the same variable x with the same exponent 2. Conversely, $6x^2$ of P and $4x$ of Q are not like terms.

Addition and Subtraction of Polynomials

We add or subtract polynomials by adding or subtracting like terms.

Example 2

(a) Add the polynomials $3x^2 + 4x + 8$ and $5x^2 + 2x - 5$.
(b) Subtract $2x^3 + 5x^2 - 3x + 2$ from $7x^3 + 9x^2 + 6x + 4$.

SOLUTION Collect like terms, using the associative and commutative properties, and then add.

(a) $$(3x^2 + 4x + 8) + (5x^2 + 2x - 5) = (3x^2 + 5x^2) + (4x + 2x) + (8 - 5)$$
$$= 8x^2 + 6x + 3$$

(b) $$(7x^3 + 9x^2 + 6x + 4) - (2x^3 + 5x^2 - 3x + 2)$$
$$= 7x^3 + 9x^2 + 6x + 4 - 2x^3 - 5x^2 + 3x - 2$$
$$= (7x^3 - 2x^3) + (9x^2 - 5x^2) + (6x + 3x) + (4 - 2)$$
$$= 5x^3 + 4x^2 + 9x + 2$$

Self-Check 1 Add the polynomials $7x^3 + 3x - 9$ and $5x^3 - 6x + 1$.

Multiplication of Polynomials

To multiply two polynomials, we multiply each term of the first polynomial by every term of the second polynomial.

Example 3 Find the product of the polynomials $2x + 3$ and $3x^2 - x + 4$.

SOLUTION First, multiply every term of the first polynomial by the second polynomial using the distributive property:

$$(2x + 3)(3x^2 - x + 4) = 2x(3x^2 - x + 4) + 3(3x^2 - x + 4)$$

Use the distributive property again, for each term.

$$= 2x(3x^2) + 2x(-x) + 2x(4) + 3(3x^2) + 3(-x) + 3(4)$$
$$= 6x^3 - 2x^2 + 8x + 9x^2 - 3x + 12 \quad \textbf{Simplify each term.}$$
$$= 6x^3 - 2x^2 + 9x^2 + 8x - 3x + 12 \quad \textbf{Collect like terms.}$$
$$= 6x^3 + 7x^2 + 5x + 12 \quad \textbf{Add like terms.}$$

Self-Check 2 Multiply $2x^2 + 3x + 1$ and $4x - 5$.

The FOIL Method

The product of two polynomials, each consisting of two terms, can be quickly computed using the so-called FOIL method.

Example 4 **The FOIL Method**

Compute $(2x + 3)(4x - 1)$.

SOLUTION The **F**irst terms ($2x$ and $4x$) are multiplied; then the **O**uter terms ($2x$ and -1), **I**nner terms (3 and $4x$) and **L**ast terms (3 and -1), are multiplied.

Finally, the results are added as follows:

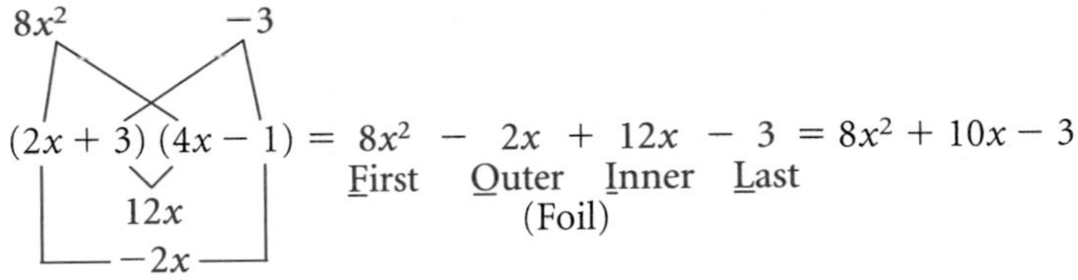

$$\underbrace{(2x+3)(4x-1)}_{\substack{8x^2,\ -3,\ 12x,\ -2x}} = \underset{\text{First}}{8x^2} - \underset{\text{Outer}}{2x} + \underset{\text{Inner}}{12x} - \underset{\text{Last}}{3} = 8x^2 + 10x - 3$$

(Foil)

We get $(2x + 3)(4x - 1) = 8x^2 + 10x - 3$.

Self-Check 3 Compute $(3x - 2)(x + 4)$.

Special Products of Polynomials

There are five special products that the reader should memorize. These results are derived using the multiplication techniques.

Special Products

Let a and b be variables or algebraic expressions.

1. $(a + b)(a - b) = a^2 - b^2$
2. $(a + b)^2 = a^2 + 2ab + b^2$
3. $(a - b)^2 = a^2 - 2ab + b^2$
4. $(a + b)^3 = a^3 + 3a^2b + 3ab^2 + b^3$
5. $(a - b)^3 = a^3 - 3a^2b + 3ab^2 - b^3$

Next we illustrate the proof of Special Product 2, leaving the remaining proofs for the student to develop in Exercises 37–40.

Proof of 2

$$\begin{aligned}(a + b)^2 &= (a + b)(a + b)\\ &= a^2 + ab + ba + b^2\\ &= a^2 + 2ab + b^2 \qquad (\text{since } ba = ab)\end{aligned}$$

The following examples illustrate the use of these products.

Example 5 **Computing a Square**

Compute $(x + 3)^2$.

SOLUTION Compare $(x + 3)^2$ with $(a + b)^2$.
Let $a = x$ and $b = 3$ in the result

$$(a + b)^2 = a^2 + 2ab + b^2$$

We get

$$(x + 3)^2 = x^2 + 2(x)(3) + (3)^2$$
$$= x^2 + 6x + 9$$

Example 6 **Computing a Cube**

Compute $(x + 2y)^3$.

SOLUTION Compare $(x + 2y)^3$ to $(a + b)^3$.
Let $a = x$ and $b = 2y$. Use these values for a and b in

$$(a + b)^3 = a^3 + 3a^2b + 3ab^2 + b^3$$

We get

$$\begin{aligned}(x + 2y)^3 &= x^3 + 3x^2(2y) + 3x(2y)^2 + (2y)^3\\ &= x^3 + 3x^2(2y) + 3x(4y^2) + 8y^3\\ &= x^3 + 6x^2y + 12xy^2 + 8y^3\end{aligned}$$

Self-Check 4 Compute $(2x - 3y)^2$ using Special Product 3.

Long Division of Polynomials

We have studied addition, subtraction, and multiplication of polynomials. We now complete this section with a discussion of division. Division of polynomials is patterned after long division of real numbers.

Example 7 Divide the polynomial $4x^3 + 8x^2 - 5x + 3$ by $2x + 1$.

SOLUTION It is important that both polynomials be written with exponents in descending order, as shown here, before starting to divide.

$2x^2 + 3x - 4$	**Divide $4x^3/2x$ to get $2x^2$.**
$2x + 1\overline{)4x^3 + 8x^2 - 5x + 3}$	
$\underline{4x^3 + 2x^2}$	**Multiply $2x^2(2x + 1)$.**
$6x^2 - 5x$	**Subtract and bring down $-5x$. Divide $6x^2/2x$ to get $3x$.**
$\underline{6x^2 + 3x}$	**Multiply $3x(2x + 1)$.**
$-8x + 3$	**Subtract and bring down 3. Divide $-8x/2x$ to get -4.**
$\underline{-8x - 4}$	**Multiply $-4(2x + 1)$.**
7	**Subtract.** **The degree of 7 is less than the degree of $2x + 1$** **This marks the end of the division.**

The **quotient** is $2x^2 + 3x - 4$ and the **remainder** is 7. The result is written

$$\frac{4x^3 + 8x^2 - 5x + 3}{2x + 1} = 2x^2 + 3x - 4 + \frac{7}{2x + 1}$$

The following example illustrates the procedure to use if certain powers of x are "missing" in the polynomial to be divided.

Example 8 **Division of Polynomials When a Term Is Missing**

Divide $2x^4 - 8x^3 - 3x + 4$ by $x^2 - 3x$.

SOLUTION Observe that there is no term in x^2 in the dividend $2x^4 - 8x^3 - 3x + 4$.

$$
\begin{array}{r|l}
 & 2x^2 - 2x - 6 \\
x^2 - 3x & 2x^4 - 8x^3 + 0x^2 - 3x + 4 \quad \textbf{Caution: No } x^2 \textbf{ term. Put in } 0x^2. \\
 & 2x^4 - 6x^3 \\
 & \quad - 2x^3 + 0x^2 \\
 & \quad - 2x^3 + 6x^2 \\
 & \qquad - 6x^2 - 3x \\
 & \qquad - 6x^2 + 18x \\
 & \qquad\qquad -21x + 4
\end{array}
$$

The degree of $-21x + 4$ is less than the degree of $x^2 - 3x$. This marks the end of the division.

Thus

$$\frac{2x^4 - 8x^3 - 3x + 4}{x^2 - 3x} = 2x^2 - 2x - 6 + \frac{-21x + 4}{x^2 - 3x}$$

Self-Check 5 Divide $3x^3 - 7x^2 + 4x + 1$ by $x - 2$.

Answers to Self-Check Exercises

1. $12x^3 - 3x - 8$
2. $8x^3 + 2x^2 - 11x - 5$
3. $3x^2 + 10x - 8$
4. $4x^2 - 12xy + 9y^2$
5. $3x^2 - x + 2 + \dfrac{5}{x - 2}$

EXERCISES R.3

In Exercises 1–6, give the degree of the polynomial, the first term, and the constant term if any.

1. $2x^2 - 3x + 1$
2. $-4x^3 + 2$
3. $7x^4 - 2x + 5$
4. $6x^3 - 2x^2 + 3x + 2$
5. $4x^8 + 9x^4$
6. $-9x + 25$

In Exercises 7–12, determine whether the algebraic expressions are polynomials.

7. $-4x^3 + 2x$
8. $5x^{-2} + 3x + 4$
9. $\dfrac{x + 1}{3x + 2}$
10. $-8x^5 + 3x - 2$
11. $7x^4$
12. $6x^2 + 3x^{-5} + 2$

In Exercises 13–20, perform the given operations of addition and subtraction of polynomials.

13. $(2x^2 + 3x + 1) + (5x^2 + x + 3)$
14. $(3x^2 - 2x + 3) + (x^2 - x + 4)$
15. $(3x^2 + 5x + 7) - (2x^2 - 2x + 4)$
16. $(4x^2 - 2x - 3) - (2x^2 + 6x - 4)$
17. $(3y^3 + 2) + (5y^2 + y + 3)$
18. $(z^4 - 2z) - (z^3 - 3z^2)$
19. $(t^2 + t + 2) + (3t^2 - 4t + 1) + (5t + 3)$
20. $(7x^2 - 2x + 1) + (4x + 3) - (x^2 - 3x + 2)$

In Exercises 21–28, multiply the polynomials.

21. $(4x + 3)(x + 2)$
22. $5x(2x - 3)$

23. $(7x + 2)(x + 3)$ **24.** $(5x - 2)6x$

25. $(x^2 + 2)(x^2 - 4)$ **26.** $(x - 2)(x^3 + 3x)$

27. $2x(x^2 - 3x + 4)$ **28.** $5x^2(x^2 - 2x + 6)$

In Exercises 29–36, multiply the polynomials.

29. $(4x^2 + 2)(x + 3)$ **30.** $(5x^3 - 2x)(3x + 4)$

31. $(x + 3)(x^2 - x + 2)$ **32.** $(x - 2)(x^2 - 2x + 3)$

33. $(2x^2 - 3x + 2)(2x^2 - x + 4)$

34. $(6x^2 + 3x - 1)(2x^2 + 6x - 1)$

35. $2y(y + 3) + 4(y + 2)$

36. $-3z(z - 1) + 2(5z - 3)$

In Exercises 37–40, derive each of the special products.

37. $(a + b)(a - b) = a^2 - b^2$

38. $(a - b)^2 = a^2 - 2ab + b^2$

39. $(a + b)^3 = a^3 + 3a^2b + 3ab^2 + b^3$

40. $(a - b)^3 = a^3 - 3a^2b + 3ab^2 - b^3$

In Exercises 41–50, use the special products to compute the algebraic expressions.

41. $(4x + 1)^2$ **42.** $(3x - 2)^2$ **43.** $(x + 4)^3$

44. $(2x - 3y)^3$ **45.** $(4x + 5y)^2$ **46.** $(6x + 2y)^2$

47. $(x - 1)^3$ **48.** $(4x^2 + 5y)^2$ **49.** $(x - y)(x + y)^3$

50. $(x + y + 2z)^2$

In Exercises 51–58, divide the first polynomial by the second using long division.

51. $x^2 + 5x + 8, \quad x + 3$

52. $4x^2 - 6x + 5, \quad 2x - 1$

53. $x^3 - 5x^2 + 4x - 7, \quad x - 3$

54. $x^3 - x + 12, \quad x - 4$

55. $6x^4 + x^2 + 1, \quad 2x^2 + 1$

56. $3x^4 + 2x^3 - 2x^2 + 1, \quad x + 1$

57. $2x^4 + 3x + 1, \quad x - 1$

58. $x^4 - 2x^3 - 5x + 3, \quad x^2 + x + 2$

In Exercises 59–64, discuss whether the statements made are true or false. Explain your answer.

59. The sum of two polynomials, each of degree 3, is a polynomial of degree 3.

60. The sum of two polynomials, each of degree 2, is a polynomial of degree less than or equal to 2.

61. The product of two polynomials, each of degree 1, is a polynomial of degree 2.

62. The product of two polynomials, each of degree 2, is a polynomial of degree less than or equal to 4.

63. Let P be an arbitrary polynomial and Q be a polynomial having constant term zero; then

(a) the polynomial $P + Q$ has constant term zero.

(b) the polynomial PQ has constant term zero.

64. Let P and Q be polynomials of unequal degrees; then

(a) $P + Q$ is of degree equal to the larger of the degrees of P and Q.

(b) the degree of PQ is larger than the degrees of either P or Q.

SECTION PROJECT

Generating Pythagorean Numbers

Each set of three natural numbers that satisfies $a^2 + b^2 = c^2$ is said to constitute a set of Pythagorean numbers. For example, $\{3, 4, 5\}$ is a set of Pythagorean numbers because $3^2 + 4^2 = 5^2$.

(a) If p and q are two natural numbers with $p > q$, one odd and one even, and are prime to one another (have no common factor other than one), then the formulas $a = p^2 - q^2, b = 2pq, c = p^2 + q^2$ give a set, $\{a, b, c\}$, of Pythagorean numbers. Prove this result.

(b) These formulas give only the so-called primitive Pythagorean numbers, where a, b, and c have no common factors. Use these formulas to generate three sets of primitive Pythagorean numbers. Verify $a^2 + b^2 = c^2$ for each set using your calculator.

(c) Let $\{a, b, c\}$ be a set of Pythagorean numbers. Using algebra, show that if n is any natural number, then $\{na, nb, nc\}$ is also a set of Pythagorean numbers. Generate three sets of Pythagorean numbers in this way from a known set of Pythagorean numbers. Verify your results using a calculator.

Factoring

- Greatest Common Factor • Factoring Polynomials of Degree Two
- Factoring by Grouping • Special Factors • Factoring Strategy
- Completing the Square

In the previous section, we discussed the multiplication of polynomials. In this section, we introduce techniques for proceeding in the reverse direction. We show how certain polynomials can be written as products of other polynomials. This process is called **factoring.** For example, it can be shown that

$$2x^3 + x^2 - 3x = x(x - 1)(2x + 3)$$

We call x, $(x - 1)$, and $(2x + 3)$ **factors** of $2x^3 + x^2 - 3x$. The simplest type of factor is called the greatest common factor.

Greatest Common Factor

Consider the polynomial $8x^3 + 2x^2 + 4x$. This polynomial consists of three terms, namely $8x^3$, $2x^2$, and $4x$. Observe that 2 is a factor of each term. x is also a factor of each term. Thus, $2x$ is a factor of the polynomial. It is the largest factor that is common to each term. This factor is called the **greatest common factor.** We can factor out $2x$ using the distributive law to get

$$8x^3 + 2x^2 + 4x = 2x(4x^2 + x + 2)$$

Example 1 Factoring Out the Greatest Common Factor

Factor out the greatest common factor of

$$2x^3y^2 + 6x^2y^2 + 8x^2y^4$$

SOLUTION 2, x^2, and y^2 are all factors of each term. $2x^2y^2$ is the greatest common factor. Factor out this expression, namely $2x^2y^2$.

$$2x^3y^2 + 6x^2y^2 + 8x^2y^4 = 2x^2y^2(x + 3 + 4y^2)$$

Example 2 Factoring Out the Greatest Common Factor

Factor out the greatest common factor of $6x^3(x + 2)^2 + 9x^2(x + 2)^4$.

SOLUTION We see that 3, x^2, and $(x + 2)^2$ are all factors of each of the terms. $3x^2(x + 2)^2$ is the greatest common factor. When we factor out this expression, we get

$$\begin{aligned} 6x^3(x + 2)^2 + 9x^2(x + 2)^4 &= 3x^2(x + 2)^2[2x + 3(x + 2)^2] \\ &= 3x^2(x + 2)^2[2x + 3(x^2 + 4x + 4)] \\ &= 3x^2(x + 2)^2(3x^2 + 14x + 12) \end{aligned}$$

Self-Check 1 Factor out the greatest common factor of $5x^4 - 15x^3 + 25x^2$.

Factoring Polynomials of Degree Two

We now discuss a method for factoring certain polynomials of degree two. Consider the polynomial:

$$3x^2 + 10x + 8$$

Factor the coefficient of x^2 and the constant term in such a way that the sum of the inner and outer products is equal to the coefficient of x as follows:

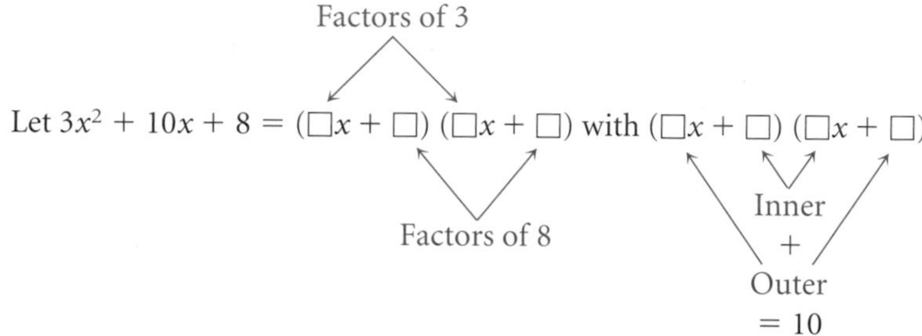

$$= (3x + 4)(x + 2)$$

The following example illustrates that, although factors exist, they may not be readily recognizable.

Example 3 Factoring a Polynomial of Degree Two

Factor $2x^2 + 11x + 12$.

SOLUTION Try $2x^2 + 11x + 12 = (2x + \square)(x + \square)$ for various factors of 12 as follows.

(a) $(2x + 6)(x + 2) = 2x^2 + 10x + 12$. No, does not work.
(b) $(2x + 2)(x + 6) = 2x^2 + 14x + 12$. No, does not work.
(c) $(2x + 4)(x + 3) = 2x^2 + 10x + 12$. No, does not work.
(d) $(2x + 3)(x + 4) = 2x^2 + 11x + 12$. Yes, does work.

Thus, $2x^2 + 11x + 12 = (2x + 3)(x + 4)$

Self-Check 2 Factor $2x^2 + 11x - 6$.

The following example illustrates that this method of factoring can be extended to polynomials in two variables.

Example 4 Factoring a Polynomial in Two Variables

Factor $2x^2 - 5xy - 3y^2$.

SOLUTION We get

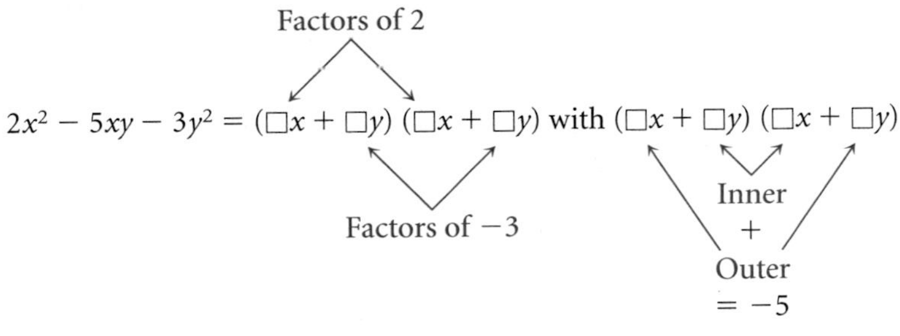

$$= (2x + y)(x - 3y)$$

Self-Check 3 Factor $6x^2 + 11xy - 10y^2$.

FACTORING BY GROUPING

Factors of some algebraic expressions may sometimes be found by grouping certain terms together. Let us factor $x^2 + xy + 4x + 4y$ using the following steps. Group together the first two terms and the second two terms such that

$$x^2 + xy + 4x + 4y = (x^2 + xy) + (4x + 4y)$$

Factor each group,

$$= x(x + y) + 4(x + y)$$

Factor out the largest common factor of each group, namely $(x + y)$,

$$= (x + y)(x + 4)$$

Thus,

$$x^2 + xy + 4x + 4y = (x + y)(x + 4)$$

Self-Check 4 Factor $xy + 2y - 6x - 12$ by grouping.

It is sometimes necessary to rearrange the terms within an expression before grouping. The following example illustrates such an expression.

Example 5 **Factoring by Grouping**

Factor $10x - 5y + xy - 50$.

SOLUTION If we group the first two terms together and the last two terms together we get

$$\begin{aligned} 10x - 5y + xy - 50 &= (10x - 5y) + (xy - 50) \\ &= 5(2x - y) + (xy - 50) \end{aligned}$$

This approach does not lead to factors. However, if we rearrange the terms of the original expression so that the two terms involving x are together, we can factor, as follows.

$$\begin{aligned} 10x - 5y + xy - 50 &= 10x + xy - 5y - 50 \\ &= x(10 + y) - 5(y + 10) \\ &= x(y + 10) - 5(y + 10) \\ &= (y + 10)(x - 5) \end{aligned}$$

Thus,

$$10x - 5y + xy - 50 = (y + 10)(x - 5)$$

SPECIAL FACTORS

Some special factors arise frequently and are listed here. The student should memorize these factors, be able to recognize situations where they arise, and be able to apply them. Some of these factors are special products from Section R.3, with the left- and right-hand sides interchanged. All these factors can be proved by multiplying out the right-hand sides.

SPECIAL FACTORS

Let a and b be variables or algebraic expressions.

1. Difference of two squares $\quad a^2 - b^2 = (a + b)(a - b)$
2. Perfect square $\quad a^2 + 2ab + b^2 = (a + b)^2$
3. Perfect square $\quad a^2 - 2ab + b^2 = (a - b)^2$
4. Sum of two cubes $\quad a^3 + b^3 = (a + b)(a^2 - ab + b^2)$
5. Difference of two cubes $\quad a^3 - b^3 = (a - b)(a^2 + ab + b^2)$

Let us look at some examples of the use of these special factors. Note that each of the expressions in Examples 6–8 can be rewritten in one of the special factor forms.

Example 6 **Difference of Two Squares**

Factor $4x^2 - 9y^2$.

SOLUTION Observe that each term in this expression is a square. We can write

$$\begin{aligned} 4x^2 - 9y^2 &= (2x)^2 - (3y)^2, \qquad \text{a difference of two squares} \\ &= (2x + 3y)(2x - 3y) \end{aligned}$$

Example 7 **Perfect Square**

Factor $x^2 + 4xy + 4y^2$.

SOLUTION Let us write

$$\begin{aligned} x^2 + 4xy + 4y^2 &= x^2 + 2x(2y) + (2y)^2, \qquad \text{a perfect square} \\ &= (x + 2y)^2 \end{aligned}$$

Example 8 **Sum of Two Cubes**

Factor $8x^3 + 27y^3$.

SOLUTION Because each term is a cube, we can write

$$\begin{aligned} 8x^3 + 27y^3 &= (2x)^3 + (3y)^3, \qquad \text{a sum of two cubes} \\ &= (2x + 3y)[(2x)^2 - (2x)(3y) + (3y)^2] \\ &= (2x + 3y)(4x^2 - 6xy + 9y^2) \end{aligned}$$

Self-Check 5 Factor $4x^2 + 12xy + 9y^2$.

Factoring Strategy

In this section, we have reviewed a number of factoring techniques. We have seen that it may be necessary to use a combination of techniques to factor an expression. Any technique could be relevant at any time, and they should all be at the student's fingertips. The more you practice, the more able you will be at picking out appropriate factoring techniques at various stages. However, the following strategy might also be helpful. Each step can apply to the whole polynomial or to certain terms within the polynomial.

Factoring Strategy

1. Factor out the greatest common factor.
2. Look for special factors.
3. Look for polynomial factors.
4. Try to group terms together for factoring.

Example 9 **Factoring Using Multiple Techniques**

Factor $2x^2 + 12xy + 18y^2 - 8$.

SOLUTION

Factor out the greatest common factor, 2:

$$2x^2 + 12xy + 18y^2 - 8 = 2(x^2 + 6xy + 9y^2 - 4)$$

Observe that $x^2 + 6xy + 9y^2$ is a perfect square:

$$= 2[(x + 3y)^2 - 4]$$

$(x + 3y)^2 - 4$ is the difference of two squares:

$$= 2(x + 3y + 2)(x + 3y - 2)$$

Thus,

$$2x^2 + 12xy + 18y^2 - 8 = 2(x + 3y + 2)(x + 3y - 2)$$

Example 10 **Factoring Using Multiple Techniques**

Factor $x^2 + 3x - y^2 - 3y$.

SOLUTION In anticipation of using the difference of two squares, group the terms together as follows:

$$x^2 + 3x - y^2 - 3y = (x^2 - y^2) + (3x - 3y)$$

Write $(x^2 - y^2)$ as the difference of two squares, and factor out the common factor in $3x - 3y$:

$$= (x - y)(x + y) + 3(x - y)$$

Factor out the common factor $(x - y)$:

$$= (x - y)(x + y + 3)$$

Thus,

$$x^2 + 3x - y^2 - 3y = (x - y)(x + y + 3)$$

Completing the Square

Completing the square is a very useful technique for changing an algebraic expression to another expression that is a perfect square. We will use it in our study of parabolas and circles. Consider the expression $x^2 + px$, where p is a real number. We can use the method of completing the square to turn this expression into a perfect square $(x + q)^2$, where q is an appropriate real number. Observe that

$$x^2 + px + \left(\frac{p}{2}\right)^2 = \left(x + \frac{p}{2}\right)^2$$

Thus, to transform $x^2 + px$ into the expression $\left(x + \frac{p}{2}\right)^2$, which is a perfect square, we add the square of half the coefficient of x.

Rule for Completing the Square

To complete the square for $x^2 + px$, add $\left(\frac{p}{2}\right)^2$ to get $\left(x + \frac{p}{2}\right)^2$.

Example 11 **Completing the Square**

Complete the square for $x^2 + 6x$.

SOLUTION Half the coefficient of x is 3. Thus, to complete the square, add $(3)^2$. We get

$$x^2 + 6x + (3)^2 = (x + 3)^2$$

Caution: Note that the final expression is not equal to the original expression. In this case, 3^2 has been added to $x^2 + 6x$ to get $(x + 3)^2$. Later we discuss how, in practice, an allowance is made for the added term.

Example 12 | Completing the Square

Complete the square for $x^2 - 2x$.

SOLUTION Half the coefficient of x is -1. Thus, to complete the square, add $(-1)^2$. We get

$$x^2 - 2x + (-1)^2 = (x - 1)^2$$

Self-Check 6 Complete the square for $x^2 - 7x$.

Answers to Self-Check Exercises

1. $5x^2(x^2 - 3x + 5)$
2. $(2x - 1)(x + 6)$
3. $(2x + 5y)(3x - 2y)$
4. $(x + 2)(y - 6)$
5. $(2x + 3y)^2$
6. $(x - \frac{7}{2})^2$

EXERCISES R.4

In Exercises 1–8, factor out the greatest common factor.

1. $4x^3 + 2x^2$ **2.** $x^3 + 2x^2 + x$
3. $3x^4 + 6x^2 + 9$ **4.** $3x^3 + 9x^2 + 6x$
5. $24x^6 + 12x^5 + 18x^4 + 30x^3$
6. $2x(x - 3)^2 + 8x^2(x - 3)$
7. $5y^2(y + 4) + 10y^3(y + 4)^2$
8. $2z^3(z + 2)^2 - 4z^2(z + 2)^3 + 12z(z + 2)^5$

In Exercises 9–18, factor the given polynomials of degree two.

9. $x^2 + x - 2$ **10.** $x^2 - 2x - 3$
11. $x^2 + 5x + 6$ **12.** $x^2 - 4$
13. $x^2 + 7x + 12$ **14.** $2x^2 + 5x - 3$
15. $3x^2 - 5x - 2$ **16.** $2y^2 + 3y - 5$
17. $10y^2 + 13y - 3$ **18.** $6z^2 - 11z - 10$

In Exercises 19–24, factor the given expressions into two factors. (See Example 3.)

19. $x^2 - xy - 2y^2$ **20.** $x^2 + xy - 2y^2$
21. $2x^2 + 5xy - 3y^2$ **22.** $3x^2 + 5xy - 2y^2$
23. $8x^2 + 2xy - y^2$ **24.** $12x^2 + xy - 6y^2$

In Exercises 25–32, factor the given expressions by grouping.

25. $x^2 - xy + 4x - 4y$ **26.** $2x^2 + xy - 6x - 3y$
27. $3x^2 + xy - 12x - 4y$ **28.** $4x - 4y + x^2 - xy$
29. $5x + 10y - yx - 2y^2$ **30.** $3u^2 - uv - 6u + 2v$
31. $xy^2 + 2y^2 - 3x - 6$ **32.** $u^3 - uv + u^2v^2 - v^3$

In Exercises 33–42, factor the given expressions using combinations of the techniques introduced in this section.

33. $x^3 + x^2 - 2x$ **34.** $x^3 + 4x^2 + 3x$
35. $x^3 - 4x^2 + 3x$ **36.** $2x^4 + 10x^3 + 8x^2$
37. $6x^2 + 27x + 12$ **38.** $8x^3 + 22x^2 - 6x$
39. $3x^2 + 3xy^3 - 9x + xy + y^4 - 3y$
40. $x^6 + x^2y^2 - 2x^2 + 3x^5 + 3xy^2 - 6x$
41. $u^2v + uv - 6v + 2u^2 + 2u - 12$
42. $12u^3v - 5u^2v - 2uv$

In Exercises 43–47, prove the results involving special factors by multiplying out the right-hand sides.

43. Difference of two squares

$$a^2 - b^2 = (a + b)(a - b)$$

44. Perfect square $a^2 + 2ab + b^2 = (a + b)^2$

45. Perfect square $a^2 - 2ab + b^2 = (a - b)^2$

46. Sum of two cubes

$$a^3 + b^3 = (a + b)(a^2 - ab + b^2)$$

47. Difference of two cubes

$$a^3 - b^3 = (a - b)(a^2 + ab + b^2)$$

In Exercises 48–54, each expression is the difference of two squares. Factor.

48. $x^2 - 4y^2$ **49.** $9x^2 - y^2$ **50.** $25x^2 - 16y^2$

51. $(3x + 2y)^2 - 4z^2$ **52.** $9x^2 - (2y + 3z)^2$

53. $16u^6 - 9v^4$ **54.** $(p + 3q)^2 - (4r - t)^2$

In Exercises 55–60, each expression is a perfect square. Factor.

55. $x^2 + 2xy + y^2$ **56.** $x^2 + 6xy + 9y^2$

57. $x^2 - 4xy + 4y^2$ **58.** $81 - 54z + 9z^2$

59. $(x + 2y)^2 - 6z(x + 2y) + 9z^2$

60. $9x^2 + 12x(3y - z) + 4(3y - z)^2$

In Exercises 61–66, each expression is the sum or difference of two cubes. Factor.

61. $8x^3 + y^3$ **62.** $8x^3 - 27y^3$

63. $125a^3 - 8b^3$ **64.** $64p^3 + 27q^3$

65. $(x + y)^3 - z^3$ **66.** $(x + 1)^3 + (x - 1)^3$

In Exercises 67–82, factor the expressions using the techniques of this section.

67. $x^2 + 2x - 3$ **68.** $4x^2 - 9y^2$

69. $x^2 + 6xy + 9y^2$ **70.** $x^2 - 36y^2$

71. $4t^2 - 4t - 8$ **72.** $(x + 2)^2 - 4y^2$

73. $u^3 - 27v^3$ **74.** $4x^6 - 9$

75. $9x^2 + 12x - 21$ **76.** $(x - 3)^3 + (y - 5)^3$

77. $9x^2 - (3y + 4z)^4$ **78.** $x^2 + 3x - 4y^2 + 6y$

79. $x^2 + 4xy + 4y^2 - 9$

80. $x^3 + 8y^3 + 2x^2y + 4xy^2$

81. $u^2 + u - 2v - 4v^2$

82. $u^2 + 5uv - 6v^2 + v - u$

In Exercises 83–88, use the method of completing the square to turn the algebraic expressions into expressions that are perfect squares.

83. $x^2 + 4x$ **84.** $x^2 - 8x$ **85.** $x^2 + 2x$

86. $x^2 + 5x$ **87.** $x^2 - 10x$ **88.** $x^2 - 3x$

R.5 Rational Expressions

- Simplification of Rational Expressions • Multiplication and Division
- Addition and Subtraction • Use of Lowest Common Denominator

So far we have introduced polynomials and defined rules for adding, subtracting, and multiplying polynomials. The rules were based on those for real numbers. This section examines algebraic expressions that are quotients of polynomials. An algebraic expression that is the quotient of two polynomials is a **rational expression.** The rules for combining rational expressions are based on the rules for working with fractions.

Let a and b represent polynomials. Consider the rational expression

$$\frac{a}{b}$$

Because the variable in the polynomial b is a real number, and division by zero is not permitted, the expression is only defined for real numbers for which b is nonzero. For example, consider

$$\frac{x^2 + 2x - 3}{x - 4}$$

This rational expression is defined when $x \neq 4$. In the following examples and exercises, we assume that all rational expressions are valid only for values of the variable for which a polynomial in a denominator is nonzero.

Simplification of Rational Expressions

Recall that a fraction can be written in lowest terms by dividing out all common factors of numerator and denominator. Rational expressions are ratios of polynomials. They can be simplified (or reduced) by dividing out common factors of the polynomials. Let the polynomials in the numerator and denominator of a rational expression have factors ac and bc. The following rule is used to simplify the expression:

$$\frac{ac}{bc} = \frac{a}{b}$$

Here we divided out the common factor c. This simplification is only valid if $c \neq 0$. As stated previously, we assume that values of x that lead to such zeros in the denominator are excluded.

Example 1 Simplifying a Rational Expression

Simplify the rational expression $\dfrac{x^2 - x - 6}{x^2 - 4x + 3}$.

SOLUTION Factor the numerator and denominator.

$$\frac{x^2 - x - 6}{x^2 - 4x + 3} = \frac{(x + 2)(x - 3)}{(x - 1)(x - 3)}$$

We now divide out the common factor $(x - 3)$. (Note that we can do this because $x - 3 \neq 0$.)

$$= \frac{(x + 2)\cancel{(x - 3)}}{(x - 1)\cancel{(x - 3)}} = \frac{(x + 2)}{(x - 1)}$$

Thus,

$$\frac{x^2 - x - 6}{x^2 - 4x + 3} = \frac{x + 2}{x - 1}$$

Self-Check 1 Simplify $\dfrac{x^2 + 3x - 10}{x^2 - 3x + 2}$.

Multiplication and Division

We now introduce the rules for multiplying and dividing rational expressions, followed by the rules for adding and subtracting. In the following examples, we illustrate the general methods and some of the techniques that can be used for special cases. For example, we show how to add rational expressions that have common factors in their denominators. The rules for multiplying and dividing rational expressions are based on those rules for fractions.

Multiplication and Division

Let a, b, and c represent polynomials.

Multiplication $\frac{a}{b} \cdot \frac{c}{d} = \frac{ac}{bd}$ Division $\frac{a}{b} \div \frac{c}{d} = \frac{a}{b} \cdot \frac{d}{c} = \frac{ad}{bc}$

Example 2 **Multiplying Rational Expressions**

Perform the multiplication $\frac{x}{x^2 - x - 2} \cdot \frac{x + 1}{x^2 + 3x}$.

SOLUTION Factor the polynomials and divide out common factors of numerators and denominators.

$$\begin{aligned}\frac{x}{x^2 - x - 2} \cdot \frac{x + 1}{x^2 + 3x} &= \frac{x}{(x + 1)(x - 2)} \cdot \frac{x + 1}{x(x + 3)} \\ &= \frac{x(x + 1)}{(x + 1)(x - 2)x(x + 3)} = \frac{1}{(x - 2)(x + 3)}\end{aligned}$$

Self-Check 2 Multiply $\frac{x^2 - x - 2}{x^2 - 2x - 3} \cdot \frac{x^2 - 9}{x^2 + 8x + 15}$.

Example 3 **Dividing Rational Expressions**

Divide $\frac{x^2 - x}{x + 2} \div \frac{x^2 + 5x}{x^2 + 3x + 2}$.

SOLUTION The rule for division of rational expressions is to convert the division into a multiplication, as in the case of fractions. We get

$$\frac{x^2 - x}{x + 2} \div \frac{x^2 + 5x}{x^2 + 3x + 2} = \frac{x^2 - x}{x + 2} \cdot \frac{x^2 + 3x + 2}{x^2 + 5x}$$

We now proceed to factor and divide out common factors:

$$\begin{aligned}&= \frac{x(x - 1)}{x + 2} \cdot \frac{(x + 1)(x + 2)}{x(x + 5)} \\ &= \frac{x(x - 1)(x + 1)(x + 2)}{x(x + 2)(x + 5)} \\ &= \frac{(x - 1)(x + 1)}{x + 5}\end{aligned}$$

Self-Check 3 Divide $\frac{2x^2 - 5x - 3}{x + 4} \div \frac{2x^2 + 3x + 1}{x^2 + 4x}$.

Addition and Subtraction

The rules for adding and subtracting rational expressions are also based on those rules for fractions.

Addition and Subtraction

Let a, b, c, and d represent polynomials.

Addition $\dfrac{a}{b} + \dfrac{c}{d} = \dfrac{ad + bc}{bd}$ Subtraction $\dfrac{a}{b} - \dfrac{c}{d} = \dfrac{ad - bc}{bd}$

Example 4 **Adding Rational Expressions**

Perform the addition $\dfrac{x+1}{x} + \dfrac{x-2}{x-3}$.

SOLUTION Using the rule for addition, we get

$$\begin{aligned}\frac{x+1}{x} + \frac{x-2}{x-3} &= \frac{(x+1)(x-3) + x(x-2)}{x(x-3)} \\ &= \frac{x^2 - 2x - 3 + x^2 - 2x}{x(x-3)} \\ &= \frac{2x^2 - 4x - 3}{x(x-3)}\end{aligned}$$

Self-Check 4 Perform the addition $\dfrac{x+1}{x-2} + \dfrac{x-3}{x+4}$.

Addition and subtraction of rational expressions that have the same denominator are straightforward.

Example 5 **Adding Rational Expressions with the Same Denominator**

Add $\dfrac{x}{x+2} + \dfrac{x^2+2x+3}{x+2}$.

SOLUTION Observe that both expressions have the same denominator, namely $x + 2$. Such additions can be performed, as for fractions, by adding the numerators:

$$\begin{aligned}\frac{x}{x+2} + \frac{x^2+2x+3}{x+2} &= \frac{x + x^2 + 2x + 3}{x+2} \\ &= \frac{x^2 + 3x + 3}{x+2}\end{aligned}$$

USE OF THE LOWEST COMMON DENOMINATOR

The addition and subtraction of rational expressions that have common factors in the denominators can be simplified using the **least common denominator** (LCD) of the expressions. The LCD is an expression of smallest degree that is a multiple of each denominator. The following examples illustrate this technique.

Example 6 **Adding Rational Expressions by Finding the LCD**

Perform the addition $\dfrac{x}{(x-1)(x+3)} + \dfrac{1}{(x-1)(x+1)}$.

SOLUTION The LCD is formed by taking each factor of the denominators to the largest power it appears in any denominator. Thus, the LCD of the two denominators in this example is

$$(x-1)(x+3)(x+1)$$

Now write each fraction in a form so that its denominator is

$$(x-1)(x+3)(x+1)$$

We get

$$\begin{aligned}\frac{x}{(x-1)(x+3)} + \frac{1}{(x-1)(x+1)} &= \frac{x(x+1)}{(x-1)(x+3)(x+1)} + \frac{(x+3)}{(x-1)(x+3)(x+1)}\\ &= \frac{x(x+1)+(x+3)}{(x-1)(x+3)(x+1)}\\ &= \frac{x^2+x+x+3}{(x-1)(x+3)(x+1)}\\ &= \frac{x^2+2x+3}{(x-1)(x+3)(x+1)}\end{aligned}$$

Self-Check 5 Compute $\dfrac{2x+1}{(x-1)(x-2)} + \dfrac{x+4}{(x-2)^2}$.

It is important to present a final answer in as simple a form as possible.

Example 7 **Adding Rational Expressions and Simplifying the Answer**

Perform the addition $\dfrac{x^2-7x+12}{2x^2+6x} + \dfrac{7}{x+3}$.

SOLUTION Factoring the denominator of the first term, we get

$$\begin{aligned}\frac{x^2-7x+12}{2x^2+6x} + \frac{7}{x+3} &= \frac{x^2-7x+12}{2x(x+3)} + \frac{7}{x+3}\\ &= \frac{x^2-7x+12+7(2x)}{2x(x+3)}\\ &= \frac{x^2+7x+12}{2x(x+3)}\end{aligned}$$

If we factor the numerator we find that we can simplify this expression. We get

$$= \frac{(x+3)(x+4)}{2x(x+3)} = \frac{x+4}{2x}$$

The next example illustrates that it can be advantageous to simplify some of the terms before adding rational expressions.

Example 8 Adding Rational Expressions

Perform the addition

$$\frac{\left(1+\frac{1}{x}\right)}{\left(1-\frac{1}{x}\right)} - \frac{1}{x-1}$$

SOLUTION First perform the additions in both the numerator and the denominator of the first term.

$$\frac{\left(1+\frac{1}{x}\right)}{\left(1-\frac{1}{x}\right)} - \frac{1}{x-1} = \frac{\left(\frac{x+1}{x}\right)}{\left(\frac{x-1}{x}\right)} - \frac{1}{x-1}$$

Write the leading term as a product and simplify.

$$= \left(\frac{x+1}{x}\right)\left(\frac{x}{x-1}\right) - \frac{1}{x-1}$$

$$= \frac{x+1}{x-1} - \frac{1}{x-1}$$

Finally add both terms. Observe that both terms have the same denominator.

$$= \frac{x+1-1}{x-1}$$

$$= \frac{x}{x-1}$$

Answers to Self-Check Exercises

1. $\frac{x+5}{x-1}$
2. $\frac{x-2}{x+5}$
3. $\frac{x(x-3)}{x+1}$
4. $\frac{2(x^2+5)}{(x-2)(x+4)}$
5. $\frac{3(x^2-2)}{(x-1)(x-2)^2}$

EXERCISES R.5

In Exercises 1–10, simplify each rational expression by dividing common factors of numerators and denominators.

1. $\dfrac{3x^2}{2x + 5x^2}$ **2.** $\dfrac{x - 2}{x^2 - 3x + 2}$

3. $\dfrac{2x + 2}{4x^2 + 6x + 2}$ **4.** $\dfrac{x^2 + x - 2}{x - 1}$

5. $\dfrac{x + 1}{x^2 + 5x + 4}$ **6.** $\dfrac{x^2 + x - 2}{x^2 + 2x - 3}$

7. $\dfrac{x^2 - 1}{x^2 + 4x + 3}$ **8.** $\dfrac{y^2 - y - 6}{y^2 - 9}$

9. $\dfrac{z^2 + 2z - 3}{z^2 + z - 6}$ **10.** $\dfrac{3t^3 - 5t^2 - 2t}{4t^3 - 7t^2 - 2t}$

In Exercises 11–16, perform the indicated multiplications and divisions.

11. $\dfrac{(x + 2)^2}{2} \cdot \dfrac{4}{x + 2}$ **12.** $\dfrac{3(x - 1)}{4(x + 3)} \cdot \dfrac{2(x + 3)^2}{x - 1}$

13. $\dfrac{2x - 1}{4x + 1} \div \dfrac{4(2x - 1)}{x}$ **14.** $\dfrac{3x}{4x + 5} \div \dfrac{6x^2}{(4x + 5)^2}$

15. $\dfrac{2u^2 + 3u - 2}{u^2 - 2u - 3} \cdot \dfrac{4u^2 + 3u - 1}{2u^2 + 9u - 5}$

16. $\dfrac{y^2 - 2y - 3}{y + 2} \div \dfrac{y + 1}{y + 2}$

In Exercises 17–24, perform the indicated additions and subtractions.

17. $\dfrac{1}{x} + \dfrac{3}{x - 1}$ **18.** $\dfrac{1}{x + 1} + \dfrac{3}{x + 2}$

19. $\dfrac{1}{x - 1} + \dfrac{4x}{x + 1}$ **20.** $\dfrac{x - 1}{x + 2} + \dfrac{x + 4}{x - 1}$

21. $\dfrac{x - 2}{x + 4} - \dfrac{x - 3}{2x - 3}$ **22.** $\dfrac{5x + 2}{2x - 3} + \dfrac{7x + 1}{2x - 3}$

23. $\dfrac{y^2 + y - 3}{y} - \dfrac{3y - 1}{3}$

24. $\dfrac{w^2 + 2w + 1}{w + 2} + \dfrac{-w^2 + 2}{w - 4}$

In Exercises 25–30, perform the indicated additions or subtractions using the LCD of the rational expressions.

25. $\dfrac{x}{(x + 1)(x + 2)} + \dfrac{2x + 1}{(x + 1)(x - 1)}$

26. $\dfrac{x - 3}{(x - 1)(2x + 1)} + \dfrac{3x}{(x + 4)(2x + 1)}$

27. $\dfrac{4x - 1}{(x + 2)(x - 1)} + \dfrac{3x + 2}{(x - 1)(x + 3)}$

28. $\dfrac{3x - 4}{(x - 1)(x + 3)^2} + \dfrac{4}{(x + 2)(x + 3)}$

29. $\dfrac{2z}{z^2 - 2z - 3} + \dfrac{3}{z^2 + 3z + 2}$

30. $\dfrac{4p + 1}{p^2 - p - 2} + \dfrac{5p - 1}{p^2 - 3p + 2}$

In Exercises 31–36, perform the indicated operations.

31. $\dfrac{x^2 + x - 2}{x^2 - x - 12} \cdot \dfrac{x^2 + 3x}{x^2 - 4}$

32. $\dfrac{2x^2 - 5x + 2}{3x^2 - 7x - 6} \cdot \dfrac{4x^2 - 11x - 3}{4x^2 - 7x - 2}$

33. $\dfrac{6x^2 - 11x - 2}{x^2 - 3x} \div \dfrac{6x^2 - 5x - 1}{x^2 - x}$

34. $\dfrac{3x^2 - 2x - 1}{x^2 + x - 2} \div \dfrac{3x^2 - 8x - 3}{(x^2 + x - 2)^2}$

35. $\dfrac{6v + 1}{v - 2} + \dfrac{2}{v} + 5v - 4$

36. $\left(\dfrac{5y + 1}{y - 2} + \dfrac{3y}{y - 1}\right) \cdot \left(\dfrac{y - 2}{y + 1}\right)$

In Exercises 37–46, simplify the fractions.

37. $\dfrac{x^2 + x}{\left(\dfrac{1}{x}\right)}$ **38.** $\dfrac{\left(x^2 + \dfrac{1}{x}\right)}{x}$

39. $\dfrac{(x - 1)}{\left(1 - \dfrac{1}{x}\right)}$ **40.** $\dfrac{(x^2 - 1)}{\left(1 + \dfrac{1}{x}\right)}$

41. $\dfrac{\left(1 - \dfrac{1}{x}\right)}{\left(1 + \dfrac{1}{x}\right)}$ **42.** $\dfrac{\left(3 + \dfrac{2}{x}\right)}{\left(3 - \dfrac{2}{x}\right)}$ **43.** $\dfrac{\left(\dfrac{1}{x} - \dfrac{3}{x^2}\right)}{\left(\dfrac{2}{x}\right)}$

44. $\dfrac{\left(\dfrac{1}{x} - x\right)}{\left(\dfrac{1 + x}{x}\right)}$ **45.** $\dfrac{\left(1 + \dfrac{1}{s}\right)}{\left(\dfrac{s + 1}{s - 1}\right)}$ **46.** $\dfrac{\left(t - \dfrac{1}{t}\right)}{\left(\dfrac{t - 1}{t}\right)}$

In Exercises 47–56, perform the indicated operations.

47. $\dfrac{1}{x-1} + \dfrac{4}{2x+3}$

48. $\dfrac{x+1}{x} \cdot \dfrac{x^2-x}{x^2-1}$

49. $\dfrac{3x}{(x-1)(x+2)} + \dfrac{x+1}{x(x-1)}$

50. $\dfrac{x}{x^2-4} \cdot \dfrac{x+2}{x^2-3x}$

51. $\dfrac{x}{(x-1)(x+2)} - \dfrac{x+1}{(x-1)(x-2)}$

52. $\dfrac{x-3}{x^2-4} \div \dfrac{x^2-3x}{x+2}$

53. $\dfrac{x+1}{x^2-2x-3} - \dfrac{x-3}{x^2+3x+2}$

54. $\dfrac{x^2+2x-8}{x^2+x-6} \cdot \dfrac{x^2+2x-3}{x^2+3x-4}$

55. $\dfrac{(x+1)}{\left(1+\dfrac{1}{x}\right)} - x + 2$

56. $\dfrac{\left(x-\dfrac{1}{x}\right)}{\left(x+\dfrac{1}{x}\right)} \cdot (x^2+1)$

Equations, Variation, Scuba Diving, and Fermat's Last Theorem

- Equations and Solutions • Equivalent Equations • Linear Equations
- Cost–Volume Analysis • Celsius and Fahrenheit Scales • Test Scores
- Variation • Scuba Diving • Electrical Resistance • Sale of Land
- Fermat's Last Theorem

In this section, we discuss equations and look at some of the ways equations can be used to analyze various situations. We discuss applications in the areas of education, business, and science. We first make clear what we mean by an equation.

Equations and Solutions

An **equation** is a statement that two algebraic expressions are equal. The following are examples of equations in the variable x:

$$2x + 4 = 10, \qquad 3x^2 - 4x + 7 = 0, \qquad \frac{x-3}{x+4} = 9x + 2$$

To **solve** such an equation is to find all values of the variable for which the equation is a true statement (the equation is satisfied). The values for which the equation is satisfied are called the **solutions** or **roots** of the equation. For example, consider the equation

$$2x + 4 = 10$$

Let us check to see whether $x = 3$ and $x = 4$ are solutions.

$x = 3$. Let $x = 3$ in the left-hand side of the equation.

$$\begin{aligned} 2x + 4 &= 2(3) + 4 \\ &= 10 \end{aligned}$$

The equation is satisfied; $x = 3$ is a solution.

$x = 4$. Let $x = 4$ in the left-hand side.

$$\begin{aligned} 2x + 4 &= 2(4) + 4 \\ &= 12 \neq 10 \end{aligned}$$

The equation is not satisfied; $x = 4$ is not a solution.

Self-Check 1

(a) Show that $x = 2$ is a solution to the equation $7x - 3 = 11$.
(b) Show that $x = -1$ and $x = 3$ are solutions to the equation $x^2 + 6x - 4 = 8x - 1$, but that $x = 2$ is not a solution.

Equivalent Equations

We now introduce a method for solving certain equations. The given equation is **transformed** into simpler equations, all of which have the same solutions, until the solutions are found. Two equations with the same solutions are called **equivalent equations.** The transformations that can be used are as follows.

Transformations for Generating Equivalent Equations

A given equation can be transformed into an equivalent equation by

1. Adding or subtracting the same real number or real expression to or from both sides of the equation.
2. Multiplying or dividing both sides of the equation by the same nonzero real number.

Let us look at some examples to see how these transformations are used to solve equations. The variable in the equation is gradually isolated, thus arriving at the solution. We often refer to the use of these transformations as writing an equation in a different form. It is advisable to check that the answer obtained is indeed the solution by substituting it back into the original equation to see that the equation is satisfied.

Example 1 Solving an Equation

Solve the equation $2x + 6 = 10$.

SOLUTION

Original equation: $2x + 6 = 10$

Subtract 6 from both sides. $2x = 4$

Divide both sides by 2. $x = 2$

Check: Substitute $x = 2$ into the original equation to see that it is satisfied.

$$\begin{aligned} 2x + 6 &= 2(2) + 6 \\ &= 10 \end{aligned}$$

The equation is satisfied; the solution is $x = 2$.

Example 2 | Solving an Equation

Solve $2(4x + 3) = (5x + 3) - 6$.

SOLUTION

Original equation:	$2(4x + 3) = (5x + 3) - 6$
Distribute.	$8x + 6 = 5x - 3$
Subtract 5*x* from both sides.	$3x + 6 = -3$
Subtract 6 from both sides.	$3x = -9$
Divide both sides by 3.	$x = -3$

Check: Let $x = -3$ in the left-hand side of the original equation.

$$\begin{aligned} 2(4x + 3) &= 2[4(-3) + 3] \\ &= -18 \end{aligned}$$

Let $x = -3$ in the right-hand side.

$$\begin{aligned} (5x + 3) - 6 &= [5(-3) + 3] - 6 \\ &= -18 \end{aligned}$$

Both sides of the equation are equal when $x = -3$. The solution is $x = -3$.

Self-Check 2 Solve $3(4 - x) = 2x - 3$.

Example 3 | Solving an Equation Using the LCD

Solve $\dfrac{4x}{3} - \dfrac{1}{2} = \dfrac{5x}{6}$.

SOLUTION This equation is first transformed by multiplying both sides by the LCD. The LCD is the smallest number that is a multiple of 3, 2, and 6. It is 6. This transformation removes the fractions.

Original equation:	$\dfrac{4x}{3} - \dfrac{1}{2} = \dfrac{5x}{6}$
Multiply both sides by LCD, 6.	$8x - 3 = 5x$
Subtract 5*x* from both sides.	$3x - 3 = 0$
Add 3 to both sides.	$3x = 3$
Divide both sides by 3.	$x = 1$

Check: If $x = 1$ is substituted into the original equation, it is found to be satisfied. The solution is $x = 1$.

Self-Check 3 Solve $\frac{7x}{8} = \frac{3x}{4} + \frac{1}{2}$.

Linear Equations

There are many types of equations. Practice identifying equations is important because various methods have been developed for solving different types of equations. Linear equations are some of the most useful equations. Let us look at linear equations in one variable.

> ***Linear Equation*** A **linear equation** in the variable x is an equation that can be written in the form
>
> $$ax + b = 0$$
>
> where a and b are real numbers, with $a \neq 0$.
>
> The equation $ax + b = 0$ is called the **standard form** of the linear equation.

Example 4 **Examples of Linear and Nonlinear Equations**

The equation $2x + 3 = 0$ is a linear equation in standard form. The equation $3x = -4$ is a linear equation because it can be written in the standard form $3x + 4 = 0$ by adding 4 to both sides.

Consider the equation $4x^2 = 3x - 7$. Rewrite the equation with all the terms on the left-hand side. We get $4x^2 - 3x + 7 = 0$. This equation is not of the form $ax + b = 0$ because of the term $4x^2$. The equation $4x^2 = 3x - 7$ is said to be **nonlinear.**

Self-Check 4 Show that the equation $8x + 2 = 3(x + 4) + 5$ is linear.

Conversion of Equations to Linear Equations

Certain equations that are not linear can be solved by changing them into linear equations. This involves multiplying both sides of the equation by an expression that contains the variable.

Caution: This technique of multiplying both sides of an equation by an expression involving a variable (unlike multiplying both sides by a constant) can lead to extraneous roots. You already know that it is advisable to check an answer by substituting it back into the original equation; however, when this method is used to solve an equation, it is *necessary* to check the result to see if it satisfies the original equation.

We now give examples to illustrate this method.

Example 5 **Converting to a Linear Equation and Then Solving**

Solve $\frac{2x + 5}{x - 1} + 3 = \frac{x - 10}{x - 1}$.

SOLUTION Multiply by the LCD $(x - 1)$ and simplify.

$$\left(\frac{2x+5}{x-1} + 3\right)(x-1) = \left(\frac{x-10}{x-1}\right)(x-1)$$

$$\left(\frac{2x+5}{x-1}\right)(x-1) + 3(x-1) = \left(\frac{x-10}{x-1}\right)(x-1)$$

$$2x + 5 + 3(x-1) = x - 10$$

$$5x + 2 = x - 10$$

$$5x - x = -10 - 2$$

$$4x = -12$$

$$x = -3$$

It can be shown by substitution that $x = -3$ satisfies the original equation. The solution is $x = -3$.

Example 6 Converting to a Linear Equation and Then Solving

Solve $3 + \dfrac{7}{x-3} = \dfrac{2x+1}{x-3}$.

SOLUTION

Multiply by the LCD $(x - 3)$ and simplify.

$$\left(3 + \frac{7}{x-3}\right)(x-3) = \left(\frac{2x+1}{x-3}\right)(x-3)$$

$$3(x-3) + \left(\frac{7}{x-3}\right)(x-3) = \left(\frac{2x+1}{x-3}\right)(x-3)$$

$$3(x-3) + 7 = 2x + 1$$

$$3x - 9 + 7 = 2x + 1$$

$$3x - 2 = 2x + 1$$

$$3x - 2x = 1 + 2$$

$$x = 3$$

$x = 3$ does not satisfy the original equation because the denominators in the equation are zero for this value of x. There are no values of x that satisfy this equation. We say that the equation has no solution.

Self-Check 5 Solve, if possible, the following equations:

(a) $\dfrac{x+3}{x-4} + 5 = \dfrac{13}{x-4}$ **(b)** $\dfrac{2x+3}{x+1} = \dfrac{1}{x+1}$

Applications . . .

Example 7 Cost–Volume Analysis

The relationship between the number of items manufactured and the total cost of production in a manufacturing process is often described by a linear equation. The equation is called a **cost–volume formula.** Let the total cost C, in dollars, of producing p items be given by

$$C = 2p + 700$$

(We shall see later in the course how such equations are constructed from data.) To find the total cost of manufacturing 500 items, let $p = 500$ in the equation. We get

$$\begin{aligned} C &= 2(500) + 700 \\ &= 1700 \end{aligned}$$

The total cost of manufacturing 500 items is \$1700.

Example 8 Celsius and Fahrenheit Scales

There are two temperature scales in common use, the Celsius and Fahrenheit scales. The relationship between the two scales is given by the linear equation

$$F = \tfrac{9}{5}C + 32$$

where C is the temperature in degrees Celsius and F is the corresponding temperature in degrees Fahrenheit. Let us find the Celsius temperatures corresponding to 50°, 68°, and 95°F. Because we are given values of F and want the corresponding values of C, we first solve the equation for C.

Original equation: $F = \tfrac{9}{5}C + 32$

Subtract 32 from both sides. $F - 32 = \tfrac{9}{5}C$

Multiply both sides by $\tfrac{5}{9}$. $\tfrac{5}{9}(F - 32) = C$

Thus, $C = \tfrac{5}{9}(F - 32)$

When $F = 50$, we get

$$\begin{aligned} C &= \tfrac{5}{9}(50 - 32) \\ &= \tfrac{5}{9}(18) = 10 \end{aligned}$$

Therefore 50°F corresponds to 10°C. Similarly, it can be shown that 68°F corresponds to 20°C, and that 95°F corresponds to 35°C.

Example 9 Test Scores

Julia Seber has test scores of 72, 82, and 75 on three tests. With a fourth test coming up, what score does she have to make on that test to get an average of 80 (B) on the first four tests?

SOLUTION It is a good idea to summarize the information in a problem like this.

Summary: Julia has scores of 72, 82, 75 on the first three tests.
She wants an average of 80 on the first four tests.
What grade must she get on the fourth test?

Let x be the necessary grade on the fourth test. The average of scores on the four tests is

$$\frac{72 + 82 + 75 + x}{4}$$

The desired average is 80. Thus,

$$\frac{72 + 82 + 74 + x}{4} = 80$$

Solve this equation for x.

Multiply both sides by 4. $$72 + 82 + 75 + x = 320$$

Simplify and solve for x. $$229 + x = 320$$
$$x = 91$$

Julia needs a score of 91 on the fourth test to raise her average to 80.

VARIATION

Many real-life situations involve one variable varying in some way with respect to other variables. Such variation leads to an equation. For example, it is known that pressure in a liquid varies directly as depth d below the surface. This statement means that there is an equation

$$P = kd$$

where k is a constant that tells us how pressure P depends on depth d.

> ***Direct Variation*** "y **varies directly** as x" means that $y = kx$ for some nonzero constant k, where k is called the **constant of variation.**

The expression "y is *directly proportional* to x" is also used for this relationship between x and y. The constant k is then referred to as the **constant of proportionality.**

Direct variation is always described by a linear equation of the form $y = kx$. To determine the equation, it is necessary to find k, the constant of variation. The following example illustrates how the constant of variation is found.

Example 10 Scuba Diving

Depth in scuba diving is measured in feet, and water pressure is measured in atmospheres. The water pressure at a depth of 20 feet is 0.6 atmospheres.* Find the

*Information from "Discover Diving Dive Center," Port Orange, Florida.

equation that relates water pressure to depth. Blue Springs in Central Florida is a popular scuba diving location. The spring is 122 feet at its deepest point. What is the maximum water pressure that a diver has to face when diving in Blue Springs?

SOLUTION Let P be the water pressure in atmospheres at a depth d feet. Recall that pressure in a liquid varies directly as depth below the surface. Thus, there is a constant k such that

$$P = kd$$

We now use the given information to find k. Substitute $d = 20$ and $P = 0.6$ into this equation

$$0.6 = 20k$$

$$k = \frac{0.6}{20} = 0.03$$

The relationship between water pressure and depth used in scuba diving is

$$P = 0.03d$$

Let us now use this equation to find the maximum pressure in Blue Springs. The depth of Blue Springs is 122 feet. When $d = 122$, $P = 0.03(122) = 3.66$. The maximum water pressure a diver faces in Blue Springs is 3.66 atmospheres.

There are other types of variation. Each variation can be written as an equation involving a constant of variation k. Some typical variations illustrating how the variation can be written as an equation are listed here.

Types of Variation

(a) y varies **inversely** as x: $y = k\left(\frac{1}{x}\right)$.

(b) u varies **directly as the square** of v: $u = k(v^2)$.

(c) t varies **directly as the cube** of s and **inversely as the square root** of h: $t = k\left(\frac{s^3}{\sqrt{h}}\right)$.

Example 11 Electrical Resistance

The electrical resistance of a wire varies directly as its length and inversely as the square of the radius of its cross section. The resistance of a wire that is 200 centimeters long with a radius of 0.5 centimeters is 40 ohms. Find the resistance of a wire of the same material, having length 160 centimeters and radius 0.4 centimeters.

SOLUTION Let R be resistance, L be length, and r be radius. We are given that R varies directly as L and inversely as the square of r. This means that there exists a constant k, such that

$$R = k\left(\frac{L}{r^2}\right)$$

We are given that $R = 40$ when $L = 200$ and $r = 0.5$. Substitute these values into the equation to find k.

$$40 = k\left(\frac{200}{0.5^2}\right)$$

$$k = \frac{(40)(0.5^2)}{200} = 0.05$$

Thus, $R = 0.05\left(\frac{L}{r^2}\right)$. Now let $L = 160$ and $r = 0.4$ to get

$$R = 0.05\left(\frac{160}{0.4^2}\right) = 50$$

The resistance of the wire is 50 ohms.

Example 12 Sale of Land

A large parcel of land is to be divided into rectangular lots and sold. The cost of a lot varies jointly as the length and width of the lot. If the cost of a lot measuring 150 feet by 100 feet is \$9000, determine the cost of a lot measuring 80 feet by 75 feet. (Figure R.21).

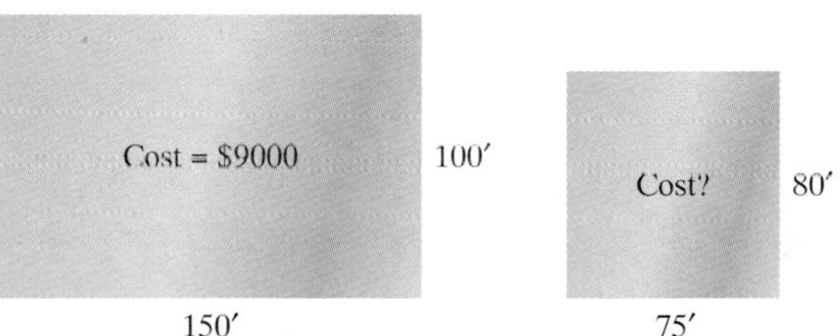

Figure R.21

SOLUTION Let C be the cost of a lot having length x and width y. C varies jointly as x and y. This means that there is a constant k such that $C = kxy$ (Figure R.22).

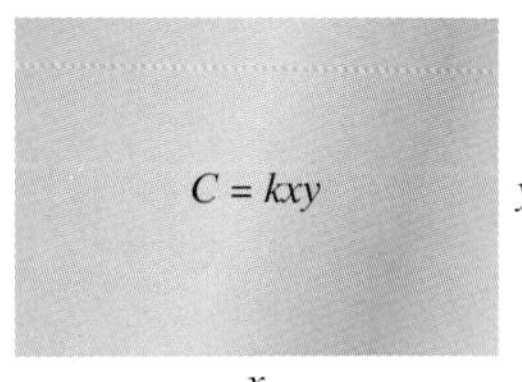

Figure R.22

We are given that $C = 9000$ when $x = 150$ and $y = 100$. Let us use this information to find the constant k. Substituting these values of C, x, and y into the equation $C = kxy$, we get

$$9000 = k(150)(100)$$

$$k = \frac{9000}{(150 \cdot 100)} = \frac{3}{5}$$

Therefore, $C = \frac{3}{5}(xy)$. When $x = 80$ and $y = 75$, we get

$$C = \frac{3}{5}(80)(75) = 3600$$

The cost of the lot is \$3600.

Example 13 Fermat's Last Theorem*

We close this section with an interesting historical tidbit about one of the most famous equations of all time. The story starts with a French mathematician, Pierre de Fermat, in approximately 1637 and ends with an English mathematician, Andrew Wiles, in 1994.

Fermat was the son of a wealthy leather merchant in the French town of Beaumont-de-Lomagne. He attended the University of Toulouse and later studied civil law at the University of Orleans. Fermat served as a lawyer and a government official in Toulouse most of his life. He became known for challenging other mathematicians to solve problems that he had already solved. These challenges and the fact that he would not reveal his own calculations infuriated others. His own countryman René Descartes called Fermat a "braggart," whereas the English mathematician John Wallis referred to him as "that damned Frenchman."

To understand the most well-known of Fermat's problems, consider the equation $x^2 + y^2 = z^2$. This equation, based on the Pythagorean Theorem, has many solutions. For example, $x = 3$, $y = 4$, $z = 5$ satisfy the equation because $3^2 + 4^2 = 5^2$. The values $x = 5$, $y = 12$, and $z = 13$ also satisfy the equation. Fermat's problem involved looking at a form of this equation that had integer powers greater than 2. In 1637, Fermat annotated a copy of Bachet's translation of Diophantus' *Arithmetika* with the following statement: "There are no positive integers such that $x^n + y^n = z^n$ for $n > 2$. I've found a remarkable proof of this fact, but there is not enough space in the margin [of the book] to write it."

Fermat claimed, for example, that the equation $x^3 + y^3 = z^3$ has no solutions that are positive integers. Neither does $x^4 + y^4 = z^4$, and so on. This result became known as Fermat's Last Theorem because it remained the only one of the results that he had posed that remained unproven throughout the years. In 1994, Andrew Wiles of Princeton University finally solved the problem that had eluded the greatest minds of the world for 358 years.

Wiles first encountered this problem as a schoolboy visiting his local library in England. He recalled that "It looked so simple, and yet all the great mathematicians in history couldn't solve it. Here was a problem, that I a ten year old could understand and I knew from that moment that I would never let go. I had to solve it." To succeed, Wiles required enormous determination to overcome the periods of self-doubt. He described his experience of attacking the problem in terms of a journey through a dark unexplored mansion: "You enter the first room of the mansion and it's completely dark. You stumble around bumping into the

*See the Williams website at www.saunderscollege.com/math/williams to link to the Internet for Simon Singh's article, "Pierre de Fermat," an excellent history of Fermat's Last Theorem.

furniture but gradually you learn where each piece of furniture is. Finally, after six months or so, you find the light switch, you turn it on and suddenly it's all illuminated. You can see exactly where you are. Then you move on to the next room and spend another six months in the dark." Wiles described the final breakthrough in his proof as follows: "I was sitting at my desk one Monday morning, when suddenly, totally unexpectedly, I had this incredible revelation. It was so indescribably beautiful, it was so simple, and so elegant. I couldn't understand how I'd missed it and I just stared at it in disbelief for twenty minutes. Then during the day I walked around the department, and I'd keep coming back to my desk looking to see if it was still there. It was still there. I couldn't contain myself, I was so excited. It was the most important moment of my working life. Nothing I ever do again will mean so much."

Even Wiles' proof had its own drama. Wiles announced his proof on June 23, 1993. A few months later, a slight but crucial flaw was found in the proof, and the mathematics world held its breath. However, on October 26, 1994, with the help of Richard Taylor of Cambridge University, Wiles corrected the original proof. The proof is extremely complex, involves mathematics that did not exist at the time of Fermat, and takes 129 pages (*Annals of Mathematics,* 1995, volume 141, pages 443–572). Did Fermat indeed have a simpler proof? Mathematicians believe that he did not.

GROUP DISCUSSION

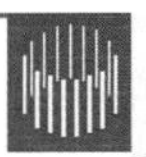

Temperature Conversion

(a) Suppose that you are going for a vacation to Europe where temperature is given in degrees Celsius. You might be in Paris, for example, and see that the temperature is 23°C. You need a quick way of finding the approximate temperature in degrees Fahrenheit. In Example 8, we saw that the exact relationship between the two scales is $F = \frac{9}{5}C + 32$. Use this equation to arrive at some rule of thumb that can be used to find the approximate temperature in degrees Fahrenheit. The emphasis is on ease and speed, not on accuracy!

(b) In July 1999, a London newspaper reported that the following were typical temperatures for some capital cities during that summer. If you were reading this article, how would you interpret these temperatures? Athens, 39°C; Geneva, 25°C; Jerusalem, 34°C; London, 20°C; Madrid, 42°C; Paris, 23°C; Rome, 31°C.

Answers to Self-Check Exercises

1. **(a)** $7(2) - 3 = 11$
   (b) $(-1)^2 + 6(-1) - 4 = 8(-1) - 1 = -9$
   $(3)^2 + 6(3) - 4 = 8(3) - 1 = 23$
   $(2)^2 + 6(2) - 4 \neq 8(2) - 1$
2. $x = 3$
3. $x = 4$
4. It can be written as $5x - 15 = 0$.
5. **(a)** $x = 5$,
   (b) No solution.

EXERCISES R.6

In Exercises 1–8, determine whether the given x values are solutions to the equations.

1. $x = 1, \quad 2x + 4 = 6$
2. $x = 3, 5x - 15 = 0$
3. $x = 1, \quad x^2 - 2x - 1 = 0$
4. $x = 1, \quad x^2 + 4x = x - 2$
5. $x = -3$ and $x = 2, \quad x^2 + x - 6 = 0$
6. $x = -1$ and $x = 3, \quad x^2 - 3x - 4 = 0$
7. $x = 2, \quad \dfrac{x - 3}{x + 1} + 4 = \dfrac{x + 9}{x + 1}$
8. $x = 3, \quad \dfrac{x + 7}{x - 2} = \dfrac{4x + 2}{11 - 3x} + 5$

In Exercises 9–16, solve the given equations for x.

9. $2x - 3 = 5$
10. $3x + 9 = 0$
11. $-2x + 9 = -11$
12. $5x + 4 = 3x - 2$
13. $7x - 1 = 12x - 3$
14. $4x - 3 = 7x + 12$
15. $3(x + 2) = 2x - 1$
16. $4(3x - 2) = 5(2x - 6)$

In Exercises 17–24, determine whether or not the given equations are linear. If an equation is linear, give its standard form.

17. $5x + 7 = 0$
18. $2x - 3 = 0$
19. $2x - 3 = 8$
20. $x^2 - 3x + 1 = 0$
21. $2(x - 3) + 4 = x + 6$
22. $\dfrac{x - 1}{x + 2} = x + 2$
23. $z + 2 = z^2 - 3z + 1$
24. $5u + 4 = 3(u - 1)$

In Exercises 25–32, solve the given equations for x, if a solution exists.

25. $5(x - 2) = 3(x + 4) - 6$
26. $9x + 3 + 4(5x - 2) = 1$
27. $8x - 1 = 2x - 3 + 9(x - 1)$
28. $5x + 4 = 3(x - 1) - 2(5 - x)$
29. $3x^2 - 2x + 4 = (x^2 - 4) + (2x^2 - x + 5)$
30. $(x - 1)(x + 1) = x^2 + 3x + 8$
31. $\dfrac{3x}{4} + \dfrac{1}{2} = \dfrac{x}{2} - 1$
32. $\dfrac{7x}{8} + \dfrac{x}{4} = \dfrac{3x}{2} + 1$

In Exercises 33–38, solve the given equations for x, if a solution exists.

33. $\dfrac{2x - 1}{x + 3} + \dfrac{3}{x + 3} = 4$
34. $\dfrac{2x - 1}{x + 4} + 5 = \dfrac{3x + 3}{x + 4}$
35. $\dfrac{x(x - 1)}{x^2 + x - 2} + 3 = \dfrac{14}{x + 2}$
36. $\dfrac{x^2 - x - 2}{x^2 + 4x + 3} + 4 = \dfrac{2x + 1}{x + 3}$
37. $\dfrac{4}{x + 2} - \dfrac{2}{x - 1} = \dfrac{3}{x + 2}$
38. $\dfrac{3}{x} - \dfrac{2x + 1}{x - 1} = \dfrac{3 - 2x}{x}$
39. **Cost–Volume Analysis** The relationship between the number of items n and the total cost C to manufacture those items is $C = 5n + 1500$. Solve this equation for n and use it to find the number of items that can be manufactured for \$5000.
40. **Depreciation of Assets** The following equation is used by accountants for straight-line depreciation of assets

$$D = \frac{C - S}{L}$$

Here, C is the initial cost of the asset, S is its salvage value, L is its estimated life, and D is depreciation per year. Solve this equation for S and use it to find the salvage value of a machine that had an initial cost of \$80,000, an annual depreciation of \$7000, and an estimated life of 10 years.

41. **Time Behavior** It has been found that there is an approximate linear relationship between a person's estimate of time lapse e and real time lapse r. Amanda Jolin (a college teacher) found that her equation is $e = 0.75r + 1.2$. If, while teaching a class, she estimates that 50 minutes have passed, while teaching a class, approximately how long will the class have been in session in reality?
42. **Grades** Jane Williams has grades of 72 and 78 on the first two tests in a class. What must she make on the third test to have an average of 80 on the three tests?
43. **Golf** Karl Alessi has rounds of 78, 74, and 68 on the first three rounds of a golf tournament. Par for the course is 72. What score does he need on the fourth and final round to average par for the tournament?
44. **Game Attendance** The attendance figures at the first four home football games of Danton College averaged 49,874. What must the attendance be at

the fifth home game (homecoming) to bring the average up to 50,000?

45. **Stocks** Daryl Anglin inherited \$62,000 and invested it in the following stocks on January 1, 1997: IBM, \$10,000; Exxon, \$8,000; GTE, \$12,000; Bristol Myers, \$7,000; Colgate Palmolive, \$10,000; Templeton Developing Markets, \$6,500; New Plan Realty, \$8,500. Because 1997 was a very good year for the stock market, the growth in the stocks for the year January 1 to December 31, 1997, was IBM, 38.1%; Exxon, 28.2%; GTE, 19.9%; Bristol Myers, 75.2%; Colgate Palmolive, 47.3%; Templeton, 4.2%; New Plan Realty, 7.2%. Determine the percentage growth in the portfolio over the year (to one decimal place, as for the given growths). Explain the steps you used to determine the answer.

46. **Production Line** The number N of plastic cups produced by a certain machine varies directly as the time t of operation of the machine. If the machine makes 12,000 cups in 6 hours, how many cups can it produce in 52 hours?

47. **Extension of a Spring** Hooke's Law for a spring implies that the length a spring stretches varies directly as the weight suspended from the spring. Find the equation that gives the extension of a spring in terms of the weight suspended if a weight of 20 pounds extends the spring 4 inches. How much will a weight of 35 pounds extend the spring?

48. **Variation** Express each of the following statements as equations.

    (a) y varies directly as the cube of x.

    (b) z varies directly as x and inversely as y.

    (c) f varies directly as the square of h and inversely as the cube root of t.

    (d) The volume of a gas V is proportional to the temperature T and inversely proportional to the pressure P.

    (e) The weight W of an object varies inversely as the square of its distance d from the center of the earth.

    (f) The horsepower H needed to run a boat varies directly as the cube of the speed.

    (g) The area A of a circle is proportional to the square of its radius.

    (h) The volume V of a sphere is proportional to the cube of its radius r.

49. **Typing Errors** Jim Kelada knows that the approximate number of typing errors he makes per page varies directly as the length of time he spends typing and the square of the typing speed. He makes two mistakes per page when typing at the rate of 50 words per minute during the first hour of typing. How many mistakes will he make when typing 55 words per minute for 3 hours?

50. **Motion of a Pendulum** The time it takes a pendulum to complete one oscillation (called its period) varies directly as the square root of its length. If the period of a pendulum 3 feet long is 1.8 seconds, what must the length be for the period to be 1 second (to two decimal places)?

51. **Planetary Motion** The Third Law of Kepler states that the time it takes for a planet to complete one revolution around the sun (called the period T of the planet) is directly proportional to the square root of the cube of the average distance d from the sun. For Earth, $T = 365$ days and $d = 93$ million miles. Mercury is the closest planet to the sun: $d = 36$ million miles. Pluto is the farthest planet from the sun: $d = 3675$ million miles. Find the duration of a "year" on Mercury and on Pluto, expressed to two decimal places.

52. **Area of a Triangle** The area of a triangle varies jointly as the height and base. If the height is doubled, how should the base be changed to triple the area?

53. **Volume of a Sphere** The volume of a sphere varies directly as the cube of its radius. How is the volume changed if the radius is tripled?

54. **Aperture of a Camera** The f-stop on a camera varies inversely as the square root of the area A of the aperture. Consider f-stops 2.8, 4, 5.6, 8, 11, and

16. Determine the ratios of the areas of the apertures of consecutive f-stops.

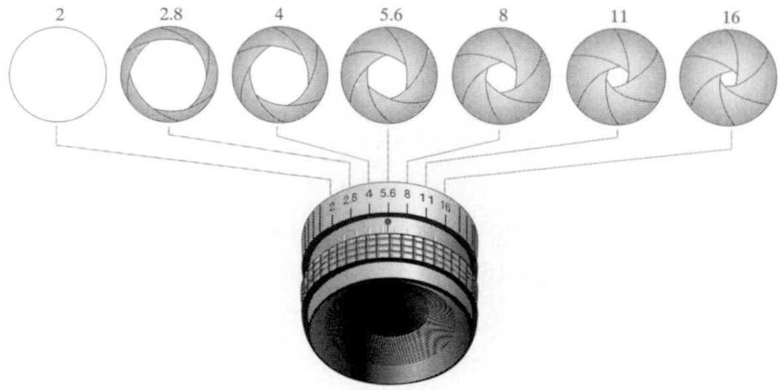

55. Sailing The wind pressure on the surface of a sail, at right angles to the direction of the wind, varies jointly as the area of the surface and the square of the velocity of the wind. What wind velocity would be necessary to cause the pressure on 80 square feet of surface to be double the pressure exerted on 10 square feet by a wind of velocity 20 miles per hour?

SECTION PROJECT

Beal's Equation

Texas millionaire and amateur mathematician Andrew Beal has produced his own generalization of Fermat's Theorem and will give an award to anyone who proves it. (See the Williams website at www.saunderscollege.com/math/williams to link to the Internet for more information.) Beal's Theorem is: "There are no positive integers x, y, and z such that $x^m + y^n = z^r$ for $m, n, r > 2$."

Give examples to illustrate Beal's Theorem.

R.7 Functions, Tables, and Skeleton Sizes

• Braking Distances • Function, Domain, and Range • Table of Function Values • Advertising Expenditures • Cases of Cheating • Maximizing Profit • Sizes of Skeletons • Difference Quotient

We live in a world where phenomena are related. For example, if the airfare on a route from New York to London is changed, the number of passengers using that route will probably change. The number of advertisements for the Ford Bronco on television affects the sales of that vehicle. The weight of a package determines the postage cost. In this section, we introduce the concept of a function. Functions are used to analyze how one variable affects another. We motivate the ideas involved with the following example describing braking distances.

BRAKING DISTANCES

The speed of a vehicle determines its braking distance. Researchers found that the relationship between speed (in miles per hour) and braking distance (in feet) for a particular model, for speeds of 30–80 mph under good dry conditions, is described by

$$\text{Braking distance} = \frac{1}{20}(\text{Speed})^2$$

For example, when the speed is 60 mph,

$$\text{Braking distance} = \frac{1}{20}60^2 = 180 \text{ feet}$$

For each speed from 30 to 80 mph, there will be a corresponding braking distance. We say that the braking distance is a function of the speed.

A convenient function notation has been developed to express this relationship. When the speed is x miles per hour, let the corresponding braking distance be $f(x)$ feet. The notation suggests that f depends on x, and we say "f of x". We write the equation

$$\text{Braking distance} = \frac{1}{20}(\text{Speed})^2$$

in the form

$$f(x) = \frac{1}{20}x^2$$

Thus, for speeds of 30, 60, and 80 mph, the corresponding braking distances $f(30)$, $f(60)$, and $f(80)$ are

$$f(30) = \frac{1}{20}30^2 = 45 \text{ feet}$$

$$f(60) = \frac{1}{20}60^2 = 180 \text{ feet}$$

$$f(80) = \frac{1}{20}80^2 = 320 \text{ feet}$$

We say that "f of 30 is 45," "f of 60 is 180," and "f of 80 is 320."

As speed increases from 30 to 80 mph, braking distance gradually increases from 45 to 320 feet. We call the set of valid x values, namely the interval [30, 80], the **domain** of the function. We call the corresponding set of $f(x)$ values, namely [45, 320], the **range** of the function.

Function, Domain, and Range

Let us formally develop some of the basic mathematical ideas of functions.

> ***Function*** Let X and Y be two sets. A **function** from X to Y is a rule that assigns to each element of X exactly one element of Y.

Let us use the notation f for the function. If x is an arbitrary element of X, let $f(x)$ be the corresponding element in Y. The set X is called the **domain** of the function. $f(x)$ is the **image** of x. The set of all images is the **range** of the function.

For convenience, we introduced function notation in terms of a function f and a variable x. Any letters can be used for the function and the variable; however, it is customary to use letters such as f, g, and h for functions and letters such as x, t, and s for variables. For example, t is often used when the variable is time.

A function f is often defined by means of an equation. For example, the equation

$$f(x) = x^2 + 1$$

where x is a real number defines a function f. Consider $x = 2$. The value of the function at $x = 2$ is

$$f(2) = 2^2 + 1 = 5$$

The image of 2 is 5. Let us find the values of the function at -3 and 4. We get

$$f(-3) = (-3)^2 + 1 = 10, \qquad f(4) = 4^2 + 1 = 17$$

Unless the domain of a function f is given explicitly, we shall take it to be the largest set of real numbers such that $f(x)$ is a real number.

Once again, consider $f(x) = x^2 + 1$. The domain of f is the set of real numbers because $f(x)$ is a real number for all real values of x. The range of f consists of real numbers that are of the form $x^2 + 1$. Because $x^2 \geq 0$ we see that $x^2 + 1 \geq 1$. The range of f is the interval $[1, \infty)$.

The results of this discussion are conveniently expressed in Figure R.23. The curved arrows indicate the relationship between the elements of the domain and the elements of the range.

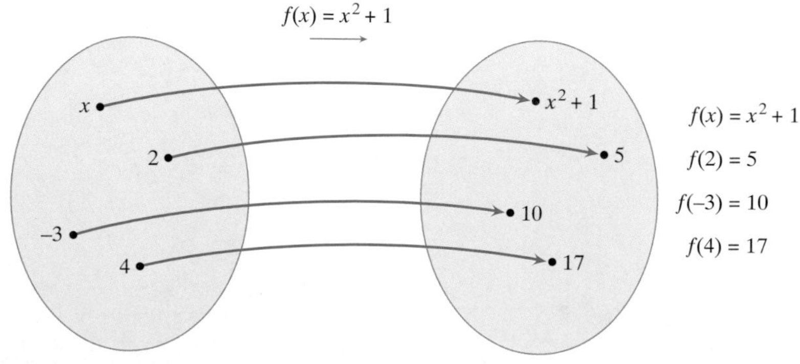

Figure R.23

Example 1 **Finding Values of a Function and Determining Its Domain and Range**

Consider the function $f(x) = \sqrt{x-4}$. Find $f(8)$ and $f(2)$, if they are defined. Find the domain and range of f.

SOLUTION Letting $x = 8$ and $x = 2$, we get

$$f(8) = \sqrt{8-4} = \sqrt{4} = 2$$
$$f(2) = \sqrt{2-4} = \sqrt{-2}$$

$\sqrt{-2}$ is not a real number. We say that $f(2)$ is **undefined.** The domain of f will consist of all values of x for which $\sqrt{x-4}$ is a real number. These are the values $x \geq 4$. Thus the domain of f is $[4, \infty)$. Observe that 2 is not in the domain of f, confirming the fact that $f(2)$ is not defined.

The range of f will consist of the values that $\sqrt{x-4}$ can take for x in the interval $[4, \infty)$. When $x = 4$, $\sqrt{x-4}$ is zero. As x increases, $\sqrt{x-4}$ gradually increases. The range of f is $[0, \infty)$.

Self-Check 1 Let $g(x) = 3x - 4$. (a) Find $g(-5)$, $g(0)$, and $g(3)$. (b) Determine the domain and range of g.

Example 2 **Values of a Function**

Consider the function $f(x) = 2x^2 - 3x + 4$. Find $f(3)$, $f(-4)$, $f(2a)$, $f(a+b)$.

SOLUTION To find the values of f, substitute for x as follows:

$$\begin{aligned}
f(3) &= 2(3)^2 - 3(3) + 4 = 18 - 9 + 4 = 13 \\
f(-4) &= 2(-4)^2 - 3(-4) + 4 = 32 + 12 + 4 = 48 \\
f(2a) &= 2(2a)^2 - 3(2a) + 4 = 8a^2 - 6a + 4 \\
f(a+b) &= 2(a+b)^2 - 3(a+b) + 4 \\
&= 2(a^2 + 2ab + b^2) - 3a - 3b + 4 \\
&= 2a^2 + 4ab + 2b^2 - 3a - 3b + 4
\end{aligned}$$

Because any real number can be used for x, the domain of this function is the set of real numbers. Generally, it is easier to find the domain of a function than the range. $f(x) = 2x^2 - 3x + 4$ is an example of what we call a quadratic function. We do not yet have the tools available to determine the range of this function. We will introduce both algebraic and geometric methods for finding the ranges of such functions in Chapter 1.

Self-Check 2 Let $g(x) = 3x^2 - 6x - 5$. Find $g(-2)$, $g(0)$, $g(-3a)$ and $g(1-2a)$.

Example 3 **Values and Domain of a Rational Function**

Let $s(t) = \dfrac{2t}{t-4}$. Determine $s(0)$, $s(4)$, and $s(8)$, if they are defined. Give the domain of s.

SOLUTION Any letter can be used for the function and the variable. In this example, s has been used for the function and t for the variable. $s(t)$ is an example of a rational function. Letting t take on the values 0, 4, and 8, we get the following:

$$s(0) = \frac{2(0)}{0 - 4} = -\frac{0}{4} = 0$$

$$s(4) = \frac{2(4)}{4 - 4} = \frac{8}{0}, \text{ which is not a real number.}$$

Thus, $s(4)$ is undefined.

$$s(8) = \frac{2(8)}{8 - 4} = \frac{16}{4} = 4$$

Thus, $s(0) = 0$, $s(4)$ is not defined, and $s(8) = 4$.

Note that $s(t)$ is defined for all real values of t except $t = 4$ because this is the only value for which the denominator is zero. The domain of s consists of the two intervals $(-\infty, 4)$ and $(4, \infty)$. This domain is said to be the **union** of these two intervals. We denote union by the symbol $\cup$. The domain is written $(-\infty, 4) \cup (4, \infty)$. (We leave the discussion of the range of a rational function until Section 3.5, where rational functions will be discussed in depth.)

Images of Elements in the Domain

Let us now look at an important property of functions. According to the definition of a function, two or more elements in the domain can have the same single image in the range. This property is illustrated in Figure R.24(a). However, two or more distinct elements in the range cannot be the images of a single element in the domain. See Figure R.24(b).

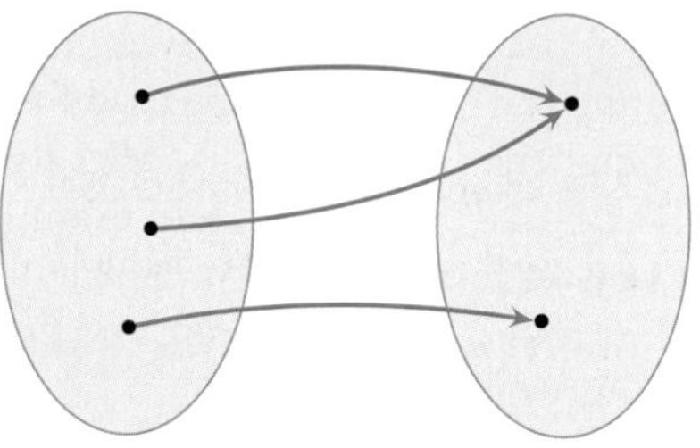

(a) Possible for functions

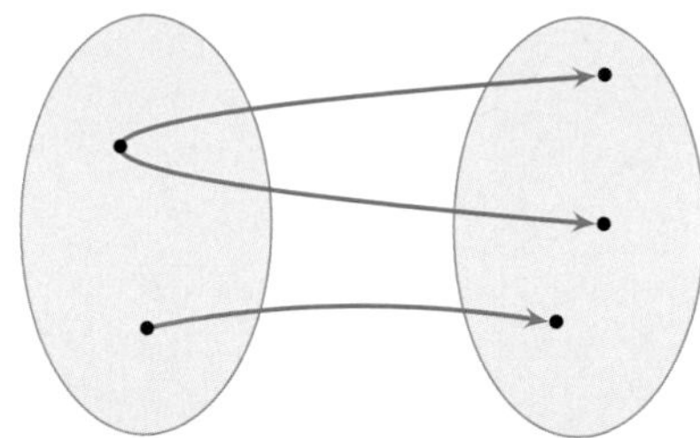

(b) Not possible for functions

Figure R.24

Example 4 Two Elements in the Domain with a Single Image

Consider the function $f(x) = 3x^2 - 5$. Show that it is possible for two or more elements in the domain to have the same single image in the range by comparing the values of $f(-2)$ and $f(2)$. Find the domain and range of f.

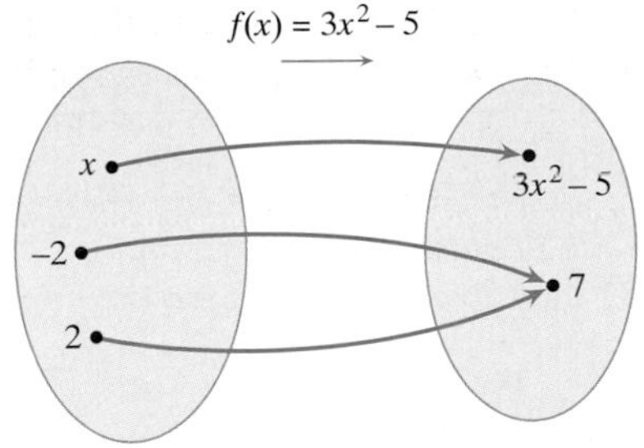

Figure R.25 Domain: Set of real numbers. Range: Set $[-5, \infty)$

SOLUTION We get

$$f(-2) = 3(-2)^2 - 5 = 12 - 5 = 7$$
$$f(2) = 3(2)^2 - 5 = 12 - 5 = 7$$

Observe that $f(-2) = f(2) = 7$. The image of both -2 and 2 is the number 7. We see that it is possible for distinct elements in the domain to have the same image. This is illustrated in Figure R.25.

Because x can be any real number, the domain of f will be the set of real numbers. The range of f will be the set of numbers of the form $3x^2 - 5$. Because $3x^2 \geq 0$, we have $3x^2 - 5 \geq -5$. The range is the set $[-5, \infty)$.

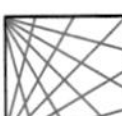

Functions with a Calculator

Some calculators can be used to compute values of a function using function notation. Consider $f(x) = x^3 - 4x$. Let us compute $f(3.4)$ (Figure R.26). $f(3.4) = 25.704$.

```
Y1=X^3-4X
Y2=
Y3=
Y4=
Y5=
Y6=
```

Enter the function.

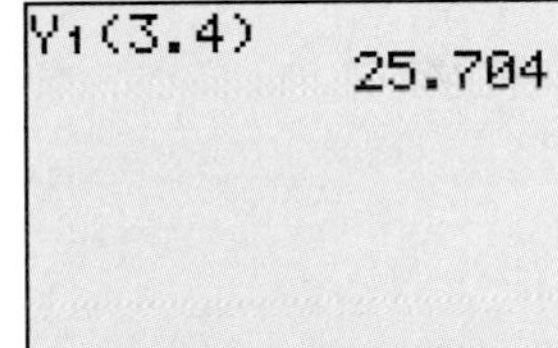

Evaluate the function at the x-value.

Figure R.26 Values of functions.

Self-Check 3 Let $g(x) = 3.8x^2 - 6.1x - 5.98$. Determine how best to use your calculator to compute values of functions. Use a calculator to find the values of $g(x)$ at $x = 2.57$, 5.6543, and -3.41.

Table of Function Values

Let us once again consider the vehicle that has a braking distance described by

$$f(x) = \frac{1}{20}x^2$$

for speeds of 30–80 miles per hour. Recall that $f(x)$ is the braking distance in feet corresponding to speed x miles per hour. The braking distances for speeds of 30, 35, . . . , 75, 80 miles per hour (in increments of 5) can be summarized as follows:

Speed (miles per hour)	30	35	40	45	50	55	60	65	70	75	80
Distance (feet)	45	61.25	80	101.25	125	151.25	180	211.25	245	281.25	320

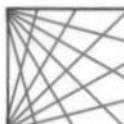

Tables with a Calculator

Many calculators have table facilities. Consider $f(x) = \frac{1}{20}x^2$. Let us construct the table of values for $f(30), f(35), \dots,$ in increments of 5. Enter the equation that describes the function (Figure R.27), and set the table minimum to 30 and the x increment to 5. Display the table of values. Scroll to see additional values.

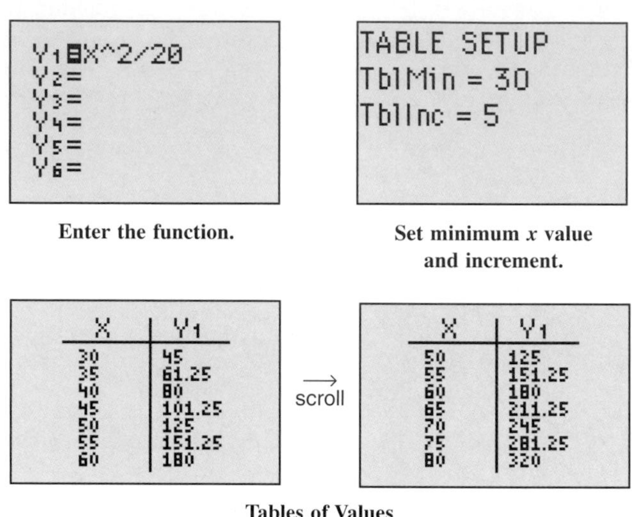

Figure R.27 Constructing a table.

Applications . . .

Example 5 Advertising Expenditures

An advertising company has studied the relationship between the amount of money $\$x$ spent on advertising a certain product and the realized sales of that product. The company found that for advertising amounts of \$1000 to \$3000, the number of items sold is closely described by the function

$$f(x) = 0.05x + 470$$

Thus, for expenditures of \$1000, \$2000, and \$3000, the corresponding sales are $f(1000), f(2000), f(3000)$, where

$$\begin{aligned} f(1000) &= 0.05(1000) + 470 = 520 \text{ items} \\ f(2000) &= 0.05(2000) + 470 = 570 \text{ items} \\ f(3000) &= 0.05(3000) + 470 = 620 \text{ items} \end{aligned}$$

As the dollar amount spent on advertising is increased from \$1000 to \$3000, sales should steadily increase from 520 to 620 items.

Note that there is a possible difference between the formal mathematical domain of a function in an application and a domain over which the function has relevance. In this example, the mathematical domain of the function

$f(x) = 0.05x + 470$ is the set of real numbers because $f(x)$ is a real number for all real values of x; however, the function is only valid for advertising expenditures of \$1000 to \$3000. The function only has physical meaning for values of x from 1000 to 3000. The meaningful domain of the function $f(x) = 0.05x + 470$ in this application is $[1000, 3000]$. When we refer to domain in applications, it will be to this meaningful set for which the function is relevant.

Further Ideas 1 When a mathematical function such as the advertising function $f(x) = 0.05x + 470$ is used in an application, we have seen that careful thought must be given to interpretation. In this application, the meaning of $f(2000) = 570$ is clear; when \$2000 is spent for advertising, the predicted sales are 570 items. What does $f(2354) = 587.7$ mean? Discuss the various possibilities.

Example 6 Cases of Cheating

The function $f(x) = -0.2x^2 + x + 11.1$ closely describes the distribution of cheating in a study involving 2500 students. In ten opportunities to cheat, $f(x)$ gives the percentage of students who cheated x times. For example, $f(4) = -0.2(4)^2 + 4 + 11.1 = 11.9$ means that the percentage of students who cheated four times was 11.9%. (We will see in Sections 1.2 and 1.5 how such functions are constructed by fitting graphs to data.)

Because x takes on the integer values 0–10 in this application, the domain of f is the set of integers $\{0, 1, 2, 3, 4, 5, 6, 7, 8, 9, 10\}$. The range consists of the corresponding values of f. The range can be found from the table of values of $f(x)$ in Figure R.28. The range is the set of real numbers $\{1.1, 3.9, 6.3, 8.3, 9.9, 11.1, 11.9, 12.3\}$. (We list each element of the set once and in ascending order). The table reveals that 11.1% of the students did not cheat. However 1.1% of the students cheated ten times, namely at every opportunity. The model provides additional information. For example, it sadly reveals that there was a tendency for the students in this sample to be honest in some situations and dishonest in others.

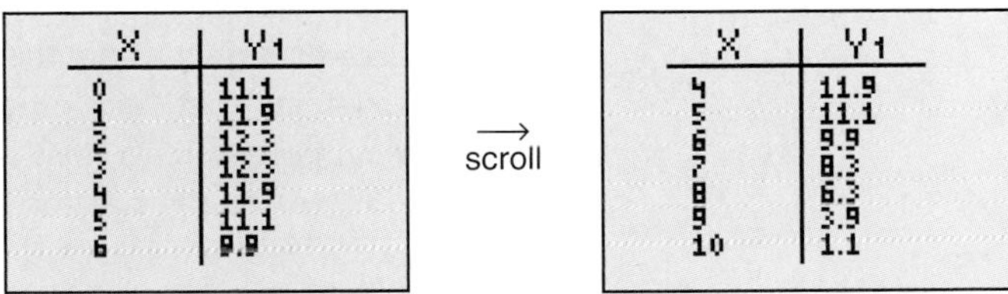

Figure R.28 Table of cases of cheating.

Further Ideas 2 Consider the study described in Example 6. What is the most common number of times that cheating occurred among these students?

Example 7 Maximizing Profit

To increase profit, manufacturing companies must decide whether to increase or decrease output of the product. In practice, profit will increase as output increases, up to a point, but then a threshold is reached, beyond which profit will gradually decrease. This is caused by such factors as more breakdowns in equipment and more energy consumption at higher production rates. This phenomenon is called the Law of Diminishing Returns. For example, a company has determined that the profit function on manufacturing n items of goods per day is $P(n) = -2n^2 + 960n$ when $200 \leq n \leq 300$, where the units of profit are dollars. The function does not accurately describe profit outside this interval. What is the domain of this profit function? What should the production level be to maximize profit and what is that profit?

SOLUTION In this example, n is the number of items manufactured. Thus, n must be a positive integer. The domain of the function $P(n)$ is the set of integers $\{200, 201, 202, \ldots, 299, 300\}$. The range of the function will be the set of corresponding values of P, namely $\{P(200), P(201), P(202), \ldots, P(299), P(300)\}$. Consider the table of P values from $P(200)$ to $P(300)$ in Figure R.29. The largest value of P occurs when $n = 240$. The production level for maximizing profit is therefore 240 units per day. The maximum profit will be \$115,200.

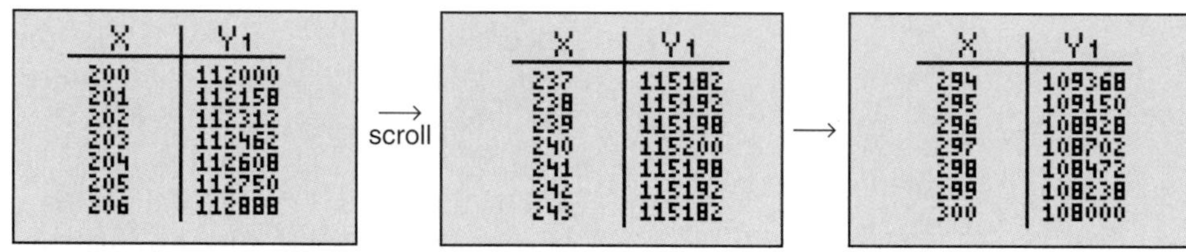

X	Y1
200	112000
201	112158
202	112312
203	112462
204	112608
205	112750
206	112888

→ scroll

X	Y1
237	115182
238	115192
239	115198
240	115200
241	115198
242	115192
243	115182

→

X	Y1
294	109368
295	109150
296	108928
297	108702
298	108472
299	108238
300	108000

Figure R.29 Table of profits.

Note that in applications it is often convenient to use letters that suggest the real variables, such as n for number of items and P for profit. However most calculators require these variables to be labeled, as here, using X and Y.

Example 8 Sizes of Skeletons

The world we live in is governed by scientific laws, and biologists attempt to discover and understand these laws. For example, they study the relationships between sizes of animals and their various biological components. Skeletons are support systems that keep animals from collapsing. Larger species must have stronger support systems than smaller species. If all linear dimensions of an animal are increased by a factor of 2, the volume will increase by a factor of 8, implying an eightfold increase in weight. The mass of the skeleton becomes an increasingly larger fraction of the weight of an animal as the size increases. For example the bones of an elephant must be proportionately heavier and stockier than those of a mouse. There is, however, a physiologic upper limit to this fraction for land animals. Beyond this weight, they cannot survive even though larger aquatic animals, such as whales, can exist because some of their weight can be supported by water. Professor Prange and his associates* have found that the skeleton/body weight ratio $f(x)$ of a land animal of weight x pounds is given by

$$f(x) = \frac{0.061x^{1.09}}{x}$$

*"Scaling of Skeletal Mass to Body Mass in Birds and Mammals," *American Naturalist,* 1979, volume 13, pages 103–22.

Using this equation, we can determine the skeleton/body weight ratio for a person who weighs 140 pounds:

$$f(140) = \frac{0.061(140^{1.09})}{140} = 0.0952 \qquad \text{to four decimal places}$$

Body weight	Skeleton weight / Body weight
X	Y1
140	.09517
160	.09632
180	.09734
200	.09827
220	.09912
240	.0999
260	.10062

Figure R.30

The table of $f(x)$ in Figure R.30 shows how the skeleton weight becomes an increasingly larger fraction of the total weight of the animal. The table shows that somewhere between 240 and 260 pounds the skeleton weight becomes a tenth of the total weight of a human being. The weight of a fully grown elephant is approximately 22,000 pounds. We get

$$f(22{,}000) = \frac{0.061(22{,}000)^{1.09}}{22{,}000} = 0.1500$$

The skeleton/body weight ratio of an elephant is 0.1500.

A brachiosaurus was approximately 70 feet long and weighed around 170,000 pounds:

$$f(170{,}000) = \frac{0.061(170{,}000)^{1.09}}{170{,}000} = 0.1803$$

The skeleton/body weight ratio of a brachiosaurus is 0.1803. Scientists believe that this number represents the largest such ratio that a land animal can sustain. Because land animals cannot sustain a skeleton that is a larger fraction of body weight, no larger land animal can exist.

Example 9 Difference Quotient

The expression $\frac{f(x+h)-f(x)}{h}$ is called a **difference quotient.** Students who go on to study calculus will find that the difference quotient plays a fundamental role in that field. It enables us to define rate of change (how rapidly things are changing). Consequently, we can discuss the rate of change of velocity (acceleration), the rate of change of profit, or the rate at which an epidemic is spreading. Let's look at the difference quotient of the function $f(x) = x^2 + 2x - 1$.

$$\frac{f(x+h)-f(x)}{h} = \frac{[(x+h)^2 + 2(x+h) - 1] - (x^2 + 2x - 1)}{h}$$

We now simplify this result.

$$\begin{aligned}\frac{f(x+h)-f(x)}{h} &= \frac{x^2 + 2xh + h^2 + 2x + 2h - 1 - x^2 - 2x + 1}{h}\\ &= \frac{2xh + h^2 + 2h}{h}\\ &= 2x + h + 2\end{aligned}$$

Self-Check 4 Determine the difference quotient $\frac{f(x+h)-f(x)}{h}$ of $f(x) = 2x^2 + 3$.

Answers to Self-Check Exercises

1. (a) $g(-5) = -19, g(0) = -4, g(3) = 5$. (b) The domain and range of g are both the set of real numbers.
2. $g(-2) = 19, g(0) = -5, g(-3a) = 27a^2 + 18a - 5$, $g(1 - 2a) = 12a^2 - 8$.
3. $g(2.57) = 3.44162$, $g(5.6543) = 81.01898226$, and $g(-3.41) = 59.00778$.
4. $4x + 2h$

Answers to Further Ideas Exercises

1. $f(2354) = 587.7$. There will be 587 items sold, but not 588 items. Thus, when $2354 is spent for advertising, the predicted sales are 587 items.
2. Largest value of f, $f(2) = f(3) = 12.3$. Thus, 12.3% of the students cheated twice, and another 12.3% cheated three times. Therefore, the most common number of times students cheated was two and three times (equal).

EXERCISES R.7

In Exercises 1–6, find the values of the functions.

1. If $f(x) = 3x - 3$, find **(a)** $f(1)$, **(b)** $f(4)$, **(c)** $f(0)$, **(d)** $f(-2)$.
2. If $f(x) = 7x + 4$, find **(a)** $f(0)$, **(b)** $f(3)$, **(c)** $f(-1)$, **(d)** $f(-5)$.
3. If $f(x) = x^2 + 2x - 3$, find **(a)** $f(2)$, **(b)** $f(5)$, **(c)** $f(0)$, **(d)** $f(-3)$.
4. If $f(x) = 3x^2 - 5x + 1$, find **(a)** $f(-3)$, **(b)** $f(-1)$, **(c)** $f(1)$, **(d)** $f(3)$.
5. If $g(x) = x^3 + 2x^2 - 4x + 3$, find **(a)** $g(-4.1)$, **(b)** $g(-2.8)$, **(c)** $g(0)$, **(d)** $g(2.3)$.
6. If $g(x) = -2x^3 + 5x - 3$, find **(a)** $g(-3.5)$, **(b)** $g(-1.26)$, **(c)** $g(0)$, **(d)** $g(4.38)$.

In Exercises 7–12, find the values of the functions if they are defined.

7. If $h(x) = \sqrt{3x - 2}$, find **(a)** $h(2)$, **(b)** $h(9)$, **(c)** $h(-1)$.
8. If $f(x) = \sqrt{2x + 5}$, find **(a)** $f(4)$ to four decimal places, **(b)** $f(0)$ to four decimal places, **(c)** $f(-3)$.
9. If $f(x) = \frac{2x + 4}{x}$, find **(a)** $f(1)$, **(b)** $f(-2)$, **(c)** $f(0)$.
10. If $f(x) = \frac{x + 3}{x - 2}$, find **(a)** $f(0)$, **(b)** $f(3)$, **(c)** $f(-1)$ to four decimal places.
11. If $f(x) = \frac{2x}{x^2 - 3x}$, find **(a)** $f(-2)$, **(b)** $f(2)$, **(c)** $f(3)$.
12. If $h(x) = \frac{x + 2}{2x^2 - 5x - 3}$, find **(a)** $h(0)$ to four decimal places, **(b)** $h(-2)$, **(c)** $h(3)$.

In Exercises 13–18, find the values of the functions.

13. If $f(x) = 4x + 1$, find $f(a)$.
14. If $f(x) = 3x - 2$, find $f(3u)$.
15. If $g(x) = 2x^2 - x + 3$, find $g(-2a)$.
16. If $f(x) = 2x$, find $f(2a + 1)$.
17. If $g(t) = 3t - 4$, find $g(3s + 2)$.
18. If $h(a) = a^2 - 2a + 3$, find $h(a + 1)$.

In Exercises 19–28, find the domain of each function.

19. $f(x) = 3x - 1$
20. $g(x) = 7x + 2$
21. $f(x) = 3x^2 + 4x - 1$
22. $f(x) = \sqrt{x}$

23. $f(x) = \sqrt{x - 3}$

24. $g(x) = \sqrt{5 - x}$

25. $f(x) = \dfrac{2x}{x - 6}$

26. $f(x) = |x + 2|$

27. $f(x) = \dfrac{1}{\sqrt{x - 4}}$

28. $f(x) = \dfrac{1}{(x + |x|)}$

In Exercises 29–36, find the domain and range of each function.

29. $f(x) = 3x - 2$

30. $f(x) = 7x - 1$

31. $g(x) = 7$

32. $f(x) = x^2$

33. $g(x) = x^2 + 8$

34. $f(x) = 2x^2 - 3$

35. $f(x) = \sqrt{x - 3}$, domain only

36. $g(x) = \sqrt{2x + 6} + 2$, domain only

In Exercises 37–44, find the domain and range of each function.

37. $f(x) = |x|$

38. $f(x) = |x - 2| - 3$

39. $g(x) = -|2x - 3|$

40. $g(x) = x^3$

41. $f(x) = x^4 - 2$

42. $f(x) = \sqrt{x - 2} + 4$

43. $f(x) = \dfrac{1}{x}$

44. $g(x) = \dfrac{1}{x - 2}$

In Exercises 45–50, use a graphing calculator to compute the values of the functions at the given points (if possible), rounded to four decimal places. What is the significance of an error message on your calculator?

45. $f(x) = 3.2x^2 + 4.1$. **(a)** $x = 2.5$, **(b)** $x = 3.92$, **(c)** $x = -7.81$.

46. $f(x) = 2.8x^2 + 3.1x - 5.67$. **(a)** $x = 3.47$, **(b)** $x = 5.983$, **(c)** $x = -4.3245$.

47. $f(x) = \sqrt{x - 2}$. **(a)** $x = 3.5$, **(b)** $x = 8.75$, **(c)** $x = 0$.

48. $f(x) = \sqrt{x^2 + 3.75}$. **(a)** $x = 9.8$, **(b)** $x = 23.3$, **(c)** $x = -3.1$.

49. $f(x) = \sqrt{x^2 - 3x - 4}$. **(a)** $x = 5$, **(b)** $x = 4$, **(c)** $x = 2$.

50. $f(x) = \dfrac{14}{x - 2.5}$. **(a)** $x = 1.4$, **(b)** $x = -5.7$, **(c)** $x = 2.5$.

In Exercises 51–54, determine the difference quotients $\dfrac{f(x + h) - f(x)}{h}$ for the given functions. Give your answers in simplified form.

51. $f(x) = x^2 + 3$

52. $f(x) = 4x - 5$

53. $f(x) = 2x^2 - x + 1$

54. $f(x) = x^3$

55. **Braking Distances** The braking distance f (in feet) of an automobile for speeds of 10–85 miles per hour under good, dry conditions is a function of speed x (in miles per hour), $f(x) = \frac{1}{25}x^2$.

(a) Determine the braking distances for speeds of 10, 15, 40, and 72 miles per hour.

(b) What is the mathematical domain of f and what is its meaningful domain in this application?

56. **Profit Function** A profit function $P(x)$ gives the total profit (in dollars) obtained by a company manufacturing x items of its goods during a certain period. A company has determined that the profit function is $P(x) = -x^2 + 850x$ when $250 \leq x \leq 550$.

(a) Determine the profits for sales of 250, 300, 350, 400, 450, 500, and 550 units.

(b) How many items should the company manufacture to get the largest profit?

57. **Vertical Motion** An object is projected vertically upward from ground level with a velocity of 160 feet per second. Its height after t seconds is a function $s(t) = -16t^2 + 160t$ of time.

(a) Determine the height every second, from 0 to 10 seconds.

(b) At what time does the object return to ground level?

(c) What is the meaningful domain of the function in this application?

58. **Vertical Motion** An object is projected vertically upward from ground level with an initial velocity of 192 feet per second. Subsequent velocities are a function of time, $v(t) = -32t + 192$.

(a) Compute velocities at 1-second intervals for 12 seconds.

(b) How do you interpret the fact that $v(6) = 0$? What information does this give?

(c) How do you interpret negative values of velocity?

59. **Spread of a Disease** A communicable disease is spreading through a certain country. The medical authorities have concluded that the disease is spreading in a manner described by the function $N(t) = 48t^2 - t^3$. Here t represents time in months from the identification of the initial case of the disease, and N represents the number of cases of the disease in thousands at time t.

(a) Determine the number of cases after 1, 2, 3, 4, 5, 6, and 7 months.

(b) Determine to the nearest month when the disease will peak.

(c) After how many months will there be no cases of the disease present?

60. Duck Population Analysis The duck population in Earl Jones Park has been estimated to have a birth pattern described by the function $N(t) = 3t^3 - 36t^2 + 108t + 750$. Here t is time in years, with $t = 0$ corresponding to 1988. $N(t)$ is the number of births that occurred during the year t.

(a) Estimate number of ducks born each year from 1990 to 1996.

(b) Estimate for the total number of ducks born from 1990 to 1996.

(c) Describe the trend in births during this period.

61. Automobile Production The number of new automobiles produced on a worldwide basis since 1986 has been estimated to be described by $N(t) = 6t^2 + 72t + 9000$, where t is time measured in years from 1986 and $N(t)$ is the number of cars in units of 100 produced during year t.

(a) Estimate the number of new cars produced annually from 1989 to 1995 and the total number of new cars manufactured during this period.

(b) If current trends continue, when will the number of new cars produced annually exceed 1,200,000?

62. Ideal Body Weight

(a) The function $M(x) = 5.5x - 220$ describes the ideal body weight of men, where x is height in inches and $M(x)$ is the corresponding weight in pounds. This equation is usually valid for heights of 60 to 84 inches. Compute M(72). What does the result mean? Give the domain and range of the function M in this application.

(b) The function $W(x) = 5.4x - 231$ describes the ideal body weight of women, where x is height in inches and $W(x)$ is the corresponding weight in pounds. This equation is usually valid for heights of 54 to 78 inches. Compute $W(66)$. What does the result mean? Give the domain and range of the function W.

63. Cable Television The function $P(t) = 0.0057t^3 - 0.302t^2 + 6.087t + 22.6$ closely describes the percentage of U.S. households with cable television during the period 1980 to 1995. t is time measured in years from 1980. For example, $P(0) = 22.6$ tells us that 22.6% of U.S. households had cable television in 1980.

(a) Determine the percentage of households with cable television in 1995.

(b) What are the predictions for the years 2000 and 2010?

(c) Compute $P(35)$. How do you interpret this number? Discuss the problem of using the function $P(t)$ to make long-term predictions.

64. Area of a Rectangle A rectangle with one side of length x units has a perimeter of length 72 units.

(a) Express the area A as a function of x.

(b) Determine $A(6)$, $A(8)$, and $A(30)$.

(c) Why do you expect $A(6)$ to be equal to $A(30)$?

(d) What is the domain of A?

65. Volume of a Box The surface area of four sides and the base of an open box having a square base of side x inches is 64 inches.

(a) Express the volume V as a function of x.

(b) Compute $V(0)$, $V(1)$, $V(2)$, …, $V(8)$.

(c) What is the domain of V in this application?

(d) For which integer value of x is the volume largest? What is that volume?

66. Housing Costs A builder constructs houses on lots that cost \$25,000.

(a) If the houses cost \$32.50 per square foot, determine the cost of houses as a function of area.

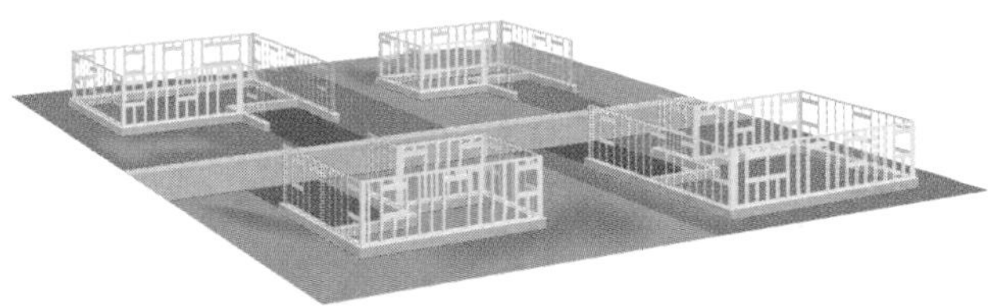

(b) Models are available with areas of 1800, 1900, 2000, 2100, 2200, 2300, and 2400 square feet. If you plan to spend \$100,000 for a house, how large a house can you afford?

CHAPTER PROJECT | Space–Time Travel

Suppose that a spaceship travels to a galaxy and then returns to Earth. If the spaceship travels a total distance d with velocity v, then we expect that the duration of the voyage will be

$$\text{Time} = \frac{\text{Distance}}{\text{Velocity}} = \frac{d}{v}$$

However, special relativity* tells us that this formula only gives the duration of the voyage for a person on Earth. Relativity tells us that the duration of the voyage for a person on the spaceship is obtained by multiplying this time by a factor of $\sqrt{1 - v^2}$, which results in a shorter time. If t_E and t_s are Earth and spaceship times, then

$$t_E = \frac{d}{v}, \qquad t_s = \frac{d}{v}\sqrt{1 - v^2}$$

Alpha Centauri is the nearest star to our solar system. *(©Harvard–Smithsonian Center for Astrophysics)*

In these formulas, the units of distance d are light years (1 light year is the distance light travels in 1 year), and the velocity v is expressed as a fraction of the velocity of light. Suppose a spaceship travels to Alpha Centauri at 0.8 the speed of light and then returns to Earth. Alpha Centauri is 4.3 light years from Earth. (It takes light 4.3 years to reach Earth from Alpha Centauri.) The total distance traveled by the spaceship, to the star and back, is 8.6 light years. Let us compute the duration of the voyage relative to Earth and the spaceship. For this voyage, we have

$$d = 8.6 \qquad \text{and} \qquad v = 0.8$$

These lead to the following values of t_E and t_s:

$$t_E = \frac{8.6}{0.8} = 10.75, \qquad t_s = \frac{8.6}{0.8}\sqrt{1 - 0.8^2} = 6.45$$

Relativity predicts that 10.75 years will have passed on Earth between the departure of the spaceship and its return. On the other hand, 6.45 years will have passed on the spaceship.

*Special relativity was developed by Albert Einstein (1879–1955) when he was 26 years of age. Einstein is considered to be one of the greatest scientists of all time. His theory of relativity revolutionized scientific thought with new ways of thinking of time, space, mass, motion, and gravitation. His famous equation, $E = mc^2$ (energy equals mass times the velocity of light squared), became a foundation stone in the development of atomic energy. Einstein received the Nobel Prize in 1921 for his work on the so-called photoelectric effect. The photoelectric cell, resulting from this work, made possible sound motion pictures and television.

Einstein lived a quiet personal life. He was fond of classical music and played the violin. He supported Zionism and was offered, but declined, the presidency of Israel in 1952. Although not associated with any orthodox religion, Einstein was deeply religious. He never believed that the universe was one of chance or chaos. He once said, "God may be sophisticated, but He is not malicious."

A person on Earth will have aged 10.75 years during the voyage, but a person on the spaceship will have aged only 6.45 years.

We do not presently have the technology to send spaceships to distant galaxies at such great velocities. Nevertheless, numerous experiments have been constructed to test this hypothesis that time on Earth varies from time recorded on an object moving relative to Earth. We mention one experiment conducted by Professor J. C. Hafele, Department of Physics, Washington University, St. Louis, MO, and Dr. Richard E. Keating, U.S. Naval Observatory, Washington, DC, reported in the journal *Science,* 1972, volume 177. During October 1971, four atomic clocks were flown around the world twice on regularly scheduled jet flights. There were slight differences in the times recorded by these clocks and the clocks on Earth. These differences were in accord with the predictions of relativity theory. This aging variation will not be realized by humans in the near future, to the extent illustrated by the trip to Alpha Centauri, because the energies required to produce such high speeds in a macroscopic body are prohibitive. However, the astronauts who went to the moon experienced this phenomenon to a very small extent.

(a) The star Sirius is 8 light years from Earth. Sirius is the nearest star other than the sun that is visible with the naked eye from most parts of North America. It is the brightest appearing of all stars. Light reaches us from the sun in 8 minutes, whereas it takes 8 years to reach us from Sirius. If a spaceship travels to Sirius and back at 0.9 the velocity of light, how many years will the voyage last for a person on Earth and for a person on the spaceship, rounded to two decimal places?

(b) A spaceship makes a round-trip flight to the bright star Capella, which is 45 light years from Earth at 0.99 the velocity of light. How many years will the voyage last for a person on Earth and for a person on the spaceship, rounded to two decimal places?

(c) Determine the distances of some other stars and compute the durations of round-trip flights to those stars at various velocities.

(d) This discussion introduced the behavior of time as described by the theory of special relativity. Are you convinced of the results? Give reasons for believing/not believing. Discuss some of the implications of this behavior of time in special relativity (e.g., the overlapping of generations).

CHAPTER R HIGHLIGHTS

Section R.1

Natural numbers: 1, 2, 3, 4, ... The numbers used for counting.

Integers: . . . , −3, −2, −1, 0, 1, 2, 3, ... The natural numbers with their negatives and zero.

Rational numbers: Numbers of the form $\frac{a}{b}$, a and b integers with $b \neq 0$.

Irrational numbers: Numbers that cannot be expressed in the form $\frac{a}{b}$. The decimal form does not terminate or have a repeating pattern.

Real numbers: The set consisting of all the rational and irrational numbers.

The real number line: The real numbers can be represented by points on this line.

Inequalities: $a < b$; a is less than b; a lies to the left of b on the real line. $a \leq b$; a is less than or equal to b; a lies to the left of b or coincides with b. $a > b$; a is greater than b; a lies to the right of b. $a \geq b$, a is greater than or equal to b; a lies to the right of b or coincides with b.

Interval: A set of numbers on the real line that forms a line segment, a half line, or an entire line. Intervals are closed, open, or half-open depending on whether the end points are included.

Rounding: Round down if the first digit discarded is less than 5. Round up if the first digit discarded is greater than or equal to 5.

Absolute value: $|a|$ is the size or magnitude of a. $|a| = a$ if $a \geq 0, |a| = -a$ if $a < 0$.

Algebraic properties: Addition and multiplication of real numbers are commutative, associative, and distributive. The additive identity is 0, and the multiplicative identity is 1. The additive inverse of a is $-a$. The multiplicative inverse of $a(\neq 0)$ is $1/a$.

Powers: $a^n = a \cdot a \cdot \ldots \cdot a$, n times. a is the base and n is the exponent.

Hierarchy of operations: Reading from left to right, all exponentiations are performed first, then multiplications and divisions, and finally additions and subtractions.

Section R.2

Exponents and radicals: $a^m \cdot a^n = a^{m+n}$. $(a^m)^n = a^{mn}$. $a^0 = 1\ (a \neq 0)$. $a^{-n} = 1/a^n$. $a^m/a^n = a^{m-n}$.

If a is a real number, m is an integer, and n is a positive integer, then

$$a^{1/n} = \sqrt[n]{a},\ a^{m/n} = (a^{1/n})^m = (\sqrt[n]{a})^m \text{ (if } \sqrt[n]{a} \text{ exists)}$$

If a is a real number, and n is a positive integer, then $\sqrt[n]{a^n} = a$ if n is odd and $|a|$ if n is even.

$$(ab)^m = a^m \cdot b^m. \qquad (a/b)^m = a^m/b^m\ (b \neq 0).$$

Scientific notation: Number is of the form $\pm a \times 10^n$, where $1 \leq a < 10$ and n is an integer. $\pm a$ is the mantissa, and n is the characteristic.

Distance, speed, time: Distance = Speed × Time.

Compound interest: Given principle P and interest r per annum, after t years amount is

$$A = P(1 + r)^t$$

Section R.3

Polynomial of degree *n* in the variable *x*: $a_n x^n + a_{n-1}x^{n-1} + \cdots + a_1 x + a_0$.

Addition of polynomials: Add like terms.

Multiplication of polynomials: Multiply each term of the first polynomial by every term of the second polynomial and simplify.

FOIL method: For multiplying two polynomials of degree 1. **F**irst terms are multiplied, then the **O**uter terms, **I**nner terms, and **L**ast terms are multiplied. Finally, all results are added.

Special products: Let a and b be variables or algebraic expressions.

$$(a+b)(a-b)=a^2-b^2, \qquad (a+b)^2=a^2+2ab+b^2,$$
$$(a-b)^2=a^2-2ab+b^2, \qquad (a+b)^3=a^3+3a^2b+3ab^2+b^3,$$
$$(a-b)^3=a^3-3a^2b+3ab^2-b^3$$

Section R.4

Special factors: Let a and b be variables or algebraic expressions.

$$a^2-b^2=(a+b)(a-b), \qquad a^2+2ab+b^2=(a+b)^2,$$
$$a^2-2ab+b^2=(a-b)^2, \qquad a^3+b^3=(a+b)(a^2-ab+b^2),$$
$$a^3-b^3=(a-b)(a^2+ab+b^2)$$

Factoring strategy: (1) Factor out the greatest common factor. (2) Look for special factors. (3) Look for polynomial factors. (4) Try to group terms together for factoring.

Completing the square: To make x^2+px into a perfect square, add $\left(\frac{p}{2}\right)^2$.

Then get

$$x^2+px+\left(\frac{p}{2}\right)^2=\left(x+\frac{p}{2}\right)^2$$

Section R.5

Rational expressions: Let a, b, c, and d represent polynomials.

$$\text{Simplification } \frac{ac}{bc}=\frac{a}{b}, \qquad \text{Multiplication } \frac{a}{b}\cdot\frac{c}{d}=\frac{ac}{bd},$$
$$\text{Division } \frac{a}{b}\div\frac{c}{d}=\frac{a}{b}\cdot\frac{d}{c}=\frac{ad}{bc}, \qquad \text{Addition } \frac{a}{b}+\frac{c}{d}=\frac{ad+bc}{bd},$$
$$\text{Subtraction } \frac{a}{b}-\frac{c}{d}=\frac{ad-bc}{bd}$$

Section R.6

Equivalent equations: Equations with the same solutions.

Generating equivalent equations: (1) Add or subtract the same real number or real expression to or from both sides of the equation. (2) Multiply or divide both sides of the equation by the same nonzero real number.

Linear equation in *x*: An equation that can be written in the form $ax+b=0$, $a\neq 0$. $ax+b=0$ is called the standard form of the linear equation.

Variation: y varies directly as x, $y=kx$; y varies inversely as x, $y=k(1/x)$; y varies directly as the square of x, $y=k(x^2)$. k is the constant of variation.

Section R.7

Function: A function f from X to Y is a rule that assigns to each element of a set X exactly one element of a set Y. X is the domain. If x is in X, $f(x)$ is the image of x in Y. The set of all images is the range.

Chapter R Review Exercises

1. Let $A = \{-7, \frac{3}{4}, 0, 2, \frac{5}{4}, 12, \sqrt{2}\}$. List all the elements of A that are **(a)** natural numbers, **(b)** integers, **(c)** integers but not natural numbers, **(d)** rational numbers, **(e)** irrational numbers.

2. Graph each of the following intervals:
   (a) $(-3, 8]$ **(b)** $[-2, \infty)$ **(c)** $(-\infty, 4)$

3. Indicate which algebraic property (or properties) of real numbers is used to justify each of the following statements.
   (a) $4 + 7 = 7 + 4$
   (b) $4 \cdot (3 + 7) = 4 \cdot 3 + 4 \cdot 7$
   (c) $3 + (4 + 7) = (3 + 4) + 7$
   (d) $(xy)z = x(yz)$
   (e) $(x + y)z = xz + yz$

4. Compute the following expressions using your calculator:
   (a) $5.793 \times (2.58 + 3.46)$
   (b) $\sqrt{37} \div (2.3 + 1.78)$, rounded to five decimal places
   (c) $(\sqrt{7.9} + 1.3^{-1}) \times 4.65$, rounded to two decimal places

5. Write the following calculator expressions in standard algebraic form and compute the result. Check your answers by performing the sequences of keystrokes on your calculator.
   (a) $8 + 9 \div 3$
   (b) $7 \times 3 + 12 \div 6$
   (c) $4 \times 3 \wedge 2 + 4 \div 2$

6. Evaluate the following by hand. Support your answers using a calculator.
   (a) $\sqrt{121}$ **(b)** $\sqrt[3]{27}$
   (c) $\sqrt[3]{\frac{-125}{8}}$ **(d)** $\left(\sqrt[4]{\frac{16}{81}}\right)^3$

7. Compute the following roots by hand. Support your answers using a calculator.
   (a) $25^{1/2}$ **(b)** $(-8)^{1/3}$
   (c) $243^{1/5}$ **(d)** $(-128)^{1/7}$

8. Compute the following (if they exist as real numbers) by hand using rules of exponents. Support your answers using a calculator.
   (a) $27^{2/3}$ **(b)** $25^{-3/2}$
   (c) $(-125)^{4/3}$ **(d)** $(-128)^{-5/4}$

9. Evaluate the following by hand using rules of exponents. Support your answers using a calculator.
   (a) $\frac{4^3}{5} \cdot 5^7$ **(b)** $(2^3) \cdot (4)^{-1}$ **(c)** $(3 \cdot 4 \cdot 5)^{-2} \cdot 45$

10. Evaluate the following by hand using rules of exponents. Support your answers using a calculator.
   (a) $2^{1/2} \cdot 2^{5/2}$ **(b)** $5^{11/4} \cdot 5^{-3/4}$ **(c)** $4^{1/3} \cdot 2^{4/3}$

11. Simplify the following algebraic expressions. Express each result in a form that involves each variable once, to a positive exponent.
   (a) x^2x^{-3} **(b)** $(x^3)^{-2}$ **(c)** $x^{2/3}x^{5/6}x^2$ **(d)** $\frac{(x^4y^{-3})^2}{(x^{-1}y^{-2})^{-3}}$

12. Simplify the following expressions by removing as many factors as possible from under the radical.
   (a) $\sqrt[3]{32x^4}$ **(b)** $\sqrt{9x^2y^7}$ **(c)** $\sqrt[3]{x^3y^8z^4}$

13. Use your calculator to compute the following to four decimal places.
   (a) 3.47^2 **(b)** $1.46^3 - 3.12^4$

14. Compute the following expressions on your calculator. Round your answers to three decimal places.
   (a) $21.67^{4/5}$ **(b)** $1.6783^{4/3}$ **(c)** $0.0007813^{-3/4}$

15. Compute the following expressions on your calculator. Round your answers to four decimal places.
   (a) $\sqrt[3]{12.78^2}$ **(b)** $\sqrt[5]{18.96^4}$ **(c)** $\sqrt[3]{-2.87^5}$

16. Express in scientific notation.
   (a) 473,000,000 **(b)** 0.0000125

17. Express in decimal notation.
   (a) 3.475×10^4 **(b)** 9.473×10^{-3}

18. Evaluate the following by hand, expressing your answers in scientific notation.
   (a) $(5.671 \times 10^3) + (4738.4 \times 10^{-4})$
   (b) $\frac{(3.15 \times 10^2) + (4.122 \times 10^3)}{(0.03 \times 10^{-2})}$

19. Use your calculator to compute the following to four decimal places.
   (a) $\frac{23.567 \times 10^6}{2.468 \times 436.768}$
   (b) $\frac{4987.3678 \times 10^{-2}}{0.023 \times 3.4527 \times 2^{-5}}$

20. **Subatomic Particles** The mass of a proton is 1.673×10^{-24} grams and the mass of a neutron is approximately the same. Most of the mass in matter comes from protons and neutrons. What is the total number of protons and neutrons in the body of a person who weighs 160 pounds, if 1 pound = 2.2×10^3 grams?

21. **Compound Interest** A sum of \$5000 is deposited in a savings account that pays an interest rate of 8%, compounded annually. Determine the amount in the account after 10 years.

22. **Compound Interest** A sum of \$8700 is deposited in a savings account that pays an interest rate of 5.5%. Determine the amount in the account after 7 years if compounding takes place (**a**) annually, (**b**) quarterly, and (**c**) monthly.

23. The volume of a sphere of radius r is $\frac{4}{3}\pi r^3$. Use your calculator to determine the volume of a sphere of radius 6.27 inches, expressed to four decimal places.

24. Perform the following operations on the polynomials.

(**a**) $(3x^2 + 4x + 6) + (5x^2 - 7x - 3)$

(**b**) $(2x + 3)(4x - 5)$

(**c**) $(x^2 - 2x + 3)(2x^2 + 3x - 1)$

25. Use the special products to compute the following.

(**a**) $(2x - 5)^2$ (**b**) $(5x^2 + y)^3$

26. Factor out the greatest common factor in the following.

(**a**) $4x^6 - 2x^3 + 12x^2$

(**b**) $3x^2(2x + 1)^3 - 12x^3(2x + 1)^2$

(**c**) $8x^3y^2z^9 - 4x^2yz^3$

27. Factor the following expressions.

(**a**) $x^2 + 6x + 5$ (**b**) $x^2 - 4x + 3$

(**c**) $2x^2 - x - 6$ (**d**) $15x^2 + 2x - 8$

28. Factor the following expressions into two factors.

(**a**) $2x^2 + 5xy - 12y^2$

(**b**) $15x^2 + 4xy - 4y^2$

(**c**) $4x^2 - 4xy + y^2$

29. Factor the following expressions by grouping.

(**a**) $2x^2 + 6xy - 4x - 12y$

(**b**) $6x^2 + 2x + 3xy + y$

(**c**) $21x^2 + 4y - 7xy - 12x$

30. Each of the following expressions is the difference of two squares. Factor each expression.

(**a**) $x^2 - 4y^2$ (**b**) $9x^2 - 16y^2$

(**c**) $4(x + 2y)^2 - 9(3x - y)^2$

31. Each of the following expressions is a perfect square. Factor each expression.

(**a**) $x^2 + 2xy + y^2$ (**b**) $9x^2 - 24xy + 16y^2$

(**c**) $9 - 12(x + 2y) + 4(x + 2y)^2$

32. Each of the following expressions is the sum or difference of two cubes. Factor each expression.

(**a**) $27x^3 + 8y^3$ (**b**) $64x^3 - y^3$

(**c**) $8x^3 - 125(2x - 1)^3$

33. Simplify the following rational expressions by dividing out common factors of numerators and denominators.

(**a**) $\dfrac{x^2 - x - 6}{x^2 + x - 2}$ (**b**) $\dfrac{x^2 - 5x + 4}{x^2 - 2x - 8}$

(**c**) $\dfrac{2x^2 + 5xy - 3y^2}{4x^2 + 17xy + 15y^2}$

34. Perform the indicated multiplications and divisions.

(**a**) $\dfrac{4(x - 2)^5}{3(x + 5y)^4} \cdot \dfrac{9(x + 5y)^3}{2(x - 2)^2}$

(**b**) $\dfrac{x^2 - 2x - 3}{x^2 + 6x + 5} \div \dfrac{2x^2 - 3x - 9}{2x^2 + 13x + 15}$

35. Perform the indicated additions and subtractions.

(**a**) $\dfrac{4}{x + 3} + \dfrac{7}{2x - 1}$ (**b**) $\dfrac{x + 2}{x + 1} - \dfrac{x - 3}{x + 5}$

(**c**) $\dfrac{2x}{(3x - 1)(x + 2)} + \dfrac{5x}{(x^2 + 3x + 2)}$

36. Determine whether the given x values are solutions to the equations.

(**a**) $x = 15,\quad 3x + 2 = 17$

(**b**) $x = 2,\quad x^2 + 3x - 10 = 0$

(**c**) $x = 3,\quad \dfrac{5x - 1}{2(x + 4)} = \dfrac{2x + 5}{3x + 2}$

37. Solve the following equations for x.

(**a**) $2x - 1 = 7$ (**b**) $5x + 2 = 12$

(**c**) $2x + 4 = x - 3$ (**d**) $9x - 3 = 3x + 15$

38. Determine whether the following equations are linear. If an equation is linear, give its standard form.

(**a**) $4x + 3 = 0$ (**b**) $3x^2 - 4x + 1$

(**c**) $5(4x - 2) = 7x + 3$

39. Solve the following equations for x.

(**a**) $(x - 1)(x + 4) = x^2 + x + 6$

(**b**) $\dfrac{3x}{2} - \dfrac{1}{2} = \dfrac{x}{4} - 4$

40. **Volume of a Cube** The volume V of a cube having length L, width W, and height H is given by $V = LWH$. Find the height of the cube having volume 50 units and a base that measures 5 by 2.

41. **Compound Interest** If a principal P is invested for t years at simple interest $r\%$, it will grow to an amount A given by $A = P + Prt$. Solve this equation for r and use it to determine the rate that will increase \$5000 to \$8000 in 10 years.

42. **Accelerated Motion** If an object starts with velocity u and accelerates with acceleration a for time t, its velocity v at the end of that time is given by $v = u + at$. Solve this equation for t. Use it to find the time it takes to accelerate with an acceleration of 8 from initial 0 to 90.

43. **Analysis of Test Scores** John Carlson has an average of 88 on four tests. What does he have to make on the fifth test to raise his average to 90?

44. **Pressure in a Liquid** The pressure at a point in a liquid varies directly as the depth below the surface. In a certain fuel tank, the pressure at a depth of 4 feet is 200 pounds per square foot.

    (a) Determine the equation that gives pressure in terms of depth.

    (b) At what depth is the pressure 300 pounds per square foot?

45. **Motion of a Ball** The distance a ball rolls down an incline varies directly as the square of its time in motion. How much farther will a ball roll in 6 seconds than in 4 seconds?

46. If $f(x) = 3x + 4$, find the following values of $f(x)$.

    (a) $f(2)$ (b) $f(5)$ (c) $f(-3)$

47. If $f(x) = 2x^2 + 4x - 3$, find the following values of $f(x)$.

    (a) $f(2)$ (b) $f(0)$ (c) $f(-3)$ (d) $f(3u)$ (e) $f(2a - 1)$

48. Find the domain and range of each of the following functions.

    (a) $f(x) = 4x + 1$ (b) $f(x) = x^2 - 6$

    (c) $f(x) = 1 - 2x^2$ (d) $f(x) = \sqrt{4 - x} + 3$

49. **Revenue Analysis** The revenue function on selling x items is $R(x) = -2x^2 + 650x$.

    (a) Determine the revenues on sales of 10, 15 and 20 items.

    (b) Why would you expect $R(0)$ to be zero?

50. **Vertical Motion** An object is projected vertically upward from ground level with velocity of v feet per second. The height s after t seconds is described by the function $s(t) = -16t^2 + vt$.

    (a) If the initial speed is 400 feet per second, determine the heights after 2, 4, and 6 seconds.

    (b) Use a table to find out how long it takes to return to ground level?

    (c) Does it take the same amount of time to come down as to go up? Investigate with a table.

51. **Computer Viruses** The number of computer viruses is increasing annually. The surge of MS-DOS viruses since 1988 is closely described by the function $n(t) = 181t^2 - 131t + 73$. t is the number of years measured from 1988 (1988 corresponds to $t = 0$). $n(t)$ is the cumulative DOS viruses for year t.

    (a) Construct a table of computer viruses for the years 1998–2005 as predicted by this function.

    (b) Is this prediction likely to be realized? Discuss.

52. **College Enrollment** The function $P(x) = 0.088x + 15.005$, which is based on statistics, has been developed to predict U.S. college enrollments for the years beginning in 1994. The units of x are years and P is millions of students. For example, $P(0) = 15.005$ tells us that the college enrollment in 1994 was approximately 15.005 million students. During which year is enrollment predicted to reach the 16 million mark?

Chapter R Test*

1. Let $A = \{-\frac{17}{9}, -1, 0, 3, \frac{5}{4}, 15, \sqrt{7}\}$. List all the elements of A that are **(a)** natural numbers **(b)** integers, but not natural numbers.

2. Indicate which algebraic property of real numbers is used to justify each of the following statements.
   (a) $(\frac{1}{4} \cdot 6) \cdot 3 = \frac{1}{4} \cdot (6 \cdot 3)$
   (b) $x(y + z) = xy + xz$

3. Compute the expression $(\sqrt{9.7} + 0.26^{-1}) \div 1.73$ to four decimal places using a calculator.

4. Write the following calculator expression in standard algebraic form and compute the result. $5 + 16 \div 2 \wedge 3 \times 4$. Check your answer by performing the sequences of keystrokes on your calculator.

5. Compute the following by hand. Support your answers using a calculator.
   (a) $\left(\sqrt[3]{\dfrac{27}{8}}\right)^2$ **(b)** $(-27)^{5/3}$

6. Simplify the following algebraic expression. Express the result in a form that involves each variable once, to a positive exponent.
$$\frac{(x^{-1}y^2)^3}{(x^{-3}y^4)^{-2}}$$

7. Simplify the following expression by removing as many factors as possible from under the radical.
$$\sqrt{48x^5y^2}$$

8. Compute the following expression on your calculator. Express the answer to four decimal places.
$$\sqrt[7]{41.82^4}$$

9. Multiply $(3x - 1)(4x + 2)$.

10. Factor **(a)** $6x^2 + 7x - 3$ **(b)** $8x^2 + 10xy - 3y^2$ **(c)** $4x^2 - 81y^2$

11. Add $\dfrac{3}{x - 1} + \dfrac{5}{2x + 1}$.

12. Solve the equation $3x + 1 = 5x + 3$ for x.

13. If $f(x) = 3x^2 + 5x - 1$, find the following values of $f(x)$.
   (a) $f(2)$ **(b)** $f(0)$ **(c)** $f(-1)$ **(d)** $f(2a - 3)$

14. Find the domains and ranges of the following functions.
   (a) $f(x) = -x^2 + 3$ **(b)** $f(x) = \sqrt{x + 2} - 1$

15. **Compound Interest** A sum of $10,000 is deposited in a savings account that pays a compound interest rate of 4.5% per annum. Determine the amount in the account after 5 years if compounding takes place monthly.

16. **GPA** Karla Friedman has a GPA of 2.95 going into the last semester. She has taken 105 hours of classes and will be taking 15 hours during the last semester. What GPA does she need during the last semester to graduate with a 3.0?

17. **Weight at Different Altitudes** The weight of an object varies inversely as the square of its distance from the center of the Earth. Tony Thomas weighs 185 pounds on Earth. The average radius of the Earth is 3956 miles. How much (expressed to two decimal places) would he weigh at an altitude of 5 miles?

18. **Motion Under Gravity** An object is thrown vertically upward from ground level at a speed of v feet per second. The height s after t seconds is described by the function $s(t) = -16t^2 + vt$. **(a)** If the initial speed is 368 feet per second, determine the height after 1, 2, and 3 seconds. **(b)** Use a table to find out how long it takes to return to ground level?

*Sample tests are given at the end of each chapter. Answers to all the questions are given in the back of the book.

Interplay Between Algebra and Geometry

René Descartes (1596 – 1650)
(©1994 Northwind Picture Archives)

In this chapter, we learn to appreciate the interplay between algebra and geometry. They complement one another in much that we do in mathematics. Much of the geometry of lines, parabolas, and circles that we introduce in this chapter was first developed by the Frenchman René Descartes (1596–1650). In fact, the Cartesian coordinate system in which all this geometry takes place is named after Descartes. Descartes grew up in a wealthy family; his father was a counselor of the Parliament of Brittany. Descartes was educated at the Jesuit College of La Flèche and studied law at Poitiers. At the age of 21, he enlisted in the army of Prince Maurice of Orange. His years of soldiering were years of leisure in which he pursued his mathematical interests, and we are told that "he studied with irresistible passion." He then settled in the Netherlands where he pursued interests in mathematics, medicine, anatomy, embryology, and meteorology. He seems to have devoted much time to dissecting many different kinds of animals. Descartes was a profound skeptic, taking nothing whatever on authority, but scrutinizing everything.

In this chapter, we derive the equations for lines, parabolas, and circles, and the student will learn how to interpret these equations. The slope and intercepts of a line, the vertex of a parabola, whether the parabola opens up or down, and the center and radius of a circle are among the algebraic topics presented.

The student will be involved in applying the mathematics. The expenses involved in scuba diving, the overall rate at which water accumulates in a reservoir when water is entering and leaving, cost–volume analysis in business, motion under gravity, and the algebra of typesetting are among the applications introduced. We explain the important technique of fitting lines and curves to data. Curve fitting is used to compare real and subjective times, to analyze the flow of a liquid, to examine expenditures on education, to investigate the growth of Intel Corporation, and so on.

We also integrate the graphing calculator into our discussions. The graphing calculator is used to clarify mathematical ideas and in applications. Techniques for using the graphing calculator to do algebra and geometry are presented. The student will graph lines, parabolas, and circles. In addition to tracing, zooming, and using the max/min finder, the zero finder, and the intersect finder, we discuss techniques for arriving at equations of curves that best fit data.

APPLICATION Understanding Subjective Time

It's easy to lose track of time. *(©Stewart Cohen/Tony Stone)*

Scientists have studied the sense of time in humans and have concluded that there is a relationship between real time and a person's estimate of time (called subjective time) that can be described by an equation. This equation varies from person to person. For example, one person may think that time goes more slowly than it really does; another may think that it goes faster.

The author determined his subjective time equation by estimating intervals of 10 minutes for an hour period and asking a coworker to record the actual times. The data were plotted and found to lie approximately on a line. Mathematics was used to find the following equation of the line that "best fits" these data:

$$t = 1.04s + 0.67$$

Suppose the author guesses at a duration of 50 minutes ($s = 50$). The real-time lapse according to this equation will be approximately 53 minutes; consequently, I would tend to keep 50-minute classes 3 minutes overtime if I did not have a watch!

How do you estimate time? In this chapter, you are asked to find your subjective time equation and arrive at some conclusions. *(This application is discussed in the Chapter Project on page 154.)*

1.1 Coordinate Systems and Graphs

- Coordinate System • The Viewing Window • Graphing with a Calculator
- Graphs • Using x- and y-Intercepts to Arrive at a Viewing Window
- Using a Table to Find a Suitable Viewing Window

Coordinate System

Coordinates enable us to describe the locations of points. In the previous chapter we saw how coordinates can be assigned to points on a line. In this chapter, we discuss how coordinates can be assigned to points in a plane. Select two real number lines in the plane, one line horizontal and the other vertical. The horizontal line is usually called the ***x*-axis,** and the vertical line is usually called the ***y*-axis.** The point at which the lines intersect is called the **origin.** On the x-axis, positive numbers fall to the right of the origin; negative numbers are located to the left of the origin. On the y-axis, positive numbers appear above the origin, and negative numbers are below the origin (Figure 1.1). The two axes, referred to as **coordinate axes**, divide the plane into four sections called *quadrants*—labeled I, II, III, and IV, counterclockwise from the upper right. Every point in the plane is either on an axis or in a quadrant.

The coordinate axes can be used to describe the location of any point in the plane. To find the location of a point P, draw vertical and horizontal lines from the point to each of the axes. Suppose that the vertical line intersects the x-axis at a and the horizontal line intersects the y-axis at b (Figure 1.2). The ordered pair (a, b) gives the location of P in the plane. The **x-coordinate** of P is a, and the **y-coordinate** of P is b. The **coordinates** of P are (a, b). Note that the x-coordinate of the point is listed first, followed by its y-coordinate. This way of describing the locations of points in a plane is called a **rectangular** or **Cartesian coordinate system.**

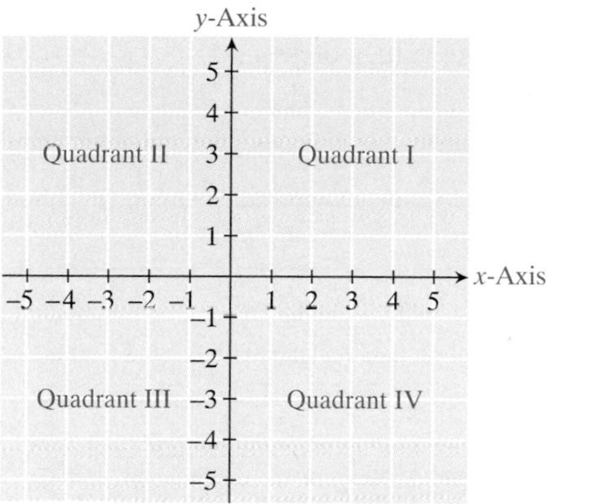

Figure 1.1 Coordinate axes and quadrants

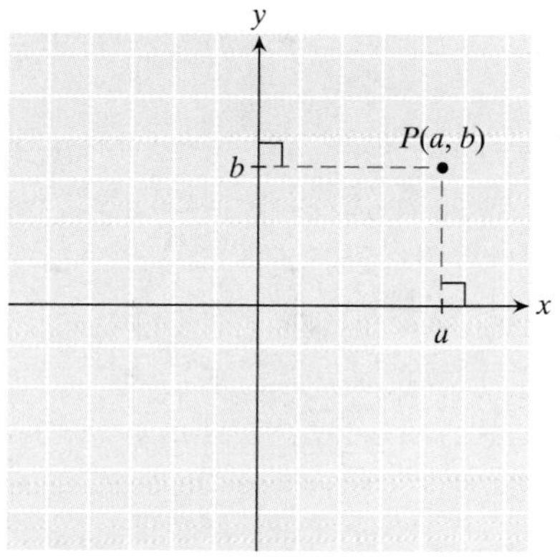

Figure 1.2 Rectangular coordinate system

Example 1 Coordinates of Points

Give the coordinates of the points P, Q, R, and S shown in Figure 1.3.

SOLUTION If we draw vertical and horizontal lines from each of the points to the coordinate axes, we see that the coordinates of the points are $P(4, 2)$, $Q(-3, 2.5)$, $R(-4, -3)$, $S(0, -4)$.

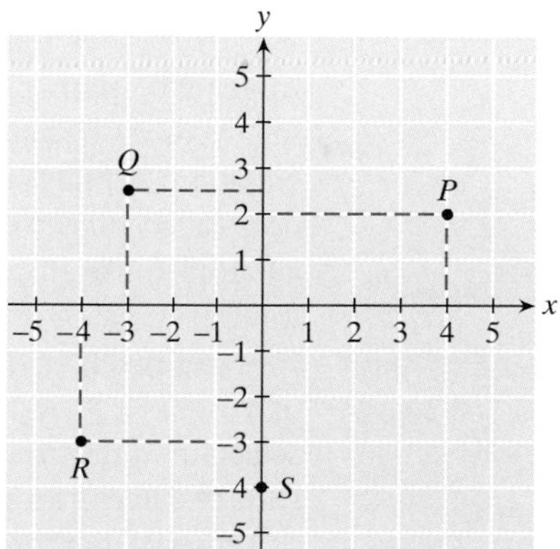

Figure 1.3

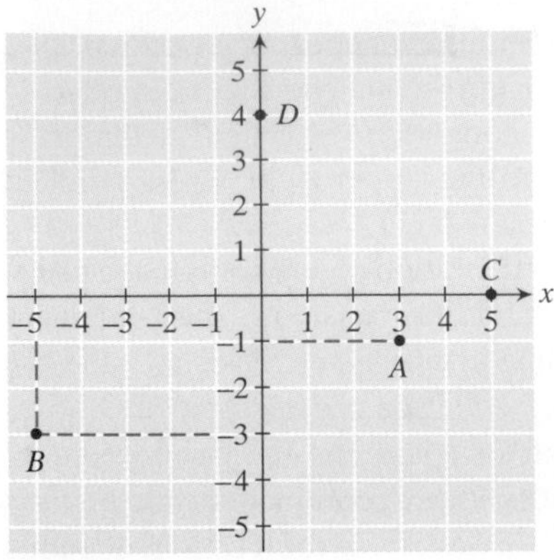

Figure 1.4

Self-Check 1 Give the coordinates of the points *A*, *B*, *C*, and *D* shown in Figure 1.4. Give the axis or quadrant in which each point lies.

Graphs

Consider the equation $y = 2x - 1$. The **graph** of this equation consists of all the points in the plane that satisfy the equation. For example, consider the point $(2, 3)$. Substituting $x = 2$ into this equation, we find that $y = 3$. The point $(2, 3)$ satisfies this equation and thus lies on its graph. Conversely, consider the point $(4, 6)$. Substituting $x = 4$ into the equation yields $y = 7$. The point $(4, 6)$ does not lie on the graph. A graph can be drawn by determining some points that lie on the graph, plotting those points, and drawing a smooth curve through the points. Enough points must be plotted to feel reasonably sure that they lead to the general shape of the graph. Let's draw the graph of $y = 2x - 1$ using this method.

Construct a table by choosing convenient values for x and finding the corresponding values of y.

x	-4	-3	-2	-1	0	1	2	3	4
y	-9	-7	-5	-3	-1	1	3	5	7

This table leads to the points $(-4, -9)$, $(-3, -7)$, $(-2, -5)$, $(-1, -3)$, $(0, -1)$, $(1, 1)$, $(2, 3)$, $(3, 5)$, and $(4, 7)$. Plot these points and draw a smooth curve through them to get the graph of $y = 2x - 1$ (Figure 1.5). For clarity, we use an ***x* scale** of 1 and a ***y* scale** of 2 on the axes. (Distances between the tick marks.)

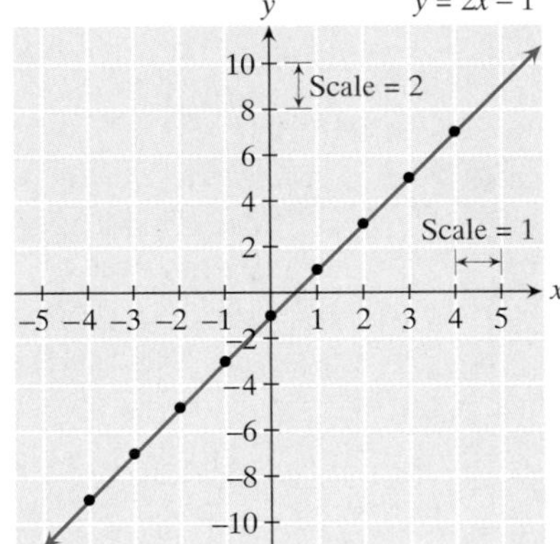

Figure 1.5 Graphing $y = 2x - 1$ by hand

The graph appears to be a straight line. We discuss straight lines in Section 1.2. We shall find that the general equation of a line is of the form $y = mx + b$.

Graphing with a Calculator

In general, many points must be plotted to get an accurate graph if you are doing this work by hand. Calculators and computers that are now available can graph equations rapidly. The graphing screen of a graphing calculator consists of small dots called **pixels,** which can be lighted. The screens of the TI-82 and TI-83 calculators, for example, have 95 pixels horizontally and 63 pixels vertically giving a total of $95 \times 63 = 5985$ pixels on the screen. A calculator displays a graph by showing pixels that correspond to points on the graph. To reproduce the graph of $y = 2x - 1$ on a calculator, one must follow three steps—enter the equation, select an appropriate graphing window, and display the graph (Figure 1.6). You should determine how to implement these steps on your calculator.

Observe that the graph appears jagged in the display. Because of the finite number of pixels on the screen, the calculator often cannot use pixels that correspond to exact points on a graph. Instead, it lights up pixels that lie as closely as possi-

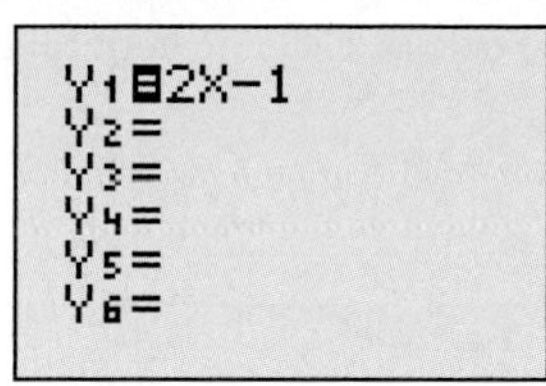

Enter the equation.

```
Xmin=-4
Xmax=4
Xscl=1
Ymin=-10
Ymax=8
Yscl=2
```

Select the window
$-4 \le x \le 4$, $-10 \le y \le 8$
with x scale 1, y scale 2.

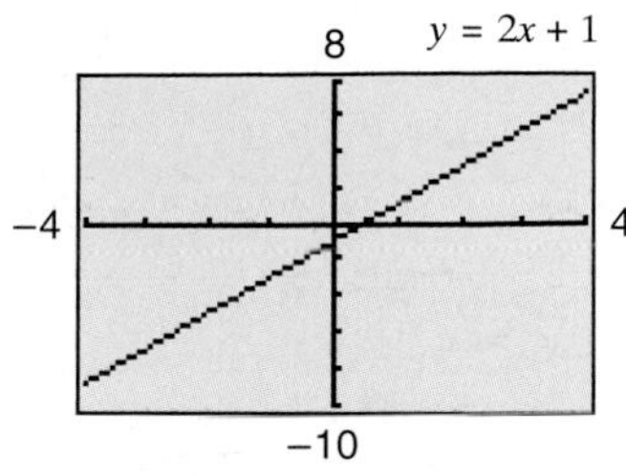

Display the graph.

Figure 1.6 Graphing $y = 2x - 1$ on a calculator

ble to points on the graph. As calculators become more sophisticated with higher-resolution screens, displays will appear smoother.

Example 2 Determining Whether Points Lie on a Graph

Determine whether the points $(-1, -5)$ and $(0, 6)$ lie on the graph of equation $y = 2x^2 + 3x - 4$.

SOLUTION Consider each point in turn. Substitute the x value into the equation to see if the corresponding y value is obtained.

Point $(-1, -5)$: When $x = -1$, $y = 2(-1)^2 + 3(-1) - 4 = -5$. The point lies on the graph.

Point $(0, 6)$: When $x = 0$, $y = 2(0)^2 + 3(0) - 4 = -4 \neq 6$. The point is not on the graph.

Example 3 Sketching a Graph by Hand and Confirming the Graph on a Calculator

Sketch the graph of $y = 3x^2$ by creating a table and plotting points; then confirm the graph on a calculator.

SOLUTION Construct the following table of convenient x and corresponding y values:

x	−4	−3	−2	−1	0	1	2	3	4
y	48	27	12	3	0	3	12	27	48

This table leads to the points $(-4, 48)$, $(-3, 27)$, $(-2, 12)$, $(-1, 3)$, $(0, 0)$, $(1, 3)$, $(2, 12)$, $(3, 27)$, and $(4, 48)$. Plot these points, and draw a smooth curve through them to get the graph of $y = 3x^2$ (Figure 1.7).

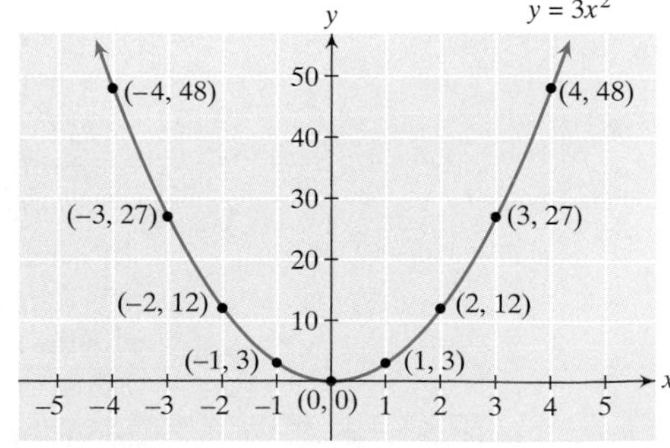

Figure 1.7

Now let us confirm the graph of $y = 3x^2$ with a calculator. Enter the equation, select an appropriate window, and display the graph (Figure 1.8). Observe that the graph is U-shaped. It is called a **parabola.** We will study parabolas in Section 1.4. The general equation of a vertical parabola is $y = ax^2 + bx + c$.

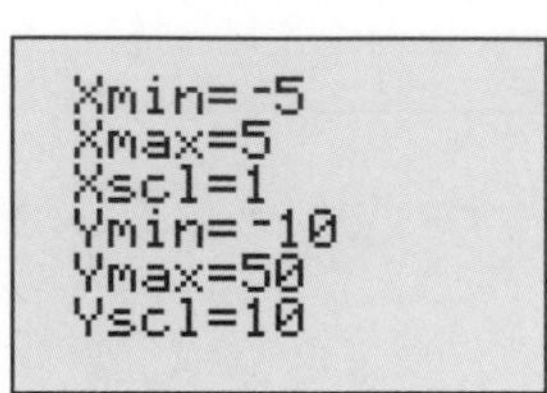

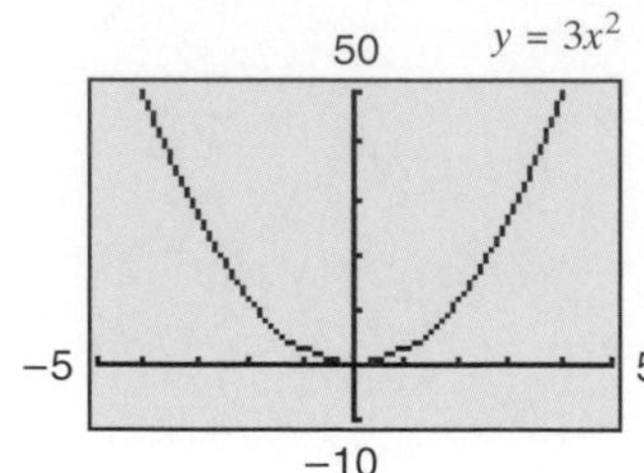

Figure 1.8 Confirmation of the graph of $y = 3x^2$

The Viewing Window

Selecting appropriate intervals of x and y for the viewing window is important because only the part of the graph that lies in the viewing window is displayed. One attempts to select intervals so the relevant part of the graph is displayed. For example, the window $-2 \leq x \leq 10$, $-10 \leq y \leq 16$ is convenient to graph $y = x^2 - 8x + 10$ (Figure 1.9). The graph is a parabola.

A point where a graph intersects the x-axis is called an ***x*-intercept.** A point where a graph intersects the y-axis is the ***y*-intercept.** The graph in Figure 1.9 has two x-intercepts and one y-intercept. The bend in the parabola is called the **vertex.** One often wants to arrive at a window that displays a part of a graph that reveals such points—these are points where "things are happening" to the graph.

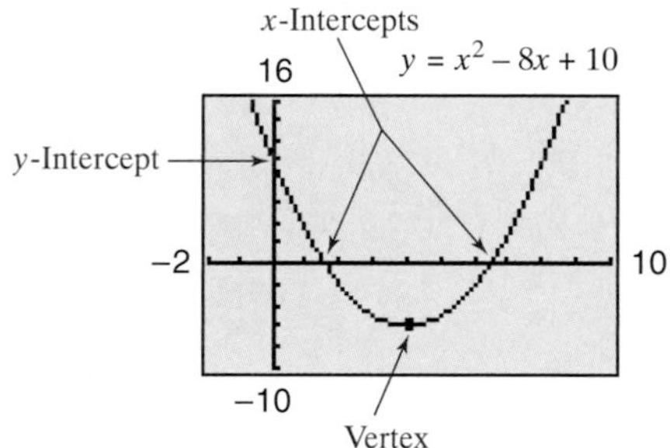

Figure 1.9 Intercepts and vertex of a parabola

Figure 1.10 shows how the graph would have appeared in other windows. Frequently, we must try windows of various sizes before we find a window that is appropriate. If in doubt about whether a window gives the complete information about the shape of a graph, try selecting a large window and working down in size until a suitable one is found. The student may also find that more than one window is needed to explore the behavior of a graph.

It is important that the student master the art of interpreting and selecting viewing windows now. These skills are fundamental to using a graphing calculator. Because we will use viewing windows extensively throughout the remainder of the course, we will look at two more examples to further this understanding.

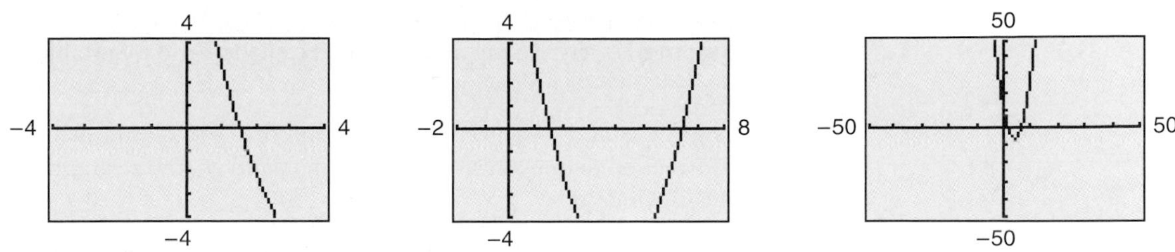

Figure 1.10 Graph of $y = x^2 - 8x + 10$ in various windows

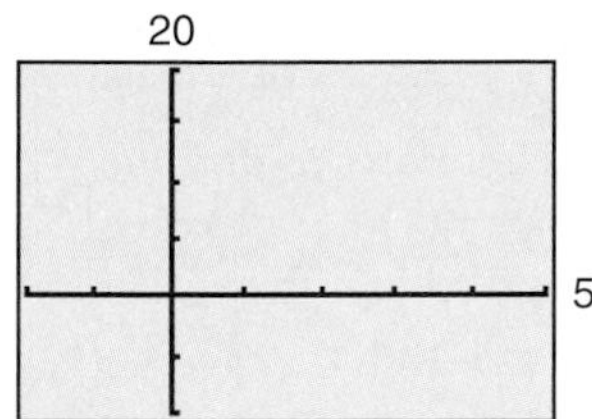

Figure 1.11

Example 4 | Description of a Viewing Window

Consider the viewing window in Figure 1.11. Partial information about the window is given. Find the complete description of the window (determine X_{min}, X_{max}, X_{scl}, Y_{min}, Y_{max}, and Y_{scl}). Confirm your answer by reproducing the window on your calculator.

SOLUTION We are given $X_{max} = 5$ and $Y_{max} = 20$. The display tells us that five tick marks lead to $X_{max} = 5$. Thus, each tick mark corresponds to 1 unit on the x-axis. We have $X_{scl} = 1$, and $X_{min} = -2$. Similarly, because four tick marks lead to $Y_{max} = 20$, each tick mark corresponds to 5 units on the y-axis. Thus, $Y_{scl} = 5$ and $Y_{min} = -10$. We verify these results on a calculator (Figure 1.12).

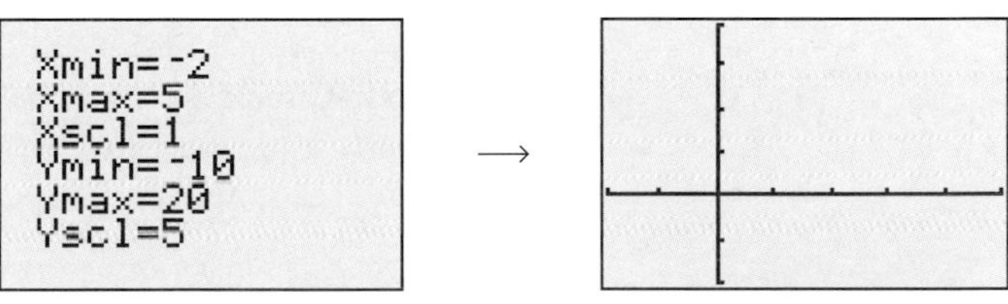

Figure 1.12

The given window

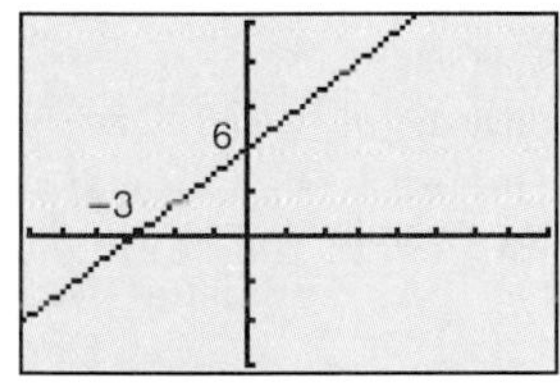

Figure 1.13

Example 5 | Description of a Viewing Window

Consider the graph of the line shown in Figure 1.13. The x-intercept and the y-intercept are displayed. Use this information to describe the viewing window (identify X_{min}, X_{max}, X_{scl}, Y_{min}, Y_{max}, and Y_{scl}).

SOLUTION Three tick marks on the x-axis lead to $x = -3$. Thus, $X_{scl} = 1$. Two tick marks on the y-axis give $y = 6$; thus, $Y_{scl} = 3$. Counting tick marks, we get $X_{min} = -6$, $X_{max} = 8$, $Y_{min} = -9$, $Y_{max} = 15$.

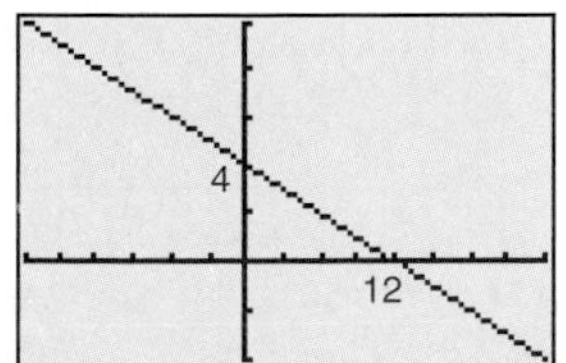

Figure 1.14

Self-Check 2 Consider the graph of the line shown in Figure 1.14. Use information from the display to describe the viewing window.

As we proceed in the course and develop more algebraic ideas, we will be able to use some of these new ideas to select suitable windows for graphs. We now introduce a method that is particularly appropriate for selecting windows for lines.

Example 6 Using x- and y-Intercepts to Find a Suitable Viewing Window

Find the x- and y-intercepts of the graph of $y = 3x + 12$ using algebra. Use this information to find a convenient viewing window for the graph. Display the graph in this window.

SOLUTION A viewing window should usually be constructed to show the x- and y-intercepts of a graph. The following methods can be used to find these intercepts.

Any point on the x-axis has y-coordinate zero. Therefore, *to find the* x*-intercept of a graph, let* y = 0 *in the equation.* Letting $y = 0$ in $y = 3x + 12$ gives $3x + 12 = 0$, leading to $x = -4$. The x-intercept is -4.

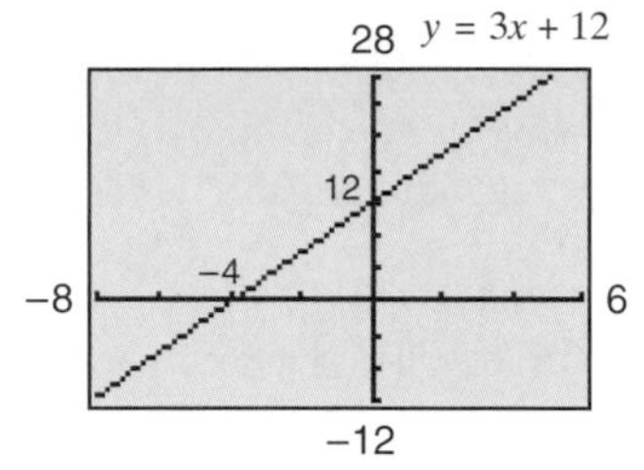

Figure 1.15

Any point on the y-axis has x-coordinate zero. Therefore, *to find the* y*-intercept of a graph, let* x = 0 *in the equation.* Letting $x = 0$ in $y = 3x + 12$ gives $y = 12$. The y-intercept is 12.

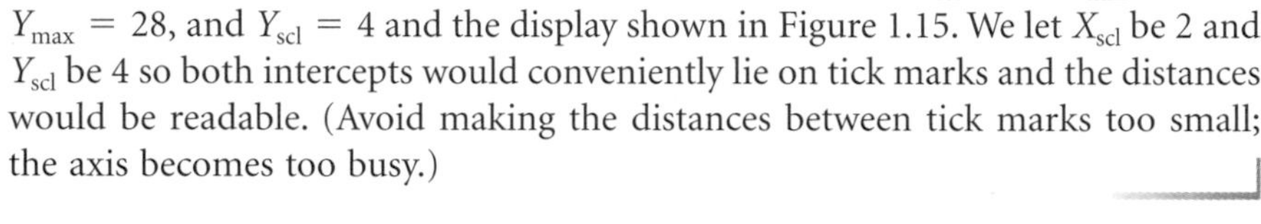

The window that we select should contain $x = -4$ and $y = 12$. This information led us to the window $X_{min} = -8$, $X_{max} = 6$, $X_{scl} = 2$, $Y_{min} = -12$, $Y_{max} = 28$, and $Y_{scl} = 4$ and the display shown in Figure 1.15. We let X_{scl} be 2 and Y_{scl} be 4 so both intercepts would conveniently lie on tick marks and the distances would be readable. (Avoid making the distances between tick marks too small; the axis becomes too busy.)

Of course, there is nothing unique about this window. The student should use these ideas as suggestions for selecting windows for graphing equations.

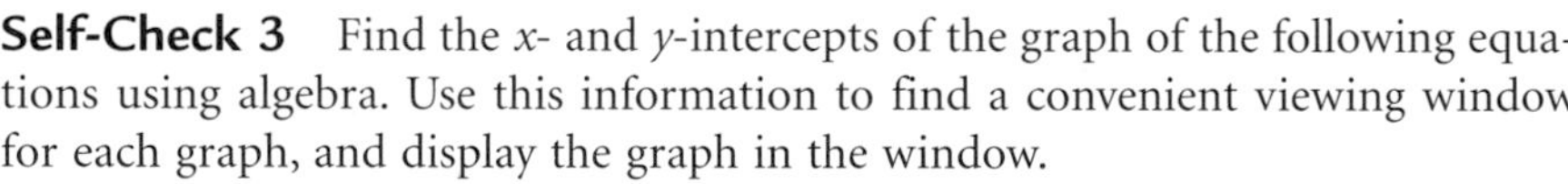

Self-Check 3 Find the x- and y-intercepts of the graph of the following equations using algebra. Use this information to find a convenient viewing window for each graph, and display the graph in the window.

(a) $y = -2x + 8$ **(b)** $y = 6x + 30$

Example 7 Using a Table to Find a Suitable Viewing Window

In this example we illustrate how a table can be used to find a suitable window for viewing a graph. In some cases, the student may find it easier to experiment with various step sizes in a table and get a feel for the behavior of the graph from the table rather than from looking at various window sizes.

Let us find a suitable window to graph the equation $y = x^2 - 12x - 7$. We arrived at the following table using a step size of 5 (Figure 1.16). Observe that as x varies from -10 to 20, y varies approximately from -50 to 200, leading to the viewing window and graph.

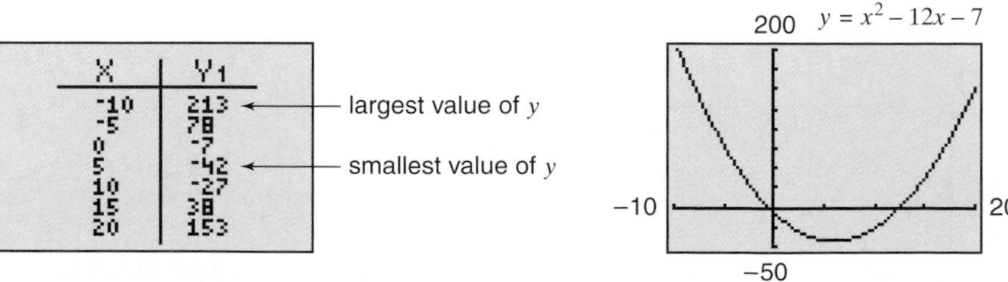

Figure 1.16 Using a table to find a suitable viewing window.

Self-Check 4 Use a graphing calculator to sketch a rough graph of each of the following parabolas. Your sketch should show the x-intercepts, y-intercepts, and vertices of the graph. Each graph has two x-intercepts. You may need to experiment with windows of various sizes before you find a suitable one.

(a) $y = x^2 - 25$ **(b)** $y = x^2 - 10x - 22$ **(c)** $y = -x^2 - 36x - 28$

We have seen three related ways of representing information in this section: namely, through **equations, tables,** and **graphs.** Throughout the course, we will see the importance of mastering all three ways, because one usually arises naturally when mathematics is used to solve a problem. Further information can then be obtained by examining the problem from the other viewpoints.

Answers to Self-Check Exercises

1. $A(3, -1)$, $B(-5, -3)$, $C(5, 0)$, and $D(0, 4)$. The point A is located in quadrant IV, B is in quadrant III, C is on the x-axis, and D is on the y-axis.
2. $X_{min} = -18$, $X_{max} = 24$, $X_{scl} = 3$, $Y_{min} = -4$, $Y_{max} = 10$, $Y_{scl} = 2$.
3. **(a)** x-intercept $= 4$, y-intercept $= 8$. **(b)** x-intercept $= -5$, y-intercept $= 30$.
4.

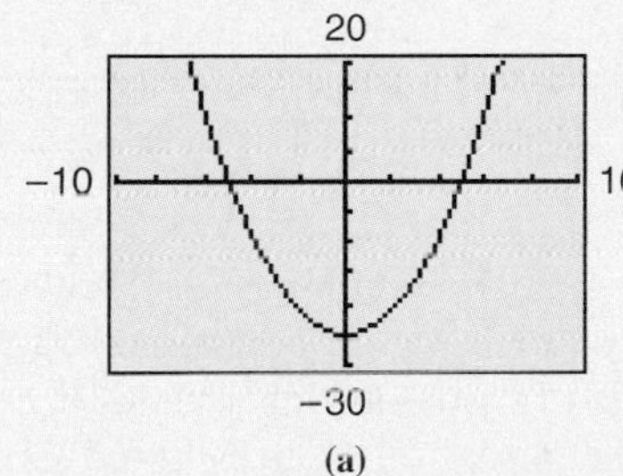

(a)

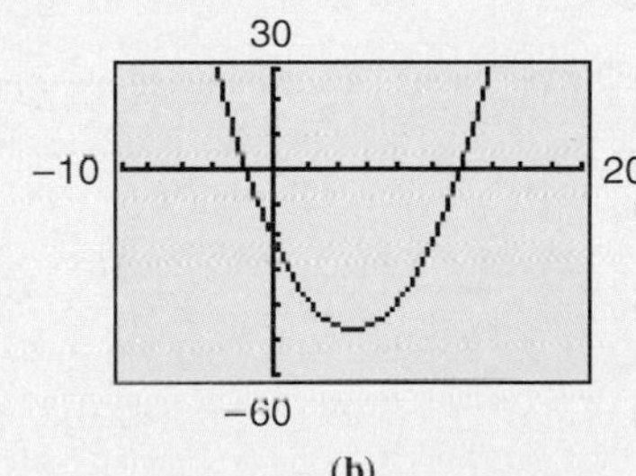

(b)

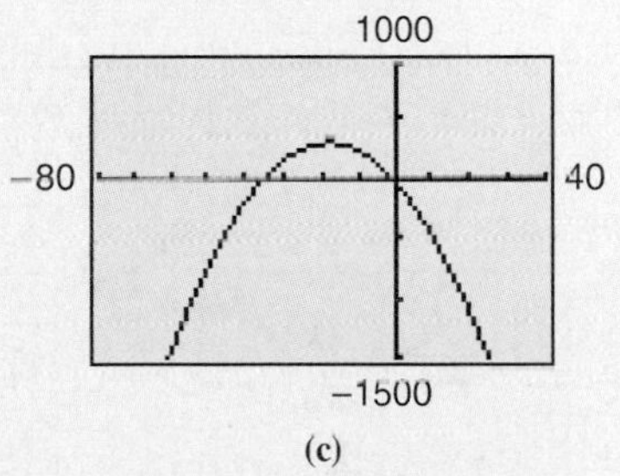

(c)

EXERCISES 1.1

In Exercises 1–4, plot the points in a rectangular coordinate system. State the axis or quadrant in which each point lies.

1. $A(2, -1)$, $B(4, 3)$, $C(6, -2)$, $D(-5, 0)$
2. $P(-5, 3)$, $Q(-3, -4)$, $R(2, 5)$, $S(6, -2)$
3. $A(3, 3)$, $B(-3, -3)$, $C(-5, -5)$, $D(-5, 5)$
4. $L(6, 0)$, $M(0, 4)$, $N(-3, 0)$, $P(0, -5)$

In Exercises 5–8, use substitution to determine whether the points lie on the graphs of the equations.

Equation	Points	
5. $y = 3x - 5$	(a) $(2, 1)$	(b) $(-3, -4)$
	(c) $(0, -5)$	
6. $y = 2x^2 - 3x + 7$	(a) $(-4, 51)$	(b) $(1, 7)$
	(c) $(3, 16)$	
7. $y = x^3$	(a) $(-1, 1)$	(b) $(3, 27)$
	(c) $(1.5, 3.375)$	
8. $y = \dfrac{1}{x - 2}$	(a) $(1, -1)$	(b) $(2, 0)$
	(c) $(4, 1)$	

In Exercises 9–14, a partial description of each viewing window is given. Give the complete description of each window (X_{min}, X_{max}, X_{scl}, Y_{min}, Y_{max}, and Y_{scl}). Confirm your answer by reproducing the window on your calculator.

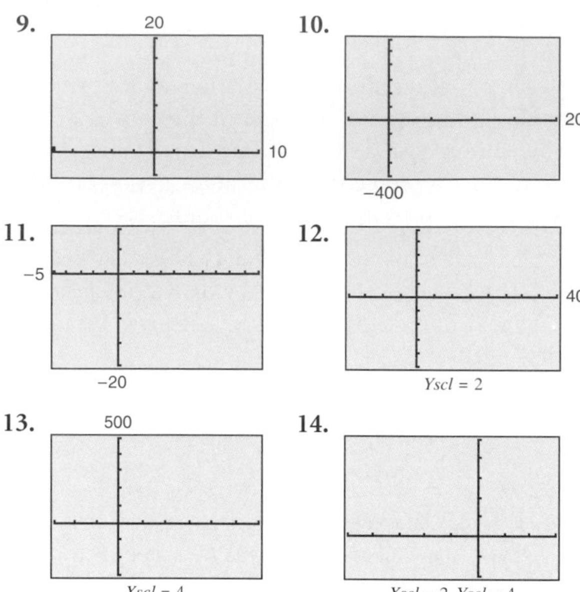

In Exercises 15–20, graphs and information about their intercepts and vertices are displayed. Use this information to describe the viewing windows (give X_{min}, X_{max}, X_{scl}, Y_{min}, Y_{max}, and Y_{scl}).

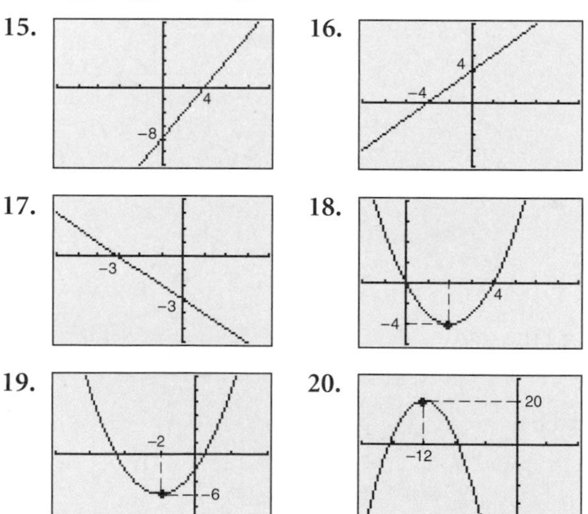

In Exercises 21–26, graphs of the equations are given. Use information from the equations and the displays to describe the viewing windows (give X_{min}, X_{max}, X_{scl}, Y_{min}, Y_{max}, and Y_{scl}). (*Hint:* The y-intercept is obtained by letting $x = 0$ in the equation. The x-intercept(s) are obtained by letting $y = 0$ in the equation.)

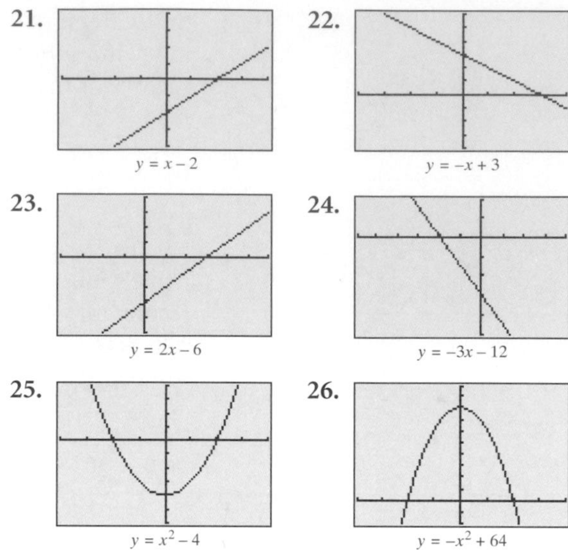

In Exercises 27–32, find the x- and y-intercepts of the graphs of the equations by letting y and then x be zero (see Example 6). Use this information to find a suitable viewing window (X_{min}, X_{max}, X_{scl}, Y_{min}, Y_{max}, and Y_{scl}) for each graph. Display the graph in the window. Use the display to sketch a rough graph.

27. $y = 2x - 16$ **28.** $y = -4x + 8$

29. $y = 5x - 15$ **30.** $y = 5x + 35$

31. $y = -3x - 27$ **32.** $y = 6x + 30$

In Exercises 33–40, use a graphing calculator to sketch a rough graph of each of the polynomials. (You will probably need to try windows of various sizes before you find a suitable one.) Your sketch should show all x-intercepts, y-intercepts, and vertices of the graph (you need not determine these points). Each graph has a single y-intercept. The polynomials of degree 2 all have two x-intercepts in these exercises, whereas the polynomials of degree 3 have three x-intercepts. State which window you used (X_{min}, X_{max}, X_{scl}, Y_{min}, Y_{max}, and Y_{scl}).

33. $y = x^2 - 64$

34. $y = -2x^2 + 32$

35. $y = -x^2 - 21x + 24$

36. $y = x^2 + 17x - 35$

37. $y = 2x^2 + 82x - 125$

38. $y = x^3 - 7x^2 + 9x + 5$

39. $y = -2x^3 + 8x^2 - 8x + 12$

40. $y = -x^3 + 9x^2 + 5x - 10$

Lines, Least-Squares Fit, and the Year 2000 Problem

• Slope • Point–Slope Form of the Equation of a Line • Slope–Intercept Form of the Equation of a Line • Horizontal and Vertical Lines • Parallel and Perpendicular Lines • Linear Equations • Least-Squares Line • Tuition Costs

SLOPE

We begin our discussion of lines by looking at the concept of **slope.** Consider the lines L_1, L_2, and L_3 in Figure 1.17. We would all agree that L_1 is steeper than L_2, and that L_2 is in turn steeper than L_3. The precise mathematical definition of slope that we adopt fits in with this intuitive idea of slope. Construct a right triangle on each line as shown and define

$$\text{Slope} = \frac{\text{Rise}}{\text{Run}}$$

Thus, according to this definition, the greater the rise corresponding to a given run, the greater the slope. We see that L_1 rises more than L_2, and that L_2 rises more than L_3 for the same run. This definition implies that L_1 is steeper than L_2, and that L_2 is steeper than L_3.

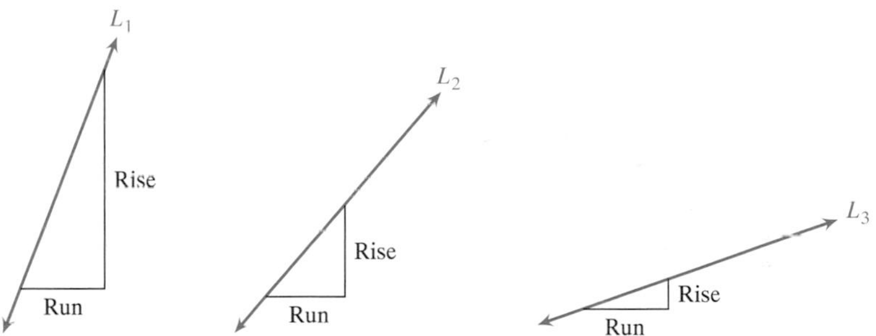

Figure 1.17 L_1 is steeper than L_2, and L_2 is steeper than L_3.

Example 1 **Sketching a Line of Given Slope Through a Given Point**

Sketch the line L of slope 2 through the point (2, 1).

SOLUTION Write slope $= 2 = \frac{2}{1}$. To sketch the line start at the point (2, 1) and use Run = 1 and Rise = 2 to arrive at the point (3, 3) (Figure 1.18). Draw the line through the points (2, 1) and (3, 3).

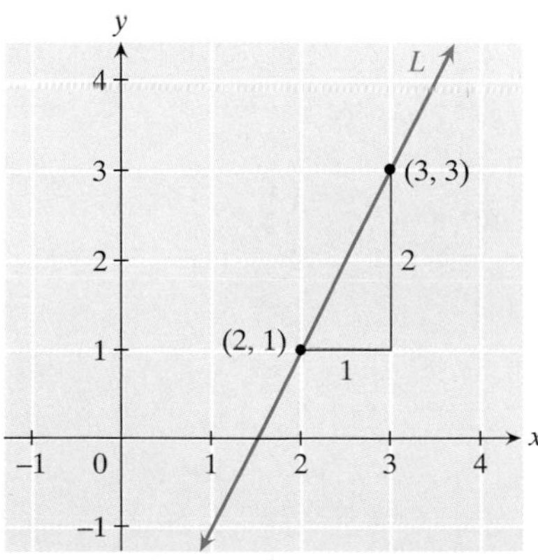

Figure 1.18

Further Ideas 1 Sketch lines of slopes 0.5, 1, and 3 through the point (3, 2) by determining another point on each line.

Example 2 Determining the Slope of a Line from Its Graph

The points at which the line in Figure 1.19(a) crosses the x- and y-axes are integers. Find the slope of the line.

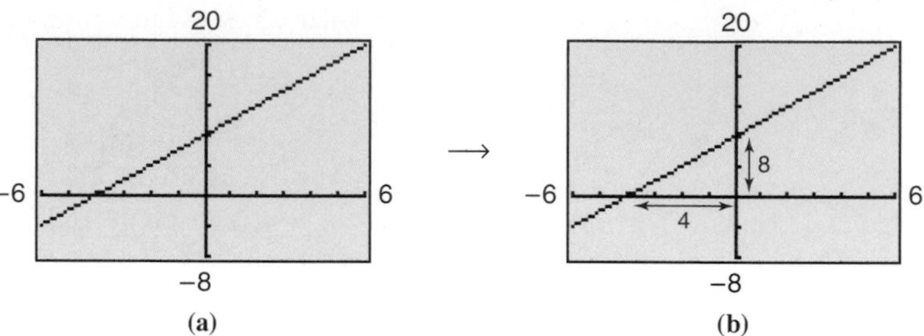

Figure 1.19

SOLUTION The viewing window has $X_{scl} = 1$ and $Y_{scl} = 4$. From Figure 1.19b we see that for Run = 4, the Rise = 8. Thus,

$$m = \frac{\text{Rise}}{\text{Run}} = \frac{8}{4} = 2$$

Self-Check 2 The points at which the line in Figure 1.20 crosses the x- and y-axes are integers. Determine the X_{scl} and Y_{scl} and then find the slope of the line.

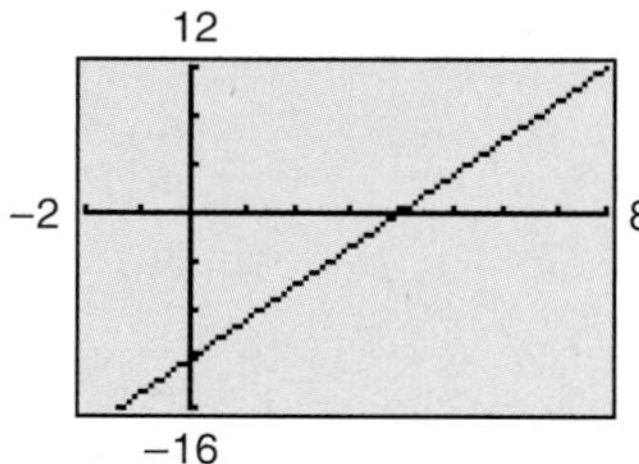

Figure 1.20

Slope of a Line Through Two Points

We now introduce the general expression for the slope of a line through two points. Let (x_1, y_1), (x_2, y_2) be two points on a line L (Figure 1.21). We see that

$$\text{Run} = x_2 - x_1 \quad \text{and} \quad \text{Rise} = y_2 - y_1$$

Thus,

$$\text{Slope} = \frac{\text{Rise}}{\text{Run}} = \frac{y_2 - y_1}{x_2 - x_1}$$

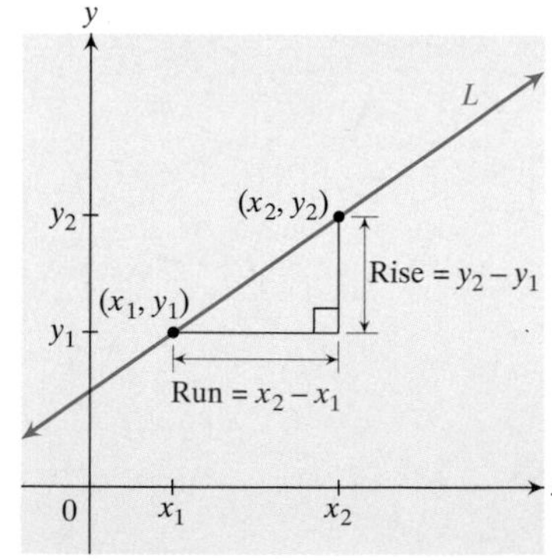

Figure 1.21

Slope The **slope** m of the line through the points (x_1, y_1) and (x_2, y_2) is

$$m = \frac{y_2 - y_1}{x_2 - x_1}, \qquad \text{provided } x_1 \neq x_2$$

Note that an implication of this definition is that a line that slopes upward from left to right has a **positive slope,** whereas a line that slopes downward from left to right has a **negative slope.** This is illustrated in Figure 1.22. The slope of a horizontal line is **zero** and the slope of a vertical line is **undefined.**

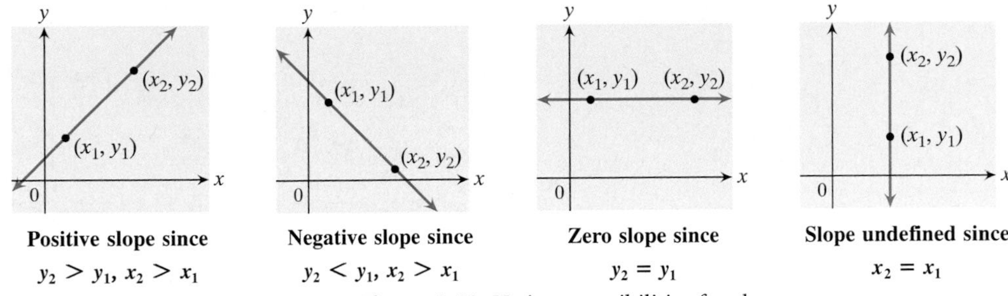

Figure 1.22 Various possibilities for slope

Example 3 Finding Slopes and Sketching Graphs

Find the slopes of the lines through the following pairs of points (if the slopes are defined). **(a)** $(-2, 5)$, $(3, -5)$, **(b)** $(-4, 3)$, $(2, 3)$, **(c)** $(2, 1)$, $(2, 4)$. Sketch the lines.

SOLUTION

(a) Let $(x_1, y_1) = (-2, 5)$ and $(x_2, y_2) = (3, -5)$; then,

$$m = \frac{y_2 - y_1}{x_2 - x_1} = \frac{-5 - 5}{3 - (-2)} = \frac{-10}{5} = -2$$

The slope is negative, indicating that the line slopes downward (Figure 1.23).

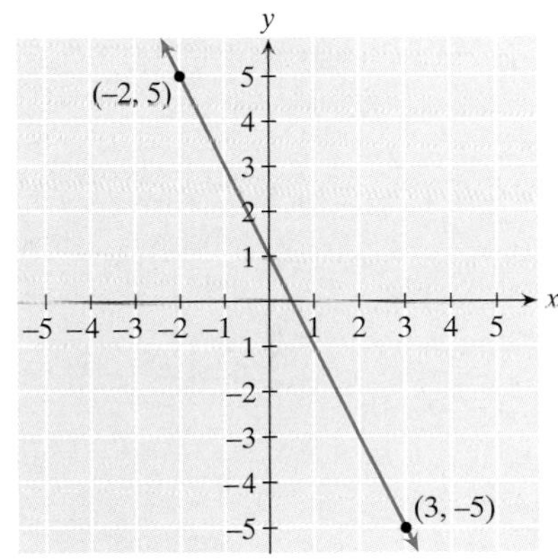

Figure 1.23

(b) Let $(x_1, y_1) = (-4, 3)$ and $(x_2, y_2) = (2, 3)$; then,

$$m = \frac{3 - 3}{2 - (-4)} = \frac{0}{6} = 0$$

The slope is zero, implying that the line is horizontal (Figure 1.24).

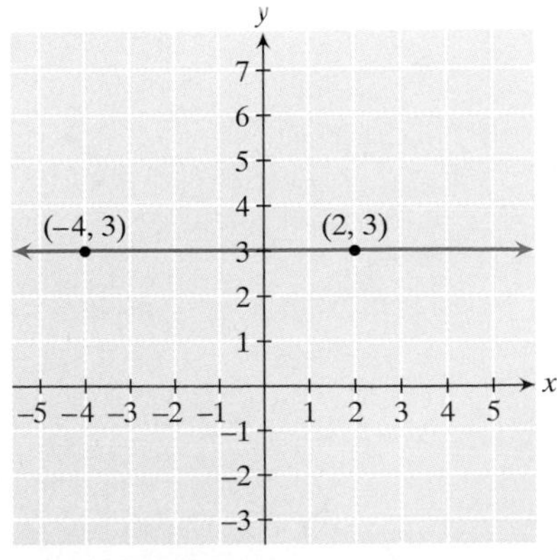

Figure 1.24

(c) Let $(x_1, y_1) = (2, 1)$ and $(x_2, y_2) = (2, 4)$. Note that $x_1 = x_2 = 2$. The line through these two points is vertical (Figure 1.25). The slope is undefined.

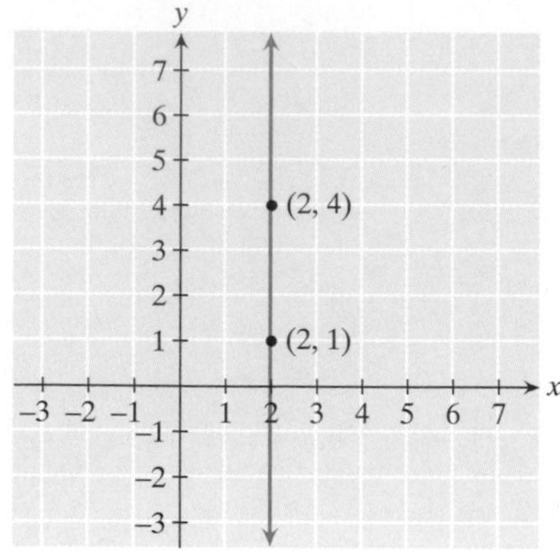

Figure 1.25

Self-Check 3 Show that the slope of the line through the points (2, 5) and (−1, −7) is 4.

The Equation of a Line

The information that defines a line can be presented in various ways. We now derive forms of the equation of a line that can be used in two different situations.

Point–Slope Form of the Equation of a Line

If we know a point on a line and the slope of the line, we have enough information to draw the line. We can find its equation from this information. Consider the line through the point (x_1, y_1) with a slope m. Let (x, y) be an arbitrary point on the line. The slope of the line through the points (x_1, y_1) and (x, y) is

$$\frac{y - y_1}{x - x_1}$$

Thus,

$$m = \frac{y - y_1}{x - x_1}$$

Rewrite the equation as $y - y_1 = m(x - x_1)$.

Point–Slope Form

The equation of the line through the point (x_1, y_1) having slope m is

$$y - y_1 = m(x - x_1)$$

Example 4 **Line Through a Given Point with a Given Slope**

Find the equation of the line through the point (2, 3) with slope 4. Graph the equation on your calculator, and use the display to support your answer.

SOLUTION Let $x_1 = 2$, $y_1 = 3$, and $m = 4$. Use the point–slope form to get the equation.

$$y - 3 = 4(x - 2)$$
$$y = 4x - 8 + 3$$
$$y = 4x - 5$$

On graphing this equation (Figure 1.26a), we see that it goes through the point (2, 3) and has slope 4.

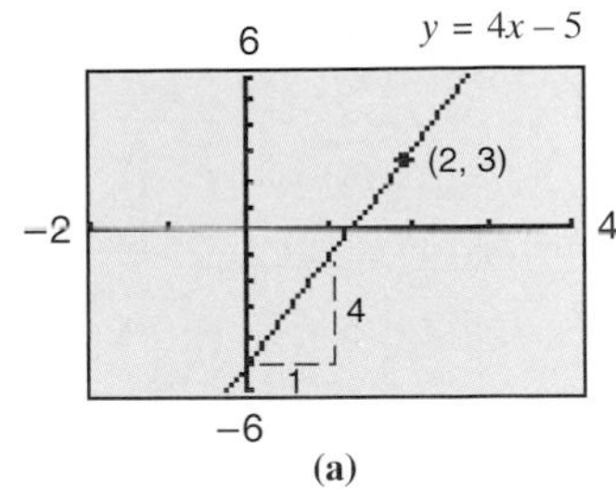

(a)

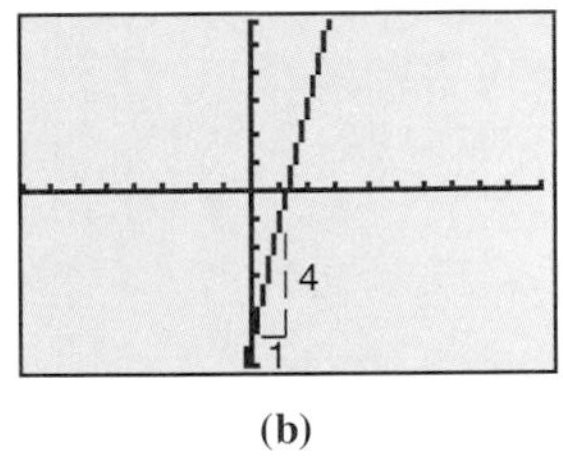

(b)

Square setting

Figure 1.26

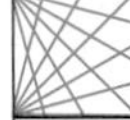

Square Setting

Note that the initial graph in Figure 1.26(a) does not appear to have slope 4. The rise does not appear to be four times the run. This is because the scales on the x- and y-axes are not the same. The scales were selected to make good use of the screen rather than to give a true scaled geometrical picture. Use the **square setting** on your calculator to see a truer display (Figure 1.26b). The horizontal and vertical tick marks have equal spacing in the square setting.

CAUTION: Entering the square mode on your calculator changes your original viewing rectangle.

Further Ideas 1 Find the equation of the line through the points (2, 3) and (5, −6). Graph the equation on your calculator, and use the display to support your answer. (*Hint:* Find the slope first and then use the point–slope form.)

Slope–Intercept Form of the Equation of a Line

The slope–intercept form of the equation of a straight line enables us to derive the equation conveniently, when given its slope and y-intercept. The y-intercept of a line is the value of y at which the line intersects the y-axis. Consider a line having slope m and y-intercept b. The line goes through the point $(0, b)$. Using the point–slope form of the equation of the line, we get

$$y - b = m(x - 0)$$
$$y = mx + b$$

Slope–Intercept Form

The equation of the straight line having slope m and y-intercept b is

$$y = mx + b$$

Example 5 **Line with a Given Slope and y-Intercept**

Find the equation of the line having slope −2 and y-intercept 3. Graph the equation on your calculator and use the display to support your answer.

SOLUTION The slope is -2, and the y-intercept is 3; thus, $m = -2$ and $b = 3$. Using the slope–intercept form, the equation is $y = -2x + 3$. Graph this equation (Figure 1.27). The line has a y-intercept of 3 and a slope of -2, confirming the result.

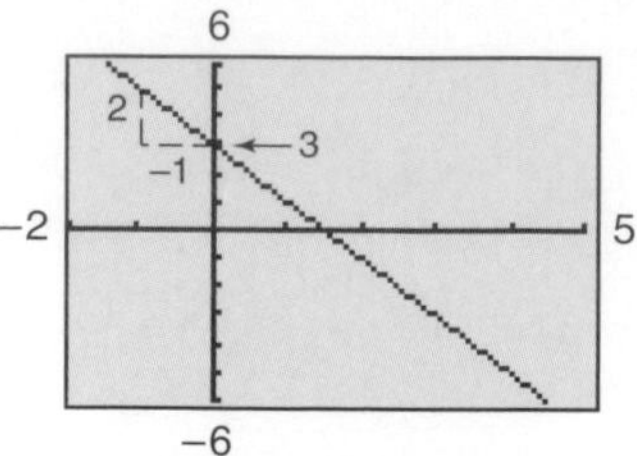

Figure 1.27 The y-intercept = 3, *rise* = 2, *run* = -1; Thus, *slope* = -2.

Self-Check 4 Find the equation of the line having slope 1 and y-intercept 6. Display the graph on your calculator in both the standard and the square settings. Use the display to check your answer.

Horizontal and Vertical Lines

Horizontal and vertical lines are special lines whose equations assume a particularly simple form. The student should be able to recognize such equations immediately. We now turn our attention to such lines.

Exploration 1

(a) Graph the lines $y = -2$, $y = 1$, $y = 3$
Find the equation and graph the horizontal line through the point (3, 4). What property do all these lines have in common?

(b) Graph the lines $x = -3$, $x = -1$, $x = 2$
Find the equation of the vertical line through the point (4, 5) and graph it. What property do all these lines have in common?

Horizontal and Vertical Lines
The horizontal line through (a, b) has equation $y = b$. The slope is zero. On the other hand, the vertical line through (a, b) has equation $x = a$. The slope is undefined.

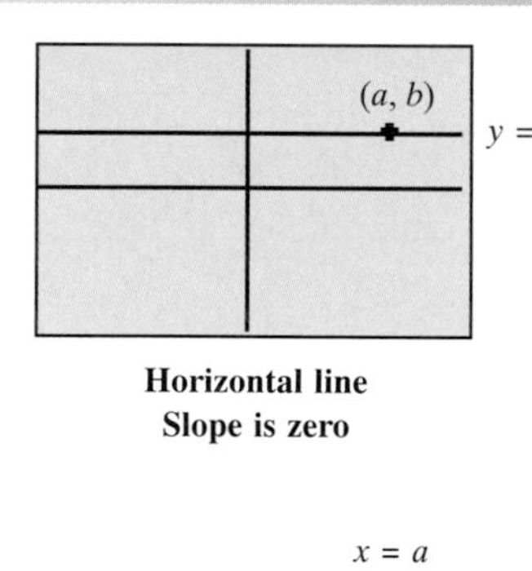

Horizontal line
Slope is zero

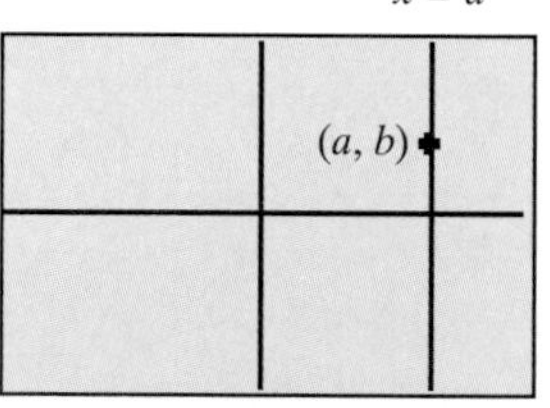

Vertical line
Slope is undefined

Vertical Line

The equation of a vertical line cannot be entered into your calculator using the standard Y = ... window. Determine how to graph a vertical line on your calculator.

Parallel and Perpendicular Lines

We think of two lines as being **parallel** if they never intersect, and two lines as being **perpendicular** if they intersect at right angles. We now make these ideas precise.

Exploration 2

(a) Graph equations $y = 2x - 3$ and $y = 2x + 1$.
Observe that these lines are parallel. Find and plot another line parallel to these two lines.

(b) Graph equations $y = 2x - 4$ and $y = -0.5x - 4$.
Observe that these lines are perpendicular. Find and graph another line that is perpendicular to $y = 2x - 4$. (Use the square setting to see that lines are perpendicular.)

Parallel and Perpendicular Lines
Let the slopes of two lines be m_1 and m_2. The lines are parallel if and only if $m_1 = m_2$. The lines are perpendicular if and only if $m_1 \cdot m_2 = -1$. (See Exercise 70 following for proof.)

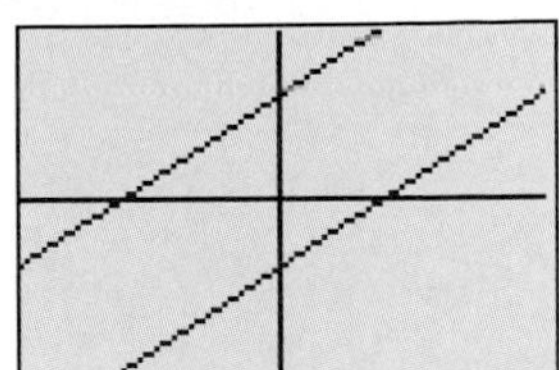

$m_1 = m_2$
Parallel lines

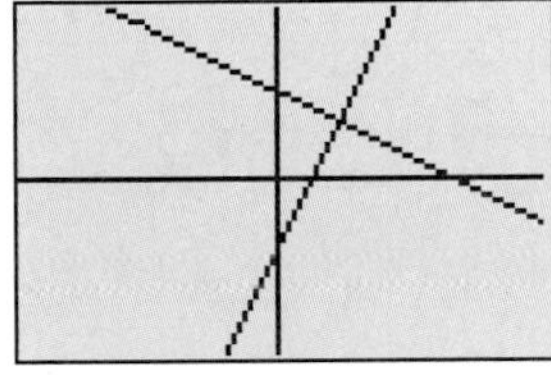

$m_1 \cdot m_2 = -1$
Perpendicular lines

Parallel and Perpendicular Lines Consider two lines with slopes m_1 and m_2.

The lines are **parallel** if and only if $m_1 = m_2$.
The lines are **perpendicular** if and only if $m_1 \cdot m_2 = -1$.

Example 6 Showing That Lines Are Parallel

Use slope to show that line L through the points $(1, 3)$ and $(3, 7)$ is parallel to line M through the points $(-4, -1)$ and $(-1, 5)$. Find the equations of these lines, and graph them on your calculator to observe that they do indeed appear to be parallel.

SOLUTION Use the formula for the slope of a line through two points to find the slopes of L and M. We get

$$\text{Slope of } L: \quad m_1 = \frac{7 - 3}{3 - 1} = \frac{4}{2} = 2$$

$$\text{Slope of } M: \quad m_2 = \frac{5 - (-1)}{-1 - (-4)} = \frac{6}{3} = 2$$

The slopes are equal; therefore the lines are parallel. We now use the point–slope form to find the equations of the lines and then graph them.

L is the line through (1, 3) with slope 2.

$$y - 3 = 2(x - 1)$$
$$y = 2x + 1$$

M is the line through $(-4, -1)$ with slope 2.

$$y - (-1) = 2[x - (-4)]$$
$$y = 2x + 7$$

We graph these lines (Figure 1.28), and they appear to be parallel.

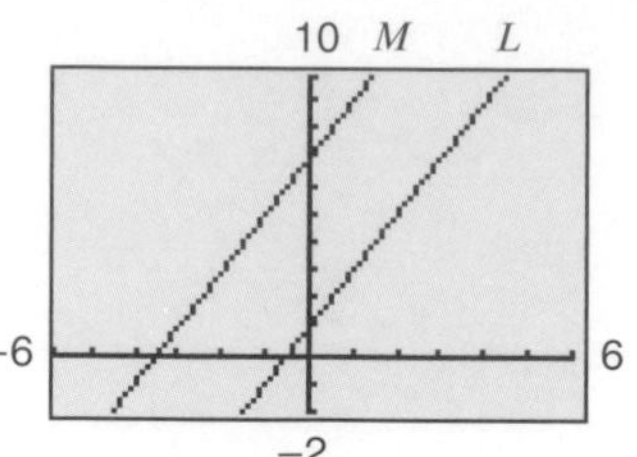

Figure 1.28 The lines M and L are parallel.

Example 7 Line Perpendicular to a Given Line

Find the equation of a line M through the point (8, 4) that is perpendicular to the line L having equation $y = -4x + 6$. Display L and M in the square setting to support your answer.

SOLUTION Comparing the given equation of L with the form $y = mx + b$, we see that its slope is -4 ($m_1 = -4$). The slope m_2 of any line that is perpendicular to this line is given by $m_1 \cdot m_2 = -1$. The slope of M is therefore

$$m_2 = \frac{-1}{m_1} = \frac{-1}{(-4)} = \frac{1}{4}$$

M is therefore the line through the point (8, 4) with slope $\frac{1}{4}$. We use the point–slope form to find the equation of M.

$$y - 4 = \frac{1}{4}(x - 8)$$
$$y = \frac{1}{4}x + 2$$

We graph L and M (Figure 1.29). The lines appear to be perpendicular.

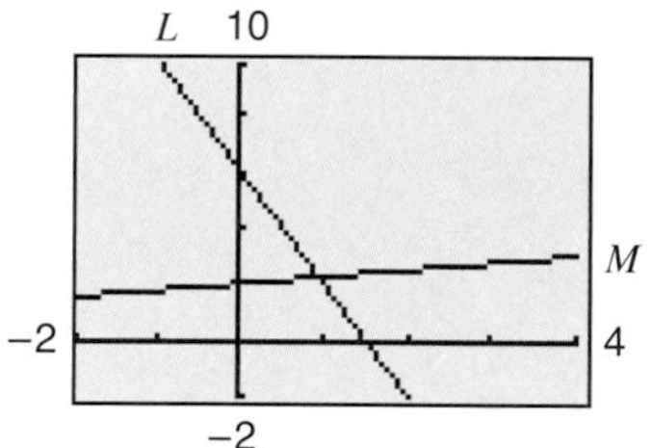

Standard setting

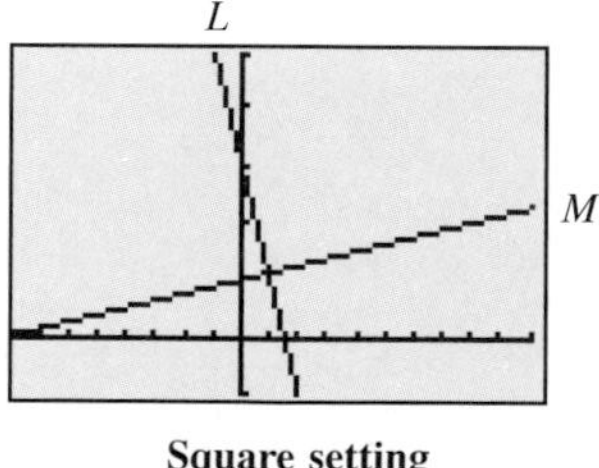

Square setting

Figure 1.29 The lines L and M are perpendicular.

Self-Check 5

(a) Find the slopes of lines that are parallel to line L through the points $(-4, 1)$ and $(-3, 6)$.
(b) Show that the line L through the points $(0, -3)$ and $(1, -5)$ and the line M through points $(1, 2)$ and $(3, 3)$ are perpendicular. Find the equations of L and M, and graph these equations on a calculator in the square setting to support your results.

CAUTION: An examination of slopes is needed to prove that two lines are parallel or perpendicular ($m_1 = m_2$ or $m_1 \cdot m_2 = -1$). A graphing calculator may be used to support such a result but not to prove it. Two lines that have almost the same slope may appear to be parallel on a calculator.

An equation of a line can be written in many forms. The slope–intercept form, $y = mx + b$, is particularly useful for giving geometrical information. This form immediately reveals the slope and y-intercept of a line. We now introduce and discuss a more general standard form of the equation of a line in two variables.

LINEAR EQUATIONS

An equation that can be written in the form $Ax + By + C = 0$, with A and B not both zero, is called a **linear equation** in the variables x and y. This form is called the **standard form** of the equation. The graph of such an equation is a straight line. We now show how to interpret the equation of a line presented in standard form.

Example 8 Sketching the Graph of a Linear Equation in Standard Form

Sketch the graph of the equation $2x - 3y - 6 = 0$. Confirm the graph on a calculator.

SOLUTION The equation is of the form $Ax + By + C = 0$; its graph will be a line. One convenient way of sketching a line is to use two points on the line. When a graph cuts the x- and y-axes, the x-intercept and y-intercept can be conveniently found and used for the two points. We use this method to sketch this graph.

Any point on the x-axis has y-coordinate zero; therefore, to find the x-intercept, let $y = 0$ in the equation:

$$2x - 3(0) - 6 = 0, \qquad \text{giving } x = 3$$

Similarly any point on the y-axis has x-coordinate zero. Letting $x = 0$ in the equation gives the y-intercept:

$$2(0) - 3y - 6 = 0, \qquad \text{giving } y = -2$$

Thus, the x-intercept is 3 and the y-intercept is -2. The graph can now be sketched (Figure 1.30a).

We write the equation in the form $y = \frac{2}{3}x - 2$ to graph on the calculator (Figure 1.30b). The calculator confirms the sketch.

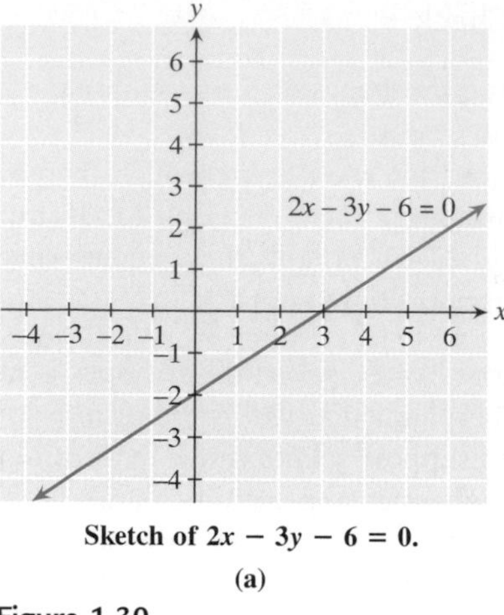

Sketch of $2x - 3y - 6 = 0$.
(a)

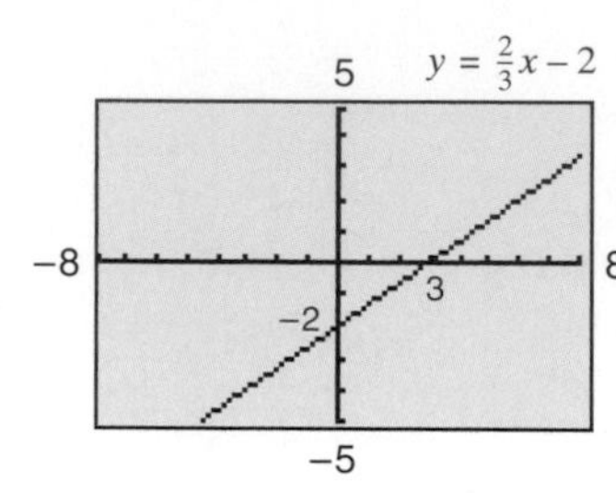

Calculator confirmation.
(b)

Figure 1.30

Further Ideas 2 Determine the slope and y-intercept of the graph of $3x - 5y + 7 = 0$. (*Hint:* Rewrite the equation in slope–intercept form.)

Example 9 Line Through a Given Point and Parallel to a Given Line

Determine the equation of the line through the point $(-4, 12)$ parallel to the line $2x + y - 10 = 0$. Graph the line on a calculator to confirm your answer.

SOLUTION Rewrite the equation $2x + y - 10 = 0$ in slope–intercept form to determine the slope.

$$y = -2x + 10$$

The slope is -2. Use the point–slope form of the equation of a line to find the line through $(-4, 12)$ with slope -2.

$$\begin{aligned} y - y_1 &= m(x - x_1) \\ y - 12 &= -2[x - (-4)] \\ y &= -2x + 4 \end{aligned}$$

The equation of the line through the point $(-4, 12)$ parallel to the line $2x + y - 10 = 0$ is $y = -2x + 4$.

The calculator graph (Figure 1.31) supports this result. The line $y = -2x + 4$ goes through the point $(-4, 12)$ and is parallel to $y = -2x + 10$.

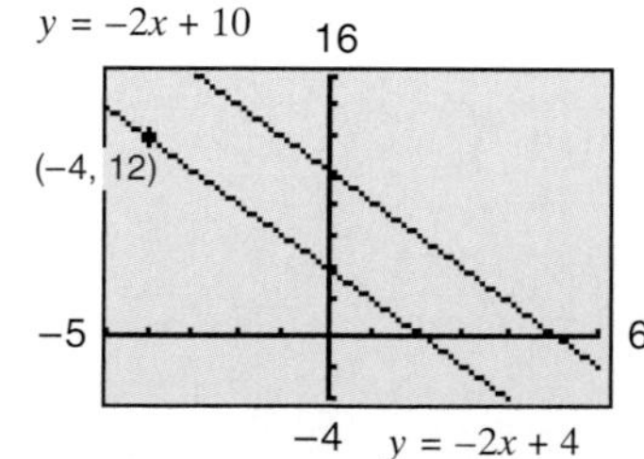

Figure 1.31

Example 10 Line Parallel to a Line Through Two Given Points

Find A so that the line $Ax - 3y - 6 = 0$ is parallel to the line passing through the points $(1, 4)$ and $(3, 12)$.

SOLUTION Rewrite the equation $Ax - 3y - 6 = 0$ in slope–intercept form to determine the slope.

$$3y = Ax - 6$$

$$y = \frac{A}{3}x - 2$$

The slope of the line $Ax - 3y + 6 = 0$ is thus $A/3$.

The slope of the line through the points (1, 4) and (3, 12) is

$$m = \frac{12 - 4}{3 - 1} = \frac{8}{2} = 4$$

For the lines to be parallel $A/3 = 4$; thus $A = 12$.

LEAST-SQUARES LINE

Many branches of science and business use equations based on data that have been determined from experimental results. Suppose that the data consist of the n data points $(x_1, y_1), (x_2, y_2), \ldots, (x_n, y_n)$. Plot these points to get a **scatter diagram** [Figure 1.32(a)]. These points lie approximately on a line. One often wants the equation of the line that comes as close as possible to passing through all the points. There are many criteria that can be used for the best fit in such cases. The one that has been found to be most satisfactory is called the **least-squares** line. Let d_1 be the vertical distance from (x_1, y_1) to a line, d_2 be the vertical distance from (x_2, y_2) to the line, and so on [Figure 1.32(b)]. The least-squares line is the line for which the sum of the squares of these vertical distances is as small as possible. It is the line that minimizes the sum $d_1^2 + d_2^2 + \cdots + d_n^2$. Calculators and computers have programs that will give the least-squares line $y = mx + b$ based on this criterion.

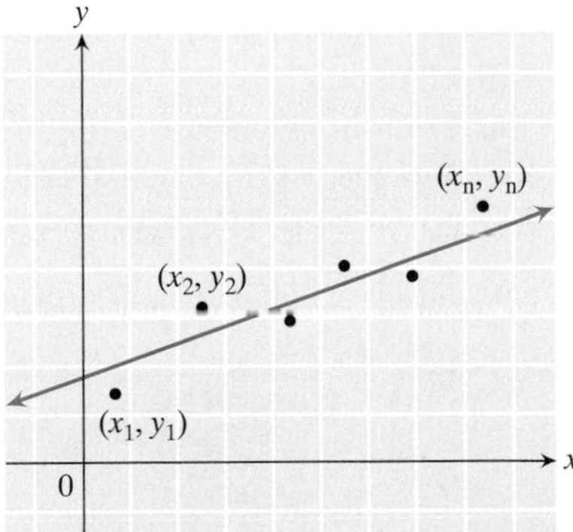

(a) Need a line that "best fits" these data.

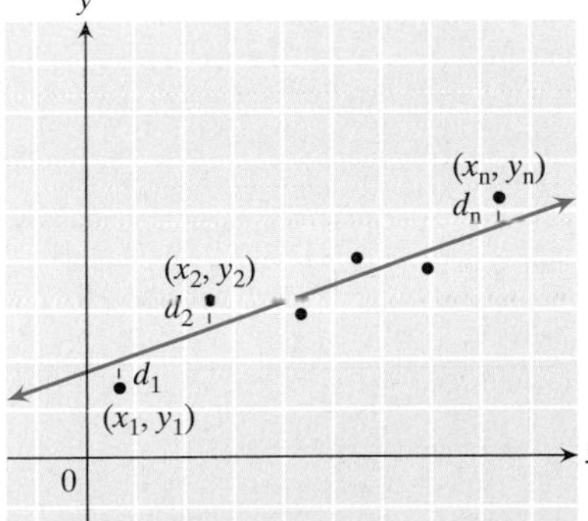

(b) The best fit is the "least-squares line." This line minimizes $d_1^2 + d_x^2 + \cdots + d_n^2$.

Figure 1.32

We now illustrate how one might estimate by hand a line of best fit for given data. We then show how to determine the least-squares line using a calculator.

Example 11 Finding the Least-Squares Line for Given Data

Consider the following data. (a) Plot these points and guess at the line that most closely fits the data. Estimate its equation. (b) Use a calculator to find the least-squares line.

$$(1, 7), (2, 8), (3, 11), (4, 17), (5, 20)$$

SOLUTION

(a) *Estimate of Best Line:* Plot the points (Figure 1.33a), and observe that the points lie approximately on a line. We draw the line L that we think best fits these points. For our line, the y-intercept might be 2 and the slope might be 3 (Figure 1.33b). Our hand estimate of the least-squares line is $y = 3x + 2$.

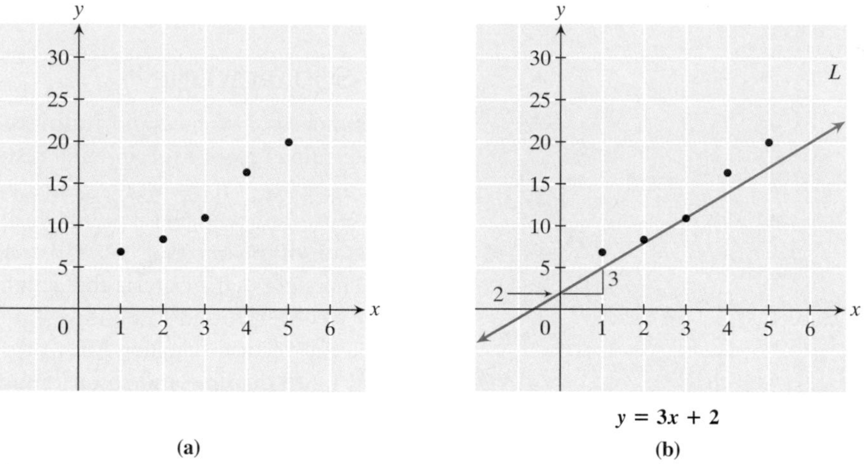

Figure 1.33 Hand estimate of a line of best fit.

(b) *Calculator Steps:* Enter and plot the data on a calculator (Figures 1.34a and 1.34b). The points lie approximately on a line. We use the calculator to determine the equation of the least-squares line for these data. The least-squares line is given to be $y = ax + b$ with $a = 3.5$ and $b = 2.1$ (Figure 1.34c). (Note that the a on our calculator is the m in the slope–intercept form.) To check the equation against the original data, enter the equation, plot the data, and graph (Figure 1.34d). The graph is seen to fit the data closely. The least-squares line is $y = 3.5x + 2.1$.

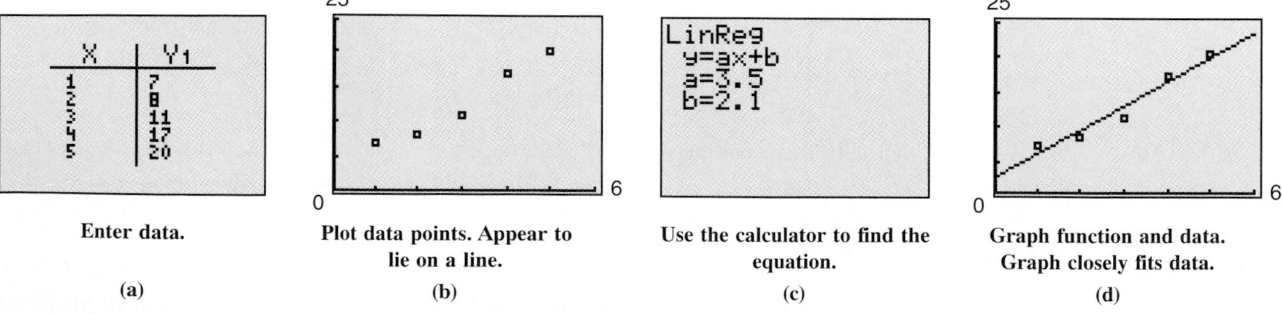

Figure 1.34 Steps in finding a least-squares line

These data give values of y for $x = 1, 2, 3, 4$, and 5. We can use this equation to predict other values of y. Suppose that we wanted to predict the value of y when $x = 2.5$. The equation gives $y = 3.5(2.5) + 2.1 = 10.85$; thus, y will be 10.85 when x is 2.5.

The area of mathematics that involves fitting equations to data is called **regression analysis.*** Because we are fitting linear equations to data in this section, we use **linear regression.** We are all aware that tuition costs are rising. Let us now use linear regression to construct a mathematical model for tuition.

Example 12 Tuition Costs

Consider the following data for in-state tuition at state universities (*World Almanac,* 1997, page 254). Numbers in parentheses give years from the base year 1990. (a) Construct the least-squares lines for these data. (b) Predict tuition costs for the year 2000. (c) Predict tuition costs for 2005.

Year (Fall)	1990 (0)	1991 (1)	1992 (2)	1993 (3)	1994 (4)	1995 (5)
Tuition ($)	2159	2410	2349	2537	2689	2811

SOLUTION

(a) We enter and plot the data on a calculator using the years 0, 1, . . . , 5, measured from 1990 for the x values on the horizontal axis of the graph (Figure 1.35). The points lie approximately on a line. We use the calculator to determine the equation of the least-squares line. The line is $y = 122.4285714x + 2186.428571$. The graph closely fits the data.

(b) Use this equation to predict tuition for the year 2000. Because the year 2000 is 10 years measured from the base year 1990, we take $x = 10$. The corresponding value of y is $y = 3410.714285$. On the basis of these statistics, using linear regression, the predicted tuition for 2000 is $3411.

(c) The year 2005 is 15 years from the base year 1990. Using $x = 15$, we get $y = 4022.857142$. Predicted tuition for 2005 is $4023.

It is important to realize that such predictions become increasingly unreliable the farther one gets from the original data. In making predictions in this fashion, we make the assumption that the current trend continues. Thus the prediction for the year 2000 is more reliable than the one for 2005. It is important to look closely at the circumstances surrounding the situation when using predictions obtained with a mathematical model.

*The method of least squares was first published by the French mathematician Adrien-Marie Legendre (1752–1833) in 1805. It at once became an important method in astronomy and in surveying and remains one of the most-used statistical tools today.

The term *regression* was introduced by Sir Francis Galton (1822–1911) in his study of heredity. He investigated data relating heights of parents and children. He concluded that the children of tall parents tend to be taller than average but not as tall as their parents. This "regression toward mediocrity" gave regression analysis its name.

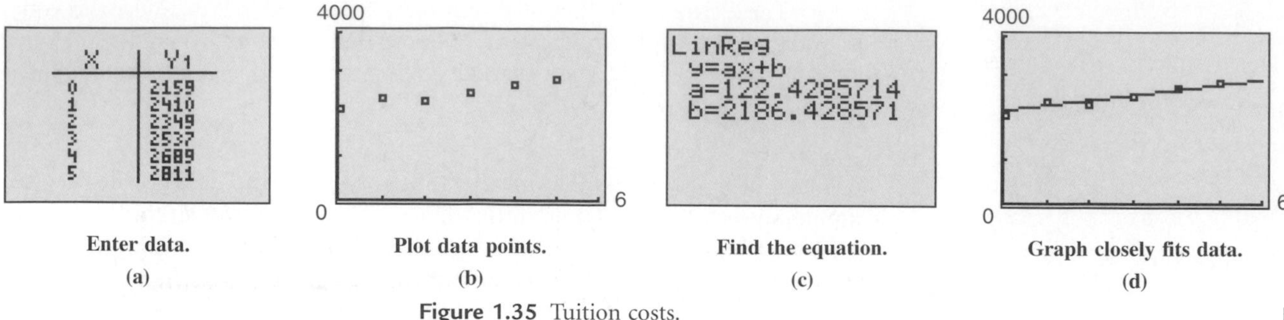

Enter data. (a) **Plot data points.** (b) **Find the equation.** (c) **Graph closely fits data.** (d)

Figure 1.35 Tuition costs.

CAUTION: It is advisable to use the number of years from the base year as the x variable in applications such as this, as we have done, rather than the absolute year ($x = 0, 1, \ldots, 5$ rather than $x = 1990, 1991, \ldots, 1995$). The x scale of 1 is of the same magnitude as the distances from the base year $0, 1, \ldots, 5$. The scale of 1 is much smaller than the year numbers $1990, \ldots, 1995$. As mentioned previously, roundoff often occurs in computations on calculators and computers. In general, the closer the numbers in a computation are in magnitude the better the accuracy.

Further Ideas 3 This discussion of time leads naturally into the so-called "Year 2000 Problem." Many computer programs involve the year. In the 1960s and 1970s, disk space was expensive, and people tried to save space by writing programs that recorded the year in terms of the last two digits of the year (1960 as 60 for example). This practice came back to haunt us with the approach of the year 2000. Such programs automatically use 00 for the year 2000, with all sorts of unforeseen repercussions. To get a feel for this problem, use your calculator to compute the least-squares line for tuition analysis using the following data. For the variable x, use the last two digits of the year. Based on these data, determine the predicted tuition for the year 2000. Be sure to use 00 for 2000. Discuss the result.

Year (Fall) (x)	1990 90	1991 91	1992 92	1993 93	1994 94	1995 95
Tuition ($)	2159	2410	2349	2537	2689	2811

GROUP DISCUSSION

Slope

In this section, we introduced the mathematical definition of slope as the ratio of rise to run. This definition was compatible with our intuitive idea of slope. Highway departments, however, use an alternative definition of slope for roads (Figure 1.36). When a road sign indicates a slope of 1 in 10, or 10%, the implication is that CB is 1 and AC is 10. Show how the highway department definition of slope is also compatible with our intuitive idea of slope. Discuss the suitability of the two definitions for the two areas. What would be the formula for the slope of the line through two given points using the highway department definition? Should highway departments change to be "in line" with mathematics?

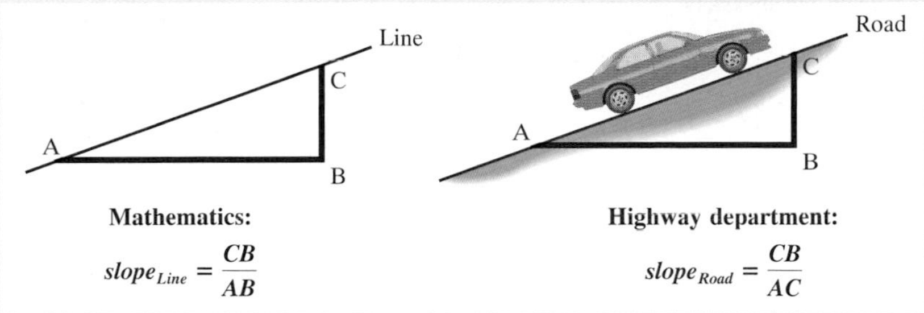

Figure 1.36

Answers to Self-Check Exercises

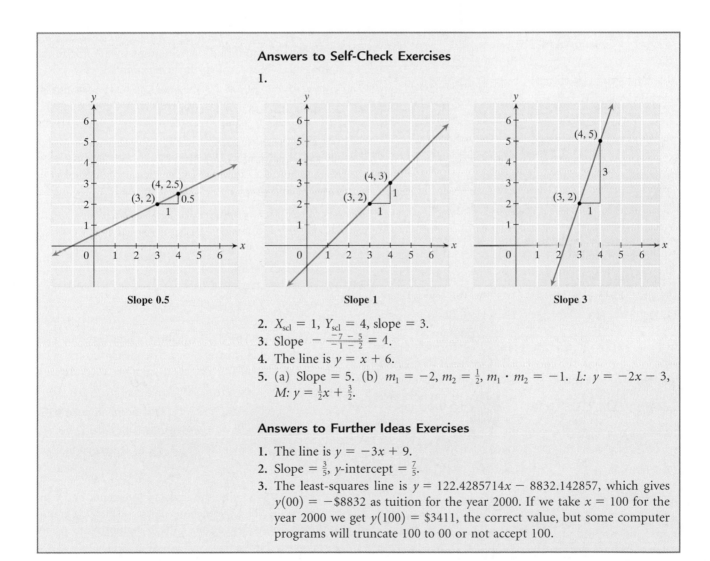

1.

Slope 0.5

Slope 1

Slope 3

2. $X_{scl} = 1$, $Y_{scl} = 4$, slope $= 3$.
3. Slope $= \frac{-7-5}{-1-2} = 4$.
4. The line is $y = x + 6$.
5. (a) Slope $= 5$. (b) $m_1 = -2$, $m_2 = \frac{1}{2}$, $m_1 \cdot m_2 = -1$. L: $y = -2x - 3$, M: $y = \frac{1}{2}x + \frac{3}{2}$.

Answers to Further Ideas Exercises

1. The line is $y = -3x + 9$.
2. Slope $= \frac{3}{5}$, y-intercept $= \frac{7}{5}$.
3. The least-squares line is $y = 122.4285714x - 8832.142857$, which gives $y(00) = -\$8832$ as tuition for the year 2000. If we take $x = 100$ for the year 2000 we get $y(100) = \$3411$, the correct value, but some computer programs will truncate 100 to 00 or not accept 100.

EXERCISES 1.2

In Exercises 1–6, the points at which the lines cross the x- and y-axes are integers. Find the slopes of the lines, if the slopes are defined.

1.

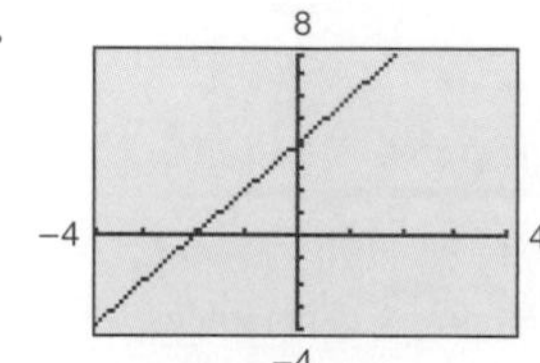

2.

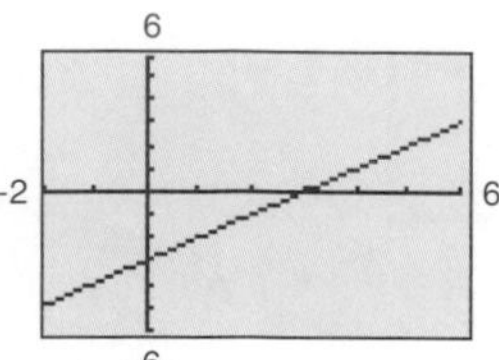

3.

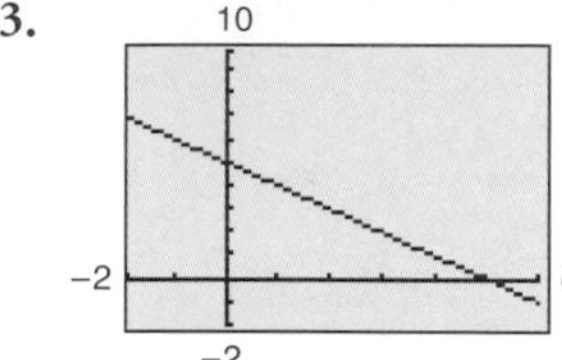

4.

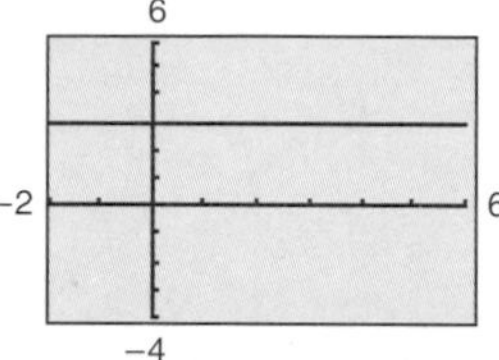

5.

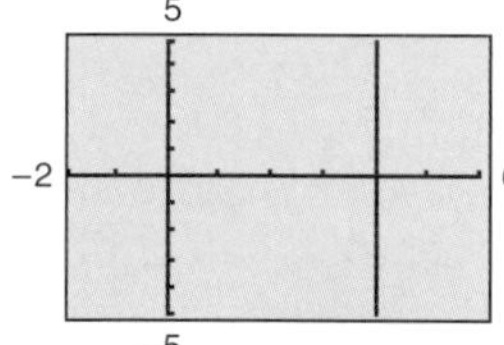

6.

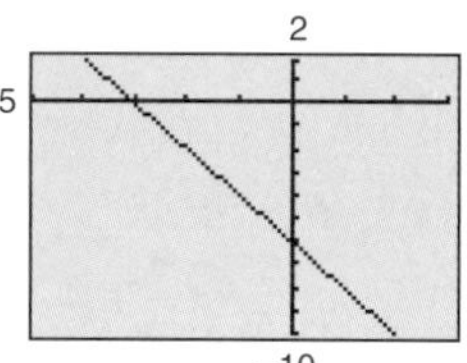

In Exercises 7–12, use the definition

$$m = \frac{y_2 - y_1}{x_2 - x_1}, \quad x_1 \neq x_2$$

to find the slopes of the lines through the given points (if the slopes are defined).

7. $(1, 1), (2, 3)$ **8.** $(-1, 0), (1, 2)$

9. $(1, -1), (1, -3)$ **10.** $(4, -1), (6, -1)$

11. $(4, 0), (2, 6)$ **12.** $(1, 9), (4, 6)$

In Exercises 13–20, use the point–slope form, $y - y_1 = m(x - x_1)$, to find the equations of the lines through the points having the given slopes. Graph the equations on your calculator, and use the displays to support your answers.

13. $(6, 8), m = 2$ **14.** $(-2, 3), m = -1$

15. $(0, 4), m = 4$ **16.** $(4, -10), m = -5$

17. $(0, 0), m = 2$ **18.** $(-1, -4), m = 0$

In Exercises 19–24, use the slope–intercept form, $y = mx + b$, to find the equations of the lines having the given slopes and y-intercepts. Graph the equations on your calculator, and use the displays to support your answers.

19. $m = 2, b = 5$ **20.** $m = -3, b = 4$

21. $m = -1, b = -6$ **22.** $m = -4, b = 0$

23. $m = 3/2, b = 5$ **24.** $m = -5/9, b = -3/5$

In Exercises 25–30, use the point–slope form,

$$y - y_1 = m(x - x_1), \text{ with } m = \frac{y_2 - y_1}{x_2 - x_1}$$

to find the equations of the lines through the points. Graph the equations on your calculator, and use the displays to support your answers.

25. $(1, 2), (2, 3)$ **26.** $(-1, 3), (1, 7)$

27. $(2, 4), (4, -2)$ **28.** $(-1, 5), (-6, -2)$

29. $(2, -3), (4, 0)$ **30.** $(0, 4), (2, 1)$

In Exercises 31–36, find the equations of the horizontal and vertical lines through the given points.

31. $(1, 4)$ **32.** $(-1, 3)$ **33.** $(6, -3)$ **34.** $(0, 5)$

35. $(-3, 0)$ **36.** $(-7, -9)$

In Exercises 37–42, determine whether the lines through the given pairs of points are parallel, perpendicular, or neither. Find the equations of the lines, and support your answers with graphs on a calculator.

37. $(0, 0), (1, 1)$ and $(1, -1), (2, 0)$

38. $(1, 2), (3, 6)$ and $(0, 1), (1, 3)$

39. $(1, -1), (4, 8)$ and $(6, 3), (9, 2)$

40. $(1, 4), (3, 9)$ and $(-4, 5), (1, 7)$

41. $(1, 1), (3, 3)$ and $(1, 0), (2, -1)$

42. $(-1, 5), (2, -1)$ and $(-2, 2), (2, -6)$

In Exercises 43–46, find the equation of line M through the given point P that is parallel or perpendicular to the given line L. Use a calculator display to support your answers.

43. L: $y = 2x - 4$. $P(6, 2)$. M is parallel to L.

44. L: $y = -3x + 5$. $P(3, -1)$. M is perpendicular to L.

45. L: $2x - 4y + 8 = 0$. $P(6, 3)$. M is parallel to L.

46. L: $5x - 2y - 3 = 0$. $P(-3, 4)$. M is perpendicular to L.

In Exercises 47–52, find the slopes, y-intercepts, and x-intercepts of the lines. Use the intercepts to sketch the graphs of the equations. Use a calculator to support your result.

47. $y = 2x + 8$ **48.** $y = -3x + 7$

49. $4x - 2y = -12$ **50.** $3x - 2y = 0$

51. $3x + 5y + 9 = 0$ **52.** $-x + 3y + 7 = 0$

In Exercises 53–58, the x- and y-intercepts of the lines are integers. Use the intercepts to find the equation of each line. Support your answer by duplicating the display on your calculator.

53. **54.**

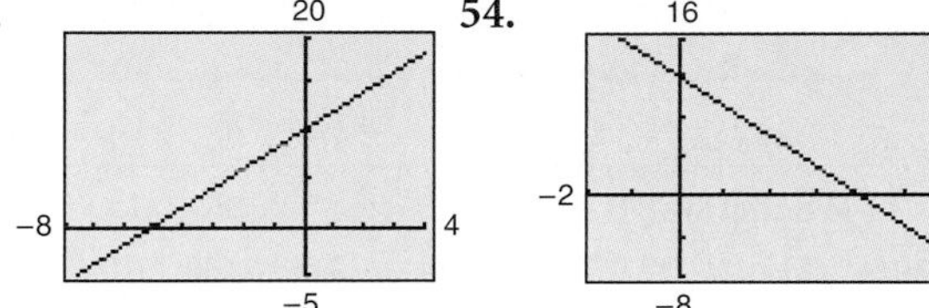

55. **56.**

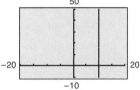

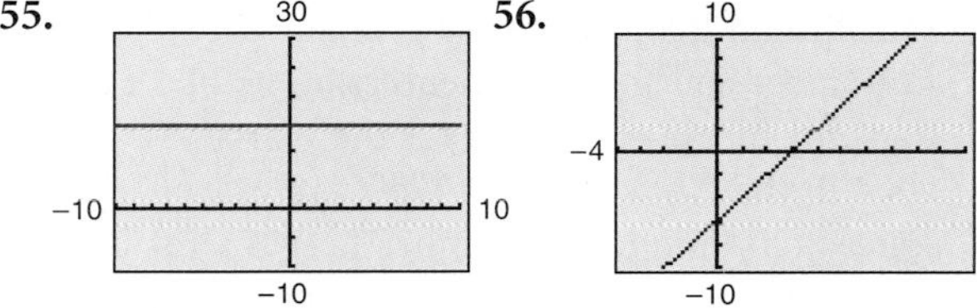

57. **58.**

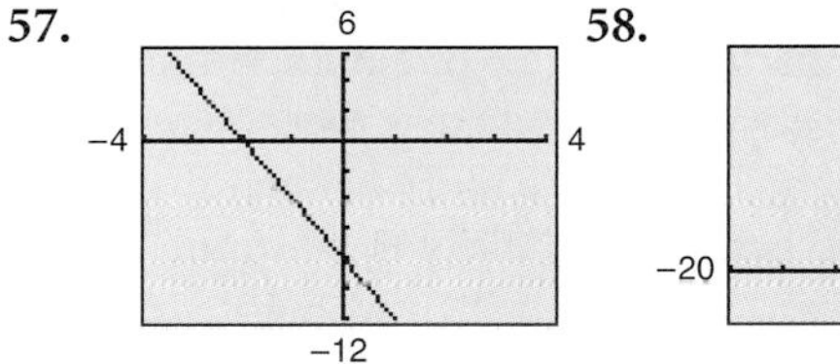

In Exercises 59–62, determine the equations of the lines satisfying the given conditions. Graph the lines on a calculator to support your answers.

59. Through the point $(1, 3)$, having y-intercept 7.

60. Through the point $(3, 5)$, parallel to the line $9x + 3y + 1 = 0$.

61. Through the point $(5, 3)$, perpendicular to the line $x - 2y + 1 = 0$.

62. With y-intercept 4, parallel to $y = 3x - 1$.

63. Find m so that the point $(-1, 4)$ lies on the line $y = mx + 5$.

64. Find b so that the point $(2, 11)$ lies on the line $y = 2x + b$.

65. Find A so that the line $Ax + 2y - 1 = 0$ is parallel to the line passing through the points $(1, 4)$ and $(3, 8)$.

66. Find B so that the line $2x + By + 5 = 0$ is perpendicular to a line having slope 4.

67. Find A so that the line $Ax + 3y + 4 = 0$ is parallel to the line $2x - y + 1 = 0$.

68. Find C so that the point $(-3, 0)$ lies on the line $5x + 4y + C = 0$.

69. Find the x-intercept of the line through the point $(-1, 6)$ perpendicular to the line $x + 2y - 3 = 0$.

70. Use similar triangles for ABC and ADC (Figure 1.37) to prove that the lines L_1 and L_2 are perpendicular if and only if $m_1 \cdot m_2 = -1$.

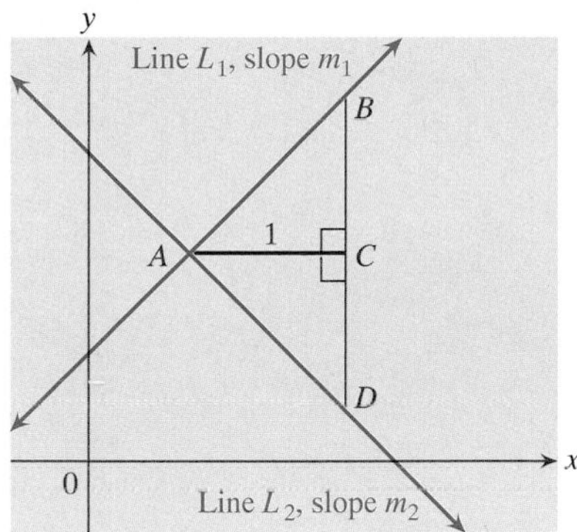

Figure 1.37

In Exercises 71–74, plot the data to see that the points lie approximately on a line. **(a)** Estimate the lines of best fit for the data points by hand. **(b)** Use a calculator to determine the least-squares lines. Use these equations to predict the values of y for the given x values.

71. $(1, 7), (2, 11), (3, 14), (4, 19)$. $x = 5$.

72. $(1, 9), (2, 11), (3, 14), (4, 16), (5, 20)$. $x = 4.25$.

73. $(1, 1), (3, 4), (4, 7), (8, 16)$. $x = 7$.

74. $(2, 20), (4, 14), (6, 13), (8, 8), (10, 1)$. $x = 7.3$.

In Exercises 75–83, plot the data to see that the relationships are approximately linear. Use a calculator to find the least-squares lines. Round answers to four decimal places unless stated otherwise.

75. Gas Consumption The rate of gasoline consumption of a car depends on its speed. To determine the effect of speed on fuel consumption, the Federal Highway Administration tested cars of various weights at various speeds. One car weighing 3980 pounds had the following results. **(a)** Determine the least-squares line corresponding to these data. **(b)** Use this equation to estimate the mileage per gallon of this car at 55 miles per hour.

Speed (miles per hour)	30	40	50	60	70
Gas used (miles per gallon)	33.1	30.3	28	26.5	24.8

76. Advertising Expenditures The Aloma Advertising Company arrived at the following statistics relating the amount of money spent on advertising the Flex Exercise Machine to the realized sale of that product. (**a**) Find the least-squares line. (**b**) Use this line to predict the sales when $4000 is spent on advertising.

Dollars spent	1000	1500	2000	2500	3000	3500
Units sold	525	540	582	600	610	634

77. GPA Analysis Winchester College compiled the following data on the verbal SAT scores of ten entering students and their graduating GPA scores 4 years later. (**a**) Determine the least-squares line. (**b**) Use it to predict the graduating GPA of a student who entered with an SAT score of 500. (**c**) Consider a student who graduates with a GPA of 3.7 but who did not take the SAT examination. What would the student probably have scored on the examination? (**d**) If you took the SAT, what would be your predicted GPA according to these statistics?

SAT	490	450	640	510	680	610	470	450	600	650
GPA	3.2	2.8	3.8	2.9	4.0	3.1	2.2	2.3	3.5	3.8

78. Mortality Analysis The Acropolis Insurance Company computed the following percentages of people dying at various ages from 20 to 65. Use linear regression to estimate the percentage of deaths at 43 years of age.

Age (years)	20	25	30	35	40	45	50	55	60	65
Deaths (%)	0.18	0.19	0.21	0.32	0.41	0.64	0.78	0.81	0.79	0.83

79. Production Costs The production numbers and costs of the Summit Company for the last 6-month period were as follows. (**a**) Use linear regression to find the best linear function for these data (called the cost–volume equation). (**b**) Use this function to predict total costs for the next month (to the nearest $10) if production was planned to rise sharply to 165 units.

	July	Aug	Sept	Oct	Nov	Dec
Units of production	120	140	155	135	110	105
Total cost ($)	6300	6500	6670	6450	6100	5950

80. Enrollment Projections The following were projections of college enrollment (in units of 1 million) for men and women for the years 1994 to 2004 in the United States (*The Chronicle of Higher Education Almanac,* September 1, 1997). (**a**) Use linear regression to find the best linear functions for the data, one for men and one for women. (**b**) Observe that there were 6.831 million men enrolled in colleges in 1994, whereas there were 8.174 million women. There were more women than men enrolled. Is this trend likely to continue, or is the trend toward more men than women? (Use the variable x, time measured from 1994, as the independent variable.)

Year (x)	1994 0	1995 1	1996 2	1997 3	1998 4	1999 5	2000 6	2001 7	2002 8	2003 9	2004 10
Men	6.831	6.781	6.752	6.765	6.811	6.892	6.970	7.041	7.116	7.161	7.216
Women	8.174	8.165	8.186	8.234	8.300	8.412	8.492	8.566	8.622	8.641	8.676

81. Analysis of Smokers The following table shows how the percentage of persons 18 years of age and older in the United States who smoke has decreased through the years. **(a)** Use linear regression to find the best linear function for these data. **(b)** What is the percentage of smokers likely to be in the year 2003? (Use 1974 as the base year with x measured from 1974.)

Year (x)	1974 0	1979 5	1985 11	1988 14	1990 16	1992 18	1993 19
Percent of smokers	37.2	33.5	30.0	27.9	25.4	26.4	25.0

(See the Williams website at www.saunderscollege.com/math/williams to link to the Internet for more information.)

82. Life Expectancy The following table shows how life expectancy (both sexes) has increased in the United States from 1940 to 1990. **(a)** Use linear regression to find the best linear function for these data. **(b)** What will the life expectancy be in the year 2000? (Use 1940 as the base year with x measured from 1940.)

Year (x)	1940 0	1950 10	1960 20	1970 30	1980 40	1990 50
Age	62.9	68.2	69.7	70.8	73.7	75.4

83. The following data for tuition at 4-year private colleges were obtained from the *World Almanac* (1997, page 254). Numbers in parentheses give years from the base year 1990. **(a)** Construct a least-squares lines from these statistics and use them to predict tuition (to the nearest dollar) for the years 2000, 2005, and 2010. **(b)** Discuss the results. How could you get more satisfactory results?

Year (Fall) (x)	1990 0	1991 1	1992 2	1993 3	1994 4	1995 5
Tuition ($)	11,379	12,192	10,294	10,952	11,522	12,216

SECTION PROJECT

Finding an Approximation for π

Use a piece of string to measure the diameters and circumferences of certain objects, such as a plate, the rim of a cup, the steering wheel of a car, or a round garbage can. Draw up a data table of points

$$\{(\text{diam}_{\text{plate}}, \text{circumf}_{\text{plate}}), (\text{diam}_{\text{wheel}}, \text{circumf}_{\text{wheel}}), \ldots\}$$

Find the least-squares line through these points. Use the line to find an approximation for π.

1.3 Linear Functions, Reservoir Monitoring, and Cost–Volume Analysis

- Linear Functions • Rate of Change • Using a Table to Get Information • CD Expenses • Scuba Diving • Reservoir Monitoring • Cost–Volume Analysis

Linear Functions

In this section we discuss functions that are defined by linear equations. Consider a linear equation

$$y = mx + b$$

As x varies, we get corresponding values of y. The equation $y = mx + b$ defines the **linear function** $f(x) = mx + b$.

For example, the linear equation $y = 2x + 5$ defines the linear function $f(x) = 2x + 5$. We can conveniently write and compute values of f at points such as $x = 3$ and $x = 7$ as follows:

$$f(3) = 2(3) + 5 = 11 \qquad \text{and} \qquad f(7) = 2(7) + 5 = 19$$

Rate of Change

The table facility of a calculator can be used to provide values of f over an interval. Let us see how the linearity of f is revealed by such a table. Consider the table of values of f over the x-interval $[-1, 5]$ (Figure 1.38a). Note how the values of f change by a constant amount 2 for each change of 1 in x. Such a constant change is the mark of a linear function. We call this constant the **rate of change of the function.** It is a measure of how rapidly the function is changing with respect to x.

The graph of $f(x) = 2x + 5$ is the straight line $y = 2x + 5$. Observe that the slope of the line, namely 2, is the change in f corresponding to a change of 1 in x (Figure 1.38b). The rate of change of f is given by the slope of its graph.

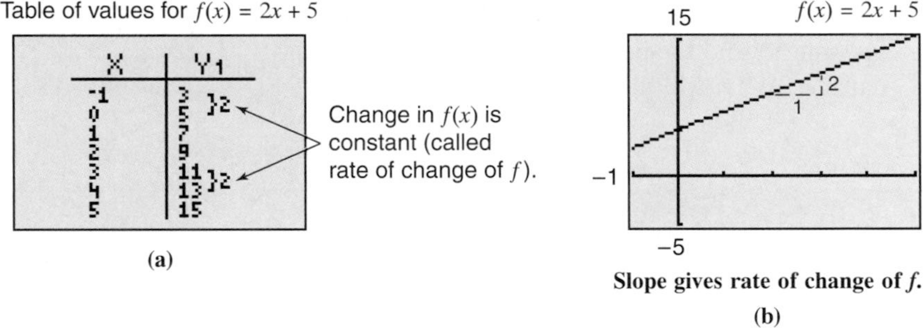

Figure 1.38 Table and graph interpretations of rate of change.

X	Y1
-4	-2
-3	.5
-2	3
-1	5.5
0	8
1	10.5
2	13

Figure 1.39

Example 1 Using a Table to Get Information

Consider the table of function values shown in Figure 1.39. Show that the function is linear, find its rate of change, find its equation, and plot its graph.

SOLUTION Observe that there is a constant change of 2.5 in function values for each change of 1 in x. The function is thus linear with a rate of change 2.5. The graph will be a line of slope 2.5. The table tells us that $f(0) = 8$. The y-intercept of the line is 8. The function is thus $f(x) = 2.5x + 8$ (Figure 1.40).

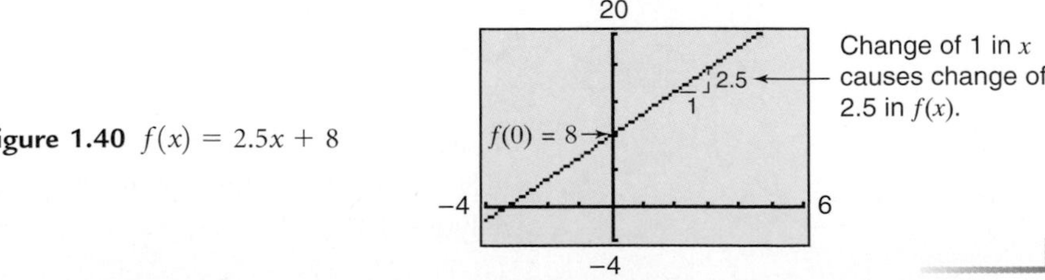

Figure 1.40 $f(x) = 2.5x + 8$

Further Ideas 1 Determine whether the functions defined by the tables in Figure 1.41 are linear. If a function is linear, find $f(0)$ and its rate of change from the table, and determine its equation.

X	Y1
-2	-2
-1	1
0	4
1	7
2	10
3	13
4	16

(a)

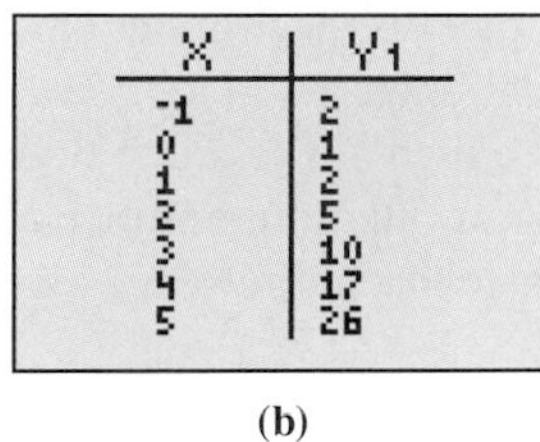

X	Y1
-1	2
0	1
1	2
2	5
3	10
4	17
5	26

(b)

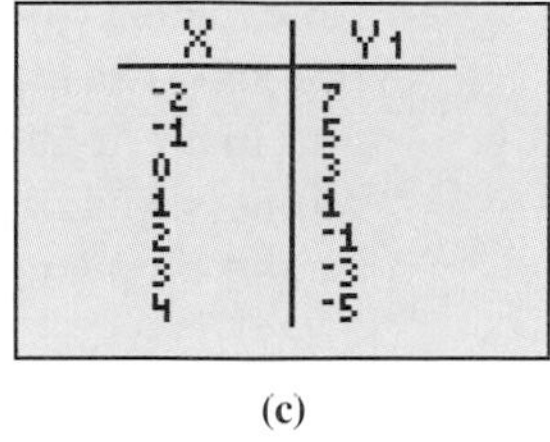

X	Y1
-2	7
-1	5
0	3
1	1
2	-1
3	-3
4	-5

(c)

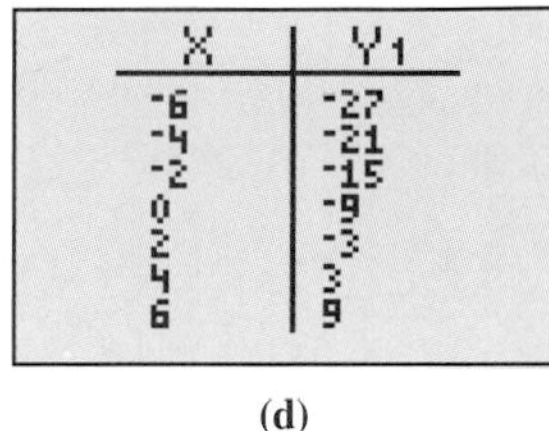

X	Y1
-6	-27
-4	-21
-2	-15
0	-9
2	-3
4	3
6	9

(d)

Figure 1.41

Applications . . .

Let us look at situations that are described by linear functions and determine information from graphs and tables. Graphs enable us to visualize these situations mathematically, giving us a feel for what is going on. It will be interesting to see the significance of slope and y-intercept in these cases. The student is encouraged to think of slope and y-intercept as having a practical as well as geometrical significance. Slope will correspond to a rate of change in each application. It will give an indication of how rapidly the function is increasing or decreasing. This section will also reenforce the natural interplay among functions, graphs, and tables.

Example 2 CD Expenses

A CD player costs \$150.75, and CDs cost \$16.96 each (including tax). Construct a function that describes the expense of buying and using the CD player. Represent this information graphically. What do the y-intercept and the slope of the graph represent in practical terms? If the cost of CDs increases, how will this affect the graph? Use the function to determine the total outlay after 10 and then 25 CDs have been purchased.

SOLUTION Suppose x CDs are purchased. Let the total expense of having the CD player up to that time be $f(x)$, then

$$f(x) = \text{Total cost of CDs} + \text{Cost of CD player}$$

$$f(x) = 16.96x + 150.75$$

The CD expenses are described by the linear function $f(x) = 16.96x + 150.75$, with the graph shown in Figure 1.42.

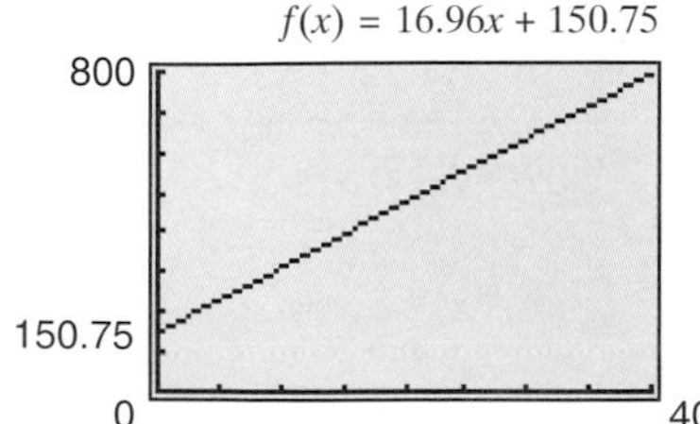

Figure 1.42 Slope = cost of a CD. y − intercept = cost of CD player.

If the cost of a CD increases, then the slope increases. The line that describes the expense of using CDs becomes steeper. The y-intercept is the cost of the CD player.

The total expenses after 10 and 25 CDs are $f(10)$ and $f(25)$. We get

$$f(10) = 16.96(10) + 150.75 = \$320.35$$
$$f(25) = 16.96(25) + 150.75 = \$574.75$$

Note that, in this example, x is the number of CDs purchased. Therefore, the function only has interpretation at $x = 1, 2, 3, \ldots$. This type of interpretation enters into many applications, such as the scuba diving analysis in the next example. In business applications, x is often the number of items manufactured. The student should be aware of this concept; however, we will not mention it every time.

Example 3 Scuba Diving

Nan Kuznicki goes scuba diving as often as she can. The air tank cost her \$257.78, and it costs \$22.50 to fill the tank each time she goes out. Construct a function that describes these expenses. Represent the information graphically.

SOLUTION Let Nan go scuba diving x times. Let the total expenses then be $f(x)$. Then

$$\begin{aligned} f(x) &= \text{Cost of } x \text{ fill-ups} + \text{Cost of the tank} \\ &= 22.50x + 257.78 \end{aligned}$$

Nan's expenses are described by the linear function $f(x) = 22.50x + 257.78$, with graph shown in Figure 1.43.

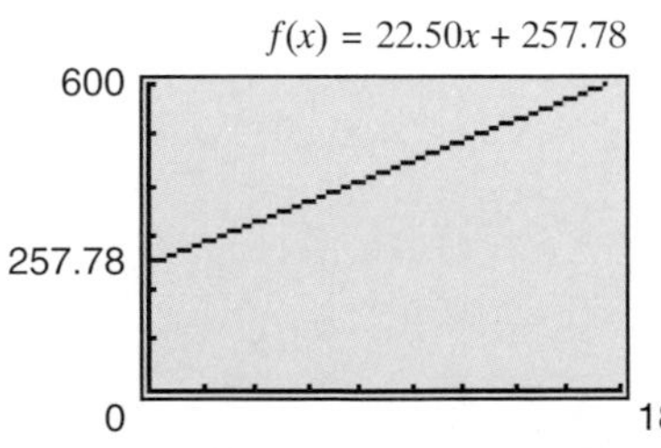

Figure 1.43 y-intercept = cost of tank. Slope = cost of an air fill-up.

Further Ideas 2

(a) Draw a table to predict Nan's costs for the first eight times she goes diving (Example 3).

(b) If Nan's friend Jane Robinson manages to buy the same air tank at a discount price of \$225, discuss how Jane's graph would differ from Nan's graph.

Example 4 | Reservoir Monitoring

Water flows into the Squaw Ridge Reservoir at the rate of 30 gallons per minute. It can be pumped out at the rate of 55 gallons per minute. The pumping process is started when there are 50,000 gallons of water in the reservoir. Construct a function that gives the number of gallons of water in the reservoir at time t minutes after the pumping starts. Use the function to determine the amount of water in the reservoir after 5 hours of pumping. What are the practical interpretations of the y-intercept and the slope of the graph of the function?

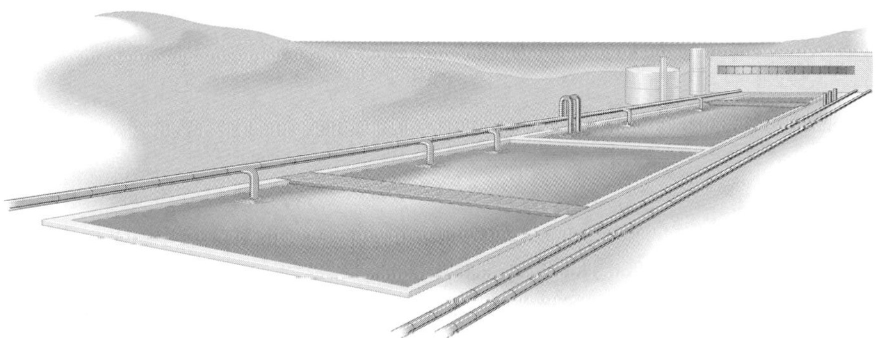

SOLUTION Let G gallons of water be in the reservoir t minutes after pumping starts, then

$$G(t) = \begin{pmatrix}\text{Amount}\\ \text{entering}\end{pmatrix} - \begin{pmatrix}\text{Amount}\\ \text{leaving}\end{pmatrix} + \begin{pmatrix}\text{Original}\\ \text{amount}\end{pmatrix}$$

$$= 30t - 55t + 50,000$$

$$= -25t + 50,000$$

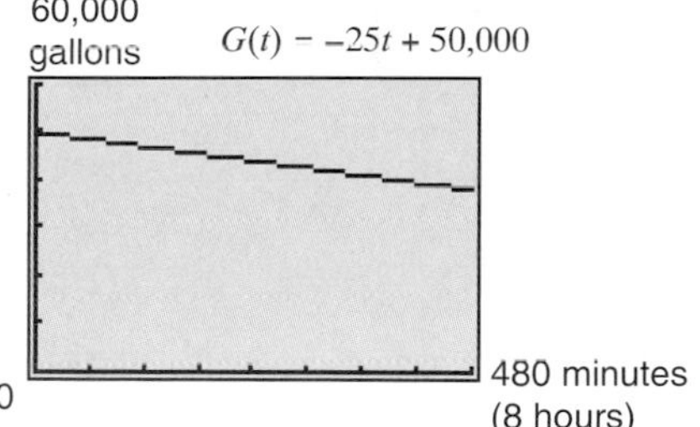

Figure 1.44 y-intercept = initial amount of water.
Slope = net rate at which water is leaving.

See the graph in Figure 1.44.

The amount of water in the reservoir after 5 hours (300 minutes) is $G(300)$.

$$G(300) = -25(300) + 50,000 = 42,500 \text{ gallons}$$

The y-intercept is the amount of water in the reservoir when pumping begins. The slope is the net rate at which water is leaving the reservoir.

What is the significance of the negative slope in this example, given that it was positive in the previous examples?

Further Ideas 3 Consider the reservoir situation in Example 4. Plot the lines $G(t) = -25t + 50{,}000$, $y = 45{,}000$ and $y = 30{,}000$ in the previous window. Determine which lines intersect in this window. What information does the fact that the lines intersect or do not intersect in this window tell you?

Example 5 Cost-Volume Analysis

Linear functions are used extensively in business analyses. The Korona Company has determined that the total cost of manufacturing 20 televisions is \$3951.20 and the total cost for 25 televisions is \$4514. Assume that the relationship between cost and volume is linear. Determine the linear equation that relates cost to volume and use it to predict the cost of manufacturing 28 televisions.

SOLUTION Let x be the number of televisions manufactured, y be the corresponding total cost, and $y = mx + b$ the **cost–volume equation.** When $x = 20$, $y = 3951.20$, and when $x = 25$, $y = 4514$. Let us use these as two points on the line to find the slope m. Let $(x_1, y_1) = (20, 3951.20)$ and $(x_2, y_2) = (25, 4514)$, then

$$m = \frac{y_2 - y_1}{x_2 - x_1} = \frac{4514 - 3951.20}{25 - 20} = 112.56$$

We now use the point–slope form to find the cost–volume equation.

$$y - 3951.20 = 112.56(x - 20)$$
$$y = 112.56x + 1700$$

Let us look at the interpretation of this equation.

Slope = Cost of producing one more television
= \$112.56, called **marginal cost**

y-intercept = Cost incurred when no televisions are produced
= \$1700, called **fixed cost**

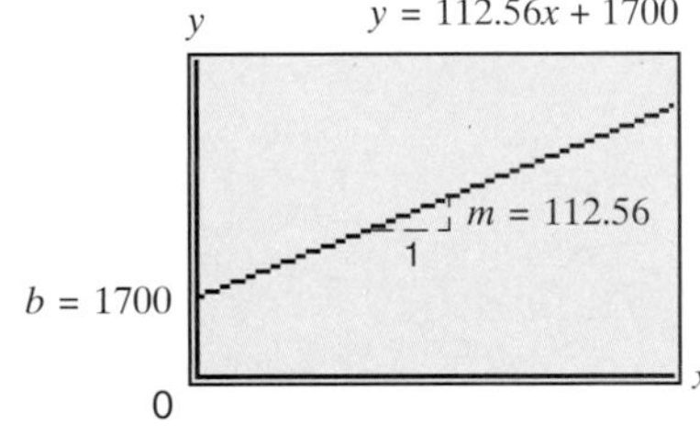

Figure 1.45 Marginal cost (slope) = cost of producing one more item. Fixed cost (y-intercept) = initial cost.

Building depreciation and property taxes are parts of the fixed cost (Figure 1.45).

We can now use this cost–volume equation to predict the cost of manufacturing 28 televisions. Let $x = 28$ in the equation.

$$y = 112.56(28) + 1700$$
$$= 4851.68$$

The predicted total cost of manufacturing 28 televisions is \$4851.68.

This function can be used to predict the cost of a range of items. Let us predict the total costs of 55, 56, . . . , 60 items. We get

x	55	56	57	58	59	60
y	7891.80	8003.36	8115.92	8228.48	8341.04	8453.60

Further Ideas 4 The PineHill Company found that the cost of manufacturing 10 wooden park tables is \$460. The cost for 15 tables is \$612.90. Find the cost–volume equation (assuming it is linear). What is the fixed cost? Predict the cost of manufacturing 20, . . . , 25 tables. Use the Table facility of a calculator to predict when the cost goes over \$15,000.

Answers to Further Ideas Exercises

1. **(a)** Linear. $f(0) = 4$, rate of change $= 3$; $f(x) = 3x + 4$. **(b)** Not linear [e.g., $f(5) - f(4) = 9$, $f(4) - f(3) = 7$]. **(c)** Linear. $f(0) = 3$, rate of change $= -2$; $f(x) = -2x + 3$. **(d)** Linear. $f(0) = -9$, rate of change $= 3$; $f(x) = 3x - 9$.
2. **(a)**

x	1	2	3	4	5	6	7	8
$f(x)$	280.28	302.78	325.28	347.78	370.28	392.78	415.28	437.78

(b) The y-intercept of Jane's graph would be the discount price of \$225. The slope of Jane's graph would be the same as Nan's, the cost of filling the tank, namely 22.5. Jane's graph would thus be parallel to Nan's graph but would lie below it.

3. Line $y = 45{,}000$ intersects G in this window. The water will have dropped to 45,000 gallons within 8 hours. Line $y = 30{,}000$ does not intersect G in this window, therefore the water will not have dropped to 30,000 within 8 hours.

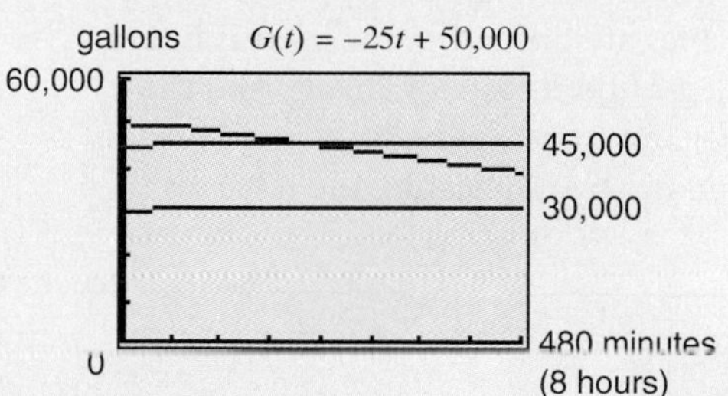

4. $y = 30.58x + 154.20$. Fixed cost $=$ \$154.20. The predicted cost of 20, . . . , 25 items is

x	20	21	22	23	24	25
y	765.80	796.38	826.96	857.54	888.12	918.70

Cost goes above \$15,000 at 45 items.

EXERCISES 1.3

1. Use a calculator to find the values of the linear function $f(x) = 3.72x - 4.3$ at **(a)** $x = -3.6$ **(b)** $x = 4.5$ **(c)** $x = 7.89$.
2. Use a calculator to find the values of the linear function $f(x) = -4.69x + 2.37$ at **(a)** $x = -9.86$ **(b)** $x = 3.24$ **(c)** $x = 35.68$.
3. Consider the linear function $f(x) = 3.5x + 9.73$. Use a table to determine the integer value of x when f first becomes greater than 120.
4. Determine whether the functions f defined by the following tables are linear. If a function is linear, use the table to find its rate of change and the value of $f(0)$ and to determine its equation. Support your answer by duplicating the table on your calculator.

(a) 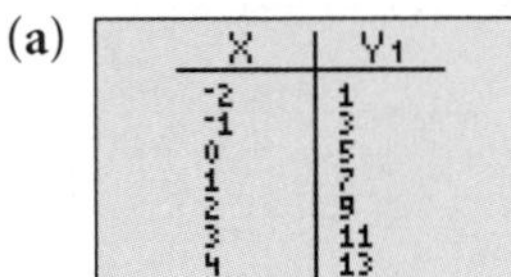

X	Y1
-2	1
-1	3
0	5
1	7
2	9
3	11
4	13

(b) 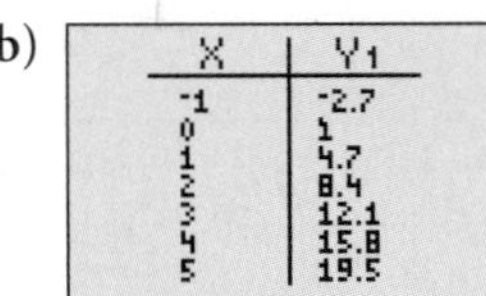

X	Y1
-1	-2.7
0	1
1	4.7
2	8.4
3	12.1
4	15.8
5	19.5

(c) 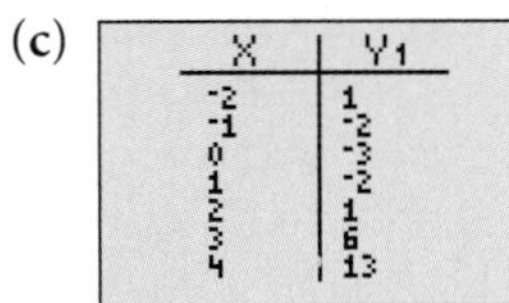

X	Y1
-2	1
-1	-2
0	-3
1	-2
2	1
3	6
4	13

(d) 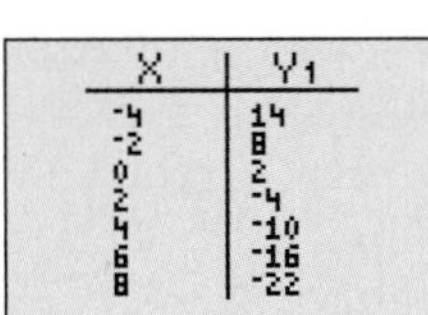

X	Y1
-4	14
-2	8
0	2
2	-4
4	-10
6	-16
8	-22

5. Determine whether the functions f defined by the following tables are linear. If a function is linear, use the table to find its rate of change and the value of $f(0)$ and to determine its equation. Support your answer by duplicating the table on your calculator.

(a) 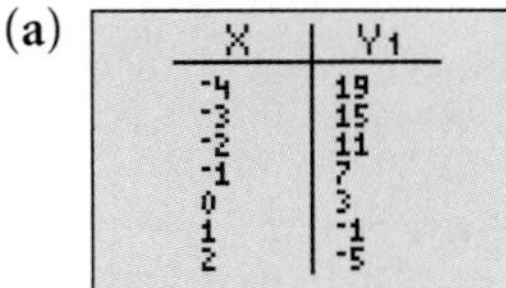

X	Y1
-4	19
-3	15
-2	11
-1	7
0	3
1	-1
2	-5

(b) 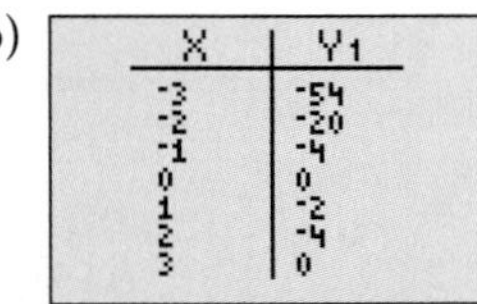

X	Y1
-3	-54
-2	-20
-1	-4
0	0
1	-2
2	-4
3	0

(c) 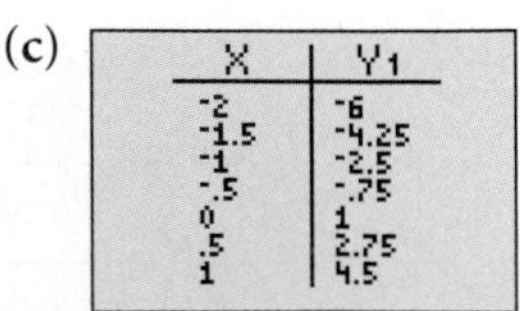

X	Y1
-2	-6
-1.5	-4.25
-1	-2.5
-.5	-.75
0	1
.5	2.75
1	4.5

(d) 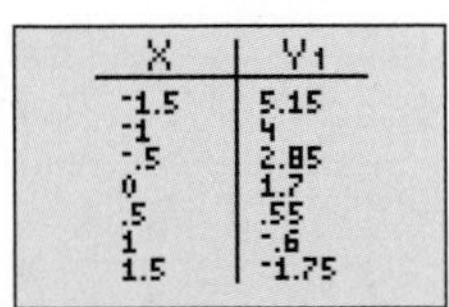

X	Y1
-1.5	5.15
-1	4
-.5	2.85
0	1.7
.5	.55
1	-.6
1.5	-1.75

6. The following tables and graphs describe linear functions. Match the tables and graphs. Find each function. Support your answer by duplicating the given tables and graphs on your calculator.

(a) 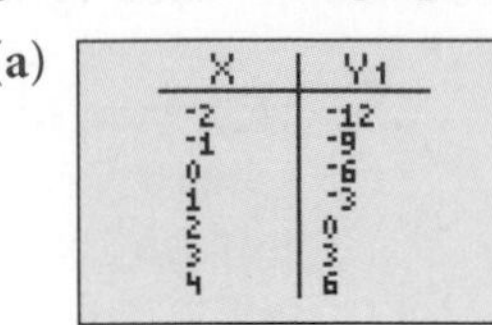

X	Y1
-2	-12
-1	-9
0	-6
1	-3
2	0
3	3
4	6

(b) 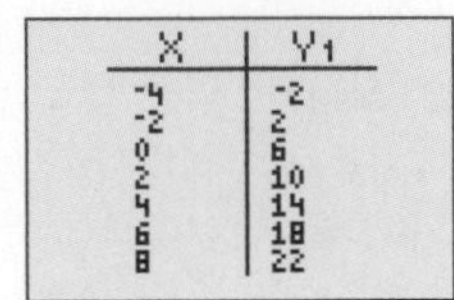

X	Y1
-4	-2
-2	2
0	6
2	10
4	14
6	18
8	22

(c) 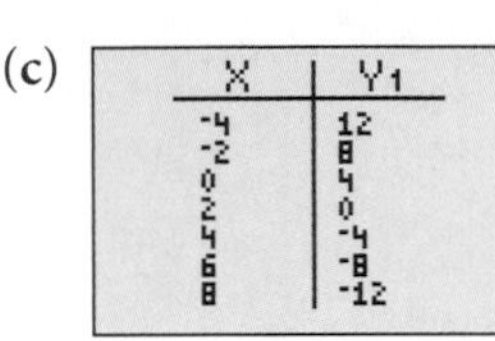

X	Y1
-4	12
-2	8
0	4
2	0
4	-4
6	-8
8	-12

(d) 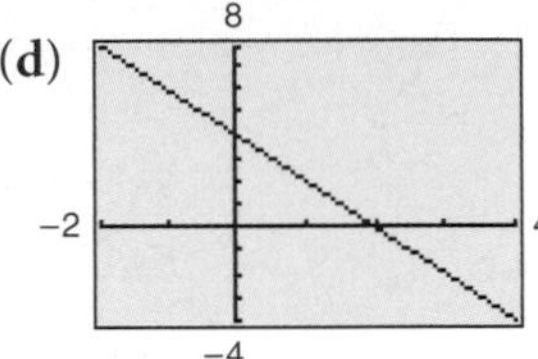

(e) 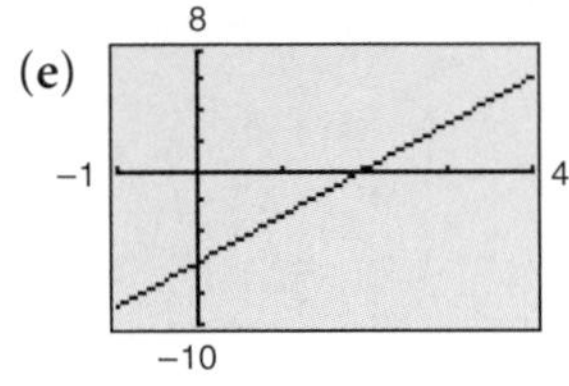

(f) 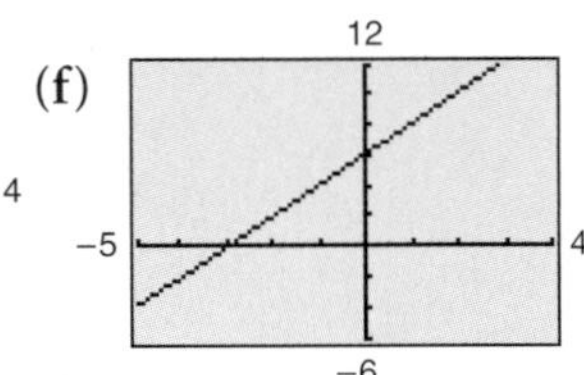

Exercises 7–12 are intended to give the student the experience of constructing functions to describe interesting situations and to think of slope and y-intercept in terms of practical concepts.

7. **Golfing Expenses** The price of a set of Kent golf clubs is \$300. It costs \$6 for regripping and maintenance per month. **(a)** Construct a function that describes the expense of buying and maintaining the golf equipment. **(b)** What do the slope and the y-intercept of the graph of this function represent physically?
8. **Computing Expenses** A RightLine computer printer costs \$250. The cost of a cartridge is \$15. The cartridge is changed once per month. **(a)** Construct a function that describes the expense of running the printer. **(b)** What do the slope and the y-intercept of the graph of this function represent physically?
9. **Keeping a Labrador Dog** A Labrador dog costs \$470. It costs \$20 a week to feed the dog. **(a)** Construct a function that describes the expense of buying and keeping the Labrador. **(b)** What do the slope and the y-intercept of the graph of this function represent physically?
10. **Running a Car** Dalmon Switzer bought a car for

\$15,000. He spends \$15 a week on gas. **(a)** Construct a function that describes Dalmon's expense of buying and running the car. (Dalmon's parents take care of maintenance and insurance.) **(b)** What do the slope and the y-intercept of the graph of this function represent physically?

11. **Television Expenses** A television costs \$250. Cable costs \$27 per month. **(a)** Construct a function that describes the expense of buying and using the TV set. **(b)** What do slope and y-intercept of the graph of this function represent physically?

12. **Golf Membership** There is an initial fee of \$250 to join the Wilson Creek Golf Club. A round of golf then costs \$15. Tim Zielske regularly plays two rounds of golf per week. **(a)** Construct a function that describes his expense of playing golf at Wilson Creek. **(b)** What do the slope and the y-intercept of the graph of this function represent physically?

13. **Nuclear Energy** The United States is very dependent on nuclear energy, with 109 commercial reactors from Maine to California. However, nuclear refuse, with deadly radiation lasting more than 1000 years, poses a grave danger to society. The chairman of the Nuclear Regulatory Commission said that storage and disposal of spent fuel will determine the future viability of nuclear power in the United States. In 1997, nuclear refuse was produced at a rate of 2000 tons a year. At that time, 34,000 tons of spent fuel had accumulated at reactor sites in 34 states. **(a)** Construct a function that describes the accumulation of nuclear refuse in the United States (assuming the situation does not change). **(b)** Predict the amount of refuse accumulated by the year 2033, when current reactor licenses will expire.

14. **Sleep** A 2-year-old child sleeps 12 hours per day on average. It has been estimated that as the child grows up the amount of daily sleep will decrease every year by 15 minutes until the age of 18, when it will level off. Construct a linear function that describes sleep as a function of age, between the ages of 2 and 18.

15. **Interstate Travel** Markers give distances every mile along Interstate 4 from Tampa on the Gulf coast of central Florida to Daytona Beach on the Atlantic coast. Judy Roberts sets off from Lakeland, which is 30 miles from Tampa, toward Daytona Beach. She travels at a steady speed of 65 miles per hour. **(a)** Find the function $s(t) = mt + b$ that gives her marker distance s from Tampa in terms of time t. **(b)** What do the slope m and s-intercept b of the graph of s correspond to in practical terms?

16. **Health Club Membership** The initial membership fee to join the Virginia Avenue Health Club is \$200. There is then a monthly fee of \$15. **(a)** Construct a function that describes the expense of being a member of the health club. **(b)** Represent the information graphically. What do the slope and the y-intercept of the graph represent practically? **(c)** Use the function to compute expenses for 24 and 36 months.

17. **Conversion Factor** **(a)** Determine the conversion factor from meters to yards, and use it to construct a function for converting meters to yards. **(b)** Two of the most important races in the Olympic Games are the 100-meter dash and the 1500-meter run. Use the conversion function to determine how these distances compare with the 100-yard sprint and the mile race.

18. **Manufacturing Costs** The Cardiff Manufacturing Company found that the cost of producing 20 electric toothbrushes is \$415. The cost for 25 toothbrushes is \$436.35. **(a)** Find the cost–volume equa-

tion. What is the fixed cost? (**b**) Use a graphing calculator to find the costs of 20, . . . , 26 toothbrushes. (**c**) Use a calculator table to determine at which item the cost reaches \$700.

19. **Production Costs** The Rodriguez Company determined that the cost of manufacturing 50 units of a product is \$190.50 and that of manufacturing 65 units is \$226.80. (**a**) Find the cost–volume equation. What is the fixed cost? (**b**) Use a graphing calculator to find the costs of 70, . . . , 76 items. (**c**) Use the calculator to determine at which item the cost reaches \$500.

20. **Production Costs** The Shah Refrigerator Company determined that the cost of manufacturing 10 refrigerators is \$3206 and the cost of 17 refrigerators is \$4977. (**a**) Find the cost–volume equation. What is the fixed cost? What is the marginal cost? (**b**) Use a graphing calculator to find the costs of 22, . . . , 28 refrigerators. (**c**) Use the calculator to determine when the cost reaches \$10,000.

In Exercises 21–26, linear graphs give information about some of the changes that have taken place in the United States in the last 20 years or so. Estimate the y-intercept and slope of each graph and determine a function that approximates the graph. Use the function to make predictions for the years 2000 and 2010 to the nearest integer (assuming that current trends continue).

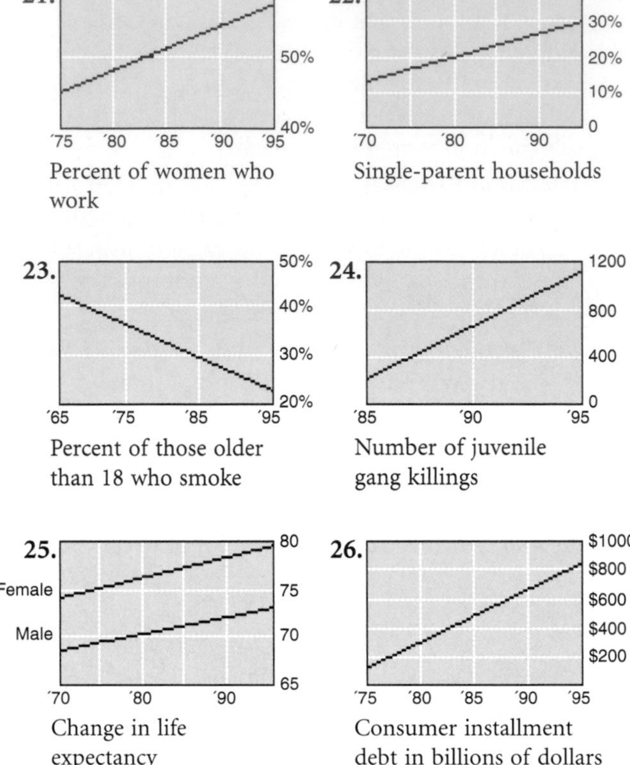

SECTION PROJECT

Constructing a Linear Model

In this section, you saw many examples of linear functions. Many of these were developed by students in a college algebra class to describe a situation in their lives. Construct and discuss such a model of a situation in your life.

1.4 Parabolas and Quadratic Equations

• Parabolas • Quadratic Equations • Vertex of a Parabola • x-Intercepts of a Parabola

In the previous sections we discussed linear equations, linear functions, and their graphs. We now turn to quadratic equations, functions, and their graphs.

Let us lay the foundation for our discussion by graphing the equation $y = x^2 + 3$ by hand. After that, we will explore the graphs of this and more-involved equations with a calculator. First, construct a table of x and y values.

x	-4	-3	-2	-1	0	1	2	3	4
y	19	12	7	4	3	4	7	12	19

This table leads to the points $(-4, 19)$, $(-3, 12)$, $(-2, 7)$, $(-1, 4)$, $(0, 3)$, $(1, 4)$, $(2, 7)$, $(3, 12)$, and $(4, 19)$. Plot these points and then draw a smooth curve through them to get the graph of $y = x^2 + 3$ (Figure 1.46).

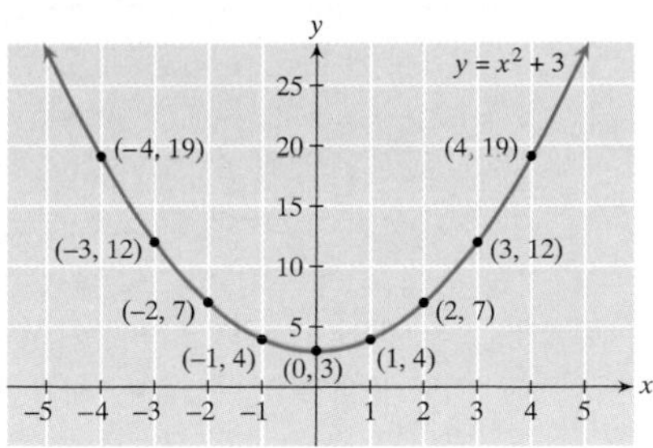

Figure 1.46 Graphing $y = x^2 + 3$ by hand.

Now let us duplicate the graph of $y = x^2 + 3$ using a calculator. Enter the equation, select an appropriate window, and display the graph (Figure 1.47). Observe that the graph is U-shaped. It is a parabola with the vertex at (0, 3).

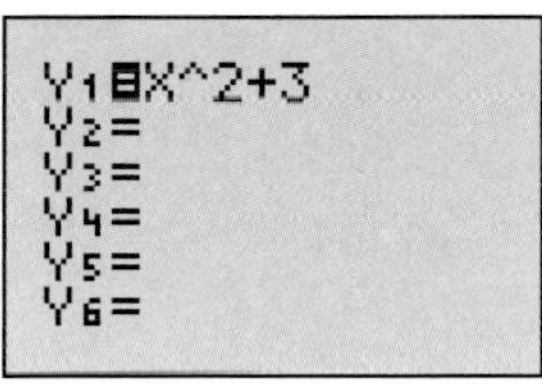

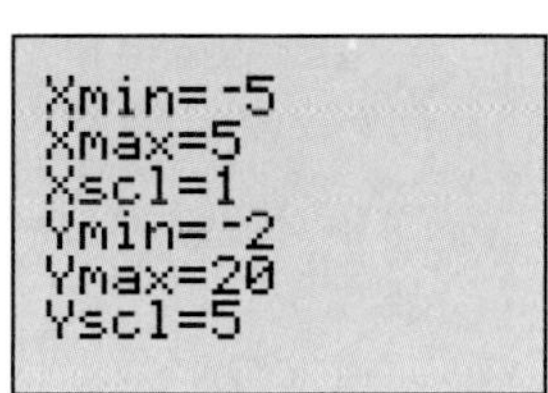

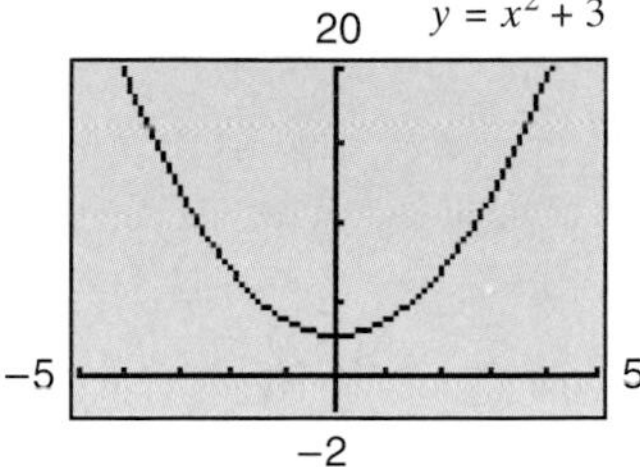

Figure 1.47 Graphing $y = x^2 + 3$ on a calculator.

The equation $y = x^2 + 3$ is a special case of $y = ax^2 + bx + c$, with $a = 1$, $b = 0$, and $c = 3$. We now give the student the opportunity to explore the graphs of such equations.

Exploration

Graph the following equations of the form $y = ax^2 + bx + c$. Observe that each graph is U-shaped. Show that each graph has a **vertex** at $x = -b/2a$. The graph is symmetrical about a certain vertical line, called its **axis.** Find the equation of each axis. How does the sign of a, the coefficient of x^2, affect the graph?

(a) $y = 2x^2 - 12x + 14$
(b) $y = -2x^2 + 8x - 5$

Parabola

An equation of the form $y = ax^2 + bx + c$, $a \neq 0$, is called a **quadratic equation** in two variables. The graph of such a quadratic equation is a **parabola.** The parabola

1. opens up if $a > 0$, down if $a < 0$,
2. has a **vertex** with x-coordinate $-b/2a$, and
3. has an **axis** with equation $x = -b/2a$.

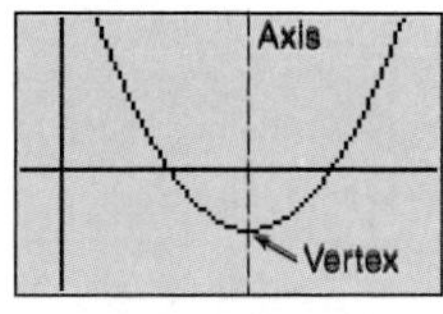

$y = ax^2 + bx + c, a > 0$

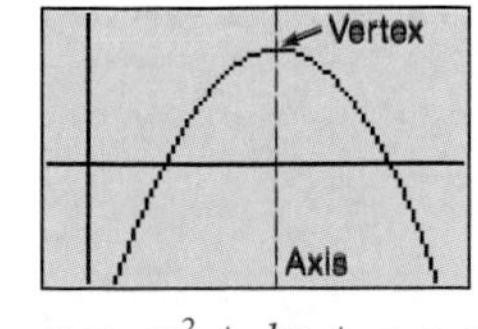

$y = ax^2 + bx + c, a < 0$

Derivation of the Formula for the Vertex of a Parabola (Optional)

We use the technique of completing the square (Section R.4) to show that the vertex is at $x = -b/2a$. First rewrite the equation $y = ax^2 + bx + c$ in a more suitable form for interpretation.

$$y = a\left(x^2 + \frac{bx}{a}\right) + c$$

Add and subtract $(b/2a)^2$ (the square of half the coefficient of x), so we can form a perfect square of the term in x.

$$y = a\left[x^2 + \frac{bx}{a} + \left(\frac{b}{2a}\right)^2 - \left(\frac{b}{2a}\right)^2\right] + c$$

Rewrite
$$y = a\left[x^2 + \frac{bx}{a} + \left(\frac{b}{2a}\right)^2\right] + c - a\left(\frac{b}{2a}\right)^2$$

Form a complete square
$$y = a\left(x + \frac{b}{2a}\right)^2 + \left(c - \frac{b^2}{4a}\right)$$

Now consider the two possibilities for this right side; namely when $a > 0$ and when $a < 0$.

If $a > 0$, $a[x + (b/2a)]^2 \geq 0$. Thus, the *smallest value* of y occurs when this term is zero; namely when $x = -b/2a$. The graph has a vertex at $x = -b/2a$ and opens upward.

If $a < 0$, $a[x + (b/2a)]^2 \leq 0$. The *largest value* of y occurs when this term is zero; namely when $x = -b/2a$. The graph has a vertex when $x = -b/2a$ and opens downward.

> ***Parabola*** A **parabola** $y = ax^2 + bx + c$ opens upward if $a > 0$ and downward if $a < 0$. The vertex has x-coordinate $-b/2a$. The line $x = -b/2a$ is an axis of symmetry.

Let us use this information to sketch the graph of a parabola.

Example 1 Sketching the Graph of a Parabola

Find the vertex and axis, and sketch a rough graph of the quadratic equation $y = 2x^2 - 8x + 4$ using algebra. Support your answer by graphing the parabola on a calculator.

SOLUTION In comparison with $y = ax^2 + bx + c$, we see that $a = 2$, $b = -8$, and $c = 4$. The x-coordinate of the vertex is

$$x = \frac{-b}{2a} = \frac{-(-8)}{2(2)} = \frac{8}{4} = 2$$

The corresponding y-coordinate is

$$y = 2(2)^2 - 8(2) + 4 = 8 - 16 + 4 = -4$$

The vertex is the point $(2, -4)$. The axis is $x = 2$. Because $a = 2 > 0$, the parabola opens up.

It is useful to find a point on the parabola in addition to the vertex. Knowing a second point gives an indication of how wide the parabola opens. Any convenient x value can be used and substituted into the equation to find a second point. If the vertex is not on the y-axis, as here, we can take $x = 0$. $x = 0$ gives $y = 4$. This value of y is the y-intercept of the graph (Section 1.1). It tells us where the graph cuts the y-axis. This information leads to the graph shown in Figure 1.48(a). The information is supported by the calculator graph in Figure 1.48(b).

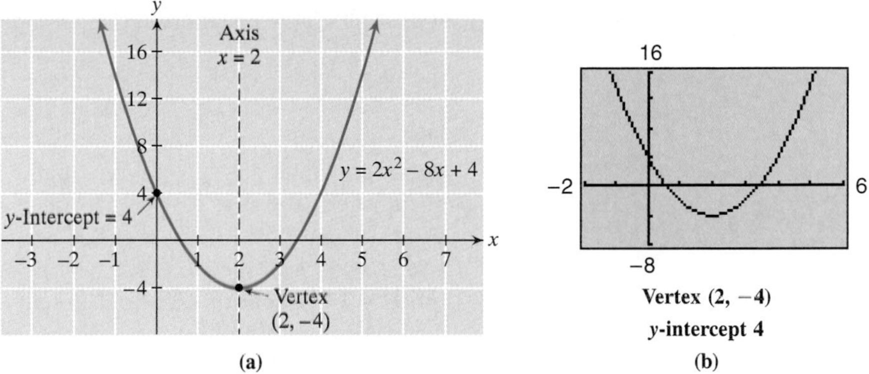

Figure 1.48 Parabola $y = 2x^2 - 8x + 4$.

Further Ideas 1

(a) Find the vertex, axis, and y-intercept, and sketch the graph of equation $y = -3x^2 + 18x - 1$. Support your answer by graphing the parabola on a calculator.

(b) Find the equation of a parabola that has vertex (3, 1) that opens downward. Corroborate your answer by graphing on a calculator.

VERTEX WITH A CALCULATOR

The vertex of a parabola is called a **maximum** or a **minimum** depending on whether the parabola opens down or up. (The vertex is the highest or lowest point on the graph.) Graphing calculators have a **max/min finder** that can be used to find the vertex. Let us graph and find the vertex of the parabola $y = -2.1x^2 + 9.7x + 5$ to four decimal places (Figure 1.49). The vertex is (2.3095, 16.2012).

Trace and *zoom* can also be used together to find the vertex but is not as efficient as using a max/min finder. We show the results of tracing and zooming three times in Figure 1.49. We still have a way to go before obtaining the vertex to four decimal places.

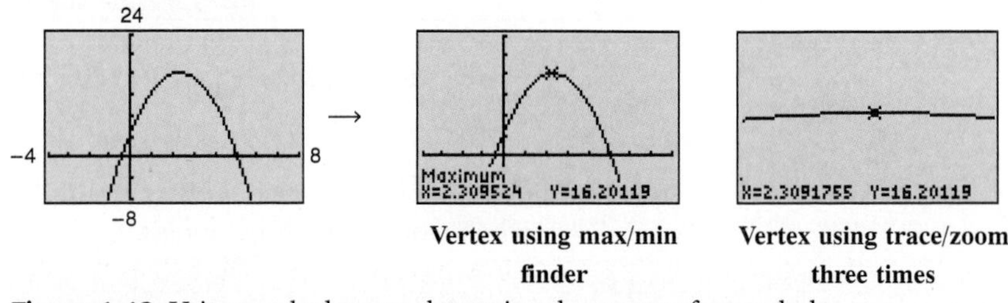

Figure 1.49 Using a calculator to determine the vertex of a parabola.

DETERMINING THE x-INTERCEPTS OF A PARABOLA

The x-intercepts are the points at which a graph crosses the x-axis. These are important points to know for any graph. We now illustrate how to find the x-intercepts of parabolas for which the equation can be factored.

Example 2 **Finding the x-Intercepts of a Parabola**

Find the x-intercepts of the parabola $y = 2x^2 - 5x - 12$.

SOLUTION The equation of the parabola can be factored as follows:

$$y = (2x + 3)(x - 4)$$

The x-intercepts are the points on the graph at which $y = 0$. We see that $y = 0$ when $2x + 3 = 0$ or $x - 4 = 0$. That is when $x = -1.5$ or $x = 4$. The x-intercepts are therefore $x = -1.5$ and $x = 4$. We confirm this result on a calculator (Figure 1.50).

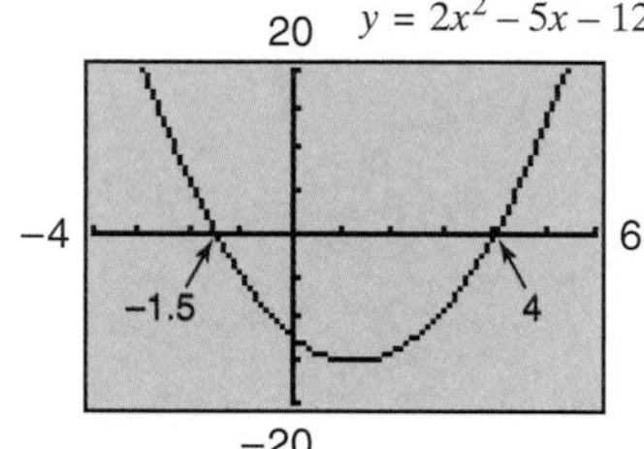

Figure 1.50 $y = 0$ when $x = -1.5, 4$

Self-Check 1 Determine the x-intercepts of the parabola $y = 2x^2 - 11x + 5$. Confirm your answer using a calculator.

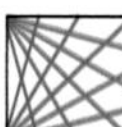

x-INTERCEPTS WITH A CALCULATOR

The x-intercepts of a parabola are also called **zeros.** (They are the points on the graph at which $y = 0$.) Graphing calculators have a **zero finder** for determining these points. Let us draw the graph of the parabola $y = x^2 - 3x - 6$ on a calculator and use the zero finder to find the x-intercepts (Figure 1.51). The x-intercepts are -1.3723 and 4.3723, rounded to four decimal places. (Trace and zoom can also be used to find x-intercepts.)

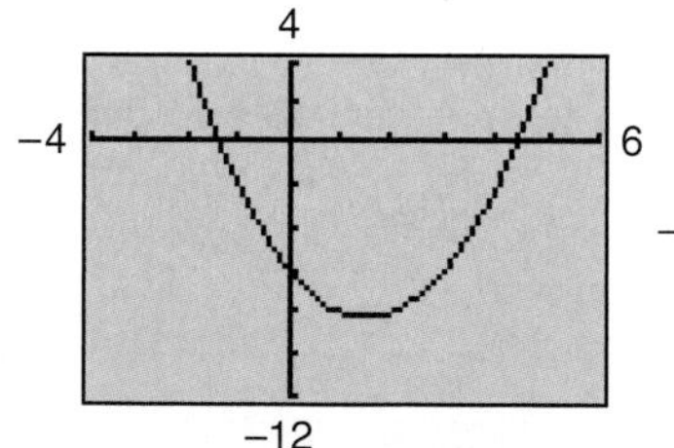

→

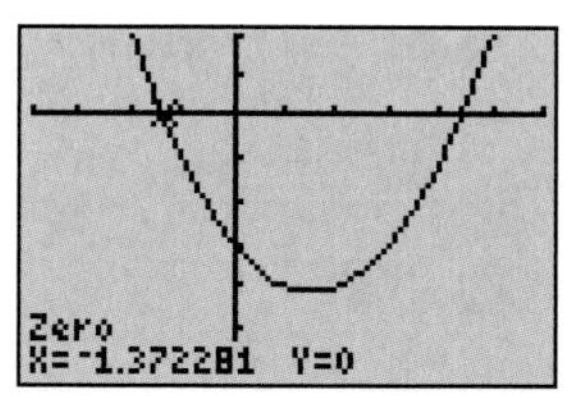

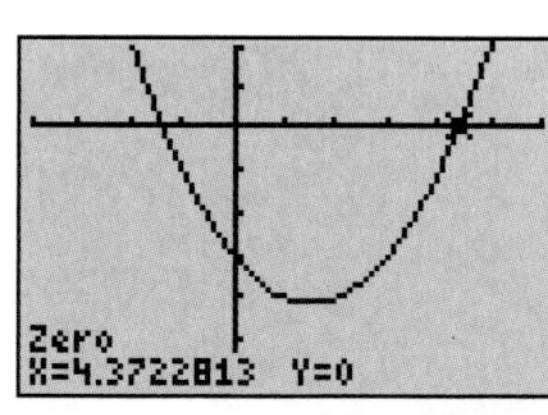

Figure 1.51 Use of zero finder to determine the x-intercepts of a parabola.

We complete this section with examples of useful algebraic techniques that arise when working with parabolas.

Example 3 **Parabola That Satisfies Certain Conditions**

Find the value of a so the point (2, 13) lies on the graph of $y = ax^2 + x - 1$.

SOLUTION For (2, 13) to lie on the graph of $y = ax^2 + x - 1$, it must satisfy this equation. Substitute $x = 2$ and $y = 13$ into this equation and solve for a.

$$13 = a(2)^2 + 2 - 1$$
$$13 = 4a + 1$$
$$a = 3$$

Thus, the point (2, 13) lies on the graph of $y = ax^2 + x - 1$ if $a = 3$.

The following example illustrates how the equation and vertex of a parabola with given x- and y-intercepts can be determined.

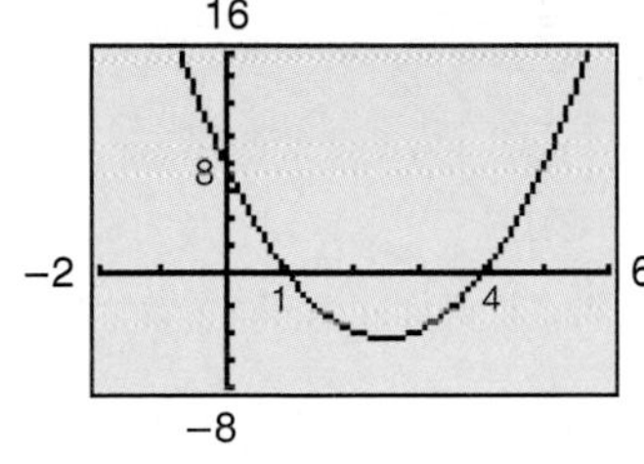

Figure 1.52

Example 4 **Equation and Vertex of a Parabola with Given Intercepts**

Consider the graph of the parabola shown in Figure 1.52 that has x-intercepts 1 and 4 and y-intercept 8.

(a) Find the equation of the parabola. Support your result by duplicating the graph on a calculator and by checking the x- and y-intercepts.
(b) Use the equation to find the vertex of the graph.

SOLUTION

(a) Consider the equation

$$y = k(x - 1)(x - 4)$$

where k is a real number. The x-intercepts are the points on the graph at which $y = 0$. Letting $y = 0$ gives $k(x - 1)(x - 4) = 0$, resulting in $x = 1$ or 4. Therefore, the x-intercepts of the graph of this equation are 1 and 4.

It remains to determine k. The y-intercept of the given graph is 8. Thus, $y = 8$ when $x = 0$. Substituting these values into the equation gives

$$8 = k(-1)(-4)$$
$$8 = 4k$$
$$k = 2$$

The equation is $y = 2(x - 1)(x - 4)$. If we multiply out, we get the equation of the parabola in standard form,

$$y = 2x^2 - 10x + 8$$

We leave it to the student to check the intercepts of this graph on a calculator.

(b) Let us use the equation to find the vertex of the parabola. Comparing the preceding equation with the standard form $y = ax^2 + bx + c$, we see that $a = 2$, $b = -10$, and $c = 8$. The x-coordinate of the vertex is

$$x = \frac{-b}{2a} = \frac{-(-10)}{2(2)} = \frac{10}{4} = 2.5$$

The corresponding y-coordinate is

$$y = 2(2.5)^2 - 10(2.5) + 8 = -4.5$$

The vertex of the parabola is $(2.5, -4.5)$.

Answer to Self-Check Exercise

1. The x-intercepts are $x = 0.5$ and $x = 5$.

Answers to Further Ideas Exercise

1. **(a)** Vertex $(3, 26)$; Axis $x = 3$; y-intercept $= -1$.

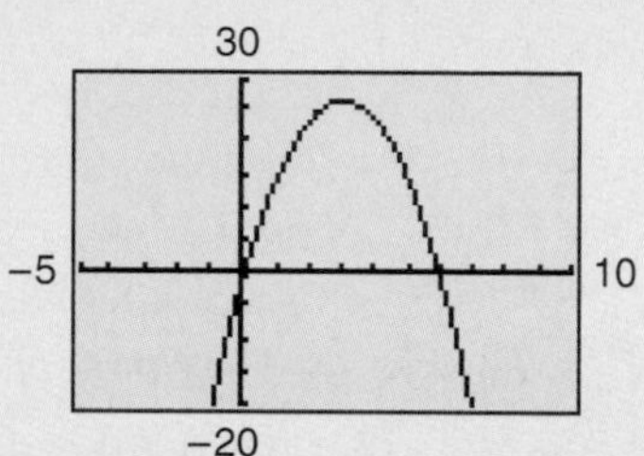

(b) $y = -x^2 + 6x - 8$. This is the parabola with $a = -1$. Get other parabolas for other values of a.

EXERCISES 1.4

In Exercises 1–6, sketch the graph of each equation over the given x-interval by plotting points and drawing a smooth curve through them. Use your graphing calculator to support your answer. (*Hint:* Use the table facility of a calculator to compute the points for Exercises 3–6.)

1. $y = x^2 + 6, [-3, 3]$
2. $y = -x^2 + 1, [-4, 4]$
3. $y = x^2 + 2x - 3, [-4, 2]$
4. $y = 2x^2 - 4x + 1, [-1, 5]$
5. $y = -3x^2 + 12x + 7, [-2, 6]$
6. $y = 4x^2 - 6x + 7, [-3, 3]$

In Exercises 7–12, find the vertex, y-intercept, and axis of symmetry of the graph of each quadratic equation. Determine whether the graph opens up or down, and sketch the graph. Check your results using a graphing calculator.

7. $y = x^2 + 8x + 15$
8. $y = 2x^2 - 4x + 7$
9. $y = 6x^2 + 36x - 5$
10. $y = -5x^2 + 20x + 3$
11. $y = 2x^2 + 5$
12. $y = -3x^2 + 18x$

In Exercises 13–18, find the vertex, y-intercept, and axis of symmetry of the graph of each quadratic equation. Determine whether the graph opens up or down, and sketch the graph. Check your results using a graphing calculator.

13. $y = -2x^2 + 10x - 1$
14. $y = 3x^2 + 12x + 4$
15. $y = 4x^2 - 14x - 3$
16. $y = -2x^2 - 8x + 3$
17. $y = 3x^2 - 7$
18. $y = 4x^2 + 20x$

In Exercises 19–26, find the vertex of each parabola (to four decimal places) using a graphing calculator.

19. $y = 2.1x^2 - 6.4x - 2.3$
20. $y = -3.3x^2 + 11.8x + 1.2$
21. $y = 6.2x^2 + 36.4x - 5.3$
22. $y = -5.7x^2 + 20.1x + 3.8$
23. $y = -0.1x^2 + 5.34x - 51$
24. $y = 0.2x^2 - 4.78x + 3$
25. $y = x^2 - 27.3x + 108$
26. $y = 0.05x^2 + 4.23x + 28.26$

In Exercises 27–34, find the x-intercepts of each parabola by factoring the equations. Use a calculator graph to support the results.

27. $y = x^2 + 2x - 15$

28. $y = x^2 - 3x + 2$

29. $y = -x^2 + 3x + 4$

30. $y = 2x^2 + x - 3$

31. $y = 2x^2 - 9x + 10$

32. $y = 6x^2 - 11x - 35$

33. $y = x^2 - 5x + 4$

34. $y = -x^2 + 7x - 10$

In Exercises 35–40, find the x-intercepts of each graph, rounded to four decimal places, using a graphing calculator.

35. $y = 2x^2 - 13x + 5$

36. $y = 2x^2 - 5x + 3$

37. $y = 6x^2 - 7x - 3$

38. $y = x^2 + x - 27$

39. $y = -x^2 + 6.3x + 5.4$

40. $y = -0.8x^2 + 10.1x - 18.4$

In Exercises 41–48, find the value of p in each situation.

41. Find p so that the point $(1, 5)$ lies on the graph of the quadratic equation $y = 2x^2 - x + p$.

42. Find p so that the point $(-2, 11)$ lies on the graph of $y = px^2 - 2x - 5$.

43. Find p so that the point $(3, 2)$ lies on the graph of $y = -x^2 + px - 1$.

44. Find p so that the point $(1, 1)$ lies on the graph of $x = py^2 + 4y - 6$.

45. Find p so that the point $(-2, p)$ lies on the graph of $y = 2x^2 + 7x - 1$.

46. Find p so that the point $(5, 3)$ lies on the graph of $y = x^2 - 10x + p$.

47. Find p so that the point $(-1, -6)$ lies on the graph of $y = 2x^2 + px - 1$.

48. Find p so that the point $(-26, 3)$ lies on the graph of $x = py^2 + 2y - 5$.

In Exercises 49–54, the graphs of parabolas, with their x- and y-intercepts, are given. **(a)** Find the equation of each parabola. Support your answers by duplicating the graphs on your calculator and checking the intercepts. **(b)** Use the equation to find the vertex of each graph.

49.

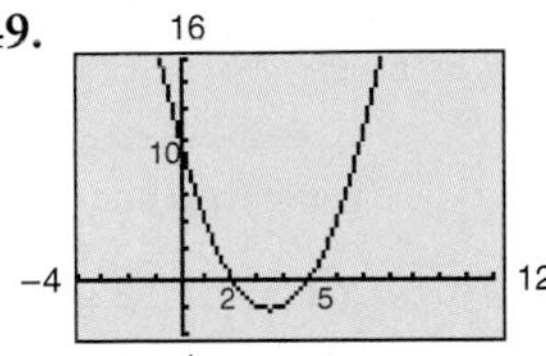

50.

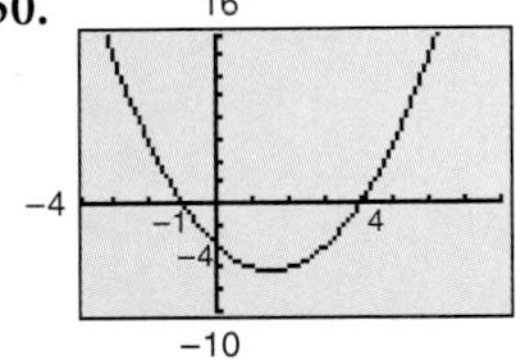

51.

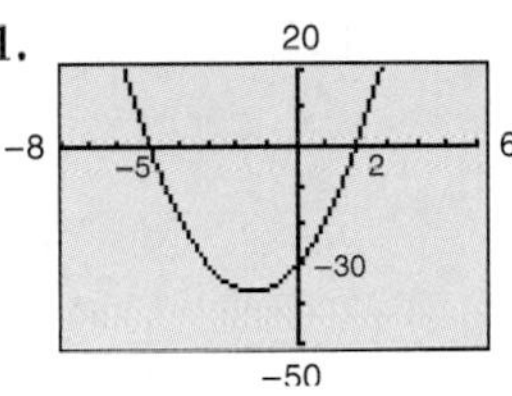

52.

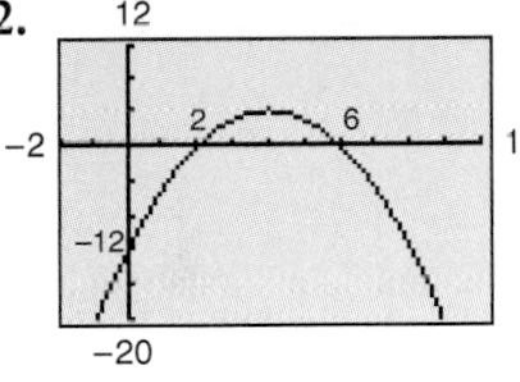

53.

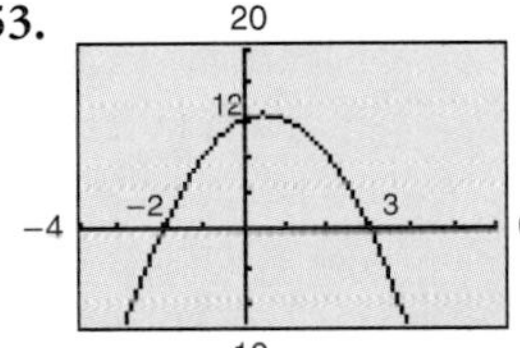

54.

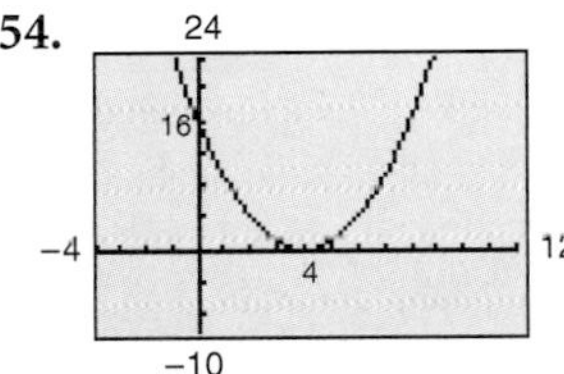

1.5 Quadratic Functions, Learning German, and Fluid Flow

- Quadratic Function • Motion Under Gravity • Learning German
- Supply and Demand • Least-Squares Parabola • The Intel Corporation
- Flow of a Fluid

QUADRATIC FUNCTIONS

In this section we discuss functions that are defined by quadratic equations. Consider a quadratic equation

$$y = ax^2 + bx + c, \qquad a \neq 0$$

As x varies, we get corresponding values of y. The equation $y = ax^2 + bx + c$ defines the **quadratic function** $f(x) = ax^2 + bx + c$.

For example, the quadratic equation $y = 2x^2 - 15x + 20$ defines the quadratic function $f(x) = 2x^2 - 15x + 20$. We can conveniently write and compute values of f at points such as $x = 3$ and $x = 8$ as follows:

$$f(3) = 2(3)^2 - 15(3) + 20 = -7 \quad \text{and} \quad f(8) = 2(8)^2 - 15(8) + 20 = 28$$

We have seen how certain problems can be solved using linear functions. We now see that other situations lend themselves to quadratic functions. We shall see how information can be obtained using algebra, a graph, or a table of values. The first example is an analysis of an object moving vertically under gravity.

Example 1 Motion Under Gravity

An object is propelled vertically from ground level with velocity v. The laws of physics tell us that the height s (in feet) is the following quadratic function of time t (in seconds):

$$s(t) = -16t^2 + vt$$

If the initial velocity is 500 feet per second, let us determine the maximum height it reaches and find out when it reaches this height and when it returns to Earth.

The actual physical motion is shown in Figure 1.53(a). The initial velocity is 500 feet per second. Thus, $v = 500$. The graph of $s(t) = -16t^2 + 500t$ is the parabola shown in Figure 1.53(b) (the mathematical view of the motion). The maximum height will be at the vertex of the parabola. The time when it returns to Earth will be a value of t such that $s(t) = 0$, a zero of the function. It will be a t-intercept of the graph.

The vertex at A is found to be the point (15.6250, 3906.25); thus, the maximum height is 3906.25 feet. This height is reached after 15.625 seconds. The t-intercept at B is found to be (31.25, 0). The object will return to Earth after 31.25 seconds.

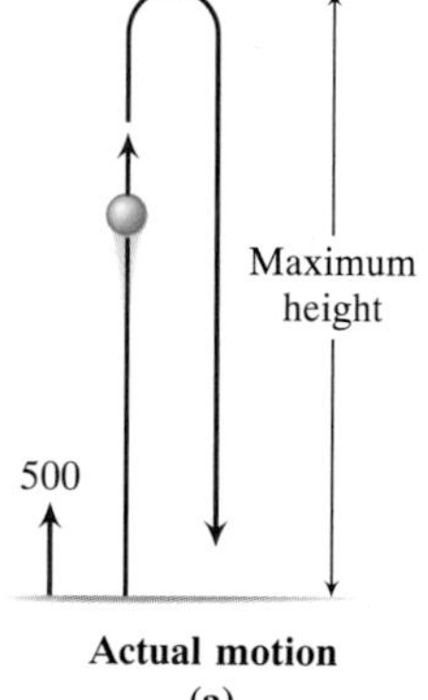

Actual motion
(a)

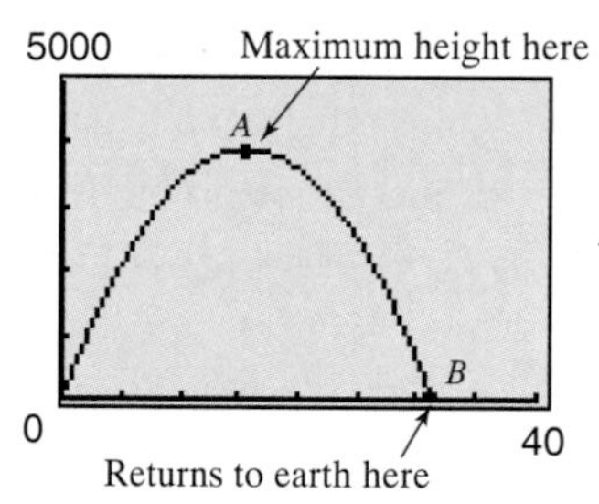

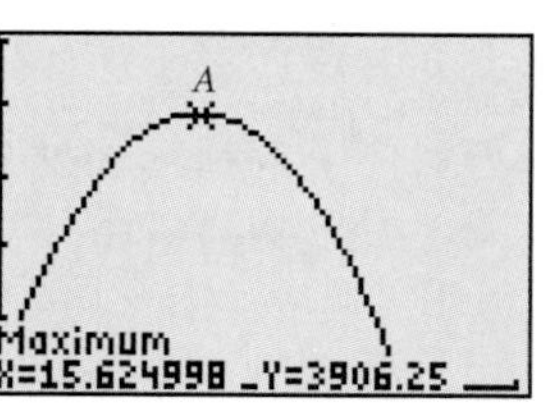

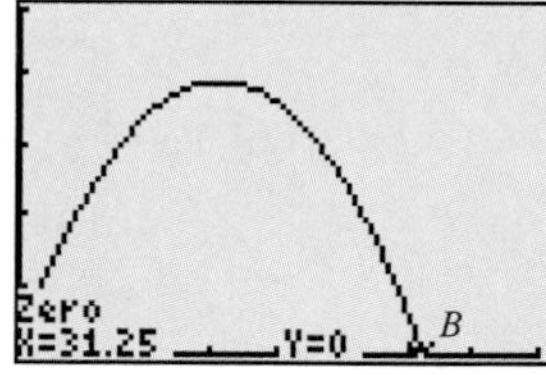

Mathematical description of the motion
(b)

Figure 1.53

Self-Check 1 Check the result in Example 1 by finding the vertex of the parabola $s(t) = -16t^2 + 500t$ using algebra. Verify that the object does return to Earth after 31.25 seconds by showing that $s(31.25) = 0$.

Exploration 1

Behavior of Object Moving Under Gravity

Look closely at the times it takes the object to reach its maximum height and the total duration of the flight in Example 1. In the light of these results, make a conjecture about a relationship between the time it takes such an object to reach its maximum height and the total time it takes to return to Earth. Gather data for other flights using various velocities to support or disprove your conjecture. Finally prove or disprove your conjecture using algebra.

Exploration 2

Looking for Symmetry

An object is propelled vertically from ground level with a velocity of 96 feet per second. Compute a table of heights every second for the duration of the flight. Observe that the table has a certain symmetry. What information does this symmetry tell you? Discuss.

A **learning function** is a convenient tool for studying learning patterns. The following example illustrates how quadratic functions are sometimes used as learning functions.

Example 2 | Learning German

An experiment involved 100 students in the learning of German words. The teacher first read one German word and then the corresponding English word. On the first trial, the teacher read the German word, and the students were asked to write down the English word. The teacher then repeated these German and English words and went on to introduce a new German word and its English counterpart. On the second trial, the teacher read these two German words, and the students were asked to write down the corresponding English words. The teacher repeated these German and English words and added yet another new German word, giving its English translation. This process was repeated until eight German words and their English counterparts had been introduced. The number of correct responses at each stage was recorded. Let us analyze the effectiveness of the learning process.

The function $f(x) = -\frac{1}{20}x^2 + x$ was found to closely describe the data found in these eight trials. x is the number of German words asked, and $f(x)$ gives the number of correct English answers. For example, $f(6) = -\frac{1}{20}(6)^2 + 6 = 4.2$ implies that out of six German words the 100 students gave an average of 4.2 correct English answers. Some of the students mastered five words, some four words, and so on. Consider the following table for $f(x)$ (Figure 1.54):

```
Y1=-X^2/20+X
Y2=
Y3=
Y4=
Y5=
Y6=
```

X	Y1
1	.95
2	1.8
3	2.55
4	3.2
5	3.75
6	4.2
7	4.55

Data on learning German

$\longrightarrow$ Scroll

X	Y1
8	4.8
9	4.95
10	5
11	4.95
12	4.8
13	4.55
14	4.2

Figure 1.54

$$f(10) = 5 \qquad \text{is seen to be the maximum value of } f$$

There would be a maximum of five correct responses when ten German words were asked at the tenth trial. It would become counterproductive to repeat the experiment any more. Fatigue and loss of concentration probably set in.

Self-Check 2 It is very natural to use a table to analyze the learning situation in Example 2. We found that the maximum value of f was 5 at $x = 10$ and interpreted this information to mean that ten trials should be conducted. Use algebra to confirm that the vertex of the parabola $f(x) = -\frac{1}{20}x^2 + x$ is indeed at the point (10, 5).

Example 3 Supply and Demand

Functions and their graphs are used extensively in business. The quantity of a good that producers are willing to **supply** at any one time depends on the **price.** The higher the price charged for an article, the greater the incentive to produce, leading to higher supplies. Thus, supply is interpreted to be an increasing function $S(p)$ of price, having a graph of the general shape indicated in Figure 1.55(a).

The **demand** for a good is also dependent on price. The higher the price charged for an article, the fewer people will buy. Thus, demand is a decreasing function $D(p)$ of price, having a graph as indicated in Figure 1.55(b).

The two sides in a market, namely supply and demand, interact to determine the price of a commodity. The price at which a commodity settles in the market is the one at which the supply and demand are equal. This price is called the **equilibrium price.** This will be the price at which the graphs of the supply and demand functions intersect (Figure 1.55c).

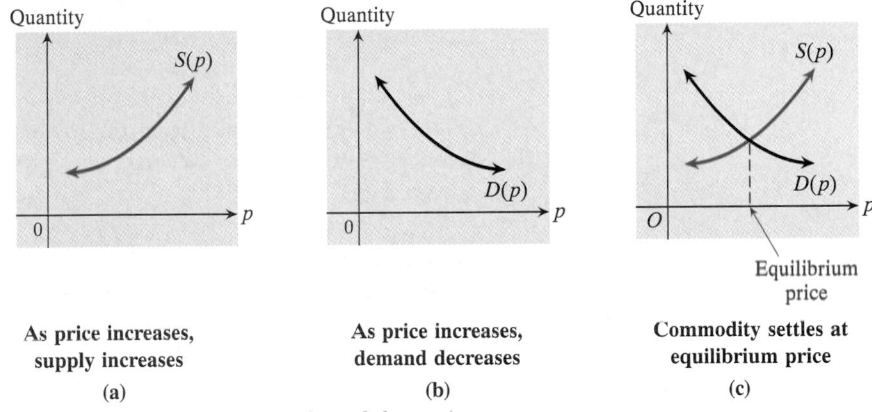

Figure 1.55 Behavior of supply and demand.

Increasing and decreasing segments of graphs of quadratic functions frequently are used to describe the supply and demand of a product. Consider the following situation. An analysis of data for a market produced the following supply and demand functions for a certain article. Let us find the equilibrium price.

$$S(p) = 0.9p^2 - 10p + 50, \qquad D(p) = 1.2p^2 - 60p + 800$$

Here p is the price of the article in dollars, S is the supply in units of one thousand, and D is the demand in units of one thousand. For example, when the price of the article is \$15, the predicted supply and demand are

$$S(15) = 0.9(15)^2 - 10(15) + 50 = 102.5 \text{ thousand items}$$
$$D(15) = 1.2(15)^2 - 60(15) + 800 = 170 \text{ thousand items}$$

Because demand is higher than supply, the market probably can stand a higher price than \$15. We will find that this is indeed so.

Let us find the equilibrium price using graphs (Figure 1.56). Using trace/zoom, we see that the equilibrium price is \$16.67. This is the price determined by the market.

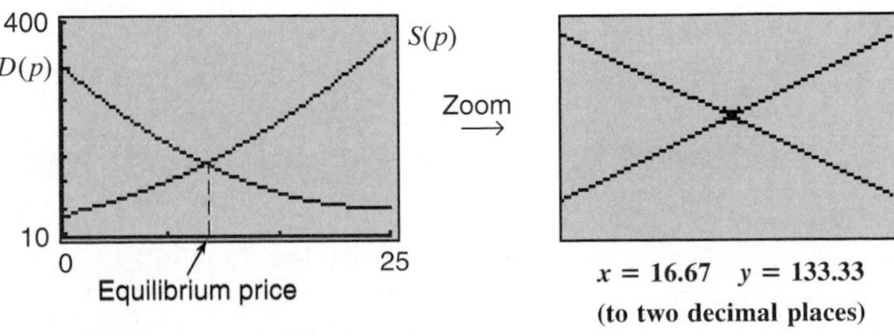

Figure 1.56

Intersect Finder

Your calculator probably has an **intersect finder.** You may prefer to use it rather than trace/zoom for finding the intersection points of graphs. The display for the intersection of $S(p)$ and $D(p)$ of Example 3 will be similar to that shown in Figure 1.57.

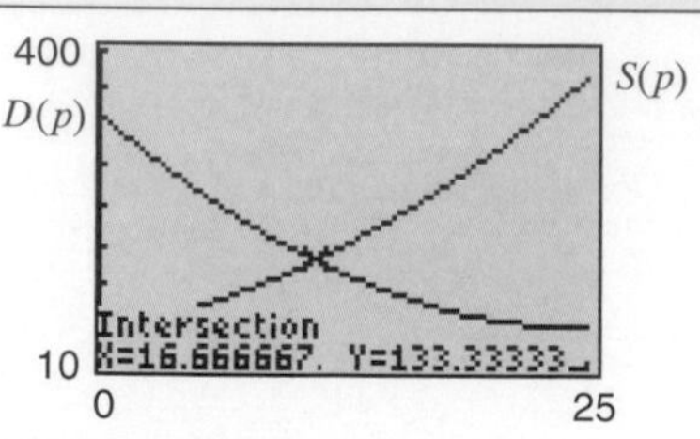

Figure 1.57

The following examples from the worlds of business and science illustrate how functions such as the ones that we used in Examples 1–3 are often constructed from data.

Example 4 The Intel Corporation

The following table shows the growth in the number of Intel Corporation employees for the period 1991 to 1997. (Intel Corporation Annual Report 1997.)

Year[a]	1991 (0)	1992 (1)	1993 (2)	1994 (3)	1995 (4)	1996 (5)	1997 (6)
Employees (at year end, in thousands)	24.6	25.8	29.5	32.6	41.6	48.5	63.7

[a]Numbers in parentheses are years measured from 1991. These are used as x values.

These figures are very interesting, in that they reveal how Intel expanded during a time when so many other companies downsized. We enter and plot these data on a calculator (Figures 1.58a and 1.58b). Observe that the points seem to lie on a parabola. In Section 1.2 we discussed a least-squares line that fitted data points that lay approximately on a line. Here we determine the equation of the least-squares parabola that best fits the data.

The least-squares parabola is found to be $y = ax^2 + bx + c$, with $a = 1.164285714$, $b = -0.7428571429$, and $c = 25.13571429$ (Figure 1.58c). To check the equation against the original data, enter the equation on a graphing calculator, plot the data, and then graph (Figure 1.58d). The graph is seen to fit the data closely. The quadratic function that best describes the number of employees at Intel over this period is

$$N(x) = 1.164285714x^2 - 0.7428571429x + 25.13571429$$

Let us use this function to predict the number of employees for the year 2000 (corresponding to $x = 9$). We get $N(9) = 112.7571428$. The predicted number of employees for the year 2000 is approximately 112.757 thousand. Will this growth in fact be sustained?

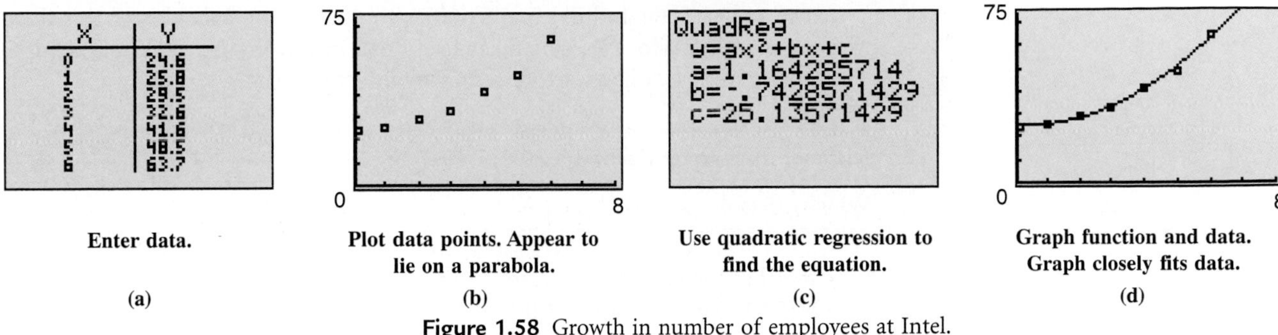

Enter data. (a)

Plot data points. Appear to lie on a parabola. (b)

Use quadratic regression to find the equation. (c)

Graph function and data. Graph closely fits data. (d)

Figure 1.58 Growth in number of employees at Intel.

Self-Check 3 The following table describes the growth in revenue of Intel Corporation for the ten years 1988 to 1997 (Intel Corporation Annual Report 1997.) Enter and plot these data. Observe that the points seem to fit a parabola closely. Determine the least-squares parabola for these data using 1988 as the base year (measure years from 1988). This function is called a **revenue function.** Predict Intel revenue for the year 2000.

Year	1988 (0)	1989 (1)	1990 (2)	1991 (3)	1992 (4)	1993 (5)	1994 (6)	1995 (7)	1996 (8)	1997 (9)
Revenue ($, in millions)	2875	3127	3921	4779	5844	8782	11,521	16,202	20,847	25,070

The following example illustrates a problem in fluid mechanics in which scientists study the behavior of flows of liquids. We again see how functions are constructed from data.

Example 5 Flow of a Fluid

A viscous fluid runs through a circular pipe with a radius of 3 inches and a length of 300 feet. The velocity of the flow gradually decreases from the center of the pipe toward the edges (Figure 1.59). This occurs because of friction at the edges. This friction causes a ripple effect throughout the pipe.

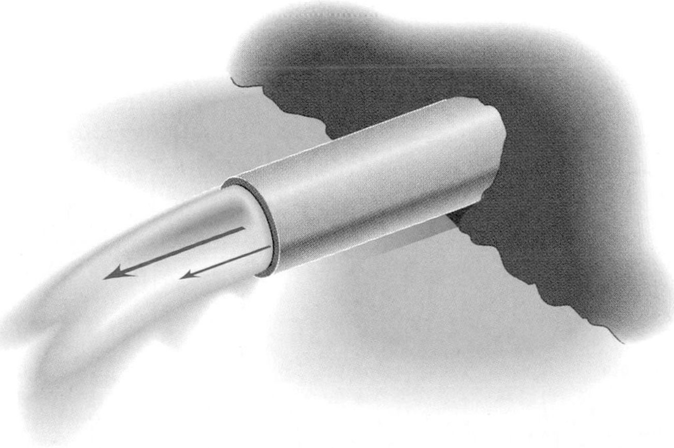

Figure 1.59 The velocity of the liquid gradually decreases from center of the pipe to the edge.

Let us collect data and use mathematics to analyze the behavior of this flow. Measurements of velocities were taken every half inch from the center of the pipe. The results were as follows, expressed to two decimal places.

Distance from center (inches)	0	0.5	1	1.5	2	2.5	3
Velocity (feet per second)	4.28	4.26	4.23	4.18	4.12	4.04	3.91

Observe that the velocity decreases from 4.28 feet per second at the center of the pipe to 3.91 feet per second at the edge. Let us determine how long it takes fluid flowing at the center of the pipe to run through the pipe and then, in contrast, the time it takes fluid at the edge to flow through the pipe.

$$\text{Center:} \qquad \text{Time} = \frac{\text{Distance}}{\text{Velocity}} = \frac{300}{4.28} = 70.09 \text{ seconds}, \qquad \text{to two decimal places}$$

$$\text{Edge:} \qquad \text{Time} = \frac{300}{3.91} = 76.73 \text{ seconds}$$

There is more than 6 seconds difference between the times.

Let us now construct a velocity function that closely fits these data. This function will enable us to determine the velocity distribution of the fluid. Enter the data into the calculator (Figure 1.60a) and plot the data points (Figure 1.60b). Observe that the points appear to lie on a downward-bending parabola, reflecting the gradual decrease in velocity from the center of the pipe to the edge. We compute the least-squares parabola for these data (Figure 1.60c).

The parabola of best fit is seen to be $y = ax^2 + bx + c$, with $a = -0.039047619$, $b = -0.0014285714$, and $c = 4.274761905$. We check the equation against the original data. Enter the equation, plot the data, and graph (Figure 1.60d). The graph is seen to fit the data points closely. The quadratic function that best describes the distribution of velocity within the pipe is therefore

$$v(x) = -0.039047619x^2 - 0.0014285714x + 4.274761905 \text{ feet per second}$$

where x is distance from the center of the pipe.

This function can be used to predict the velocity at any distance from the center of the pipe. Let us predict the velocity 1.75 inches from the center of the pipe. We get $v(1.75) = 4.15375$. The velocity 1.75 inches from the center of the pipe is estimated to be 4.15 feet per second, expressed to two decimal places. Observe that we have rounded this velocity to two decimal places because the original data were given to two decimal places; we cannot claim more accuracy than this.

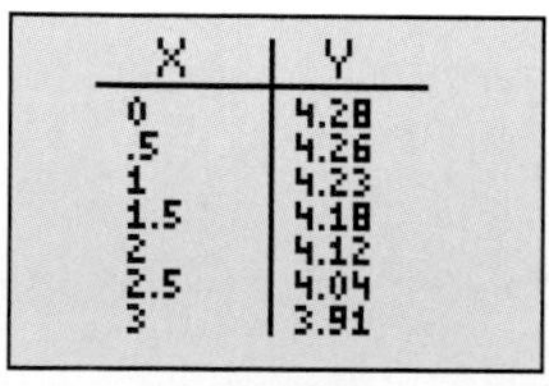

Enter data.

(a)

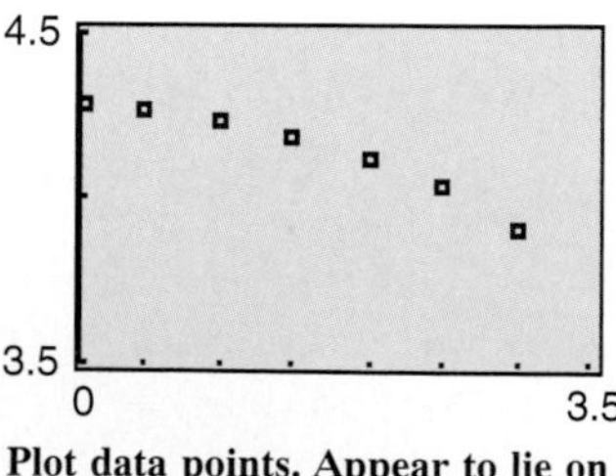

Plot data points. Appear to lie on a parabola.

(b)

Use quadratic regression to find the equation.

(c)

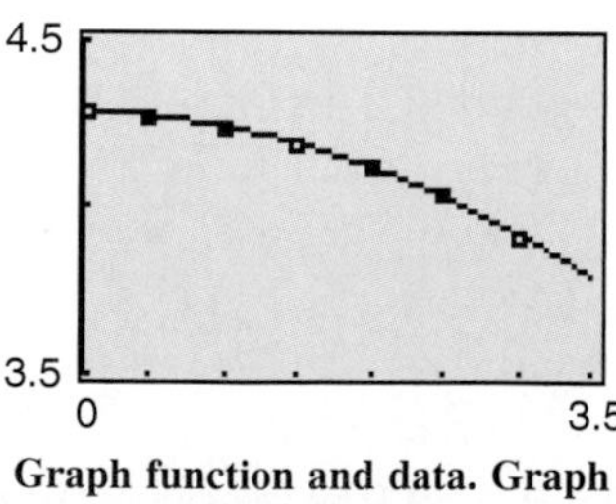

Graph function and data. Graph closely fits data.

(d)

Figure 1.60 Distribution of velocity within a pipe.

As mentioned previously, the area of mathematics that involves fitting equations to data is called regression analysis. In these examples we used **quadratic regression** to determine quadratic equations. The first step in finding an equation for given data is to plot the data, as here, and then to decide on the class of equations (quadratic equations here). It is sometimes known in advance which type of equation fits a given situation. For example, the dissipation of a drug in the body is often described by a polynomial of degree 4 (this application will be discussed in Section 3.2). Once the general class of equations has been determined, regression is used to find the specific equation that best describes the situation.

Answer to Self-Check Exercise

3. $R(t) = 333.6780303t^2 - 528.1689394t + 3163.736364$. The year 2000 corresponds to $t = 12$. $R(12) = 44875.34545$. According to these data, the revenue for the year 2000 will be approximately $44,875 million!

EXERCISES 1.5

1. **Motion Under Gravity** An object is projected vertically upward from ground level with a velocity of 384 feet per second. The height s after t seconds is given by the quadratic function $s(t) = -16t^2 + 384t$. Find its maximum height and the time at which it reaches that height.
2. **Motion Under Gravity** If an object is projected vertically from height s_0 with velocity v, its motion is described by the quadratic function $s(t) = -16t^2 + vt + s_0$. An object is projected from a height of 9920 feet with a velocity of 832 feet per second. Use a graphing calculator to determine its maximum height and the duration of the flight. Confirm your answer using algebra.
3. **Motion Under Gravity** If an object is projected vertically from ground level with velocity v, its motion is described by the quadratic function $s(t) = -16t^2 + vt$. An object that is projected vertically from ground level reaches a maximum height after 19.5 seconds. What was its initial velocity?
4. **Area of a Circle** The area A of a circle is a function of its radius, $A(r) = \pi r^2$. What is the domain of this function? Use a table to determine the first integer radius for which the area is greater than 800.
5. **Revenue Analysis** In a certain market area, it has been estimated that the total revenue R from selling a certain television is related to the price p of the set as follows: $R(p) = 553p - p^2$. What price should be charged to obtain the largest revenue possible? **(a)** Use algebra. **(b)** Use a table. **(c)** Use a graph.
6. **Revenue Analysis** In a certain market area, it has been estimated that the total revenue R in dollars from selling a certain refrigerator is related to the price p of the set as follows: $R(p) = 862p - 1.25p^2$.
   (a) Use algebra to determine the price that should be charged to maximize revenue.
   (b) Check your answer with a graph.
7. **Revenue Analysis** The revenue $R(\$)$ of the Canberra Company is the following function of units produced, x.

$$R(x) = \left(-\frac{1}{38.32}\right)x^2 + 90.45x$$

   (a) Use algebra to determine the production level that gives the largest revenue.
   (b) Check your answer with a graph.
   (c) Use a table to determine the production interval for which the revenue is above $60,000.
8. **Points of Intersection of Graphs** **(a)** Graph $f(x) = x^2 - 4$ and $g(x) = -x + 5$ on your calculator. **(b)** Determine the points of intersection of these graphs, rounded to two decimal places.
9. **Points of Intersection of Graphs** **(a)** Graph $f(x) = 0.8x^2 + x - 3$ and $g(x) = 1.3x^2 - 3x - 8$ on your calculator. **(b)** Determine the points of intersection of these graphs, rounded to two decimal places.
10. **Supply and Demand Model** An analysis of a certain market produced the following supply and demand functions: $S(p) = p^2 - 15p + 285$ and

$D(p) = p^2 - 90p + 2136$. The units of p are dollars, and the units of S and D are thousands of articles. Determine the equilibrium price of the commodity in the market.

11. **Supply and Demand Model** An analysis of a certain market produced the following supply and demand functions: $S(p) = 1.4p^2 - 10p + 150$ and $D(p) = 0.9p^2 - 60p + 1500$. The units of p are dollars, and the units of S and D are thousands of articles. Use your graphing calculator to determine the equilibrium price of the commodity in the market.

12. **Supply and Demand Model** An analysis of a certain market produced the following supply and demand functions: $S_1(p) = 0.75p^2 - 8p + 130$ and $D(p) = 0.8p^2 - 50p + 800$. The units of p are dollars, and the units of S and D are thousands of articles.

    (a) Use your graphing calculator to determine the approximate equilibrium price of the commodity in the market.

    (b) Suppose shortages of material disrupt the supply so the function becomes $S_2(p) = 0.75p^2 - 8p + 80$. How does this affect the equilibrium price of the commodity? Discuss.

13. **Dow Jones Industrial Average** The following table gives the annual highs for the Dow Jones for the years 1990 to 1998.

    (a) Using 1990 as year 0, 1991 as year 1, and so on as x values and the corresponding Dow Jones numbers as y values, plot these data and show that the points lie approximately on a parabola. Find the quadratic function that best fits these data, and use it to predict the value of the Dow for the year 2002.

    (b) When is the Dow predicted to reach 15,000 based on these data?

Year	1990	1991	1992	1993	1994	1995	1996	1997	1998
Dow	2999	3168	3413	3794	3978	5216	6560	8254	9337

14. **Flow of a Fluid** Fluid runs through a circular pipe with a radius of 4 inches and a length of 200 feet. Experiments have shown that the velocity of the flow gradually decreases from the center of the pipe toward the edges because of friction at the edges. Measurements of velocities were taken every half inch from the center of the pipe. The results, expressed to two decimal places, were as follows.

Distance from center (inches)	0	0.5	1	1.5	2	2.5	3	3.5	4
Velocity (feet per second)	6.42	6.40	6.35	6.30	6.22	6.13	6.03	5.91	5.78

(a) Determine how long it takes fluid flowing at the center of the pipe to run through the pipe and then, in contrast, the time it takes liquid at the edge to flow through the pipe, rounded to two decimal places.

(b) Use quadratic regression to find a quadratic function that fits these data.

(c) Use the function to predict the velocity of the fluid at a distance of 3.2 inches from the center of the pipe, rounded to two decimal places.

15. **Revenue Analysis** Vespa Acoustic Systems has the following data on sales and corresponding revenue:

Sales, x (units)	500	1,000	1,500	2,000	2,500	3,000
Revenue, R ($)	67,500	125,000	172,500	210,000	237,500	255,000

(a) Find the revenue function $R(x)$ (a quadratic function) that best fits these data.

(b) What would be the expected revenue from sales of 2400 items?

(c) What would be the predicted sales from 3500 items?

16. **Braking Distances** *Kar Magazine* investigated the braking distances (in feet) of a certain model of automobile for speeds of 30 to 70 miles per hour under good, dry conditions. The data collected are shown here.

    (a) Plot these points to show that a quadratic function can be used to describe the relationship between braking distance and speed. Use quadratic regression to find a function for braking distance.

    (b) Determine the braking distances for speeds of 55 and 65 miles per hour.

    (c) Use the function to predict the braking distance for 75 miles per hour.

Speed (mph)	30	40	50	60	70
Braking distance (feet)	109.5	162	222.5	291	367.5

17. **Production of Automobiles** Data for the number of new automobiles produced on a worldwide basis are shown here. Time t is measured in years from 1986, and the number of cars N is in units of 100. Plot these points to show that a quadratic function can be used to approximate the relationship between N and t.

    (a) Use quadratic regression to find a suitable function $N(t)$.

    (b) Use this function to predict the approximate number of cars produced in the year 2000.

Years from 1986, t	0	2	4	6	8	10
Number of cars, N (units of 100)	9,000	9,168	9,384	9,648	9,960	10,320

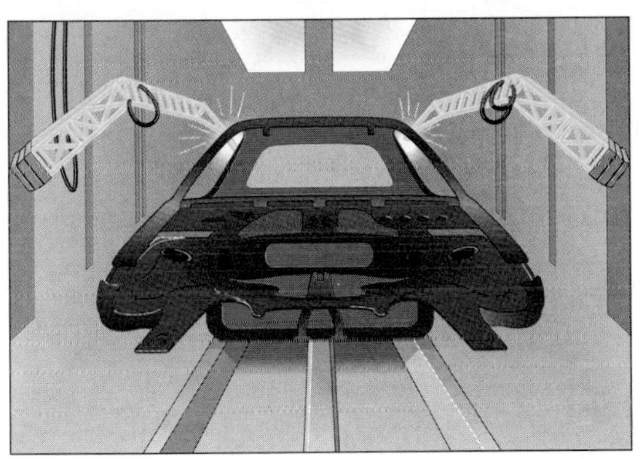

18. Consider the following data for the average size of a farm in the United States from 1950 to 1995 (*The World Almanac,* 1997, page 157).

    (a) Plot the data, show that they closely fit a parabola, and find the equation of the parabola of best fit.

    (b) Use the equation to predict the average farm size for 2002.

    (c) How do you view the short-term trend?

    (d) How do you view the long-term prediction according to these data?

Year	1950	1960	1970	1980	1995
Size of farm (acres)	213	297	374	426	469

19. **U.S. Trade Balance with Japan** The following table gives the U.S. trade balances with Japan for 1987 through 1994. Plot the three sets of data to see that they lie approximately on parabolas.

    (a) Use quadratic regression to find three quadratic functions for these data.

    (b) Predict the export, import, and difference figures for the year 2000 based on these data. (See the Williams Web site at

    www.saunderscollege.com/math/williams

    to link to the Internet for more information.)

Years from 1987	0	1	2	3	4	5	6	7
Exports	28.2	37.7	44.5	48.6	48.1	47.8	47.9	53.5
Imports	84.6	89.5	93.6	89.7	91.5	97.4	107.2	119.1
Difference ($, billions)	−56.4	−51.8	−49.1	−41.1	−43.4	−49.6	−59.3	−65.5

1.6 Distances, Circles, and Typesetting

• The Distance Formula • The Midpoint Formula • Circle, Radius, and Center • Standard Form of the Equation of a Circle • General Forms of the Equation of a Circle • Algebra in Typesetting

In this section, we derive formulas for the distance between points and for the midpoint of a line segment. These discussions lead naturally to the equation of a circle. As an application of this area of mathematics, we see how a professional typesetter uses these mathematical tools to lay out the graphics in a text.

Many of the results of this section are based on the Pythagorean Theorem. The Pythagorean Theorem gives the relationship among the lengths of the sides of a right triangle. In this section, we will use the notation $d(P, Q)$ for the distance between the points P and Q. The distance between the points P and R would be $d(P, R)$ and so on. We remind the student of the Pythagorean Theorem, writing it in terms of this notation. The side opposite the right angle of a right triangle is called the **hypotenuse.**

THE PYTHAGOREAN THEOREM

Consider a right triangle PQR (Figure 1.61). The hypotenuse is PQ, and the other two sides are PR and QR; then $d(P, Q)^2 = d(P, R)^2 + d(Q, R)^2$.

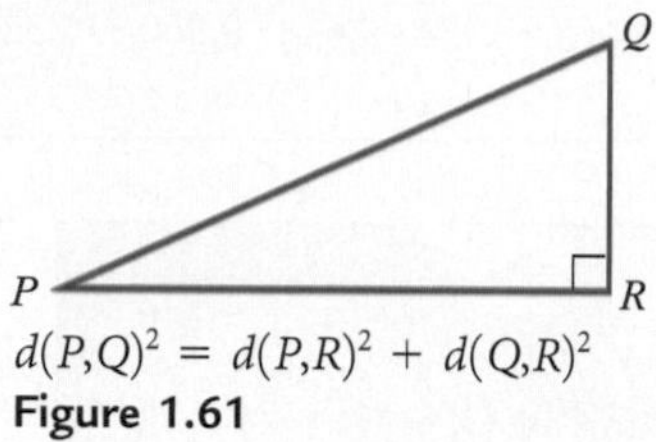

$d(P,Q)^2 = d(P,R)^2 + d(Q,R)^2$

Figure 1.61

THE DISTANCE FORMULA

Let us derive a formula for the distance $d(P, Q)$ between two points $P(x_1, y_1)$ and $Q(x_2, y_2)$. First, form the right triangle PQR, shown in Figure 1.62. Observe that R has the same x value as Q and the same y value as P. R is thus the point (x_2, y_1). We get

$$d(Q, R) = d(A, B) = y_2 - y_1$$
$$d(P, R) = d(C, D) = x_2 - x_1$$

The Pythagorean Theorem gives

$$d(P, Q)^2 = d(P, R)^2 + d(Q, R)^2$$

Taking the principal square root because distance must be positive, we get

$$d(P, Q) = \sqrt{d(P, R)^2 + d(Q, R)^2}$$
$$= \sqrt{(x_2 - x_1)^2 + (y_2 - y_1)^2}$$

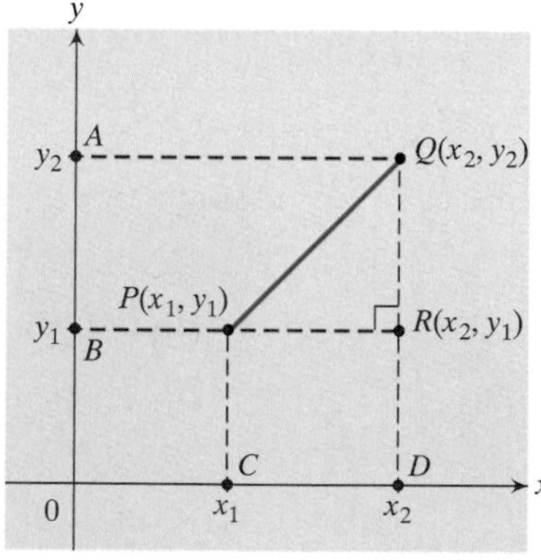

Figure 1.62

We have arrived at the formula for the distance between two points, both located in the first quadrant. The result holds also if the points are located anywhere in the plane. Let us summarize this result.

The Distance Formula

The distance between the points $P(x_1, y_1)$ and $Q(x_2, y_2)$ is

$$d(P, Q) = \sqrt{(x_2 - x_1)^2 + (y_2 - y_1)^2}$$

Example 1 Distance Between Two Points

Find the distance between the points (2, 8) and (6, 3), rounded to four decimal places.

SOLUTION Let $(x_1, y_1) = (2, 8)$ and $(x_2, y_2) = (6, 3)$ (Figure 1.63). We get

$$\begin{aligned} d[(2, 8), (6, 3)] &= \sqrt{(6-2)^2 + (3-8)^2} \\ &= \sqrt{(4)^2 + (-5)^2} \\ &= \sqrt{16 + 25} = \sqrt{41} \\ &= 6.403124237 \\ \text{Distance} &= 6.4031, \quad \text{to four decimal places} \end{aligned}$$

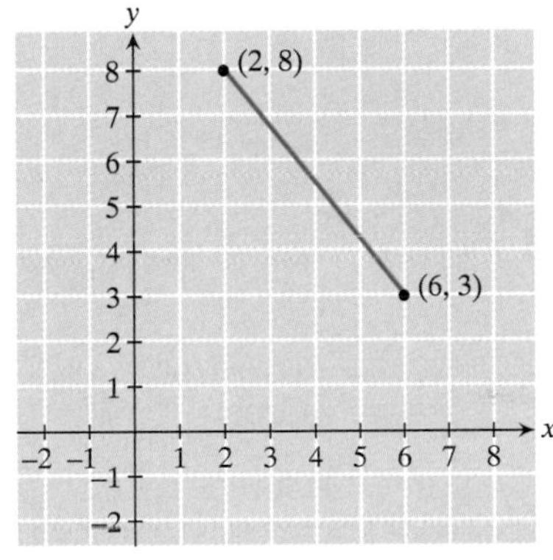

Figure 1.63

Note that it does not matter which points we select as (x_1, y_1) and (x_2, y_2) in the distance formula because the quantities $(x_2 - x_1)$ and $(y_2 - y_1)$ are squared.

Self-Check 1 Find the distance between the points $(-2, 6)$ and $(4, -1)$, rounded to four decimal places.

Example 2 Points on the y-Axis a Certain Distance from a Given Point P

Find all the points on the y-axis that are distance 5 from the point $P(4, 9)$.

SOLUTION Let a point on the y-axis distance 5 from $P(4, 9)$ be denoted $(0, y)$. Draw a rough sketch (Figure 1.64). There appear to be two such points, Q and R. We therefore expect two values of y. Apply the distance formula to the points (4, 9) and $(0, y)$.

$$\sqrt{(4-0)^2 + (9-y)^2} = 5$$

Square both sides and solve for y.

$$\begin{aligned} 16 + (9-y)^2 &= 25 \\ (9-y)^2 - 9 &= 0 \\ y^2 - 18y + 72 &= 0 \\ (y-6)(y-12) &= 0 \\ y &= 6 \text{ or } 12 \end{aligned}$$

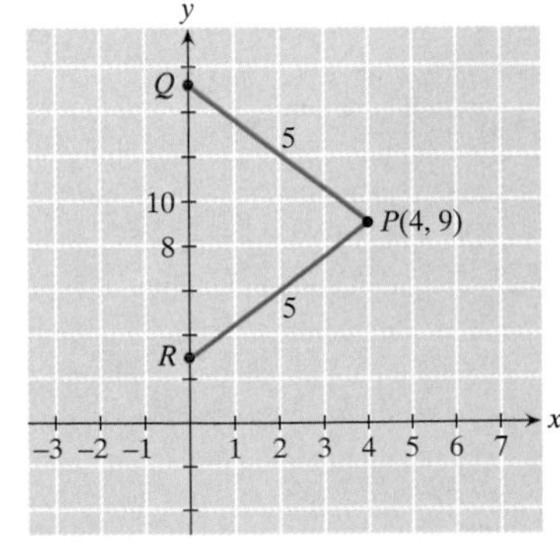

Figure 1.64

The points on the y-axis distance 5 from (4, 9) are therefore (0, 6) and (0, 12).

Further Ideas 1 Find all the points on the line $y = x$ that are distance 5 from the point $P(2, 3)$. Express your results to two decimal places.

The Midpoint Formula

Let us now derive a formula for the midpoint $R(x, y)$ of the line segment joining the points $P(x_1, y_1)$ and $Q(x_2, y_2)$. Because R is midway between P and Q, the x is located midway between x_1 and x_2 (Figure 1.65). Thus, $x - x_1 = x_2 - x$.

Rearrange to get

$$2x = x_1 + x_2$$

$$x = \frac{x_1 + x_2}{2}$$

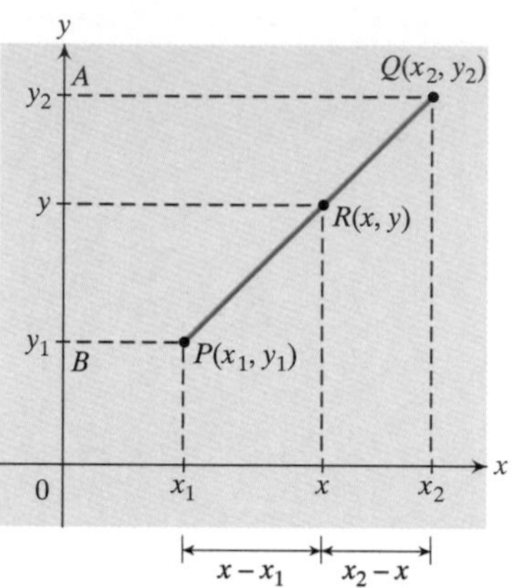

Figure 1.65

Similarly, the y-coordinates of P, Q, and R lead to

$$y = \frac{y_1 + y_2}{2}$$

We have found formulas for the coordinates of the midpoint of PQ when P and Q both are located in the first quadrant. Again, these formulas also hold if the points are located anywhere in the plane. We summarize the result.

Midpoint Formula

The midpoint of the line segment joining the points (x_1, y_1) and (x_2, y_2) is

$$\left(\frac{x_1 + x_2}{2}, \frac{y_1 + y_2}{2}\right)$$

Example 3 Midpoint of a Line Segment

Find the midpoint of the line segment joining the points $(-4, 5)$ and $(2, 1)$.

SOLUTION Let $(x_1, y_1) = (-4, 5)$ and $(x_2, y_2) = (2, 1)$. The midpoint formula gives

$$\left(\frac{x_1 + x_2}{2}, \frac{y_1 + y_2}{2}\right) = \left(\frac{-4 + 2}{2}, \frac{5 + 1}{2}\right)$$

$$= (-1, 3)$$

The midpoint is $(-1, 3)$ (Figure 1.66).

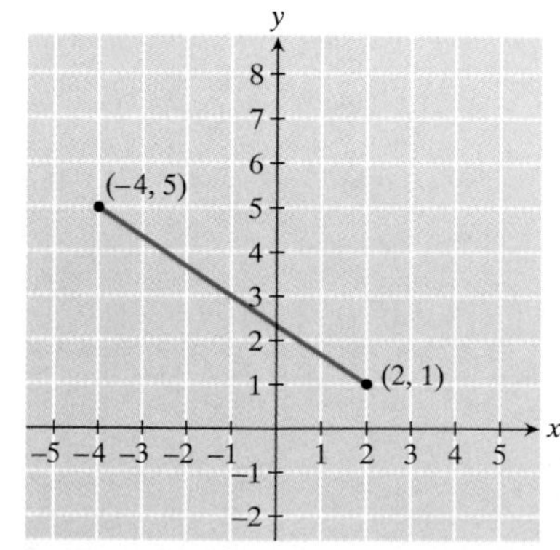

Figure 1.66

Self-Check 2 Find the midpoint of the line segment joining the points $(5, -6)$ and $(7, 2)$.

Example 4 End Point of a Line Segment

The midpoint of a line segment is (3.8, 7.2). If one of the end points of the line segment is (1, 4), find the other end point.

SOLUTION Let the other end point be (x, y). Draw a rough sketch (Figure 1.67). The midpoint of the line segment joining the points (1, 4) and (x, y) is

$$\left(\frac{1+x}{2}, \frac{4+y}{2}\right)$$

This is the point (3.8, 7.2). Thus,

$$\frac{1+x}{2} = 3.8 \quad \text{and} \quad \frac{4+y}{2} = 7.2$$

Solve these equations for x and y.

$$1 + x = 7.6 \quad \text{and} \quad 4 + y = 14.4$$
$$x = 6.6 \quad \text{and} \quad y = 10.4$$

The other end point of the line segment is (6.6, 10.4).

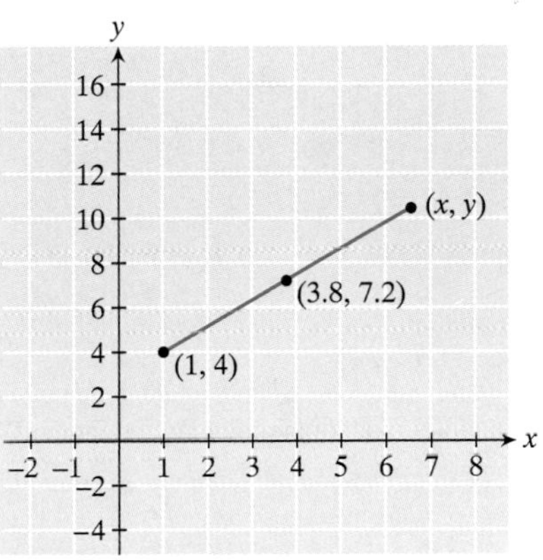

Figure 1.67

CIRCLE

We are now in a position to discuss circles. We define a circle and use the distance formula to arrive at the equation.

> ***Circle*** A **circle** is a set of points in the plane that are fixed distance from a given point. This distance is called the **radius** and the given point is called the **center.**

Consider a circle with center $C(h, k)$ and radius r (Figure 1.68). Let $P(x, y)$ be an arbitrary point on the circle. Apply the distance formula to CP.

$$d(C, P) = \sqrt{(x-h)^2 + (y-k)^2}$$

Thus,

$$\sqrt{(x-h)^2 + (y-k)^2} = r$$

Square both sides of the equation to get the equation of the circle

$$(x-h)^2 + (y-k)^2 = r^2$$

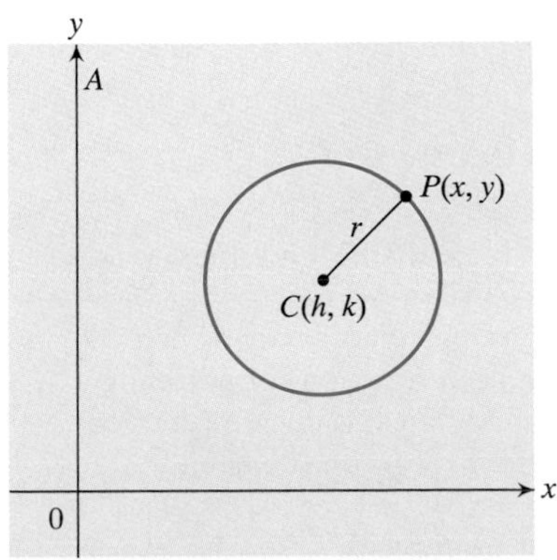

Figure 1.68

Standard Form of the Equation of a Circle

The **standard form** of the equation of a circle, with center (h, k) and radius r, is

$$(x - h)^2 + (y - k)^2 = r^2$$

Example 5 Center, Radius, and Graph of a Circle

Find the center and radius and graph the following circle on a calculator. Use the graph to confirm the center and radius.

$$(x - 6)^2 + (y + 4)^2 = 9$$

SOLUTION First, write this equation in standard form

$$(x - h)^2 + (y - k)^2 = r^2$$

Note that this involves writing $+4$ as $-(-4)$ in the second term as follows:

$$(x - 6)^2 + [y - (-4)]^2 = 9$$

Compare with the standard form. We see that $h = 6$, $k = -4$, and $r^2 = 9$. The equation therefore represents a circle having center $(6, -4)$ and radius 3.

To graph the circle $(x - 6)^2 + (y + 4)^2 = 9$ on a calculator, we must solve this equation for y. First, rewrite the equation as

$$(y + 4)^2 = 9 - (x - 6)^2$$

Take the square root of each side of the equation. Remember that there are two roots: one positive and one negative.

$$y + 4 = \sqrt{9 - (x - 6)^2} \quad \text{or} \quad y + 4 = -\sqrt{9 - (x - 6)^2}$$

$$y = \sqrt{9 - (x - 6)^2} - 4 \quad \text{or} \quad y = -\sqrt{9 - (x - 6)^2} - 4$$

Enter the equations $Y_1 = \sqrt{9 - (X - 6)^2} - 4$ and $Y_2 = -\sqrt{9 - (X - 6)^2} - 4$ into the calculator and graph (Figure 1.69). (The first equation represents the top semicircle, and the second equation represents the bottom semicircle.) The center and radius are confirmed. Observe that the graph has gaps because of the limited resolution of the calculator screen, as discussed previously.

Note that in the standard setting the graph does not appear to form a circle due to different scales on the x- and y-axes. Select the square setting to get a true circle.

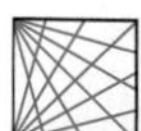

CIRCLE

Your calculator may have a built-in function for graphing a circle. If it has such a function, use it to graph the circle $(x - 6)^2 + (y + 4)^2 = 9$ of Example 5.

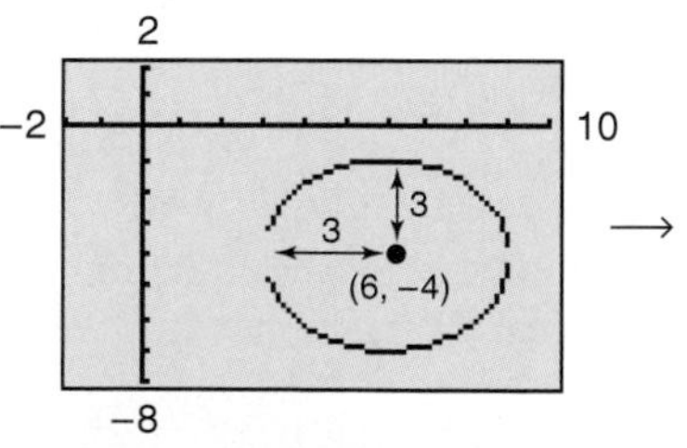

Standard setting

→

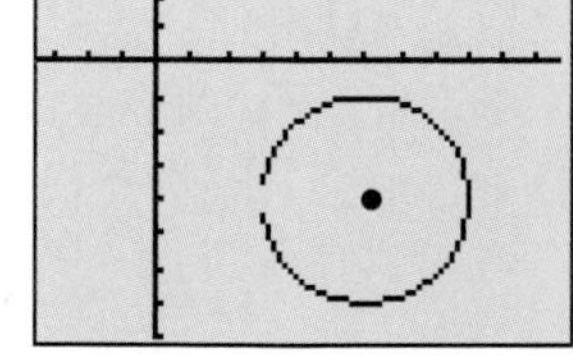

Square setting

Figure 1.69 Circle $(x - 6)^2 + (y + 4)^2 = 9$

Example 6 **Circle with a Given Center and Radius**

Find the equation of the circle with center $(-4, 5)$ and radius 8.

SOLUTION Substitute $h = -4$, $k = 5$, and $r = 8$ into the standard form of the equation of a circle, $(x - h)^2 + (y - k)^2 = r^2$. We get

$$[x - (-4)]^2 + (y - 5)^2 = 8^2$$

The equation of the circle is

$$(x + 4)^2 + (y - 5)^2 = 64$$

We graph this equation on a calculator (Figure 1.70). It does indeed correspond to a circle with center $(-4, 5)$ and radius 8.

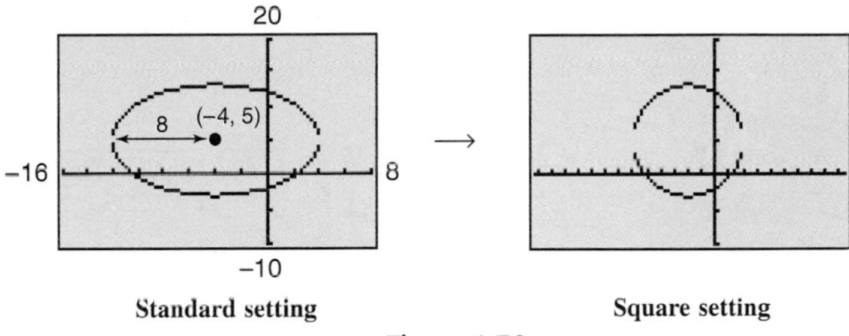

Figure 1.70

Self-Check 3 Find the equation of the circle having center $(6, -2)$ and radius 7. Graph the equation on your calculator, and use the display to check your answer.

Example 7 **Circle with End Points of a Diameter Given**

Find the equation of the circle having the points $A(4, 5)$ and $B(10, 5)$ as end points of a diameter. Graph the equation on your calculator, and use the display to confirm your answer.

SOLUTION First, we find the center of the circle and then its radius. This information can then be used to write the standard form of the equation of the circle.

Draw a rough sketch of the circle (Figure 1.71a). The center C of the circle will be the midpoint of the line segment AB.

$$C = \left(\frac{4 + 10}{2}, \frac{5 + 5}{2}\right) = (7, 5)$$

The radius of the circle is

$$d(A, C) = \sqrt{(7 - 4)^2 + (5 - 5)^2} = 3$$

The equation of the circle is thus

$$(x - 7)^2 + (y - 5)^2 = 9$$

We graph this equation on a calculator (Figure 1.71b). It is a circle with diameter $A(4, 5)$ $B(10, 5)$.

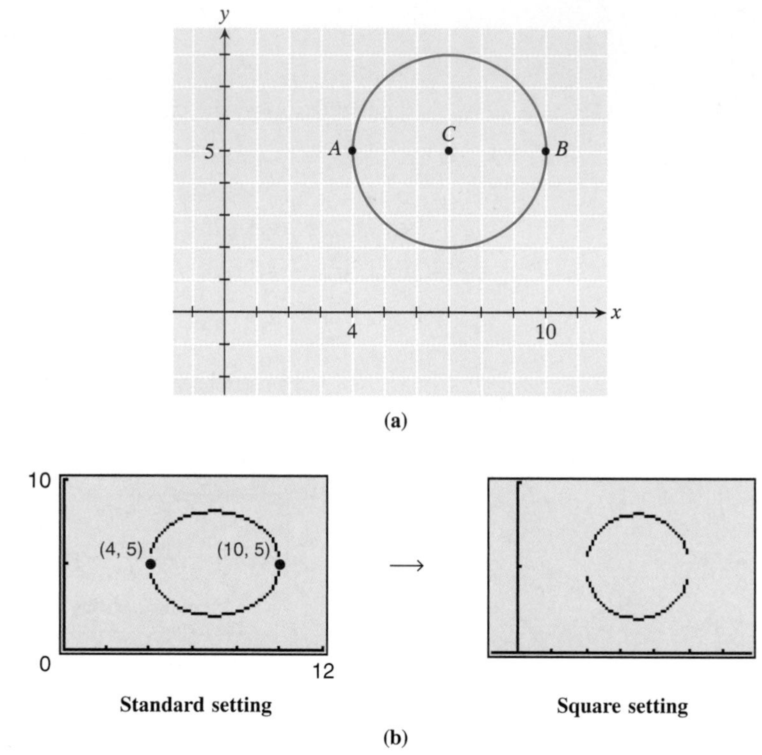

Figure 1.71

Up to this time we have only considered equations of circles in the standard form $(x - h)^2 + (y - k)^2 = r^2$. Let us now look at more general forms of the equation of a circle.

General Forms of the Equation of a Circle

The standard form of the equation of a circle center (h, k) and radius r is $(x - h)^2 + (y - k)^2 = r^2$. Let us expand the squares in the first two terms.

$$x^2 - 2xh + h^2 + y^2 - 2yk + k^2 = r^2$$

Letting $a = -2h$, $b = -2k$, $c = h^2 + k^2 + r^2$, this equation can be written

$$x^2 + y^2 + ax + by + c = 0$$

Some equations of circles may appear in this form. Such equations must be rewritten in standard form to determine the center and radius. The algebraic technique of completing the square that we introduced in Section R.4 can be used to do this.

Example 8 **Rewriting the Equation of Circle in Standard Form**

Find the center and radius of the circle having equation

$$x^2 + 4x + y^2 - 8y - 5 = 0$$

SOLUTION This equation is not in standard form. Let us use the method of completing the square to rewrite the equation in standard form. The center and radius can then be read. Separate the x and y terms as follows:

$$(x^2 + 4x) + (y^2 - 8y) - 5 = 0$$

Form complete squares of both the x and the y terms by adding appropriate numbers within each parenthesis. For the x term, *add the square of half the coefficient of* x, namely 2^2. For the y term, *add the square of half the coefficient of* y, namely $(-4)^2$. Compensate for the numbers added in this manner by adding them also to the right side of the equation.

$$(x^2 + 4x + 2^2) + [y^2 - 8y + (-4)^2] - 5 = 2^2 + (-4)^2$$

We are now able to write the x and y terms as complete squares.

$$(x + 2)^2 + (y - 4)^2 - 5 = 20$$
$$(x + 2)^2 + (y - 4)^2 = 25$$

In comparison with the standard form, we see that $h = -2$, $k = 4$, and $r^2 = 25$. Thus the center of the circle is $(-2, 4)$, and the radius is 5 (Figure 1.72).

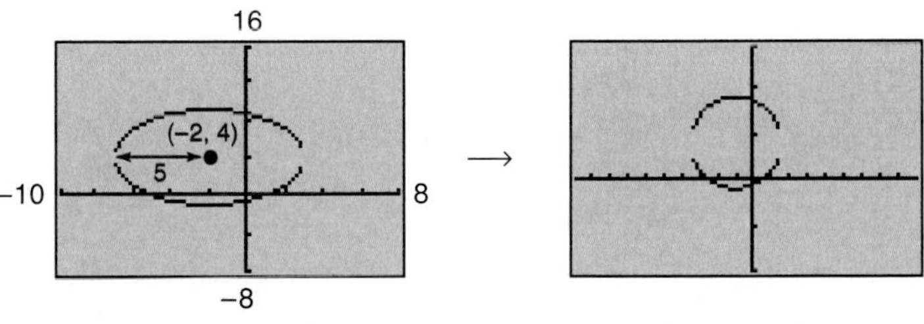

Standard setting　　Square setting

Figure 1.72

Self-Check 4 Find the center and radius of the circle having equation

$$x^2 + 2x + y^2 - 6y = 26$$

Graph the circle on your calculator, and use the display to check your answer.

We now look at the problem of determining whether a given equation represents a circle. Sometimes it is readily apparent that the equation is not of the appropriate form. At other times, the equation has to be rewritten before it is known.

Example 9 **Determining Whether an Equation Is That of a Circle**

Determine whether the following equation is that of a circle:

$$x^2 + 4x + 8y - 7 = 0$$

SOLUTION The equation is not of the form $x^2 + y^2 + ax + by + c = 0$. (There is no y^2 term.) It cannot be rewritten in the standard form of the equation of a circle; thus the equation is not that of a circle.

The following example illustrates that not all equations that are of the form $x^2 + y^2 + ax + by + c = 0$ are circles.

Example 10 **Determining Whether an Equation Is That of a Circle**

Determine whether the following equation is that of a circle:

$$x^2 + y^2 + 6x + 8y + 28 = 0$$

SOLUTION The test for determining whether an equation represents a circle is to see whether it can be rewritten in the standard form of the equation of a circle. Separate the x and y terms.

$$(x^2 + 6x) + (y^2 + 8y) + 28 = 0$$

Form complete squares of both the x and the y terms. For the x term, add the square of half the coefficient of x, namely 3^2. For the y term, add the square of half the coefficient of y, namely 4^2. Compensate for the numbers added in this manner by adding them also to the right side of the equation. We get

$$\begin{aligned}(x^2 + 6x + 3^2) + (y^2 + 8y + 4^2) + 28 &= 3^2 + 4^2\\ (x + 3)^2 + (y + 4)^2 + 28 &= 25\\ (x + 3)^2 + (y + 4)^2 &= -3\end{aligned}$$

There is no real number r such that $r^2 = -3$; therefore, this equation is not that of a circle. There are no points that satisfy this equation because the left side $(x + 3)^2 + (y + 4)^2$ is never negative.

Algebra in Typesetting

Typesetters use the algebraic formulas for distance, midpoint, and circle to produce figures in overhead slides, manuscripts, and textbooks. For example, a graphics typesetter using the popular typesetting software $\TeX$ must enter the locations of the vertices of a figure and commands to draw lines between those vertices into a program to draw a figure. We illustrate the steps by drawing an equilateral triangle. Suppose the typesetter wants to draw an equilateral triangle at the location where the vertices of the base are $A(1, 1)$ and $C(3, 1)$ (Figure 1.73).

The location of the third vertex B is needed. By symmetry, B has x-coordinate 2. Let B be the point $(2, y)$, where y is such that $BC = AC$.

The distance formula gives

$$BC = \sqrt{(2-3)^2 + (y-1)^2} \quad \text{and} \quad AC = 2$$

Thus,

$$\sqrt{(2-3)^2 + (y-1)^2} = 2$$

Let us solve this equation for y.

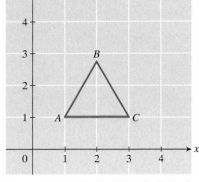

Figure 1.73 $A(1, 1)$ $C(3, 1)$ $B(2, y)$
Triangle to be typeset

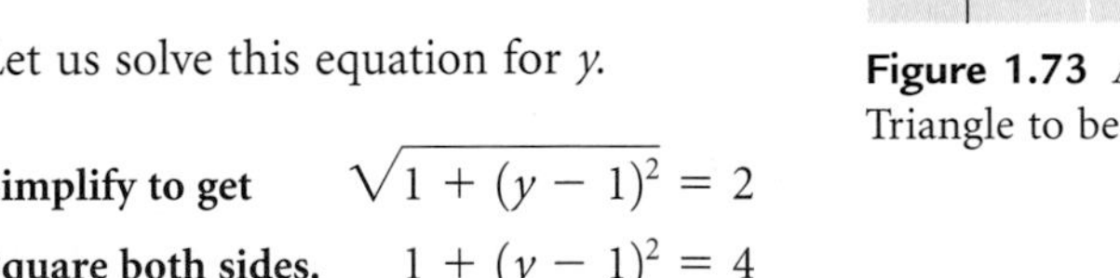

Simplify to get $\sqrt{1 + (y-1)^2} = 2$

Square both sides. $1 + (y-1)^2 = 4$

Simplify. $(y-1)^2 = 3$

Take the square root of both sides.

$$y - 1 = 1.7321 \quad \text{or} \quad y - 1 = -1.7321, \quad \text{rounded to four decimal places}$$
$$y = 2.7321 \quad \text{or} \quad y = -0.7321$$

From Figure 1.73, we see that the first value of y is the one needed here (the second value of y leads to a point equidistant from A and C that is below the x-axis). B is the point $(2, 2.7321)$.

We can check what the triangle will look like by displaying it on a calculator. Using $A(1, 1)$, $B(2, 2.7321)$, and $C(3, 1)$, we get the display shown in Figure 1.74.

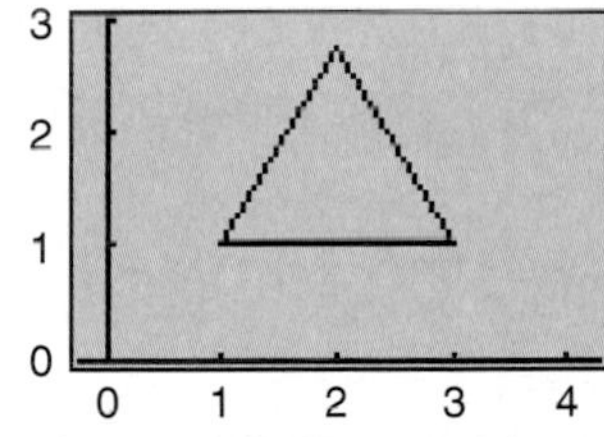

Figure 1.74 Square setting

The graphics typesetter can thus use the vertices $(1, 1)$, $(3, 1)$, $(2, 2.7321)$ to obtain the equilateral triangle at the desired location.

Group Discussion

Planning Artwork

Construct some geometrical artwork that might appear in a book you are writing. A graphics typesetter is to set the figures for this artwork. Calculate the data the typesetter needs and, if possible, graph the figures on your calculator to ensure that you are going to give the typesetter the correct data.

Answers to Self-Check Exercises

1. Distance is 9.2195.
2. Midpoint is $(6, -2)$.
3. The circle is $(x - 6)^2 + (y + 2)^2 = 49$. See graph.

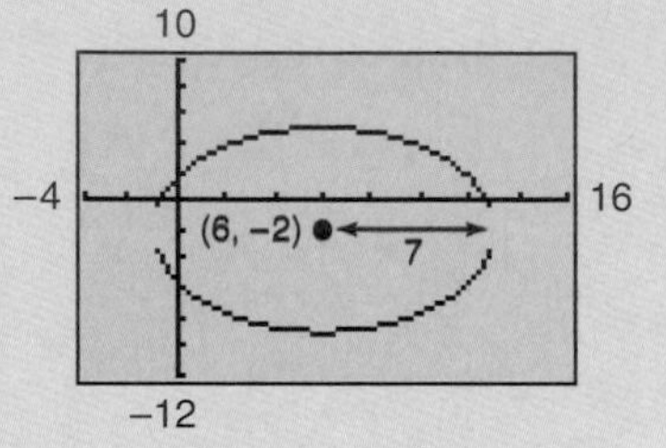

Standard setting

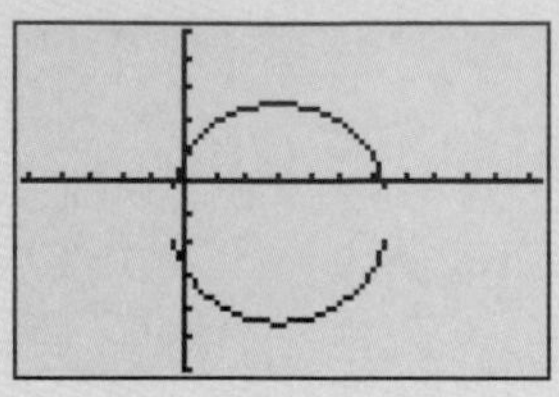

Square setting

4. The circle is $(x + 1)^2 + (y - 3)^2 = 36$, the center is $(-1, 3)$, and the radius is 6.

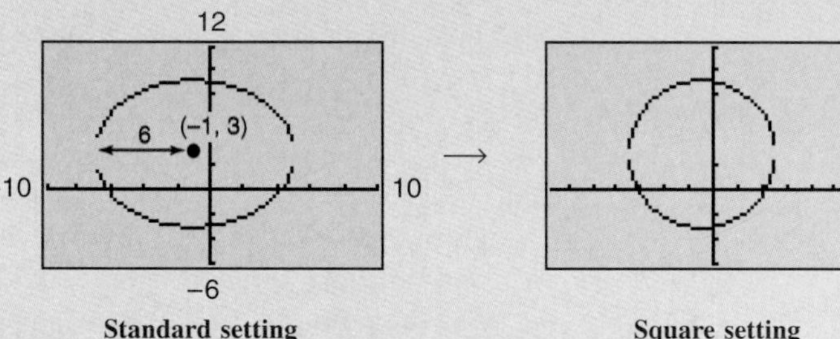

Standard setting Square setting

Answer to Further Ideas Exercise

1. The points are of the form (x, x). $d((x, x), (2,3)) = 5$ leads to $x = -1$ or $x = 6$. The points are $(-1, -1)$ and $(6, 6)$.

EXERCISES 1.6

In Exercises 1–6, use the distance formula to compute the distances between the given points, rounded to four decimal places.

1. $(1, 3), (2, 7)$ **2.** $(-2, 4), (5, 6)$
3. $(1, 1), (3, -6)$ **4.** $(2, 4), (-3, 6)$
5. $(-3, 0), (-2, 1)$ **6.** $(0, 0), (-2, 5)$

In Exercises 7–12, use the midpoint formula to find the midpoint of the line segment joining the given points.

7. $(1, 2), (3, 6)$ **8.** $(2, 5), (4, 1)$
9. $(3, 2), (6, 8)$ **10.** $(-1.25, 6.97), (3.72, -2.43)$
11. $(0, 0), (5, -2)$ **12.** $(1, -3), (0, -7)$

In Exercises 13–18, find the centers and radii of the circles having the given equations. Graph the equations on your calculator, and use the displays to support your answers.

13. $(x + 1)^2 + (y - 2)^2 = 4$
14. $(x - 3)^2 + (y + 5)^2 = 25$
15. $(x + 5)^2 + (y - 2)^2 = 17$
16. $(x + 6)^2 + (y - 8)^2 = 12$
17. $(x - 1)^2 + (y - 3)^2 = 8$
18. $(x - 2)^2 + (y + 7)^2 = 19$

In Exercises 19–24, determine the equations of the circles having the given centers and radii (in standard form).

Graph the equations on your calculator, and use the displays to support your answers.

19. Center (3, 1), radius 4
20. Center (5, −1), radius 7
21. Center (0, 4), radius 5
22. Center (−2, 0), radius 3
23. Center (1.5, 6), radius 3.2
24. Center (−3, −12), radius 4.1

In Exercises 25–32, find the centers and radii of the circles having the given equations.

25. $x^2 + 4x + y^2 - 6y = 23$
26. $x^2 - 8x + y^2 - 10y - 8 = 0$
27. $x^2 + y^2 - 6x - 2y = -9$
28. $x^2 + y^2 + 4x + 6y = 54$
29. $4x^2 - 8x + 4y^2 - 24y + 39 = 0$
30. $2x^2 + 4x + 2y^2 - 4y = -3$
31. $9x^2 + 9y^2 - 72y + 108 = 0$
32. $2x^2 + 2y^2 + 8x - 12y = -10$

In Exercises 33–40, determine whether the equations are equations of circles. (Can the equations be written in the standard form of an equation of a circle?) If the equation is that of a circle, give its center and radius.

33. $x^2 - 2x + y^2 - 10y - 10 = 0$
34. $x^2 + 6x + y^2 + 4y = 20$
35. $x^2 - 2x + y^2 + 10y + 33 = 0$
36. $x^2 - 14x + y^2 - 2y + 53 = 0$
37. $x^2 - 6x + y^2 + 7 = 0$
38. $x^2 - 2x + 2y = 3$
39. $x^2 - 6x - 4y^2 - 8y + 9 = 0$
40. $4x^2 + 8x + 4y^2 - 40y = 40$
41. Find the equations of the concentric circles displayed here. Check your answers by duplicating the graphs on your calculator.

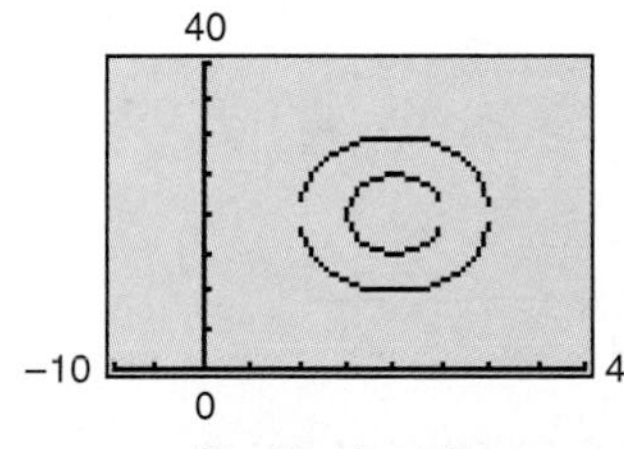

Standard setting

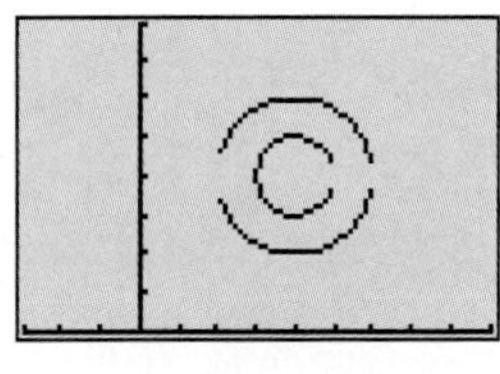
Square setting

In Exercises 42–44, find the equations of the circles displayed. Support your answers by duplicating the graphs on your calculator. (All displays are in standard setting.)

42.
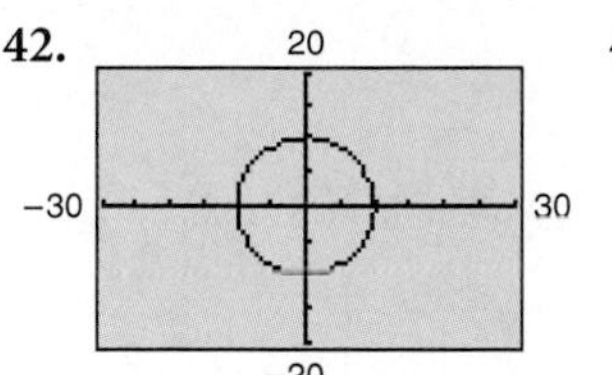

43.
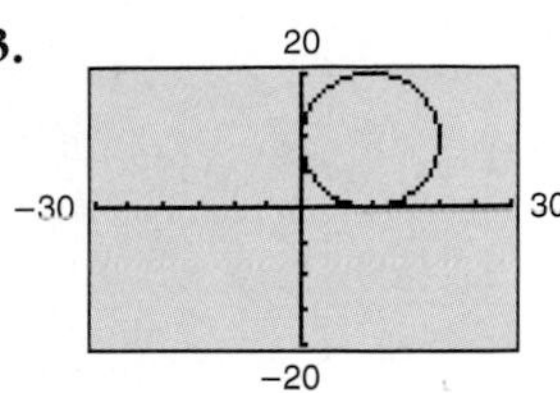

44.
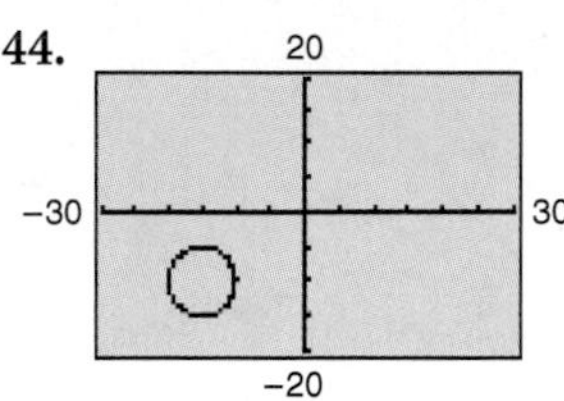

In Exercises 45–47, find the equations of the semicircles. Support your answers by duplicating the graphs on your calculator. (All displays are in standard setting.)

45.
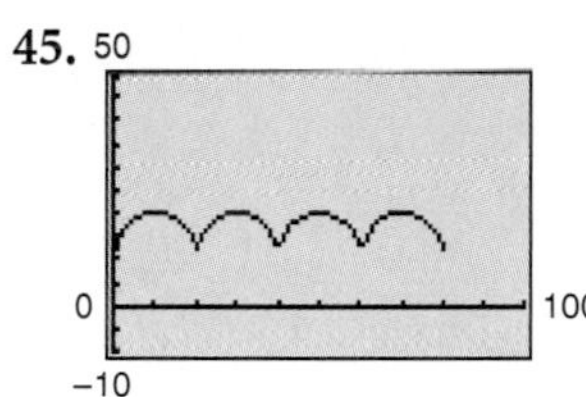

46.
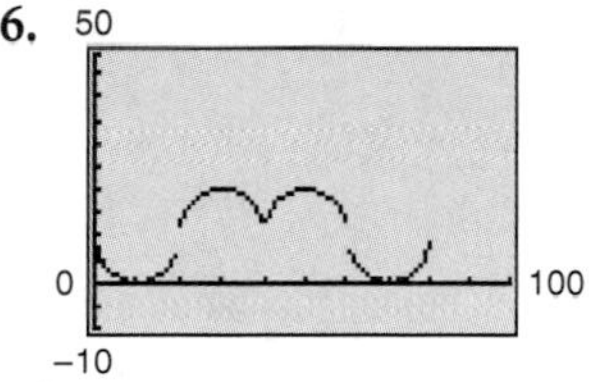

47.
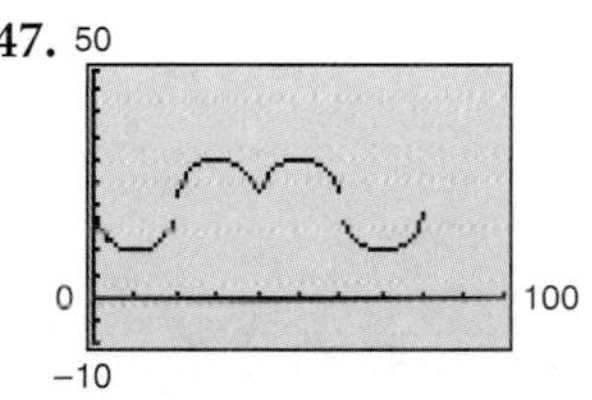

Typesetting A graphics typesetter of a geometry book wants to set the figures in Exercises 48–51. He contracts the computation of the vertices of the figures to you (a mathematician). Calculate these data; then, if possible, graph the figures on your calculator to ensure that you are going to give the typesetter the correct data.

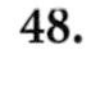
48.

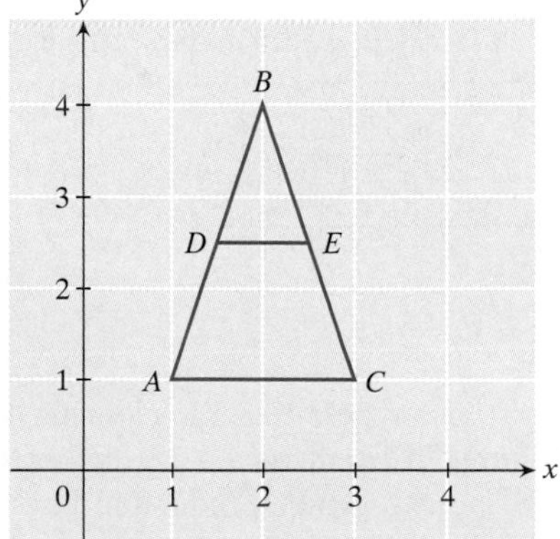

$A(1, 1)$ $B(2, 4)$ $C(3, 1)$
D and E are midpoints of the sides.

49.

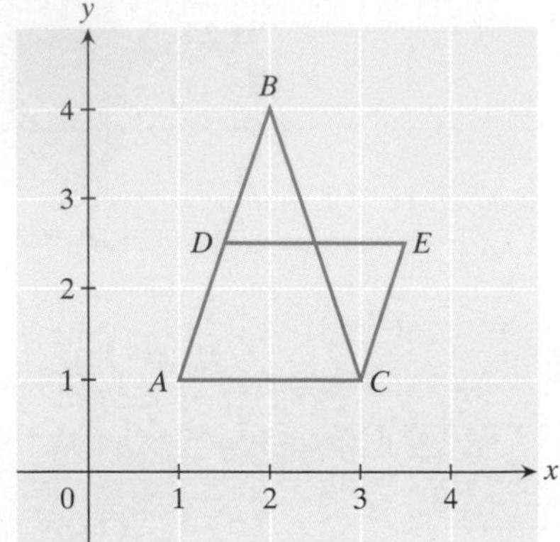

$A(1, 1)$ $B(2, 4)$ $C(3, 1)$
D is midpoint of the side.
$ACED$ is a parallelogram.

50.

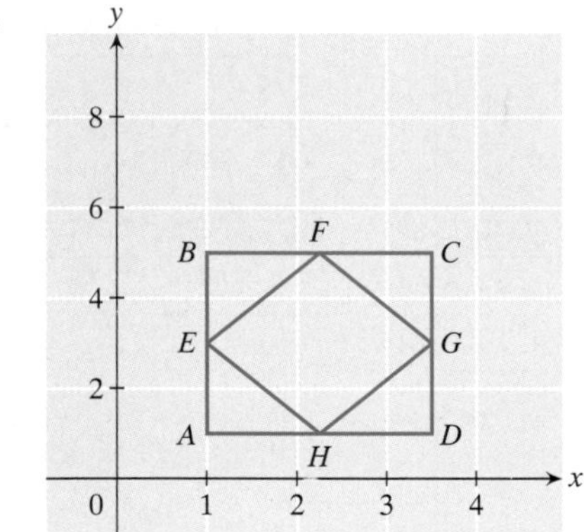

$A(1, 1)$ $B(1, 5.3)$ $C(3.4, 5.3)$ $D(3.4, 1)$
$E, F, G, H,$ are midpoints of the sides.

51.

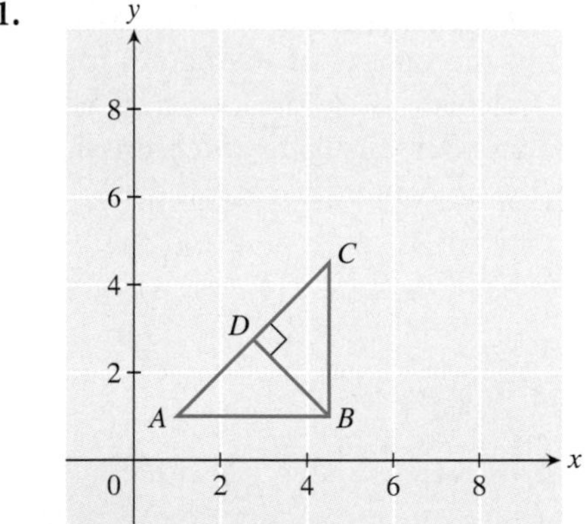

$A(1, 1)$ $AB = BC = 3.5$

Typesetting A graphics typesetter of a geometry book wants to set the figures in Exercises 52–55. He contracts the computation of points, centers, radii, and lines that define the figures to you. Determine the data the typesetter needs. Check the data by duplicating the figures on your calculator.

52.

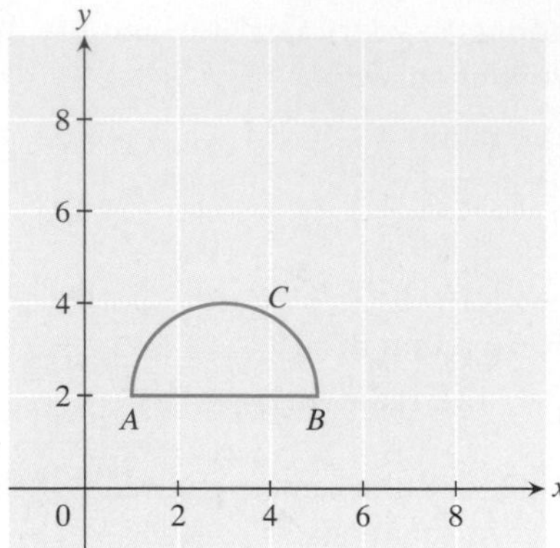

Typesetter needs points A, B, and equation of semicircle C.

53.

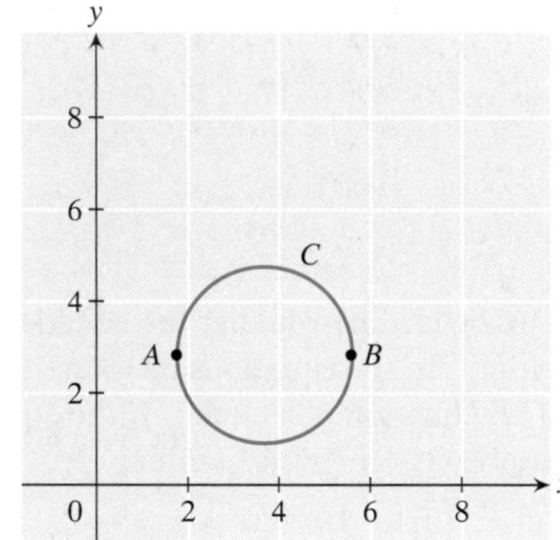

C is a circle with diameter AB.
$A(1.8, 2.8)$, $B(5.6, 2.8)$.
Typesetter needs center, radius, and equation of circle.

54.

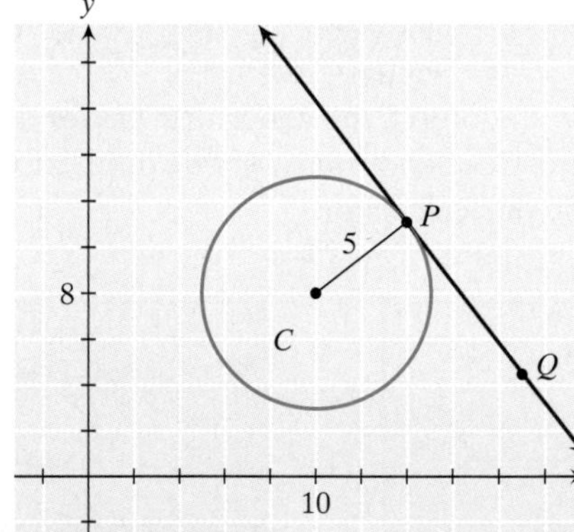

$C(10, 8)$, $P(13, 12)$. PQ is perpendicular to CP.
Typesetter needs equations of circle and PQ.

55. 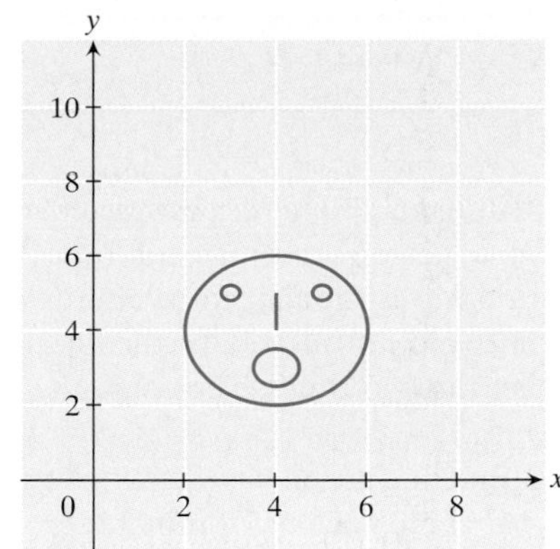

Suggest data.

56. Find the lengths of the sides of the triangle having vertices $A(2, 1)$, $B(4, 0)$, and $C(-1, -5)$, and then use the Pythagorean Theorem to prove the triangle is a right triangle. Check that it is a right triangle geometrically by drawing the triangle in the square setting on your calculator.

57. Are the points $A(1, 5)$, $B(1, 9)$, and $C(-2, -5)$ the vertices of a right triangle? Test using algebra and a graphics display.

58. Show by displaying on a calculator that the points $A(1, -5)$, $B(5, -6)$, and $C(4, -2)$ are the vertices of an isosceles triangle (two of the sides have equal length). Verify the result also using algebra.

59. Show that the points $A(3, -1)$, $B(5, 4)$, $C(-2, 1)$, and $D(0, 6)$ are the vertices of a square.

60. Find all the points on the y-axis that are distance 5 from the point $(3, 1)$.

61. Find all the points on the x-axis that are distance 13 from the point $(-4, 12)$.

62. Find the values of x for which the distance between the points $(x, 1)$ and $(6, 4)$ is 5.

63. Find the values of y for which the distance between the points $(-3, 4)$ and $(9, y)$ is 13.

64. The midpoint of a line segment is $(1, 3)$. If one of the end points of the line segment is $(3, 7)$, find the other end point.

65. If P is the point $(1, 2)$ and Q is the point $(5, 6)$, find the coordinates of the point R that is three fourths of the way from P to Q.

66. If $P(x_1, y_1)$ and $Q(x_2, y_2)$ are both in the first quadrant with $x_2 > x_1$ and $y_2 > y_1$, find a formula for the coordinates of the point R that is three fourths of the way from P to Q.

67. Find the x- and y-intercepts of the circle $(x - 3)^2 + (y - 1)^2 = 25$ using a calculator. Confirm your answer using algebra. (The x-intercept is the x-coordinate of the point at which the circle intersects the x-axis. The y-intercept is the y-coordinate of the point at which the circle cuts the y-axis.)

68. Find the x- and y-intercepts, if there are any, of the circle $(x - 5)^2 + (y - 1)^2 = 4$ using a calculator. Confirm your answer using algebra.

69. Find the equation of the circle having the points $A(3, 1)$ and $B(3, 7)$ as end points of a diameter. Graph the equation on your calculator, and use the display to support your answer.

70. Find the equation of the circle with the points $A(2, 1)$ and $B(10, 7)$ as end points of a diameter.

71. Find the equation of the circle having center $(5, 2)$ that touches the x-axis at the point $(5, 0)$. Graph the equation on your calculator, and use the display to support your answer.

72. Find the equation of the circle having center $(-4, -3)$ that touches the y-axis at the point $(0, -3)$. Graph the equation on your calculator, and use the display to support your answer.

73. Find the equation of the circle that touches the x-axis at $x = 2$ and touches the y-axis at $y = 2$. Graph the equation on your calculator, and use the display to support your answer.

74. Find the equation of a circle with center $(4, 2)$, passing through the point $(9, -3)$. Graph the equation on your calculator, and use the display to support your answer.

75. Find the circumference and area of the circle having equation $x^2 + 4x + y^2 - 14y - 11 = 0$.

76. Find the equations of four circles that do not intersect one another, all of which lie inside the rectangle defined by the four lines $x = 1$, $x = 8$, $y = 2$, $y = 9$. Graph the circles and lines on your calculator, and use the display to support your answers.

77. Find the equation of the circle that touches the lines $x = 5$, $y = 4$, and $y = -2$. Graph the circle on your calculator, and use the display to support your answers.

78. Consider the circle $x^2 - 4x + y^2 - 2y - 11 = 0$. Find the other end point of the diameter through the point $(-2, 1)$. Graph the circle on your calculator, and use the display to support your answers.

CHAPTER 1 PROJECT Real and Subjective Times

Scientists have studied the sense of time in humans and have concluded that there is a relationship between real time and a person's estimate of time (called subjective time). In most cases, this relationship is thought to be approximately linear. Let t be real time, s be subjective time, and $t = ms + b$ be the linear equation that describes the connection between the two times. The constants m and b vary from person to person.

The author determined his subjective time line as follows. Starting at a given time (while watching television), he estimated intervals of 10 minutes for a total of 60 minutes. A coworker recorded the actual times. The results were as follows:

Subjective time, s (minutes)	10	20	30	40	50	60
Real time, t (minutes)	8.5	22	32	38	50	63

These data were plotted and found to lie approximately on a line. The graph of the data points is called a **scatter diagram.** The least-squares line was then computed (Figure 1.75).

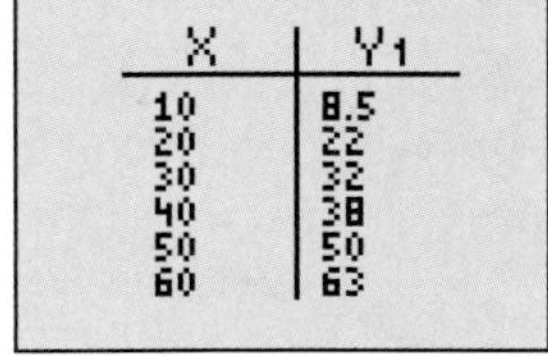

Enter data.

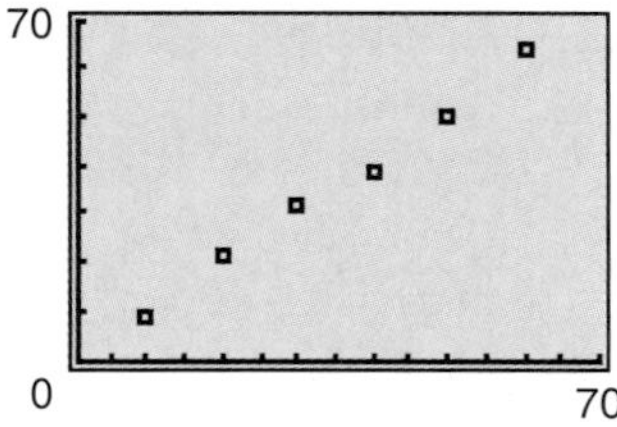

Plot data on a calculator or by hand. Points lie approximately on a line.

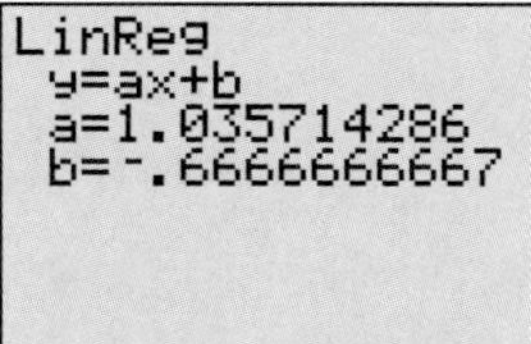

Compute least-squares line. $y = 1.04x + 0.67$ (to two decimal places).

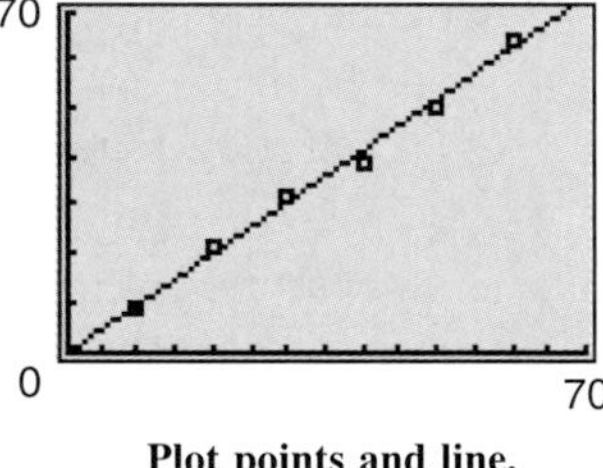

Plot points and line. Line is a good fit.

Figure 1.75 Relationship between real time and subjective time.

As a result, there is the following approximate relationship between real time t and the author's estimate of time s:

$$t = 1.04s + 0.67$$

We round the numbers to only two decimal places because we cannot claim too much accuracy in our experiment. The line is seen to fit the data closely.

With the aid of another person, take real and subjective time data, and compute your subjective time equation. Describe the conditions under which the experiment took place. Make a case for the conditions being suitable for the experiment. If you guess at 1 hour how long will the actual time be? Do you think that time goes faster or slower than it actually does? Are you satisfied with your concept of time? If not, why not? How could you change it?

WEB PROJECT 1

To explore the concepts of this chapter further, try the interesting WebProject. Each activity relates the ideas and skills you have learned in this chapter to a real-world applied situation. Visit the Williams Web site at

www.saunderscollege.com/math/williams

to find the complete WebProject, related questions, and live links to the Internet.

CHAPTER 1 HIGHLIGHTS

Section 1.1

Rectangular Coordinate System: Way of describing locations of points. In a plane the x-axis intersects the y-axis at right angles, and the point of intersection is the origin. Every point has a location (x, y).
Graph: The graph of an equation is the set of points that satisfy the equation.
Scales: x scale and y scale are the distances between tick marks on the x- and y-axes.
Pixel: Small dots on graphing calculator screen that can be lighted.
Viewing Window: The portion of the coordinate plane that appears in the display.
Intercepts: x- and y-intercepts are the points at which a graph cuts the x- and y-axes.

Section 1.2

Slope: A measure of how steep a line is. The slope of a line through the points (x_1, y_1) and (x_2, y_2) is $(y_2 - y_1)/(x_2 - x_1)$ if $x_1 \neq x_2$. The slope of a vertical line is undefined.
Point–Slope Form: Equation of line through (x_1, y_1) with slope m: $y - y_1 = m(x - x_1)$.
Slope–Intercept Form: Equation of line having slope m and y-intercept b: $y = mx + b$.
Square Setting: Calculator window in which horizontal and vertical tick marks have equal spacing.
Parallel and Perpendicular Lines: Let slopes be m_1 and m_2. Lines are parallel if $m_1 = m_2$. Lines are perpendicular if $m_1 \cdot m_2 = -1$.
Linear Equation: An equation that can be written in the form $Ax + By = C$, with A and B not both zero, is a linear equation in x and y. Its graph is a line.
Least-Squares Line: Line of best fit through given data points.

Section 1.3

Linear Function: $f(x) = mx + b$.
Rate of Change: Rate of change of $f(x) = mx + b$ is m.
Cost–Volume Equation: $y = mx + b$, where x is the number of items manufactured, y is the corresponding cost, m is the marginal cost, and b is the fixed cost.

Section 1.4

Parabola: Graph of $y = ax^2 + bx + c$. The vertex is at $x = -b/2a$. The axis is $x = -b/2a$. Opens up if $a > 0$ and down if $a < 0$. The y-intercept is c.
Quadratic Function: $f(x) = ax^2 + bx + c, a \neq 0$.

Section 1.5

Equilibrium Price: Price at which a commodity settles in the market, when supply is equal to demand.

Section 1.6

Distance Formula: The distance between $P(x_1, y_1)$ and $Q(x_2, y_2)$ is $d(P, Q) = \sqrt{(x_2 - x_1)^2 + (y_2 - y_1)^2}$.
Midpoint: Point on a line, midway between points (x_1, y_1) and (x_2, y_2), is

$$\left(\frac{x_1 + x_2}{2}, \frac{y_1 + y_2}{2}\right)$$

Circle: Set of points in the plane that are a fixed distance from a given point. This distance is called the radius, and the point is called the center. Standard form of circle with center (h, k) and radius r is $(x - h)^2 + (y - k)^2 = r^2$.

CHAPTER 1 REVIEW EXERCISES

1. Plot the following points in a rectangular coordinate system.
   (a) $A(3, 5)$, $B(-2, 4)$, $C(-6, -4)$, $D(8, -1)$
   (b) $P(-5, -4)$, $Q(7, -3)$, $R(-8, 2)$, $S(4,3)$
2. Use algebra to determine whether the following points lie on the graphs of the equations.

	Equation	Points
(a)	$y = 2x - 1$	$(-1, -3), (0, -2), (3, 5)$
(b)	$y = 3x^2 - 2x + 1$	$(-2, 17), (1, 2), (6, 98)$
(c)	$y = \dfrac{1}{x-6} + 3\sqrt{x-2}$	$(6, 0), (3, \frac{11}{3}), (11, \frac{46}{5})$

3. Consider the following viewing windows. Only a partial description of each window is given. Give the complete description of each window (X_{min}, X_{max}, X_{scl}, Y_{min}, Y_{max}, and Y_{scl}). Support your answer by reproducing the window on your calculator.

(a) 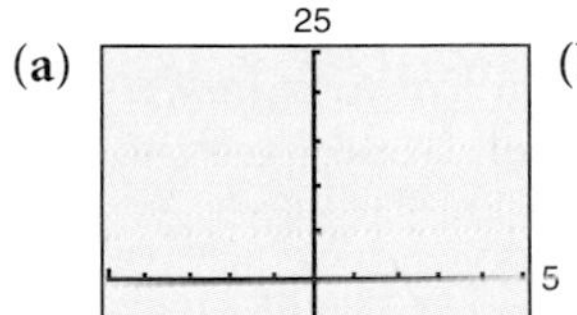

(b) 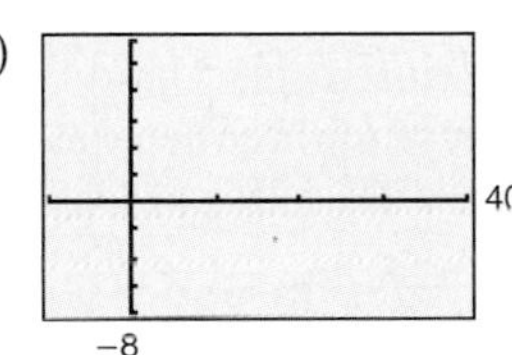

(c) 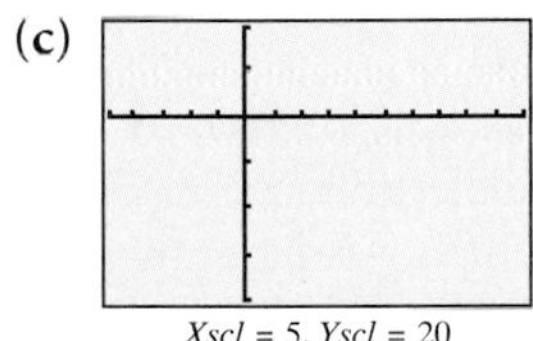

$Xscl = 5$, $Yscl = 20$

4. Find the x- and y-intercepts of the graphs of the following linear equations using algebra. Use this information to find a convenient viewing window (X_{min}, X_{max}, X_{scl}, Y_{min}, Y_{max}, and Y_{scl}) for each graph. Display the graph in the window. Use the display to sketch a rough graph.
   (a) $y = x - 7$ (b) $y = -2x + 5$
   (c) $y = 12x - 36$
5. Sketch the lines through the following points. Determine the slopes of the lines, if they are defined.
   (a) $(2, 6), (7, 1)$ (b) $(-3, 6), (7, -3)$
   (c) $(-3, -8), (-1, -2)$ (d) $(2, 8), (2, -9)$
6. Use the point–slope form to find the equations of the lines through the following points having the given slopes. Graph the equations on your calculator, and use the displays to support your answers.
   (a) $(3, 5)$, $m = 4$ (b) $(-2, 7)$, $m = -2$
   (c) $(-6, -2)$, $m = 0$
   (d) $(5, -3)$, slope does not exist
7. Find the slope of a line that is parallel to the line through the following pairs of points.
   (a) $(-1, 5)$ and $(7, 17)$ (b) $(2, 9)$ and $(-3, 1)$
   (c) $(-5, 1)$ and $(-3, -6)$
8. Find the slope of a line that is perpendicular to the line through the following pairs of points.
   (a) $(3, -6), (-1, 12)$ (b) $(1, 1), (7, 7)$
   (c) $(-3, 8), (-2, -4)$
9. Use slopes to determine whether the three points $(1, -3)$, $(3, 1)$, and $(5, 5)$ lie on a line.
10. Use the point–slope form of the equation of a line to find the equations of the lines through the following points. Graph the equations on your calculator, and use the displays to support your answers.
    (a) $(1, 4), (7, -8)$ (b) $(-2, -3), (2, 5)$
    (c) $(5, 5), (-3, 9)$
11. Use the slope–intercept form of the equation of a line to find the equations of the lines having the following slopes and y-intercepts. Graph the lines on a calculator in the square setting, and use the displays to support your answers.
    (a) $m = 3, b = 4$ (b) $m = -1, b = 7$
    (c) $m = 6, b = -2$
12. Find the equation of the line that has slope 5 and x-intercept -2. Graph the line on a calculator in the square setting, and use the display to support your answer.
13. Find the equation of the line through the point $(2, -5)$ perpendicular to the line $y = -2x + 4$. Graph the lines on a calculator in the square setting, and use the display to support your answer.
14. Find the equation of the line through the point $(2, 7)$ perpendicular to the line $-6x + 2y - 9 = 0$. Graph the lines on a calculator in the square setting, and use the display to support your answer.
15. Find the equation of the line with x-intercept -2, parallel to the line $-3x + y - 7 = 0$. Graph the lines on a calculator in the square setting, and use the display to support your answer.

16. Find A so that the line $Ax + 3y - 7 = 0$ is parallel to the line $6x - 2y = 5$. Graph the lines on a calculator in the square setting, and use the display to support your answer.

17. Find the x-intercept of the line through the point $(1, 5)$ perpendicular to the line $4x - 2y - 6 = 0$.

18. Use a calculator to find the values of the linear function $f(x) = 2.47x - 3.251$ at $x = -3.45, 2.79$, and 23.72.

19. Determine whether the functions $f(x)$ defined by the following tables are linear. If a function is linear, use the table to find its rate of change and the value of $f(0)$, and arrive at its equation.

(a)

X	Y1
-3	-14
-2	-10
-1	-6
0	-2
1	2
2	6
3	10

(b)

X	Y1
-2	5
-1	2
0	1
1	2
2	5
3	10
4	17

(c)

X	Y1
-1.5	8.3
-1	6.9
-.5	5.5
0	4.1
.5	2.7
1	1.3
1.5	-.1

20. Soccer Cleats A pair of Adidas "World Cup" cleats costs \$105. *A* set of replaceable screw-in studs costs \$15. Studs are replaced every 4 months.

(a) Find an equation that describes the expense of buying and maintaining the cleats.

(b) Represent this information graphically. What do the y-intercept and the slope represent physically?

21. Cost–Volume Analysis A company found that the cost of manufacturing 20 items of a certain product is \$525.50. The cost of manufacturing 30 items is \$560.80.

(a) Find the cost–volume formula.

(b) What is the cost of manufacturing 50 items?

(c) Use a graphing calculator to find the costs of 70, . . . , 76 items.

(d) Use the calculator to determine when the cost increases above \$800.

22. Automobile Production The following table gives the number of passenger cars, in units of 1000, in various countries (*Statistical Yearbook of the United Nations*). Use a least-squares line to predict the number of passenger cars in these countries in the year 2005 (to the nearest thousand).

	1988	1989	1990	1991	1992	1993
Canada	9,745	9,985	10,255	10,199	10,530	10,731
Japan	21,280	22,667	23,659	24,613	25,539	26,386
USA	116,575	116,573	118,459	123,461	123,698	126,728

23. Find the vertices, y-intercepts, and axes, and sketch the graphs of the following quadratic equations, using algebra. Verify your graphs using a graphing calculator.

(a) $y = x^2 - 4x + 9$ **(b)** $y = x^2 + 6x + 1$

(c) $y = 2x^2 - 8x - 3$ **(d)** $y = -3x^2 + 18x - 7$

24. Find a so that the point $(1, 6)$ lies on the graph of $y = 2x^2 - 3x + a$. Check your answer by graphing the equation on a calculator and confirming that the graph does pass through the point $(1, 6)$.

25. Find the equation of a parabola that has x-intercepts -2 and 3 and opens up. Confirm your answer by graphing the equation on a calculator and observing that the properties are satisfied.

26. Construct two quadratic equations that have x-intercepts 1 and -4. Confirm your answers by graphing the equations on a calculator and examining the x-intercepts.

27. Motion Under Gravity The height s feet after t seconds of an object projected vertically from ground level with velocity v feet per second is $s = -16t^2 + vt$. An object is propelled vertically from ground level with velocity of 287 feet per second. Use a calculator to determine its maximum height and the time at which it reaches this height (rounded to two decimal places). Use algebra to confirm your answer.

28. Supply and Demand Model An analysis of a certain market produced the following supply and demand functions: $S(p) = 0.64p^2 - 7p + 124$, $D(p) = 0.72p^2 - 48p + 970$. The units of p are dollars, and the units of S and D are thousands of articles. Use your graphing calculator to determine the approximate equilibrium price of the commodity in the market.

29. Flow of a Glacier A glacier in Switzerland is 7 miles long. The average depth of the glacier is 250 feet. Measurements of the velocity of the glacier were taken every 50 feet from the surface, down to a depth of 200 feet. The results were as follows:

Depth (feet from the surface)	0	50	100	150	200
Velocity (feet/year)	316	344.5	378	416.5	460

(a) Use quadratic regression to find a quadratic function that fits these data.

(b) Use the function to predict the velocity of the glacier at a depth of 250 feet?

(c) At the present rate, how many years will it take ice at the surface, middle, and base of the glacier to travel from the source to the end of the glacier?

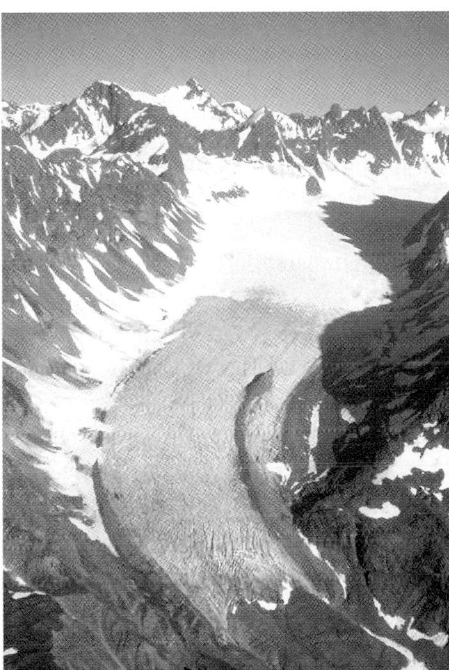

Ice moves more slowly at the surface of a glacier than at the base. *(U.S. Geological Survey)*

30. Use the distance formula to compute the distances between the following points, rounded to four decimal places.

(a) $(1, 5), (3, 9)$ (b) $(-2, 6), (2, 5)$

(c) $(3, -4), (-7, -1)$

31. Find the midpoints of the line segments joining the following points.

(a) $(1, 3), (7, 9)$ (b) $(-4, -3), (-8, 7)$

(c) $(8, 0), (-9, 3)$

32. The midpoint of a line segment is $(2, 5)$. If one end point is $(-1, 2)$, find the other end point.

33. Find all the points on the y-axis that are distance 5 from the point $(3, -1)$.

34. Find all the points $(x, 3x)$ that are distance 2 from the point $(1, 5)$.

35. Find the centers and radii of the circles having the following equations. Graph the equations on your calculator, and use the displays to support your answers.

(a) $(x - 3)^2 + (y + 4)^2 = 9$

(b) $(x + 1)^2 + (y + 7)^2 = 25$

(c) $x^2 + y^2 - 4x - 6y + 9 = 0$

(d) $x^2 + y^2 + 6x + 2y = -4$

36. Determine the equations of the circles having the following centers and radii. Graph the equations on your calculator, and use the displays to support your answers.

(a) Center $(3, 6)$, radius 3

(b) Center $(-1, 4)$, radius 6

37. Find the equation of the circle having the points $A(1, 5)$ and $B(5, 7)$ as ends of a diameter. Give its center and radius. Graph the equation on your calculator, and use the display to support your answer.

38. Find the equation of the circle having center $(-3, 1)$ that touches the x-axis at the point $(-3, 0)$. Graph the equation on your calculator, and use the display to support your answer.

39. Find the circumference and area of the circle $x^2 + y^2 - 2x - 8y + 13 = 0$.

40. Find the equations of the following graphs. Confirm your answers by duplicating the graphs on your calculator. (All displays are in standard setting.)

(a)

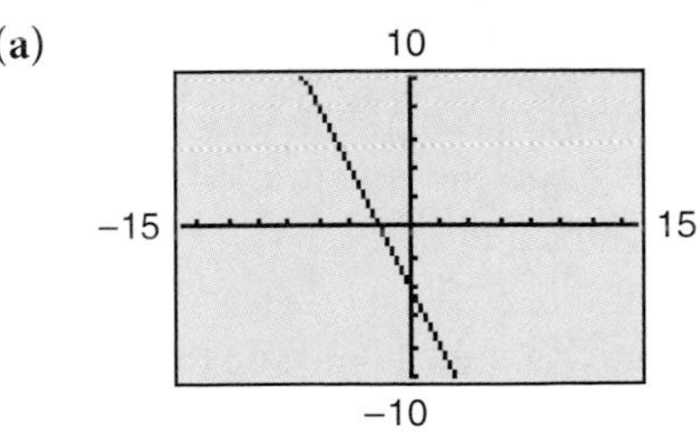

(b)

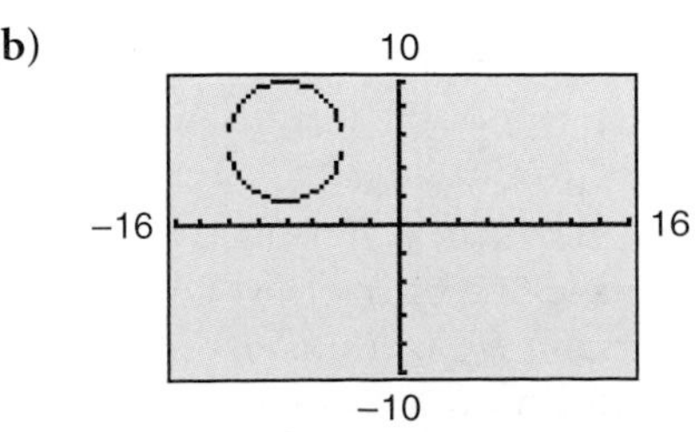

Chapter 1 Test

1. Use algebra to determine whether the following points lie on the graph of the equation.

Equation	Points
$y = 2x^2 - 4x + 3$	$(-2, 19), (1, 1), (5, 31)$

2. Consider the following viewing window. Only a partial description of the window is given. Give the complete description of each window (X_{min}, X_{max}, X_{scl}, Y_{min}, Y_{max}, and Y_{scl}). Support your answer by reproducing the window on your calculator.

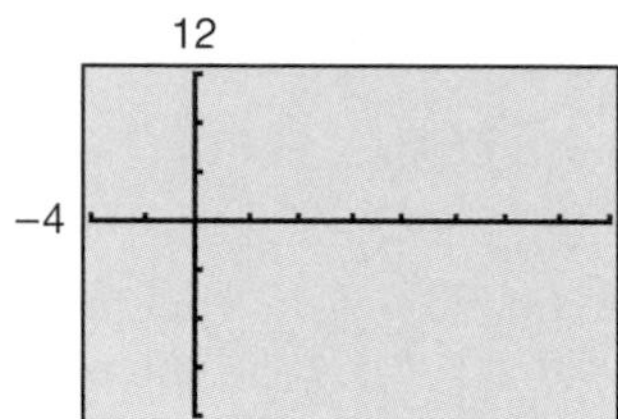

3. Find the equation of the line through the point $(-2, 3)$ having slope -1. Graph the equation on your calculator, and use the display to support your answer.
4. Find the slope of the lines that are parallel to the line through the points $(2, -3)$ and $(-1, 6)$.
5. Find the slope of the lines that are perpendicular to the line through the points $(1, 1)$ and $(3, 8)$.
6. Find the equation of the line through the points $(-2, 3)$ and $(-5, 9)$. Graph the equation on your calculator, and use the display to support your answers.
7. Find the equation of the line having slope -2 and y-intercept 3. Graph the equation on a calculator in the square setting, and use the display to support your answer.
8. Find the equation of the line through the point $(-3, 4)$ perpendicular to the line $-2x + y - 6 = 0$. Graph the lines on a calculator in the square setting, and use the display to support your answer.
9. Find the x-intercept of the line through the point $(2, -8)$ parallel to the line $12x - 3y - 3 = 0$.
10. Show that the function $f(x)$ defined by the following table is linear. Use the table to find $f(0)$ and its rate of change and to determine its equation.

X	Y1
-3	9
-2	7
-1	5
0	3
1	1
2	-1
3	-3

11. Find the vertex, y-intercept, and axis, and sketch the graph of $y = x^2 - 6x - 7$ using algebra. Verify your graph using a graphing calculator.
12. Construct a quadratic equation that has x-intercepts -3 and 5 and y-intercept 15.
13. Use the distance formula to compute the distance between the points $(2, 4)$ and $(-3, 12)$. Express the answer to four decimal places.
14. Find all the points on the line $y = x$ that are distance 9 from the point $(6, 3)$. Give the answers to two decimal places.
15. Find the centers and radii of the circles having the following equations. Graph the equations on your calculator, and use the displays to support your answers.
    (a) $(x + 5)^2 + (y - 1)^2 = 36$
    (b) $x^2 + y^2 - 6x - 2y + 6 = 0$
16. Determine the equation of the circle having center $(2, -7)$ and radius 9. Graph the equation on your calculator, and use the display to support your answer.
17. Find the equation of the circle having center $(2, 5)$ that touches the y-axis at the point $(0, 5)$. Graph the equation on your calculator, and use the display to support your answer.
18. **Manufacturing Costs** A company found that the cost of manufacturing 15 items of a certain product is \$631.50. The cost of manufacturing 30 items is \$669.
    (a) Find the cost–volume formula.
    (b) What is the cost of manufacturing 22 items?
19. **Populations Densities** The following table gives the population per square mile of land of the United States every decade for the years 1900 through 1990 (*The World Almanac,* 1997, page 382). Use a least-squares line to predict the population per square mile for the year 2010, expressed to one decimal place.

Year	1900	1910	1920	1930	1940	1950	1960	1970	1980	1990
Population	21.5	26.0	29.9	34.7	37.2	42.6	50.6	57.5	64.0	70.3

20. Supply and Demand An analysis of a certain market has produced the following supply and demand functions. $S(p) = 0.58p^2 - 6p + 121$, $D(p) = 0.83p^2 - 52p + 890$. The units of p are dollars, and the units of S and D are thousands of articles. Use your graphing calculator to determine the approximate equilibrium price of the commodity in the market.

2 Solving Equations and Inequalities

Sir Isaac Newton (1642 – 1727)
(© 1992 Northwind Picture Archives)

Solving equations is at the heart of doing and applying mathematics. Algebraic techniques have been developed to solve various types of equations. In this chapter, we introduce techniques for solving some of the most important types of equations. Methods for solving quadratic equations and equations that can be changed into quadratic equations are introduced, and geometrical ways of viewing the equations and the solutions are presented. The geometrical approach leads to numerical methods that can be applied on a graphing calculator for solving equations that cannot be solved algebraically.

Methods for solving inequalities arise naturally from methods for solving equations. Equations and inequalities are used in the study of the motion of an object under gravity. The original research in this area was done by Sir Isaac Newton (1642–1727) in Cambridge, England. Newton once said that the concept of the universal force of gravity came to him while he was drinking tea in the garden and saw an apple fall. Newton has been described as "one of the greatest names in the history of human thought" because of his significant contributions to mathematics, physics, and astronomy. For example, he explained the decomposition of light and the theory of the rainbow, invented the reflecting telescope and the sextant, and created the mathematical field of calculus. Unfortunately this last achievement involved Newton in much controversy. The question arose as to whether the German mathematician Gottfried Wilhelm Leibnitz also developed calculus independently of Newton. It is now believed that Newton and Leibnitz both developed calculus independently of one another. This controversy is regrettable because of the long and bitter alienation that it produced between English and Continental mathematicians. Newton was one of the intellectual giants of the ages, and his body was interred in Westminster Abbey in London where a magnificent monument was erected in his memory.

APPLICATION | Motion Under Gravity

Future rocket scientist.
(©Bushnell/Soifer/Tony Stone)

An object is fired vertically from ground level with velocity v feet per second. From the laws of physics, we know that the height s (in feet) is a quadratic function of time t (in seconds), $s(t) = -16t^2 + vt$. This function gives us much information about the motion. For example, suppose that the initial velocity is 256 feet per second, and we want to find when the object is at a height of 768 feet, we solve the equation $-16t^2 + 256t = 768$. The solutions to this equation tell us that the object is at a height of 768 feet 4 seconds and 12 seconds into the flight. The graph in Figure 2.1 supports this conclusion. *(This application is discussed in Section 2.2, Example 1, on page 175.)*

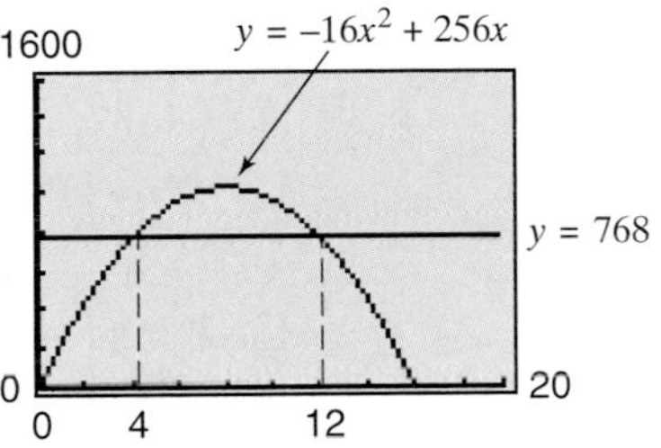

Figure 2.1 Projectile is at a height of 768 feet after 4 and 12 seconds into the flight.

2.1 Quadratic Equations

• Graphical Interpretation of Solutions • Solution by Factoring • Solution by Extracting Square Roots • Solution by Completing the Square • The Quadratic Formula • Discriminant

We introduced quadratic equations in the previous chapter and discussed their parabola graphs. In this section, we take an in-depth look at this important class of equations. We discuss various ways of solving quadratic equations and the geometrical interpretation of those solutions.

Quadratic Equation A **quadratic equation** in x is an equation that can be written in the form

$$ax^2 + bx + c = 0$$

where a, b, and c are real numbers with $a \neq 0$. $ax^2 + bx + c = 0$ is said to be the *standard form* of the quadratic equation.

Consider the following examples:

$3x^2 - 5x + 6 = 0$ is a quadratic equation in standard form.
$2x^2 + 3x = 7$ is a quadratic equation because it can be written in the form $2x^2 + 3x - 7 = 0$.
$2x^3 + x^2 + 7x - 1 = 0$ is not a quadratic equation because of the $2x^3$ term.

We introduce a number of algebraic methods for solving quadratic equations that are in standard form. If a quadratic equation is not in standard form, it should first be transformed to this form.

Graphical Interpretation of Solutions

The graphical view that we have of the solutions of a quadratic equation comes from our knowledge of parabolas. Observe that the solutions to the equation $ax^2 + bx + c = 0$ are the points on the graph of the parabola $y = ax^2 + bx + c$, where $y = 0$. These are the x-intercepts of the graph.

The solutions to the quadratic equation $ax^2 + bx + c = 0$ are the x-intercepts of the graph of the parabola $y = ax^2 + bx + c$ (Figure 2.2).

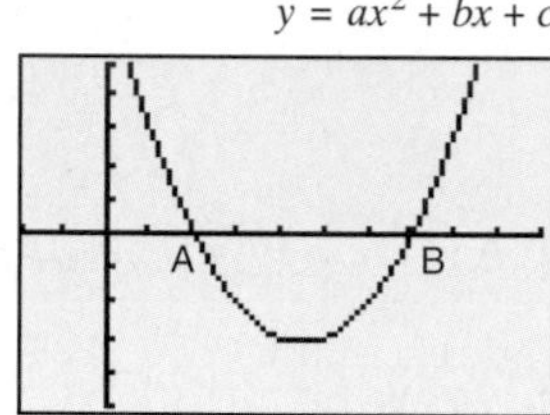

Figure 2.2 The solutions to $ax^2 + bx + c = 0$ are A and B.

We present four methods for solving quadratic equations in this section.

Solution by Factoring

Factoring can be used to solve $ax^2 + bx + c = 0$ when it is possible to factor the polynomial on the left side of the equation as the product of two first-degree polynomials. Let us solve the quadratic equation

$$x^2 - 2x - 3 = 0$$

and look at the geometric interpretation of the solutions.

Factor the polynomial to get

$$(x + 1)(x - 3) = 0$$

The product of the two real numbers $(x + 1)$ and $(x - 3)$ is zero if and only if at least one of these numbers is zero. This implies that either

$$x + 1 = 0 \quad \text{or} \quad x - 3 = 0$$
$$x = -1 \quad \text{or} \quad x = 3$$

There are two solutions (or **roots**):

$$x = -1 \quad \text{and} \quad x = 3$$

(Figure 2.3).

The solutions of $x^2 - 2x - 3 = 0$ are the points on the graph of $y = x^2 - 2x - 3$ where $y = 0$; the x-intercepts of the parabola.

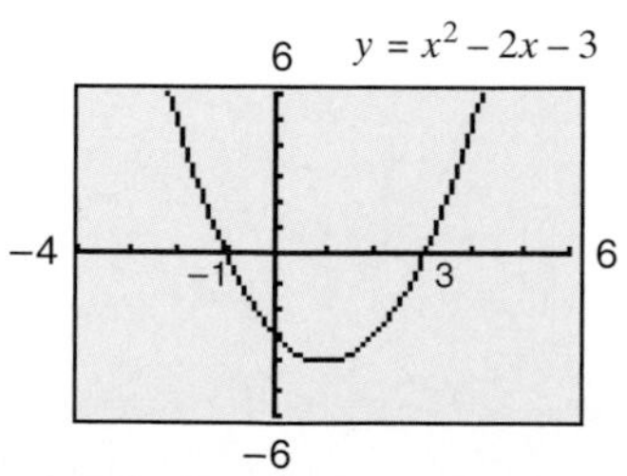

Figure 2.3 The solutions are -1 and 3.

Example 1 **Solving by Factoring**

Solve the equation $2x^2 = x + 3$ using algebra. Use a graph to support your answer.

SOLUTION Rewrite the equation in standard form $2x^2 - x - 3 = 0$.

Factor the polynomial to get

$$(2x - 3)(x + 1) = 0$$

$$2x - 3 = 0 \quad \text{or} \quad x + 1 = 0$$

$$2x = 3 \quad \text{or} \quad x = -1$$

$$x = \tfrac{3}{2} \quad \text{or} \quad x = -1$$

There are two solutions:

$x = \frac{3}{2}$ and $x = -1$ (Figure 2.4).

The solutions of $2x^2 = x + 3$ are the x-intercepts of $y = 2x^2 - x - 3$.

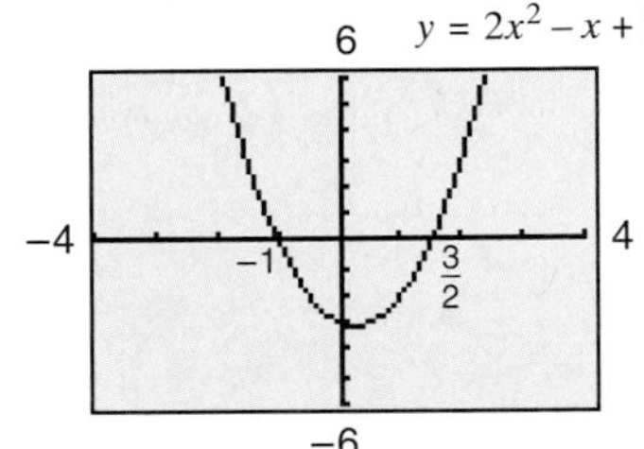

Figure 2.4 The solutions are $x = -1$ and $x = \frac{3}{2}$.

Self-Check 1 Solve the quadratic equation $2x^2 - x = 10$ using algebra. Check your answer with a graph.

Example 2 **Solving by Factoring**

Solve $x^2 - 6x + 9 = 0$.

SOLUTION Factor the polynomial.

$$(x - 3)(x - 3) = 0$$

The repeated factor leads to the single equation

$$x - 3 = 0$$
$$x = 3$$

The quadratic equation has the single solution $x = 3$. Because the factor $(x - 3)$ is a repeated factor, the root $x = 3$ is referred to as a **repeated root.**

Further Ideas 1

(a) Draw the graph of the parabola $y = x^2 - 6x + 9$ on your calculator. How is the fact that the equation $x^2 - 6x + 9 = 0$ has a repeated root (Example 2) revealed by this graph?

(b) Solve the equation $x^2 - 8x + 16 = 0$ using a graph. Confirm your answer using algebra.

Solution by Extracting Square Roots

This method gives the roots of a quadratic equation that is in the form $x^2 = k$, with $k > 0$. Rewrite the equation in standard form and then factor as follows:

$$x^2 - k = 0$$
$$(x - \sqrt{k})(x + \sqrt{k}) = 0$$
$$x - \sqrt{k} = 0 \quad \text{or} \quad x + \sqrt{k} = 0$$
$$x = \sqrt{k} \quad \text{or} \quad x = -\sqrt{k}$$

The equation $x^2 = k$ has two solutions,

$$x = \sqrt{k} \quad \text{and} \quad x = -\sqrt{k}$$

We write these two solutions $x = \pm\sqrt{k}$.

THE SOLUTIONS OF $x^2 = k$

The solutions of $x^2 = k$, $k > 0$, are $x = \pm\sqrt{k}$.

We can conveniently interpret this method as taking the square root of both sides of the equation as long as we remember that there are two roots: one positive and one negative. For example, the equation $x^2 = 14$ has the two solutions $x = \pm\sqrt{14}$.

Further Ideas 2 Solve $4x^2 = 25$ using algebra. *Hint:* Rewrite in the form $x^2 = k$.

Example 3 Solving by Extracting Roots

Solve $(x + 2)^2 = 64$.

SOLUTION The method of extracting square roots can be used for solving this equation. We can take the square root of both sides of the equation, remembering that there are two roots, one positive and one negative.

$$x + 2 = \sqrt{64} \quad \text{or} \quad x + 2 = -\sqrt{64}$$
$$x + 2 = 8 \quad \text{or} \quad x + 2 = -8$$
$$x = 6 \quad \text{or} \quad x = -10$$

The solutions are $x = 6$ and $x = -10$ (Figure 2.5).

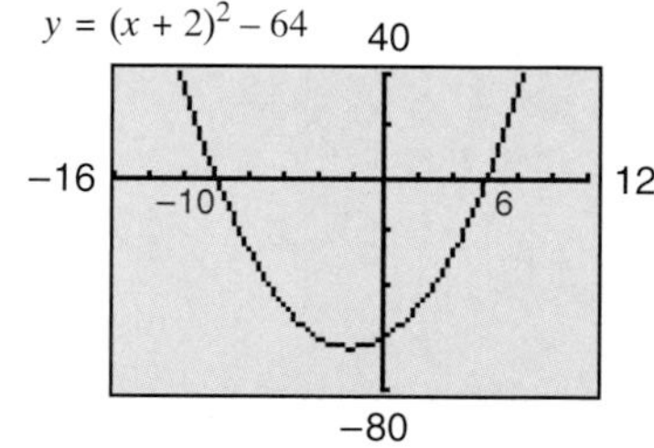

Figure 2.5

Solution by Completing the Square

The equation in the previous example, namely $(x + 2)^2 = 64$, lends itself to solution by extracting roots because the term involving x is a perfect square. The technique of completing the square (Section R.4) can be used to rewrite a quadratic equation in a form in which the left side becomes a perfect square. The equation can then be solved by extracting roots.

Example 4 **Solving by Completing the Square**

Solve $x^2 - 4x - 5 = 0$ by first completing the square and then extracting the roots.

SOLUTION

Rewrite the equation with the constant term on the right.

$$x^2 - 4x = 5$$

Complete the square for the left side. Add $(-2)^2$, the square of half the coefficient of x, to both sides.

$$x^2 - 4x + (-2)^2 = 5 + (-2)^2$$

The left side is now a complete square.

$$(x - 2)^2 = 9$$

Solve this equation by extracting roots.

$$x - 2 = \sqrt{9} \quad \text{or} \quad x - 2 = -\sqrt{9}$$

$$x - 2 = 3 \quad \text{or} \quad x - 2 = -3$$

$$x = 5 \quad \text{or} \quad x = -1$$

The solutions are $x = 5$ and $x = -1$ (Figure 2.6).

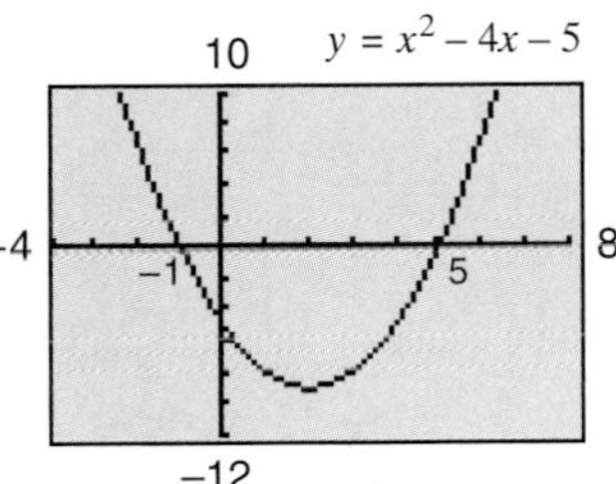

Figure 2.6

Although this equation provided a suitable introduction to the method of completing the square, the equation can be solved more easily by factoring. The following equation cannot be solved by factoring.

Example 5 **Solving by Completing the Square**

Solve $2x^2 - 4x - 1 = 0$.

SOLUTION This example illustrates the method of completing the square when the coefficient of x^2 is not 1.

Divide by 2 to make the coefficient of x^2 equal to 1, and then proceed as in Example 4.

$$x^2 - 2x - \tfrac{1}{2} = 0$$

Rewrite the equation with the constant term on the right.

$$x^2 - 2x = \tfrac{1}{2}$$

Complete the square for the left side. Add $(-1)^2$, the square of half the coefficient of x, to both sides.

$$x^2 - 2x + (-1)^2 = \tfrac{1}{2} + (-1)^2$$

The left side is now a complete square.

$$(x - 1)^2 = \tfrac{3}{2}$$

Solve this equation by extracting roots.

$$x - 1 = \sqrt{\tfrac{3}{2}} \quad \text{or} \quad x - 1 = -\sqrt{\tfrac{3}{2}}$$

$$x = 1 + \sqrt{\tfrac{3}{2}} \quad \text{or} \quad x = 1 - \sqrt{\tfrac{3}{2}}$$

The solutions are $x = 1 + \sqrt{\frac{3}{2}}$ and $x = 1 - \sqrt{\frac{3}{2}}$. These results can be checked using a graph or by direct substitution into the original equation. We illustrate the check by substitution using a calculator (Figure 2.7). Note that the calculator gives 10^{-13} (a very small number), not zero, when substituting the latter root. This is due to roundoff error. The solutions are correct.

Figure 2.7 Verifying the solutions to $2x^2 - 4x - 1 = 0$.

Self-Check 2 Solve $3x^2 - 12x + 7 = 0$ using the method of completing the square.

Up to this time we have solved specific quadratic equations by various methods. Now let us use this method of completing the square to derive a **general formula** for solving quadratic equations. This formula will enable us not only to solve equations, but also to arrive at general results about solutions to quadratic equations.

The Quadratic Formula

Consider the quadratic equation in standard form $ax^2 + bx + c = 0$, $a \neq 0$. First, divide by a.

$$x^2 + \frac{b}{a}x + \frac{c}{a} = 0$$

Rewrite the equation with the constant term on right side.

$$x^2 + \frac{b}{a}x = -\frac{c}{a}$$

Complete the square for the left side by adding $(\frac{b}{2a})^2$ to both sides.

$$x^2 + \frac{b}{a}x + \left(\frac{b}{2a}\right)^2 = \left(\frac{b}{2a}\right)^2 - \frac{c}{a}$$

Write the left side as a square, and simplify the right side.

$$\left(x + \frac{b}{2a}\right)^2 = \frac{b^2}{4a^2} - \frac{c}{a}$$

$$\left(x + \frac{b}{2a}\right)^2 = \frac{b^2 - 4ac}{4a^2}$$

Take the square root of each side of this equation.

$$x + \frac{b}{2a} = \pm\sqrt{\frac{b^2 - 4ac}{4a^2}}$$

Thus,

$$x = -\frac{b}{2a} \pm \frac{\sqrt{b^2 - 4ac}}{2a}$$

$$= \frac{-b \pm \sqrt{b^2 - 4ac}}{2a}$$

Let us summarize this result.

THE QUADRATIC FORMULA

The solution to the quadratic equation $ax^2 + bx + c = 0$, $a \neq 0$ is

$$x = \frac{-b \pm \sqrt{b^2 - 4ac}}{2a}$$

The next example illustrates the use of this formula for a quadratic equation that has two solutions. The two solutions correspond to the two x-intercepts of the graph of the equation.

Example 6 Solving by Using the Quadratic Formula

Determine the solutions of $2x^2 - 3x - 7 = 0$, rounded to four decimal places.

SOLUTION In the standard form $ax^2 + bx + c = 0$, we see that $a = 2$, $b = -3$, and $c = -7$. (Always include the negative sign with the coefficient or constant term.) Substitute these values of a, b, and c into the quadratic formula to get the solutions.

$$x = \frac{-(-3) \pm \sqrt{(-3)^2 - 4(2)(-7)}}{2(2)}$$

Using a calculator we get

$$x_1 = \frac{-(-3) + \sqrt{(-3)^2 - 4(2)(-7)}}{2(2)} = 2.7656$$

$$x_2 = \frac{-(-3) - \sqrt{(-3)^2 - 4(2)(-7)}}{2(2)} = -1.2656$$

There are two solutions, $x = -1.2656$ and $x = 2.7656$ (Figure 2.8).

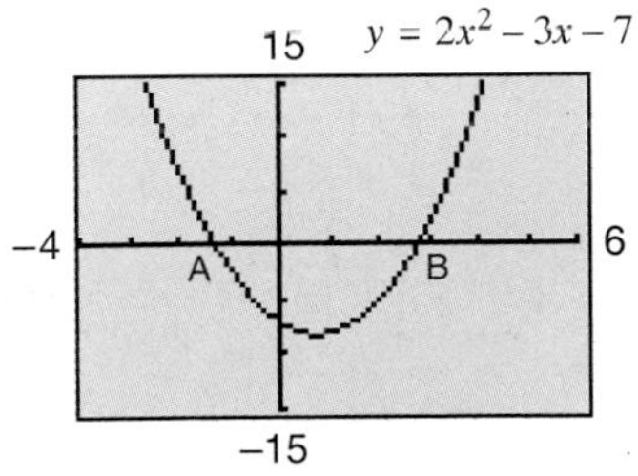

15 $y = 2x^2 - 3x - 7$
-4
6
A
B
-15

Figure 2.8 The two solutions are A and B. A = −1.2656, B = 2.7656.

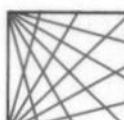

Solutions of Quadratic Equations with a Calculator

The solutions to a quadratic equation $ax^2 + bx + c = 0$ are the x-intercepts of the graph of the parabola $y = ax^2 + bx + c$. We can find the solutions on a calculator using the zero finder technique introduced in Section 1.4 (or trace/zoom). We illustrate the use of a zero finder to confirm the solutions $x = -1.2656$ and $x = 2.7656$ of the quadratic equation $2x^2 - 3x - 7 = 0$ expressed to four decimal places (Figure 2.9).

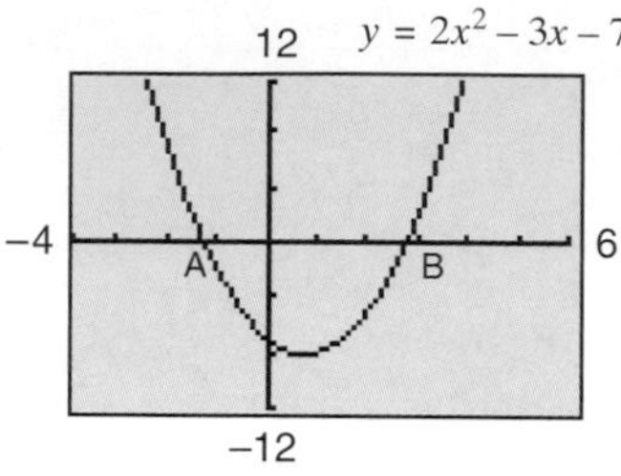

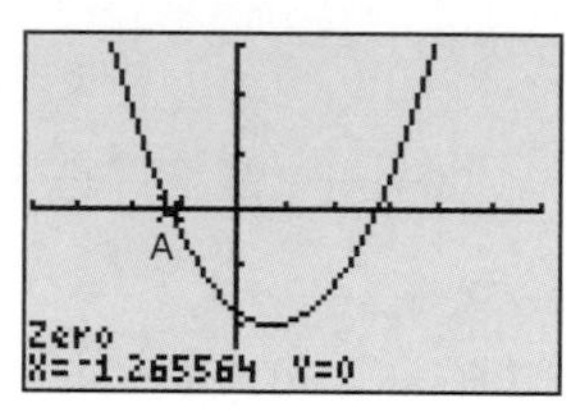

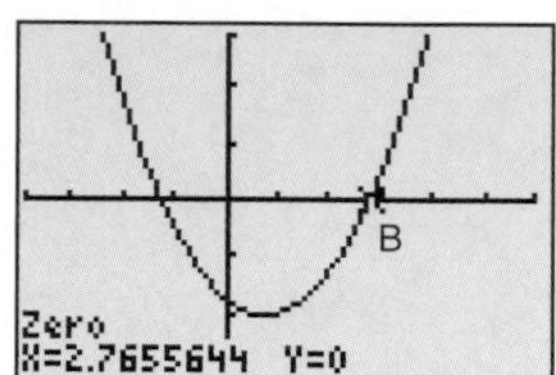

Figure 2.9 Solutions to $2x^2 - 3x - 7 = 0$ using a zero finder.

The quadratic formula gives an algebraic way of solving a quadratic equation. The calculator provides a geometric way using a zero finder or trace/zoom. The algebraic and geometric methods complement each other.

Self-Check 3 Solve $3x^2 - 5x - 7 = 0$ to four decimal places, using the quadratic formula. Confirm the results using a calculator.

The following example illustrates that the quadratic formula can lead to a single solution. A single solution arises when the graph of the equation touches the x-axis.

Example 7 Solving by Using the Quadratic Formula

Solve $x^2 - 6x + 9 = 0$.

SOLUTION In comparison with the standard form, we see that $a = 1$, $b = -6$, and $c = 9$. The quadratic formula gives

$$x = \frac{6 \pm \sqrt{(-6)^2 - 4(1)(9)}}{2(1)}$$

$$= \frac{6 \pm \sqrt{36 - 36}}{2} = \frac{6}{2} = 3$$

The equation has the single solution $x = 3$ (Figure 2.10).

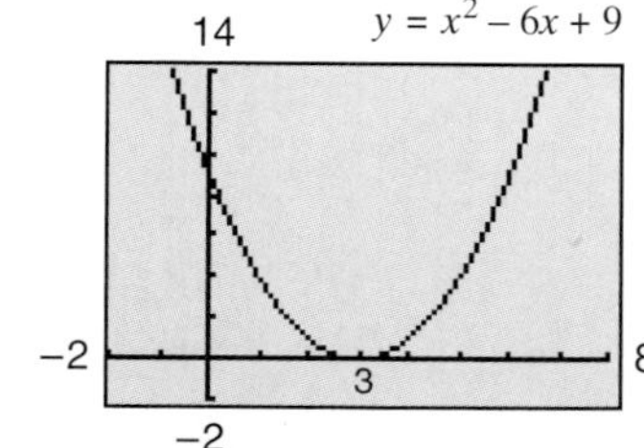

Figure 2.10 The single solution is A.

Next we will see that a quadratic equation may have no real number solutions. This case arises when the graph has no x-intercepts.

Example 8 Solving by Using the Quadratic Formula

Solve $x^2 + 4x + 5 = 0$.

SOLUTION With $a = 1$, $b = 4$, and $c = 5$, the quadratic formula yields

$$x = \frac{-4 \pm \sqrt{4^2 - 4(1)(5)}}{2(1)} = \frac{-4 \pm \sqrt{-4}}{2}$$
$$= -2 \pm \sqrt{-1}$$

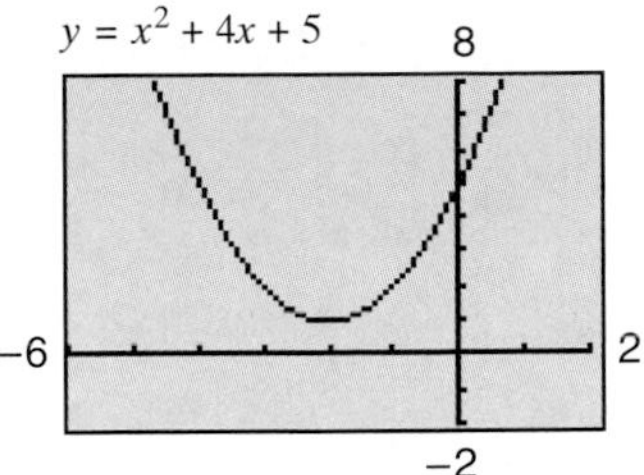

Figure 2.11 No x-intercepts. No real solutions.

The equation has no real number solutions because $-2 + \sqrt{-1}$ and $-2 - \sqrt{-1}$ are not real numbers. They are called **complex numbers**. We shall study such numbers in Chapter 4. (See Figure 2.11.)

Self-Check 4 Use the quadratic formula to show that the equation $2x^2 - 3x + 4 = 0$ has no real solutions. Check this result using a graph.

THE DISCRIMINANT

We have seen examples of quadratic equations that have two, one, and no solutions. Now let us examine the conditions under which these three classes of solutions arise in terms of algebra and geometry.

Consider the quadratic equation $ax^2 + bx + c = 0$. We know that the solutions are given by the quadratic formula

$$x = \frac{-b \pm \sqrt{b^2 - 4ac}}{2a}$$

The quantity under the radical sign, namely $b^2 - 4ac$, is called the **discriminant.** The sign of the discriminant determines the type of solutions.

1. If $b^2 - 4ac > 0$, the equation has two real solutions:

$$x = \frac{-b + \sqrt{b^2 - 4ac}}{2a} \quad \text{and} \quad x = \frac{-b - \sqrt{b^2 - 4ac}}{2a}.$$

2. If $b^2 - 4ac = 0$, the equation has one real solution: $x = -\dfrac{b}{2a}$.

3. If $b^2 - 4ac < 0$, the equation has no real solutions.

The solutions of $ax^2 + bx + c = 0$ correspond to the x-intercepts of the graph of the parabola $y = ax^2 + bx + c$. These three algebraic possibilities arise from the fact that the graph can have two, one, or no x-intercepts (Figure 2.12).

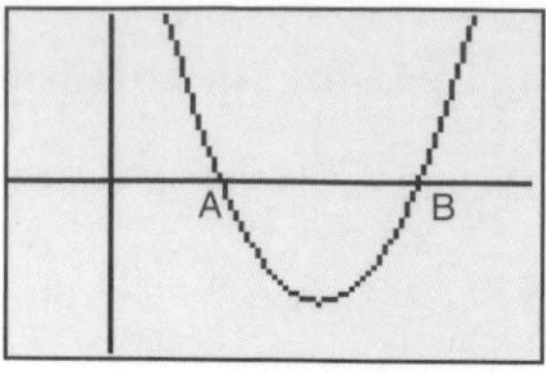

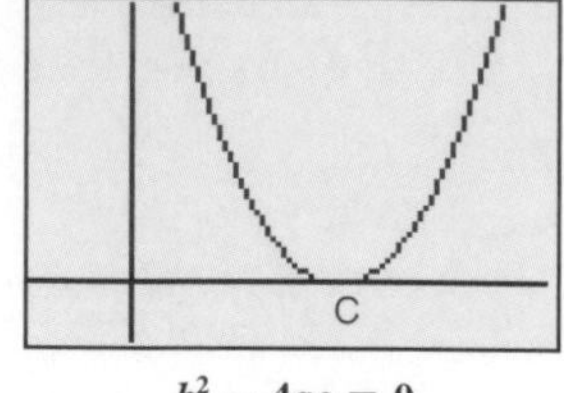

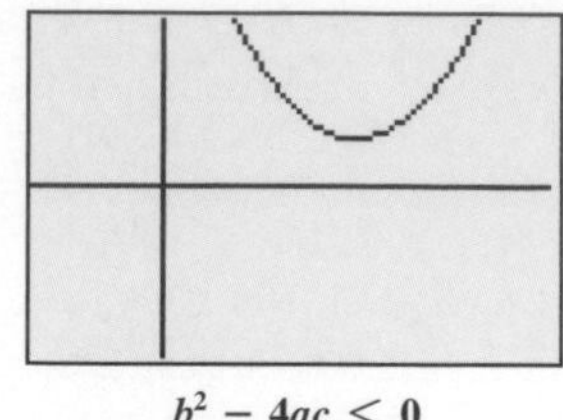

$b^2 - 4ac > 0$
two real solutions: A, B

$b^2 - 4ac = 0$
one real solution: C

$b^2 - 4ac < 0$
no real solutions

Figure 2.12 Information from the discriminant.

Example 9 Using the Discriminant to Check Type of Solutions

Use the discriminant to show that the equation $3x^2 - 4x + 7 = 0$ has no real solutions. Check your answer by graphing on a calculator.

SOLUTION For this equation, $a = 3$, $b = -4$, and $c = 7$.

$$\begin{aligned} b^2 - 4ac &= (-4)^2 - 4(3)(7) \\ &= -68 < 0 \end{aligned}$$

The discriminant is negative; thus the equation has no real solutions (Figure 2.13).

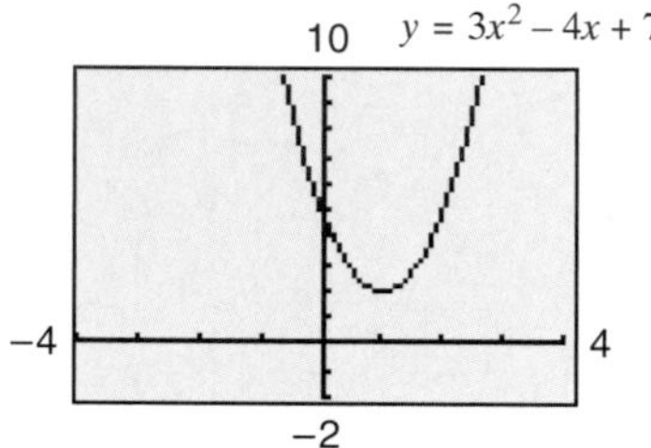

Figure 2.13 No x-intercepts; therefore, no real solutions.

The following example illustrates how the discriminant can be used to construct equations with certain properties.

Example 10 Using the Discriminant to Get a Single Solution

Find the values of k for which the following equation has only one solution. Check your answers using a graph.

$$x^2 + kx + 4 = 0$$

SOLUTION The equation will have one solution if the discriminant is zero. Therefore, equate the discriminant to zero and solve for k. In comparison with the standard form, $a = 1$, $b = k$, and $c = 4$. The discriminant is

$$b^2 - 4ac = k^2 - 4(1)(4) = k^2 - 16$$

The discriminant is zero if

$$\begin{aligned} k^2 - 16 &= 0 \\ k^2 &= 16 \\ k &= \pm 4 \end{aligned}$$

The equation will have a single solution if $k = 4$ or $k = -4$ (Figure 2.14).

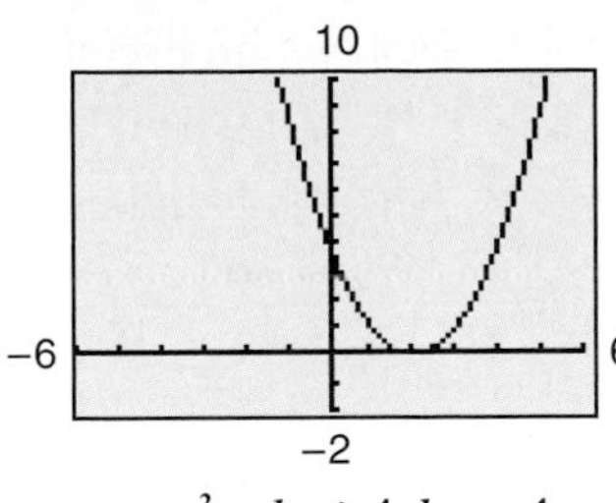

$y = x^2 - kx + 4, k = -4$

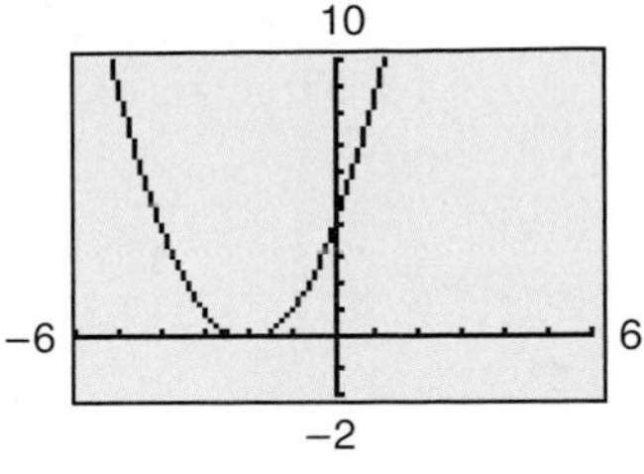

$y = x^2 - kx + 4, k = 4$

Figure 2.14

Observe that the graphs for $k = -4$ and $k = 4$ both intersect the x-axis in a single point, indicating one solution in each case. These solutions are seen to be 2 (corresponding to $k = -4$) and -2 (corresponding to $k = 4$).

Answers to Self-Check Exercises

1. $x = -2, x = \frac{5}{2}$. $-5 \le x \le 5$, $-15 \le y \le 5$ is an appropriate window.
2. $x^2 - 4x + \frac{7}{3} = 0, (x-2)^2 = -\frac{7}{3} + 4, x = 2 \pm \sqrt{\frac{5}{3}}$.
3. $x = -0.9067, 2.5734$.
4. $x = \dfrac{3 \pm \sqrt{-23}}{4}$. The graph has no x-intercept, thus no solutions.

Answers to Further Ideas Exercises

1. **(a)** The graph has a single x-intercept. Window $-5 \le x \le 10$, $-5 \le y \le 15$.
   (b) $x = 4$, repeated root. Window $-5 \le x \le 10$, $-5 \le y \le 15$.
2. $x = \pm\frac{5}{2}$.

EXERCISES 2.1

In Exercises 1–8, solve the quadratic equations by factoring. Use graphs to support your answers.

1. $x^2 - 3x + 2 = 0$ **2.** $x^2 + 4x + 3 = 0$

3. $x^2 - 4x = 0$ **4.** $x^2 - 6x + 9 = 0$

5. $-x^2 - 2x + 8 = 0$ **6.** $x^2 - 1 = 0$

7. $2x^2 + 5x - 3 = 0$ **8.** $-3x^2 + x + 2 = 0$

In Exercises 9–16, solve the equations by factoring. Use graphs to support your answers.

9. $4x^2 - 12x + 9 = 0$ **10.** $6x^2 + 13x - 5 = 0$

11. $7x^2 + 13x - 2 = 0$ **12.** $2x^2 + 7x - 15 = 0$

13. $x^2 - 2x = 2x$ **14.** $9x^2 = 48x - 64$

15. $16z^2 - 8z + 1 = 0$

16. $6u^2 + 10u = 3 - 4u^2 - 3u$

In Exercises 17–26, solve the equations by extracting square roots. Use graphs to support your answers.

17. $x^2 = 9$ **18.** $4x^2 - 1 = 0$ **19.** $3x^2 - 27 = 0$

20. $x^2 = \frac{16}{9}$ **21.** $x^2 - \frac{1}{25} = 0$ **22.** $(2x - 1)^2 = 9$

23. $(5x + 2)^2 = 4$ **24.** $9x^2 + 3 = 28$

25. $(3a + 4)^2 = 25$ **26.** $(2t - 1)^2 - 16 = 0$

In Exercises 27–32, solve the quadratic equations to four decimal places, by extracting square roots. Use graphs to support your answers.

27. $x^2 = 17$ **28.** $x^2 = 14.28$ **29.** $(x - 2)^2 = 8.6$

30. $(5x + 2)^2 = 7$ **31.** $(3x + 4)^2 = 5.6$

32. $(7x - 9)^2 = 12.98$

In Exercises 33–38, solve the quadratic equations by completing the square. Check your answers using substitution.

33. $x^2 + 2x - 24 = 0$ **34.** $x^2 - 6x - 27 = 0$

35. $x^2 - 4x - 12 = 0$ **36.** $x^2 - 2x - 3 = 0$

37. $x^2 + 8x + 7 = 0$ **38.** $x^2 + 6x - 7 = 0$

In Exercises 39–44, solve the quadratic equations by completing the square. You may leave your answers in forms involving radicals. Check your answers using substitution.

39. $2x^2 - 12x + 17 = 0$ **40.** $3x^2 + 12x + 5 = 0$

41. $2x^2 + 4x - 3 = 0$ **42.** $4x^2 - 8x + 1 = 0$

43. $3x^2 + 6x + 2 = 0$ **44.** $5x^2 - 20x + 17 = 0$

In Exercises 45–50, solve the equations using a method of your choice. You may leave your answers in forms involving radicals.

45. $x^2 + 6x - 7 = 0$ **46.** $(x - 3)^2 = 9$

47. $x^2 + 5x - 3 = 0$ **48.** $10x^2 - x - 3 = 0$

49. $(2x - 4)^2 - 11 = 0$ **50.** $2x^2 - 2x - 5 = 0$

In Exercises 51–58, use the quadratic formula to solve (if possible) the equations. Express answers to four decimal places. Use graphs to support your answers.

51. $x^2 + 5x + 1 = 0$ **52.** $x^2 - 3x + 4 = 0$

53. $2x^2 + 4x - 1 = 0$ **54.** $x^2 + 2x + 4 = 0$

55. $x^2 + 4x + 4 = 0$ **56.** $4x^2 - 9x - 9 = 0$

57. $2s^2 - 3s + 4 = 0$ **58.** $t^2 + 4t + 3 = 0$

In Exercises 59–64, use the quadratic formula to solve (if possible) the equations. Express answers to four decimal places. Use graphs to support your answers.

59. $4x^2 - 2x - 1.1 = 0$ **60.** $3x^2 - x - 2 = 0$

61. $-2t^2 + 4t - 7 = 0$ **62.** $p^2 + 2.1p - 1.9 = 0$

63. $3x^2 - 4.8x + 1.92 = 0$

64. $7.1x^2 - 3.2x + 1.8 = 0$

In Exercises 65–68, use the quadratic formula to solve the quadratic equations.

65. $ax^2 - x - 2.5 = 0$, for $a = 1$, 2.1, and 4.3.

66. $5.29x^2 - 14.26x + c = 0$, for $c = 3.21$, 9.61, and 12.73.

67. $2.7x^2 + bx + 3.1 = 0$, for $b = 2.3$, 7.3, and -9.2.

68. $ax^2 + bx - 4.6 = 0$, for $a = 1.2$ and $b = 7.6$, and then for $a = 1.5$ and $b = 6.4$.

In Exercises 69–76, solve the quadratic equations graphically to four decimal places, using either a zero finder or trace/zoom.

69. $1.7x^2 - 3.8x - 4 = 0$

70. $-3.2x^2 + 1.9x + 6.1 = 0$

71. $1.7x^2 - 4.3x - 5.1 = 0$

72. $2.5x^2 + 3.6x - 7.1 = 0$

73. $-5.3x^2 + 7.2x + 2.3 = 0$

74. $-8.3x^2 + 29.7x - 7.3 = 0$

75. $3.6x^2 - 42.8x + 12.7 = 0$

76. $-5.4x^2 - 31.3x + 5.6 = 0$

In Exercises 77–84, use the discriminant to determine whether the equations have two real solutions, a single real solution, or no real solutions. Use graphs to support your answers.

77. $x^2 + x + 3 = 0$ **78.** $x^2 + 3x - 1 = 0$

79. $4x^2 + x + 1 = 0$ **80.** $3x^2 + x - 2 = 0$

81. $2x^2 + 4x + 2 = 0$ **82.** $4.1x^2 - 1.9x + 1.2 = 0$

83. $3x^2 - 4.8x + 1.92 = 0$

84. $3.1x^2 + 0.8x - 2.3 = 0$

In Exercises 85–90, find the value(s) of k for which each equation has a single solution. Use graphs to support your answers.

85. $2x^2 + kx + 8 = 0$ **86.** $4x^2 + kx + 1 = 0$

87. $2x^2 + 2x + k = 0$ **88.** $kx^2 - 10x + 5 = 0$

89. $2x^2 + kx + k = 0$ **90.** $kx^2 + 4x + k = 0$

91. Consider the equation $ax^2 - 5.7x - 6.3 = 0$, where a is a real number. Find values of a that give two real solutions, a single solution, and no real solutions.

Applications of Quadratic Equations Including Motion and Architecture

- Motion under Gravity • Guidelines for Applying Mathematics
- Architecture • Number Puzzle • Distance Between Ships • Canoeing

We have seen how certain problems can be solved using linear equations. Other situations require quadratic equations. In this section, we now gain further experience describing situations in mathematical terms and then using mathematics to arrive at results. Let us consider a familiar application; namely, motion under gravity, which is described by a quadratic equation.

Example 1 **Motion Under Gravity**

An object is propelled vertically from ground level with velocity v. We have already seen that the height s at time t is given by the equation $s = -16t^2 + vt$. If the initial velocity is 256 feet per second, when will the object be at a height of 768 feet ?

SOLUTION Because of initial velocity, $v = 256$, the height at time t is given by

$$s = -16t^2 + 256t$$

We suspect that the object will reach 768 feet on the way up and on the way down (Figure 2.15).

The object will be at height 768 feet when

$$-16t^2 + 256t = 768$$

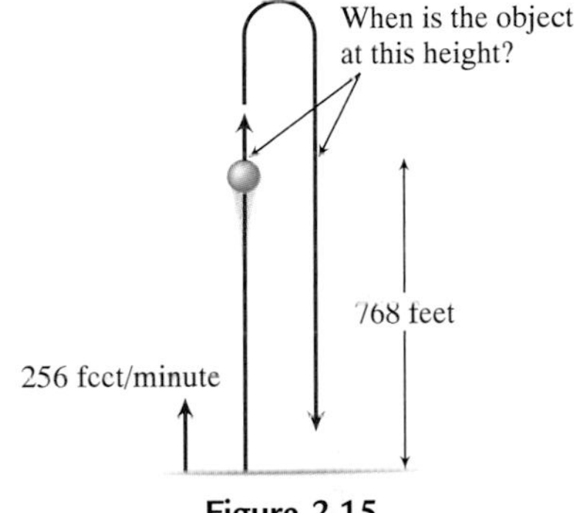

Figure 2.15

Let us solve this quadratic equation and then look at the graphical interpretation of the result.

Write the equation in standard form, simplify, and solve.

$$16t^2 - 256t + 768 = 0$$
$$16(t^2 - 16t + 48) = 0$$
$$16(t - 4)(t - 12) = 0$$
$$t = 4 \quad \text{or} \quad t = 12$$

The object will reach a height of 768 feet after 4 seconds on the way up and after 12 seconds on the way down (Figure 2.16).

Graph $y = -16t^2 + 256t$ and $y = 768$. Then, $-16t^2 + 256t = 768$ where the graphs intersect.

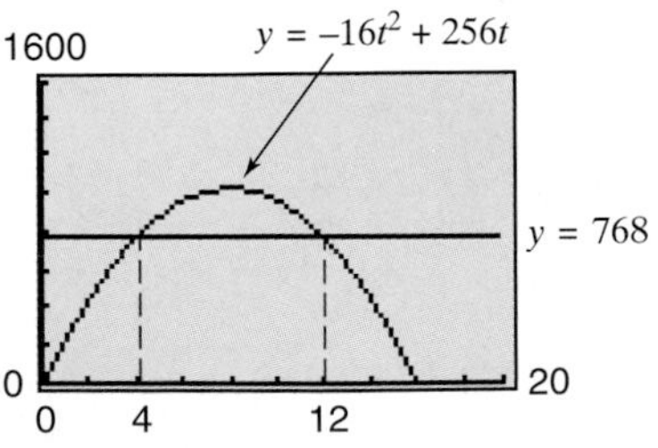

Figure 2.16 Graphs intersect at $t = 4$ and $t = 12$. Object will be at a height of 768 feet after 4 and 12 seconds.

Further Ideas 1 Using a graph, show that the object in Example 1 never reaches a height of 1200 feet.

In this example we started our discussion with the equation that described the physical situation. Very often the initial equation must be constructed. This can sometimes be done from data, as we saw in Sections 1.2 and 1.5. However, applications often are presented initially in words. The equation must be developed from a verbal description of the problem before it can be solved to give required information. Developing the initial mathematical "picture" and then proceeding step by step to arrive at the answer, an approach to problem solving called **mathematical modeling,** is an extremely important art. Although there are no rules that work for all occasions, the following guidelines can be helpful in this process.

Guidelines for Applying Mathematics

(a) Summarize the situation. Pinpoint the unknown—we often use the letter x for it.
(b) Draw a sketch (if possible).
(c) Decide on a strategy, and find an equation in x.
(d) Solve this equation for x.
(e) Interpret the result.

Using these guidelines, we now give examples of ways to translate a given problem into mathematics and determine a solution.

Example 2 Planning a Library Extension

Henry Mason University is planning to extend the library building. The current building is rectangular and measures 325 feet by 210 feet. The planned extension will double this area and is to be carried out uniformly on the north and east sides (architect's plan in Figure 2.17). The architects, Roe and Brown, used the following steps to find the width of the extension.

SOLUTION Summarize the situation. The current library is 325 feet by 210 feet.

Area of current library $= 325 \times 210$
$= 68{,}250$ square feet

Area of extension $= 68{,}250$ square feet

Area of new library $= 2 \times 68{,}250$ square feet
$= 136{,}500$ square feet

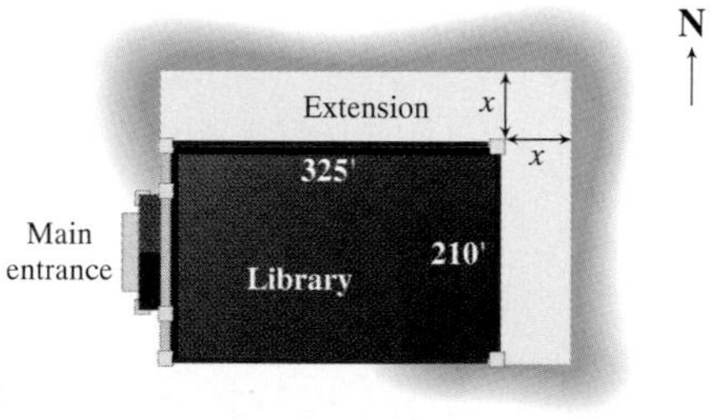

Figure 2.17

We need to find the width of the extension. Let x be the width of the extension (Figure 2.17).

Decide on a strategy for finding x. Use Figure 2.17 to find an expression for the area of the new library in terms of x, and then equate this expression to the known area of the new library. Solving this equation will lead to the required width of the extension.

$$\text{Length of new library} = (325 + x)$$
$$\text{Width of new library} = (210 + x)$$

Thus,

$$\begin{aligned} \text{Area} &= (325 + x)(210 + x) \\ &= 68{,}250 + 325x + 210x + x^2 \\ &= 68{,}250 + 535x + x^2 \end{aligned}$$

Equating this expression to the known area 136,500, gives the equation

$$x^2 + 535x + 68{,}250 = 136{,}500$$
$$x^2 + 535x - 68{,}250 = 0$$

Solving this equation using the quadratic formula gives $x = -213.62$ and $x = 70.87$ (rounded to two decimal places). The negative number has no practical significance in this problem. The width of the extension should be 70.87 feet.

The next example involves a number puzzle. Although such puzzles can in themselves be interesting to solve, they also give excellent practice at developing a mathematical description of a problem. A sketch is not used in this example.

Example 3 Number Puzzle

Find two positive integers that have sum 10 and product 21.

SOLUTION Let one integer be x. The sum of the two integers is 10. Thus, the second integer is $(10 - x)$. Let us use the fact that the product of the two numbers is 21 to determine an equation in x.

$$x(10 - x) = 21$$
$$10x - x^2 = 21$$
$$x^2 - 10x + 21 = 0$$

Solve this equation to get x and thus the two numbers.

$$(x - 3)(x - 7) = 0$$
$$x = 3 \quad \text{or} \quad x = 7$$

The first number is 3 or 7. The second number will be $(10 - x)$. The second number is therefore 7 or 3. The two numbers are 3 and 7.

Example 4 Distance Between Ships

The ship *Cardonia* sails from Miami traveling due north for Boston at 15 knots. An hour later the *Mars* sails east from Miami at 20 knots, with destination Gibraltar. When will the ships be 100 nautical miles apart?

SOLUTION Let the ships be 100 nautical miles apart t hours after the *Cardonia* leaves port. The *Cardonia* will then have been sailing for t hours and the *Mars* will have been sailing for $t - 1$ hours. Use the equation Distance = Speed × Time to arrive at the distance traveled by each ship.

$$\text{Distance traveled by the } \textit{Cardonia} = 15t$$

$$\text{Distance traveled by the } \textit{Mars} = 20(t - 1)$$

We are now able to construct a sketch of the situation (Figure 2.18). It remains to determine the value of t for which the distance BC between the ships is 100.

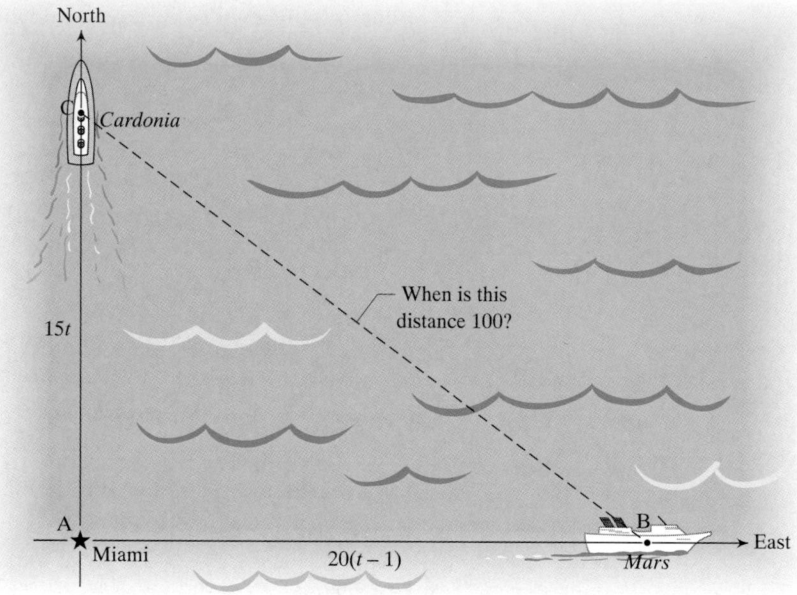

Figure 2.18

Let us use the Pythagorean Theorem for the triangle ABC, with $BC = 100$, to arrive at an equation in t. The Pythagorean Theorem gives

$$(15t)^2 + [20(t - 1)]^2 = 100^2$$

We now proceed to simplify and solve this equation.

$$225t^2 + 400(t^2 - 2t + 1) = 10{,}000$$

$$625t^2 - 800t - 9600 = 0$$

Using the quadratic formula, we find the solutions to be $t = 4.611095567$ and $t = -3.331095567$. The negative number has no significance in this problem. The ships will be 100 nautical miles apart after approximately 4.6 hours (4 hours 36 minutes) after the *Cardonia* leaves Miami.

Example 5 Canoeing

Jenna Mathias and Jeff Hernandez paddled a canoe 6 miles downstream from its source at Alexander Springs in The Ocala National Forest and then rowed back upstream to their starting point. The return journey, upstream, took 1 hour more than the downstream trip. The park brochure describes the river as flowing at a leisurely 1.5 miles per hour. What was the average speed of the students for the round-trip (in still water)? Verify your answer graphically.

SOLUTION Let us summarize the situation.

Distance down river = 6 miles.
It took 1 hour more to paddle upriver.
Speed of current = 1.5 miles per hour.
What was students' average speed in still water ? Let it be x.

Let us find expressions for the time going downstream, and then for the time to come back upstream. The difference will be 1 hour. This will give us an equation in x.

Because the canoe was going with the current downstream and against the current upstream,

$$\text{Speed of canoe downstream} = x + 1.5$$
$$\text{Speed of canoe upstream} = x - 1.5$$

We can now use the Time, Distance, Speed formula to determine the downstream and upstream times.

$$\text{Time} = \frac{\text{Distance}}{\text{Speed}}$$

We get

$$\text{Time going downstream} = \frac{6}{x + 1.5}$$
$$\text{Time going upstream} = \frac{6}{x - 1.5}$$

The return journey upstream took 1 hour more than the downstream trip. Thus,

$$\frac{6}{x - 1.5} - \frac{6}{x + 1.5} = 1 \qquad \text{(a)}$$

To solve this equation, first multiply both sides by $(x - 1.5)(x + 1.5)$ to eliminate denominators. We get

$$6(x + 1.5) - 6(x - 1.5) = (x - 1.5)(x + 1.5)$$
$$6x + 9 - 6x + 9 = x^2 - 2.25$$
$$x^2 = 20.25$$
$$x = 4.5 \quad \text{or} \quad x = -4.5$$

The speed in still water should be positive. Thus, $x = 4.5$. Check $x = 4.5$ in Equation (a). It is satisfied.

Thus, the students averaged 4.5 miles per hour in the canoe (Figure 2.19 for the graphical check).

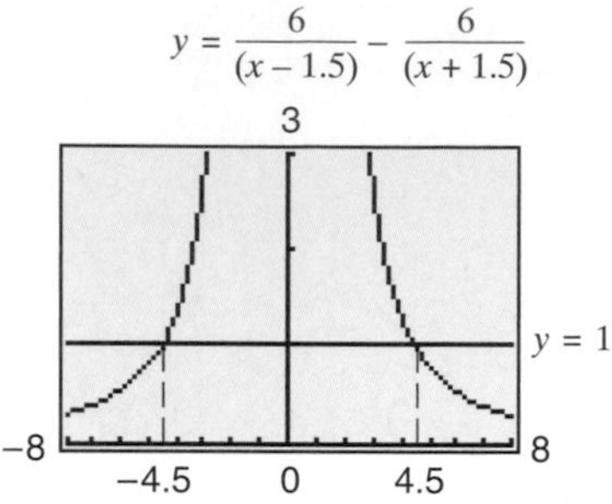

Figure 2.19 $\frac{6}{x - 1.5} - \frac{6}{x + 1.5} = 1$ when $x = -4.5$ and $x = 4.5$. The speed of the canoe is 4.5 miles per hour.

Caution: Multiplying both sides of an equation by an expression involving x, such as multiplying by $(x - 1.5)(x + 1.5)$ in Example 5, can lead to extraneous roots. When this method is used to solve an equation, it is necessary to check the result to see whether it satisfies the original equation.

Further Ideas 2 Suppose that the return journey upstream took 2 hours more than the downstream journey in this canoeing trip. Modify the graphs in an appropriate way, and use the intersect of the graphs to determine the new average speed in still water.

Answers to Further Ideas Exercises

1. Graphs of $y = -16t^2 + 256t$ and $y = 1200$ do not intersect. (See graph below left.) Therefore, $-16t^2 + 256t \neq 1200$ for any value of t. The object never reaches 1200 feet.
2. $\frac{6}{x - 1.5} - \frac{6}{x + 1.5} = 2$ when $x = 3.354102$ (See graph below right.) The average speed in still water would be 3.35 miles per hour.

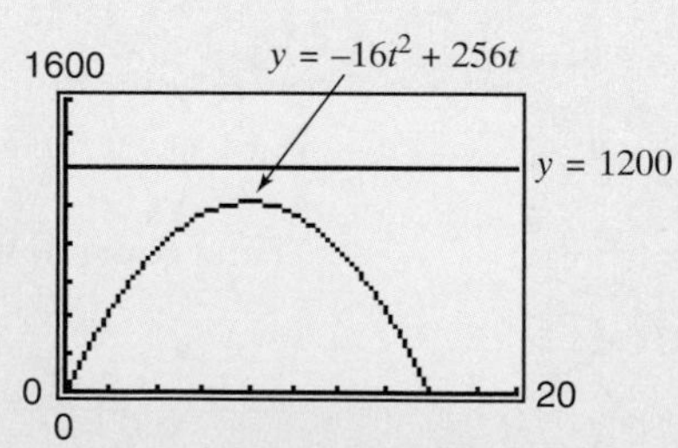

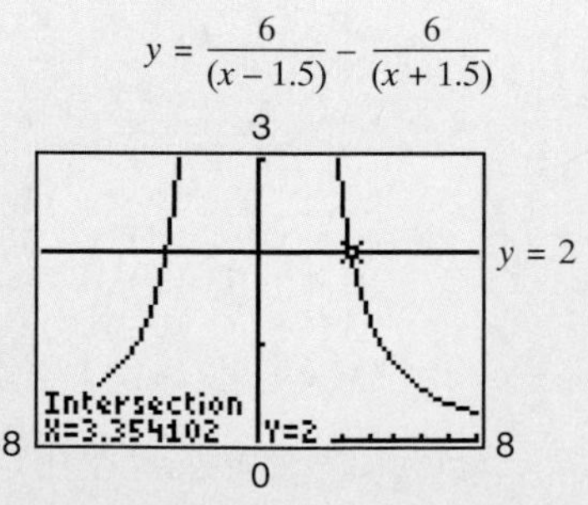

EXERCISES 2.2

MOTION UNDER GRAVITY

In Exercises 1–6, use the equation $s = -16t^2 + vt$, which gives the height s feet at time t seconds for an object fired upward with velocity v feet per second.

1. An object is propelled vertically with a velocity of 320 feet per second. Use algebra to determine when it will be at a height of 1200 feet. Confirm your result using a graph.
2. An object is fired vertically with a velocity of 500 feet per second. When will it be at a height of 1000 feet (to two decimal places)?
3. An object is fired vertically with a velocity of 160 feet per second. When will it be at height 400 feet? *Hint:* The quadratic equation has a repeated root; what does this mean?
4. An object is propelled vertically with a velocity of 192 feet per second. Using both algebra and a graph, show that it never reaches a height of 600 feet.
5. An object is fired vertically from ground level with a velocity of 192 feet per second. For how long is it in the air?
6. An object is fired vertically from ground level with an initial velocity v. Prove that it takes the same time on the way up as it does to come down again to ground level.
7. **University Parking** The administration at Bangor University has decided to extend the parking lot at Raichel Hall to meet increased student parking needs. The dormitory is rectangular, measuring 500 feet by 200 feet, with a 50-foot wide parking lot on three sides (Figure 2.20). The plan is to add half as much again in parking space, uniformly around the present area. What must the width of the added parking space be?

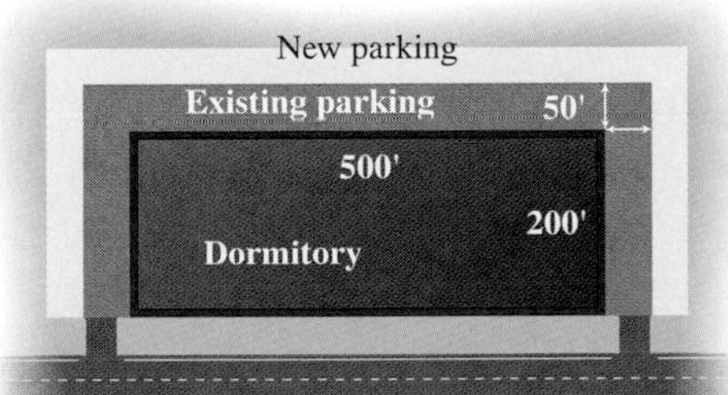

Figure 2.20

8. **Shopping Center Parking** The Tomoka shopping mall is rectangular, measuring 600 feet by 300 feet. It is surrounded by a parking area of uniform width, 75 feet (Figure 2.21). The plan is to double the parking area, adding new parking uniformly around the present area. What should the width of the new area be?

Figure 2.21

9. **Landscaping** The Gonzalez family wants to sod a fringe area around a 20-foot by 12-foot swimming pool (Figure 2.22). A nursery has 228 square feet of sod available. If the area is a uniform width around the pool, how wide will it be with this amount of sod?

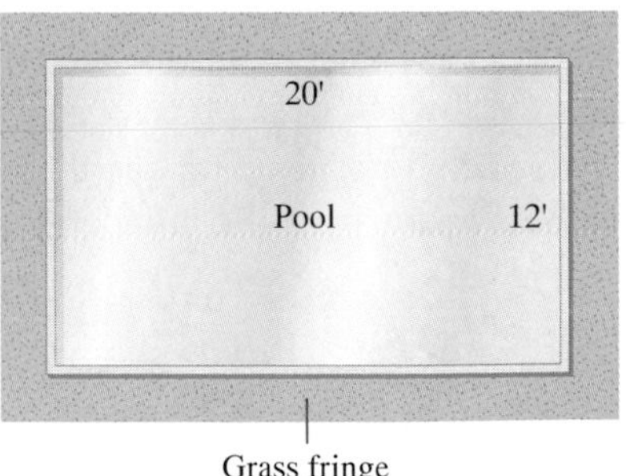

Figure 2.22

10. **Extending a Patio** The Gittens family plans to enlarge a 20-foot by 8-foot concrete patio that borders the back of the house. The plan is to extend the patio a uniform width on all three sides (Figure 2.23). Eighty square feet of concrete is available for the extension. What will the width of the extension be?

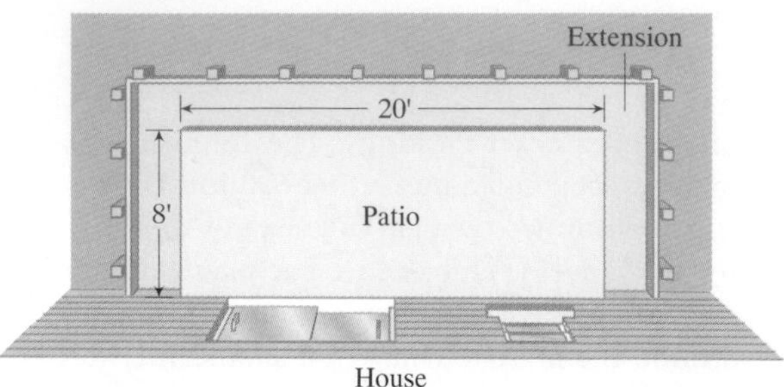

Figure 2.23

11. **Laying a Driveway** Tom Gehrke wants to lay down a concrete driveway. The length of the driveway is to be 10 feet greater than the width. If 264 square feet of concrete is available, what will be the dimensions of the driveway?
12. **Landscaping** Dimitri Carlino wants to sod a rectangular region. The region is 6 feet longer one way than the other. 135 square feet of sod is available. What are the dimensions of the region that can be sodded?

PUZZLES

13. Find two positive integers that have a sum of 20 and a product of 91.
14. Find two consecutive positive integers for which the product is 240.
15. Find two consecutive odd numbers for which the product is 143.
16. The sum of an integer and its reciprocal is $\frac{17}{4}$. Find the integer.
17. The difference between an integer and its reciprocal is $\frac{35}{6}$. Find the integer.
18. If 12 is added to a positive number, the result is the same as that for squaring the number. Find the number.
19. If 5 is added to a number, the result is the same as that obtained by multiplying its reciprocal by 14. What is the number?

JOURNEYS

20. **Distance Between Ships** The ship *Archimedes* leaves San Francisco sailing South at 12 knots. Two hours later the *Lauderdale* leaves San Francisco sailing west at 15 knots. When will the ships be 50 nautical miles apart? (Measure time from the departure of the *Archimedes.*)
21. **Distance Between Planes** Two planes depart from Dallas airport at the same time from different runways, one flies east at 500 miles per hour, the other flies north at 425 miles per hour. How long will it take the planes to be 200 miles apart?

22. **Speed of a Boat** The cruise boat *Sweet Orleans* travels to a point 45 miles upstream on the Mississippi from New Orleans and then returns. The outward journey, against the current takes $1\frac{1}{2}$ hours more than the return journey with the current. If the river flows at 4 miles per hour, what is the average speed of the boat in still water?
23. **Flying Speed** A light Cesna plane travels with constant speed from Phoenix to Juniper, a distance of 1170 miles, and then returns. There was a tailwind of 40 miles per hour going and a headwind of 40 miles per hour on the way back. The difference in times between the outward and return journeys was 1 hour and 45 minutes. Use a graph to determine the speed of the plane in still air.

2.3 Further Types of Equations

• Quadratic-Type Equations • Equations with Radicals • Equations with a Rational Exponent • Solutions to Nonlinear Equations Using Graphs

In Sections R.6 and 2.1, we saw how certain equations can be changed into linear or quadratic equations. Solutions of these equations then led to solutions of the original equations. We now introduce further classes of equations that can be solved by converting the equations into linear or quadratic equations.

QUADRATIC-TYPE EQUATIONS

Quadratic-Type Equations An equation is said to be of **quadratic type** if it can be written in the form

$$au^2 + bu + c = 0, \qquad a \neq 0$$

where u is an expression in some variable.

We now give examples of quadratic-type equations and show how quadratic equations can be used to solve such equations.

Example 1 **Changing to a Quadratic Equation and Solving**

Solve $x^4 - 10x^2 + 9 = 0$, and check the solutions graphically.

SOLUTION We make the substitution $u = x^2$. The equation then becomes a quadratic equation

$$u^2 - 10u + 9 = 0$$

Now let us proceed to factor and solve this quadratic equation in the variable u.

$$(u - 1)(u - 9) = 0$$
$$u - 1 = 0 \quad \text{or} \quad u - 9 = 0$$
$$u = 1 \quad \text{or} \quad u = 9$$

The solutions of $x^4 - 10x^2 + 9 = 0$ are the points on the graph of $y = x^4 - 10x^2 + 9$ where $y = 0$. They are the x-intercepts of the graph.

Because $u = x^2$, we get

$$x^2 = 1 \quad \text{or} \quad x^2 = 9$$
$$x = \pm 1 \quad \text{or} \quad x = \pm 3$$

There are thus four solutions to this equation: $x = -1, 1, -3, 3$ (Figure 2.24).

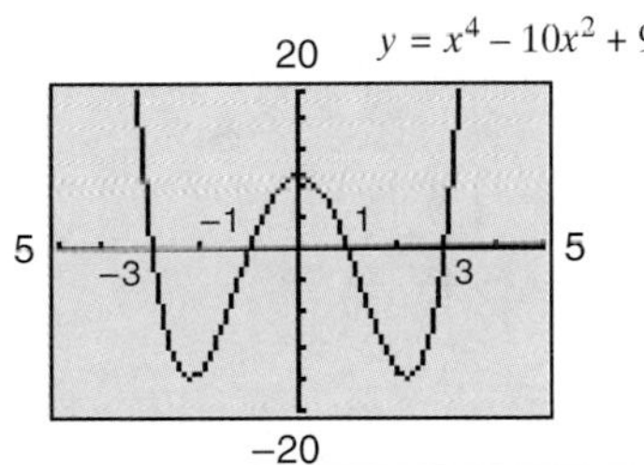

Figure 2.24

Example 2 **Changing to a Quadratic Equation and Solving**

Solve $(x - 1)^2 - (x - 1) - 6 = 0$.

SOLUTION Let $u = (x - 1)$. This substitution gives a quadratic equation in u.

$$u^2 - u - 6 = 0$$
$$(u + 2)(u - 3) = 0$$
$$u = -2 \quad \text{or} \quad u = 3$$

Because $u = x - 1$, $x = u + 1$. Thus, $x = -1$ or $x = 4$. There are two solutions to this equation: $x = -1$ and $x = 4$ (Figure 2.25).

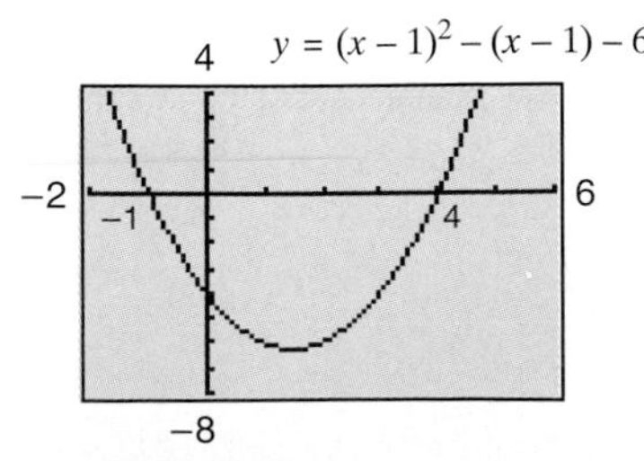

Figure 2.25

Example 3 **Changing to a Quadratic Equation and Solving**

Solve $2x^{-2} - x^{-1} - 1 = 0$.

SOLUTION Let $u = x^{-1}$. Thus, $x^{-2} = u^2$. The equation becomes a quadratic equation when written in terms of u.

$$2u^2 - u - 1 = 0$$
$$(2u + 1)(u - 1) = 0$$
$$u = -\tfrac{1}{2} \quad \text{or} \quad u = 1$$

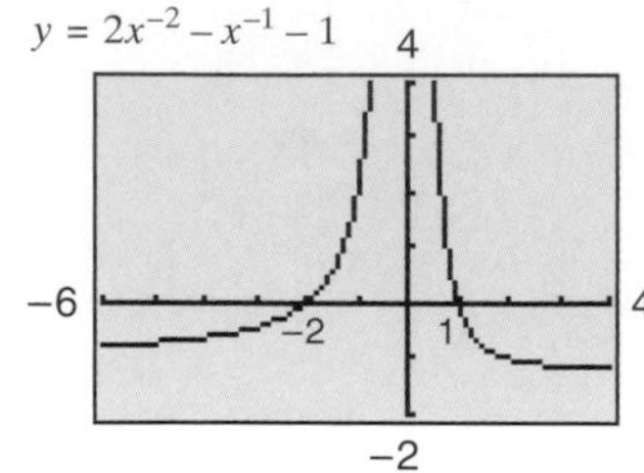

Figure 2.26

Because $u = x^{-1}$, we get

$$x^{-1} = -\tfrac{1}{2} \quad \text{or} \quad x^{-1} = 1$$
$$x = -2 \quad \text{or} \quad x = 1$$

The solutions are $x = -2$ and $x = 1$ (Figure 2.26).

Self-Check 1 Solve the following equations using algebra. Check your answer using a graph. **(a)** $x^4 - 5x^2 + 6 = 0$ **(b)** $x^{-2} - 4x^{-1} + 4 = 0$

Equations with Radicals

An equation with a radical may sometimes be converted into a linear or a quadratic equation by raising both sides to the power of the index of the radical. It is essential to check all answers when this method is used to solve an equation because extraneous roots may be introduced in the process.

Example 4 **Solving an Equation with a Radical**

Solve $\sqrt{2x + 1} = 3$.

SOLUTION Eliminate the radical by squaring both sides of the equation and solve.

$$2x + 1 = 9$$
$$2x = 8$$
$$x = 4$$

Graph $y = \sqrt{2x + 1}$ and $y = 3$. Then $\sqrt{2x + 1} = 3$ where the graphs intersect.

Check in the original equation to see that $x = 4$ satisfies it.

$$\sqrt{2x + 1} = \sqrt{2(4) + 1} = 3$$

The equation is satisfied. There is a unique solution $x = 4$ (Figure 2.27).

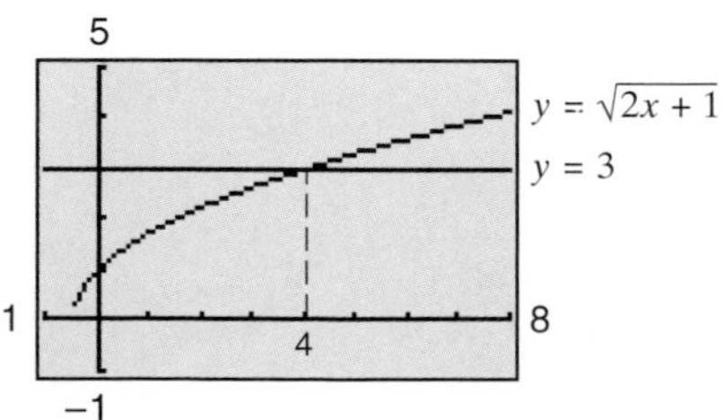

Figure 2.27 Graphs intersect when $x = 4$.

Example 5 Solving an Equation with a Radical

Solve $x = \sqrt{4 - 3x}$.

SOLUTION Eliminate the radical by squaring both sides of the equation. We get a quadratic equation:

$$x^2 = 4 - 3x$$

Rewrite this quadratic equation in standard form and solve.

$$\begin{aligned} x^2 + 3x - 4 &= 0 \\ (x - 1)(x + 4) &= 0 \\ x = 1 \quad \text{or} \quad x &= -4 \end{aligned}$$

There are two possible solutions, namely $x = 1$ and $x = -4$.

It is now necessary to check both of these values in the original equation. Substituting $x = 1$ into the right side of the equation gives

$$\begin{aligned} \sqrt{4 - 3x} &= \sqrt{4 - 3(1)} \\ &= 1 \\ &= x \end{aligned}$$

$x = 1$ satisfies the equation; it is a solution.

Substituting $x = -4$ in the right side gives

$$\begin{aligned} \sqrt{4 - 3x} &= \sqrt{4 - 3(-4)} \\ &= \sqrt{4 + 12} \\ &= \sqrt{16} \\ &= 4 \\ &\neq x \end{aligned}$$

$x = -4$ does not satisfy the equation; it is not a solution. The equation has the single solution $x = 1$ (Figure 2.28).

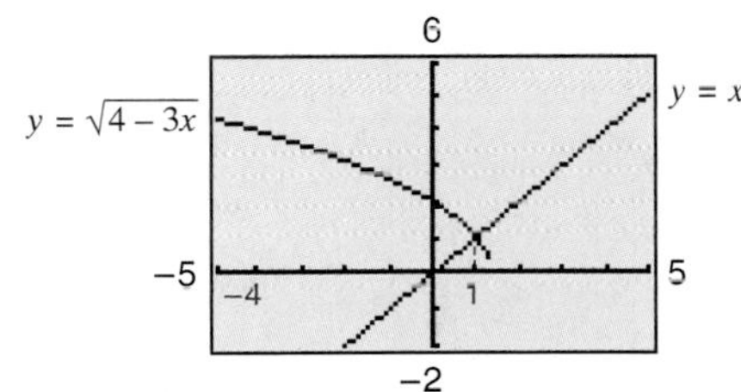

Figure 2.28 The graphs of $y = x$ and $y = \sqrt{4 - 3x}$ intersect at $x = 1$ but not at $x = -4$. $x = 1$ is a solution but $x = -4$ is not.

Further Ideas 1

(a) Using a graph, verify that $x = 1$ is the only solution to $x = \sqrt{4 - 3x}$.

(b) Solve $x = \sqrt{2x + 3}$. Use a graph to support your answer.

Example 6 Solving an Equation with a Radical

Solve the equation $\sqrt[3]{x^2 + 2x} + 1 = 0$.

SOLUTION When possible, as here, isolate the radical on one side of the equation. Write the equation as

$$\sqrt[3]{x^2 + 2x} = -1$$

Cube both sides of the equation to eliminate the radical and solve.

$$x^2 + 2x = -1$$
$$x^2 + 2x + 1 = 0$$
$$(x + 1)(x + 1) = 0$$
$$x = -1$$

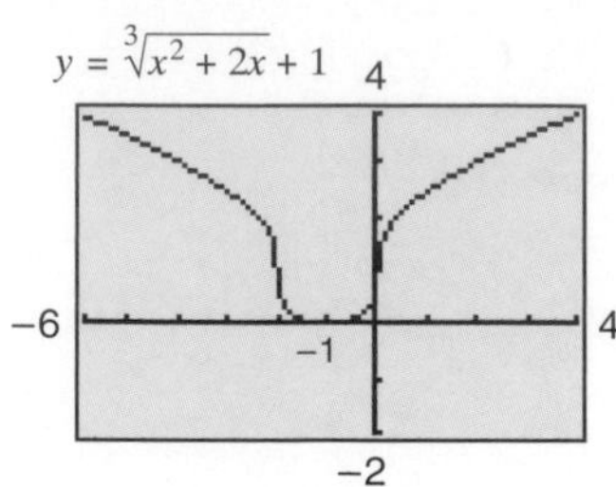

Figure 2.29 $\sqrt[3]{x^2 + 2x} + 1 = 0$ when $x = -1$.

Check in the original equation.

$$\sqrt[3]{x^2 + 2x} + 1 = \sqrt[3]{(-1)^2 + 2(-1)} + 1$$
$$= \sqrt[3]{-1} + 1 = 0$$

The equation is satisfied when $x = -1$ (Figure 2.29).

The next example illustrates that it may be necessary to repeat the process of raising both sides of an equation to a power to solve the equation.

Example 7 **Solving an Equation with Two Radicals**

Solve $\sqrt{x + 2} - \sqrt{2x + 5} + 1 = 0$.

SOLUTION When an equation involves two radicals, rewrite the equation with a radical on each side and then square both sides as follows:

$$\sqrt{x + 2} = \sqrt{2x + 5} - 1$$

Square both sides to eliminate the left radical.

$$(\sqrt{x + 2})^2 = (\sqrt{2x + 5})^2 + 2(-1)(\sqrt{2x + 5}) + (-1)^2$$
$$x + 2 = 2x + 5 - 2\sqrt{2x + 5} + 1$$

Isolate the radical.

$$x + 2 - 2x - 5 - 1 = 2x + 5 - 2\sqrt{2x + 5} + 1$$
$$-x - 4 = -2\sqrt{2x + 5}$$

Square both sides again to eliminate the radical, and solve.

$$x^2 + 8x + 16 = 4(2x + 5)$$
$$x^2 + 8x + 16 = 8x + 20$$
$$x^2 - 4 = 0$$
$$(x + 2)(x - 2) = 0$$
$$x = -2 \quad \text{or} \quad x = 2$$

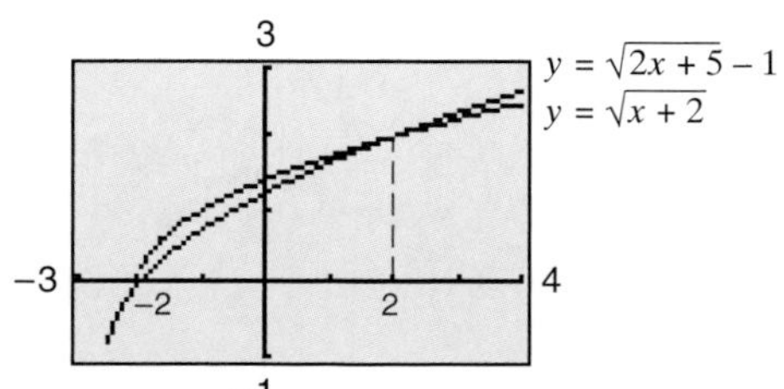

Figure 2.30 $\sqrt{x + 2} = \sqrt{2x + 5} - 1$ when $x = -2, 2$.

These values are found to satisfy the original equation. There are two solutions $x = -2$ and $x = 2$ (Figure 2.30).

Self-Check 2 Solve $\sqrt{2x-1} - \sqrt{x-4} = 2$. Check your answer using a graph.

EQUATIONS WITH A RATIONAL EXPONENT

The following examples illustrate ways of solving certain equations with rational exponents.

Example 8 Solving an Equation with a Rational Exponent

Solve $x^{3/5} = 8$.

SOLUTION To solve this equation, it is necessary to eliminate the exponent $\frac{3}{5}$. This can be accomplished by first raising both sides of the equation to the power 5 and then taking the cube root of each side of the resulting equation.

Raise to power 5.

$$(x^{3/5})^5 = 8^5$$
$$x^3 = 32{,}768$$

Take the cube root.

$$x = 32{,}768^{1/3}$$
$$x = 32$$

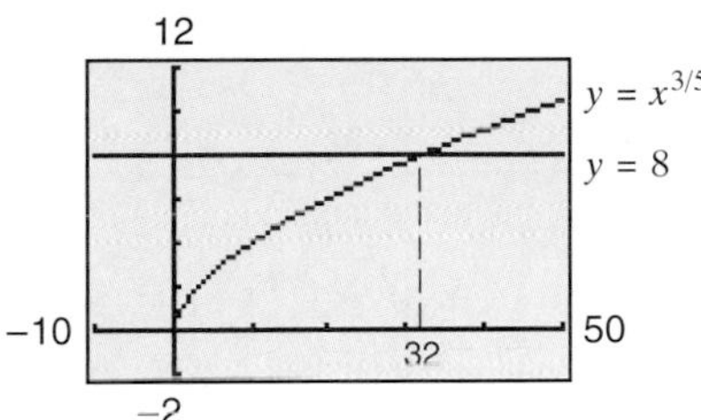

Figure 2.31 $x^{\frac{3}{5}} = 8$ when $x = 32$.

Check in the original equation.

$$x^{3/5} = 32^{3/5} = (32^{1/5})^3 = 2^3 = 8$$

The equation is satisfied. The solution is $x = 32$ (Figure 2.31).

Example 9 Solving an Equation with a Rational Exponent

Solve $(x-1)^{2/3} = 25$.

SOLUTION To solve this equation, it is necessary to eliminate the exponent $\frac{2}{3}$. This can be done by first cubing both sides of the equation and then taking the square root of each side of the resulting equation. Remember that there are two roots, one positive and one negative, when taking the square root.

Raise to power 3.

$$[(x-1)^{2/3}]^3 = 25^3$$
$$(x-1)^2 = 15{,}625$$

Take the square root.

$$x - 1 = \pm\sqrt{15{,}625}$$
$$x = 1 \pm \sqrt{15{,}625}$$
$$x = 1 \pm 125$$
$$x = -124 \quad \text{or} \quad 126$$

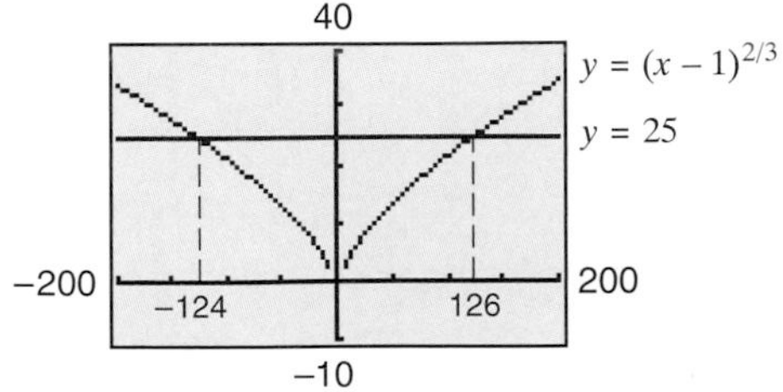

Figure 2.32 $(x-1)^{\frac{2}{3}} = 25$ when $x = -124$ or 126.

Both values are found to satisfy the original equation. See Figure 2.32 for the graphical check.

CAUTION: Taking an odd-number root of both sides of an equation, such as the cube root in Example 8, will result in a single solution. Taking an even-number root, such as the square root in Example 9, will produce a positive and a negative number, resulting in two solutions.

Self-Check 3 Solve $(x - 3)^{3/5} = 8$. Use a graph to support your answer.

SOLUTIONS TO NONLINEAR EQUATIONS USING GRAPHS

We have discussed algebraic methods for finding exact solutions to certain types of equations and the use of graphs to confirm and visualize the results. Many equations, however, cannot be solved using algebraic techniques. The graph then gives us a way of solving the equation. Computers or calculators are used to find solutions to equations by finding the x-intercepts of their graphs.

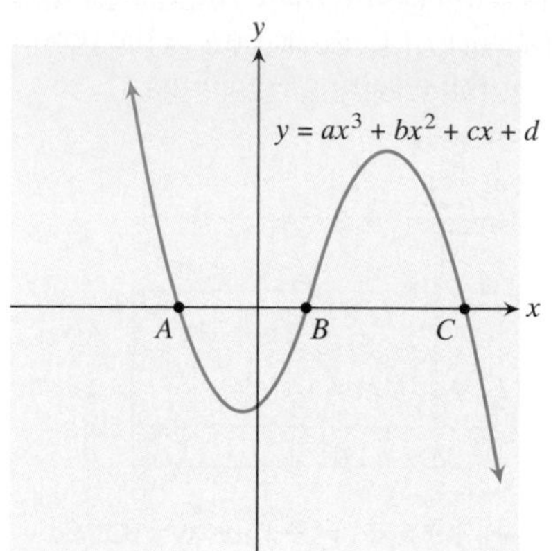

Figure 2.33 A, B, and C are the solutions to $ax^3 + bx^2 + cx + d = 0$.

Consider a nonlinear equation, such as $ax^3 + bx^2 + cx + d = 0$. Suppose the graph of $y = ax^3 + bx^2 + cx + d$ is as shown in Figure 2.33. We see that $y = 0$ at A, B, and C. These x-intercepts of the graph are the solutions* to $ax^3 + bx^2 + cx + d = 0$. These intercepts are found using a zero finder or trace/zoom. The solutions obtained in this manner are usually approximate solutions.

To arrive at all the solutions of an equation using this graphical method, it is helpful to know in advance how many solutions to expect. An equation of the form $a_nx^n + a_{n-1}x^{n-1} + \cdots + a_1x + a_0 = 0$, $a_n \neq 0$, is called a polynomial equation of degree n. The following result gives useful information about the number of solutions of polynomial equations.

NUMBER OF SOLUTIONS

A polynomial equation of degree n has at most n solutions.

We now use this result to arrive at all the solutions to a polynomial equation of degree 3.

Example 10 Solving an Equation Using a Graph

Use a graph to solve $x^3 - x^2 - 6x + 4 = 0$. Express the solutions to four decimal places,

SOLUTION Because this equation is a polynomial equation of degree 3, it has at most three solutions. Graph $y = x^3 - x^2 - 6x + 4$ in a suitable window (Figure 2.34). We see that the equation has three solutions, corresponding to the points A, B, and C. The solutions are $x = -2.3234$, 0.6421, and 2.6813, rounded to four decimal places.

*Whenever we refer to solutions in this discussion, we mean real number solutions. An equation may have solutions that are complex numbers (Section 2.1). We shall discuss these in Chapter 4.

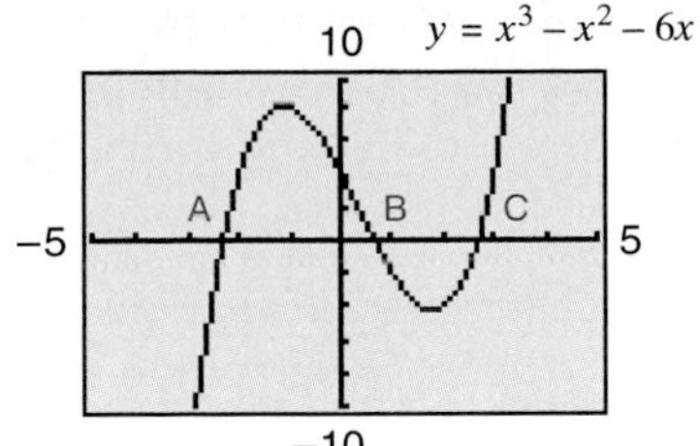

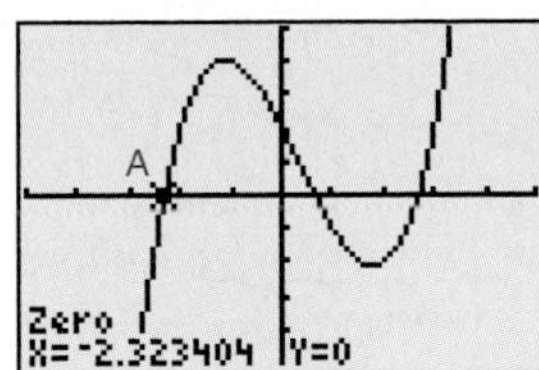

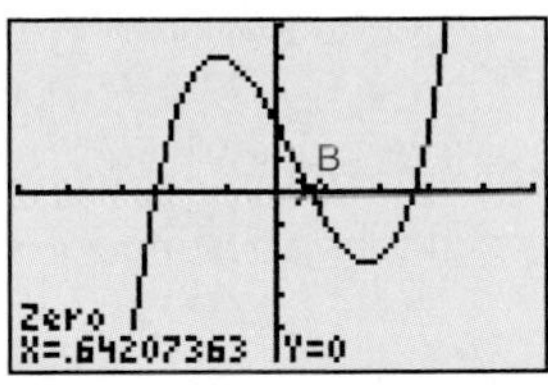

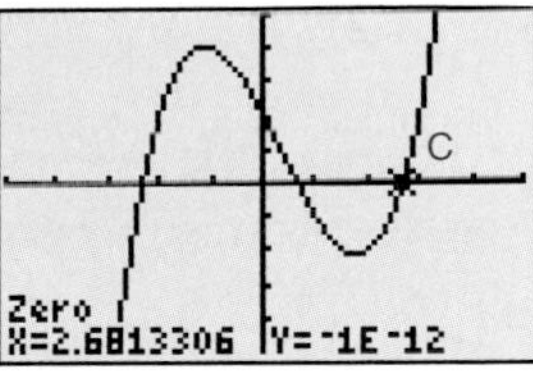

Figure 2.34 The solutions of $x^3 - x^2 - 6x + 4 = 0$ are A, B, and C.

CAUTION: A polynomial equation of degree n has at most n solutions. This result gives us a useful upper bound on the number of solutions. There could, however, be fewer than n solutions. The following Self-Check exercise illustrates a polynomial equation of degree 3 that has only one solution.

Self-Check 4 Solve the equation $x^3 - 5x^2 = -2x + 4$, to four decimal places. *Hint:* First write the equation in the standard form $x^3 - 5x^2 + 2x - 4 = 0$.

Exploration

Controlling the Solutions to a Cubic

Use a graphing calculator to find values of k for which the polynomial equation $x^3 - x^2 - 8x + k = 0$ has (a) three solutions, (b) two solutions, and (c) one solution. Explain the method that you used. *Hint:* 12 is an interesting value of k.

Answers to Self-Check Exercises

1. **(a)** $x = \pm\sqrt{2}$ and $x = \pm\sqrt{3}$ **(b)** $x = \frac{1}{2}$.
2. Two solutions, $x = 5, 13$.
3. Single solution, $x = 35$.
4. Single solution, $x = 4.7563$.

Answers to Further Ideas Exercise

1. **(a)** The graph has a single x-intercept. **(b)** $x = 3$.

EXERCISES 2.3

In Exercises 1–8, solve the equations (if solutions exist) by using an appropriate substitution. Use graphs to support your answers.

1. $x^4 + 2x^2 - 3 = 0$
2. $x^4 - x^2 - 2 = 0$
3. $x^4 - 5x^2 + 4 = 0$
4. $x^4 - 1 = 0$
5. $x^4 - 6x^2 + 8 = 0$
6. $2x^4 + 5x^2 + 3 = 0$
7. $6x^4 + x^2 - 2 = 0$
8. $3x^4 + 10x^2 + 3 = 0$

In Exercises 9–14, solve the equations by using an appropriate substitution. Use graphs to support your answers.

9. $x^4 - 2x^2 - 15 = 0$
10. $x^4 + 5x^2 - 14 = 0$
11. $(x - 1)^2 + 2(x - 1) - 3 = 0$
12. $(x + 2)^2 + 3(x + 2) - 10 = 0$

13. $(p - 3)^2 - 5(p - 3) + 4 = 0$

14. $(v + 1)^2 - 4(v + 1) - 32 = 0$

In Exercises 15–20, solve the equations by using an appropriate substitution. Use graphs to support your answers.

15. $x^6 - 9x^3 + 8 = 0$ **16.** $x^6 - 28x^3 + 27 = 0$

17. $x^{-2} - 4x^{-1} + 3 = 0$ **18.** $x^{-2} - 2x^{-1} - 8 = 0$

19. $x^{-2} - 3x^{-1} - 10 = 0$ **20.** $x^{-2} - 7x^{-1} + 12 = 0$

In Exercises 21–28, solve the equations (if solutions exist) by using an appropriate substitution. Use graphs to support your answers.

21. $8(x - 1)^2 - 6(x - 1) + 1 = 0$

22. $x + 2\sqrt{x} - 3 = 0$ **23.** $x - \sqrt{x} - 6 = 0$

24. $3x + 14\sqrt{x} + 8 = 0$ **25.** $x^{1/2} + 3x^{1/4} - 4 = 0$

26. $x^{1/2} - 3x^{1/4} - 10 = 0$

27. $\dfrac{1}{(x + 1)^2} + \dfrac{3}{(x + 1)} + 2 = 0$

28. $\dfrac{1}{(x - 2)^2} + \dfrac{2}{(x - 2)} - 3 = 0$

In Exercises 29–38, solve the equations. Use graphs to support your answers.

29. $\sqrt{2x - 1} = 3$ **30.** $\sqrt{3x + 1} = 4$

31. $\sqrt{7x + 11} = 9$ **32.** $x = \sqrt{x + 2}$

33. $x = \sqrt{x + 6}$ **34.** $x = \sqrt{3x}$

35. $\sqrt{2t + 8} = t$ **36.** $r = -\sqrt{r + 6}$

37. $\sqrt[3]{x + 1} = 2$ **38.** $\sqrt[3]{2x - 3} - 1 = 0$

In Exercises 39–44, solve the equations. Use graphs to support your answers.

39. $\sqrt{2x - 3} = \sqrt{x + 3}$ **40.** $\sqrt{3x - 2} = \sqrt{x + 4}$

41. $2\sqrt{x + 1} = \sqrt{2x + 3} + 1$

42. $\sqrt{x + 5} - \sqrt{x - 1} = 2$

43. $\sqrt{2x + 1} + \sqrt{x - 3} = 4$

44. $5x - \sqrt{2 - 23x} = 0$

In Exercises 45–54, solve the equations (if solutions exist). Use graphs to support your answers.

45. $x^{3/5} = 27$ **46.** $x^{2/5} = 16$

47. $(x - 1)^{1/3} = -2$ **48.** $(3x + 1)^{3/4} = 8$

49. $(x + 2)^{2/3} = 4$ **50.** $(x - 3)^{1/4} = -1$

51. $(x + 2)^{-2/5} = \frac{1}{4}$ **52.** $(4x^2 + 1)^{1/2} = 2x + 3$

53. $(x + 5)^{-5/3} = 32$ **54.** $(x - 3)^{2/3} = 9$

In Exercises 55–64, solve the equations (if solutions exist) using the methods of this section. Use graphs to support your answers.

55. $2x^4 + 3x^2 - 5 = 0$ **56.** $x^4 - 3x^2 - 18 = 0$

57. $x^{-2} - 8x^{-1} + 12 = 0$ **58.** $\sqrt{7 - 6x} - x = 0$

59. $\sqrt{5x + 6} - 1 = \sqrt{x + 1}$

60. $(2x - 3)^2 - 5(2x - 3) + 4 = 0$

61. $x^{2/3} - 4 = 0$ **62.** $(x - 4)^{3/5} = 64$

63. $x^6 - 4x^3 + 3 = 0$ **64.** $x^{-4} + 6x^{-2} + 8 = 0$

In Exercises 65–70, solve the quadratic equations (if solutions exist) to two decimal places using a zero finder or trace/zoom.

65. $x^2 + 8x - 2 = 0$

66. $-2x^2 - 10x + 1 = 0$

67. $-3x^2 + 7x + 15 = 0$

68. $7.2x^2 + 37.8x + 17.4 = 0$

69. $5.76x^2 + 17.28x - 51.84 = 0$

70. $8x^2 - 8x + 23 = 0$

In Exercises 71–78, solve the following equations (if solutions exist), to four decimal places, using a zero finder or trace/zoom.

71. $10x^3 + 17x^2 - 35x - 12 = 0$

72. $-4x^3 + 16x^2 - 9x - 5 = 0$

73. $5x^3 - 2x^2 - 17x + 7 = 0$

74. $2x^3 - 5x^2 - 28x + 15 = 0$

75. $x^4 + x^3 - 14x^2 - x + 12 = 0$

76. $6x^4 + x^3 + 21x^2 + 6x - 5 = 0$

77. $-x^3 + 2x^2 + 3x = 5$

78. $x^4 + 5x = 13x - x^3 - 7$

79. Use the discriminant to determine the values of k for which the equation $y = x^2 - 10x + k$ has **(a)** two real solutions, **(b)** one real solution, and **(c)** no real solutions. Demonstrate your answers using graphs on a calculator.

80. Investigate graphs of $y = x^3 - 6x^2 + 9x + k$ for various values of k on a calculator to find values of k for which the polynomial equation $x^3 - 6x^2 + 9x + k = 0$ has **(a)** one real solution, **(b)** two real solutions, and (**c**) three real solutions. (**d**) Can the equation ever have no solutions? *Hint:* Start with $k = 2$.

81. Explore the graph of $y = x^4 - 9x^2 + 3x + k$ for various values of k on a calculator. Find values of k for which the polynomial equation $x^4 - 8x^2 + 3x + k = 0$ has (**a**) four real solutions, (**b**) three real solutions, and (**c**) two real solutions. (**d**) Can the equation ever have one real solution or no real solution?

2.4 Inequalities, Temperature Range of a Computer, and Spread of a Disease

• Inequalities • Properties of Inequalities • Double Inequalities • Quadratic Inequalities • Rational Inequalities • Solving Inequalities Graphically • Operating Temperature Range of a Computer • Spread of a Disease

Inequalities

So far in this chapter, we have discussed equations. Recall that an equation is a statement that two algebraic expressions are equal. We now extend our discussion to inequalities. An **inequality** is a statement that one expression is greater than, greater than or equal to, less than, or less than or equal to another statement. The following are examples of inequalities and the symbols used:

greater than ($>$)	$2x > 8$
greater than or equal to ($\geq$)	$4x - 2 \geq 3$
less than ($<$)	$5x + 2 < 9$
less than or equal to ($\leq$)	$3x - 7 \leq x + 2$

The properties of inequalities follow. These properties follow from the definitions of inequalities introduced in Section R.1.

Properties of Inequalities

Let a, b, c, and d be real numbers.

1. Transitive property: If $a < b$ and $b < d$, then $a < d$.
2. Addition of inequalities: If $a < b$ and $c < d$, then $a + c < b + d$.
3. Addition of a constant: If $a < b$, then $a + c < b + c$.
4. Multiplication by a nonzero constant:
   (**a**) If $a < b$ and $c > 0$, then $ac < bc$.
   (**b**) If $a < b$ and $c < 0$, then $ac > bc$. (Note that the inequality sign is reversed when c is negative.)

Analogous properties hold for $>$, $\leq$, and $\geq$.

Many of the terms and concepts used with equations apply to inequalities. The **solutions** of an inequality are the values of the variable that satisfy the inequality. Two inequalities that have the same solutions are **equivalent.** These properties of inequalities are valid for algebraic expressions because the expressions that we use represent real numbers. We can solve certain inequalities by gradually isolating the variable, as with equations, using these properties. Once again, care must be taken when multiplying both sides of an inequality by a negative number. The sign of the inequality is then reversed. The following examples illustrate the method.

Example 1 Solving an Inequality

Let us solve the inequality

$$-2x + 1 < 7$$

Gradually isolate x on the left side of the inequality, using the properties of inequalities.

State the original inequality.	$-2x + 1 < 7$
Add -1 to both sides.	$-2x < 6$
Divide both sides by -2.	$x > -3$

(Dividing by -2 is the same as multiplying by $-\frac{1}{2}$, a negative number; therefore, the sign of the inequality is reversed.)

The inequality has many solutions; namely, all x values greater than -3. We can also express these solutions using interval notation as $(-3, \infty)$.

We can check this result graphically. First, graph the lines $y = -2x + 1$ and $y = 7$ on a calculator (Figure 2.35). The inequality $-2x + 1 < 7$ occurs when the graph of $y = -2x + 1$ lies below the line $y = 7$. This is when $x > -3$.

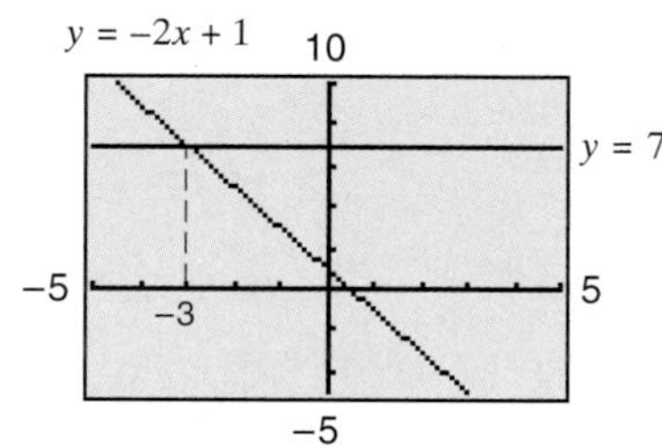

Figure 2.35 Graph of $y = -2x + 1$ lies below $y = 7$ when $x > -3$. Thus, $-2x + 1 < 7$ when $x > -3$.

Example 2 Solving an Inequality

Solve $5x - 16 \geq 2x - 4$.

SOLUTION Let us collect all the x terms on the left of the inequality sign and the constant terms on the right.

State the original inequality.	$5x - 16 \geq 2x - 4$
Subtract $2x$ from both sides.	$3x - 16 \geq -4$
Add 16 to both sides.	$3x \geq 12$
Divide both sides by 3.	$x \geq 4$

The set of solutions is $[4, \infty)$.

Now let us check this result graphically. Graph $y = 5x - 16$ and $y = 2x - 4$ (Figure 2.36). The inequality $5x - 16 \geq 2x - 4$ occurs when the first graph lies on or above the second graph. We see that this is over the interval $[4, \infty)$.

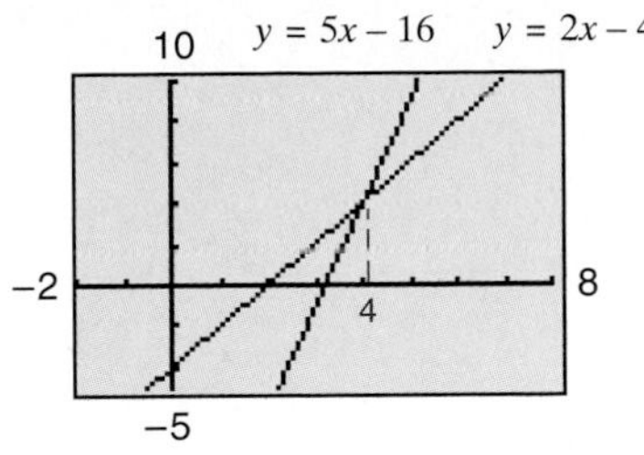

Figure 2.36 Graph of $y = 5x - 16$ lies on or above $y = 2x - 4$ when $x \geq 4$. Thus, $5x - 16 \geq 2x - 4$ when $x \geq 4$.

Self-Check 1 Solve $3x - 1 > -2x + 9$. Check your answer using a graph.

Double Inequalities

A statement may involve a pair of inequalities. For example, consider

$$-5 < 2x + 3 < 7$$

This is called a **double inequality.** It means

$$-5 < 2x + 3 \quad \text{and} \quad 2x + 3 < 7$$

A double inequality often may be solved by operating on both inequalities at the same time, as illustrated in Example 3.

Example 3 Solving a Double Inequality

Solve $-5 < 2x + 3 < 7$.

SOLUTION Gradually isolate the variable x as follows:

State the original inequality.	$-5 < 2x + 3 < 7$
Subtract 3 throughout.	$-8 < 2x < 4$
Divide by 2 throughout.	$-4 < x < 2$

The solution is made up of all the real numbers that are greater than -4 and less than 2. The set of solutions is $(-4, 2)$.

Let us look at the graphic interpretation. Graph $y = 2x + 3$, $y = -5$, and $y = 7$ (Figure 2.37). The inequality $-5 < 2x + 3 < 7$ will occur when the graph of $y = 2x + 3$ lies between the lines $y = -5$ and $y = 7$. We see that $-5 < 2x + 3 < 7$ over the interval $(-4, 2)$.

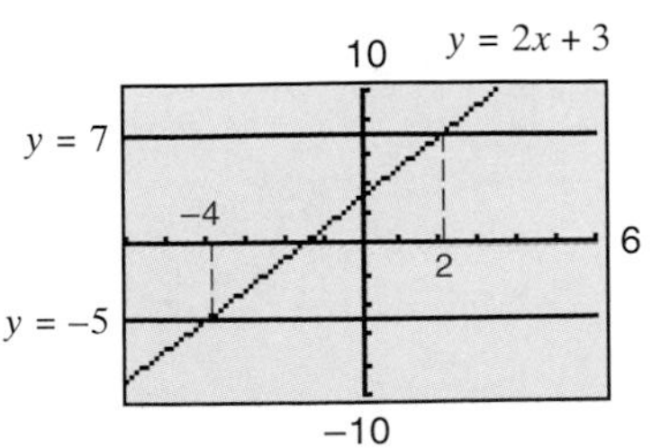

Figure 2.37 $-5 < 2x + 3 < 7$ when $-4 < x < 2$.

Further Ideas 1

(a) Solve $-3 \leq 4x + 1 \leq 9$. Check your answer using a graph.
(b) Use the display in Figure 2.38 to estimate the values of x for which $-3 < 2x + 1 < 13$. Confirm your answer using algebra.

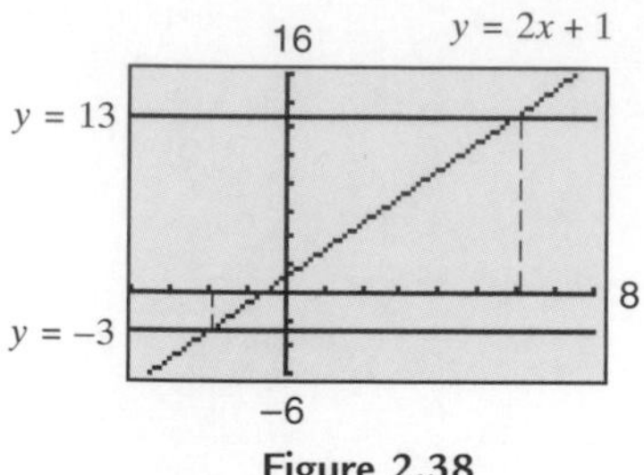

Figure 2.38

QUADRATIC INEQUALITIES

Inequalities may involve polynomials. For example, consider $x^2 - 2x - 3 < 0$. Such an inequality is called a **quadratic inequality.** We now give a method that can be used to solve a quadratic inequality in which the polynomial can be factored.

Example 4 Solving a Quadratic Inequality

Solve $x^2 - 2x - 3 < 0$.

SOLUTION Factor the polynomial and write the inequality in factored form:

$$(x + 1)(x - 3) < 0$$

To solve this inequality, we first determine where $(x + 1)(x - 3)$ is zero.

$$(x + 1)(x - 3) = 0 \qquad \text{when } x = -1 \quad \text{or} \quad x = 3$$

These values of x are called **critical values.** They divide the real line into three intervals, namely $x < -1$, $-1 < x < 3$, and $x > 3$ (Figure 2.39).

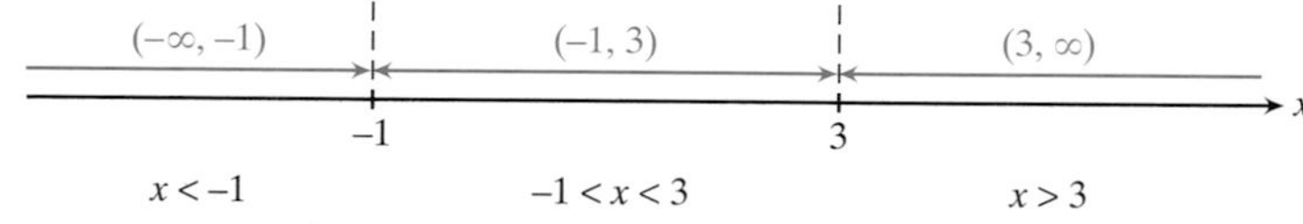

Figure 2.39

We now summarize where the factors $(x + 1)$ and $(x - 3)$ are negative, where they are zero, and where they are positive over this partition of the line. This information will tell us where the product $(x + 1)(x - 3)$ is negative, zero and positive.

$(x + 1)$: $x + 1 < 0$ when $x < -1$, $x + 1 = 0$ when $x = 1$, $x + 1 > 0$ when $x > -1$.
$(x - 3)$: $x - 3 < 0$ when $x < 3$, $x - 3 = 0$ when $x = 3$, $x - 3 > 0$ when $x > 3$.

Write this information as the first two rows of the following **sign diagram.** Multiply the signs together over the intervals to get the signs (and zeros) of the product $(x + 1)(x - 3)$ over these intervals (Figure 2.40).

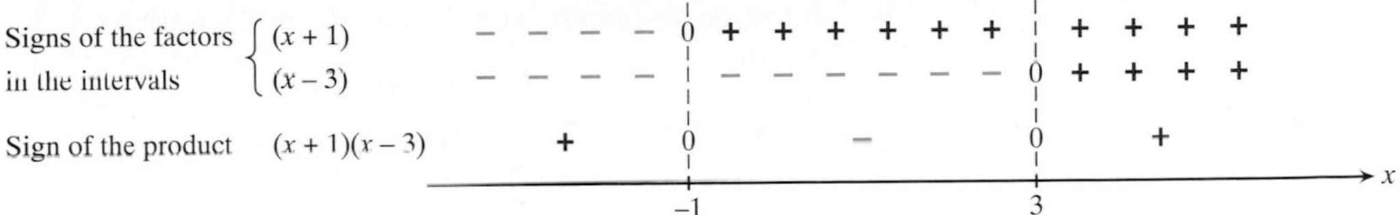

Figure 2.40

The last row of this diagram shows that

$$\begin{aligned}(x+1)(x-3) &> 0 \quad \text{over } (-\infty, -1) \text{ and } (3, \infty)\\(x+1)(x-3) &= 0 \quad \text{at } x=-1 \text{ and } x=3\\(x+1)(x-3) &< 0 \quad \text{over } (-1, 3)\end{aligned}$$

The set of solutions is thus $(-1, 3)$.

We found that the solutions to a quadratic equation could be viewed geometrically. Let us look at the solutions of the quadratic inequality $x^2 - 2x - 3 < 0$ from the geometrical point of view.

Graph $y = x^2 - 2x - 3$ (Figure 2.41). The inequality $x^2 - 2x - 3 < 0$ will occur when the graph lies below the x-axis. The graph is below the x-axis over the interval $(-1, 3)$. The set of solutions is thus $(-1, 3)$.

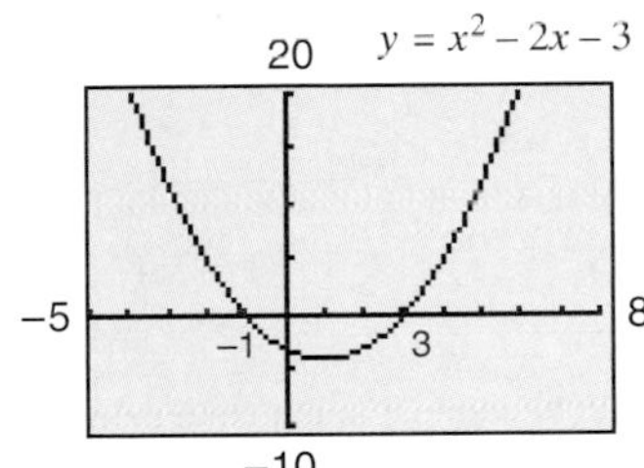

Figure 2.41 $x^2 - 2x - 3 < 0$ when $-1 < x < 3$.

Let us look at a few more examples to see how this method is used to solve quadratic inequalities.

Example 5 Solving a Quadratic Inequality

Solve $x^2 + x - 2 \geq 0$.

SOLUTION Factor the polynomial and write the inequality in factored form.

$$(x+2)(x-1) \geq 0$$

First determine the critical values.

$$(x+2)(x-1) = 0 \quad \text{when } x = -2 \quad \text{or} \quad x = 1$$

These points divide the real line into the three intervals $(-\infty, -2)$, $(-2, 1)$, and $(1, \infty)$ (Figure 2.42).

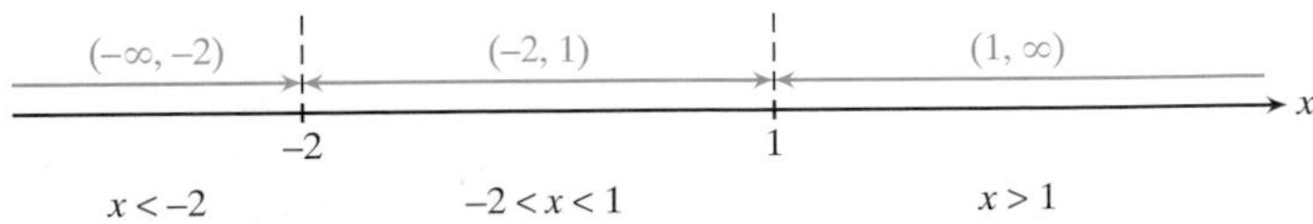

Figure 2.42

Summarize where the factors $(x + 2)$ and $(x - 1)$ are negative, zero, and positive.

$(x + 2)$: $x + 2 < 0$ when $x < -2$, $x + 2 = 0$ when $x = -2$, $x + 2 > 0$ when $x > -2$.
$(x - 1)$: $x - 1 < 0$ when $x < 1$, $x - 1 = 0$ when $x = 1$, $x - 1 > 0$ when $x > 1$.

Write this information as the first two rows of a sign diagram. Multiply the signs together over the intervals to get the signs (and zeros) of the product $(x + 2)(x - 1)$ over these intervals (Figure 2.43).

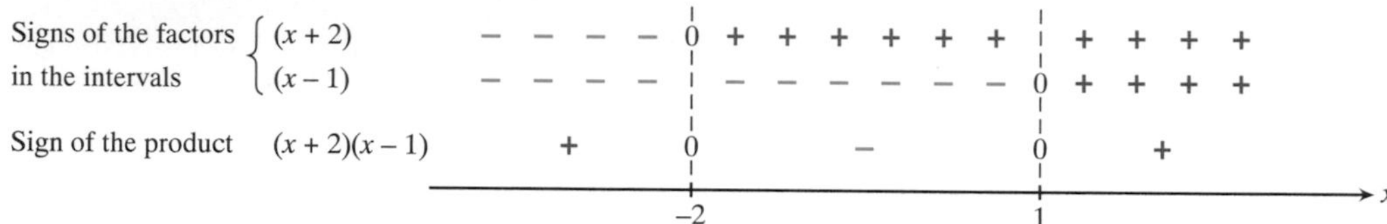

Figure 2.43

We see that $(x + 2)(x - 1) \geq 0$ for values of x in the interval $(-\infty, -2]$ and for those in the interval $[1, \infty)$. The set of solutions is said to be the **union** of these two intervals. We denote union by the symbol $\cup$. We write the set of solutions as $(-\infty, -2] \cup [1, \infty)$ (Figure 2.44).

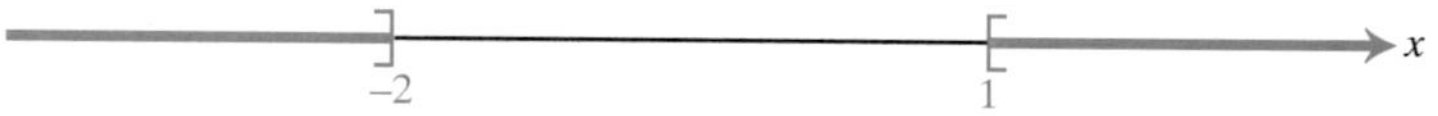

Figure 2.44 $(-\infty, -2]\cup[1, \infty)$

Let us check this result using geometry. Graph $y = x^2 + x - 2$ (Figure 2.45). The inequality $x^2 + x - 2 \geq 0$ will be satisfied when the graph lies on or above the x-axis. We see that the graph is on or above the x-axis over the interval $(-\infty, -2]$ and also over the interval $[1, \infty)$. The set of solutions is thus $(-\infty, -2] \cup [1, \infty)$.

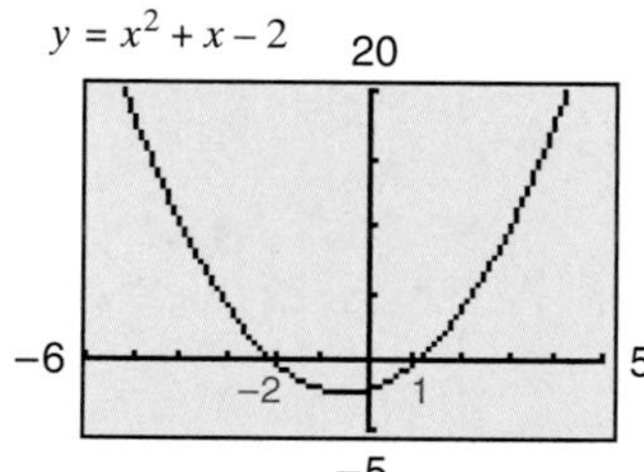

Figure 2.45 $x^2 + x - 2 \geq 0$ when $x \leq -2$ and $x \geq 1$.

Further Ideas 2

(a) Solve $x^2 - x - 6 > 0$. Support your answer using a graph.

(b) Use the display in Figure 2.46 to estimate the values of x for which $-x^2 + 8x - 12 \geq 0$. Confirm your answer using algebra.

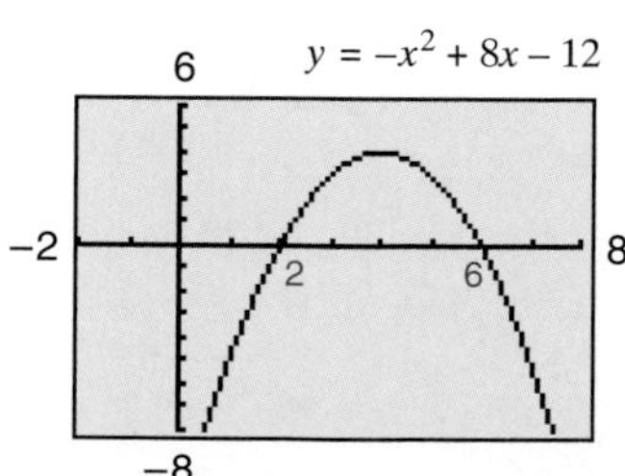

Figure 2.46

RATIONAL INEQUALITIES

The terms in an inequality may be rational expressions. These inequalities are solved by combining all the nonzero terms into one rational expression and using the critical values of this expression. The method is illustrated in Example 6.

Example 6 Solving a Rational Inequality

Solve the rational inequality $\frac{2x}{x-5} < -3$.

SOLUTION Simplify into one rational expression step by step as follows.

State the original inequality. $$\frac{2x}{x-5} < -3$$

Add 3 to both sides. $$\frac{2x}{x-5} + 3 < 0$$

Place over a common denominator. $$\frac{2x + 3(x-5)}{x-5} < 0$$

Simplify the numerator in stages. $$\frac{5x-15}{x-5} < 0, \frac{5(x-3)}{x-5} < 0$$

This inequality will be true when

$$\frac{x-3}{x-5} < 0$$

Thus the original inequality is equivalent to

$$\frac{x-3}{x-5} < 0$$

We now solve this inequality. The **critical values** of a rational expression such as $(x-3)/(x-5)$ are the values of x that make the numerator or denominator zero. The critical values are thus $x = 3$ and $x = 5$. These points divide the real line into the intervals $(-\infty, 3)$, $(3, 5)$, and $(5, \infty)$, as seen in Figure 2.47.

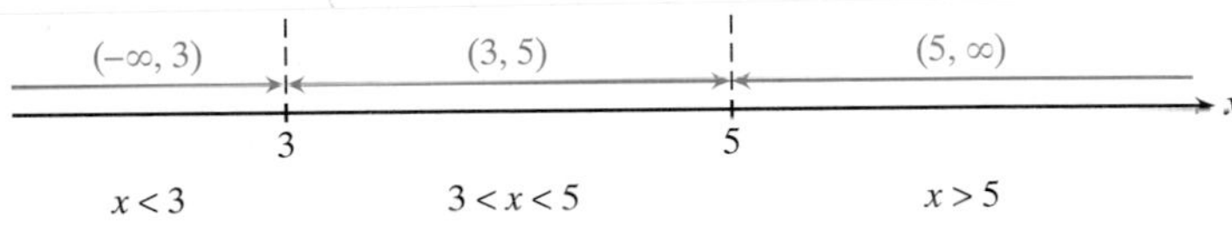

Figure 2.47

The signs and zeros of the rational expression are given by the signs and zeros of the numerator and the signs of the denominator over these intervals. Summarize these.

$(x-3)$: $x-3 < 0$ when $x < 3$, $x-3 = 0$ when $x = 3$, $x-3 > 0$ when $x > 3$.

$(x-5)$: $x-5 < 0$ when $x < 5$, $x-5 > 0$ when $x > 5$.

Represent this information in a sign diagram (Figure 2.48) and determine the signs of the quotient $(x-3)/(x-5)$.

		$x<3$	3	$3<x<5$	5	$x>5$
Signs of the factors in the intervals	$(x-3)$	− − − −	0	+ + + + + +		+ + + +
	$(x-5)$	− − − −		− − − − − −	0	+ + + +
Sign of the quotient	$\frac{(x-3)}{(x-5)}$	+	0	−		+

Figure 2.48

It can be seen that $(x-3)/(x-5) < 0$ for values of x in the interval (3, 5). The set of solutions is thus (3, 5).

We check this result using geometry. Graph $y = 2x/(x-5)$ (Figure 2.49). Observe that the graph is below the line $y = -3$ over the interval (3, 5). Thus, $2x/(x-5) < -3$ over (3, 5). The set of solutions is (3, 5).

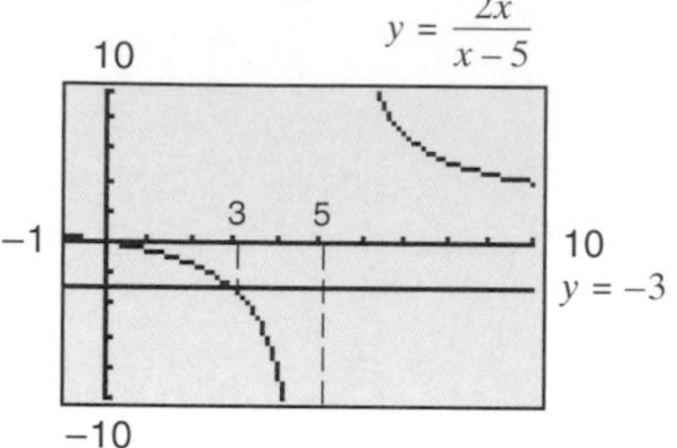

Figure 2.49 $\frac{2x}{x-5} < -3$ when $3 < x < 5$.

Self-Check 2 Solve $\frac{6x+2}{x-3} \geq 1$. Support your answer using a graph.

Solving Inequalities Graphically

Inequalities are solved graphically in very much the same way as equations. The solutions of an inequality are the values of x that satisfy the inequality. Let us solve the inequality $-x^3 + 2x^2 + 3x - 4 > 0$, to two decimal places.

We graph the equation $y = -x^3 + 3x^2 + 2x - 4$ (Figure 2.50). The solutions of the inequality will be the values of x over which $y > 0$. These will be the values of x for which the graph lies above the x-axis.

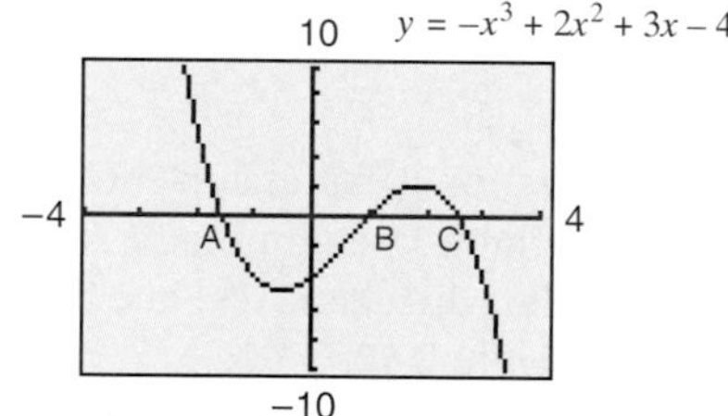

Figure 2.50 $-x^3 + 2x^2 + 3x - 4 > 0$, when $x <$ A and B $< x <$ C

The points A, B, and C are found on a calculator (using trace/zoom or a zero finder) to be

$$A = -1.56, \quad B = 1, \quad C = 2.56 \quad \text{(rounded to two decimal places)}$$

The graph is above the x-axis on the intervals $(-\infty, -1.56)$ and $(1, 2.56)$. Thus, the set of solutions to the inequality is $(-\infty, -1.56) \cup (1, 2.56)$

Further Ideas 3 Solve the inequality $x^3 - 5x^2 > 3x - 12$ using a graph. Express the solution to two decimal places. *Hint:* It is easiest to solve such an inequality by rewriting it in the form $x^3 - 5x^2 - 3x + 12 > 0$.

Applications . . .

Example 7 Operating Temperature Range of a Computer

The relationship between degrees Fahrenheit and degrees Celsius is described by the equation $F = \frac{9}{5}C + 32$. A supercomputer can operate between 41°F and 95°F. What is its Celsius range of operation?

SOLUTION We can describe the range of operation of the computer in degrees Fahrenheit by the inequality

$$41 < F < 95$$

Substitute $F = \frac{9}{5}C + 32$ into this inequality to get the restrictions on C.

$$41 < \frac{9}{5}C + 32 < 95$$

Solve this inequality.

$$41 - 32 < \frac{9}{5}C < 95 - 32$$

$$9 < \frac{9}{5}C < 63$$

$$45 < 9C < 315$$

$$5 < C < 35$$

The computer will thus operate between 5°C and 35°C.

We check this result using a graph. Graph $y = \frac{9}{5}x + 32$ (Figure 2.51). The inequality $41 < \frac{9}{5}x + 32 < 95$ will be satisfied when the graph of $y = \frac{9}{5}x + 32$ lies between the lines $y = 41$ and $y = 95$. We see that this holds over the interval (5, 35).

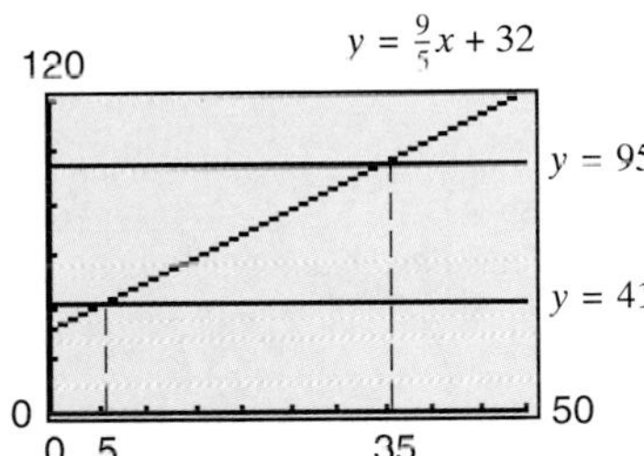

Figure 2.51 $41 < \frac{9}{5}x + 32 < 95$ when $5 < x < 95$.

Example 8 Spread of a Disease

Medical authorities have estimated that a communicable disease is spreading through a country in a manner described by the equation $N = -t^2 + 12t$, where t is the time in months from the identification of the disease and N is the number (in thousands) of new cases at time t. How long will it take the number of new cases to finally decrease below 20,000? Support your answer using a graph.

SOLUTION Let us find the interval of time for which $N < 20$ (thousand). This will be given by

$$-t^2 + 12t < 20$$

Solve this inequality.

$$-t^2 + 12t - 20 < 0$$

$$(-t + 2)(t - 10) < 0$$

Find the critical values.

$$(-t + 2)(t - 10) = 0$$

The critical values are $t = 2$ and $t = 10$. These points define the intervals $t < 2$, $2 < t < 10$, and $t > 10$. We now determine where the factors $(-t + 2)$ and $(t - 10)$ are negative, zero, and positive.

$(-t + 2)$: $-t + 2 < 0$ when $t > 2$, $-t + 2 = 0$ when $t = 2$,
$-t + 2 > 0$ when $t < 2$.
$(t - 10)$: $t - 10 < 0$ when $t < 10$, $t - 10 = 0$ when $t = 10$,
$t - 10 > 0$ when $t > 10$.

Represent this information in a sign diagram (Figure 2.52), and compute the signs of the product $(-t + 2)(t - 10)$.

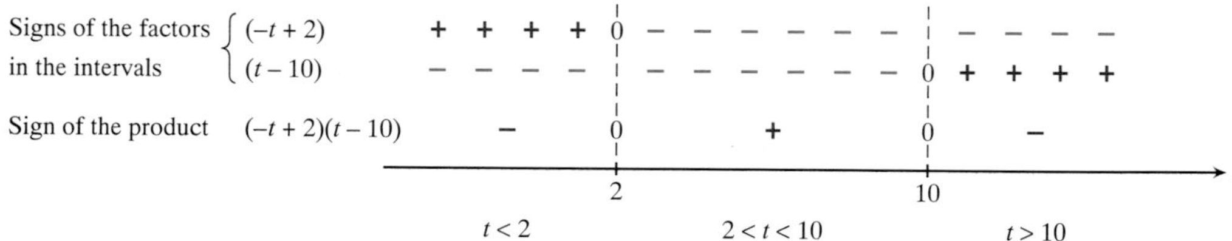

Figure 2.52

Thus, $(-t + 2)(t - 10) < 0$ for $t < 2$ and $t > 10$. The values of t that satisfy the inequality $-t^2 + 12t < 20$ are those in the intervals $t < 2$ and $t > 10$.

Physically, this result means that the number of new cases of the disease, described by $-t^2 + 12t$, will be less than 20,000 when time is less than 2 months or more than 10 months. Ten months after the initial outbreaks, the number of new cases of the disease will finally decrease to less than 20,000.

We check this result using a graph. Graph $N = -t^2 + 12t$ (Figure 2.53). $N < 20$ when the graph lies below the line $t = 20$. Observe that $-t^2 + 12t < 20$ over the intervals $[0, 2)$ and $(10, 12]$. Ten months after the initial outbreak, the number of new cases of the disease will decrease to less than 20,000. Note that $-t^2 + 12t = 0$ when $t = 12$. There will be no new cases of the disease after 12 months.

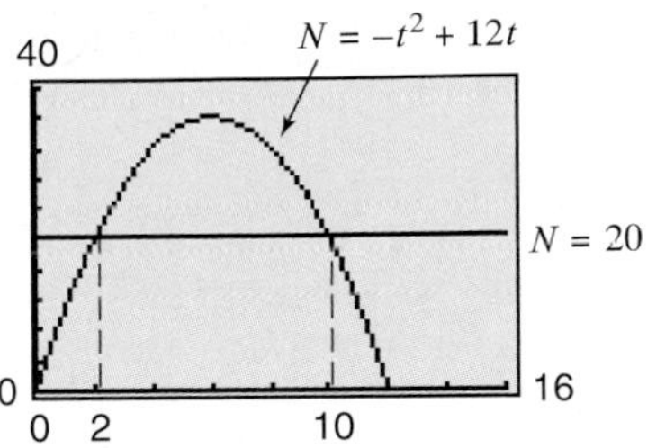

Figure 2.53 $-t^2 + 12t < 20$ when $t < 2$ and $t > 10$.

Group Discussion

Solutions to Quadratic Inequalities

All the quadratic inequalities that we discussed have had many solutions. On the other hand, we have seen that quadratic equations can have a single solution or no solutions. This prompts us to ask whether quadratic inequalities that have a single solution or no solution can exist. Investigate this. We suggest that you be guided by geometry. If you are convinced that such inequalities can exist, use geometry to construct examples. Can questions about the motion of an object fired vertically from ground level be expressed in terms of inequalities that turn out to have a single solution or no solutions? Discuss.

Answers to Self-Check Exercises

1. $(2, \infty)$.
2. $(-\infty, -1] \cup (3, \infty)$.

Answers to Further Ideas Exercises

1. **(a)** $[-1, 2]$, **(b)** $(-2, 6)$.
2. **(a)** $(-\infty, -2) \cup (3, \infty)$, **(b)** $x = [2, 6]$.
3. $(-1.60, 1.47) \cup (5.13, \infty)$.

EXERCISES 2.4

In Exercises 1–6, solve the linear inequalities. Write the solutions in interval notation. Check your answer graphically.

1. $x + 2 \le 7$ **2.** $3x - 1 < 8$ **3.** $-2x + 1 < 11$

4. $4x + 1 > 7$ **5.** $5x \le x + 12$ **6.** $3x \ge 10 - 2x$

In Exercises 7–12, solve the linear inequalities. Write the solutions in interval notation. Check your answer graphically.

7. $5x - 2 \le x + 2$ **8.** $3x - 1 \ge 5$

9. $19 - 3x > 5x + 1$ **10.** $3x + 7 \le x + 1$

11. $2z + 5 > -3z + 2$ **12.** $7t - 2 \ge 4t + 7$

In Exercises 13–18, solve the double inequalities. Write the solutions in interval notation. Support your answers using graphs.

13. $1 < 3x - 2 < 7$ **14.** $5 \le 2x + 1 \le 9$

15. $5 < x + 3 \le 7$ **16.** $-4 \le -x + 2 < 6$

17. $-9 < 4x - 5 < 15$ **18.** $-4 < -x + 2 < 8$

In Exercises 19–24, solve the double inequalities. Write the solutions in interval notation. Support your answers using graphs.

19. $2 < -2x + 4 \le 12$ **20.** $7 > x - 3 \ge 4$

21. $13 \ge -3x + 1 \ge 4$ **22.** $-3 \le 4t - 1 < 7$

23. $1 \le 2x + 3 \le 5$ **24.** $-4 < -x - 3 < 1$

In Exercises 25–28, use the displays to estimate the values of x for which the inequalities are true. Confirm your answers using algebra.

25. $-4 < x + 2 < 8$ **26.** $-5 \le 2x - 1 \le 11$

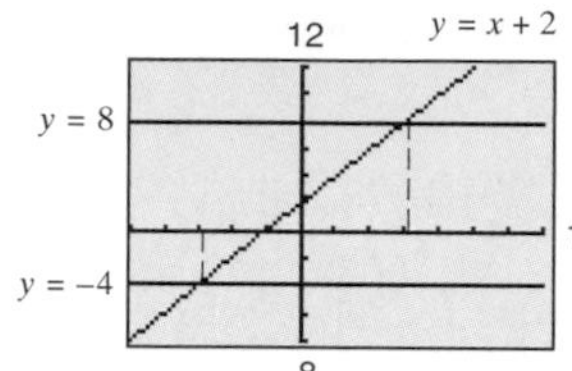

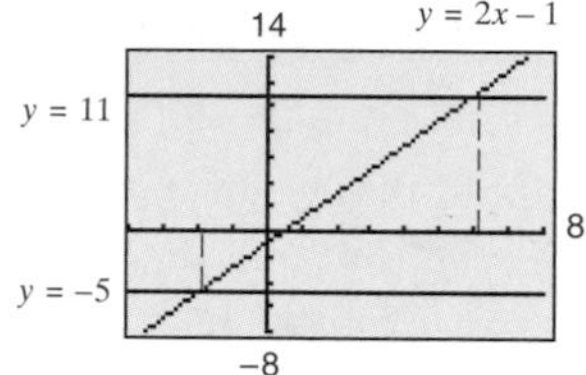

27. $2 < -x + 4 < 8$ **28.** $-5 \le -2x + 3 < 13$

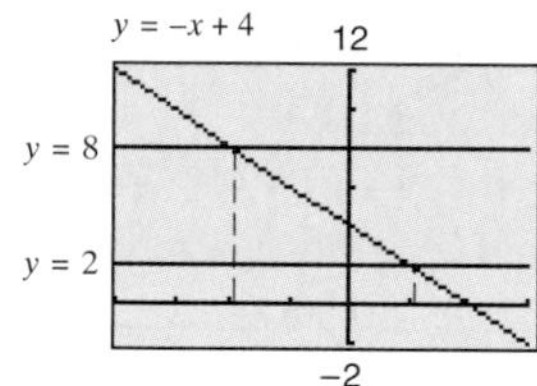

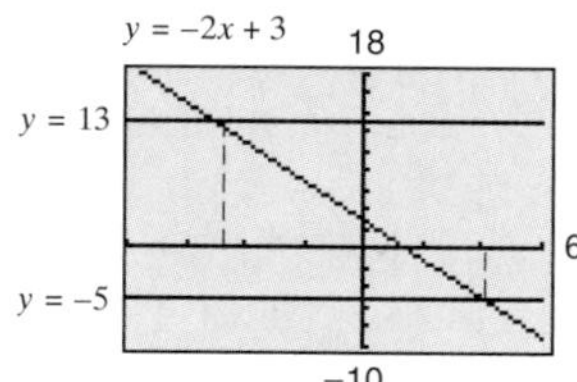

In Exercises 29–34, solve the quadratic inequalities. Support your answers using a graph.

29. $x^2 + x - 2 < 0$ **30.** $x^2 - 4x + 3 \ge 0$

31. $-x^2 - x + 6 < 0$ **32.** $x^2 - 6x + 8 \le 0$

33. $x^2 - 1 > 0$ **34.** $-x^2 + 5x - 4 \ge 0$

In Exercises 35–40, solve the quadratic inequalities. Support your answers using graphs.

35. $x^2 - 9x - 7 < 3$ **36.** $x^2 - 4x - 3 \le 2$

37. $x^2 - 8 > 1$ **38.** $-2x^2 + x + 1 > 6x - 2$

39. $x^2 - 5x - 9 \le -2x + 1$

40. $x^2 + x - 3 < 2x^2 + 4x - 7$

In Exercises 41–44, use the displays to estimate the values of x for which the inequalities are true. Confirm your answers using algebra.

41. $x^2 - 5x + 4 < 0$ **42.** $x^2 + 2x - 15 \ge 0$

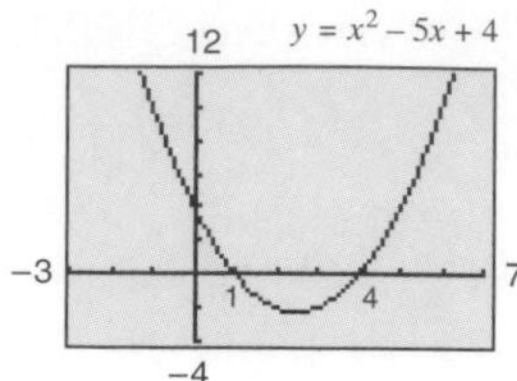

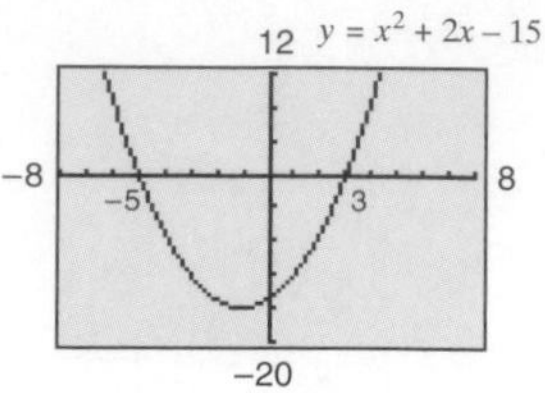

43. $-x^2 + 8x - 12 < 0$ **44.** $-x^2 - 2x + 48 \ge 0$

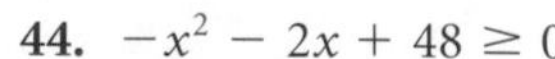

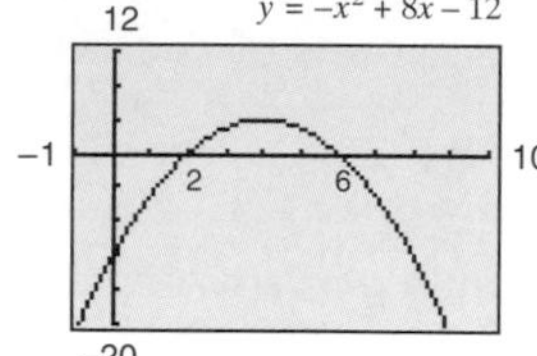

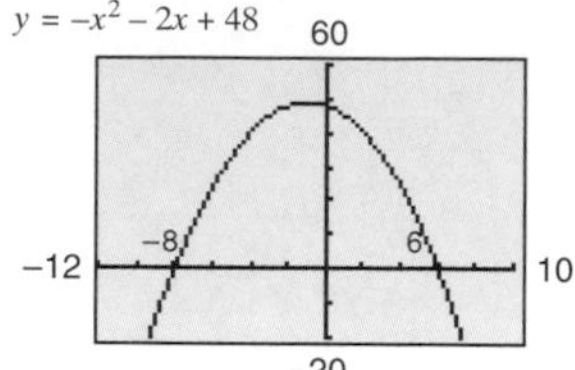

In Exercises 45–50, solve the rational inequalities. Support your answers using graphs.

45. $\dfrac{3x + 1}{x - 1} < 2$ **46.** $\dfrac{2x - 1}{x - 3} \ge 1$

47. $\dfrac{-x + 8}{x - 2} < 1$ **48.** $\dfrac{5x + 2}{x + 2} \le 3$

49. $\dfrac{6x}{x - 4} > -6$ **50.** $\dfrac{-4x + 2}{x + 3} \le -2$

In Exercises 51–54, use the given displays to solve the inequalities. The x-intercepts of the graphs are given.

51. $x^3 - 2x^2 - 5x + 6 > 0$

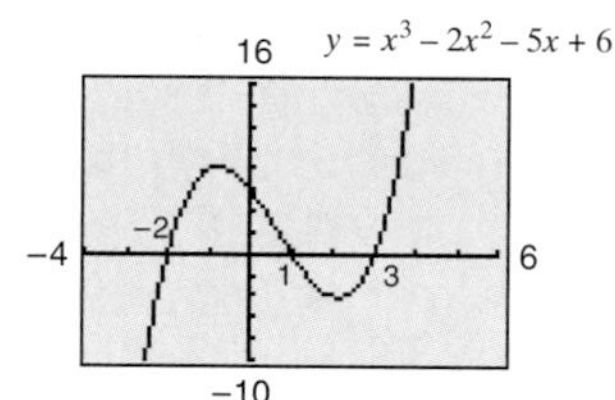

52. $-x^3 + 15x^2 - 68x + 96 \le 0$

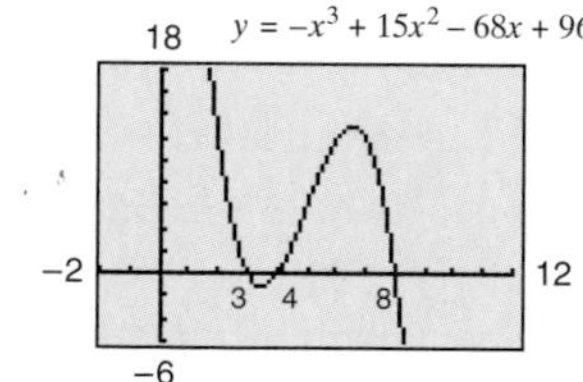

53. $x^4 + 5x^3 + 5x^2 - 5x - 6 \geq 0$

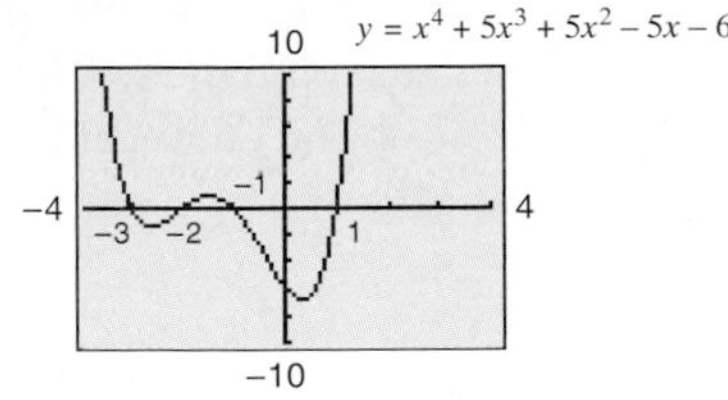

54. $-x^4 + 3x^3 + 15x^2 - 19x - 30 < 0$

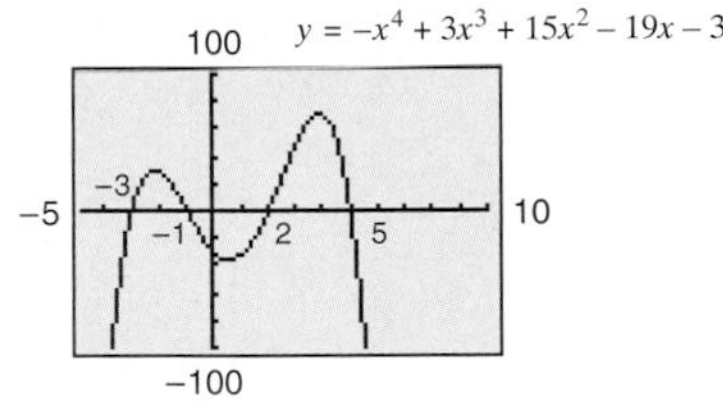

In Exercises 55–60, solve the inequalities using graphing techniques. Round the solutions to two decimal places.

55. $x^2 - 4x - 6 < 0$

56. $x^2 - 11x + 23 \geq 0$

57. $-x^2 + 6x + 15 > 0$

58. $-2x^2 + 7x - 11 > 0$

59. $x^3 - 12x + 9 > 0$

60. $x^3 - 3x^2 - 35x + 69 \leq 0$

In Exercises 61–66, solve the inequalities using graphing techniques. Round the solutions to two decimal places. *Hint:* Rewrite each inequality with all terms on the left and find the zeros.

61. $-x^3 + 3x^2 + 4x \leq 7$

62. $x^3 > 5x^2 + 2x + 3$

63. $x^3 - 2x^2 - 5x + 6 > 0$

64. $-x^3 + 15x^2 - 68x + 123 \geq 0$

65. $\dfrac{6x + 2}{x - 2} < 3$

66. $\dfrac{3x - 11}{x + 4} \geq -6$

67. There is only one real number that does not satisfy the inequality $x^2 > 6x - 9$. Solve the inequality and determine that number. Use graphs of $y = x^2$ and $y = 6x - 9$ to explain this result.

68. Use graphs to show that the set of solutions to the inequality $x^2 > x - 6$ is the set of real numbers.

TEMPERATURE

In Exercises 69–72, use the equation $F = \frac{9}{5}C + 32$ or its equivalent form $C = \frac{5}{9}(F - 32)$ to find the Fahrenheit or Celsius interval.

69. Celsius interval corresponding to temperatures greater than or equal to 59°F.

70. Celsius interval corresponding to Fahrenheit temperatures lying between 50°F and 86°F (inclusive).

71. Fahrenheit interval corresponding to temperatures greater than 20°C.

72. Fahrenheit interval corresponding to temperatures greater than or equal to 0°C and less than or equal to 40°C.

73. **Manufacturing** The relationship between total cost C (in dollars) and number of items n produced in a manufacturing process is $C = 16n + 1200$. If the total cost has to be kept between \$32,000 and \$38,000 (inclusive), what limitation does this imply on the number of items produced? Support your answer using a graph.

74. **Simple Interest** The total interest I that accumulates on a principal P invested at a rate r per annum simple interest for t years is $I = Prt$. Anita Oyler invests \$1000 at 5% simple interest. If she wants the interest to be more than \$200, how long must she leave the money in the account? Support your answer using a graph.

75. **Simple Interest** A principal P invested at a rate r per annum simple interest increases to $A = P + Prt$ in t years. (a) If \$5000 is to increase to more than \$10,000 at 4% per annum, how long must it be left in the account? (b) If the money is to increase to more than \$6000 at 6% in 10 years, what must the initial investment be?

76. **Fencing a Field** The perimeter P of a rectangle of length L and width W is $P = 2L + 2W$. The width of a rectangular field is to be 150 feet. What are the limitations on the length of the field if a maximum of 750 feet of fencing is available to surround it? Support your answer using a graph.

77. **Triangle** The area A of a triangle having base b and height h is $A = \frac{1}{2}bh$. If the area of a triangle with base 6 inches is to be greater than 24 square inches, how large must the height be? Support your answer using a graph.

In Exercises 78 and 79, the equation $s = -16t^2 + vt$ gives the height s feet at time t seconds of an object fired upward with a velocity of v feet per second. Support your answers using a graph.

78. **Motion Under Gravity** If the initial velocity is 128 feet per second, when is the object 112 feet high or higher?

79. **Motion Under Gravity** If the initial velocity is 240 feet per second, when is the object under 576 feet?

80. **Advertising Strategy** The Gwynt Company has determined that there is a relationship $y = 0.04x + 500$ between amount $\$x$ spent on advertising and realized sales y items of its perfume. If the company wants to sell more than 10,000 items, how much money must it spend on advertising? Support your answer using a graph.

81. **Television Sales** The Ulrich Household Goods Company has estimated that the demand D (in thousands) for a type of television is related to the price P of the television (in dollars) according to the equation $D = 900P - P^2$. Use a graph to determine the interval of prices that leads to a demand of more than 100,000 televisions.

82. **Miles per Gallon** The *Northern California Car Magazine* reports that, for cars, there is a relationship

$$y = -(0.83 \times 10^{-6})x^2 - (0.71 \times 10^{-3})x + 45$$

between weights of cars (x pounds) and miles per gallon (y). Use a graph to determine the weight interval of cars (to the nearest pound) that get between 20 and 30 miles per gallon?

2.5 Equations and Inequalities with Absolute Values

- Equations with Absolute Values • Inequalities with Absolute Values

We complete the chapter by discussing methods for solving equations and inequalities that involve absolute values.

Equations with Absolute Values

To motivate ideas let us solve the following equation:

$$|x| = 2$$

There are two numbers whose magnitudes are 2; namely, 2 and -2. There are therefore two solutions to this equation: $x = 2$ and $x = -2$. Observe that we can view the solutions as the points on the real line distance 2 from the origin (Figure 2.54).

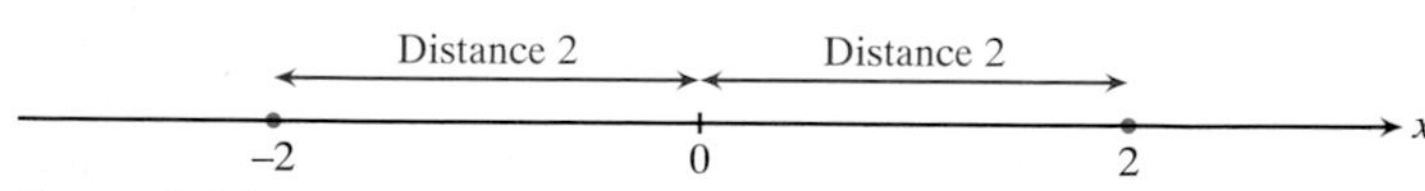

Figure 2.54

We can generalize this result as follows.

THE EQUATION $|a| = b$

If a and b are real numbers, with b positive, then

$$|a| = b \quad \text{if and only if} \quad a = b \quad \text{or} \quad a = -b$$

We now see how this result leads to methods for solving more general equations that involve absolute values. The result is valid when the numbers a and b are replaced by algebraic expressions because the expressions we use represent real numbers.

Example 1 Solving an Equation with an Absolute Value

Solve the equation $|2x - 5| = 3$.

SOLUTION The expression $2x - 5$ is a real number whose magnitude is 3. Thus, $2x - 5$ is equal to 3 or to -3. We get

$$\begin{aligned} 2x - 5 &= 3 \quad &\text{or}\quad 2x - 5 &= -3 \\ 2x &= 8 \quad &\text{or}\quad 2x &= 2 \\ x &= 4 \quad &\text{or}\quad x &= 1 \end{aligned}$$

There are two solutions: $x = 4$ and $x = 1$.

Self-Check 1 Solve $|3x + 1| - 5$.

Consider now the equation $|a| = |b|$ for two numbers a and b. This equation implies that the magnitude of a is equal to that of b. Thus, $a = b$ or $a = -b$.

THE EQUATION $|a| = |b|$

If a and b are real numbers, then

$$|a| = |b| \quad \text{if and only if} \quad a = b \quad \text{or} \quad a = -b$$

Example 2 Solving an Equation with Two Absolute Values

Solve $|3x + 4| = |2x + 6|$.

SOLUTION The expressions $3x + 4$ and $2x + 6$ are real numbers whose magnitudes are equal. Thus,

$$\begin{aligned} 3x + 4 &= 2x + 6 && \text{or} & 3x + 4 &= -(2x + 6) \\ 3x - 2x &= 6 - 4 && \text{or} & 3x + 4 &= -2x - 6 \\ x &= 2 && \text{or} & 3x + 2x &= -6 - 4 \\ &&&& 5x &= -10 \\ &&&& x &= -2 \end{aligned}$$

The solutions are $x = 2$ and $x = -2$.

Self-Check 2 Solve $|4x - 5| = |2x - 7|$.

Inequalities with Absolute Values

Let us solve the inequality

$$|x| < 5$$

This inequality is satisfied by all the real numbers whose magnitudes are less than 5. Naturally, all the real numbers in the interval $(0, 5)$ satisfy this inequality. Note, however, that the negatives of these numbers also have magnitudes less than 5 and so also does zero. For example, $|-4| < 5$. The set of solutions is therefore the open interval $(-5, 5)$. We can also write $-5 < x < 5$ for this set of solutions. Observe that the solutions are all the points with a distance less than 5 from the origin (Figure 2.55).

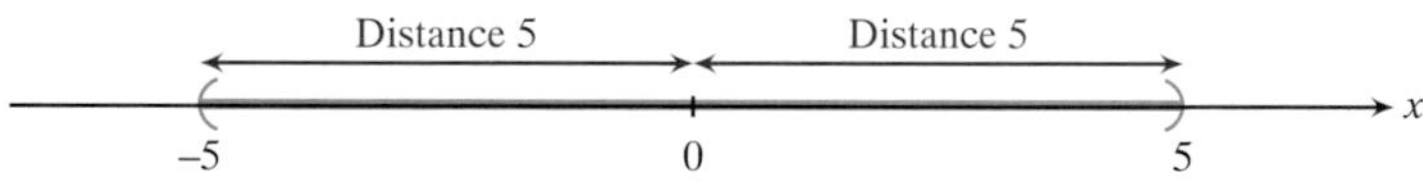

Figure 2.55

We can generalize these ideas. Consider two real numbers a and b, where b is positive. If $|a| < b$, then the magnitude of a is less than that of b, implying that $-b < a < b$. Conversely, if $|a| > b$, then the magnitude of a is greater than that of b, implying that $a < -b$ or $a > b$. Let us summarize these observations.

Inequalities with Absolute Values

If a and b are real numbers, with b being positive, then

$$\begin{aligned} |a| < b \quad &\text{if and only if} \quad -b < a < b \\ |a| > b \quad &\text{if and only if} \quad a < -b \ \text{ or } \ a > b \end{aligned}$$

These results also hold if $<$ is replaced with $\leq$ and $>$ is replaced with $\geq$.

Now let us use these results to solve inequalities that involve absolute values. As in the case of equations, these results hold when the real numbers a and b are replaced with algebraic expressions.

Example 3 **Solving an Inequality with an Absolute Value**

Solve $|x| > 3$.

SOLUTION This inequality is satisfied by all real numbers whose magnitudes are greater than 3. These are the numbers that are greater than 3 and the numbers that are less than -3. They are the numbers in the intervals $(-\infty, -3)$ and $(3, \infty)$. The set of solutions is $(-\infty, -3) \cup (3, \infty)$. We can also write $x < -3$ or $x > 3$.

Observe that we can view the solutions as all numbers that lie more than 3 units from the origin (Figure 2.56).

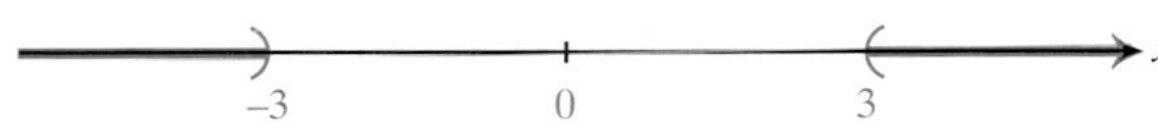

Figure 2.56

Example 4 **Solving an Inequality with an Absolute Value**

Solve $|2x - 1| < 5$.

SOLUTION The expression $2x - 1$ is a number whose magnitude is less than 5. Thus, $2x - 1$ lies between -5 and 5. We get

$$-5 < 2x - 1 < 5$$
$$-4 < 2x < 6$$
$$-2 < x < 3$$

The set of solutions is $(-2, 3)$, as shown in Figure 2.57.

Figure 2.57

We can check this result graphically. Graph $y = |2x - 1|$ and $y = 5$ (Figure 2.58). The inequality $|2x - 1| < 5$ is true when the graph of $y = |2x - 1|$ lies below the line $y = 5$. This is true when $-2 < x < 3$. The set of solutions is thus $(-2, 3)$.

Note the interesting V shape of the graph of $y = |2x - 1|$. We will understand why this is so in the next chapter.

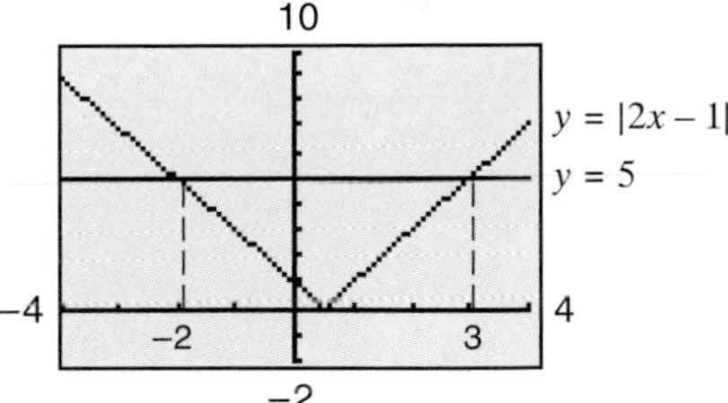

Figure 2.58 $|2x - 1| < 5$ when $-2 < x < 3$.

Example 5 **Solving an Inequality with an Absolute Value**

Solve $|2x + 3| \geq 7$.

SOLUTION The expression $2x + 3$ is a number whose magnitude is greater than or equal to 7. Thus $2x + 3$ is less than or equal to -7 or greater than or equal to 7.

$$\begin{aligned} 2x + 3 &\le -7 \quad &\text{or}\quad 2x + 3 &\ge 7 \\ 2x &\le -10 \quad &\text{or}\quad 2x &\ge 4 \\ x &\le -5 \quad &\text{or}\quad x &\ge 2 \end{aligned}$$

The solutions are the points in the intervals $(-\infty, -5]$ and $[2, \infty)$. The set of solutions is $(-\infty, -5] \cup [2, \infty)$, as shown in Figure 2.59.

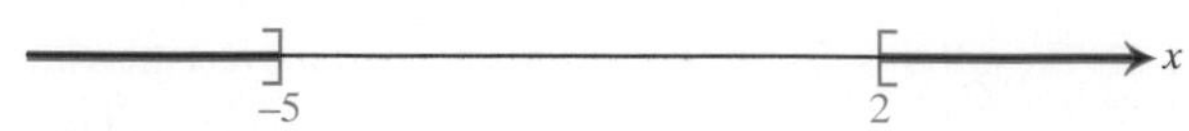

Figure 2.59

We can check this result graphically. Graph $y = |2x + 3|$ and $y = 7$ (Figure 2.60). The inequality $|2x + 3| = 7$ occurs when the graph of $y = |2x + 3|$ lies on or above the line $y = 7$. This occurs when $x \le -5$ or $x \ge 2$. The set of solutions is thus $(-\infty, -5] \cup [2, \infty)$.

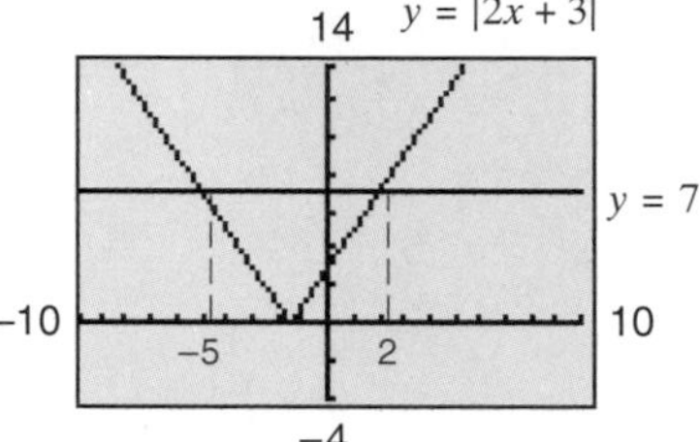

Figure 2.60 $|2x + 3| \ge 7$ when $x \le -5$ or $x \ge 2$.

Further Ideas 1

(a) Solve $|2x - 3| \le 5$. Check your answer graphically.

(b) Use the display of Figure 2.61 to solve $|x - 3| \ge 2$. Check your answer using algebra.

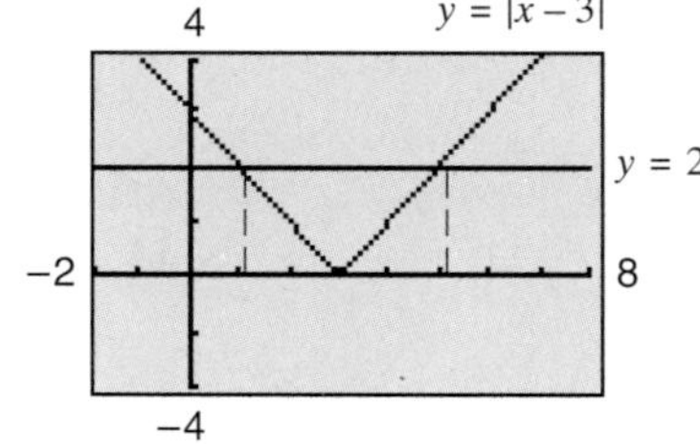

Figure 2.61

We solved a double inequality in Section 2.4. Now let us solve a double inequality that includes an absolute value.

Example 6 Solving a Double Inequality with an Absolute Value

Solve $2 < |x + 3| < 9$.

SOLUTION We solve this inequality in two parts. Solve $2 < |x + 3|$ and $|x + 3| < 9$ separately and then bring the results together. Using the properties of inequalities discussed in this section we get the following:

(a) $2 < |x + 3|$: $x + 3 < -2$ or $x + 3 > 2$, $x < -5$ or $x > -1$ (Figure 2.62a).

(b) $|x + 3| < 9$: $-9 < x + 3 < 9$, $-12 < x < 6$ (Figure 2.62b).

We now find the condition on x for it to satisfy both (a) and (b). x must be less than -5, or x must be greater than -1. Further, x must lie between -12 and 6. These conditions are all satisfied if x lies between -12 and -5 or between -1 and 6 (Figure 2.62c). These are the shaded parts common to the graphs in Figure 2.62a and 2.62b. The solution set is $(-12, -5) \cup (-1, 6)$.

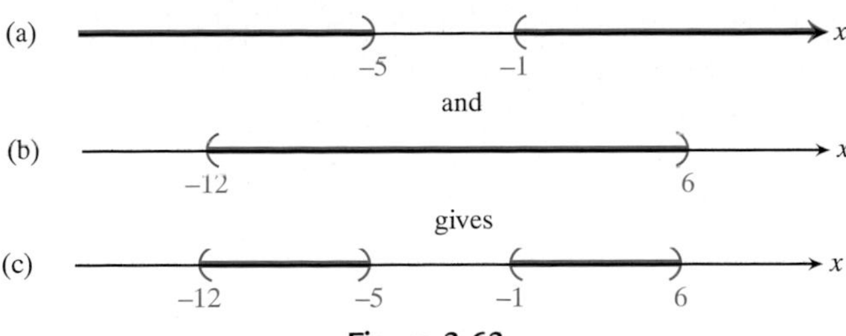

Figure 2.62

Let us check this result graphically. Graph $y = |x + 3|$, $y = 2$, and $y = 9$ (Figure 2.63). $2 < |x + 3| < 9$ holds when the graph of $y = |x + 3|$ lies between the lines $y = 2$ and $y = 9$. This occurs over the intervals $(-12, 5)$ and $(-1, 6)$. The set of solutions is thus $(-12, -5) \cup (-1, 6)$.

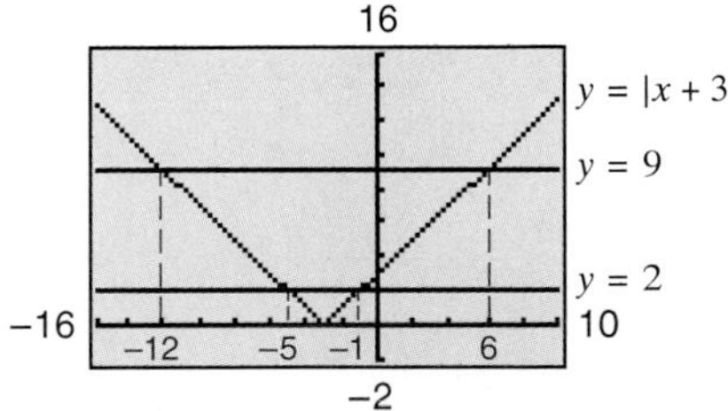

Figure 2.63 $2 < |x + 3| < 9$ when $-12 < x < -5$ or $-1 < x < 6$.

Answers to Self-Check Exercises

1. $x = -2$ and $x = \frac{4}{3}$.
2. $x = -1$ and $x = 2$.

Answer to Further Ideas Exercise

1. **(a)** $[-1, 4]$, **(b)** $(-\infty, 1] \cup [5, \infty)$.

EXERCISES 2.5

In Exercises 1–8, solve the equations.

1. $|x| = 5$ **2.** $|x - 3| = 7$ **3.** $|x - 4| = 6$
4. $|2x - 1| = 6$ **5.** $|3x + 4| = 9$ **6.** $|2x - 3| = 7$
7. $|3z - 1| = 4$ **8.** $|t - 3| = 0$

In Exercises 9–16, solve the equations.

9. $|2x - 4| = |x - 2|$ **10.** $|4x + 1| = |3x + 6|$
11. $|3x + 2| = |x - 6|$ **12.** $|2 - 3x| = |2x + 1|$
13. $|5x - 2| = |2 - 3x|$ **14.** $|w - 3| = |w + 1|$
15. $|-5x| = 20$ **16.** $|2 + x| = |2 - x|$

In Exercises 17–22, solve the equations.

17. $|2x - 3| = |-9|$ **18.** $|3x| - |-2| = 4$
19. $|3x| = 15$ **20.** $\dfrac{1}{|x + 2|} = \dfrac{1}{4}$
21. $\dfrac{3}{|2x - 1|} = \dfrac{6}{7}$ **22.** $\dfrac{1}{|x|} = 5$

In Exercises 23–34, solve the inequalities. Write the solutions in interval notation. Support your answers using a graph.

23. $|x| < 4$ **24.** $|x| \le 6$
25. $|x| > 5$ **26.** $|x + 2| \le 7$
27. $|x - 3| > 8$ **28.** $|2x + 1| < 5$
29. $|2x - 3| \ge 7$ **30.** $|3x + 4| < 8$
31. $|1 - 2x| > 7$ **32.** $|1 - v| \le 6$
33. $|9 - 3t| \ge 6$ **34.** $|5 - 3x| < 9$

In Exercises 35–38, use the given displays to estimate the values of x for which the inequalities are valid. Confirm your answers using algebra.

35. $|2x + 1| < 7$

36. $|2x - 1| \geq 3$

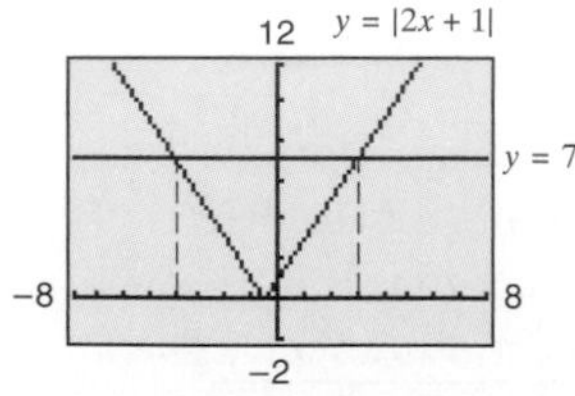

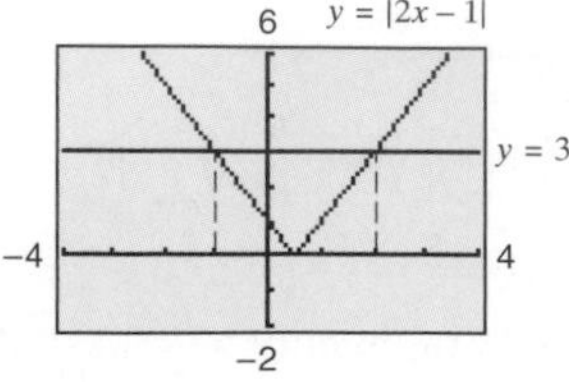

37. $|2x - 5| \leq 7$

38. $|1 - x| > 3$

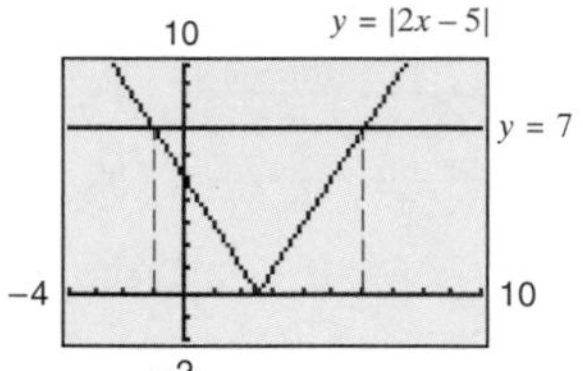

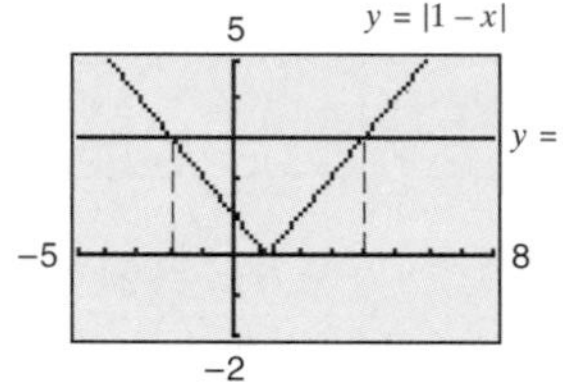

In Exercises 39–42, use a graph to solve the inequalities. Use algebra to confirm your answer.

39. $|x + 1| < 5$

40. $|x - 1| > 8$

41. $|2x - 7| \leq 7$

42. $|5 - 2x| \geq 9$

In Exercises 43–52, solve the inequalities. Support your answers using a graph.

43. $|5x - 7| \geq |-12|$

44. $|x| > -4$

45. $|x| + 2 < 8$

46. $-|2x| + 3 > 5$

47. $\left|\dfrac{x - 1}{3}\right| \leq 5$

48. $\left|\dfrac{2x + 1}{5}\right| > 3$

49. $2 < |x| < 5$

50. $1 \leq |x - 2| < 6$

51. $9 > |2x - 5| > 0$

52. $-7 < |2x + 1| < 9$

CHAPTER 2 PROJECT — Algorithms for Finding Roots and Solving Equations

Today we do so much of our mathematics using calculators and computers. This project consists of two numerical methods that let you catch a glimpse of the ways calculators and computers execute some of the mathematics.

Computing a Square Root

Evaluating a square root is basic to mathematics. We have seen the need to take the square root of a number on numerous occasions in this course. For example, the extracting square roots and completing the square methods introduced in this chapter depend on being able to take square roots. Today we have calculators that enable us to approximate square roots at the press of a few keys, but this has not always been the case! This project shows us how square roots were found before the age of technology and then demonstrates how roots can now be found using technology. It gives us an idea of how a square root is possibly computed on your calculator.

Heron of Alexandria was an Egyptian mathematician of the first century BCE. His most important work was his *Metrica,* written in three volumes. They were discovered in Constantinople in 1896. Of particular interest in this work is his method of approximating the square root of a nonsquare natural number. He proposed that we let n be the number and a and b, the two factors. That is, $n = ab$ (a and b need not be natural numbers). Then $(a + b)/2$ would be an approximation for $\sqrt{n}$. This approximation improves with the closeness of a and b. The step can be repeated to get better and better approximations on a calculator using $(\text{Ans} + n/\text{Ans})/2$. Such a sequence of steps that tells us how to solve a problem is called an **algorithm.** In practice calculators are programmed to execute such algorithms at the push of a key.

(a) Use this method to find $\sqrt{10}$ as accurately as you possibly can on your calculator. (Determine the best way to implement the algorithm on your calculator.) Record your answers at each stage. Check your answer by comparing it with the value of $\sqrt{10}$ given by your calculator using the square root key.
(b) Explain why the method works.

SOLUTIONS OF A QUADRATIC EQUATION

We now provide an algorithm called **Newton's Method** (named after Sir Isaac Newton) for finding the roots of a quadratic equation. Let $ax^2 + bx + c = 0$ be the quadratic equation. Let x_0 be a guess at a root (any real number). Then

$$x_1 = x_0 - \frac{ax_0^2 + bx_0 + c}{2ax_0 + b}$$

will be closer to a root. The step can be repeated to get better and better approximations using the previous answer repeatedly as follows:

$$\text{Ans} - \frac{a\text{Ans}^2 + b\text{Ans} + c}{2a\text{Ans} + b}$$

The method leads to an approximation of the root closest to the original guess, if there are two roots, and to the single root, if there is only one root. The zero finder of your calculator executes an algorithm such as this.

(a) Construct a quadratic equation that has two known solutions and test the algorithm.
(b) Construct a quadratic equation that has a single solution and test the algorithm.
(c) Investigate what happens if the algorithm is used for a quadratic equation that has no real solutions.
(d) The algorithm will not usually work for $ax^2 + bx + c = 0$ when the initial guess is $x = -b/2a$. Test this on your calculator. This guess corresponds to the vertex of the parabola. It is halfway between the solutions—the calculator does not know which solution to converge to! Stay away from this number as a guess.
(e) Try to find out from the manufacturer of your calculator which algorithms are used to compute square roots and zeros.

(Readers who go on to calculus will derive Newton's Method in that course.)

WEB PROJECT 2

To explore the concepts of this chapter further, try the interesting WebProject. Each activity relates the ideas and skills you have learned in this chapter to a real-world applied situation. Visit the Williams Web site at

www.saunderscollege.com/math/williams

to find the complete WebProject, related questions, and live links to the Internet.

Chapter 2 Highlights

Section 2.1

Quadratic Equation in x: An equation that can be written in the form $ax^2 + bx + c = 0$, where a, b, and c are real numbers with $a \neq 0$. This is the standard form.

Graphical Representation of Solutions to a Quadratic Equation: They are the x-intercepts of the graph of the parabola $y = ax^2 + bx + c$.

Roots of a Quadratic Equation: Another way of talking about its solution.

Roots by Factoring: If $ax^2 + bx + c = 0$ can be written $(px + q)(rx + t) = 0$ the roots (or solutions) are $x = -q/p$ and $x = -t/r$.

Repeated Root: If $ax^2 + bx + c = 0$ can be written $(px + q)(px + q) = 0$, $x = -q/p$ is a repeated root. There is a single solution, $x = -q/p$.

Solutions by Extracting Square Roots: The solutions of $x^2 = k$, where $k > 0$, are $x = \pm\sqrt{k}$.

Solutions by Completing the Square: $ax^2 + bx + c = 0$. Write as $ax^2 + bx = -c$. Make $ax^2 + bx$ into a perfect square and then extract square roots.

Quadratic Formula: Solutions to $ax^2 + bx + c = 0$, $a \neq 0$, are

$$x = \frac{-b \pm \sqrt{b^2 - 4ac}}{2a}.$$

Discriminant: Quadratic equation $ax^2 + bx + c = 0$, $a \neq 0$. Discriminant is $b^2 - 4ac$. If $b^2 - 4ac > 0$, there are two solutions, if $b^2 - 4ac = 0$, there is one solution; if $b^2 - 4ac < 0$, there are no real solutions.

Section 2.3

Quadratic-Type Equation: An equation that can be written in the form

$$au^2 + bu + c = 0, \quad a \neq 0.$$

Number of Roots: A polynomial equation of degree n has at most n solutions.

Section 2.4

Inequality: A statement that one expression is greater than, greater than or equal to, less than, or less than or equal to another statement.

Properties of Inequalities: If $a < b$ and $b < d$ then $a < d$. If $a < b$ and $c < d$ then $a + c < b + d$. If $a < b$, then $a + c < b + c$. If $a < b$ and $c > 0$, then $ac < bc$. If $a < b$ and $c < 0$, then $ac > bc$.

Section 2.5

Equations with Absolute Values:

$|a| = b$ if and only if $a = b$ or $a = -b$.

$|a| < b$ if and only if $-b < a < b$. $|a| > b$ if and only if $a < -b$ or $a > b$.

$|a| \leq b$ if and only if $-b \leq a \leq b$. $|a| \geq b$ if and only if $a \leq -b$ or $a \geq b$.

CHAPTER 2 REVIEW EXERCISES

1. Solve the following equations by factoring. Use graphs to support your answers.
   (a) $x^2 + x - 2 = 0$ (b) $x^2 + 10x + 21 = 0$
   (c) $2x^2 + 9x - 5 = 0$ (d) $6x^2 - 17x - 14 = 0$
2. Solve the following equations by extracting square roots. Round answers to four decimal places. Use graphs to support your answers.
   (a) $x^2 = 16$ (b) $9x^2 - 1 = 0$ (c) $2x^2 - 8 = 0$
   (d) $(x - 3)^2 = 25$ (e) $(4x + 1)^2 = 7$
3. Solve the following quadratic equations by completing the square. Check your answers using substitution.
   (a) $x^2 - 10x + 21 = 0$ (b) $x^2 + 14x + 24 = 0$
   (c) $2x^2 - 16x + 31 = 0$ (d) $3x^2 - 6x - 8 = 0$
4. Solve (if possible) the following quadratic equations using the quadratic formula. Express answers to four decimal places. Use graphs to support your answers.
   (a) $x^2 + 4x - 3 = 0$ (b) $x^2 - x - 9 = 0$
   (c) $2x^2 + 8x - 7 = 0$ (d) $3x^2 + 9x + 1 = 0$
   (e) $4x^2 + x + 2 = 0$ (f) $x^2 + 6x + 9 = 0$
5. Use trace/zoom or a zero finder with a graph on your calculator to solve the quadratic equation $2.3x^2 - 3.5x - 6.8 = 0$. Express the answer to four decimal places.
6. Use the discriminant to determine whether the following equations have two real solutions, a single real solution, or no real solutions. Use graphs to support your answers.
   (a) $x^2 + 3x - 5 = 0$ (b) $3x^2 + x - 2 = 0$
   (c) $7x^2 + 2x + 3 = 0$ (d) $9x^2 - 6x + 1 = 0$
7. Construct a quadratic equation that has solutions $x = 1$ and $x = -4$. Use a graph to support your answer.
8. **Speed of a Boat** The tourist boat *Sunrise Queen* travels to a point 40 miles upstream on the Ohio River from its home port near Cairo and then returns. The upstream journey takes 2 hours more than the return journey. If the river flows at 5 miles per hour, what is the speed of the boat in still water? Use graphs to support your answers.
9. **Motion Under Gravity** An object is propelled vertically from ground level with a velocity of 224 feet per second. When will it be at a height of 720 feet? (The height s feet after t seconds of an object propelled vertically from ground level with velocity v feet per second is $s = -16t^2 + vt$.)
10. Solve the following equations by using an appropriate substitution. Use graphs to support your answers.
    (a) $x^4 - x^2 - 6 = 0$ (b) $2x^4 + 5x^2 - 3 = 0$
    (c) $15(x + 3)^4 - 7(x + 3)^2 - 2 = 0$
    (d) $x - x^{1/2} - 12 = 0$
11. Solve the following equations (if possible) by taking powers of sides of the equations. Check your answers using algebra and using a graph.
    (a) $\sqrt{4x - 3} = 1$ (b) $\sqrt{3x + 7} = 7$
    (c) $\sqrt{3x} = \sqrt{6x - 1}$ (d) $\sqrt{4x + 3} = \sqrt{2x - 1}$
    (e) $\sqrt{x + 2} + \sqrt{x - 3} = 5$
    (f) $(x + 3)^{2/3} = 9$
    (g) $x^{-4/5} = 16$
12. Solve the following equations (if solutions exist), to four decimal places, using trace/zoom or a zero finder with a graph.
    (a) $3x^3 - 7x^2 - 22x + 19 = 0$
    (b) $x^3 + 2x^2 + 5x + 16 = 0$
    (c) $4x^4 + 20x^3 + 3x^2 - 47x + 40 = 0$
13. Solve each of the following inequalities for x. Write the answers in interval notation. Support your answers using a graph.
    (a) $x - 3 \le 7$ (b) $2x - 7 < 15$
    (c) $3x + 2 \ge -x$ (d) $4x - 5 < 7x + 1$
14. Solve the following double inequalities. Support your answers using a graph.
    (a) $5 \le 2x + 1 \le 7$ (b) $-3 < 4x - 3 < 9$
    (c) $22 \ge 3x + 1 \ge 5$ (d) $-3 < 4x + 1 \le 7$
15. Solve the following quadratic inequalities. Support your answers using a graph.
    (a) $x^2 - x - 6 \le 0$ (b) $x^2 - 5x + 4 > 0$
    (c) $x^2 - 5x - 1 < -1$
16. Solve the following rational inequalities. Support your answers using a graph.
    (a) $\dfrac{5x - 1}{2x + 1} \ge 3$ (b) $\dfrac{6x + 3}{x + 4} \le 5$
    (c) $\dfrac{-x + 1}{x - 5} > -3$

17. **Motion Under Gravity** If an object is fired vertically with an initial velocity of 288 feet per second, when is the object higher than 1232 feet? (The height s feet after t seconds of an object propelled vertically from ground level with velocity v feet per second is $s = -16t^2 + vt$.)
18. **Simple Interest** If a principal P is invested at a rate r per annum simple interest, it will grow to an amount $A = P + Prt$ in t years. Scott Dimick invests \$1000 at 6% per annum simple interest. How long will it take to get more than \$1480?
19. Solve the following equations (if possible).
    (a) $|x + 1| = 7$ (b) $|x - 4| = 9$
    (c) $|2x - 3| = 5$ (d) $|2x + 1| = |x - 3|$
    (e) $|3x - 2| = |4x + 1|$ (f) $-|2x| + 4 = 7$
20. Solve the following inequalities. Write your answers in interval notation. Check your answers using a graph.
    (a) $|x| \leq 7$ (b) $|x + 3| \leq 4$
    (c) $|3x - 5| < 6$ (d) $|2x - 5| \geq 13$
    (e) $|7x - 3| > 6$ (f) $1 \leq |x| \leq 7$
    (g) $-3 < |x + 4| < 5$
21. Use the following displays to estimate the values of x for which the inequalities are valid. Verify your answers using algebra.
    (a) $5 < 4x + 1 < 21$ (b) $x^2 + 2x - 24 \geq 0$

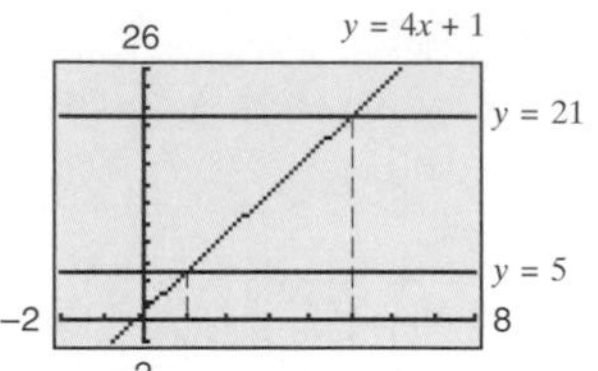

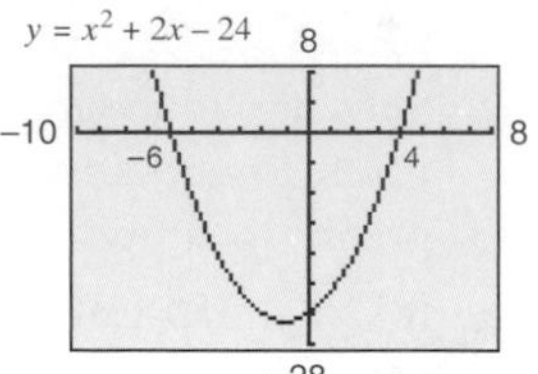

    (c) $|2x + 3| < 9$

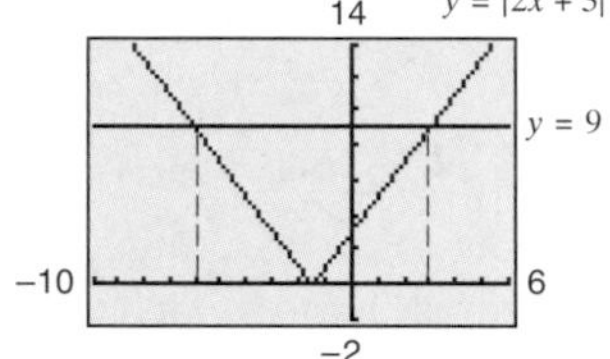

CHAPTER 2 TEST

1. Solve the equation $2x^2 - 17x + 35 = 0$ by factoring. Use a graph to support your answer.
2. Solve the equation $6x^2 - 54 = 0$ by extracting square roots. Use a graph to support your answer.
3. Solve the equation $2x^2 - 8x + 3 = 0$ by completing the square. Check your answer using substitution.
4. Solve (if possible) the following quadratic equations using the quadratic formula. Round answers to four decimal places. Use graphs to support your answers.
    (a) $x^2 + 3x - 5 = 0$ (b) $x^2 - 2x + 7 = 0$
5. Use a zero finder or trace/zoom with a graph on your calculator to solve the quadratic equation $-2.8x^2 - 4.7x + 9.3 = 0$. Round the answer to four decimal places.
6. An object is propelled vertically from ground level with a velocity of 224 feet per second. When will it be at a height of 640 feet? (The height s feet after t seconds of an object propelled vertically from ground level with velocity v feet per second is $s = -16t^2 + vt$.)
7. Solve $x^4 - 2x^2 - 15 = 0$ by using an appropriate substitution. Use a graph to support your answer.
8. Solve the equation $\sqrt{3x + 7} = 5$ by taking powers of both sides of the equations. Check your answers using algebra and using a graph.
9. Solve the equation $2x^3 - 5x^2 - 19x + 62 = 0$, to four decimal places, using trace/zoom or a zero finder with a graph.
10. Solve the inequality $3x - 4 < 7x + 8$. Write the answer in interval notation. Support your answer using a graph.
11. Solve the quadratic inequality $x^2 - 6x + 5 < 0$. Support your answer using a graph.

12. Solve the absolute value equation $|2x - 1| = |3x + 7|$.
13. Solve the inequality $-2 < |x + 3| < 7$. Write your answer in interval notation. Check your answer using a graph.
14. **Planning a Driveway** A man wants to lay down a concrete driveway. The length of the driveway is to be 25 feet greater than the width. If 350 square feet of concrete is available, what will be the dimensions of the driveway?
15. **Motion Under Gravity** An object is fired vertically with an initial velocity of 272 feet per second. Use inequalities to determine when the object is higher than 960 feet. (The height s feet after t seconds of an object propelled vertically from ground level with velocity v feet per second is $s = -16t^2 + vt$.)

CUMULATIVE TEST* CHAPTERS R, 1, AND 2

1. Compute the following expression, rounded to four decimal places, using a calculator.

   (a) $(\sqrt{8.3} + 0.57^{-1}) \div 2.91$ (b) $\sqrt[6]{27.38^5}$
2. Compute the following by hand. Support your answers using a calculator.

   (a) $\left(\sqrt[4]{\dfrac{16}{81}}\right)^3$ (b) $(-125)^{2/3}$
3. Simplify the following algebraic expression. Express the result in a form that involves each variable once, to a positive exponent.
$$\frac{(x^{-2}y^3)^4}{(x^{-2}y^5)^{-2}}$$
4. Multiply $(2x - 3)(3x + 5)$.
5. Factor. **(a)** $12x^2 - 5x - 2$ **(b)** $9x^2 - 25y^2$
6. Find the domains and ranges of the following functions.

   (a) $f(x) = -2x^2 + 4$ (b) $f(x) = \sqrt{x - 1} + 3$
7. Find the equation of the line through the point $(3, -1)$ having slope 2. Graph the equation on your calculator, and use the display to support your answer.
8. Find the equation of the line through the points $(-1, 5)$ and $(4, -5)$. Graph the equation on your calculator and use the display to support your answers.
9. Find the x-intercept of the line through the point $(-1, 5)$ parallel to the line $10x - 2y + 4 = 0$.
10. Find the vertex, y-intercept, and axis, and sketch the graph of $y = x^2 - 8x - 4$ using algebra. Verify your graph using a graphing calculator.
11. Use the distance formula to compute the distance between the points $(1, 5)$ and $(-2, 8)$. Round the answer to four decimal places.
12. Find the centers and radii of the circles having the following equations. Graph the equations on your calculator, and use the displays to support your answers.

    (a) $(x - 3)^2 + (y + 7)^2 = 25$

    (b) $x^2 + y^2 + 2x - 6y + 1 = 0$
13. Solve the equation $2x^2 + 5x - 12 = 0$ by factoring. Use a graph to support your answer.
14. Solve (if possible) the following quadratic equations using the quadratic formula. Round answers to four decimal places. Use graphs to support your answers.

    (a) $x^2 + 4x - 7 = 0$ (b) $x^2 - 3x + 8 = 0$
15. Solve $x^4 - 2x^2 - 8 = 0$ by using an appropriate substitution. Use a graph to support your answer.

*Cumulative Tests are given after every three or four chapters. Answers to all the questions are given in the back of the book.

16. Solve the equation $2x^3 - 4x^2 - 18x + 57 = 0$, to four decimal places, using trace/zoom or a zero finder with a graph.
17. Solve the inequality $x - 4 < 3x + 2$. Write the answer in interval notation. Support your answer using a graph.
18. **Analysis of Test Scores** A student has an average of 78 after three tests. What does she need on the fourth test to bring the average up to 80?
19. **Cost–Volume Analysis** A company has found that the cost of manufacturing 10 items of a certain product is \$520. The cost of manufacturing 20 items is \$550. Find the cost–volume formula. What is the cost of manufacturing 17 items?
20. **Motion Under Gravity** An object is propelled vertically from ground level with velocity of 256 feet per second. When will it be at height of 880 feet? (The height s feet after t seconds of an object propelled vertically from ground level with velocity v feet per second is $s = -16t^2 + vt$.)

3 Further Development of Functions

Leonhard Euler (1707 – 1783)
(©1994 Northwind Picture Archives)

The concept of a function permeates much of mathematics and is an important tool in applying mathematics. We have seen that a function can be used to describe how one variable varies with respect to another. The function notation $f(x)$ (we say "f is a function of x") suggests this dependency on x. This notation was first used by the Swiss mathematician Leonhard Euler. Euler was born in Basel, Switzerland, in 1707. His father, who was a minister, gave him his first instruction in mathematics and later sent him to the university in Basel. Euler wrote his dissertation at the age of 19 on the masting of ships. He served as professor in St. Petersburg, Russia, and later as head of the Prussian Academy in Berlin. His amazing productivity was not in the least impaired when he had the misfortune to become totally blind.

In this chapter, we look into algebraic and geometrical techniques for working with functions. We will see how different functions are related and how the graphs of functions can be shifted vertically and horizontally, reflected in the x- and y-axes, and stretched. Through such operations as addition, multiplication, and composition, functions will be combined to produce new functions that are useful in applications.

We will see how functions can be used to help a cosmetics company decide how often to advertise its product on television, and by scientists to learn more about animals.

Often we need to know where functions are increasing in value, where they are decreasing, and when they reach their maximum or minimum values. For example, the spread of an epidemic may be described by a function, allowing authorities to estimate its duration and intensity. Functions are also used to tell doctors how rapidly a drug dissipates in the body. These applications and the following analysis of the path of an asteroid are discussed in this chapter.

APPLICATION | Tracking an Asteroid

Asteroid 243 Ida gathered by the Galileo spacecraft on its way to Jupiter. *(©NASA/The National Audubon Society Collection/Photo Researchers)*

Recently there has been much interest in the paths of asteroids. In March 1998, for example, a false alarm about asteroid 1997XF11 coming dangerously close to Earth created quite a stir (see "Spaceguard Survey of Asteroid and Comet Impact Hazards" at http://www.saunderscollege.com/math/williams for work being done by NASA on tracking asteroids). In this chapter, we see how a segment of the orbit of an asteroid can be approximated by a parabola. Using a coordinate system having Earth at the origin, astronomers have arrived at the following coordinates of an asteroid as it approaches the vicinity of Earth.

x coordinate (million miles)	40	36	32	28	24	20
y coordinate (million miles)	9.00	9.88	10.92	12.12	13.48	15.00

The path of the asteroid in the neighborhood of Earth is predicted from these data to be approximately $y = 0.005x^2 - 0.6x + 25$. We get the following picture of the path of the asteroid (A) relative to Earth (E) (Figure 3.1). The asteroid is nearest to Earth somewhere around point **x**. We are able to predict that the closest the asteroid gets to Earth is about 21.91 million miles. (*This application is discussed in Section 3.2, Example 7, on pages 234–235.*)

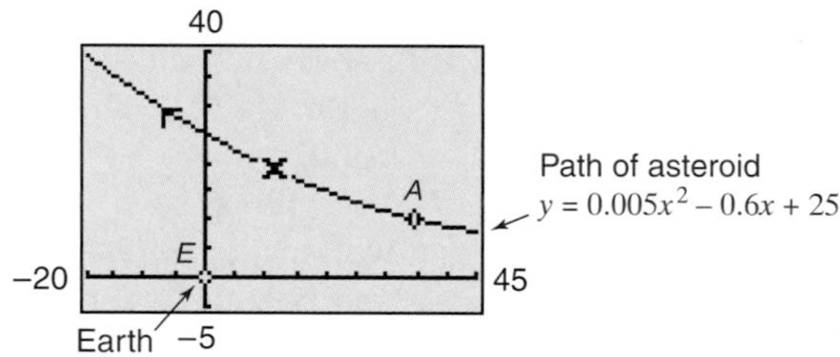

Figure 3.1 Modeling the path of an asteroid.

3.1 Polynomial Functions

- The Vertical Line Test • Linear Functions • Quadratic Functions
- Polynomial Functions • The Graphs of Polynomial Functions

We have seen various graphs in this course. Some are graphs of functions; some are not graphs of functions. There is a convenient way of telling whether a graph is that of a function.

THE VERTICAL LINE TEST

A function f assigns to each element x of its domain exactly one element $f(x)$ in the range (Section R.7). The graph of a function f is made up of all points $(x, f(x))$ in the xy-plane, where x is in the domain of f. This means that there can be only one point $(x, f(x))$ on the graph that corresponds to each x in the domain. This observation leads to the vertical line test for determining whether a given set of points is the graph of a function.

THE VERTICAL LINE TEST

A set of points in the plane is the graph of a function if and only if every vertical line intersects the graph in at most one point.

Example 1 **Recognizing the Graphs of Functions**

Determine which of the curves in Figure 3.2 are graphs of functions.

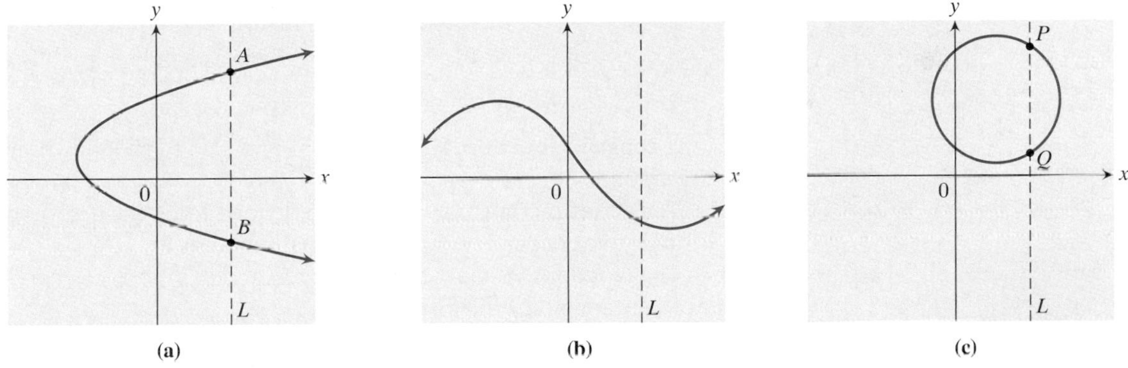

Figure 3.2

SOLUTION

(a) Not the graph of a function, because there is a vertical line L that intersects the curve at the two points A and B.
(b) The graph of a function. Every vertical line intersects the curve at one point.
(c) Not the graph of a function, because there is a vertical line L that intersects the curve at the two points P and Q.

The two classes of functions that we have paid closest attention to up to the present have been linear functions and quadratic functions. In this section, we extend the discussion to the more general class of polynomial functions.

LINEAR FUNCTIONS

A function of the form $f(x) = ax + b$, where a and b are real numbers, is called a **linear function.** The graph of f is the line $y = ax + b$. For example, the function $f(x) = 2x + 7$ is a linear function. The graph of f is the line $y = 2x + 7$.

Quadratic Functions

A function of the form $f(x) = ax^2 + bx + c$, where a, b, and c are real numbers with $a \neq 0$, is called a **quadratic function.** The graph of f is the parabola $y = ax^2 + bx + c$. For example, the function $f(x) = 2x^2 - 12x + 8$ is a quadratic function. The graph of f is the parabola $y = 2x^2 - 12x + 8$.

Linear and quadratic functions are special cases of **polynomial functions.** We find that some of the ideas that we introduced for linear and quadratic functions apply to polynomial functions in general. For example, it is useful to know the x- and y-intercepts of the graphs of polynomial functions. The idea of the vertex of a parabola generalizes to points called maxima and minima of graphs of polynomials.

Polynomial Functions

A function of the form $f(x) = a_nx^n + a_{n-1}x^{n-1} + \cdots + a_1x + a_0$, where $a_n, \ldots, a_0$ are real numbers, with $a_n \neq 0$, and n is a nonnegative integer is called a **polynomial function of degree *n* in the variable *x*.**

The following are examples of polynomial functions.

$f(x) = 2x - 3$ is a polynomial function of degree 1.
$f(x) = -4x^2 + 5x + 1$ is a polynomial function of degree 2.
$f(x) = 0.67x^4 + 0.14x^3 - 1.83x^2 + 7.26x$ is a polynomial function of degree 4.

A **constant function,** such as $f(x) = 6$, is said to be a polynomial function of degree 0.

The **zero function** $f(x) = 0$ is also defined to be the zero polynomial. There is no degree assigned to this polynomial function.

The functions $f(x) = (x + 1)/(x - 5)$ and $f(x) = 2^x + 3$ are not polynomials. They are not of the form $f(x) = a_nx^n + \cdots + a_1x + a_0$. We will study these types of functions in Sections 3.5 and 5.1.

Polynomials That Are the Product of Linear Factors

Many polynomial functions that we use in algebra are the products of linear factors; that is, they can be written in the form $f(x) = (px + q)(rx + s)\cdots$. These functions deserve special attention.

Example 2 Finding the Degree of a Polynomial Function

Show that $f(x) = (2x - 3)(x - 1)(x + 4)$ is a polynomial function of degree 3.

SOLUTION We multiply out the expression on the right.

$$\begin{aligned} f(x) &= (2x - 3)(x - 1)(x + 4) \\ &= (2x - 3)(x^2 + 3x - 4) \\ &= 2x(x^2 + 3x - 4) - 3(x^2 + 3x - 4) \\ &= 2x^3 + 6x^2 - 8x - 3x^2 - 9x + 12 \\ &= 2x^3 + 3x^2 - 17x + 12 \end{aligned}$$

f is a polynomial function of degree 3.

The result of this example can be generalized as follows.

A Polynomial Expressed as Factors

A function $f(x) = (px + q)(rx + s), \ldots, (p \neq 0, r \neq 0, \text{etc.})$ that is the product of n linear factors is a polynomial function of degree n.

Graphs of Polynomial Functions

We have seen that the graph of a quadratic function (a parabola) is a smooth curve. This characteristic applies to all polynomial functions. The graph of every polynomial function is a smooth curve with no sharp corners, jumps, or breaks. Let $f(x)$ be a typical polynomial function, with the graph shown in Figure 3.3. The **y-intercept** of a graph is the value of y at which the graph crosses the y-axis. It is the point on the graph at which $x = 0$. The y-intercept is thus $f(0)$. The **x-intercepts** of a graph are the values of x at which the graph crosses the x-axis. They are values of x for which $f(x) = 0$. Because the x-intercepts are points at which $f(x) = 0$, they are also called **zeros** of the function. The y-intercept and x-intercepts are valuable points to know in sketching the graph of a function.

Intercepts of a Graph

The y-intercept of the graph of $y = f(x)$ is $f(0)$.
The x intercepts of the graph of $y = f(x)$ are the roots of $f(x) = 0$.

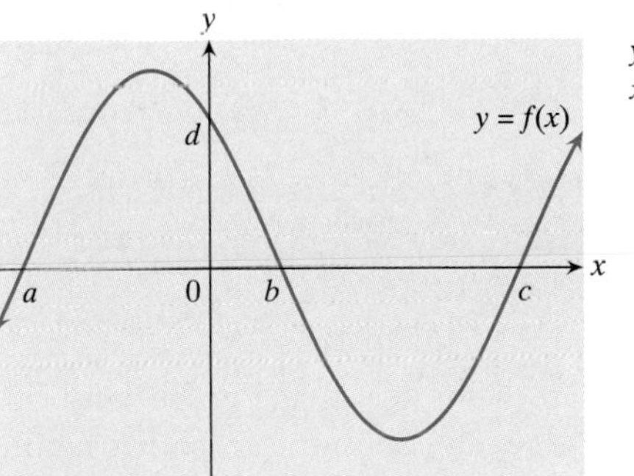

y-Intercept: $d = f(0)$
x-Intercepts: a, b, and c, are the roots of $f(x) = 0$ (called zeros of f)

Figure 3.3 The x- and y-intercepts of a graph.

Example 3 Finding x- and y-Intercepts

Find the x- and y-intercepts of the graph of $f(x) = x^3 - x^2 - 6x$.

SOLUTION The x-intercepts are the values of x for which $f(x) = 0$. They are the roots of the equation

$$x^3 - x^2 - 6x = 0$$

Factor in stages to get the roots.

$$x(x^2 - x - 6) = 0$$
$$x(x + 2)(x - 3) = 0$$

The roots are $x = 0$, $x = -2$ and $x = 3$. The x-intercepts of the graph are 0, -2, and 3.

Further, $f(0) = 0$; thus the y-intercept is 0. The graph goes through the origin.

Self-Check 1 Determine the x- and y-intercepts of the graph of

$$f(x) = x^3 + 5x^2 + 4x$$

Let us look at a method of sketching the graphs of certain polynomial functions next. We will use this technique to sketch graphs of other functions, such as rational functions, later in the course. It is an important mathematical technique that the student will meet in other areas of mathematics.

Example 4 Sketching the Graph of a Polynomial Function

Sketch the graph of $f(x) = (x + 4)(x - 1)(x - 3)$.

SOLUTION f is a polynomial function of degree 3. Let us first determine the y-intercept. It is given by $f(0)$. We get

$$f(0) = (4)(-1)(-3) = 12$$
$$y\text{-intercept} = 12$$

Now let us find the x-intercepts. They are the roots of the equation $f(x) = 0$. We get

$$(x + 4)(x - 1)(x - 3) = 0$$
$$x\text{-intercepts} = -4, 1, 3$$

The graph thus crosses the x-axis at $x = -4$, $x = 1$, and $x = 3$. These three points divide the real line into the intervals

$$(-\infty, -4), \quad (-4, 1), \quad (1, 3), \quad (3, \infty)$$

On each interval, the graph must be completely above the x-axis or completely below it because the graph has no jumps or breaks and does not cross the x-axis on the interval. To determine whether the graph lies above or below the x-axis on a given interval, compute the sign of $f(x)$ at a convenient test point in the interval. If the sign of f is positive, the graph lies above the x-axis on that interval; if the sign is negative, it lies below the x-axis. Tabulate the results as follows.

Interval	Test Point	$f(x) = (x + 4)(x - 1)(x - 3)$	Above/Below x-Axis
$(-\infty, -4)$	$x = -5$	$f(-5) = -48 < 0$	Below
$(-4, 1)$	$x = 0$	$f(0) = 12 > 0$	Above
$(1, 3)$	$x = 2$	$f(2) = -6 < 0$	Below
$(3, \infty)$	$x = 4$	$f(4) = 24 > 0$	Above

Summarize this information graphically (Figure 3.4).

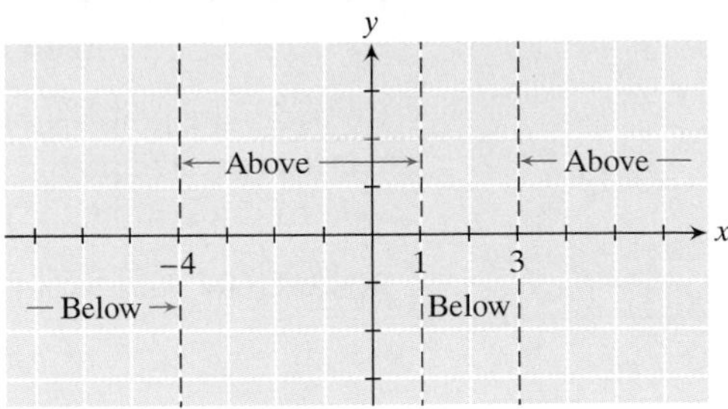

Figure 3.4

This information, together with y-intercept $= 12$ and x-intercepts $= -4, 1, 3$, leads to the rough graph shown in Figure 3.5(a). We confirm this graph on a calculator in Figure 3.5(b).

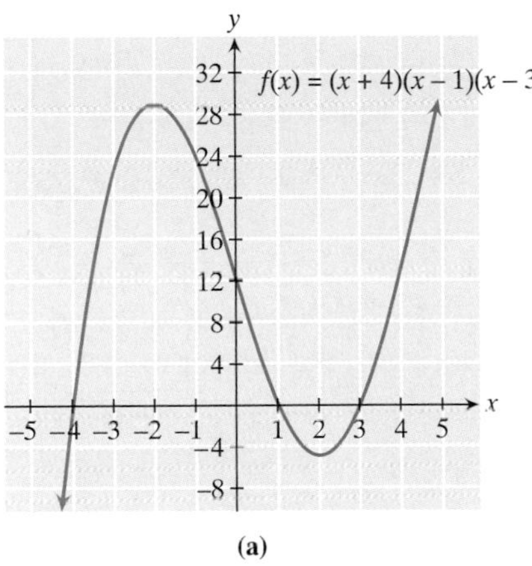

(a)

40

−5

4

−8

(b)

Figure 3.5

Exploration

Consider the graph of

$$f(x) = k(x - a)(x - b)(x - c)$$

Prove that the x-intercepts are a, b, and c.

Use a calculator to explore the graph of

$$f(x) = k(x - 1)(x - 2)(x - 5)$$

for various values of k. How does the value of k affect the graph?

$\boldsymbol{f(x) = k(x - a)(x - b)(x - c)}$

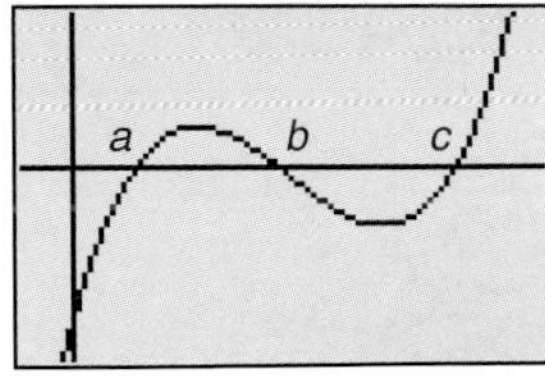

x-Intercepts are a, b, and c.

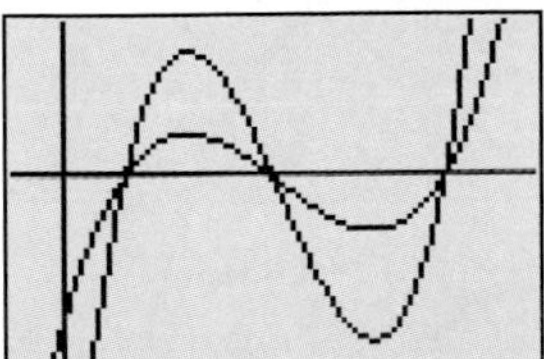

k is a stretching/compressing factor.

Example 5 The Equation of a Polynomial Function with Given Intercepts

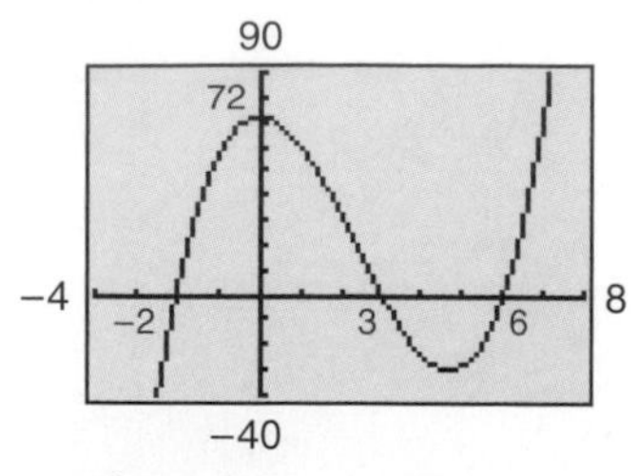

Figure 3.6

Consider the graph shown in Figure 3.6, which has x-intercepts -2, 3, and 6 and y-intercept 72. Find the polynomial function that has this graph. Support your answer by duplicating the graph on a calculator and checking the x- and y-intercepts.

SOLUTION Guided by the preceding exploration, let the function be

$$f(x) = k(x - a)(x - b)(x - c)$$

with x-intercepts a, b, and c. Because the x-intercepts of our graph are -2, 3, and 6, then $a = -2$, $b = 3$, and $c = 6$. The function can thus be written

$$\begin{aligned} f(x) &= k(x - (-2))(x - 3)(x - 6) \\ &= k(x + 2)(x - 3)(x - 6) \end{aligned}$$

It remains to find the value of k. The y-intercept is 72; thus, $f(0) = 72$. Letting $x = 0$ in this equation for $f(x)$ gives

$$\begin{aligned} 72 &= k(2)(-3)(-6) \\ &= 36k \end{aligned}$$

Therefore, $k = 2$. The function is $f(x) = 2(x + 2)(x - 3)(x - 6)$.

We leave it to the student to draw the graph of this function on a calculator to see that the shape is correct and that the intercepts are the specified ones.

Answer to Self-Check Exercise

1. $f(x) = x^3 + 5x^2 + 4x = x(x^2 + 5x + 4) = x(x + 1)(x + 4)$.
   $f(x) = 0$ when $x = 0, -1, -4$. The x-intercepts are $0, -1, -4$.
   $f(0) = 0$. The y-intercept is 0.

EXERCISES 3.1

In Exercises 1–6, use the vertical line test to determine whether the given graphs are those of functions.

1.

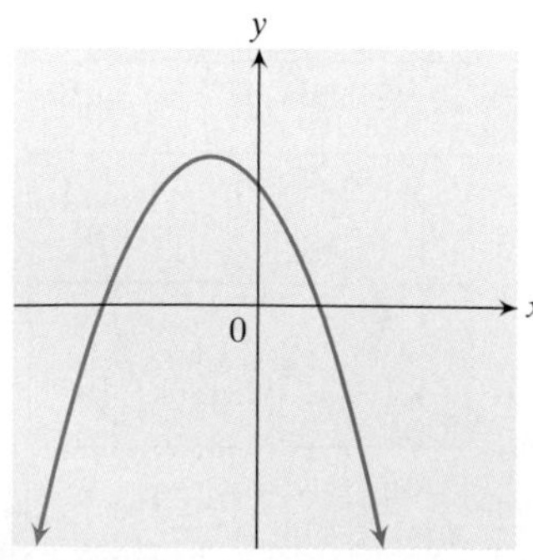

2.

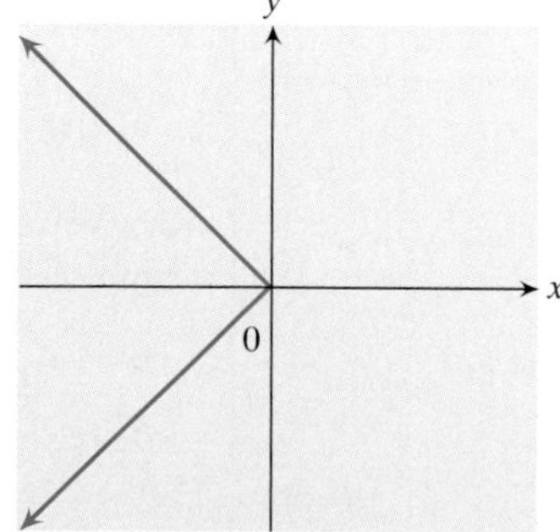

3.

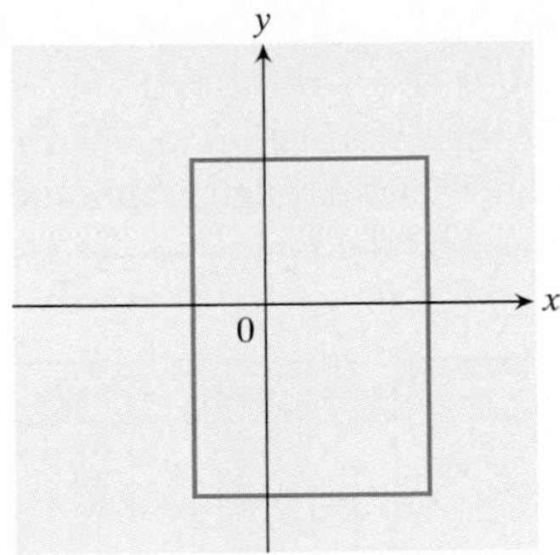

4.

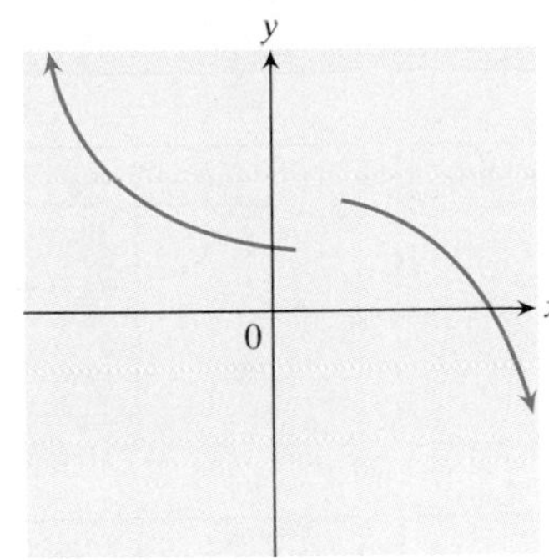

5.

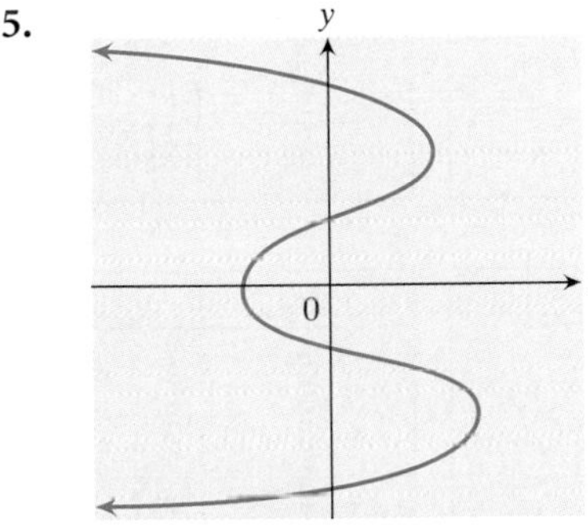

6.

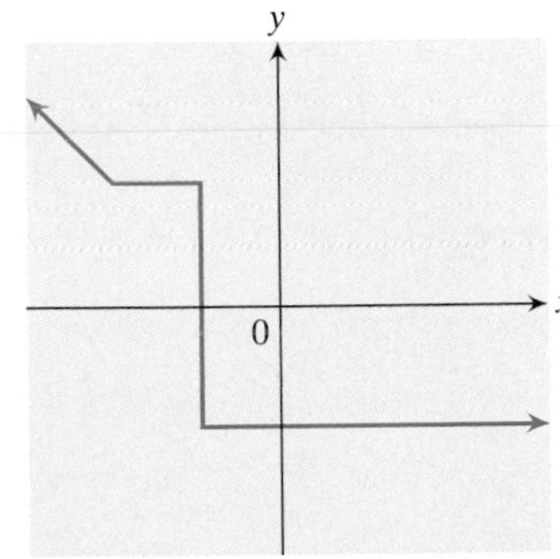

In Exercises 7–10, confirm the degrees of the polynomial functions.

7. Show that $f(x) = (x - 3)(x + 5)$ is a polynomial function of degree 2.
8. Show that $f(x) = (3x - 2)(x + 4)$ is a polynomial function of degree 2.
9. Show that $f(x) = (x + 1)(x - 2)(x + 5)$ is a polynomial function of degree 3.
10. Show that $f(x) = (2x + 1)(x - 3)(x + 4)$ is a polynomial function of degree 3.

In Exercises 11–16, determine the x-intercepts of the graphs of the polynomial functions. *Hint:* First factor; then solve $f(x) = 0$.

11. $f(x) = x^2 - 2x - 3$
12. $f(x) = x^2 - 2x + 1$
13. $f(x) = 2x^2 - 4x - 6$
14. $f(x) = 6x^2 + x - 1$
15. $f(x) = x^2 - 4x - 5$
16. $f(x) = 3x^2 + 7x + 4$

In Exercises 17–22, determine the x-intercepts of the graphs of the polynomial functions.

17. $f(x) = x^3 - 3x^2 + 2x$
18. $f(x) = 2x^3 - 5x^2 - 3x$
19. $f(x) = x(x - 2)(x + 5)$
20. $f(x) = x(2x - 3)(5x + 1)$
21. $f(x) = x^2(x + 3)(3x - 7)$
22. $f(x) = (x - 2)(x + 3)(2x - 4)$

In Exercises 23–28, determine the x- and y-intercepts of the graphs of the functions, and sketch the graphs using algebraic techniques. Support your answers by duplicating the graphs on your calculator.

23. $f(x) = (x - 1)(x - 3)(x - 6)$
24. $f(x) = x(x + 3)(x - 5)$
25. $f(x) = (x + 4)(x + 1)(2 - x)$
26. $f(x) = x^2(x - 3)(x + 4)$
27. $f(x) = x^2(x - 2)^2(x + 3)$
28. $f(x) = x^3(x + 5)(x - 7)$

In Exercises 29–34, determine the x- and y-intercepts of the graphs of the functions, and sketch the graphs using algebraic techniques. Support your answers by duplicating the graphs on your calculator.

29. $f(x) = x^4 - 4x^2$
30. $f(x) = x^4 + 3x^3$
31. $f(x) = x^3 - 7x^2 + 10x$
32. $f(x) = x^4 + 2x^3$
33. $f(x) = x(x^2 - 1)(x + 3)$
34. $f(x) = (x^2 - 4)(x^2 - 9)$

In Exercises 35–38, find the polynomial functions having the given graphs. The x- and y-intercepts of the graphs are given. Support your answers by duplicating the graphs on your calculator and checking the intercepts. *Hint:* Let

$f(x) = k(x - a)(x - b)(x - c)$. Find the values of a, b, c, and k.

35.

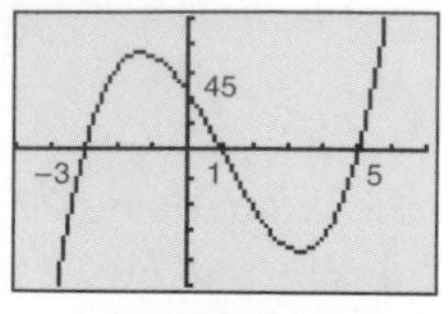

36.

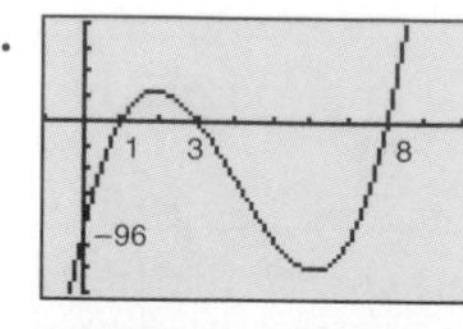

37.

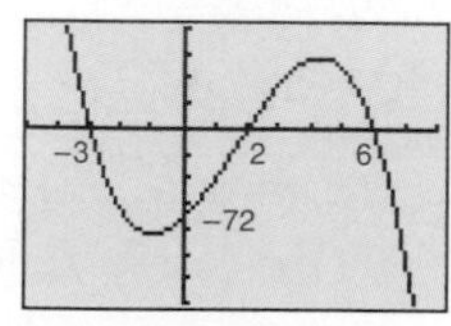

38.

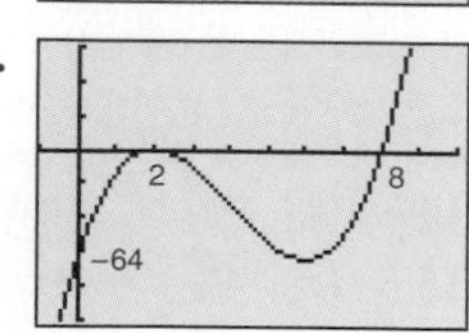

In Exercises 39–42, construct a polynomial function with a graph having the given properties. Check your answers with a calculator display.

39. Degree 2 with two x-intercepts at $x = -2$ and $x = 3$, with f being positive on $(-\infty, -2)$ and $(3, \infty)$ and negative on $(-2, 3)$.

40. Degree 2 with two x-intercepts at $x = 1$ and $x = 5$, with f being negative on $(-\infty, 1)$ and $(5, \infty)$ and positive on $(1, 5)$.

41. Degree 3 with three x-intercepts at $x = -2$, 1, and 4, with f being negative on $(-\infty, -2)$ and $(1, 4)$ and positive on $(-2, 1)$ and $(4, \infty)$.

42. Degree 3 with three x-intercepts at $x = -3$, 0, and 6, with f being positive on $(-\infty, -3)$, and $(0, 6)$, and negative on $(-3, 0)$ and $(6, \infty)$.

In Exercises 43–46, the tables describe the behavior of polynomial functions of the form

$$f(x) = k(x - a)(x - b)(x - c)$$

Use the tables to find the x- and y-intercepts of each graph, where the graph is above the x-axis and where it is below the x-axis. Use this information to determine the equation of the function. Sketch a rough graph, and check your answers by duplicating the table and the graph on your calculator.

43.

X	Y1
-7	-198
-6	-80
-5	0
-4	48
-3	70
-2	72
-1	60

X	Y1
0	40
1	18
2	0
3	-8
4	0
5	30
6	88

44.

X	Y1
-4	55
-3	0
-2	-27
-1	-32
0	-21
1	0
2	25

X	Y1
3	48
4	63
5	64
6	45
7	0
8	-77
9	-192

45.

X	Y1
-4	-270
-3	-96
-2	0
-1	36
0	30
1	0
2	-36

X	Y1
1	0
2	-36
3	-60
4	-54
5	0
6	120
7	324

46.

X	Y1
-1	24
0	5
1	0
2	3
3	8
4	9
5	0

X	Y1
1	0
2	3
3	8
4	9
5	0
6	-25
7	-72

Construction of Functions, Optimization, and Drug Administration

- Increasing and Decreasing Functions • Maxima and Minima of Functions
- Immigrants to the United States • Oral Administration of Drugs
- Laying Down Fiber Optic Cables • Tracking an Asteroid

So far, we have seen how functions are used to describe many phenomena. In applications one is often interested in knowing where functions are increasing in value, where they are decreasing, and where they reach their maximum or minimum values. For example, when tracking an epidemic, authorities can use a function to describe how long the epidemic will spread, when it will peak, and when it will probably be over. Astronomers who have computed the path of an asteroid will want to know how close it will get to the Earth. After a drug is administered orally, the concentration in the blood gradually increases to a maximum and then drops off with time. Physicians need to know when the concentration peaks. In this section, we develop the mathematics for constructing functions to describe situations such as these and show how to use the functions to arrive at answers to the questions posed. We have already laid the mathematical groundwork for discussing such problems. The concepts of maxima and minima* introduced in the discussion of vertices of quadratic functions in Section 1.4 will be extended to other functions in this section.

Increasing and Decreasing Functions and Maxima and Minima of Functions

Consider the graph of a function $f(x)$ shown in Figure 3.7. Let us clarify the ideas of **increasing, decreasing,** and **constant** behavior over intervals. We read this graph from left to right and say that f is

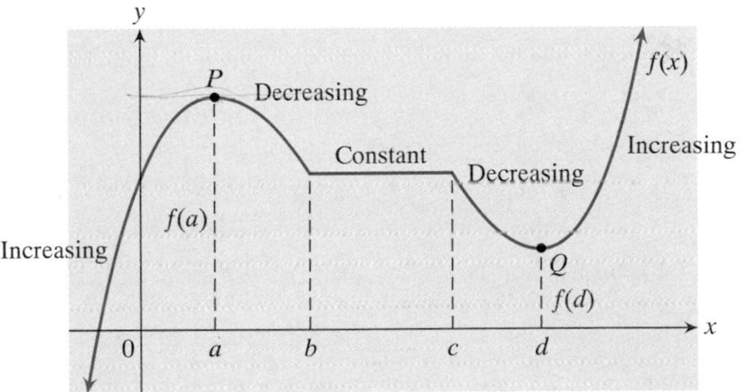

Figure 3.7

increasing over $(-\infty, a]$
decreasing over $[a, b]$
constant over $[b, c]$
decreasing over $[c, d]$
increasing over $[d, \infty)$

We also introduce the concepts of **local maxima** and **local minima.** Observe that $f(a)$ is greater than the value of f at neighboring points. Further, $f(d)$ is smaller than the value of f at neighboring points. We say that

f has a local maximum of $f(a)$ at $x = a$
f has a local minimum of $f(d)$ at $x = d$

We also say that the graph is increasing, decreasing, or constant over these intervals and that the graph has a local maximum at the point P and a local minimum at the point Q.

*The plural of maximum is maxima, and the plural of minimum is minima.

Let us look at these concepts within a familiar setting, namely the parabola. Recall that we discussed the vertex of a parabola in Section 1.4. There is a maximum or a minimum at the vertex, and the function is increasing on one side of the vertex and decreasing on the other.

Example 1 | Finding the Maxima and Minima of a Function

Sketch the graph of $f(x) = x^2 - 4x + 7$. Determine any maxima and minima and where the function is increasing and decreasing.

SOLUTION The graph of the standard quadratic function $f(x) = ax^2 + bx + c$ is a parabola with vertex at $x = -b/2a$. The graph opens up if $a > 0$ and down if $a < 0$. Compare $f(x) = x^2 - 4x + 7$ with this form; we get $a = 1$, $b = -4$, and $c = 7$. The vertex of f is at

$$x = \frac{-b}{2a} = \frac{-(-4)}{2} = 2$$

The corresponding value of f is

$$f(2) = (2)^2 - 4(2) + 7 = 3$$

The vertex is thus (2, 3).

Because $a = 1 > 0$, the graph opens up. The y-intercept is given by $f(0)$. The y-intercept is 7. This information leads to the graph shown in Figure 3.8. The function f has the following properties:

minimum of 3 at $x = 2$
decreasing over $(-\infty, 2]$
increasing over $[2, \infty)$

The minimum of 3 in this instance is the smallest value of f for the whole graph. We call it the minimum value of f. The function does not have a maximum.

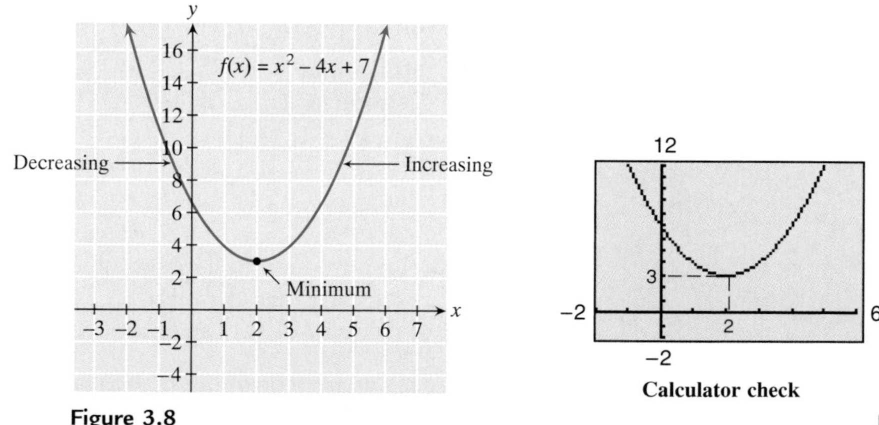

Figure 3.8

The Number of Maxima and Minima of a Function

The maxima and minima are often the most important points on the graph of a function. In displaying the graph of a given function on a calculator it is often important to know that one has a viewing window that displays all these points. The following result is helpful.

Number of Maxima and Minima

A polynomial function of degree n has at most $n - 1$ local maxima and minima.

Caution: This result puts an upper limit on the number of maxima and minima. The actual number of maxima and minima combined could be less than $n - 1$.

Example 2 Finding Maxima and Minima

Find the maxima and minima of $f(x) = x^3 - 7x^2 + 9x + 4$, rounded to four decimal places.

SOLUTION We graph the function in a suitable window (Figure 3.9). Note that this window displays a local maximum and a local minimum. The function is of degree 3. It thus has at most two maxima and minima. The window therefore shows all the maxima and minima. Use a max/min finder (or trace/zoom) to find the maximum and minimum points, as for a parabola in Section 1.4. We get

local maximum of 7.2362 at $x = 0.7699$
local minimum of -8.0510 at $x = 3.8968$

with $f(x)$ increasing over $(-\infty, 0.7699]$, decreasing over $[0.7699, 3.8968]$, and increasing over $[3.8968, \infty)$.

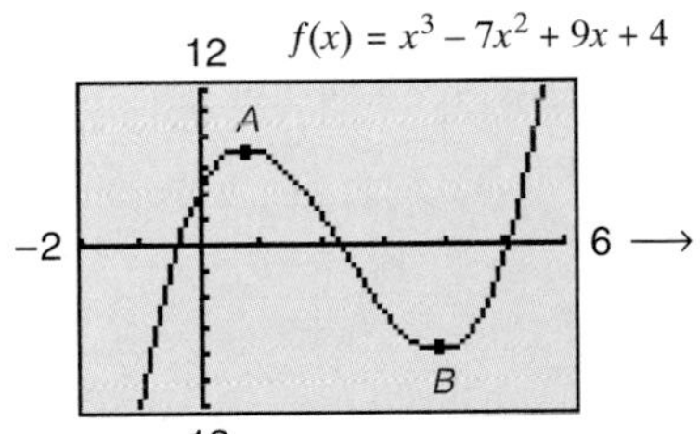

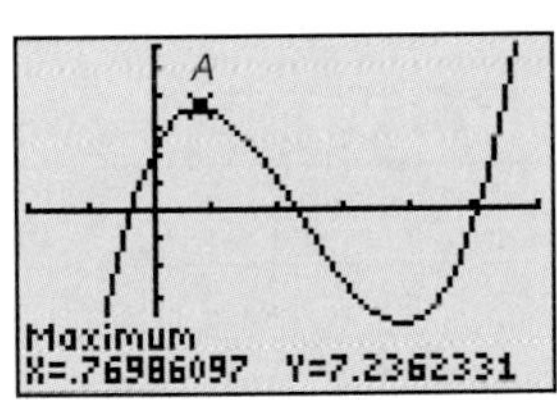

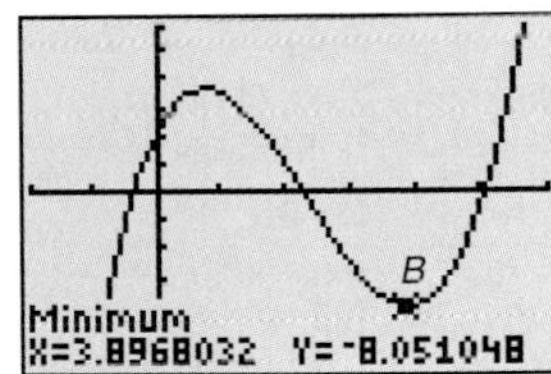

Figure 3.9

Self-Check 1 Sketch the graph of $f(x) = -x^3 + 6x^2 - 10x + 9$ on a graphing calculator. Determine the maxima and minima of f, where it is increasing, and where it is decreasing, to four decimal places.

Now let us see how these ideas are implemented. We shall construct functions that describe real situations and find where those functions increase, where they decrease, and where they have maxima and minima. Sometimes we will arrive at results using algebra, and then use a graph to confirm the results. At other times, we will arrive at the initial result using a graph. Very often the student has a choice of which method to use. We encourage familiarity with both approaches. As we have seen so often, algebra and geometry complement one another and give deeper insight into a problem.

Applications . . .

The fact that a polynomial function of degree n has at most $n - 1$ local maxima and minima means that a polynomial of degree 3 often has one maximum and one minimum. A polynomial of degree 4 often has two maxima and one minimum or one maximum and two minima. We can use this information to construct polynomial functions that closely fit given data. The following example illustrates how we are led to a polynomial of degree 3 to describe the percentage of foreign-born people living in the United States.

Example 3 Immigrants to the United States

The following table gives the percentage of the U.S. population that was foreign-born for the years 1900 through 1995. (*The World Almanac,* 1997, p. 383) The author is one of these immigrants having come to the United States from Great Britain in 1962!

Year	1900	1910	1920	1930	1940	1950	1960	1970	1980	1990	1995
Percent foreign-born	13.6	14.7	13.2	11.6	8.8	6.9	5.4	4.8	6.2	7.9	8.8

Let us find a least-squares function that best fits these data and use it to make predictions. We enter and plot these data on a calculator using the years as measured from 1900 for the horizontal axis (Figure 3.10) so that 1900 corresponds to $x = 0$, 1910 to $x = 10$, and so on. Observe that the data seem to follow a graph that has one maximum and one minimum, suggesting a polynomial of degree 3. The least-squares polynomial of degree 3 is computed and graphed. The fit is good. The relationship between the time (as measured from 1900) and the percentage of foreign-born people living in the United States is closely described by the following function.

$$P(t) = 0.000065681704t^3 - 0.0075753924t^2 + 0.0779685048t + 14.03222399$$

Let us use this function to predict the percentage of foreign-born people in the year 2000, 100 years after the base year 1900. We get $P(100) = 11.75685447$. On the basis of these statistics, the foreign-born people living in the United States in the year 2000 will be about 11.76%, which is about the same as in 1930.

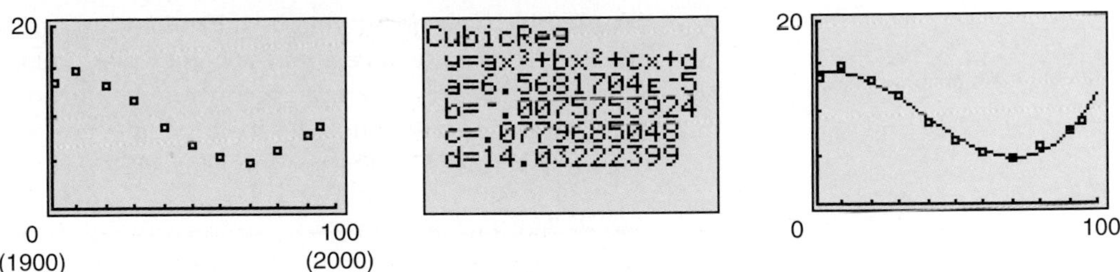

Figure 3.10 Foreign-born living in the United States.

Example 4 Oral Administration of Drugs

After a drug is administered orally, the concentration in the blood rises fairly sharply to a maximum and then drops off more slowly. This behavior can be described by part of the graph of a polynomial of degree 4. The following function describes the typical behavior of concentration as a function of time:

$$C(t) = -0.0067t^4 + 0.1794t^3 - 1.8393t^2 + 7.2226t$$

The units of time are hours, and the units of drug are micrograms per milliliter (μg/mL). This function was derived by fitting a polynomial of degree 4 to data, as in the previous example. We first graph the function (Figure 3.11). Observe that the maximum concentration occurs after 3.28 hours.

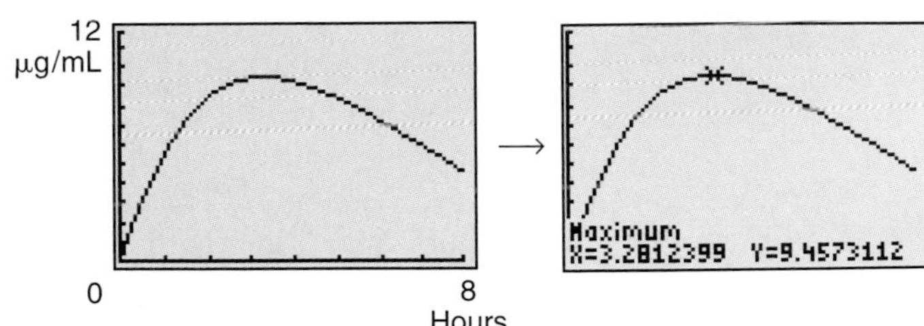

Figure 3.11 Drug concentration.

Suppose that a patient must be monitored carefully while the level of this drug in the blood is more than 9.0 μg/mL. We leave it to you to show that this period is from 2.40 hours after administration to 4.33 hours.

Further Ideas 1

The National Center for Health Statistics has released the following information regarding AIDS for the period 1990 to 1996. (See the Williams Web site at www.saunderscollege.com/math/williams to link to the Inernet for more information.) These statistics are for the highest risk group, namely adults between 25 and 44 years of age. Determine a polynomial function that closely fits these data, and discuss the graph. Numbers in parentheses give years from the base year 1990. Use these numbers for the horizontal axis.

Year	1990 (0)	1991 (1)	1992 (2)	1993 (3)	1994 (4)	1995 (5)	1996 (6)
Deaths per 100,000	23.3	26.5	29.9	32.9	36.7	36.9	27.2

Although the statistics are definitely encouraging, one naturally must be guarded in drawing conclusions from such limited data.

Of course, the method of constructing functions varies from situation to situation. We have seen examples in which regression analysis was used. We discussed guidelines for constructing mathematical models in Section 2.2. The following examples build on these ideas and illustrate further techniques for constructing functions.

Example 5 Enclosing a Paddock

Lance Williams has 800 feet of fencing to enclose a rectangular paddock. Prove that the maximum area he can enclose is a square with sides of 200 feet.

SOLUTION Let the length of one side of the paddock be x and the length of the other side be y (Figure 3.12). Area $A = xy$.

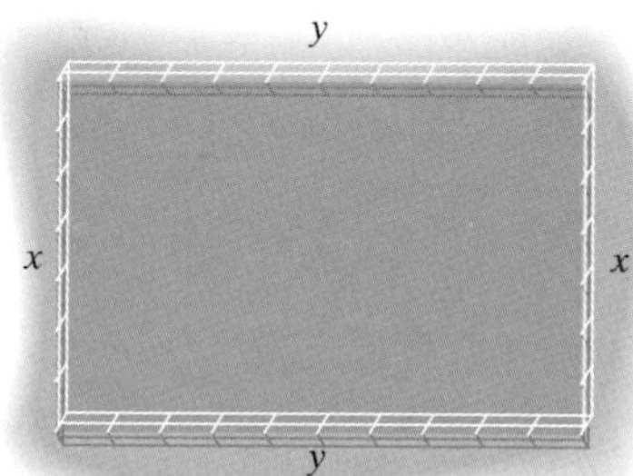

Figure 3.12

We want to find the values of x and y that maximize A. We find that x and y are not independent. This fact is extremely important because it enables us to express A in terms of the single variable x. We then have the tools to find the maximum value. The length of the perimeter is 800 feet. Thus,

$$2x + 2y = 800$$
$$x + y = 400$$

Solve this equation for y, and substitute the value of y into the expression $A = xy$.

$$y = 400 - x$$
$$A = x(400 - x)$$

A is now a function of a single variable. We graph $A(x)$ and find its maximum value.

$A(x)$ has a maximum at the point P (Figure 3.13). We see that the maximum area is $A = 40{,}000$ when $x = 200$. The corresponding value of y is $y = 400 - x = 200$. Thus, the area of the paddock is a maximum when the region is a square with sides of 200 feet.

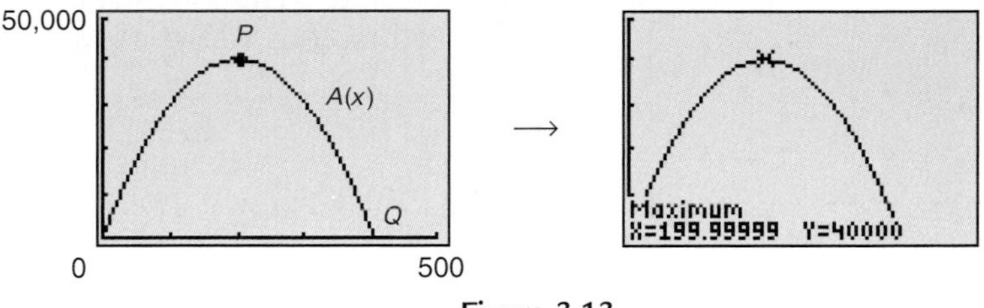

Figure 3.13

Further Ideas 2

(a) In Example 5, we used a graphing calculator to find the value of x that led to the maximum value A. Observe that the equation $A = x(400 - x)$ is a quadratic equation. Its graph is thus a parabola. Use algebra to show that the vertex of this parabola is (200, 40,000), confirming the result already obtained.

(b) Consider the graph of $A(x)$ in Example 5. Determine the intervals of x over which x is increasing and where it is decreasing. What does this gradual increase and decrease mean in terms of area? To what do the points O and Q in Figure 3.13 correspond in terms of the physical situation?

Example 6 Laying Down Fiber Optic Cables

Computer terminals are to be installed in Elizabeth Hall of Stetson University. The building is across a street 50 feet wide, and is 400 feet north of the Computer Center (Figure 3.14). The terminals are to be hard wired with fiber optic cables. The cost of laying the cables under the street is \$15 per foot and under the grass south of Elizabeth Hall is \$10 per foot. How should the lines be laid to minimize cost to the university?

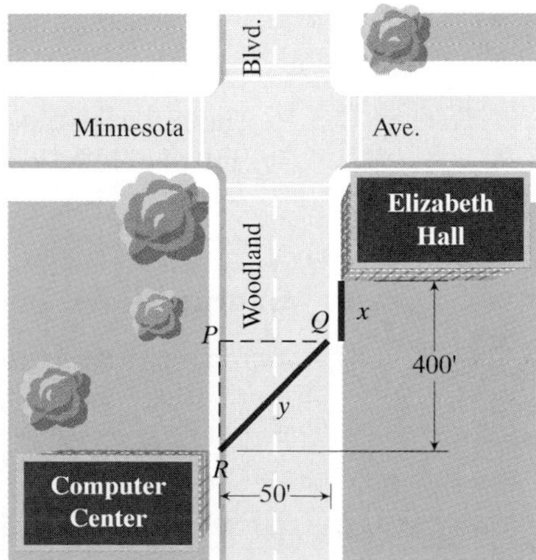

Figure 3.14 Fiber optic plan for Stetson University.

SOLUTION Let x feet of cable be laid under the grass and y feet under the road (see Figure 3.14). The total cost is

$$C = 10x + 15y$$

The variables x and y are not independent, however. Consider the triangle PQR. The Pythagorean Theorem gives

$$QR^2 = PR^2 + PQ^2$$

We have that $QR = y$, $PQ = 50$, $PR = 400 - x$. Thus,

$$y^2 = (400 - x)^2 + 50^2$$

Take the square root of both sides and simplify.

$$y = \sqrt{x^2 - 800x + 162{,}500}$$

Substitute this expression for y into $C = 10x + 15y$ to get

$$C(x) = 10x + 15\sqrt{x^2 - 800x + 162{,}500}$$

We graph $C(x)$ using a calculator and determine the minimum (Figure 3.15). The minimum value is $C = 4559$ when $x = 355$.

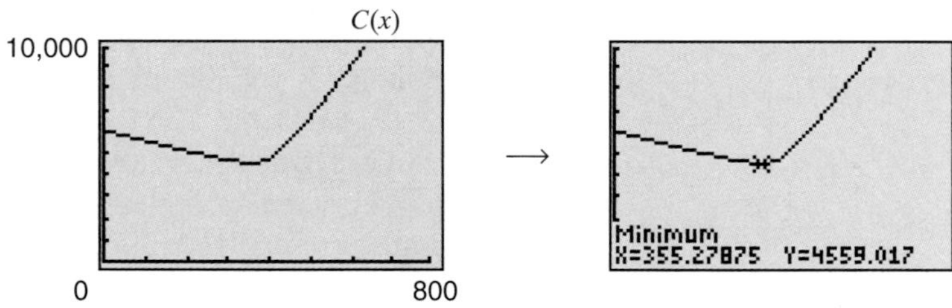

Figure 3.15

The corresponding value of y is $\sqrt{x^2 - 800x + 162{,}500} = 67$. Thus, 355 feet of cable should be laid under the grass and 67 feet should be laid under the road.

Example 7 Tracking an Asteroid

Astronomers have determined that the path of a certain asteroid as it approaches Earth is closely described in a convenient coordinate system by the equation $y = 0.005x^2 - 0.6x + 25$. Earth is at the origin of this coordinate system and x and y are measured in millions of miles (see Exercise 27 for the derivation of this equation). Let us draw the path of the asteroid and then determine how close it gets to Earth.

We get the trajectory shown in Figure 3.16. The path of the asteroid A shows that it is nearest to Earth somewhere around the point **x**.

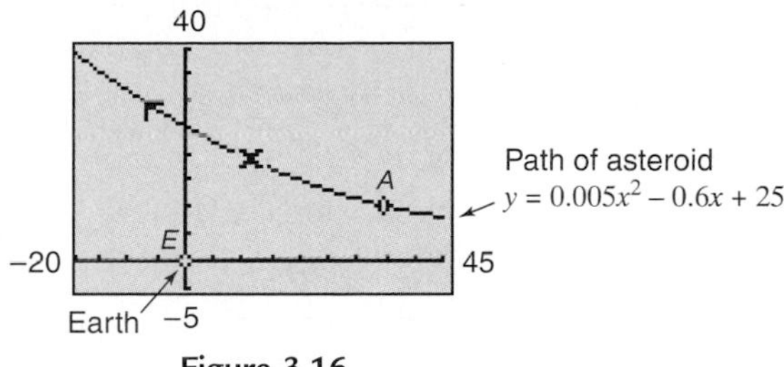

Figure 3.16

Let us derive a function for the distance d between Earth E and the asteroid A. The minimum value of this function tells us how close the asteroid comes to Earth. The point A lies on the curve $y = 0.005x^2 - 0.6x + 25$. It is thus a point of the form

$$A(x, 0.005x^2 - 0.6x + 25)$$

d is the distance between the points $E(0, 0)$ and $A(x, 0.005x^2 - 0.6x + 25)$. Using the distance formula, we get

$$\begin{aligned} d(x) &= \sqrt{(x-0)^2 + ((0.005x^2 - 0.6x + 25) - 0)^2} \\ &= \sqrt{x^2 + (0.005x^2 - 0.6x + 25)^2} \end{aligned}$$

We plot the graph of this function and find that it has a minimum value of 21.913621 (Figure 3.17). The closest the asteroid gets to Earth is thus approximately 21.91 million miles.

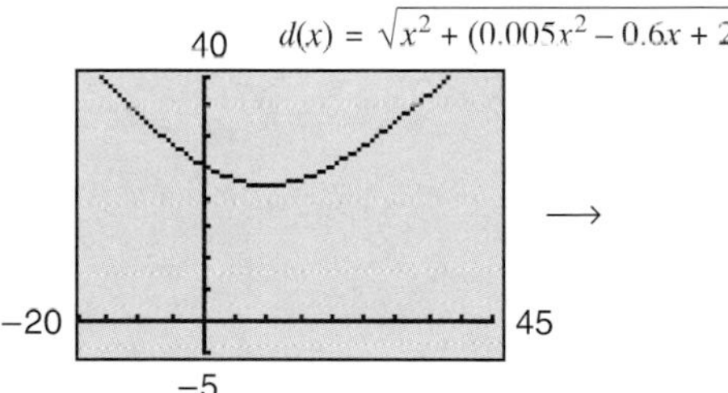

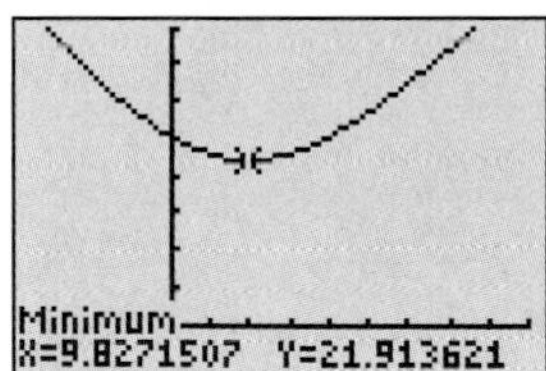

Figure 3.17

Answers to Self-Check Exercise

1. Local minimum of 3.9113 at $x = 1.1835$; local maximum of 6.0887 at $x = 2.8165$. Decreasing over $(-\infty, 1.1835]$, increasing over $[1.1835, 2.8165]$, and decreasing over $[2.8165, \infty)$.

Answers to Further Ideas Exercises

1. The data rise slowly then drop off sharply (see screen below). It is similar to the behavior of drug dissipation in Example 4, with the drop coming at the end. Part of a polynomial of degree 4 fits these data. However, note

that this function predicts that AIDS will soon disappear. Although the statistic of 27.2 for 1996 is encouraging, in that it indicates a downward trend, this single large drop probably skews the data too much. One will have to wait for more data to find a function that better describes a downward trend in the number of cases of AIDS.

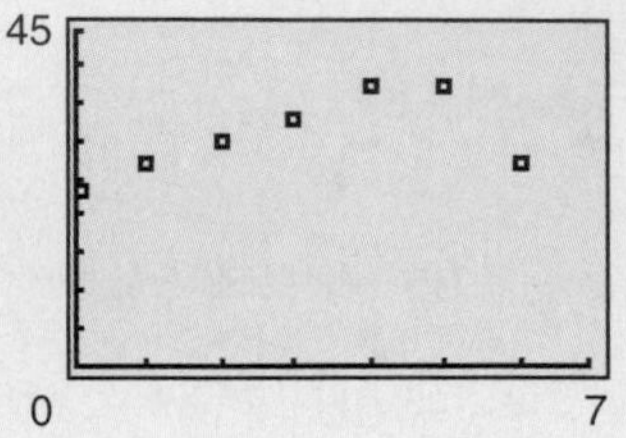

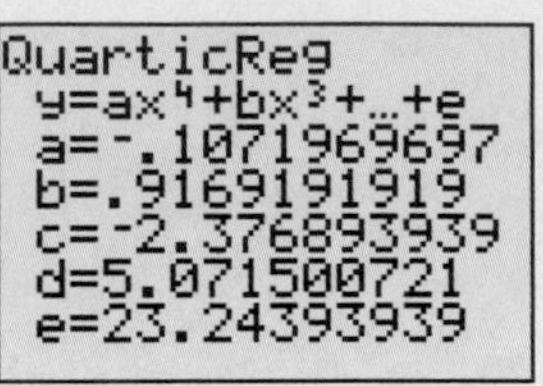

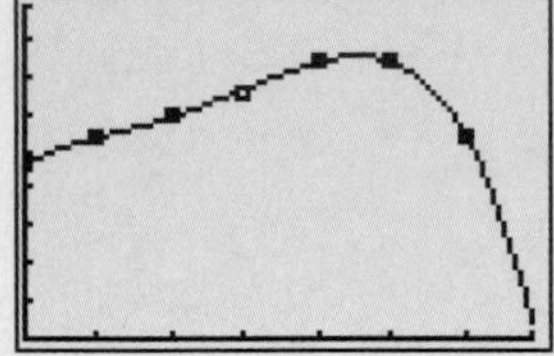

2. **(b)** A is increasing over $[0, 200]$ and decreasing over $[200, 400]$. At $O(0, 0)$ the fence is all used up in one direction, enclosing no area. At $Q(400, 0)$, the fence is all used up in the other direction, enclosing no area. The area gets bigger and bigger as the region gets closer and closer to being a square. It peaks at square shape and then decreases again.

EXERCISES 3.2

In Exercises 1–6, graphs of functions are given. Determine the x-intervals over which the functions are increasing, decreasing, and constant, and find the x values of any local maxima or minima.

1.

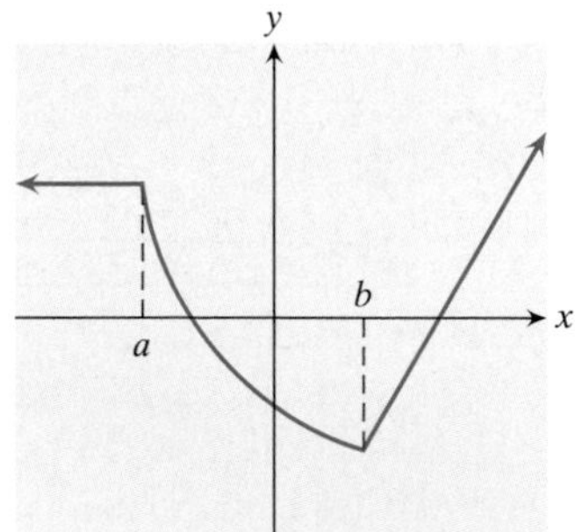

2.

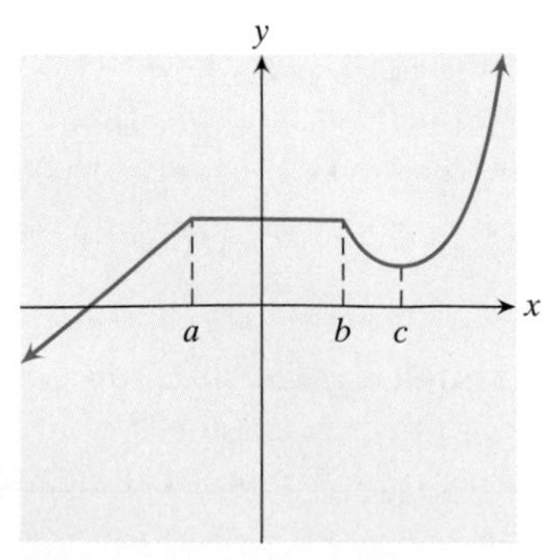

3.

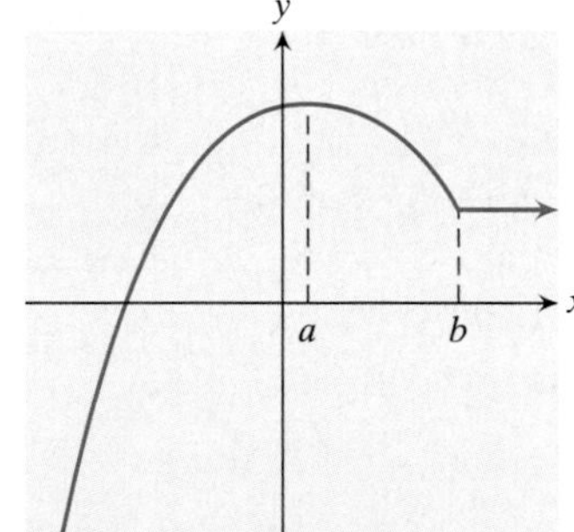

4.

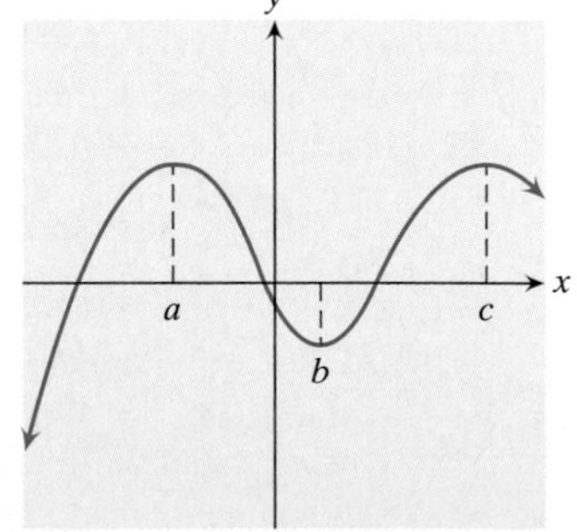

5.

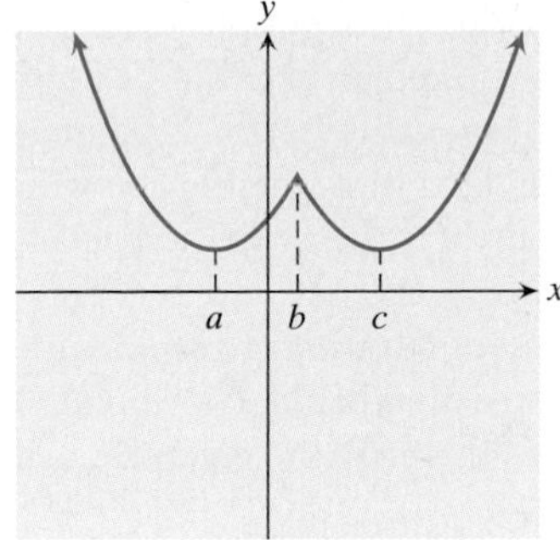

6.

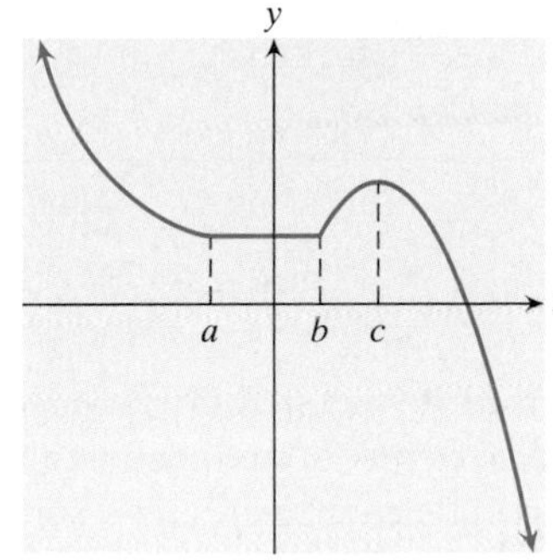

In Exercises 7–12, sketch a graph of the quadratic functions using algebra (see Example 1). Find the local maxima and minima and the intervals over which the quadratic functions are increasing and decreasing. Check your results using a graphing calculator.

7. $f(x) = x^2 + 8x + 10$ **8.** $f(x) = -x^2 + 4x + 4$

9. $f(x) = -4x^2 + 8x + 7$

10. $f(x) = 2x^2 - 16x + 35$

11. $f(x) = x^2 - 6x + 4$ **12.** $f(x) = -2x^2 - 8x - 5$

In Exercises 13–20, graph the functions on a graphing calculator. Find the local maxima and minima and the intervals over which the functions are increasing and decreasing, rounded to four decimal places.

13. $f(x) = 4x^3 - 3x^2 - 8x + 5$

14. $f(x) = -x^3 + x^2 + x + 7$

15. $f(x) = -5x^3 + 7x^2 + 3x + 9$

16. $f(x) = 2x^3 + 5x^2 + 3x + 1$

17. $f(x) = -2x^3 + 7x^2 + 9x - 4$

18. $f(x) = x^3 - 4x^2 - 53x + 168$

19. $f(x) = x^3 + 34x^2 + 40x - 3000$

20. $f(x) = -x^3 + 37x^2 - 320x - 52$

21. Analyzing a Duck Population The number of rare ducks in the Wilmot Zoo varies. The population between 1990 and 1998 is described by the function

$$N(t) = 2t^3 - 27t^2 + 84t + 416$$

where $t = 0$ corresponds to 1990. When in this 8-year period did the population peak, and when was it at its lowest?

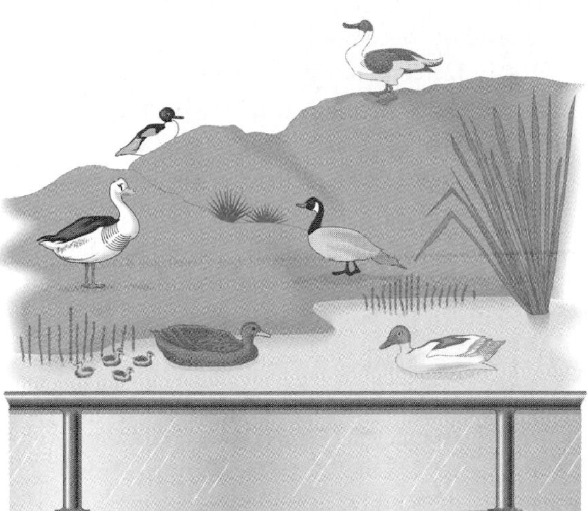

22. Motion Under Gravity An object is projected vertically upward from ground level with a velocity of 384 feet per second. The height s after t seconds is given by the quadratic function

$$s(t) = -16t^2 + 384t$$

(a) Find when the object reaches maximum height and determine that height.

(b) Determine the intervals of time over which the height is increasing and when it is decreasing.

23. Planning Book Production Bryce and Rand Book Publishers have undertaken market research to help decide how many copies of the first edition of its college physics book to produce. Printing too many copies would mean that a number would remain unsold for a long period, whereas printing too few would mean that the company could not take full advantage of the available market. The research group arrived at the following function relating profit P (in thousands of dollars) to number of books produced N (in thousands of books). How many copies of the first edition should the publisher produce to achieve the largest profit?

$$P(N) = -\frac{N^3}{15} + \frac{4N^2}{5} + 4N$$

24. Oral Administration of Drugs The following function C describes the concentration of a drug in the body t hours after it has been administered orally to a patient:

$$C(t) = -0.0082t^4 + 0.1932t^3 - 1.7651t^2 + 6.4283t$$

(a) Determine when the amount of the drug in the blood peaks and what this amount is.

(b) The patient must be carefully monitored while the level of this drug in the blood is above 7.3 μg/mL. Determine when this is.

25. **Popularity of Brand Names** The following function is based on market research carried out annually for the years 1978 to 1998 to gauge public opinion regarding brand names. The function reflects the people in a sample size of 3500 each year who agreed with the statement "I try to stick to well-known brand names."

$$f(x) = -0.0004044x^4 + 0.01745x^3 - 0.1505x^2 - 1.4929x + 82.19$$

The domain of f is $[0, 20]$, with 0 corresponding to the year 1978 and 20 corresponding to 1998. The range is $[0, 100]$, with the units giving the percentage of people who agreed with this statement. Graph the function, and use it to describe the public attitude toward brand names during this period. Is the interest in brand names on the rise or has it peaked and is decreasing?

26. **Long Distance Trucking** The Delta Long Distance Trucking Company has data indicating that the cost in dollars C of operating a truck that averages v miles per hour for a trip of distance d miles is

$$C(v) = \frac{22d}{v} + 307 + \frac{680v}{87}$$

(a) Use a graphing calculator to determine the velocity (to the nearest mile per hour) at which the trucks should travel to minimize costs on 1000-mile trips?

(b) Examine graphs for various values of d to determine whether the optimal velocity will increase or decrease as the distance of the trips increases.

(c) At what distance will the driver be tempted to average 75 miles per hour?

27. **Tracking an Asteroid** In Example 7, we discussed how astronomers tracked an asteroid as it approached the vicinity of the Earth. The equation of the path of the asteroid in a suitable coordinate system with the Earth at the origin was given to be $y = 0.005x^2 - 0.6x + 25$. Astronomers derived this equation using quadratic regression with the following data, consisting of the coordinates of the approaching asteroid. Verify the equation using these data.

x coordinate (million miles)	40	36	32	28	24	20
y coordinate (million miles)	9.00	9.88	10.92	12.12	13.48	15.00

28. **Tracking an Asteroid** The asteroid Hannaford (named after its discoverer) is approaching the vicinity of the Earth. Its coordinates, in a suitable system with the Earth at the origin, are as follows. Use quadratic regression to show that the equation of its path is approximately

$$y = -0.02x^2 + 2.6x - 34.5$$

x coordinate (million miles)	40	35	30	25	20
y coordinate (million miles)	37.5	32.0	25.5	18.0	9.5

29. **Tracking an Asteroid** The path of the asteroid Hannaford as viewed from Earth is approximately $y = -0.02x^2 + 2.6x - 34.5$, where x and y are measured in millions of miles (Exercise 28). Determine how close the asteroid gets to Earth.

30. **Spread of an Epidemic** An epidemic is spreading through a certain county. The medical authorities have determined the following data on the total number of cases every 5 months since the initial outbreak. It is now 30 months since the initial outbreak was recorded.

(a) Use cubic regression to find the function $N(t)$ that describes the behavior of this epidemic.

(b) Show, using a graph and data points, that your function closely fits the data points.

(c) Use the function to predict when the disease will peak (to the nearest month).

(d) After how many months will there be no cases of the epidemic if the current trend continues?

Time in months from outbreak, t	5	10	15
Total number of cases of epidemic, N	16,375	59,000	118,125

Time in months from outbreak, t	20	25	30
Total number of cases of epidemic, N	184,000	246,875	297,000

31. Nuclear Warheads It has been shown that the total number of nuclear warheads in the world during the 50-year period from 1945 to 1995 is closely described by the function

$$N(t) = -0.075t^4 + 3.8t^3 + 11.65t^2$$

t is time measured in years from 1945 when the first atomic bomb was developed, and $N(t)$ is the number of nuclear warheads at time t. The super powers are now reducing their nuclear arsenals.

(a) Determine when the number of nuclear warheads peaked and what that number was.

(b) What was the number of nuclear warheads in 1995 according to this function?

(c) Are nuclear weapons ever likely to be completely abolished? Discuss. What sort of graph do you think will describe $N(t)$ after 1995?

32. Maximizing Volume A box with an open top is to be made from a square piece of cardboard having sides 3 feet long. Square pieces are to be cut from each corner and the sides folded up. Let the length of each of the square pieces be x (Figure 3.18).

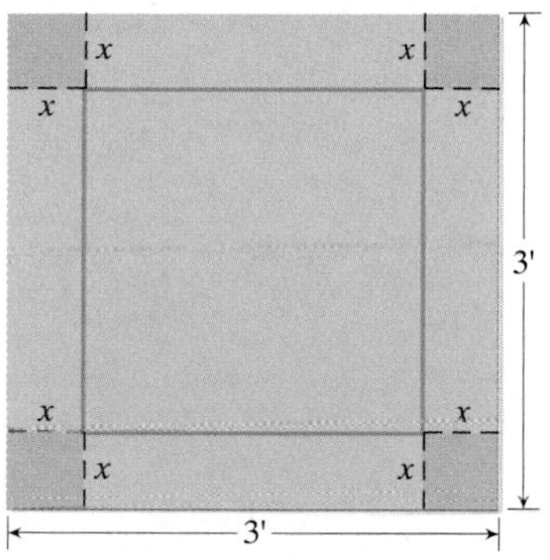

Figure 3.18

(a) Express the volume as a function of x.

(b) Find the dimensions of the box that yield maximum volume. What is that maximum volume?

(c) What values of x, to two decimal places, would yield a volume of 1.5 cubic feet?

33. Laying Power Lines The Ranford Power Company has two plants, one on each side of the River Dee, which is 100 yards wide. It wants to lay a power line between the two buildings. Plant 1 is 500 yards north of Plant 2 (Figure 3.19). The cost of laying the line along the bank is \$20 per yard and under the river is \$50 per yard. How much line should go along the bank and how much should go under the river to minimize cost?

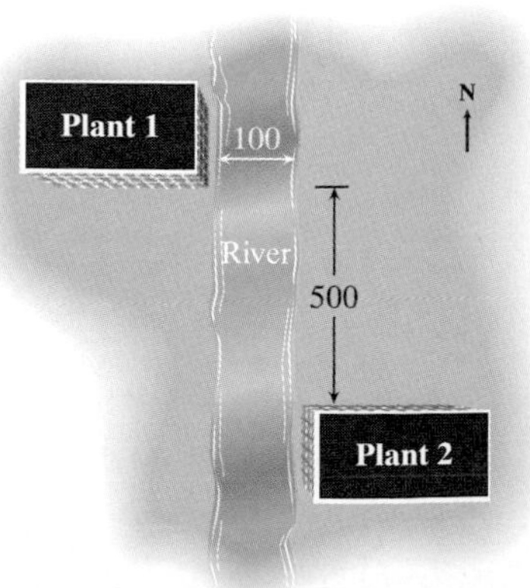

Figure 3.19

34. Traffic Tolls A mathematical analysis based on the results of a questionnaire gives the number N of people using a stretch of the Simon Grant Toll road daily to be the following function of the toll p in dollars:

$$N(p) = \frac{2500}{p^2 + 2}$$

The current toll is 75 cents.

(a) How many people now use the road daily?

(b) If the toll is raised to \$1, how many people daily will stop using the road?

(c) What is the significance of the function $R = Np$? What information can you get from the graph of this function?

3.3 Special Functions, Symmetry, and Consumer Awareness

• Domains, Ranges, and Graphs • The Square Root Function and the Absolute Value Function • The Square Root Function • Symmetry • Piecewise-Defined Functions • Discounting in Bulk Buying • Consumer Awareness • Sizes of Brains in Mammals

In this section, we return to the concepts of domain and range introduced in Section R.7. We introduce ways of graphing functions and show how information about the domains and ranges of functions can be obtained from their graphs.

We remind the student that a function f is a rule that assigns to each element x of a set X exactly one element $f(x)$ of a set Y. The set X is called the **domain** of the function. $f(x)$ is the **image** of x. The set of all images is called the **range** of the function. The **graph** of f is the graph of the equation $y = f(x)$.

Domains, Ranges, and Graphs

The following examples illustrate how to sketch a rough graph of a function using algebra, and how to find the range from the graph. Once again, we see how algebra and geometry go hand in hand. At this time, we stress the value of arriving at rough sketches of graphs using algebraic techniques, in addition to determining graphs using calculators. Both methods are important and complement one another.

Example 1 Finding the Range of a Function from Its Graph

Determine the domain of the function $f(x) = x^2 - 6x + 13$ and sketch its graph. Confirm the graph with a graphing calculator. Use the graph to find the range of f.

SOLUTION We remind the student that, unless the domain of a function f is given explicitly, we will take it to be the largest set of real numbers such that $f(x)$ is a real number. Because any real number can be used for x in

$$f(x) = x^2 - 6x + 13$$

the domain of f is the set of real numbers, $(-\infty, \infty)$.

f is a quadratic function. The graph of f is the parabola $y = x^2 - 6x + 13$, where x is a real number. To sketch the graph of a parabola, we find its vertex, determine whether it opens up or down, and find a second point on the graph (usually the y-intercept) to indicate how wide the parabola opens. On comparing this equation with the standard form $ax^2 + bx + c$, we see that $a = 1$, $b = -6$, and $c = 13$. The vertex is at

$$x = \frac{-b}{2a} = \frac{-(-6)}{2(1)} = 3$$

The corresponding y value is $y = (3^2) - 6(3) + 13 = 4$. The vertex is $(3, 4)$. Because $a = 1 > 0$, the parabola opens upward. Furthermore, when $x = 0$, $y = 13$, telling us that the y-intercept is 13. We use this information to sketch the graph in Figure 3.20 and the calculator to confirm the graph.

The graph of f is the set of points (x, y) that satisfy the equation $y = x^2 - 6x + 13$. Let us look at the geometrical interpretation of the domain and range. The domain of f consists of the x-coordinates of points on the graph. It can be seen that the domain is the whole of the x-axis, namely the interval $(-\infty, \infty)$. The range of f will be the set of y-coordinates of the points on the graph. Points on the graph have y-values $y \geq 4$. The range of f is thus the interval $[4, \infty)$ (Figure 3.20).

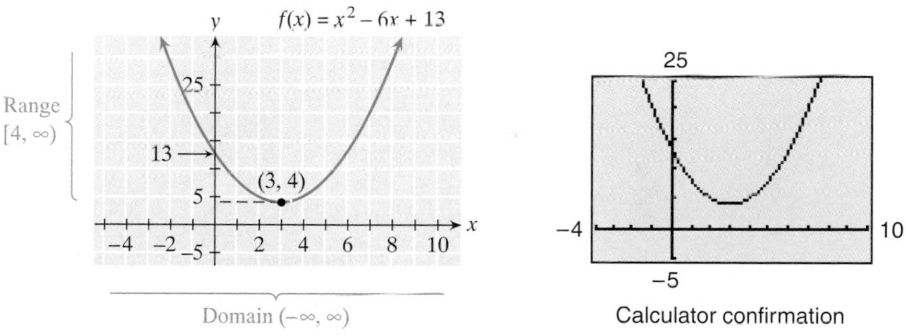

Figure 3.20 Domain and range of $f(x) = x^2 - 6x + 13$.

Example 2 Finding the Range of a Function from Its Graph

Consider the function $f(x) = 2x^2 - 16x + 14$, with domain restricted to the interval $[2, 9]$. Sketch the graph of f. Use the graph to find the range of f. Confirm the results with a graphing calculator.

SOLUTION The graph of f is the part of the parabola $y = 2x^2 - 16x + 14$ for which x lies in the interval $[2, 9]$. The vertex is found using algebra to be the point $(4, -18)$. Because $a = 2 > 0$, the parabola opens upward.

The points on the graph corresponding to $x = 2$ and $x = 9$ are called **end points.** Let us find these points.

When $x = 2$, $y = 2(2^2) - 16(2) + 14 = -10$.
When $x = 9$, $y = 2(9^2) - 16(9) + 14 = 32$.

The end points are $(2, -10)$ and $(9, 32)$. The graph of f is the segment of the parabola $y = 2x^2 - 16x + 20$ that extends from the point $(2, -10)$ to the point $(9, 32)$. Plot the end points and the vertex and sketch this segment of the parabola (Figure 3.21).

Now let us use this graph to find the range of f. The graph of f is the set of points (x, y) that satisfy the equation $y = 2x^2 - 16x + 14$ for values of x in the interval $[2, 9]$. The range of f will consist of the y-coordinates of the points on the graph. We see that points on the graph have y values from -18 to 32. The range of f is thus the interval $[-18, 32]$.

We can conveniently confirm the graph and range on a graphing calculator using the window defined by the domain and range, namely $2 \le x \le 9$ and $-18 \le y \le 32$ (Figure 3.21). The shape of our graph is confirmed. Observe that the graph is perfectly framed in this window.

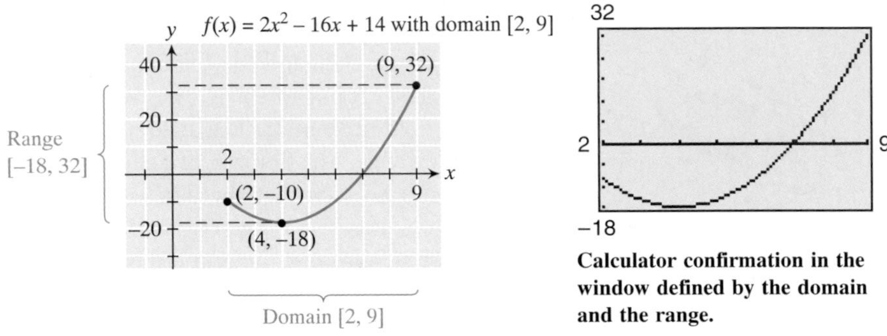

Figure 3.21 Range and domain of $f(x) = 2x^2 - 16x + 14, 2 \le x \le 9$.

Self-Check 1 Sketch the graph of $f(x) = -x^2 + 4x - 1$ when the domain of f is restricted to the interval $[0, 5]$. Determine the range of f. Confirm your answers with a graphing calculator.

Let us use these techniques to sketch the graph and discuss the domain and range of the important function $f(x) = \sqrt{x}$.

THE SQUARE ROOT FUNCTION $f(x) = \sqrt{x}$

The domain of the **square-root function** $f(x) = \sqrt{x}$ is the set of all possible x values. $\sqrt{x}$ is only defined if $x \ge 0$. Thus, the domain of f is $[0, \infty)$.

Let us sketch the graph of f by plotting points. For convenience, select values of x in the domain that have integer roots (Figure 3.22).

x	0	1	4	9	16	25
$f(x)$	0	1	2	3	4	5

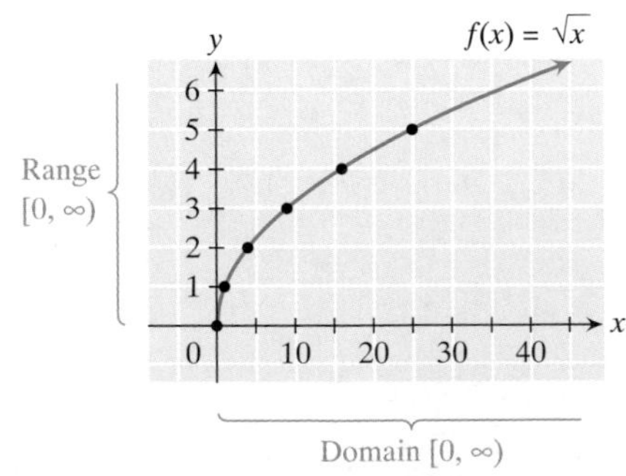

Figure 3.22

The domain consists of the origin and points on the positive x-axis. The range is the set of y-values for points on the graph. We see that $y \geq 0$ for points on the graph. Thus, the range is $[0, \infty)$.

We confirm the graph using a calculator. Observe that none of the graph lies to the left of the y-axis or below the x-axis, supporting the results for domain and range.

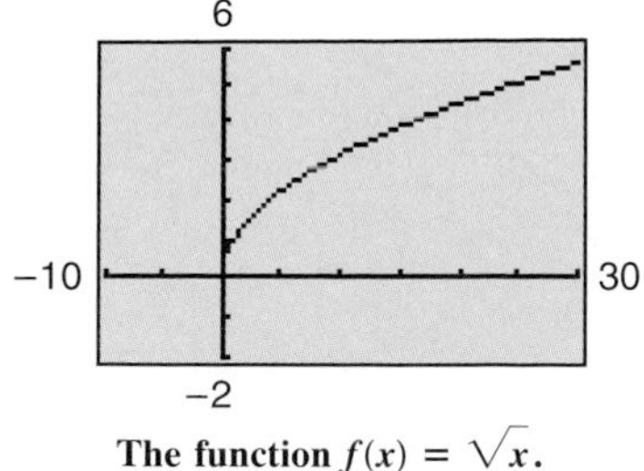

The function $f(x) = \sqrt{x}$.

Further Ideas 1 The following function gives the relationship between the swimming speed V (centimeters per second) and the length L (centimeters) of a salmon. (J. R. Brett, "The Respiratory Metabolism and Swimming Performance of Young Sockeye Salmon," *Journal of the Fisheries Research Board of Canada* **21:** 1183–1226):

$$V = 19.5\sqrt{L}$$

Use a graph to determine how large a salmon must be (to the nearest tenth of a centimeter) to be able to swim 75 centimeters per second.

SYMMETRY

The graphs of certain functions display symmetries. These symmetries enable us to better understand the behavior of the functions and also the phenomena described by those functions. We now introduce symmetries with respect to the y-axis and with respect to the origin and illustrate these symmetries with the graphs of $f(x) = x^2$ and $f(x) = x^3$.

Symmetry with Respect to the y-Axis

Consider the graph of the function f in Figure 3.23(a). Note the symmetry of points such as A and A' of the graph about the y-axis. The part of the graph of f for $x < 0$ is the mirror image in the y-axis of the part for $x > 0$. We say that

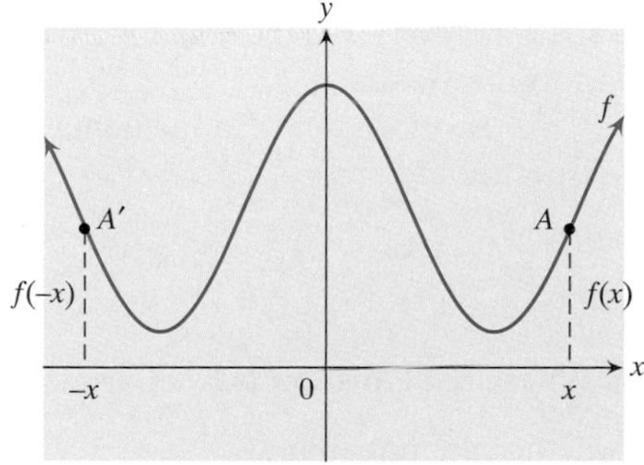

Symmetry about y-axis.
This symmetry happens when $f(-x) = f(x)$.

(a)

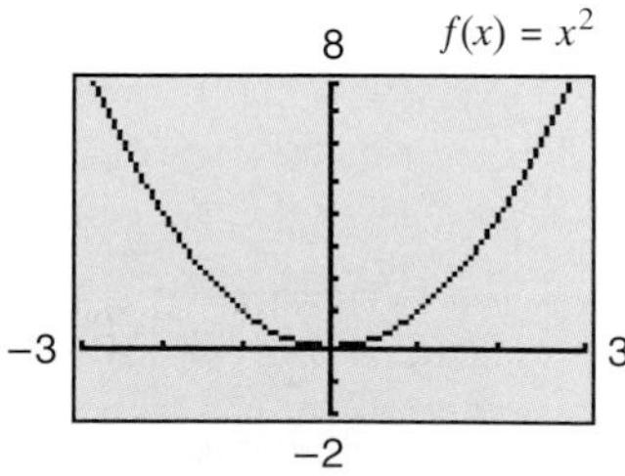

Symmetry about y-axis.

(b)

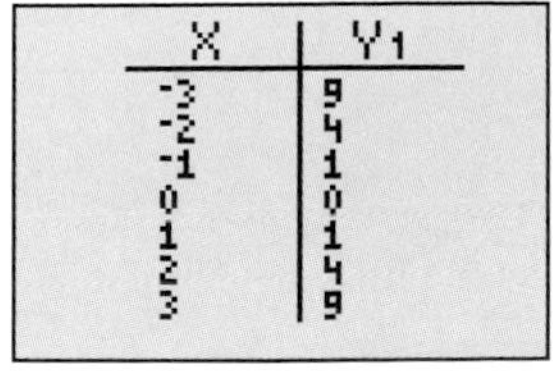

X	Y1
-3	9
-2	4
-1	1
0	0
1	1
2	4
3	9

$f(-3) = f(3)$, $f(-2) = f(2)$ . . . .

(c)

Figure 3.23

the graph is symmetric with respect to the y-axis. The symmetry occurs because the values of f at any x and $-x$ are identical. The graph of $f(x) = x^2$ is such a graph (Figure 3.23b). The table reveals this symmetry (Figure 3.23c). For example $f(-3) = f(3)$ and $f(-2) = f(2)$. A function that is symmetric with respect to the y-axis is called an **even function.**

Symmetry with Respect to the Origin

Consider the graph of f in Figure 3.24(a). Note the symmetry of points such as P and P' on the graph about the origin. This graph is said to be symmetric with respect to the origin. The symmetry occurs because the value of f at $-x$ is the negative of the value of f at x. The graph of $f(x) = x^3$ is such a graph (Figure 3.24b).The table reveals this symmetry in Figure 3.24(c). For example $f(-3) = -f(3)$ and $f(-2) = -f(2)$. A function that is symmetric with respect to the origin is called an **odd function**.

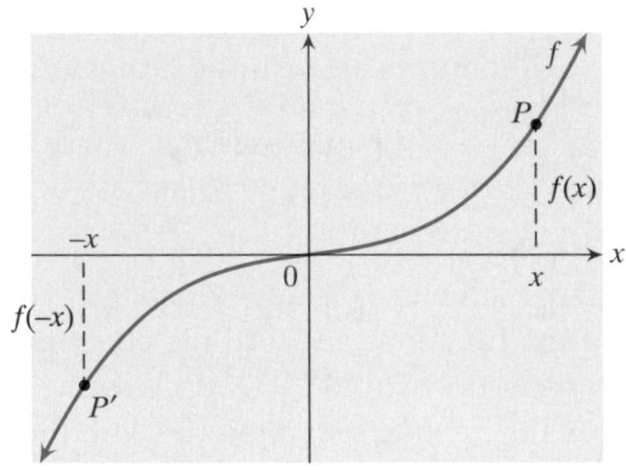

Symmetry about origin.
This symmetry happens when $f(-x) = -f(x)$.
(a)

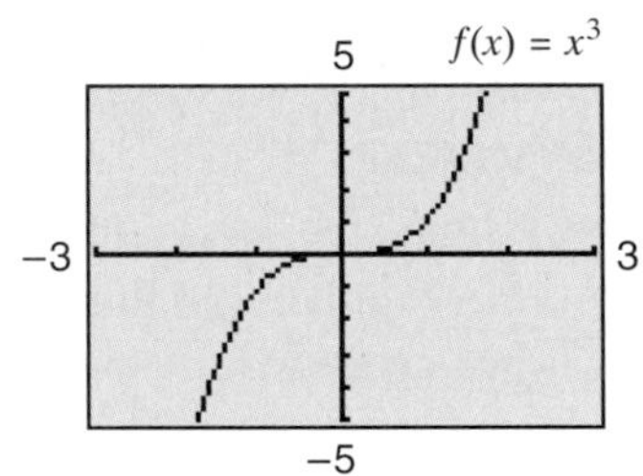

Symmetry about origin.
(b)

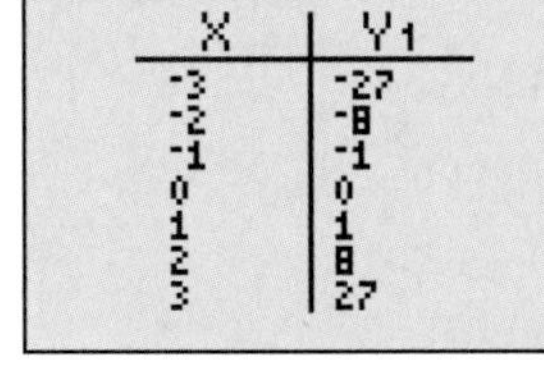

$f(-3) = -f(3), f(-2) = -f(2) \ldots.$
(c)

Figure 3.24

Even and Odd Functions

A function is **even** if $f(-x) = f(x)$ for every x in the domain. An even function is symmetric with respect to the y-axis.

A function is **odd** if $f(-x) = -f(x)$ for every x in the domain. An odd function is symmetric with respect to the origin.

Example 3 Determining Whether a Function is Even, Odd, or Neither

(a) Use algebra to determine whether the function $f(x) = x^4 + 2$ is even, odd, or neither.

(b) Draw the graph of f with a graphing calculator. Indicate any symmetry in the graph that supports your conclusion to part (a).

SOLUTION

(a) We find an expression for $f(-x)$. To do this, replace x with $-x$ in the equation of $f(x)$.

$$f(-x) = (-x)^4 + 2 = x^4 + 2$$
$$= f(x)$$

Thus, $f(-x) = f(x)$. f is an even function.

(b) We graph $f(x) = x^4 + 2$ (Figure 3.25). The graph is seen to be symmetric about the y-axis, confirming that it is an even function.

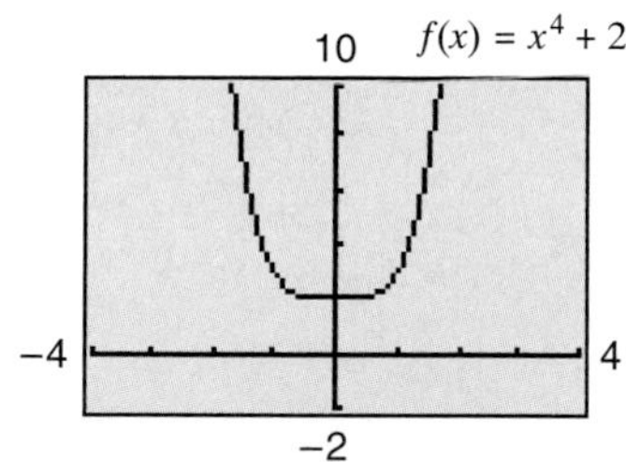

Figure 3.25

Self-Check 2 Use algebra to determine whether the following functions are even, odd, or neither. Draw the graphs of the functions with a graphing calculator. Indicate any symmetry in the graphs that supports your conclusions about whether the functions are even, odd, or neither.

(a) $f(x) = 3x^4 - 5$ **(b)** $f(x) = 5x + 7$ **(c)** $f(x) = 2x^3 + 4x$

Exploration

Consider the tables shown in Figure 3.26. Determine whether the corresponding functions $f(x)$, $g(x)$, $h(x)$, and $k(x)$ are likely to be even, odd, or neither based on this information.

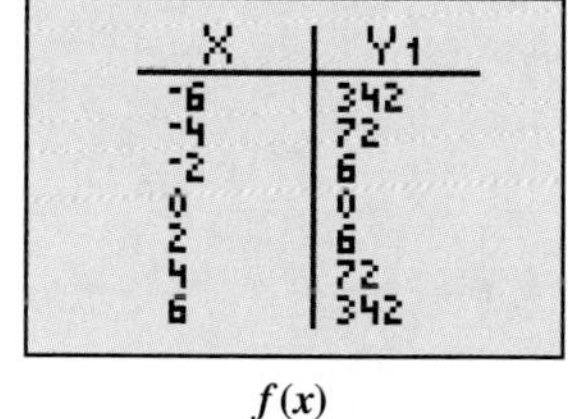

X	Y1
-6	342
-4	72
-2	6
0	0
2	6
4	72
6	342

$f(x)$

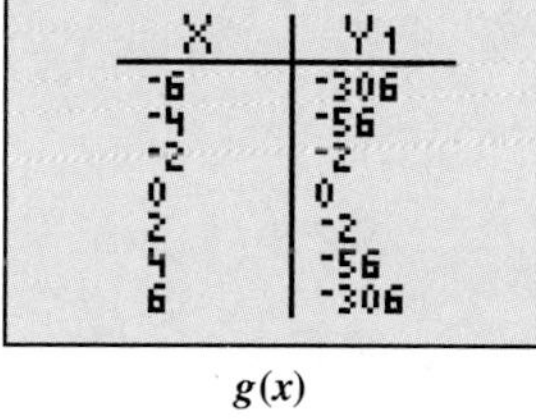

X	Y1
-6	-306
-4	-56
-2	-2
0	0
2	-2
4	-56
6	-306

$g(x)$

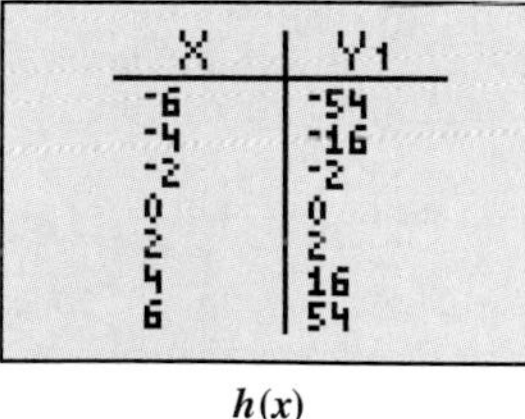

X	Y1
-6	-54
-4	-16
-2	-2
0	0
2	2
4	16
6	54

$h(x)$

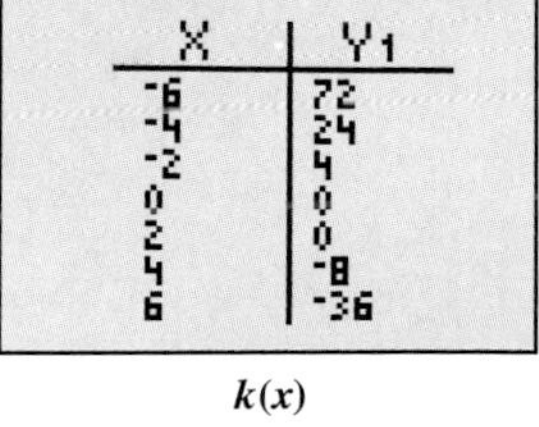

X	Y1
-6	72
-4	24
-2	4
0	0
2	0
4	-8
6	-36

$k(x)$

Figure 3.26

We have seen various ways of graphing functions. We now illustrate how to make use of symmetry to sketch the graph of a function. We use symmetry to draw the graph and to discuss the geometrical interpretation of domain and range of the absolute value function.

THE ABSOLUTE VALUE FUNCTION $f(x) = |x|$

Let us show that the **absolute value function** $f(x) = |x|$ is an even function and then use this information to sketch its graph. Substitute $-x$ for x in $f(x) = |x|$. We get

$$\begin{aligned} f(-x) &= |-x| \\ &= |x| \\ &= f(x) \end{aligned}$$

Thus, $f(-x) = f(x)$. f is an even function.

$f(x)$ is defined for all real values of x. Thus, the domain of f is $(-\infty, \infty)$. Because f is an even function, its graph is symmetric about the y-axis. Let us draw the part of the graph that corresponds to values of $x \geq 0$. The remainder of the graph can then be obtained by drawing the mirror image of this part in the y-axis. For $x \geq 0$, $f(x) = x$. Thus, the graph of f for these values of x is a straight line through the origin with slope 1 (Figure 3.27a). Add the mirror image of this part in the y-axis to get the complete graph of $f(x) = |x|$ (Figure 3.27b). The range of f is seen to be $[0, \infty)$.

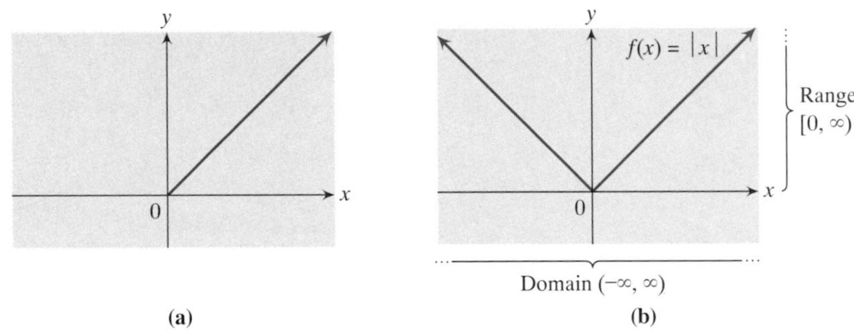

Figure 3.27 The absolute value function.

ABSOLUTE VALUE FUNCTION WITH A CALCULATOR

The function $f(x) = |x|$ can be entered into a calculator using the built-in absolute value (Figure 3.28). We confirm the graph of $f(x) = |x|$ in Figure 3.27.

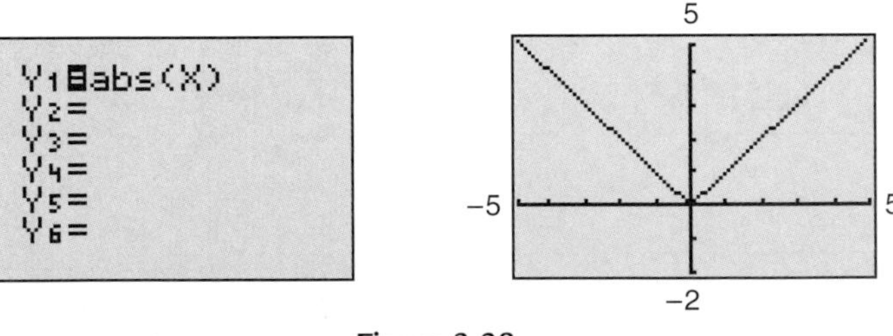

Figure 3.28

PIECEWISE-DEFINED FUNCTIONS

We have seen how functions are constructed to describe various situations. Sometimes a quantity will behave in a particular manner over a certain interval and in a different manner over another interval. The following example illustrates how equations can be brought together in a piecewise-defined function for a whole interval.

Example 4 A Piecewise-Defined Function

Consider the following function, which is defined by different equations over two intervals of the domain. We call such a function a **piecewise-defined function.**

$$f(x) = \begin{cases} x^2 & \text{if } x \leq 2 \\ x + 4 & \text{if } x > 2 \end{cases}$$

If $x \leq 2$, then $f(x) = x^2$. The graph of f over $(-\infty, 2]$ is the part of the parabola that lies over this interval. Because $x = 2$ is in this interval, the end point $(2, 4)$ will be part of the graph. We indicate this by means of a solid circle. See part **a** of the graph in Figure 3.29.

When $x > 2$, then $f(x) = x + 4$. The graph of f over $(2, \infty)$ is the part of the line that lies over this interval. Because $x = 2$ is not in this interval, the point $(2, 6)$ is not part of the graph. We indicate this by means of an open circle. See part **b** of the graph in Figure 3.29.

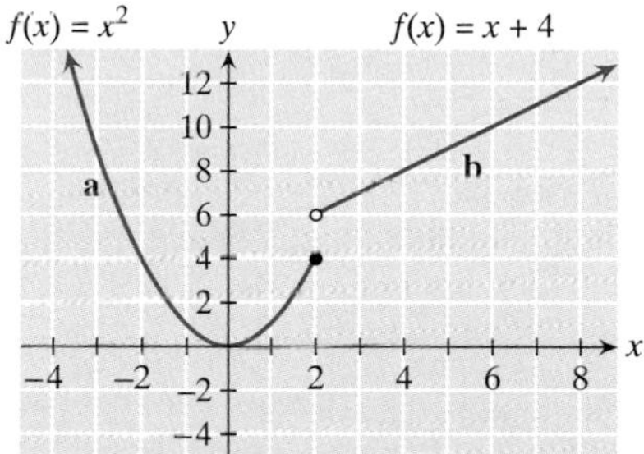

Figure 3.29 Graph of a piecewise-defined function.

PIECEWISE-DEFINED FUNCTIONS WITH A CALCULATOR

Piecewise-defined functions can be graphed on some calculators. We enter and graph the function of Example 4 in Figure 3.30.

$$f(x) = \begin{cases} x^2 & \text{if } x \leq 2 \\ x + 4 & \text{if } x > 2 \end{cases}$$

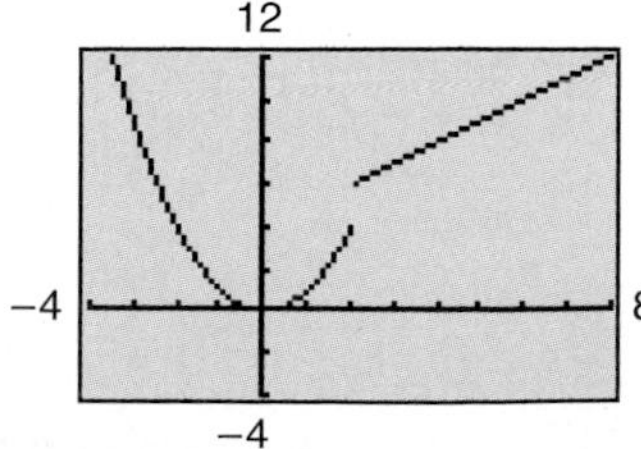

Figure 3.30

Applications . . .

The following example illustrates how mathematics can be used to model a very common situation in business, namely discounting when buying in bulk.

Example 5 Discounting in Bulk Buying

The list price of a certain textbook is \$47.00. For an order of 25 or more books, there is a discount of \$4 per book.

(a) Construct a function that expresses the total cost of books in terms of the number of books ordered.
(b) Use the function to find the costs of 12 books and of 30 books.
(c) Find the costs of 23 and 25 books and comment on these figures.

SOLUTION

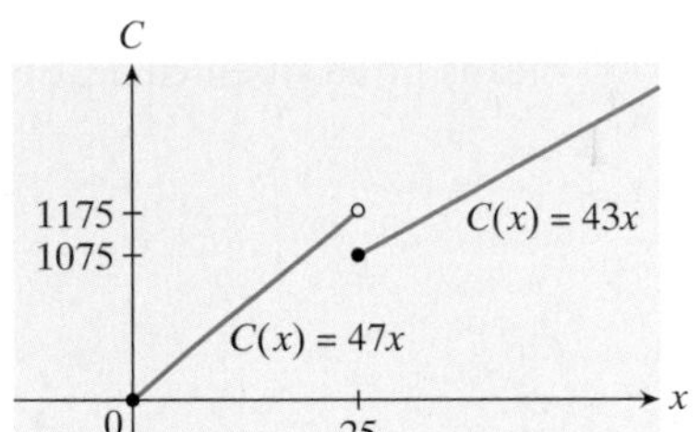

Figure 3.31 Cost function for bulk buying.

(a) The list price of the book is \$47. There is a discount of \$4 for an order of 25 or more books. Thus, the cost is \$43 if the order is 25 or more.

Let x books be ordered. If $x < 25$, the total cost is $47x$. If $x \geq 25$, the total cost is $43x$. Let $C(x)$ be the function that gives the cost of x books. The domain of C is divided naturally into the two parts $0 \leq x < 25$ and $x \geq 25$, and C is defined as follows over those parts.

$$C(x) = \begin{cases} 47x & \text{if } 0 \leq x < 25 \\ 43x & \text{if } x \geq 25 \end{cases}$$

When $x = 25$, $47x = 1175$. When $x = 25$, $43x = 1075$. The end points of the graph are (25, 1175) and (25, 1075). The first end point is not included in the graph; the second one is included (Figure 3.31).

(b) We use this function to find the costs of 12 and 30 books. Substituting $x = 12$ into the first part of the function and $x = 30$ into the second part, we get $C(12) = 564$ and $C(30) = 1290$. The cost of 12 books is \$564, and the cost of 30 books is \$1290.

(c) We now find the costs of 23 and 25 books. Using the first part of C, we find that $C(23) = 1081$. Using the second part of C, we find that $C(25) = 1075$. The cost of 23 books is \$1081, whereas the cost of 25 books is \$1075. It is cheaper to buy 25 books than 23, not an unusual outcome with this type of discount structure.

Example 6 Consumer Awareness

A cosmetics company has undertaken a study of the effectiveness of a certain television commercial in advertising a product. Consumer awareness of the product was estimated for periods after the running of the commercial. Awareness was evaluated on a scale of 0 through 5, with 0 indicating that a person had little awareness of the product and 5 indicating that a person was very aware of the product. The results were as follows.

Let t be time in days measured from the showing of the commercial. Awareness was found to be closely described by the function $A(t) = -t/2 + 5$ for up to 10 days after the showing. Observe that $f(0) = 5$ and $f(10) = 0$. The initial awareness when the commercial is shown is 5. The awareness becomes 0 when $t = 10$; that is, after 10 days.

We can construct and graph the following piecewise-defined function for describing awareness (Figure 3.32).

$$A(t) = \begin{cases} -\dfrac{t}{2} + 5 & \text{if } t \leq 10 \\ 0 & \text{if } t > 10 \end{cases}$$

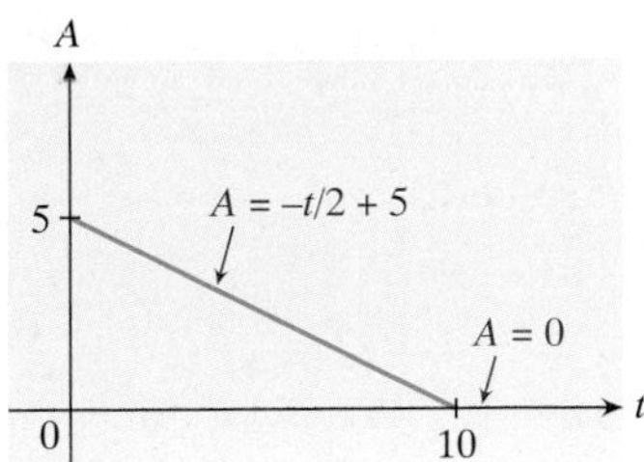

Figure 3.32 Graph of consumer awareness.

Suppose that the company does not want awareness to drop below the level 3. How often should the commercial be shown? Let us determine when awareness drops to 3. Solve $-t/2 + 5 = 3$.

$$-\frac{t}{2} + 5 = 3$$

$$-\frac{t}{2} = -2$$

$$t = 4$$

Four days after the showing of the commercial, consumer awareness drops to 3. The company should show the commercial every 4 days to ensure that awareness does not drop below 3. The desired advertising graph would then be as shown in Figure 3.33. Observe that this graph is obtained by giving awareness a boost of 2 every 4 days to bring the rating back up from 3 to 5. The new awareness function $D(t)$ is thus

$$D(t) = \begin{cases} -\dfrac{t}{2} + 5 & \text{if } 0 \leq t < 4 \\ -\dfrac{t}{2} + 7 & \text{if } 4 \leq t < 8 \\ -\dfrac{t}{2} + 9 & \text{if } 8 \leq t < 12 \\ \ldots & \end{cases}$$

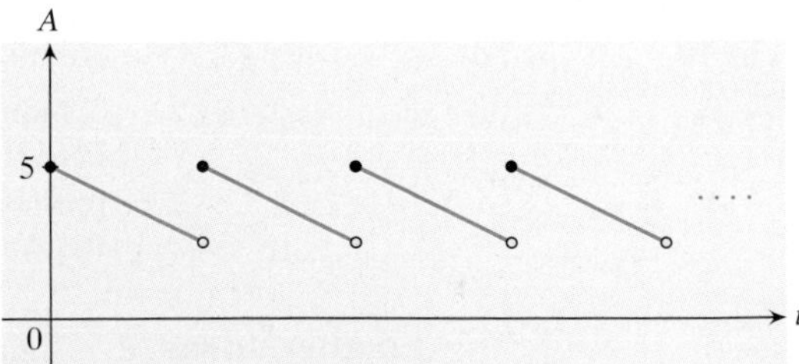

Figure 3.33 Desired advertising graph.

Example 7 Sizes of Brains in Mammals

We discussed previously how the world we live in is governed by the laws of chemistry and physics and that animals must live within the bounds of these laws. Biologists attempt to discover and understand these laws. They study the relationships between sizes of animals and their various biological components. For example, they have studied the mental capabilities of mammals in relation to their brain sizes. Knut Schmidt-Nielson in his book *Scaling—Why Is Animal Size So Important?* (Cambridge University Press, 1985) discusses how the mass of the brain (M_{brain}, in kilograms) is a function of body mass (M_{body}, in kilograms) for nonhuman mammals,

$$M_{\text{brain}} = 0.01(M_{\text{body}})^{0.7}$$

Modern reptiles weigh up to about 10^2 kilograms. Their brain sizes are therefore, according to this function, up to $0.01(10^2)^{0.7}$ kilograms, or approximately 0.25 kilograms. This brain size as a fraction of body size is approximately $0.25/10^2$, or 0.0025. It has often been stated that the giant dinosaurs had very small brains relative to their body size and that competition from brainier mammals led to their ultimate extinction. However, recent research with brain cavities in fossil dinosaur skulls has shown that the brain size of dinosaurs as a fraction of body size is of magnitude 0.0025. Thus, although dinosaurs did have smaller brains than modern mammals, their brain size was within the range of the body size that could be expected for reptiles. Furthermore, modern reptiles with their relatively small brains have survived competition with mammals for hundreds of millions of years. It is therefore difficult to accept the argument that dinosaurs became extinct because of their disproportionately small brains.

Tyrannosaurus Rex (or "tyrant lizard") had a head the size of an easy chair and up to 60 teeth, some the size and shape of bananas. (*©John Sibbick*)

Further Ideas 2 Brain sizes of humans are approximately described by the equation $M_{\text{brain}} = 0.085(M_{\text{body}})^{0.66}$ (all weights in kilograms). What is your approximate brain size, given that 1 kilogram is 2.2046 pounds? (Einstein was a man of relatively small stature!)

GROUP DISCUSSION

Symmetry

We have seen functions that have graphs that are symmetric with respect to the y-axis (even functions) and functions that have graphs that are symmetric with respect to the origin (odd functions). It is very natural to ask whether there are functions that have graphs that are symmetric with respect to the x-axis. Discuss.

Answers to Self-Check Exercises

1. The vertex is $(2, 3)$, and it opens down. $f(0) = -1$, $f(5) = -6$. The range is $[-6, 3]$.
2. **(a)** even **(b)** neither **(c)** odd

Answers to Further Ideas Exercises

1. Salmon must be at least 14.8 centimeters long.

EXERCISES 3.3

In Exercises 1–10, sketch the graph of each function. Determine the domain and range of each function. Confirm your results using a graphing calculator, with the window $X_{\min} = -5$, $X_{\max} = 5$, $X_{\text{scl}} = 1$, $Y_{\min} = -10$, $Y_{\max} = 10$, and $Y_{\text{scl}} = 2$.

1. $f(x) = 2x - 4$
2. $f(x) = -2x + 8$
3. $f(x) = x$
4. $g(x) = -2x$
5. $f(x) = x^2 + 2x + 3$
6. $f(x) = -x^2 + 4x - 5$
7. $g(x) = 2x^2 - 8x + 5$
8. $f(x) = -3x^2 + 6x - 1$
9. $g(x) = x^2 + 2$
10. $f(x) = -x^2 + 5$

In Exercises 11–16, sketch the graph of each function. Determine the domain and range of each function. Confirm your results using a graphing calculator, with the window $X_{\min} = -5$, $X_{\max} = 5$, $X_{\text{scl}} = 1$, $Y_{\min} = -10$, $Y_{\max} = 10$, and $Y_{\text{scl}} = 2$.

11. $f(x) = x^4$
12. $g(x) = x^5$
13. $g(x) = -x^2$
14. $f(x) = 2\sqrt{x}$
15. $f(x) = ax^2$, for $a = 2$, 3, and -5, in the same coordinate system.
16. $f(x) = a\sqrt{x}$, for $a = 2$, 3, and -5 in the same coordinate system.

In Exercises 17–22, functions and domains are given. Sketch the graph of each function, and find its range. Confirm your results using a graphing calculator. Use the domain and the range as the x and y intervals of your window.

17. $f(x) = 2x - 1$ with domain $[-2, 2]$.
18. $f(x) = 3x + 5$ with domain $[-3, 4]$.
19. $g(x) = -2x + 8$ with domain $(0, 5)$. Observe that the domain is an open interval. What effect does this have on the range?
20. $f(x) = x^2 - 4x - 1$ with domain $[0, 5]$.
21. $f(x) = 2x^2 + 4x + 1$ with domain $[-3, 2]$.
22. $f(x) = -4x^2 + 24x - 8$ with domain $(0, 6)$ (the domain is an open interval).

In Exercises 23–26, functions and domains are given. Sketch the graph of each function and find its range. Confirm your results using a graphing calculator. Use the domain and the range as the x and y intervals of your window. (See Example 2.)

23. $f(x) = x^3$ with domain $[-2, 2]$.
24. $f(x) = 2\sqrt{x}$ with domain $[4, 25]$.
25. $h(x) = -x^2$ with domain $[-4, 3]$.
26. $f(x) = 2x - 3$ with domain $[-2, 5)$.

In Exercises 27–36, **(a)** use algebra to determine whether the functions are even, odd, or neither by computing

$f(-x)$. **(b)** Draw the graph of each function on a graphing calculator, using the window $X_{\min} = -5$, $X_{\max} = 5$, $X_{\text{scl}} = 1$, $Y_{\min} = -10$, $Y_{\max} = 10$, and $Y_{\text{scl}} = 2$. Discuss any symmetry in the graph that supports your conclusion to part (a).

27. $f(x) = x^2 - 4$
28. $f(x) = x^3$
29. $f(x) = x + 4$
30. $f(x) = x^2$
31. $f(x) = 4x^3 + 2$
32. $f(x) = 7x^4 - 2$
33. $f(x) = \dfrac{1}{x^2 - 4}$
34. $f(x) = \sqrt{x}$
35. $f(x) = 4$
36. $f(x) = 4x - 3x^3$

In Exercises 37–46, **(a)** draw the graphs of the functions with the aid of a graphing calculator, using the window $X_{\min} = -5$, $X_{\max} = 5$, $X_{\text{scl}} = 1$, $Y_{\min} = -10$, $Y_{\max} = 10$, and $Y_{\text{scl}} = 2$. Indicate any symmetry about either the y-axis or the origin in the graphs. Use any such symmetry to decide whether the functions are even, odd, or neither. **(b)** Determine $f(-x)$ and $-f(x)$ for each function. Use these to confirm your answer to part (a).

37. $f(x) = \dfrac{6}{x}$
38. $f(x) = \dfrac{4}{x^2}$
39. $f(x) = \dfrac{8}{x^3}$
40. $f(x) = 2x^2 + 4x - 6$
41. $f(x) = 3x^3 - x + 5$
42. $f(x) = \sqrt{x^3 + 4x}$
43. $f(x) = |x| + 4$
44. $f(x) = |x - 4|$
45. $f(x) = |x^2 - 4|$
46. $f(x) = |x| + x^2$

In Exercises 47–52, determine whether the functions that correspond to the given tables are likely to be even, odd, or neither based on this information. Give reasons for your answers.

47.

X	Y1
-6	288
-4	48
-2	0
0	0
2	0
4	48
6	288

48.

X	Y1
-15	-336
-10	-99
-5	-12
0	0
5	12
10	99
15	336

49.

X	Y1
-6	-45
-4	-16
-2	-3
0	0
2	-1
4	0
6	9

50.

X	Y1
-3	30
-2	10
-1	2
0	0
1	-2
2	-10
3	-30

51.

X	Y1
-9	-1320
-6	-262.8
-3	-17.1
0	0
3	-17.1
6	-262.8
9	-1320

52.

X	Y1
-1.5	.825
-1	.3
-.5	.075
0	0
.5	-.075
1	-.3
1.5	-.825

In Exercises 53–58, sketch the graphs of the piecewise-defined functions. Confirm your graphs, if possible, using a calculator with the window $X_{\min} = -5$, $X_{\max} = 5$, $X_{\text{scl}} = 1$, $Y_{\min} = -10$, $Y_{\max} = 10$, and $Y_{\text{scl}} = 2$.

53. $f(x) = \begin{cases} 2x + 1 & \text{if } x \le 4 \\ 2 & \text{if } x > 4 \end{cases}$

54. $f(x) = \begin{cases} x - 3 & \text{if } x < 2 \\ 3x - 5 & \text{if } x \ge 2 \end{cases}$

55. $f(x) = \begin{cases} -4x + 5 & \text{if } x \le 3 \\ -7 & \text{if } x > 3 \end{cases}$

56. $f(x) = \begin{cases} 5 & \text{if } x \le -6 \\ -8 & \text{if } x > 6 \end{cases}$

57. $f(x) = \begin{cases} 6 - x & \text{if } x < 2 \\ x - 6 & \text{if } x \ge 2 \end{cases}$

58. $f(x) = \begin{cases} x & \text{if } x \le -2 \\ -2x + 4 & \text{if } x > -2 \end{cases}$

In Exercises 59–64, sketch the graphs of the piecewise-defined functions. Confirm your graphs, if possible, using a calculator with the window $X_{\min} = -5$, $X_{\max} = 5$, $X_{\text{scl}} = 1$, $Y_{\min} = -10$, $Y_{\max} = 10$, and $Y_{\text{scl}} = 2$.

59. $f(x) = \begin{cases} x^2 & \text{if } x \le 0 \\ x/2 & \text{if } x > 0 \end{cases}$

60. $f(x) = \begin{cases} x^2 - 3 & \text{if } x \le 1 \\ -x - 1 & \text{if } x > 1 \end{cases}$

61. $f(x) = \begin{cases} -x + 1 & \text{if } x < -2 \\ x^3 & \text{if } x \ge -2 \end{cases}$

62. $f(x) = \begin{cases} x & \text{if } x \le 3 \\ -4 & \text{if } x > 3 \end{cases}$

63. $f(x) = \begin{cases} -3 & \text{if } x \le -2 \\ x^2 & \text{if } -2 < x \le 3 \\ 2 & \text{if } x > 3 \end{cases}$

64. $f(x) = \begin{cases} 5 & \text{if } x \le -2 \\ -x + 1 & \text{if } -2 < x \le 3 \\ x^2/4 & \text{if } x > 3 \end{cases}$

65. Use algebra to find the value of k that will make the following function f continuous; that is, have no jumps in its graph. Check your answer by drawing a graph of f for that value of k.

$$f(x) = \begin{cases} x + 1 & \text{if } x \le 3 \\ -x - k & \text{if } x > 3 \end{cases}$$

66. Use algebra to find the value of k that will make the following function f continuous; that is, have no jumps in its graph. Check your answer by drawing a graph of f for that value of k.

$$f(x) = \begin{cases} x^2 & \text{if } x \le 2 \\ -x + k & \text{if } x > 2 \end{cases}$$

67. Consider the function $f(x) = x^3 - 9x^2 + 15x$, with domain $[0, 8]$. Use a graphing calculator (with max/min finder) to determine its range.

68. Consider the function $f(x) = x^3 - 15x^2 + 48x + 12$, with domain $[1, 9]$. Use a graphing calculator (with max/min finder) to determine its range.

69. Growth in a Child's Vocabulary Researchers studying the increase in a child's vocabulary during the second year of life found that the number of words in a child's vocabulary at age t months is closely described by the function

$$N(t) = t^2 - 10t - 20$$

where $12 \le t \le 24$. For example,

$$N(16) = (16)^2 - 10(16) - 20 = 76$$

means that a 16-month-old child has a vocabulary of about 76 words.

(a) Determine how many words a child might know at 12 months and at 24 months.

(b) What is the vocabulary increase during the first 6 months of this period and during the second 6 months.

(c) It is clear that this increase cannot continue indefinitely. Graph $N(t)$. What do you think the general shape of the learning curve will look like over the first 16 years of a child's life? Give reasons.

70. Discount on Buying in Bulk Prindle and Jones Book Publishers sell a certain accounting textbook for \$35.00. For an order of 10 or more books, there is a discount of \$2 per book.

(a) Construct a function that expresses the total cost of books in terms of the number of books ordered.

(b) Use the function to find the costs of 8 books and of 22 books.

71. Discount in Printing The print shop at Winford University makes copies at a rate of 5 cents per page for the first 50 pages, 3 cents per page for the next 50 copies, and 2 cents per page for over 100 pages.

(a) Construct a function that describes the total cost of having material copied at the print shop.

(b) Use the function to find the costs of 75 and 120 pages.

72. Cost of Modernizing Equipment The Coburn Engineering Company has equipment that costs \$40 a day to run. They decide to modernize the equipment at a cost of \$400. The equipment will then cost \$35 a day to run.

(a) Construct two cost functions: one to describe the old situation and one to describe the new.

(b) Use the graphs of these functions to determine how long it will take for the modernized equipment to make up the \$400.

(c) Use algebra to confirm your answer.

73. Consumer Awareness A company has found that, after its product is advertised on television, consumer awareness of the product is described by the following function of time t (in days).

$$A(t) = \begin{cases} -t/4 + 20 & \text{if } 0 \le t \le 80 \\ 0 & \text{if } t > 80 \end{cases}$$

(a) Sketch the graph of $A(t)$.

(b) The company does not want awareness to drop below a level of 14. Determine and sketch the graph of the function that describes the desired advertising procedure. How often should the commercial be shown?

74. Monthly Gas Charges The monthly charges of the Central Texas Gas Company are as follows:

Gas Used	Charge
First 75 cubic meters	\$23.00 minimum cost
Next 100 cubic meters	12¢ per cubic meter
Next 250 cubic meters	9¢ per cubic meter
Above 425 cubic meters	6¢ per cubic meter

(a) Construct a function that can be used to compute the monthly gas bill.

(b) Use the function to find the costs of 110, 260, and 830 cubic meters of gas.

75. Sizes of Brains The average weights of the brains of humans and nonhuman mammals, expressed as functions of body weight (all in kilograms), are

Humans: $M_{\text{brain}} = 0.085(M_{\text{body}})^{0.66}$

Mammals (nonhuman): $M_{\text{brain}} = 0.01(M_{\text{body}})^{0.70}$

The elephant is the largest land animal and can weigh about 6000 kilograms.

(a) Find the weight of the brain of a 6000-kilogram elephant, to two decimal places.

(b) Estimate the brain weight of a 140-pound person (1 kilogram = 2.2046 pounds). (The functions are taken from *Scaling—Why Is Animal Size So Important?* by Knut Schmidt-Nielson, Cambridge University Press, 1985.)

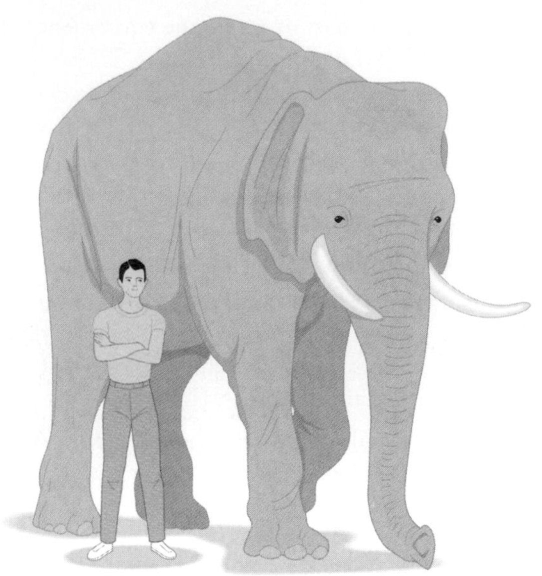

76. Sizes of Eggs Data on the weights of the eggs of 800 species of birds were collected and analyzed. It was found that egg weight was the following function of body weight (all in grams):

$$M_{egg} = 0.277(M_{body})^{0.77}$$

(a) A hummingbird weighs 2.5 grams. What is its egg weight, rounded to two decimal places?

(b) The largest bird that ever lived was the elephant bird from Madagascar. It weighed about 500 kilograms. What was the approximate weight of its egg both in grams and pounds

$$(1 \text{ kilogram} = 2.2046 \text{ pounds})?$$

(c) An ostrich egg weighs 1 kilogram. Use a graph to determine the weight of the ostrich in kilograms, to two decimal places. (Research on these topics was carried out by H. Rahn and his associates and is reported in "Relation of Avian Egg Weight to Body Weight," *Auk* **92:** 750–765, 1975).

77. Metabolic Rates Physiologists have given much study to the metabolic rates of mammals and birds. The metabolic rate R (in appropriate units) has been found to be the following function of body mass M (in kilograms):

$$R = 73.3(M)^{0.74}$$

(a) Find the metabolic rates of a 20-gram mouse and a 300-kilogram cow, rounded to two decimal places.

(b) The metabolic rate per unit mass is called the specific metabolic rate (Royal Society Symbols Committee, 1975). Find the function that describes specific metabolic rate as a function of body weight. Plot the graph of this function. Discuss why the graph tells you that the metabolic rate per unit mass decreases with body size. What does this imply in terms of the relative restlessness of mice, sheep, horses, and elephants?

SECTION PROJECT

Adding to the Library of Functions

In this chapter, we discussed the graphs of linear functions, quadratic functions, polynomial functions, the square root function, and the absolute value function. As we mentioned, students should build up a library of functions so that they quickly recognize these functions and sketch their graphs. This project gives students the opportunity to add to this collection.

Graph the following functions on your calculator and arrive at the general shapes of graphs of functions of the form $f(x) = x^n$ and $f(x) = x^{1/n}$, where n is an integer, even and odd. These are the classes of functions that contain the functions $f(x) = x^2$, $f(x) = x^3$, and $f(x) = \sqrt{x}$ of this section. Investigate the symmetries of these graphs. Use a calculator to find the points of intersection of the graphs.

(a) $f(x) = x^2$, $g(x) = x^4$, $h(x) = x^6$

(b) $f(x) = x^3$, $g(x) = x^5$, $h(x) = x^7$

(c) $f(x) = x^{1/2}$, $g(x) = x^{1/4}$, $h(x) = x^{1/6}$

(d) $f(x) = x^{1/3}$, $g(x) = x^{1/5}$, $h(x) = x^{1/7}$

Shifting and Stretching Graphs, Supply and Demand for Wheat

• Vertical and Horizontal Shifts • Reflections • Stretching and Compressing • Motion Under Gravity • Supply and Demand for Wheat • Jogging • Modeling with Graphs

We have introduced various classes of functions so far and have encouraged students to build up a collection of graphs of useful functions. We now find out how graphs of certain other functions can be obtained from these known graphs through horizontal shifts, vertical shifts, and reflections and by stretching. We illustrate some of these transformations and the way that they can be combined in Figure 3.34.

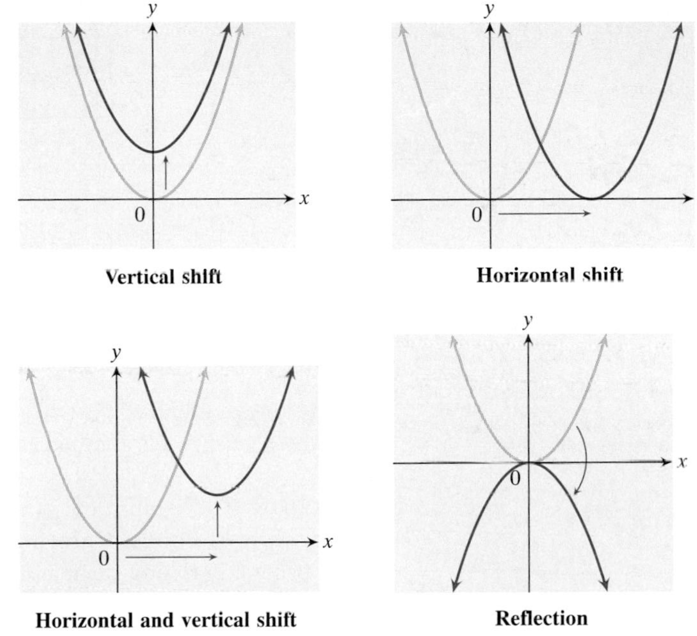

Figure 3.34 Shifting and reflecting graphs.

Let us now explore the relationships between the forms of the functions that have graphs that are related in these ways.

VERTICAL AND HORIZONTAL SHIFTS

Exploration 1

Vertical Shifts

Graph $f(x) = x^2$, $g(x) = x^2 + 2$, and $h(x) = x^2 - 3$ on your calculator in the window $-4 \le x \le 4$, $-4 \le y \le 8$. Observe that the graph of $g(x)$ can be obtained from the graph of $f(x)$ by shifting it upward 2 units. Discuss how the graph of $h(x)$ can be obtained from that of $f(x)$.

Observe that $g(x)$ can be interpreted as a function of the form $f(x) + c$ for the value $c = 2$. $h(x)$ is a function of the form $f(x) + c$ for $c = -3$. These observations lead to the following results.

Vertical Shifts of Graphs

Let c be a positive real number. The graph of $g(x) = f(x) + c$ can be obtained from that of $f(x)$ by an upward shift of c units. The graph of $h(x) = f(x) - c$ can be obtained from that of $f(x)$ by a downward shift of c units.

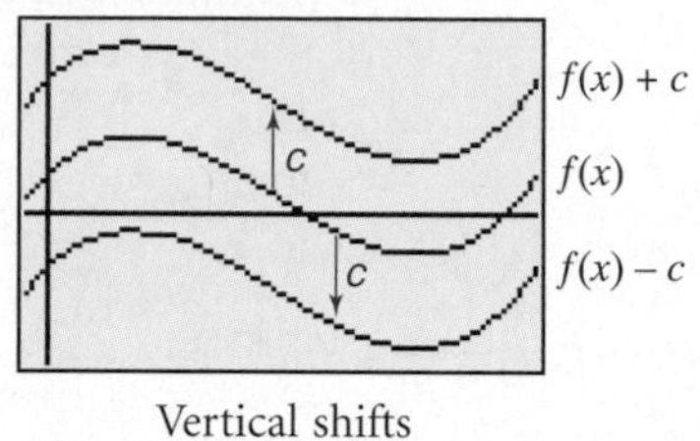

Vertical shifts

We now illustrate how this result can be used to sketch graphs.

Example 1 Using a Vertical Shift to Sketch a Graph

Use the known graph of $f(x) = |x|$ to sketch the graph of $g(x) = |x| - 2$. Check your answer using a graphing calculator.

SOLUTION The graph of $g(x) = |x| - 2$ can be obtained from that of $f(x) = |x|$ by shifting the graph downward 2 units (Figure 3.35). We check this result using a graphing calculator. The domain of g is seen to be $(-\infty, \infty)$ and the range is $[-2, \infty)$.

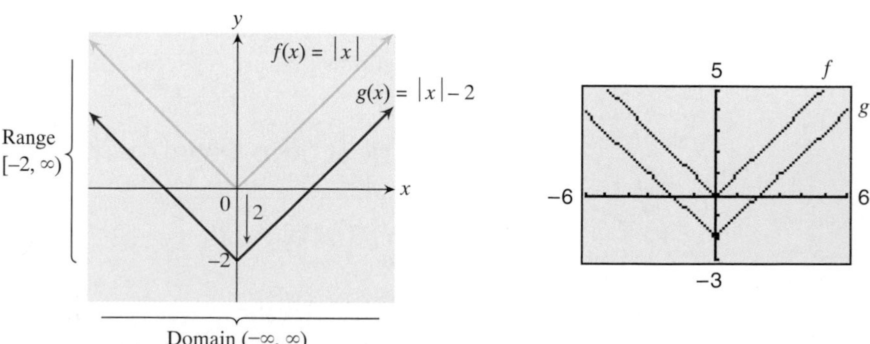

Figure 3.35 Vertical shift of $f(x) = |x|$.

Exploration 2

Horizontal Shifts

Graph $f(x) = x^2$, $g(x) = (x + 2)^2$, and $h(x) = (x - 4)^2$ on your calculator in the window $-6 \leq x \leq 6$, $-1 \leq y \leq 4$. Discuss how the graphs of $g(x)$ and $h(x)$ can be obtained from the graph of $f(x)$.

Observe that $g(x)$ can be interpreted as a function of the form $f(x + c)$ for the value $c = 2$. $h(x)$ is a function of the form $f(x + c)$ for $c = -4$. These observations lead to the following result.

HORIZONTAL SHIFTS OF GRAPHS

Let c be a positive real number. The graph of $g(x) = f(x + c)$ can be obtained from that of $f(x)$ by a horizontal shift of c units to the left. The graph of $h(x) = f(x - c)$ can be obtained from that of $f(x)$ by a horizontal shift of c units to the right. *Note:* **plus** is a shift to the **left; minus** is a shift to the **right.**

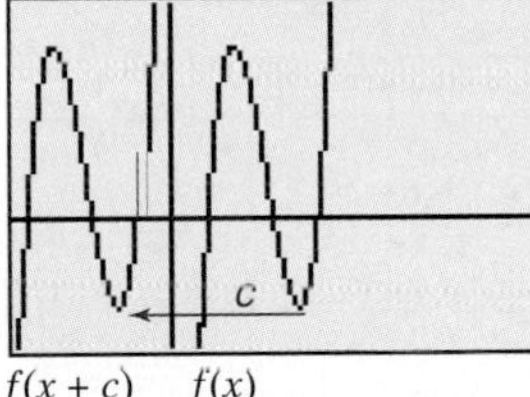

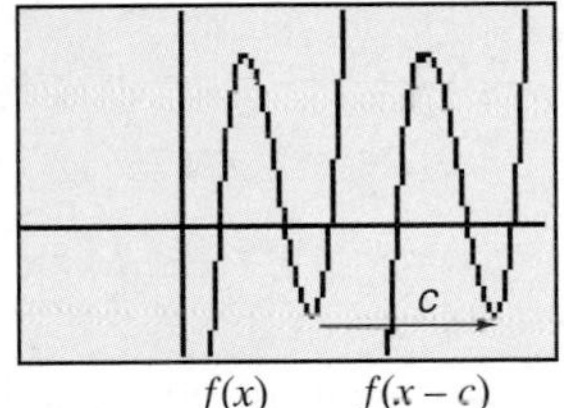

Horizontal shifts.

Example 2 Using a Horizontal Shift to Sketch a Graph

Sketch the graph of $g(x) = (x + 4)^3$. Check your answer using a graphing calculator.

SOLUTION The graph of $g(x) = (x + 4)^3$ can be obtained from the graph of $f(x) = x^3$ by shifting the graph horizontally 4 units to the left (Figure 3.36). We check the graph using a calculator. The domain of g is $(-\infty, \infty)$. The range is also $(-\infty, \infty)$.

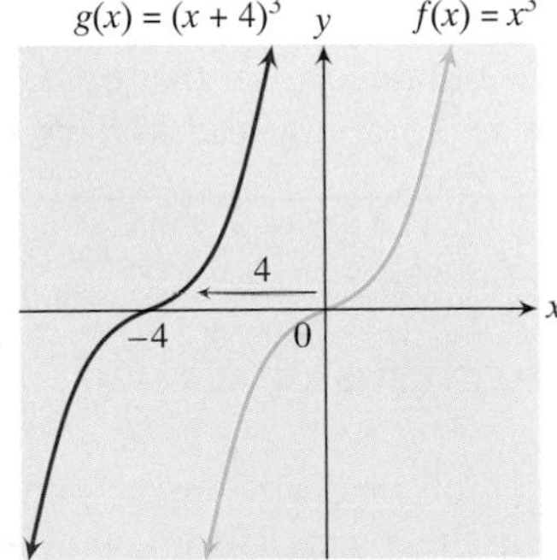

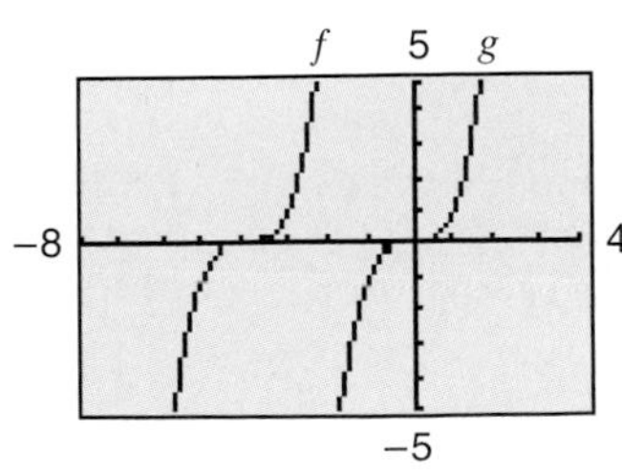

Figure 3.36 Horizontal shift of $f(x) = x^3$

The techniques of vertical and horizontal shifts can be combined to sketch graphs. The following example illustrates this.

Example 3 **Using Shifts to Sketch a Graph**

Sketch the graph of $g(x) = (x - 3)^2 + 1$. Support your answer using a graphing calculator.

SOLUTION Let us start with the known graph of $f(x) = x^2$. We perform a horizontal shift of 3 units to the right followed by a vertical shift of 1 unit up (Figure 3.37). The domain of g is $(-\infty, \infty)$. The range is $[1, \infty)$.

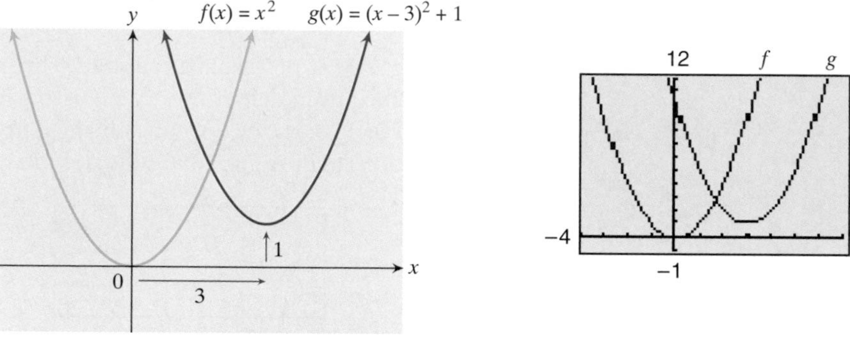

Figure 3.37 Horizontal and vertical shifts of $f(x) = x^2$.

REFLECTIONS

Exploration 3

Reflections

(a) Graph $f(x) = x^2 + 1$ and $g(x) = -x^2 - 1$ on your calculator in the window $-3 \leq x \leq 3$, $-5 \leq y \leq 5$. Discuss how the graph of $g(x)$ can be obtained from that of $f(x)$.

(b) Graph $f(x) = x^3 + 2$ and $h(x) = (-x)^3 + 2$ on your calculator in the window $-3 \leq x \leq 3$, $-10 \leq y \leq 10$. Discuss how the graph of $h(x)$ can be obtained from that of $f(x)$.

These explorations lead to the following general results.

REFLECTIONS OF GRAPHS

The graph of $g(x) = -f(x)$ can be obtained from that of $f(x)$ by a reflection in the x-axis. The graph of $h(x) = f(-x)$ can be obtained from that of $f(x)$ by a reflection in the y-axis.

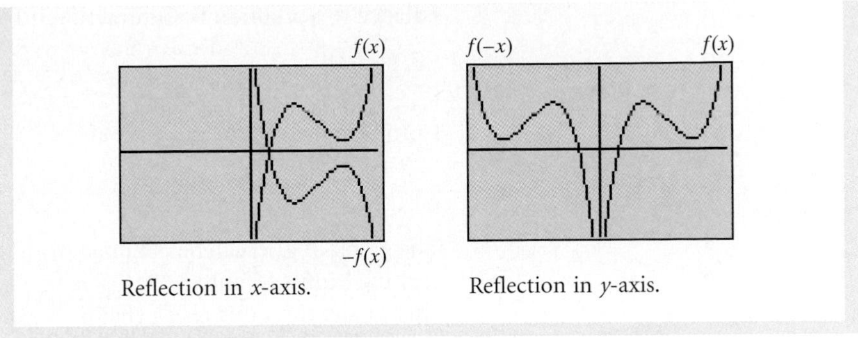

Reflection in x-axis. Reflection in y-axis.

Example 4 Using a Reflection and a Shift to Sketch a Graph

Sketch the graph of $g(x) = -|x| + 4$. Support your answer using a graphing calculator.

SOLUTION Sketch the graph in two steps. Start with the known graph of $f(x) = |x|$ and perform a reflection in the x-axis followed by a vertical shift of 4 units upward (Figure 3.38). The domain of g is $(-\infty, \infty)$. The range of g is $(-\infty, 4]$.

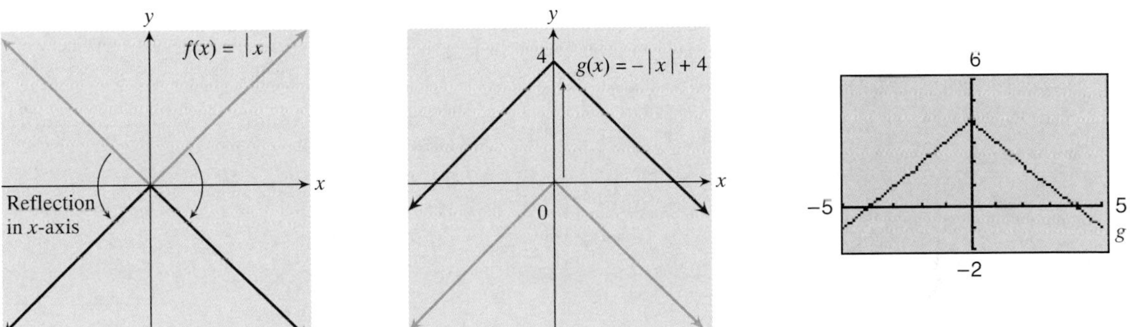

Figure 3.38 Reflection and vertical shift of $f(x) = |x|$.

We saw in Section 1.4 how the graph of a parabola can be sketched by finding its vertex and determining whether it opens upward or downward. The following example illustrates how the graph of a parabola can also be sketched using shifts.

Example 5 Using Shifts to Sketch the Graph of a Parabola

Rewrite the quadratic function $g(x) = x^2 - 6x + 11$ in a form that shows shifts, and use this information to sketch its graph. Use a graphing calculator to support the graph.

SOLUTION We use the technique of completing the square to rewrite $g(x)$ in a form that reveals shifts. Separate the terms involving x.

$$g(x) = (x^2 - 6x) + 11$$

Add the square of half the coefficient of x to complete the square of the first part.

Balance this addition by subtracting it from the second part.

$$g(x) = (x^2 - 6x + (-3)^2) + 11 - (-3)^2$$

Complete the square and simplify.

$$g(x) = (x - 3)^2 + 2$$

The graph of $g(x)$ can be obtained from that of $f(x) = x^2$ by shifting right 3 units and up 2 units (Figure 3.39a). The vertex of the graph of $g(x)$ is (3, 2). The calculator supports this result (Figure 3.39b).

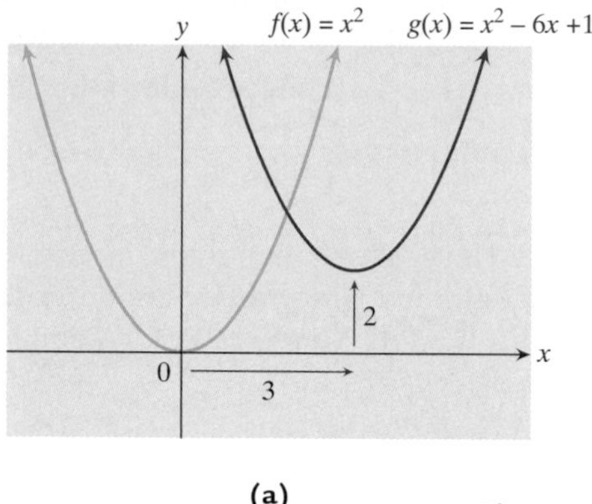

(a) (b)

Figure 3.39

Further Ideas 1 The graph of $g(x)$ in Figure 3.40 can be obtained from that of $f(x) = x^2$ through a reflection and shifts. Find the function $g(x)$. Check your answer by duplicating the graph on a calculator.

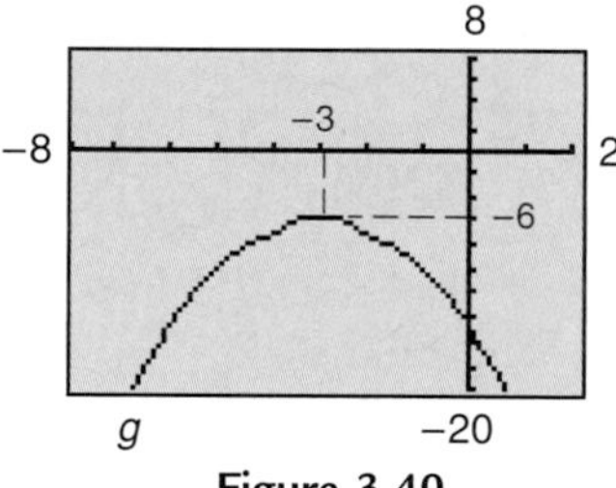

Figure 3.40

Stretching and Compressing

Exploration 4

Stretching and Compressing

(a) Graph $f(x) = x^2$, $g(x) = 2x^2$, and $h(x) = (\frac{1}{2})x^2$ on your calculator in the window $-4 \leq x \leq 4$, $0 \leq y \leq 8$.

(b) Graph $f(x) = x^3$, $g(x) = 2x^3$, and $h(x) = (\frac{1}{2})x^3$ on your calculator in the window $-2 \leq x \leq 2$, $-8 \leq y \leq 8$.

Discuss the effect of the factors 2 and $\frac{1}{2}$ on the graph of $f(x)$ in each case.

These explorations lead to the following results about stretching and compressing graphs.

Vertically Stretching and Compressing

Let c be a real number greater than 1. The graph of $g(x) = cf(x)$ is the graph of $f(x)$ stretched vertically by a factor c. The graph of $h(x) = (1/c)f(x)$ is the graph of $f(x)$ compressed vertically by a factor c.

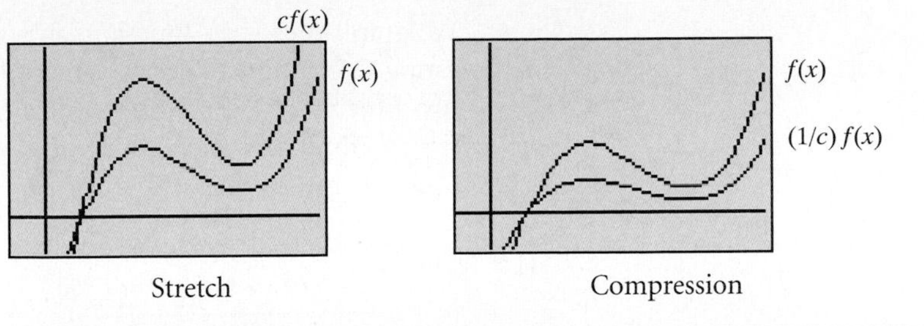

We now see how these transformations arise naturally in applications.

Applications . . .

Example 6 Motion Under Gravity

In Section 1.5, we introduced the function that describes the motion of an object propelled vertically under gravity. If an object is propelled at time $t = 0$ from ground level ($s = 0$) with velocity v, then height s at time t is described by

$$s(t) = -16t^2 + vt \qquad (1)$$

Suppose that an object is propelled vertically now, not from ground level but from an initial height h. The value of s at each t will now be h greater than previously. Thus,

$$s(t) = -16t^2 + vt + h \qquad (2)$$

Observe that Equation (2) is obtained from Equation (1) through a vertical shift of h. Thus, the physical raising of the starting point up a distance h is described mathematically as a vertical shift of h. We illustrate these physical and mathematical perspectives in Figure 3.41.

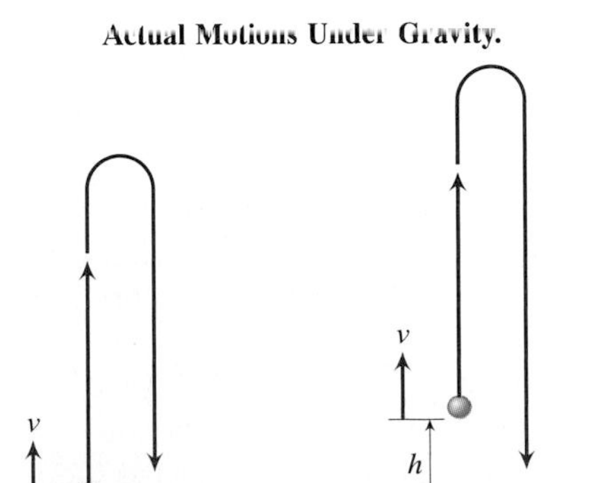

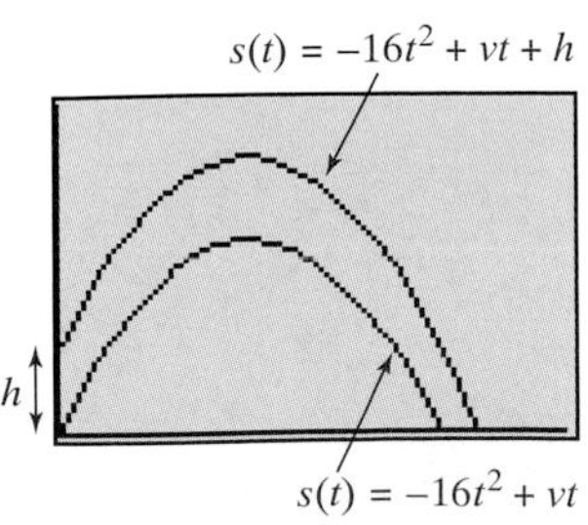

Figure 3.41

Let us compare the motions of two objects: one projected upward with a velocity of 256 feet per second from ground level and the other projected upward with a velocity of 256 feet per second from a height of 300 feet at the same instant. The functions that describe the two motions are

$$s_1(t) = -16t^2 + 256t \qquad \text{and} \qquad s_2(t) = -16t^2 + 256t + 300$$

We graph these two functions in a convenient window (Figure 3.42). It is interesting to examine tables of values of s_1 and s_2. The tables show the heights of the two objects at 1-second intervals. Both are propelled upward at the same time $t = 0$. Observe how the value of s_2 is always 300 more than s_1, demonstrating the upward shift of 300.

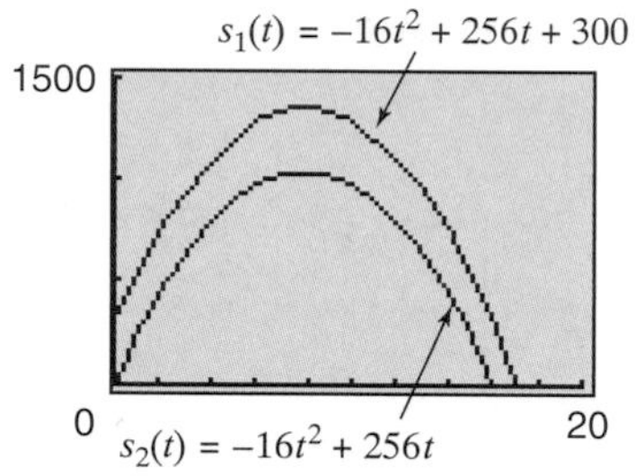

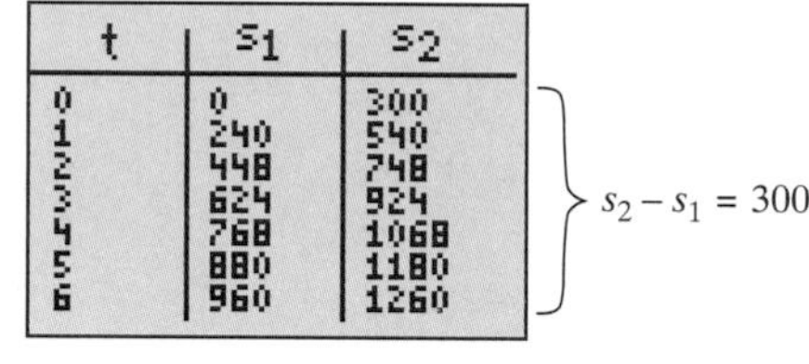

t	s_1	s_2
0	0	300
1	240	540
2	448	748
3	624	924
4	768	1068
5	880	1180
6	960	1260

Figure 3.42 Graph and table views of motion.

Exploration 5

Interpretation of a Horizontal Shift

We saw in Example 6 how vertical shifts can arise in investigating motions. We now give the student the opportunity to determine how horizontal shifts arise in analyzing motions. Consider two objects: one projected upward with a velocity of 256 feet per second from ground level and the other projected upward with a velocity of 256 feet per second from ground level 6 seconds after the first. Find the functions that describe each motion. Explain how this physical situation corresponds to a horizontal shift.

Example 7 | Supply and Demand for Wheat

Economists slide graphs a great deal in their analyses! We now illustrate such shifting for a supply function. We introduced supply and demand functions in Example 3 in Section 1.5. The two sides in a market: namely supply and demand, interact to determine the price of a commodity. The price at which a commodity settles in a market is the one at which supply is equal to demand. This price is called the **equilibrium price.** This will be the price at which the graphs of the supply and demand functions cross (Figure 3.43a).

Let $S(p)$ be the predicted supply function for wheat based on normal weather. Let us see how economists modify the supply–demand model to allow for bad weather. The effect of bad weather will be to reduce the amount of wheat the

Wheat production is sensitive to weather. Bad weather reduces the amount of wheat available, driving up the price. (*© David Austen/Tony Stone*)

farmers will supply at every market price by a constant factor c. The actual supply function becomes $S(p) - c$. Note the implication of this supply shift. The equilibrium price is increased from the expected p_1 to p_2 (Figure 3.43b). We pay more for our wheat.

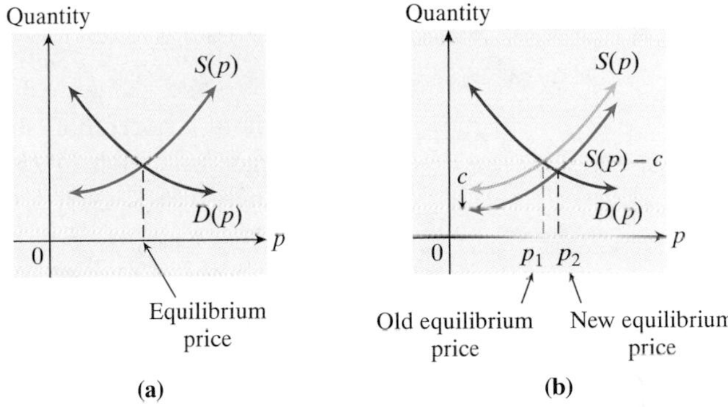

Figure 3.43 Effect of bad weather on the price of wheat.

Further Ideas 2 Suppose that rising family incomes make everyone want more wheat. The economic effect is that at each price there will be more demand. The demand function becomes $D(p) + c$. Build this function into Figure 3.43(a) to show how this vertical shift causes the equilibrium price to increase.

Example 8 Jogging

Jeff Williams starts out jogging at a steady 6 miles per hour, and 12 minutes later Michelle Williams starts out from the same place running at 8 miles per hour. Find functions that describe both motions, expressing distance as a function of the time measured from when the first jogger started out. Plot the graphs of these functions on your calculator. Use the graphs to determine when Michelle catches up with Jeff.

SOLUTION Let t be time measured in hours from when the first jogger (Jeff) starts out. The first jogger runs at 6 miles per hour. Thus, the distance the first jogger will have run at time t hours is $f(t) = 6t$. The second jogger (Michelle) starts 12 minutes later. The 12 minutes can be written as 12/60 hours, or 0.2 hours.

At time t, the second jogger will have been running for $(t - 0.2)$ hours. The second jogger runs at 8 miles per hour. The distance that the second jogger will have completed at time t is therefore $g(t) = 8(t - 0.2)$.

We graph f and g (Figure 3.44). Note that the later start of the second jogger corresponds to a horizontal shift of 0.2 from the origin in the graph of g. The second jogger will catch up with the first jogger when they have both run the same distance, that is when $f(t) = g(t)$. This corresponds to the point A, the intersection of the graphs. A is found (using an intersect finder or trace/zoom) to be the point (0.8, 4.8). The second jogger thus catches up with the first jogger 0.8 hours (or 45 minutes) after the first jogger sets out. They will then both be 4.8 miles from the start.

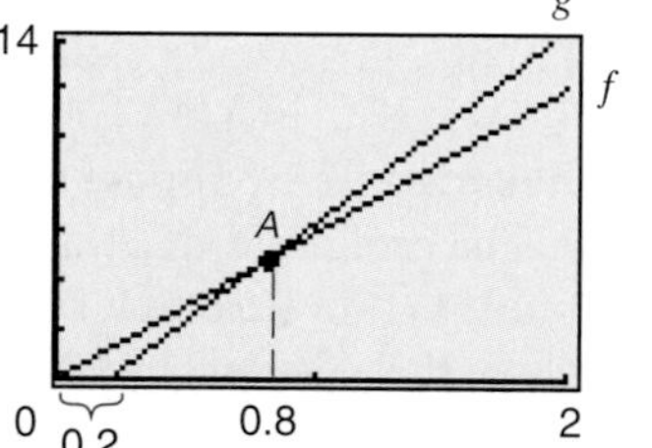

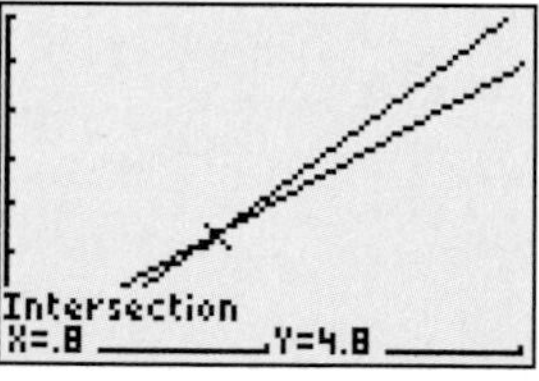

Figure 3.44 The second jogger catches up with the first.

Further Ideas 3 Verify, using algebra, that the second jogger catches up with the first jogger 0.8 hours after the first jogger sets out and that they will then both be 4.8 miles from the start.

Modeling with Graphs

Information is often given in a form that lends itself to sketching a rough graph. The general shape of a graph can convey much information about a situation.

Example 9 Temperature Graph

From the time the author got up until noon it gradually warmed up. At noon, a cold front came through. The temperature dropped rapidly until 2 PM. It then slowly warmed up again until 8 PM, from which time the temperature remained pretty constant until the author went to bed. Draw a graph of temperature as a function of time.

SOLUTION Let time be the horizontal axis and temperature be the vertical axis. Figure 3.45 shows a possible graph of the situation. The first part of the graph, labeled *a*, represents the part of the day from getting up to noon, when the temperature gradually increased. Part *b* of the graph represents the rapid decrease in temperature until 2 PM. Part *c* shows the slow warming until 8 PM. Finally, part *d* depicts the constant temperature from 8 PM to bedtime.

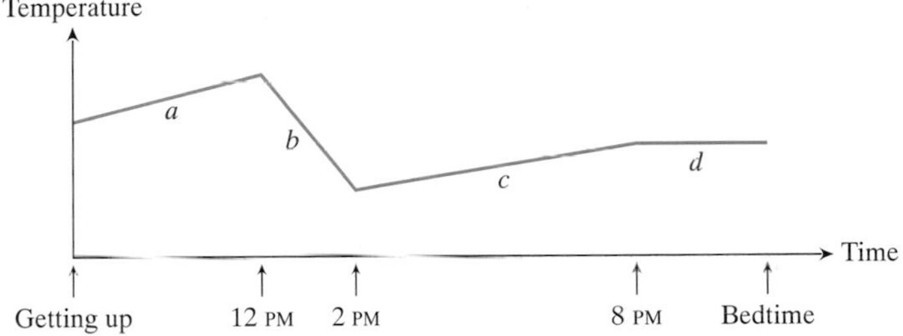

Figure 3.45 Daytime temperature.

Example 10 Going Skiing

Gerhard Wulf and Denise Ogden left Denver for Vail for some skiing. A few miles outside Denver, Gerhard realized that he had forgotten his ski boots. They had to return to Denver to get the boots before continuing on to Vail. Sketch a graph of distance from Denver as a function of time.

SOLUTION Let time be the horizontal axis and distance from Denver be the vertical axis. We get the graph shown in Figure 3.46. The first part of the graph, labeled *a*, represents the first part of the journey, when the distance from Denver is gradually increasing with time. Then comes the instance of realization that the boots had been forgotten. Part *b* of the graph represents the return to Denver, when the distance from Denver is gradually decreasing. Part *c* then represents the uninterrupted journey to Vail.

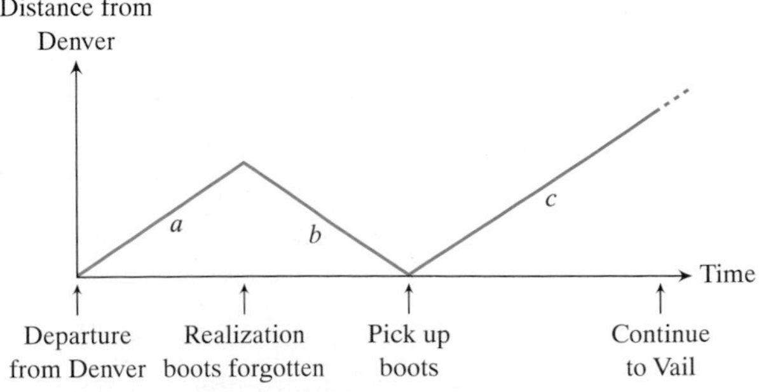

Figure 3.46 Going skiing.

If specific information is given, the rough graphs can be used to develop a function that describes the situation. Such functions are often piecewise-defined functions. We illustrate these steps in constructing a function with the following example.

Example 11 Driving from New York to Washington, DC

Cynthia Kubic drove from New York City to Washington, DC. She stopped and had lunch in Philadelphia on the way. We can represent this information with the graph shown in Figure 3.47. The first part of the graph, labeled a, represents the first part of the journey, when the distance from New York is gradually increasing with time. The horizontal part b represents the lunch stop, when the distance from New York remains constant while time is increasing. The final part c represents the final stage of the journey from Philadelphia to Washington, DC, with the distance from New York gradually increasing with time until Washington, DC is reached.

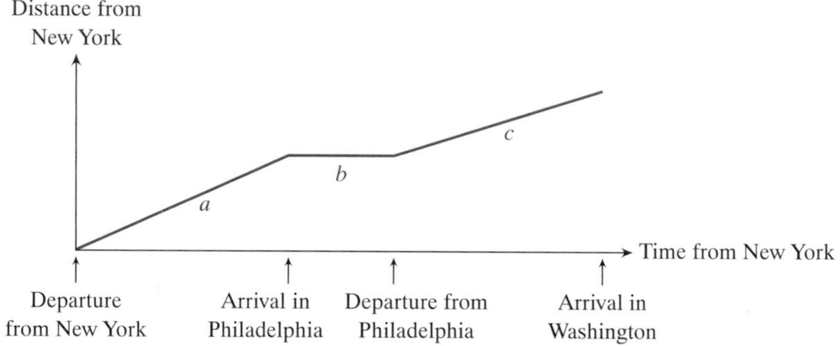

Figure 3.47 Traveling from New York to Washington.

Suppose Cynthia drove at an average speed of 60 miles per hour from New York to Philadelphia and at 65 miles per hour from Philadelphia to Washington, DC. The distance from New York to Philadelphia is 90 miles and from Philadelphia to Washington, DC is 130 miles. Thus, using Time = Distance/Speed, she took $90/60 = 1.5$ hours to travel from New York to Philadelphia. She took $130/65 = 2$ hours to travel from Philadelphia to Washington, DC. Suppose she stopped for 1 hour for lunch in Philadelphia. This information now leads to the more complete graph shown in Figure 3.48.

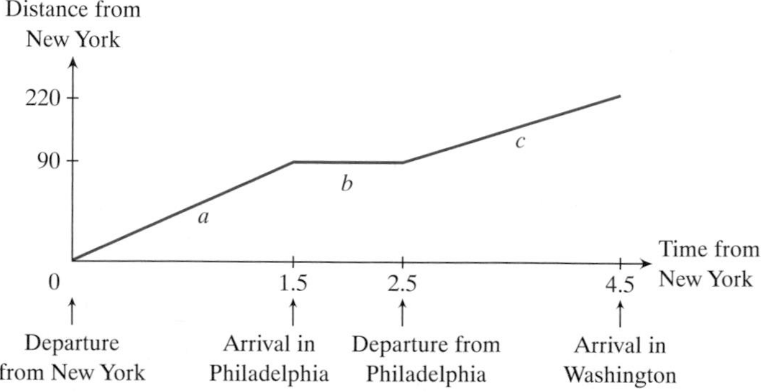

Figure 3.48 Breakdown of journey from New York to Washington.

We can use the relationship Distance = Speed $\times$ Time and shifts to get the piecewise-defined function $d(t)$ that describes this graph.

$$d(t) = \begin{cases} 60t & 0 \le t \le 1.5 & \text{New York to Philadelphia} \\ 90 & 1.5 \le t \le 2.5 & \text{Lunch break} \\ 65(t - 2.5) + 90 & 2.5 \le t \le 4.5 & \text{Philadelphia to Washington, DC} \end{cases}$$

Self-Check 1 Arrive at the various parts of the function $d(t)$ in Example 11.

Example 12 **Modeling a Kickoff in the NFL**

A football field has the dimensions shown in Figure 3.49.

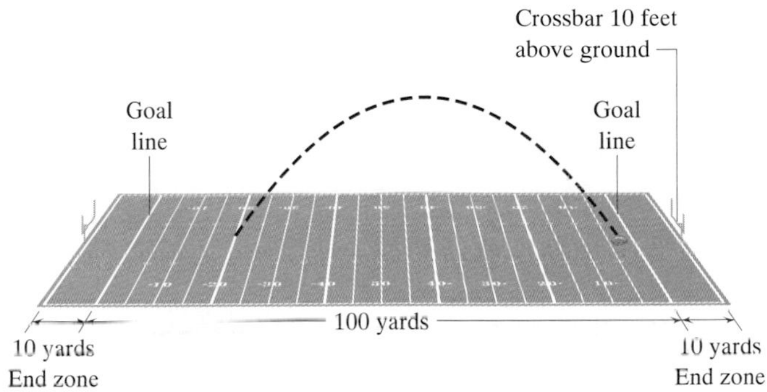

Figure 3.49 Football field.

Let us construct a mathematical model of the kickoff. The trajectory of the ball is a parabola. Kickers aim for an initial trajectory angle of 45°. This gives maximum distance for the kick. The function for such a trajectory is known to be

$$f(x) = -\frac{32}{v^2}x^2 + x$$

where $f(x)$ is the height in feet at distance x feet measured from the goal line (own) and v is the velocity of the ball in feet per second immediately after the kick. In the NFL, the kickoff takes place from the 35-yard line (105 feet). We allow for this with a horizontal shift. The function that describes the kickoff from the 35-yard line is

$$f(x) = -\frac{32}{v^2}(x - 105)^2 + (x - 105)$$

Let us determine whether a kick with an initial velocity of 75 feet per second will reach the end zone. Using a graphing calculator, enter the equation

Y = [(-)] 32(X [−] 105) [^] 2 [÷] 75 [^] 2 [+] X [−] 105

Select a convenient window that extends the length of the field, 0–100 yards (300 feet) with tick marks every 10 yards (30 feet):

$$X_{\min} = 0,\ X_{\max} = 300,\ X_{\text{scl}} = 30,\ Y_{\min} = 0,\ Y_{\max} = 100,\ Y_{\text{scl}} = 10$$

The graph in Figure 3.50 shows that the ball lands short of the end zone.

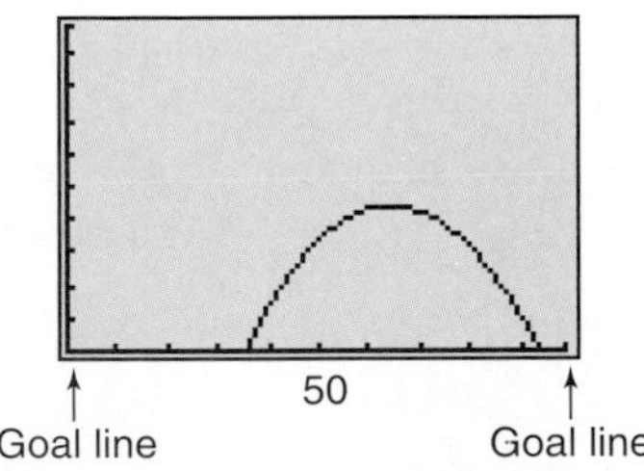

Figure 3.50 The kickoff lands short of the end zone.

Further Ideas 4

(a) Use algebra to find the minimum velocity that will ensure that the ball reaches the end zone. Check the result using a graph.

(b) Suppose that the kick goes out of bounds; it has to be taken again from the 30-yard line. What is the new minimum velocity for the ball to reach the end zone?

Group Discussion

Constructing a Graph

Locate a central building on campus, such as a tower, library, or student union. Consider a student living in a certain dormitory on campus. Draw a rough graph of the distance of the student from the building during a 24-hour period, midnight to midnight. *Hint:* Horizontal axis is time; vertical axis is distance from the central building.

Answer to Self-Check Exercise

1. Part a. Distance = Speed × Time gives $d(t) = 60t$. Part b. Distance from New York is constant, $d(t) = 90$. Part c. Travel time out of Philadelphia is $t - 2.5$. Thus, distance from Philadelphia is $d(t) = 65(t - 2.5)$. Distance from New York is given by $d(t) = 65(t - 2.5) + 90$.

Answers to Further Ideas Exercises

1. Reflect $f(x) = x^2$ in the x-axis; shift left 3 and down 6.

$$g(x) = -(x + 3)^2 - 6$$

2. Equilibrium price increases from p_1 to p_2 (see below).

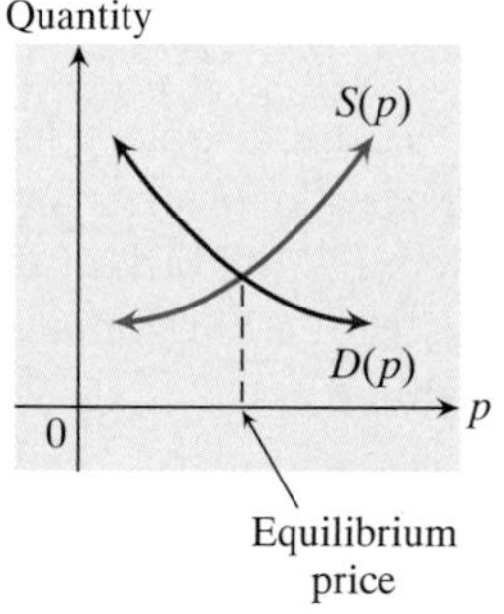

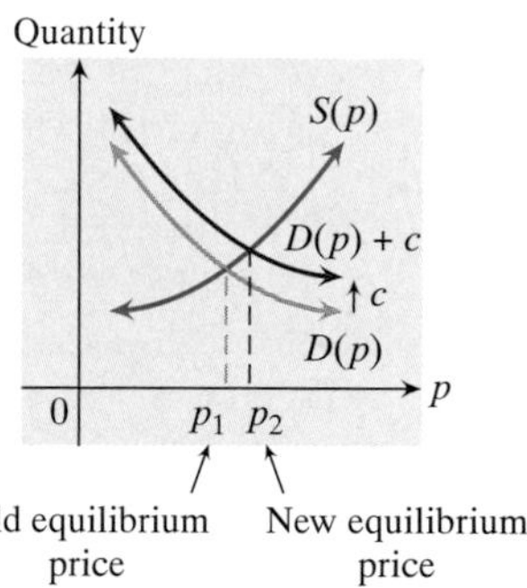

3. The second jogger catches up with the first when $g(t) = f(t)$, $8(t - 0.2) = 6t$, $2t = 1.6$, $t = 0.8$ hours. At that time $f(t) = 6(0.8) = 4.8$ miles.

4. **(a)** If the ball reaches the end zone, $x = 300$ feet. We want f to be zero when $x = 300$. Thus, $0 = -(32/v^2)(300 - 105)^2 + (300 - 105)$, which gives $v = 78.99$ feet per second.

   (b) Motion now described by

$$f(x) = -\frac{32}{v^2}(x - 90)^2 + (x - 90)$$

Again, we want f to be zero when $x = 300$. Thus

$$0 = -\frac{32}{v^2}(300 - 90)^2 + (300 - 90)$$

which gives $v = 81.97$ feet per second.

EXERCISES 3.4

In Exercises 1–10, (a) use vertical shifts, horizontal shifts, and reflections of the graphs of $f(x) = x^2$ or $f(x) = x^3$ to sketch the graphs of the given functions. Describe the transformations used. (b) Check your results using a graphing calculator, with the window $X_{min} = -10$, $X_{max} = 10$, $X_{scl} = 2$, $Y_{min} = -20$, $Y_{max} = 20$, and $Y_{scl} = 4$.

1. $g(x) = x^2 + 4$
2. $g(x) = x^2 - 7$
3. $g(x) = (x + 3)^2$
4. $g(x) = (x - 6)^2$
5. $g(x) = x^3 - 4$
6. $g(x) = -x^3$
7. $g(x) = -x^3 + 6$
8. $g(x) = (x + 3)^2 + 4$
9. $g(x) = -(x - 4)^7$
10. $g(x) = -(-x + 6)^3$

In Exercises 11–18, (a) use transformations of the graphs of $f(x) = cx^n$, $f(x) = |x|$, or $f(x) = c\sqrt{x}$ to sketch the graphs of the given functions. Describe the transformations used. (b) Check your results using a graphing calculator, with the window $X_{min} = -10$, $X_{max} = 10$, $X_{scl} = 2$, $Y_{min} = -20$, $Y_{max} = 20$, and $Y_{scl} = 4$.

11. $g(x) = (x - 3)^2 - 8$
12. $g(x) = -(x - 6)^3 + 4$
13. $g(x) = |x + 3| - 5$
14. $g(x) = -|x - 2| + 3$
15. $g(x) = -2x^4 + 10$
16. $g(x) = -3(x + 4)^5 - 2$
17. $g(x) = \sqrt{x + 6} - 8$
18. $g(x) = -\sqrt{x - 1} + 6$

In Exercises 19–24, rewrite each quadratic function $g(x)$ in a form that shows shifts and use this information to sketch its graph (see Example 5). Use a graphing calculator to support your answer in the window $X_{min} = -5$, $X_{max} = 5$, $X_{scl} = 1$, $Y_{min} = -5$, $Y_{max} = 5$, and $Y_{scl} = 1$.

19. $g(x) = x^2 - 4x + 2$
20. $g(x) = x^2 - 4x$
21. $g(x) = x^2 + 2x + 4$
22. $g(x) = x^2 + 8x + 14$
23. $g(x) = -x^2 + 4x - 1$
24. $g(x) = -x^2 - 8x - 19$

In each of Exercises 25–30, sketch the graphs of g for the given values of c, in one coordinate system. Check your answers using a graphing calculator in the window $X_{min} = -5$, $X_{max} = 5$, $X_{scl} = 1$, $Y_{min} = -5$, $Y_{max} = 5$, and $Y_{scl} = 1$.

25. $g(x) = x^2 + c$, $c = -2, 0, 2$
26. $g(x) = (x - c)^2 + 2$, $c = -3, 0, 3$
27. $g(x) = -(x + 2)^3 + c$, $c = -3, 0, 3$
28. $g(x) = |x + c| - 4$, $c = -2, 0, 2$
29. $g(x) = -\sqrt{x + c} + c$, $c = -3, 0, 3$
30. $g(x) = -(x + c)^2 + c$, $c = -4, 0, 4$

In Exercises 31–33, find the functions having the given graphs using transformations of $f(x) = x^2$. Each window is $X_{min} = -10$, $X_{max} = 10$, $X_{scl} = 1$, $Y_{min} = -20$, $Y_{max} = 20$, and $Y_{scl} = 5$. Check your answer by duplicating the display on your calculator.

31.

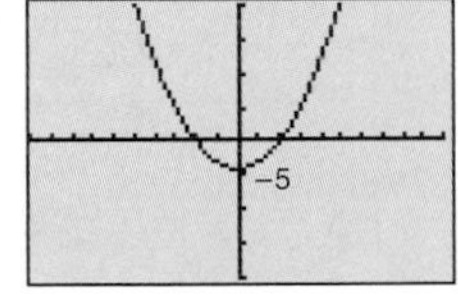

32.

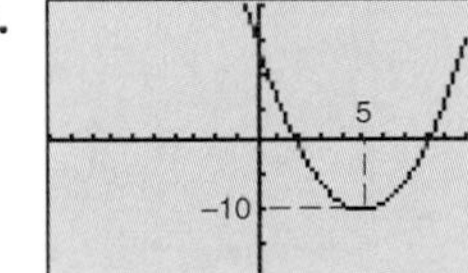

33.

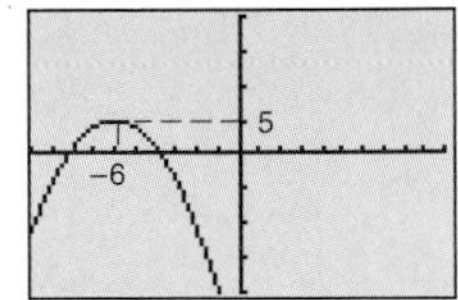

In Exercises 34–36, find the functions having the given graphs using transformations of $f(x) = x^3$. Each window is $X_{min} = -10$, $X_{max} = 10$, $X_{scl} = 1$, $Y_{min} = -30$, $Y_{max} = 30$, and $Y_{scl} = 5$. Check your answer by duplicating the display on your calculator.

34.

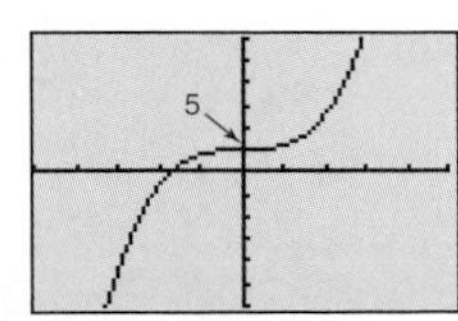

35.

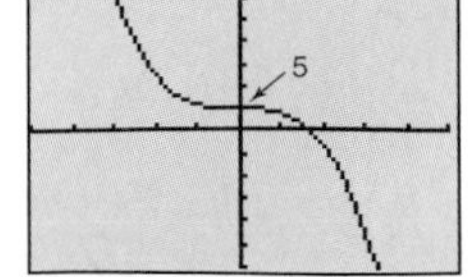

36.

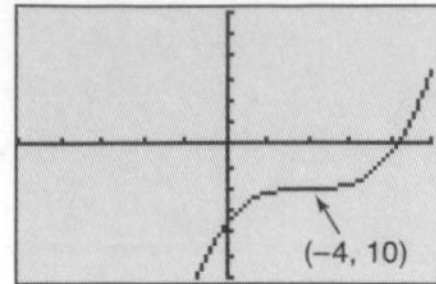

In Exercises 37–39, find the functions having the given graphs using transformations of $f(x) = |x|$. Each window is $X_{\text{min}} = -10$, $X_{\text{max}} = 10$, $X_{\text{scl}} = 1$, $Y_{\text{min}} = -10$, $Y_{\text{max}} = 10$, and $Y_{\text{scl}} = 1$. Check your answer by duplicating the display on your calculator.

37.

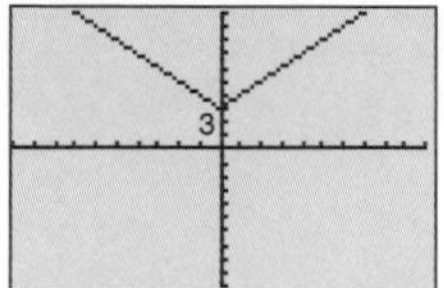

38.

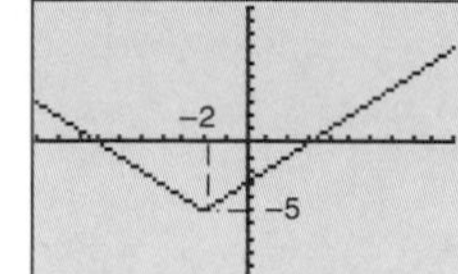

39.

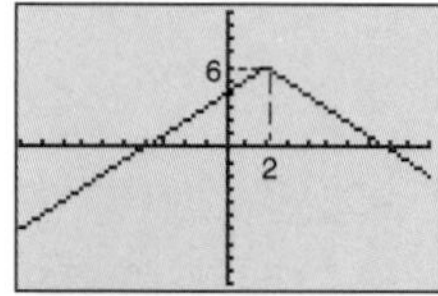

In Exercises 40–42, find the functions having the given graphs using transformations of $f(x) = \sqrt{x}$. Each window is $X_{\text{min}} = -10$, $X_{\text{max}} = 10$, $X_{\text{scl}} = 1$, $Y_{\text{min}} = -4$, $Y_{\text{max}} = 4$, and $Y_{\text{scl}} = 1$. Check your answer by duplicating the display on your calculator.

40.

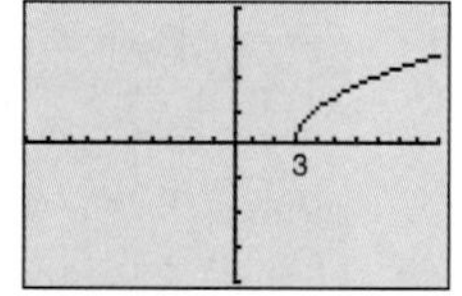

41.

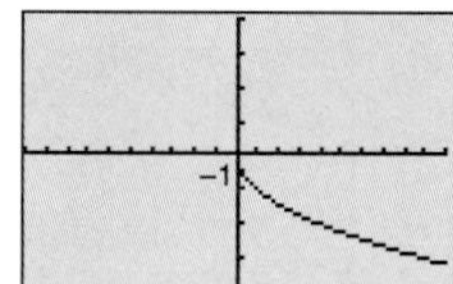

42.

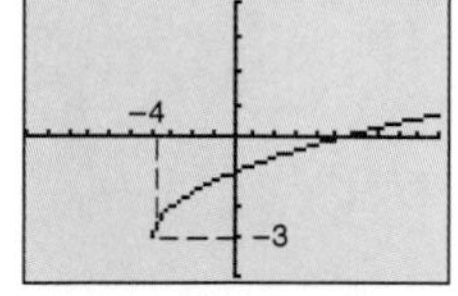

In Exercises 43–45, duplicate the graphs of the parabolas on your graphing calculator using the window $X_{\text{min}} = -10$, $X_{\text{max}} = 10$, $X_{\text{scl}} = 1$, $Y_{\text{min}} = -20$, $Y_{\text{max}} = 20$, and $Y_{\text{scl}} = 5$. Use translations and reflections of known functions. Give each function. (The coordinates of all the vertices have integer values.)

43.

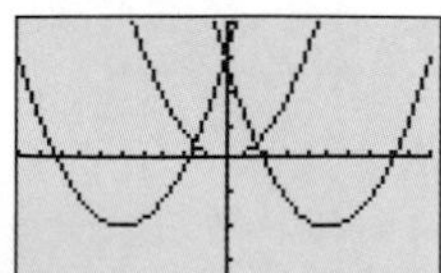

44.

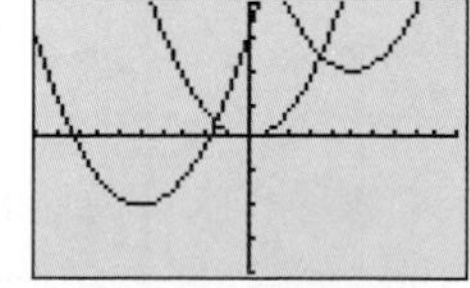

45.

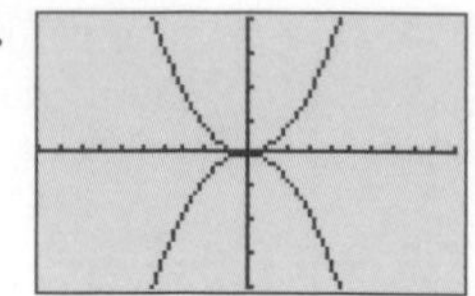

In Exercises 46–48, duplicate the graphs on your graphing calculator using the window $X_{\text{min}} = -10$, $X_{\text{max}} = 10$, $X_{\text{scl}} = 1$, $Y_{\text{min}} = -10$, $Y_{\text{max}} = 10$, and $Y_{\text{scl}} = 2$. Use translations and reflections of known functions. Give each function. (The coordinates of all the "corners" have integer values.)

46.

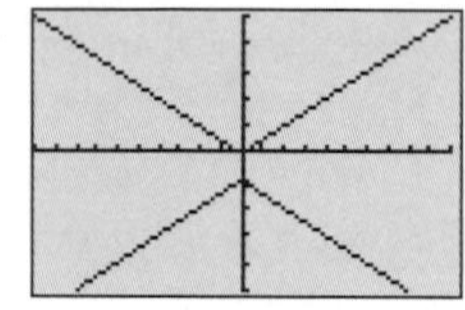

47.

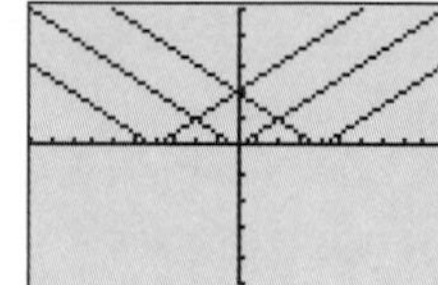

48.

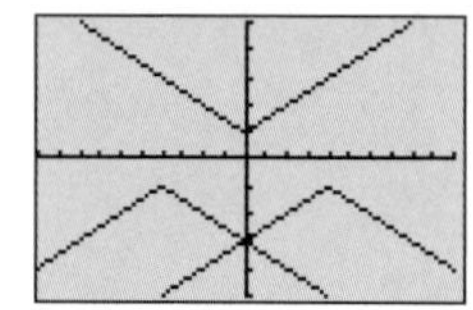

49. **(a)** Consider the following tables of $f(x) = x^2$ and $g(x) = x^2 + 2$. Explain how these tables are related.

(b) Consider the tables of $f(x) = x^2$ and $h(x)$. Find a function $h(x)$ that has these values.

(c) Consider the tables of $f(x) = x^3$ and $k(x)$. Find a function $k(x)$ that has these values.

Verify your answers by duplicating the tables on your calculator.

X	f	g
-3	9	11
-2	4	6
-1	1	3
0	0	2
1	1	3
2	4	6
3	9	11

(a)

X	f	h
-3	9	6
-2	4	1
-1	1	-2
0	0	-3
1	1	-2
2	4	1
3	9	6

(b)

X	f	k
-3	-27	-22
-2	-8	-3
-1	-1	4
0	0	5
1	1	6
2	8	13
3	27	32

(c)

50. **(a)** Consider the following tables of $f(x) = x^2$ and $g(x) = (x - 3)^2$. Explain how these tables are related.

(b) Consider the tables of $f(x) = x^2$ and $h(x)$. Find a function $h(x)$ that has these values.

(c) Consider the tables of $f(x) = x^3$ and $k(x)$. Find a function $k(x)$ that has these values.

Verify your answers by duplicating the tables on your calculator.

X	f	g
-9	81	144
-6	36	81
-3	9	36
0	0	9
3	9	0
6	36	9
9	81	36

(a)

X	f	h
-9	81	169
-5	25	81
-1	1	25
3	9	1
7	49	9
11	121	49
15	225	121

(b)

X	f	k
-5	-125	-27
-3	-27	-1
-1	-1	1
1	1	27
3	27	125
5	125	343
7	343	729

(c)

51. Match the following tables and graphs, giving explanations. Determine the equation of each graph. Check your answers by duplicating the tables and graphs on your calculator.

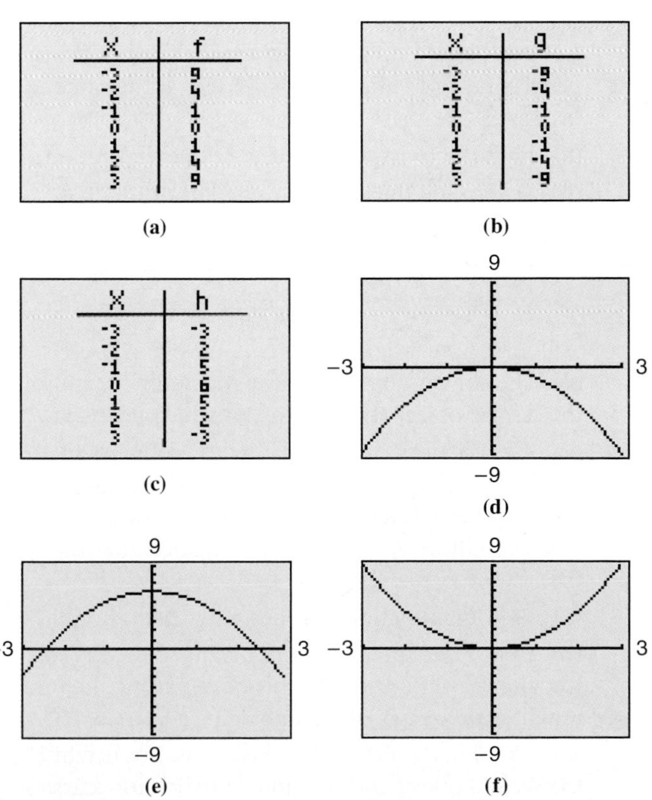

X	f
-3	9
-2	4
-1	1
0	0
1	1
2	4
3	9

(a)

X	g
-3	-9
-2	-4
-1	-1
0	0
1	-1
2	-4
3	-9

(b)

X	h
-3	-3
-2	2
-1	5
0	6
1	5
2	2
3	-3

(c)

52. The following tables are of functions of the form $f(x) = x^2 + c$ or $f(x) = -x^2 + c$ for various values of c. Find the functions. Check your answers by duplicating the tables on your calculators.

X	f
-3	12
-2	7
-1	4
0	3
1	4
2	7
3	12

(a)

X	f
-3	5
-2	0
-1	-3
0	-4
1	-3
2	0
3	5

(b)

X	f
-3	-4
-2	1
-1	4
0	5
1	4
2	1
3	-4

(c)

53. The following tables are of functions of the form $f(x) = (x - c)^2$ or $f(x) = -(x - c)^2$ for various values of c. Find the functions. Check your answers by duplicating the tables on your calculators.

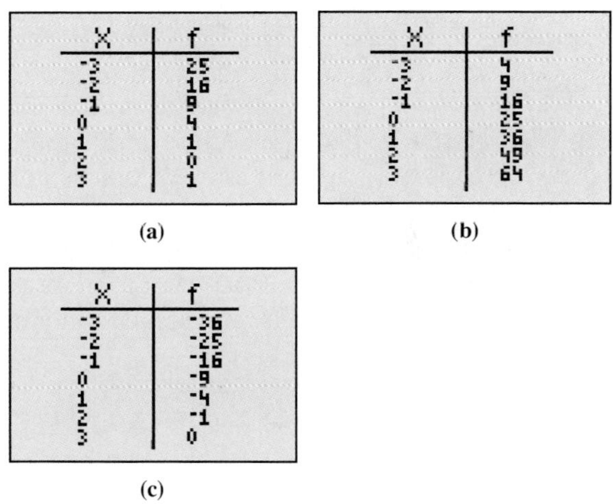

X	f
-3	25
-2	16
-1	9
0	4
1	1
2	0
3	1

(a)

X	f
-3	4
-2	9
-1	16
0	25
1	36
2	49
3	64

(b)

X	f
-3	-36
-2	-25
-1	-16
0	-9
1	-4
2	-1
3	0

(c)

In Exercises 54–62, draw rough graphs to describe the behaviors.

54. **Distance in Terms of Time** Don Viosca drove at a constant speed on Interstate 4. At a certain point in time he realized that he was running behind schedule and increased his speed to a new constant. Sketch a graph of distance as a function of time.

55. **Flying from Miami to London** A flight starting in Miami, Florida, has a stop in Orlando, Florida, and then continues to London, England.

(a) Sketch a graph of distance from Miami as a function of time.

(b) Sketch a graph of distance from Orlando as a function of time.

(c) Sketch a graph of distance from London as a function of time.

56. **Driving from San Diego to Los Angeles** Cynthia Heskin drove from San Diego to Los Angeles, stayed in Los Angeles for a day, and then drove on to San Francisco. She stayed in San Francisco for a few days and then drove back to San Diego, stopping once again in Los Angeles for a few hours on the way. Draw a graph of distance from San Diego as a function of time.

57. **Miles per Gallon and Speed** As an automobile increases in speed from rest it gets more and more miles per gallon until it reaches a certain speed. From then on the miles per gallon gradually decrease. Sketch a graph of miles per gallon as a function of speed.

58. **Water Pressure and Depth** As one dives deeper into water, the pressure increases. Sketch a graph of pressure as a function of depth.

59. **Airline Profits and Fares** An airline company decides to increase air fares on its route from Gainesville to Chicago in order to increase profit. It is found that profits increase up to a certain fare but then start to decrease because fewer people are flying. Sketch a graph of profit as a function of air fare.

60. **Terminal Velocity** An object falling freely under gravity will fall faster and faster until it reaches its so-called terminal velocity. It will then continue to fall at this fixed velocity. The terminal velocity varies from body to body depending on its size. Sketch a graph of velocity as a function of time for a body falling freely under gravity. (The terminal velocity of a sky diver is around 120 miles per hour, based on information in the movie *Terminal Velocity.*)

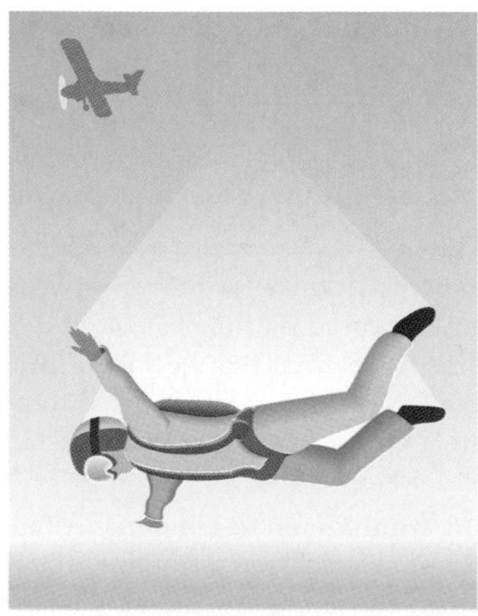

61. **Motion of a Pendulum** Sketch a graph of the speed of the bob of a pendulum as a function of time as it moves through one complete oscillation starting at an extreme position.

62. **Motion of a Spring** A weight at the end of a spring oscillates from its highest position to its lowest and then back up again, through two complete oscillations. Sketch the graph of displacement from the original position as a function of time.

Motion Under Gravity

Exercises 63–65 involve the use of vertical and horizontal translations in describing vertical motion under gravity (see Example 6). In these exercises, use the result that the height of an object projected upward with initial velocity v from ground level at time $t = 0$ is $s(t) = -16t^2 + vt$ at time t.

63. An object is projected upward with a velocity of 288 feet per second from ground level. Another object is projected upward with a velocity of 288 feet per second from a height of 352 feet at the same instant.

(a) Find the function that describes each motion.

(b) Determine the maximum height of the first object and the time at which this occurs using the formula that the vertex of a parabola

$$y = ax^2 + bx + c$$

is at $x = -b/2a$.

(c) Use shifts to determine the maximum height of the second object and the time at which it occurs.

64. An object is projected upward with a velocity of 288 feet per second from ground level. Another object is projected upward with a velocity of 288 feet per second from ground level 5 seconds after the first.

(a) Find the function that describes each motion.

(b) Determine the maximum height of the first object and the time at which this occurs using the formula for the vertex of a parabola (see Exercise 63).

(c) Use shifts to determine the maximum height of the second object and the time at which it occurs.

65. An object is projected upward with a velocity of 288 feet per second from ground level. Another object is projected upward with a velocity of 288 feet per second from a height of 352 feet, 5 seconds after the first.

(a) Find the function that describes each motion.

(b) Determine the maximum height of the first object and the time at which this occurs using the formula for the vertex of a parabola (See Exercise 63).

(c) Use shifts to determine the maximum height of the second object and the time at which it occurs.

66. Interstate Travel Debbie Pringle leaves Denver for Kansas City on Interstate 70 at 8:00 AM. She travels at an average speed of 60 miles per hour (allowing for stops). Brian Williams leaves Denver for Kansas City at 8:30 AM and travels at 68 miles per hour. Kansas City is 600 miles from Denver. Construct functions that describe both trips. Use your graphing calculator to find out when, and how far from Denver, Brian catches up with Debbie. Confirm your answer using algebra.

In Exercises 67 and 68, draw rough graphs to describe the situations; then draw more detailed graphs using the given information. Find the function that describes the graph.

67. Travel from Cleveland to Chicago Tessa Senez drove from Cleveland to Chicago. On the way she stopped in Toledo for a meal. The distance from Cleveland to Toledo is 110 miles, and that from Toledo to Chicago is 245 miles. The average speed from Cleveland to Toledo was 55 miles per hour. The average speed from Toledo to Chicago was 70 miles per hour. Tessa spent an hour eating her meal in Toledo. Draw a graph of the distance from Cleveland as a function of time. Give the piecewise-defined function that describes this graph.

68. Travel from Louisville to Kansas City Carlos Perez drove from Louisville to Kansas City. On the way, he stopped in St. Louis. The distance from Louisville to St. Louis is 240 miles and that from St. Louis to Kansas City is 285 miles. The average speed from Louisville to St. Louis was 60 miles per hour. The average speed from St. Louis to Kansas City was 75 miles per hour. Carlos took a half-hour break in St. Louis before continuing to Kansas City. Draw a graph of the distance from Louisville as a function of time. Give the piecewise-defined function that describes this graph.

69. Breaking the Mile Barrier Dr. Roger Bannister broke the 4-minute-mile barrier with a time of 3 minutes 59.4 seconds for the mile on the Iffley Road track, Oxford, England, at 6:10 PM on May 6, 1954.

(a) How fast would Bannister have been running (in miles per hour) if he had run the race at a constant speed?

(b) Find the function that describes his distance from the starting line as a function of time, assuming he ran the race at a constant speed.

(c) Sketch the graph of the function.

(d) How would the actual graph of his run probably have differed from this graph? Discuss.

70. How would you expect good weather and unexpectedly cheap fertilizer to affect the supply of wheat? Illustrate geometrically why you would expect the equilibrium price of wheat to decrease under such conditions. *Hint:* See Example 7.

In Exercises 71–73, you are asked to construct and investigate models for various sports. Can these models be used for coaching strategy/advice? The trajectory of an object is determined by its initial velocity and initial direction (Figure 3.51).

The **trajectory** of an object having initial velocity v measured in feet per second and initial direction m is

$$f(x) = -\frac{16(1 + m^2)}{v^2}x^2 + mx$$

The **initial direction** is the slope of the tangent line. ($m = 1$ corresponds to 45°, see Example 12.) x is the horizontal distance from the origin (in feet), and $f(x)$ is the corresponding height.

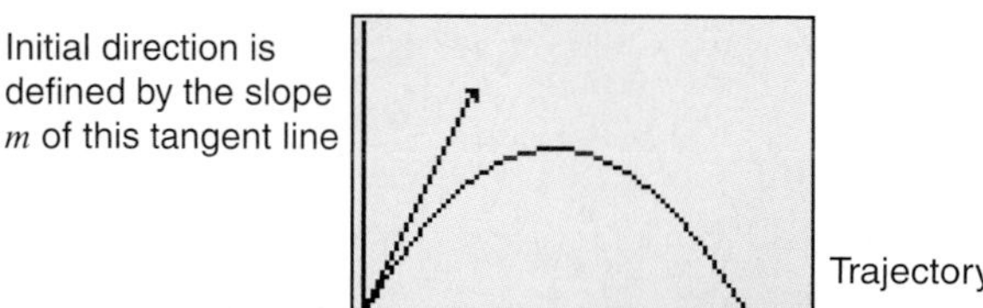
Initial direction is defined by the slope m of this tangent line
Trajectory

Figure 3.51 Trajectory of an object moving under gravity.

71. Football The height of the crossbar of the goal post is 10 feet. Does a field goal from the opponent's 40-yard line kicked with initial velocity of 75 feet per second at an angle of 45° go over the crossbar?

72. Baseball It is 400 feet from home plate of a baseball field to the center field fence. Assume that the bat connects with the ball at height of 3 feet and that the fence is 8 feet high. Does a hit down center field with an initial velocity of 115 feet per second at an angle of 45° go over the fence?

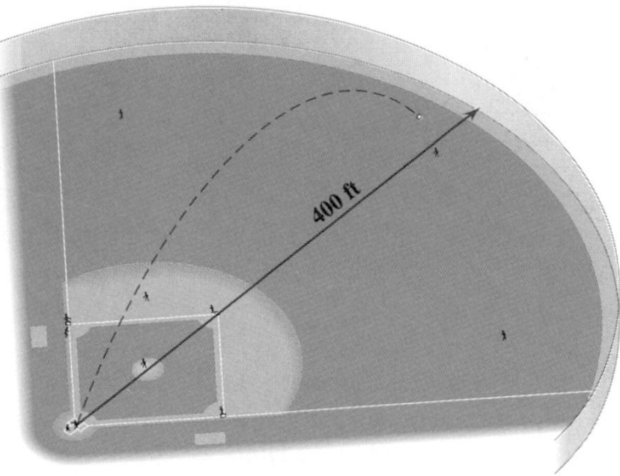

73. Golf A par 3 hole is 160 yards long. The tee is 21 feet higher than the green. A golfer hits the ball in the direction of the hole with initial velocity of 105 feet per second at an angle of 45°. The green is circular (radius 45 feet) with the pin at the center of the green. Does the ball land on the green?

74. Artillery Range Artillery guns are located on cliffs, overlooking settlements in a valley 2000 feet below. The guns can fire shells at 500 feet per second, at varying angles. Using the function given for Exercises 71–73, find the range of the guns.

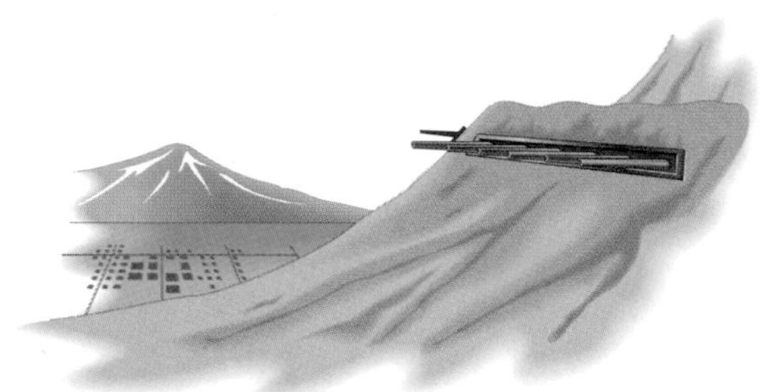

SECTION PROJECT
Half-Timber Designs

Houses and churches in many towns and villages in Europe are decorated with *fachwerk* (Germany) or *half-timber* (England). Fine examples are to be found in the university town of Tübingen in Germany and in the cathedral city of Chester in England. Figure 3.52 shows popular fachwerk designs. Find the equations that give these designs and reproduce them on your calculator using the graphing window $X_{\min} = 0$, $X_{\max} = 5$, $X_{\text{scl}} = 2$, $Y_{\min} = 0$, $Y_{\max} = 5$, and

$Y_{scl} = 2$. All designs are constructed using linear functions, absolute value functions, and "circles." Construct your own interesting fachwerk design, and give the equations for reproducing it on a graphing calculator.

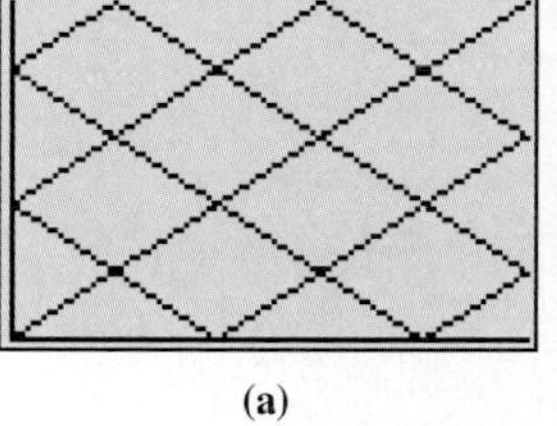
(a)

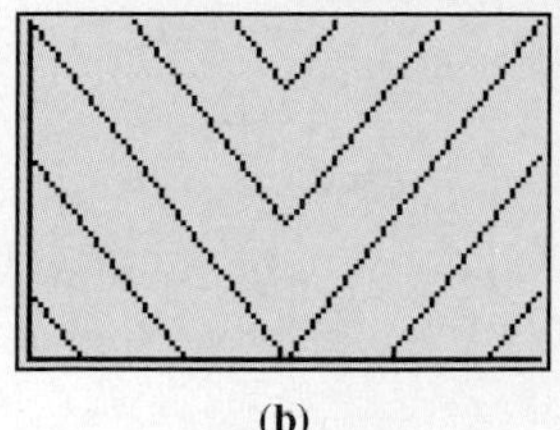
(b)

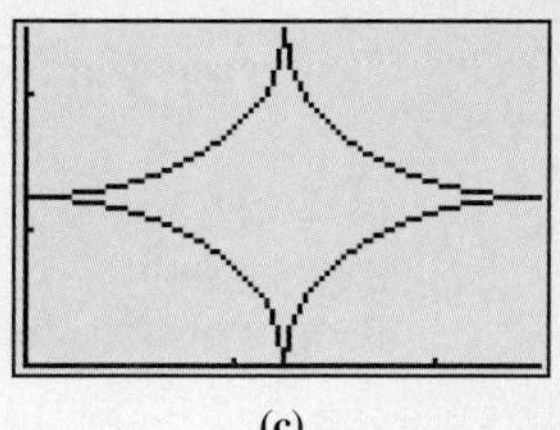
(c)

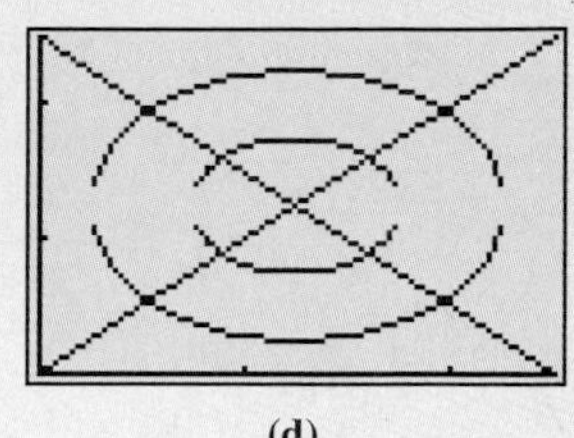
(d)

Figure 3.52

3.5 Rational Functions

• Rational Functions • Asymptotes • Sketching the Graph of a Rational Function

We discussed rational expressions in Section R.5. With this background, we now investigate functions that are defined in terms of rational expressions. The graphs of these functions exhibit interesting phenomena unlike any we have discussed to date. The graphs get closer and closer to certain lines called asymptotes. Such functions often arise in the discussion of the long-term behavior of phenomena.

Rational Functions

A function f of the form

$$f(x) = \frac{h(x)}{g(x)}$$

where $g(x)$ and $h(x)$ are polynomial functions is called a **rational function.** Note that $f(x)$ is not defined if $g(x) = 0$. Thus, the domain of a rational function is the set of all real numbers except the zeros of the denominator function.

Example 1 Determine the domain of each of the following rational functions.

(a) $f(x) = \dfrac{x^2 - 7x + 5}{x - 3}$ **(b)** $g(x) = \dfrac{x + 1}{(x + 1)(x - 5)}$

SOLUTION

(a) The domain of f is the set of all real numbers except 3.

(b) The domain of g is the set of all real numbers except -1 and 5.

We introduce the concepts involved in graphing a rational function with the following examples.

Example 2 **Sketching the Graph of a Rational Function**

Sketch the graph of $f(x) = \dfrac{x^2 - x - 6}{x - 3}$.

SOLUTION The domain of f is the set of all real numbers except 3. Before starting to graph f, let us express it in its **simplest form** by dividing out any common factors of the numerator and denominator. We get

$$f(x) = \frac{(x + 2)(x - 3)}{x - 3}$$

Divide out the common factor $(x - 3)$, indicating that x cannot take on the value 3.

$$f(x) = x + 2, \qquad x \neq 3$$

The graph of f is the straight line $y = x + 2$ with a gap at the point where $x = 3$, that is at the point (3, 5) (Figure 3.53a). Note that $x \neq 3$ is part of the simplified form of the function in this example. It is important to make sure that the information about the domain of the function is not lost when simplifying. The graph is verified in the calculator display of Figure 3.53(b); however, the resolution of your calculator screen is probably not sufficient to show the gap. The table in Figure 3.53(c) further supports the linear relationship $f(x) = x + 2$, with f undefined at $x = 3$.

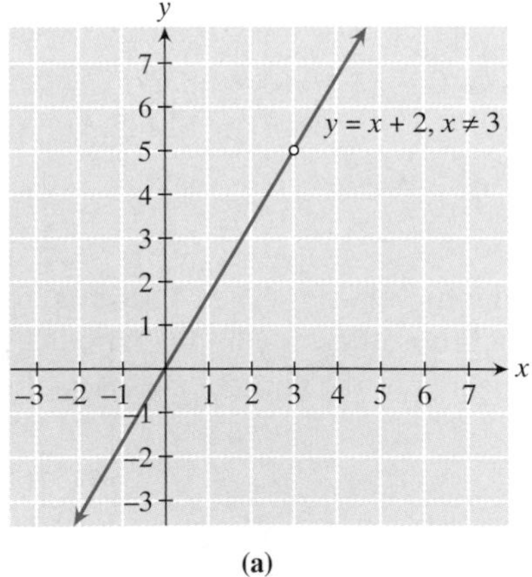

(a)

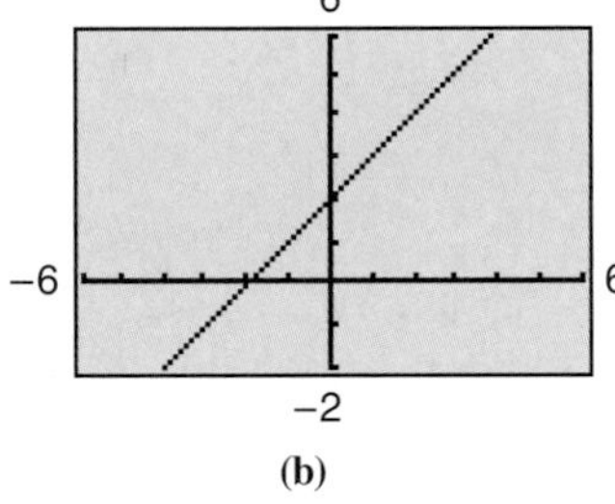

(b)

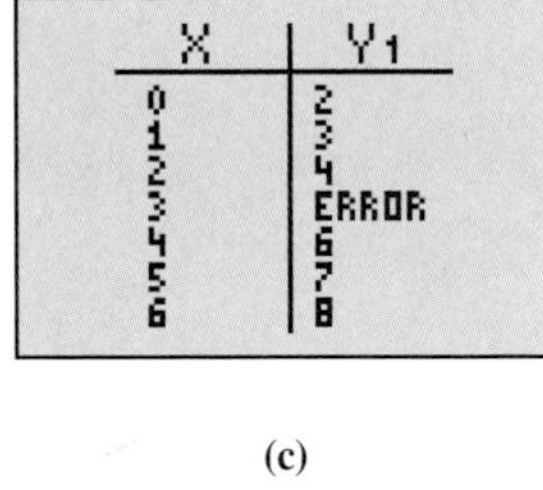

X	Y1
0	2
1	3
2	4
3	ERROR
4	6
5	7
6	8

(c)

Figure 3.53 The rational function $f(x) = \dfrac{x^2 - x - 6}{x - 3}$.

Horizontal and Vertical Asymptotes

Not all rational functions simplify to polynomials. We now illustrate the behavior of such a rational function.

Example 3 Sketching the Graph of a Rational Function

Sketch the graph of $f(x) = \frac{1}{x}$.

SOLUTION f is a rational function in which the numerator is constant. It is already in simplified form. f is defined for all values of x except $x = 0$.

Let us sketch the graph of f by plotting points. To understand the behavior of f, we selected the following x values on either side of $x = 0$ and computed the corresponding y values. We are particularly interested in the behavior of the function around $x = 0$ and have thus selected a number of values close to 0.

$x < 0$:	x	-4	-2	-1	$-1/2$	$-1/4$	$-1/10$	$-1/100$	$-1/1000$
	$f(x) = 1/x$	$-1/4$	$-1/2$	-1	-2	-4	-10	-100	-1000

$x > 0$:	x	$1/1000$	$1/100$	$1/10$	$1/4$	$1/2$	1	2	4
	$f(x) = 1/x$	1000	100	10	4	2	1	$1/2$	$1/4$

These points lead to the graph shown in Figure 3.54. We confirm the graph on a calculator. Observe that as x gets closer to 0 from the left side, the values of $f(x)$ are negative, with increasing magnitude. On the other hand, as x gets closer and closer to 0 from the right side, the values of $f(x)$ become larger positive numbers. We say that $f(x)$ approaches $-\infty$ as x approaches 0 from the left, and $f(x)$ approaches ∞ as x approaches 0 from the right. We write

$f(x) \to -\infty$ as $x \to 0$ from the left
$f(x) \to \infty$ as $x \to 0$ from the right

The graph of f gets closer and closer to the y-axis, the line $x = 0$, but never meets it. The line $x = 0$ is called a **vertical asymptote** of the graph.

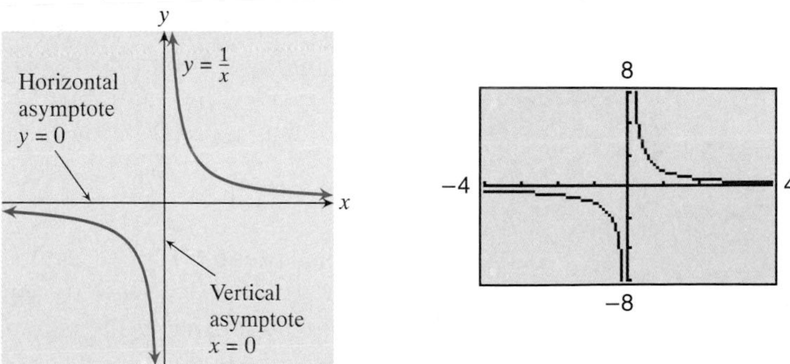

Figure 3.54 Graph and asymptotes of $f(x) = \frac{1}{x}$.

Observe also that as the values of x increase or decrease, the values of $f(x)$ get closer and closer to zero. We say that $f(x)$ approaches zero as x approaches ∞.

$f(x)$ also approaches zero as x approaches $-\infty$. This is expressed as

$$f(x) \to 0 \text{ as } x \to \infty$$
$$f(x) \to 0 \text{ as } x \to -\infty$$

The x-axis is called a **horizontal asymptote** of the graph. Asymptotes play a central role in the behavior of rational functions.

Example 4 **Sketching the Graph of a Rational Function**

Sketch the graph and find the vertical and horizontal asymptotes of the function $f(x) = 1/(x - 4)$. Support your answers using a graphing calculator.

SOLUTION The graph of $f(x) = 1/(x - 4)$ can be obtained from the graph of $g(x) = 1/x$ by shifting the graph horizontally 4 units to the right. This shift leads to the graph shown in Figure 3.55. The graph is confirmed in the calculator display.

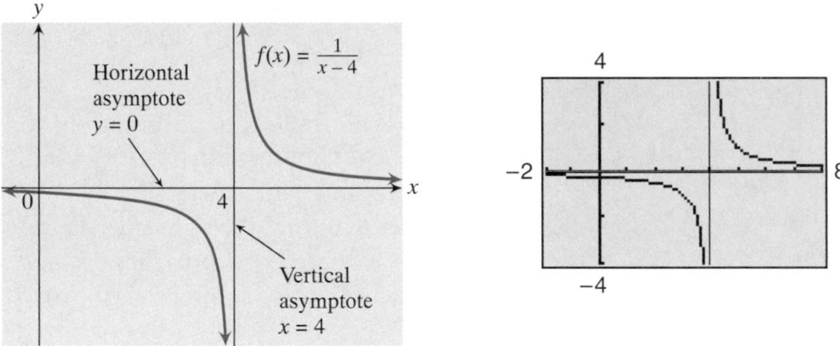

Figure 3.55 Graph and asymptotes of $f(x) = \dfrac{1}{x - 4}$.

The function f has a vertical asymptote $x = 4$, and a horizontal asymptote $y = 0$. The vertical asymptote is at the zero of the function in this denominator. Observe the vertical line drawn by this calculator at $x = 4$. Your calculator may or may not draw such a line. This calculator has computed points on the graph on both sides of the asymptotes and joined them. This line may be interpreted as the vertical asymptote! Be aware that it is not part of the graph.

Further Ideas 1 Use shifts of $g(x) = 1/x$ to sketch the graph of $f(x) = 1/(x - 2) + 4$. Give the equations of vertical and horizontal asymptotes. Support your answers using a graphing calculator.

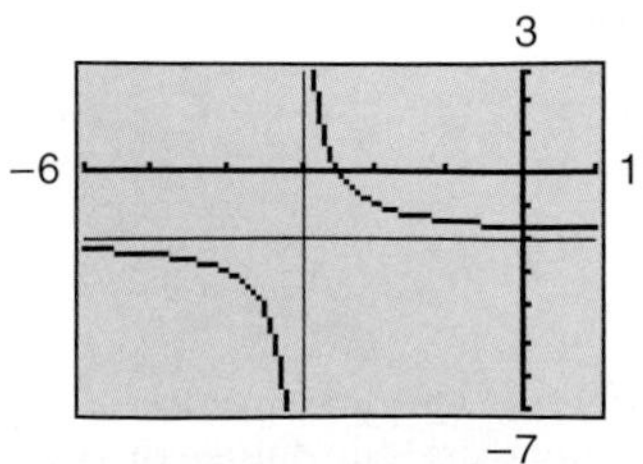

Figure 3.56

Example 5 **Finding a Rational Function from Its Graph**

Consider the graph of Figure 3.56 in which asymptotes have been added (solid lines). Give the equations of the asymptotes. The graph is that of a function $f(x)$ obtained from $g(x) = 1/x$ using shifts. Find the function $f(x)$. Support your answer by duplicating the graph on your calculator.

SOLUTION There is a single vertical asymptote that intersects the x-axis at -3. The vertical asymptote is thus $x = -3$. There is a single horizontal asymptote that intersects the y-axis at -2. The horizontal asymptote is $y = -2$.

The graph can be obtained from that of $g(x) = 1/x$ by a horizontal shift of 3 units to the left and a vertical shift of 2 units down. The function is thus

$$f(x) = \frac{1}{x + 3} - 2$$

On graphing f in the given window, we get the display shown in Figure 3.57, supporting the answer. (Your calculator may or may not produce a vertical line at $x = -3$ for the reason discussed in Example 4.)

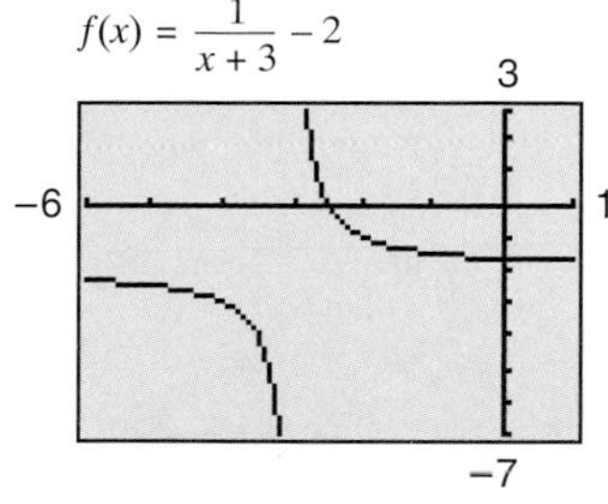

Figure 3.57

We now give formal definitions of vertical and horizontal asymptotes.

> ***Vertical and Horizontal Asymptotes*** Let f be a rational function.
>
> The line $x = a$ is a **vertical asymptote** of the graph of f if $f(x) \to \infty$ or $f(x) \to -\infty$ as $x \to a$ from the left or right.
>
> The line $y = b$ is a **horizontal asymptote** of the graph of f if $f(x) \to b$ as $x \to \infty$ or as $x \to -\infty$.

These concepts are illustrated in Figure 3.58.

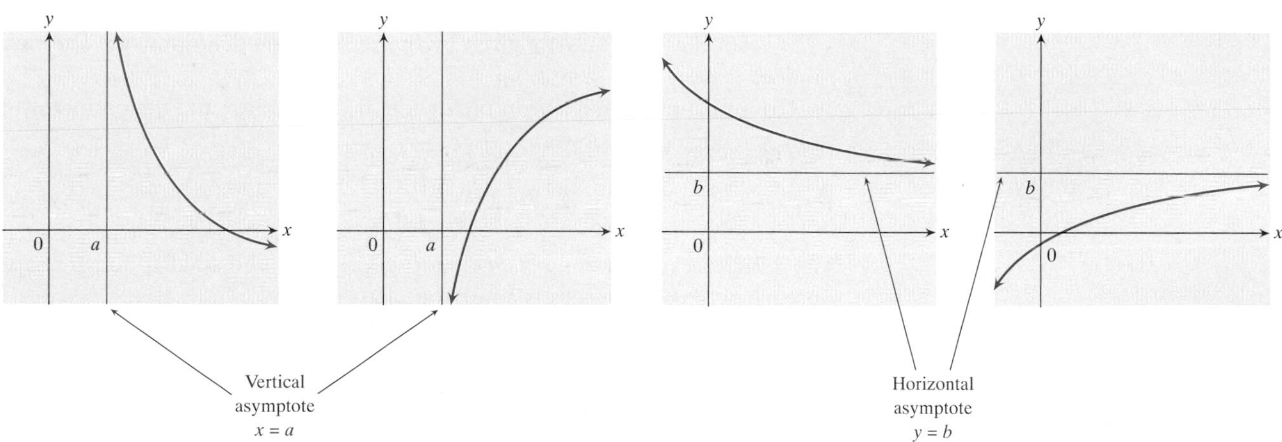

Figure 3.58 Horizontal and vertical asymptotes.

Now let us look at further types of rational functions. The following example leads to ways of determining vertical and horizontal asymptotes of rational functions for which the degree of the numerator is either less than the degree of the denominator or equal to it.

Example 6 Finding the Equations of Asymptotes

Determine the vertical and horizontal asymptotes of the following functions.

(a) $f(x) = \dfrac{x + 3}{x^2 - x - 2}$
Degree of numerator less than degree of denominator.

(b) $g(x) = \dfrac{2x^2 - 5x - 3}{x^2 - x - 12}$
Degree of numerator equal to degree of denominator.

SOLUTION Let us graph the functions using a calculator to get an idea of where asymptotes lie (Figure 3.59). The vertical asymptotes of f appear to be in the vicinity of the lines $x = -1$ and $x = 2$. The x-axis appears to be a horizontal asymptote. The vertical asymptotes of g seem to be in the vicinity of $x = -3$ and $x = 4$, whereas $y = 2$ might be a horizontal asymptote.

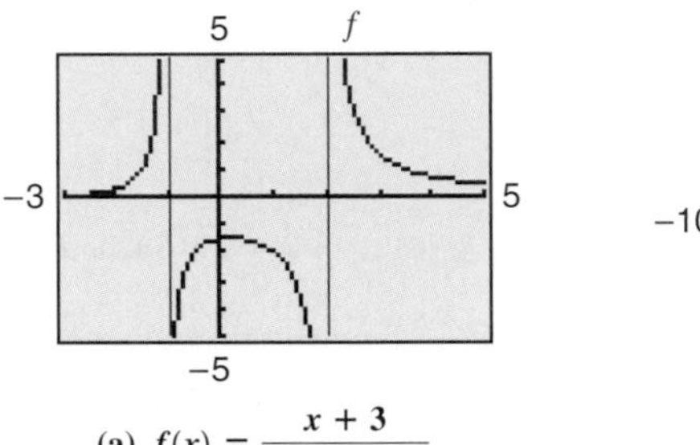

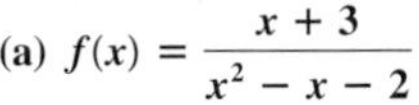

(a) $f(x) = \dfrac{x + 3}{x^2 - x - 2}$

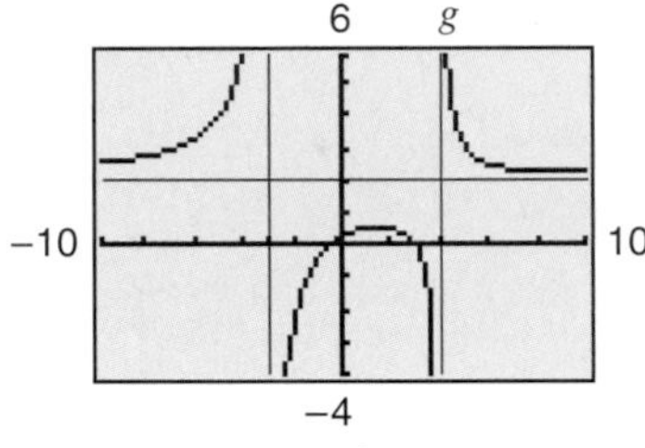

(b) $g(x) = \dfrac{2x^2 - 5x - 3}{x^2 - x - 12}$

Figure 3.59

Let us now confirm these results using algebra.

(a) Factor the denominator of f.

$$f(x) = \frac{x + 3}{x^2 - x - 2} = \frac{x + 3}{(x + 1)(x - 2)}$$

The vertical asymptotes are given by the zeros of the denominator. They are indeed $x = -1$ and $x = 2$.

To find a horizontal asymptote divide the numerator and denominator of f by x^2 (the highest power of x).

$$f(x) = \frac{1/x + 3/x^2}{1 - 1/x - 2/x^2}$$

As x increases the terms $1/x$, $3/x^2$, $2/x^2$ get smaller and smaller and $f(x)$ approaches zero. The x-axis is a horizontal asymptote.

(b) Factor the denominator of g.

$$g(x) = \frac{2x^2 - 5x - 3}{x^2 - x - 12} = \frac{2x^2 - 5x - 3}{(x + 3)(x - 4)}$$

The vertical asymptotes are given by the zeros of the denominator. They are $x = -3$ and $x = 4$.

To find the horizontal asymptote, divide the numerator and denominator of g by x^2 (the highest power of x).

$$g(x) = \frac{2 - 5/x - 3/x^2}{1 - 1/x - 12/x^2}$$

As x increases, the terms $5/x$, $3/x^2$, $1/x$, and $12/x^2$ get smaller and smaller and $g(x)$ approaches 2. The line $y = 2$ is a horizontal asymptote. Observe that the horizontal asymptote can be obtained by taking the ratio of the coefficients of x^2 in the numerator and denominator.

These examples lead to the following rules for quickly determining the vertical and horizontal asymptotes of rational functions of this type.

Determining Vertical and Horizontal Asymptotes

Let

$$f(x) = \frac{h(x)}{g(x)}, \quad \text{where} \quad h(x) = a_n x^n + \cdots + a_1 x + a_0$$

and

$$g(x) = b_n x^n + \cdots + b_1 x + b_0$$

Assume that h and g have no common factors.

Vertical Asymptotes

The vertical asymptotes of $f(x)$ are given by the zeros of $g(x)$.

Horizontal Asymptote

(i) If the degree of h is less than the degree of g, the x-axis is the only horizontal asymptote.

(ii) If the degree of h is equal to the degree of g, then the line $y = a_n/b_n$ is the only horizontal asymptote.

(iii) If the degree of h is greater than the degree of g, there is no horizontal asymptote. (We illustrate this case later by means of an example.)

Self-Check 1 Arrive at the equations of the vertical and horizontal asymptotes of the following functions using the preceding rules. Use a graph drawn on a calculator to check your answer.

(a) $f(x) = \dfrac{x - 1}{x^2 + x - 6}$ **(b)** $g(x) = \dfrac{-3x^2 - 4x + 1}{x^2 - x - 2}$

Example 7 Finding Asymptotes and Sketching a Graph

Find the asymptotes and sketch the graph of

$$f(x) = \frac{x - 2}{x^2 - x - 12}$$

Support your graph using a calculator.

SOLUTION We use four steps to sketch the graph. We determine the vertical asymptotes, the horizontal asymptote, the zeros of the function, and where it is above and where it is below the x-axis.

1. Vertical Asymptotes: Factor the denominator of f. We get
$$f(x) = \frac{x-2}{(x+3)(x-4)}$$
The vertical asymptotes are given by the zeros of the denominator. We have that $(x+3)(x-4) = 0$ when $x = -3$ or $x = 4$. The vertical asymptotes are the lines $x = -3$ and $x = 4$.
2. Horizontal Asymptote: The degree of $h(x) = x - 2$ is one, and the degree of $g(x) = x^2 - x - 12$ is two. The degree of the numerator is less than that of the denominator. Whenever this is the case the horizontal asymptote is the x-axis.
3. x-Intercepts: The x-intercepts are the zeros of the numerator. We get $x - 2 = 0$ when $x = 2$. There is a single x-intercept, $x = 2$.
4. Above or Below the x-Axis: Consider the intervals defined by the vertical asymptotes $x = -3$, $x = 4$ and the x-intercept 2. We determine whether the graph lies above or below the x-axis on these intervals.

Interval	**Test Point**	$f(x) = \frac{x-2}{x^2-x-12}$	**Above/Below x-Axis**
$(-\infty, -3)$	$x = -4$	$f(-4) = -3/4 < 0$	Below
$(-3, 2)$	$x = 0$	$f(0) = 1/6 > 0$	Above
$(2, 4)$	$x = 3$	$f(3) = -1/6 < 0$	Below
$(4, \infty)$	$x = 5$	$f(5) = 3/8 > 0$	Above

This information is summarized in Figure 3.60a and leads to the graph shown in Figure 3.60b.

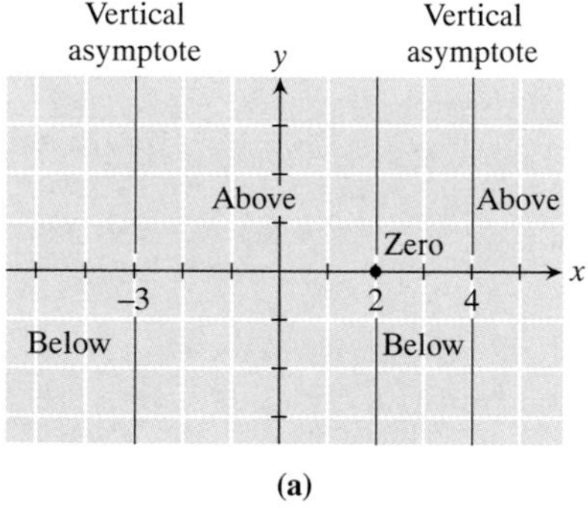

(a)

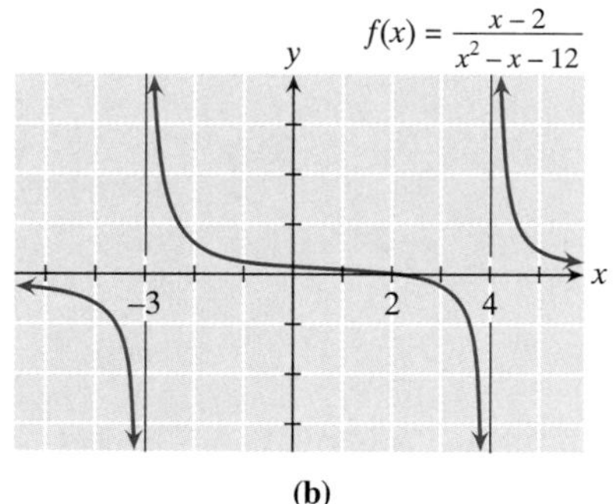

(b)

Figure 3.60 Graph of $f(x) = \frac{x-2}{x^2-x-12}$

We support this graph, using a calculator (Figure 3.61). (Your calculator may or may not produce vertical lines at $x = -3$ and $x = 4$ for the reason discussed in Example 4.)

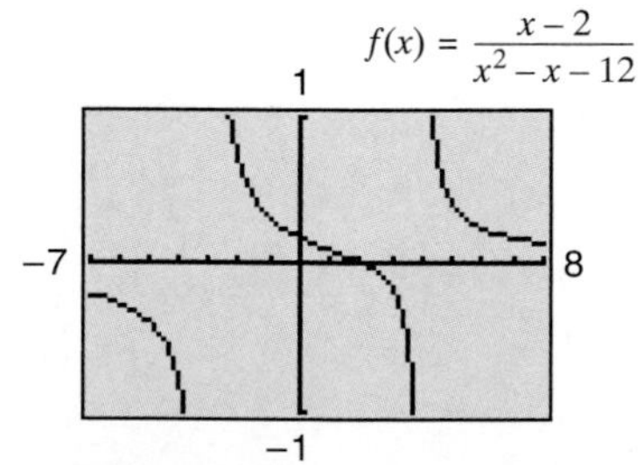

Figure 3.61

Self-Check 2 Find the asymptotes and sketch the graph of

$$f(x) = \frac{2x^2 - x - 15}{x^2 - 2x - 24}$$

Support your graph using a calculator.

Slant Asymptotes

When the degree of the numerator of a rational function is one more than the degree of the denominator, there will be a slant asymptote. The following example illustrates this situation.

Example 8 Graphing a Function with a Slant Asymptote

Find the asymptotes, and analyze the behavior of

$$f(x) = \frac{x^2 - 7x + 17}{x - 4}$$

Confirm your answers by graphing f and its asymptotes on a calculator.

SOLUTION Divide $x^2 - 7x + 17$ by $x - 4$ as follows.

$$\begin{array}{r l}
x \phantom{{}+{}} - 3 \phantom{{}+ 17} & \textbf{Divide } x^2/x \textbf{ to get } x. \\
x - 4\overline{)x^2 - 7x + 17} & \\
\underline{x^2 - 4x \phantom{{}+ 17}} & \textbf{Multiply } x(x - 4). \\
-3x + 17 & \textbf{Subtract and bring down 17. Divide } -3x/x \textbf{ to get } -3. \\
\underline{-3x + 12} & \textbf{Multiply } -3(x - 4). \\
5 & \textbf{Subtract.}
\end{array}$$

This means that we can write

$$f(x) = x - 3 + \frac{5}{x - 4}$$

As x gets closer and closer to 4 from either side, the $5/(x - 4)$ part of $f(x)$ gets larger and larger in magnitude. The line $x = 4$ is a vertical asymptote.

On the other hand, as x increases or decreases indefinitely, the term $5/(x - 4)$ becomes smaller and smaller. $f(x)$ gets to resemble the line $y = x - 3$. The line $y = x - 3$ is a **slant asymptote.**

We graph $f(x)$ and the lines $x = 4$ and $y = x - 3$ on a calculator (Figure 3.62). It can be seen that the line $x = 4$ is a vertical asymptote. Furthermore, the graph of f gets closer to the line $y = x - 3$ as x increases and also as x decreases, confirming that the line $y = x - 3$ is an asymptote.

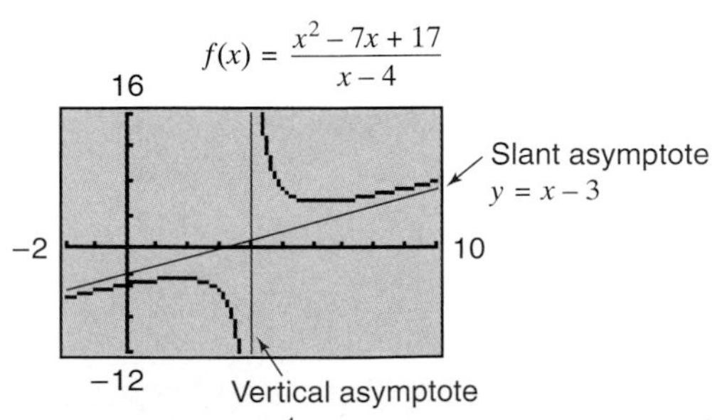

Figure 3.62 Graph and asymptotes $f(x) = \dfrac{x^2 - 7x + 17}{x - 4}$

General Result for Slant Asymptotes

We can generalize the result of Example 8 as follows. Let $f(x)$ be a rational function for which the degree of the numerator is one more than the degree of the denominator. Division can be used to write f in the form $f(x) = q(x) + r(x)$, where $q(x)$ is a linear function and $r(x)$ is a rational function. The degree of the numerator of $r(x)$ is smaller than the degree of the denominator. The function $y = q(x)$ is a slant asymptote of f. The vertical asymptotes of $f(x)$ are those of $r(x)$. There is no horizontal asymptote.

Self-Check 3 Find the equations of the asymptotes of

$$f(x) = x + 5 + \frac{3}{x - 4}$$

using algebra. Confirm your answers by graphing f and its asymptotes on a calculator.

Answers to Self-Check Exercises

1. **(a)** Vertical asymptotes $x = -3$, $x = 2$. Horizontal asymptote x-axis. **(b)** Vertical asymptotes $x = -1$, $x = 2$. Horizontal asymptote $y = -3$.
2. Vertical asymptotes $x = -4$ and $x = 6$. Horizontal asymptote $y = 2$.
3. One slant asymptote, $y = x + 5$. One vertical asymptote, $x = 4$.

Answer to Further Ideas Exercise

1. Shift graph of $g(x)$ 2 units right and 4 units up. Vertical asymptote $x = 2$. Horizontal asymptote $y = 4$.

EXERCISES 3.5

In Exercises 1–6, determine the domains of the rational functions.

1. $f(x) = \dfrac{2x + 1}{x}$ **2.** $f(x) = \dfrac{3x - 2}{x - 4}$

3. $f(x) = \dfrac{4x + 2}{3x + 1}$ **4.** $f(x) = \dfrac{2x^2 - 3x}{x}$

5. $f(x) = \dfrac{x - 3}{(x + 2)(x - 1)}$ **6.** $f(x) = \dfrac{(x + 2)(x - 1)}{(x + 2)(x + 4)}$

In Exercises 7–12, determine the domains of the rational functions.

7. $f(x) = \dfrac{(x - 8)(x + 4)}{x(x - 3)(x + 7)}$

8. $f(x) = \dfrac{x}{x^2 - 3x + 2}$

9. $f(x) = \dfrac{1}{x^2 - 4x}$ **10.** $f(x) = \dfrac{x + 3}{x^2 - 10x + 21}$

11. $f(x) = \dfrac{x + 5}{x^2 + 3x - 4}$ **12.** $f(x) = \dfrac{x^2 + 1}{x^2 - 1}$

In Exercises 13–18, simplify each rational function by dividing out the common factors of the numerators and denominators. Sketch the graphs. Support your answers by duplicating the graphs on your calculator. (The calculator will probably not show the gaps in the graphs.)

13. $f(x) = \dfrac{x(x - 2)}{x - 2}$ **14.** $f(x) = \dfrac{(x - 4)(x + 1)}{x + 1}$

15. $f(x) = \dfrac{(2x + 5)(3x - 2)}{3x - 2}$

16. $f(x) = \dfrac{2x^2 - 3x}{x}$

17. $f(x) = \dfrac{x^2 + 3x - 4}{x - 1}$ **18.** $f(x) = \dfrac{3x - 6}{x - 2}$

In Exercises 19–22, simplify each rational function by canceling out common factors of numerators and denominators. Sketch the graphs. Support your answers by duplicating the graphs on your calculator.

19. $f(x) = \dfrac{x^2 + 3x - 4}{x + 4}$ **20.** $f(x) = \dfrac{x^3 - 3x^2}{x - 3}$

21. $f(x) = \dfrac{2x^2 - 2x - 4}{2x - 4}$

22. $f(x) = \dfrac{12x^2 - 5x - 2}{4x + 1}$

In Exercises 23–28, give the equations of all vertical and horizontal asymptotes of the graphs of the functions. Sketch the graphs of these functions using shifts, reflection, and stretches of the known graph of $g(x) = 1/x$. Support your answers by duplicating the graphs on your calculator.

23. $f(x) = \dfrac{1}{x - 2}$ **24.** $f(x) = \dfrac{1}{x + 2} + 6$

25. $f(x) = \dfrac{3}{x} - 4$ **26.** $f(x) = \dfrac{2}{x - 1} - 4$

27. $f(x) = -\dfrac{1}{x + 3} + 2$ **28.** $f(x) = -\dfrac{1}{(x - 2)} - 5$

In Exercises 29–34, give the equations of all vertical and horizontal asymptotes, find the zeros, and sketch the graphs of the rational functions. Support your answers by duplicating the graphs on your calculator.

29. $f(x) = \dfrac{x - 1}{x^2 - 2x - 3}$ **30.** $f(x) = \dfrac{x + 1}{x^2 + x - 6}$

31. $f(x) = \dfrac{2}{x^2 - 5x + 4}$ **32.** $f(x) = \dfrac{3}{x^2 - 7x + 10}$

33. $f(x) = \dfrac{2x^2 - 5x - 25}{x^2 - x - 12}$

34. $f(x) = \dfrac{3x^2 + 14x - 24}{x^2 + 5x - 24}$

In Exercises 35–38, give the equations of all vertical and horizontal asymptotes, find the zeros, and sketch the graphs of the rational functions. Support your answers by duplicating the graphs on your calculator.

35. $f(x) = \dfrac{x^2 - 2x - 3}{x^2 + 2x - 8}$

36. $f(x) = \dfrac{3x^2 - 11x - 20}{x^2 - x - 20}$

37. $f(x) = \dfrac{3x + 5}{x - 2}$ **38.** $f(x) = \dfrac{4x + 7}{x + 4}$

In Exercises 39–42, $X_{scl} = 1$ and $Y_{scl} = 1$ in each display. Give the equations of all vertical and horizontal asymptotes of the given graphs. Use this information to find equations for the graphs. Support your answers by duplicating the graphs on your calculator. *Hint:* Some graphs may be obtained from others through shifts and reflections.

39. 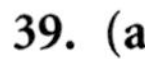

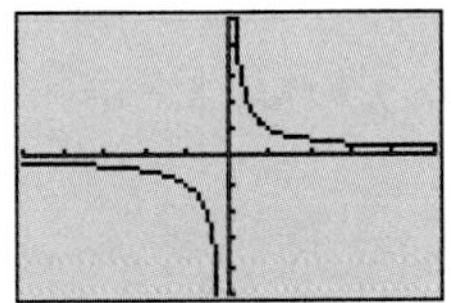

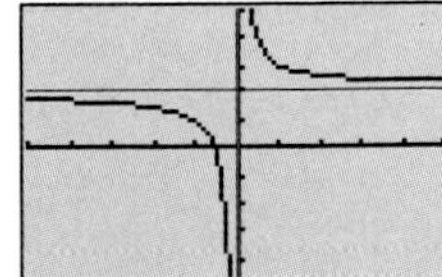

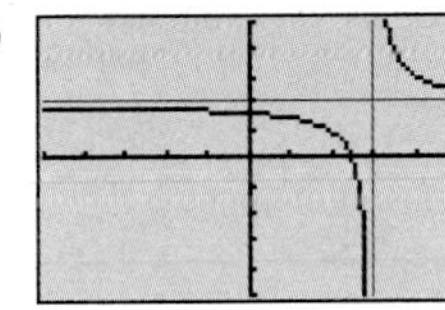

40. (a)

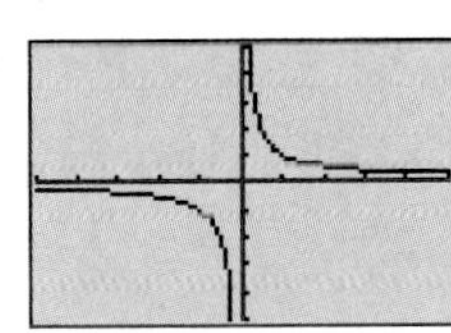

(b)

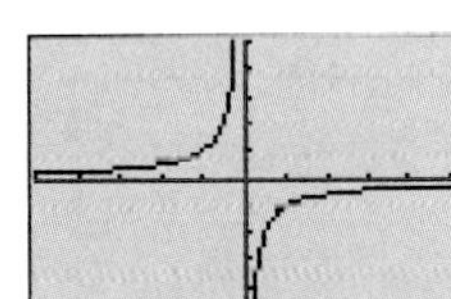

(c)

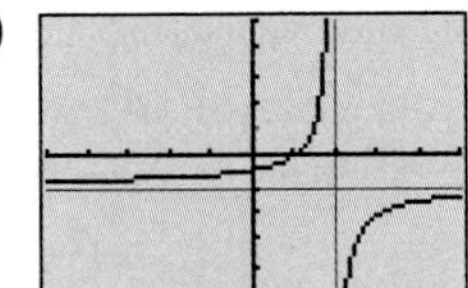

41. (a)

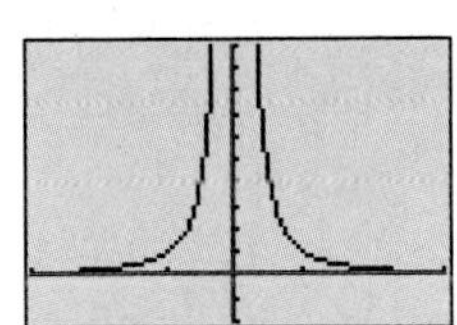

(b)

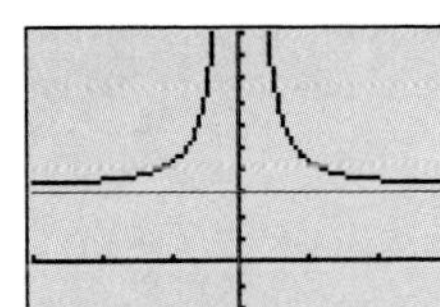

(c)

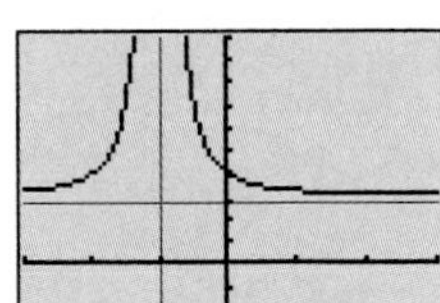

42. (a)

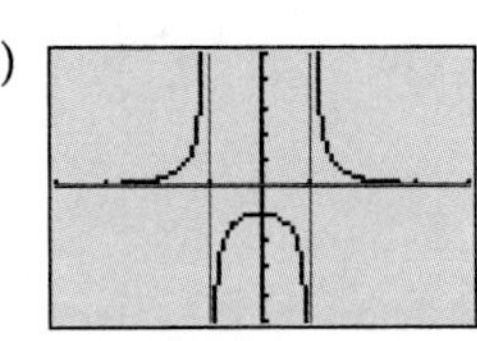

(b)

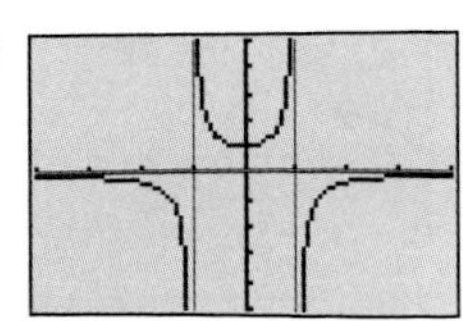

(c) 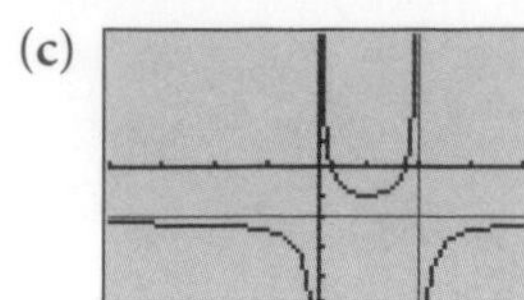

In Exercises 43–50, find the equations of all the asymptotes using algebra. Confirm your answers by graphing f and its asymptotes on a calculator.

43. $f(x) = x + 3 + \dfrac{7}{x - 2}$

44. $f(x) = x + 4 + \dfrac{1}{x}$

45. $f(x) = \dfrac{x^2 - 4x + 5}{x - 3}$

46. $f(x) = \dfrac{x^2 + 7x + 10}{x + 3}$

47. $f(x) = \dfrac{x^2 + x + 17}{x - 4}$

48. $f(x) = \dfrac{2x^2 + 9x - 3}{x + 5}$

49. $f(x) = \dfrac{-3x^3 + x^2 + 4}{x^2}$

50. $f(x) = \dfrac{x^3 - 2x^2 + x - 3}{x^2 - 16}$

51. Draw the graphs of $f(x) = x^{-1/2}$, $g(x) = x^{-1/4}$, and $h(x) = x^{-1/6}$, all in the same calculator screen. From these graphs, deduce the shapes of the graphs of the family of functions of the form $f(x) = x^{-1/n}$, where n is a positive even integer.

52. Draw the graphs of $f(x) = x^{-1/3}$, $g(x) = x^{-1/5}$, and $h(x) = x^{-1/7}$, all in the same calculator screen. From these graphs, deduce the shapes of the graphs of the family of functions of the form $f(x) = x^{-1/n}$, where n is a positive odd integer.

3.6 Operations on Functions and an Introduction to the Field of Chaos

• Sum, Difference, Product, and Quotient of Functions • Composition of Functions • Cost, Revenue, and Profit Functions • The Effect of the Middle Man • An Introduction to the Theory of Chaos • Modeling a Population of Trout

Sum, Difference, Product, and Quotient of Functions

We discussed operations on real numbers and on polynomials in Sections R.1 and R.3. There were four basic operations; namely, addition, subtraction, multiplication, and division. We now extend these operations to functions. We will see how functions can be added, subtracted, multiplied, and divided. An additional operation called composition will also be introduced.

Let us begin by looking at the operation of addition. Consider the functions f and g, defined as follows.

$$f(x) = 7x + 4, \qquad g(x) = 3x - 2$$

Add these functions to get a new function denoted $f + g$. We get

$$\begin{aligned}(f + g)(x) &= f(x) + g(x)\\ &= (7x + 4) + (3x - 2)\\ &= 10x + 2\end{aligned}$$

The notations $f - g$, fg, and f/g are used respectively for the difference, product, and quotient of f and g. For the preceding two functions f and g, we have

$$(f - g)(x) = (7x + 4) - (3x - 2) = 4x + 6$$
$$(fg)(x) = (7x + 4)(3x - 2) = 21x^2 - 2x - 8$$
$$\left(\frac{f}{g}\right)(x) = \frac{7x + 4}{3x - 2}, \qquad x \neq \frac{2}{3}$$

Caution: Note that $(f/g)(x)$ is not defined at the point where the denominator $g(x)$ is zero. We now make these ideas precise with the following definitions.

Operations on Functions

Let f and g be functions of a variable x. Then

Sum	$(f + g)(x) = f(x) + g(x)$
Difference	$(f - g)(x) = f(x) - g(x)$
Product	$(fg)(x) = f(x)g(x)$
Quotient	$\left(\frac{f}{g}\right)(x) = \frac{f(x)}{g(x)}, \ g(x) \neq 0$

It is important when working with a function to be aware of the domain of that function. A function will only be defined for values of the variable that are in its domain. These operations introduce four new functions: the sum, difference, product, and quotient. Each of these functions will have its own domain. These domains will naturally depend on the domains of f and g. These definitions imply that $f + g$, $f - g$, and fg all exist at values of x, where both f and g are defined. We must leave out the values of x at which $g(x)$ is zero from this set for the domain of f/g.

Domains

The domains of $f + g$, $f - g$, and fg will be the values of x common to the domains of both f and g. This is the intersection of the domains of f and g.

The domain of f/g will be those numbers x in the intersection of the domains of f and g for which $g(x) \neq 0$.

Example 1 Combining Functions

Let $f(x) = \sqrt{x - 4}$, $g(x) = x - 6$. Determine $(f + g)(x)$, $(f - g)(x)$, $(fg)(x)$, and $(f/g)(x)$. Give the domain of each function.

SOLUTION We need the domains of f and g to determine the domains of the other functions. The domain of f will consist of all values of x for which $\sqrt{x - 4}$ is a real number. These are the values of x for which $x - 4 \geq 0$. Thus, the domain of f is $[4, \infty)$. The domain of g is the set of real numbers, or $(-\infty, \infty)$. The intersection of these two intervals is $[4, \infty)$. We get

$(f + g)(x) = \sqrt{x - 4} + x - 6$. Domain $[4,\infty)$.
$(f - g)(x) = \sqrt{x - 4} - x + 6$. Domain $[4,\infty)$.
$(fg)(x) = (\sqrt{x - 4})(x - 6)$. Domain $[4,\infty)$.
$(f/g)(x) = (\sqrt{x - 4})/(x - 6)$. This is a rational function. We must be careful in determining the domain of f/g. It will be the points in the domains of both f and g for which $g(x) = x - 6$ is not zero. It will be the points in $[4, \infty)$ for which $x \neq 6$. The domain of f/g is thus $[4,6) \cup (6,\infty)$.

We draw the graph of f/g on a calculator (Figure 3.63). The graph does not exist to the left of $x = 4$ and has a vertical asymptote at $x = 6$. This confirms that the domain is $[4, 6) \cup (6, \infty)$.

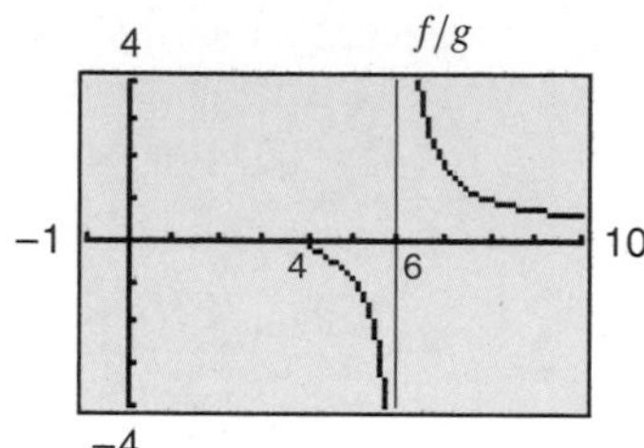

Figure 3.63

Further Ideas 1 Consider the functions $f(x) = \sqrt{x - 4}$ and $g(x) = x - 6$ of Example 1. Display the graphs of $f + g$, $f - g$, and fg on your calculator (Figure 3.64). Verify that the domains of each of these functions is $[4, \infty)$.

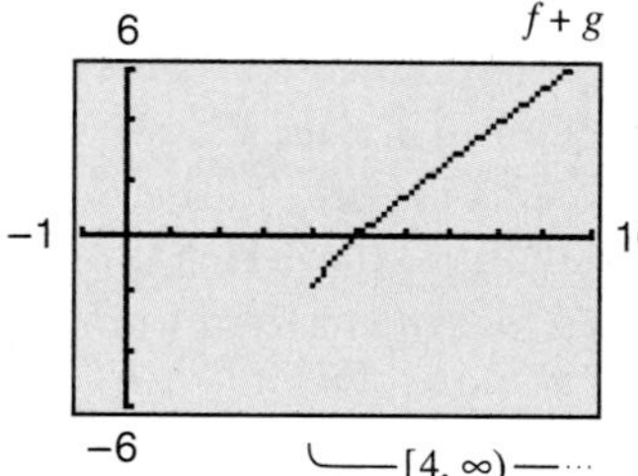

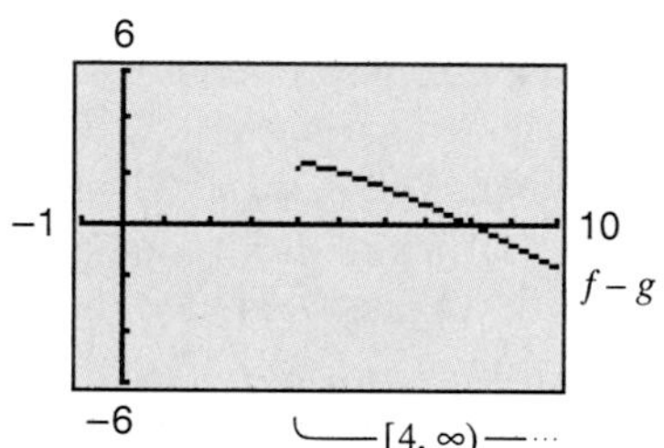

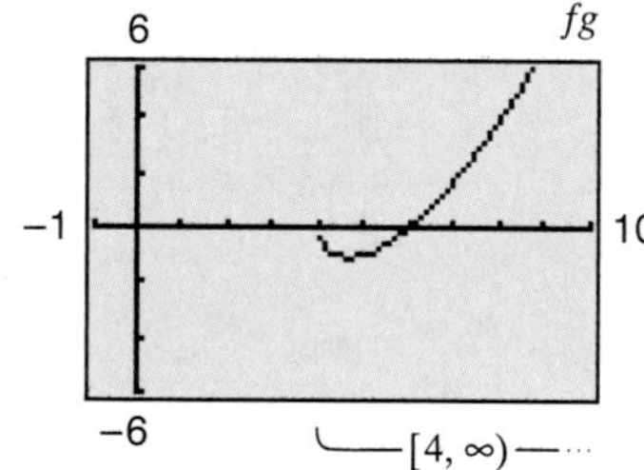

Figure 3.64 You may be able to enter these functions into your calculator as Y1 = √(X − 4), Y2 = X − 6, Y3 ■ Y1 + Y2, Y4 ■ Y1 − Y2, Y5 ■ Y1 * Y2.

Example 2 Finding the Values of Combinations of Functions

If $f(x) = x - 5$ and $g(x) = 2x^2 + 3$, find the following function values, if they are defined. **(a)** $(f + g)(3)$ **(b)** $(fg)(2)$ **(c)** $(f/g)(5)$ **(d)** $(g/f)(5)$

SOLUTION We compute each function first; then we evaluate it at the point.

(a) $(f + g)(x) = 2x^2 + x - 2$. $(f + g)(3) = 2(3)^2 + 3 - 2 = 19$.
(b) $(fg)(x) = (x - 5)(2x^2 + 3) = 2x^3 - 10x^2 + 3x - 15$.
$(fg)(2) = 2(2)^3 - 10(2)^2 + 3(2) - 15 = -33$.
(c) $(f/g)(x) = \dfrac{x - 5}{2x^2 + 3}$. $(f/g)(5) = \dfrac{0}{53} = 0$.
(d) $(g/f)(x) = \dfrac{2x^2 + 3}{x - 5}$. $(g/f)(5)$ is not defined since $f(5) = 0$.

Further Ideas 2 Let $f(x) = \sqrt{x^2 - 9}$, $g(x) = 2x - 8$. Display the graph of f/g on a calculator. Use the display to determine the domain of f/g. Verify your answer using algebra.

We have discussed ways of adding, subtracting, multiplying, and dividing functions. We now introduce yet another important way of combining functions.

Composition of Functions

Let f be a function that assigns to each element x of a set X the element y of a set Y (Figure 3.65). Let g be a function that assigns to each y of the set Y the element z in a set Z. The effect of f followed by g is to assign to x the element z. We can interpret this sequence as defining a new function that assigns to each element x in X an element z in Z. This function is called the composite function of g and f and is denoted $g \circ f$.

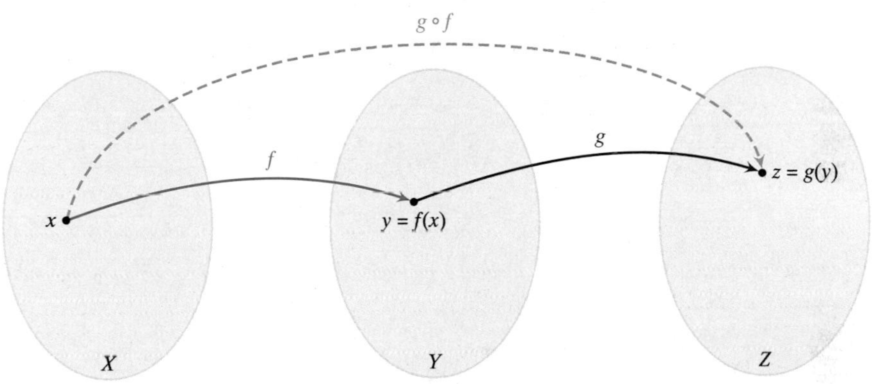

Figure 3.65 Composition of two functions.

> ***Composite Function*** Let f and g be functions. The **composite function** of g and f is denoted $g \circ f$ and is defined by
>
> $$(g \circ f)(x) = g(f(x))$$
>
> for all x in the domain of f such that $f(x)$ is in the domain of g.

Example 3 **Constructing a Composite Function**

Let $f(x) = 4x^2$ and $g(x) = 3x - 1$. Find $(g \circ f)(x)$.

SOLUTION According to our definition,

$$\begin{aligned}(g \circ f)(x) &= g(f(x)) \\ &= g(4x^2)\end{aligned}$$

This means that $4x^2$ should be substituted for x in the preceding expression for $g(x)$. This gives

$$\begin{aligned}(g \circ f)(x) &= 3(4x^2) - 1\\ &= 12x^2 - 1\end{aligned}$$

Thus, $(g \circ f)(x) = 12x^2 - 1$.

Example 4 **Finding the Domain of a Composite Function**

Let $f(x) = x^2 - x - 6$ and $g(x) = 1/x$. Construct $(g \circ f)(x)$. Compute $(g \circ f)(4)$. Give the domain of the function $g \circ f$. Display the graph of $g \circ f$ on a calculator and confirm the domain.

SOLUTION We first construct $g \circ f$ using the definition of function composition.

$$\begin{aligned}(g \circ f)(x) &= g(f(x))\\ &= g(x^2 - x - 6)\\ &= \frac{1}{x^2 - x - 6}\end{aligned}$$

Thus, $(g \circ f)(x) = \dfrac{1}{x^2 - x - 6}$.

Let us now compute $(g \circ f)(4)$. Let $x = 4$ in the above expression.

$$(g \circ f)(4) = \frac{1}{4^2 - 4 - 6} = \frac{1}{6}$$

To determine the domain of $g \circ f$ factor the denominator as follows.

$$(g \circ f)(x) = \frac{1}{(x + 2)(x - 3)}$$

The domain of $g \circ f$ is the set of all real numbers except -2 and 3.

We draw the graph of $g \circ f$ on a calculator and confirm the domain. The graph has vertical asymptotes at $x = -2$ and $x = 3$. $g \circ f$ is defined at all values of x except $x = -2$ and $x = 3$ (Figure 3.66).

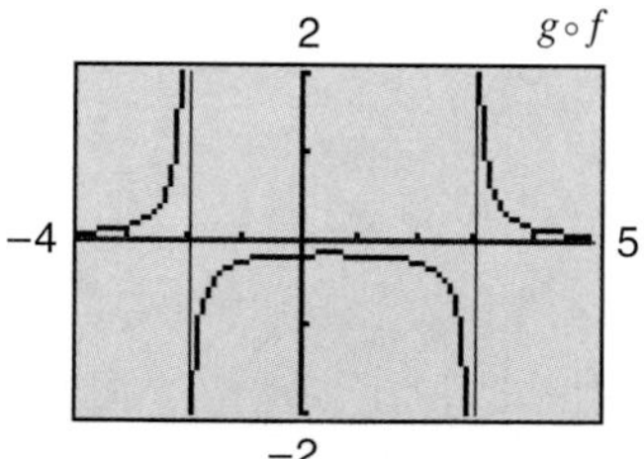

Figure 3.66 You may be able to enter Y1 = X^2 − X − 6, Y2 = 1/X, Y3 ■ Y2(Y1).

Self-Check 1 1. Let $f(x) = x^2 - 4$ and $g(x) = \sqrt{x}$. Determine $(g \circ f)(x)$ and $(f \circ g)(x)$ and their domains. Compute $(g \circ f)(2)$ and $(f \circ g)(-3)$. Verify your results from graphs of $g \circ f$ and $f \circ g$ on a calculator.

Note that in general $g \circ f$ and $f \circ g$ are different functions. $g \circ f \neq f \circ g$. This implies that the operation of function composition is not commutative.

Applications . . .

Example 5 Cost, Revenue, and Profit Functions

A total cost function $C(x)$ gives a manufacturing company the total cost of producing x units of a commodity over a given period, such as a week or a month. A total revenue function $R(x)$ gives the total revenue for selling x units. Companies construct such functions from data and use them to plan production. In practice, a company wants to decide on the production level necessary to maximize its profits. This example illustrates how this is achieved.

A company that manufactures motor boats is operating with weekly cost and revenue functions

$$C(x) = \frac{x^3}{120} - 6x^2 + 2370x + 4500 \qquad \text{and} \qquad R(x) = -x^2 + 2000x$$

The company is producing 340 boats per week. Should it change production to increase the weekly profit?

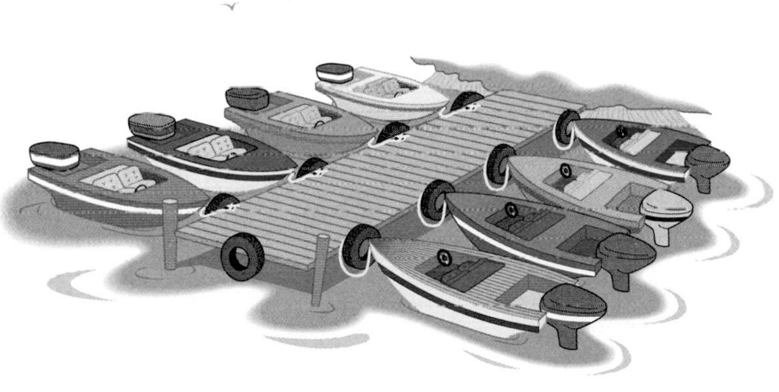

Profit is defined to be revenue minus cost.

$$\text{Profit} = \text{Revenue} - \text{Cost}$$

Let $P(x)$ be the weekly profit. Thus,

$$P(x) = R(x) - C(x)$$

$$P(x) = (-x^2 + 2000x) - \left(\frac{x^3}{120} - 6x^2 + 2370x + 4500\right)$$

$$= -\frac{x^3}{120} + 5x^2 - 370x - 4500$$

The current production of 340 yields a profit $P(340) = \$120{,}166.67$. The aim is to find the value of x for which P is a maximum. Weekly production should be changed from 340 boats to this number to ensure maximum profit.

The graph of P is drawn in a suitable window (Figure 3.67). The maximum value of the function $P(x)$ is seen to be $P = 121{,}506.27$ at $x = 358.74509$.

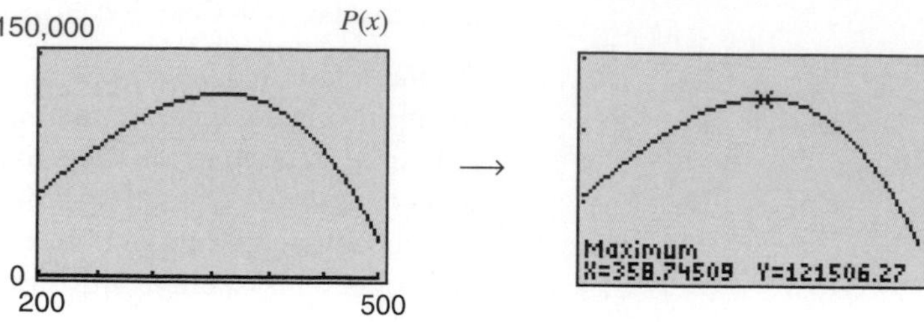

Figure 3.67

Because it is not meaningful to discuss the production of parts of boats, we examine the values of $P(x)$ at $x = 358$ and $x = 359$. We get $P(358) = \$121{,}504.07$ and $P(359) = \$121{,}506.00$. The maximum profit therefore occurs when the company manufactures 359 boats weekly. The company should increase boat production from 340 boats per week to 359 boats per week. There would be a predicted increase of profit from $120,166.67 to $121,506.

Example 6 The Effect of the Middleman

The price of a manufactured item increases in stages from the factory cost of manufacturing the item to the price that the customer pays. The final price of a product naturally depends on the number of middlemen. Let us model this process for a situation involving the production and sale of software when there are two intermediate levels between manufacturer and customer. This model can then be modified to take in any number of middlemen. Let the factory cost of a software product be $$x$. Suppose that the following increases take place at the following stages.

1. *Wholesaler* increases the price by 25% and adds $12 for shipping and handling.
2. *Retailer* adds 40% to the wholesale price and adds $5 for customer service.

Let us determine the final price that the customer pays for the software as a function of the manufacturing price $$x$. As an example of the use of this function, let us find what the consumer pays for software that costs $78.86 to produce.

The wholesale price $$w$ of the software is given by the function

$$\begin{aligned} w(x) &= x + (25\% \text{ of } x) + 12 = x + \frac{25x}{100} + 12 \\ &= 1.25x + 12 \end{aligned}$$

The retail price $$r$ is a function of the wholesale price.

$$\begin{aligned} r(w) &= w + (40\% \text{ of } w) + 5 = w + \frac{40w}{100} + 5 \\ &= 1.4w + 5 \end{aligned}$$

We form the composite function r ∘ w to express the retail price that the customer pays as a function of the original manufacturing price x. We get

$$\begin{aligned} r \circ w\ (x) &= r(w(x)) \\ &= 1.4(1.25x + 12) + 5 \\ &= 1.75x + 21.8 \end{aligned}$$

Now let us look at the situation when the factory cost of the software is \$78.86. When $x = 78.86$, we get

$$\begin{aligned} r(78.86) &= 1.75(78.86) + 21.8 \\ &= 159.81 \end{aligned}$$

Thus, when the factory price of the software is \$78.86, the price to the customer is \$159.81.

Example 7 **An Introduction to the Theory of Chaos**

Algebra enables us to discover and describe a great deal about the world around us. We now give you a glimpse of the power of the mathematics that you are mastering. You will see how mathematics is being used in many disciplines to understand what has heretofore been thought of as erratic, unpredictable behavior. We introduce you to the science of **chaos.**

Functions and the operation of function composition, together with the idea of proportionality, are used by scientists in mathematical models to describe the growth and decline of populations and to predict their long-term behavior. These can be populations of mammals, birds, insects, or fish. In these models, the independent variable x describes the population as a fraction, with 0 representing extinction and 1 representing the greatest conceivable population. $f(x)$ gives the population one unit of time later. This unit could be, for example, one day, one year, or one generation. The population one generation later (this is the term used to cover all these units) will depend on the current population x and the room for future population growth, $1 - x$. $f(x)$ is thus proportional to $x(1 - x)$. We have arrived at the **logistic function,** which is the basic function used in many sciences in models to describe growth,

$$f(x) = cx(1 - x), \qquad \text{with domain } 0 \leq x \leq 1$$

c is the constant of proportionality. It is selected so that the results of the model fit known data. It is usually a number between 1 and 4.

For example, the function $f(x) = 2x(1 - x)$ might describe the behavior of a colony of moths. If the population at a certain time is 0.75, the population of the next generation will be

$$f(0.75) = 2(0.75)(1 - 0.75) = 0.375$$

The population has decreased from 0.75 to 0.375, that is, from three quarters of the largest conceivable population to three-eights of it. The population will fluctuate, possibly decreasing for a number of generations, then increasing, and so on. The mathematics is used to study these fluctuations.

Observe how naturally the use of function composition arises in this application. The populations of the first and second generations are x and $f(x)$, whereas the population of the third generation will be $f(f(x))$ or $f \circ f(x)$. The population of the fourth generation will be $f \circ f \circ f(x)$, and so on. The populations of the first four generations of this colony of moths are

First generation:	$x = 0.75$
Second generation:	$f(x) = 0.375$
Third generation:	$f \circ f(x) = f(f(x)) = f(0.375) = 0.46875$
Fourth generation:	$f \circ f \circ f(x) = f(f(f(x)) = f(0.46875) = 0.498046875$

The populations of the first four generations written to two decimal places are

First	Second	Third	Fourth
0.75,	0.38,	0.47,	0.50

We see that this population first decreases sharply and then increases.

This process of repeated composition of a single function is called **iteration.** We now illustrate how to use the calculator to compute the numbers in an iterative process such as this.

Example 8 Modeling a Population of Trout

Let $f(x) = 2.5x(1 - x)$ describe the behavior of the trout population of a mountain lake. If the population at a certain time is 0.72, let us predict the populations of the next three generations using a calculator.

We start with the initial value of x, namely 0.72, and compute $f(0.72)$. This answer then becomes the new x, and the new value of f is computed and so on. To carry out this process on a calculator, we write the equation $f(x) = 2.5x(1 - x)$ in the form

$$\text{New result} = 2.5(\text{Previous answer})[1 - (\text{Previous answer})]$$

The iterative process can then be implemented as shown in Figure 3.68. We see that the population fluctuates. The first four generations (rounded to two decimal places) are 0.72, 0.50, 0.62, and 0.59.

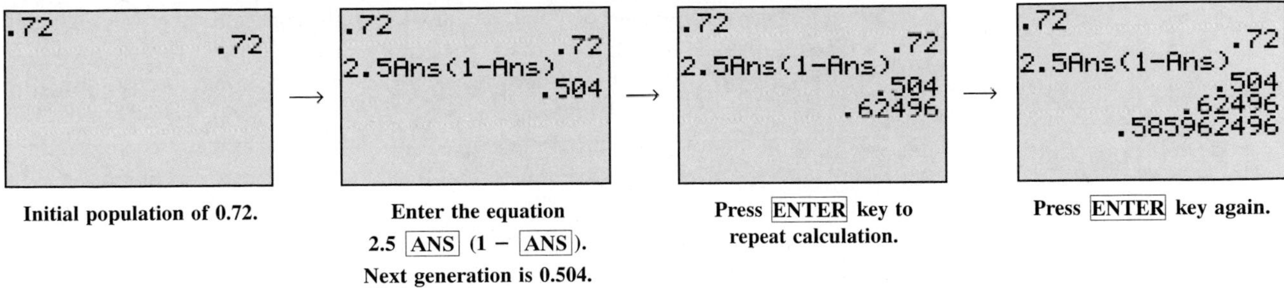

Figure 3.68 Generations of a population of trout.

Further Ideas 3 Determine how to implement the sequence of steps in Example 8 on your calculator. Determine the populations of the first ten generations of trout, writing the results to two decimal places.

Example 9 Illustration of Chaos

To understand the situations described by logistic functions, we need to know how to interpret iterations of functions $f(x) = cx(1 - x)$ for various values of c. Consider the previous trout population. Current populations are described by $f(x) = 2.5x(1 - x)$. What are the implications if conditions change and behavior is described by $f(x) = 3.2x(1 - x)$? The behavior of the trout population could change as a result of pollution or the influx of some other kinds of fish. We shall find some surprising results—phenomena that scientists have only recently become aware of and as yet do not fully understand. Let us look at the long-term behavior of an initial population of $x = 0.72$ under the various conditions of $c = 2.5$, $c = 3.2$, $c = 3.5$, and $c = 3.7$. Using a calculator or computer, we arrive at the following results. We display these results to two decimal places so that the student can clearly see the concepts involved.

$\mathbf{f(x) = 2.5x(1 - x)}$**:** 0.72, 0.50, 0.62, 0.59, 0.61, 0.60, 0.60, 0.60, . . .
The population reaches a steady state of 0.60.

$\mathbf{f(x) = 3.2x(1 - x)}$**:** 0.72, 0.65, 0.73, 0.63, . . . , 0.80, 0.51, . . . , 0.80, 0.51, . . . , 0.80, 0.51, . . .
In the long-term, the population oscillates between 0.80 and 0.51. It is said to be of period two.

$\mathbf{f(x) = 3.5x(1 - x)}$**:** 0.72, 0.71, 0.73, 0.69, . . . , 0.87, 0.38, 0.83, 0.50, . . . , 0.87, 0.38, 0.83, 0.50, . . . , 0.87, 0.38, 0.83, 0.50, . . .
The population oscillates among 0.87, 0.38, 0.83, and 0.50. It is of period four.

$\mathbf{f(x) = 3.7x(1 - x)}$**:** 0.72, 0.75, 0.70, 0.78, . . . , 0.65, 0.85, 0.48, 0.92, . . . , 0.38, 0.87, 0.42, 0.90, . . .
There is no pattern. This is an example of **chaos.** The population appears to fluctuate at random.

Each of these models is an example of a **dynamical system,** a process that changes in time. Pioneering work in the field of chaos was done by mathematician James Yorke of the University of Maryland and biologist Robert May of Princeton University in the mid 1970s. Yorke coined the term *chaos* in his paper "Period Three Implies Chaos," *American Mathematical Monthly,* **82,** 985–992, 1975. Interestingly enough, when Yorke attended an international conference a number of years later in East Berlin, he met Russian mathematician A. N. Sarkovski, who claimed that he had arrived at the same results earlier. In fact, Sarkovski had indeed done the research, unknown to scientists in the West! In this discussion, we introduced the concept of chaos within the framework of models in biology and ecology. However chaos is a newly emerging field that explains the behavior of phenomena in many scientific disciplines. It offers a way of seeing order and pattern where before only disorder had been observed. For example, the techniques of chaos are being used to attempt to understand the behavior of the stock market, changing weather patterns, and the motions of planets and galaxies. Chaos is being used to explain the complicated rhythms of the human heart, the clustering of cars on expressways, the swirling of cigarette smoke, and the formation of clouds. Readers who are interested in finding out more about the fascinating field of chaos should read *Chaos—Making a New Science* by James Gleick (Viking, 1987).

Further Ideas 4

(a) Determine the first five generations (rounded to two decimal places) of the population described by the function $f(x) = 3x(1 - x)$ if the initial population is 0.8. What are the long-term predictions of this model?

(b) Show that the initial population of 0.5 with growth described by $f(x) = 4x(1 - x)$ reaches a steady state of 0 (becomes extinct).

Group Discussion

Graphs of f(x) and |f(x)|

(a) Graph $f(x)$ and $|f(x)|$, in turn, for $f(x) = x^2 - 4x + 3$ on your calculator.

(b) Graph of $f(x)$ and $|f(x)|$ for $f(x) = x^3 - 10x^2 + 27x - 18$ on your calculator.

(c) Graph $f(x)$ and $|f(x)|$ for $f(x) = (x - a)(x - b)(x - c)$ for various values of a, b, and c. Use the graphs from parts a and b and these graphs to describe how the graphs of $f(x)$ and $|f(x)|$ are related.

(d) Give an algebraic explanation of the relationship between the graphs of $f(x)$ and $|f(x)|$ for the preceding functions.

Answer to Self-Check Exercise

1. $(g \circ f)(x) = \sqrt{x^2 - 4}$, with domain $(-\infty, -2] \cup [2, \infty)$. $(f \circ g)(x) = |x| - 4$, with domain $[0, \infty)$. $(g \circ f)(2) = 0$, $(f \circ g)(-3)$ does not exist.

Answers to Further Ideas Exercises

2. $\left(\frac{f}{g}\right)(x) = \frac{\sqrt{x^2 - 9}}{2x - 8}$. Domain $(-\infty, -3] \cup [3, 4) \cup (4, \infty)$.

3. First ten generations of trout, rounded to two decimal places: 0.72, 0.50, 0.62, 0.59, 0.61, 0.60, 0.60, 0.60, 0.60, 0.60. The population becomes steady at 0.60, that is at three fifths the greatest conceivable population.
4. **(a)** First five generations, rounded to two decimal places: 0.80, 0.48, 0.75, 0.56, 0.74. The long-term population oscillates between 0.65 and 0.68. **(b)** Populations are 0.5, 1, 0, 0, 0, . . . ; they become extinct in three generations.

EXERCISES 3.6

In Exercises 1–6, construct the functions $(f+g)(x)$, $(f-g)(x)$, $(fg)(x)$, and $(f/g)(x)$ for the given functions. Give the domain of each function.

1. $f(x) = 2x + 4,\ g(x) = x - 3$
2. $f(x) = -3x + 5,\ g(x) = 4x + 2$
3. $f(x) = 3x^2 - 2x + 1,\ g(x) = 8x^2 + 2x - 1$
4. $f(x) = 2x + 1,\ g(x) = 3x^2 - x - 2$
5. $f(x) = 2|x| + 3,\ g(x) = |x| - 3$
6. $f(x) = x - 3,\ g(x) = 2\sqrt{x + 4}$

In Exercises 7–16, construct $(f+g)(x)$, $(f-g)(x)$, $(fg)(x)$, and $(f/g)(x)$ for the given functions. Give the domain of each function. Use graphs to support your answers.

7. $f(x) = 2,\ g(x) = 1/(x - 4)$
8. $f(x) = 3x + 2,\ g(x) = 1/x^2$
9. $f(x) = x + 3,\ g(x) = 1/(x + 3)$
10. $f(x) = x + 3,\ g(x) = \sqrt[4]{x - 4}$
11. $f(x) = 7,\ g(x) = -7$
12. $f(x) = x - 3,\ g(x) = x - 3$
13. $f(x) = 5x^2,\ g(x) = \sqrt{x - 4}$
14. $f(x) = x^2,\ g(x) = 1/x^2$
15. $f(x) = \sqrt{x - 3} + 4,\ g(x) = \sqrt{x - 4}$
16. $f(x) = \sqrt{x - 3} + 2,\ g(x) = \sqrt{x - 3}$

In Exercises 17–24, $f(x) = 2x - 1$, $g(x) = 5x + 2$, and $h(x) = -x + 3$. Find the value of each function, if it is defined.

17. $(f+g)(3)$ **18.** $(f-g)(1)$ **19.** $(fg)(-2)$
20. $(f-h)(4)$ **21.** $\left(\dfrac{f}{g}\right)(4)$ **22.** $(fh)(-3)$
23. $\left(\dfrac{g}{h}\right)(3)$ **24.** $\left(\dfrac{g}{f}\right)(0.5)$

In Exercises 25–32, find $(g \circ f)(x)$ and $(f \circ g)(x)$. Give the domain of each function.

25. $f(x) = 2x + 7,\ g(x) = 3x + 2$
26. $f(x) = 1 - 2x,\ g(x) = 3x - 5$
27. $f(x) = x^2,\ g(x) = 3x + 1$
28. $f(x) = 2x^2 + 1,\ g(x) = 4x - 2$
29. $f(x) = 3x^2 - x + 5,\ g(x) = 4$
30. $f(x) = x + 5,\ g(x) = 1/(1 - x)$
31. $f(x) = x,\ g(x) = 1/x$
32. $f(x) = 4,\ g(x) = -2$

In Exercises 33–38, find $(g \circ f)(x)$ and $(f \circ g)(x)$. Give the domain of each function.

33. $f(x) = |x|,\ g(x) = -3$ **34.** $f(x) = |x|,\ g(x) = 1/x$
35. $f(x) = 2x^2 - x + 4,\ g(x) = -3x + 2$
36. $f(x) = (x + 2)/(2x - 3),\ g(x) = 1/x$
37. $f(x) = |x + 1|,\ g(x) = x^2 - 4x + 3$
38. $f(x) = x^4 + 2x^2 + 1,\ g(x) = 2$

In Exercises 39–42, find $(g \circ f)(x)$ and $(f \circ g)(x)$. Determine the domain of each function using algebra. Use graphs to support your answers.

39. $f(x) = \sqrt{x - 3},\ g(x) = x^2 + 3$
40. $f(x) = 2x + 3,\ g(x) = 1/(x - 1)$
41. $f(x) = \sqrt{x - 3},\ g(x) = \sqrt{x + 5}$
42. $f(x) = \sqrt{x + 2},\ g(x) = \sqrt{x - 1}$

In Exercises 43–50, $f(x) = 2x + 1$, $g(x) = 3x - 2$, and $h(x) = x^2$.

43. Find $(f \circ g)(2)$ **44.** Find $(g \circ f)(-2)$
45. Find $(f \circ h)(1)$ **46.** Find $(h \circ g)(-3)$
47. Find $(f \circ f)(a)$ **48.** Find $(h \circ h)(a)$
49. Find $(g \circ g)(3)$ **50.** Find $(f \circ g)(h(x))$

In Exercises 51–56, $f(x) = 2x + 3$, $g(x) = \sqrt{x + 2}$, and $h(x) = x^2$. Express each function as a composite of functions chosen from *f*, *g*, and *h*.

51. $F(x) = 2\sqrt{x + 2} + 3$ **52.** $G(x) = 2x^2 + 3$

53. $H(x) = \sqrt{2x + 5}$ **54.** $F(x) = \sqrt{x^2 + 2}$

55. $G(x) = x + 2$ **56.** $H(x) = 4x^2 + 12x + 9$

In Exercises 57–62, $f(x) = 2x + 1$, $g(x) = 3x - 2$, and $h(x) = x^2 + 1$. Express each function as a composite function of functions chosen from *f*, *g*, and *h*.

57. $F(x) = 6x + 1$ **58.** $G(x) = 9x^2 - 12x + 5$

59. $H(x) = 6x - 3$ **60.** $F(x) = x^4 + 2x^2 + 2$

61. $H(x) = 3x^2 + 1$ **62.** $G(x) = 4x + 3$

63. Production of Microcomputers The Omega Company manufactures and sells a certain type of microcomputer for \$400. The total cost *C* of producing *x* computers per week is described by the function

$$C(x) = \frac{x^3}{275} - 2x^2 + 658x + 2400$$

The company total revenue *R* from these *x* computers is

$$R(x) = 400x$$

The company is manufacturing 250 computers per week. Should the company change its production rate to maximize profit? *Hint:* Profit is

$$P(x) = R(x) - C(x)$$

64. Manufacturing Televisions The Victor Company manufactures televisions. The cost of manufacturing *x* televisions per day is

$$C(x) = \frac{x^3}{40} - 5x^2 + 350x + 1920$$

The total revenue for selling these *x* televisions is $R(x) = -x^2 + 260x$. The company is manufacturing 100 televisions per day. Should the company change its production rate to maximize profit?

65. Average Cost Function Let $C(x)$ be the total cost function of a business for producing *x* units of an item in a certain period. An average function that is of considerable importance in business is associated with this cost function; it is called the **average cost function** and is denoted $C_A(x)$. $C_A(x)$ is the average cost of manufacturing each of the *x* units and is defined to be the following quotient:

$$C_A(x) = \frac{C(x)}{x}$$

A company has determined that the total cost of manufacturing *x* units per day of a certain commodity is

$$C(x) = \frac{x^3}{10} - 47x^2 + 8{,}000x + 57{,}600$$

(a) Determine the average cost of producing 150 units.

(b) Graph the function $C_A(x)$. Where does the graph increase and where does it decrease? What does the shape of the graph tell you about the production?

(c) Find the daily production that minimizes average cost.

(d) What is that minimum average cost?

66. A company has determined that the total cost of manufacturing motorcycles is

$$C(x) = \frac{x^3}{4} - 49x^2 + 4{,}000x + 10{,}000$$

when *x* cycles per week are produced.

(a) Determine the average cost function

$$C_A(x) = C(x)/x$$

(b) Compute $C_A(41) - C_A(40)$. What does this number mean?

(c) At what production level does the average cost of each motorcycle start increasing?

67. Numbers of State Park Visitors The number of people visiting Grinwald State Park per annum in the 50-year period from 1945 to 1995 is closely described by the following function:

$$N(t) = 2t^3 - 156t^2 + 2{,}530t + 66{,}800$$

t is time in years from 1945, with $t = 0$ corresponding to 1945. $N(t)$ is the number of visitors during the year t.

(a) Construct the difference function

$$D(t) = N(t + 1) - N(t), \text{ with } 0 \leq t \leq 49$$

(b) What information does the function $D(t)$ give you?

(c) Compute $D(5)$, $D(20)$, and $D(45)$. What do these values mean?

(d) Graph $D(t)$. Determine the minimum value of this function. What information does this give?

68. **Calculator Costs** Let the factory cost of a calculator be $\$x$. Suppose the following increases take place. First, the wholesaler increases the price by 20% and adds \$7 for shipping and handling. The retailer then adds 35% to the wholesale price and adds \$9 for customer service.

(a) Determine the final price that the customer pays for the calculator as a function of the manufacturing price $\$x$.

(b) If the factory cost of the calculator is \$35, what will the consumer pay?

DYNAMICAL SYSTEMS

69. Consider the logistic function $f(x) = cx(1 - x)$ for $c = 2.5$, $c = 3.2$, $c = 3.5$, and $c = 4$. Let the initial population be 0.5. Determine whether the populations reach a steady state, become periodic, or are chaotic, rounded to six decimal places.

70. Consider the logistic function $f(x) = 3.55x(1 - x)$ with initial population 0.72. Show that the long-term populations have period eight, to six decimal places.

71. Consider the logistic function $f(x) = 3.8397x(1 - x)$ with initial population 0.72. Show that the long-term populations have period three, to six decimal places.

72. **Analysis of a Deer Population** The deer population of the Pasco State Forest was 0.6 in 1990 (0.6 of the estimated maximum population that the area could support). The populations in the years following were found to be very closely described by the logistic function $f(x) = 3.35x(1 - x)$. [If x is the population one year, $f(x)$ is the population the following year.]

(a) Predict the populations for the years 1997–2000, rounded to four decimal places.

(b) What are the long-term predicted minimum and maximum sizes of the deer population?

73. **Predicting a University Library Budget** Eight percent of the annual budget of Dortmal University was devoted to the library in 1998. The percentage during the years following was described by the logistic function $f(x) = 1.2x(1 - x)$. How high will the percentage become if things continue in this manner? *Hint:* Express percentage as a decimal. The initial value of x is 0.08.

74. **Equilibrium Levels** (a) Consider the logistic function $f(x) = cx(1 - x)$. Suppose a population reaches the equilibrium level k, k being nonzero. Show that $k = 1 - 1/c$. *Hint:* a population reaches equilibrium level k when $k = ck(1 - k)$.

In this section we saw that the population of 0.72 for $c = 2.5$ settles down to an equilibrium level of 0.6. We see that $0.6 = 1 - 1/2.5$, illustrating the result.

(b) We can use the result from part a to construct a logistic function that has a desired equilibrium level. Construct a logistic function $f(x)$ that has equilibrium level 0.5. Observe that this equilibrium level is independent of the initial value of an orbit. The initial value affects the number of iterations it takes to converge to the equilibrium level. Show that the initial populations of 0.4 and 0.8 both converge to the equilibrium level 0.5 for this function.

(c) Is it possible to construct a logistic equation having equilibrium level 0.8? Investigate.

SECTION PROJECT

Constructing a Logistic Function to Describe a Situation

Describe a situation that could, in your opinion, be described by a logistic function. Give a realistic logistic equation that would describe this situation and discuss some of the implications obtained by applying the equation.

3.7 Inverse Functions, One-to-One, and Cryptography

- Inverse Function • One-to-one Function • The Horizontal Line Test
- The Graph of a Function and Its Inverse • Predicting Sales
- Cryptography

When we have used functions, it has usually been in the context of determining values of f corresponding to certain points in the domain of f. However, occasions arise when we are interested in doing things the other way around. We want to know the points in the domain that lead to given values of f. In Section R.7, we used a model based on an automobile's braking distance to introduce the concept of a function. Let us use another such application to introduce the idea of the inverse of a function.

Probable Speed in an Accident

Let the braking distance of a certain model of car be described by the function $f(x) = 4x - 80$ for speeds of between 30 and 110 miles per hour, where x is the speed of a vehicle, and $f(x)$ is the corresponding braking distance in feet. For example, when the speed is 60 miles per hour, the braking distance is

$$f(60) = 4(60) - 80 = 160 \text{ feet}$$

It is also useful to be able to do things the other way around, namely to compute the speeds corresponding to given braking distances. Suppose such a car was involved in an accident. The skid marks stretched for 300 feet. What was the probable speed of the car? To solve this equation for speed, we would add 80 and then divide by 4. We arrive at the inverse of f; namely, the function $g(x) = (x/4) + 20$ (still keeping the letter x for the independent variable). The notation f^{-1} is used for the inverse function. Thus, $f^{-1}(x) = (x/4) + 20$. We can now use this function to determine the speeds that correspond to various braking distances. For example, the speed for a braking distance of 300 feet is $f^{-1}(300) = 95$ miles per hour. On the basis of this information, the vehicle involved in the accident was doing 95 miles per hour.

Caution: Note that the inverse of the function f is denoted f^{-1}. The -1 is part of the notation for the inverse function. It should not be confused with a negative exponent.

With this informal background, let us discuss the concept of an inverse function.

Inverse Function

Let f be a function having domain A and range B. Let a be any element in A with $f(a) = b$. If there is a function having domain B and range A that inverts the effect of f in that it brings b back to a, then such a function is called the **inverse** of f and is denoted f^{-1} (Figure 3.69).

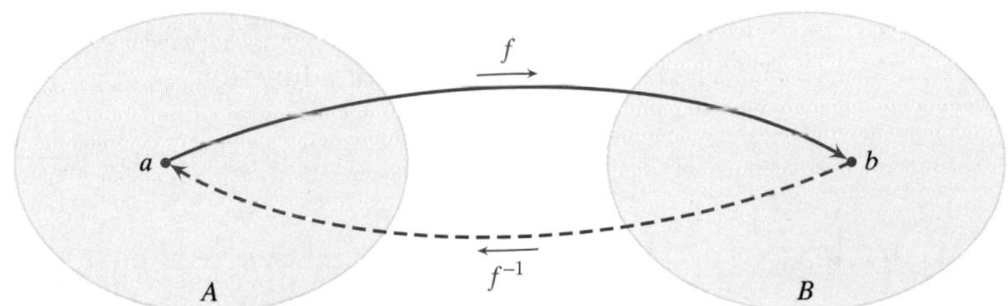

Figure 3.69 Inverse function. Note that the range of f is the domain of f^{-1} and vice versa.

Now let us investigate the relationship between f and f^{-1}. We have that

$$f(a) = b \quad \text{and} \quad f^{-1}(b) = a$$

Substitute for b from the first equation into the second to get

$$f^{-1}(f(a)) = a$$

Similarly, substitute for a from the second equation into the first to get

$$f(f^{-1}(b)) = b$$

Thus,

$$f^{-1}(f(a)) = a \quad \text{and} \quad f(f^{-1}(b)) = b$$

This discussion leads to the following definition.

> ***Inverse Function*** The function g is the **inverse** of f if
>
> $$g(f(x)) = x \text{ for every } x \text{ in the domain of } f$$
>
> and
>
> $$f(g(x)) = x \text{ for every } x \text{ in the domain of } g$$
>
> The domain of g is the range of f. We write f^{-1} for g.

This definition implies that if the function g is the inverse of f, then it is also true that f is the inverse of g.

Example 1 Confirming an Inverse

Let $f(x) = 3x + 7$. Show that $g(x) = (x - 7)/3$ is the inverse function of $f(x)$ using algebra.

SOLUTION According to this definition, we need to show that

$$g(f(x)) = x \qquad \text{and} \qquad f(g(x)) = x$$

We get

$$g(f(x)) = g(3x + 7) = \frac{(3x + 7) - 7}{3} = x$$

$$f(g(x)) = f\left(\frac{x - 7}{3}\right) = 3\left(\frac{x - 7}{3}\right) + 7 = x$$

Thus, g is the inverse of f. We write $f^{-1}(x) = (x - 7)/3$. Because the range of f is the set of real numbers, the domain of f^{-1} is the set of real numbers.

Self-Check 1 Let $f(x) = 5x - 1$. Show that $g(x) = (x + 1)/5$ is the inverse of $f(x)$.

The Graph of a Function and Its Inverse

There is an interesting relationship between the graph of a function and that of its inverse. Consider again the function $f(x) = 3x + 7$ and its inverse

$$f^{-1}(x) = \frac{x - 7}{3}$$

If we graph these functions, together with the line $y = x$, we see that the graph of f is the mirror image of the graph of f in the line $y = x$ (Figure 3.70). The graphs of f and f^{-1} are symmetric about the line $y = x$.

Now let us show that we can always expect the graph of a function and its inverse to be symmetric about the line $y = x$.

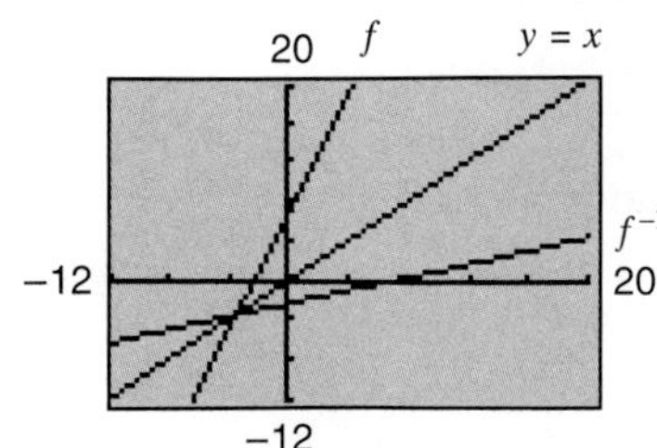

Figure 3.70 Graphs of $f = 3x + 7$ and $f^{-1} = (x - 7)/3$ are symmetric about line $y = x$.

Symmetry Between the Graphs of a Function and Its Inverse

Let a be a number in the domain of f and let $f(a) = b$. By the definition of f^{-1}, we know that $f^{-1}(b) = a$. Thus, if the point (a, b) is on the graph of f, then the point (b, a) is on the graph of f^{-1}. The points (a, b) and (b, a) are symmetric with respect to the line $y = x$, as shown in Figure 3.71. Thus, the graphs of f and f^{-1} are symmetric about the line $y = x$.

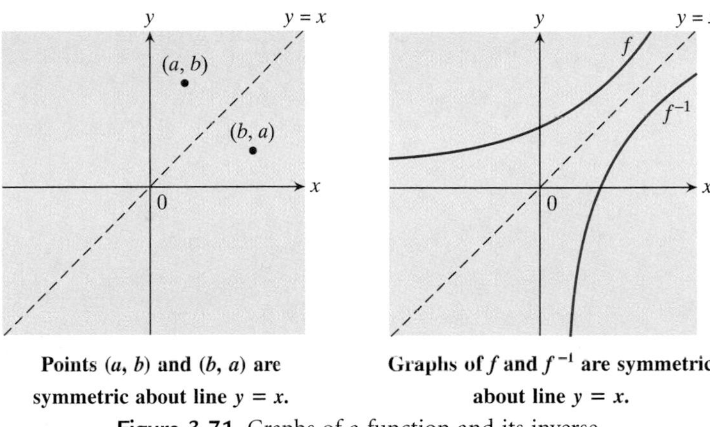

Points (a, b) and (b, a) are symmetric about line $y = x$.

Graphs of f and f^{-1} are symmetric about line $y = x$.

Figure 3.71 Graphs of a function and its inverse.

GRAPHS OF f AND f^{-1}

The graphs of f and f^{-1} are symmetric about the line $y = x$.

We now give a method for finding the inverse of a function. Consider the function $y = f(x)$. It is customary to use the letter x for the independent variable and y for the dependent variable of a function. Because the range of f will become the domain of f^{-1} and *vice versa,* we interchange the letters x and y in the equation $y = f(x)$. We then solve this equation for y to get the function f^{-1}. Note that not every function has an inverse. We shall discuss the criterion for an inverse in detail later in this section.

FINDING THE INVERSE OF A FUNCTION f

1. Write the function as $y = f(x)$.
2. Interchange x and y.
3. Solve the equation for y in terms of x. If y defines a function, then this is the inverse of f. Replace y by the symbol $f^{-1}(x)$. If y does not define a function, then f does not have an inverse.
4. The domain of f^{-1} is the range of f.

We now give an example to illustrate this method of arriving at an inverse. In Example 1, we gave an algebraic method of confirming an inverse. We will support the inverse in this example using the symmetry of the graphs about the line $y = x$.

Example 2 Finding the Inverse of a Function

Find the inverse of $f(x) = 5x - 3$. Support your answer with a graph.

SOLUTION We use the steps for finding the inverse.

1. Write the function as $y = f(x)$.
$$y = 5x - 3$$
2. Interchange x and y.
$$x = 5y - 3.$$
3. Solve for y.
$$5y = x + 3$$
$$y = \frac{x+3}{5}$$
Replace y by f^{-1}.
$$f^{-1}(x) = \frac{x+3}{5}$$
4. The range of f is $(-\infty, \infty)$; thus the domain of f^{-1} is $(-\infty, \infty)$. See Figure 3.72 for the graphs of f and f^{-1}.

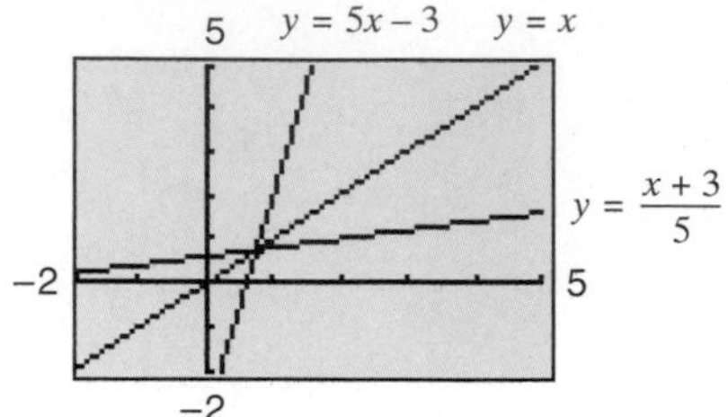

Figure 3.72 Graphs of $y = 5x - 3$ and $y = (x + 3)/5$ are symmetric about the line $y = x$. Thus, if $f(x) = 5x + 3$, then $f^{-1}(x) = (x + 3)/5$.

The following example illustrates the care that must be taken in determining the domain of the inverse function. The domain of the inverse is the range of the original function.

Example 3 Finding the Inverse of a Function

Find the inverse of the function $f(x) = \sqrt{x - 4}$. Support your answer with a graph.

SOLUTION We use the steps for finding an inverse.

1. Write the function as $y = f(x)$.
$$y = \sqrt{x - 4}$$
2. Interchange x and y.
$$x = \sqrt{y - 4}$$
3. Solve for y.
$$x^2 = y - 4$$
$$y = x^2 + 4$$
Replace y by f^{-1}.
$$f^{-1}(x) = x^2 + 4$$
4. The range of f is $[0, \infty)$; thus, the domain of f^{-1} is $[0, \infty)$.

See Figure 3.73.

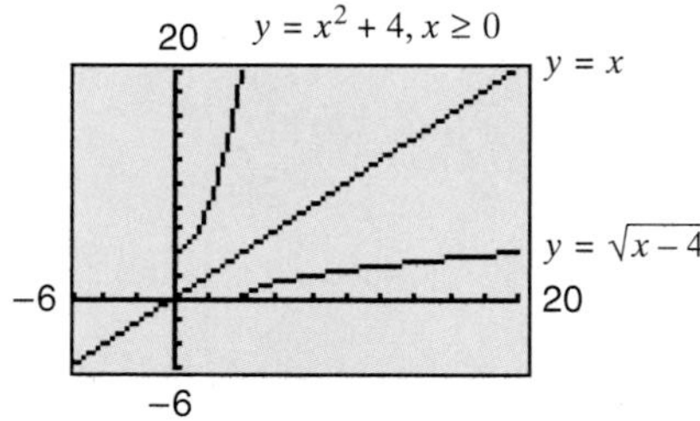

Figure 3.73 Graphs of $y = \sqrt{x - 4}$ and $y = x^2 + 4, x \geq 0$, are symmetric about the line $y = x$. Thus, if $f(x) = \sqrt{x - 4}$, then $f^{-1}(x) = x^2 + 4, x \geq 0$.

CAUTION: The domain of a function normally depends solely on that function. However, the function f^{-1}, as the notation implies, is closely related to f. It does not have the freedom of a normal function. The domain of the inverse function f^{-1} is the range of f.

Self-Check 2 Find the inverse of $f(x) = \sqrt{x + 3} + 2$. Verify your answer by graphing f, f^{-1}, and the line $y = x$ on your calculator and checking for symmetry. Give the domain of f^{-1}.

The following example illustrates a function that does not have an inverse function.

Example 4 Finding That an Inverse Function Does Not Exist

Show that function $f(x) = x^2$ does not have an inverse. Let us go through the steps for finding the inverse of a function.

1. Write the function as $y = f(x)$. $\quad y = x^2$
2. Interchange x and y. $\quad x = y^2$
3. Solve for y. $\quad y = \pm\sqrt{x}$

Observe that y does not define a function because there are two distinct values of y corresponding to a single value of x. Thus, f does not have an inverse. If we return to the original function $f(x) = x^2$, we can see why this situation arises. There are distinct values of x in the domain that have the same image $f(x)$. For example, $x = -2$ and $x = 2$ have the same image 4. The inverse would have to "bring 4 back to -2 and 2." This violates the definition of a function. There can be no inverse.

These observations lead to the following conclusions concerning the existence of the inverse of a function. If distinct elements of the domain have distinct images, the inverse exists. Such a function is said to be one-to-one. Conversely, if there are certain elements in the domain with the same image, there will not be an inverse. This function is not one-to-one.

One-to-One Function

Consider the function $f(x) = 4x - 1$. Take two points in the domain, such as $x = 1$ and $x = 2$. We get

$$f(1) = 4(1) - 1 = 3 \qquad \text{and} \qquad f(2) = 4(2) - 3 = 5$$

Thus, for the two points $x = 1$ and $x = 2$, we get $f(1) \neq f(2)$. Similarly, any two distinct points in the domain of f will lead to different points in the range. Such a function is said to be **one-to-one**. It will have an inverse.

On the other hand, consider the function $g(x) = x^4$. At $x = -2$ and at $x = 2$

$$g(-2) = (-2)^4 = 16 \qquad \text{and} \qquad g(2) = (2)^4 = 16$$

Thus, although $-2 \neq 2$, we get $g(-2) = g(2)$. Such a function is not one-to-one. It will not have an inverse.

One-to-One Function A function f is **one-to-one** if for any two elements a and b in the domain of f, whenever $a \neq b$ then $f(a) \neq f(b)$ (Figure 3.74).

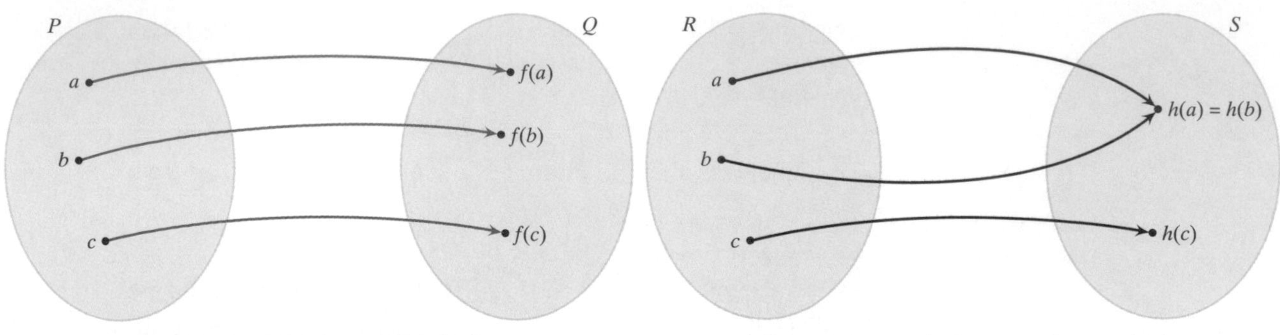

f is one-to-one (distinct elements in the domain have distinct images).

h is not one-to-one (at least two elements in the domain have the same image).

Figure 3.74 Concept of one-to-one.

Example 5 Using Algebra to Check Whether a Function Is One-to-One

(a) Prove that $f(x) = 5x + 2$ is one-to-one.
(b) Prove that $g(x) = \sqrt{x^2 - 5}$ is not one-to-one.

SOLUTION We use the definition to prove these results.

(a) Let a and b be distinct elements in the domain of f. Because $a \neq b$, then

$$5a \neq 5b$$

$$5a + 2 \neq 5b + 2$$

Thus, $f(a) \neq f(b)$, f is one-to-one.

(b) A way to show that a function is not one-to-one is to find two distinct numbers in the domain that have the same function value. Consider $x = -3$ and $x = 3$. We get

$$g(-3) = \sqrt{(-3)^2 - 5} = \sqrt{9 - 5} = \sqrt{4} = 2$$

and

$$g(3) = \sqrt{(3)^2 - 5} = \sqrt{9 - 5} = \sqrt{4} = 2$$

Because $-3 \neq 3$ but $g(-3) = g(3)$, the function g is not one-to-one.

Self-Check 3

(a) Show that $f(x) = 2x^3 + 1$ is one-to-one.
(b) Show that $g(x) = 3x^2 - 7$ is not one-to-one.

The Horizontal Line Test

There is a convenient graphical test that can be used to determine whether a function is one-to-one. Suppose that the horizontal line $y = k$ cuts the graph of $y = f(x)$ at a number of points, such as (a, k) and (b, k). Thus, $f(a) = f(b) = k$. The implication is that f is not one-to-one. Conversely, if every horizontal line cuts the graph in one or no points, then f is one-to-one.

THE HORIZONTAL LINE TEST

A function is one-to-one if and only if every horizontal line cuts its graph in at most one point.

Figure 3.75 illustrates this test. In Figure 3.75(a), f is one-to-one because every horizontal line, such as L, cuts the graph in one point, such as A. In Figure 3.75(b), g is not one-to-one because there is a horizontal line L that cuts the graph in two points B and C.

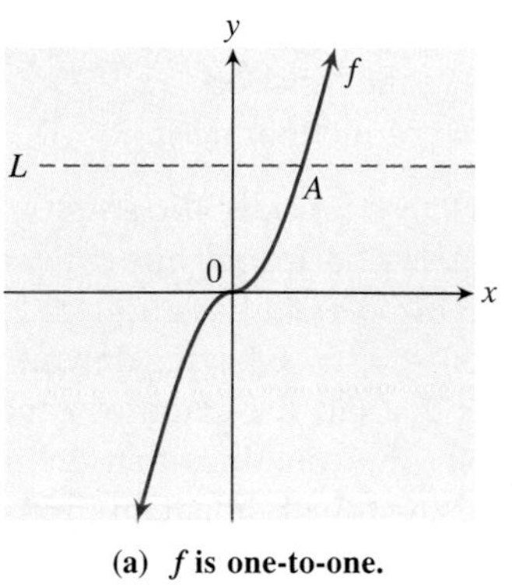

(a) f is one-to-one.

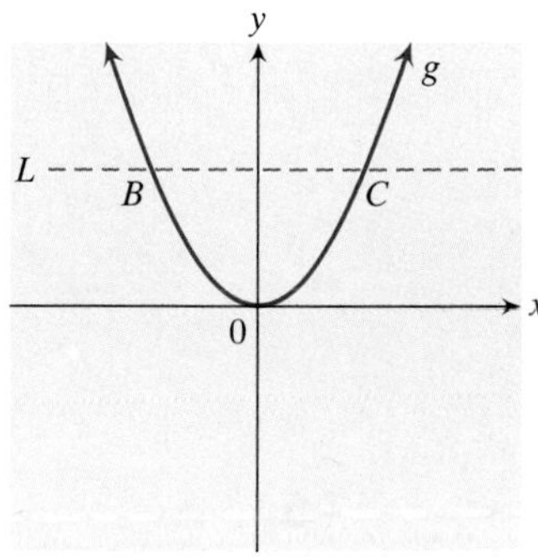

(b) g is not one-to-one.

Figure 3.75 The horizontal line test.

Example 6 Using a Graph to Check Whether a Function Is One-to-One

Use graphs to show that
(a) $f(x) = 2x - 3$ is one-to-one.
(b) $g(x) = x^2 - 4x + 1$ is not one-to-one.

SOLUTION

(a) The graph of $f(x) = 2x - 3$ is a straight line. Observe that every horizontal line, such as $y = 2$, cuts the graph in one point (Figure 3.76). Thus f is one-to-one.

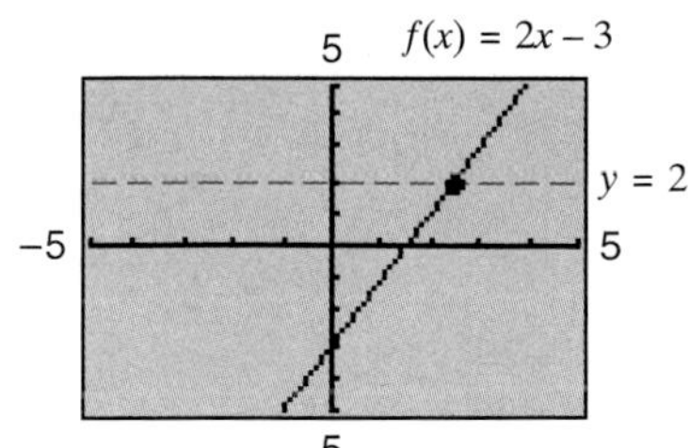

Figure 3.76

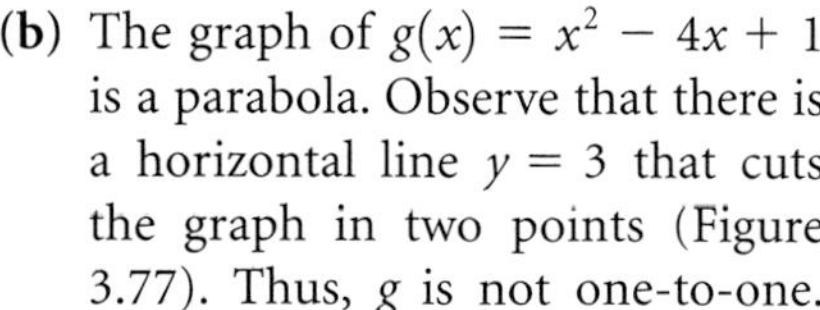

(b) The graph of $g(x) = x^2 - 4x + 1$ is a parabola. Observe that there is a horizontal line $y = 3$ that cuts the graph in two points (Figure 3.77). Thus, g is not one-to-one.

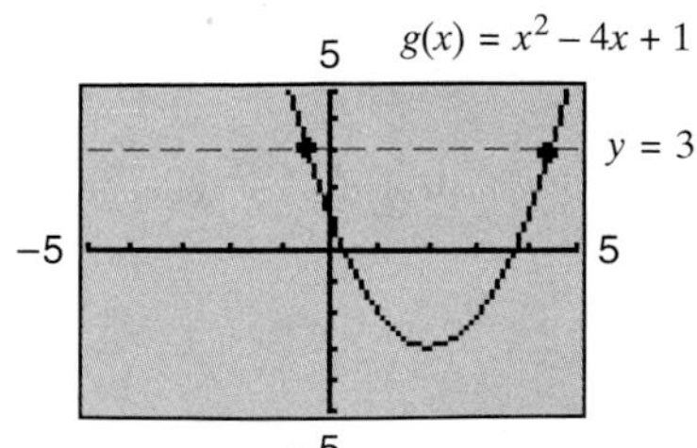

Figure 3.77

Further Ideas 1

(a) Use a graph to show that the function $f(x) = x^3$ is one-to-one.

(b) Is every polynomial function of degree 3 one-to-one? If not, use your calculator to find a polynomial function of degree 3 that is not one-to-one. Give its graph and the equation of a specific horizontal line that cuts the graph in more than one point.

The following example illustrates the important role the domain plays in determining whether a function has an inverse.

Example 7 Finding the Inverse of a Function

Show that the function $f(x) = x^2 - 6$ with domain $[0, \infty)$ has an inverse function. Determine that inverse.

SOLUTION Consider the graph of $f(x) = x^2 - 6$ with domain $[0, \infty)$ (Figure 3.78a). Every horizontal line either cuts this graph in one point or not at all. Thus, f is one-to-one and has an inverse. It is instructive to compare f with the function $g(x) = x^2 - 6$ having domain $(-\infty, \infty)$ (Figure 3.78b). There are horizontal lines that cut the graph of g in two points. Thus, g does not have an inverse function. The equations that define f and g are the same. The domain decided whether the function had an inverse.

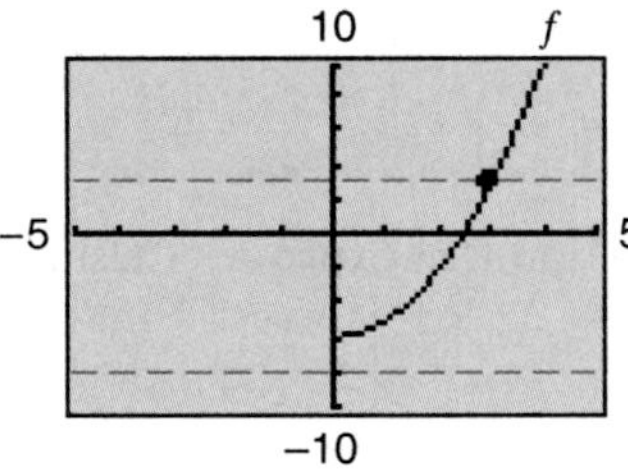

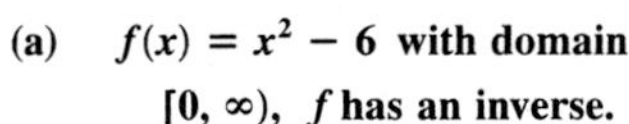

(a) $f(x) = x^2 - 6$ **with domain** $[0, \infty)$**,** f **has an inverse.**

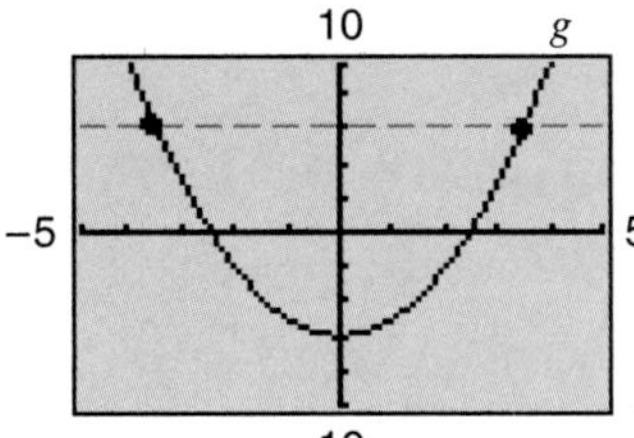

(b) $g(x) = x^2 - 6$ **with domain** $(-\infty, \infty)$**,** g **does not have an inverse.**

Figure 3.78

We now compute the inverse of f, remembering that its domain is $[0, \infty)$. Write the function as $y = f(x)$.

$$y = x^2 - 6, \quad \text{with } x \geq 0 \text{ since the domain is } [0, \infty)$$

Interchange x and y.

$$x = y^2 - 6, \quad \text{with } y \geq 0$$

Solve for y.

$$y^2 = x + 6$$

$$y = \sqrt{x + 6} \quad \text{(take the positive root because } y \geq 0\text{)}$$

Thus,

$$f^{-1}(x) = \sqrt{x + 6}$$

Let us consider the range of $f(x) = x^2 - 6$ for values of x in $[0, \infty)$. Because $x^2 \geq 0$, $x^2 - 6 \geq -6$. The range of f is $[-6, \infty)$. (The graph of f in Figure 3.78a confirms this.) The domain of f^{-1} is thus $[-6, \infty)$.

In applications, the inverse of a function often arises when one wants to interchange independent and dependent variables. The letters of the variables can carry significance, and it may not be desirable to interchange these letters. We now present a familiar application within this setting.

Applications . . .

Example 8 Converting Temperature Scales

The function that expresses Fahrenheit temperature in terms of Celsius temperature is $F(C) = \frac{9}{5}C + 32$. To express Celsius in terms of Fahrenheit, we solve this equation for C as follows:

$$F = \frac{9}{5}C + 32$$

$$\frac{9}{5}C = F - 32$$

$$C = \frac{5}{9}(F - 32)$$

The function that expresses Celsius in terms of Fahrenheit is $C(F) = \frac{5}{9}(F - 32)$. This is the inverse of $F(C)$.

Further Ideas 2 ***Motion of a Pendulum*** The period of a pendulum is the time it takes to perform one complete oscillation. The period T seconds for a pendulum of length L feet is given by the formula $T(L) = 2\pi\sqrt{L/32}$. Determine the inverse of T, and use this inverse to find the length of a pendulum that would have a period of 1 second.

Example 9 Predicting Sales

Consider the following function, which describes the sales s of a certain item as a function of price p:

$$s(p) = -100p + 13{,}600$$

This function can be used to determine the predicted sales when price p is charged for the item. For example, when the price is \$125.95, the predicted sales are

$$s(39.23) = -100(124.95) + 13{,}600 = 1{,}105 \text{ units}$$

However, the company may also want to know how much it should charge to get certain sales (the inverse problem). For example, what is the price that leads to sales of 1500 units? In such instances, we are given the value of s, and we need to compute the corresponding value of p. We use the inverse function. We solve the equation $s = -100p + 13{,}600$ for p.

$$-100p = s - 13{,}600$$
$$100p = 13{,}600 - s$$
$$p = \frac{1}{100}(13{,}600 - s)$$

The function that expresses price p in terms of sales s is thus

$$p(s) = \frac{1}{100}(13{,}600 - s)$$
$$p(s) = 136 - 0.01s$$

This function can now be used to give the prices that are predicted to lead to various sales. For example, when $s = 1{,}500$ units,

$$p(1{,}500) = 136 - 0.01(1{,}500)$$
$$= 121$$

The price that should lead to sales of 1500 items is \$121.

Example 10 Cryptography

Cryptography is the process of coding and decoding messages. The word comes from the Greek *kryptos* meaning "hidden." The technique can be traced back to the ancient Greeks. Today governments use sophisticated methods of coding and decoding messages. One type of coding technique makes use of a one-to-one function to encode a message. We shall illustrate the method using a simple encoding function. Let the message be

MEET AT ONE

and the encoding function be

$$f(x) = 3x + 7$$

We assign a number to each letter of the alphabet. Let us associate each letter with its position in the alphabet. A is 1, B is 2, and so on, as follows. Let a space between words be denoted by the number 27.

A	B	C	D	E	F	G	H	I	J	K	L	M
1	2	3	4	5	6	7	8	9	10	11	12	13

N	O	P	Q	R	S	T	U	V	W	X	Y	Z	-
14	15	16	17	18	19	20	21	22	23	24	25	26	27

According to this convention, the digital form of the message is

M	E	E	T	-	A	T	-	O	N	E
13	5	5	20	27	1	20	27	15	14	5

We now code this digital form of the message. Enter the function into the calculator as $Y = 3X + 7$ and arrive at the table of codes for the numbers 1 through 27 shown in Figure 3.79.

X	Y1
1	10
2	13
3	16
4	19
5	22
6	25
7	28

X	Y1
8	31
9	34
10	37
11	40
12	43
13	46
14	49

X	Y1
15	52
16	55
17	58
18	61
19	64
20	67
21	70

X	Y1
22	73
23	76
24	79
25	82
26	85
27	88
28	91

Figure 3.79 Table of codes.

This scheme results in the following coded message.

Message:	M	E	E	T	-	A	T	-	O	N	E
Digital Form:	13	5	5	20	27	1	20	27	15	14	5
Coded Form:	46	22	22	67	88	10	67	88	52	49	22

The message would be transmitted as 46, 22, 22, 67, 88, 10, 67, 88, 52, 49, 22.

To decode this message, the receiver would reverse this procedure. The receiver would be given the encoding function, reproduce the table in Figure 3.79, and use it in the opposite direction to decode the numbers 46, . . . , 22, to arrive back at original numbers 13, . . . , 5 (in effect using the inverse function). These numbers would be associated with their position in the alphabet, resulting in the message MEET-AT-ONE, as follows:

46	22	22	67	88	10	67	88	52	49	22
13	5	5	20	27	1	20	27	15	14	5
M	E	E	T	-	A	T	-	O	N	E

In this example, we have associated letters with their position number in the alphabet. Any one-to-one numerical correspondence can be used. In practice, the encoding function is complex. Naturally, the more complex the function, the more difficult it is to break the code. The only requirement is that the function be one-to-one for the 27 integers that make up the domain; otherwise, one would not know which of several numbers to choose in decoding.

Answers to Self-Check Exercises

1. $g(f(x)) = \dfrac{(5x - 1) + 1}{5} = x$, and $f(g(x)) = 5\left(\dfrac{x + 1}{5}\right) - 1 = x$, thus, g is the inverse of f.

2. $f^{-1}(x) = (x - 2)^2 - 3$, domain $[2, \infty)$. Graphs of f and f^{-1} are symmetric about $y = x$.

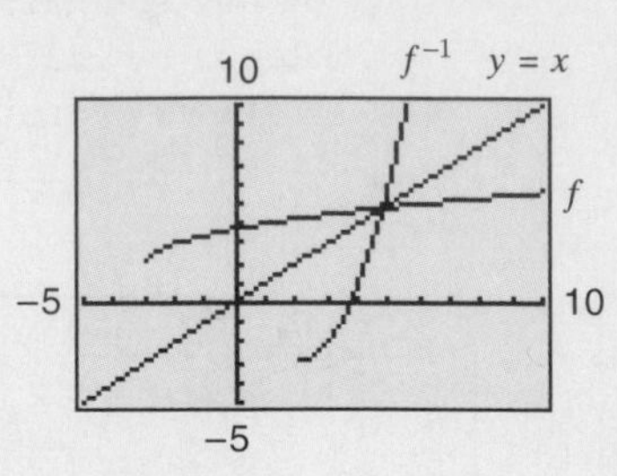

3. **(a)** If $a \neq b$, $a^3 \neq b^3$, $2a^3 \neq 2b^3$, $2a^3 + 1 \neq 2b^3 + 1$, $f(a) \neq f(b)$; thus, f is one-to-one.
   (b) $g(-1) = g(1) = -4$; thus g is not one-to-one.

Answers to Further Ideas Exercises

1. **(a)** Every horizontal line cuts the graph of $f(x) = x^3$ at one point (see graph on the left below); thus f is one-to-one.
   (b) Consider graphs of $f(x) = ax^3 + bx^2 + cx + d$ for various values of a, b, c, and d. Soon find a graph for which a horizontal line cuts the graph in three places [e.g., $f(x) = x^3 - 5x^2 + 2x + 15$ (see graph on the right below). This function is not one-to-one.

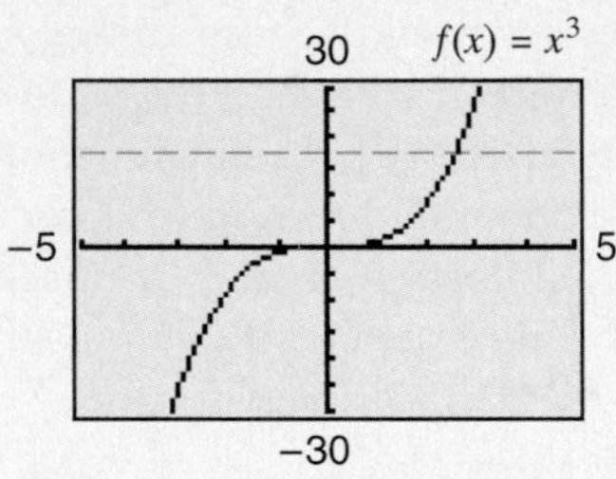

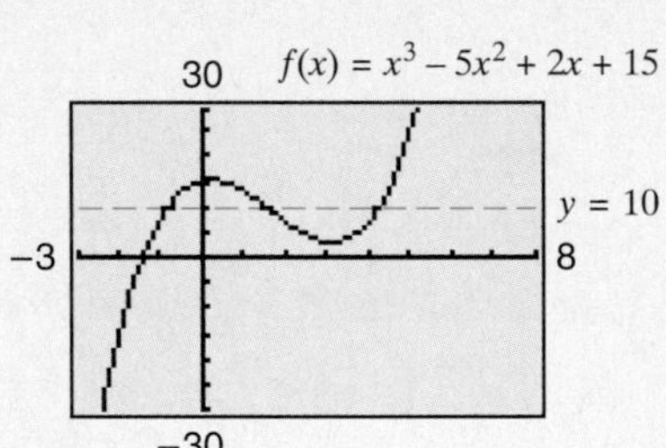

2. Inverse is $L(T) = 8T^2/\pi^2$. $L(1) = 0.81$ feet, rounded to two decimal places.

EXERCISES 3.7

In Exercises 1–6, show that g is the inverse of f using algebra.

1. $f(x) = 2x - 1$, $g(x) = (x + 1)/2$
2. $f(x) = 3x + 4$, $g(x) = (x - 4)/3$
3. $f(x) = x/4$, $g(x) = 4x$
4. $f(x) = (x - 1)/5$, $g(x) = 5x + 1$
5. $f(x) = x^3$, $g(x) = \sqrt[3]{x}$
6. $f(x) = 5x$, $g(x) = x/5$

In Exercises 7–12, show that g is the inverse of f using algebra.

7. $f(x) = 3 - 2x$, $g(x) = (3 - x)/2$
8. $f(x) = 1 - x^3$, $g(x) = \sqrt[3]{1 - x}$
9. $f(x) = \sqrt{x - 5}$, $g(x) = x^2 + 5$ with domain $x \geq 0$
10. $f(x) = \sqrt{2 - x}$, $g(x) = 2 - x^2$ with domain $x \geq 0$
11. $f(x) = x/(x + 1)$, $x \neq -1$;
    $g(x) = x/(1 - x)$, $x \neq 1$
12. $f(x) = (x + 1)/(x - 2)$, $x \neq 2$;
    $g(x) = (2x + 1)/(x - 1)$, $x \neq 1$

In Exercises 13–18, each function has an inverse. Find the inverse. Verify the inverse using graphs.

13. $f(x) = 5x + 2$
14. $f(x) = -2x + 3$
15. $f(x) = 2x$
16. $f(x) = 7 - 3x$

17. $f(x) = (x + 3)/4$ **18.** $f(x) = (5 - 2x)/7$

In Exercises 19–24, each function has an inverse. Find the inverse. Verify the inverse using graphs.

19. $f(x) = x^3 + 3$ **20.** $f(x) = \sqrt{x - 5}$

21. $f(x) = \dfrac{1}{\sqrt{x}}$ **22.** $f(x) = \sqrt[3]{x} - 2$

23. $f(x) = x^2$, with domain $x \geq 0$

24. $f(x) = 3 - x^2$, with domain $x \leq 0$

In Exercises 25–30, graphs of functions are given. Use the horizontal line test to determine whether the functions are one-to-one.

25.

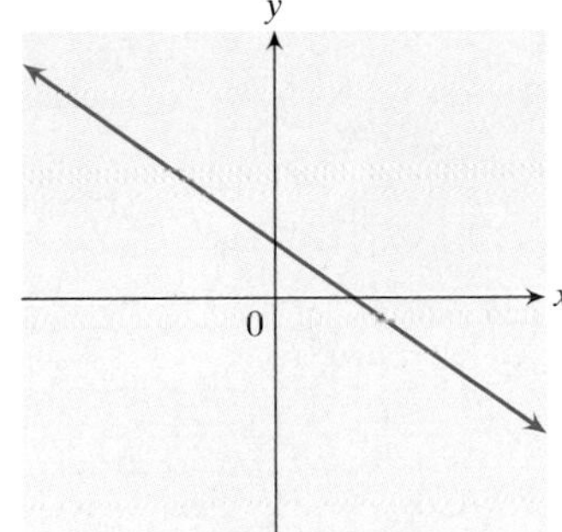

26.

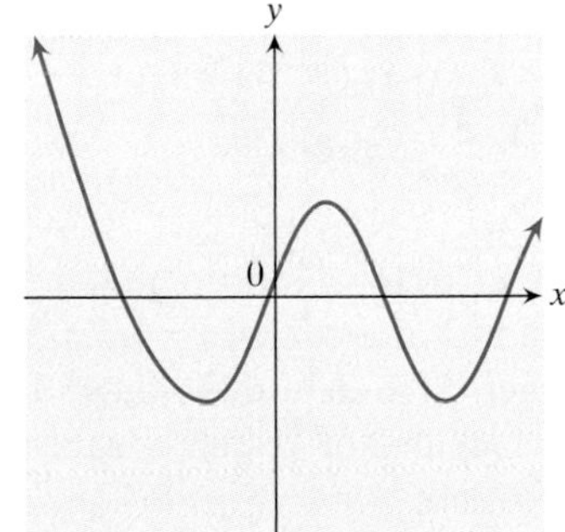

27.

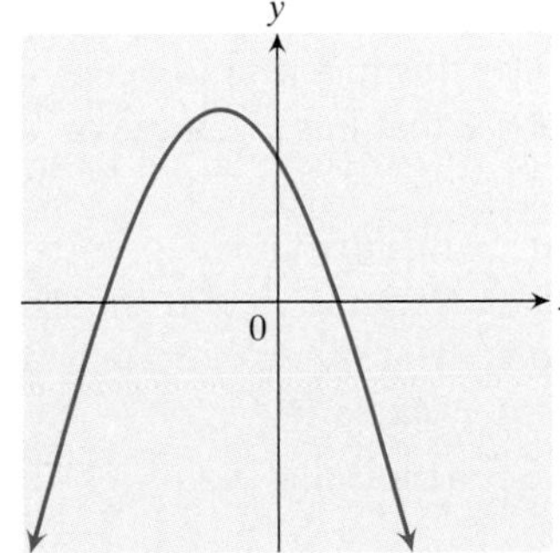

28.

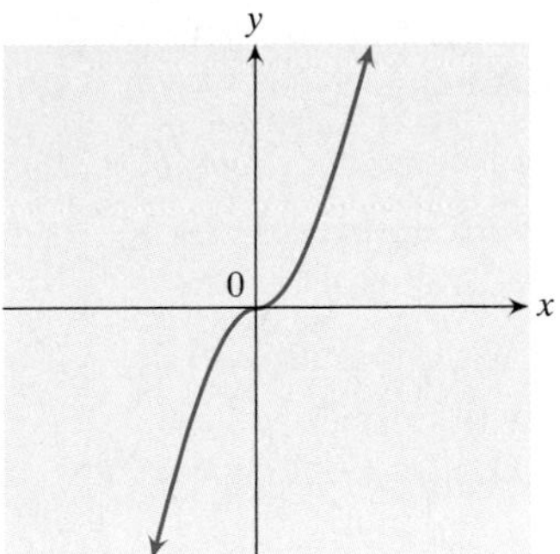

29.

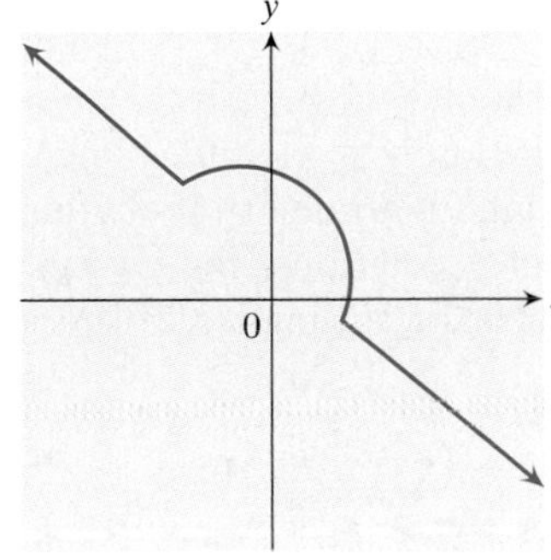

30.

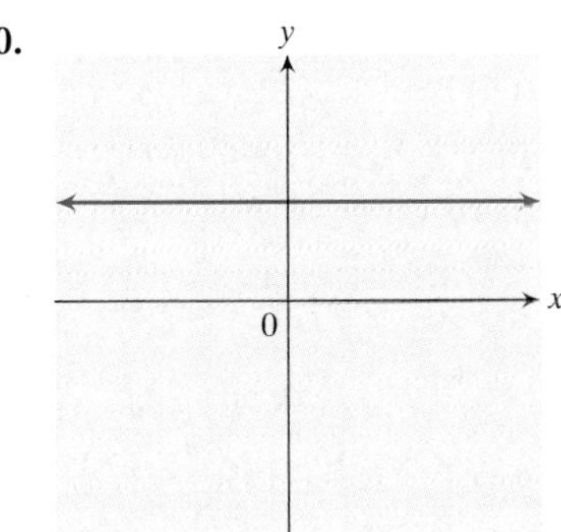

In Exercises 31–36, determine whether the functions are one-to-one (a) by using algebra and (b) by sketching a rough graph and using the horizontal line test.

31. $f(x) = 2x + 3$ **32.** $f(x) = -3x + 7$

33. $f(x) = 4$ **34.** $f(x) = x^2 - 3$

35. $f(x) = -x^2$ **36.** $f(x) = x^2 - 4x + 5$

In Exercises 37-44, determine whether the functions are one-to-one (a) by using algebra and (b) by sketching a rough graph and using the horizontal line test.

37. $f(x) = x^3$ **38.** $f(x) = x^4$ **39.** $f(x) = |x|$

40. $f(x) = \sqrt{x}$ **41.** $f(x) = 1/x$ **42.** $f(x) = x^2 - 4$

43. $g(x) = (x - 3)^2$ **44.** $h(x) = 1/(x - 3)$

In Exercises 45–52, graph each function on your calculator, and use the horizontal line test to show that the function is one-to-one and hence has an inverse function. Find the inverse and give its domain.

45. $f(x) = \sqrt[3]{x + 4}$
46. $f(x) = \sqrt{9 - x^2}$, with domain $0 \leq x \leq 3$
47. $f(x) = (\sqrt{x} - 5)^3$
48. $f(x) = (x^2 + 7)/3$, with domain $x \geq 0$
49. $f(x) = 1/(x - 4)$, with domain $x > 4$
50. $f(x) = 1/(2x + 3)$, with $x < -3/2$
51. $f(x) = |x + 3|$, with domain $x \geq -3$
52. $f(x) = |2x - 5| + 2$, with domain $x \geq 5/2$

In Exercises 53–60, determine whether g is the inverse of f by graphing f, g, and the line $y = x$ on the same calculator screen and by looking for symmetry about this line. If g is not the inverse of f, determine the inverse algebraically, and check your answer using a calculator.

53. $f(x) = 3x - 2$, $g(x) = (x + 2)/3$
54. $f(x) = 2x + 4$, $g(x) = (x - 2)/4$
55. $f(x) = 2x - 1$, $g(x) = 0.5x + 1$
56. $f(x) = x^3 + 1$, $g(x) = \sqrt[3]{x - 1}$
57. $f(x) = x^2 + 3$ with domain $x \geq 0$,
$g(x) = \sqrt{x - 3}$ with $x \geq 0$
58. $f(x) = (x + 2)^2 - 4$ with domain $x \leq -2$,
$g(x) = \sqrt{x + 4} - 2$ with $x \geq -4$
59. $f(x) = \sqrt{x - 5}$, $g(x) = x^2 + 5$
60. $f(x) = \sqrt{x - 4} + 2$, $g(x) = (x - 2)^2 + 4$

61. Consider the function $f(x) = x^3 + kx^2 + 3x - 4$, for various integer values of k. Use your graphing calculator to find the integer values of k for which this function is one-to-one, and therefore has an inverse function. *Hint:* Apply the horizontal line test for graphs for $k = -5, -4, -3, -2, \ldots$.
62. Consider the function $f(x) = |x| + kx - 1$ for various integer values of k. Use your graphing calculator to find the integer values of k for which this function is one-to-one and therefore has an inverse function. *Hint:* Apply the horizontal line test for various values of k.
63. Prove that every linear function $f(x) = ax + b$, $a \neq 0$, has an inverse function. Find the inverse. Observe that the inverse is also a linear function.
64. Prove that a constant function $f(x) = c$ does not have an inverse function.
65. Prove that the quadratic function
$$f(x) = ax^2 + bx + c,\ a \neq 0$$
with domain the set of real numbers, does not have an inverse function.
66. Prove that the quadratic function
$$f(x) = ax^2 + bx + c,\ a \neq 0$$
with domain $[-b/(2a), \infty)$, has an inverse function. Find the inverse.
67. Show that the function $f(x) = ax^3 + b$, $a \neq 0$, has an inverse function. Find the inverse.
68. **Illumination** The intensity of illumination from a source of light is a function of the distance from that source. At a distance of d feet, the intensity is $I = 240{,}000/d^2$ candle power. Determine the inverse of $I(d)$, and use this inverse to find the distance at which the intensity is 100 candle power, rounded to four decimal places.

69. **Radius and Area of a Circle** Find the function that expresses the radius of a circle as a function of its area. Use this function to determine the radius of the circle that has an area of 1 square inch, rounded to four decimal places.
70. **Radius and Volume of a Sphere** Find the function that expresses the radius of a sphere as a function of its volume. Use this function to determine the radius of the sphere that has a volume of 16 cubic inches.
71. **Accelerated Motion** Consider an object moving with nonzero constant acceleration a. If the initial velocity is u, its velocity v at time t is described by the function $v(t) = u + at$.

    (**a**) Find the inverse function that expresses t as a function of v.

    (**b**) If $u = 20$ feet per second and $a = 25$ feet per second^2, find the time at which the object is moving at 95 feet per second.

72. Motion Under Gravity The distance s feet that a freely falling object drops from rest in time t seconds is given by the function $s(t) = 16t^2$.

(a) Find the inverse function.

(b) The Sears Tower in Chicago is 1454 feet high. Use the inverse function to find how long it would take an object dropped from the top of the tower to reach the ground below.

73. Manufacturing Analysis General Products Ltd. has estimated that the total cost C in dollars for manufacturing x units of a certain product is described by the function $C(x) = 2x + 700$.

(a) Find the inverse function that expresses x in terms of C.

(b) Use the inverse function to find the number of items that can be manufactured for $1230.

74. Simple Interest Let an initial sum P be deposited in an account that pays simple interest at an annual rate r. At the end of t years, this will have grown to an amount A given by $A(t) = P(1 + rt)$.

(a) Find the inverse function of A.

(b) Use the inverse to find out how long it will take $2000 to increase to $2400 at 5%.

75. Weights at Different Altitudes The average radius of the Earth is 3956 miles. If the weight of a person on Earth is W_E pounds, then that person's weight $W(d)$ at a distance d miles above the Earth is known to be given by the function

$$W(d) = W_E\left(\frac{3956}{3956 + d}\right)^2$$

The Earth's gravity is considered to be effective to about 1 million miles from Earth. This function is valid for this region, which is called translunar space.

(a) Amy Cowan weighs 140 pounds. Sketch a rough graph of the function $W(d)$ for Amy. Give an explanation of how you arrived at the shape of the graph. Verify the shape of the graph on a calculator.

(b) Consider the graph of W for Amy. Use the graph to explain that the function $W(d)$ is one-to-one over the interval $[0, 1{,}000{,}000]$. Determine the inverse function. Use this inverse to find the distance from Earth at which Amy would weigh 1 pound.

Cryptography

76. Use the function $f(x) = 3x + 7$ of Example 10 to encode the message RETREAT.

77. Use the function $f(x) = 3x + 7$ of Example 10 to decode the message 13, 70, 82, 88, 67, 34, 88, 64, 67, 52, 16, 40.

78. Use the function $f(x) = 2x + 5$ to encode the message COME-HOME.

79. Use the function $f(x) = 2x + 5$ to decode the message 43, 11, 7, 45, 45, 15, 41.

CHAPTER 3 PROJECT | Recalling Ability

An experiment was conducted to determine whether one recalls better after a period of sleep or after a period of being awake. Subjects were given ten nonsense syllables to remember, either before going to bed or before normal daytime activity. The following functions were found to closely describe recall for periods of up to 8 hours for these subjects.

After a period of sleep: $f(x) = 0.04x^2 - x + 10$
After a period of being awake: $g(x) = 0.12x^2 - 2x + 10$

For example, $f(4) = 6.64$ and $g(4) = 3.92$ mean that on being awakened after 4 hours of sleep subjects remembered 6.64 syllables on average, whereas they remembered 3.92 syllables after 4 hours of normal daytime activity.

(a) Determine retention in each case after 8 hours.

(b) After some duration of time, the number of syllables retained during sleep is twice the number retained during daytime activities. Use tables for the functions to determine this period.

(c) Construct the function $h(x) = f(x) - g(x)$. What is the domain of this function? Graph this function on your calculator, and determine the value of x at which the maximum occurs. What information does this value of x tell you?

(d) Do these functions fit your retention patterns? Select partners in the class and investigate. Record data, and if these functions do not fit your patterns, find appropriate functions using linear regression. Compare your retention powers with others in the class. How can you improve your retention powers?

WEB PROJECT 3

To explore the concepts of this chapter further, try the interesting WebProject. Each activity relates the ideas and skills you have learned in this chapter to a real-world applied situation. Visit the Williams Web site at

www.saunderscollege.com/math/williams

to find the complete WebProject, related questions, and live links to the Internet.

Chapter 3 Highlights

Section 3.1

Vertical Line Test: A set of points in the plane is the graph of a function if and only if every vertical line intersects the graph in at most one point.

Linear Function: $f(x) = ax + b$, where a and b are real numbers. The graph is a line.

Quadratic Function: $f(x) = ax^2 + bx + c$, where a, b, and c are real numbers with $a \neq 0$. The graph is a parabola.

Polynomial Function of Degree n: $f(x) = a_n x^n + a_{n-1}x^{n-1} + \cdots + a_1 x + a_0$, with $a_n \neq 0$, and n is a nonnegative integer.

Intercepts: The y-intercept of the graph of $y = f(x)$ is $f(0)$. The x-intercepts of the graph of $y = f(x)$ are the roots of $f(x) = 0$. The x-intercepts are also called zeros of f.

Section 3.2

Increasing (Decreasing): f is increasing (decreasing) over an interval if graph is rising (descending) in going left to right over the interval.

Local Maximum (Minimum): $f(a)$ is greater (smaller) than the value of f at neighboring points.

Max/Min of a Polynomial Function of Degree n: The graph has at most $n - 1$ local maxima and minima.

Section 3.3

Square Root Function: $f(x) = \sqrt{x}$. The domain and range are both $[0, \infty)$.

Symmetry: Function is even if $f(-x) = f(x)$ for every x in the domain. An even function is symmetric with respect to the y-axis; e.g. $f(x) = x^2$. Function is odd if $f(-x) = -f(x)$ for every x in the domain. An odd function is symmetric with respect to the origin; e.g., $f(x) = x^3$.

Absolute Value Function: $f(x) = |x|$. The domain is $(-\infty, \infty)$, and the range is $[0, \infty)$.

Piecewise-Defined Function: Function is defined by different equations over different parts of its domain.

Section 3.4

Vertical Shifts: The graph of $g(x) = f(x) + c$, $c > 0$, can be obtained from that of $f(x)$ by an upward shift of c units. The graph of $g(x) = f(x) - c$, $c > 0$, can be obtained from that of $f(x)$ by a downward shift of c units.

Horizontal Shifts: The graph of $g(x) = f(x + c)$, $c > 0$, can be obtained from that of $f(x)$ by a horizontal shift of c units to the left. The graph of $h(x) = f(x - c)$, $c > 0$, can be obtained from that of $f(x)$ by a horizontal shift of c units to the right.

Reflections: The graph of $g(x) = -f(x)$ can be obtained from that of $f(x)$ by a reflection in the x-axis. Graph of $h(x) = f(-x)$ can be obtained from that of $f(x)$ by a reflection in the y-axis.

Vertically Stretching and Compressing: The graph of $g(x) = cf(x)$, $c > 1$, is the graph of $f(x)$ stretched vertically by a factor c. The graph of $h(x) = (1/c)f(x)$, $c > 1$, is the graph of $f(x)$ compressed vertically by a factor c.

Section 3.5

Rational Function: $f(x) = h(x)/g(x)$. f is in simplest form when any common factors of numerator and denominator have been divided out.

Vertical and Horizontal Asymptotes: Let f be a rational function. Line $x = a$ is a vertical asymptote of the graph of f if $f(x) \to \infty$ or $f(x) \to -\infty$ as $x \to a$

from the left or the right. Line $y = b$ is a horizontal asymptote if $f(x) \to b$ as $x \to \infty$ or as $x \to -\infty$.

Section 3.6

Operations on Functions: Sum $(f + g)(x) = f(x) + g(x)$. Difference $(f - g)(x) = f(x) - g(x)$. Product $(fg)(x) = f(x)g(x)$. Quotient $(f/g)(x) = f(x)/g(x)$, $g(x) \neq 0$. Composition $(g \circ f)(x) = g(f(x))$.

Logistic Function: $f(x) = cx(1 - x)$, with domain $0 \leq x \leq 1$, c constant. Such functions are used to describe population dynamics in the field of chaos. An orbit of populations may be periodic, approach a steady state, or be chaotic.

Section 3.7

Inverse Function: The function g is the inverse of f if $g(f(x)) = x$ for every x in the domain of f and $f(g(x)) = x$ for every x in the domain of g. We write f^{-1} for g. The range of f is the domain of g and *vice versa.*

Graphs of f and f^{-1}: The graphs of f and f^{-1} are symmetric about the line $y = x$.

Finding the Inverse of a Function:

(1) Write the function as $y = f(x)$.
(2) Interchange x and y.
(3) Solve the equation for y in terms of x. Replace y by the symbol $f^{-1}(x)$.
(4) The domain of f^{-1} is the range of f.

One-to-One Function: f is one-to-one if, for elements a and b in the domain of f, whenever $a \neq b$, then $f(a) \neq f(b)$.

Horizontal Line Test: A function is one-to-one if and only if every horizontal line cuts its graph in at most one point.

CHAPTER 3 REVIEW EXERCISES

1. Determine the x- and y-intercepts of the graphs of the following functions, and sketch the graphs using algebraic techniques. Support your answers by duplicating the graphs on your calculator.

(a) $f(x) = (x + 1)(x - 2)(x - 5)$
(b) $f(x) = x^2(x - 3)(x + 2)$
(c) $g(x) = x^3 + 3x^2 - 4x$

2. Construct a polynomial function of degree 3 with a graph having the following properties. Three x-intercepts at $x = -3$, 2, and 4, y-intercept $y = 48$, with f being negative on $(-\infty, -3)$ and $(2, 4)$ and positive on $(-3, 2)$ and $(4, \infty)$. Check your answers with a calculator display.

3. Graph the following functions on your graphing calculator. Find the intervals over which the functions are increasing, where they are decreasing, and the values (rounded to two decimal places) of the local maxima, minima, and zeros.

(a) $f(x) = 2x^3 + 5x^2 + 3x + 1$
(b) $f(x) = (x - 3)^4 + 5x - 17$

4. Advertising and Sales An advertising agency found that the sales of a certain product are described by the function

$$s(t) = t^3 - 89t^2 + 1{,}600t + 10{,}000 \text{ units}$$

t days after the end of an advertising campaign. The function is valid for $t \leq 40$. Graph the function and determine when sales (number of units of the product sold) will peak.

5. **Motion of a Rocket** A rocket is propelled upward. The height of the rocket in feet at 5-second intervals during the first 25 seconds is recorded here.

(a) Use cubic regression to find a function $s(t)$ that fits these data.

(b) Determine when the rocket reaches its maximum altitude and when it falls back to Earth (rounded to two decimal places).

(c) Give the intervals of time over which the height is increasing and when it is decreasing.

Time	5	10	15	20	25
Altitude	1952.5	3180.0	3757.5	3760.0	3262.5

6. Sketch the graph of each of the following functions. Give the range of each function. Support your results using a graphing calculator. Use the domain and the range as the x and y intervals of your window.

(a) $f(x) = 3x - 2$, with domain $[-3, 5]$

(b) $f(x) = -2x + 4$, with domain $[-5, 5]$

(c) $f(x) = 2x^2 + 8x + 3$, with domain $[-3, 1]$

7. Sketch the graphs of the following functions.

(a) $f(x) = \begin{cases} x + 3 & \text{if } x \leq 2 \\ -x - 1 & \text{if } x > 2 \end{cases}$

(b) $g(x) = \begin{cases} x^2 + 1 & \text{if } x < 2 \\ -x - 1 & \text{if } x \geq 2 \end{cases}$

8. **Pay of an Employee** Tompkins and Associates pays its employees \$8 an hour during a regular work day, 8 AM to 5 PM. The rate rises to overtime pay of \$12 an hour between 5 PM and midnight.

(a) Construct a function that describes the pay of an employee who works from 8 AM to midnight.

(b) Use the function to find the pay of someone who works from 8 AM to 7 PM.

9. **Incubation Times** H. Rahn and his associates ("Relation of Avian Egg Weight to Body Weight," *Auk* **92**: 750–765, 1975) have found that the incubation time (t_{inc}, in days) for birds has been found to be the following function of the body mass of the parent bird (M_b, in grams).

$$t_{\text{inc}} = 9.105(M_b)^{0.167}$$

The body sizes of birds are from about 2.5 grams for hummingbirds to 500 kilograms for the largest elephant bird. What is the range of incubation times (in days), to one decimal place?

10. (i) Use algebra to determine whether the following functions are even, odd, or neither by computing $f(-x)$. (ii) Sketch the graphs of the functions with the aid of a graphing calculator, using the window $X_{\min} = -4$, $X_{\max} = 4$, $X_{\text{scl}} = 1$, $Y_{\min} = -4$, $Y_{\max} = 4$, and $Y_{\text{scl}} = 1$. Indicate any symmetry in the graphs that supports your conclusion to part i.

(a) $f(x) = -2/x^2$ (b) $f(x) = 3$

(c) $g(x) = -4x^3$ (d) $h(x) = x^4 + x$

11. Use vertical and horizontal shifts, and reflections of known graphs to sketch the graphs of the following functions. State the known graph and transformations used. Check your results using a graphing calculator, with the window $X_{\min} = -10$, $X_{\max} = 10$, $Y_{\min} = -20$, and $Y_{\max} = 20$.

(a) $g(x) = x^2 - 3$ (b) $g(x) = (x - 2)^2 + 4$

(c) $g(x) = |x - 3| - 5$

(d) $g(x) = -(x - 3)^2 - 4$

(e) $g(x) = \sqrt{x - 1} + 3$

12. Graph the functions $f(x) = x^2$, $g(x) = x^2 - 4$, $h(x) = (x + 3)^2 + 2$ in the same coordinate system. Explain how the graphs of g and h can be obtained from the graph of f.

13. In the following exercises find the equations of the functions having the given graphs using transformations of $f(x) = x^3$. Each window is $X_{\min} = -5$, $X_{\max} = 5$, $X_{\text{scl}} = 1$, $Y_{\min} = -30$, $Y_{\max} = 30$, and $Y_{\text{scl}} = 5$. Check your answer by duplicating the display on your calculator.

(a)
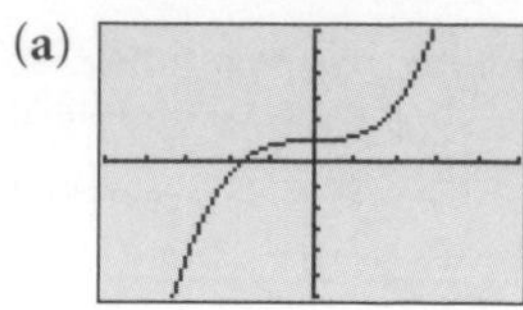

(b)
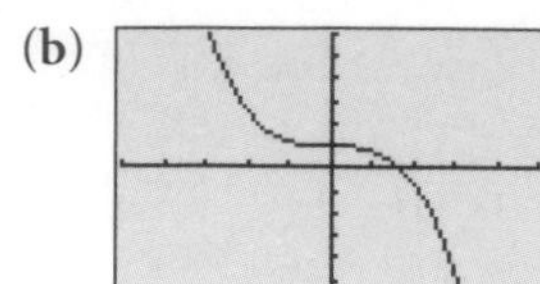

(c)
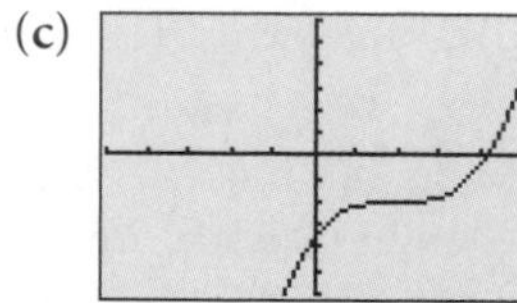

14. **(a)** Consider the following tables of $f(x) = x^2$ and $g(x) = (x - 1)^2$. Explain how these tables are related.

(b) Consider the tables of $f(x) = x^2$ and $h(x)$. What is the function $h(x)$?

(c) Consider the tables of $f(x) = x^2$ and $k(x)$. What is the function $k(x)$? Check your answers by duplicating the tables on your calculator.

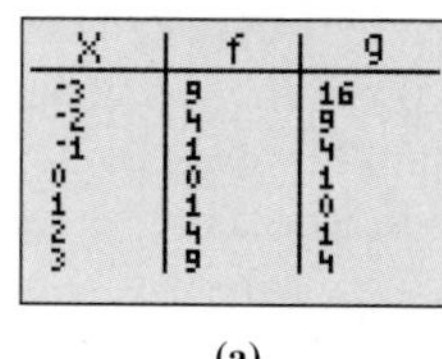

X	f	g
-3	9	16
-2	4	9
-1	1	4
0	0	1
1	1	0
2	4	1
3	9	4

(a)

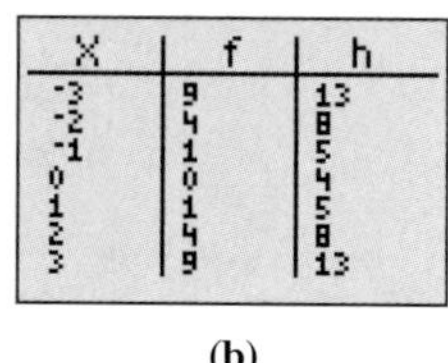

X	f	h
-3	9	13
-2	4	8
-1	1	5
0	0	4
1	1	5
2	4	8
3	9	13

(b)

X	f	k
-3	9	20
-2	4	13
-1	1	8
0	0	5
1	1	4
2	4	5
3	9	8

(c)

15. Relative Motion As you drive on the interstate, you spot a police car in your mirror. The police car gradually catches up with you, overtakes you, and disappears ahead of you. Sketch a graph of the distance between you and the police car as a function of time, from the moment it appears to the time it disappears.

16. Sales and Prices As the price of an item increases, fewer and fewer items are bought by consumers. Sketch a graph of number of items purchased as a function of price.

17. Parking Fees A college has a parking garage that accomodates 150 cars. The garage was full when no parking fee was charged. The administration decided to charge for parking. Naturally, the number of people using the garage decreased as the price increased. The number of people using the garage is a function of the parking fee. Construct a function that possibly describes the situation.

18. Traveling on an Interstate Highway Tom Smith leaves Dallas for Houston on Interstate 45 at 9:00 AM. He travels at an average speed of 60 miles per hour. Sally Jones leaves Dallas for Houston at 9:15 AM and travels at 75 miles per hour. Houston is 243 miles from Dallas. Construct functions that describe both trips. Use a graphing calculator to find out when Sally catches up with Tom. How far were they from Dallas, when Sally caught up with Tom?

19. Sketch the graphs of the following functions. Determine the intervals over which the functions are increasing, decreasing, or constant.

(a) $f(x) = x^2 + 1$ **(b)** $f(x) = -x^3 + 3$

(c) $g(x) = |x - 4|$

(d) $f(x) = \begin{cases} 5 & \text{if } x \leq 4 \\ x - 2 & \text{if } x > 4 \end{cases}$

(e) $g(x) = \begin{cases} -x + 1 & \text{if } x \leq -4 \\ x^2 + 1 & \text{if } x \geq -2 \end{cases}$

20. Determine the domains of the following rational functions.

(a) $f(x) = \dfrac{3x^2 - 4x}{x}$ **(b)** $f(x) = \dfrac{x - 2}{x^2 + 2x - 8}$

21. Give equations of any vertical or horizontal asymptotes and the domains, and sketch the graphs of the following rational functions. Support your answers by duplicating the graphs on your calculator.

(a) $f(x) = \dfrac{x^2 + 2x - 3}{x - 1}$

(b) $f(x) = \dfrac{2}{x - 4} + 3$

(c) $f(x) = \dfrac{5}{x^2 + x - 6}$

(d) $g(x) = \dfrac{x - 1}{(x + 2)(x - 4)} + 4$

22. Find the sum, difference, product, and quotient of each of the following pairs of functions. Give the domain of each function. Verify the domains by graphing the functions on a calculator.

(a) $f(x) = 3x - 1,\ g(x) = -4x + 5$

(b) $f(x) = 4x + 1,\ g(x) = 2\sqrt{3x + 4}$

(c) $f(x) = x^2 + 3x + 1,\ g(x) = 1/x^2$

23. Find $(f \circ g)(x)$ and $(g \circ f)(x)$ for the following pairs of functions. Give the domain of each function.

Verify the domains by graphing the functions on a calculator.

(a) $f(x) = 3x + 5$, $g(x) = 2x - 7$

(b) $f(x) = 4x + 1$, $g(x) = 3x^2 + 4x - 1$

(c) $f(x) = \sqrt{x + 4}$, $g(x) = x^2 - 6x - 11$

24. Express $F(x) = -12x + 7$ as a composite of the functions $f(x) = 3x - 1$ and $g(x) = -4x + 3$.

25. Express $H(x) = 20x^2 + 8x - 2$ as a composite of the functions $f(x) = 4x + 2$ and

$$g(x) = 5x^2 + 2x - 1$$

26. Show that $g(x) = (x - 3)/4$ is the inverse of $f(x) = 4x + 3$ using algebra.

27. Determine whether the following functions are one-to-one using algebra. Support your answer by drawing a graph and using the horizontal line test.

(a) $f(x) = 5x - 1$ (b) $f(x) = x^3 + 1$

(c) $f(x) = 2x^2 - 4x - 1$

28. Each of the following functions is one-to-one. Find the inverse. Give its domain. Check your answer by graphing f, its inverse, and the line $y = x$ on a calculator and looking for symmetry about line $y = x$.

(a) $f(x) = 7x - 4$ (b) $f(x) = 4 - x^2$, $x \geq 0$

CHAPTER 3 TEST

1. Determine the x- and y-intercepts of

$$f(x) = (x + 2)(x - 4)(x - 6)$$

and sketch the graph using algebraic techniques. Support your answer by duplicating the graph on a calculator.

2. Sketch the graph of the function

$$f(x) = x^2 + 6x + 1$$

with domain $[-9, 4]$. Give the range of f. Support your results using a graphing calculator. Use the domain and the range as the x and y intervals of your window.

3. Use algebra to show that the function

$$f(x) = x^4 - 3x^2$$

is an even function. Sketch the graph of f with the aid of a graphing calculator, using the window $X_{min} = -4$, $X_{max} = 4$, $X_{scl} = 1$, $Y_{min} = -4$, $Y_{max} = 4$, and $Y_{scl} = 1$. Indicate any symmetry in the graph that supports your conclusion that f is even.

4. Graph the functions

$$f(x) = x^2, \qquad g(x) = (x - 4)^2 + 3$$

in the same coordinate system. Explain how the graph of g can be obtained from the graph of f using translations.

5. Consider the following tables of $f(x) = x^2$ and $g(x)$. What is the function $g(x)$? Check your answer by duplicating the table on your calculator.

X	f	g
-3	9	1
-2	4	0
-1	1	1
0	0	4
1	1	9
2	4	16
3	9	25

6. Give the equations of any vertical or horizontal asymptotes, identify the domain, and sketch the graph of the function

$$f(x) = \frac{2}{x^2 - 3x - 4}$$

Support your answer by duplicating the graph on your calculator.

7. Graph $f(x) = 2x^3 + 3x^2 - 5x + 4$ on your graphing calculator. Find the intervals over which f is increasing, where it is decreasing, the values of the local maxima and minima, and the zeros, rounded to two decimal places.

8. Find the sum, difference, product, and quotient of the following pair of functions: $f(x) = 3x - 1$, $g(x) = \sqrt{x - 2}$. Give the domain of each function.

9. Find the composite function $(g \circ f)(x)$ of $f(x) = \sqrt{x - 3}$, $g(x) = x^2 + 5x + 7$. Give the domain of this function.

10. Show that the function $f(x) = 3x + 2$ is one-to-one. Find its inverse.

11. Consumer Awareness A company has found that, after its product is advertised on television, consumer awareness of the product is described by the following function of time t (in days):

$$A(t) = \begin{cases} -t/3 + 25 & \text{if } 0 \leq t \leq 75 \\ 0 & \text{if } t > 75 \end{cases}$$

(a) Sketch the graph of $A(t)$.

(b) The company does not want awareness to drop below a level of 18. Determine the function $D(t)$ that describes the desired advertising procedure and sketch the graph.

(c) How often should the commercial be shown?

12. Deer Population The number of deer in the Ocala National Forest in Florida varies in population. It has been estimated that the population between 1990 and 2000 was described by the function

$$N(t) = 2t^3 - 32t^2 + 83t + 1000$$

where $t = 0$ corresponds to 1990. When in this period did the population peak, and when was it at its lowest?

13. Manufacturing Costs A company has estimated that the total cost C in dollars for manufacturing x units of a certain product is described by the function $C(x) = 6.5x + 520$. Find the inverse function that expresses x in terms of C. Use it to find the number of items that can be manufactured for \$17,225.

Further Theory of Polynomials

Carl Friedrich Gauss (1777–1855) *(© Library of Congress)*

Polynomial functions are among the most useful functions in mathematics. For example, we have already seen that the motion of an object under gravity is described by a polynomial function. It behooves us to know as much as we can about polynomials. This chapter is devoted primarily to furthering the student's understanding of polynomials.

So far the numbers we have used in this course have been real numbers. However, we have occasionally seen situations in which the set of real numbers was inadequate. Recall that, in the discussion of roots to a quadratic equation, we were confronted by $\sqrt{-1}$. This is not a real number. It is called an imaginary number and is denoted i. In this chapter, we formally extend the set of real numbers to a larger set of numbers of the form $a + bi$, where a and b are both real numbers. These numbers are called complex numbers. They were studied and used by one of the greatest mathematicians of all time, Carl Friedrich Gauss (1777–1855). One of the legends about Gauss is that he detected an error in his father's bookkeeping at the age of 3! It is recorded that he had a strong will, and that his character showed a curious mixture of self-conscious dignity and childlike simplicity. He was not very communicative and at times was morose. Gauss spent his professional life as director of the observatory at the University of Göttingen, Germany.

We shall use complex numbers to gain insight into roots of polynomials. Complex numbers are used extensively in fields such as electrical engineering and physics.

APPLICATION | Fractal Geometry

A fractal. Such beautiful images are generated on a computer using a mathematical procedure called iteration. *(© Bruce Bauslaugh, University of Calgary)*

Computer graphics systems based on traditional Euclidean geometry are suitable for creating pictures of man-made objects, such as machinery, buildings, and airplanes. Images of such objects can be created using lines, circles, and other geometric figures. However, these techniques are not appropriate when it comes to constructing images of natural objects, such as animals, trees, and landscapes. In 1975, Benoit B. Mandelbrot, a research fellow at IBM, introduced a new geometry, which he called *fractal geometry,* that can be used to describe natural phenomena. Fractal is a convenient label for irregular and fragmented self-similar shapes. The story behind the word *fractal* is interesting. Mandelbrot came across the Latin adjective *fractus,* from the verb *frangere,* to break, in his son's Latin book. The resonance of the main English cognates *fracture* and *fraction* seemed appropriate; consequently, he coined the word *fractal.* Fractal objects contain structures nested within one another. Each smaller structure is a miniature, although not necessarily identical, version of the larger form. In this section the student is introduced to the Mandelbrot set, a fractal named after Benoit B. Mandelbrot. *(This application is discussed in Section 4.2, pages 337–339, and in the Chapter Project, page 354.)*

4.1 Synthetic Division, Zeros, and Factors

- Long Division, Quotient, and Remainder • The Division Algorithm
- The Remainder Theorem • Synthetic Division • Zeros of Polynomials
- Factors of Polynomials and the Factor Theorem

LONG DIVISION, QUOTIENT, AND REMAINDER

In this section, we analyze the method of long division of polynomials. This analysis leads to a shortcut of this method called synthetic division. Synthetic division is used to gain further insight into the zeros and factors of polynomials.

Let us divide the polynomial $2x^3 - 7x^2 + 5x + 2$ by $x - 3$ and look closely at the form the result takes. It is important that both polynomials be written with exponents in descending order, as here, before starting to divide.

$$
\begin{array}{r l}
2x^2 - x + 2 & \textbf{Divide } 2x^3/x \textbf{ to get } 2x^2. \\
x - 3\overline{)2x^3 - 7x^2 + 5x + 2} & \\
\underline{2x^3 - 6x^2} & \textbf{Multiply } 2x^2(x-3). \\
-\ x^2 + 5x & \textbf{Subtract and bring down } 5x. \textbf{ Divide } -x^2/x \textbf{ to get } -x. \\
\underline{-\ x^2 + 3x} & \textbf{Multiply } -x(x-3). \\
2x + 2 & \textbf{Subtract and bring down 2. Divide } 2x/x \textbf{ to get 2.} \\
\underline{2x - 6} & \textbf{Multiply } 2(x-3). \\
8 & \textbf{Subtract.} \\
 & \textbf{The degree of 8 is less than the degree of } x - 3. \\
 & \textbf{This marks the end of the division.}
\end{array}
$$

The quotient is $2x^2 - x + 2$ and the remainder is 8. The result is written

$$\frac{2x^3 - 7x^2 + 5x + 2}{x - 3} = 2x^2 - x + 2 + \frac{8}{x - 3}$$

THE DIVISION ALGORITHM

Let us generalize this result. Let $f(x)$ be a polynomial, and let us divide $f(x)$ by another nonzero polynomial $g(x)$, which is of degree less than $f(x)$, using long division. The process of division continues until a quotient $q(x)$ is obtained with remainder $r(x)$, which is either zero or is a polynomial of degree less than $g(x)$, as previously. We can write

$$\frac{f(x)}{g(x)} = q(x) + \frac{r(x)}{g(x)}$$

Multiply both sides of this identity by $g(x)$.

$$f(x) = g(x)q(x) + r(x)$$

We have arrived at a result that is called the **Division Algorithm.**

DIVISION ALGORITHM

Let $f(x)$ and $g(x)$ be two polynomials with $g(x)$ being a nonzero polynomial of degree less than $f(x)$. Then there exist unique polynomials $q(x)$ and $r(x)$ such that

$$f(x) = g(x)q(x) + r(x)$$

where $r(x)$ is either the zero polynomial or a polynomial of degree less than $g(x)$. $f(x)$ is the **dividend,** $g(x)$ is the **divisor,** $q(x)$ is the **quotient,** and $r(x)$ is the **remainder.**

THE REMAINDER THEOREM

In this section, we will be primarily concerned with applying the Division Algorithm when the divisor is a polynomial of the form $x - k$, where k is a constant. Because the remainder will be zero or of degree less than the divisor, the remainder will be a constant. From the Division Algorithm

$$f(k) = (x - k)q(x) + \text{constant.}$$

Let $x = k$ in this equation.

$$f(k) = (k - k)q(k) + \text{constant}$$
$$f(k) = \text{constant}$$

This proves the following result:

Remainder Theorem

If a polynomial $f(x)$ is divided by $x - k$, then the remainder is $f(k)$.

$$f(x) = (x - k)q(x) + f(k)$$

There is a shortcut of the long division method called synthetic division that can be used when the divisor is a function of the form $x - k$. The method can be justified by comparing it with long division.

Synthetic Division

We explain the method of synthetic division in terms of polynomials of degree 3. The pattern can then be extended to polynomials of higher degree. Consider the polynomial $f(x) = ax^3 + bx^2 + cx + d$ and the divisor $g(x) = x - k$. Display the polynomial in terms of its coefficients a, b, c, and d and the divisor as k. Perform the following steps, in the sequence indicated by the arrows.

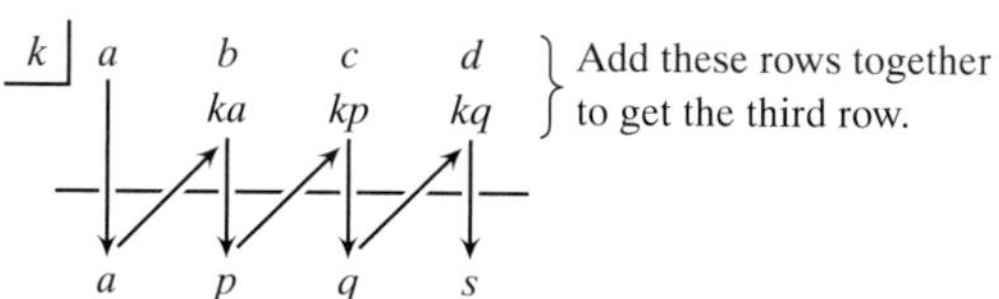

The first three numbers in the last row will be the coefficients of the quotient and the last number is the remainder $f(k)$.

Quotient: $q(x) = ax^2 + px + q$

Remainder: $f(k) = s$

Example 1 Quotient, Remainder, and Function Value by Synthetic Division

Use synthetic division to find the quotient and remainder when

$$f(x) = 2x^3 - 7x^2 + 5x + 2$$

is divided by $g(x) = x - 3$. Use the remainder to compute $f(3)$.

SOLUTION We get

$$\begin{array}{r|rrrr} 3 & 2 & -7 & 5 & 2 \\ & & 6 & -3 & 6 \\ \hline & 2 & -1 & 2 & 8 \end{array}$$

The first three numbers in the last line are the coefficients of the quotient $q(x)$, and the last number is the remainder. The quotient is thus $q(x) = 2x^2 - x + 2$, and the remainder is 8. By the Remainder Theorem $f(3) = 8$.

Example 2 **Quotient, Remainder, and Function Value by Synthetic Division**

(a) Use synthetic division to find the quotient and remainder when

$$f(x) = 3x^4 - 2x^2 + 4x - 3$$

is divided by $x + 2$. Use the remainder to compute $f(-2)$.

(b) Use substitution to verify the value of $f(-2)$.

SOLUTION

(a) In this example, we see how synthetic division can be extended to polynomials of higher degree than 3. Observe also that there is no x^3 in

$$f(x) = 3x^4 - 2x^2 + 4x - 3$$

The coefficient of x^3 is entered as being zero.

$$\begin{array}{r|rrrrr} -2 & 3 & 0 & -2 & 4 & -3 \\ & & -6 & 12 & -20 & 32 \\ \hline & 3 & -6 & 10 & -16 & 29 \end{array}$$

The first four numbers in the last line are the coefficients of the quotient, and the last number is the remainder; thus, the quotient is

$$q(x) = 3x^3 - 6x^2 + 10x - 16$$

and the remainder is 29. The Remainder Theorem gives $f(-2) = 29$.

(b) Using substitution,

$$\begin{aligned} f(-2) &= 3(-2)^4 - 2(-2)^2 + 4(-2) - 3 \\ &= 3(16) - 2(4) - 8 - 3 \\ &= 48 - 8 - 8 - 3 \\ &= 29 \end{aligned}$$

We see that $f(-2) = 29$, confirming the result obtained by synthetic division.

Self-Check 1 Use synthetic division to find the quotient and remainder when $f(x) = 2x^4 - 3x^3 + 4x - 7$ is divided by $x + 1$. Determine $f(-1)$.

ZEROS OF A POLYNOMIAL

We remind the student that the number k is a **zero** of a polynomial function $f(x)$ if $f(k) = 0$. The real-number zeros of $f(x)$ are the points at which the graph of $y = f(x)$ cuts the x-axis. Because $f(x)$ is the remainder after dividing $f(x)$ by $x - k$, we get the following result.

ZEROS OF A POLYNOMIAL

The number k is a zero of a polynomial $f(x)$ if and only if the remainder on dividing $f(x)$ by $x - k$ is zero.

Example 3 **Verifying a Zero Using Synthetic Division**

Show that 5 is a zero of $f(x) = x^5 - 4x^4 + x^3 - 26x^2 - 22x + 10$. Use a graphing display on a calculator to support your algebraic conclusions.

SOLUTION Divide $f(x)$ by $x - 5$ using synthetic division.

$$\begin{array}{r|rrrrrr} 5 & 1 & -4 & 1 & -26 & -22 & 10 \\ & & 5 & 5 & 30 & 20 & -10 \\ \hline & 1 & 1 & 6 & 4 & -2 & 0 \end{array}$$

The remainder on dividing $f(x)$ by $x - 5$ is zero. Thus, 5 is a zero of $f(x)$. The graph in Figure 4.1 confirms this result.

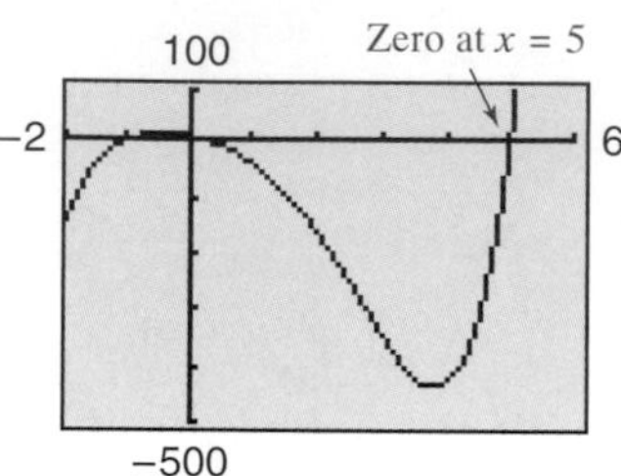

Figure 4.1
$f(x) = x^5 - 4x^4 + x^3 - 26x^2 - 22x + 10$

Self-Check 2 Show that -2 is a zero of $f(x) = 3x^4 + 2x^3 - 7x^2 - x - 6$. Use a graphing display on a calculator to support your algebraic conclusions.

FACTORS OF A POLYNOMIAL

$x - k$ is a **factor** of the polynomial $f(x)$ if there exists a polynomial $q(x)$ such that $f(x) = (x - k)q(x)$. The Remainder Theorem leads to the following result about factors.

FACTOR THEOREM

The polynomial $x - k$ is a factor of the polynomial $f(x)$ if and only if $f(k) = 0$.

Thus, $x - k$ is a factor of $f(x)$ if and only if k is a zero of $f(x)$. Factors lead to zeros and vice versa. Synthetic division can be used to determine whether certain polynomials are factors of others.

Example 4 **Checking for a Factor Using Synthetic Division**

Determine whether $x - 7$ is a factor of $f(x) = 2x^3 - 14x^2 + 2x - 14$.

SOLUTION Divide $2x^3 - 14x^2 + 2x - 14$ by $x - 7$ to see whether the remainder is zero.

$$\begin{array}{r|rrrr} 7 & 2 & -14 & 2 & -14 \\ & & 14 & 0 & 14 \\ \hline & 2 & 0 & 2 & 0 \end{array}$$

The remainder is zero; thus, $f(7) = 0$, telling us that $x - 7$ is a factor of f. The graph in Figure 4.2 confirms this result.

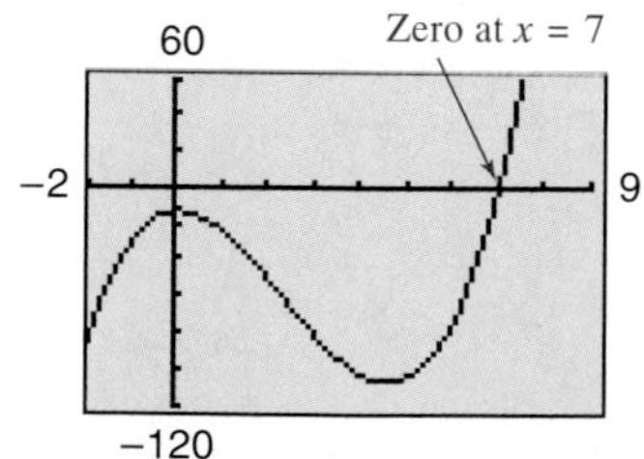

Figure 4.2
$f(x) = 2x^3 - 14x^2 + 2x - 14$

Self-Check 3 Determine whether $x + 4$ is a factor of

$$f(x) = x^4 + 5x^3 + 3x^2 - 6x - 7$$

Use a graph to support your algebraic conclusions.

Answers to Self-Check Exercises

1. Note that the coefficient of x^2 is zero. Write

$$f(x) = 2x^4 - 3x^3 + 0x^2 + 4x - 7$$

$q(x) = 2x^3 - 5x^2 + 5x - 1$, remainder $= -6$; thus $f(-1) = -6$.

3. $x + 4$ is not a factor; there is a remainder of 1 after dividing. The graph shows that $x = -4$ is not a zero of f.

EXERCISES 4.1

In Exercises 1–8, use synthetic division to find the quotient and remainder when the polynomial $f(x)$ is divided by the polynomial $g(x)$.

1. $f(x) = 2x^3 + 3x^2 + x + 3$, $g(x) = x - 2$
2. $f(x) = x^3 - 3x^2 + 2x + 1$, $g(x) = x - 4$
3. $f(x) = -2x^3 + x^2 + 4x - 2$, $g(x) = x + 1$
4. $f(x) = 4x^3 + 2x^2 + 5x + 8$, $g(x) = x + 4$
5. $f(x) = x^4 + 2x^3 - 3x^2 + 8x - 7$, $g(x) = x - 5$
6. $f(x) = -3x^3 + 5x$, $g(x) = x - 7$
7. $f(x) = x^4 + 2x^3 + 7x + 3$, $g(x) = x + 6$
8. $f(x) = 2x^5 + 5x^4 - 3x^3 - 4x^2 + x + 1$, $g(x) = x - 1$

In Exercises 9–16, use synthetic division to find the quotient and remainder when the polynomial $f(x)$ is divided by the polynomial $g(x) = x - k$. Use the remainder to compute $f(k)$.

9. $f(x) = 3x^3 - 4x^2 + x + 1$, $g(x) = x - 1$
10. $f(x) = x^3 + 2x + 3$, $g(x) = x - 3$
11. $f(x) = 4x^3 + 2x^2 + 3x + 8$, $g(x) = x + 2$
12. $f(x) = x^3 + 3x^2 - 5x - 2$, $g(x) = x - 2$
13. $f(x) = x^4 + 2x^3 - 2x^2 + 6x + 7$, $g(x) = x + 1$
14. $f(x) = -x^3 + 2x^2 + 3$, $g\ (x) = x - 6$
15. $f(x) = 3x^4 - 4x^3 - x^2 + 2x + 4$, $g(x) = x + 8$
16. $f(x) = x^5 - 6x^4 + 5x^3 + x^2 + 3x - 9$, $g(x) = x - 9$

In Exercises 17–24, use synthetic division with an appropriate divisor to compute the value of f.

17. $f(x) = 4x^3 - x^2 + 2x + 4$. Compute $f(1)$.
18. $f(x) = 4x^3 - x^2 + 2x + 4$. Compute $f(3)$.
19. $f(x) = 4x^3 - x^2 + 6x + 3$. Compute $f(-2)$.
20. $f(x) = x^3 - 2x^2 + 3x + 1$. Compute $f(-1)$.
21. $f(x) = -2x^3 + 3x^2 + 2x - 2$. Compute $f(-2)$.
22. $f(x) = 2x^4 - 3x^3 + x^2 - 2x - 4$. Compute $f(4)$.
23. $f(x) = 7x^3 + x$. Compute $f(-2)$.
24. $f(x) = 3x^5 + 2x^4 - 4x^3 - x^2 + 5x + 12$. Compute $f(6)$.

In Exercises 25–32, use synthetic division to determine whether r is a zero of the polynomial $f(x)$. Use a graphing display on a calculator to support your algebraic conclusions.

25. $f(x) = x^3 - x^2 - 2x + 8$, $r = -2$
26. $f(x) = -x^3 - x^2 + 7x - 20$, $r = -4$
27. $f(x) = x^3 - 4x^2 + 3x + 2$, $r = 2$
28. $f(x) = 7x^3 + x^2 + 5x + 1$, $r = 5$
29. $f(x) = 4x^3 - 9x^2 - 8x + 3$, $r = 3$
30. $f(x) = x^3 - 8x^2 + 21x - 20$, $r = 4$
31. $f(x) = 3x^5 - 8x^3 - 6x^2 + x - 10$, $r = 2$
32. $f(x) = 7x^4 + 2x^2 - 3x + 7$, $r = -5$

In Exercises 33–40, use synthetic division to determine whether the second polynomial is a factor of $f(x)$. Use a graph to support your algebraic conclusions.

33. $f(x) = x^3 - 4x^2 + 9x - 10$, $x - 2$
34. $f(x) = x^3 - 7x^2 + 4x + 24$, $x - 3$

35. $f(x) = 5x^3 - 20x^2 + 5x + 30,\ x - 3$

36. $f(x) = x^4 + 7x^3 + 12x^2 - 3x - 18,\ x + 3$

37. $f(x) = 4x^3 - 4x^2 + 7x + 1,\ x + 4$

38. $f(x) = x^3 - 8x^2 + 17x - 10,\ x - 5$

39. $f(x) = 8x^4 - 2x^3 + 7x^2 - 2x - 1,\ x - \frac{1}{2}$

40. $f(x) = 2x^3 + 9x^2 - 17x + 6,\ x + 6$

41. Use synthetic division to determine a value of k for which $x + 1$ is a factor of $kx^3 + 2x^2 - 3x + 4$.

42. Use synthetic division to determine a value of k for which $x - 3$ is a factor of $2x^3 - kx^2 + x - 3$.

43. Use synthetic division to determine a value of k for which $x - 5$ is a factor of $kx^4 + 2x^3 - 20kx^2 + 25x$.

44. Determine a value of k for which 3 is a zero of $2x^3 - kx^2 + 9x - 24$.

45. Determine a value of k for which -2 is a zero of $kx^4 + x^3 - 3x^2 + 5x - 5$.

46. Compute $f(2)$ for the polynomial

$$f(x) = 3x^2 - 4x + 1$$

(a) using synthetic division, and **(b)** using substitution.

47. Compute $f(-3)$ for the polynomial

$$f(x) = 2x^3 + 5x^2 - 2x - 4$$

(a) using synthetic division and **(b)** using substitution.

48. Compute $f(4)$ for the polynomial

$$f(x) = 3x^4 - 2x^3 + 7x^2 - 3x - 1$$

(a) using synthetic division and **(b)** using substitution.

4.2 Complex Numbers and the Mandelbrot Set

• Imaginary Numbers • Powers of i • Complex Numbers • Complex Roots of a Polynomial • Addition, Subtraction, Multiplication, and Division of Complex Numbers • Magnitude of a Complex Number • Fractals

Imaginary Numbers

The numbers we have used in this course thus far have been real numbers. However, we have occasionally seen situations in which the set of real numbers was inadequate. For example, $\sqrt{-1}$ arose in the discussion of roots to a quadratic equation. This is not a real number because there is no real number x such that $x^2 = -1$. In this section, we will formally extend the set of real numbers to a larger set called the complex numbers. We shall use complex numbers to discuss roots of polynomials.

We commence the construction of complex numbers by looking at $\sqrt{-1}$. Let us introduce a new number, denoted i, to stand for $\sqrt{-1}$. The French mathematician René Descartes called this number an imaginary number in the 17th century. The name has remained. The letter i for $\sqrt{-1}$ goes back to the Swiss mathematician Leonhard Euler in the 18th century.

> ***The Imaginary Number i*** The **imaginary number *i*** is defined by $i = \sqrt{-1}$, with $i^2 = -1$.

We can add this imaginary number to itself any number of times. For example,

$$i + i + i = 3i$$

We can multiply i by any real number. For example, we write

$$7 \cdot i = 7i$$

Any such number is also called an imaginary number.

Imaginary Numbers If c is a nonzero real number, then any number of the form ci is called an **imaginary number.**

For example, $4i$, $5.7i$, $-3.96i$, and πi are all imaginary numbers. We are now in a position to define the square roots of negative real numbers in terms of imaginary numbers.

Radicals and Imaginary Numbers

Let a be a positive number. Then $\sqrt{-a}$ is not a real number because there is no real number whose square is $-a$. However, there is an imaginary number whose square is $-a$. Consider the imaginary number $\sqrt{a}\,i$. Using the properties of exponents we get

$$\begin{aligned}(\sqrt{a}\,i)^2 &= (\sqrt{a})^2(i)^2\\ &= (a)(-1)\\ &= -a\end{aligned}$$

Thus, $\sqrt{a}\,i$ is a square root of $-a$. It is called the **principal square root** of $-a$. Additionally, $-\sqrt{a}\,i$ is also a square root of $-a$ since $(-\sqrt{a}\,i)^2$ is also equal to $-a$.

This discussion leads to the following definition for the square root of a negative number.

Square Root of a Negative Number

Let a be a positive real number. Then

$$\sqrt{-a} = \sqrt{a}\,i.$$

Example 1 Simplifying Radicals

Simplify each of the following: **(a)** $\sqrt{-9}$ **(b)** $\sqrt{-144}$

SOLUTION

(a) $\sqrt{-9} = \sqrt{9}i = 3i$
(b) $\sqrt{-144} = \sqrt{144}\,i = 12i$

Caution: Not all the rules for radicals when roots are real extend to imaginary numbers. The rule $\sqrt{a}\sqrt{b} = \sqrt{ab}$, which may be used for positive real numbers

a and b, does not apply if a and b are negative. We illustrate the correct and incorrect ways of computing such a product:

Correct $\sqrt{-2} \cdot \sqrt{-8} = \sqrt{2}i \cdot \sqrt{8}i = (\sqrt{2} \cdot \sqrt{8})i^2 = \sqrt{16}(-1) = -4$
Incorrect $\sqrt{-2} \cdot \sqrt{-8} = \sqrt{(-2)(-8)} = \sqrt{16} = 4$

POWERS OF i

If we compute powers of i up to the eighth power, we observe that a pattern develops.

$i^1 = i$ $\qquad i^5 = i^4 \cdot i = 1 \cdot i = i$

$i^2 = -1$ $\qquad i^6 = i^5 \cdot i = i \cdot i = -1$

$i^3 = i^2 \cdot i = (-1)i = -i$ $\qquad i^7 = i^6 \cdot i = (-1)i = -i$

$i^4 = i^2 \cdot i^2 = (-1)(-1) = 1$ $\qquad i^8 = i^7 \cdot i = (-i)i = -i^2 = 1$

Observe that the value of i^5 is the same as that of i^1, that the value of i^6 is the same as that of i^2 and so on. Any positive integer power i^n of i is given by one of i, i^2, i^3, or i^4. Define $i^0 = 1$ to fit into this scheme. To find which one, divide n by 4 and use the remainder as the exponent. Let us see why this works.

Let m be the divisor and r the remainder after dividing n by 4. Thus, $n = 4 \cdot m + r$. We now get

$$i^n = i^{(4 \cdot m)+r} = i^{(4 \cdot m)} \cdot i^r = (i^4)^m \cdot i^r = 1 \cdot i^r = i^r$$

As an example, let us compute i^{14}. After dividing 14 by 4, we get

$$14 = (3 \cdot 4) + 2$$

The remainder is 2, thus, $i^{14} = i^2 = -1$.

POWERS OF i

Let n be a positive integer. Let r be the remainder on dividing n by 4. Then $i^n = i^r$.

Self-Check 1 Determine i^{37}.

COMPLEX NUMBERS

We are now in a position to combine the set of imaginary numbers with the set of real numbers to form a set of numbers called complex numbers.

Complex Numbers Let a and b be real numbers. A number of the form

$$a + bi$$

is called a **complex number.** a is called the **real part,** and b the **imaginary part** of the complex number.

Note that any real number is of the form $a + bi$, with $b = 0$. Thus, the real numbers form a subset of the set of complex numbers. Furthermore, any imaginary number is of this form, with $a = 0$ and $b \neq 0$. The imaginary numbers also form a subset of the set of complex numbers.

Example 2 **Analyzing Complex Numbers**

(a) $2 + 3i$ is a complex number. The real part is 2 and the imaginary part is 3.

(b) 6 is a real number. It is also a complex number having real part 6 and imaginary part 0.

(c) $4i$ is an imaginary number. It is also a complex number having real part 0 and imaginary part 4.

Example 3 **Complex Roots of a Polynomial**

Let us return to the quadratic equation $x^2 + 4x + 5 = 0$ of Example 8, Section 2.1. On comparison with the standard form of a quadratic equation

$$ax^2 + bx + c = 0$$

we see that $a = 1$, $b = 4$, and $c = 5$. The quadratic formula gives

$$\begin{aligned} x &= \frac{-b \pm \sqrt{b^2 - 4ac}}{2a} \\ &= \frac{-4 \pm \sqrt{4^2 - 4(1)(5)}}{2(1)} \\ &= -2 \pm \sqrt{-1} \end{aligned}$$

At the time in Section 2.2, we concluded correctly that the equation has no real number solutions because $\sqrt{-1}$ is not a real number. We can now, however, write the roots of the quadratic equation in the form $x = -2 \pm i$. The quadratic equation has two complex roots.

Further Ideas 1

(a) Determine the roots of the quadratic equation $x^2 + 2x + 26 = 0$.

(b) Use the quadratic formula to prove that if a quadratic equation has complex roots, then they occur in pairs of the form $x = p + qi$ and $x = p - qi$, where p and q are real numbers. ($x = p - qi$ is called the complex conjugate of $x = p + qi$. We discuss complex conjugates later in this section.)

Addition, Subtraction, Multiplication, and Division of Complex Numbers

We introduced operations of addition, subtraction, and division of real numbers in Chapter R and have used them throughout the course. It is natural at this time to introduce operations of addition, subtraction, multiplication, and division for complex numbers. Because real numbers and imaginary numbers are subsets of the set of complex numbers, the operations that we introduce are consistent with the operations that we have already discussed for those sets. We first give a concept of equality of complex numbers that leads to definitions of operations.

Equality of Complex Numbers

Let $a + bi$ and $c + di$ be two complex numbers. Then

$$a + bi = c + di \text{ if and only if } a = c \text{ and } b = d$$

Thus, two complex numbers are equal if and only if their real parts are equal and their imaginary parts are equal. For example, if

$$a + bi = 2 + 4i \qquad \text{then} \qquad a = 2 \text{ and } b = 4$$

Addition and Subtraction of Complex Numbers

Let $a + bi$ and $c + di$ be two complex numbers. Then

$$(a + bi) + (c + di) = (a + c) + (b + d)i$$
$$(a + bi) - (c + di) = (a - c) + (b - d)i$$

Thus, to add two complex numbers we add their real parts and add their imaginary parts.

Example 4 **Adding and Subtracting Complex Numbers**

Perform the indicated operations:

(a) $(3 + 2i) + (5 - 6i)$ **(b)** $(-2 + 4i) - (3 + 7i)$

SOLUTION

(a) $(3 + 2i) + (5 - 6i) = (3 + 5) + (2 - 6)i = 8 - 4i$
(b) $(-2 + 4i) - (3 + 7i) = (-2 - 3) + (4 - 7)i = -5 - 3i$

Self-Check 2 Add and subtract as indicated.

(a) $(4 + 9i) + 6i$ **(b)** $(7 - 2i) - (3 + 8i)$

Let us now develop a rule for multiplying two complex numbers. Let $a + bi$ and $c + di$ be two complex numbers. Multiply these two numbers using the distributive rule for multiplying polynomials.

$$\begin{aligned}(a + bi)(c + di) &= a(c + di) + bi(c + di)\\ &= ac + adi + bci + bdi^2\\ &= ac + (ad + bc)i - bd, \qquad \text{because } i^2 = -1\\ &= (ac - bd) + (ad + bc)i\end{aligned}$$

Thus, we define multiplication as follows.

MULTIPLICATION OF TWO COMPLEX NUMBERS

Let $a + bi$ and $c + di$ be two complex numbers. Then

$$(a + bi)(c + di) = (ac - bd) + (ad + bc)i$$

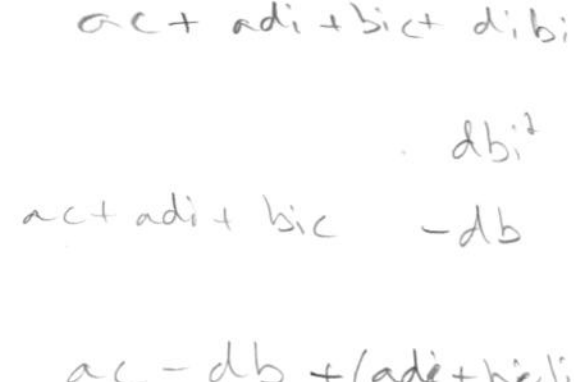

We recommend that the student not memorize this rule but rather be able to multiply complex numbers using the rules for multiplying polynomials. Results should be simplified using $i^2 = -1$, as illustrated previously.

Example 5 **Multiplying Complex Numbers**

Find the product $(2 - 5i)(4 + i)$.

SOLUTION Using the preceding rule, we get

$$\begin{aligned}(2 - 5i)(4 + i) &= 2(4 + i) - 5i(4 + i)\\ &= 8 + 2i - 20i - 5i^2\\ &= 8 - 18i + 5\\ &= 13 - 18i\end{aligned}$$

Self-Check 3 Find the product $(-2 + 4i)(1 + 3i)$.

Associated with every complex number $a + bi$ is another complex number $a - bi$, called its **complex conjugate.** For example, the complex conjugate of $2 + 3i$ is $2 - 3i$. The concept of the complex conjugate is used in dividing one complex number by another. We now illustrate division of complex numbers.

Example 6 **Dividing Complex Numbers**

The method of division is to multiply both the numerator and the denominator by the complex conjugate of the denominator and simplify. This technique will always create a real number in the denominator. To illustrate the method let us compute $(2 - 3i)/(1 + 2i)$.

$$\begin{aligned}\frac{2 - 3i}{1 + 2i} &= \frac{2 - 3i}{1 + 2i} \cdot \frac{1 - 2i}{1 - 2i} = \frac{(2 - 3i)(1 - 2i)}{(1 + 2i)(1 - 2i)}\\ &= \frac{2(1 - 2i) - 3i(1 - 2i)}{1(1 - 2i) + 2i(1 - 2i)} = \frac{2 - 4i - 3i + 6i^2}{1 - 2i + 2i - 4i^2}\\ &= \frac{2 - 7i - 6}{1 + 4} = \frac{-4 - 7i}{5} = -\frac{4}{5} - \frac{7}{5i}\end{aligned}$$

Thus, $\dfrac{2 - 3i}{1 + 2i} = -\dfrac{4}{5} - \dfrac{7}{5}i.$

Note that the final quotient of two complex numbers is written in the form $a + bi$. Such a way of writing complex numbers is called the **standard form.**

Self-Check 4 Determine $(1 - 2i)/(3 + 4i)$. Write the answer in standard form.

MAGNITUDE OF A COMPLEX NUMBER

We have discussed the concept of magnitude or size of a real number in Section R.1. Complex numbers also have magnitudes. Let $z = a + bi$ be a complex number. Because real numbers are a subset of complex numbers, it is quite natural that we use the same symbol $|z|$ for the magnitude of a complex number as for that of a real number. We also call $|z|$ the **absolute value** of z and define it as follows.

Magnitude of a Complex Number The **magnitude** of the complex number $z = a + bi$ is denoted $|z|$ and is defined

$$|z| = \sqrt{a^2 + b^2}$$

We discussed how real numbers can be represented as points on a line and how the magnitude of a real number corresponded to its distance from the origin. We can represent a complex number $a + bi$ as a point (a, b) in the plane. Observe that the magnitude of $a + bi$ corresponds to the distance of the point (a, b) from the origin Figure 4.3.

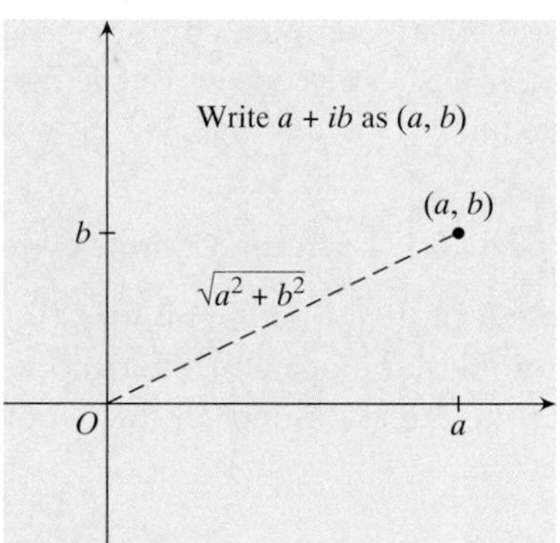

Figure 4.3 Magnitude of a complex number. $|a + bi| = \sqrt{a^2 + b^2}$

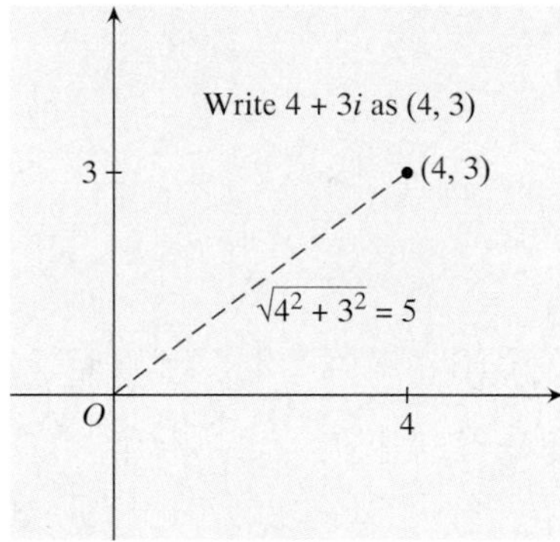

Figure 4.4 Magnitude of a complex number. $|4 + 3i| = 5$

For example,

$$|4 + 3i| = \sqrt{4^2 + 3^2} = 5, \qquad \text{(see Figure 4.4)}$$

$$|2 + 3i| = \sqrt{2^2 + 3^2} = 3.6056, \qquad \text{rounded to four decimal places}$$

$$|i| = \sqrt{0^2 + 1^2} = 1$$

$$|1| = \sqrt{1^2 + 0^2} = 1$$

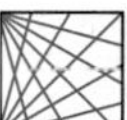

Complex Number Arithmetic Using a Calculator

Addition, subtraction, multiplication, and division of complex numbers can be carried out on some calculators. Ordered pair notation is used for entering and interpreting results on calculators. The complex number $2 + 3i$, for example, is entered as $(2, 3)$. The reverse procedure is used in interpreting the calculator results. The result of $(-4, 5)$, for example, would correspond to the complex number $-4 + 5i$. We compute $(3 + 2i) + (5 - 6i)$, $(2 - 5i)(4 + i)$, $(2 - 3i)/(1 + 2i)$, and $|2 + 3i|$ on a Texas Instruments TI-85 calculator (Figure 4.5).

```
(3,2)+(5,-6)
              (8,-4)
(2,-5)(4,1)
            (13,-18)
(2,-3)/(1,2)
          (-.8,-1.4)
```

$\longrightarrow (3 + 2i) + (5 - 6i) = 8 - 4i$

$\longrightarrow (2 - 5i)(4 + i) = 13 - 18i$

$\longrightarrow \dfrac{(2 - 3i)}{(1 + 2i)} = -0.8 - 1.4i$

```
abs (2,3)
      3.60555127546
```

$\longrightarrow |2 + 3i| = 3.6056$, rounded to four decimal places

Figure 4.5 Complex arithmetic on a calculator.

Determine whether your calculator can carry out complex arithmetic. If so, verify these results.

Self-Check 5 If your calculator can perform complex arithmetic, compute the following.

(a) $(2 - 6i)(5 + 3i)$ **(b)** $\dfrac{2 + 4i}{1 + 3i}$ **(c)** $|5 + 7i|$ **(d)** $\dfrac{(3 + 4i)(-1 + 2i)}{1 + i}$

Fractals

The process of repeated composition of a single function $f(x)$ to get $f \circ f(x)$, $f \circ f \circ f(x)$, and so on is called **iteration.** We introduced iteration of functions in Section 3.6 to study fluctuations in populations in the field of chaos. Iteration, in which the independent variables are complex numbers, is being used in a relatively new area of mathematics called **fractal geometry.** We now discuss the ideas involved and introduce the most well-known fractal called the **Mandelbrot set.**

Consider the function $f(z) = z^2 + c$, where c is a complex variable for which $|c| \leq 2$. We start with the initial value zero for z, $z = 0 + 0i$. A calculator or computer can be used to find the **orbits** (values of $f(0)$, $f \circ f(0)$, $f \circ f \circ f(0), \ldots$) for various values of c, and to decide whether the orbit "escapes" to infinity or not; that is, whether the points in an orbit become increasingly large in magnitude or whether they remain finite. Complex numbers such as $a + bi$ can be plotted on the screen as points (a, b). The calculator or computer is programmed to color black those points whose orbits remain finite. The region that results is called the **Mandelbrot set** (named after the Polish-born French mathematician Benoit Mandelbrot). See Figure 4.6 (different colors have been used for surrounding points that increase in magnitude at various rates).

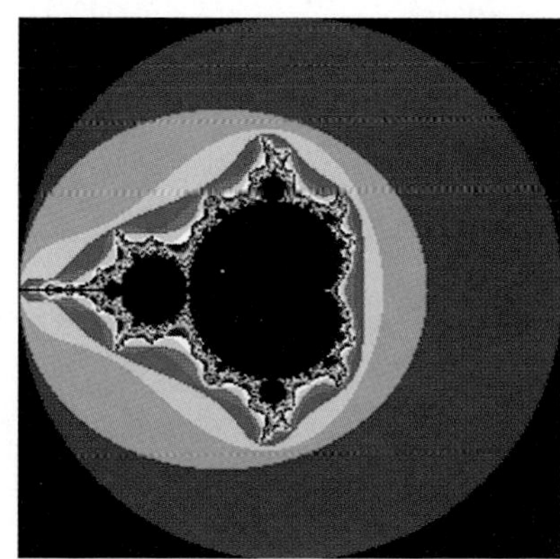

Figure 4.6 The Mandelbrot set (colored in black). *(© Bruce Bauslaugh, University of Calgary)*

Let us consider the orbits for $c = 1 - i$ and $c = -1 + 0.2i$.

$\boldsymbol{f(z) = z^2 + (1 - i)}$:

$$0 + 0i \rightarrow \underset{f(0)}{1 - i} \rightarrow \underset{f(1-i)}{1 - 3i} \rightarrow \underset{f(1-3i)}{-7 - 7i} \rightarrow \underset{f(-7-7i)}{1 + 97i} \rightarrow \underset{f(1+97i)}{-9407 + 193i} \rightarrow \cdots$$

The numbers become increasingly larger in magnitude. The number $1 - i$ escapes to infinity; the point $(1, -1)$ is not in the Mandelbrot set.

$\boldsymbol{f(z) = z^2 + (-1 + 0.2i)}$:

$$0 + 0i \rightarrow \underset{f(0)}{-1 + 0.2i} \rightarrow \underset{f(-1+2i)}{-0.04 - 0.2i} \rightarrow \underset{f(-0.04-0.2i)}{-1.0384 + 0.216i} \rightarrow \underset{f(-1.0384+0.216i)}{0.0316186 - 0.248589i} \rightarrow \cdots$$

The numbers do not become increasingly larger in magnitude. The number $-1 + 2i$ does not escape; the point $(-1, 2)$ is in the Mandelbrot set. This point would be colored as being part of the Mandelbrot set.

Students, who have a calculator that handles complex numbers can compute these iterations, as shown in Figure 4.7, which is similar to the method used for computing iterations of real numbers in Section 3.6.

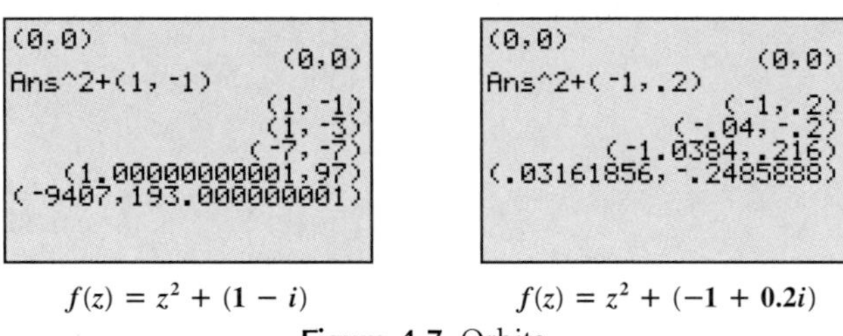

$f(z) = z^2 + (1 - i)$ $\qquad$ $f(z) = z^2 + (-1 + 0.2i)$

Figure 4.7 Orbits.

Further Ideas 2

(a) Verify the preceding orbits that correspond to $c = 1 - i$ and $c = -1 + 0.2i$.

(b) Show that the point corresponding to i lies inside the Mandelbrot set, whereas that corresponding to 1 lies outside the Mandelbrot set.

It is interesting to look at enlargements of certain areas of the Mandelbrot set (Figure 4.8). This can be done by "zooming in" on these regions using a computer. Observe that each enlargement bears a strong resemblance to part of the

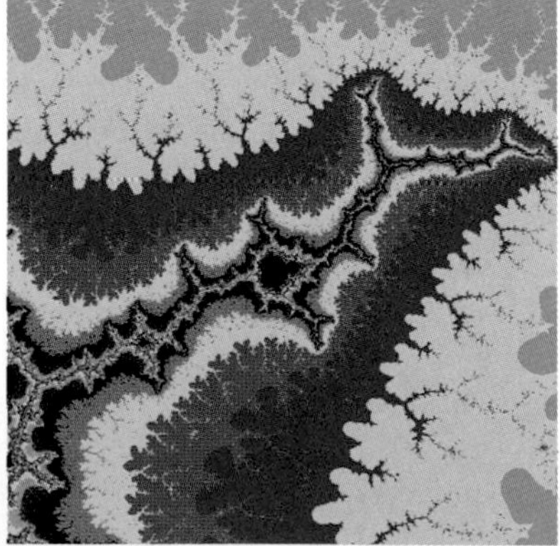

Figure 4.8 Magnification reveals that buried deep within the Mandelbrot set are baby Mandelbrot sets. *(© Bruce Bauslaugh, University of Calgary)*

original picture. This feature of the picture is called **self-similarity.** Objects that have this characteristic are called **fractals.** The Mandelbrot set is unimaginably complex and beautiful. Fractal geometry has indeed introduced a new art form. Colorful fractal images now decorate walls of offices and appear in calendars and on covers of magazines.

Answers to Self-Check Exercises

1. $37 = (9 \cdot 4) + 1$. Thus $i^{37} = i^1 = i$.
2. **(a)** $4 + 15i$ **(b)** $4 - 10i$
3. $-14 - 2i$
4. $-\frac{1}{5} - \frac{2}{5}i$
5. **(a)** $28 - 24i$ **(b)** $1.4 - 0.2i$ **(c)** 8.6023, to four decimal places **(d)** $-4.5 + 6.5i$

Answers to Further Ideas Exercises

1. **(a)** $x = -1 \pm 5i$
   (b) Consider $ax^2 + bx + c = 0, a \neq 0$.

$$x = \frac{-b \pm \sqrt{b^2 - 4ac}}{2a}$$

Because the equation has complex roots, let $\sqrt{b^2 - 4ac} = ri$. Then

$$x = \frac{-b \pm ri}{2a} = p + qi \text{ or } p - qi$$

2. **(b)** Orbit of i: $i \to -1 + i \to -i \to -1 + i \to -i \to -1 + i \cdots$. Lies in Mandelbrot set.
   Orbit of 1: $1 \to 2 \to 5 \to 26 \to 677 \to 458{,}330 \cdots$. Lies outside Mandelbrot set.

EXERCISES 4.2

In Exercises 1–8, identify each number as real, imaginary, or complex.

1. 4 **2.** $6i$ **3.** 7.2 **4.** $2 + 3i$
5. $-8i$ **6.** $-4 - 6i$ **7.** -9.56 **8.** $7 - 3.2i$

In Exercises 9–16, express each of the roots in terms of i.

9. $\sqrt{-4}$ **10.** $\sqrt{25}$ **11.** $\sqrt{-36}$ **12.** $-\sqrt{-81}$
13. $\sqrt{121}$ **14.** $\sqrt{-100}$ **15.** $\sqrt{-49}$ **16.** $-\sqrt{64}$

In Exercises 17–22, compute the powers of i.

17. i^{19} **18.** i^{36} **19.** i^{41} **20.** i^{100}
21. i^{10} **22.** i^{74}

In Exercises 23–28, give the real part and the imaginary part of each complex number.

23. $2 + 3i$ **24.** $5 - 6i$ **25.** $7i$ **26.** $-3 + 7i$
27. $-8i$ **28.** 17

In Exercises 29–34, give the complex conjugate of each complex number.

29. $3 + 2i$ **30.** $-6 - 7i$ **31.** $3 - 5i$ **32.** $-2 + 8i$
33. $9i$ **34.** 12

In Exercises 35–44, solve the equations. The roots may be real or complex. You may leave the solutions in a form involving radicals.

35. $x^2 + 2x + 3 = 0$ **36.** $x^2 - x - 2 = 0$

37. $x^2 - 2x + 1 = 0$ **38.** $x^2 + 2x + 4 = 0$

39. $x^2 + 4x + 4 = 0$ **40.** $4x^2 - 9x - 9 = 0$

41. $x^2 + 4 = 0$ **42.** $x^2 + x + 3 = 0$

43. $-x^2 + 2x - 5 = 0$ **44.** $3x^2 - 2x + 5 = 0$

In Exercises 45–54, add and subtract as indicated. Write each answer in the standard form $a + bi$.

45. $(2 + 3i) + (5 + 2i)$ **46.** $(7 + 2i) + (-1 + i)$

47. $(9 - 2i) - (7 + 3i)$ **48.** $(-2 - 4i) - 6$

49. $(7 - 3i) + (-4 + 2i)$ **50.** $(3 + 6i) - 7i$

51. $(-3 + 6i) + (-3 - 8i)$

52. $(4 + 2i) + (3 + 2i) + (4 - 5i)$

53. $(3 + i) + (9 - 2i) + i$

54. $(4 - 7i) + (2 + 5i) - (7 + 4i)$

In Exercises 55–64, compute the products. Write each answer in the standard form $a + bi$. If you have a calculator that handles complex numbers, check your answers on the calculator.

55. $(1 + 2i)(3 + i)$ **56.** $(5 - 6i)(2 + i)$

57. $(6 + i)(2 - i)$ **58.** $(-6 - 2i)(-4 + 3i)$

59. $-4(3 + 2i)$ **60.** $(2 + i)(2 - i)$

61. $(2 + 5i)^2$ **62.** $(-7 + 2i)(6 + 2i)$

63. $(3 - 4i)^2$ **64.** $(3 + i)^4$

In Exercises 65–74, divide as indicated. If you have a calculator that handles complex numbers, check your answers on the calculator.

65. $\dfrac{2 + 3i}{5 + i}$ **66.** $\dfrac{4 - i}{2 + 3i}$ **67.** $\dfrac{2 - 3i}{1 + i}$ **68.** $\dfrac{1 - 2i}{-3 + 2i}$

69. $\dfrac{5 + i}{5 - i}$ **70.** $\dfrac{2 - 3i}{-1 + 4i}$ **71.** $\dfrac{i}{2 + i}$ **72.** $\dfrac{3}{i}$

73. $\dfrac{5 - 3i}{2i}$ **74.** $\dfrac{2i}{4i}$

In Exercises 75–81, compute the magnitude of each complex number. Round the answers to four decimal places.

75. $12 + 5i$ **76.** $-2 + 4i$ **77.** $3 - 5i$

78. $-1 + 7i$ **79.** 9 **80.** $4i$

81. $-6 - 3i$

82. Show that the sum of a complex number and its complex conjugate is always a real number.

83. Show that the difference of a complex number and its complex conjugate is always an imaginary number.

84. Find a complex number $a + bi$ such that

$$(a + bi)^2 = i$$

Fractals

Determine whether the numbers in Exercises 85–92 correspond to points that lie inside or outside the Mandelbrot set.

85. $0.5 + i$ **86.** $0.2 + 0.4i$ **87.** -1

88. $-1.2 + 0.25i$ **89.** $-1 + 0.5i$ **90.** $-2 + 0.1i$

91. $0.1 + 0.5i$ **92.** $0.4i$

93. Investigate whether the points corresponding to $-2, -3, -2.000001$ lie inside or outside the Mandelbrot set. Determine whether points that correspond to numbers of the form $-2.00\ldots001$ lie in the Mandelbrot set. What does your investigation tell you about the probable location of the point $(2, 0)$ for the Mandelbrot set? Discuss.

4.3 Complex Zeros, Complex Factors, and the Fundamental Theorem of Algebra

- Complex Factorization Theorem • Fundamental Theorem of Algebra
- Constructing a Polynomial with Given Zeros • Complex Zeros Theorem
- Conjugate Zeros Theorem

In Section 4.1, we discussed zeros and factors of polynomials, but we restricted the discussion to working with real numbers. In this section, we extend the discussion to the complex number system. We will refer to zeros and factors that are complex numbers as **complex zeros** and **complex factors.** The Remainder Theorem and the Factor Theorem, introduced in Section 4.1, are also valid when the numbers are complex numbers.

The following example illustrates the application of the Factor Theorem to a polynomial with complex coefficients.

Example 1 Checking for a Factor Using Synthetic Division

Determine whether $x - 2i$ is a factor of

$$f(x) = 2x^3 + (1 - 3i)x^2 + (2 - i)x + 2$$

SOLUTION By the Factor Theorem, $x - 2i$ will be a factor of $f(x)$ if $f(2i) = 0$. Divide $f(x)$ by $x - 2i$ using synthetic division to see whether the remainder $f(2i)$ is zero.

$$\begin{array}{r|rrrr} 2i & 2 & 1 - 3i & 2 - i & 2 \\ & & 4i & -2 + 2i & -2 \\ \hline & 2 & 1 + i & i & 0 \end{array}$$

The remainder is zero. Thus, $f(2i) = 0$, implying that $x - 2i$ is a factor of $f(x)$.

Further Ideas 1 Let $f(x) = x^3 + (2 + i)x^2 - (3 - 4i)x - 1 + 5i$. $f(i)$ will be the remainder after dividing $f(x)$ by $x - i$ (by the Remainder Theorem). Show that $f(i) = 1 + 8i$.

We now give some important theoretical results, results that the student should be aware of even though the proofs of some of these results are beyond the scope of this course.

FUNDAMENTAL THEOREM OF ALGEBRA

Every polynomial of degree $n > 1$ has at least one complex zero.

The first proof of this theorem was given by the German mathematician Carl Friedrich Gauss in his doctoral dissertation when he was 22 years old.

The Fundamental Theorem of Algebra leads to the following more general result. We are able to provide the proof of this result.

Complete Factorization Theorem

A polynomial $f(x)$ of degree $n > 1$ has n linear complex factors. $f(x)$ may be written

$$f(x) = a(x - r_1)(x - r_2) \ldots (x - r_n)$$

where $a, r_1, r_2, \ldots, r_n$ are complex numbers.

Proof By the Fundamental Theorem of Algebra, we know that $f(x)$ has at least one complex zero. Let it be r_1. Thus, $x - r_1$ is a root of $f(x)$. Therefore, there exists a polynomial $q_1(x)$ of degree $n - 1$ such that

$$f(x) = (x - r_1)q_1(x).$$

By the Fundamental Theorem of Algebra, $q_1(x)$ has at least one complex root. Let it be r_2. Thus, there exists a polynomial $q_2(x)$ such that

$$q_1(x) = (x - r_2)q_2(x).$$

This implies that

$$f(x) = (x - r_1)(x - r_2)q_2(x).$$

Continuing thus, after n steps, we arrive at a polynomial $q_n(x)$ of degree 0. This polynomial is a nonzero complex number a. Thus,

$$f(x) = a(x - r_1)(x - r_2)\cdots(x - r_n),$$

proving the theorem.

Notes on the Complete Factorization Theorem

1. a is the coefficient of x^n in $f(x)$.
2. In practice, some or all of the numbers $a, r_1, r_2, \ldots, r_n$ may be real numbers, because the real numbers are a subset of the set of complex numbers.
3. Not all the $r_1, r_2, \ldots, r_n$ may be distinct; that is, some of the factors may be **repeated factors.**

We now give a number of examples to illustrate these two theorems.

Example 2 | Constructing a Polynomial with Given Zeros

Construct a polynomial of degree 3 that has zeros 1, 2, and -4.

SOLUTION By the Factor Theorem, the polynomial will have factors $(x - 1)$, $(x - 2)$, and $(x - (-4))$. Complete Factorization then implies that the polynomial will be of the form

$$f(x) = a(x - 1)(x - 2)(x + 4)$$

There will be a polynomial of degree 3 that has zeros 1, 2, and -4 for each value of the complex number a. For example, letting $a = 1$, we get

$$\begin{aligned} f(x) &= (x - 1)(x - 2)(x + 4) = (x - 1)(x^2 + 2x - 8) \\ &= x^3 + x^2 - 10x + 8 \end{aligned}$$

The polynomial $f(x) = x^3 + x^2 - 10x + 8$ has zeros 1, 2, and -4.

Self-Check 1 Construct a polynomial of degree 3 that has zeros 1, 3, and i.

Example 3 Constructing a Polynomial with Given Zeros

Construct a polynomial $f(x)$ of degree 3 that has zeros 0, -1, and $1 - i$ and is such that $f(1) = i$.

SOLUTION In Example 2, we had the freedom to select any value for the number a. Here the additional restriction $f(1) = i$ forces a specific value for a. Because $f(x)$ has zeros 0, -1, and $1 - i$, it must be of the form

$$\begin{aligned} f(x) &= a(x - 0)(x + 1)(x - 1 + i) \\ &= ax(x + 1)(x - 1 + i) \end{aligned}$$

Because $f(1) = i$, on substituting $x = 1$ we get,

$$\begin{aligned} i &= a(1)(1 + 1)(1 - 1 + i) \\ i &= 2ai \\ a &= \frac{1}{2} \end{aligned}$$

Thus

$$\begin{aligned} f(x) &= \frac{1}{2}x(x + 1)(x - 1 + i) \\ &= \frac{1}{2}x^3 + \frac{i}{2}x^2 + \left(-\frac{1}{2} + \frac{i}{2}\right)x \end{aligned}$$

Example 4 Determining Further Zeros, Given One Zero

If 2 is a zero of the polynomial $f(x) = x^3 - 4x^2 + x + 6$, determine the other zeros.

SOLUTION Since 2 is a zero of $f(x)$, then $(x - 2)$ is a factor. Thus, there exists a polynomial $q(x)$ such that

$$f(x) = (x - 2)q(x)$$

Divide $f(x)$ by $x - 2$ to get $q(x)$. Using synthetic division, we get

$$\begin{array}{r|rrrr} 2 & 1 & -4 & 1 & 6 \\ & & 2 & -4 & -6 \\ \hline & 1 & -2 & -3 & 0 \end{array}$$

The last row gives $q(x) = x^2 - 2x - 3$. Thus,

$$\begin{aligned} f(x) &= (x-2)(x^2 - 2x - 3) \\ &= (x-2)(x+1)(x-3) \end{aligned}$$

The other zeros of $f(x)$ are seen to be -1 and 3.

Self-Check 2 If 1 is a zero of the polynomial $f(x) = x^3 + x^2 - 17x + 15$, determine the other zeros.

The following theorem gives a maximum to the number of complex zeros for a polynomial function.

Complex Zeros Theorem

A polynomial $f(x)$ of degree $n > 1$ has at most n distinct complex zeros.

Proof We know that $f(x)$ can be written $f(x) = a(x - r_1)(x - r_2) \cdots (x - r_n)$. The zeros of $f(x)$ are the numbers $r_1, r_2, \ldots, r_n$. If these numbers are all distinct, there will be n zeros; otherwise, the number of zeros will be less than n.

If the factor $(x - r_i)$ appears m times in the polynomial $f(x)$, we say that r_i is a **zero of multiplicity** m. For example, consider the polynomial

$$f(x) = 2(x+5)(x-4)^2(x-7)^3$$

There are three zeros; namely -5, 4, and 7. -5 occurs once, 4 occurs twice, and 7 occurs three times. We say that -5 is of multiplicity one, and 4 is of multiplicity two, whereas 7 is of multiplicity three.

Example 5 Constructing a Polynomial with Given Characteristics

Construct a polynomial $f(x)$ of degree 3 that has a zero -1 of multiplicity one and a zero 4 of multiplicity two and is such that $f(5) = 24$.

SOLUTION Because -1 is a zero of multiplicity one and 4 is a zero of multiplicity two, $f(x)$ will be of the form

$$f(x) = a(x+1)(x-4)^2$$

Because $f(5) = 24$, on substituting $x = 5$, we get

$$\begin{aligned} 24 &= a(5+1)(5-4)^2 \\ 24 &= 6a \\ a &= 4 \end{aligned}$$

Thus

$$f(x) = 4(x + 1)(x - 4)^2 = 4(x + 1)(x^2 - 8x + 16)$$
$$= 4x^3 - 28x^2 + 32x + 64$$

Self-Check 3 Construct a polynomial of degree 3 that has two zeros, namely 1 of multiplicity two and 5 of multiplicity one, and is such that $f(2) = -6$.

The following theorem (which we state without proof) tells us that if a certain type of polynomial has a complex zero, then its complex conjugate will also be a zero.

Conjugate Zeros Theorem

Let $f(x)$ be a polynomial of degree >1 with real coefficients. If $a + bi$ is a zero of $f(x)$, then the complex conjugate $a - bi$ is also a zero.

Let us illustrate the Conjugate Zeros Theorem for a polynomial of degree 2.

Example 6 Determining Zeros of a Polynomial

Determine the zeros of $f(x) = x^2 - 4x + 20$.

SOLUTION $f(x)$ is a polynomial of degree >1 having real coefficients. Thus, any complex zeros should occur in pairs: a complex number and its conjugate. The zeros of $f(x)$ are the numbers for which $f(x) = 0$. The zeros are thus the solutions of

$$x^2 - 4x + 20 = 0$$

Let us solve this equation using the quadratic formula. We get

$$x = \frac{-(-4) \pm \sqrt{(-4)^2 - 4(1)(20)}}{2} = \frac{-(-4) \pm \sqrt{-64}}{2}$$
$$= \frac{4 \pm 8i}{2} = 2 \pm 4i$$

The zeros of the polynomial are $2 + 4i$ and its conjugate $2 - 4i$, confirming the statement of the theorem.

Example 7 Constructing a Polynomial with Given Characteristics

Construct a polynomial $f(x)$ of degree 3 with real coefficients and zeros 3 and $1 + 2i$.

SOLUTION Because $f(x)$ has real coefficients and $1 + 2i$ is a zero, its conjugate $1 - 2i$ must also be a zero. Because $f(x)$ is of degree 3, it can have no more than three zeros. The zeros of $f(x)$ are thus 3, $1 + 2i$, and $1 - 2i$. This implies that the factors of $f(x)$ are $(x - 3)$, $(x - (1 + 2i))$, and $(x - (1 - 2i))$. $f(x)$ will be of the form

$$f(x) = a(x - 3)(x - (1 + 2i))(x - (1 - 2i))$$

For example, let $a = 1$ (the most convenient value). Multiply out to get

$$f(x) = (x - 3)(x - 1 - 2i)(x - 1 + 2i) = (x - 3)(x^2 - 2x + 5)$$
$$= x^3 - 5x^2 + 11x - 15$$

Answers to Self-Check Exercises

1. Let $f(x) = (x - 1)(x - 3)(x - i) = (x - 1)(x^2 - 3x - ix + 3i)$. $f(x) = x^3 - (4 + i)x^2 + (3 + 4i)x - 3i$.
2. Other zeros are -5 and 3.
3. Let

$f(x) = a(x - 1)^2(x - 5) = a(x^2 - 2x + 1)(x - 5) = ax^3 - 7ax^2 + 11ax - 5a$
$f(2) = 8a - 28a + 22a - 5a = -3a$. But $f(2) = -6$. Thus $a = 2$.
$f(x) = 2x^3 - 14x^2 + 22x - 10$.

EXERCISES 4.3

In Exercises 1–6, use synthetic division to determine whether $g(x)$ is a factor of $f(x)$.

1. $f(x) = x^3 - 4ix^2 + (3 + i)x + 3 - 18i$, $g(x) = x - 3i$
2. $f(x) = 2x^3 + (1 + 5i)x^2 - 3x + 1$, $g(x) = x + i$
3. $f(x) = 3x^3 - 5ix^2 + (3 + 4i)x + 3 - 2i$, $g(x) = x - 2i$
4. $f(x) = (3 + i)x^3 + 4x^2 + (2 - 3i)x + 1 - 2i$, $g(x) = x + 1$
5. $f(x) = 2ix^3 + (5 - 6i)x^2 - (23 + 7i)x + 12 - 4i$, $g(x) = x - 4$
6. $f(x) = 4x^3 - 15ix^2 + (2 - 6i)x + 1 + 7i$, $g(x) = x - 4i$

In Exercises 7–12, use synthetic division with an appropriate divisor to compute the value of f.

7. $f(x) = 3x^3 - 3ix^2 + (2 - 4i)x + 1 + 2i$, $f(2)$
8. $f(x) = -2x^3 + (1 - 2i)x^2 + (1 + 4i)x + 4 - 3i$, $f(3i)$
9. $f(x) = 2ix^3 - 4ix^2 + (7 + 3i)x + 2 + 5i$, $f(-2i)$
10. $f(x) = (1 + 2i)x^3 - 4ix^2 + (1 + 4i)x - 1 - 4i$, $f(1 + 2i)$
11. $f(x) = (2 - 4i)x^3 + (4 - 2i)x^2 + (3 + 4i)x + 7 + 2i$, $f(2 - 3i)$
12. $f(x) = (4 - i)x^4 + (14 - 16i)x^3 + (4 - 2i)x^2 + (2 + 3i)x - 7$, $f(-4 + 3i)$

In Exercises 13–20, determine the zeros of the polynomials and state the multiplicity of each zero.

13. $f(x) = (x - 2)(x + 3)(x + 5)$
14. $f(x) = (x - 4)(x - 3)(x + 7)$
15. $f(x) = (x - 4)(x + 2)^2$
16. $f(x) = (x + 1)(x - 2)(x + 5)^3$
17. $f(x) = (x - 5)^3(x + 7)^4$
18. $f(x) = (x - 3)(x + 3)^2(x + 5)^3$
19. $f(x) = (x - 2i)(x + 3 - 4i)^2$
20. $f(x) = (x + 2 - 5i)^3(x + 7 - 3i)^4$

In Exercises 21–30, construct polynomials of degree 3 that have the indicated zeros.

21. 1, 3, 5
22. -2, 1, 7
23. -2, 0, 6
24. -2, 3, 3 (3 is of multiplicity two)
25. -3, 2, 2
26. 1, 5, $-3 - 2i$
27. 2, $1 - 2i$, $3 + i$
28. -1, $2 + i$, $3 - 4i$
29. 2, i, i
30. $1 - i$, $3 + i$, $4 - i$

In Exercises 31–40, one zero of $f(x)$ is given. Determine all the zeros.

31. $f(x) = x^3 - 6x^2 + 11x - 6$, 1
32. $f(x) = x^3 - 3x - 2$, -1
33. $f(x) = x^3 - 7x + 6$, -3
34. $f(x) = x^3 - 3x^2 - 18x + 40$, 5

35. $f(x) = x^3 - 6x^2 + 12x - 8$, 2
36. $f(x) = x^3 + x^2 - 8x - 12$, 3
37. $f(x) = 3x^3 - 14x^2 + 13x + 6$, 2
38. $f(x) - x^3 - 2x^2 - 3x$, -1
39. $f(x) = x^3 + 2x^2 + x + 2$, $-i$
40. $f(x) = x^3 - 3x^2 + 4x - 12$, $2i$

In Exercises 41–50, construct polynomials $f(x)$ that have the given properties.

41. Degree 3 with zeros -1, 2, and 2 (i.e., 2 of multiplicity two) such that $f(3) = 8$
42. Degree 3 with zeros 1, -3, and 3 such that $f(4) = 63$
43. Degree 3 with a zero 2 of multiplicity three, such that $f(3) = -4$
44. Degree 3 with zeros 1 and 4 (of multiplicity two), such that $f(7) = 270$
45. Degree 3 with zeros -2 (of multiplicity two) and -4, such that $f(-5) = 18$
46. Degree 4 with zeros -3 (of multiplicity two) and 4 (of multiplicity two), such that $f(3) = 36$
47. Degree 4 with zeros 1 and 3 (of multiplicity three), such that $f(5) = 16$
48. Degree 3 with zeros i, $-1 + 3i$ (of multiplicity two) such that $f(3i) = 6i$
49. Degree 3 with zeros 2, $1 + i$, and $3 - 2i$ such that $f(2 + i) = 3 + i$
50. Degree 3 with zeros i, $-2 + i$, and $1 - i$ such that $f(1 + 2i) = -2 + i$

In Exercises 51–57, construct polynomials $f(x)$ having real coefficients that have the given numbers among their zeros.

51. Degree 3 with zeros 4 and $2 - i$
52. Degree 3 with zeros 3 and $-2 + 3i$
53. Degree 3 with zeros -2 and $3 - 5i$
54. Degree 3 with zeros and -1 and $5i$
55. Degree 4 with zeros 1, 3, and $5 - i$
56. Degree 4 with zeros $2 - i$, and $3 + 2i$
57. Degree 4 with zeros $-3 + i$ and $2 - 4i$

4.4 Computing Zeros of Polynomials

- Rational Zero Theorem • Finding the Rational Zeros of a Polynomial
- Finding All the Zeros of a Polynomial

In the preceding sections we introduced theorems that give information about zeros of polynomials and discussed methods of checking whether certain numbers are zeros of given polynomials. We now give a method for finding zeros of polynomials. The method is based on the following theorem.

Rational Zero Theorem

If the polynomial

$$f(x) = a_nx^n + a_{n-1}x^{n-1} + \cdots + a_1x + a_0, \qquad a_n \neq 0$$

has integer coefficients and if the rational number p/q, written in lowest terms, is a zero of $f(x)$, then

1. p is a factor of the constant term a_0.
2. q is a factor of the leading coefficient a_n.

Proof p/q is a zero of $f(x)$; thus, $f(p/q) = 0$. This gives

$$a_n\left(\frac{p}{q}\right)^n + a_{n-1}\left(\frac{p}{q}\right)^{n-1} + \cdots + a_1\left(\frac{p}{q}\right) + a_0 = 0$$

Multiply both sides by q^n, and add $-a_0q^n$ to both sides:

$$a_np^n + a_{n-1}p^{n-1}q + \cdots + a_1pq^{n-1} = -a_0q^n$$

Factor out the common factor p on the left side:

$$p(a_np^{n-1} + a_{n-1}p^{n-2}q + \cdots + a_1q^{n-1}) = -a_0q^n$$

This implies that p is a factor of $-a_0q^n$. Because p/q is in lowest terms, p and q have no common factors. Thus, p is a factor of a_0, proving part 1 of the theorem. Part 2 is proved in a similar manner.

In applying this theorem to find zeros, we will use the relationship between zeros and factors repeatedly. We remind the student that r is a zero of a polynomial $f(x)$ if and only if $x - r$ is a factor of $f(x)$. We now give a number of examples to illustrate how the Rational Zero Theorem is used.

Example 1 Finding the Rational Zeros of a Polynomial

Find all the rational zeros of the polynomial

$$f(x) = x^3 + 4x^2 + x - 6$$

SOLUTION Let p/q be a rational zero. The Rational Zero Theorem tells us that p is a factor of the constant term -6 and that q is a factor of the leading coefficient 1. The possible values of p are thus the factors of -6. They are ±1, ±2, ±3, and ±6. The possible values of q are ±1. Possible values of p/q the rational zeros of $f(x)$, are thus ±1, ±2, ±3, and ±6.

Let us check to see whether 1 is a zero. Divide $f(x)$ by $x - 1$.

$$\begin{array}{r|rrrr} 1 & 1 & 4 & 1 & -6 \\ & & 1 & 5 & 6 \\ \hline & 1 & 5 & 6 & 0 \end{array}$$

The remainder is zero; thus, $x - 1$ is a factor, and 1 is a zero.

The last line of the synthetic division tells us that

$$f(x) = (x - 1)(x^2 + 5x + 6)$$

Factor the last expression to get

$$f(x) = (x - 1)(x + 2)(x + 3)$$

The zeros of $f(x)$ are 1, -2, and -3.

We verify the result with a graphing calculator (Figure 4.9). The x-intercepts of the graph are seen to be 1, -2, and -3.

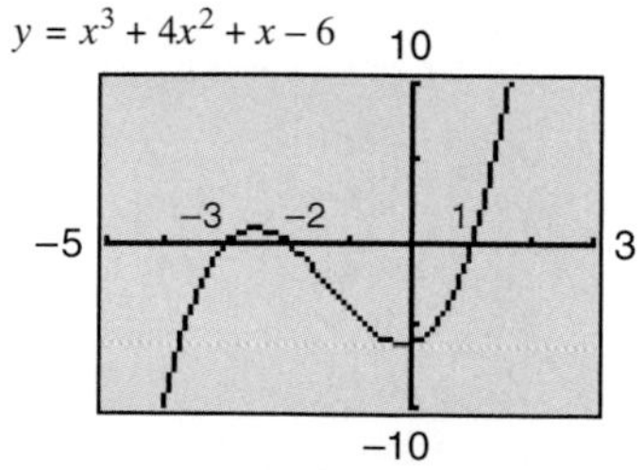

Figure 4.9

Self-Check 1 Find all the rational zeros of the polynomial

$$f(x) = x^3 + 2x^2 - 11x - 12$$

Verify your answer with a graphing calculator.

Example 2 | Finding the Rational Zeros of a Polynomial

Find all the rational zeros of the polynomial

$$f(x) = x^4 + 6x^3 - 7x^2 - 48x - 36$$

SOLUTION Let p/q be a rational zero. The Rational Zero Theorem tells us that p is a factor of the constant term -36 and that q is a factor of the leading coefficient 1. Possible values of p are $\pm 1, \pm 2, \pm 3, \pm 4, \pm 6, \pm 9, \pm 12, \pm 18$, and ± 36. The possible values of q are ± 1. The possible zeros are the quotients p/q. Likely zeros are thus $\pm 1, \pm 2, \pm 3, \pm 4, \pm 6, \pm 9, \pm 12, \pm 18$, and ± 36. (Note that all zeros must be factors of the constant term if the leading coefficient is 1, as here.)

Let us check to see whether 1 is a zero. Divide $f(x)$ by $x - 1$.

$$\begin{array}{r|rrrrr} 1 & 1 & 6 & -7 & -48 & -36 \\ & & 1 & 7 & 0 & -48 \\ \hline & 1 & 7 & 0 & -48 & -84 \end{array}$$

The remainder is not zero; thus 1 is not a zero. Let us check to see whether -1 is a zero. Divide $f(x)$ by $x + 1$.

$$\begin{array}{r|rrrrr} -1 & 1 & 6 & -7 & -48 & -36 \\ & & -1 & -5 & 12 & 36 \\ \hline & 1 & 5 & -12 & -36 & 0 \end{array}$$

The remainder is zero; thus, $x + 1$ is a factor, and -1 is a zero.

The last line of the division gives

$$f(x) = (x + 1)(x^3 + 5x^2 - 12x - 36)$$

The trial-and-error process is now continued. At this stage, use the polynomial $x^3 + 5x^2 - 12x - 36$ rather than $f(x)$ because it is of smaller degree. Any remaining rational zeros of $f(x)$ must also be the rational zeros of this polynomial. Test $-1, \pm 2, \pm 3$, and so on. (Note that -1 could be of multiplicity more than one, so it still must be examined.) It is found that -1 and 2 are not zeros. On testing -2 by dividing by $x + 2$, we get

$$\begin{array}{r|rrrr} -2 & 1 & 5 & -12 & -36 \\ & & -2 & -6 & 36 \\ \hline & 1 & 3 & -18 & 0 \end{array}$$

The remainder is zero; thus, $x + 2$ is a factor, and -2 is a zero.

The last line of the division leads to

$$f(x) = (x + 1)(x + 2)(x^2 + 3x - 18)$$

Factor the last expression to get

$$f(x) = (x + 1)(x + 2)(x - 3)(x + 6)$$

The zeros of $f(x)$ are $-1, -2, -6$, and 3.

We verify the result with a graphing calculator (Figure 4.10).

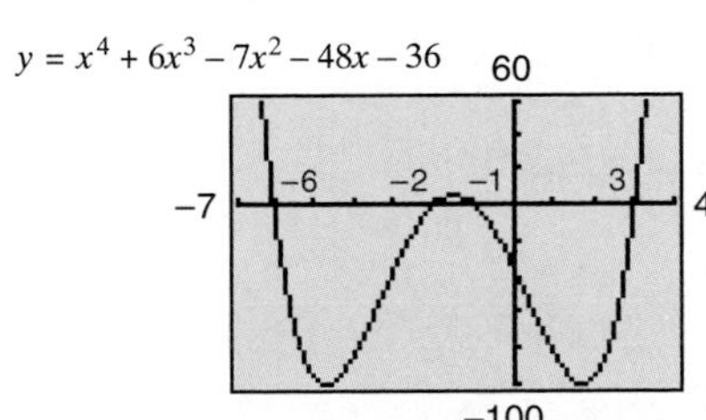

Figure 4.10

Self-Check 2 Find all the rational zeros of the polynomial

$$f(x) = x^4 + 3x^3 - 3x^2 - 7x + 6$$

Verify your answer with a graphing calculator.

The following example illustrates the use of the Rational Zero Theorem when the leading coefficient is a number other than 1.

Example 3 Finding the Rational Zeros of a Polynomial

Find all the rational zeros of the polynomial

$$f(x) = 4x^3 - 5x^2 - 23x + 6$$

SOLUTION Let p/q be a rational zero; p is a factor of 6, and q is a factor of 4. The possible values of p are thus ± 1, ± 2, ± 3, and ± 6. The possible values of q are ± 1, ± 2, and ± 4. Possible rational zeros are formed by taking all possible quotients of the form p/q. Thus, possible rational zeros are ± 1, ± 2, ± 3, ± 6, $\pm \frac{1}{2}$, $\pm \frac{3}{2}$, $\pm \frac{1}{4}$, and $\pm \frac{3}{4}$. It is found using synthetic division that ± 1, $+2$ are not zeros. We next test -2. Divide $f(x)$ by $(x + 2)$.

$$\begin{array}{r|rrrr} -2 & 4 & -5 & -23 & 6 \\ & & -8 & 26 & -6 \\ \hline & 4 & -13 & 3 & 0 \end{array}$$

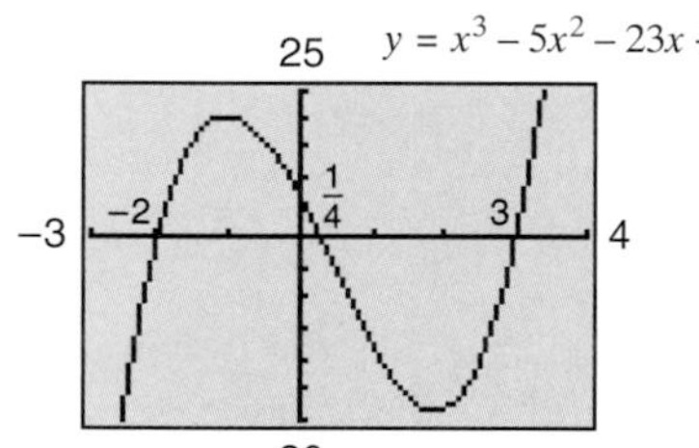

Figure 4.11

The remainder is zero; thus, $x + 2$ is a factor. The division gives

$$f(x) = (x + 2)(4x^2 - 13x + 3)$$

Factor further to get

$$f(x) = (x + 2)(4x - 1)(x - 3)$$

The zeros are -2, $\frac{1}{4}$, and 3. We verify these zeros with a graphing calculator (Figure 4.11).

Self-Check 3 Find all the rational zeros of the polynomial

$$f(x) = 6x^3 + 17x^2 - 4x - 3$$

Verify your answer with a graphing calculator.

The following example illustrates how this method of finding zeros can be adapted to polynomials with rational coefficients.

Example 4 Finding the Rational Zeros of a Polynomial

Find all the rational zeros of the polynomial

$$f(x) = x^4 + \frac{5}{4}x^3 - \frac{15}{8}x^2 - \frac{5}{8}x + \frac{1}{4}$$

SOLUTION $f(x)$ has rational coefficients; thus, the Rational Zero Theorem cannot be applied directly. The zeros of $f(x)$ will be the values of x that satisfy

$$x^4 + \frac{5}{4}x^3 - \frac{15}{8}x^2 - \frac{5}{8}x + \frac{1}{4} = 0$$

Multiply both sides of this equation by 8 to convert the coefficients to integers. We get

$$8x^4 + 10x^3 - 15x^2 - 5x + 2 = 0$$

The zeros of $g(x) = 8x^4 + 10x^3 - 15x^2 - 5x + 2$, which have integer coefficients, will be the same as the zeros of $f(x)$. The Rational Zero Theorem can be applied to this polynomial.

Let p/q be a rational zero; p is a factor of 2, and q is a factor of 8. The possible values of p are ± 1, and ± 2. The possible values of q are ± 1, ± 2, ± 4, and ± 8. Thus, the possible values of p/q are ± 1, ± 2, $\pm\frac{1}{2}$, $\pm\frac{1}{4}$, and $\pm\frac{1}{8}$.

Let us test 1. Divide $g(x)$ by $x - 1$.

$$\begin{array}{r|rrrrr} 1 & 8 & 10 & -15 & -5 & 2 \\ & & 8 & 18 & 3 & -2 \\ \hline & 8 & 18 & 3 & -2 & 0 \end{array}$$

The remainder is 0; thus, $x - 1$ is a factor. The division gives

$$g(x) = (x - 1)(8x^3 + 18x^2 + 3x - 2)$$

Continue testing the possible zeros in the polynomial $8x^3 + 18x^2 + 3x - 2$. It is found that 1, 1, and 2 are not factors. On testing -2 by dividing by $(x + 2)$, we get

$$\begin{array}{r|rrrr} -2 & 8 & 18 & 3 & -2 \\ & & -16 & -4 & 2 \\ \hline & 8 & 2 & -1 & 0 \end{array}$$

The remainder is zero; thus, $x + 2$ is a factor. We now get

$$\begin{aligned} g(x) &= (x - 1)(x + 2)(8x^2 + 2x - 1) \\ &= (x - 1)(x + 2)(4x - 1)(2x + 1) \end{aligned}$$

The factors of $g(x)$, and thus of $f(x)$, are $(x - 1)$, $(x + 2)$, $(4x - 1)$, and $(2x + 1)$. These factors give the zeros 1, -2, $\frac{1}{4}$, and $-\frac{1}{2}$.

We verify these results graphically (Figure 4.12).

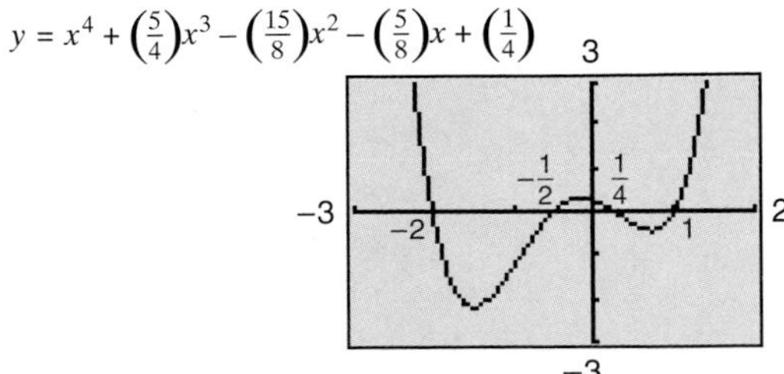

Figure 4.12

Self-Check 4 Find all the rational zeros of the polynomial

$$f(x) = x^4 + \frac{1}{6}x^3 - \frac{38}{3}x^2 + \frac{61}{6}x - 2$$

Verify your answer graphically.

The final example of this section illustrates how complex zeros of a polynomial can sometimes be found after all the rational zeros have been determined.

Example 5 Finding All the Zeros of a Polynomial

Determine all the zeros of the polynomial

$$f(x) = x^4 - 3x^3 + 4x^2 - 6x + 4$$

SOLUTION Let p/q be a rational zero; p is a factor of 4, and q is a factor of 1. The possible rational zeros are thus ±1, ±2, and ±4. Let us test 1. Divide $f(x)$ by $x - 1$.

$$\begin{array}{r|rrrrr} 1 & 1 & -3 & 4 & -6 & 4 \\ & & 1 & -2 & 2 & -4 \\ \hline & 1 & -2 & 2 & -4 & 0 \end{array}$$

The remainder is 0; thus, $x - 1$ is a factor. We get

$$f(x) = (x - 1)(x^3 - 2x^2 + 2x - 4)$$

We now use the polynomial $x^3 - 2x^2 + 2x - 4$ to search for remaining zeros. It is found that 1 and -1 are not zeros. On testing 2 by dividing by $x - 2$, we get

$$\begin{array}{r|rrrr} 2 & 1 & -2 & 2 & -4 \\ & & 2 & 0 & 4 \\ \hline & 1 & 0 & 2 & 0 \end{array}$$

The remainder is 0; thus, $x - 2$ is a factor. The last line of the division leads to

$$f(x) = (x - 1)(x - 2)(x^2 + 2)$$

$x^2 + 2$ is now factored. We can write

$$x^2 + 2 = x^2 - (-2) = x^2 - (\sqrt{2}i)^2 = (x + \sqrt{2}i)(x - \sqrt{2}i)$$

Thus,

$$f(x) = (x - 1)(x - 2)(x + \sqrt{2}i)(x - \sqrt{2}i)$$

The zeros of $f(x)$ are thus 1, 2, $-\sqrt{2}i$, and $\sqrt{2}i$.

Self-Check 5 Find all the zeros of $f(x) = x^4 - x^3 - x^2 - x - 2$.

We have given a method here for determining zeros of certain polynomials. Although this method is satisfactory for polynomials of small degree, it is not an efficient method for polynomials of high degree. The need to find zeros of polynomials is an important one in the application of mathematics. Much research has gone into the area, and computer methods that lead to good approximations of zeros can be implemented. These methods are presented in more advanced classes.

Answers to Self-Check Exercises

1. Zeros are 3, -1, and -4.
2. Zeros are 1, 1, -2, and -3.
3. Zeros are $\frac{1}{2}$, $-\frac{1}{3}$, and -3.
4. Zeros are -4, $\frac{1}{3}$, $\frac{1}{2}$, and 3.
5. Zeros are -1, 2, i, and $-i$.

EXERCISES 4.4

In Exercises 1–10, find all the zeros of the polynomials. The zeros are all integers for these polynomials. Verify your answer with a graphing calculator.

1. $f(x) = x^3 - 6x^2 + 11x - 6$
2. $f(x) = x^3 + 3x^2 - 6x - 8$
3. $f(x) = x^3 - 7x + 6$
4. $f(x) = x^3 - 2x^2 - 11x + 12$
5. $f(x) = x^3 + 5x^2 - 2x - 24$
6. $f(x) = x^3 - x^2 - 17x - 15$
7. $f(x) = x^4 - 2x^3 - 9x^2 + 2x + 8$
8. $f(x) = x^4 - 4x^3 + 3x^2 + 4x - 4$
9. $f(x) = x^4 - 2x^3 - 4x^2 + 2x + 3$
10. $f(x) = x^4 - 8x^3 + 22x^2 - 24x + 9$

In Exercises 11–20, find all the zeros of the polynomials. The zeros are all rational numbers. Verify your answer with a graphing calculator.

11. $f(x) = 3x^3 + x^2 - 8x + 4$
12. $f(x) = 2x^3 - 3x^2 - 11x + 6$
13. $f(x) = 2x^3 - 5x^2 + 4x - 1$
14. $f(x) = 3x^3 + 4x^2 - 13x + 6$
15. $f(x) = 4x^3 - 11x^2 + 5x + 2$
16. $f(x) = 2x^3 + 7x^2 + 8x + 3$
17. $f(x) = 6x^3 + 7x^2 - 9x + 2$
18. $f(x) = 6x^3 - 11x^2 + x + 4$
19. $f(x) = 2x^4 - 3x^3 - x^2 + 3x - 1$
20. $f(x) = 16x^4 - 28x^3 - 12x^2 + 7x + 2$

In Exercises 21–26, find all the zeros of the polynomials. Note that the polynomials have rational coefficients. Verify your answer with a graphing calculator.

21. $f(x) = x^3 + \frac{5}{2}x^2 + \frac{1}{2}x - 1$
22. $f(x) = x^3 + \frac{15}{4}x^2 - \frac{7}{4}x - 3$
23. $f(x) = x^3 + \frac{19}{4}x^2 + \frac{11}{4}x - 1$
24. $f(x) = x^3 + \frac{3}{4}x^2 - \frac{3}{8}x - \frac{1}{8}$
25. $f(x) = 3x^4 - \frac{7}{2}x^3 - \frac{23}{2}x^2 + 14x - 2$
26. $f(x) = x^4 - \frac{5}{6}x^3 - \frac{5}{6}x^2 + \frac{5}{6}x - \frac{1}{6}$

In Exercises 27–32, find all the zeros of the polynomials. The polynomials have complex zeros.

27. $f(x) = x^4 - x^3 - 2x - 4$
28. $f(x) = 2x^4 - 2x^3 - 11x^2 - x - 6$
29. $f(x) = 3x^4 + x^3 + x^2 + x - 2$
30. $f(x) = x^3 + x^2 - 2$
31. $f(x) = x^3 - 2x^2 + 3x - 6$
32. $f(x) = x^3 + 5x^2 + 17x + 13$

CHAPTER 4 PROJECT | Julia Sets

We introduced the Mandelbrot set in Section 4.2. This project introduces you to further fractals called Julia sets. Consider the function $f(z) = z^2 - 1$, where z is a complex number with $|z| \le 1$. A calculator or computer can be used to find the orbit of each point in this region and to decide whether the orbit escapes to infinity or not; that is, whether the points in an orbit become increasingly larger in magnitude or whether they remain finite. Complex numbers can be plotted on a screen, with the real part being represented by the x-axis and the imaginary part by the y-axis. A calculator or computer can be programmed to plot those points for which orbits do not escape. The picture obtained is called the **Julia set** of $f(z) = z^2 - 1$. There is a Julia set for each function of the form $f(z) = z^2 + c$, where c is a parameter such that $|c| \le 2$.

Use your library or other resource to find information about Julia sets. Write a brief report of Julia sets, giving examples of points that lie within and without those sets.

WEB PROJECT 4

To explore the concepts of this chapter further, try the interesting WebProject. Each activity relates the ideas and skills you have learned in this chapter to a real-world applied situation. Visit the Williams Web site at

www.saunderscollege.com/math/williams

to find the complete WebProject, related questions, and live links to the Internet.

Chapter 4 Highlights

Section 4.1

The Division Algorithm: Let $f(x)$ and $g(x)$ be two polynomials with $g(x)$ being a nonzero polynomial of degree less than $f(x)$. There are unique polynomials $q(x)$ and $r(x)$ such that $f(x) = g(x)q(x) + r(x)$, where $r(x)$ is either the zero polynomial or a polynomial of degree less than $g(x)$. $f(x)$ is the dividend, $g(x)$ is the divisor, $q(x)$ is the quotient, and $r(x)$ is the remainder.

The Remainder Theorem: If a polynomial $f(x)$ is divided by $x - k$, then the remainder is $f(k)$. $f(x) = (x - k)q(x) + f(k)$.

Synthetic Division: A shortcut of the long division method.

Zeros of Polynomials: The number k is a zero of a polynomial $f(x)$ if and only if the remainder after dividing $f(x)$ by $x - k$ is zero.

Factor Theorem: The polynomial $x - k$ is a factor of the polynomial $f(x)$ if and only if $f(k) = 0$.

Section 4.2

Imaginary Number i: $i = \sqrt{-1}$, with $i^2 = -1$.

Square Root of a Negative Number, $\sqrt{-a}$: If a is a real positive number, $\sqrt{-a} = \sqrt{a}i$.

Complex Number: $a + bi$, where a and b are real numbers; a is the real part, and b is the imaginary part of the complex number.

Equality of Complex Numbers: $a + bi = c + di$ if and only if $a = c$ and $b = d$.

Addition and Subtraction of Complex Numbers:
$(a + bi) + (c + di) = (a + c) + (b + d)i$,
$(a + bi) - (c + di) = (a - c) + (b - d)i$.

Multiplication of Complex Numbers:
$(a + bi)(c + di) = (ac - bd) + (ad + bc)i$.

Complex Conjugate: The complex conjugate of $a + bi$ is $a - bi$.

Division of Complex Numbers: Multiply both the numerator and the denominator by the complex conjugate of the denominator and simplify.

Magnitude of a Complex Number: If $z = a + bi$, then $|z| = \sqrt{a^2 + b^2}$.

Mandelbrot Set: A fractal generated by the repeated composition of the function $f(z) = z^2 + c$ for $|c| \leq 2$. The points whose orbits remain finite make up the Mandelbrot set.

Section 4.3

Fundamental Theorem of Algebra: Every polynomial of degree $n > 1$ has at least one complex zero.

Complete Factorization Theorem: A polynomial $f(x)$ of degree $n > 1$ has n linear complex factors. $f(x)$ may be written

$$f(x) = a(x - r_1)(x - r_2)\cdots(x - r_n)$$

where $a, r_1, r_2, \ldots, r_n$ are complex numbers.

Complex Zeros Theorem: A polynomial $f(x)$ of degree $n > 1$ has at most n distinct complex zeros.

Conjugate Zeros Theorem: Let $f(x)$ be a polynomial of degree >1 with real coefficients. If $a + bi$ is a zero of $f(x)$, then the complex conjugate $a - bi$ is also a zero.

Section 4.4

Rational Zero Theorem: If the polynomial $f(x) = a_nx^n + \cdots + a_1x + a_0$, $a_n \neq 0$, has integer coefficients, and if the rational number p/q, written in lowest terms, is a zero of $f(x)$, then (1) p is a factor of the constant term a_0 and (2) q is a factor of the leading coefficient a_n.

Chapter 4 Review Exercises

1. Use synthetic division to find the quotient and remainder when $f(x)$ is divided by $g(x)$.
   (a) $f(x) = x^3 - 3x^2 + 4x - 1$, $g(x) = x - 3$
   (b) $f(x) = 2x^3 + 5x^2 + 2x + 6$, $g(x) = x + 2$
   (c) $f(x) = 2x^4 - 3x^3 - 4x^2 + 6x - 5$, $g(x) = x - 4$
2. Use synthetic division with an appropriate divisor to compute the values of $f(x)$.
   (a) $f(x) = x^3 + 3x^2 - 5x - 1$, $f(2)$
   (b) $f(x) = 4x^3 - 3x^2 - 7x + 3$, $f(5)$
   (c) $f(x) = -x^4 + 2x^3 - 5x^2 + 4x - 6$, $f(-3)$
3. Use synthetic division to determine whether r is a zero of $f(x)$.
   (a) $f(x) = x^3 - 7x^2 + 14x - 8$, $r = 4$
   (b) $f(x) = 2x^3 + 7x^2 - 3x - 2$, $r = 2$
   (c) $f(x) = 3x^4 + 4x^3 - 8x^2 + 16x - 15$, $r = -3$
4. Use synthetic division to determine whether the second polynomial is a factor of $f(x)$.
   (a) $f(x) = x^3 - 2x^2 - 2x + 1$, $x - 1$
   (b) $f(x) = 2x^3 - 3x^2 + 4x + 36$, $x + 2$
   (c) $f(x) = -2x^4 + x^3 + 12x^2 - 23x + 12$, $x + 3$
5. Compute each of the following roots in terms of i.
   (a) $\sqrt{-16}$ (b) $\sqrt{-25}$ (c) $\sqrt{36}$ (d) $-\sqrt{-169}$
6. Compute the following powers of i.
   (a) i^7 (b) i^{21} (c) i^{35} (d) i^{82}
7. Solve the following equations. You may leave the solutions in a form with radicals.
   (a) $x^2 + 2x + 5 = 0$ (b) $2x^2 - 3x + 4 = 0$
8. Perform the following additions, subtractions and multiplications. If you have a calculator that handles complex numbers, check your answers on the calculator.
   (a) $(3 - 4i) + (7 + 6i)$ (b) $(1 + 4i) - (8 + 5i)$
   (c) $(8 - 2i) + (3 - 9i)$ (d) $(1 - 2i)(7 + 5i)$
   (e) $(6 + 4i)(7 + i)$ (f) $(2 - 3i)^2$
9. Divide as indicated and write the answer in the standard form $a + bi$. If you have a calculator that handles complex numbers, check your answers on the calculator.
   (a) $\dfrac{1 + 4i}{-2 - 3i}$ (b) $\dfrac{6 - 2i}{1 + i}$ (c) $\dfrac{5 - 3i}{8 + 3i}$
10. Use synthetic division to find the value of k for which $x - 2$ is a factor of $x^3 - 6x^2 + kx - 10$.
11. Use synthetic division to determine whether $x + 2i$ is a factor of
   $$f(x) = x^3 + (-4 + 2i)x^2 + (1 - 8i)x + 2i$$
12. Use synthetic division with an appropriate divisor to compute the value of $f(-3i)$ for
   $$f(x) = 2x^3 + (2 - 3i)x^2 + (4 - i)x + 5i$$
13. Construct a polynomial of degree 3 with zeros -3, 2, and 5.
14. Given that 3 is a zero of
   $$f(x) = x^3 + x^2 - 17x + 15$$
   find the other zeros.
15. Construct a polynomial $f(x)$ with zeros 1, 2, and -3 such that $f(3) = 36$.
16. Find all the zeros of the following polynomials.
   (a) $f(x) = 2x^3 - 3x^2 - 11x + 6$
   (b) $f(x) = x^3 + \frac{5}{2}x^2 + \frac{1}{2}x - 1$
   (c) $f(x) = x^3 - 2x^2 + 3x - 6$
   (d) $f(x) = x^3 + 5x^2 + 9x + 5$

CHAPTER 4 TEST

1. Use synthetic division to find the quotient and remainder when $f(x) = x^3 - 2x^2 + 5x - 2$ is divided by $g(x) = x - 3$.
2. Use synthetic division with an appropriate divisor to compute the value of $f(3)$ for

$$f(x) = x^3 + 5x^2 - 4x - 3$$

Check your answer by calculating $f(3)$ with a calculator.

3. Use synthetic division to determine whether 2 is a zero of $f(x) = 2x^3 + 6x^2 - 4x - 5$. Check your answer by calculating $f(2)$ with a calculator.
4. Compute each of the following roots in terms of i.

(a) $\sqrt{-144}$ (b) $-\sqrt{-81}$

5. Compute the following powers of i.

(a) i^9 (b) i^{23}

6. Solve the equation $2x^2 + 3x + 7 = 0$. You may leave the solutions in a form involving radicals.
7. Add and multiply as indicated. If you have a calculator that handles complex numbers, check your answers on the calculator.

(a) $(2 - 5i) + (3 + 7i)$

(b) $(1 + 6i)(3 - 4i)$

8. Divide $(1 + 3i)/(2 - 5i)$. Write the answer in the standard form $a + bi$. If you have a calculator that handles complex numbers, check your answers on the calculator.
9. Use synthetic division to find the value of k for which $x - 2$ is a factor of $x^3 - 3x^2 + kx - 8$.
10. Use synthetic division to determine whether $x + 3i$ is a factor of

$$f(x) = x^3 + (-4 + 3i)x^2 + (1 - 8i)x - 12 + 3i$$

11. Given that 2 is a zero of

$$f(x) = x^3 - 4x^2 - 11x + 30$$

find the other zeros.

12. Find all the zeros of the polynomial

$$f(x) = 2x^3 + 3x^2 - 9x - 10$$

5 Exponential and Logarithmic Functions

John Napier (1550–1617)
(© Library of Congress)

The function $f(x) = k(2^{x/c})$, where c and k are constants, is an example of an exponential function. Observe that the variable x appears as an exponent, hence the name exponential function. The inverse of an exponential function is called a logarithmic function. Logarithms were invented by John Napier (1550–1617) of Merchiston, Scotland. One of the greatest curiosities of the history of science is that logarithms were investigated before exponential functions. Their invention was timed admirably however. Logarithms were used by Kepler in his study of planetary orbits and by Galileo in his study of the stars. It has been said that "the invention of logarithms, by shortening the labors, doubled the life of the astronomer."

Exponential functions are used to describe the growth and decay of many natural phenomena. The growths of population, industrialization, food production, consumption of nonrenewable resources, and pollution are all described by exponential functions. The graphs in Figures 5.1 and 5.2 show the exponential growth of world population and the exponential decay of the radioactive substance carbon-14. Figure 5.1 shows how world population is exploding. It is predicted to grow from 5.94 billion in 1998 to more than 20 billion in less than 80 years. Figure 5.2 shows how the decay of carbon-14 is used to date artifacts. The funeral boat of the Egyptian Sesostris III was found to be about 3600 years old. Dr. Willard Libby of the University of Chicago received a Nobel Prize in 1960 for developing this method of carbon dating.

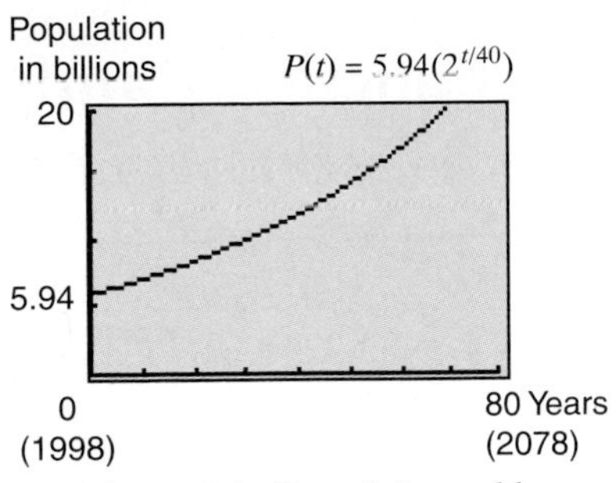

Figure 5.1 Growth in world population.

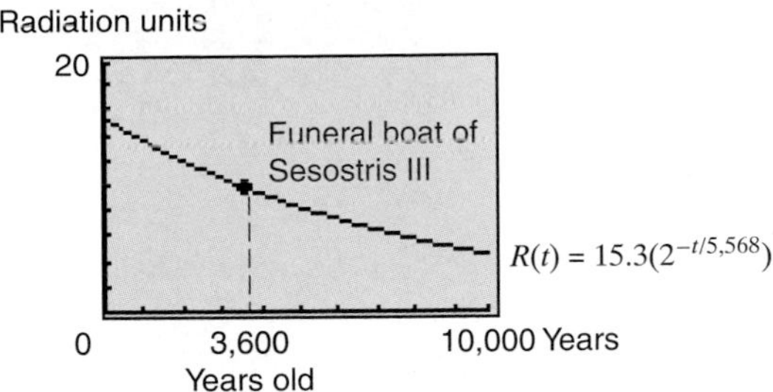

Figure 5.2 Decay of carbon-14–used to date artifacts

APPLICATION The Bacterium *Escherichia coli*

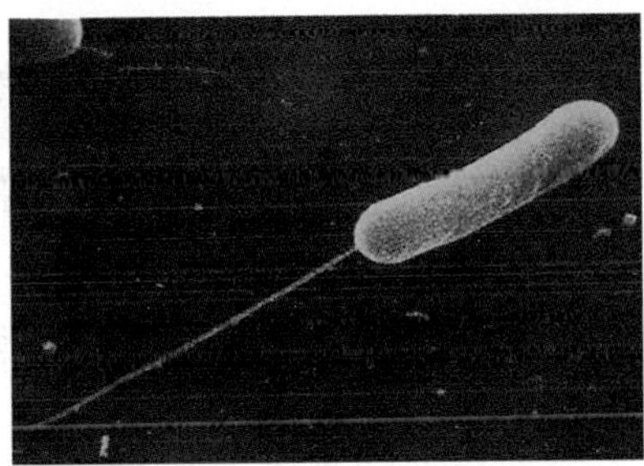

Escherichia coli are rod-shaped bacteria found in our intestines. Most strains are harmless, but 0157:H7 recently caused hamburger contamination in some regions. *(© David M. Phillips/ Visuals Unlimited)*

The bacterium *Escherichia coli* (commonly referred to as *E. coli*) not only causes milk to sour but also sickened 500 Washington State residents in 1993 after they ate hamburgers. Supermarkets recalled hamburgers from 124 stores across New England in the summer of 1998 after tests confirmed the presence of *E. coli* in meat samples from two stores. An outbreak of *E. coli* at a county fair in northern New York in August 1999 killed a 3-year-old girl and sickened about 85 other people. The bacterium made the August 3, 1998, cover of *TIME Magazine* under the heading "The Killer Germ." It is believed that *E. coli* is most commonly spread through undercooked meat and unwashed fruits and vegetables. The study of bacteria such as this is carried out in The Federal Centers for Disease Control in Atlanta. In this chapter, we study the growth of *E. coli* in laboratory conditions at 37°C. The graph of the rapid growth is shown in Figure 5.3. The algebraic analysis tells us that an initial count of 1000 of this dangerous bacterium grows to more than 40,000 in less than 100 minutes. (*This application is discussed in Section 5.1, Example 5, pages 367–368.*)

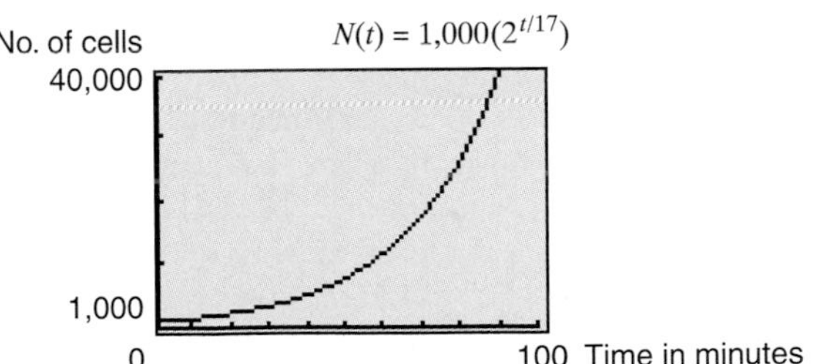

Figure 5.3 Growth of *Escherichia coli* in the laboratory at 37°C.

5.1 Exponential Functions and the Growth of *Escherichia coli*

- Exponential Functions • Exponential Growth • Exponential Decay
- Shifts and Reflections • Doubling Time of an Exponential Function
- Population Growth • Bacterial Growth

EXPONENTIAL FUNCTIONS

Let us make the distinction between algebraic functions and exponential functions clear by looking at examples. Consider the following algebraic functions of various types:

$$f(x) = 4x^3 + 5x - 3, \qquad g(x) = (7x^2 - 2x + 3)(x^7 - 5),$$

$$h(x) = \frac{2x^2 + x - 5}{x^2 + 1}, \qquad k(x) = (2x^2 + 3x + 1)^{3/2}$$

Observe that the variable x enters into each of these functions as a base and that each exponent is constant. In contrast, consider the function

$$f(x) = 2^x$$

The variable x is an exponent in this function, and the base is the constant 2. Because the variable is an exponent, the function is called an **exponential function.**

We now lay the mathematical groundwork for using such functions. The most basic types of exponential functions are of the form $f(x) = a^x$, where a is a positive real number not equal to one.

Exponential Function The function $f(x) = a^x$, where $a > 0$, $a \neq 1$, and x is a real number, is called the **exponential function** with **base** a.

For example, $f(x) = 2^x$ is the exponential function with base 2, and $g(x) = (\frac{1}{3})^x$ is the exponential function with base $\frac{1}{3}$.

GRAPH OF $f(x) = 2^x$

We now look at graphs of exponential functions. Let us sketch the graph of $f(x) = 2^x$ by plotting points. This will lead to graphs of more general exponential functions. Construct a table.

x	-3	-2	-1	0	1	2	3
$f(x) = 2^x$	$2^{-3} = \frac{1}{8}$	$2^{-2} = \frac{1}{4}$	$2^{-1} = \frac{1}{2}$	$2^0 = 1$	$2^1 = 2$	$2^2 = 4$	$2^3 = 8$

This table leads to the points $(-3, \frac{1}{8})$, $(-2, \frac{1}{4})$, $(-1, \frac{1}{2})$, $(0, 1)$, $(1, 2)$, $(2, 4)$, $(3, 8)$. Plot these points and draw a smooth curve through them to get the graph of $f(x) = 2^x$ (Figure 5.4).

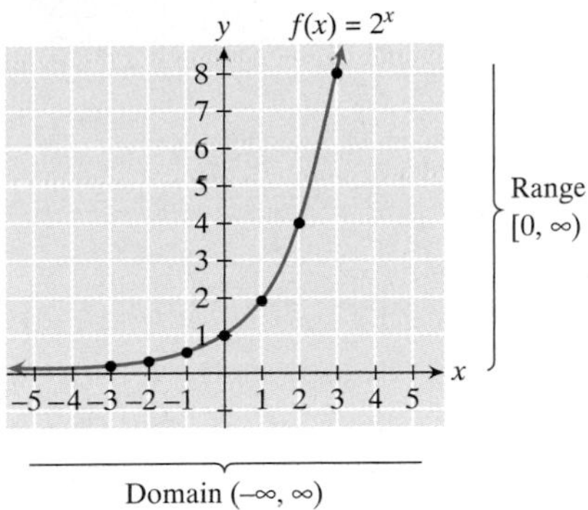

Figure 5.4 The function $f(x) = 2^x$.

Note that we have only discussed rational exponents so far. We have not, for example, discussed values of 2^x where x is an irrational number, such as 2^{π}. We shall assume that the values of exponential functions, such as 2^x, at irrational values of x are such that the graph is a continuous smooth curve. The domain of $f(x) = 2^x$ is the set of real numbers and the range is $(0, \infty)$.

Observe that, as x decreases, the graph gets closer and closer to the x-axis but never meets the axis. In symbols, $f(x) \to 0$ as $x \to -\infty$. The x-axis is a horizontal asymptote. On the other hand, $f(x)$ increases indefinitely as x increases. We write $f(x) \to \infty$ as $x \to \infty$.

Graphing Exponential Functions with a Calculator

We illustrate the graph of $f(x) = 2^x$ on a calculator (Figure 5.5). Note that the graph appears to touch the negative x-axis in this window. The resolution of the calculator screen is not sufficient to reveal that the curve does get closer and closer to the axis without meeting it.

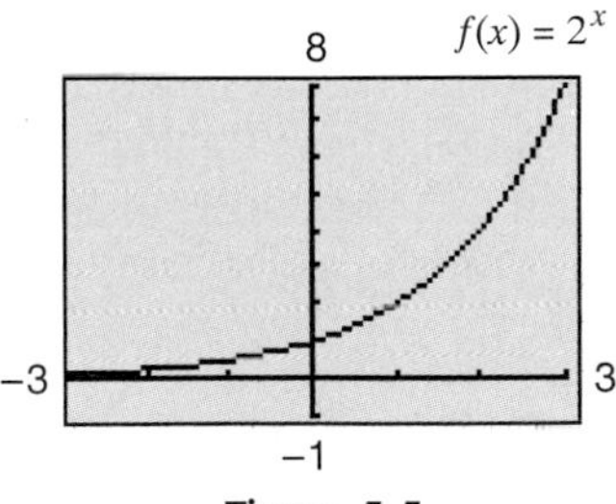

Figure 5.5

We can generalize the preceding observations to any exponential function of the same form as $f(x) = 2^x$ as follows.

Behavior of $f(x) = a^x$ where $a > 1$

The domain of f is $(-\infty, \infty)$. The range of f is $(0, \infty)$. The x-axis is a horizontal asymptote. $f(x)$ increases indefinitely as x increases (see Figure 5.6).

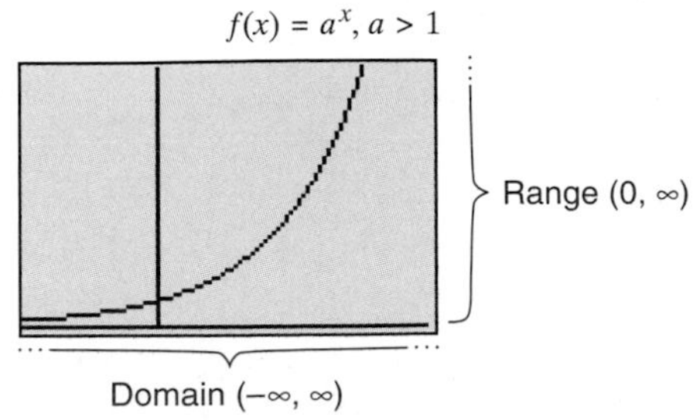

Figure 5.6

Let us now determine the relative locations of the graphs of exponential functions of the type $f(x) = a^x$ where $a > 1$.

Exploration 1

Graph the functions $f(x) = 2^x$, $g(x) = 3^x$ on a calculator. Do these graphs intersect at a point? Record the intervals of x where $f(x) > g(x)$ and where $f(x) < g(x)$.

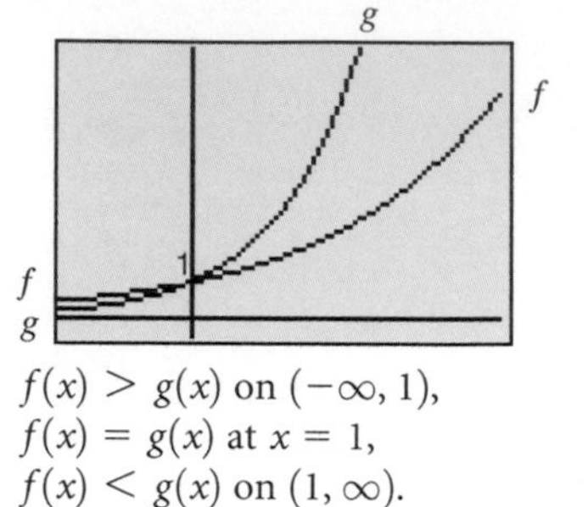

$f(x) > g(x)$ on $(-\infty, 1)$,
$f(x) = g(x)$ at $x = 1$,
$f(x) < g(x)$ on $(1, \infty)$.

Behavior of $g(x) = a^x$ where $0 < a < 1$

Let us now look at exponential functions of the form $g(x) = a^x$ where $0 < a < 1$. $g(x) = (\frac{1}{2})^x$ is such an exponential function. Because $\frac{1}{2}$ can be written 2^{-1}, we can write $g(x) = (\frac{1}{2})^x$ in the form $g(x) = 2^{-x}$. It is customary to write exponential functions where $0 < a < 1$ in this form, with a negative exponent. We now ask the student to investigate the shape of the graph of such an exponential function.

Exploration 2

Graph the functions $f(x) = 2^x$ and $g(x) = 2^{-x}$ on a calculator. How would you describe the relationship between the graphs?
The important features of these graphs are as follows.

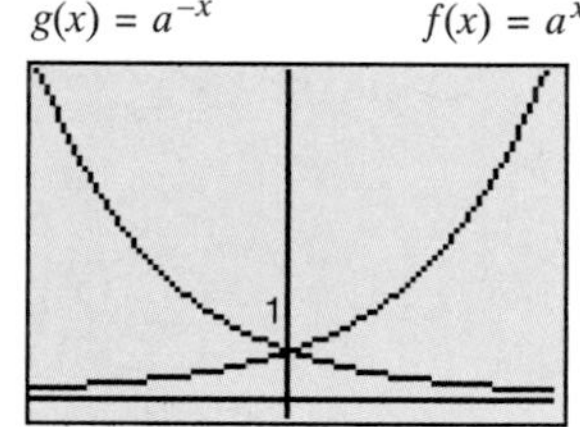

Graph of $g(x)$ is mirror image of $f(x)$ in y-axis (because $g(x) = f(-x)$).

CHARACTERISTICS OF GRAPHS OF EXPONENTIAL FUNCTIONS

$f(x) = a^x$, with $a > 1$: $f(x)$ is always positive. As x decreases, $f(x)$ decreases. The graph gets closer and closer to the *x-axis* but never meets it. The x-axis is a horizontal asymptote. As x increases, $f(x)$ increases indefinitely. This type of function is called one of **exponential growth.**

$g(x) = a^{-x}$, with $a > 1$: $g(x)$ is always positive. As x decreases, $g(x)$ increases indefinitely. As x increases, $g(x)$ decreases. The graph gets closer and closer to the x-axis but never meets it. The x axis is a horizontal asymptote. This type of function is called one of **exponential decay.**

The y-intercept of each graph is 1, because $f(0) = g(0) = a^0 = 1$.

Vertical and Horizontal Shifts and Reflections

Graphs of certain functions involving exponential expressions can be sketched from these graphs through the use of shifts and reflections or by stretching or compressing, as in Section 3.4.

Example 1 Using a Shift to Sketch the Graph of an Exponential Function

Use the known graph of $f(x) = 2^x$ to sketch the graph of $g(x) = 2^x + 3$. Check your answer using a graphing calculator. Give the domain and range of f.

SOLUTION The graph of $g(x) = 2^x + 3$ can be obtained from that of $f(x) = 2^x$ by shifting the graph upward 3 units (Figure 5.7a). We verify this graph using a graphing calculator (Figure 5.7b). The domain of g is the same as that of f; namely, $(-\infty, \infty)$. The range of f is $(0, \infty)$. The upward shift of f means that the range of g is $(3, \infty)$.

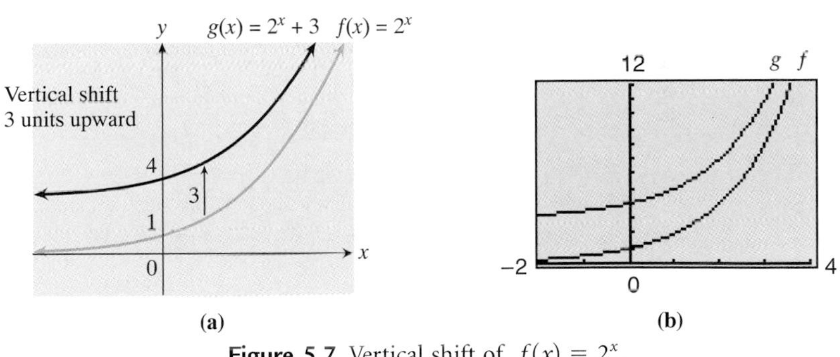

Figure 5.7 Vertical shift of $f(x) = 2^x$.

Self-Check 1 Sketch the graph of $g(x) = 3^{x+2} - 4$ using horizontal and vertical shifts of $f(x) = 3^x$. State the shifts used. Check your answer using a graphing calculator. Give the domain and range of g.

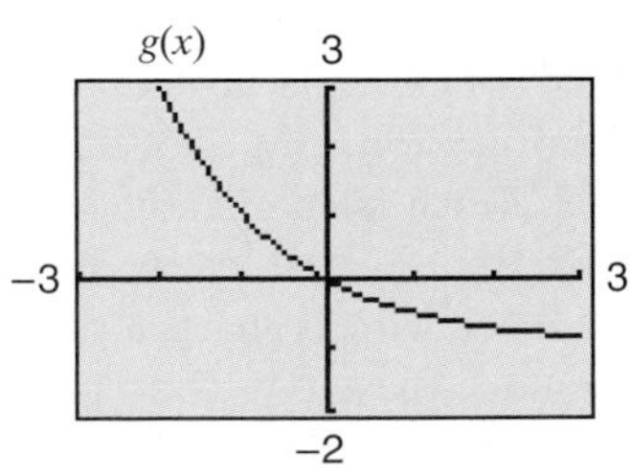

Figure 5.8

Example 2 Using a Reflection and a Shift to Sketch a Graph

The graph of $g(x)$ in Figure 5.8 can be obtained from the known graph of $f(x) = 2^x$ by a reflection and a vertical shift. Find the function $g(x)$. Check your answer by duplicating the given graph of $g(x)$ on your calculator.

SOLUTION We graph $f(x) = 2^x$ as our starting graph (Figure 5.9). A reflection in the y-axis gives $h(x) = 2^{-x}$, a decreasing curve. A vertical downward shift of 1 gives $g(x) = 2^{-x} - 1$, a graph that passes through the origin. This is the given graph. Thus, $g(x) = 2^{-x} - 1$.

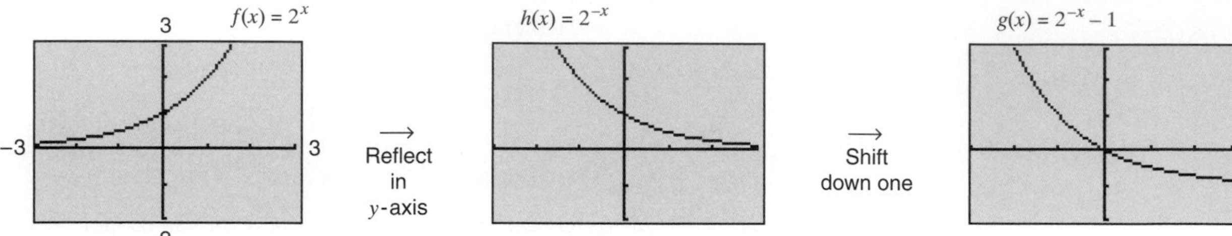

Figure 5.9 Reflection and shift of $f(x) = 2^x$.

Example 3 Stretching a Graph

Use the known graph of $f(x) = 2^x$ to sketch the graph of $g(x) = 3(2^x)$. Check your answer using a graphing calculator.

SOLUTION The graph of $g(x) = 3(2^x)$ can be obtained from that of $f(x) = 2^x$ by stretching it vertically by a factor of 3 (Figure 5.10a). We verify this graph using a graphing calculator (Figure 5.10b).

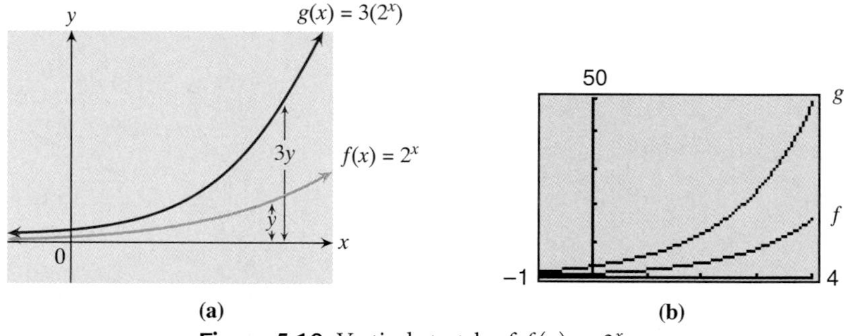

Figure 5.10 Vertical stretch of $f(x) = 2^x$.

Applications . . .

The growths of many phenomena are exponential in nature; they can be approximately described by exponential functions. For example, growth of population, bacteria, industrialization, food production, consumption of nonrenewable resources, and pollution are exponential. Much human activity, from the use of fertilizer to the expansion of cities, can be represented by exponential growth. An exponential function that describes such behavior is characterized by the initial value and doubling time. We now introduce these two concepts and then look at specific applications in the fields of population and bacterial growth, and computing.

DOUBLING TIME OF AN EXPONENTIAL FUNCTION

Consider the exponential function

$$f(x) = k(2^{x/c})$$

where k and c are constants and c is positive. Let us interpret k and c. We see that

$$f(0) = k(2^0) = k$$

Thus, k is the value of $f(x)$ *at* $x = 0$. k is called the **initial value** of f.

Let us now interpret the constant c. Using the properties of exponents,

$$f(x + c) = k[2^{(x+c)/c}] = k[2^{(x/c+1)}] = k(2^{x/c}2^1) = 2k(2^{x/c}) = 2f(x)$$

Thus, an increase of c in x doubles the value of f. c is called the **doubling time** of f. The doubling time is an indication of how rapidly the growth is taking place.

EXPONENTIAL GROWTH

Exponential growth can be described by the function

$$f(x) = k(2^{x/c}), \qquad c > 0$$

where k is the initial value, and c is the doubling time.

Consider the exponential function $f(x) = 5(2^{x/3})$. The initial value of this function is 5 and its doubling time is 3. Let us examine its graph (Figure 5.11a). We see that $f(0) = 5, f(3) = 10$, and $f(6) = 20$. The initial value is 5, and the value of f is doubling every 3 units.

The table of f in Figure 5.11b further illustrates the doubling aspect of this exponential function every 3 units. As x increases 0, 3, 6, 9, . . . units, $f(x)$ increases 5, 10, 20, 40, . . . units. The real impact of exponential growth lies in its doubling characteristic. A few doublings can lead quickly to very large numbers. Predict the values of $f(21)$ and $f(24)$. Check your answers using a graph.

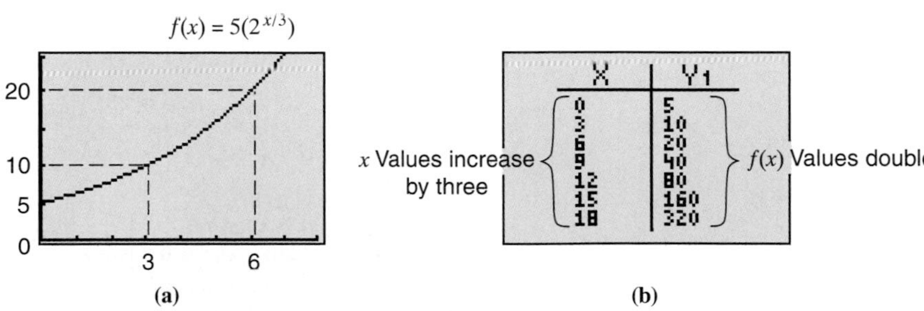

Figure 5.11 Graph and table views of doubling time.

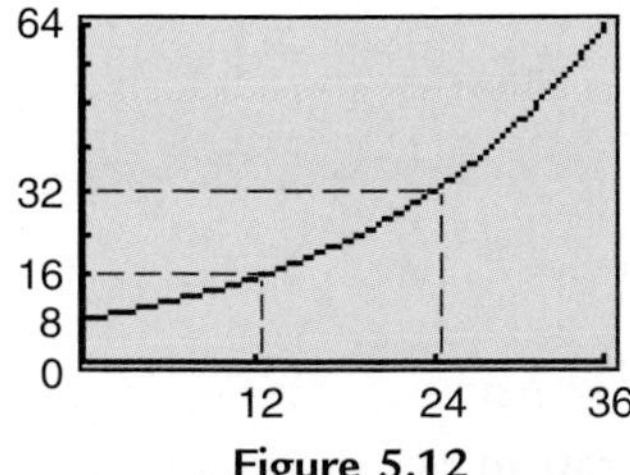

Figure 5.12

Further Ideas 1

(a) The graph of Figure 5.12 describes an exponential function $f(x) = k(2^{x/c})$. Find the function and give its initial value and its doubling time.

(**b**) Consider the exponential decay $g(x) = k(2^{-x/c})$, $c > 0$. Show that an increase of c in x halves the value of g. It is characteristic of all exponential decay that the time it takes a quantity to halve is constant. c is called the **half-life** of the function $g(x)$. Graph the function $g(x) = 128(2^{-x/4})$ and use the graph to demonstrate, in the manner of Example 3, that the half-life of g is 4. Construct a table for values of $g(x)$ for $x = 0, 4, 8, 12, 16, 20, 24$. Discuss how this table reveals that the half-life of $g(x)$ is 4.

Example 4 Population Growth

The world population in August 1998 was approximately 5.94 billion. World population is currently observed to be doubling approximately every 40 years. Let us construct a function that describes world population growth and use it to predict the population for the year 2050.

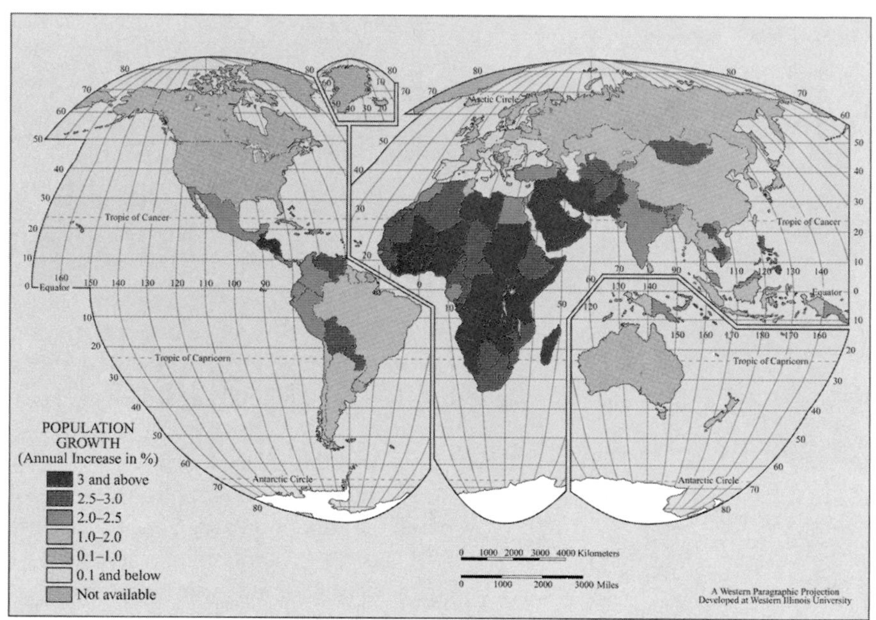

Because the time it takes world population to double is constant, the population is growing exponentially. Thus, population P at time t can be described by the function

$$P(t) = k(2^{t/c})$$

for suitable values of k and c. Let t be time in years measured from 1998. World population in 1998 was 5.94 billion. Thus, $k = 5.94$. The doubling time is 40 years, giving $c = 40$. Therefore,

$$P(t) = 5.94(2^{t/40})$$

The year 2050 is 52 years after 1998, corresponding to $t = 52$.

$$P(52) = 5.94(2^{52/40}) = 14.63, \qquad \text{rounded rate to two decimal places}$$

According to our model, if the current growth rate continues, the world population can be expected to be up from 5.94 billion in 1998 to about 14.63 billion by the year 2050.

We graph the predicted world population for the next 80 years (Figure 5.13). Observe how the population increases from 5.94 billion to more than 20 billion in less than 80 years. We see how the doubling aspect of exponential growth leads rapidly to very large numbers.

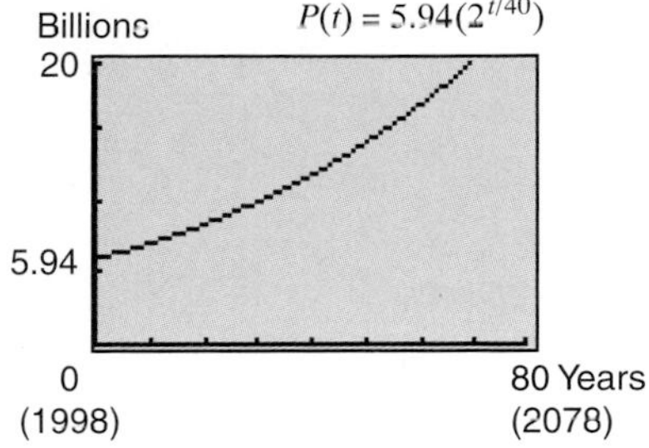

Figure 5.13 World population growth.

Further Ideas 2 In 1991 there were 400,000 programmers using the C++ computer language. The number was doubling on the average every 7.5 months.* Show that if this rate of growth had prevailed, then every person on the planet would have become a C++ programmer before the year 2000. This illustrates the care that must be taken in interpreting predictions made with mathematical models. There is always the stipulation "if things remain unchanged." As was mentioned in Section 1.2 regarding the use of linear regression, one must always look closely at the circumstances surrounding the situation. In this case, the model revealed the phenomenal growth that took place in the number of C++ users, but for a limited time. The programming language Java, which is suitable for programming for the Internet, then came along.

Example 5 Bacterial Growth

Bacterial growth is described by an exponential function. Let us first see why this is to be expected—it gives us insights into why many natural phenomena can be expected to grow exponentially. We shall then look at the growth of the bacterium *Escherichia coli.* Numerous chemical reactions are involved in the multiplication of bacteria; cells divide, a single cell giving rise to two cells. Those two cells each divide, giving rise to four cells. The four cells give rise to eight cells and so on. If we follow the growth from one cell we get the following progression:

$$1 \to 2 \to 4 \to 8 \to 16 \to 32 \to \cdots$$

This growth can be expressed

$$2^0 \to 2^1 \to 2^2 \to 2^4 \to 2^5 \to \cdots \to 2^n \to \cdots$$

The exponents in this sequence represent the number of generations in the growth. For example, after four generations, the single cell will have given rise to 2^4 cells.

It is known that the time it takes a cell of a given bacterium to divide under given conditions is constant; it is the same for all cells. This is the doubling time c of the bacterium. The time required for bacteria to develop through n generations is $t = nc$. Thus, $n = t/c$. If we start with a single bacterium, then it will grow

*Chuck Allison "Jumping into Java", *C/C++ Users Journal.* 1999, 17(1):69.

to 2^n bacteria in time t; that is, $2^{t/c}$ bacteria. If we start with k bacteria, they will grow to $k(2^{t/c})$ in time t. Thus, the growth of bacteria is described by the exponential function

$$N(t) = k(2^{t/c})$$

where t is time, $N(t)$ is the number of cells of the bacterium at time t, k is the initial count, and c is the doubling time (called the **generation time** in bacteriology).

Study has determined that the important bacterium *E. coli* has a doubling time of 17 minutes at 37°C in a laboratory; $c = 17$. This bacterium is also used by molecular biologists in experiments involving genetic manipulation. Consider an initial count of 1000 bacteria; $k = 1000$. Thus, the growth of the bacterium can be described under these conditions by the function

$$N(t) = 1000(2^{t/17})$$

where t is time in minutes, and $N(t)$ is the number of cells of the bacterium at time t. Let us compute N at convenient values of t (multiples of 17).

t	0	17	34	51	68	85	102
$N(t)$	1,000	2,000	4,000	8,000	16,000	32,000	64,000

This table confirms that the count of bacteria doubles every 17 minutes.

Let us use the graphing calculator to determine how long it takes the population to grow to 20,000 (Figure 5.14). After tracing and zooming a number of times, we find that the population is 20,000 after 73.47 minutes.

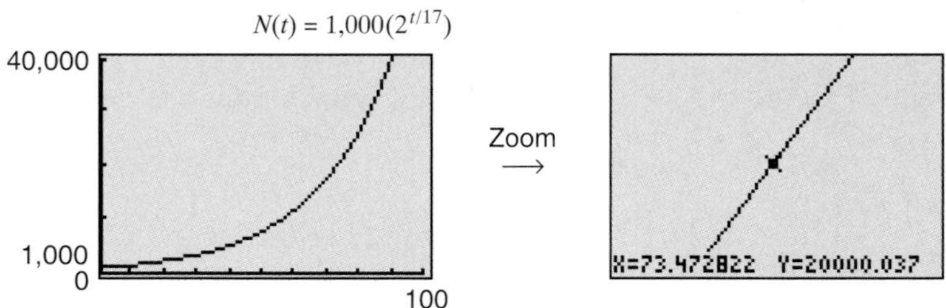

Figure 5.14 Growth of *Escherichia coli.*

Further Ideas 3 The bacterium *Bacillus subtilis* can be found in soil and in the contents of intestines. Its doubling time is 26 minutes. Give the function that describes the growth of this bacterium for an initial count of 200 bacteria. Use a table to determine when the population will reach 3200? Verify your answer using a graph.

GROUP DISCUSSION

Good Grades

You are given a dollar from a benevolent aunt at the start of your university education. She promises that she will double it every month that you are in school for the period that you are making good progress on your studies, keeping your grade point average 3.0 or higher. If it drops below 3.0, it will discontinue, but will start again once it reaches 3.0. How much could it amount to in 4 years? Discuss.

Answer to Self-Check Exercise

1. Horizontal shift 2 units to the left, 4 units down. Domain of g is $(-\infty, \infty)$ and the range is $(-4, \infty)$. See graph below.

Answers to Further Ideas Exercises

1. **(a)** $f(x) = 8(2^{x/12})$. Initial value $= 8$, doubling time $= 12$.
   (b) $g(x + c) = k(2^{-(x+c)/c}) = k(2^{(-x/c-1)}) = k(2^{-x/c}2^{-1}) = \frac{1}{2}k(2^{-x/c}) = \frac{1}{2}g(x)$.

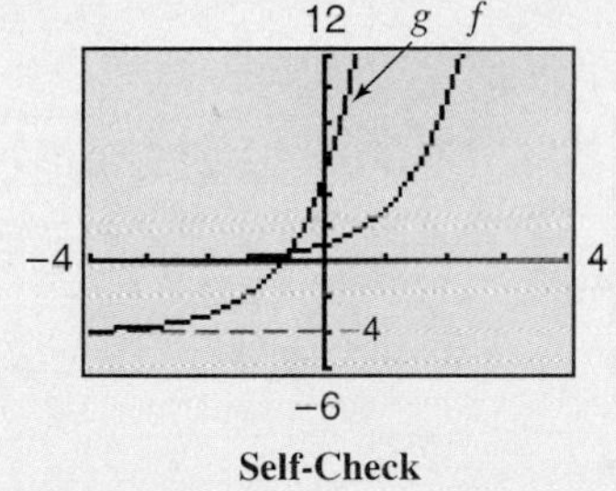

Self-Check

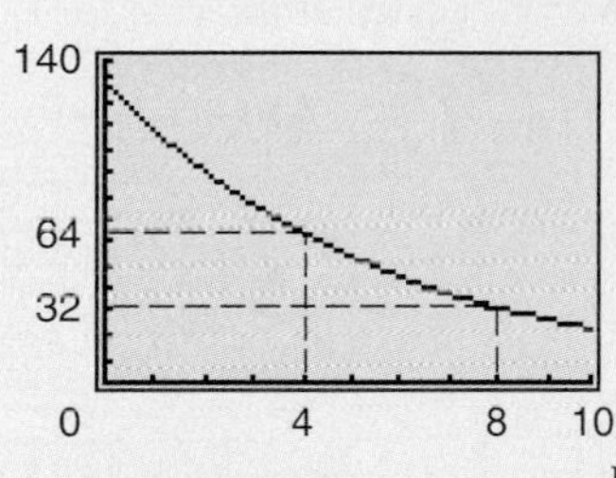

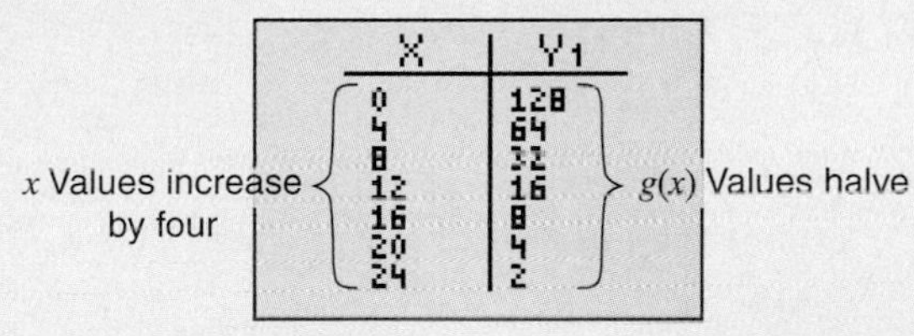

X	Y1
0	128
4	64
8	32
12	16
16	8
20	4
24	2

Further Ideas 1(b)

2. 400,000 is 0.0004 billion. 7.5 months is 0.625 years. $N(t) = 0.0004(2^{t/0.625})$. $t = 0$ in 1991, $t = 9$ in the year 2000. $N(9) = 8.65$ billion programmers, to two decimal places. The model of Example 4 predicts world population for 2000 to be $P(t) = 5.94(2^{t/40})$ with $t = 2$. $P(2) = 6.15$ billion.
3. $N(t) = (200)2^{t/26}$. 104 minutes.

EXERCISES 5.1

In Exercises 1–4, sketch the graphs of the functions by plotting the points corresponding to $x = -3, -2, -1, 0, 1, 2, 3$. Check your answers by drawing the graphs on a calculator in the window $-3 \le x \le 3$, $0 \le x \le 10$.

1. $f(x) = 3^x$
2. $f(x) = 2^{-x}$
3. $f(x) = 2^{3x}$
4. $f(x) = (\frac{1}{3})^x$

In Exercises 5–8, give the results, rounded to four decimal places.

5. Let $f(x) = 2^x$. Compute **(a)** $f(1.7)$, **(b)** $f(3.42)$, **(c)** $f(-2.1)$.

6. Let $f(x) = 3.7^x$. Compute **(a)** $f(1.5)$, **(b)** $f(0.61)$, **(c)** $f(-4.8)$.

7. Let $g(x) = 4^{-x}$. Compute **(a)** $g(1.6)$, **(b)** $g(3.75)$, **(c)** $g(-5.2)$.

8. Let $h(x) = 2.5^{-x}$. Compute **(a)** $h(2.34)$, **(b)** $h(-1.23)$, **(c)** $h(-3.2473)$.

In Exercises 9–14, (a) use vertical and horizontal shifts of the known graph of $f(x)$ to sketch the graph of $g(x)$. Describe the shifts used. (b) Check your results using a graphing calculator. (c) Give the domain and range of g.

9. $g(x) = 2^x + 3, f(x) = 2^x$
10. $g(x) = 4^x - 2, f(x) = 4^x$
11. $g(x) = 3^{-x} + 5, f(x) = 3^{-x}$
12. $g(x) = 2^{x-4}, f(x) = 2^x$
13. $g(x) = 3^{x+1}, f(x) = 3^x$
14. $g(x) = 6^{-(x-2)}, f(x) = 6^{-x}$

In Exercises 15–20, **(a)** use transformations of the known graph of $f(x)$ to sketch the graph of $g(x)$. Describe the transformations used. **(b)** Check your results using a graphing calculator. **(c)** Give the domain and range of g.

15. $g(x) = 5^{-(x+3)}, f(x) = 5^{-x}$
16. $g(x) = 2^{x-4} + 3, f(x) = 2^x$
17. $g(x) = 4^{-(x-5)} - 2, f(x) = 4^{-x}$
18. $g(x) = 2^{x+3} + 1, f(x) = 2^x$
19. $g(x) = 3(2^x), f(x) = 2^x$
20. $g(x) = 3^{2x}, f(x) = 3^x$

21. Graph $f(x) = 3^x$ and $g(x) = 4^x$ on a calculator. Determine the intervals of x over which $f > g$, $f = g$, $f < g$.

22. Let $f(x) = 3^x$. Show that $f(2 + 3) = f(2) \cdot f(3)$. Is it true that $f(a + b) = f(a) \cdot f(b)$ for any two real numbers a and b?

The graphs of $g(x)$ in Exercises 23–28 can be obtained from the known graph of $f(x) = 2^x$ by shifts and reflections. The viewing window is $-4 \le x \le 4$, $-3 \le y \le 3$. Find the functions $g(x)$ having these graphs and describe the transformations used. Check your answer by duplicating the given graph on your calculator.

23.

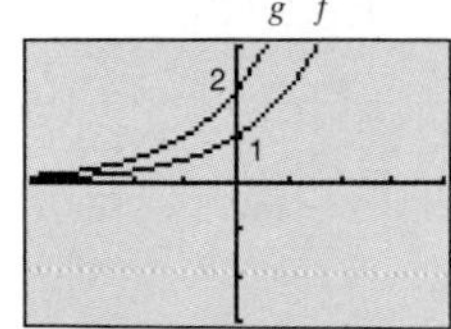

24.

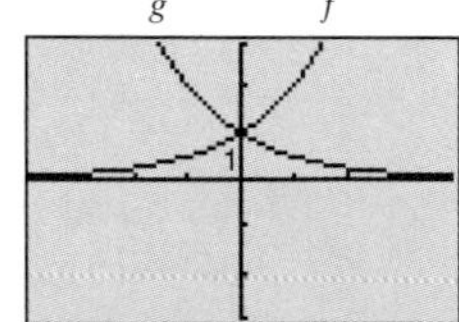

25.

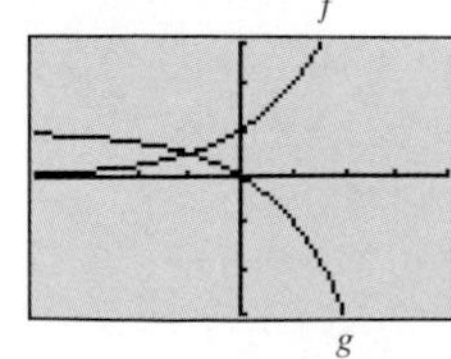

26.

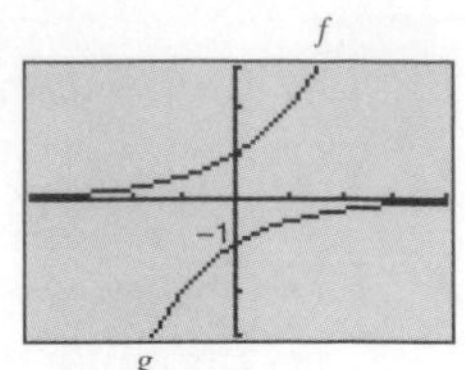

27.

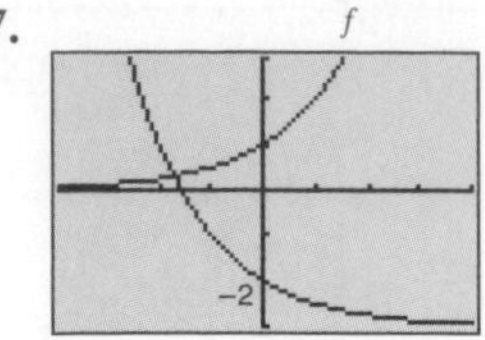

28.

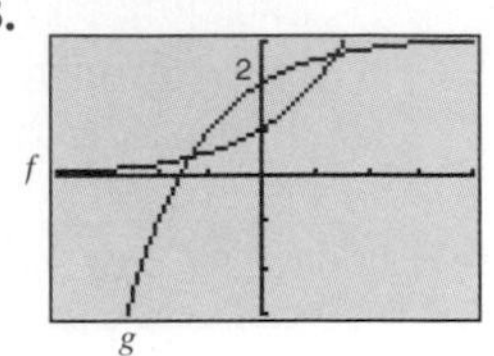

In Exercises 29–32, graphs of $f(x) = 2^x$ and a second function $g(x)$ are given in the viewing window $-6 \le x \le 6$, $0 \le y \le 8$. Each function g is of the form $g(x) = \pm 2^{\pm x + a} + b$, where a and b are integers. Find $g(x)$ using algebra. To aid in your investigation, the points of intersection of the graphs are given. Check your answer by duplicating the given graph of $g(x)$ on your calculator.

29.

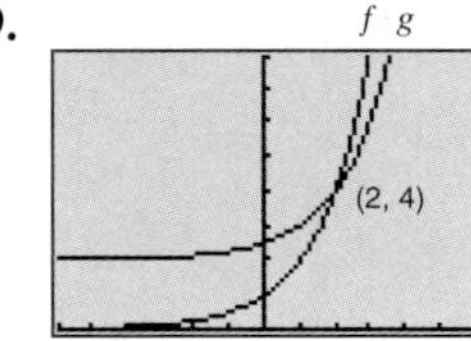

30.

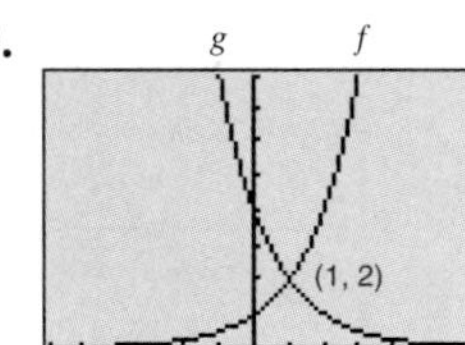

31.

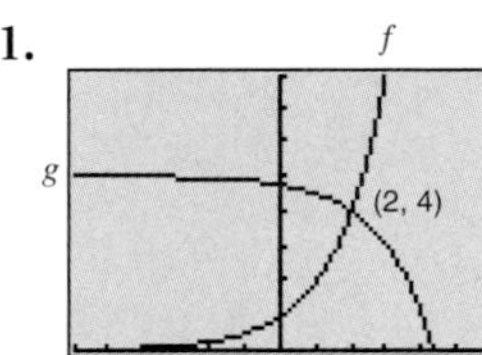

32.

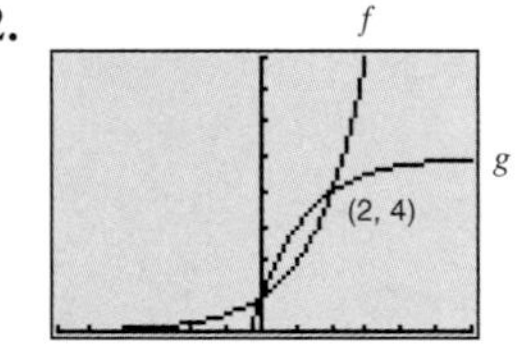

33. The following graph describes an exponential function $f(x) = k(2^{x/c})$. Give the initial value and doubling time of the function. Find the function. Verify your answer by duplicating the given graph on your calculator.

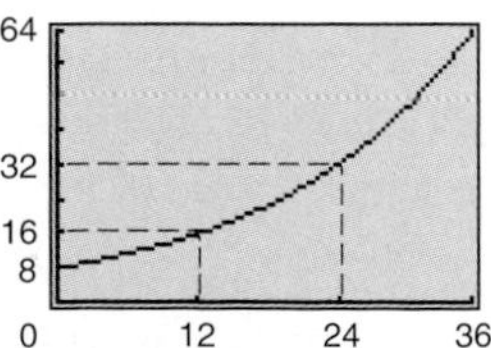

34. Match the following functions and graphs, giving an explanation for your choices. Verify your answers by duplicating the graphs on your calculator.

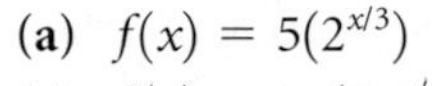

(a) $f(x) = 5(2^{x/3})$ (b) $f(x) = 7(2^{x/4})$

(c) $f(x) = 72(2^{-x/11})$ (d) $f(x) = 48(2^{-x/7})$

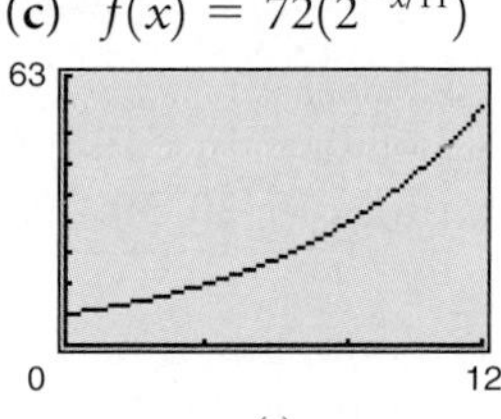

(e)

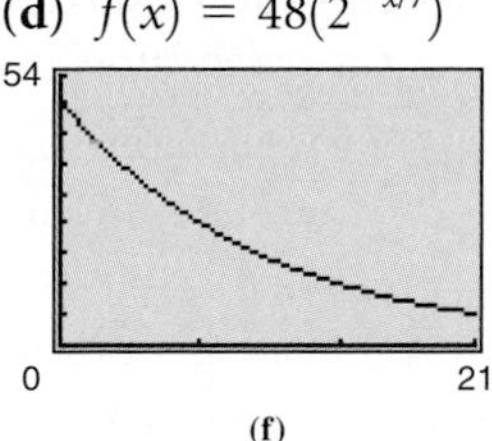

(f)

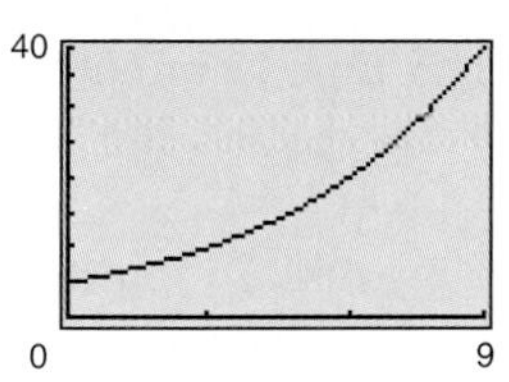

(g)

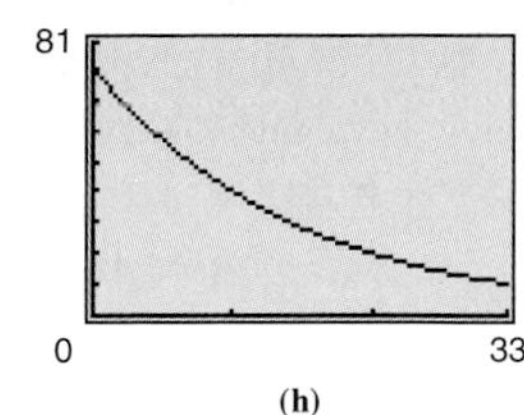

(h)

35. The following table describes an exponential function $f(x) = k(2^{x/c})$. Find the function. Verify your answer by duplicating the given table on your calculator. Give the initial value and doubling time of the function.

X	Y1
0	6
3	7.5595
6	9.5244
9	12
12	15.119
15	19.049
18	24

36. The following tables and graphs describe growth or decay of three exponential functions of the form $f(x) = k(2^{x/c})$. Give the initial value and doubling time (or halving time) of each function. Match the tables and graphs. Find each exponential function. Verify your answer by duplicating the given table and graph on your calculator.

(a)

X	Y1
0	8
1	11.314
2	16
3	22.627
4	32
5	45.255
6	64

(b)

X	Y1
0	7
3	8.8194
6	11.112
9	14
12	17.639
15	22.224
18	28

(c)

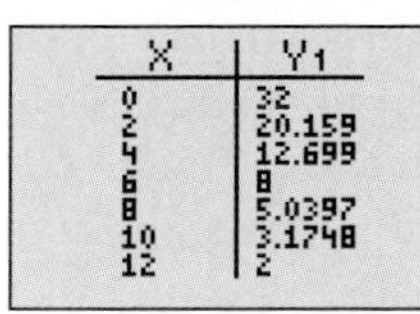

X	Y1
0	32
2	20.159
4	12.699
6	8
8	5.0397
10	3.1748
12	2

(d)

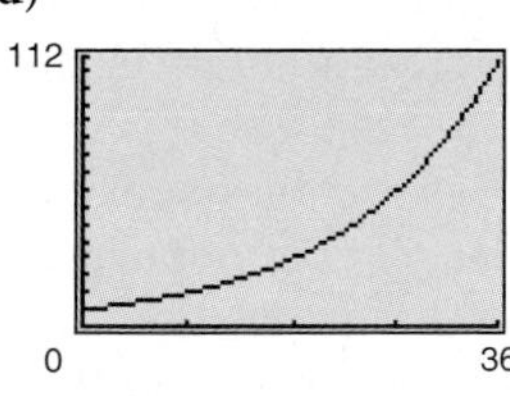

(e)

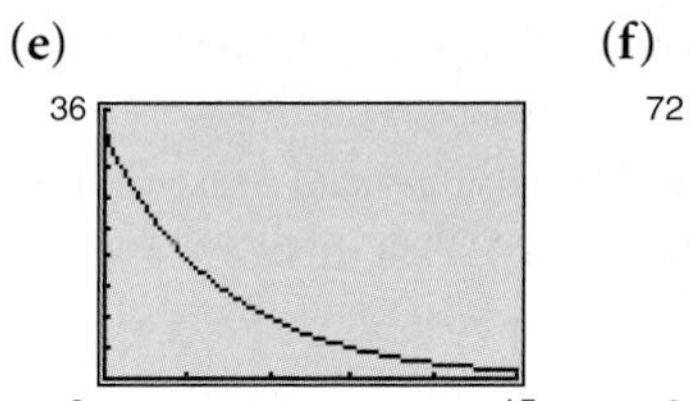

(f)

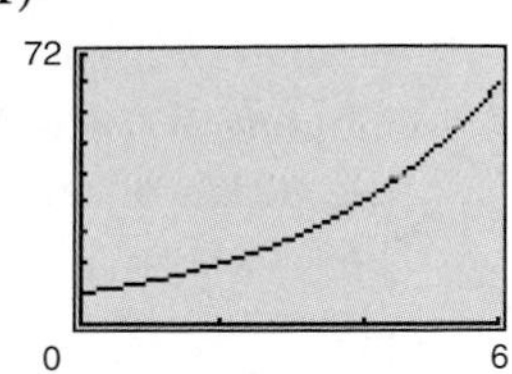

37. World Population World population is approximately described by the exponential function $P(t) = 5.94(2^{t/40})$ (see Example 4). $P(t)$ is the population in billions, and t is time in years, measured from 1998. **(a)** Predict world population for the year 2020, in billions, rounded to two decimal places. **(b)** What was the population in 1970?

38. Consumption of Crude Oil The worldwide consumption of crude oil has been growing exponentially with a doubling time of 19.8 years. The amount of oil consumed in 1995 was 7,841,777 thousand metric tons. **(a)** Construct the exponential function that describes the consumption of oil. **(b)** Predict the consumption of oil in the year 2005.

39. Moore's Law Since integrated circuits were developed, the number of transistors that can be packed on a computer chip has increased at a phenomenal rate. This rate, at which the transistor count doubles approximately every 18 months, has become an axiom known as **Moore's Law.** It is named after Gordon Moore, who first noticed the trend in the early 1960s. In 1995 there were 5.5 million transistors in the Pentium Pro chip. Predict, to the nearest thousand, the number of transistors in the "chip of 2002."

40. Population Growth The population growth of Torbay City is described by the function $P(t) = 20(2^{t/10})$. $P(t)$ is the population in thousands at time t in years, measured from 1980. If the current growth continues in this manner, when will the population reach 160 thousand?

41. Population Growth The cities of Penton and Trent have populations of 150 thousand and 200 thousand, respectively. They are growing with doubling times of 20 and 30 years, respectively. **(a)** After how many years (to the nearest year) will the population of Penton equal that of Trent, if current trends continue? **(b)** What will the populations then be, to the nearest thousand?

42. Population Growth The population growth of a certain country is described by $P(t) = 30(2^{t/45})$. t is the time measured from 1985, and P is the population in units of one million. In which year is the population predicted to reach 40 million?

43. **Population Growth of the United States** The population of the United States in 1997 was 268.5 million. (See the Williams website at www.saunders.college.com/math/williams to link to the Internet for the world and U.S. popclocks.) The population is growing exponentially, with a doubling time of about 65 years. If the current trend continues, when is the population predicted to reach 300 million?

44. **World Industrial Production** World industrial production is growing exponentially according to the equation $y = 100(2^{t/10})$. Production is measured on a scale, with the 1963 production equal to 100. Will production surpass 1000 by the year 2000?

45. **Doubling Time** The current population of a certain country is 8 million. What is the smallest doubling time (to one decimal place) for the population to reach 12 million in 15 years?

46. **Population Growth of Mexico** The population of Mexico had a doubling time of 12 years for the period 1975 to 1985. The population of Mexico in 1975 was 58.4 million. **(a)** Construct the exponential function that describes the growth in the population of Mexico based on these statistics. **(b)** Compute the predicted population for the year 1992. **(c)** Use your library or other resources to determine the actual population in 1992. Compare the two numbers. How would you modify the model in light of this recent number? What is your best estimate for the population in the year 2005?

BACTERIAL GROWTH

In Exercises 47–50, determine the functions that describe the growths of the bacteria, given the initial counts. The doubling times (in minutes) of the various bacteria are in laboratory surroundings (37°C). Compute the counts at the given times without using a calculator.

47. *Pseudomonas natriegens*—doubling time 10 minutes and initial count 100: times 10, 20, 30, 40, 50, and 60 minutes.

48. *Bacillus stearothermophilus*—doubling time 11 minutes and initial count 500: times 11, 22, 33, 44, 55, and 66 minutes.

49. *Bacillus coagulans*—doubling time 13 minutes and initial count 600: times 26, 52, 78, 104, and 130 minutes.

50. *Rhizobium melitoli*—doubling time 75 minutes and initial count 50: times 225, 450, 675, and 900 minutes.

51. **Timing Bacterial Growth** The growth of a certain bacterium is described by $N(t) = 1000\,(2^{t/25})$. How long (to the nearest minute) will it take the bacterium to reach 5000?

52. **Timing Bacterial Growth** The growth of the bacterium *Pseudomonas natriegens* is described by the equation $N(t) = k(2^{t/10})$. How long (to the nearest minute) will it take an initial count of 500 to grow to 2500?

53. **Determining Doubling Time** An initial count of 1000 of the bacterium *Bacillus subtilis* grew to approximately 1500 in 15 minutes. Find the doubling time (to the nearest minute) of this bacterium.

54. **Comparing Growths** An initial population of 100 of bacterium A, having a doubling time of 12 minutes, is isolated and studied in a laboratory. Then 20 minutes later, a population of 100 of bacterium B, having a doubling time of 8 minutes, is isolated and studied. When will the population of bacterium B equal that of bacterium A?

SECTION PROJECT

Populations of China and India

(a) The populations of China and India are increasing exponentially. China now has the world's largest population. However the population of India is catching up fast. Visit your library or the Web and find data on the populations of China and India. Use this information to construct exponential functions that describe the population growths of China and India. Give the doubling time of each population. Use the graphs of these functions to predict when the population of India is likely to surpass that of China. What is the political reason for the population of India catching up with that of China?

Section Project *(continued)*

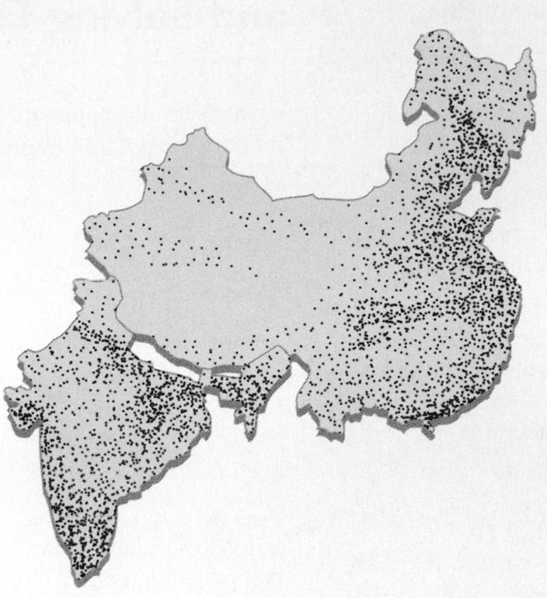

(b) We have used doubling time as the measure of exponential growth in this chapter. **Rate of growth** is also used as a measure of exponential growth. The rate of growth is expressed as a percentage per unit of time. In practice, models used are very approximate, and the following general rule of thumb is used to relate the rate of growth (r) to the doubling time (c):

$$r \approx \frac{70}{c}$$

For example, we saw that the doubling time of world population is currently 40 years in Example 4 of Section 5.1. The corresponding rate of growth is (70/40)%, or 1.75% per annum. Use your previous results to determine the growth rates of China and India. Use library or Web resources to search for statistics on the growth rates of countries. Record the growth rates of some countries that interest you. Which is the country with the largest growth rate, and how high is that rate?

(c) Show that the exact relationship between rate of growth and doubling time of the exponential function $f(t) = k2^{t/c}$ is

$$r = 100(2^{1/c} - 1)$$

Graph the functions $r(c) = 70/c$ and $r(c) = 100(2^{1/c} - 1)$. Use the graphs to show the validity of the rule of thumb $r \approx 70/c$. Support the approximation also using tables.

5.2 The Natural Exponential Function and Solving Exponential Equations

• The Natural Exponential Function • Solving Exponential Equations • Continuous Compounding • Color Control on a TV Screen • Learning Curve

The Natural Exponential Function

An important irrational number, designated by the letter e, is frequently used as a base for exponential functions. The number e is the value approached by

$$\left(1 + \frac{1}{n}\right)^n$$

as n becomes larger and larger. We give a few of the values of this expression in the following table. We see that $e \approx 2.71828$.

n	1	2	10	100	1,000	10,000	100,000	1,000,000
$\left(1 + \frac{1}{n}\right)^n$	2	2.25	2.59374	2.70481	2.71692	2.71815	2.71827	2.71828

Further Ideas 1 Graph the function $f(x) = \left(1 + \frac{1}{x}\right)^x$ on your calculator. What is the equation of the horizontal asymptote?

In Section 5.1, we discussed exponential functions $f(x) = a^x$, where a is a positive real number (not equal to one) called the base of the exponential function. We now introduce an exponential function where the base is this irrational number e. This exponential function plays a key role in much of mathematics. In this section, we will see the role it plays in the world of finance and in television technology.

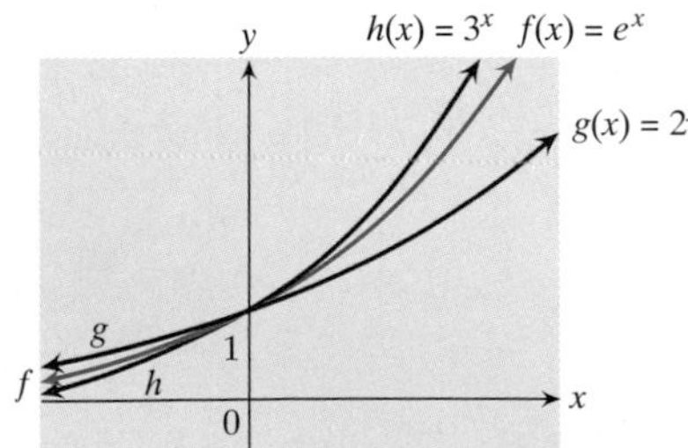

Figure 5.15 Position of the graph of $f(x) = e^x$

> ***The Natural Exponential Function*** The exponential function $f(x) = e^x$, to the base e, is called the **natural exponential function.**

The Graph of the Natural Exponential Function

Because $2 < e < 3$, the graph of $f(x) = e^x$ will lie between the graphs of $g(x) = 2^x$ and $h(x) = 3^x$ (Figure 5.15).

Computing Values and Graphing the Natural Exponential Function

The exponential key on a calculator is usually $\boxed{e^x}$. It can be used to compute values of the natural exponential function and also to enter the function into the calculator. We find the value of $e^{2.73}$ to illustrate computation. We enter the function and plot $f(x) = e^x$ to illustrate graphing (Figure 5.16).

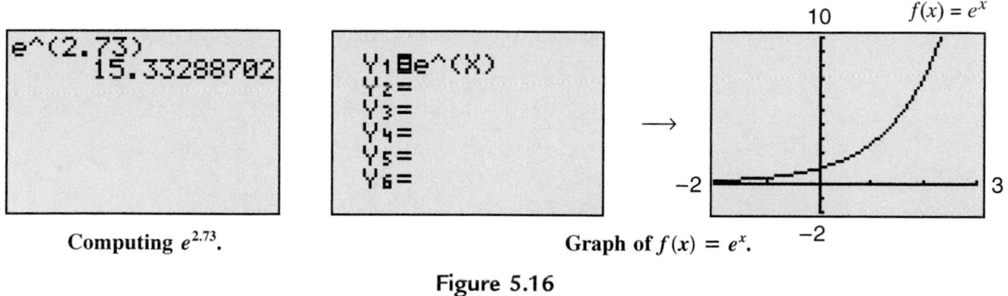

Computing $e^{2.73}$. Graph of $f(x) = e^x$.

Figure 5.16

Further Ideas 2 Use the exponential key on your calculator to find the value of e. To how many decimal places of accuracy does your calculator display e? Compare the accuracies of the value of e on the various models of calculators in your class. Which calculator gives the most accurate value of e?

Reflections and Shifts

Graphs of certain exponential functions to the base e can be sketched from that of $f(x) = e^x$ by using reflections and shifts. We now give an example to illustrate the ideas.

Example 1 **Using a Reflection and a Shift to Sketch a Graph**

Use the known graph of $f(x) = e^x$ to sketch the graph of $g(x) = e^{-x} - 2$. Check your answer using a graphing calculator. Give the domain and range of g.

SOLUTION We arrive at the graph of $g(x) = e^{-x} - 2$ from the graph of $f(x) = e^x$, using a reflection in the y-axis and a vertical shift of 2 down (Figure 5.17). The domain of g is the same as that of f; namely $(-\infty, \infty)$. The range of f is $(0, \infty)$. The downward shift means that the range of g is $(-2, \infty)$.

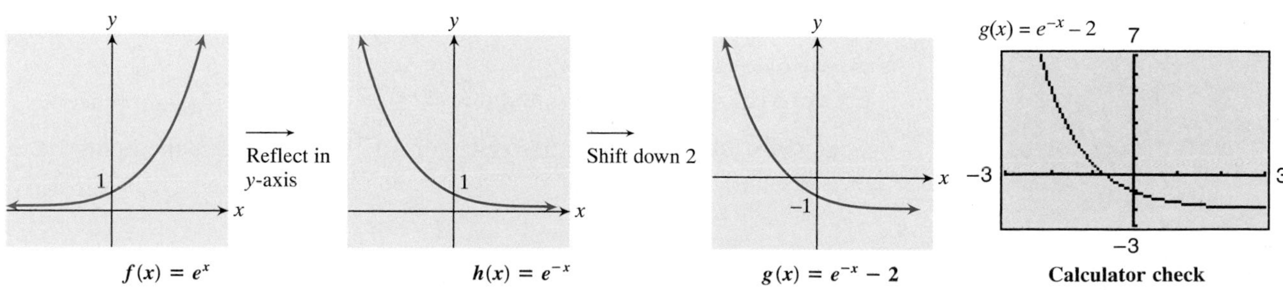

Figure 5.17 Reflection and horizontal shift of $f(x) = e^x$.

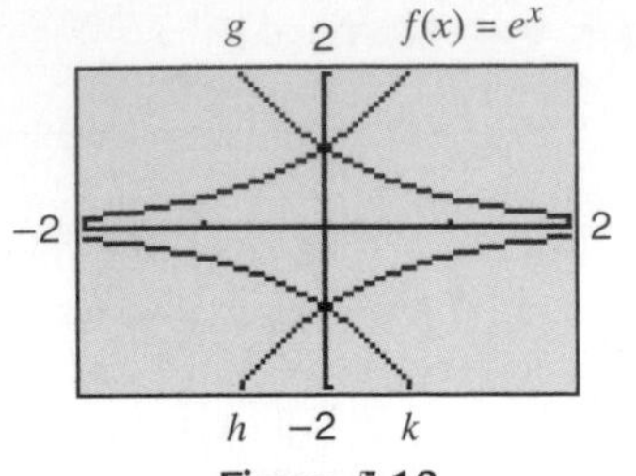

Figure 5.18

Further Ideas 3

(a) Sketch the graph of $g(x) = e^{x-2} + 3$ using horizontal and vertical shifts of $f(x) = e^x$. State the shifts used. Check your answer using a graphing calculator. Give the domain and range of f.

(b) The graph of the function $f(x) = e^x$ is displayed in Figure 5.18. Find the other functions: $g(x)$, $h(x)$, and $k(x)$. Check your answers by duplicating the graphs on your calculator.

Solving Exponential Equations

In previous chapters, we solved equations that involved the variable x as a base. Let us now look at equations in which the variable appears as an exponent.

Example 2 Solve $2^{x-1} = 2^5$.

SOLUTION When the bases of the left and right sides of an equation are equal, as here, the exponents must be equal. We get

$$x - 1 = 5$$
$$x = 6$$

The solution is $x = 3$.

Example 3 Solve $3^{2x+4} = 9^4$.

SOLUTION The left side of this equation is to base 3, whereas the right side is to base 9. Rewrite the equation with both sides having the same base; then equate exponents.

$$3^{2x+4} = (3^2)^4$$
$$3^{2x+4} = (3^8)$$
$$2x + 4 = 8$$
$$x = 2$$

The solution is $x = 2$.

Self-Check 1 Solve the equation $2^{3x+7} = 4^5$.

Example 4 If $2^n = 16$, compute n^3.

SOLUTION The left side of this equation is to base 2; express the right side to the same base. We get

$$2^n = 2^4$$
$$n = 4$$

Using this value for n, we now get $n^3 = 4^3 = 64$.

SOLUTIONS TO EXPONENTIAL EQUATIONS USING GRAPHS

In the preceding examples, we presented algebraic methods for solving certain types of exponential equations. Many exponential equations, however, cannot be solved (or are solved with difficulty) using algebraic techniques. The graphing approach can then be used to find an approximate solution.

Example 5 Solving an Exponential Equation Using Graphs

Solve $3^{(2x-5)} = 7$ to four decimal places using graphs.

SOLUTION Graph $y = 3^{(2x-5)}$ and $y = 7$ (Figure 5.19). The point of intersection A of the two graphs will be the solution to this equation. The solution rounded to four decimal places is found to be $x = 3.3856$ using an intersect finder (or trace/zoom).

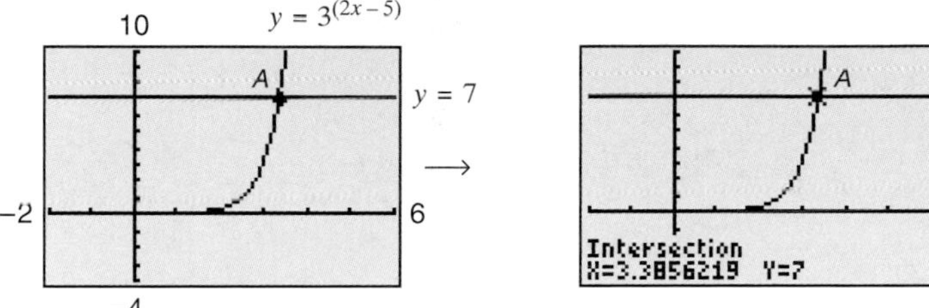

Figure 5.19 The solution to $3^{(2x-5)} = 7$ expressed to four decimal places is $x = 3.3856$.

Self-Check 2 Solve $2^{(x^2-2x+1)} = 5$ to four decimal places.

Example 6 Solving an Exponential Equation Using a Graph

Solve $e^{x^2} - e^{2x} - 5 = 0$ to four decimal places.

SOLUTION Graph $y = e^{x^2} - e^{2x} - 5$ (Figure 5.20). The x-intercepts will be the solutions to this equation. The solutions, to four decimal places, are seen to be $x = -1.2747$ and $x = 2.0398$ using a zero finder (or trace/zoom).

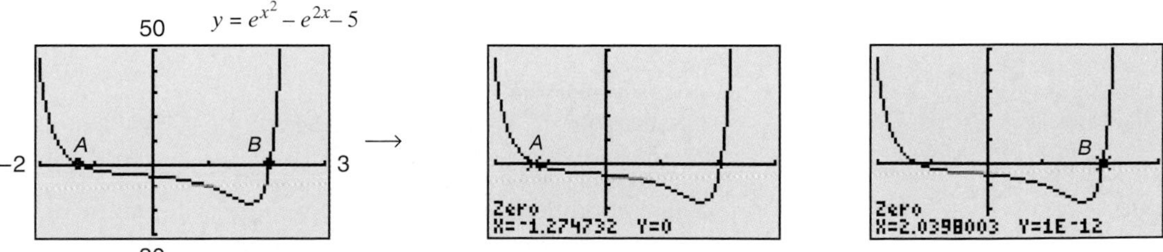

Figure 5.20 The solutions of $e^{x^2} - e^{2x} - 5 = 0$ expressed to four decimal places are $x = -1.2747$ and $x = 2.0398$.

2 Applications . . .

Our first application illustrates how the natural exponential function arises in the world of finance.

Compound Interest

Let a sum of money P be deposited in an account that pays interest at an annual rate r, compounded n times per year. There will be n occasions each year when interest is calculated and added to the principal. The interest rate that is used each time is r/n.

The amount A at the end of the first interval is given by

$$A = \underbrace{P}_{\text{Principal}} + \underbrace{P\left(\frac{r}{n}\right)}_{\substack{\text{Interest on } P \\ \text{during first interval}}} = P\left(1 + \frac{r}{n}\right)$$

The amount A at the end of the second interval is given by

$$A = P\left(1 + \frac{r}{n}\right) + \underbrace{P\left(1 + \frac{r}{n}\right)\left(\frac{r}{n}\right)}_{\substack{\text{Interest on } P(1 + r/n) \\ \text{during second interval}}} = P\left(1 + \frac{r}{n}\right)^2$$

During t years, there are nt such intervals. The amount A in the account at the end of t years according to this pattern will be

$$A = P\left(1 + \frac{r}{n}\right)^{nt}$$

Let us summarize this important result.

> **Compound Interest**
>
> If a sum of money P is deposited in an account that pays interest at an annual rate r, compounded n times per year, the amount A in the account at the end of t years will be
>
> $$A = P\left(1 + \frac{r}{n}\right)^{nt}.$$

Example 7 Compound Interest

A sum of \$1000 is deposited in an account that pays interest at an annual rate of 8%, compounded quarterly. The account will mature at the end of 3 years. Determine the amount in the account at that time.

SOLUTION Let us summarize the given information.

Initial sum: $P = 1000$.

Interest rate: $r = 8\% = 8/100 = 0.08$.

Time: $t = 3$.
Compounded quarterly: $n = 4$ (four times per year).

The formula $A = P\left(1 + \frac{r}{n}\right)^{nt}$ gives

$$A = 1000\left(1 + \frac{0.08}{4}\right)^{(4\cdot 3)} = 1000(1 + 0.02)^{12}$$
$$= 1000(1.02)^{12} = 1268.241795$$

The amount in the account at the end of 3 years will be \$1268.24 to the nearest cent.

Self-Check 3 Gerald Rhodes deposits \$2500 in an account that pays interest at an annual rate of 10%, compounded semiannually. Determine the amount in the account after 4 years.

Continuous Compounding

We have developed a formula for determining the amount that accumulates in t years when compounding takes place n times per year. Let us now proceed to find the formula for **continuous compounding.**

The amount in an account at the end of t years, compounded at an annual rate r, n times per year is

$$A = P\left(1 + \frac{r}{n}\right)^{nt}$$

If the number of times n the money is compounded per year is gradually increased, it will approach continuous compounding. Let $k = n/r$, then $n = kr$. We can write

$$A = P\left(1 + \frac{1}{k}\right)^{krt} = P\left[\left(1 + \frac{1}{k}\right)^{k}\right]^{rt}$$

As n increases, k will increase. The term $\left(1 + \frac{1}{k}\right)^{k}$ will approach e, and A will approach Pe^{rt}. The amount at the end of t years with continuous compounding is thus

$$A = Pe^{rt}$$

Continuous Compounding

If a sum of money P in an account that pays interest at an annual rate r is compounded continuously, the amount in the account at the end of t years will be

$$A = Pe^{rt}$$

Example 8 Continuous Compounding

Arwel Lloyd deposits \$5000 in an account that pays interest at an annual rate of 9%, compounded continuously. Determine the amount in the account at the end of 5 years.

SOLUTION

Initial sum: $P = 5000$.
Interest rate: $r = 9\% = 9/100 = 0.09$.
Time: $t = 5$.

The formula for continuous compounding gives

$$A = Pe^{rt} = 5000e^{(0.09 \cdot 5)} = 5000e^{0.45} = 7841.560927$$

The amount in the account after 5 years is \$7841.56.

Self-Check 4 Determine the growths in \$5000 deposited at an annual rate of 4% for 10 years, compounded (**a**) annually, (**b**) quarterly, (**c**) monthly, (**d**) daily, and (**e**) continuously.

Example 9 Color Control on a Television Screen

Color can be thought of as being made up of three primary colors; namely red, blue, and green because there are three types of cones in the retina of the eye, each sensitive to light of one of red, green, or blue hues. Experiments have measured the response of the cones to various wavelengths of light. The responses are described by the curves and their functions shown in Figure 5.21.

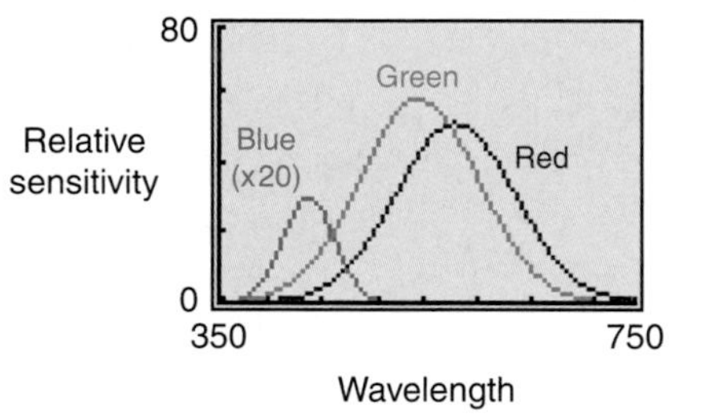

Red: $r(w) = 51e^{-((w-580)/80)^2}$
Green: $g(w) = 58e^{-((w-545)/80)^2}$
Blue: $b(w) = 1.5e^{-((w-440)/35)^2}$

Figure 5.21 Eye response to red, green, and blue.

The functions and curves indicate that the blue cones are far less sensitive than the red and green cones. In fact, the graph for blue sensitivity would not appear in Figure 5.21 unless we had followed the customary practice of displaying the graph of $20 \times b(w)$. The functions show, for instance, that when the wavelength of light is 500 nanometers (nm), the relative sensitivities of the eye to the various primary colors are

$$r(500) = 18.76,\ g(500) = 42.27,\ b(500) = 0.08 \text{ (to two decimal places)}$$

Because $g(500)$ is much larger than $r(500)$ or $b(500)$, light of wavelength 500 nanometers appears green to the eye.

The eye sees yellow when the responses of the red and green cones are approximately equal. The point of intersection of the graphs of $r(w)$ and $g(w)$ is found to be at the wavelength of 574 nanometers. The wavelength of yellow light is therefore about 574 nanometers.

A television screen consists of pixels made up of red, green, and blue phosphor dots. These are programmed to simulate the response of the eye to specific wavelengths. For example, pixels that look yellow would have a combination of red and green dots that would produce the same response in the red and green cones of the eye as light of approximately 574 nanometers.

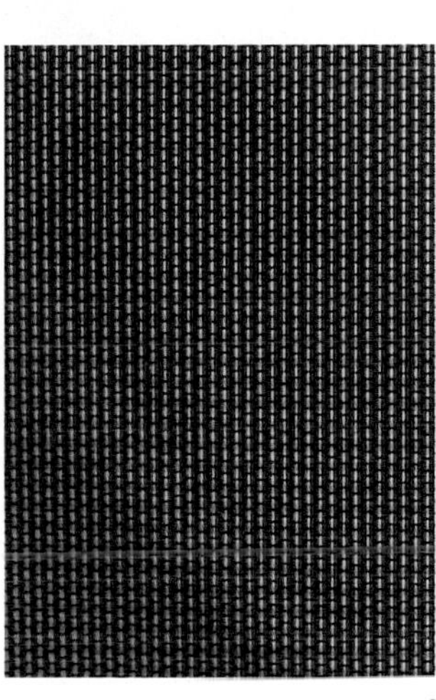

A television screen consists of pixels made up of red, green, and blue phosphor dots. The dots are activated so that a pixel appears to be the desired color. (*© George Semple*)

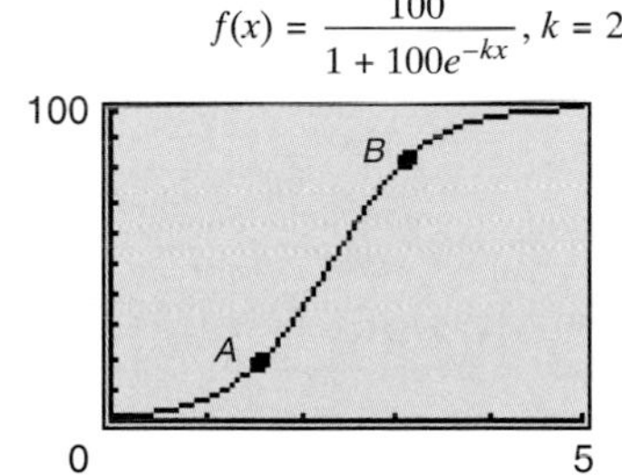

Figure 5.22 A learning curve.

Example 10 **Learning Curve**

The function $f(x) = 100/(1 + 100e^{-kx})$ represents a class of functions that can be used to model learning. k is called a **parameter.** Different values of k will give different functions that are appropriate for describing various situations. We graph f for $k = 2$ in the window $0 \leq x \leq 5$, $0 \leq y \leq 100$ (Figure 5.22). Such a graph is called a **learning curve** because at first one learns slowly (O to A); the material and the method of teaching may be new. One then goes through a more rapid phase of learning (A to B) as one settles in to the approach and the material. Finally the learning slows down (after B) as the material becomes more difficult. This graph might describe a 5-hour workshop of the use of spreadsheets or wordprocessing, for example, with $f(x)$ giving the percentage of material learned at time x.

The student is asked to use this function to model learning in a 50-minute class in the group discussion at the end of this section.

Group Discussion

Learning Model

Use the function $f(x) = \dfrac{100}{1 + 100e^{-kx}}$ to construct a model of the learning process in a typical 50-minute university class (see Example 10). Use the window $0 \leq x \leq 1$ (1 hour), $0 \leq y \leq 100(\%)$. Try various values of k to find the one that you think best fits the learning pattern of your class. Explain the various parts of the learning curve.

Answers to Self-Check Exercises

1. $2^{3x+7} = 4^5$, $2^{3x+7} = 2^{2(5)}$, $3x + 7 = 10$, $x = 1$.
2. The graphs of $y = 2^{(x^2-2x+1)}$ and $y = 5$ intersect at $x = -0.5238, 2.5238$, to four decimal places.
3. $A = P\left(1 + \dfrac{r}{n}\right)^{nt} = 2500\left(1 + \dfrac{0.1}{2}\right)^{(2)(4)} = \$3693.64.$
4. $P = 5000.$ $r = 0.04.$ $t = 10.$ **(a)** $A = 5000(1 + 0.04)^{10} = \$7401.22.$

   (b) $A = 5000\left(1 + \dfrac{0.04}{4}\right)^{(4)(10)} = \$7444.32.$

   (c) $A = 5000\left(1 + \dfrac{0.04}{12}\right)^{(12)(10)} = \$7454.16.$

   (d) $A = 5000\left(1 + \dfrac{0.04}{365}\right)^{(365)(10)} = \$7458.96.$

   (e) $A = 5000e^{(0.04)(10)} = \$7459.12.$

Answers to Further Ideas Exercises

1. See the graph of function $f(x) = \left(1 + \dfrac{1}{x}\right)^x$ below. The horizontal asymptote is $y = e$ because f gets closer to e as x increases.

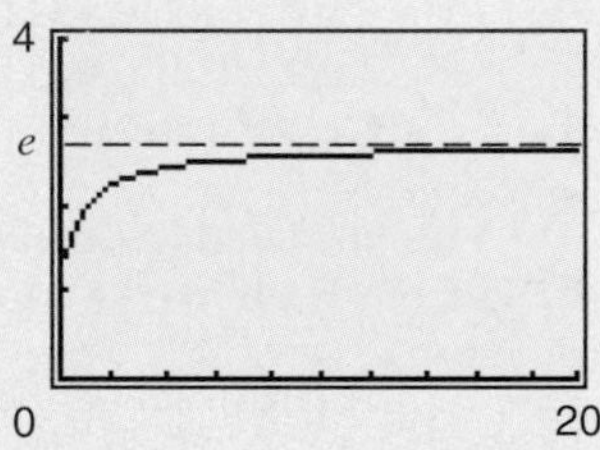

2. Evaluate e^1. We got 2.718281828, the value to nine decimal places.
3. **(a)** A horizontal shift of 2 units to the right, 3 units up. The domain of g is $(-\infty, \infty)$, and the range is $(3, \infty)$.
   (b) $g(x) = e^{-x}$. $h(x) = -e^{-x}$. $k(x) = -e^x$.

EXERCISES 5.2

In Exercises 1–4, use a calculator to arrive at the results, rounding to four decimal places.

1. Compute the values of e^x at **(a)** $x = 0.25$ **(b)** $x = 0.85$ **(c)** $x = -1.6$ **(d)** $x = -5.4$
2. Compute the values of e^x at **(a)** $x = 0.65$ **(b)** $x = 4.7$ **(c)** $x = -0.40$ **(d)** $x = -2$
3. **(a)** Compute the values of e^x at $x = -3, -2, -1, 0, 1, 2$, and 3. **(b)** Use these values to plot the graph of $f(x) = e^x$. Confirm the graph with a calculator.
4. **(a)** Compute the values of e^{-x} at $x = -3, -2, -1, 0, 1, 2, 3$. **(b)** Use these values to plot the graph of $g(x) = e^{-x}$. Confirm the graph with a calculator. **(c)** Discuss the relationship between the graphs of $f(x) = e^x$ and $g(x) = e^{-x}$.

In Exercises 5–10, use shifts, reflections, and stretches of the known graph of $f(x) = e^x$ to sketch the graphs of the functions. Describe the transformations used. Check your results using a graphing calculator.

5. $g(x) = -e^x$
6. $g(x) = e^x + 2$
7. $g(x) = e^{x-3}$
8. $g(x) = e^{x+1} - 2$
9. $g(x) = e^{1-x} + 3$
10. $g(x) = 2e^x$

In Exercises 11–22, solve the given equations (if possible) using algebraic methods. Check your solution using graphs.

11. $3^{x+1} = 3^7$
12. $4^{3x-1} = 4^{x+5}$
13. $7^{5x+2} = 7^{3x-5}$
14. $2^{2x+4} = 8^x$
15. $3^{1-2x} = 9^{x-1}$
16. $4^{x+3} = 8^{x+1}$
17. $9^{2x+3} = 27$
18. $2^{4x+1} = 1/32$
19. $8^{2x+1} = 4^{3x+1}$
20. $e^{3x+2} = e^{x-1}$
21. $e^{5-x} = e^{x-1}$
22. $e^{x^2-3x+5} = e^{x+2}$

23. If $2^n = 32$, compute n^3.
24. If $3^n = 81$, compute n^{-1}.
25. If $n^4 = 16$, compute n^6.
26. If $n^{-2} + 1 = 50$, compute 2^n to four decimal places.

In Exercises 27–38, solve the given equations (if possible) to four decimal places using graphs.

27. $3^{(x+2)} = 12$
28. $2^{(3x-4)} = 9$
29. $5^{(4x+9)} = 15$
30. $3^{(x^2+x-3)} = 5$
31. $2^{(3x^2+2x-5)} = 11$
32. $5^{(x^2-7x+4)} = -13$
33. $5^{x^3} = 13$
34. $2^{(x^3+2x-1)} = 14$
35. $e^{x^2} - e^{2x} - 31 = 0$
36. $e^{(2x^2-3x)} - 2e^{2.4x} = -376$
37. $e^x + e^{-x} = 12$
38. $e^x - e^{-x} = 10$

Compound Interest

In Exercises 39–42, the sums of money are deposited in accounts at the given interest rate for the given periods. Compute the amounts in the accounts under the various compounding plans (to the nearest cent).

39. \$8000 at 6% for 4 years compounded **(a)** annually, **(b)** continuously.
40. \$3000 at 5% for 2 years compounded **(a)** annually, **(b)** continuously.
41. \$12,000 at 9% for 3 years compounded **(a)** semiannually, **(b)** continuously.
42. \$42,500 at 8% for 5 years 6 months compounded **(a)** quarterly, **(b)** continuously.

Compound Interest

In Exercises 43–46, the sums of money are deposited in accounts at the given interest rate, for the given periods. Compute the amounts in the accounts under the various compounding plans (to the nearest cent).

43. \$5325 at 10% for 7 years 3 months compounded **(a)** quarterly, **(b)** continuously.
44. \$25,000 at 7.5% for 5 years compounded **(a)** monthly, **(b)** continuously.
45. \$50,250 at 7% for 3 years compounded **(a)** quarterly, **(b)** continuously.
46. \$7500 at 7.5% for 8 years 6 months compounded **(a)** monthly, **(b)** continuously.

Continuous Compounding

In Exercises 47–50, use the graph of the function $A = Pe^{rt}$ on your calculator to compute times (to nearest month).

47. Determine the time it would take \$5000 to increase to \$6000 at 5% compounded continuously.
48. Determine the time it would take \$8000 to increase to \$10,000 at 7% compounded continuously.
49. Determine (to two decimal places) the time it would take \$4000 to double at 6% compounded continuously. Prove that the doubling time is the same at this interest rate, no matter what the initial principle.
50. Determine the doubling time for 10% compounded continuously.

Continuous Compounding

In Exercises 51–54, use the graph of the function $A = Pe^{rt}$ on your calculator to compute interest rates (to two decimal places).

51. Determine the interest rate that would enable \$1000 to increase to \$1522 (to the nearest dollar) in 7 years, compounded continuously.
52. Determine the interest rate that would enable \$5000 to increase to \$5810 (to the nearest dollar) in 3 years, compounded continuously.
53. Determine the interest rate that would enable \$8000 to increase to \$12,560 in 10 years, compounded continuously.
54. Determine the interest rate that would enable \$5000 to double in 10 years, compounded continuously.
55. **Luminosity Function** The sum of the three functions that describe the sensitivity of the eye to the primary colors, namely,

$$f(w) = 51e^{-((w-580)/80)^2} + 58e^{-((w-545)/80)^2} + 1.5e^{-((w-440)/35)^2}$$

is called the **luminosity function.** It shows the eye's response to light of constant luminance as the dominant wavelength is varied. (See Example 9 in this section.) Graph $f(w)$ and determine the wavelength of the light to which the eye is most sensitive.

SECTION PROJECT

Cooling Experiment

The cooling of a hot liquid is described by the equation

$$T = be^{-at} + c$$

where T is the temperature at time t, and a, b, and c are constants. Heat water in a pan and record its temperature at different times during the cooling period. Use the information you get in the preceding equation to determine a, b, and c, and thus the equation for describing the cooling of water. The constant a varies from liquid to liquid, but the constants b and c depend on the initial temperature of the liquid and the room temperature. Interpret the b and c that you have obtained in this manner.

5.3 Logarithmic Functions and Seismography

- Logarithmic Function • Common and Natural Logarithmic Functions
- Graphs of Logarithmic Functions • Earthquake Measurements

In Section 5.1, we saw that world population growth is defined by the exponential function $P(t) = 5.94(2^{t/40})$, where t is time measured in years from 1998 and $P(t)$ is the population in billions at time t. We can use this function to predict the population P at any future time t. For example, if we want to predict the population for the year 2010, we use $t = 12$ and arrive at $P(12) = 7.26$ billion, rounded to two decimal places. Demographers, however, often want to do the converse; namely, estimate the times when populations will reach certain values. For example, suppose we want to know when the population reaches 8 billion; the time t would be given by $8 = 5.94(2^{t/40})$. The problem of finding this value of t should remind the student of the previous discussion of inverse functions. The inverse of the exponential function $P(t) = 5.94(2^{t/40})$ will express t as a function of P, enabling us to answer such questions. The inverse of an exponential function is called a **logarithmic function.**

Logarithmic Function

Let us first show that an exponential function $g(x) = a^x$ $(a > 0,\ a \neq 1)$ has an inverse. A horizontal line cuts the graph of $g(x) = a^x$ in at most one point (Figure 5.23). By the horizontal line test, the exponential function is one-to-one. It thus has an inverse.

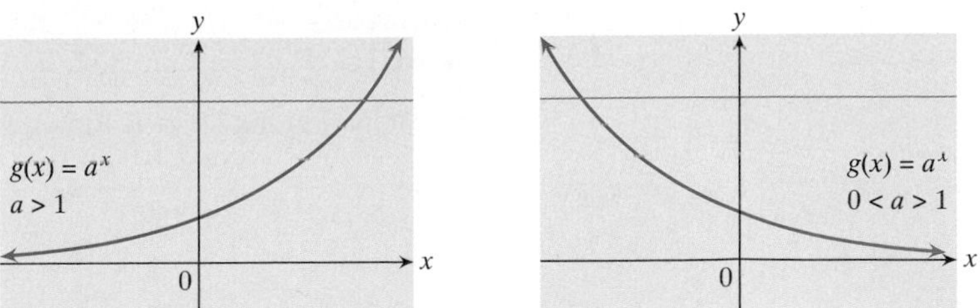

Figure 5.23 Horizontal line cuts the graph of $g(x) = a^x$ in, at most, one point; thus, g has an inverse.

Denote the inverse of $g(x) = a^x$ by $f(x) = \log_a x$. f is called the **logarithmic function with base *a*.**

Exponential and Logarithmic Functions

The exponential function $g(x) = a^x$ and the logarithmic function $f(x) = \log_a x$, where $a > 0$, $a \neq 1$, are inverses of one another.

Example 1 **Inverses**

(a) The inverse of the exponential function $g(x) = 2^x$ is $f(x) = \log_2 x$.
(b) The inverse of the logarithmic function $f(x) = \log_5 x$ is $g(x) = 5^x$.

Conversion Between Exponential and Logarithmic Forms

Consider the exponential function $g(x) = a^x$ and its inverse $f(x) = \log_a(x)$. Let y be the image of x under g; thus $y = a^x$. By the definition of inverse function, x will be the image of y under f; thus $x = \log_a y$. This means that we can write an exponential equation in logarithmic form and vice versa.

$$x = \log_a y \quad \text{if and only if} \quad y = a^x$$

Certain equations can be changed from exponential form to logarithmic form; other equations can be changed from logarithmic to exponential form using the preceding result. The student should become proficient at changing from one form to the other.

Example 2

(a) Write the equation $y = 7^x$ in logarithmic form.
(b) Write the equation $y = \log_3 x$ in exponential form.

SOLUTION

(a) Use the definition of logarithm. We get $y = 7^x$ is equivalent to $x = \log_7 y$.
(b) The equation $y = \log_3 x$ is equivalent to $x = 3^y$.

Certain logarithmic values can be computed by rewriting in exponential form.

Example 3 Compute $\log_2 8$.

SOLUTION Let $\log_2 8 = y$. Rewrite this equation in exponential form and solve for y.

$$8 = 2^y$$
$$2^3 = 2^y$$
$$y = 3$$

Thus

$$\log_2 8 = 3$$

Self-Check 1 Compute $\log_3 9$.

Common and Natural Logarithmic Functions

Logarithmic functions to base e and base 10 are the most commonly used logarithmic functions. They are given special notations and names.

$f(x) = \log_e x$ is written $f(x) = \ln x$ and called the **natural logarithm.**
$f(x) = \log_{10} x$ is written $f(x) = \log x$ and called the **common logarithm.**

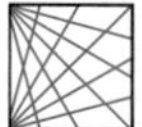

Computing Logarithms with a Calculator

A calculator can be used to find values of logarithms. The key for ln is usually LN and the key for log is LOG. Let us compute ln 4.5673 and log 4.5673 (Figure 5.24). (We show how to compute logarithms to other bases in Section 5.4.)

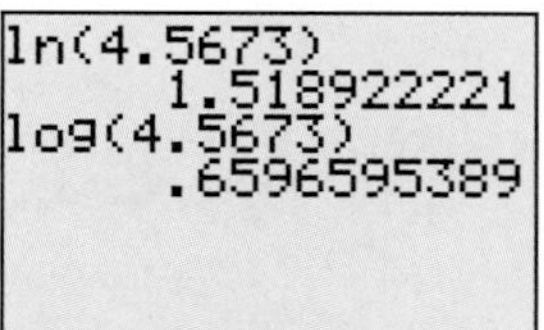

Figure 5.24

Exploration

Enter $f(x) = \ln(e^x)$ and $g(x) = e^{\ln x}$ in your calculator and examine the tables of values for $x = 1, 2, 3, \ldots$. What information do the tables tell you?

We now look at graphs of logarithmic functions. We first graph $f(x) = \log_2 x$ by hand. This discussion leads to the graphs of general logarithmic functions.

GRAPH $f(x) = \log_2 x$

Let us sketch the graph of $f(x) = \log_2 x$ by plotting points. The inverse of $f(x)$ is the function $g(x) = 2^x$. We find points that lie on the graph of $g(x)$ and then use these to find points that lie on the graph of $f(x)$. Construct a table of convenient values for the exponential function $g(x)$.

x	-2	-1	0	1	2	3
$g(x)$	$\frac{1}{4}$	$\frac{1}{2}$	1	2	4	8

This table leads to the following points on the graph of $g(x)$: $(-2, \frac{1}{4})$, $(-1, \frac{1}{2})$, $(0, 1)$, $(1, 2)$, $(2, 4)$, and $(3, 8)$.

The graphs of $g(x)$ and its inverse $f(x)$ are symmetric about the line $y = x$. Thus, if a point (a, b) is on the graph of $g(x)$, then (b, a) will lie on the graph of $f(x)$. The graph of $f(x)$ thus passes through the points $(\frac{1}{4}, -2)$, $(\frac{1}{2}, -1)$, $(1, 0)$, $(2, 1)$, $(4, 2)$, and $(8, 3)$. These points are plotted, and the graph of $f(x) = \log_2 x$ is drawn (Figure 5.25).

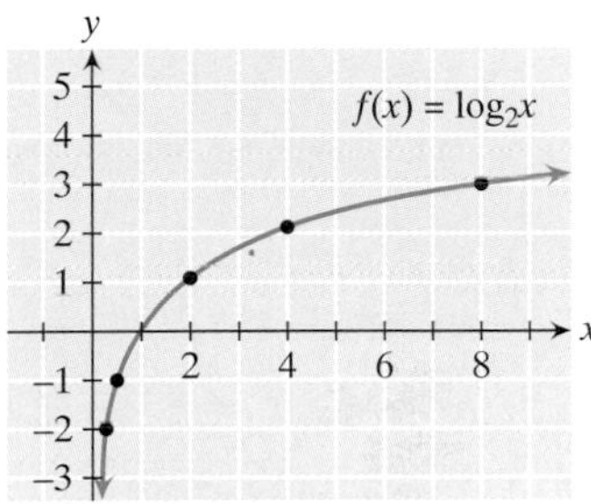

Figure 5.25 The function $f(x) = \log_2 x$

SHAPE OF THE GRAPH OF A LOGARITHMIC FUNCTION

The preceding discussion leads to the shapes of the graphs of all logarithmic functions of the form $f(x) = \log_a x$ with $a > 1$. Because the graphs of a function and their inverse are symmetric about the line $y = x$, we can sketch the graph of a logarithmic function from the graph of its inverse exponential function. In Figure 5.26, the known graph of $g(x) = a^x$ $(a > 1)$ is used to sketch the graph of $f(x) = \log_a x$. Observe that the x-intercept of this logarithmic function is 1; there is no y-intercept. The domain of $f(x) = \log_a x$ is $(0, \infty)$ and the range is $(-\infty, \infty)$. The y-axis is a vertical asymptote.

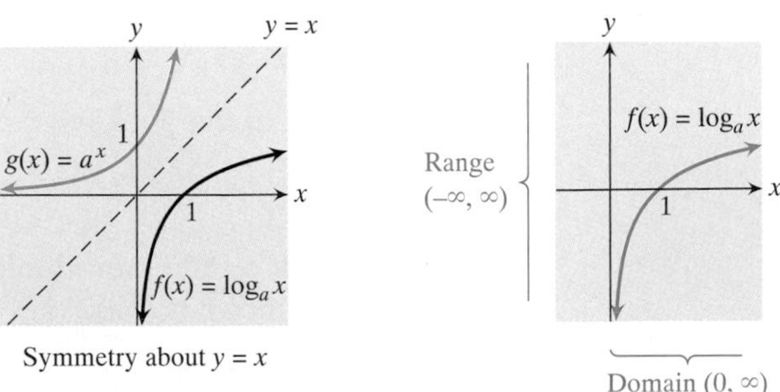

Figure 5.26 Graphs of $f(x) = \log_a x$ and $g(x) = a^x$.

Let us now determine the relative locations of the graphs of logarithmic functions of the type $f(x) = \log_a x$ where $a > 1$.

Exploration

Sketch the graphs of $F(x) = 2^x$ and $G(x) = 3^x$. Use these two graphs and symmetry about the line $y = x$ to sketch the graphs of $f(x) = \log_2 x$ and $g(x) = \log_3 x$. Do the graphs of $f(x)$ and $g(x)$ intersect at a point? Record the intervals of x where $f(x) > g(x)$ and where $f(x) < g(x)$.

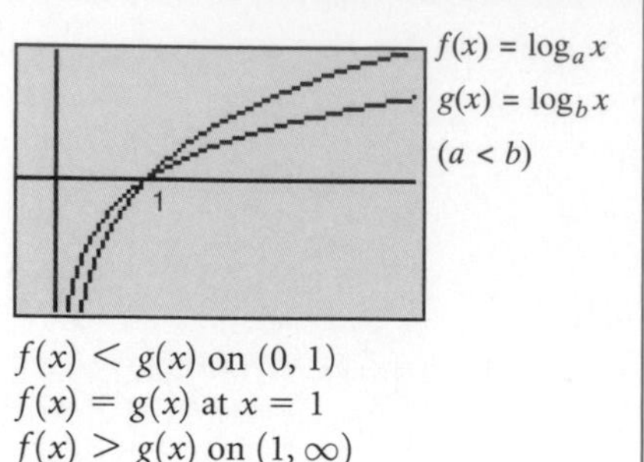

$f(x) < g(x)$ on $(0, 1)$
$f(x) = g(x)$ at $x = 1$
$f(x) > g(x)$ on $(1, \infty)$

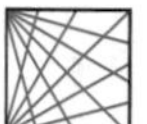

Graphing Logarithmic Functions with a Calculator

Graphing calculators can be used to graph both natural and common logarithmic functions and to understand the behavior of these functions. Let us sketch the graphs of $g(x) = e^x$, $f(x) = \ln x$, and $y = x$ in the same coordinate system (Figure 5.27). Observe that $f(x)$ and $g(x)$ are symmetric about the line $y = x$, demonstrating that $f(x)$ is $g^{-1}(x)$.

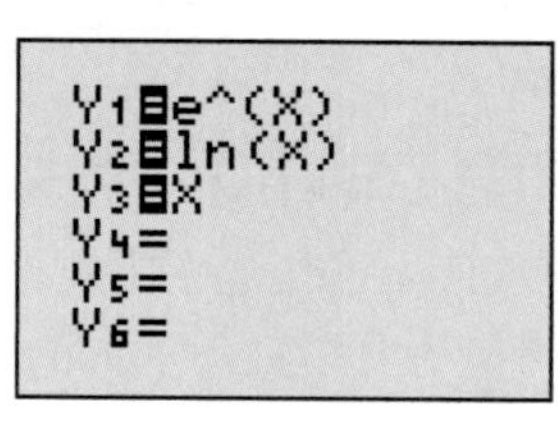

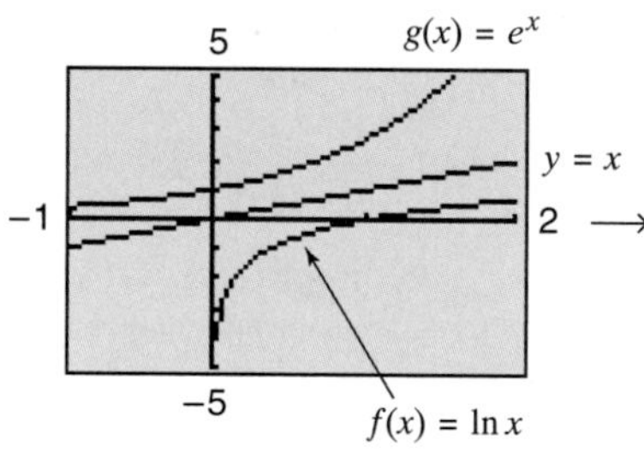

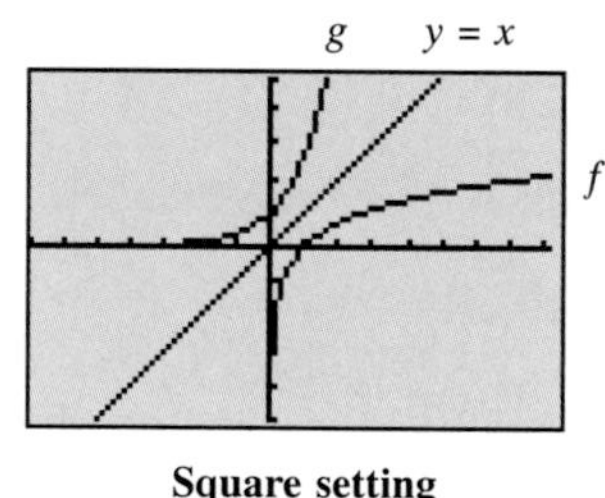

Square setting

Figure 5.27

Note that the x-axis is a horizontal asymptote for $g(x) = e^x$ and that the y-axis is a vertical asymptote for $f(x) = \ln(x)$. The resolution of the calculator screen is not sufficient to reveal that the curves get closer to these axes without meeting them.

Further Ideas 1 Use graphs on your calculator to show that $f(x) = \log x$ is the inverse of $g(x) = 10^x$.

Graphs of certain logarithmic functions can be sketched using shifts and reflections of known graphs.

Example 4 **Using Shifts to Sketch Graphs**

Use shifts of the known graph of $f(x) = \ln x$ to sketch the graph of

$$g(x) = \ln(x - 2) + 4$$

Check your answer using a graphing calculator.

SOLUTION We can arrive at the graph of $g(x) = \ln(x - 2) + 4$ from the graph of $f(x) = \ln x$ using a shift of 2 to the right and a shift of 4 up (Figure 5.28). The graph is confirmed by the calculator display. Note that the line $x = 2$ is a vertical asymptote of $g(x) = \ln(x - 2) + 4$. The calculator graph of $g(x)$ stops in midair. This is because the screen resolution is not sufficient to display any other pixels closer to the line $x = 2$ in this window.

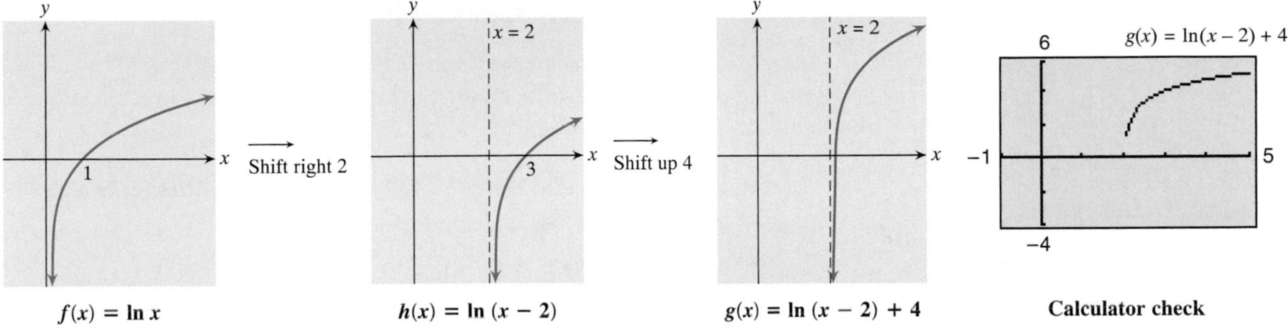

Figure 5.28 Horizontal and vertical shifts of $f(x) = \ln x$

Example 5 Finding the Inverse of an Exponential Function

Find the inverse of the function $f(x) = e^{x+1} - 3$ using algebra. Verify your answer by graphing f and this new function on a calculator and checking that they are symmetric about the line $y = x$.

SOLUTION We use the method of Section 3.7 for finding the inverse of a function. Write as $y = f(x)$.

$$y = e^{x+1} - 3$$

Interchange x and y.

$$x = e^{y+1} - 3$$

Solve for y in stages.

$$e^{y+1} = x + 3$$

Write in logarithmic form.

$$y + 1 = \ln(x + 3)$$

$$y = \ln(x + 3) - 1$$

Thus, $f^{-1}(x) = \ln(x + 3) - 1$ (Figure 5.29).

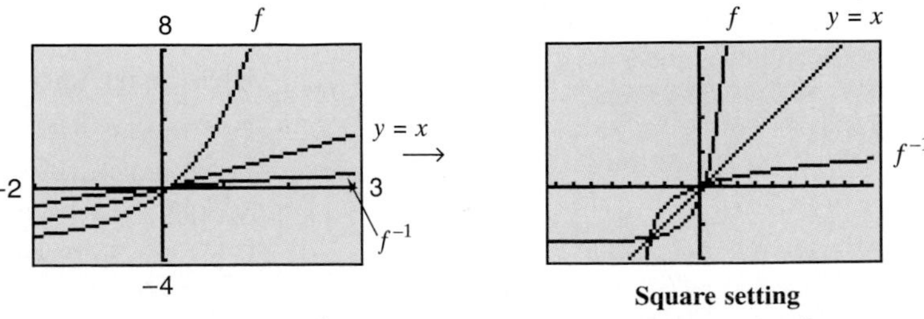

Figure 5.29 f and f^{-1} are symmetric about the line $y = x$; thus, $f^{-1}(x) = \ln(x + 3) - 1$ is indeed the inverse of $f(x) = e^{x+1} + 3$.

Further Ideas 2 Show that the inverse of $f(x) = e^{x+a} + b$ is

$$g(x) = \ln(x - b) - a$$

Discuss how the graphs of f and g are related to the graphs of $F(x) = e^x$ and $G(x) = \ln x$ through horizontal and vertical shifts if a and b are both positive.

Example 6 | Earthquake Measurement

The magnitude R of an earthquake is measured on the Richter scale. The following were the magnitudes of some major earthquakes that caused much loss of life and damage.

Earthquake Location and Date	Magnitude on the Richter Scale
Los Angeles, 1994	6.7
Kobe, Japan, 1995	7.2
Mexico City, 1978	8.6
Lebu, Chile, 1960	8.9

Severe earthquakes, such as the one in Kobe, Japan, are rare. Annually, we experience one great earthquake (8.0 points or higher on the Richter scale), 18 major (7–7.9 points), 120 strong (6–6.9 points), and 800 moderate (5–5.9) ones. *(© CORBIS/Michael S. Yamashita)*

The Richter scale, although convenient for seismologists to work with, is misleading for the layperson. We are accustomed to thinking in terms of linear scales. The Richter scale is not linear. We shall see that the Kobe earthquake was three times as severe as the Los Angeles earthquake, a fact that is not apparent from the table. Let us generate a table that is more meaningful.

The formula that defines the Richter scale follows. An earthquake of intensity I has a Richter scale reading of R, where

$$R = \log \frac{I}{I_0}$$

I_0 is a minimum perceptible intensity. A comparison of I values gives us more information than does a comparison of R values. Let us solve this equation for I. Write the equation in exponential form (remembering that log is to base 10)

$$10^R = \frac{I}{I_0}$$

Thus,

$$I = I_0 \times 10^R$$

We now get the following table that gives the earthquakes as multiples of the minimum intensity I_0.

Earthquake Location and Date	Intensity
Los Angeles, 1994	$I_0 \times 10^{6.7}$
Kobe, Japan, 1995	$I_0 \times 10^{7.2}$
Mexico City, 1978	$I_0 \times 10^{8.6}$
Lebu, Chile, 1960	$I_0 \times 10^{8.9}$

If we take ratios, we get the following table of the sizes of the earthquakes as multiples of the intensity of the Los Angeles earthquake.

Location and Date	Number of Times More Intense Than the LA Earthquake
Kobe, Japan, 1995	3.16
Mexico City, 1978	79.43
Lebu, Chile, 1960	158.50

This table gives us a much better understanding of the relative intensities of the earthquakes than does the scientific Richter scale.

Further Ideas 3

(a) The San Francisco earthquake of 1989, which took place during the World Series, had a Richter reading of 7.1. How much more intense than the Los Angeles earthquake of 1994 was this earthquake?

(b) Prove that every increase of 1 in the Richter scale implies a tenfold increase in earthquake intensity.

Answer to Self-Check Exercise

1. $\log_3 9 = 2$.

Answers to Further Ideas Exercises

1. Graph $g(x) = 10^x, f(x) = \log x$ and $h(x) = x$ on the calculator, using appropriate x and y intervals (say $-2 \le x \le 2, -2 \le y \le 2$). The graphs of f and g are symmetric about the line $y = x$, showing that f is the inverse of g.
2. $y = e^{x+a} + b, x = e^{y+a} + b, x - b = e^{y+a}, y = \ln(x - b) - a$. The inverse of $f(x) = e^{x+a} + b$ is $g(x) = \ln(x - b) - a$. f is obtained from F by a horizontal shift of a to the left and an upward shift of b. g is obtained from G by a horizontal shift of b to the right and downward shift of a.
3. **(a)** $10^{7.1}/10^{6.7} = 2.51$. The San Francisco earthquake was 2.51 times the intensity of the Los Angeles earthquake.
   (b) $I_{R+1} = I_0 10^{(R+1)} = I_0 10(10^R) = 10I_0(10^R) = 10I_R$.

EXERCISES 5.3

In Exercises 1–6, find the inverse of each function.

1. $g(x) = 4^x$ **2.** $g(x) = 7^x$ **3.** $f(x) = 12^x$
4. $f(x) = \log_5 x$ **5.** $f(x) = \log_8 x$ **6.** $h(x) = \log_{27} x$

In Exercises 7–14, write each equation in logarithmic form.

7. $y = 4^x$ **8.** $y = 9^x$ **9.** $y = 13^x$
10. $5^3 = 125$ **11.** $2^5 = 32$ **12.** $10^{-2} = 0.01$
13. $2^{-3} = 0.125$ **14.** $(1/4)^2 = 0.0625$

In Exercises 15–22, write each equation in exponential form.

15. $y = \log_5 x$ **16.** $y = \log_7 x$ **17.** $y = \log_{12} x$
18. $\log_3 81 = 4$ **19.** $\log_2 8 = 3$ **20.** $\log_6 6 = 1$
21. $\log_7 1 = 0$ **22.** $\log_2(1/8) = -3$

In Exercises 23–34, compute the value of each logarithm.

23. $\log_3 9$ **24.** $\log_2 16$ **25.** $\log_6 1$
26. $\log_5 125$ **27.** $\log_7 7$ **28.** $\log_4(1/16)$

29. $\log_{10} 1000$ **30.** $\log_{100} 0.01$ **31.** $\log_6(1/216)$
32. $\log_{125}(1/5)$ **33.** $\log_{16} 2$ **34.** $\log_{64} 4$

In Exercises 35–40, use a calculator to compute the logarithmic functions, rounding to four decimal places.

35. Compute $\ln x$ at the following x values.
(a) 2.13 **(b)** 6.78 **(c)** 0.41 **(d)** 0.07823

36. Determine $\log x$ at the following x values.
(a) 0.72 **(b)** 3.82 **(c)** 12.93 **(d)** 645.27

37. Determine $\ln x$ at the following x values.
(a) 0.52 **(b)** 5.92 **(c)** 16.93 **(d)** 478.29

38. Determine $\ln x$ at the following x values.
(a) 23.56 **(b)** 251 **(c)** 0.45 **(d)** 0.0000578

39. Compute $\log x$ at the following x values.
(a) 0.361 **(b)** 4.982 **(c)** 16.032 **(d)** 543.21

40. Compute $\ln x$ at the following x values.
(a) 0.562 **(b)** 7.984 **(c)** 34.2546 **(d)** 762.9086

In Exercises 41–48, use shifts and reflections of the known graphs of $f(x) = \ln x$ and $f(x) = \log x$ to sketch a graph of each function. Describe the transformations used. Verify the graph using a calculator.

41. $g(x) = \ln x + 3$ **42.** $g(x) = \ln x - 2$
43. $g(x) = \log(x - 4)$ **44.** $g(x) = \ln(x - 1) + 2$
45. $g(x) = \log(x + 1) - 3$
46. $g(x) = -\log x$ **47.** $g(x) = \ln(-x)$
48. $g(x) = -\ln(x + 3) - 2$

In Exercises 49–54, find the inverse of each function using algebra.

49. $f(x) = 2^x - 3$ **50.** $f(x) = 3^{(x-4)}$
51. $f(x) = 7^{(2x+5)} + 3$ **52.** $f(x) = \log_2 3x - 1$
53. $f(x) = \log_6(4x + 7)$ **54.** $f(x) = \log_8(2x - 3) + 5$

In Exercises 55–60, find the inverse of each function using algebra. Use your graphing calculator to draw the graphs of f, f^{-1}, and $y = x$ in square setting. Verify your answers by checking whether the graphs of f and f^{-1} are symmetric about the line $y = x$.

55. $f(x) = e^x + 1$ **56.** $f(x) = e^{(x-1)}$
57. $f(x) = e^{(2x+4)} + 1$ **58.** $f(x) = \ln 3x - 1$
59. $f(x) = \ln(3x + 6)$ **60.** $f(x) = \ln(2x - 4) + 2$

The graphs of $g(x)$ in Exercises 61–66 can be obtained from the known graph of $f(x) = \ln x$ by horizontal shifts and reflections. The viewing window is

$$-4 \le x \le 4, -3 \le y \le 3$$

Find the functions $g(x)$ having these graphs and describe the transformations used. Check your answer by duplicating the given graph on your calculator.

61.

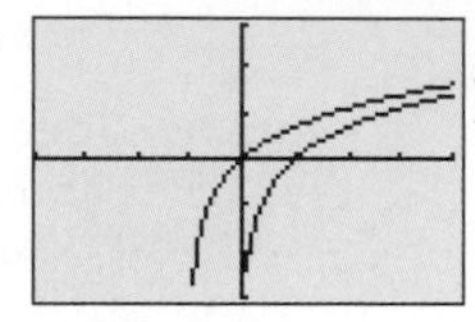

62.

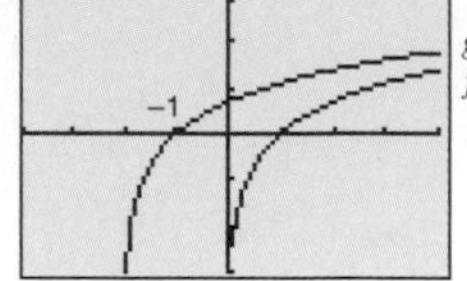

63.

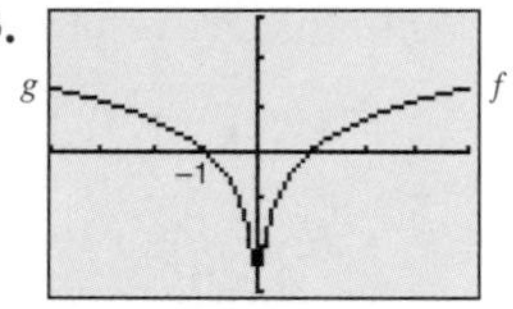

64.

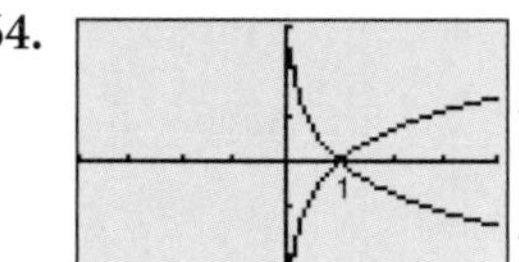

65.

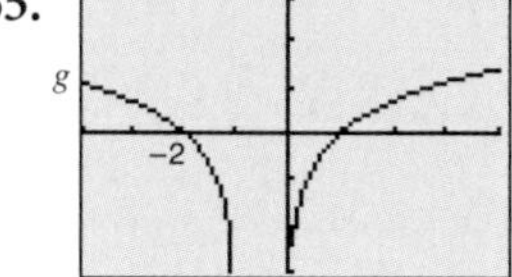

66.

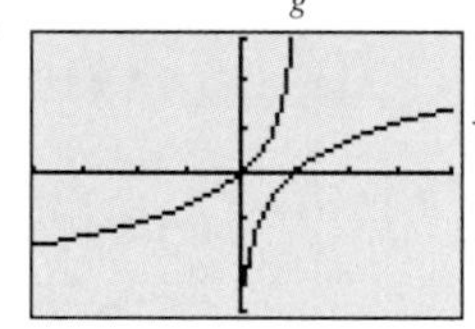

In Exercises 67–69, draw the graphs of the functions on your calculator. Give the equations of the asymptotes, and describe any symmetries in the graphs. Explain why the symmetries occur in the graphs.

67. $f(x) = \ln x^2$ **68.** $f(x) = \ln\left(\frac{1}{x}\right)$

69. $f(x) = \ln|x|$

70. Determine how many times more powerful each of the following earthquakes were than the Los Angeles earthquake of 1994 (to two decimal places). **(a)** The San Francisco earthquake of 1906 had a Richter scale reading of 6.9. **(b)** An earthquake of 8.2 on the Richter scale occurred in Tangshan, China, on July 28, 1976. This earthquake caused 242,000 deaths. **(c)** The earthquake that caused havoc in Izmit, Turkey, on August 16, 1999 had a Richter-scale reading of 7.8.

Properties of Logarithms and Solving Logarithmic Equations

• Solving Logarithmic Equations • Logarithmic Identities • Properties of Logarithms • Change of Base in a Logarithm • World Population Growth • Dating of Artifacts

In Section 5.3, we saw how equations involving logarithms arise in applications. In this section, we look at the algebraic properties of logarithms. These properties will enable us to manipulate equations containing logarithms. For example, we will use these properties in a mathematical model that is used to date artifacts.

We remind the student of an important result from Section 5.3, which is used to convert between logarithmic equations and exponential equations and vice versa.

CONVERSION BETWEEN EXPONENTIAL AND LOGARITHMIC FORMS

$$\log_a y = x \quad \text{if and only if} \quad y = a^x \quad (a > 0, a \neq 1).$$

SOLVING LOGARITHMIC EQUATIONS

We now use this result to solve an equation involving a logarithm by converting it into an equation involving an exponent.

Example 1 **Rewriting a Logarithmic Equation in Exponential Form**

Solve the equation $\log_2(x + 3) = 4$.

SOLUTION We know that

$$\log_2(x + 3) = 4 \quad \text{if and only if} \quad x + 3 = 2^4$$

Thus,

$$x + 3 = 2^4$$
$$x + 3 = 16$$
$$x = 13$$

Self-Check 1 Solve $\log_9(x - 4) = 1/2$.

The following example illustrates the reverse procedure. We rewrite an exponential equation in logarithmic form.

Example 2 **Rewriting an Exponential Equation in Logarithmic Form**

Rewrite the equation $y = 2^{(3x-6)}$ to express x in terms of y.

SOLUTION Write the equation in logarithmic form and then solve for x.

$$\log_2 y = 3x - 6$$
$$\log_2 y + 6 = 3x$$
$$x = \left(\frac{1}{3}\right)\log_2 y - 2$$

LOGARITHMIC IDENTITIES

There are two identities that are useful for simplifying expressions involving logarithms.

LOGARITHMIC IDENTITIES

1. $a^{\log_a x} = x$, for all $x > 0$.
2. $\log_a a^x = x$, for any real number x.

Proof Define a function $f(x) = a^x$ and its inverse $f^{-1}(x) = \log_a(x)$. The properties $f(f^{-1}(x)) = x$ and $f^{-1}(f(x)) = x$ of inverses are now used to prove the above identities. $f(f^{-1}(x)) = x$ gives $f(\log_a(x)) = x$, leading to $a^{\log_a x} = x$.

Conversely, $f^{-1}(f(x)) = x$ gives $f^{-1}(a^x) = x$, leading to $\log_a a^x = x$.

We now show how these identities are used to evaluate expressions and to solve equations.

Example 3 Evaluate $3(\log_5 5^2)$.

SOLUTION The first logarithmic identity implies that $\log_5 5^2 = 2$, thus,

$$3(\log_5 5^2) = 3(2)$$
$$= 6$$

Example 4 Solve $5y = 4^{\log_4(x-6)}$ for x.

SOLUTION The second logarithmic identity implies that $4^{\log_4(x-6)} = x - 6$, thus,

$$5y = x - 6$$
$$x = 5y + 6$$

Self-Check 2

(a) Evaluate $-4(7^{\log_7 5})$.

(b) Solve $7y = \log_3(2x - 5)$ for x.

Properties of Logarithms

The following are fundamental properties of logarithms that are used in the computation and the algebraic manipulation of logarithms.

> **Properties of Logarithms**
>
> 1. $\log_a(xy) = \log_a x + \log_a y$
> 2. $\log_a\left(\dfrac{x}{y}\right) = \log_a x - \log_a y$
> 3. $\log_a(x^r) = r \log_a x$
>
> where x and y are positive numbers, r is a real number, and a is a positive number not equal to 1.

Proof of Property 1. Let $p = \log_a x$ and $q = \log_a y$. Write these equations in exponential form.

$$x = a^p \qquad \text{and} \qquad y = a^q$$

Multiply these and use the properties of exponents.

$$xy = a^p \cdot a^q$$
$$xy = a^{p+q}$$

Write this equation in logarithmic form.

$$\log_a(xy) - p + q$$
$$= \log_a x + \log_a y, \qquad \text{by definition of } p \text{ and } q$$

Thus, $\log_a(xy) = \log_a x + \log_a y$

The proofs of Properties 2 and 3 are similar (see Exercise 87).

Example 5 | Using the Properties of Logarithms

If $\log_a 3 = 1.0986$ and $\log_a 5 = 1.6094$, find **(a)** $\log_a 15$, **(b)** $\log_a(5/3)$, and **(c)** $\log_a 9$.

SOLUTION

(a) $\log_a 15 = \log_a(3 \cdot 5)$

$= \log_a 3 + \log_a 5$ Property 1

$= 1.0986 + 1.6094 = 2.708$

(b) $\log_a\left(\dfrac{5}{3}\right) = \log_a 5 - \log_a 3$ Property 2

$= 1.6094 - 1.0986 = 0.5108$

(c) $\log_a 9 = \log_a 3^2$

$= 2 \log_a 3$ Property 3

$= 2(1.0986) = 2.1972$

Self-Check 3 If $\log_a 3 = 0.4771$ and $\log_a 7 = 0.8451$, find (**a**) $\log_a 21$, (**b**) $\log_a(3/7)$, and (**c**) $\log_a 343$.

Rewriting Expressions

Properties of logarithms often are used to rewrite expressions.

Example 6 Express $\log_a(x^4y^{1/2}/z^3)$ in terms of logarithms of x, y, and z.

SOLUTION

$$\begin{aligned}\log_a(x^4y^{1/2}/z^3) &= \log_a(x^4y^{1/2}) - \log_a z^3 && \text{Property 2}\\ &= \log_a x^4 + \log_a y^{1/2} - \log_a z^3 && \text{Property 1}\\ &= 4\log_a x + \left(\frac{1}{2}\right)\log_a y - 3\log_a z && \text{Property 3}\end{aligned}$$

Example 7 **Combining Logarithmic Terms**

Write the following expression as a single logarithm:

$$5\log_a x + \log_a(x-7) - 2\log_a x$$

SOLUTION

$$\begin{aligned}5\log_a x + \log_a(x-7) - 2\log_a x &= \log_a x^5 + \log_a(x-7) - \log_a x^2 && \text{Property 3}\\ &= \log_a x^5(x-7) - \log_a x^2 && \text{Property 1}\\ &= \log_a x^5(x-7)/x^2 && \text{Property 2}\\ &= \log_a x^3(x-7)\end{aligned}$$

Self-Check 4 Write the following expression as a single logarithm:

$$\log_a(x^2-4) - \log_a(x+2) + 3\log_a 2$$

Solving Logarithmic Equations

We now give examples to illustrate how the properties of logarithms can be used to solve equations.

Example 8 Solve the equation $\log_3 4 + \log_3 x = 2$ for x.

SOLUTION Using Property 1, the terms on the left-hand side of the equation may be combined.

$$\log_3(4x) = 2$$

Write this equation in exponential form and solve.

$$\begin{aligned}4x &= 3^2\\ 4x &= 9\\ x &= 2.25\end{aligned}$$

Self-Check 5 Solve the equation $4 \log_2 x - 3 \log_2 x = 5$ for x.

CAUTION: Care must be taken when roots are involved in solving an equation. The following example illustrates that extraneous roots may be introduced.

Example 9 **Extraneous Roots**

Solve the equation $2 \log_6 x = \log_6 25$.

SOLUTION Use Property 2 to get

$$\log_6 x^2 = \log_6 25$$

When the bases are equal, as here (base 6), we equate the arguments

$$x^2 = 25$$

$$x = 5 \qquad \text{or} \qquad x = -5$$

Note that $x = -5$ is not a solution of the original equation because $\log_6(-5)$ does not exist. $x = -5$ is not in the domain of the logarithmic function. The only solution is $x = 5$.

Solutions to Logarithmic Equations Using Graphs

In the previous examples, we presented algebraic methods for solving certain types of logarithmic equations. However, many logarithmic equations cannot be solved using algebraic techniques. Graphs are then used to find approximate solutions, as for such exponential equations.

Example 10 **Solving a Logarithmic Equation with a Graph**

Solve the equation $\ln(x^2 + 2x + 1) = -3x + 7$ to four decimal places.

SOLUTION Graph the equations $y = \ln(x^2 + 2x + 1)$ and $y = -3x + 7$ (Figure 5.30). The point of intersection of the graphs will be the solution. The solution to four decimal places is seen to be $x = 1.6769$ using an intersect finder (or trace/zoom).

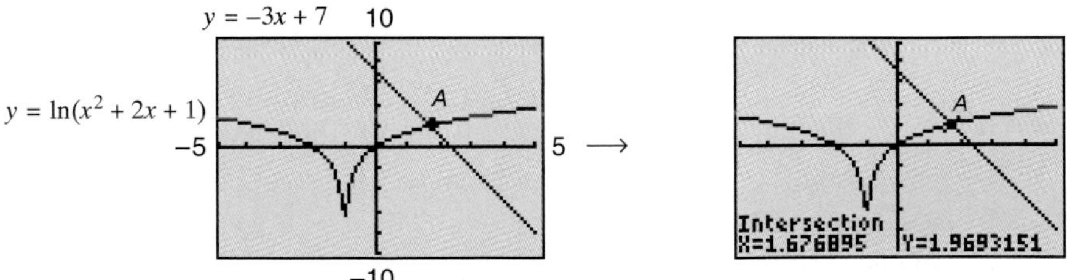

Figure 5.30 The solution to $\ln(x^2 + 2x + 1) = -3x + 7$ expressed to four decimal places is $x = 1.6769$.

Change of Base in a Logarithm

It is sometimes desirable to change the base of a logarithm. Calculators, for example, have keys for computing logarithms to base e or to base 10. To compute a logarithm to any other base using a calculator, one has to convert that logarithm to base e or to base 10. The formula for changing base follows:

CHANGE OF BASE

$$\log_b x = \frac{\log_a x}{\log_a b}$$

Proof Let $p = \log_b x$. Write this equation in exponential form.

$$b^p = x$$

Take the logarithm (base a) of both sides of this equation.

$$\log_a b^p = \log_a x$$

Use Property 3 of logarithms.

$$p \log_a b = \log_a x$$

Solve this equation for p.

$$p = \frac{\log_a x}{\log_a b}$$

Substitute back for p to get the desired result.

$$\log_b x = \frac{\log_a x}{\log_a b}$$

In Section 5.3, we saw how to use a calculator to compute values of natural and common logarithms. Let us now see how a calculator can be used to compute the logarithm to a base other than e or 10.

Example 11 **Computing a Logarithm to Base 6**

Compute $\log_6 4$ to four decimal places.

SOLUTION Express this logarithm in terms of natural logarithms and then use a calculator. The formula for change of base gives

$$\log_6 4 = \frac{\ln 4}{\ln 6}$$

A calculator then gives

$$\log_6 4 = 0.7737, \text{ rounded to four decimal places}$$

Self-Check 6 Compute $\log_5 7$ to four decimal places (a) using natural logarithms and (b) using common logarithms. (You should, of course, get the same result both ways.)

In Section 5.3, we saw how to use a graphing calculator to draw graphs of natural and common logarithms. Let us now see how a calculator can be used to sketch the graph of a logarithm to a base other than e or 10.

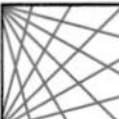

Graphing General Logarithmic Functions with a Calculator

Let us sketch the graphs of $f(x) = 5^x$, $g(x) = \log_5 x$, and $y(x) = x$ in the same coordinate system. Enter $g(x) = \log_5 x$ into the calculator as $\ln x/\ln 5$. We see that f and g are symmetric about the line $y = x$, demonstrating that g is f^{-1} (Figure 5.31).

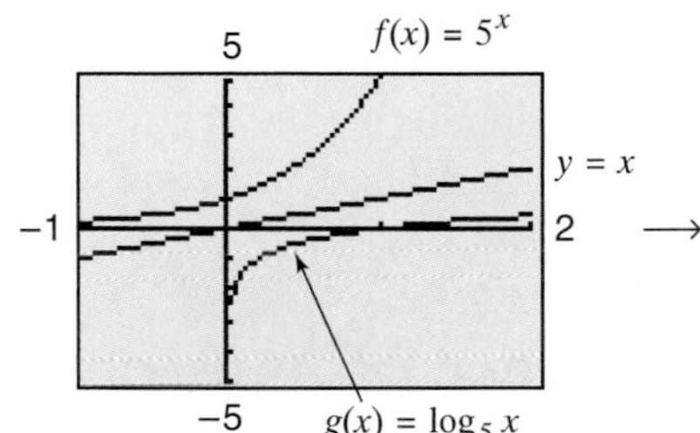

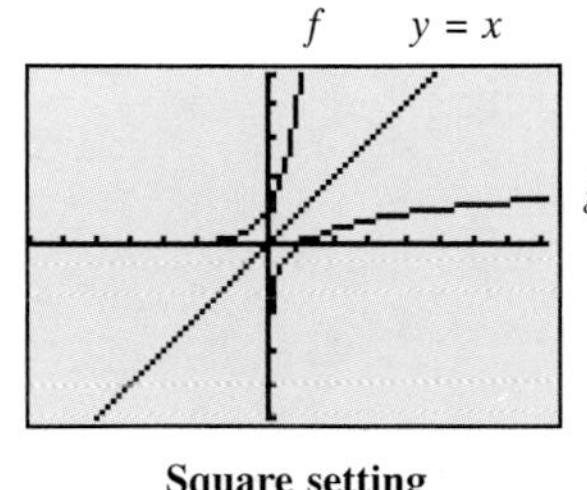

Square setting

Figure 5.31

Applications . . .

World Population Growth

Example 12 **World Population Growth**

World population growth is defined by the exponential function $P(t) = 5.94(2^{t/40})$, where t is time measured in years from 1998 and $P(t)$ is the population in billions at time t (Section 5.1). (a) Determine the inverse of this function. (b) Use the inverse to predict when the population can be expected to reach 8 billion.

SOLUTION

(a) To arrive at the inverse of P, we solve $P(t) = 5.94(2^{t/40})$ for t.

$$2^{t/40} = \frac{P}{5.94}$$

We rewrite it in logarithmic form and then solve for t.

$$t/40 = \log_2\left(\frac{P}{5.94}\right)$$

$$t = 40\log_2\left(\frac{P}{5.94}\right)$$

We write this function in terms of natural logarithms as follows. The inverse function is

$$t(P) = \frac{40 \ln\left(\frac{P}{5.94}\right)}{\ln 2}$$

(b) We can use this function to predict the number of years (measured from 1998) that it will take for the world population to reach various sizes. The time it will take for the population to reach 8 billion is

$$t(8) = \frac{40 \ln\left(\frac{8}{5.94}\right)}{\ln 2}$$

$$= 17.18, \qquad \text{to two decimal places}$$

The world population was 5.94 billion in 1998. It is predicted to exceed 8 billion during the 17th year after 1998; namely, in the year 2015!

Dating of Artifacts

The carbon dating of artifacts is an example of the use of exponential decay. Logarithms to base 2 arise naturally in this application. We have seen how exponential growth is described in terms of doubling time: the time it takes a quantity to double. Exponential decay is described in terms of **half-life:** the time it takes a quantity to halve.

Example 13 Carbon Dating of Artifacts

Carbon-14 is a radioactive substance; it gives off radiant energy. Carbon-14 atoms in the atmosphere are taken in by plants. Animals either eat plants or eat other animals that eat plants; thus, they also contain carbon-14. When a plant or animal dies it ceases to take in carbon-14, and the amount of carbon-14 within it decays without being replenished.

Scientists have found that if the source of radioactivity is not replenished, the radioactivity decreases exponentially. The half-life of carbon-14 is 5568 years. The amount of carbon-14 radiation given off by 1 gram of living substance is constant at 15.3 units. After t years of decay, the amount of radioactivity $R(t)$ given off by 1 gram is

$$R(t) = 15.3\,(2^{-t/5568})$$

The technique of carbon dating involves determining the age of an artifact by measuring the amount of carbon-14 left in a sample. In this application, the radiation R is known, and the corresponding time t is wanted. We therefore determine the inverse function. We solve for t.

$$2^{-t/5568} = \frac{R}{15.3}$$

Rewrite this equation in logarithmic form.

$$-\frac{t}{5568} = \log_2\left(\frac{R}{15.3}\right)$$

$$t = -5568 \log_2\left(\frac{R}{15.3}\right)$$

$$t = -\frac{5568 \ln(R/15.3)}{\ln 2}$$

Let us summarize this important result.

Function for Carbon Dating

If the carbon-14 measurement for an artifact is R units, the following function can be used to give the age of the artifact in years:

$$t(R) = -\frac{5568 \ln(R/15.3)}{\ln 2}$$

We now give a specific example. Wood from the deck of the funeral boat of the Egyptian Sesostris III was measured for radioactivity. The amount of radioactivity given off by 1 gram was estimated to be 9.75; thus $R = 9.75$. Let us determine the age of the boat. Using the preceding function, the age t is

$$t(9.75) = -\frac{5568 \ln\left(\frac{9.75}{15.3}\right)}{\ln 2}$$

$$= 3619.520321$$

The boat is approximately 3600 years old—it dates back to about 1600 BCE.

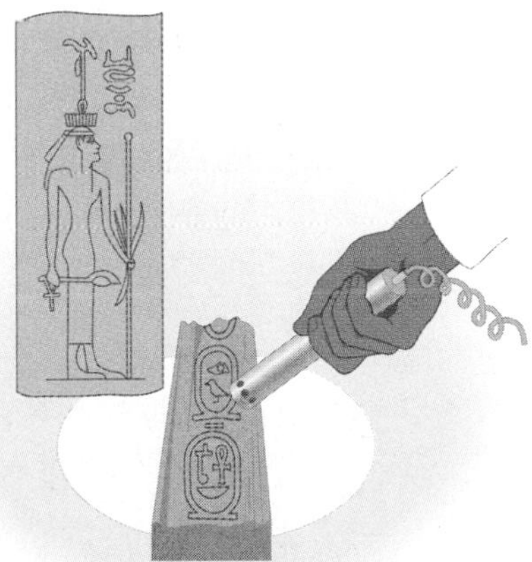

We conclude this example by mentioning one of the most recent important results of carbon dating: On December 15, 1998, Dr. Gunter Dreyer, head of the German Archaeological Institute, announced that clay tablets uncovered in southern Egypt from the tomb of King Scorpion I probably represent the earliest known writing by humankind. The subject these tablets deal with may be of no surprise. It is taxes. They record linen and oil deliveries made about 5300 years ago as tithes to the king. The discovery throws open for debate a widely held belief among historians that the first people to write were the Sumerians of the Mesopotamian civilization sometime before 3000 BCE.

Self-Check 7 The radioactivity of wood from the foundation cribbing for a fortification wall in a fort at Alisher Huyuh, Turkey, was found to be 10.26 units. How old is the fort?

Answers to Self-Check Exercises

1. $\log_9(x-4) = 1/2$. $(x-4) = 9^{1/2}$, $x = 4+3$, $x = 7$.
2. **(a)** $-4(7^{\log_7 5}) = -4(5) = -20$.
   (b) $7y = \log_3(2x-5)$, $3^{7y} = 2x-5$, $x = (3^{7y})/2 + 5/2$.
3. **(a)** $\log_a 21 = \log_a(3 \cdot 7) = \log_a 3 + \log_a 7 = 1.3222$.
   (b) $\log_a(3/7) = \log_a 3 - \log_a 7 = -0.3680$.
   (c) $\log_a 343 = \log_a(7^3) = 3\log_a(7) = 2.5353$.
4. $\log_a(x^2-4) - \log_a(x+2) + 3\log_a 2 = \log_a(x^2-4) - \log_a(x+2) + \log_a 2^3 = \log_a[2^3(x^2-4)/(x+2)] = \log_a[8(x-2)]$.
5. $4\log_2 x - 3\log_2 x = 5$, $\log_2 x^4 - \log_2 x^3 = 5$, $\log_2(x^4/x^3) = 5$, $\log_2(x) = 5$, $x = 2^5 = 32$.
6. $\log_5 7 = \dfrac{\ln 7}{\ln 5} = 1.2091$. $\log_5 7 = \dfrac{\log 7}{\log 5} = 1.2091$.
7. $t(R) = \dfrac{-5568\ln(R/15.3)}{\ln 2} = \dfrac{-5568\ln(10.26/15.3)}{\ln 2} = 3210$ years.

EXERCISES 5.4

In Exercises 1–12, solve each logarithmic equation for x.

1. $\log_3(x-4) = 2$
2. $\log_7(x+5) = 2$
3. $\log_2(2x-3) = 4$
4. $\log_4(3x+5) = 2$
5. $\log_5(3x-2) = 3$
6. $\log_7(4x-1) = -1$
7. $\log_x(3-2x) = 2$
8. $\log_x(5x-6) = 2$
9. $\log_x 81 = 4$
10. $\log_{x+1} 9 = 2$
11. $\log_2 16 = x$
12. $\log_3 9 = x+4$

In Exercises 13–20, rewrite each equation to express x in terms of y.

13. $2y = 3^{(x+1)}$
14. $y = 5^{(2x+6)}$
15. $4y = 2^{(x-5)}$
16. $5y = \log_2(x-6)$
17. $2y = \log_3(4x+1)$
18. $7y = \log_{10}(3x-8)$
19. $\log_2 18y = \log_2(3x+6)$
20. $\log_7(4y+5) = \log_7(2x-1)$

In Exercises 21–26, use the logarithmic identities $a^{\log_a x} = x$ and $\log_a a^x = x$ to evaluate each expression.

21. $4(\log_3 3^2)$
22. $-6(\log_7 7^{-2})$
23. $5(\log_4 4^2)$
24. $7(4^{\log_4 3})$
25. $-2(5^{\log_5 8})$
26. $7(3^{\log_3 9})$

In Exercises 27–30, use the properties of logarithms to compute the logarithms.

27. If $\log_a 6 = 1.7918$ and $\log_a 2 = 0.6931$, find
(a) $\log_a 12$ **(b)** $\log_a 3$ **(c)** $\log_a 36$ **(d)** $\log_a(1/3)$

28. If $\log_a 5 = 0.699$ and $\log_a 3 = 0.4771$, find
(a) $\log_a 15$ **(b)** $\log_a 0.6$ **(c)** $\log_a 125$ **(d)** $\log_a(5/3)$

29. If $\log_a 7 = 1.9459$ and $\log_a 4 = 1.3863$, find

(a) $\log_a 1.75$ (b) $\log_a 2401$ (c) $\log_a(4/7)$
(d) $\log_a 16$

30. If $\log_a 4 = 0.6021$ and $\log_a 3 = 0.4771$, find
(a) $\log_a 27$ (b) $\log_a 12$ (c) $\log_u 144$
(d) $\log_a 0.1875$

In Exercises 31–36, write the expressions in terms of logarithms of x, y, and z.

31. $\log_a(xy/z)$ **32.** $\log_a(x^2y^3z^{-4})$
33. $\log_b(x^3y^2/z^{-1})$ **34.** $\log_7(x^2yz^{-3}) + \log_7[2x/(y^{-1}z)]$
35. $\log_3(25yz/x) - \log_3(5x^3y/z^{-2})$
36. $\log_4(2xz/y) + \log_4(x^2yz^3) - 3\log_4(4x^{-1}y^{-2}z^{-3})$

In Exercises 37–42, write the expressions as a single logarithm.

37. $3\log_a 2x + \log_a x - 2\log_a 4x$
38. $-2\log_b(4x - 3) + 3\log_b x + 2\log_b x(4x - 3)$
39. $7\log_c x + 3\log_c x(5x + 2)^{-1} - \log_c x^2(5x + 2)^{-2}$
40. $2\log_7 3(4x - 1) + 3\log_7 x(4x - 1) - \log_7 9x$
41. $\log_{10} x(x - 2) + 2\log_{10} y(x - 2) - 3\log_{10} xy(x - 2) + 4\log_{10}(x - 2)$
42. $-\log_2 9xy + 4\log_2 3x^{-1}y - 3\log_2 2xy^2 + 2\log_2 4x^3y^{-4}$

In Exercises 43–50, use the properties of logarithms to solve the equations for x.

43. $\log_2 3 + \log_2 x = 4$
44. $\log_3 5 - \log_3 2x = 2$
45. $\log_2(x - 4) + \log_2 6 = 3$
46. $\log_4(x + 3) - \log_4 x = 1$
47. $\log_3(3x + 6) - \log_3(x - 2) = 2$
48. $\log_7(x - 4) + \log_7 3 = \log_7 2x$
49. $\log_8(5x + 2) = \log_8 6 - \log_8 3$
50. $\log_4 8x - \log_4 x = 2$

In Exercises 51–58, use the properties of logarithms to solve the equations for x.

51. $2\log_7 3x - \log_7 2x = 1$
52. $\log_3 x + \log_3 6x - \log_3 2x = 4$
53. $2\log_9 3x - \log_9 x = 1$
54. $3\log_2 2x - 2\log_2 x = 4$
55. $3\log_6 x = \log_6 125$
56. $2\log_7(x - 1) = \log_7 4$
57. $\log_2(x^2 - 7x + 10) = 2$
58. $\log_4(x - 3)^2 = 2$

In Exercises 59–66, solve the equations (if possible), rounding to four decimal places, using graphs.

59. $\ln(2x - 5) = -2x + 13$
60. $\ln(3x + 7) - 2x - 8$
61. $\ln(x^2 + 2x + 7) = 3x - 7$
62. $\ln(x^2 - 3x + 4) = -2x - 5$
63. $\ln(x^3 + 2x) = 2x - 8$
64. $\ln(3x - 5) = e^{2x-4} + 1$
65. $\ln(2x + 1) = 3^{4-x} - 2$
66. $\ln(4x + 9) = x^2 - 8x + 12$

In Exercises 67–72, compute the logarithms to four decimal places by converting the given logarithms to natural logarithms and then using a calculator.

67. $\log_3 5$ **68.** $\log_5 7$ **69.** $\log_7 2.4$
70. $\log_2 3.95$ **71.** $\log_6 75.43$ **72.** $\log_8 473.62$

In Exercises 73–78, compute the expressions to four decimal places, by using the properties of logarithms and a calculator.

73. $\log_3 5 + \log_3 7$ **74.** $2\log_5 4 - \log_5 8$
75. $3\log 2 - 2\log 10$ **76.** $3\log_7 4 - 5\log_7 2$
77. $\log_8 820 + 3\log_8 13 - \log_8 5$
78. $5\log_3 120 - 2\log_3 99.3 + 2.3\log_3 7.6$

In Exercises 79–86, solve the equations for x to four decimal places by writing them in logarithmic form and then using a calculator.

79. $5 = e^{3x}$ **80.** $6 = e^{-2x}$ **81.** $72 = 10^{5x}$
82. $12 = 10^{-x/4}$ **83.** $4 = 2^x$ **84.** $3.4 = 3^{2x}$
85. $5.1 = 7^{(2x-1)}$ **86.** $7.8 = 2^{(3x+5)}$

87. Prove the following two properties of logarithms introduced in this section.
(a) $\log_a(x/y) = \log_a x - \log_a y$
(b) $\log_a(x^r) = r\log_a x$

88. (a) Prove that the function $g(x) = \log_{0.5} x$ is the inverse of $f(x) = 2^{-x}$. (b) Use this knowledge to draw a rough graph of $g(x) = \log_{0.5} x$. (c) Verify your result by drawing the graph of g, f, and $y = x$ on a calculator and observing symmetry. (d) Why is the graph of $g(x) = \log_{0.5} x$ concave up, not concave down, as other logarithmic functions we have discussed? (e) Arrive at results for the general shapes of graphs of functions of the form $f(x) = \log_a x$, $0 < a < 1$.

Carbon Dating

In Exercises 89–93, determine the age of each artifact to the nearest decade, given the radioactivity.

89. Wood from the floor of a central room in a large *hilani* (palace) of the Syro-Hittite period in the city of Tayinat in northwest Syria was found to have radioactivity of 11.17 units.

90. A slab of wood from a roof beam of the tomb of Viziar Hemaka, contemporaneous with King Udima, First Dynasty, in Sakkara, Egypt, was found to have radioactivity of 8.33 units.

91. Samples of basketry from Lovelock Cave, Nevada, were found to have radioactivity of 12.40 units.

92. Charcoal from a temple at Tilantonga, Oaxaca, Mexico, was found to have radioactivity of 11.07 units.

93. Lake mud from Neasham near Darlington in the extreme north of England was found to have radioactivity of 3.96 units.

94. Predicting World Population We have seen that world population growth is defined by the exponential function $P(t) = 5.94(2^{t/40})$, where t is time in years measured from 1998 and $P(t)$ is the population in billions at time t. The inverse of this function is

$$t(P) = \frac{40 \ln(P/5.94)}{\ln 2}$$

We used this function in this section to show that the world population is predicted to exceed 8 billion in the year 2015. Use $t(P)$ to predict when the world population will reach 10 billion.

95. Predicting Growth of *Escherichia coli* In Section 5.1, we saw how the growth of bacteria is described by the exponential function $N(t) = k(2^{t/c})$, where t is time, $N(t)$ is the number of cells of the bacterium at time t, k is the initial count, and c is the doubling (or generation) time. **(a)** Determine $t(N)$, the inverse of this function. **(b)** The bacterium *Escherichia coli* has a doubling time of 17 minutes at 37°C in a laboratory; $c = 17$. Use $t(N)$ to compute how long (to the nearest tenth of a minute) it will take a single bacterium of *Escherichia coli* to grow to 1000 bacteria.

96. Determining Doubling Time The growth of bacteria is described by the exponential function $N(t) = k(2^{t/c})$ where t is time, $N(t)$ is the number of cells of the bacterium at time t, k is the initial count, and c is the doubling (or generation) time. In this exercise, we see how counts of bacteria are used in the laboratory to determine the doubling time c. **(a)** Show that the preceding equation can be rewritten in the logarithmic form

$$c = \frac{t(\ln 2)}{\ln(N/k)}$$

If k and a further count N at a time t can be measured in the laboratory, this equation can be used to compute the doubling time c for the bacterium to two decimal places.

(b) An initial count for a culture of *Bacillus thermophilus* in the laboratory is 1,000 cells. The count 60 minutes later is 13,454. Determine the doubling time and hence the function that describes the growth. How long does it take the initial count of 1,000 bacteria to reach 50,000, to the nearest second? This bacterium is of the *Bacillus* genus—it can lead to tetanus, tuberculosis, or whooping cough.

Exponential Regression and Intravenous Drug Administration

- Exponential Regression • Intravenous Drug Administration
- Mainstreaming in the Computer World

Exponential Regression

We have used doubling time (or half-life) to construct exponential functions. The doubling time of something that is growing exponentially is not always readily available for constructing the exponential function. One often needs to construct an appropriate exponential function from data. We saw in Chapters 1 and 3 how functions that describe real situations can be constructed from data by using regression. We now illustrate how **exponential regression** can be used to determine exponential functions from data.

Example 1 College Education

Since 1940, the number of people in the United States with 4 years of college education has been increasing dramatically. The following statistics were obtained from the web. (See the Williams website at www.saunderscollege.com/math/williams to link to the Internet for more information.) Numbers in parentheses give years from the base year 1940. Let us construct a function $N(t)$ that closely fits these data.

Year, t	1940 (0)	1950 (10)	1960 (20)	1970 (30)	1980 (40)	1990 (50)
Number of people with 4 years of college (in millions), N	3.407	5.272	7.617	12.062	22.193	33.291

We enter and plot the data on a calculator (Figure 5.32a), using the years measured from 1940 for the horizontal axis. Observe that the growth appears to be exponential (Figure 5.32b). Use exponential regression on a calculator to find the function that best fits these data (Figure 5.32c). The function is

$$N(t) = 3.256628189(1.047281015)^t$$

Check the function against the original data. Enter the function, plot the data, and graph. The graph is seen to fit the data points closely (Figure 5.32d).

This function can now be used to predict future numbers. For example, the prediction for the year 2000, when t will be 60, is

$$N(60) = 3.256628189(1.047281015)^{60} = 52.06693624$$

We round this number to three decimal places (like the original data). The prediction is that 52.067 million people will have 4 years of college education in the year 2000. (See Exercise 9 for a discussion of the significance of this base number 1.047281015.)

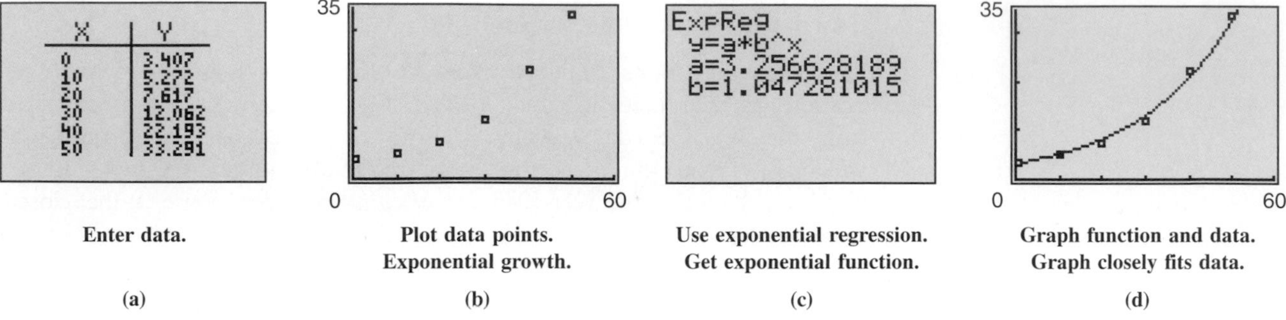

Enter data. (a)

Plot data points. Exponential growth. (b)

Use exponential regression. Get exponential function. (c)

Graph function and data. Graph closely fits data. (d)

Figure 5.32 People with a college education.

Further Ideas 1 *Microsoft Corporation Assets* The total assets of Microsoft Corporation for the years 1993 to 1997 are given in the following table (*Microsoft Annual Report,* 1997). Numbers in parentheses give years from the base year 1993. Plot the data (using number of years measured from 1993 on the horizontal axis) to show that assets are growing exponentially. Use exponential regression on a calculator to find the function that best fits these data; then predict the total assets of Microsoft for the year 2001.

Year, t	1993 (0)	1994 (1)	1995 (2)	1996 (3)	1997 (4)
Assets, A (millions of dollars)	3,805	5,363	7,210	10,093	14,387

We introduced piecewise-defined functions in Section 3.3 and have seen the usefulness of such functions in applications on numerous occasions. The next example illustrates how such a function is needed to plan the repeated application of a drug.

Example 2 Intravenous Drug Administration

After the intravenous administration of a drug, the drug is distributed rapidly in the blood stream (unlike the oral administration that was discussed in Section 3.2) and then decreases exponentially through biotransformation. The quantity $Q(t)$ of the drug remaining in the body at time t after intravenous administration is given by

$$Q(t) = k(2^{-t/c})$$

where k is the initial concentration of the drug, and c is a positive constant. This function is one of exponential decay. Its interpretation is analogous to the previous examples of exponential decay. Here, c is the time it takes the amount of drug in the body to halve. This time is the half-life of the drug (or in pharmaceutical language its **mean residence time** [MRT]). Let us look at the application of a spe-

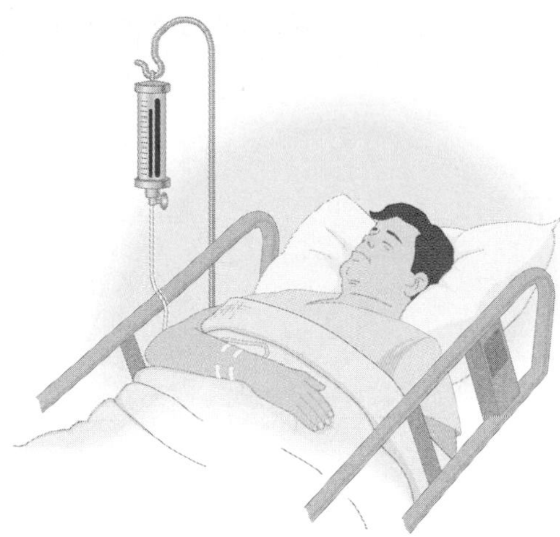

cific drug, sodium phenobarbital. This drug is used to treat epilepsy and cirrhosis of the liver.

A patient initially is given 300 milligrams sodium phenobarbital. The half-life of the sodium phenobarbital is approximately 60 hours. The initial amount of the drug in the body will thus halve after 60 hours. Let us model this decrease. The quantity of sodium phenobarbital in the body after t hours is described by the function

$$Q(t) = 300(2^{-t/60})$$

We graph the decay of the drug over a 180-hour period ($7\frac{1}{2}$ days) and display a table of values (Figure 5.33).

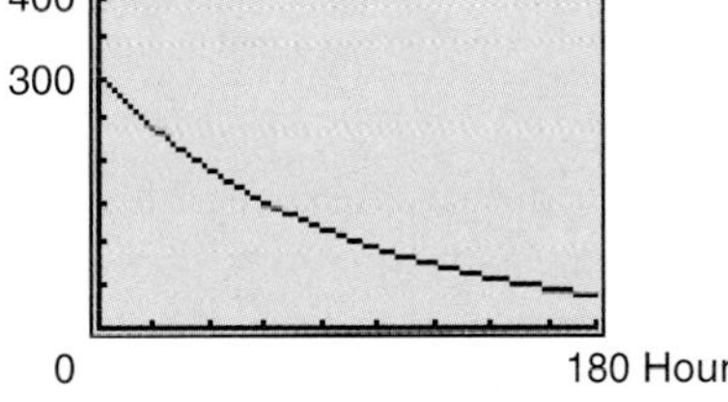

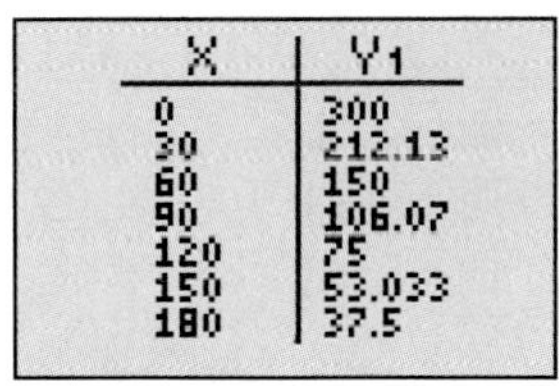

X	Y1
0	300
30	212.13
60	150
90	106.07
120	75
150	53.033
180	37.5

Figure 5.33 Decay of sodium phenobarbital during a week.

The table tells us that there should be 37.5 milligrams sodium phenobarbital remaining in the body after 180 hours. Half-lives of drugs vary greatly. For instance, aspirin has a half-life of 15 minutes, penicillin has a half-life of approximately 30 minutes, and digitoxin, which is used to control hepatitis, has a half-life of about 10 days. Some drugs can persist in the blood for years.

Most drugs are administered in a constant dosage, given over regular intervals for prolonged periods of time. When appropriate, it has become common to administer a drug once every half-life. Because the half-life of sodium phenobarbital is 60 hours, half the original 300 milligrams will have decayed in this time. Therefore, this drug could be administered over a period of time with a dose of 150 milligrams every 60 hours. The level of sodium phenobarbital in the body would then be kept between 150 milligrams and 300 milligrams. The graph and table of such a prescription over the first three doses are shown in Figure 5.34.

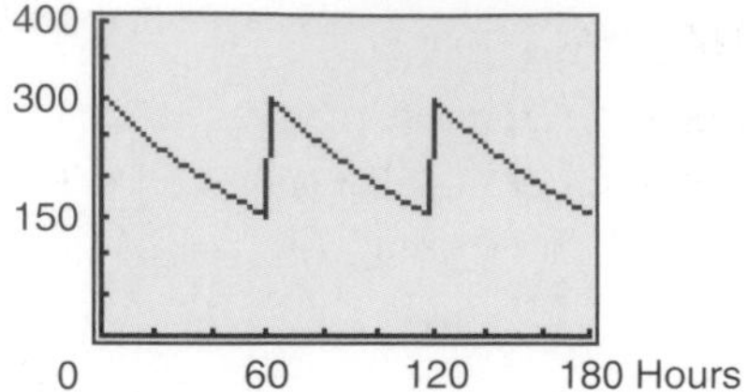

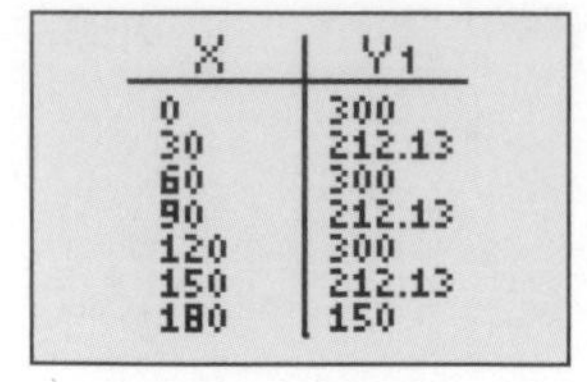

X	Y1
0	300
30	212.13
60	300
90	212.13
120	300
150	212.13
180	150

Figure 5.34 Administration of sodium phenobarbital with a dose every 60 hours.

Observe that the graph is duplicated over the intervals [0, 60), [60, 120), and [120, 180). We can get the piecewise-defined function that describes this graph from horizontal translations of 60 and 120 of the original function.

$$Q(t) = \begin{cases} 300(2^{-t/60}) & \text{if} \quad 0 \le t < 60 \\ 300(2^{-(t-60)/60}) & \text{if} \quad 60 \le t < 120 \\ 300(2^{-(t-120)/60}) & \text{if} \quad 120 \le t < 180 \end{cases}$$

We have modeled the administration over the first three doses. The function that describes the behavior of the drug over a longer period would naturally involve further pieces described by translations of 180, 240, and so forth.

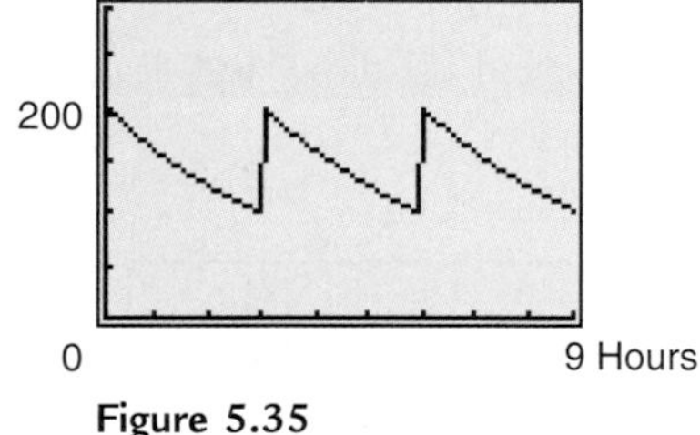

Figure 5.35

Further Ideas 2 The drug meperidine has a half-life of 3 hours. An initial dose of 200 milligrams of this drug was administered intravenously to a patient to treat acute viral hepatitis. Doses of 100 milligrams were then given every 3 hours. The graph of Figure 5.35 describes the quantity of the drug in the body over the first 9 hours. Determine the function that describes this behavior over 9 hours. How much meperidine is in the body after 8 hours?

We have discussed learning functions and curves on numerous occasions. There are many such functions that are used to model a learning experience. Our last example in this section illustrates learning curves for hardware and software productivity.

Example 3 | Mainstreaming in the Computer World

It is a well-known fact that if new technology does not achieve mainstream status quickly enough it dies. During the Industrial Age, "quick" meant about 50 to 100 years (e.g., invention of the telephone). In the Postindustrial age (about 1945 to 1990), "quick" meant 10 to 20 years (television, business personal computers). In the current Information Age, "quick" means 5 to 10 years. In an article entitled "The Big Software Chill", Ted Lewis of the Naval Postgraduate School claims that current software development is not taking place at a rate fast enough to attain mainstream status, whereas hardware is.* The implication is that software tools will continuously disappear to be replaced by new tools, whereas hardware will remain in a more permanent state. In this example, we give the student the

*Ted Lewis, "The Big Software Chill", *Computer.* 1996; 29(3)12.

opportunity to follow the argument presented by Lewis, using the mathematics that we developed, but couched in the jargon of the field.

The learning curves of hardware and software productivity are computed as $P(t) = B^t$, where t is time in years from 1960 and P is the cost per unit of performance. The values of B are 0.675 for the hardware industry and 0.955 for the software industry. The learning curves for the two sides of the computing world, namely $P(t) = 0.675^t$ and $P(t) = 0.955^t$, are shown in Figure 5.36a.

The **mainstream curves** are obtained from $M = 2/(P + 1) - 1$ for the appropriate P. M is the market share. The mainstream curves for the hardware and software industries are $M(t) = 2/(0.675^t + 1) - 1$ and $M(t) = 2/(0.955^t + 1) - 1$. These curves for the time interval $t = 0$ (1960) to $t = 40$ (2000) (Figure 5.36b) reveal how quickly technology attains mainstream status. Lewis is honest enough to say, "Don't ask why this works, because there is no sound theory behind forecasting." Note how these curves indicate that hardware attains mainstream status within 10 years, whereas software does not, thereby supporting Lewis' argument. Lewis goes on to discuss how a computer language such as C++ is under siege because its early momentum has not been sustained. "Until someone discovers or invents a breakthrough software technology that puts us on a faster learning curve, the big software chill will get bigger."

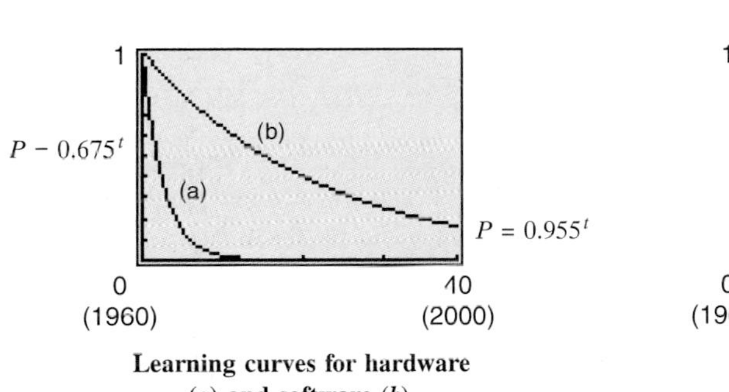

Learning curves for hardware (*a*) and software (*b*).

(a)

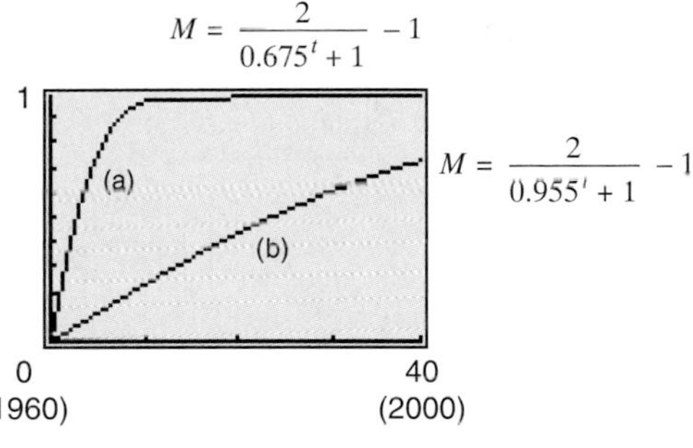

Mainstreaming curves for hardware (*a*) and software (*b*).

(b)

Figure 5.36

Further Ideas 3 The learning curve for software errors per thousand lines of code (KLOC) at NASA Goddard Space Flight Center is given by $P(t) = B^t$ with $B = 0.969$, for the period 1976 to 1996. This learning curve is typical of the results returned by error-detecting software. Use a graphing calculator to sketch the learning and mainstream curves for software errors. Is the software for detecting errors ever likely to achieve mainstream according to this evidence?

Answers to Further Ideas Exercises

1. $A(t) = 3801.62189(1.38990259)^t$. In 2001, $t = 8$. $A(8) = \$52,947.30$ million.
2. See $Q(t)$ below. $Q(8) = 125.992105$. There would be 126 milligrams of meperidine in the body after 8 hours.

$$Q(t) = \begin{cases} 200(2^{-t/3}) & \text{if} \quad 0 \le t < 3 \\ 200(2^{-(t-3)/3}) & \text{if} \quad 3 \le t < 6 \\ 200(2^{-(t-6)/3}) & \text{if} \quad 6 \le t < 9 \end{cases}$$

3. $P(t) = 0.969^t, \qquad M(t) = \dfrac{2}{0.969^t + 1} - 1$

See following graphs. $M(24) = 0.5024$. This is well below 1. Software for detecting errors is not likely to achieve mainstream.

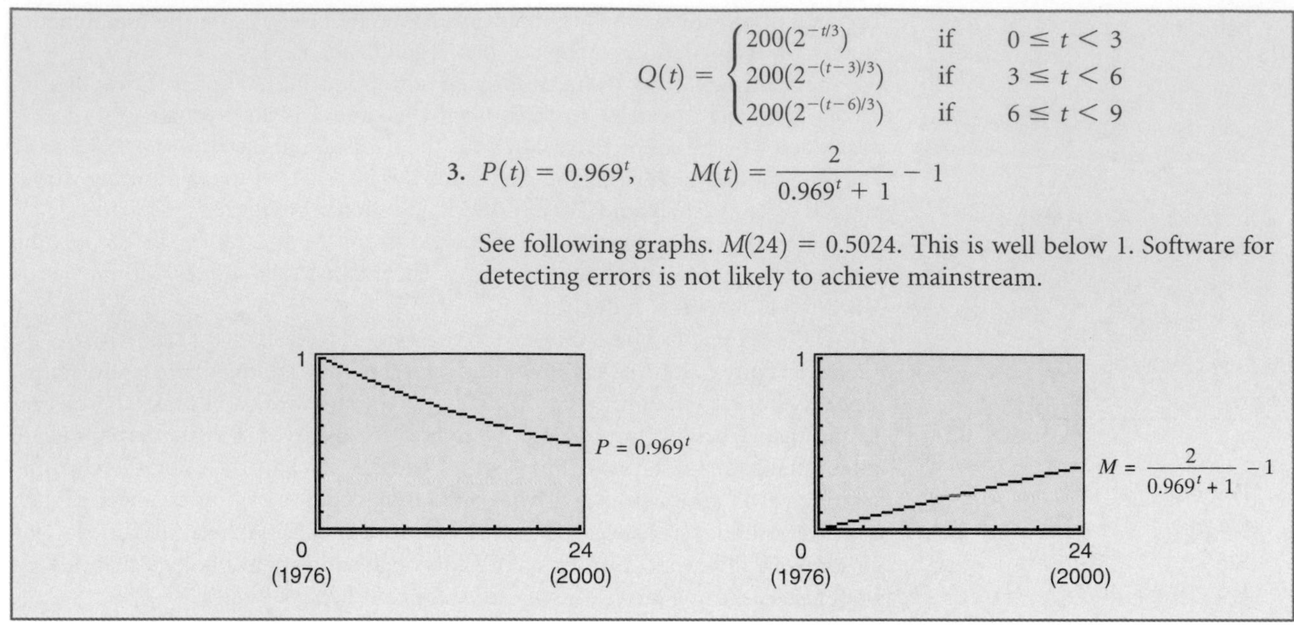

EXERCISES 5.5

In Exercises 1–7,* use exponential regression to analyze the situations. Numbers in parentheses give years from the base year. Plot the data using these numbers for the horizontal axis.

1. **High School Education** The following table gives the number of people in the United States with 4 years of high school education. (**a**) Plot the data to see that the number of people with high school educations is apparently increasing exponentially. Use regression to find the best exponential function for these data. (**b**) Predict the number of people in the United States with 4 years of high school education in the year 2000, rounding to three decimal places.

Year	1940 (0)	1950 (10)	1960 (20)	1970 (30)	1980 (40)	1990 (50)
Number of people with 4 years of high school (millions)	10.552	17.624	28.477	37.134	47.934	60.119

*For exercises 1–5, see the Williams Web site at www.saunderscollege.com/math/williams to link to the Internet for more information.

2. **Minimum Wage** The following table describes the increase in minimum wage in the United States at certain times since 1945. (**a**) Plot the data to see that the minimum wage is increasing exponentially. Use exponential regression to find the best exponential function for these data. (**b**) Predict the minimum wage for the year 2010.

Year	1945 (0)	1961 (16)	1967 (22)	1974 (29)	1980 (35)	1989 (44)	1997 (52)
Minimum wage ($)	0.40	1.15	1.40	2.00	3.10	3.35	5.15

3. **Personal Earnings** The following table describes the growth in median annual personal earnings. (**a**) Plot the data to see that this number is increasing exponentially. Use exponential regression to find the best exponential function for these data. (**b**) Predict when median annual earnings will reach $25,000.

Year	1940 (0)	1950 (10)	1960 (20)	1970 (30)	1980 (40)	1990 (50)
Annual earnings ($)	746	1,926	2,894	4,375	8,549	13,898

4. **Self-employment** The following table describes the growth in the number of self-employed workers in the United States (in millions). (**a**) Plot the data to see that this number is increasing exponentially. Use exponential regression to find the best exponential function for these data. (**b**) Predict the number of self-employed workers in the year 2020.

Year	1950 (0)	1960 (10)	1970 (20)	1980 (30)	1990 (40)
Number of workers	4.19	6.87	6.27	8.20	12.50

5. **Medical Research** The following data give government expenditure (in millions of dollars) on medical research in the United States since 1960. (**a**) Plot the data to see that this number is increasing exponentially. Use exponential regression to find the best exponential function for these data. (**b**) Predict the expenditure on medical research in the year 2005.

Year	1960 (0)	1965 (5)	1970 (10)	1975 (15)	1980 (20)	1985 (25)	1990 (30)	1993 (33)
Expenditure on medical research (millions of dollars)	449	1,227	1,684	2,648	4,924	6,903	11,312	12,780

6. **Workman's Compensation** The following data give government expenditure (in millions of dollars) on workman's compensation in the United States in 5-year periods since 1960. **(a)** Plot the data to see that this number is increasing exponentially. Use exponential regression to find the best exponential function for these data. **(b)** When is expenditure likely to reach $100 billion?

Year	1960 (0)	1965 (5)	1970 (10)	1975 (15)	1980 (20)	1985 (25)	1990 (30)	1993 (33)
Expenditure on workman's compensation (millions of dollars)	1,308.5	1,859.4	2,960.4	6,479.1	13,457.2	22,263.6	41,703.4	43,376.2

7. **CREF Stock** The CREF Stock Account is the largest singly managed equity account in the world. Its assets in 1997 totaled $100 billion. Most university professors have their retirement stocks in CREF. The account was established in 1952. Its growth since 1952 is given by the following chart (*Investment Forum,* **1**(2): 1997). **(a)** Is this account growing exponentially over this whole period? **(b)** If not, how can the data be adjusted to predict likely assets for the year 2007? **(c)** Predict the CREF assets for 2007.

Year	1952 (0)	1957 (5)	1962 (10)	1967 (15)	1972 (20)	1977 (25)	1982 (30)	1987 (35)	1992 (40)	1997 (45)
Assets ($)	1 million	49 million	239 million	1 billion	3 billion	4 billion	11 billion	26 billion	47 billion	100 billion

8. **Growth of a Tree** For a period of its life, the growth of a tree follows the equation $d = k(2^{t/c})$, where d is the diameter in inches, t is the number of years from the beginning of this period, and k and c are constants. At the beginning of this period, d is 5 inches, and after 8 years, d is 10 inches. **(a)** Use this information to determine k and c. **(b)** What is the significance of the value of c? **(c)** What will the diameter of the tree be after 16 years, assuming that the growth will still be described by this law at that time?

9. **Analyzing Exponential Growth Functions**
(**a**) The growth of a certain population is described by the function $f(t) = k(3^{t/p})$. Show that p is the time it takes the population to triple.
(**b**) The growth of a certain population is described by the function $f(t) = k(a^{t/q})$, where $a > 0$. Show that q is the time it takes the population to multiply by a.
(**c**) In Example 1, we saw that the number of people with 4 years of college education is closely described by $N(t) = 3.256628189(1.047281015)^t$. N is measured in millions, and t is the time in years measured from 1940. This function was found on a calculator using exponential regression. The calculator gives the exponential function to base 1.047281015. What is the significance of this number? A calculator will give an exponential function of best fit in the form $f(t) = k(a^t)$. What is the significance of a?

10. **Growth of a Bacterium** The doubling time of a bacterium is c minutes. Determine the time it will take a population of this bacterium to grow eightfold. Note that the time it takes to multiply by eight is less than four times the doubling time. This is characteristic of exponential growth. Things get better or worse very quickly.

Exercises 11–14 involve the dissipation of drugs in the body.

11. **Dissipation of Penicillin** The half life of penicillin is 30 minutes. The initial amount of penicillin in the body will therefore have halved in 30 minutes. A patient is given 250 milligrams intravenously. (**a**) Give the function that describes the dissipation of the drug in the body. (**b**) How much penicillin will be left in the body after 90 minutes?

12. **Dissipation of Prednisolone** The drug prednisolone has a half-life of 4 hours. An initial dose of 100 milligrams of this drug was administered intravenously to a patient to treat acute viral hepatitis. Doses of 50 milligrams were then given every 4 hours. The graph in Figure 5.37 describes the quantity of the drug in the body over the first 12 hours. (**a**) Determine the function that describes this behavior over 12 hours. (**b**) How much prednisolone is in the body after 11 hours (rounded to two decimal places)?

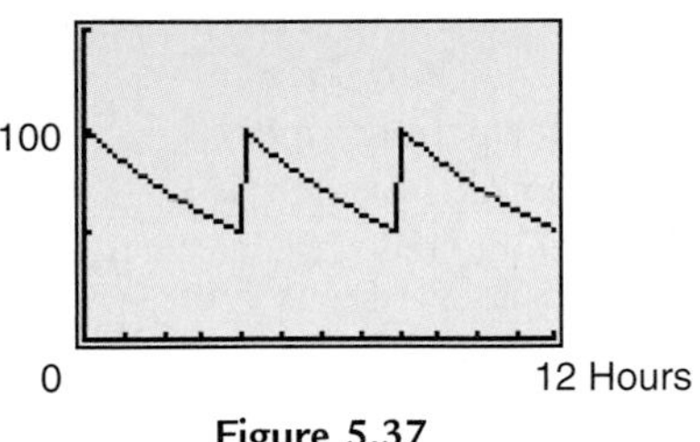

Figure 5.37

13. **Dissipation of Procainamide** Procainamide is used to treat cardiac arrhythmia and has a half-life of 3 hours. First, 80 milligrams procainamide is administered intravenously to a patient, followed by 40 milligrams every 3 hours for a 12-hour period. At this point, the treatment ceases. (**a**) Sketch a graph of the quantity in the body for the 18 hours after the drug was first administered. Determine the function that describes the dissipation of this drug in the body. (**b**) How much procainamide remains after 20 hours (to two decimal places)?

14. **Dissipation of Guanethidine** Some drugs are given in small amounts more frequently than once every half-life. Guanethidine has a half-life of about 10 days and is used to treat hypertension. Consider a case of outpatient therapy. An initial dose of 20 milligrams is given; then a dose is given every day for each of the following 3 days to bring the level up to 20 milligrams (there are four doses, including the first). The treatment is then discontinued but monitored. (**a**) Define and graph the function that describes this schedule. (**b**) What quantity of guanethidine is in the body at the end of a week (to two decimal places)?

CHAPTER 5 PROJECT | Simulating the Government Model for World Population

Information about world and United States populations can be found on the Internet (see the Williams Web site at www.saunderscollege.com/math/williams for more information). The numbers are continually updated. It is frightening to actually see the increases taking place. The world population on 08/11/98 at 15:58:22 GMT was 5,935,428,096. We based our population number of 5.94 billion in Section 5.1 on this number. Look up this site and discover what the world population is currently. The numbers given on the Internet are being produced by some mathematical model similar to ours, based on figures like our 5.94 billion and 40-year doubling time. We have seen models that use initial value and doubling time in Section 5.1 and models that use exponential regression in this section. Record data produced by the government model for world population. Construct a model that will closely predict these government figures.

WEB PROJECT 5

To explore the concepts of this chapter further, try the interesting WebProject. Each activity relates the ideas and skills you have learned in this chapter to a real-world applied situation. Visit the Williams Web site at

www.saunderscollege.com/math/williams

to find the complete WebProject, related questions, and live links to the Internet.

Chapter 5 Highlights

Section 5.1

Exponential Function: $f(x) = a^x$ where $a > 0$, $a \neq 1$. a is called the base.
Exponential Growth: Described by the function $f(x) = k(2^{x/c})$, $c > 0$, where k is the initial value, and c is the doubling time.
Exponential Decay: Described by the function $f(x) = k(2^{-x/c})$, $c > 0$, where k is the initial value, and c is the halving time.

Section 5.2

The Number *e*: The value approached by $\left(1 + \dfrac{1}{n}\right)^n$ as n becomes larger and larger. e is an irrational number. $e \approx 2.71828$.
Natural Exponential Function: $f(x) = e^x$.
Compound Interest: If a sum of money P is deposited in an account that pays interest at an annual rate r, compounded n times per year, amount A in the account after t years is $A = P(1 + r/n)^{nt}$.
Continuous Compounding: If a sum of money P in an account that pays interest at an annual rate r is compounded continuously, the amount in the account at the end of t years is $A = Pe^{rt}$.

Section 5.3

Logarithmic Function: Denote the inverse of $g(x) = a^x$ by $f(x) = \log_a x$. f is called the logarithmic function with base a.
Inverse Functions: $g(x) = a^x$ and $f(x) = \log_a x$, where $a > 0$, $a \neq 1$, are inverses of one another.
Conversion Rule: $x = \log_a y$ if and only if $y = a^x$.
Natural Logarithmic Functions: $f(x) = \log_e x$, written $f(x) = \ln x$.
Common Logarithmic Function: $f(x) = \log_{10} x$, written $f(x) = \log x$.

Section 5.4

Logarithmic Identities: $a^{\log_a x} = x$, for all $x > 0$, $\log_a a^x = x$, for any real number x.
Properties of Logarithms: $\log_a(xy) = \log_a x + \log_a y$, $\log_a(x/y) = \log_a x - \log_a y$, $\log_a(x^r) = r\log_a x$, where x, y are positive numbers, r is a real number, and a is a positive number not equal to 1.
Change of Base: $\log_b x = \dfrac{\log_a x}{\log_a b}$.

Chapter 5 Review Exercises

1. Sketch the graphs of the following functions. Check your results using a graphing calculator.
(a) $f(x) = 6^x$ (b) $f(x) = 3^{-x}$

2. Use standard graphs of $f(x) = a^x$ for an appropriate a to sketch the graphs of the following functions. Describe any shifts or reflections used. Check your results using a graphing calculator.
(a) $g(x) = 3^x + 2$ (b) $g(x) = 2^{x+1}$
(c) $g(x) = 6^{x-2} - 3$ (d) $g(x) = -3^{x+2}$

3. Use the known graph of $f(x) = e^x$ to sketch the graphs of the following functions. Describe the shifts or reflections used. Check your results using a graphing calculator.
(a) $g(x) = e^x - 3$ (b) $g(x) = e^{-x} + 1$
(c) $g(x) = e^{x+1} + 2$ (d) $g(x) = -e^x + 6$

4. Solve the following equations for x. Check your solutions using a graph.
(a) $4^{x-3} = 4^5$ (b) $2^{4x+1} = 2^{7x-3}$
(c) $4^{7x-2} = 8^3$ (d) $9^{4x-3} = 27$
(e) $e^{3x+2} = e^{5x-4}$

5. Solve the following equations, rounding to two decimal places, using graphs.
(a) $3^{(x-4)} = 5$ (b) $2^{(x^2+2x-4)} = 9$
(c) $2^{(x^3+3x-2)} = 7$

6. Write each of the following equations in logarithmic form.
(a) $y = 3^x$ (b) $y = 2^{4x}$ (c) $3^{-2} = \frac{1}{9}$

7. Write each of the following equations in exponential form.
(a) $y = \log_3 x$ (b) $y = \log_7 5x$
(c) $\log_3\left(\frac{1}{27}\right) = -3$

8. Compute the value of each of the following logarithms.
(a) $\log_2 4$ (b) $\log_3 27$ (c) $\log_5\left(\frac{1}{125}\right)$
(d) $\log_{27}\left(\frac{1}{3}\right)$

9. Find the inverse of each of the following functions using algebra.
(a) $f(x) = 3^x - 5$
(b) $f(x) = 5^{(3x+5)} - 7$
(c) $f(x) = \log_4 5x + 8$
(d) $f(x) = \log_3(2x - 5) + 9$

10. Solve the following equations for x.
(a) $\log_2(2x + 4) = 3$ (b) $\ln(3x - 2) = 5$
(c) $\log_x 8 = 3$ (d) $3y = 4^{(7x+1)}$

11. Use the properties of logarithms to write each of the following expressions as a single logarithm.
(a) $4 \log_a x + 2 \log_a(x - 3) - \log_a 5x$
(b) $4 \log_7(x + 1) - 3 \log_7 x(x + 1)^2 + \log_7 5x^3$

12. Use the properties of logarithms to solve the following equations for x.
(a) $\log_3 x + \log_3 2 = 4$
(b) $\log_2(7x - 4) - \log_2 2 = 3$
(c) $\log_2(x^2 + 4x - 1) = 2$
(d) $\log_3(x - 1)^2 - 2 \log_3(x + 4) = 4$

13. Solve the following equations (if possible) to two decimal places using graphs.
(a) $\ln(3x + 1) = -2x + 5$
(b) $\ln(x^2 + 3x + 5) = 3x^2 - 7$

In Exercises 14 and 15, use a calculator to find the values of the functions, rounded to four decimal places.

14. $f(x) = e^x$ and $g(x) = e^{-x}$ at $x = -3.2, -1.8, 0,$ 4.3, and 5.7.

15. (a) $f(x) = \ln x$ and $g(x) = \log x$ at $x = 0.00038$, 0.75, 1, 3.45, 12.78. (b) Compare the values of $\ln x$ and $\log x$ for $0 < x < 1$, $x = 1$, and $x > 1$. What does this suggest about the graphs of $f(x) = \ln x$ and $g(x) = \log x$? Verify your conjecture by drawing the graphs of $f(x) = \ln x$ and $g(x) = \log x$ on your calculator.

16. Determine the following logarithms to four decimal places by converting to natural logarithms and then using a calculator. Use the properties of logarithms if necessary. (a) $\log_2 6$ (b) $\log_5 3.5$
(c) $\log_4 5.8 - \log_4 2$ (d) $\log_5 6.7 + \log_5 2.3$

17. Use shifts and reflections of the graph of $f(x) = \ln(x)$ to sketch the graphs of the following functions. Describe any shifts or reflections used. Check your results using a graphing calculator.
(a) $g(x) = \ln x + 3$ (b) $g(x) = \ln x - 1$
(c) $g(x) = \ln(x - 1)$ (d) $g(x) = \ln(x + 1) - 2$
(e) $g(x) = -\ln(x - 2)$ (f) $g(x) = \ln(3 - x)$

18. **Population Growth** The population growth of Century City is described by the function $P(t) = 35(2^{t/25})$. $P(t)$ is the population in thousands at time t in years, measured from 1990. **(a)** If the current growth continues, determine in which year the population will reach 50 thousand, using a graph of $P(t)$. **(b)** Verify your result by expressing t as a function of P and determining its value when $P = 50$ thousand.

19. **Growth of Bacteria** The initial count of a bacterium is 500. The doubling time is 15 minutes. **(a)** Determine the function that describes the growth of the bacteria. **(b)** Compute the counts every 15 minutes for 1 hour by hand.

20. **Growth of Bacteria** The doubling time of the bacterium *Vibrio marinus* is 80 minutes. **(a)** Determine the function that describes the growth of this bacterium. **(b)** If the initial count is 200, find the counts after 4, 8, 12, and 16 hours by hand.

21. **Growth of Bacteria** The doubling time for the bacterium *Rhizobium melitoli* is 75 minutes. Determine the time (to the nearest tenth of a minute) that it will take a population of this bacterium to grow to ten times its original size.

22. **Carbon Dating** Glacial wood collected in a glacial till in Story County, Iowa, had radioactivity of 1.99 units. The till is thought to be Mankato. How old is the till, to the nearest decade?

23. The graph in Figure 5.38 describes an exponential function $f(x) = k(2^{x/c})$. Use the graph to find the initial value, the doubling time, and then the function. Verify your answer by duplicating the given graph on your calculator.

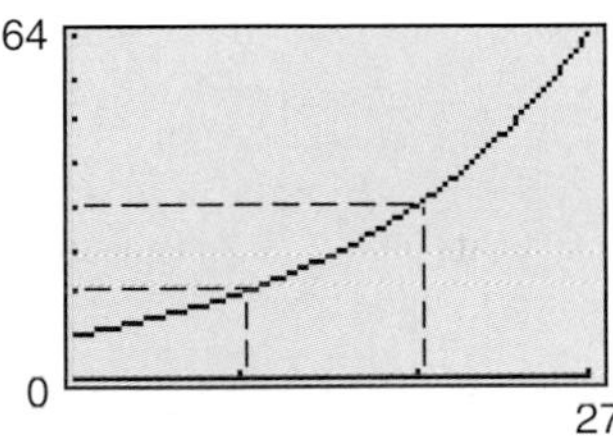

Figure 5.38

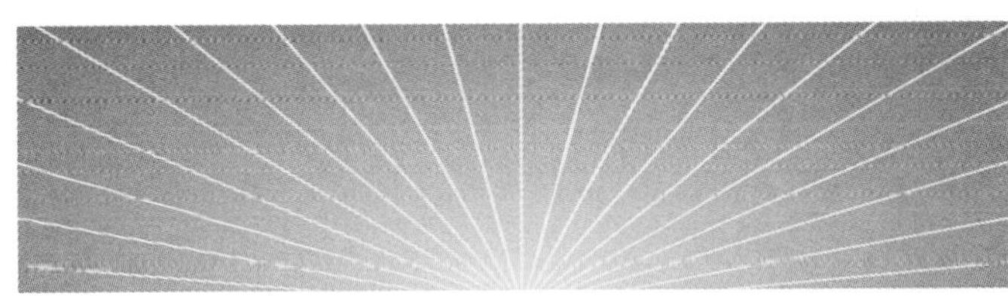

CHAPTER 5 TEST

1. Use the known graph of $f(x) = 2^x$ and shifts to sketch the graph of $g(x) = 2^{x+4} - 5$. Describe the shifts used. Check your results using a graphing calculator.

2. Solve the following equations for x. Check your solutions using a graph.
   (a) $2^{x-4} = 4^3$ **(b)** $9^{2x-3} = 81$ **(c)** $e^{-x+4} = e^{3x-5}$

3. Solve the following equations for x. Check your results using graphs.
   (a) $2^{4x/3} = 16$ **(b)** $3^{-2x/5} = 9$

4. Solve the equation $4^{(2x-5)} = 3$, rounding to two decimal places, using graphs.

5. Write each of the following equations in logarithmic form.
   (a) $y = 5^x$ **(b)** $y = 3^{7x}$

6. Write each of the following equations in exponential form.
   (a) $y = \log_6 x$ **(b)** $y = \log_3 4x$

7. Compute $\log_3(1/81)$.

8. Find the inverse of the function $f(x) = \log_2(x - 3) + 7$ using algebra.

9. Solve the following equations for x using algebra.
   (a) $\ln(4x - 3) = 8$, to four decimal places
   (b) $2y = 3^{(5x-1)}$

10. Use the properties of logarithms to solve the following equations for x.
    (a) $\log_4 x + \log_4 2 = 1$
    (b) $\log_2(x + 1)^2 - 2\log_2(x - 3) = 6$

11. Solve the equation $\ln(2x - 3) = -2x + 9$ to two decimal places using graphs.
12. Determine the following logarithms to four decimal places by converting to natural logarithms and then using a calculator. (**a**) $\log_3 4.7$ (**b**) $\log_2 8.3 + \log_2 5.8$ *Hint*: Use the properties of logarithms in part (**b**).
13. **City Population Growth** The population growth of a certain city has a doubling time of 53 years. The population was 20 thousand in 1995. (**a**) Construct an exponential function $P(t)$ that describes population as a function of time. (**b**) If the current growth continues, determine when the population is predicted to reach 30 thousand, using a graph of $P(t)$. (**c**) Verify your result by expressing t as a function of P and determining its value when $P = 30$.
14. **Carbon Dating of Fortification Wall** Wood from the foundation of a fortification wall in a mound at Alisher Huyuh, Turkey, was found to have radioactivity $R = 10.26$ units. It is known that after t years of decay the amount of radioactivity $R(t)$ given off by 1 gram of a radioactive substance is $R = (15.3)(2^{-t/5568})$. Solve this equation for t and determine the approximate age of the wall.

Cumulative Test Chapters 3, 4, and 5

1. Sketch the graph of the function $f(x) = x^2 - 4x + 2$, with domain $[-3, 8]$. Give the range of f. Support your results using a graphing calculator. Use the domain and the range as the x and y intervals of your window.
2. Graph the functions $f(x) = x^2$ and $g(x) = (x - 3)^2 - 5$ in the same coordinate system. Explain how the graph of g can be obtained from the graph of f using translations.
3. Give the equations of any vertical or horizontal asymptotes and the domain, and sketch the graph of the function

$$f(x) = \frac{2}{x^2 - x - 2}$$

Support your answer by duplicating the graph on your calculator.
4. Graph $f(x) = 2x^3 + 4x^2 - 6x + 3$ on your graphing calculator. Find the intervals over which f is increasing, where it is decreasing, and the values of the local and maxima and minima and the zeros (round to two decimal places).
5. Find the composite function $(g \circ f)(x)$ of $f(x) = \sqrt{x - 4}$ and $g(x) = x^2 + 3x + 5$. Give the domain of this function.
6. Show that the function $f(x) = -2x + 6$ is one-to-one. Find its inverse.
7. Use synthetic division to find the quotient and remainder when $f(x) = x^3 + 3x^2 - 4x + 1$ is divided by $g(x) = x - 2$.
8. Use synthetic division with an appropriate divisor to compute the value of $f(4)$ for $f(x) = x^3 + 2x^2 - 8x - 5$. Check your answer by calculating $f(4)$ with a calculator.
9. Compute each of the following roots in terms of i.
   (**a**) $\sqrt{-81}$ (**b**) $-\sqrt{-36}$
10. Compute the following powers of i.
   (**a**) i^{14} (**b**) i^{35}
11. Perform the following additions and multiplications. If you have a calculator that handles complex numbers, check your answers on the calculator.
   (**a**) $(3 - i) + (4 + 9i)$ (**b**) $(2 + 3i)(5 - 7i)$

12. Perform the division $(2 + 7i)/(1 - i)$. Write the answers in the standard form $a + bi$. If you have a calculator that handles complex numbers, check your answers on the calculator.

13. Use the known graph of $f(x) = 2^x$ and shifts to sketch the graph of $g(x) = 2^{x-3} + 7$. Describe the shifts used. Check your results using a graphing calculator.

14. Solve the equation $3^{(-2x+7)} = 4$ to two decimal places using graphs.

15. Compute $\log_2(1/32)$.

16. Solve the following equations for x using algebra.

(**a**) $\ln(5x + 3) = 7$, to four decimal places

(**b**) $y = 2^{(2-3x)}$

17. Spread of a Disease The spread of a disease in a community is approximately described by the function $N(t) = 1.8t^3 - 28t^2 + 90t$. t is the time measured in years from the initial outbreak. $N(t)$ is the number of new cases at time t. (**a**) If the current trend continues, when will the numbers start decreasing. (**b**) How long after the initial case will there be no more new cases.

18. City Population Growth The population growth of a certain city has a doubling time of 34 years. The population was 27 thousand in 1995. Construct an exponential function $P(t)$ that describes population as a function of time. (**a**) If the current growth continues, determine when the population is predicted to reach 35 thousand, using a graph of $P(t)$. (**b**) Verify your result by expressing t as a function of P and determining its value when $P = 35$.

6 Systems of Equations and Inequalities

George B. Dantzig (1914–)
(© Edward W. Souza News Service, Stanford University)

A linear equation is an equation such as $x + 4y = 7$ or $2x + 3y - z = 6$, where all the variables are of degree one. Linear equations are the most basic of all equations. In this section, we look at systems of more than one such equation. Such systems often arise when modeling real situations. For example, we will see how currents through an electrical circuit are determined using systems of linear equations. A linear inequality is an expression such as $3x - y \leq 7$. Methods for solving systems of linear inequalities arise naturally from methods for solving systems of linear equations. Linear inequalities are used to analyze the optimal use of resources in an important area of applied mathematics called linear programming.

Historically, linear programming was first developed by George B. Dantzig and his associates at the U.S. Department of the Air Force. However, the emphasis in applications has now moved to the general industrial area. Linear programming is used by Exxon Corporation to determine the optimal blend of gasoline and by the H. J. Heinz Company to determine the specifications for a processed cheese spread. Linear programming is such an important technique that the 1975 Nobel Prize in economics was awarded to Professors Leonid Kantorovich of the former Soviet Union and Tjalling C. Koopmans of the United States for their work in linear programming. Kantorovich was also awarded the Stalin and Lenin prizes for his work in this field. He showed how linear programming could be used to improve economic planning in the former Soviet Union. Koopmans developed his linear programming theory while planning the optimal movement of ships back and forth across the Atlantic during World War II.

The following example from this chapter illustrates how constraints (often in the form of limited money and time) arise on production in industry and how optimal values (often of profit) are found under these constraints using linear programming.

APPLICATION Manufacturing Calculators

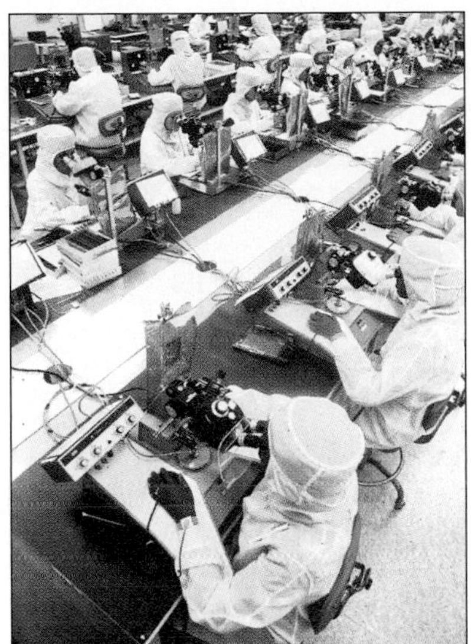

Producing semi-conductor chips for calculators, in a clean room. *(© Chip Henderson/Tony Stone)*

A company manufactures two types of calculators, the Calc1 and the Calc2. The manufacturing times and costs of these calculators are Calc1: 5 hours, \$8 and Calc2: 2 hours, \$10. The profits on the calculators are Calc1: \$3 and Calc2: \$2. How many of each type of calculator should be manufactured weekly to maximize profit?

Let x and y be the numbers of Calc1 and Calc2 manufactured. We find that the constraints on the production process are described by the following system of linear inequalities:

$$5x + 2y \leq 900$$
$$8x + 10y \leq 2800$$
$$x \geq 0, y \geq 0$$

The shaded region in Figure 6.1 represents the values of x and y that satisfy this system of inequalities.

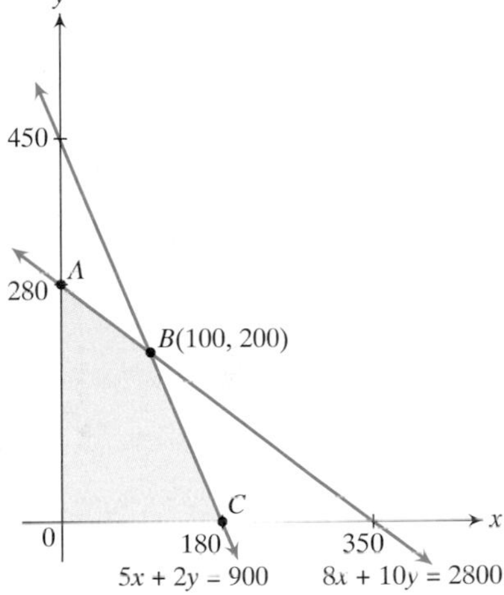

Figure 6.1

The profit is described by the function $f = 3x + 2y$. The maximum value of the function under these constraints is found to be 700 at B, when $x = 100$ and $y = 200$. The interpretation of these results is that the maximum weekly profit is \$700, when 100 Calc1 calculators and 200 Calc2 calculators are manufactured. *(This application is discussed in Section 6.6, Example 3, pages 465–467.)*

6.1 Systems of Linear Equations in Two Variables

• Systems of Linear Equations in Two Variables • Method of Substitution • Method of Elimination • Unique Solution • No Solution • Many Solutions • Solutions Using Graphs • Textbook Purchasing • Mixing Acids

Systems of Linear Equations in Two Variables

Consider the equations

$$\begin{aligned} x + 2y &= 8 \\ 2x - y &= 6 \end{aligned}$$

Each equation is a linear equation. The set of equations is called a **system of linear equations** in two variables. Values of x and y that satisfy both equations are called **solutions** to the system. Let us now discuss the geometrical interpretation of such a system and its solutions.

The graph of a linear equation is a straight line. A system of two linear equations in two variables can correspond to two lines that intersect, two lines that are parallel, or a single line. (Two distinct equations, such as $x + 2y = 3$ and $2x + 4y = 6$, can have the same graph.) Any solution corresponds to a point that lies on all lines. If the lines intersect, there will be a unique solution. If the lines are parallel, there will be no solution. If both equations correspond to a single line, there will be infinitely many solutions; namely, any point on the line will be a solution. These various possibilities are illustrated in Figure 6.2.

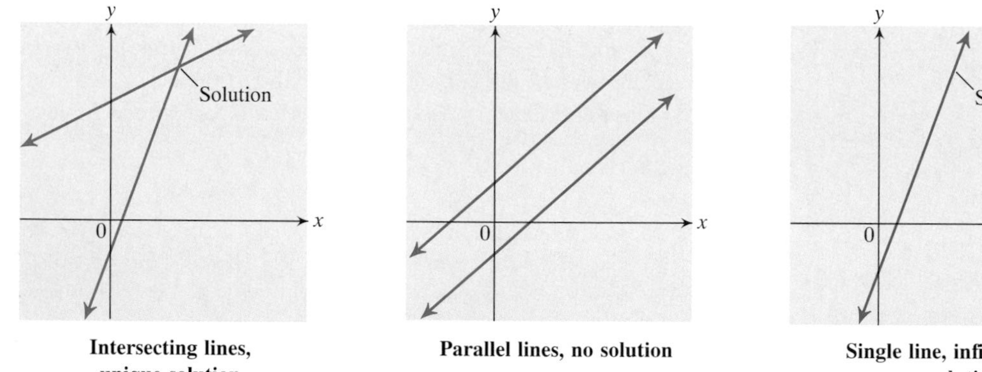

Figure 6.2 Unique, no, or many solutions.

If a system has a unique solution or many solutions it is said to be **consistent.** If a system has no solutions it is said to be **inconsistent.** We now introduce a systematic way of solving a system of two linear equations in two variables.

Method of Substitution

We illustrate the method by solving the following system of equations, and verify the answer using both algebra and graphs.

$$x + 2y = 8 \qquad (1)$$
$$2x - \quad y = 6 \qquad (2)$$

Select one of the equations and solve for one of the variables in terms of the other.

Let us select equation 1 and solve for x.

$$x = -2y + 8 \qquad (3)$$

Substitute for x into equation 2 and solve for y.

$$2(-2y + 8) - y = 6$$
$$-4y + 16 - y = 6$$
$$-5y = -10$$
$$y = 2$$

Substitute $y = 2$ into equation 3 to get

$$x = -2(2) + 8 = 4$$

The system has the unique solution

$$x = 4, \qquad y = 2$$

These values are seen to satisfy the equations $x + 2y = 8$ and $2x \quad y = 6$. The graphs of Figure 6.3 also support this result.

Solve each equation for y and plot lines.

$$y = -x/2 + 4$$
and
$$y = 2x - 6$$

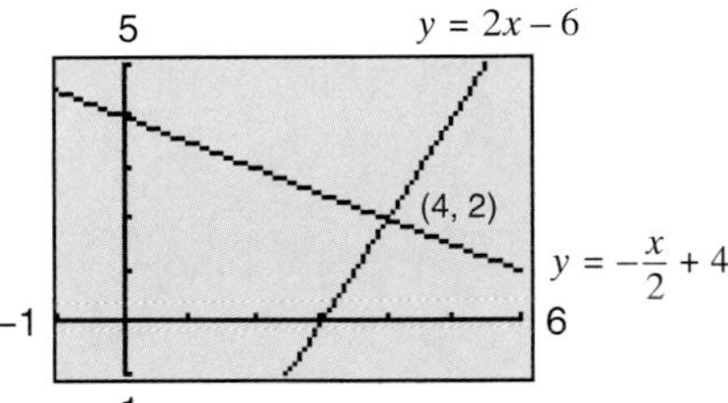

Figure 6.3 The solution is $x = 4$, $y = 2$.

Self-Check 1 Solve the following system using the method of substitution, and verify your answer using algebra and graphs.

$$2x + 3y = 3$$
$$3x - \quad y = 10$$

We now introduce a second method of solving systems of linear equations, called the method of elimination. Although the method of substitution is convenient for solving systems of two equations, the method of elimination leads to a way of solving systems of many equations.

Method of Elimination

This method involves multiplying one or both of the equations by suitable constants so that when the equations are added, one of the variables is eliminated. The resulting equation in one variable is then solved. The value of the second variable is then obtained by substitution into one of the equations. We illustrate the method by solving the previous system.

$$x + 2y = 8 \qquad (1)$$
$$2x - \quad y = 6 \qquad (2)$$

Multiply both sides of equation 1 by -2 to get the system

$$\begin{aligned} -2x - 4y &= -16 \\ 2x - y &= 6 \end{aligned}$$

This system is **equivalent** to the original system in that both systems have the same solution. Add these equations to eliminate x.

$$\begin{aligned} -2x - 4y &= -16 \\ 2x - y &= 6 \\ \hline - 5y &= -10 \\ y &= 2 \end{aligned}$$

The y value of the solution is 2. Substitute this value of y into equation 1 to find x. (Any equation can be used. Select the least complicated.)

$$\begin{aligned} x + 2(2) &= 8 \\ x &= 4 \end{aligned}$$

The system has the unique solution $x = 4$ and $y = 2$.

Note that, in this example, we multiplied equation 1 by -2 to get an equation that could be added to equation 2 to eliminate x. The selection of such numbers is at the heart of this method. The following example illustrates that it may be necessary to multiply both equations by suitable numbers.

Example 1 Solving Using Elimination

Solve the following system and verify the answer using algebra and graphs.

$$\begin{aligned} 2x + 3y &= 3 \qquad (1) \\ -3x + 4y &= -13 \qquad (2) \end{aligned}$$

SOLUTION Multiply both sides of equation 1 by 3 and both sides of equation 2 by 2. This enables us to add the resulting equations to eliminate x.

$$\begin{aligned} 6x + 9y &= 9 \\ -6x + 8y &= -26 \\ \hline 17y &= -17 \\ y &= -1 \end{aligned}$$

Substitute y back into (1) to get x.

$$\begin{aligned} 2x + 3(-1) &= 3 \\ 2x &= 6 \\ x &= 3 \end{aligned}$$

The system has the unique solution

$$x = 3, \qquad y = -1$$

These values satisfy the equations $2x + 3y = 3$ and $-3x + 4y = -13$. The graphs of Figure 6.4 also support this result.

Solve each equation for y and plot lines.

$$y = -2x/3 + 1$$
and
$$y = 3x/4 - 13/4$$

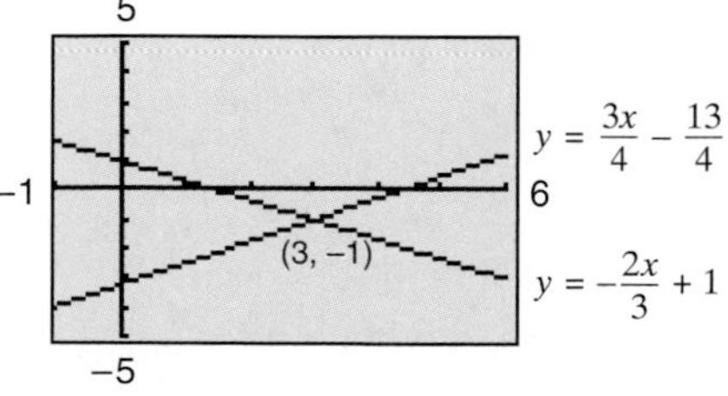

Figure 6.4 The solution is $x = 3$, $y = -1$.

Self-Check 2 Solve the following system using the method of elimination, and verify your answer using algebra and graphs.

$$\begin{aligned} 3x - 2y &= 0 \\ 2x + y &= 7 \end{aligned}$$

We now give an example of a system that has no solution.

Example 2 System with No Solution

Solve if possible the following system and verify the answer using graphs.

$$\begin{aligned} x - 2y &= 4 \qquad (1) \\ -2x + 4y &= 6 \qquad (2) \end{aligned}$$

SOLUTION Multiply equation 1 by 2 and add the resulting equations to eliminate x.

$$\begin{aligned} 2x - 4y &= 8 \\ -2x + 4y &= 6 \\ \hline 0 &= 14 \end{aligned}$$

$0 = 14$ is false. It is an indication that the system is inconsistent. The system has no solution (Figure 6.5).

Solve each equation for y and plot lines.

$$y = x/2 - 2$$

and

$$y = x/2 + 3/2$$

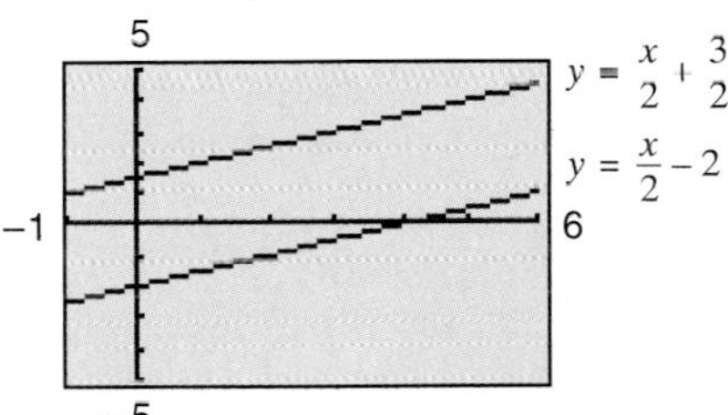

Figure 6.5 Lines are parallel. There is no point of intersection and no solution.

Self-Check 3 Show that the following system has no solution. Verify your answer using graphs.

$$\begin{aligned} -2x + 4y &= 12 \\ x - 2y &= -14 \end{aligned}$$

The following example illustrates a system of equations that has many solutions.

Example 3 System with Many Solutions

Solve the following system and verify the answer using a graph.

$$\begin{aligned} 4x - 2y &= 6 \qquad (1) \\ -6x + 3y &= -9 \qquad (2) \end{aligned}$$

SOLUTION Multiply equation 1 by 6 and equation 2 by 4. Add the resulting equations.

$$\begin{aligned} 24x - 12y &= 36 \\ -24x + 12y &= -36 \\ \hline 0 &= 0 \end{aligned}$$

The true statement $0 = 0$ is an indication that the equations are equivalent. There are infinitely many solutions (Figure 6.6).

Solve each equation for y and plot lines.

$$y = 2x - 3$$
(only one equation)

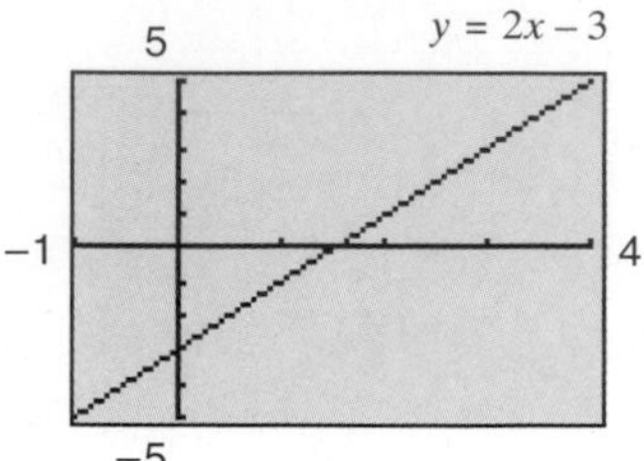

Figure 6.6 Any point on the line is a solution.

We now illustrate how to express these many solutions. Select one of the equations and solve for y in terms of x. Let us select equation 1.

$$-2y = -4x + 6$$
$$y = 2x - 3$$

The many solutions are of the form $(x, 2x - 3)$, where x can take on any value. We can generate any number of specific solutions by giving x various values. For example, $x = 1$ leads to the solution $(1, -1)$; $x = 2$ gives the solution $(2, 1)$; and so on.

Self-Check 4 Solve the following system of equations, and verify the answer using a graph.

$$\begin{aligned} 2x - 4y &= 8 \\ -3x + 6y &= -12 \end{aligned}$$

Solutions to Systems of Linear Equations Using Graphs

In the preceding examples, we used algebra to solve systems of equations that had "clean" solutions, and we used graphs to confirm those solutions. Graphs can also be used to actually find approximate solutions to systems of linear equations in two variables. We now illustrate the graphing approach.

Example 4 Solving a System of Equations using Graphs

Solve the following system of linear equations to four decimal places using graphs.

$$\begin{aligned} 5x - 3y &= 7 \\ -3x + 7y &= 8 \end{aligned}$$

SOLUTION Solve each equation for y. This puts the equations in a form that can be graphed.

$$y = \frac{5x}{3} - \frac{7}{3}$$

$$y = \frac{3x}{7} + \frac{8}{7}$$

Graph the lines (Figure 6.7). The point of intersection A of the graphs will be the solution to this equation. The solution to four decimal places is found to be $x = 2.8077$, $y = 2.3462$ using an intersect finder (or trace/zoom).

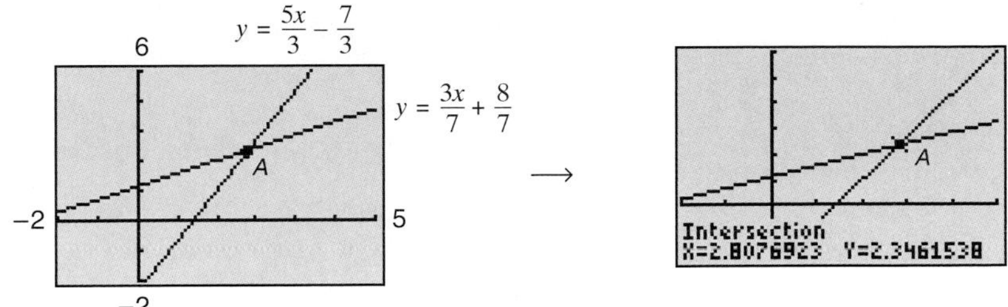

Figure 6.7 The solution, expressed to four decimal places, is $x = 2.8077$, $y = 2.3462$.

We now give examples to illustrate how systems of linear equations can arise in practice.

Applications . . .

Example 5 Equation of a Line Through Two Points

Find the equation of the line that passes through the points (1, 2), and (4, 11).

SOLUTION We discussed the use of the point–slope form of the equation to find the equation of a line through two given points in Section 1.2. We now give a second method based on the elimination method. The general equation of a line is $y = mx + b$. There will be a unique line that passes through the points (1, 2) and (4, 11). The points can be used to find the values of m and b and thus the equation of the line. Because the points lie on the line, they must satisfy its equation. Substitute each point into the equation $y = mx + b$.

$$\begin{aligned} &\text{Point } (1,2) \text{ gives} \qquad 2 = m(1) + b \\ &\text{Point } (4,11) \text{ gives} \qquad 11 = m(4) + b \end{aligned}$$

Rewrite these equations in standard form with the variables on the left sides of the equations.

$$\begin{aligned} m + b &= 2 \qquad (1) \\ 4m + b &= 11 \qquad (2) \end{aligned}$$

Values for m and b, and thus the equation of the line, are found by solving this system of equations. In this system, it is more straightforward to eliminate b than m.

Multiply equation 1 by -1 and add to equation 2.

$$\begin{aligned} -m - b &= -2 \\ 4m + b &= 11 \\ \hline 3m \qquad &= 9 \\ m &= 3 \end{aligned}$$

Substitute this value of m into equation 1 to get b.

$$\begin{aligned} 3 + b &= 2 \\ b &= -1 \end{aligned}$$

The equation of the line is thus $y = 3x - 1$.

Example 6 Textbook Purchasing

The Littleton College Bookstore bought 500 algebra books and 200 trigonometry books. The total cost of the books was $20,000. Later the store bought an additional 30 algebra books and 40 trigonometry books for $1900. Determine the cost to the bookstore of both an algebra book and a trigonometry book.

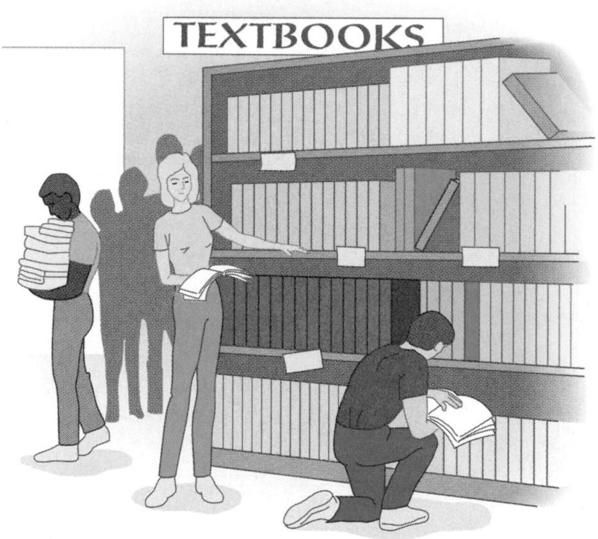

SOLUTION Let the cost of each algebra book be $\$x$ and that of each trigonometry book be $\$y$.

The cost of 500 algebra books at $\$x$ each is $\$500x$.
The cost of 200 trigonometry books at $\$y$ each is $\$200y$.

Because the total cost of these books was $20,000, we get

$$500x + 200y = 20{,}000$$

Let us now look at the second purchase.

The cost of 30 algebra books at $\$x$ each is $\$30x$.
The cost of 40 trigonometry books at $\$y$ each is $\$40y$.

Because the total cost of the second purchase was $1900, we get

$$30x + 40y = 1900$$

The values of x and y can be found by solving the system of equations

$$\begin{aligned} 500x + 200y &= 20{,}000 \quad (1) \\ 30x + 40y &= 1{,}900 \quad (2) \end{aligned}$$

Simplify equation 1 by dividing throughout by 100, and simplify equation 2 by dividing by 10.

$$\begin{aligned} 5x + 2y &= 200 \quad (3) \\ 3x + 4y &= 190 \quad (4) \end{aligned}$$

Multiply equation 3 by -3 and equation 4 by 5 and add to eliminate x.

$$\begin{aligned} -15x - 6y &= -600 \\ 15x + 20y &= 950 \\ \hline 14y &= 350 \\ y &= 25 \end{aligned}$$

Substitute this value of y back into equation 3 to get x. (Any equation can be used.)

$$\begin{aligned} 5x + 2(25) &= 200 \\ x &= 30 \end{aligned}$$

The system has the unique solution $x = 30$ and $y = 25$. Thus, the price of each algebra book is \$30, and the price of each trigonometry book is \$25.

Example 7 Mixing Acids

A chemist has two concentrates of acid: a 20% solution and a 25% solution. How much of each solution should be mixed to get 80 milliliters of a 23.75% solution?

SOLUTION Let x milliliters of the 20% acid and y milliliters of the 25% acid be used. Because 80 milliliters of the mixture is needed,

$$x + y = 80 \qquad (1)$$

The amount of acid in x milliliters at 20% is $\frac{20x}{100}$, and the amount of acid in y milliliters at 25% is $\frac{25y}{100}$. Because the mixture is to be 80 milliliters at 23.75%,

$$\frac{20x}{100} + \frac{25y}{100} = \frac{23.27 \times 80}{100}$$

Simplify this equation.

$$20x + 25y = 1900 \qquad (2)$$

We solve the system of equations 1 and 2 to determine x and y.

$$\begin{aligned} x + \quad y &= \quad 80 \qquad (1) \\ 20x + 25y &= 1900 \qquad (2) \end{aligned}$$

Multiply equation 1 by -20 and add to equation 2 to eliminate x.

$$\begin{aligned} -20x - 20y &= -1600 \\ 20x + 25y &= \quad 1900 \\ \hline 5y &= \quad 300 \\ y &= \quad 60 \end{aligned}$$

Substitute this value of y into equation 1 to get x.

$$\begin{aligned} x + 60 &= 80 \\ x &= 20 \end{aligned}$$

Thus, the chemist should mix 20 milliliters of the 20% acid and 60 milliliters of the 25% acid to get 80 milliliters of the 23.75% solution.

Answers to Self-Check Exercises

1. Solution is $x = 3$ and $y = -1$.
2. Solution is $x = 2$ and $y = 3$.
4. Infinitely many solutions $\left(x, \frac{x}{2} - 2\right)$, x any value. Both equations correspond to the line $y = \frac{x}{2} - 2$.

EXERCISES 6.1

In Exercises 1–6, solve the systems of equations using the method of substitution, and verify the answers using graphs.

1. $\begin{aligned} x + 2y &= 5 \\ -x + y &= 1 \end{aligned}$ **2.** $\begin{aligned} x + y &= 2 \\ 3x - y &= -6 \end{aligned}$ **3.** $\begin{aligned} 2x + 3y &= 9 \\ -x + 2y &= 6 \end{aligned}$

4. $\begin{aligned} 2x - y &= 1 \\ 4x + 5y &= 9 \end{aligned}$ **5.** $\begin{aligned} 3x + 2y &= 1 \\ 4x + y &= 2 \end{aligned}$ **6.** $\begin{aligned} 2x - y &= 6 \\ x - 5y &= 2 \end{aligned}$

In Exercises 7–12, solve the systems of equations using the method of substitution, and verify the answers using graphs.

7. $\begin{aligned} 2x + y &= 8 \\ 3x - 2y &= 12 \end{aligned}$ **8.** $\begin{aligned} -3x + 2y &= 2 \\ 2x + y &= -6 \end{aligned}$

9. $\begin{aligned} x + 2y &= 6 \\ -2x + y &= -2 \end{aligned}$ **10.** $\begin{aligned} 4x + 7y &= 10 \\ 2x - 3y &= 8 \end{aligned}$

11. $\begin{aligned} x + 2y &= 5 \\ -x + 2y &= -1 \end{aligned}$ **12.** $\begin{aligned} x + y &= 3 \\ 3x - 4y &= 16 \end{aligned}$

In Exercises 13–20, solve the given systems using the method of elimination, and verify the answers using graphs.

13. $\begin{aligned} x + 2y &= 3 \\ x - y &= 0 \end{aligned}$ **14.** $\begin{aligned} x + 2y &= 5 \\ 2x - y &= 0 \end{aligned}$

15. $\begin{aligned} x + 4y &= -1 \\ 3x - 2y &= 11 \end{aligned}$ **16.** $\begin{aligned} 3x - y &= 11 \\ 2x + y &= 9 \end{aligned}$

17. $\begin{aligned} 3x + y &= -4 \\ 6x + 2y &= -8 \end{aligned}$ **18.** $\begin{aligned} 2x + 3y &= -10 \\ 5x + 2y &= -3 \end{aligned}$

19. $\begin{aligned} 2x + y &= 8 \\ 3x - 2y &= 12 \end{aligned}$ **20.** $\begin{aligned} x - 2y &= 3 \\ -2x + 4y &= 12 \end{aligned}$

In Exercises 21–28, solve the given systems using the method of elimination, and verify the answers using graphs.

21. $\begin{aligned} 4x + 2y &= 16 \\ 2x - 3y &= 0 \end{aligned}$ **22.** $\begin{aligned} -3x + 2y &= 2 \\ 2x + y &= -6 \end{aligned}$

23. $\begin{aligned} 3x + 6y &= 3 \\ 2x + 4y &= 2 \end{aligned}$ **24.** $\begin{aligned} 4x - y &= 7 \\ -8x + 2y &= 6 \end{aligned}$

25. $\begin{aligned} 3x \phantom{{}-y} &= 12 \\ 2x - y &= 7 \end{aligned}$ **26.** $\begin{aligned} y &= -3 \\ x + 2y &= 1 \end{aligned}$

27. $\begin{aligned} 2x + y &= 4 \\ -2x - y &= 3 \end{aligned}$ **28.** $\begin{aligned} 4x - y &= 2 \\ -8x + 2y &= -4 \end{aligned}$

In Exercises 29–37, find solutions (if they exist) of the systems of equations using graphs, to four decimal places. Use an intersect finder or trace/zoom.

29. $\begin{aligned} x + 2y &= 3 \\ x - y &= 0 \end{aligned}$ **30.** $\begin{aligned} x + 2y &= 4 \\ 3x - y &= 0 \end{aligned}$

31. $\begin{aligned} x + 4y &= -2 \\ 2x - 3y &= 10 \end{aligned}$ **32.** $\begin{aligned} 3x - y &= 11 \\ x + y &= 8 \end{aligned}$

33. $\begin{aligned} x + 3y &= -4 \\ 2x + 6y &= -8 \end{aligned}$ **34.** $\begin{aligned} 2x + 3y &= -10 \\ 5x + 2y &= 11 \end{aligned}$

35. $\begin{aligned} 2.1x + y &= 8.7 \\ 5.3x - 2y &= 13.2 \end{aligned}$ **36.** $\begin{aligned} x - 2.3y &= 5 \\ -3.4x + 6y &= 11 \end{aligned}$

37. $\begin{aligned} 0.8x - 5.2y &= 1.8 \\ 4.2x - 1.1y &= 9.3 \end{aligned}$

In Exercises 38–48, construct a system of two linear equations in two variables having the given solutions.

38. $x = 2, y = 3$ **39.** $x = -1, y = 2$

40. $x = 3, y = 1$ **41.** $x = 5, y = -3$

42. $x = 0, y = 4$ **43.** $x = 6, y = 0$

44. $x = -2, y = -6$ **45.** $x = 3, y = -7$

46. Many solutions of the form $(x, 3x + 2)$.

47. Many solutions of the form $(x, x - 3)$.

48. Many solutions of the form $(x, 2x - 1)$.

49. Consider the following system of equations for various values of c.

$$\begin{aligned} x + y - 4 \\ cx + 2y = 2 \end{aligned}$$

Which values of c result in a system that has **(a)** a unique solution and, **(b)** no solution? **(c)** Show that the system cannot have many solutions.

50. Consider the following system of equations for various values of c.

$$\begin{aligned} x + y &= 2 \\ 2x - cy &= 2 \end{aligned}$$

Which values of c result in a system with **(a)** a unique solution **(b)** no solutions. **(c)** Prove that the system cannot have many solutions.

In Exercises 51–58, find equations of the lines that pass through the given pairs of points.

51. $(2, 3), (1, 1)$ **52.** $(-1, -2), (1, 6)$

53. $(0, 1), (2, -2)$ **54.** $(4, 27), (2, 9)$

55. $(-2, -8), (0, -2)$ **56.** $(-2, 5), (-1, 3)$

57. $(1, -1), (-1, -7)$ **58.** $(1, 2), (4, 2)$

59. Texbook Purchasing The Brunswick College Bookstore buys 400 algebra books and 200 trigonometry books. The total cost was \$23,800.

Later the bookstore buys an additional 52 algebra books and 11 trigonometry books for \$2569. Determine the cost to the bookstore of an algebra book and of a trigonometry book.

60. **Car Buying** Main Street Cars buys 23 new cars and 5 new trucks for \$404,000. They later buy 7 cars and 2 trucks for \$133,000. Find the cost of a car and of a truck.

61. **Buying Fans** Ted Griffey buys four Roma ceiling fans and two Seabreeze ceiling fans for a total cost of \$256. He later buys one of each fan for \$96. Find the cost of each fan.

62. **Purchasing Computers** Stetson University buys 80 microcomputers and 6 printers for a total cost of \$99,156. It later buys an additional 15 microcomputers and 2 printers for \$19,052. Find the price of each microcomputer and printer.

63. **Small Change** Gerallt has \$2.30 in his pocket, made up of quarters and dimes. If he has a total of 14 coins, how many of each type does he have?

64. **Coins in a Parking Meter** A parking meter contains \$6.95 in quarters and dimes. There are 17 more dimes than quarters. How many of each type of coin is in the meter?

65. **Age Puzzle** The sum of the ages of Alison and Jane is 30, and the difference in their ages is 2. How old are the children?

66. **Age Puzzle** A father is 22 years older than his daughter. The sum of their ages is 58. How old are the father and the daughter?

67. **Investment Analysis** The Rodriguez family invested \$5000 split between a bond that paid 8% interest per year and a mutual fund that paid 10%. The total return on the two investments at the end of the year is \$420. How much was invested in each?

68. **Investment Analysis** Tonya Bond invested \$9000 split between a bond that paid 7.5% interest per year and a mutual fund that paid 9.5%. The total return on the two investments at the end of the year was \$840 How much was invested in each?

69. **Number Puzzle** Three times a certain number when added to four times another number gives 70. Two times the first number when added to twice the second gives 16. What are the two numbers?

70. **Number Puzzle** Four times a certain number when added to five times another number gives 33. Three times the first added to twice the second gives 16. What are the two numbers?

71. **Size of a Rectangle** The perimeter of a rectangle is 200 feet. The length is 20 feet longer than the width. Find the length and width of the rectangle.

72. **Size of a Rectangle** The perimeter of a rectangle is 172 feet. The length is 22 feet longer than the width. Find the length and width of the rectangle.

73. **Mixing Nuts** Pickwick Health Food Store sells nuts at \$2.50 per pound and raisins at \$1.75 per pound. The store wishes to make 25 pounds of a mixture of nuts and raisins, to be sold at \$2.20 per pound. How many pounds of nuts and how many pounds of raisins should go into the mixture?

74. **Mixing Coffee Types** The Coffee Place wishes to mix two types of coffee, one that sells at \$1.50 per pound and another that sells at \$2.00 per pound, to get 20 pounds selling at \$1.80 per pound. How much of each type of coffee should go into the mixture?

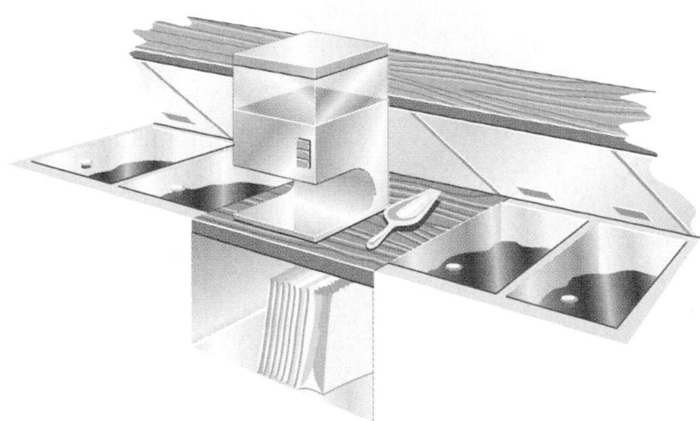

75. **Transporting Refrigerators** Ashley Appliances has manufacturing plants at Kent and Denton. Its refrigerators are sold in Sanford. It costs \$25 to transport each refrigerator from Kent to Sanford and \$30 to transport each refrigerator from Denton to Sanford. A total of 50 refrigerators are to be transported at a cost of \$1340. How many refrigerators are transported from each location?

76. **Manufacturing Fertilizer** The Schmidt Company makes two types of fertilizer, Growfast and Weedkill, using chemicals alpha and zeta. Fertilizer Growfast is made up of 80% chemical alpha and 20% chemical zeta. Fertilizer Weedkill is made up of 60% chemical alpha and 40% chemical zeta. The manufacturer has 34 tons of chemical alpha avail-

able and 16 tons of chemical zeta. How much of each type of fertilizer can be manufactured?

77. **Motion Down a Plane** The function

$$v(t) = v_0 + at$$

describes the motion of an object sliding down an inclined plane. t is time, a is acceleration, v_0 is the initial velocity, and $v(t)$ is the velocity at time t. If $v(3) = 29$ and $v(7) = 61$, find the initial velocity and the acceleration.

78. **Motion Under Gravity** The motion of an object fired vertically upward under gravity is described by $s(t) = -16t^2 + v_0t + s_0$. t is time, s_0 is initial height, v_0 is initial velocity, and $s(t)$ is the height of the object at time t. If $s(1) = 234$ and $s(3) = 506$, find the initial velocity and the initial height.

79. **Manufacturing Cost** The relationship between the total number of items x manufactured by a company and the total cost involved c is described by the equation $c = mx + b$, called the cost–volume formula. If 500 units cost $1700 and 1000 units cost $2820, find m and b. Use this equation to predict the cost of manufacturing 1200 units.

6.2 Systems of Linear Equations in Three Variables

• Systems of Linear Equations in Three Variables • Elementary Transformations • Gauss–Jordan Elimination for Unique Solutions • Matrices and Systems of Linear Equations • Elementary Row Operations

In Section 6.1, we solved systems of two linear equations in two variables using the method of substitution and the method of elimination. In this section and the next, we extend the method of elimination to systems of linear equations having three or more variables. We live in a world that is very dependent on electrical networks. Our computers, televisions, automobiles, and planes are only a few items that use electrical networks. These networks are planed using mathematics. We will see how systems of linear equations are used in this planing stage to determine currents through electrical networks.

Systems of Linear Equations in Three Variables

Consider the system of equations

$$\begin{aligned} x + y + z &= 2 \\ 2x + 3y + z &= 3 \\ x - y - 2z &= -6 \end{aligned}$$

Each equation in this system is of degree 1. Such an equation is called a **linear equation** and describes a plane in three-dimensional space. Values of x, y, and z that satisfy all the equations are called **solutions.** Solutions correspond to points that lie on all three planes. There can be a unique solution, many solutions, or no solutions, depending on whether the planes intersect at a point or at many points or have no points in common. We illustrate some of these possibilities in Figure 6.8. We give a systematic way of solving such systems, called the **method of Gauss–Jordan elimination.** In this section, we discuss systems with unique solutions; we discuss systems with many solutions and no solutions in the next section.

Unique solution.

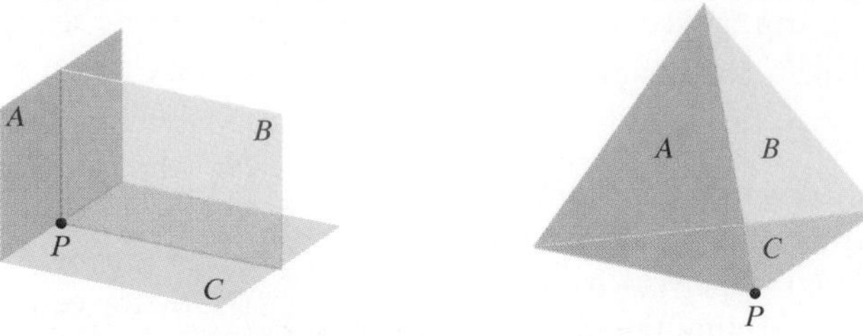

Three planes—*A*, *B*, and *C*—intersect at a single point *P*. *P* corresponds to a unique solution.

No solutions.

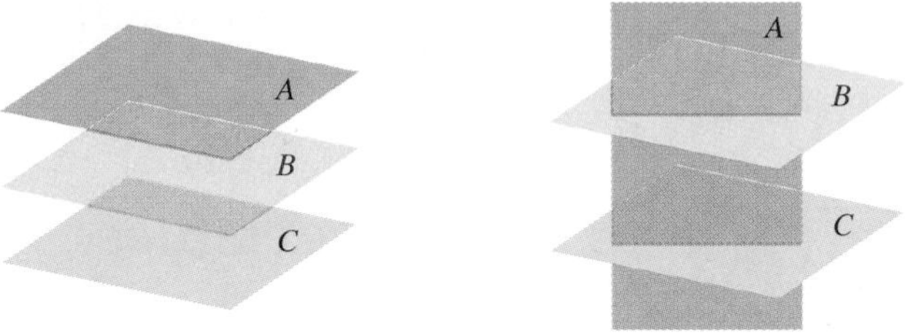

Planes *A*, *B*, and *C* have no points in common. There are no solutions.

Many solutions.

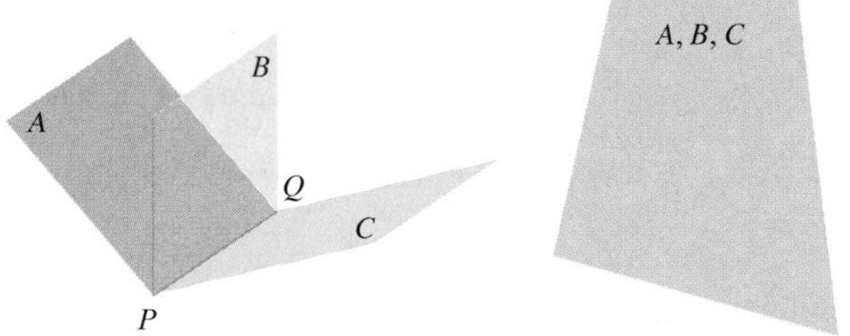

Three planes—*A*, *B*, and *C*—intersect in a line *PQ*. Any point on the line is a solution.

Three equations represent the same plane. Any point on the plane is a solution.

Figure 6.8

GAUSS–JORDAN ELIMINATION FOR UNIQUE SOLUTIONS

We now extend the method of elimination to a system of three linear equations in three variables. The following transformations do not change the solutions of a system of equations.

ELEMENTARY TRANSFORMATIONS

1. Interchange the positions of two equations.
2. Multiply both sides of an equation by a nonzero constant.
3. Add a multiple of one equation to another equation.

Elementary transformations preserve solutions because the order of the equations does not affect the solution; multiplying an equation throughout by a nonzero constant does not change the truth of the equality; and adding equal quantities to both sides of an equality results in an equality.

Systems of equations that are related through elementary transformations, and thus have the same solutions, are called **equivalent systems.** The symbol $\approx$ is used to indicate equivalent systems of equations.

The method of Gauss–Jordan* elimination uses elementary transformations to eliminate variables in a systematic manner, until we arrive at a system that gives the solution. We illustrate the method with the following example.

Example 1 Solving a System of Three Linear Equations in Three Variables

Solve the system of linear equations

$$\begin{aligned} x + y + z &= 2 \\ 2x + 3y + z &= 3 \\ x - y - 2z &= -6 \end{aligned}$$

SOLUTION We use the first equation to eliminate x from the second and third equations.

$x + y + z = 2$	$\approx$	$x + y + z = 2$
$2x + 3y + z = 3$	**Add (-2) Equation 1 to Equation 2**	$y - z = -1$
$x - y - 2z = -6$	**Add (-1) Equation 1 to Equation 3**	$-2y - 3z = -8$

Next use the second equation to eliminate y from the first and third equations.

$\approx$	$x + 2z = 3$
Add (-1) Equation 2 to Equation 1	$y - z = -1$
Add (2) Equation 2 to Equation 3	$-5z = -10$

Multiply the last equation by $-\frac{1}{5}$ to get z.

	$x + 2z = 3$
$\approx$	$y - z = -1$
$(-\frac{1}{5})$ Equation 3	$z = 2$

Finally use the third equation to eliminate z from the first and second equations.

*Carl Friedrich Gauss (1777–1855) was one of the greatest mathematical scientists of all time. He taught for 47 years at the University of Göttingen, Germany, and contributed to many areas of mathematics, such as algebra, probability, and statistics. He discovered a way to calculate the orbits of asteroids. Gauss was not really a physicist in the sense of searching for new phenomena, but rather a mathematician who attempted to formulate in exact mathematical terms the experimental results of others. His personal life was tragic; he suffered from political turmoil and financial problems associated with the French Revolution and democratic revolutions in Germany.

Wilhelm Jordan (1842–1899) taught geodesy at the Technical College of Karlsruhe, Germany. His major work was a handbook on geodesy that contained his work on systems of equations. Jordan was considered a fine teacher and writer.

$$\approx \qquad x = -1$$

Add (−2) Equation 3 to Equation 1 $\quad y = 1$

Add Equation 3 to Equation 2 $\quad z = 2$

The solution to this system, and therefore also the solution to the original system is $x = -1$, $y = 1$, and $z = 2$.

Geometrically, each of the three original equations represents a plane in three-dimensional space. The fact that there is a unique solution means that these three planes intersect at a point. The solution $(-1, 1, 2)$ gives the coordinates of this point.

We now show that it is unnecessary to carry the variables x, y, and z at all stages in this process. The coefficients of the variables give us all the information at each stage. To be able to do this, we need the concept of a matrix.

Matrices and Systems of Linear Equations

We introduce rectangular arrays of numbers, called **matrices,** to describe systems of linear equations.

> ***Matrix*** A *matrix* is a rectangular array of numbers. The numbers in the array are called the **elements** of the matrix.

Examples of matrices, in the standard notation are

$$\begin{bmatrix} 1 & -3 & 5 \\ 2 & -7 & 4 \end{bmatrix} \quad \begin{bmatrix} 2 & -1 \\ 0 & 7 \\ 9 & 5 \end{bmatrix} \quad \begin{bmatrix} 1 & 3 & -2 \\ 4 & 7 & 8 \\ -3 & 0 & 5 \end{bmatrix}$$

The location of an element in a matrix is described by stating the row and column in which it lies. Rows are labeled starting from the top of the matrix, columns are labeled starting from the left. For example, the element 4 of the first matrix lies in row 2, column 3. The element 9 of the second matrix is in row 3, column 1.

There are two important matrices associated with every system of linear equations. The coefficients of the variables form the **matrix of coefficients** of the system. The coefficients, together with the constant terms, form the **augmented matrix** of the system. Consider the system

$$\begin{aligned} x + y + z &= 2 \\ 2x + 3y + z &= 3 \\ x - y - 2z &= -6 \end{aligned}$$

The matrix of coefficients and the augmented matrix for this system are as follows:

$$\begin{bmatrix} 1 & 1 & 1 \\ 2 & 3 & 1 \\ 1 & -1 & -2 \end{bmatrix} \quad \begin{bmatrix} 1 & 1 & 1 & 2 \\ 2 & 3 & 1 & 3 \\ 1 & -1 & -2 & -6 \end{bmatrix}$$

Matrix of coefficients   Augmented matrix

Observe that the matrix of coefficients is embedded in the augmented matrix. We say that it is a **submatrix** of the augmented matrix. The augmented matrix completely describes the system. In solving a system of linear equations, each equivalent system in the sequence can be represented by an augmented matrix. It becomes unnecessary to write down the variables $x, y, z, \ldots$. Instead of performing elementary transformations on the systems of equations, we can perform analogous transformations called **elementary row operations** on the augmented matrices. One important implication is that we can perform these transformations and arrive at solutions using calculators and computers. We now summarize these row operations.

Elementary Row Operations on Matrices

1. Interchange two rows.
2. Multiply the elements of a row by a nonzero constant.
3. Add a multiple of the elements of one row to the corresponding elements of another row.

Two matrices are said to be **row equivalent** if one can be obtained from the other by using a sequence of elementary row operations. We now show how these row operations are used to solve a system of linear equations.

Example 2 Solving a System of Linear Equations Using Matrices

Solve the following system of linear equations:

$$\begin{aligned} x - 2y + 4z &= 12 \\ 2x - y + 5z &= 18 \\ -x + 3y - 3z &= -8 \end{aligned}$$

SOLUTION Start with the augmented matrix and use the first row to create zeros in the first column. (This corresponds to using the first equation to eliminate x from the second and third equations.) We use the letter "R" to denote row.

$$\begin{bmatrix} 1 & -2 & 4 & 12 \\ 2 & -1 & 5 & 18 \\ -1 & 3 & -3 & -8 \end{bmatrix} \begin{matrix} \approx \\ \textbf{Add } (-2) \textbf{ R1 to R2} \\ \textbf{Add R1 to R3} \end{matrix} \begin{bmatrix} 1 & -2 & 4 & 12 \\ 0 & 3 & -3 & -6 \\ 0 & 1 & 1 & 4 \end{bmatrix}$$

Now multiply row 2 by $\frac{1}{3}$. (This corresponds to making the coefficient of y in the second equation 1.)

$$\begin{matrix} \approx \\ \\ \left(\frac{1}{3}\right)\textbf{R2} \end{matrix} \begin{bmatrix} 1 & -2 & 4 & 12 \\ 0 & 1 & -1 & -2 \\ 0 & 1 & 1 & 4 \end{bmatrix}$$

Next, create zeros in the second column as follows. (This corresponds to using the second equation to eliminate y from the first and third equations.)

$$\begin{matrix} \approx \\ \textbf{Add (2) R2 to R1} \\ \textbf{Add } (-1) \textbf{ R2 to R3} \end{matrix} \begin{bmatrix} 1 & 0 & 2 & 8 \\ 0 & 1 & -1 & -2 \\ 0 & 0 & 2 & 6 \end{bmatrix}$$

Multiply row 3 by $\frac{1}{2}$. (This corresponds to making the coefficient of z in the third equation 1.)

$$\underset{\left(\frac{1}{2}\right)\textbf{R3}}{\approx} \begin{bmatrix} 1 & 0 & 2 & 8 \\ 0 & 1 & -1 & -2 \\ 0 & 0 & 1 & 3 \end{bmatrix}$$

Finally, create zeros in the third column. (This corresponds to using the third equation to eliminate z from the first and second equations.)

$$\underset{\substack{\textbf{Add }(-2)\textbf{ R3 to R1} \\ \textbf{Add R3 to R2}}}{\approx} \begin{bmatrix} 1 & 0 & 0 & 2 \\ 0 & 1 & 0 & 1 \\ 0 & 0 & 1 & 3 \end{bmatrix}$$

This matrix corresponds to the system

$$\begin{aligned} x \qquad &= 2 \\ y \quad &= 1 \\ z &= 3 \end{aligned}$$

The solution is $x = 2$, $y = 1$, $z = 3$.

Overview of Gauss–Jordan Elimination

This method of reducing a system of linear equations to an equivalent simpler system, using matrices, involves creating ones and zeros in certain locations of matrices. These numbers are created in a systematic manner, column by column, according to the following pattern, where * indicates a possible nonzero element.

$$\begin{bmatrix} * & * & * & * \\ * & * & * & * \\ * & * & * & * \end{bmatrix} \approx \begin{bmatrix} 1 & * & * & * \\ * & * & * & * \\ * & * & * & * \end{bmatrix} \approx \begin{bmatrix} 1 & * & * & * \\ 0 & * & * & * \\ 0 & * & * & * \end{bmatrix} \approx \begin{bmatrix} 1 & * & * & * \\ 0 & 1 & * & * \\ 0 & * & * & * \end{bmatrix}$$

$$\approx \begin{bmatrix} 1 & 0 & * & * \\ 0 & 1 & * & * \\ 0 & 0 & * & * \end{bmatrix} \approx \begin{bmatrix} 1 & 0 & * & * \\ 0 & 1 & * & * \\ 0 & 0 & 1 & * \end{bmatrix} \approx \underset{\text{Reduced form}}{\begin{bmatrix} 1 & 0 & 0 & * \\ 0 & 1 & 0 & * \\ 0 & 0 & 1 & * \end{bmatrix}} \begin{matrix} \leftarrow x \\ \leftarrow y \\ \leftarrow z \end{matrix}$$

The ones are called **leading ones.** This final matrix is called the **reduced form** of the original matrix. The numbers in the last column give the solution to the system.

Self-Check 1 Solve the following system of equations.

$$\begin{aligned} x - y + 2z &= 5 \\ -2x + 3y - 5z &= -12 \\ x + y + 2z &= 3 \end{aligned}$$

The following example illustrates that it may be necessary to interchange two rows at some stage to proceed as in Example 2.

Example 3 Gauss–Jordan Elimination with a Row Interchange

Solve the system

$$\begin{aligned} 4x + 8y - 12z &= 44 \\ 3x + 6y - 8z &= 32 \\ -2x - y &= -7 \end{aligned}$$

SOLUTION We start with the augmented matrix and proceed as follows. (Note the use of zero in the augmented matrix for the coefficient of the missing variable z in the third equation.)

$$\begin{bmatrix} 4 & 8 & -12 & 44 \\ 3 & 6 & -8 & 32 \\ -2 & -1 & 0 & -7 \end{bmatrix} \underset{\left(\frac{1}{4}\right)\text{R1}}{\approx} \begin{bmatrix} 1 & 2 & -3 & 11 \\ 3 & 6 & -8 & 32 \\ -2 & -1 & 0 & -7 \end{bmatrix}$$

$$\underset{\substack{\text{Add } (-3)\text{ R1 to R2} \\ \text{Add } (2)\text{ R1 to R3}}}{\approx} \begin{bmatrix} 1 & 2 & -3 & 11 \\ 0 & 0 & 1 & -1 \\ 0 & 3 & -6 & 15 \end{bmatrix}$$

At this stage, we need a nonzero element in location row 2, column 2 to continue. To achieve this, we interchange the second row with the third row (a **later** row) and then proceed.

$$\underset{\text{R2} \leftrightarrow \text{R3}}{\approx} \begin{bmatrix} 1 & 2 & -3 & 11 \\ 0 & 3 & -6 & 15 \\ 0 & 0 & 1 & -1 \end{bmatrix} \underset{\left(\frac{1}{3}\right)\text{R2}}{\approx} \begin{bmatrix} 1 & 2 & -3 & 11 \\ 0 & 1 & -2 & 5 \\ 0 & 0 & 1 & -1 \end{bmatrix}$$

$$\underset{\text{Add } (-2)\text{ R2 to R1}}{\approx} \begin{bmatrix} 1 & 0 & 1 & 1 \\ 0 & 1 & -2 & 5 \\ 0 & 0 & 1 & -1 \end{bmatrix} \underset{\substack{\text{Add } (-1)\text{ R3 to R1} \\ \text{Add } (2)\text{ R3 to R2}}}{\approx} \begin{bmatrix} 1 & 0 & 0 & 2 \\ 0 & 1 & 0 & 3 \\ 0 & 0 & 1 & -1 \end{bmatrix}$$

We have arrived at the reduced form. This matrix corresponds to the system

$$\begin{aligned} x &= 2 \\ y &= 3 \\ z &= -1 \end{aligned}$$

The solution is $x = 2$, $y = 3$, $z = -1$.

Self-Check 2 Solve the following system of equations.

$$\begin{aligned} x - y - 2z &= -5 \\ -2x + 2y + 5z &= 13 \\ -x + 2y + 3z &= 9 \end{aligned}$$

We now give examples of applications of systems of linear equations in three variables. In these examples, we focus on showing how a given problem gives rise to a system of linear equations; we then supply the solution and interpret it. We leave it as an exercise for the student to use Gauss–Jordan elimination to determine the solution.

Example 4

Find an equation of a parabola that passes through the three points $(-1, 4)$, $(2, 1)$, and $(3, 8)$. Verify your answer using a graph.

SOLUTION The general equation of a parabola is $y = ax^2 + bx + c$. There will be a unique parabola that passes through the three given points. The points can be used to find the values of a, b, and c, and thus the equation of the parabola.

Because the points lie on the parabola, they must satisfy its equation. Substitute each point into the equation $y = ax^2 + bx + c$.

$$\begin{aligned}
&\text{Point } (-1, 4): && 4 = a(-1)^2 + b(-1) + c \\
&\text{Point } (2, 1): && 1 = a(2)^2 + b(2) + c \\
&\text{Point } (3, 8): && 8 = a(3)^2 + b(3) + c
\end{aligned}$$

Simplify these three equations in the variables a, b, and c, and write them in standard form, with the variables on the left sides of the equations.

$$\begin{aligned}
a - b + c &= 4 \\
4a + 2b + c &= 1 \\
9a + 3b + c &= 8
\end{aligned}$$

This system of equations is solved using Gauss–Jordan elimination. The solution is found to be $a = 2$, $b = -3$, and $c = -1$. The equation of the parabola is thus

$$y = 2x^2 - 3x - 1$$

(Figure 6.9).

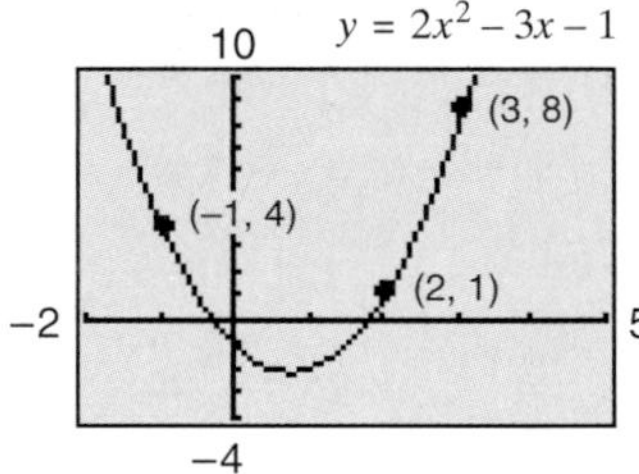

Figure 6.9 This parabola passes through the points $(-1, 4)$, $(2, 1)$, $(3, 8)$.

Self-Check 3 Determine the equation of the parabola that passes through the three points $(1, 0)$, $(2, 5)$, and $(3, 12)$. Verify your answer using a graph.

Example 5 Analyzing Money

Jose Torres has 18 bills, made up of ones, fives, and tens. The total value of the money is \$52. If he has two more fives than tens, how many of each type of bill are there?

SOLUTION Let Jose have x one-dollar bills, y fives, and z tens. We shall use the given information to construct three linear equations in x, y, and z.

The total number of bills is 18; thus,

$$x + y + z = 18$$

The total value of the money is \$52; thus,

$$x + 5y + 10z = 52$$

Finally, Jose has two more fives than tens; thus,

$$y - z = 2$$

To find the number of bills, x, y, and z, we must solve the following system of three equations in three variables.

$$\begin{aligned} x + \ y + \quad z &= 18 \\ x + 5y + 10z &= 52 \\ y - \quad z &= \ \ 2 \end{aligned}$$

This system is solved using Gauss–Jordan elimination. The solution is $x = 12$, $y = 4$, and $z = 2$. Therefore, Jose has 12 one-dollar bills, 4 five-dollar bills, and 2 ten-dollar bills.

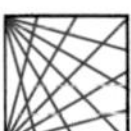

Row Operations Using Calculators

Elementary row operations on matrices can be performed using calculators. Calculators vary considerably in the way matrices are entered and displayed and the operations are implemented. We discuss the steps involved in solving a system of linear equations. We will use the standard matrix display here because it can be understood by all calculator users. You should determine how to implement each of these steps on your particular calculator and interpret the way a matrix is displayed.

Let us discuss the system of linear equations of Example 3. We solve the system

$$\begin{aligned} 4x + 8y - 12z &= 44 \\ 3x + 6y - \ 8z &= 32 \\ -2x - \ y \qquad &= -7 \end{aligned} \quad \text{with augmented matrix} \quad A = \begin{bmatrix} 4 & 8 & -12 & 44 \\ 3 & 6 & -8 & 32 \\ -2 & -1 & 0 & -7 \end{bmatrix}$$

The following steps are implemented on a calculator.

Enter and display Matrix A

$$\begin{bmatrix} 4 & 8 & -12 & 44 \\ 3 & 6 & -8 & 32 \\ -2 & 1 & 0 & -7 \end{bmatrix}$$

$\longrightarrow$ Perform $\left(\frac{1}{4}\right)$ Row 1

$$\begin{bmatrix} 1 & 2 & -3 & 11 \\ 3 & 6 & -8 & 32 \\ -2 & 1 & 0 & -7 \end{bmatrix}$$

$\longrightarrow$ Add (−3)Row 1 to Row 2

$$\begin{bmatrix} 1 & 2 & -3 & 11 \\ 0 & 0 & 1 & -1 \\ -2 & 1 & 0 & -7 \end{bmatrix}$$

$\longrightarrow$ Add (2)Row 1 to Row 3

$$\begin{bmatrix} 1 & 2 & -3 & 11 \\ 0 & 0 & 1 & -1 \\ 0 & 3 & -6 & 15 \end{bmatrix}$$

$\longrightarrow$ Swap Row 2, Row 3

$$\begin{bmatrix} 1 & 2 & -3 & 11 \\ 0 & 3 & -6 & 15 \\ 0 & 0 & 1 & -1 \end{bmatrix}$$

$\cdots\longrightarrow$ Arrive at reduced form

$$\begin{bmatrix} 1 & 0 & 0 & 2 \\ 0 & 1 & 0 & 3 \\ 0 & 0 & 1 & -1 \end{bmatrix}$$

This matrix corresponds to the system

$$\begin{aligned} x \quad &= 2 \\ y \quad &= 3 \\ z &= -1 \end{aligned}$$

The solution is $x = 2$, $y = 3$, and $z = -1$.

Some calculators have a **simultaneous equations solver** that can be used to solve systems of linear equations. Such a solver will give you the solution immediately but will not display the transformations used. Your calculator may also give the **reduced form** of a matrix (sometimes called **reduced row echelon form,** abbreviated **rref**). Investigate these possibilities. Use your calculator to check the answers that you have obtained using algebra.

Answers to Self-Check Exercises

1. Solution is $x = 2$, $y = -1$, and $z = 1$.
2. Solution is $x = 2$, $y = 1$, and $z = 3$.
3. Parabola is $y = x^2 + 2x - 3$.

EXERCISES 6.2

In Exercises 1–6, determine the matrix of coefficients and the augmented matrix of each system of equations.

1. $\begin{aligned} x + 3y &= 7 \\ 2x - 5y &= -3 \end{aligned}$

2. $\begin{aligned} 5x + 2y - 4z &= 8 \\ x + 3y + 6z &= 4 \\ 4x + 6y - 9z &= 7 \end{aligned}$

3. $\begin{aligned} -x + 3y - 5z &= -3 \\ 2x - 2y + 4z &= 8 \\ x + 3y \quad &= 6 \end{aligned}$

4. $\begin{aligned} 5x + 4y &= 9 \\ 2x - 8y &= -4 \\ x + 2y &= 3 \end{aligned}$

5. $\begin{aligned} x \quad &= 8 \\ y \quad &= 2 \\ z &= -7 \end{aligned}$

6. $\begin{aligned} -x + 3y &= 9 \\ 3x - 2y &= 11 \\ x + 8y &= 1 \\ -x - 5y &= 4 \end{aligned}$

In Exercises 7–14, interpret the matrices as augmented matrices of systems of equations. Write down each system of equations.

7. $\begin{bmatrix} 1 & 2 & 3 \\ 4 & 5 & 6 \end{bmatrix}$

8. $\begin{bmatrix} 7 & 9 & 8 \\ 6 & 4 & -3 \end{bmatrix}$

9. $\begin{bmatrix} 1 & 9 & -3 \\ 5 & 0 & 2 \end{bmatrix}$

10. $\begin{bmatrix} 8 & 7 & 5 & -1 \\ 4 & 6 & 2 & 4 \\ 9 & 3 & 7 & 6 \end{bmatrix}$

11. $\begin{bmatrix} 2 & -3 & 6 & 4 \\ 7 & -5 & -2 & 3 \\ 0 & 2 & 4 & 0 \end{bmatrix}$

12. $\begin{bmatrix} 0 & -2 & 4 \\ 5 & 7 & -3 \\ 6 & 0 & 8 \end{bmatrix}$

13. $\begin{bmatrix} 1 & 0 & 0 & 3 \\ 0 & 1 & 0 & 8 \\ 0 & 0 & 1 & 4 \end{bmatrix}$

14. $\begin{bmatrix} 1 & 2 & -1 & 6 \\ 0 & 1 & 4 & 5 \\ 0 & 0 & 1 & -2 \end{bmatrix}$

In Exercises 15–20, you are given a matrix followed by an elementary row operation. **(a)** Determine each resulting matrix by hand. **(b)** Perform these transformations using a calculator.

15. $\begin{bmatrix} 2 & 6 & -4 & 0 \\ 1 & 2 & -3 & 6 \\ 8 & 3 & 2 & 5 \end{bmatrix} \approx$ $\left(\frac{1}{2}\right)$**R1**

16. $\begin{bmatrix} 0 & -8 & 4 & 3 \\ 2 & 7 & 5 & 1 \\ 3 & -5 & 8 & 9 \end{bmatrix} \approx$ **R1 ↔ R2**

17. $\begin{bmatrix} 1 & 2 & 3 & -1 \\ -1 & 1 & 7 & 1 \\ 2 & -4 & 5 & -3 \end{bmatrix} \approx$ **Add R1 to R2**, **Add (−2) R1 to R3**

18. $\begin{bmatrix} 1 & 2 & 3 & -4 \\ 0 & 1 & 2 & 1 \\ 0 & -4 & 3 & -5 \end{bmatrix} \approx$ **Add (−2) R2 to R1**, **Add (4) R2 to R3**

19. $\begin{bmatrix} 1 & 0 & 4 & -3 \\ 0 & 1 & -3 & 2 \\ 0 & 0 & 1 & 5 \end{bmatrix} \approx$ **Add (−4) R3 to R1**, **Add (3) R3 to R2**

20. $\begin{bmatrix} 1 & 0 & 2 & 7 \\ 0 & 1 & 5 & -3 \\ 0 & 0 & -2 & 8 \end{bmatrix} \underset{\left(-\frac{1}{2}\right)\text{R3}}{\approx}$

In Exercises 21–24, interpret each row operation as a stage in determining the reduced form of a matrix. Why has the indicated operation been selected? What particular aim does it accomplish in terms of the systems of linear equations that are described by the matrices?

21. $\begin{bmatrix} 1 & -4 & 3 & 5 \\ -2 & 1 & 7 & 5 \\ 4 & 0 & -3 & 6 \end{bmatrix} \underset{\substack{\text{Add (2) R1 to R2} \\ \text{Add (}-4\text{) R1 to R3}}}{\approx} \begin{bmatrix} 1 & -4 & 3 & 5 \\ 0 & -7 & 13 & 15 \\ 0 & 16 & -15 & -14 \end{bmatrix}$

22. $\begin{bmatrix} 1 & 2 & -4 & 7 \\ 0 & 3 & 9 & -6 \\ 0 & 4 & 7 & -8 \end{bmatrix} \underset{\left(\frac{1}{3}\right)\text{R2}}{\approx} \begin{bmatrix} 1 & 2 & -4 & 7 \\ 0 & 1 & 3 & -2 \\ 0 & 4 & 7 & -8 \end{bmatrix}$

23. $\begin{bmatrix} 1 & 3 & -4 & 5 \\ 0 & 0 & -2 & 6 \\ 0 & 1 & 3 & -8 \end{bmatrix} \underset{\text{R2} \leftrightarrow \text{R3}}{\approx} \begin{bmatrix} 1 & 3 & -4 & 5 \\ 0 & 1 & 3 & -8 \\ 0 & 0 & -2 & 6 \end{bmatrix}$

24. $\begin{bmatrix} 1 & 2 & 5 & 0 \\ 0 & 1 & 2 & -3 \\ 0 & -3 & 1 & -2 \end{bmatrix} \underset{\substack{\text{Add (}-2\text{) R2 to R1} \\ \text{Add (3) R2 to R3}}}{\approx} \begin{bmatrix} 1 & 0 & 1 & 6 \\ 0 & 1 & 2 & -3 \\ 0 & 0 & 7 & -11 \end{bmatrix}$

In Exercises 25–28, interpret each row operation as a stage in determining the reduced form of a matrix. Why has the operation been selected?

25. $\begin{bmatrix} 1 & 0 & 2 & 6 \\ 0 & 1 & -1 & 3 \\ 0 & 0 & 1 & 2 \end{bmatrix} \underset{\substack{\text{Add (}-2\text{) R3 to R1} \\ \text{Add R3 to R2}}}{\approx} \begin{bmatrix} 1 & 0 & 0 & 2 \\ 0 & 1 & 0 & 5 \\ 0 & 0 & 1 & 2 \end{bmatrix}$

26. $\begin{bmatrix} 0 & 2 & 4 & -1 \\ 4 & 3 & 2 & -8 \\ 5 & -7 & 1 & 2 \end{bmatrix} \underset{\text{R1} \leftrightarrow \text{R2}}{\approx} \begin{bmatrix} 4 & 3 & 2 & -8 \\ 0 & 2 & 4 & -1 \\ 5 & -7 & 1 & 2 \end{bmatrix}$

27. $\begin{bmatrix} 1 & 0 & 3 & 7 \\ 0 & 1 & 4 & 2 \\ 0 & 0 & -2 & 6 \end{bmatrix} \underset{\left(-\frac{1}{2}\right)\text{R3}}{\approx} \begin{bmatrix} 1 & 0 & 3 & 7 \\ 0 & 1 & 4 & 2 \\ 0 & 0 & 1 & -3 \end{bmatrix}$

28. $\begin{bmatrix} 1 & 0 & -2 & 4 \\ 0 & 1 & 3 & -4 \\ 0 & 0 & 1 & -3 \end{bmatrix} \underset{\substack{\text{Add (2) R3 to R1} \\ \text{Add (}-3\text{) R3 to R2}}}{\approx} \begin{bmatrix} 1 & 0 & 0 & -2 \\ 0 & 1 & 0 & 5 \\ 0 & 0 & 1 & -3 \end{bmatrix}$

In Exercises 29–32, the systems of equations all have unique solutions. Solve the systems by hand using the method of Gauss–Jordan elimination.

29. $\begin{aligned} x + z &= 3 \\ 2y - 2z &= -4 \\ y - 2z &= 5 \end{aligned}$

30. $\begin{aligned} x + y + 3z &= 6 \\ x + 2y + 4z &= 9 \\ 2x + y + 6z &= 11 \end{aligned}$

31. $\begin{aligned} x - y + 3z &= 3 \\ 2x - y + 2z &= 2 \\ 3x + y - 2z &= 3 \end{aligned}$

32. $\begin{aligned} -x + y - z &= -2 \\ 3x + y + z &= 10 \\ 4x + 2y + 3z &= 14 \end{aligned}$

In Exercises 33–38, the systems of equations all have unique solutions. **(a)** Solve the systems by hand using the method of Gauss–Jordan elimination. **(b)** Solve with a calculator using the method of Gauss–Jordan elimination.

33. $\begin{aligned} 2y + 4z &= 8 \\ 2x + 2y &= 6 \\ x + y + z &= 5 \end{aligned}$

34. $\begin{aligned} x - 2y - 4z &= -9 \\ x + 5y + 10z &= 19 \\ 2x - 3y - 5z &= -13 \end{aligned}$

35. $\begin{aligned} x + 2y + 3z &= 14 \\ 2x + 5y + 8z &= 36 \\ x - y &= -4 \end{aligned}$

36. $\begin{aligned} x - y - z &= -1 \\ -2x + 6y + 10z &= 14 \\ 2x + y + 6z &= 9 \end{aligned}$

37. $\begin{aligned} 2x + 2y - 4z &= 14 \\ 3x + y + z &= 8 \\ 2x - y + 2z &= -1 \end{aligned}$

38. $\begin{aligned} x + 2y - z &= 3 \\ x + 3y - 2z &= -6 \\ -x - y + 3z &= 6 \end{aligned}$

In Exercises 39 and 40, the systems of equations have unique solutions. Solve the systems using a calculator.

39. $\begin{aligned} 1.5x + 3z &= 15 \\ -x + 7y - 9z &= -45 \\ 2x + 5z &= 22 \end{aligned}$

40. $\begin{aligned} -3x - 6y - 15z &= -3 \\ x + 1.5y + 4.5z &= 0.5 \\ -2x - 3.5y - 8.5z &= -2 \end{aligned}$

In Exercises 41–48, find the equations of the parabolas through the given points. Verify your answer using a graph.

41. $(1, 4), (2, 11), (3, 22)$

42. $(1, 4), (2, 3), (3, 0)$

43. $(-1, 0), (1, 6), (2, 12)$

44. $(1, 1), (3, -15), (5, -47)$

45. $(-1, 1), (1, 5), (2, 19)$

46. $(0, -7), (1, -3), (3, 23)$

47. $(2, 8), (4, 6), (6, -4)$

48. $(3, -13), (4, -21), (5, -31)$

49. **Analyzing Money** Daphne DeLand has ten bills made up of ones, fives, and tens. The total value of the money is \$44. If she has two more fives than tens, how many of each type of bill are there?

50. **Analyzing Money** Enrico Sanchez has 11 bills made up of ones, fives, and tens. The total value of the money is \$36. If he has five more ones than tens, how many of each type of bill are there?

51. **Analyzing Money** Carol McDonald has 13 bills made up of ones, fives, and tens. If she has four more ones than fives and three more fives than tens, what is the total value of the money?

52. **Analyzing Money** Janice Sylvia has 15 bills, made up of ones, fives, and tens. If she has one more ten than ones and two more tens than fives, what is the total value of the money?

53. **Money in a Vending Machine** A vending machine at the Jasper University student center contains nickels, dimes, and quarters. At the end of the day, the machine contains 780 coins. The total value of the money is \$137. If the machine has 80 more nickels than dimes, how many of each type of coin is in the machine?

54. **Money in a Vending Machine** A vending machine contains nickels, dimes, and quarters. At the end of the day the machine contains \$104.50. It has 180 more dimes than nickels and 110 more quarters than dimes. How many of each type of coin is in the machine?

55. **Investment** Timothy Marpole invested \$8000 in three types of bonds: one at 7% interest per annum, one at 8%, and one at 10%. The total return on these bonds at the end of the first year was \$730. If \$3000 more was invested at 10% than at 8%, how much was invested in each bond?

56. **Investment** Donna Grabowski invested \$13,250 in three types of bonds: one at 7%, one at 9%, and one at 12%. The total return on the bonds at the end of the first year was \$1345. If \$1500 more was invested at 9% than at 7%, how much was invested in each bond?

57. **Manufacturing** The Designer's Choice Company manufactures tables, chairs, and cupboards using metal, wood, and plastic. The following chart gives the amounts of these items (in convenient units) that go into each product.

	Metal	Wood	Plastic
Table	2	6	1
Chair	1	2	1
Cupboard	3	5	2

If the company has 150 units of metal, 330 units of wood, and 120 units of plastic available, how many tables, chairs, and cupboards can be manufactured?

58. **Manufacturing** The Furniture Emporium manufactures desks, cabinets, and chairs. These items are made out of metal, wood, and plastic. The following chart represents, in convenient units, the amounts of these items that go into each product.

	Metal	Wood	Plastic
Desk	3	4	2
Cabinet	6	1	1
Chair	1	2	2

If the company has 460 units of metal, 165 units of wood, and 125 units of plastic, how many desks, cabinets, and chairs can be manufactured?

59. **Linear Motion** The motion of an object moving along a straight line is described by the equation $s(t) = (1/2)at^2 + v_0t + s_0$. In this equation, t is time, s_0 is the initial distance from a fixed point 0 on the line, v_0 is the initial velocity, a is constant acceleration, and $s(t)$ is the distance from 0 at time t. Find a, v_0, and s_0 if **(a)** $s(1) = 30$, $s(2) = 56$, and $s(3) = 86$. **(b)** $s(2) = 49$, $s(4) = 73$, and $s(6) = 93$.

6.3 Gauss–Jordan Elimination and Electrical Networks

• Reduced Form • Gauss–Jordan Elimination, the General Case • Unique, Many, and No Solutions • Electrical Networks

In Section 6.2, we used the method of Gauss–Jordan elimination to solve systems of three equations in three variables that had a unique solution. We shall now discuss the method in its more general setting, where there can be a unique solution, many solutions, or no solutions. Our approach again will be to start from the augmented matrix of the given system and to perform a sequence of elementary row operations that will result in a simpler matrix (the reduced form), which leads directly to the solution.

Reduced Form

We now give the general definition of reduced form. The student will observe that the reduced matrices discussed in Section 6.2 all conform to this definition.

Reduced Form A matrix is in **reduced form** if

1. Any rows consisting entirely of zeros are grouped at the bottom of the matrix.
2. The first nonzero element of each other row is 1. This element is called a **leading one.**
3. The leading one of each row after the first is positioned to the right of the leading one of the previous row.
4. All other elements in a column that contains a leading one are zero.

The following matrices are all in reduced form.

$$\begin{bmatrix} 1 & 0 & 8 \\ 0 & 1 & 2 \\ 0 & 0 & 0 \end{bmatrix} \quad \begin{bmatrix} 1 & 0 & 0 & 7 \\ 0 & 1 & 0 & 3 \\ 0 & 0 & 1 & 9 \end{bmatrix} \quad \begin{bmatrix} 1 & 4 & 0 & 0 \\ 0 & 0 & 1 & 0 \\ 0 & 0 & 0 & 1 \end{bmatrix} \quad \begin{bmatrix} 1 & 2 & 3 & 0 \\ 0 & 0 & 0 & 1 \\ 0 & 0 & 0 & 0 \end{bmatrix}$$

The following matrices are not in reduced form for the reasons given:

$$\begin{bmatrix} 1 & 2 & 0 & 4 \\ 0 & 0 & 0 & 0 \\ 0 & 0 & 1 & 3 \end{bmatrix}$$

Row of zeros not at bottom of matrix

$$\begin{bmatrix} 1 & 2 & 0 & 3 & 0 \\ 0 & 0 & 3 & 4 & 0 \\ 0 & 0 & 0 & 0 & 1 \end{bmatrix}$$

First nonzero element in row 2 is not 1

$$\begin{bmatrix} 1 & 0 & 0 & 2 \\ 0 & 0 & 1 & 4 \\ 0 & 1 & 0 & 3 \end{bmatrix}$$

Leading one in row 3 not to the right of leading one in row 2

$$\begin{bmatrix} 1 & 7 & 0 & 8 \\ 0 & 1 & 0 & 3 \\ 0 & 0 & 1 & 2 \end{bmatrix}$$

Nonzero element above leading one in row 2

There are usually many sequences of row operations that can be used to transform a given matrix to reduced form; however, they all lead to the same reduced form. We say that *the reduced form of a matrix is unique.*

GAUSS–JORDAN ELIMINATION, THE GENERAL CASE

The method of Gauss–Jordan elimination is an important systematic way (called an algorithm) for arriving at the reduced form. It can be programmed on a calculator or a computer. We now summarize the method and then give examples of its implementation.

GAUSS–JORDAN ELIMINATION

1. Write down the augmented matrix of the system of linear equations.
2. Derive the reduced form of the augmented matrix using elementary row operations. This is done by creating leading ones, then zeros above and below each leading one, column by column starting with the first column.
3. Write down the system of equations corresponding to the reduced echelon form. This system gives the solution.

We now show how this method is used to solve various systems of equations. The following example illustrates how to solve a system of linear equations that has many solutions. The reduced form is derived. It then becomes necessary to interpret the reduced form, expressing the many solutions in a clear manner.

Example 1 **System with Many Solutions**

Solve, if possible, the following system of equations:

$$\begin{aligned} 3x - 3y + 3z &= 9 \\ 2x - y + 4z &= 7 \\ 3x - 5y - z &= 7 \end{aligned}$$

SOLUTION Start with the augmented matrix and follow the Gauss–Jordan algorithm.

$$\begin{bmatrix} 3 & -3 & 3 & 9 \\ 2 & -1 & 4 & 7 \\ 3 & -5 & -1 & 7 \end{bmatrix} \underset{\left(\frac{1}{3}\right)\text{R1}}{\approx} \begin{bmatrix} 1 & -1 & 1 & 3 \\ 2 & -1 & 4 & 7 \\ 3 & -5 & -1 & 7 \end{bmatrix}$$

$$\underset{\substack{\textbf{Add } (-2)\textbf{ R1 to R2} \\ \textbf{Add } (-3)\textbf{ R1 to R3}}}{\approx} \begin{bmatrix} 1 & -1 & 1 & 3 \\ 0 & 1 & 2 & 1 \\ 0 & -2 & -4 & -2 \end{bmatrix} \underset{\substack{\textbf{Add R2 to R1} \\ \textbf{Add } (2)\textbf{ R2 to R3}}}{\approx} \begin{bmatrix} 1 & 0 & 3 & 4 \\ 0 & 1 & 2 & 1 \\ 0 & 0 & 0 & 0 \end{bmatrix}$$

We have arrived at the reduced form. The corresponding system of equations is

$$\begin{aligned} x + \phantom{y + {}} 3z &= 4 \\ y + 2z &= 1 \end{aligned}$$

There are many values of x, y, and z that satisfy these equations. This is a system of equations that has many solutions. x is called the **leading variable** of the first equation, and y is the leading variable of the second equation. To express these many solutions, we write the leading variables in each equation in terms of the remaining variables. We get

$$x = -3z + 4$$
$$y = -2z + 1$$

Let us assign the arbitrary value r to z. The **general solution** to the system is

$$x = -3r + 4, \qquad y = -2r + 1, \qquad z = r$$

As r ranges over the set of real numbers, we get infinitely many solutions. r is called a **parameter.** We can get specific solutions by giving r different values. For example,

$$r = 1 \quad \text{gives} \quad x = 1, \quad y = -1, \quad z = 1$$
$$r = -2 \quad \text{gives} \quad x = 10, \quad y = 5, \quad z = -2$$

Self-Check 1 Solve the following system using Gauss–Jordan elimination.

$$\begin{aligned} x + y + 2z &= 2 \\ x + 2y + 5z &= 0 \\ 2x + y + z &= 6 \end{aligned}$$

In Section 6.2, we saw that it sometimes becomes necessary to interchange rows during Gauss–Jordan elimination. The following example further illustrates this situation.

Example 2 Gauss–Jordan Elimination with a Row Interchange

Solve the following system using Gauss–Jordan elimination.

$$\begin{aligned} x + 4y + 3z &= 1 \\ 2x + 8y + 11z &= 7 \\ x + 6y + 7z &= 3 \end{aligned}$$

SOLUTION Start with the augmented matrix and create zeros in the first column.

$$\begin{bmatrix} 1 & 4 & 3 & 1 \\ 2 & 8 & 11 & 7 \\ 1 & 6 & 7 & 3 \end{bmatrix} \underset{\substack{\text{Add } (-2) \text{ R1 to R2} \\ \text{Add } (-1) \text{ R1 to R3}}}{\approx} \begin{bmatrix} 1 & 4 & 3 & 1 \\ 0 & 0 & 5 & 5 \\ 0 & 2 & 4 & 2 \end{bmatrix}$$

At this stage, we need a nonzero element in location row 2, column 2 to continue. We interchange the second row with the third row (a *later* row) and then proceed.

$$\underset{\text{R2} \leftrightarrow \text{R3}}{\approx} \begin{bmatrix} 1 & 4 & 3 & 1 \\ 0 & 2 & 4 & 2 \\ 0 & 0 & 5 & 5 \end{bmatrix} \underset{\left(\frac{1}{2}\right)\text{R2}}{\approx} \begin{bmatrix} 1 & 4 & 3 & 1 \\ 0 & 1 & 2 & 1 \\ 0 & 0 & 5 & 5 \end{bmatrix}$$

$$\underset{\text{Add } (-4) \text{ R2 to R1}}{\approx} \begin{bmatrix} 1 & 0 & -5 & -3 \\ 0 & 1 & 2 & 1 \\ 0 & 0 & 5 & 5 \end{bmatrix} \underset{\left(\frac{1}{5}\right)\text{R2}}{\approx} \begin{bmatrix} 1 & 0 & -5 & -3 \\ 0 & 1 & 2 & 1 \\ 0 & 0 & 1 & 1 \end{bmatrix}$$

$$\underset{\substack{\text{Add } (5) \text{ R3 to R1} \\ \text{Add } (-2) \text{ R3 to R2}}}{\approx} \begin{bmatrix} 1 & 0 & 0 & 2 \\ 0 & 1 & 0 & -1 \\ 0 & 0 & 1 & 1 \end{bmatrix}$$

This matrix is in reduced form. The corresponding system of equations is

$$\begin{aligned} x \qquad\quad &= 2 \\ y \quad &= -1 \\ z &= 1 \end{aligned}$$

There is a unique solution $x = 2$, $y = -1$, and $z = 1$.

Self-Check 2 Solve the following system using Gauss–Jordan elimination.

$$\begin{aligned} x + 2y + 4z &= 15 \\ 2x + 4y + 9z &= 33 \\ x + 3y + 5z &= 20 \end{aligned}$$

The next example illustrates the method for a system that has no solution.

Example 3 System with No Solution

Solve, if possible, the system

$$\begin{aligned} x + 4y + z &= 2 \\ x + 2y - z &= 0 \\ 2x + 6y \qquad &= 3 \end{aligned}$$

SOLUTION We get the following sequence of matrices.

$$\begin{bmatrix} 1 & 4 & 1 & 2 \\ 1 & 2 & -1 & 0 \\ 2 & 6 & 0 & 3 \end{bmatrix} \underset{\substack{\textbf{Add } (-1)\textbf{ R1 to R2} \\ \textbf{Add } (-2)\textbf{ R1 to R3}}}{\approx} \begin{bmatrix} 1 & 4 & 1 & 2 \\ 0 & -2 & -2 & -2 \\ 0 & -2 & -2 & -1 \end{bmatrix}$$

$$\underset{\left(-\frac{1}{2}\right)\textbf{R2}}{\approx} \begin{bmatrix} 1 & 4 & 1 & 2 \\ 0 & 1 & 1 & 1 \\ 0 & -2 & -2 & -1 \end{bmatrix} \underset{\substack{\textbf{Add } (-4)\textbf{ R2 to R1} \\ \textbf{Add } (2)\textbf{ R2 to R3}}}{\approx} \begin{bmatrix} 1 & 0 & -3 & -2 \\ 0 & 1 & 1 & 1 \\ 0 & 0 & 0 & 1 \end{bmatrix}$$

$$\underset{\substack{\textbf{Add } (2)\textbf{ R3 to R1} \\ \textbf{Add } (-1)\textbf{ R3 to R2}}}{\approx} \begin{bmatrix} 1 & 0 & -3 & 0 \\ 0 & 1 & 1 & 0 \\ 0 & 0 & 0 & 1 \end{bmatrix}$$

This matrix is in reduced form. The corresponding system of equations is

$$\begin{aligned} x \qquad - 3z &= 0 \\ y + z &= 0 \\ 0 &= 1 \end{aligned}$$

The last statement is false, indicating that the system of equations is inconsistent. The system has no solution.

Self-Check 3 Show that the following system has no solution.

$$\begin{aligned} x + y + z &= 7 \\ 2x + y + 3z &= 10 \\ -x + y - 3z &= 0 \end{aligned}$$

Application . . .

Example 4 Electrical Networks

Systems of linear equations are used to determine the currents through various branches of electrical networks. The following two laws, which are based on experimental verification in the laboratory, lead to the equations.

Kirchhoff's Laws*

1. *Junctions.* All the current flowing into a junction must flow out of it.
2. *Paths.* The sum of the *IR* terms (*I* denotes current, and *R* stands for resistance) in any direction around a closed path is equal to the total voltage in the path in that direction.

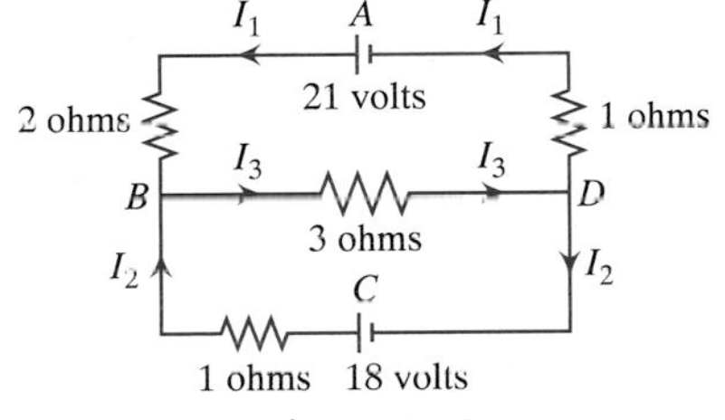

Figure 6.10

Consider the electrical network shown in Figure 6.10. The batteries (labeled |⊢) are 21 and 18 volts. The resistances (labeled ⩘⩘) are one 2-ohm, one 3-ohm, and two 1-ohm. Let us determine the currents I_1, I_2, and I_3 through the branches of this network.

Kirchhoff's Laws refer to junctions and closed paths. There are two junctions in this circuit, namely the points *B* and *D*. There are three closed paths, namely *ABDA*, *CBDC*, and *ABCDA*. Apply the laws to the junctions and paths.

Junctions

Junction *B*, $I_1 + I_2 = I_3$

Junction *D*, $I_3 = I_1 + I_2$

These two equations result in a single linear equation

$$I_1 + I_2 - I_3 = 0$$

Paths

Path *ABDA* $2I_1 + 3I_3 + 1I_1 = 21$

Simplify $3I_1 + 3I_3 = 21$

$I_1 + I_3 = 7$

Path *CBDC* $1I_2 + 3I_3 = 18$

*Gustav Robert Kirchhoff (1824–1887) was educated at the University of Königsberg and did most of his teaching at the University of Heidelberg. His major contributions were in the experimental discovery and theoretical analysis of the laws of electromagnetic radiation. Kirchhoff was a master teacher whose texts set a standard for the teaching of theoretical physics in German universities. He was described as "not easily drawn out but of a cheerful and obliging disposition."

It is not necessary to look further at path *ABCDA*. We now have a system of three linear equations in three unknowns, I_1, I_2, and I_3. Path *ABCDA* in fact leads to an equation that is a combination of the last two equations; there is no new information.

The problem thus reduces to solving the following system of three linear equations in three variables:

$$\begin{aligned} I_1 + I_2 - I_3 &= 0 \\ I_1 + I_3 &= 7 \\ I_2 + 3I_3 &= 18 \end{aligned}$$

Using the method of Gauss–Jordan elimination, we get

$$\begin{bmatrix} 1 & 1 & -1 & 0 \\ 1 & 0 & 1 & 7 \\ 0 & 1 & 3 & 18 \end{bmatrix} \underset{(-1)\,\text{R1} + \text{R2}}{\approx} \begin{bmatrix} 1 & 1 & -1 & 0 \\ 0 & -1 & 2 & 7 \\ 0 & 1 & 3 & 18 \end{bmatrix}$$

$$\underset{(-1)\,\text{R2}}{\approx} \begin{bmatrix} 1 & 1 & -1 & 0 \\ 0 & 1 & -2 & -7 \\ 0 & 1 & 3 & 18 \end{bmatrix} \underset{\substack{(-1)\,\text{R2} + \text{R1} \\ (-1)\,\text{R2} + \text{R3}}}{\approx} \begin{bmatrix} 1 & 0 & 1 & 7 \\ 0 & 1 & -2 & -7 \\ 0 & 0 & 5 & 25 \end{bmatrix}$$

$$\underset{\left(\frac{1}{5}\right)\text{R3}}{\approx} \begin{bmatrix} 1 & 0 & 1 & 7 \\ 0 & 1 & -2 & -7 \\ 0 & 0 & 1 & 5 \end{bmatrix} \underset{\substack{(-1)\,\text{R3} + \text{R1} \\ (2)\,\text{R3} + \text{R2}}}{\approx} \begin{bmatrix} 1 & 0 & 0 & 2 \\ 0 & 1 & 0 & 3 \\ 0 & 0 & 1 & 5 \end{bmatrix}$$

The currents are $I_1 = 2$ amperes, $I_2 = 3$ amperes, and $I_3 = 5$ amperes. The solution is unique, as is to be expected in this physical situation.

Group Discussion

Systems with the Same Matrix of Coefficients

Consider two systems of three linear equations in three variables having the same matrix of coefficients. Is it possible for one system to have a unique solution and the other to have many solutions? Discuss.

Answers to Self-Check Exercises

1. Many solutions are $x = r + 4$, $y = -3r - 2$, and $z = r$.
2. Unique solution $x = -1$, $y = 2$, and $z = 3$.

EXERCISES 6.3

In Exercises 1–9, determine whether the matrices are in reduced form. If a matrix is not in reduced form, give a reason.

1. $\begin{bmatrix} 1 & 0 & 2 \\ 0 & 1 & 3 \end{bmatrix}$

2. $\begin{bmatrix} 1 & 2 & 0 & 4 \\ 0 & 0 & 1 & 7 \end{bmatrix}$

3. $\begin{bmatrix} 1 & 2 & 5 & 6 \\ 0 & 1 & 3 & -7 \end{bmatrix}$

4. $\begin{bmatrix} 1 & 4 & 0 & 5 \\ 0 & 0 & 2 & 9 \end{bmatrix}$

5. $\begin{bmatrix} 1 & 0 & 0 \\ 0 & 1 & 0 \\ 0 & 0 & 1 \end{bmatrix}$

6. $\begin{bmatrix} 1 & 5 & 0 \\ 0 & 0 & 1 \\ 0 & 0 & 0 \end{bmatrix}$

7. $\begin{bmatrix} 1 & 0 & 0 & 4 \\ 0 & 1 & 0 & 5 \\ 0 & 0 & 1 & 9 \end{bmatrix}$

8. $\begin{bmatrix} 1 & 0 & 0 & 3 & 2 \\ 0 & 2 & 0 & 6 & 1 \\ 0 & 0 & 1 & 2 & 3 \end{bmatrix}$

9. $\begin{bmatrix} 1 & 0 & 3 & 0 \\ 0 & 1 & 6 & 0 \\ 0 & 0 & 0 & 1 \end{bmatrix}$

In Exercises 10–18, determine whether the matrices are in reduced form. If a matrix is not in reduced form, give a reason.

10. $\begin{bmatrix} 1 & 0 & 3 & -2 \\ 0 & 0 & 1 & 8 \\ 0 & 1 & 4 & 9 \end{bmatrix}$

11. $\begin{bmatrix} 1 & 2 & 0 & 0 & 4 \\ 0 & 0 & 1 & 0 & 6 \\ 0 & 0 & 0 & 1 & 5 \end{bmatrix}$

12. $\begin{bmatrix} 1 & 5 & 0 & 2 & 0 \\ 0 & 0 & 1 & 9 & 0 \\ 0 & 0 & 0 & 0 & 1 \\ 0 & 0 & 0 & 0 & 0 \end{bmatrix}$

13. $\begin{bmatrix} 1 & 0 & 4 & 2 & 6 \\ 0 & 1 & 2 & 3 & 4 \\ 0 & 0 & 0 & 1 & 2 \\ 0 & 0 & 0 & 0 & 1 \end{bmatrix}$

14. $\begin{bmatrix} 1 & 0 & 2 & 0 & 3 \\ 0 & 0 & 0 & 0 & 0 \\ 0 & 1 & 2 & 0 & 7 \\ 0 & 0 & 0 & 1 & 3 \end{bmatrix}$

15. $\begin{bmatrix} 1 & 0 & 4 & 0 & 0 \\ 0 & 1 & 2 & 0 & 0 \\ 0 & 0 & 0 & 1 & 0 \\ 0 & 0 & 0 & 0 & 1 \end{bmatrix}$

16. $\begin{bmatrix} 1 & 0 & 0 & 5 & 3 \\ 0 & 0 & 0 & 0 & 0 \\ 0 & 1 & 2 & 3 & 7 \end{bmatrix}$

17. $\begin{bmatrix} 0 & 0 & 1 & 0 & 4 \\ 0 & 0 & 0 & 1 & 5 \\ 0 & 1 & 0 & 0 & 3 \end{bmatrix}$

18. $\begin{bmatrix} 1 & 5 & -3 & 0 & 7 \\ 0 & 0 & 0 & 1 & 4 \\ 0 & 0 & 0 & 0 & 0 \end{bmatrix}$

In Exercises 19–24, each matrix is the reduced form of the augmented matrix of a system of linear equations. Give the solution (if it exists) to each system of equations.

19. $\begin{bmatrix} 1 & 0 & 0 & 2 \\ 0 & 1 & 0 & 4 \\ 0 & 0 & 1 & -3 \end{bmatrix}$

20. $\begin{bmatrix} 1 & 0 & -3 & 4 \\ 0 & 1 & 2 & 8 \\ 0 & 0 & 0 & 0 \end{bmatrix}$

21. $\begin{bmatrix} 1 & 3 & 0 & 6 \\ 0 & 0 & 1 & -2 \\ 0 & 0 & 0 & 0 \end{bmatrix}$

22. $\begin{bmatrix} 1 & 0 & 5 & 0 \\ 0 & 1 & -7 & 0 \\ 0 & 0 & 0 & 1 \end{bmatrix}$

23. $\begin{bmatrix} 1 & 0 & 2 & 3 \\ 0 & 1 & 0 & 5 \\ 0 & 0 & 0 & 0 \end{bmatrix}$

24. $\begin{bmatrix} 1 & 0 & 0 & 7 \\ 0 & 1 & -3 & 2 \\ 0 & 0 & 0 & 0 \end{bmatrix}$

In Exercises 25–30, solve (if possible) each of the systems of three equations in three variables, using the method of Gauss–Jordan elimination.

25. $\begin{aligned} x + 4y + 3z &= 1 \\ 2x + 8y + 11z &= 7 \\ x + 6y + 7z &= 3 \end{aligned}$

26. $\begin{aligned} x + 2y + 4z &= 15 \\ 2x + 4y + 9z &= 33 \\ x + 3y + 5z &= 20 \end{aligned}$

27. $\begin{aligned} x + y + z &= 7 \\ 2x + 3y + z &= 18 \\ -x + y - 3z &= 1 \end{aligned}$

28. $\begin{aligned} x + 4y + z &= 2 \\ x + 2y - z &= 0 \\ 2x + 6y \phantom{{}+z} &= 3 \end{aligned}$

29. $\begin{aligned} x - y + z &= 3 \\ 2x - y + 4z &= 7 \\ 3x - 5y - z &= 7 \end{aligned}$

30. $\begin{aligned} x + y + 6z &= 2 \\ -x \phantom{{}+y} - 2z &= 1 \\ 2x + 3y + 16z &= 8 \end{aligned}$

In Exercises 31–38, solve (if possible) each of the systems of three equations in three variables using the method of Gauss–Jordan elimination.

31. $\begin{aligned} 3x + 6y - 3z &= 6 \\ -2x - 4y - 3z &= -1 \\ 3x + 6y - 2z &= 10 \end{aligned}$

32. $\begin{aligned} x + 2y + z &= 7 \\ x + 2y + 2z &= 11 \\ 2x + 4y + 3z &= 18 \end{aligned}$

33. $\begin{aligned} x + 2y - z &= 3 \\ 2x + 4y - 2z &= 6 \\ 3x + 6y + 2z &= -1 \end{aligned}$

34. $\begin{aligned} x + 2y + 3z &= 8 \\ 3x + 7y + 9z &= 26 \\ 2x + \phantom{7y+{}} 6z &= 11 \end{aligned}$

35. $\begin{aligned} y + 2z &= 5 \\ x + 2y + 5z &= 13 \\ x + \phantom{2y+{}} 2z &= 4 \end{aligned}$

36. $\begin{aligned} x + 2y + 8z &= 7 \\ 2x + 4y + 16z &= 14 \\ y + 3z &= 4 \end{aligned}$

37. $\begin{aligned} x + y - 3z &= 10 \\ -3x - 2y + 4z &= -24 \end{aligned}$

38. $\begin{aligned} 2x - 6y - 14z &= 38 \\ -3x + 7y + 15z &= -37 \end{aligned}$

39. Construct examples of the following: **(a)** a system of linear equations with more variables than equations, having no solution. **(b)** a system of linear equations with more equations than variables, having a unique solution.

40. If a is non-zero, determine the conditions on b, c, and d for the matrix

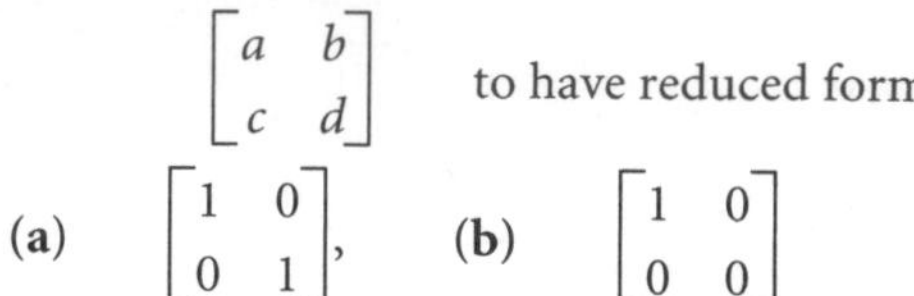

$\begin{bmatrix} a & b \\ c & d \end{bmatrix}$ to have reduced form
(a) $\begin{bmatrix} 1 & 0 \\ 0 & 1 \end{bmatrix}$, (b) $\begin{bmatrix} 1 & 0 \\ 0 & 0 \end{bmatrix}$

In Exercises 41–46, determine the currents in the various branches of the electrical networks. The units of current are amperes, and the units of resistance are ohms.

41.

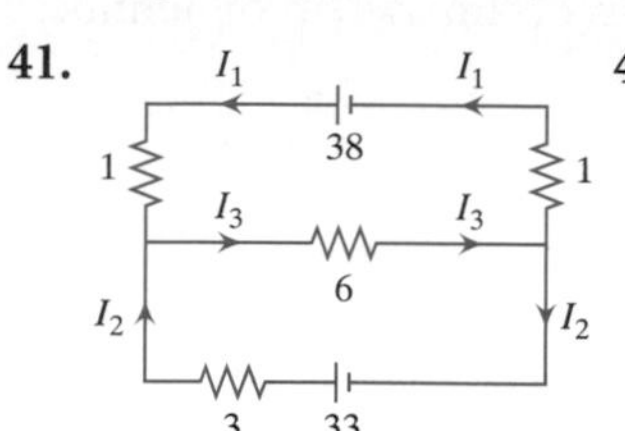

42.

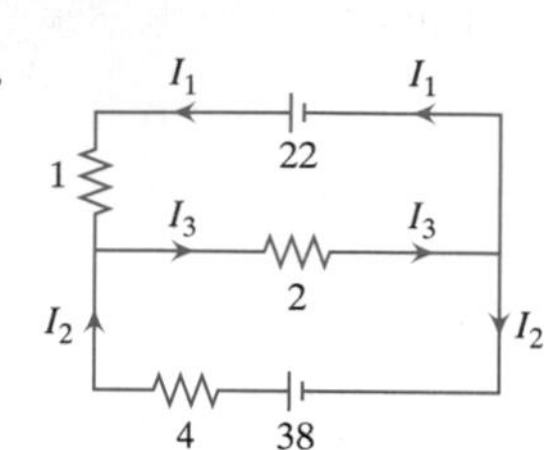

43.

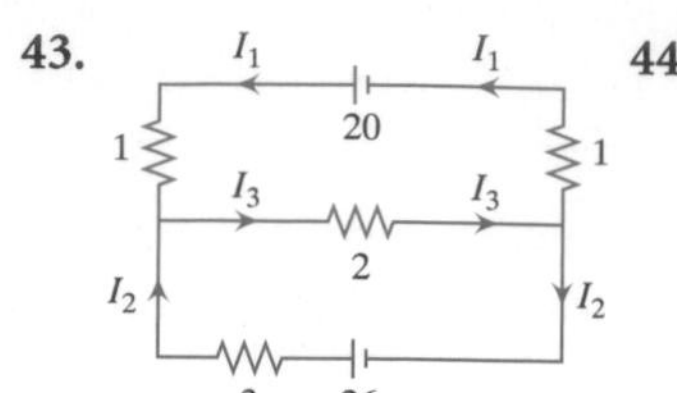

44.

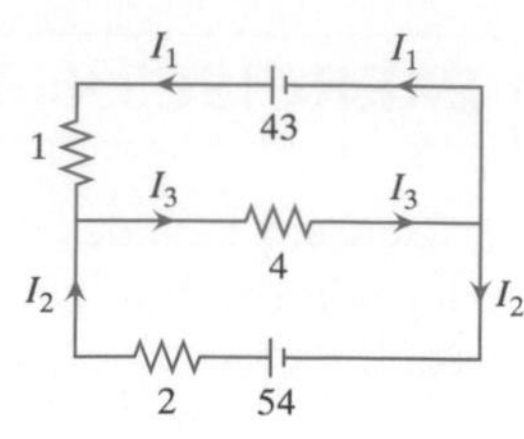

45.

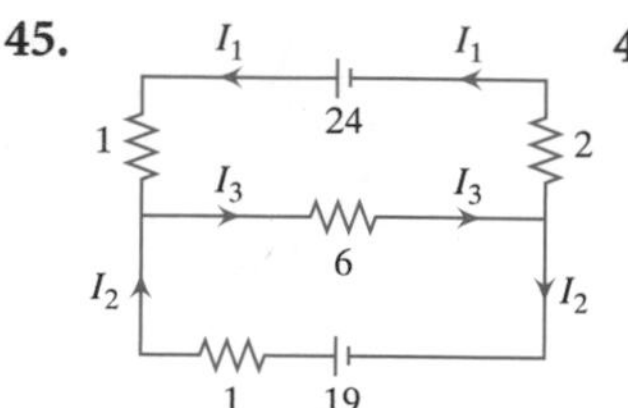

46.

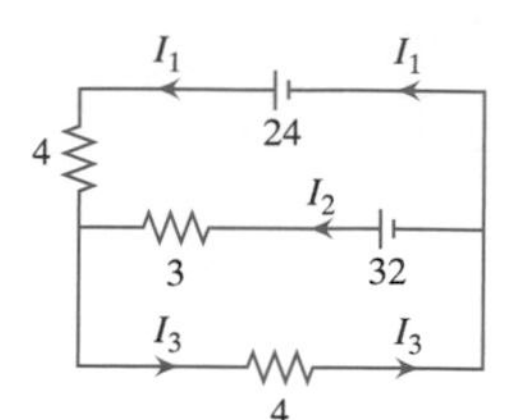

6.4 Systems of Nonlinear Equations

• Systems of Nonlinear Equations • Method of Substitution • Method of Elimination • Solutions Using Graphs

A **system of nonlinear equations** is a system of equations in which at least one of the equations is not of degree 1. In general, such systems can be difficult to solve. We will illustrate a method of substitution, a method of elimination, and a graphing method for solving certain systems of nonlinear equations.

Method of Substitution

As for linear systems of equations, this method involves substituting for one variable from one equation into the other equation.

Example 1 **Solving Using Substitution**

Solve the nonlinear system

$$\frac{x^2}{9} + \frac{y^2}{4} = 1$$
$$x = 2$$

SOLUTION Substitute for x from the second equation into the first equation.

$$\frac{2^2}{9} + \frac{y^2}{4} = 1$$
$$\frac{y^2}{4} = \frac{5}{9}$$

$$y^2 = \frac{20}{9}$$

$$y = \pm\sqrt{\frac{20}{9}} = \pm\sqrt{\frac{4 \times 5}{9}} = \pm\frac{2\sqrt{5}}{3}$$

There are two solutions to this system:

$$x = 2,\ y = \frac{2\sqrt{5}}{3} \qquad \text{and} \qquad x = 2,\ y = -\frac{2\sqrt{5}}{3}$$

There are geometrical interpretations to the solutions of this system of nonlinear equations.

The solutions are the points $A\left(2, \frac{2\sqrt{5}}{3}\right)$, $B\left(2, -\frac{2\sqrt{5}}{3}\right)$ of intersection of the graph of $\frac{x^2}{9} + \frac{y^2}{4} = 1$, an ellipse, and the graph of $x = 2$, a vertical line (Figure 6.11).

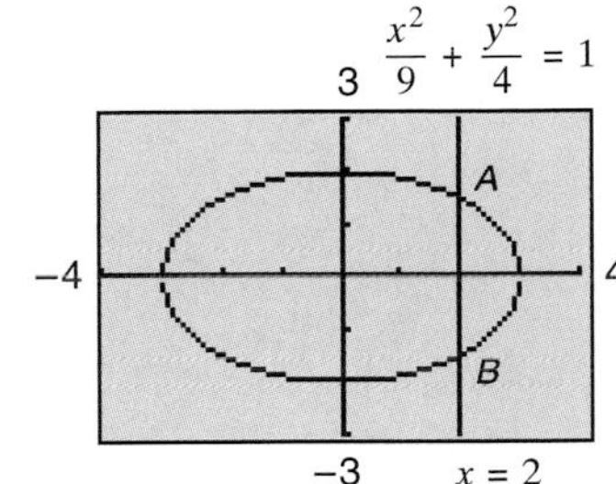

Figure 6.11

Further Ideas 1 Solve the following system of nonlinear equations and verify your answer using graphs.

$$x^2 - 3y^2 = 24$$
$$y = 2$$

Example 2 Solving Using Substitution

Solve the following system of equations and verify the answer using graphs.

$$\frac{xy}{2} - 7x = -6 \qquad (1)$$

$$-2x + y = 4 \qquad (2)$$

SOLUTION Solve equation 2 for y in terms of x.

$$y = 2x + 4 \qquad (3)$$

Substitute this value of y into equation 1

$$\frac{x(2x + 4)}{2} - 7x = -6$$

$$x^2 + 2x - 7x = -6$$

$$x^2 - 5x + 6 = 0$$

Factor the equation

$$(x - 2)(x - 3) = 0$$

Thus, $x = 2$, or $x = 3$. Substitute these values of x into (3) to get the corresponding values of y.

$$x = 2 \quad \text{gives} \quad y = 8$$
$$x = 3 \quad \text{gives} \quad y = 10$$

There are two solutions to the system: (2, 8) and (3, 10). Let us verify these solutions graphically.

Solve each equation for y and plot graphs.

$$y = \frac{2(7x - 6)}{x} \quad \text{and} \quad y = 2x + 4$$

The solutions are the points $A(2, 8)$ and $B(3, 10)$ (Figure 6.12).

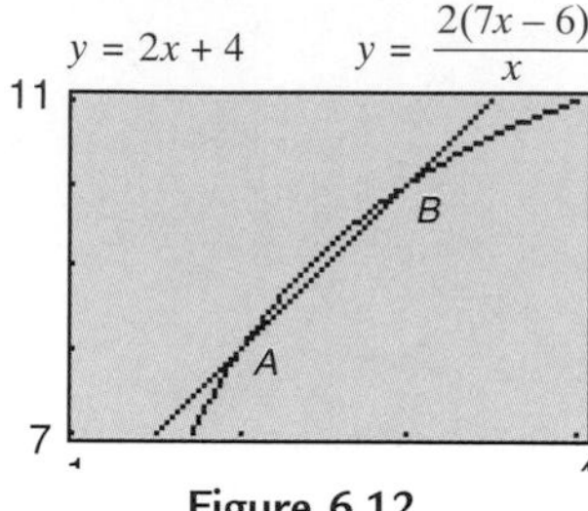

Figure 6.12

Further Ideas 2 Solve the following system of nonlinear equations and verify your answer using graphs.

$$xy - 7y = -20$$
$$3x - y = -2$$

METHOD OF ELIMINATION

As in the case of linear systems of equations, this method involves eliminating one variable by adding a multiple of one equation to the other.

Example 3 Solving Using Elimination

Solve the nonlinear system

$$x^2 + y^2 = 13 \qquad (1)$$
$$3x^2 - y^2 = 23 \qquad (2)$$

SOLUTION Observe that if we add the equations, we can eliminate the variable y. This system lends itself better to an elimination of a variable by adding than it does to the method of substitution.

Add equations 1 and 2.

$$\begin{aligned} x^2 + y^2 &= 13 \\ 3x^2 - y^2 &= 23 \\ \hline 4x^2 \quad &= 36 \\ x^2 &= 9 \\ x &= \pm 3 \end{aligned}$$

There are two possible values of x; namely, $x = -3$ and $x = 3$. Substitute these values of x into equation 1 to get the corresponding values of y.

$$\begin{aligned} (\pm 3)^2 + y^2 &= 13 \\ y^2 &= 4 \\ y &= \pm 2 \end{aligned}$$

Each of these values of y correspond to each value of x. There are thus four solutions: $(-3, -2)$, $(-3, 2)$, $(3, -2)$, and $(3, 2)$. Let us verify these solutions graphically.

Solve each equation for y and plot graphs.

$$y = \pm\sqrt{13 - x^2}$$

and

$$y = \pm\sqrt{3x^2 - 23}$$

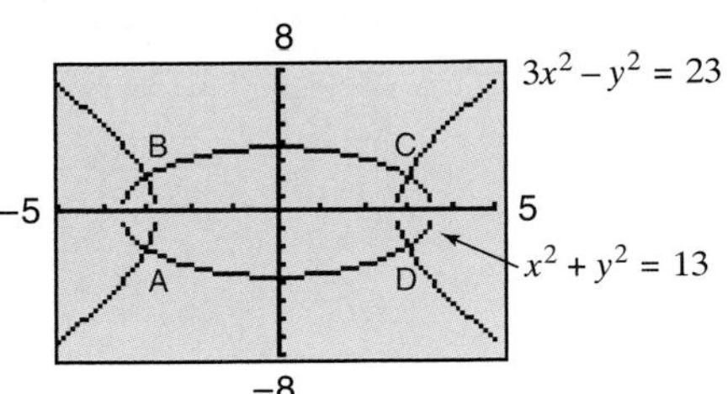

Figure 6.13

The solutions are the points

$A(-3, -2)$, $B(-3, 2)$, $C(3, 2)$, $D(3, -2)$

(Figure 6.13).

Further Ideas 3 Solve the following system of nonlinear equations and verify your answer using graphs.

$$2x^2 - y^2 = -1$$
$$3x^2 - 2y^2 = -6$$

Example 4 Solving Using Elimination

Solve the system

$$x^2 + 2xy + y^2 = 12 \qquad (1)$$
$$-x^2 + xy - y^2 = -3 \qquad (2)$$

SOLUTION Add equations 1 and 2 to eliminate x^2 and y^2.

$$\begin{aligned} x^2 + 2xy + y^2 &= 12 \\ -x^2 + xy - y^2 &= -3 \\ \hline 3xy &= 9 \\ xy &= 3 \end{aligned}$$

Solve this equation for y in terms of x.

$$y = \frac{3}{x} \qquad \text{if } x \neq 0 \qquad (3)$$

Substitute for y into equation 1 to get an equation involving only x.

$$x^2 + 2x\left(\frac{3}{x}\right) + \left(\frac{3}{x}\right)^2 = 12$$

$$x^2 + 6 + \frac{9}{x^2} = 12$$

Multiply both sides by x^2.

$$x^4 + 6x^2 + 9 = 12x^2$$

$$x^4 - 6x^2 + 9 = 0$$

Factor.

$$(x^2 - 3)(x^2 - 3) = 0$$
$$x^2 = 3$$
$$x = \pm\sqrt{3}$$

There are two values of x. The corresponding values of y can be found by substituting these values of x into equation 3.

$$x = -\sqrt{3} \quad \text{gives} \quad y = \frac{3}{x} = -\frac{3}{\sqrt{3}} = -\sqrt{3}$$

$$x = \sqrt{3} \quad \text{gives} \quad y = \frac{3}{x} = \frac{3}{\sqrt{3}} = \sqrt{3}$$

The solutions are $(-\sqrt{3}, -\sqrt{3})$, and $(\sqrt{3}, \sqrt{3})$.

In arriving at these solutions, we made the assumption $x \neq 0$. The possibility of $x = 0$ leading to a solution must be investigated. Let $x = 0$ in the original system. Equation 1 gives $y^2 = 12$, whereas equation 2 gives $-y^2 = -3$. Because y^2 cannot equal both 12 and 3, there is no solution with $x = 0$. We have arrived at the complete set of solutions.

We can only verify solutions graphically if we can plot the equations. This can be done if y can be isolated in the equations. y cannot be easily isolated in these equations; therefore, we do not verify the solutions graphically. The solutions are verified algebraically by substituting into the original equations. The results $x = -\sqrt{3}$, $y = -\sqrt{3}$ and $x = \sqrt{3}$, $y = \sqrt{3}$ both satisfy these equations.

Further Ideas 4 Solve the following system of equations:

$$x^2 - xy + y^2 = 4$$
$$-x^2 + 2xy - y^2 = 0$$

Solutions to Systems of Nonlinear Equations Using Graphs

Graphs can be used to find approximate solutions to nonlinear equations that can be written in the form $y = \ldots$. We illustrate the method with the following example.

Example 5 Solving a System of Nonlinear Equations Using Graphs

Solve the following system of equations to four decimal places using graphs.

$$y = 2x^3 - 6x^2 + 4$$
$$y = x^2 - 7x + 8$$

SOLUTION Graph the equations (Figure 6.14). The point of intersection A of the graphs will be the solution to the system of equations. The solution to four decimal places is $x = 2.3837$ and $y = -3.0038$ using an intersect finder (or trace/zoom).

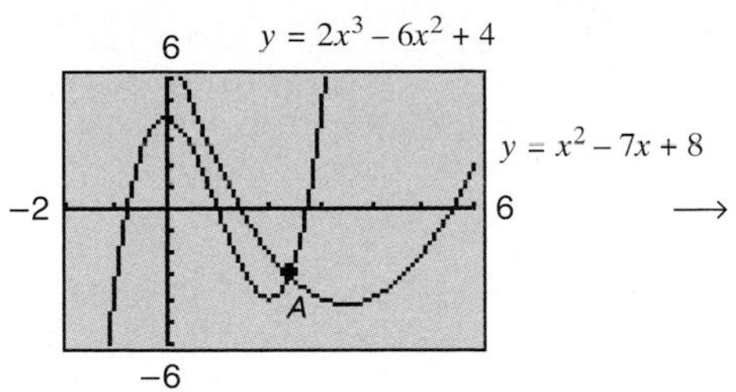

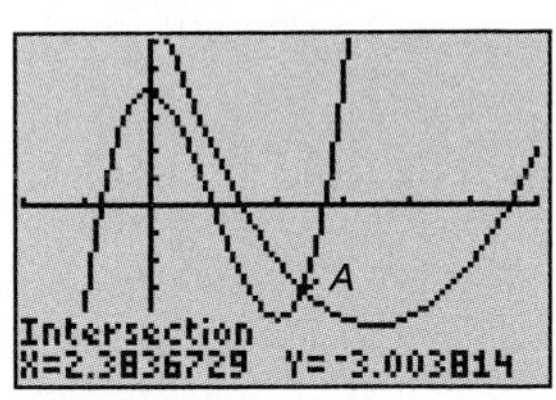

Figure 6.14 The solution, expressed to four decimal places, is $x = 2.3837$, $y = -3.0038$.

Answers to Further Ideas Exercises

1. Solutions are $(-6, 2)$ and $(6, 2)$.
2. Solutions are $(\frac{1}{3}, 3)$ and $(6, 20)$.
3. Solutions are $(-2, -3)$, $(-2, 3)$, $(2, -3)$, and $(2, 3)$.
4. Solutions are $(2, 2)$ and $(-2, -2)$.

EXERCISES 6.4

In Exercises 1–6, solve the systems of nonlinear equations using the method of substitution. Verify your answers using graphs.

1. $\begin{aligned} x^2 + y^2 &= 26 \\ x &= 1 \end{aligned}$
2. $\begin{aligned} x^2 + y^2 &= 13 \\ x &= -3 \end{aligned}$
3. $\begin{aligned} x^2 + 9y^2 &= 97 \\ x &= 4 \end{aligned}$
4. $\begin{aligned} 3x^2 - y^2 &= 5 \\ y &= -2 \end{aligned}$
5. $\begin{aligned} 5x^2 + 2y^2 &= 28 \\ y &= 3 \end{aligned}$
6. $\begin{aligned} 8x^2 - 23y^2 &= 36 \\ x &= 2y \end{aligned}$

In Exercises 7–12, solve the systems of nonlinear equations using the method of substitution. Verify your answers using graphs.

7. $\begin{aligned} 2x^2 + 3y^2 &= 5 \\ x + y &= 0 \end{aligned}$
8. $\begin{aligned} x^2 + y^2 &= 20 \\ x + 2y &= 0 \end{aligned}$
9. $\begin{aligned} -3x^2 - 4x + y^2 &= 26 \\ 2x - y &= -1 \end{aligned}$
10. $\begin{aligned} 10x^2 - 13x - y^2 &= -2 \\ 3x - y &= 2 \end{aligned}$
11. $\begin{aligned} x^2 + 12x + 2y^2 &= 30 \\ x + y &= 3 \end{aligned}$
12. $\begin{aligned} -3x^2 - 4x + y^2 &= 2 \\ 2x + y &= -1 \end{aligned}$

In Exercises 13–18, solve the systems of nonlinear equations using the method of elimination. Verify your answers using graphs.

13. $\begin{aligned} x^2 + y^2 &= 5 \\ 5x^2 - y^2 &= 1 \end{aligned}$
14. $\begin{aligned} x^2 + y^2 &= 10 \\ 3x^2 - y^2 &= -6 \end{aligned}$
15. $\begin{aligned} 2x^2 + 3y^2 &= 12 \\ 3x^2 - y^2 &= -4 \end{aligned}$
16. $\begin{aligned} 2x^2 + y^2 &= 3 \\ -x^2 + 3y^2 &= 2 \end{aligned}$
17. $\begin{aligned} 2x^2 + y^2 &= 33 \\ 3x^2 + 4y^2 &= 52 \end{aligned}$
18. $\begin{aligned} 3x^2 - y^2 &= 3 \\ -x^2 + 4y^2 &= 32 \end{aligned}$

In Exercises 19–24, solve the systems of nonlinear equations using the method of elimination. Verify your answers using graphs.

19. $\begin{aligned} 2x^2 + 3y^2 &= 7 \\ -4x^2 + 2y^2 &= 2 \end{aligned}$
20. $\begin{aligned} 2x^2 - 6y^2 &= 26 \\ 3x^2 - 4y^2 &= 59 \end{aligned}$
21. $\begin{aligned} 2x^2 + 3y^2 &= 32 \\ 3x^2 + 7y^2 &= 48 \end{aligned}$
22. $\begin{aligned} 7x^2 + 2y^2 &= 18 \\ 4x^2 + 5y^2 &= 45 \end{aligned}$
23. $\begin{aligned} 2x^2 + y^2 &= 27 \\ -x^2 + 3y^2 &= 74 \end{aligned}$
24. $\begin{aligned} x^2 - 2y^2 &= 7 \\ 2x^2 + 3y^2 &= 35 \end{aligned}$

In Exercises 25–30, solve the systems of equations by either the method of substitution or the method of elimination. Verify your answers using graphs.

25. $x^2 - y^2 = 7$
$2x - y = 5$

26. $2x^2 + y^2 = 3$
$-x^2 + 3y^2 = 2$

27. $2x^2 + 4x - y^2 = 0$
$-x + y = 2$

28. $2x^2 - y^2 = -8$
$2x + y = 0$

29. $4x^2 + 2y^2 = 22$
$2x^2 - y^2 = -7$

30. $3x^2 - y^2 = -13$
$2x^2 + y^2 = 18$

In Exercises 31–36, solve the systems of equations using algebraic techniques.

31. $3xy - 2x = -2$
$x - 3y = 1$

32. $xy - 5x = 6$
$x - y = -4$

33. $xy/4 - x = 3$
$4x - y = -12$

34. $y^2 + xy = 3$
$x^2 = 4$

35. $y^2 + 3xy = 18$
$y^2 = 9$

36. $y^2x + 4y^4x^2 + x = 42$
$y^2x = 3$

In Exercises 37–44, solve the systems of equations using algebraic techniques (if solutions exist).

37. $x - y = -1$
$xy = 2$

38. $x^2 - 2xy + y^2 = 1$
$-x^2 + 3xy - y^2 = 5$

39. $x^2 - 2xy + y^2 = 1$
$x^2 - 3xy + y^2 = -1$

40. $x^2 + 2y^2 = 6$
$xy = 2$

41. $x^2 + 3xy + (4/9)y^2 = 4$
$xy = 3$

42. $1/x + 2/y = 5$
$-1/x + 3/y = 5$

43. $2/x - 3/y = -1$
$5/x + 2/y = 26$

44. $2/x + 3/y = 7$
$3/x - 1/y = -6$

In Exercises 45–50, find the solutions (if they exist) of the systems of equations using graphs, to four decimal places.

45. $y = x^2 - 3$
$y = 2x + 3$

46. $y = 2x^2 - 7x + 2$
$y = -x + 2$

47. $y = -3x^2 + x + 7$
$y = x + 3$

48. $y = -2x^2 - x + 4$
$y = x^2 - 2x - 7$

49. $y = -x^2 - x + 1$
$y = x^2 - 6x + 8$

50. $y = x^3 + 1$
$y = x^2 + 2x - 4$

In Exercises 51–56, find the solutions (if they exist) of the systems of equations using graphs, to four decimal places.

51. $y = -2x + 1$
$y = e^x + 2$

52. $y = x^2 + 3x + 1$
$y = e^x + 1$

53. $y = 2e^{x-2} - 3$
$y = \ln(2x - 3)$

54. $y = x^2 - 4$
$y = \ln x - 5$

55. $y = x^3 + 2$
$y = -x^2 - 1$

56. $y = x^2 - 3x - 1$
$y = -x^2 + 4x + 5$

6.5 Systems of Linear Inequalities

• Linear Inequalities • The Graph of an Inequality • Systems of Linear Inequalities • Solutions to Systems of Inequalities

We have discussed systems of linear and nonlinear equations. We now turn our attention to a related topic, **systems of linear inequalities.** In this section, we discuss solutions to systems of linear inequalities. In the next section, we shall see how systems of linear inequalities are used to make decisions about how to use limited resources effectively. For example, we will see how a company can decide on a strategy for manufacturing calculators, given certain monetary and time constraints.

Linear Inequalities

An expression such as $x + 2y = 6$ is called a linear equation in the variables x and y. The graph of such an equation is a straight line, representing all the points in the plane that satisfy the equation.

An expression such as $x + 2y \leq 6$ is called a **linear inequality.** It will also have a graph; namely, the set of all points in the plane that satisfy the inequality. For example, consider the point (2, 1). Let us see if this point lies on the graph of $x + 2y \le 6$. Let $x = 2$ and $y = 1$ on the left side of the inequality.

$$\begin{aligned} x + 2y &= 2 + 2(1) \\ &= 4 \le 6 \end{aligned}$$

The point (2, 1) satisfies the inequality and thus lies on the graph of $x + 2y \le 6$.

Consider the point (2, 4). Let $x = 2$ and $y = 4$.

$$\begin{aligned} x + 2y &= 2 + 2(4) \\ &= 10 \not\le 6 \end{aligned}$$

The point (2, 4) does not lie on the graph of $x + 2y \le 6$.

Example 1 **Checking Whether a Point Lies on the Graph of an Inequality**

Determine whether the point (1, 4) lies on the graph of $x - 3y \le 2$.

SOLUTION Let $x = 1$ and $y = 4$ on the left side of the inequality.

$$\begin{aligned} x - 3y &= 1 - 3(4) \\ &= -11 \le 2 \end{aligned}$$

The point (1, 4) lies on the graph of $x - 3y \le 2$.

Self-Check 1 Determine whether the point (−3, 2) lies on the graph of $2x + 6y \le 3$.

The Graph of an Inequality

The graph of an inequality, such as $x + 2y \le 6$, is made up of many points. It consists of all the points on the line $x + 2y = 6$ and all the points on one side of this line. To find the relevant side, select a suitable point not on the line to see whether it satisfies the inequality. (0, 0) is a convenient point to use if the line does not go through the origin.

Let us draw the graph of $x + 2y \le 6$. First draw the line $x + 2y = 6$ (Figure 6.15). Let us determine whether (0, 0) lies on the graph. Let $x = 0$, and $y = 0$. We get

$$\begin{aligned} x + 2y &= 0 + 2(0) \\ &= 0 \le 6 \end{aligned}$$

The point (0, 0) satisfies the inequality and thus lies on the graph of $x + 2y \le 6$.

The graph of $x + 2y \le 6$ consists of all points on the line $x + 2y = 6$ and points on the side of the line containing the origin. Such a region is called a **half-plane.** The graph of $x + 2y \le 6$ is the line and the shaded region in Figure 6.15.

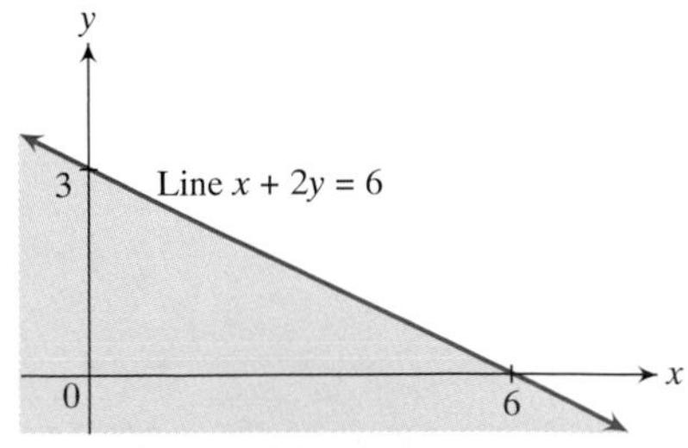

Figure 6.15

Example 2 Graphing an Inequality

Sketch the graph of the inequality $3x - 4y \leq 0$.

SOLUTION Draw the line $3x - 4y = 0$ (see Figure 6.16). Let us select (2, 1) as a convenient point not on the line $3x - 4y = 0$ to determine the half-plane that lies on the graph. Note that the origin cannot be used for this inequality because it lies on the line. Let $x = 2$ and $y = 1$ on the left side of the inequality. We get

$$\begin{aligned} 3x - 4y &= 3(2) - 4(1) \\ &= 2 \not\leq 0 \end{aligned}$$

The point (2, 1) does not satisfy the inequality. It does not lie on the graph. The graph of $3x - 4y \leq 0$ thus contains the half-plane that does not include the point (2, 1). The graph is made up of the line and the shaded region in Figure 6.16.

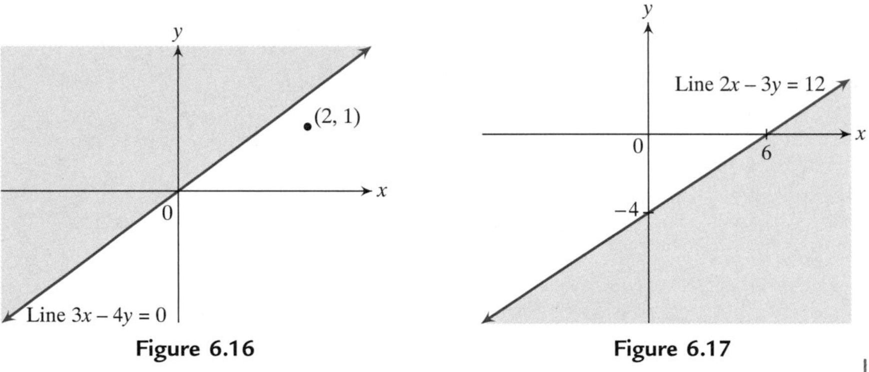

Figure 6.16 Figure 6.17

Self-Check 2 Show that the graph of $2x - 3y \geq 12$ is the shaded region in Figure 6.17.

Example 3 Graphing an Inequality

Sketch the graph of $4x - y > 8$.

SOLUTION Observe that this is a strict inequality. The points on the line $4x - y = 8$ do not lie on the graph. The line will form a boundary to the graph. It is convenient to use a dotted line to indicate that the line is not part of the graph. Sketch the dotted line $4x - y = 8$. Select (0, 0) as the test point. Let $x = 0$ and $y = 0$. We get

$$\begin{aligned} 4x - y &= 4(0) - 0 \\ &= 0 \not> 8 \end{aligned}$$

The origin does not lie on the graph. The graph consists of the half-plane not containing the origin. The graph is the shaded region (Figure 6.18).

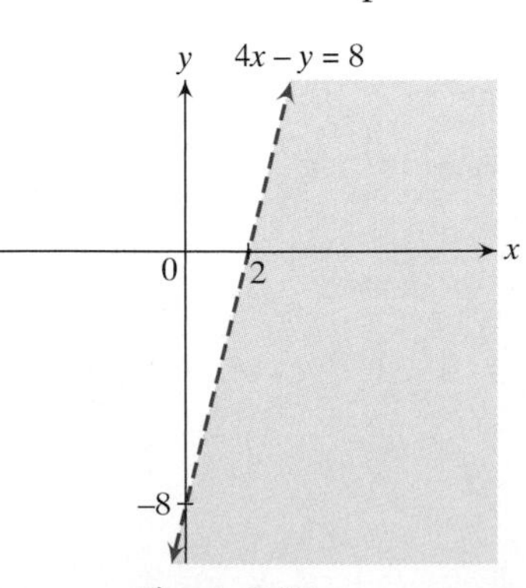

Figure 6.18

Systems of Linear Inequalities

The solution to a system of inequalities consists of the points that satisfy all the inequalities. It is the intersection of the graphs of the inequalities.

Example 4 Solutions to a System of Inequalities

Graph the set of solutions to the following system of inequalities.

$$\begin{aligned} x + y &\le 5 \\ x - 2y &\le -1 \end{aligned}$$

SOLUTION The graph of the set of solutions will be the points that satisfy both inequalities. It can be found by first graphing $x + y \le 5$, and $x - 2y \le -1$. The graph of the solution set will be the set of points common to both graphs.

In Figure 6.19, the graph of $x + y \le 5$ is the region on and below the line. The graph of $x - 2y \le -1$ is on and above the line. (We let you verify this.) The intersection of these two regions is shaded green. This is the graph of the set of solutions to the system of inequalities.

The point A is called a **vertex** of the graph. In applications it is usually necessary to find the vertices. We shall see the significance of a vertex in the following section. A is the point of intersection of the lines $x + y = 5$ and $x - 2y = -1$. A can be found by solving the system

$$\begin{aligned} x + y &= 5 \\ x - 2y &= -1 \end{aligned}$$

A is the point $(3, 2)$.

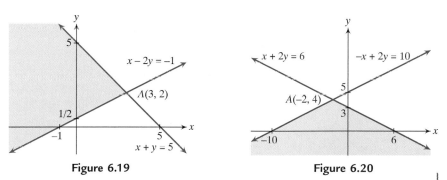

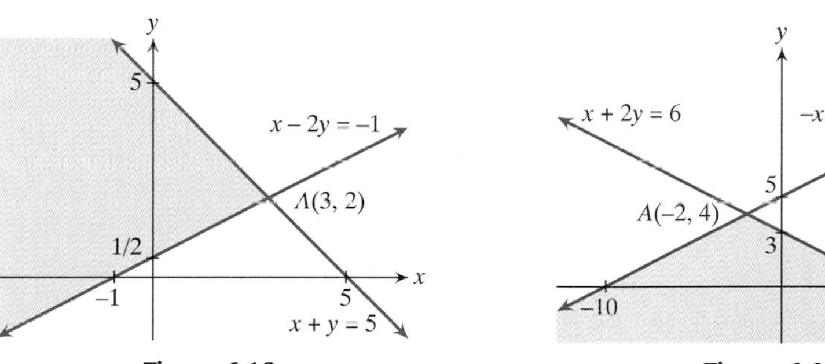

Figure 6.19

Figure 6.20

Self-Check 3 Show that the shaded region and vertex in Figure 6.20 are the solutions and the vertex of the following system of inequalities.

$$\begin{aligned} x + 2y &\le 6 \\ -x + 2y &\le 10 \end{aligned}$$

Example 5 Solutions to a System of Inequalities

Graph the set of solutions to the following system of inequalities. Find the vertices of the graph.

$$\begin{aligned} x - y &\le 2 \\ 2x + y &\le 10 \\ x &\ge 0 \\ y &\ge 0 \end{aligned}$$

SOLUTION The graph of the set of solutions will be the intersection of the graphs of the four separate inequalities.

The graph of $x - y \leq 2$ is found to be the region on and above the line $x - y = 2$.
The graph of $2x + y \leq 10$ is the region on and below the line $2x + y = 10$.
The graph of $x \geq 0$ is the region on and to the right of the y-axis.
The graph of $y \geq 0$ is the region on and above the x-axis.

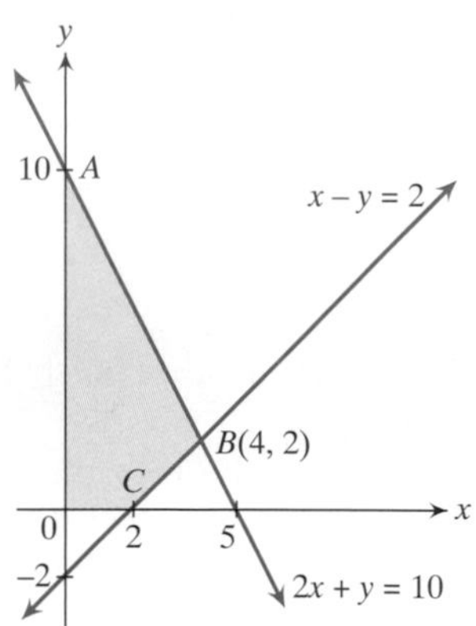

Figure 6.21

The graph of the set of solutions is the shaded region in Figure 6.21. The vertices are the points A, B, C, and O.

A is the y-intercept of $2x + y = 10$; $A(0, 10)$.
B is the intersection of $x - y = 2$ and $2x + y = 10$; $B(4, 2)$.
C is the x-intercept of $x - y = 2$; $C(2, 0)$.
O is the origin, $(0, 0)$.

The vertices of the graph are $A(0, 10)$, $B(4, 2)$, $C(2, 0)$, and $O(0, 0)$.

Self-Check 4 Graph the set of solutions to the following system of inequalities.

$$x + y \leq 5$$
$$2x + y \leq 8$$
$$x \geq 0$$
$$y \geq 0$$

Answers to Self-Check Exercises

1. The point does not lie on the graph.
2. Take a point in the shaded region, say $(6, -4)$. We have $2(6) - 3(-4) = 24 \geq 12$; the inequality is satisfied. Thus, the solution is the shaded region.
3. The origin satisfies both inequalities. The shaded region is the intersection of the two solution sets. The lines intersect at $(-2, 4)$.
4. See Figure 6.22. Vertices are $A(0, 5)$, $B(3, 2)$, $C(4, 0)$, and $O(0, 0)$.

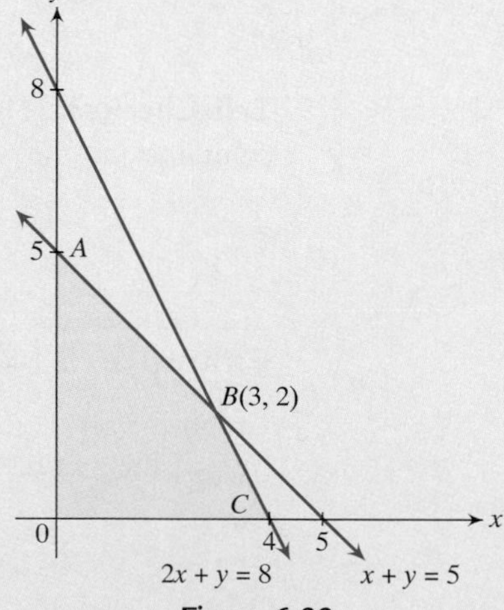

Figure 6.22

EXERCISES 6.5

In Exercises 1–8, determine whether the given points lie on the graphs of the inequalities.

1. $(1, 2)$, $2x + y \le 6$
2. $(3, 1)$, $4x + 3y \le 9$
3. $(-1, 3)$, $2x + 5y \le 17$
4. $(-2, 4)$, $x + y \ge 2$
5. $(7, 1)$, $3x + y \le 22$
6. $(-3, 4)$, $-x + 4y > 20$
7. $(4, 1)$, $2x + y < 12$
8. $(2, -3)$, $6x - 2y < -12$

In Exercises 9–14, determine whether the given points lie on the graphs of the inequalities.

9. $(2, 4)$, $2x + 3y \le 10$
10. $(-3, 1)$, $4x + 2y > -10$
11. $(2, 4)$, $2x + y \ge 6$
12. $(1, 1)$, $-3x + y > -1$
13. $(-2, 3)$, $4x - 6y \le 6$
14. $(1, -1)$, $x - 3y \ge 0$

In Exercises 15–24, sketch the graphs of the inequalities.

15. $x + y \le 3$
16. $2x - y \le 3$
17. $4x - y \le -2$
18. $4x - 2y < 4$
19. $5x + y < 4$
20. $2x + y \ge 3$
21. $-2x + y \ge 3$
22. $-x + 2y < -6$
23. $x \ge 3$
24. $y \le 2$

In Exercises 25–34, sketch the graphs of the inequalities.

25. $x \le -2$
26. $y > 3$
27. $x > -3$
28. $2x + 5y \le 1$
29. $4x + y \le 3$
30. $2x + y \ge 4$
31. $3x - y \le 2$
32. $-2x + y < 3$
33. $3x + 4y < -2$
34. $7x - 3y \ge 6$

In Exercises 35–42, graph the sets of solutions to the systems of inequalities.

35. $2x - y \le 5$, $x + y \le 4$
36. $2x - y \le -5$, $x + y \le 1$
37. $-x + 2y \ge -2$, $6x - 2y \ge -5$
38. $2x + 3y \le 13$, $4x + y \le 8$
39. $3x - y \le -9$, $2x + 4y \le 8$
40. $4x - y \le 7$, $2x + 5y \le -13$
41. $2x - y \le 1$, $9x + 2y \le 16$
42. $x + 2y \le 6$, $2x + y \ge 6$

In Exercises 43–52, graph the sets of solutions to the systems of inequalities.

43. $2x + y \le 3$, $4x + y \le 5$, $x \ge 0$
44. $-x + 2y \le 6$, $2x + y \le 7$, $x \ge 0$
45. $x + 2y \le 6$, $4x - 3y \le 12$, $y \ge 0$
46. $2x + 3y \le 12$, $2x + y \le 5$, $y \ge 0$
47. $x + 3y \le 12$, $2x + y \le 6$, $x \ge 0$, $y \ge 0$
48. $x + 2y \le 8$, $5x + 3y \le 21$, $x \ge 0$, $y \ge 0$
49. $2x + 5y \le 20$, $3x + y \le 10$, $x \ge 0$, $y \ge 0$
50. $3x + y \le 5$, $-2x + 3y \le -1$, $x \ge 0$, $y \ge 0$
51. $2x + 3y \le 12$, $2x - y \le 4$, $x \ge 1$, $y \ge 0$
52. $x + 2y \le 14$, $2x + y \le 9$, $x \ge 1$, $y \ge 2$

In Exercises 53–58, graph the sets of solutions to the systems of inequalities. Find the vertices.

53. $x + 3y \le 7$, $2x - y \le 7$
54. $2x + 5y \le 14$, $-x + 3y \le 4$
55. $x - y \le 2$, $4x + 5y \ge 17$
56. $2x - y \ge -1$, $x + y \ge 4$
57. $5x + 2y \le 16$, $3x + 4y \ge 4$
58. $x + 2y \le 8$, $-x + 3y \ge 2$

In Exercises 59–64, graph the sets of solutions to the systems of inequalities. Find the vertices.

59. $x + 2y \le 9$, $2x + y \le 12$, $x \ge 0$, $y \ge 0$
60. $2x + y \le 9$, $x + 3y \le 17$, $x \ge 0$, $y \ge 0$
61. $3x + y \le 11$, $4x + 2y \le 16$, $x \ge 0$, $y \ge 0$
62. $x + 2y \le 8$, $3x + y \le 14$, $x \ge 1$, $y \ge 0$
63. $2x + y \le 9$, $x + y \le 7$, $x \ge 0$, $y \ge 2$
64. $x + y \le 9$, $2x + y \le 12$, $x \ge 2$, $y \ge 3$

6.6 Linear Programming and Optimal Use of Resources

• Monetary Constraint • Time Constraint • A Linear Programming Problem—Manufacturing Calculators • Linear Programming—General Geometrical Method • Optimizing the Production of Fertilizer

Linear programming is a branch of mathematics that is used to plan the distribution of limited resources. It was first developed during World War II to plan transportation across the Atlantic. It is now used widely in industry and in business.

The constraints that arise from limited resources are described in linear programming by systems of linear inequalities. In applying the method, it becomes necessary to solve systems of linear inequalities. We now give examples to illustrate how monetary and time constraints are described by linear inequalities.

Example 1 Monetary Constraint

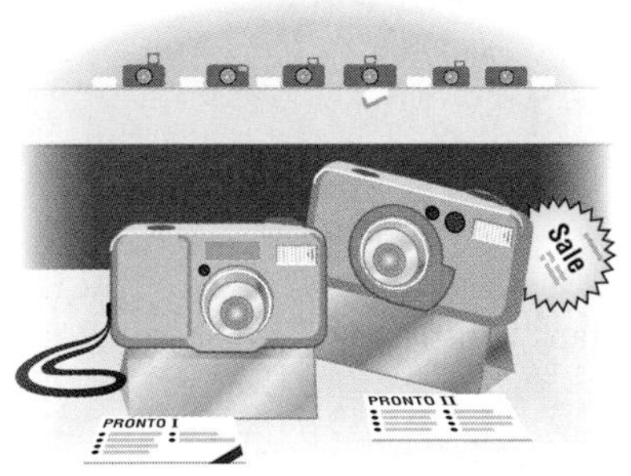

A company manufactures two types of cameras, the Pronto I and the Pronto II. The Pronto I costs \$55 to manufacture, whereas the Pronto II costs \$78. The total funds available for the production of these cameras are \$235,000. We can describe this monetary constraint on the production of cameras by a linear inequality. Let the company manufacture x of Pronto I and y of Pronto II. We get

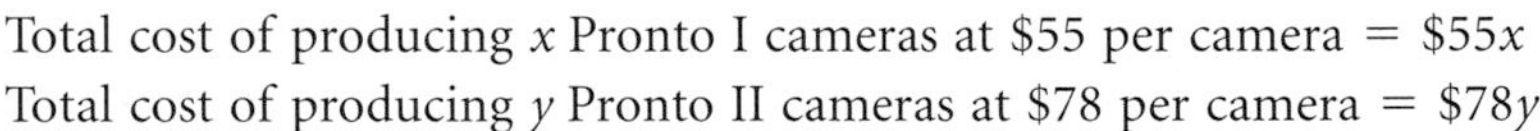

Total cost of producing x Pronto I cameras at \$55 per camera $= \$55x$
Total cost of producing y Pronto II cameras at \$78 per camera $= \$78y$

Thus,

$$\text{Total manufacturing cost} = \$(55x + 78y)$$

Because the funds available are \$235,000, we get the constraint

$$55x + 78y \leq 235{,}000$$

Self-Check 1 A company manufactures two types of office desks, the Officemate and the Executive. The Officemate costs \$195 to manufacture and the Executive costs \$263. The company manufactures x of the Officemate and y of the Executive. The total funds available are \$32,000. Describe this monetary constraint by means of an inequality.

Example 2 Time Constraint

In Example 1, we saw how a monetary constraint was represented by an inequality. Another constraint that often arises is time. Only a certain amount of time is available for production. This example illustrates a time constraint.

A company makes two types of personal computers, the Jupiter and the Cosmos. It takes 27 hours to assemble a Jupiter computer, and it takes 34 hours to assemble a Cosmos computer. The total labor time allotted for this work is 800 hours. Let us describe this time constraint by means of an inequality. Let the company assemble x Jupiter and y Cosmos computers.

Total time to assemble x Jupiters at 27 hours per computer $= 27x$ hours
Total time to assemble y Cosmos at 34 hours per computer $= 34y$ hours

Thus,

$$\text{Total time to assemble computers} = (27x + 34y) \text{ hours}$$

Because the allotted time is 800 hours, we get the constraint

$$27x + 34y \leq 800$$

Let us now use these concepts to solve a linear programming problem.

Example 3 A Linear Programming Problem: Manufacturing Calculators

A company manufactures two types of calculators, model Calc1 and model Calc2. It takes 5 hours and 2 hours to manufacture a Calc1 and a Calc2, respectively. The company has 900 hours available per week for the production of calculators. The manufacturing cost of each Calc1 is \$8 and the manufacturing cost of each Calc2 is \$10. The total funds available per week for production are \$2800. The profit on each Calc1 is \$3, and the profit on each Calc2 is \$2. How many of each type of calculator should be manufactured weekly to obtain maximum profit?

SOLUTION In this problem there are two constraints: time and money. The aim of the company is to maximize profit under these constraints. We first find the inequalities that describe the time and monetary constraints. Let the company manufacture x of Calc1 and y of Calc2.

Time Constraint

Time to manufacture x Calc1 at 5 hours each $= 5x$ hours
Time to manufacture y Calc2 at 2 hours each $= 2y$ hours
Total manufacturing time $= (5x + 2y)$ hours

There are 900 hours available; therefore,

$$5x + 2y \leq 900$$

Monetary Constraint

Total cost of manufacturing x Calc1 at \$8 each $= \$8x$
Total cost of manufacturing y Calc2 at \$10 each $= \$10y$
Total production costs $= \$(8x + 10y)$

There is \$2800 available for production of calculators; therefore,

$$8x + 10y \leq 2800$$

Furthermore, x and y represent numbers of calculators manufactured. These numbers cannot be negative. We therefore get two more constraints:

$$x \geq 0, \qquad y \geq 0$$

The constraints on the production process are thus described by the following system of linear inequalities:

$$\begin{aligned} 5x + \ \ 2y &\leq 900 \\ 8x + 10y &\leq 2800 \\ x &\geq 0 \\ y &\geq 0 \end{aligned}$$

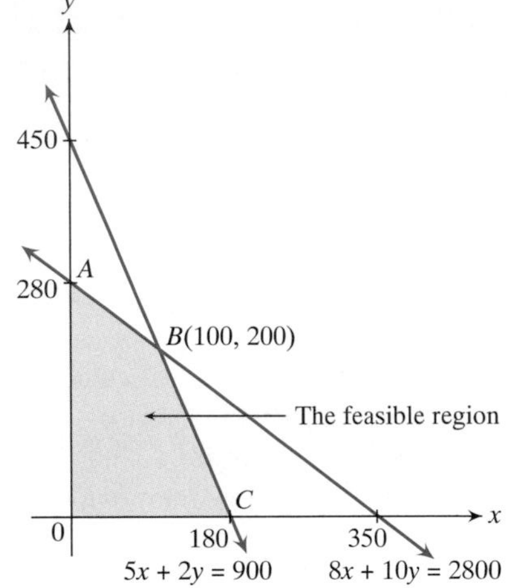

Figure 6.23

The graph of the set of solutions to this system is sketched in Figure 6.23. The vertices are $A(0, 280)$, $B(100, 200)$, $C(180, 0)$, and $O(0, 0)$. (We leave the details of sketching the graph and finding the vertices to the student as these were discussed in Section 6.5.) The graph of the set of solutions is called the **feasible region** of the linear programming problem. Any point in this region satisfies all the inequalities and is thus a possible schedule.

We next find an expression for profit.

Profit

Weekly profit on x Calc1s at \$3 per calculator = \$$3x$
Weekly profit on y Calc2s at \$2 per calculator = \$$2y$
Total weekly profit = \$$(3x + 2y)$

Let us introduce a **profit function:**

$$f = 3x + 2y$$

The problem thus reduces to mathematically finding values of x and y that give the maximum value of f under these constraints. Such a function f is called the **objective function** of the linear programming problem. It would be an endless task to examine the value of the function $f = 3x + 2y$ at all points in the feasible region, looking for the maximum. It has been proved mathematically that the maximum value of f will occur at a vertex of the feasible region; namely, at one of the points A, B, C, or O; or, if there is more than one point at which the maximum occurs, it will be along one edge of the region, such as AB or BC. Hence, we have only to examine the values of $f = 3x + 2y$ at the vertices A, B, C, and O. We get

Vertex	$f = 3x + 2y$
$A(0, 280)$	$f_A = 3(0) + 2(280) = 560$
$B(100, 200)$	$f_B = 3(100) + 2(200) = 700$
$C(180, 0)$	$f_C = 3(180) + 2(0) = 540$
$O(0, 0)$	$f_O = 3(0) + 2(0) = 0$

The maximum value of f is 700, and it occurs at B, namely when $x = 100$ and $y = 200$. Because f is profit, x is the number of Calc1 calculators manufactured, and y is the number of Calc2 calculators manufactured; the maximum weekly profit is \$700, and this occurs when 100 Calc1 calculators and 200 Calc2 calculators are manufactured.

We now summarize the steps involved in solving a linear programming problem.

SOLVING A LINEAR PROGRAMMING PROBLEM

1. Determine the system of linear inequalities that describes the constraints.
2. Sketch the graph of the set of solutions to this system of inequalities.
3. Find the vertices of the graph.
4. Determine the function to be maximized.
5. Find the value of the function at each vertex of the graph.
6. Select the vertex (or edge) at which the function is a maximum.
7. Interpret the result.

In the following example we reinforce the mathematical aspects of solving linear programming problems.

Example 4 Find the maximum value of the function $f = x + 3y$ and the point at which it occurs, subject to the following constraints:

$$\begin{aligned} x + 2y &\leq 8 \\ 4x + 4y &\leq 24 \\ x &\geq 0 \\ y &\geq 0 \end{aligned}$$

SOLUTION The graph of the set of solutions is sketched, and the vertices are found (see Figure 6.24). Compute the value of f at each vertex.

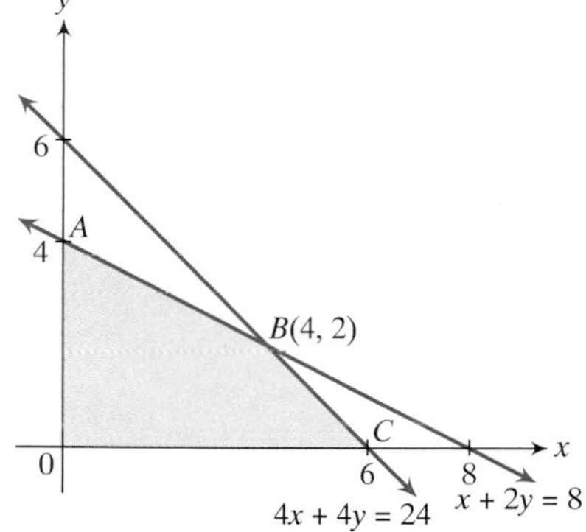

Figure 6.24

Vertex	$f = x + 3y$
$A(0, 4)$	$f_A = 0 + 3(4) = 12$
$B(4, 2)$	$f_B = 4 + 3(2) = 10$
$C(6, 0)$	$f_C = 6 + 3(0) = 6$
$O(0, 0)$	$f_O = 0 + 3(0) = 0$

The maximum value of f is 12 at vertex A, when $x = 0$ and $y = 4$.

Self-Check 2 Find the maximum value of the function $f = 3x + y$ and the point at which it occurs, subject to the constraints

$$\begin{aligned} 5x + 2y &\leq 20 \\ 5x + y &\leq 15 \\ x &\geq 0 \\ y &\geq 0 \end{aligned}$$

Example 5 Optimizing the Production of Fertilizer

The Central Ohio Agricultural Company manufactures two types of fertilizer, Fastgrow and Suregrow, using chemicals A and B. Fastgrow is made up of 10% chemical A and 90% chemical B. Suregrow is made up of 70% chemical A and 30% chemical B. There are 14 tons of chemical A available and 36 tons of chemical B available. What quantities of Fastgrow and Suregrow should be produced to obtain as much fertilizer as possible?

SOLUTION Let the manufacturer produce x tons of Fastgrow and y tons of Suregrow. There are two restrictions in this problem: the amounts of chemicals A and B that are available.

Chemical A

$$\text{Amount of chemical A in Fastgrow} = \frac{10}{100}x \text{ tons}$$

$$\text{Amount of chemical A in Suregrow} = \frac{70}{100}y \text{ tons}$$

Because 14 tons of chemical A are available,

$$\frac{10}{100}x + \frac{70}{100}y \leq 14$$

that is

$$10x + 70y \leq 1400$$
$$x + 7y \leq 140$$

Chemical B

$$\text{Amount of chemical B in Fastgrow} = \frac{90}{100}x \text{ tons}$$

$$\text{Amount of chemical B in Suregrow} = \frac{30}{100}y \text{ tons}$$

Because 36 tons of chemical B are available,

$$\frac{90}{100}x + \frac{30}{100}y \leq 36$$

that is

$$90x + 30y \leq 3600$$
$$3x + y \leq 120$$

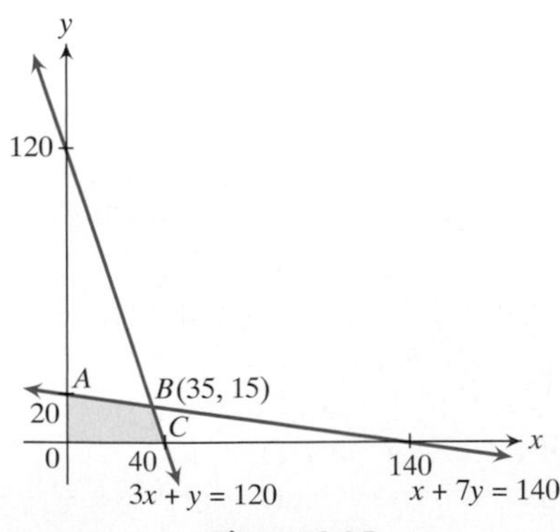

Figure 6.25

Because the amount of fertilizer cannot be negative, we also have the constraints $x \geq 0$ and $y \geq 0$. The constraints on the fertilizer are thus

$$x + 7y \leq 140$$
$$3x + y \leq 120$$
$$x \geq 0$$
$$y \geq 0$$

The graph of this system of inequalities is given in Figure 6.25. The vertices of the graph are found to be $A(0, 20)$, $B(35, 15)$, $C(40, 0)$, and $O(0, 0)$.

The aim of the company is to produce as much fertilizer as possible. The total amount of fertilizer produced is $(x + y)$ tons. Let $f = x + y$. We compute the values of f at the vertices of the graph

Vertex	$f = x + y$
$A(0, 20)$	$f_A = 0 + 20 = 20$
$B(35, 15)$	$f_B = 35 + 15 = 50$
$C(40, 0)$	$f_C = 40 + 0 = 40$
$O(0, 0)$	$f_O = 0 + 0 = 0$

The maximum value of f is 50 at vertex $B(35, 15)$; therefore, $x = 35$ and $y = 15$. The manufacturer should produce 35 tons of Fastgrow and 15 tons of Suregrow to obtain as much fertilizer as possible.

Answers to Self-Check Exercises

1. Constraint is $195x + 263y \le 32{,}000$.
2. Maximum of 11 at $(2, 5)$.

EXERCISES 6.6

In Exercises 1–6, find the maximum value of the function and the point at which it occurs, subject to the given constraints.

1. $f = 5x + 2y$
   $2x + y \le 11$
   $3x + y \le 15$
   $x \ge 0, y \ge 0$
2. $f = 2x + y$
   $x + y \le 5$
   $2x + 3y \le 13$
   $x \ge 0, y \ge 0$
3. $f = 4x + y$
   $2x + y \le 4$
   $6x + y \le 8$
   $x \ge 0, y \ge 0$
4. $f = 2x + y$
   $4x + y \le 16$
   $x + y \le 7$
   $x \ge 0, y \ge 0$
5. $f = 4x + 3y$
   $x + 3y \le 15$
   $2x + y \le 10$
   $x \ge 0, y \ge 0$
6. $f = 2x + 3y$
   $2x + 4y \le 16$
   $3x + 2y \le 12$
   $x \ge 0, y \ge 0$

In Exercises 7–12, find the maximum value of the function and the point at which it occurs, subject to the given constraints.

7. $f = 4x - y$
   $x + 2y \le 8$
   $3x + 2y \le 16$
   $x \ge 0, y \ge 0$
8. $f = x + 5y$
   $4x + y \le 12$
   $x + y \le 6$
   $x \ge 0, y \ge 0$
9. $f = 3x + 3y$
   $x + y \le 4$
   $3x + y \le 6$
   $x \ge 0, y \ge 0$
10. $f = x + 2y$
    $2x + y \le 6$
    $x + 3y \le 8$
    $x \ge 0, y \ge 0$
11. $f = 2x + 6y$
    $3x + 2y \le 16$
    $x + 2y \le 11$
    $x \ge 2, y \ge 1$
12. $f = x + y$
    $2x + y \le 10$
    $3x + 4y \le 25$
    $0 \le x \le 4, 0 \le y \le 5$

13. **Manufacturing Calculators** The Kansas Office Supply manufactures two types of calculators, model KC-1 and model KC-2. It takes 1 hour and 4 hours in labor time to manufacture each KC-1 and KC-2, respectively. The cost of manufacturing the KC-1 is \$30 and that of manufacturing a KC-2 is \$20. The company has 1600 hours of labor time available and \$18,000 in running costs. The profit on each KC-1 is \$10 and on each KC-2 is \$8. What should the production schedule be to ensure maximum profit?
14. **Manufacturing Fans** The Corfu Company manufactures two types of electric fans, the Cooler and the Comforter. Two machines, I and II, are needed to manufacture each product. It takes 3 minutes on

each machine to produce a Cooler. Producing a Comforter takes 1 minute on machine I and 2 minutes on machine II. The total time available on machine I is 3000 minutes, and on machine II the time available is 4500 minutes. The company realizes a profit of \$15 on each Cooler and \$7 on each Comforter. How many of each type of fan should the company manufacture to obtain the largest profit?

15. **Transporting Refrigerators** The Widmaer Refrigerator Company has two plants in towns Chester and Crewe. Its refrigerators are sold in a certain town, Flint. It takes 20 hours (packing, transportation, and so on) to transport a refrigerator from Chester to Flint and 10 hours to move it from Crewe to Flint. It costs \$60 to transport each refrigerator from Chester to Flint and \$10 per refrigerator for delivery from Crewe to Flint. There is a total of 1200 hours of labor time available, and a total of \$2400 is budgeted for transportation. The profit on each refrigerator from Chester is \$40, and the profit on each refrigerator from Crewe is \$20. How many refrigerators should the company transport from Chester and how many from Crewe to maximize profit?

16. **Oil Production** The maximum daily production of the Ponca City Oil Refinery is 1400 barrels. The refinery can produce two types of oil: gasoline for automobiles and heating oil for domestic purposes. The production costs per barrel are \$6 for gasoline and \$8 for heating oil. The daily production budget is \$9600. The profit per barrel is \$3.50 on gasoline and \$4 on heating oil. What is the maximum profit that can be realized daily, and what quantities of each type of oil are then produced?

17. **Tailoring** A tailor has 80 square yards of cotton material and 120 square yards of woolen material. A suit requires 2 square yards of cotton and 1 square yard of wool. A dress requires 1 square yard of cotton and 3 square yards of wool. How many of each garment should the tailor make to maximize income, if a suit and a dress each sell for \$90? What is the maximum income?

18. **Crop Allocation** A farmer has to decide how many acres of a 40-acre plot are to be devoted to growing strawberries and how many to growing tomatoes. It takes 8 hours to pick an acre of strawberries and 6 hours to pick an acre of tomatoes. There will be 300 hours of labor available for the picking. The profit per acre is \$700 for the strawberries and \$600 for the tomatoes. How many acres of each should be grown to maximize profit?

19. **Purchasing School Buses** The Volusia County School District is buying new buses. It has a choice of two types. The Torro costs \$18,000 and holds 25 passengers. The Sprite costs \$22,000 and holds 30 passengers. An amount of \$572,000 has been budgeted for the new buses. A maximum of 30 drivers will be available to drive the buses. At least 17 Sprite buses must be ordered because of the desirability of having a certain number of large-capacity buses. How many of each type should be purchased to carry a maximum number of students?

20. **Producing Fertilizer** Scott's makes two types of fertilizer, Fastgrow and Nourisher, using chemicals chA and chB. Fastgrow is made up of 80% chemical chA and 20% chemical chB. Nourisher is made up of 60% chemical chA and 40% chemical chB. The manufacturer requires at least 30 tons of Fastgrow and at least 50 tons of Nourisher. There are available 100 tons of chemical chA and 50 tons of chemical chB. If Scott's wants to make as much fertilizer as possible, what quantities of Fastgrow and Nourisher should be produced?

21. **Constructing a Menu** West Essex Hospital wants to design a dinner menu containing two items, M and N. Each ounce of M provides 1 unit of vitamin A and 2 units of vitamin B. Each ounce of N provides 1 unit of vitamin A and 1 unit of vitamin B. The two dishes must provide at least 7 units of vitamin A and at least 10 units of vitamin B. If each ounce of M costs 8 cents and each ounce of N costs 12 cents, how many ounces of each item should the hospital serve to minimize cost? *Hint:* The minimum value of the objective function will also occur at the vertex, or along the side of the feasible region.

22. **Clothes Manufacturing** Tom Underwood has 10 square yards of cotton, 10 square yards of wool, and 6 square yards of silk. A pair of slacks requires 1 square yard of cotton, 2 square yards of wool, and 1 square yard of silk. A skirt requires 2 square yards of cotton, 1 square yard of wool, and 1 square yard of silk. The net profit on a pair of slacks is \$3 and the net profit on a skirt is \$4. How many skirts and how many slacks should be made to maximize profit?

23. **Shipping** (**a**) Makai Shipping has trucks that can carry a maximum of 12,000 pounds of cargo with a maximum volume of 9000 cubic feet. They ship for two companies: Pringle Company packages weigh 5 pounds each and have a volume of 5 cubic feet, and Williams Company packages weigh 6 pounds, with a volume of 3 cubic feet. By contract, the shipper makes a profit of \$3 on each package from Pringle and \$4 on each package from Williams. How many packages from each company should the shipper carry? (**b**) A lawyer points out the fine print at the bottom of the Pringle contract, which says that the shipper must carry at least 240 packages from Pringle. How should the shipper now divide the work? How much did the clause cost the shipper in profit?

CHAPTER 6 PROJECT | Linear Programming

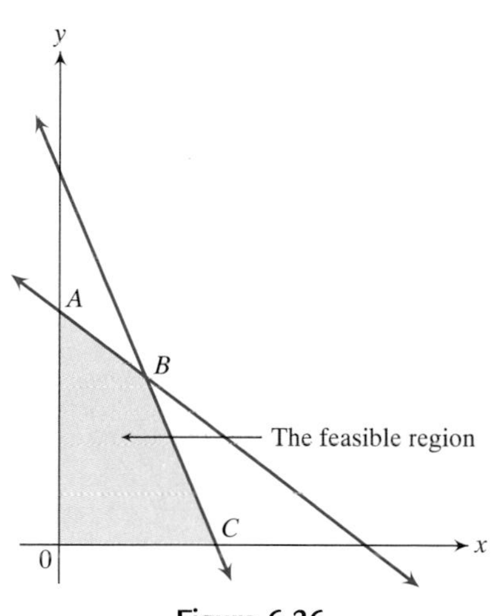

Figure 6.26

We have applied the method of linear programming to a number of situations in this chapter. In this project, the student will discover why the method of linear programming works. Recall that the region of points that satisfies the system of inequalities in a linear programming is called the feasible region. A typical feasible region *ABCO* is shown in Figure 6.26. The linear programming problems that we have discussed involve finding the maximum value of a function *f*, called the objective function, over this feasible region. It would be an endless task to examine the value of the function at all points in the feasible region, looking for the maximum. We have used the result that the maximum value of *f* will occur at a vertex of the feasible region, namely at one of the points *A*, *B*, *C*, or *O*; or if there is more than one point at which the maximum occurs, it will be along one edge of the region, such as *AB* or *BC*. Use a geometrical argument to show why the maximum occurs at a vertex or an edge in this manner. *Hint:* You may look up this explanation in a book that discusses linear programming. However, present the argument in your own words, using notation that we have used in this course.

WEB PROJECT 6

To explore the concepts of this chapter further, try the interesting WebProject. Each activity relates the ideas and skills you have learned in this chapter to a real-world applied situation. Visit the Williams Web site at

www.saunderscollege.com/math/williams

to find the complete WebProject, related questions, and live links to the Internet.

Chapter 6 Highlights

Section 6.1

Consistent System of Linear Equations: Has a unique or many solutions.
Inconsistent System of Linear Equations: Has no solutions.
Method of Substitution: Solve for one variable in terms of the other. Use substitution to then get the solution.
Method of Elimination: Eliminate one variable to get the value of other variable. Substitute to get complete solution.

Section 6.2

Gauss–Jordan elimination: A systematic method of solving systems of linear equations.
Elementary Transformations: (1) Interchange the positions of two equations. (2) Multiply both sides of an equation by a nonzero constant. (3) Add a multiple of one equation to another equation.
Equivalent Systems: Systems of linear equations that are related through elementary transformations. They have the same solution(s).
Matrix: A rectangular array of numbers. The numbers in the array are called the elements of the matrix.
Matrix of Coefficients: A matrix made up of the coefficients in a system of linear equations.
Augmented Matrix: A matrix made up of the coefficients together with the constant terms of a system of linear equations. This matrix describes the system.
Elementary Row Operations on Matrices: (1) Interchange two rows. (2) Multiply the elements of a row by a nonzero constant. (3) Add a multiple of the elements of one row to the corresponding elements of another row.
Two Matrices Are Row Equivalent: One can be obtained from the other by using a sequence of elementary row operations.

Section 6.3

Matrix in Reduced Form: (1) Any rows with all zeros are grouped at the bottom of the matrix. (2) The first nonzero element of each other row is 1. (called a **leading one.**) (3) The leading one of each row after the first is positioned to the right of the leading one of the previous row. (4) All other elements in a column that contains a leading one are zero. The reduced form of a matrix is unique.
Gauss–Jordan Elimination: (1) Write down the augmented matrix of the system of linear equations. (2) Derive the reduced form of the augmented matrix using elementary row operations. This is done by creating leading ones, then zeros above and below each leading one, column by column starting with the first column. (3) Write down the system of equations corresponding to the reduced echelon form. This system gives the solution.
General Solution: Final form of expressing many solutions.
Kirchhoff's Laws: Laws for flow of current through a network.

1. *Junctions.* All the current flowing into a junction must flow out of it.
2. *Paths.* The sum of the *IR* terms (*I* denotes current and *R* denotes resistance) in any direction around a closed path is equal to the total voltage in the path in that direction.

Section 6.4

System of Nonlinear Equations: System of equations in which at least one of the equations is not of degree 1. Can be solved using substitution or elimination or by finding points of intersection of the graphs using a calculator.

Section 6.5

System of Linear Inequalities: Each expression in the system is a linear inequality. The solution to each inequality will be all points on one side of a line (a half-plane). The solution to the system will be the intersection of these regions.

Section 6.6

Linear Programming: A branch of mathematics that is used to plan the distribution of limited resources. The limitation in resources is described by a system of linear inequalities.

Solving a Linear Programming Problem:

1. Determine the system of linear inequalities that describes the constraints.
2. Sketch the graph of the solutions to the system of inequalities (feasible region).
3. Find the vertices of the set of the graph.
4. Determine the function to be maximized (objective function).
5. Find the value of the function at each vertex of the graph.
6. Select the vertex (or edge) where the function is a maximum.
7. Interpret the result.

CHAPTER 6 REVIEW EXERCISES

1. Solve the following systems of equations using the method of substitution, and verify the answers using graphs.

(a) $x - 2y = 3$
$3x + 5y = 20$

(b) $3x - 2y = 12$
$2x + y = 1$

(c) $-x + 3y = 1$
$4x - 5y = 10$

2. Solve the following systems using the method of elimination, and verify the answers using graphs.

(a) $x + 2y = 8$
$3x + 5y = 20$

(b) $3x - 4y = -13$
$2x + y = -5$

(c) $5x + y = 14$
$3x - 4y = -10$

3. Construct a system of two linear equations in two variables that has the unique solution $(2, -5)$.

4. Construct a system of two linear equations in two variables that has many solutions of the form $(x, 3x - 7)$.

5. Find the equations of the lines passing through the following pairs of points.

(a) $(1, 3), (3, -1)$ (b) $(4, 2), (3, 0)$

(c) $(-3, 6), (2, -5)$

6. A man has \$1.55 in his pocket, made up of quarters and dimes. If he has a total of eight coins how many of each type of coin does he have?

7. A college book store buys 400 mathematics books and 300 history books for a total cost of \$25,300. It later buys 20 mathematics books and 25 history books for a total cost of \$1615. Determine the price of each mathematics and each history book.

8. Solve the following systems of equations using the method of Gauss–Jordan elimination.

(a) $\begin{aligned} x + y + 3z &= 4 \\ x + 2y + 4z &= 7 \\ 2x + y + 4z &= 8 \end{aligned}$ **(b)** $\begin{aligned} 2x + 4y + 14z &= -4 \\ 2x + 5y + 17z &= -5 \\ x - y &= -1 \end{aligned}$

(c) $\begin{aligned} x + 3y - z &= 4 \\ 2x + 5y + 4z &= 3 \\ 3x + 8y + 3z &= 5 \end{aligned}$

9. Find the equations of the parabolas through the following points. Verify your answer using a graph.

(a) $(1, 0), (2, 7), (-1, -2)$

(b) $(-2, 5), (0, 1), (3, 40)$

10. Solve the following systems of equations using the method of Gauss–Jordan elimination.

(a) $\begin{aligned} x + y + 3z &= 4 \\ x + 2y + 4z &= 7 \\ 2x + y + 4z &= 8 \end{aligned}$

(b) $\begin{aligned} 2x + 4y + 14z &= -4 \\ 2x + 5y + 17z &= -5 \\ x - y &= -1 \end{aligned}$

(c) $\begin{aligned} x - y + z &= -4 \\ -x + 2y + z &= 7 \\ x + y + 5z &= 2 \end{aligned}$

(d) $\begin{aligned} x - 2y - 4z &= 7 \\ x - y - z &= 5 \\ 3x - 4y - 6z &= 18 \end{aligned}$

(e) $\begin{aligned} x - y - 3z &= 2 \\ 2x - y - 2z &= 5 \\ x + y + 5z &= 4 \end{aligned}$

(f) $\begin{aligned} x + y + 6z &= 4 \\ -x + y + 2z &= -2 \\ 3x + 4y + 22z &= 14 \end{aligned}$

11. Solve (if possible) the following systems of nonlinear equations. Verify your answers (if possible) using graphs.

(a) $\begin{aligned} x^2 - y^2 &= 15 \\ x + y &= 5 \end{aligned}$ **(b)** $\begin{aligned} 2x^2 + y^2 &= 9 \\ 2x - y &= 3 \end{aligned}$

(c) $\begin{aligned} 2x^2 + 3y^2 &= 35 \\ x^2 + y^2 &= 13 \end{aligned}$ **(d)** $\begin{aligned} 2x^2 + y^2 &= 8 \\ x + y &= 4 \end{aligned}$

(e) $\begin{aligned} y^2 + 2xy &= 33 \\ y^2 &= 9 \end{aligned}$ **(f)** $\begin{aligned} \frac{1}{x} + \frac{5}{y} &= \frac{3}{2} \\ \frac{2}{x} - \frac{1}{y} &= \frac{4}{5} \end{aligned}$

12. Find solutions of the following systems of equations to two decimal places using graphs. Use an intersect finder or trace/zoom.

(a) $\begin{aligned} y &= x^2 + 1 \\ y &= x + 2 \end{aligned}$ **(b)** $\begin{aligned} y &= x^2 + 2x + 3 \\ y &= -x^3 - 3 \end{aligned}$

(c) $\begin{aligned} y &= e^{x-1} + 1 \\ y &= \ln(x - 1) + 4 \end{aligned}$

13. Determine whether the following points lie on the graphs of the inequalities.

(a) $(1, 5), 3x + y \le 7$

(b) $(-2, 3), 5x + 2y \le 9$

(c) $(4, 7), -2x + 7y \ge 6$

(d) $(0, 0), x + y \le 4$

(e) $(2, -1), 3x - 2y \ge 10$

(f) $(4, -3), 3x + 5y \le -4$

14. Sketch the graphs of the following inequalities.

(a) $x + 2y \le 4$ **(b)** $3x - y \le 4$

(c) $-4x + 5y \le 3$ **(d)** $7x + 12y \ge 28$

(e) $x \ge 2$ **(f)** $y \ge -3$

(g) $7x - 3y \ge 5$ **(h)** $3x - 6y \ge 3$

15. Graph the sets of solutions to the following systems of inequalities. Find the vertices of the graphs.

(a) $\begin{aligned} x + 4y &\le 8 \\ -x + 2y &\le -2 \end{aligned}$ **(b)** $\begin{aligned} 2x + 3y &\le 11 \\ 4x - 7y &\ge -43 \end{aligned}$

(c) $\begin{aligned} x + y &\le 7 \\ 3x - 2y &\le 1 \end{aligned}$ **(d)** $\begin{aligned} 3x + 2y &\le 13 \\ x + y &\le 5 \\ x \ge 0, y &\ge 0 \end{aligned}$

(e) $\begin{aligned} 2x + 3y &\le 14 \\ 4x + y &\le 8 \\ x \ge 0, y &\ge 0 \end{aligned}$ **(f)** $\begin{aligned} 4x + 2y &\le 16 \\ 3x + y &\le 10 \\ x \ge 1, y &\ge 2 \end{aligned}$

16. Find the maximum values of the following functions and the points at which those maxima occur, under the given constraints.

(a) $\begin{aligned} x + 2y &\le 12, f = x + 4y \\ 3x + y &\le 11 \\ x \ge 0, y &\ge 0 \end{aligned}$

(b) $\begin{aligned} 2x + 3y &\le 12, f = 4x + 3y \\ 3x + y &\le 11 \\ x \ge 0, y &\ge 0 \end{aligned}$

(c) $\begin{aligned} x + 2y &\le 8, f = 2x + 4y \\ 2x + y &\le 10 \\ x \ge 0, y &\ge 0 \end{aligned}$

(d) $\begin{aligned} x + 3y &\le 19, f = 4x + 3y \\ 4x + y &\le 21 \\ x \ge 1, y &\ge 2 \end{aligned}$

17. Production of Television Sets Royal Electronics manufactures two types of television sets, the Panorama and the Vision. It takes 4 hours and 1 hour in labor time to manufacture the Panorama and the Vision, respectively. The cost of manufacturing a Panorama set is \$200, whereas that of manufacturing a Vision set is \$300. The company has 2200 hours of labor and \$150,000 in manufacturing costs available each week. The profit on each Panorama is $40 and on each Vision it is $35. How many of each type of television set should be produced each week to ensure maximum profit?

18. **Furniture Manufacturing** The Bee-Line Manufacturing Company makes cabinets and tables. The material for each cabinet costs $20, and the material for each table costs $30. The funds available daily for the production are $1900. It takes 1 hour to assemble a cabinet and 2 hours to assemble a table. There are 110 hours of labor available daily. The profit on each cabinet is $7, and the profit on each table is $12. How many cabinets and how many tables should be manufactured daily to achieve maximum profit? What is the maximum possible daily profit?

CHAPTER 6 TEST

1. Solve the following system of equations using the method of substitution, and verify the answers using a graph.

$$x + 3y = 14$$
$$2x + y = 8$$

2. Solve the following system of equations using the method of elimination, and verify the answers using a graph.

$$x + 2y = 1$$
$$2x + 3y = 3$$

3. Use a system of two equations to find the equation of the line passing through the points (1, 4) and (2, 7).

4. Solve the following system of equations using the method of Gauss–Jordan elimination.

$$x + y + z = 2$$
$$2x + y - z = 5$$
$$x + 3y + 4z = 3$$

5. Find the equation of the parabola through the points (1, 4), (3, 14), and (−1, 10). Verify your answer using a graph.

6. Solve the following system of nonlinear equations. Verify your answer using a graph.

$$2x^2 - y^2 = -8$$
$$x + y = 2$$

7. Determine whether the following points lie on the graphs of the inequalities.
   (a) $(2, 4)$, $3x + 2y \le 16$
   (b) $(3, -4)$, $-2x + y \le -11$

8. Graph the sets of solutions to the following system of inequalities. Find the vertices of the graph. Find the maximum value of $f = 2x + 3y$ under these constraints.

$$x + 3y \le 9$$
$$2x - y \le 4$$
$$x \ge 0, y \ge 0$$

9. **Manufacturing Bicycles** A company manufactures two types of bicycles, the Cruiser and the Jet. It takes 5 hours and 4 hours in labor time to manufacture the Cruiser and the Jet, respectively. The cost of manufacturing a Cruiser is $600, whereas that of manufacturing a Jet is $800. The company has 3620 hours of labor and $556,000 in manufacturing costs available each week. The profit on each Cruiser is $70, and on each Jet it is $65. How many Cruisers and Jets should be produced each week to ensure maximum profit?

10. **Purchasing Textbooks** A college book store buys 600 English books and 320 mathematics books for a total cost of $36,240. It later buys 50 English books and 20 mathematics books for a total cost of $2740. Determine the price of each English and each mathematics book.

7 Matrices

Gabriel Cramer (1704–1752)
(The Granger Collection, New York)

A matrix is a rectangular array of numbers. Matrices are useful tools in applications. For example, origin–destination matrices are used by highway departments to analyze traffic patterns. Einstein used a matrix to describe the force of gravity in his Theory of General Relativity.

One of the contributors to our knowledge of matrices was Gabriel Cramer (1704–1752). Cramer was educated and lived in Geneva, Switzerland, and traveled widely in Europe. He contributed to the fields of geometry and probability and was described as "a great poser of stimulating problems." He is best known for Cramer's rule for solving systems of linear equations. We introduce this rule in Section 7.4; however, it is not his original work! Instead, he explained it and encouraged its use. His greatest contribution was probably in the thankless (but important) task of editing and distributing the works of others. Cramer's interests were broad, being involved in civic life, the restoration of cathedrals, and excavations. He was said to be "friendly, good-humored, pleasant in voice and appearance, possessing a good memory, judgment, and health."

In this chapter, we develop a theory of matrices motivated by the structure that we have for real numbers. For example, because we can add and multiply real numbers, we introduce operations of addition and multiplication for matrices. We shall see how data are correlated and manipulated (often on computers) using matrices. As an application, we look at how doctors' offices can use matrices in their accounting departments. Matrix multiplication is used to develop a model that describes population movement between suburbs and the centers of American cities.

APPLICATION | Group Relationships in Sociology

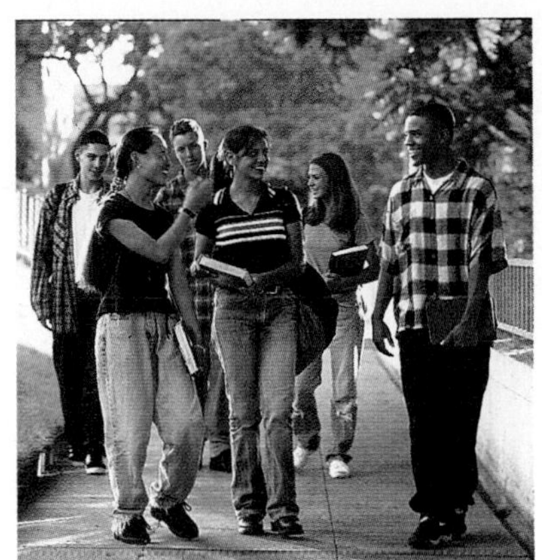

Social interaction among students.
(© David Young Wolff/Tony Stone)

Sociologists use matrices to analyze interaction within groups of people. Consider a group of five people. A sociologist might be interested in finding out which one of the five has the most influence over the other members. The group is asked to fill out a short questionnaire, and then the researchers tabulate the answers. We label the group members $M_1, M_2, \ldots, M_5$ and represent the results in the form of the graph shown in Figure 7.1. The ensuing analysis of this graph using a distance matrix leads to the conclusion that M_3 is the most influential person in the group, followed by M_5, M_2, and M_1. If you want to start a rumor you should tell it to M_3! *(This application is discussed in Example 11 on pages 486–487.)*

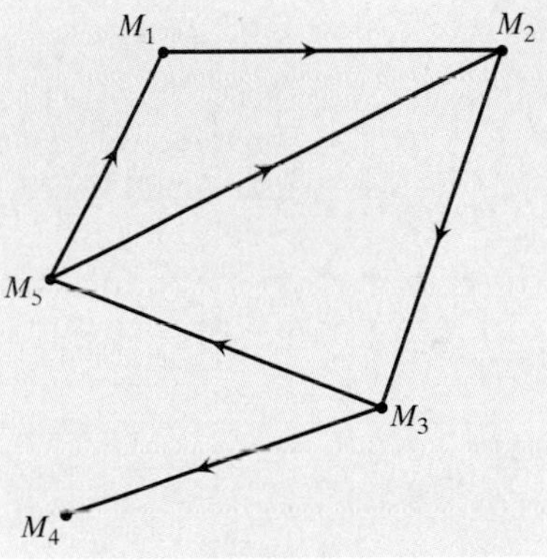

Figure 7.1

7.1 Matrices and Group Relationships in Sociology

- Matrix Concepts • Addition and Subtraction • Scalar Multiplication
- Origin–Destination Matrix • Medical Accounting • Trade Figures
- Group Relationships in Sociology

We remind the student that a **matrix** is a rectangular array of numbers. The numbers in the array are called the **elements** of the matrix. We now introduce some of the ways of interpreting matrices by means of an application.

Origin–Destination Matrix

Highway departments use matrices to analyze traffic patterns. Such matrices are called origin–destination matrices (O–D matrices). An O–D matrix may be used, for example, to estimate the likely impact of a new shopping area. A region of interest is divided into zones, and an estimate of daily traffic between zones is made.

The following matrix represents daily traffic in a small city that is divided into four zones: 1, 2, 3, and 4.

		Destination Zone			
		1	2	3	4
	1	1500	1200	300	400
Origin	2	1000	5000	700	550
Zone	3	340	900	4000	2500
	4	750	550	2750	1400

The interpretation of the elements is as follows. Consider the 700 in row 2, column 3 (blue). It indicates that the daily traffic from zone 2 to zone 3 is 700 vehicles. The largest element in the matrix is 5000 in row 2, column 2. This tells us that the largest traffic flow is the internal flow of 5000 vehicles per day in zone 2.

Rows, Columns, and Size of a Matrix

This example illustrates the usefulness of interpreting matrices in terms of rows and columns. It is very natural to label rows from top to bottom and columns from left to right. The location of each element in the matrix is given by the row and column in which it lies. We now formally define these important matrix ideas.

Each matrix has a certain number of rows and a certain number of columns. For example, the following matrix has two rows and three columns.

Three columns

$$\begin{array}{lc} & \downarrow \quad \downarrow \quad \downarrow \\ \text{Two} \rightarrow & \begin{bmatrix} 5 & 6 & 2 \\ 9 & 3 & 1 \end{bmatrix} \\ \text{rows} \rightarrow & \end{array}$$

The **size** of a matrix is given in terms of the number of rows and the number of columns in the matrix. The size of the preceding matrix is said to be 2×3 and is read "two by three." The number of rows is given first, followed by the number of columns.

When the number of rows is equal to the number of columns, the matrix is said to be a **square matrix.** If a matrix contains only one row, it is said to be a **row matrix.** If a matrix contains only one column, it is said to be a **column matrix.** The following are examples of various types of matrices.

$$\begin{bmatrix} 3 & 1 & 0 & 7 \\ -5 & -3 & 9 & 8 \\ 1 & -4 & 2 & -7 \end{bmatrix} \quad \begin{bmatrix} 4 & -2 & 6 \\ 5 & 9 & 1 \\ 3 & 8 & 2 \end{bmatrix} \quad \begin{bmatrix} 0 & -1 & 5 & 3 \end{bmatrix} \quad \begin{bmatrix} 9 \\ 4 \\ 1 \end{bmatrix}$$

3 × 4 matrix, 3 rows, 4 columns; 3 × 3 matrix, square matrix; 1 × 4 matrix, row matrix; 3 × 1 matrix, column matrix

Self-Check 1 Give the size of the matrix $\begin{bmatrix} 4 & 2 \\ -6 & -8 \\ 2 & 1 \end{bmatrix}$.

Location of an Element in a Matrix

Each element in a matrix has a certain **location.** The location is described by giving the row and the column in which the element lies. Rows are labeled from the top of the matrix, and columns are labeled from the left. Consider the following matrix:

$$\begin{array}{r} \\ \\ \text{Row 1} \rightarrow \\ \text{Row 2} \rightarrow \\ \text{Row 3} \rightarrow \end{array} \begin{array}{cccc} \text{Col 1} & \text{Col 2} & \text{Col 3} & \text{Col 4} \\ \downarrow & \downarrow & \downarrow & \downarrow \\ -2 & 5 & 7 & 6 \\ 1 & -9 & 3 & 2 \\ 5 & 4 & 8 & -4 \end{array}$$

The element 3 (red) is in row 2 and column 3. This element is in the (2, 3) location. The element −4 (blue) is in row 3, column 4. It is in the (3, 4) location. Observe that the convention is to first give the row in which the element lies, followed by the column.

Example 1 Locating Elements in a Matrix

Consider the matrix $\begin{bmatrix} 1 & 4 & -1 \\ 2 & 5 & 9 \\ 1 & 7 & 6 \end{bmatrix}$.

(a) Give the location of the element 7 (red).
(b) Give the element in location (2, 3).

SOLUTION

(a) The element 7 (red) is in row 3 and column 2. It is thus in the (3, 2) location.
(b) The element in location (2, 3) lies in row 2, column 3; it is 9.

Self-Check 2 Give the location of the element 8 (red) in the following matrix:

$$\begin{bmatrix} 1 & 2 & -1 & 8 & 9 \end{bmatrix}$$

Matrix Notation

Matrices are usually denoted by capital letters and their elements by the corresponding lowercase letters. Every element of a matrix has a location, specified by its row and column. Let A be a matrix. Then a_{ij} is the element in row i, column j of A.

$$a_{ij}$$

First subscript indicates row — Second subscript indicates column

Example 2 Determining Elements of a Matrix

Determine a_{12}, a_{22}, and a_{34} for the following matrix A.

$$A = \begin{bmatrix} 1 & 2 & -3 & 5 \\ -1 & 4 & 7 & 9 \\ -2 & 6 & 11 & 0 \end{bmatrix}$$

SOLUTION

a_{12} is the element in row 1, column 2. $a_{12} = 2$.
a_{22} is the element in row 2, column 2. $a_{22} = 4$.
a_{34} is the element in row 3, column 4. $a_{34} = 0$.

Every matrix has a certain number of rows and columns. We refer to an **arbitrary matrix** having m rows and n columns as being an $m \times n$ matrix. We can visualize an $m \times n$ matrix A as follows:

$$A = \begin{bmatrix} a_{11} & a_{12} & \cdots & a_{1n} \\ a_{21} & a_{22} & \cdots & a_{2n} \\ \vdots & \vdots & \vdots & \vdots \\ a_{m1} & a_{m2} & \cdots & a_{mn} \end{bmatrix}$$

We now formally define the concept of equality for matrices. It enables us to discuss matrix equations.

Equality of Matrices Two matrices are said to be equal if they are of the same size, and if their corresponding elements are equal.

We have used equations for much of this course. This definition of equality now enables us to extend this important concept to matrices. We illustrate some matrix equations and methods of solving them.

Example 3 Solving a Matrix Equation

Solve the following matrix equation:

$$\begin{bmatrix} x & x + y \\ 3z & 5 \end{bmatrix} = \begin{bmatrix} 2 & 9 \\ -12 & 5 \end{bmatrix}$$

SOLUTION A solution of this matrix equation will consist of values of x, y, and z that satisfy the equation. The matrices are both 2×2 matrices. They are of the same size. Comparing corresponding elements we get the following system of equations:

$$\begin{aligned} x &= 2 \\ x + y &= 9 \\ 3z &= -12 \\ 5 &= 5 \end{aligned}$$

This system has the unique solution $x = 2$, $y = 7$, and $z = -4$.

Example 4 **Solving a Matrix Equation**

Solve the matrix equation

$$\begin{bmatrix} x & x - y \\ x - z & x + y + z \end{bmatrix} = \begin{bmatrix} 5 & 3 \\ 8 & 1 \end{bmatrix}$$

SOLUTION Comparing corresponding elements of the matrices gives the following system of equations:

$$\begin{aligned} x &= 5 \\ x - y &= 3 \\ x - z &= 8 \\ x + y + z &= 1 \end{aligned}$$

The first equations give $x = 5$, $y = 2$, and $z = -3$. These values do not satisfy the fourth equation. The system has no solution, therefore, the matrix equation has no solution.

Self-Check 3 Solve the following matrix equation:

$$\begin{bmatrix} x & x + y \\ x + y + z & x - z \end{bmatrix} = \begin{bmatrix} 1 & 2 \\ 4 & -1 \end{bmatrix}$$

Addition and Subtraction of Matrices

We now develop an algebra of matrices that is motivated by the structure that we have for real numbers. We can add and multiply real numbers. These operations are of course used daily. We introduce operations of addition and multiplication for matrices that will enhance their usefulness.

Addition and Subtraction of Matrices

The sum of two matrices of the same size is obtained by adding corresponding elements.

Subtraction is performed on matrices of the same size by subtracting corresponding elements.

If the matrices are not of the same size, they cannot be added or subtracted. We then say that the sum or difference does not exist.

Example 5 Adding and Subtracting Matrices

Let $A = \begin{bmatrix} 1 & 2 & -1 \\ 3 & 0 & 4 \end{bmatrix}$, $B = \begin{bmatrix} 7 & 1 & 2 \\ 3 & 1 & 5 \end{bmatrix}$, $C = \begin{bmatrix} 1 & 2 \\ 4 & 1 \end{bmatrix}$

Compute the matrices $A + B$, $A - B$, and $A + C$, if they exist.

SOLUTION

$$A + B = \begin{bmatrix} 1 & 2 & -1 \\ 3 & 0 & 4 \end{bmatrix} + \begin{bmatrix} 7 & 1 & 2 \\ 3 & 1 & 5 \end{bmatrix}$$
$$= \begin{bmatrix} 1+7 & 2+1 & -1+2 \\ 3+3 & 0+1 & 4+5 \end{bmatrix} = \begin{bmatrix} 8 & 3 & 1 \\ 6 & 1 & 9 \end{bmatrix}$$

$$A - B = \begin{bmatrix} 1 & 2 & -1 \\ 3 & 0 & 4 \end{bmatrix} - \begin{bmatrix} 7 & 1 & 2 \\ 3 & 1 & 5 \end{bmatrix} = \begin{bmatrix} 1-7 & 2-1 & -1-2 \\ 3-3 & 0-1 & 4-5 \end{bmatrix}$$
$$= \begin{bmatrix} -6 & 1 & -3 \\ 0 & -1 & -1 \end{bmatrix}$$

$A + C$ does not exist because A is a 2×3 matrix and C is a 2×2 matrix. They are not the same size.

Self-Check 4

Let $A = \begin{bmatrix} 1 & -1 \\ 2 & 4 \end{bmatrix}$, $B = \begin{bmatrix} 3 & 2 \\ -1 & 6 \end{bmatrix}$, $C = \begin{bmatrix} 5 & 1 & 2 & 4 \\ 3 & 0 & -1 & 7 \end{bmatrix}$

Compute $A + B$ and $A - C$, if they exist.

This rule for adding and subtracting two matrices is extended in a natural way to a number of matrices. A number of matrices that are of the same size are added by adding corresponding elements.

Example 6 Adding Three Matrices

Let $A = \begin{bmatrix} 1 & 3 \\ -1 & 4 \end{bmatrix}$, $B = \begin{bmatrix} 7 & 9 \\ -2 & 6 \end{bmatrix}$, $C = \begin{bmatrix} -2 & 5 \\ 1 & 3 \end{bmatrix}$

Compute $A + B + C$.

SOLUTION

$$A + B + C = \begin{bmatrix} 1 & 3 \\ -1 & 4 \end{bmatrix} + \begin{bmatrix} 7 & 9 \\ -2 & 6 \end{bmatrix} + \begin{bmatrix} -2 & 5 \\ 1 & 3 \end{bmatrix}$$
$$= \begin{bmatrix} 1+7-2 & 3+9+5 \\ -1-2+1 & 4+6+3 \end{bmatrix} = \begin{bmatrix} 6 & 17 \\ -2 & 13 \end{bmatrix}$$

Scalar Multiplication

Let us now turn our attention to multiplying matrices by numbers. When working with matrices, it is customary to refer to numbers as **scalars.** The rule for multiplying a matrix by a scalar follows.

Scalar Multiplication

The product of a scalar and a matrix is obtained by multiplying every element of the matrix by the scalar.

Example 7 Multiplying a Matrix by a Scalar

Multiply the matrix $\begin{bmatrix} 1 & 2 & 4 \\ 3 & -1 & 6 \end{bmatrix}$ by the scalar 6.

SOLUTION $6\begin{bmatrix} 1 & 2 & 4 \\ 3 & -1 & 6 \end{bmatrix} = \begin{bmatrix} 6 & 12 & 24 \\ 18 & -6 & 36 \end{bmatrix}$

The following example illustrates how expressions can involve both addition and scalar multiplication of matrices.

Example 8 Computing a Matrix Expression

If $\quad A = \begin{bmatrix} 1 & 4 & 3 \\ -1 & 2 & 1 \end{bmatrix}, \quad B = \begin{bmatrix} 0 & -5 & 3 \\ 1 & 6 & 4 \end{bmatrix}, \quad C = \begin{bmatrix} 6 & 1 & 4 \\ 2 & 1 & 8 \end{bmatrix}$

and $p = 4$, $q = -2$, compute $pA + qB + C$.

SOLUTION Observe the use of uppercase letters for matrices and lowercase letters for scalars in this example. This is standard practice when working with matrices and scalars. We get

$$\begin{aligned} pA + qB + C &= 4\begin{bmatrix} 1 & 4 & 3 \\ -1 & 2 & 1 \end{bmatrix} + (-2)\begin{bmatrix} 0 & -5 & 3 \\ 1 & 6 & 4 \end{bmatrix} + \begin{bmatrix} 6 & 1 & 4 \\ 2 & 1 & 8 \end{bmatrix} \\ &= \begin{bmatrix} 4 & 16 & 12 \\ -4 & 8 & 4 \end{bmatrix} + \begin{bmatrix} 0 & 10 & -6 \\ -2 & -12 & -8 \end{bmatrix} + \begin{bmatrix} 6 & 1 & 4 \\ 2 & 1 & 8 \end{bmatrix} \\ &= \begin{bmatrix} 10 & 27 & 10 \\ -4 & -3 & 4 \end{bmatrix} \end{aligned}$$

Applications . . .

Example 9 Medical Accounting

Data are often correlated and manipulated using matrices. This example illustrates the natural use of matrix addition in handling data. A clinic has three doctors, each with a specialty. Patients attending the clinic see more than one

doctor. The accounts are drawn up monthly and handled systematically through the use of matrices. We illustrate this accounting with four patients. The monthly bills are summarized in the following 4 × 3 matrix, for which entries are in dollars.

$$\text{Patient}\quad \begin{array}{c} \\ A \\ B \\ C \\ D \end{array} \overset{\text{Doctor}}{\begin{array}{c} \begin{array}{ccc} I & \quad II & \quad III \end{array} \\ \begin{bmatrix} 95 & 40 & 0 \\ 0 & 25 & 75 \\ 80 & 120 & 0 \\ 0 & 0 & 60 \end{bmatrix} \end{array}}$$

The patients are denoted *A*, *B*, *C*, and *D*. The doctors are *I*, *II*, and *III*. The bill of patient *A*, for example, from doctor *I* is \$95. The bill from doctor *II* is \$40, and so on. During a given 3-month period, there will be three such matrices. The quarterly bills are given by adding these matrices. Assume that the monthly bills are as follows:

$$\begin{array}{c} \\ A \\ B \\ C \\ D \end{array} \overset{\text{First Month}}{\begin{array}{c} \begin{array}{ccc} I & \quad II & \quad III \end{array} \\ \begin{bmatrix} 95 & 40 & 0 \\ 0 & 25 & 75 \\ 80 & 120 & 0 \\ 0 & 0 & 60 \end{bmatrix} \end{array}} + \overset{\text{Second Month}}{\begin{array}{c} \begin{array}{ccc} I & \quad II & \quad III \end{array} \\ \begin{bmatrix} 50 & 0 & 0 \\ 0 & 0 & 80 \\ 100 & 0 & 0 \\ 50 & 100 & 0 \end{bmatrix} \end{array}} + \overset{\text{Third Month}}{\begin{array}{c} \begin{array}{ccc} I & \quad II & \quad III \end{array} \\ \begin{bmatrix} 35 & 40 & 120 \\ 0 & 60 & 0 \\ 200 & 0 & 0 \\ 100 & 20 & 0 \end{bmatrix} \end{array}}$$

$$= \overset{\text{Quarter}}{\begin{array}{c} \begin{array}{ccc} I & \quad II & \quad III \end{array} \\ \begin{bmatrix} 180 & 80 & 120 \\ 0 & 85 & 155 \\ 380 & 120 & 0 \\ 150 & 120 & 60 \end{bmatrix} \end{array}}$$

Thus, A's bill from doctor I during the 3 months are \$95, \$50, and \$35. The quarterly bill will be $\$(95 + 50 + 35) = \$180$, and so on.

Although this analysis and others like it can be carried out without the use of matrices, the handling of large amounts of data is now often done on computers using matrix techniques.

Example 10 Trade Figures

This example illustrates the natural use of scalar multiplication in handling data. The following matrix represents the total trade figure between countries for a recent 5-year period. The numbers are values in billions of U.S. dollars.

		To			
		Canada	Japan	Britain	United States
From	Canada	0	5	6.2	65
	Japan	4.2	0	15	54
	Britain	9	4	0	22
	United States	540	20.5	60	0

The total exports from Britain to the United States, for example, were valued at \$22,000,000,000. To obtain the annual average over this 5-year period, we multiply this matrix by $\frac{1}{5}$. The average figures would be

		To			
		Canada	Japan	Britain	United States
From	Canada	0	1	1.24	13
	Japan	0.84	0	3	10.8
	Britain	1.8	0.8	0	4.4
	United States	10.8	4.1	12	0

Example 11 Group Relationships in Sociology

Matrices are used in many areas of sociology. We give an example that involves analyzing a group of people. Consider a group of five people. A sociologist is interested in finding out which one of the five has the most influence over, or dominates, the other members. The group is asked to fill out the following short questionnaire:

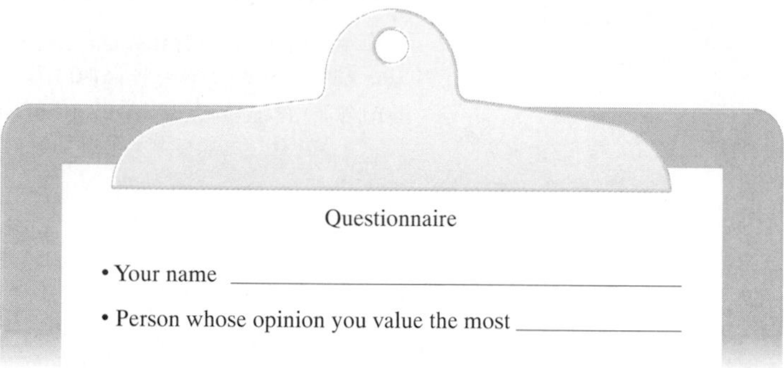
Questionnaire

• Your name ______

• Person whose opinion you value the most ______

These answers are then tabulated. Let us for convenience label the group members $M_1, M_2, \ldots, M_5$. Suppose the results are as given in the table of Figure 7.2a.

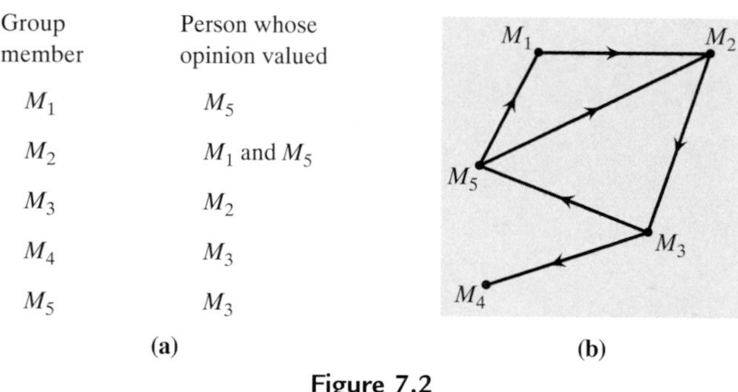

Group member	Person whose opinion valued
M_1	M_5
M_2	M_1 and M_5
M_3	M_2
M_4	M_3
M_5	M_3

(a) (b)

Figure 7.2

The sociologist makes the assumption that the person whose opinion a member values most is the person who influences that member most. Thus influence goes from the right column to the left column in the table in Figure 7.2a. We can represent these results by the diagram in Figure 7.2b. Such a diagram is called a **digraph.** The group members are represented by the **vertices** and direct influence is represented by a line segment (called an **arc** of the digraph), with the direction of influence being indicated on the line. For example, group members M_1 and M_2 are represented by the vertices M_1 and M_2 in the digraph. M_1 influences M_2; therefore, there is a line segment from M_1 to M_2.

Let us say that the **distance** from vertex M_1 to M_3 is 2 because we can get from M_1 to M_3 in two stages, namely along the line segment M_1M_2 then M_2M_3. The distance from M_1 to M_5, on the other hand, is 3. Construct the following **distance matrix** D for this digraph, where d_{ij} = distance from M_i to M_j. Because the distance from M_1 to M_1 is 0, we make $d_{11} = 0$. The distance from M_1 to M_2 is 1; therefore, $d_{12} = 1$. The distance from M_1 to M_3 is 2; therefore, $d_{13} = 2$, and so on. Let $d_{ij} = \infty$ when one cannot get from one vertex to another. For example,

$d_{45} = \infty$. Add up all the elements in each row and exhibit the sums as shown here.

$$D = \begin{bmatrix} 0 & 1 & 2 & 3 & 3 \\ 3 & 0 & 1 & 2 & 2 \\ 2 & 2 & 0 & 1 & 1 \\ \infty & \infty & \infty & 0 & \infty \\ 1 & 1 & 2 & 3 & 0 \end{bmatrix} \quad \begin{matrix} 9 \\ 8 \\ 6 \\ \infty \\ 7 \end{matrix}$$

Distance Matrix — Row Sums

In this graph, arcs correspond to direct influence; 2-paths, 3-paths, and so on, correspond to indirect influence. Thus, presumably, the smaller the distance from M_i to M_j, the greater the influence M_i has on M_j. The sum of the elements in row i gives the total distance of M_i to the other vertices. This leads to the following interpretation of row sums:

The smaller row sum i, *the greater the influence of person* M_i *on the group.*

We see that the smallest row sum is 6; for row 3. Thus, M_3 is the most influential person in the group, followed by M_5, M_2 and then M_1. If you want to start a rumor you should tell it to M_3!

The area of mathematics that deals with digraphs is called graph theory. Readers who are interested in learning more about graph theory are referred to *Introduction to Graph Theory* by Robin J. Wilson (John Wiley and Sons, 1987) and *Discrete Mathematical Structures* by Fred S. Roberts (Prentice Hall, 1976). The former book is a beautiful introduction to the mathematics of graph theory; the latter has a splendid collection of applications. The article "Predicting Chemistry by Topology" by Dennis H. Rouvray (*Scientific American*, **40,** September 1986) contains a fascinating account of how graph theoretical methods are being used to predict chemical properties of molecules that have not yet been synthesized.

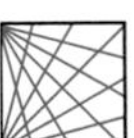

MATRIX OPERATIONS USING CALCULATORS

Matrix computations can be performed with calculators. Calculators vary considerably by the way in which matrices are entered and displayed and operations implemented. We illustrate the ideas involved. You should determine how to implement each of the steps and interpret the displays on your particular calculator.

Let us compute $A + B$, and $3A$ for the following matrices A and B.

$$A = \begin{bmatrix} 2 & 4 \\ 1 & 3 \end{bmatrix}, \quad B = \begin{bmatrix} 5 & -2 \\ 0 & 3 \end{bmatrix}$$

[A]
[2 4]
[1 3]

Enter and display *A*.

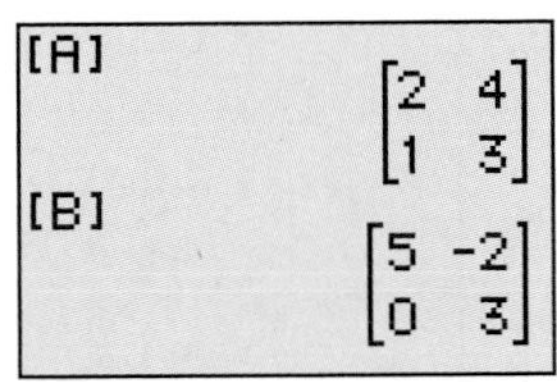

Enter and display *B*.

[A]+[B]
[7 2]
[1 6]

Compute *A* + *B*.

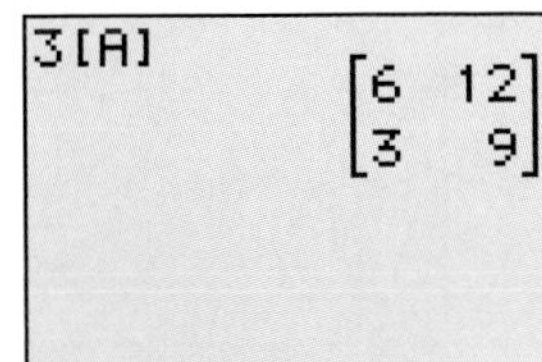

Compute 3*A*.

Self-Check 5 Use a calculator to compute $3.2A + 5.4B$, where

$$A = \begin{bmatrix} 1 & 2.7 & -3.6 \\ 8.2 & 4 & 8 \end{bmatrix}, \quad B = \begin{bmatrix} 4.3 & 3.9 & 2.5 \\ -6.7 & 0 & 3 \end{bmatrix}$$

Answers to Self-Check Exercises

1. Size of the matrix is 3×2.
2. Location is (1, 4).
3. Solution is $x = 1$, $y = 1$, and $z = 2$.
4. $A + B = \begin{bmatrix} 4 & 1 \\ 1 & 10 \end{bmatrix}$, $A - C$ does not exist.
5. $3.2A + 5.4B = \begin{bmatrix} 26.42 & 29.7 & 1.98 \\ -9.94 & 12.8 & 41.8 \end{bmatrix}$

EXERCISES 7.1

In Exercises 1–10, give the sizes of the matrices. If the matrix is square, a row matrix, or a column matrix, indicate so.

1. $\begin{bmatrix} 1 & 2 \\ -1 & 3 \end{bmatrix}$

2. $\begin{bmatrix} 1 & 2 & 4 \\ -1 & 3 & 6 \end{bmatrix}$

3. $\begin{bmatrix} 1 & 3 \\ 4 & 6 \\ 2 & 1 \end{bmatrix}$

4. $\begin{bmatrix} -1 & 2 & 3 \end{bmatrix}$

5. $\begin{bmatrix} 1 \\ 3 \\ 4 \end{bmatrix}$

6. $\begin{bmatrix} 6 \end{bmatrix}$

7. $\begin{bmatrix} 5 \\ 9 \end{bmatrix}$

8. $\begin{bmatrix} 1 & -2 & 5 \\ 6 & 1 & 3 \\ 8 & 1 & 2 \end{bmatrix}$

9. $\begin{bmatrix} 1 & 6 & -1 & 5 \\ 6 & 1 & 3 & 9 \end{bmatrix}$

10. $\begin{bmatrix} 2 & 4 & 8 & 1 & 6 & -12 \end{bmatrix}$

In Exercises 11–22, give the indicated elements of the matrices A, B, C, and D.

$$A = \begin{bmatrix} 5 & 4 \\ 1 & 3 \end{bmatrix}, \quad B = \begin{bmatrix} 2 & -1 \\ 6 & 3 \end{bmatrix}, \quad C = \begin{bmatrix} 1 & 2 & 4 \\ 7 & 8 & 3 \end{bmatrix},$$

$$D = \begin{bmatrix} 5 & 6 & 1 & 9 \\ 3 & 4 & -2 & -1 \\ 7 & 1 & 5 & 6 \end{bmatrix}$$

11. a_{12} **12.** b_{22} **13.** c_{23} **14.** a_{21} **15.** c_{11} **16.** d_{24}
17. d_{32} **18.** c_{21} **19.** d_{33} **20.** d_{14} **21.** d_{34} **22.** c_{13}

In Exercises 23–34, give the indicated elements of the matrices P and Q.

$$P = \begin{bmatrix} 1 & 2 & -1 & 3 \\ 4 & -4 & 6 & 2 \\ 0 & 9 & 8 & 1 \end{bmatrix},$$

$$Q = \begin{bmatrix} 2 & -1 & 0 & 6 & 9 & 8 \\ 1 & 2 & 5 & 3 & 7 & 2 \\ 8 & 2 & 1 & 11 & -6 & 3 \\ 4 & 3 & 7 & 1 & 5 & -12 \end{bmatrix}$$

23. p_{14} **24.** p_{33} **25.** q_{24} **26.** q_{45} **27.** q_{46} **28.** p_{32}
29. q_{35} **30.** p_{24} **31.** p_{13} **32.** p_{31} **33.** q_{26} **34.** q_{36}

In Exercises 35–44, solve the matrix equations.

35. $\begin{bmatrix} x & x + y \end{bmatrix} = \begin{bmatrix} 2 & 5 \end{bmatrix}$

36. $\begin{bmatrix} x & y \\ x + y & z \end{bmatrix} = \begin{bmatrix} -1 & 3 \\ 2 & 4 \end{bmatrix}$

37. $\begin{bmatrix} x & x + y \\ x - y & x + z \end{bmatrix} = \begin{bmatrix} 2 & 3 \\ 1 & 4 \end{bmatrix}$

38. $\begin{bmatrix} x & x + y + z \\ x - y & z \end{bmatrix} = \begin{bmatrix} 3 & 5 \\ 2 & 4 \end{bmatrix}$

39. $\begin{bmatrix} x & x - y \\ x + y & x + z \end{bmatrix} = \begin{bmatrix} 3 & 4 \\ 6 & 8 \end{bmatrix}$

40. $\begin{bmatrix} x & x+y \\ x-y & 3 \end{bmatrix} = \begin{bmatrix} -2 & 0 \\ -4 & z \end{bmatrix}$

41. $\begin{bmatrix} x+y & x+z \\ x+y+z & x-1 \end{bmatrix} = \begin{bmatrix} 3 & 4 \\ 6 & 0 \end{bmatrix}$

42. $\begin{bmatrix} x+y+z & x-z \\ 3 & -1 \end{bmatrix} = \begin{bmatrix} 8 & 3 \\ y+z & y-z \end{bmatrix}$

43. $\begin{bmatrix} x-y+z & y \\ x+z & 7 \end{bmatrix} = \begin{bmatrix} 0 & 3 \\ 3 & 7 \end{bmatrix}$

44. $\begin{bmatrix} x & x+y \\ x+z & z \end{bmatrix} = \begin{bmatrix} 1 & -1 \\ 5 & 6 \end{bmatrix}$

In Exercises 45–54, perform the indicated operations, if possible, using the matrices A, B, C, and D. Check your answers using a calculator.

$$A = \begin{bmatrix} 1 & 2 \\ -1 & 3 \end{bmatrix}, \quad B = \begin{bmatrix} 4 & 2 \\ 5 & -3 \end{bmatrix}, \quad C = \begin{bmatrix} 5 & 1 \\ 2 & 0 \end{bmatrix},$$

$$D = \begin{bmatrix} 4 & 6 & -1 \\ 2 & 3 & 4 \end{bmatrix}, \quad E = \begin{bmatrix} 2 & -1 & 0 \\ 5 & 1 & 6 \end{bmatrix}$$

45. $A + B$
46. $A - B$
47. $A + D$
48. $D + E$
49. $A + 2B$
50. $2A - B$
51. $A + B + C$
52. $4A - B + 3C$
53. $A + 5E$
54. $A + 2B + 3E$

In Exercises 55–64, perform the indicated operations, if possible, using the matrices P, Q, R, S, and T. Check your answers using a calculator.

$$P = \begin{bmatrix} 1 & 2 \\ 3 & 4 \\ -1 & 2 \end{bmatrix}, \quad Q = \begin{bmatrix} 4 & 5 \\ 1 & 2 \\ 0 & 6 \end{bmatrix},$$

$$R = \begin{bmatrix} 1 & 2 & -1 \\ 3 & 4 & 2 \end{bmatrix}, \quad S = \begin{bmatrix} 5 & 2 & 6 \\ 3 & 1 & 0 \end{bmatrix},$$

$$T = \begin{bmatrix} 1 & 4 & 0 \\ -2 & 3 & -5 \end{bmatrix}$$

55. $P + Q$
56. $P + 2Q$
57. $Q + 3S$
58. $5R - 3S$
59. $P - Q$
60. $2P - 3Q$
61. $R + 2S - 3T$
62. $-3S + R$
63. $P + 2Q + 3R - 4T$
64. $2P - 4Q + 3P$

In Exercises 65–70, solve the matrix equations.

65. $\begin{bmatrix} x+y+z & y-z \\ z & x^2 \end{bmatrix} = \begin{bmatrix} 8 & 1 \\ 1 & 25 \end{bmatrix}$

66. $\begin{bmatrix} x^2 & x+y \\ x-y & 3 \end{bmatrix} = \begin{bmatrix} 4 & 0 \\ -4 & z \end{bmatrix}$

67. $\begin{bmatrix} x^2 & x+y \\ 4 & y \end{bmatrix} = \begin{bmatrix} 9 & 3 \\ 4 & -3 \end{bmatrix}$

68. $\begin{bmatrix} x^2-2x & x+y \\ x+z & x+y+z \end{bmatrix} = \begin{bmatrix} -1 & 3 \\ 4 & 6 \end{bmatrix}$

69. $\begin{bmatrix} x^2-3x & 2x \\ y & z \end{bmatrix} = \begin{bmatrix} -2 & 6 \\ 4 & 3 \end{bmatrix}$

70. $\begin{bmatrix} x^2-4x & x+y \\ y-z & z \end{bmatrix} = \begin{bmatrix} 12 & 5 \\ 4 & 3 \end{bmatrix}$

71. If $A = \begin{bmatrix} 1 & 2 \\ 3 & -1 \end{bmatrix}$, $B = \begin{bmatrix} 5 & 1 \\ 3 & 2 \end{bmatrix}$, and $C = \begin{bmatrix} -2 & 7 \\ 1 & 4 \end{bmatrix}$, and $D = 2A - 3B + 4C$, find **(a)** d_{12} **(b)** d_{22} **(c)** $3d_{21}$ **(d)** $d_{12} + 2d_{22}$.

72. If $A = \begin{bmatrix} -1 & 2 & 6 \\ 0 & 1 & 4 \end{bmatrix}$, $B = \begin{bmatrix} 5 & -1 & 2 \\ 0 & 3 & 1 \end{bmatrix}$, and $C = \begin{bmatrix} 5 & 1 & 6 \\ 2 & 1 & 4 \end{bmatrix}$, and $D = 4A - 2B + 4C$, determine **(a)** d_{23} **(b)** d_{13} **(c)** $a_{11} + b_{11} - d_{11}$ **(d)** $2a_{12} - 3b_{22} + c_{13} - 4d_{22}$.

73. If $A = \begin{bmatrix} -1 & 2 & 1 & 0 & 6 \\ 4 & -3 & 7 & 0 & -3 \\ 5 & 7 & 4 & -2 & 5 \\ 7 & 0 & 1 & 2 & -3 \end{bmatrix}$, answer the following.

(a) If $a_{ij} = 2$, what are the possible values of i and j?

(b) If $a_{i3} = 1$, what are the possible values of i?

(c) If $a_{2j} = -3$, what are the possible values of j?

(d) If $a_{ij} = 4$ and $i = j$, what are the possible values of i and j?

(e) If $a_{ij} = 0$ and $i < j$, what are the possible values of i and j?

(f) If $a_{ij} = 7$ and $j = i - 1$, what are the possible values of i and j?

74. **NFL Tables** The following matrix gives the statistics at a certain point of the season in the Western Division of the National Football League.

	Won	Lost	Tied
Denver	4	1	1
Oakland	4	0	1
Seattle	2	3	0
Kansas City	2	2	0
San Diego	1	4	0

How many games has (a) Oakland won (b) Kansas City lost?

75. **Traffic Analysis** We have seen how highway departments use O–D matrices to analyze traffic patterns. A region of interest is divided into zones and an estimate of daily traffic between zones is made. The following matrix represents daily traffic in a small city that is divided into four zones: I, II, III, and IV.

Destination Zone

$$\begin{array}{cc} & \begin{array}{cccc} \text{I} & \text{II} & \text{III} & \text{IV} \end{array} \\ \begin{array}{c} \text{Origin} \\ \text{Zone} \end{array} \begin{array}{c} \text{I} \\ \text{II} \\ \text{III} \\ \text{IV} \end{array} & \begin{bmatrix} 1000 & 1700 & 300 & 200 \\ 1400 & 6000 & 900 & 600 \\ 400 & 900 & 2000 & 1200 \\ 700 & 460 & 2400 & 1800 \end{bmatrix} \end{array}$$

For example, the daily traffic from zone II to zone III is 900 vehicles [the element in location (2,3)]. Using this O–D matrix, determine the following:

(a) the daily traffic from zone I to zone IV
(b) the daily traffic from zone III to zone II
(c) internal traffic in zone II
(d) internal traffic in zone IV

Will such an O–D matrix always be square?

76. **Medical Accounting** A clinic has three doctors, I, II, and III. The accounts of three patients, A, B, and C, over 3 consecutive months are given by the following matrices:

First month, Second month

$$\begin{array}{c} \\ A \\ B \\ C \end{array} \begin{array}{c} \begin{array}{ccc} \text{I} & \text{II} & \text{III} \end{array} \\ \begin{bmatrix} 420 & 0 & 32 \\ 0 & 60 & 45 \\ 125 & 0 & 160 \end{bmatrix} \end{array}, \quad \begin{array}{c} \begin{array}{ccc} \text{I} & \text{II} & \text{III} \end{array} \\ \begin{bmatrix} 0 & 70 & 48 \\ 32 & 45 & 80 \\ 0 & 400 & 30 \end{bmatrix} \end{array},$$

Third month

$$\begin{array}{c} \begin{array}{ccc} \text{I} & \text{II} & \text{III} \end{array} \\ \begin{bmatrix} 0 & 65 & 70 \\ 40 & 127 & 0 \\ 85 & 0 & 38 \end{bmatrix} \end{array}$$

(a) Determine the matrix that gives the quarterly accounts. What is B's quarterly bill from doctor III?

(b) Determine the matrix that gives the average monthly accounts. What is C's average monthly bill from doctor II?

77. **Group Analysis** The following tables represent information obtained from questionnaires given to groups of people. In each case, construct the digraph that describes the influence structure within the group. Use the distance matrix to rank the members according to their influence within the group.

(a)

Group Member	*Whose Opinion Is Valued*
M_1	M_2
M_2	M_5
M_3	M_2
M_4	M_3
M_5	M_1 and M_4

(b)

Group Member	*Whose Opinion Is Valued*
M_1	M_2
M_2	M_3
M_3	M_1 and M_4
M_4	M_5
M_5	M_1

(c)

Group Member	*Whose Opinion Is Valued*
M_1	M_2 and M_4
M_2	M_3
M_3	M_5
M_4	M_5
M_5	M_1

(d)

Group Member	*Whose Opinion Is Valued*
M_1	M_4
M_2	M_1
M_3	M_2
M_4	M_2 and M_5
M_5	M_1

78. **Street System** The network shown in Figure 7.3 describes a system of streets in a city downtown area. Many of the streets are one-way. Interpret the network as a digraph. Find its distance matrix.

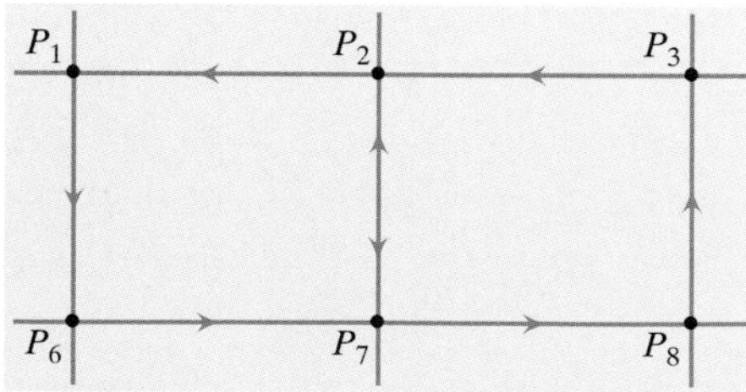

Figure 7.3

79. **Street System** The following distance matrix D describes the distances connecting four junctions, P, Q, R, and S, in a small city. Draw the street map.

$$D = \begin{array}{c} \\ P \\ Q \\ R \\ S \end{array} \begin{array}{c} \begin{array}{cccc} P & Q & R & S \end{array} \\ \begin{bmatrix} 0 & 1 & 2 & 3 \\ 2 & 0 & 1 & 2 \\ 1 & 1 & 0 & 1 \\ 1 & 2 & 1 & 0 \end{bmatrix} \end{array}$$

80. Digraphs Consider a digraph with vertices $P_1, \ldots, P_n$. The **adjacency matrix** A of the digraph is a matrix of zeros and ones, such that

$$a_{ij} = \begin{cases} 1 & \text{if there is an arc from vertex } P_i \text{ to } P_j \\ 0 & \text{otherwise} \end{cases}$$

Determine the adjacency matrix of each of the following digraphs.

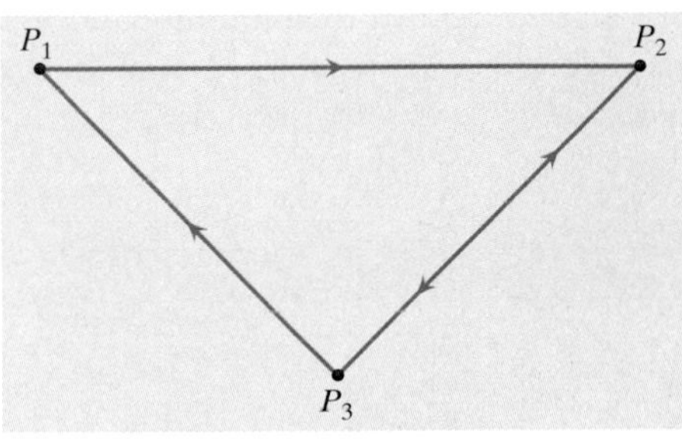

(a)

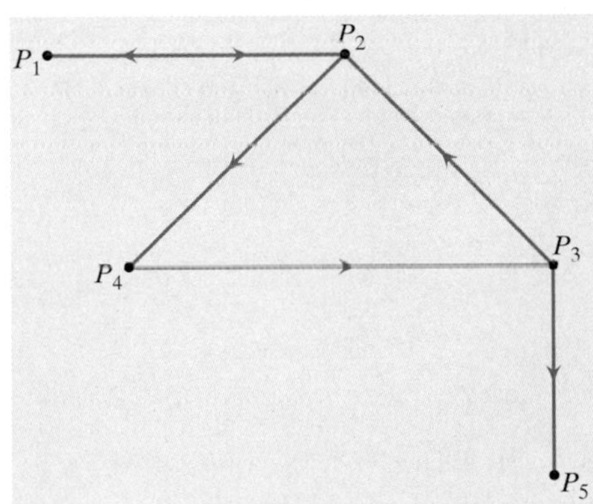

(b)

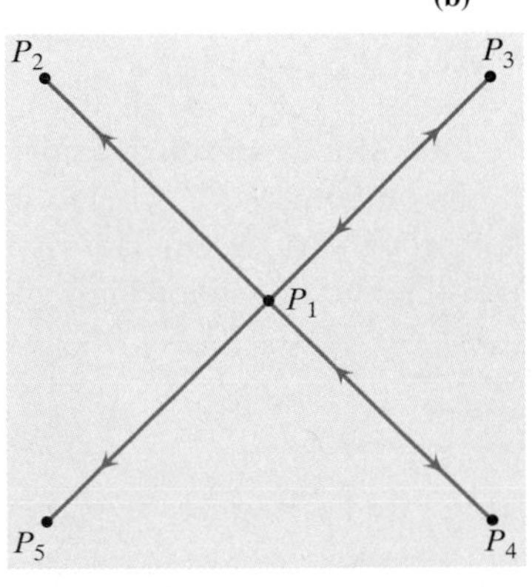

(c)

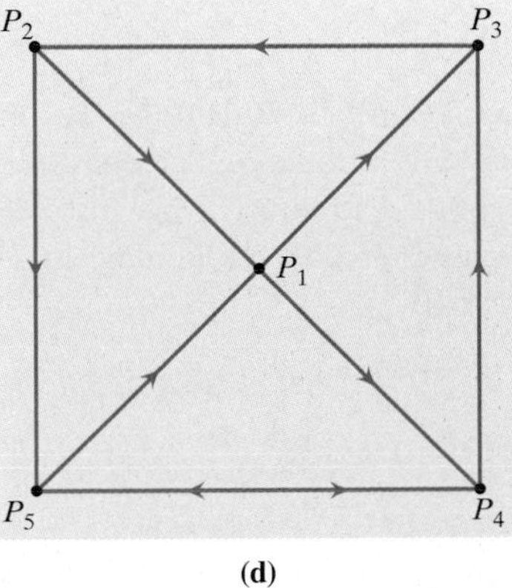

(d)

81. Digraphs Sketch the digraphs that have the following adjacency matrices.

(a) $\begin{bmatrix} 0 & 1 & 1 & 1 \\ 1 & 0 & 0 & 0 \\ 1 & 0 & 0 & 0 \\ 1 & 0 & 0 & 0 \end{bmatrix}$ (b) $\begin{bmatrix} 0 & 1 & 1 & 0 \\ 0 & 0 & 1 & 1 \\ 0 & 0 & 0 & 1 \\ 1 & 0 & 0 & 0 \end{bmatrix}$

(c) $\begin{bmatrix} 0 & 1 & 1 & 0 \\ 1 & 0 & 1 & 0 \\ 0 & 0 & 0 & 1 \\ 1 & 1 & 0 & 0 \end{bmatrix}$

82. Living Location A preacher has five churches under his care at towns *A*, *B*, *C*, *D*, and *E* (Figure 7.4). The line segments between the towns indicate roads, and the numbers are the distances in miles between the towns along those roads. He has to preach in a different town on each of five consecutive Sundays. (**a**) Construct a distance matrix for this graph. Use the results from the row sums of the distance matrix to decide where the preacher should live. (**b**) Do you observe a certain symmetry in the distance matrix? Describe this symmetry in your own words. Does this symmetry arise by chance in this particular case or would you expect the distance matrix for all such situations (with different numbers of towns and distances) to always exhibit such symmetry? Such a matrix is called a **symmetric matrix.**

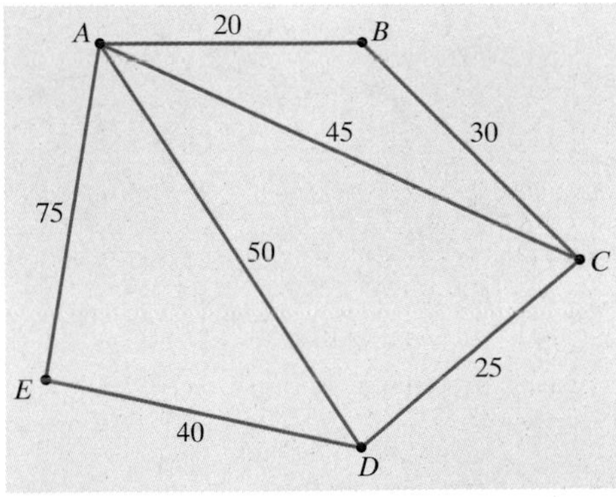

Figure 7.4

7.2 Multiplication of Matrices and Population Movements

• Matrix Multiplication • Size of a Product Matrix • Powers of a Matrix
• Population Movement Model

Matrix Multiplication

In Section 7.1, we defined addition, subtraction, and scalar multiplication of matrices. We now introduce a rule for multiplying matrices by first looking at an application.

Let us use matrices to analyze the wages of three people working in a restaurant. (The model that we develop can be extended to accommodate any number of workers.) The following matrix A describes a typical work week for these three people.

Hours of Work Per Week

$$\begin{array}{r c} & \begin{array}{ccc} \text{M–F} & \text{Sat} & \text{Sun} \end{array} \\ \begin{array}{r} \text{Person 1} \\ \text{Person 2} \\ \text{Person 3} \end{array} & \begin{bmatrix} 40 & 4 & 2 \\ 20 & 2 & 3 \\ 10 & 5 & 6 \end{bmatrix} \end{array}$$

Matrix A

For example, person 3, a student, works 10 hours total Monday through Friday, but is free to work 5 hours on Saturdays and 6 hours on Sundays.

Workers are paid an hourly wage of \$5 Monday through Friday. On Saturdays the hourly wage is \$6 and on Sundays it is \$8. We construct the following wage column matrix B:

Wages Per Hour (\$)

$$\begin{array}{r c} \begin{array}{r} \text{M–F} \\ \text{Sat} \\ \text{Sun} \end{array} & \begin{bmatrix} 5 \\ 6 \\ 8 \end{bmatrix} \end{array}$$

Matrix B

The total weekly wages of these three people are obtained by multiplying hours by wages as follows:

	M–F		Sat		Sun
Person 1	40×5	+	4×6	+	$2 \times 8 = \$240$
Person 2	20×5	+	2×6	+	$3 \times 8 = \$136$
Person 3	10×5	+	5×6	+	$6 \times 8 = \$128$

We can arrive at these numbers if we multiply matrix A by matrix B as follows. We multiply each row of A times the column of B as follows.

$$\begin{bmatrix} 40 & 4 & 2 \\ 20 & 2 & 3 \\ 10 & 5 & 6 \end{bmatrix} \begin{bmatrix} 5 \\ 6 \\ 8 \end{bmatrix} = \begin{bmatrix} (40 \quad 4 \quad 2)\begin{pmatrix} 5 \\ 6 \\ 8 \end{pmatrix} \\ (20 \quad 2 \quad 3)\begin{pmatrix} 5 \\ 6 \\ 8 \end{pmatrix} \\ (10 \quad 5 \quad 6)\begin{pmatrix} 5 \\ 6 \\ 8 \end{pmatrix} \end{bmatrix}$$

Multiply a row by a column by multiplying corresponding elements and adding.

$$= \begin{bmatrix} 40 \times 5 & + & 4 \times 6 & + & 2 \times 8 \\ 20 \times 5 & + & 2 \times 6 & + & 3 \times 8 \\ 10 \times 5 & + & 5 \times 6 & + & 6 \times 8 \end{bmatrix}$$

$$= \begin{bmatrix} 240 \\ 136 \\ 128 \end{bmatrix} \begin{matrix} \textbf{Wages of Person 1} \\ \textbf{Wages of Person 2} \\ \textbf{Wages of Person 3} \end{matrix}$$

The handling of large amounts of data is now often done on computers using these matrix techniques. Let us look at another example.

Example 1 Multiplying Matrices

Consider the following matrices A and B:

$$A = \begin{bmatrix} 1 & 3 \\ 2 & 0 \end{bmatrix}, \qquad B = \begin{bmatrix} 5 & 0 & 1 \\ 3 & -2 & 6 \end{bmatrix}$$

We compute the product AB by interpreting A in terms of its rows and B in terms of its columns and multiplying the rows by the columns in the following systematic manner:

$$AB = \begin{bmatrix} 1 & 3 \\ 2 & 0 \end{bmatrix} \begin{bmatrix} 5 & 0 & 1 \\ 3 & -2 & 6 \end{bmatrix}$$

$$= \begin{bmatrix} (1 \quad 3)\begin{pmatrix} 5 \\ 3 \end{pmatrix} & (1 \quad 3)\begin{pmatrix} 0 \\ -2 \end{pmatrix} & (1 \quad 3)\begin{pmatrix} 1 \\ 6 \end{pmatrix} \\ (2 \quad 0)\begin{pmatrix} 5 \\ 3 \end{pmatrix} & (2 \quad 0)\begin{pmatrix} 0 \\ -2 \end{pmatrix} & (2 \quad 0)\begin{pmatrix} 1 \\ 6 \end{pmatrix} \end{bmatrix} \begin{matrix} \leftarrow \textbf{First row of } A \textbf{ times each column of } B \textbf{ in turn} \\ \leftarrow \textbf{Second row of } A \textbf{ times each column of } B \textbf{ in turn} \end{matrix}$$

Multiply a row by a column by multiplying corresponding elements and adding,

$$= \begin{bmatrix} (1 \times 5) + (3 \times 3) & (1 \times 0) + (3 \times -2) & (1 \times 1) + (3 \times 6) \\ (2 \times 5) + (0 \times 3) & (2 \times 0) + (0 \times -2) & (2 \times 1) + (0 \times 6) \end{bmatrix}$$

$$= \begin{bmatrix} 14 & -6 & 19 \\ 10 & 0 & 2 \end{bmatrix}$$

Example 2 Multiplying Matrices

If $A = \begin{bmatrix} 2 & 1 \\ 7 & 0 \\ -3 & -2 \end{bmatrix}$ and $B = \begin{bmatrix} -1 & 0 \\ 3 & 5 \end{bmatrix}$, determine AB.

SOLUTION Multiplying the rows of A by the columns of B in the appropriate manner we get

$$AB = \begin{bmatrix} 2 & 1 \\ 7 & 0 \\ -3 & -2 \end{bmatrix} \begin{bmatrix} -1 & 0 \\ 3 & 5 \end{bmatrix}$$

$$= \begin{bmatrix} (2 \quad 1)\begin{pmatrix} -1 \\ 3 \end{pmatrix} & (2 \quad 1)\begin{pmatrix} 0 \\ 5 \end{pmatrix} \\ (7 \quad 0)\begin{pmatrix} -1 \\ 3 \end{pmatrix} & (7 \quad 0)\begin{pmatrix} 0 \\ 5 \end{pmatrix} \\ (-3 \quad -2)\begin{pmatrix} -1 \\ 3 \end{pmatrix} & (-3 \quad -2)\begin{pmatrix} 0 \\ 5 \end{pmatrix} \end{bmatrix}$$

← **First row of A times each column of B in turn**

← **Second row of A times each column of B in turn**

← **Third row of A times each column of B in turn**

$$= \begin{bmatrix} -2 + 3 & 0 + 5 \\ -7 + 0 & 0 + 0 \\ 3 - 6 & 0 - 10 \end{bmatrix} = \begin{bmatrix} 1 & 5 \\ -7 & 0 \\ -3 & -10 \end{bmatrix}$$

Self-Check 1 Compute the product AB of the following matrices:

$$A = \begin{bmatrix} 1 & 2 \\ -1 & 3 \end{bmatrix}, \quad B = \begin{bmatrix} 0 & 4 & 1 \\ 2 & 1 & 5 \end{bmatrix}$$

Exploration

Consider the following matrices A and B; let $C = AB$. Each element of C is found by multiplying a certain row of A by a certain column of B. Find the elements c_{12} and c_{34} of C without computing the whole of the matrix C.

$$A = \begin{bmatrix} 1 & 0 & -3 \\ 4 & 5 & -2 \\ 0 & 7 & -5 \end{bmatrix}, \quad B = \begin{bmatrix} 2 & 0 & -4 & 5 \\ 1 & -3 & 2 & 5 \\ 0 & -2 & 5 & 1 \end{bmatrix}$$

Describe how you have arrived at these elements ($c_{12} = 6$, $c_{34} = 30$).

We now give the formal definition of matrix multiplication. We define the product by giving the rule for arriving at an arbitrary element of the product matrix in the preceding manner.

MATRIX MULTIPLICATION

The element in row i and column j of the product matrix AB is obtained by multiplying the corresponding elements of row i of A with column j of matrix B and adding the products.

Having defined a rule for multiplying matrices, let us now look at some implications of the rule. Consider the following matrices A and B.

$$A = \begin{bmatrix} 3 & -1 & 2 \\ 4 & 1 & 5 \end{bmatrix}, \qquad B = \begin{bmatrix} 2 & 1 \\ 3 & 4 \end{bmatrix}$$

Let us attempt to compute AB using the matrix multiplication rule. We get

$$AB = \begin{bmatrix} 3 & -1 & 2 \\ 4 & 1 & 5 \end{bmatrix}\begin{bmatrix} 2 & 1 \\ 3 & 4 \end{bmatrix} = \begin{bmatrix} (3 \quad -1 \quad 2)\begin{pmatrix} 2 \\ 3 \end{pmatrix} & (3 \quad -1 \quad 2)\begin{pmatrix} 1 \\ 4 \end{pmatrix} \\ (4 \quad 1 \quad 5)\begin{pmatrix} 2 \\ 3 \end{pmatrix} & (4 \quad 1 \quad 5)\begin{pmatrix} 1 \\ 4 \end{pmatrix} \end{bmatrix}$$

We cannot compute $(3 \quad -1 \quad 2)\begin{pmatrix} 2 \\ 3 \end{pmatrix}$ because there is no element in B to match the 2 of A; $(3 \times 2) + (-1 \times 3) + (2 \times ?)$. The same shortcoming applies to all the other elements of AB. We say that the product AB does not exist. There would be matching elements and the product would then exist if the number of columns in A was equal to the number of rows in B. We arrive at the following result:

EXISTENCE OF A PRODUCT

The product AB of two matrices A and B exists if and only if the number of columns in A is equal to the number of rows in B.

Self-Check 2

Let $A = \begin{bmatrix} 1 & 2 \\ -1 & 3 \end{bmatrix}$, $B = \begin{bmatrix} 1 & 2 & -1 \\ 3 & 4 & 9 \end{bmatrix}$, and $C = \begin{bmatrix} 0 & 4 & 1 \\ 2 & 1 & 5 \\ 4 & 2 & 8 \end{bmatrix}$

Determine whether AB, AC, and BC exist. (You need not compute the products if they exist.)

Example 3 Investigating Whether Matrix Multiplication Is Commutative

Consider the following matrices A and B:

$$A = \begin{bmatrix} 8 & 3 \\ -9 & 0 \end{bmatrix}, \qquad B = \begin{bmatrix} 5 & 3 & 2 \\ -1 & 5 & 7 \end{bmatrix}$$

Observe that AB exists because A has two columns and B has two rows; however, BA does not exist because B has three columns and A has two rows. This example illustrates that the order in which one multiplies matrices is important.

Matrices are not commutative under multiplication. In general,

$$AB \neq BA.$$

Size of a Product Matrix

Let us now investigate the size of a matrix that results from multiplying two matrices. Because the product AB of two matrices A and B only exists if the number of columns in A is equal to the number of rows in B, let A be an $m \times n$ matrix and B be an $n \times r$ matrix. A has n columns and B has n rows; therefore, AB exists.

The first row of AB is obtained by multiplying the first row of A by each column of B in turn; thus, the number of columns in AB is equal to the number of columns in B. The first column of AB results from multiplying each row of A in turn by the first column of B; thus, the number of rows in AB is equal to the number of rows in A.

Size of a Product Matrix

If A is an $m \times n$ matrix and B is an $n \times r$ matrix, then the product AB is an $m \times r$ matrix.

$$\underset{m \times n}{A} \times \underset{n \times r}{B} = \underset{m \times r}{AB}$$

Inside match

Outside give size of AB

We can illustrate this result as follows:

Example 4 Determining the Size of a Product Matrix

If A is a 4×3 matrix and B is a 3×5 matrix, what is the size of the product AB?

SOLUTION A has three columns, and B has three rows; thus, AB exists. It will be a 4×5 matrix.

$$\underset{4 \times 3}{A} \times \underset{3 \times 5}{B} = \underset{4 \times 5}{AB}$$

Inside match

Outside give size of AB

Self-Check 3 If A is a 2×5 matrix and B is a 5×6 matrix, what is the size of AB?

Multiplying Many Matrices

We can extend the concept of the product of two matrices to the product of a number of matrices. A product ABC will only exist if both AB and BC exist. The following example illustrates the ideas involved.

Example 5 Multiplying Three Matrices

$$\text{Let} \quad A = \begin{bmatrix} 1 & 2 \\ -1 & 3 \end{bmatrix}, \quad B = \begin{bmatrix} 3 & 4 & -1 \\ 0 & 1 & 2 \end{bmatrix}, \quad C = \begin{bmatrix} 2 \\ 3 \\ -1 \end{bmatrix}$$

Compute ABC, if it exists.

SOLUTION The rows and columns of A, B, and C match up as shown in the figure. Thus, the product exists and will be a 2×1 matrix.

$$\begin{array}{ccccccc} A & \times & B & \times & C & = & ABC \\ 2 \times 2 & & 2 \times 3 & & 3 \times 1 & & 2 \times 1 \end{array}$$

Match Match

ABC is 2×1

We can compute the product ABC in stages as follows:

$$AB = \begin{bmatrix} 1 & 2 \\ -1 & 3 \end{bmatrix} \begin{bmatrix} 3 & 4 & -1 \\ 0 & 1 & 2 \end{bmatrix} = \begin{bmatrix} 3 & 6 & 3 \\ -3 & -1 & 7 \end{bmatrix}$$

$$ABC = (AB)C = \begin{bmatrix} 3 & 6 & 3 \\ -3 & -1 & 7 \end{bmatrix} \begin{bmatrix} 2 \\ 3 \\ -1 \end{bmatrix} = \begin{bmatrix} 21 \\ -16 \end{bmatrix}$$

The matrices in such a product can be grouped in any manner for actual computation, as long as the order is maintained. For example, we could have computed BC first, to get $A(BC)$.

Matrix multiplication is associative.

$$(AB)C = A(BC)$$

Powers of a Matrix

We can multiply a square matrix by itself any number of times; that is, we can take powers of square matrices. The notation for powers of a matrix is similar to the notation for powers of a real number. If A is a square matrix, then A multiplied by itself k times is written A^k.

$$A^k = \underbrace{AA \ldots A}_{k \text{ Times}}$$

Example 6 Computing the Cube of a Matrix

If $A = \begin{bmatrix} 1 & -2 \\ -1 & 0 \end{bmatrix}$, compute A^3.

SOLUTION We get

$$A^2 = \begin{bmatrix} 1 & -2 \\ -1 & 0 \end{bmatrix}\begin{bmatrix} 1 & -2 \\ -1 & 0 \end{bmatrix} = \begin{bmatrix} 3 & -2 \\ -1 & 2 \end{bmatrix}$$

$$A^3 = (A^2)A = \begin{bmatrix} 3 & -2 \\ -1 & 2 \end{bmatrix}\begin{bmatrix} 1 & -2 \\ -1 & 0 \end{bmatrix} = \begin{bmatrix} 5 & -6 \\ -3 & 2 \end{bmatrix}$$

Further Ideas 1 If $A = \begin{bmatrix} 2 & -3 \\ -1 & 0 \end{bmatrix}$, compute A^4 (using two matrix multiplications).

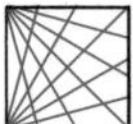

Matrix Multiplication Using a Calculator

Calculators can be used to multiply matrices. We illustrate the steps. You should determine how to implement these steps on your calculator. Let us compute the product AB for the following matrices A and B.

$$A = \begin{bmatrix} 2 & 4 \\ 1 & 3 \end{bmatrix}, \quad B = \begin{bmatrix} 5 & -2 \\ 0 & 3 \end{bmatrix}$$

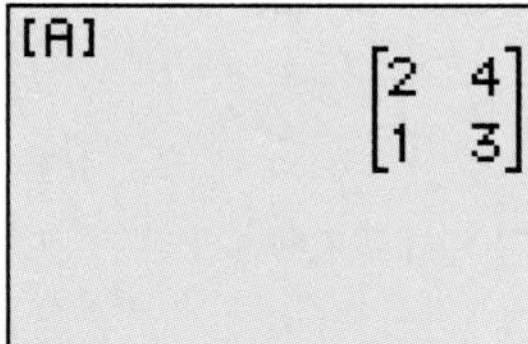

Enter *A*.

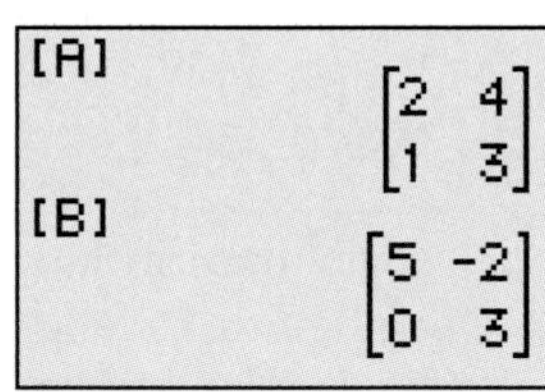

Enter *B*.

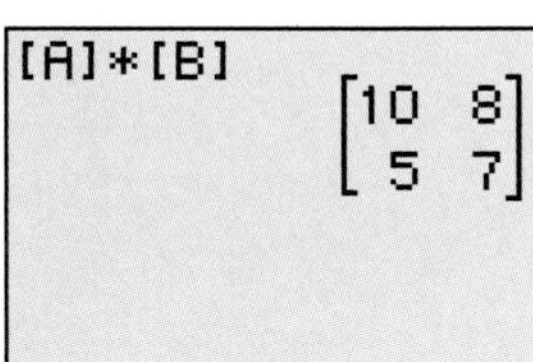

Compute *AB*.

Example 7 Powers of a Matrix with a Calculator

Consider the matrix

$$A = \begin{bmatrix} 1 & 2 \\ 3 & 4 \end{bmatrix}.$$

Determine how to compute A^2, A^3, and A^4, on your calculator. Does your calculator have a way of using the previous answer and multiplying it by A as follows to get the next power? Press the Enter key to repeat the previous calculation of Ans * A to get further powers.

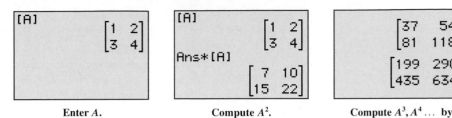

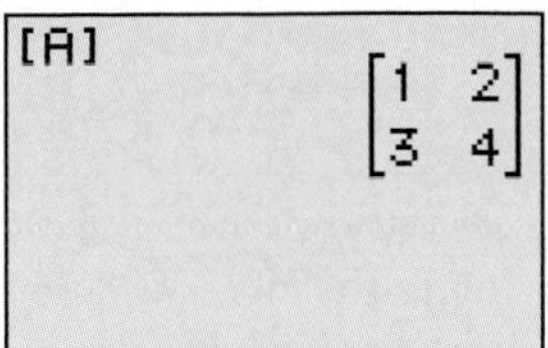

Enter A.

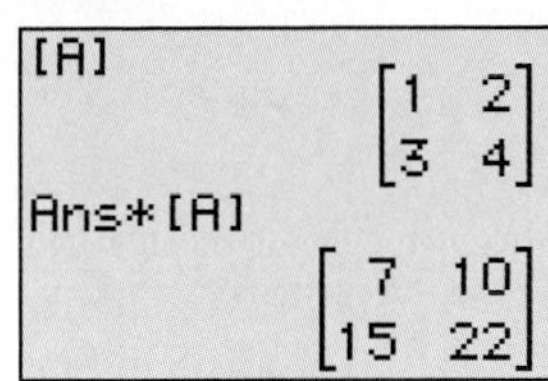

Compute A^2.

[37 54]
[81 118]
[199 290]
[435 634]

Compute $A^3, A^4 \ldots$ by pressing ENTER.

At times one is only interested in a single power of a matrix. Some calculators have the facility to compute a single power. Consider the matrix

$$A = \begin{bmatrix} 1 & 2 \\ 3 & 4 \end{bmatrix}.$$

Determine whether it is possible to arrive at the value of A^4 directly as follows.

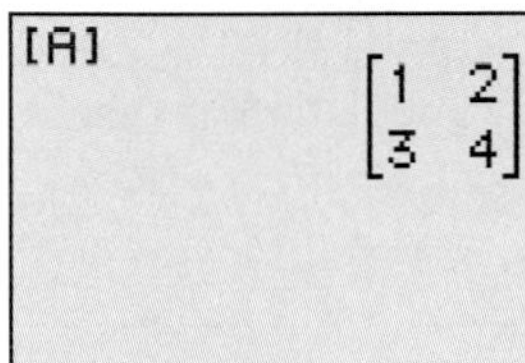

Enter A.

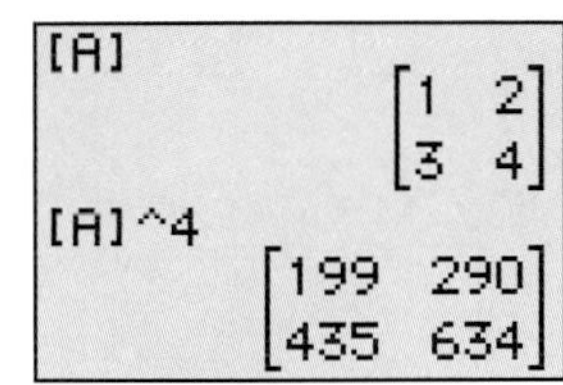

Compute A^4.

Self-Check 4

(a) Compute AB using a calculator, where

$$A = \begin{bmatrix} 2.1 & 3.4 & 9 \\ 3.2 & 5 & -3.7 \end{bmatrix} \quad \text{and} \quad B = \begin{bmatrix} 4.1 & -5.8 \\ 2.3 & 9.8 \\ 0 & 1.5 \end{bmatrix}$$

(b) Compute A^5 where $A = \begin{bmatrix} 2 & -3 & 1 \\ 4 & 0 & -2 \\ 1 & 3 & -1 \end{bmatrix}$.

Further Application . . .

We have discussed how the rule for multiplying matrices is motivated by applications. We now give an example of how matrix multiplication arises in demography.

Example 8 A Population Movement Model

In this example, we develop a model of population movement between cities and surrounding suburbs in the United States.

The number of people living in cities in the United States during 1995 was 60 million, whereas the number of people living in the surrounding suburbs was 140 million. Let us represent this information by the matrix

$$X_0 = \begin{bmatrix} 60 \\ 140 \end{bmatrix}.$$

Consider the population flow from cities to suburbs. During 1995, the probability of a person staying in the city was 0.96 (96% chance of staying in the city). The probability of moving to the suburbs was 0.04 (4% chance). For the reverse flow, the probability of a person moving to the city was 0.01 and the probability of remaining in suburbia was 0.99. These probabilities can be written as the elements of a matrix A called a **transition matrix.**

$$\begin{array}{cccl} & \multicolumn{2}{c}{\text{(From)}} & \\ & \text{City} & \text{Suburb} & \text{(To)} \\ A = & \left[\begin{matrix} 0.96 \\ 0.04 \end{matrix}\right. & \left.\begin{matrix} 0.01 \\ 0.99 \end{matrix}\right] & \begin{matrix} \text{City} \\ \text{Suburb} \end{matrix} \end{array}$$

The probability of moving from location a to location b is given by the element in column a and row b. Consider now the population distribution in 1996, one year later. We can reason as follows:

$$\begin{aligned} \text{City population in 1996} &= \begin{matrix}\text{People who remained}\\ \text{from 1995}\end{matrix} + \begin{matrix}\text{People who moved in}\\ \text{from the suburbs}\end{matrix} \\ &= (0.96 \times 60) + (0.01 \times 140) \\ &= 59 \text{ million} \end{aligned}$$

$$\begin{aligned} \text{Suburban population in 1996} &= \begin{matrix}\text{People who moved in}\\ \text{from the city}\end{matrix} + \begin{matrix}\text{People who stayed}\\ \text{from 1995}\end{matrix} \\ &= (0.04 \times 60) + (0.99 \times 140) \\ &= 141 \text{ million} \end{aligned}$$

Observe that we can arrive at these numbers using matrix multiplication.

$$\begin{bmatrix} 0.96 & 0.01 \\ 0.04 & 0.99 \end{bmatrix} \begin{bmatrix} 60 \\ 140 \end{bmatrix} = \begin{bmatrix} 59 \\ 141 \end{bmatrix}$$

Using 1995 as the base year, let X_1 be the population in 1996, one year later. We can write

$$X_1 = AX_0$$

Assume that the population flow represented by the matrix A is unchanged over the years. The population distributions after 2, 3, . . . , n years are given by

$$X_2 = AX_1, X_3 = AX_2, \ldots, X_n = AX_{n-1}$$

Using these figures for X_0 and A, the predictions of this model (using a calculator, see Further Ideas 2) are

$$\begin{array}{cccc} 1995 & 1996 & 1997 & 1998 \\ X_0 = \begin{bmatrix} 60 \\ 140 \end{bmatrix} \begin{matrix} \text{City} \\ \text{Suburb} \end{matrix}, & X_1 = \begin{bmatrix} 59 \\ 141 \end{bmatrix}, & X_2 = \begin{bmatrix} 58.05 \\ 141.95 \end{bmatrix}, & X_3 = \begin{bmatrix} 57.15 \\ 142.85 \end{bmatrix}, \quad \ldots \end{array}$$

rounded to two decimal places.

Observe how the city population is decreasing annually, whereas that of the suburbs is increasing. We can also use this model to make a prediction for the population at any time in the future. The preceding steps imply that the population X_n, n years in the future, is given by

$$X_n = A^n X_0$$

For example, the predicted distribution for the year 2010, 15 years after 1995, is (using a calculator)

$$X_{15} = A^{15}X_0 = \begin{bmatrix} 0.96 & 0.01 \\ 0.04 & 0.99 \end{bmatrix}^{15} \begin{bmatrix} 60 \\ 140 \end{bmatrix} = \begin{bmatrix} 49.27 \\ 150.73 \end{bmatrix} \begin{matrix} \text{City} \\ \text{Suburb} \end{matrix}$$

A number of such long-term predictions reveal that the population distribution approaches

$$\begin{bmatrix} 40 \\ 160 \end{bmatrix}.$$

Thus, if conditions do not change, the model predicts that city population will gradually approach 40 million whereas the population of suburbia will approach 160 million.

Further Ideas 2 To calculate the population distributions $X_1, X_2, X_3, \ldots$ of Example 8 on a graphing calculator, enter matrix A, and X_0 as matrix B. Compute $X_1 = A * B$. Repeated use of $A *$ Ans then gives further population distributions. Verify the values of

$$X_1 = \begin{bmatrix} 59 \\ 141 \end{bmatrix}, X_2 = \begin{bmatrix} 58.05 \\ 141.95 \end{bmatrix}, X_3 = \begin{bmatrix} 57.15 \\ 142.85 \end{bmatrix}, \text{ and } X_{15} = \begin{bmatrix} 49.27 \\ 150.73 \end{bmatrix}$$

of Example 8 using your graphing calculator. Additionally, verify that the long-term prediction is $\begin{bmatrix} 40 \\ 160 \end{bmatrix}$.

Answers to Self-Check Exercises

1. $AB = \begin{bmatrix} 4 & 6 & 11 \\ 6 & -1 & 14 \end{bmatrix}$.
2. AB exists, AC does not exist, BC exists.
3. AB is a 2×6 matrix.
4. **(a)** $AB = \begin{bmatrix} 16.43 & 34.64 \\ 24.62 & 24.89 \end{bmatrix}$, **(b)** $A^5 = \begin{bmatrix} 281 & -651 & 151 \\ 846 & -54 & -522 \\ 85 & 750 & -436 \end{bmatrix}$.

Answers to Further Ideas Exercises

1. $A^4 = \begin{bmatrix} 61 & -60 \\ -20 & 21 \end{bmatrix}$.

EXERCISES 7.2

In Exercises 1–16, perform the indicated computations, if possible, using the matrices, *A*, *B*, *C*, and *D*. Check your answers using a calculator.

$$A = \begin{bmatrix} 2 & 0 \\ -1 & 3 \end{bmatrix}, \quad B = \begin{bmatrix} 1 & 2 \\ 3 & 4 \end{bmatrix}, \quad C = \begin{bmatrix} 1 & 0 & 2 \\ -1 & 3 & 4 \end{bmatrix},$$

$$D = \begin{bmatrix} 3 & 0 \\ -2 & 4 \end{bmatrix}$$

1. AB **2.** BA **3.** AC **4.** CA

5. AD **6.** CD **7.** BC **8.** DC

9. ABC **10.** A^2 **11.** B^2 **12.** C^2

13. $AB + A$ **14.** $2AB$ **15.** DB **16.** DAB

In Exercises 17–26, perform the indicated computations, if possible, using the matrices, *A*, *B*, *C*, *D*, and *E*. Check your answers using a calculator.

$$A = \begin{bmatrix} 1 & 2 \\ -1 & 3 \end{bmatrix}, \quad B = \begin{bmatrix} 0 & 4 \\ 6 & 1 \end{bmatrix}, \quad C = \begin{bmatrix} 1 \\ 2 \end{bmatrix},$$

$$D = \begin{bmatrix} 1 & 3 \end{bmatrix}, \quad E = \begin{bmatrix} 4 & -1 \\ 2 & 1 \end{bmatrix}$$

17. AB **18.** BA **19.** AC

20. CD **21.** DE **22.** $2AE$

23. $2AB - 3AE$ **24.** CB **25.** $A^2 + B$

26. $3AB + 2C^2$

In Exercises 27–40, *A*, *B*, *C*, *D*, *E*, and *F* are matrices of the given sizes. Give the sizes of the matrix expressions, if they exist.

$A\ 2 \times 2$, $B\ 2 \times 2$, $C\ 2 \times 3$, $D\ 3 \times 3$, $E\ 2 \times 1$, $F\ 1 \times 3$

27. AB **28.** AC **29.** CA

30. AE **31.** AF **32.** $(FD)^2$

33. EA **34.** CD **35.** $2DA$

36. ABE **37.** EFD **38.** AB^2C

39. $A^4 + 3B$ **40.** $CD^2 + 4CD$

In Exercises 41–50, *P*, *Q*, *R*, *S*, and *T* are matrices of the given sizes. Give the sizes of the matrix expressions, if they exist.

$P\ 2 \times 2$, $Q\ 2 \times 4$, $R\ 4 \times 4$, $S\ 4 \times 2$, $T\ 4 \times 2$

41. PQ **42.** PR **43.** QRT

44. $(SP)^2$ **45.** QP **46.** $S + 3T$

47. $RS + T$ **48.** $QT - 5P^3$ **49.** $6TQ + 8RS$

50. $R - 4ST$

In Exercises 51–62, perform the indicated computations, if possible, using the matrices *A*, *B*, *C*, and *D*. Check your answers using a calculator.

$$A = \begin{bmatrix} 1 & -1 \\ 2 & 0 \end{bmatrix}, \quad B = \begin{bmatrix} 3 & -2 \\ 4 & 1 \end{bmatrix}, \quad C = \begin{bmatrix} 1 & 0 \\ 2 & 1 \\ -1 & 3 \end{bmatrix},$$

$$D = \begin{bmatrix} 2 & 0 & 1 \\ 1 & 4 & -3 \end{bmatrix}$$

51. ABD **52.** CAD

53. $AC + BC$ **54.** A^3

55. A^2BC **56.** $AD + 5BD + 4D$

57. $CBAB$ **58.** AD^2

59. DA **60.** $DC + 2A^2$

61. $(DC)^2 + A^2$ **62.** $CAB + 2C$

In Exercises 63–72, *A*, *B*, *C*, *D*, and *E* are matrices of the given sizes. Give the sizes of the matrix expressions, if they exist.

$A\ 2 \times 2$, $B\ 2 \times 4$, $C\ 4 \times 4$, $D\ 4 \times 2$, $E\ 4 \times 2$

63. A^2 **64.** AB^2 **65.** $ABCD$

66. BC^2D **67.** $EBCD$ **68.** $(EBCD)^3$

69. $(AB)^4$ **70.** $CD + CEA^2$ **71.** $BCD + 2A^2$

72. $(C^3DB)^5$

In Exercises 73–80, perform the indicated operations on a calculator using the matrices *A* and *B*.

$$A = \begin{bmatrix} 3.2 & 5.8 \\ 0 & -2.9 \end{bmatrix}, \quad B = \begin{bmatrix} 1.4 & 7 \\ -3.8 & 9 \end{bmatrix}$$

73. AB **74.** BA

75. $3.5AB$ **76.** $3.4AB + 2.1A$

77. $AB - BA$ **78.** A^4

79. $0.0007AB^5$ **80.** $0.3ABA - 0.1A^2B^3$

In Exercises 81–86, perform the indicated operations on a calculator, if possible, using the matrices *P*, *Q*, and *R*.

$$P = \begin{bmatrix} 0 & -2.1 \\ 4.3 & 7 \end{bmatrix}, \quad Q = \begin{bmatrix} 1 & 2.8 \\ -5.1 & 6.8 \end{bmatrix},$$

$$R = \begin{bmatrix} 1.1 & 4 & -1.3 \\ 2 & 3.8 & -5 \end{bmatrix}$$

81. PQ **82.** PQR **83.** $3.7RP$

84. $3PR + 4QR$ **85.** R^2 **86.** $4.5P^2R - 3.8Q^3R$

POPULATION MOVEMENT

87. Construct a mathematical model of population flow between metropolitan and nonmetropolitan areas of the United States, given that their respective populations in 1995 were 200 million and 65 million. The probabilities are given by the matrix

$$\begin{array}{c} \text{(From)} \\ \begin{array}{cc} \text{Metro} & \text{Nonmetro} \end{array} \quad \text{(To)} \\ \begin{bmatrix} 0.99 & 0.02 \\ 0.01 & 0.98 \end{bmatrix} \begin{array}{l} \text{Metro} \\ \text{Nonmetro} \end{array} \end{array}$$

Predict the population distributions of metropolitan and nonmetropolitan areas for the years 1996 through 2000 (in millions, rounded to two decimal places). Predict also the population for the year 2010 and the long-term prediction.

88. In the period 1990 to 1995, the total population of the United States increased by 1% per annum. Assume that the population increases annually by 1% during the years immediately following. Build this factor into the model of the Exercise 87, and predict the populations of metropolitan and nonmetropolitan areas for the years 2000 and 2010 (in millions, to two decimal places).

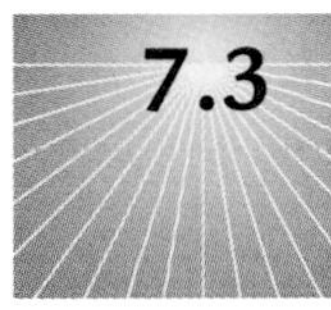

The Inverse of a Matrix and the Interdependence of Industries

- Identity Matrix • Inverse of a Matrix • Solving Systems of Equations
- The Leontief Model from Economics

In this section, we introduce the concept of matrix inverse. We shall see how an inverse can be used to solve certain systems of linear equations and to analyze the interdependence of industries.

IDENTITY MATRIX

We first introduce matrices that play a role in matrix theory that is similar to the role played by 1 for real numbers. The characteristic of 1 is that $a \times 1 = 1 \times a = a$ for any real number a. A square matrix that has 1s in the diagonal locations from the upper-left corner to the lower-right corner and zeros elsewhere is called an **identity matrix.** The notation for such an $n \times n$ matrix is I_n, and the 1s are called the **main diagonal** of the matrix. If A is an arbitrary matrix the same size as I_n, then it is easy to see that

$$AI_n = I_nA = A$$

The following are examples of identity matrices:

$$I_2 = \begin{bmatrix} 1 & 0 \\ 0 & 1 \end{bmatrix}, \quad I_3 = \begin{bmatrix} 1 & 0 & 0 \\ 0 & 1 & 0 \\ 0 & 0 & 1 \end{bmatrix}, \quad I_4 = \begin{bmatrix} 1 & 0 & 0 & 0 \\ 0 & 1 & 0 & 0 \\ 0 & 0 & 1 & 0 \\ 0 & 0 & 0 & 1 \end{bmatrix}$$

Example 1 **Illustrating the Characteristics of an Identity Matrix**

Let $A = \begin{bmatrix} 5 & 0 \\ 3 & -2 \end{bmatrix}$. Show that $AI_2 = I_2A = A$.

SOLUTION If we multiply, we get

$$AI_2 = \begin{bmatrix} 5 & 0 \\ 3 & -2 \end{bmatrix}\begin{bmatrix} 1 & 0 \\ 0 & 1 \end{bmatrix} = \begin{bmatrix} 5 & 0 \\ 3 & -2 \end{bmatrix} = A$$

$$I_2A = \begin{bmatrix} 1 & 0 \\ 0 & 1 \end{bmatrix}\begin{bmatrix} 5 & 0 \\ 3 & -2 \end{bmatrix} = \begin{bmatrix} 5 & 0 \\ 3 & -2 \end{bmatrix} = A$$

Inverse of a Matrix

To motivate the concept of the inverse of a matrix, let us look at the multiplicative inverse of a real number. If a is a nonzero real number, then we say that b is the multiplicative inverse of a if

$$ab = ba = 1$$

For example, $\frac{1}{4}$ is the inverse of 4 because

$$4 \times \tfrac{1}{4} = \tfrac{1}{4} \times 4 = 1$$

We now give the definition of inverse of a square matrix based on this observation.

> ***Inverse of a Matrix*** An $n \times n$ matrix A has an inverse B if and only if $AB = BA = I_n$.

It is customary to denote the inverse of a matrix A by A^{-1}.

A nonsquare matrix does not have an inverse. Neither, however, does every square matrix have an inverse. We shall see an example of such a matrix later in this section. A matrix is said to be **invertible** if it has an inverse.

Example 2 **Showing that a Matrix Is the Inverse of a Given Matrix**

Prove that the matrix $A = \begin{bmatrix} 1 & 2 \\ 3 & 4 \end{bmatrix}$ has inverse $B = \begin{bmatrix} -2 & 1 \\ \frac{3}{2} & -\frac{1}{2} \end{bmatrix}$.

SOLUTION We have that

$$AB = \begin{bmatrix} 1 & 2 \\ 3 & 4 \end{bmatrix}\begin{bmatrix} -2 & 1 \\ \frac{3}{2} & -\frac{1}{2} \end{bmatrix} = \begin{bmatrix} 1 & 0 \\ 0 & 1 \end{bmatrix} = I_2$$

and

$$BA = \begin{bmatrix} -2 & 1 \\ \frac{3}{2} & -\frac{1}{2} \end{bmatrix}\begin{bmatrix} 1 & 2 \\ 3 & 4 \end{bmatrix} = \begin{bmatrix} 1 & 0 \\ 0 & 1 \end{bmatrix} = I_2$$

Thus $AB = BA = I_2$, proving that the matrix A has inverse B. We write

$$A^{-1} = \begin{bmatrix} -2 & 1 \\ \frac{3}{2} & -\frac{1}{2} \end{bmatrix}$$

We now present a method (without proof) for finding the inverse of a matrix. The method involves finding the reduced form of a matrix using the method of Gauss–Jordan elimination.

Gauss–Jordan Elimination for Finding the Inverse of a Matrix

Let A be an $n \times n$ matrix.

1. Adjoin the identity $n \times n$ matrix I_n to A to form the matrix $[A : I_n]$.
2. Compute the reduced form of $[A : I_n]$ using Gauss–Jordan elimination. If the reduced form is of the form $[I_n : B]$, then B is the inverse of A. If the reduced echelon form is not of this form, then A has no inverse.

Example 3 Finding the Inverse of a 2 × 2 Matrix

Determine the inverse of the matrix $\begin{bmatrix} 1 & 2 \\ 3 & 4 \end{bmatrix}$.

SOLUTION In this example, we derive the inverse discussed in Example 2.

Adjoin I_2 to A and compute the reduced form of the resulting matrix.

$$[A : I_2] = \begin{bmatrix} 1 & 2 & 1 & 0 \\ 3 & 4 & 0 & 1 \end{bmatrix} \underset{\textbf{Add } (-3) \textbf{ R1 to R2}}{\approx} \begin{bmatrix} 1 & 2 & 1 & 0 \\ 0 & -2 & -3 & 1 \end{bmatrix}$$

$$\underset{(-\frac{1}{2}) \textbf{ R2}}{\approx} \begin{bmatrix} 1 & 2 & 1 & 0 \\ 0 & 1 & \frac{3}{2} & -\frac{1}{2} \end{bmatrix} \underset{\textbf{Add } (-2) \textbf{ R2 to R1}}{\approx} \begin{bmatrix} 1 & 0 & -2 & 1 \\ 0 & 1 & \frac{3}{2} & -\frac{1}{2} \end{bmatrix}$$

Thus, $A^{-1} = \begin{bmatrix} -2 & 1 \\ \frac{3}{2} & -\frac{1}{2} \end{bmatrix}$

Example 4 Finding the Inverse of a 3 × 3 Matrix

Determine the inverse of the matrix

$$A = \begin{bmatrix} 1 & -1 & -2 \\ 2 & -3 & -5 \\ -1 & 3 & 5 \end{bmatrix}$$

SOLUTION Applying the method of Gauss–Jordan elimination, we get

$$[A:I_3] = \begin{bmatrix} 1 & -1 & -2 & 1 & 0 & 0 \\ 2 & -3 & -5 & 0 & 1 & 0 \\ -1 & 3 & 5 & 0 & 0 & 1 \end{bmatrix}$$

$$\underset{\substack{\textbf{Add } (-2)\textbf{ R1 to R2} \\ \textbf{Add R1 to R3}}}{\approx} \begin{bmatrix} 1 & -1 & -2 & 1 & 0 & 0 \\ 0 & -1 & -1 & -2 & 1 & 0 \\ 0 & 2 & 3 & 1 & 0 & 1 \end{bmatrix}$$

$$\underset{(-1)\textbf{ R2}}{\approx} \begin{bmatrix} 1 & -1 & -2 & 1 & 0 & 0 \\ 0 & 1 & 1 & 2 & -1 & 0 \\ 0 & 2 & 3 & 1 & 0 & 1 \end{bmatrix}$$

$$\underset{\substack{\textbf{Add R2 to R1} \\ \textbf{Add } (-2)\textbf{ R2 to R3}}}{\approx} \begin{bmatrix} 1 & 0 & -1 & 3 & -1 & 0 \\ 0 & 1 & 1 & 2 & -1 & 0 \\ 0 & 0 & 1 & -3 & 2 & 1 \end{bmatrix}$$

$$\underset{\substack{\textbf{Add R3 to R1} \\ \textbf{Add } (-1)\textbf{ R3 to R2}}}{\approx} \begin{bmatrix} 1 & 0 & 0 & 0 & 1 & 1 \\ 0 & 1 & 0 & 5 & -3 & -1 \\ 0 & 0 & 1 & -3 & 2 & 1 \end{bmatrix}$$

Thus, $A^{-1} = \begin{bmatrix} 0 & 1 & 1 \\ 5 & -3 & -1 \\ -3 & 2 & 1 \end{bmatrix}$

Self-Check 1 Find the inverse of the matrix $A = \begin{bmatrix} 2 & -1 & 1 \\ -5 & 2 & 1 \\ 3 & -1 & -1 \end{bmatrix}$.

The following example illustrates a square matrix that does not have an inverse.

Example 5 Determine, if possible, the inverse of the matrix $A = \begin{bmatrix} 1 & 2 \\ 2 & 4 \end{bmatrix}$.

SOLUTION

$$[A:I_2] = \begin{bmatrix} 1 & 2 & 1 & 0 \\ 2 & 4 & 0 & 1 \end{bmatrix} \underset{\textbf{Add } (-2)\textbf{ R1 to R2}}{\approx} \begin{bmatrix} 1 & 2 & 1 & 0 \\ 0 & 0 & -2 & 1 \end{bmatrix}$$

If we proceed from this stage to the reduced form, the first matrix (the leftmost two columns) will not change. This part will not become I_2. Thus, A does not have an inverse.

Self-Check 2 Show that the following matrix A does not have an inverse.

$$A = \begin{bmatrix} 1 & 2 & 0 \\ 4 & 5 & 1 \\ 2 & 1 & 1 \end{bmatrix}$$

Matrix Inverse Using a Calculator

Calculators can be used to find the inverses of matrices. We illustrate the steps. You should determine how to implement these steps on your calculator. Let us compute A^{-1} for the following matrix A:

$$A = \begin{bmatrix} 2 & 4 \\ 1 & 3 \end{bmatrix}$$

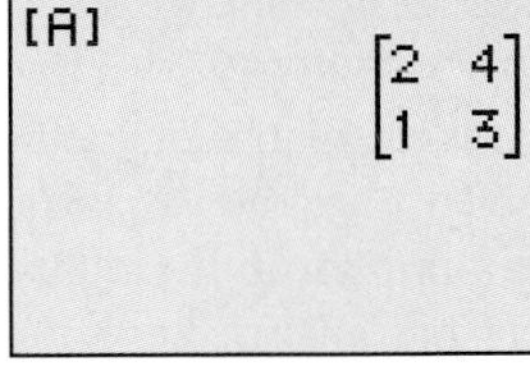

Enter A.

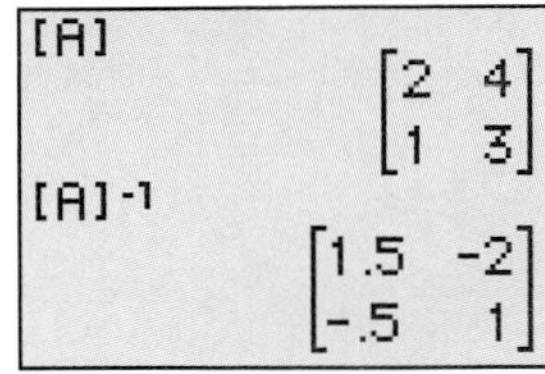

Compute A^{-1}.

Self-Check 3 Use your calculator to compute the inverse of the following matrix A. Give the elements of the inverse, rounded to two decimal places.

$$A = \begin{bmatrix} 0.5 & 1 & 0 \\ 3 & -0.4 & 1 \\ -0.6 & 5 & -2 \end{bmatrix}$$

Solving Systems of Equations

A matrix inverse can be used to solve certain systems of equations. Consider the following system of three linear equations in three variables:

$$\begin{aligned} x - y - 2z &= 1 \\ 2x - 3y - 5z &= 3 \\ -x + 3y + 5z &= -2 \end{aligned}$$

This system can be written in the following matrix form, using the matrix of coefficients.

$$\begin{bmatrix} 1 & -1 & -2 \\ 2 & -3 & -5 \\ -1 & 3 & 5 \end{bmatrix} \begin{bmatrix} x \\ y \\ z \end{bmatrix} = \begin{bmatrix} 1 \\ 3 \\ -2 \end{bmatrix}$$

If the matrix of coefficients is invertible, the system has a unique solution given by

$$\begin{bmatrix} x \\ y \\ z \end{bmatrix} = \begin{bmatrix} 1 & -1 & -2 \\ 2 & -3 & -5 \\ -1 & 3 & 5 \end{bmatrix}^{-1} \begin{bmatrix} 1 \\ 3 \\ -2 \end{bmatrix}$$

This inverse was already found in Example 4. Using that result, we get

$$\begin{bmatrix} x \\ y \\ z \end{bmatrix} = \begin{bmatrix} 0 & 1 & 1 \\ 5 & -3 & -1 \\ -3 & 2 & 1 \end{bmatrix} \begin{bmatrix} 1 \\ 3 \\ -2 \end{bmatrix} = \begin{bmatrix} 1 \\ -2 \\ 1 \end{bmatrix}$$

The unique solution is $x = 1$, $y = -2$, and $z = 1$.

The previous example illustrates a general method for solving certain systems of linear equations. A system of linear equations can be written in the form

$$AX = B$$

where A is the matrix of coefficients of the system, X is a column matrix in which the elements are the variables, and B is a column matrix in which the elements are the numbers on the right-hand sides of the equations. If A is an $n \times n$ matrix that has an inverse, both sides of this equation can be multiplied by A^{-1} to get

$$\begin{aligned} A^{-1}AX &= A^{-1}B \\ I_nX &= A^{-1}B \\ X &= A^{-1}B \end{aligned}$$

yielding the solution.

A system of linear equations can be written in matrix form $AX = B$. If A has an inverse, the system has a unique solution given by $X = A^{-1}B$.

Self-Check 4

(a) Use the matrix inverse method to solve the system of equations.

$$\begin{aligned} 2x - y + z &= 1 \\ -5x + 2y + z &= -2 \\ 3x - y - z &= 3 \end{aligned}$$

Hint: The inverse of the matrix of coefficients of this system was determined in Self-Check 1.

(b) Use the matrix inverse method on a calculator to solve the system of equations. Express the solution to two decimal places.

$$\begin{aligned} 3x - y + z &= 4 \\ 2x + 3y - z &= -2 \\ x + 3y - 4z &= 9 \end{aligned}$$

Hint: Enter $A = \begin{bmatrix} 3 & -1 & 1 \\ 2 & 3 & -1 \\ 1 & 3 & -4 \end{bmatrix}$ and $B = \begin{bmatrix} 4 \\ -2 \\ 9 \end{bmatrix}$ and use keystrokes such as

[MATRX] 1 [x⁻¹] [×] [MATRX] 2 [ENTER].

Application . . .

Example 6 The Leontief Model from Economics

The matrix concepts introduced in this section are used to analyze the interdependence of economies in a mathematical model called the Leontief input–output model. Wassily Leontief received a Nobel Prize in 1973 for his work in this field. The practical applications of this model have proliferated until it has now become a standard tool for investigating economic structures ranging from cities and corporations to states and countries.

Consider an economic situation that involves a number of interdependent industries. The output of any one industry is needed as input by the other industries and also by nonproducing parts of the economy, such as consumers and governments, called the **open sector.** The Leontief model is used to determine the production required of each industry to meet industrial and open-sector requirements.

The interdependence of the industries is described by a matrix A called the **input–output matrix** and the demands of the open sector are described by a column matrix D. Let X be a column matrix that describes the outputs of the industries necessary to meet the demands of all the industries and the open sector. It can be shown that X is given by

$$X = (I - A)^{-1}D$$

where I is the appropriate identity matrix.

Let us see how this model works. Consider an economy consisting of three industries having the following input–output matrix A and open-sector demand matrix D.

$$A = \begin{bmatrix} 0.20 & 0.20 & 0.30 \\ 0.50 & 0.50 & 0 \\ 0 & 0 & 0.20 \end{bmatrix}, \qquad D = \begin{bmatrix} 9 \\ 12 \\ 16 \end{bmatrix}$$

The element a_{ij} of A is the amount from industry i required to produce one unit from industry j. For example $a_{12} = 0.20$ means that 20 cents' worth of the product of industry 1 goes into every dollar of the product of industry 2. The fact that $a_{22} = 0.50$ means 50 cents' worth of the product of industry 2 goes into the production of every dollar of its own product. The elements of D imply that the demands of the open sector from industries 1, 2, and 3 are \$9 million, \$12 million, and \$16 million, respectively.

Let us determine the output levels required of the industries to meet the demands of the other industries and of the open sector. Compute $X = (I - A)^{-1}D$ using a calculator. Enter matrices I_3, A, and D into the calculator, and use such keystrokes as

(MATRX 1 − MATRX 2) x^{-1} × MATRX 3 ENTER.

We get

$$X = \begin{bmatrix} 33 \\ 57 \\ 20 \end{bmatrix}$$

The output levels necessary from industries 1, 2, and 3 to meet all the demands are \$33 million, \$57 million, and \$20 million, respectively.

Readers who are interested in finding out more about applications of this model should read "The World Economy of the Year 2000" by Wassily W. Leontief, (*Scientific American* 1980; September: 166). The article describes the application of this model to a world economy. The model was commissioned by the United Nations, with special financial support from the Netherlands. In the model, the world is divided into 15 distinct geographic regions, each one described by an individual input–output matrix. The regions are then linked by a larger matrix, which is used in an input–output model. Overall, more than 200 economic sectors are included in the model. By feeding in various values, economists use the model to create scenarios of future world economic conditions.

Answers to Self-Check Exercises

1. $A^{-1} = \begin{bmatrix} 1 & 2 & 3 \\ 2 & 5 & 7 \\ 1 & 1 & 1 \end{bmatrix}$

2. $A : I_3 \approx \cdots \approx \begin{bmatrix} 1 & 0 & \frac{2}{3} & -\frac{5}{3} & \frac{2}{3} & 0 \\ 0 & 1 & -\frac{1}{3} & \frac{4}{3} & -\frac{1}{3} & 0 \\ 0 & 0 & 0 & 2 & -1 & 1 \end{bmatrix}$

Thus, the reduced form of $[A : I_3]$ will not be of the type $[I_3 : B]$.

3. $A^{-1} = \begin{bmatrix} -1.27 & 0.61 & 0.30 \\ 1.64 & -0.30 & -0.15 \\ 4.47 & -0.94 & -0.97 \end{bmatrix}$

4. **(a)** Solution is $x = 6$, $y = 13$, and $z = 2$.
(b) Solution is $x = 1.68$, $y = -3.19$, and $z = -4.23$.

EXERCISES 7.3

In Exercises 1–8, select the appropriate I_n for each matrix, and show that $AI_n = I_nA$.

1. $\begin{bmatrix} 3 & 6 \\ 1 & 0 \end{bmatrix}$ **2.** $\begin{bmatrix} -1 & 5 \\ 2 & 3 \end{bmatrix}$

3. $\begin{bmatrix} 4 & -5 \\ 2 & 3 \end{bmatrix}$ **4.** $\begin{bmatrix} 1 & 2 & -1 \\ 3 & 0 & 4 \\ 1 & 6 & 2 \end{bmatrix}$

5. $\begin{bmatrix} 0 & 7 & 6 \\ -2 & 4 & 3 \\ 1 & 9 & 8 \end{bmatrix}$ **6.** $\begin{bmatrix} 1 & 5 & 7 \\ -4 & 3 & 2 \\ 0 & 6 & 1 \end{bmatrix}$

7. $\begin{bmatrix} 1 & 2 & -1 & 3 \\ 4 & 6 & 0 & 5 \\ 7 & 1 & 2 & 9 \\ 12 & 3 & 4 & 1 \end{bmatrix}$ **8.** $\begin{bmatrix} 5 & 0 & 0 & 1 \\ 2 & 6 & -5 & 9 \\ 9 & 8 & 7 & 1 \\ 2 & 3 & 5 & 4 \end{bmatrix}$

In Exercises 9–20, determine whether the given pairs of matrices are inverses of each other by checking to see whether their product is the identity matrix.

9. $\begin{bmatrix} 3 & 7 \\ 2 & 5 \end{bmatrix}, \begin{bmatrix} 5 & -7 \\ -2 & 3 \end{bmatrix}$ **10.** $\begin{bmatrix} 2 & -1 \\ 3 & 0 \end{bmatrix}, \begin{bmatrix} 1 & 3 \\ 1 & 6 \end{bmatrix}$

11. $\begin{bmatrix} 5 & -2 \\ -7 & 3 \end{bmatrix}, \begin{bmatrix} 3 & 2 \\ 7 & 5 \end{bmatrix}$ **12.** $\begin{bmatrix} -1 & 1 \\ 3 & -2 \end{bmatrix}, \begin{bmatrix} 2 & 1 \\ 3 & 1 \end{bmatrix}$

13. $\begin{bmatrix} 3 & 2 \\ 4 & 3 \end{bmatrix}, \begin{bmatrix} 3 & -2 \\ -4 & 3 \end{bmatrix}$

14. $\begin{bmatrix} 2 & 3 \\ -1 & 6 \end{bmatrix}, \begin{bmatrix} 5 & 6 \\ -3 & -4 \end{bmatrix}$

15. $\begin{bmatrix} 1 & 0 & 2 \\ -1 & 2 & 3 \\ 1 & -1 & 0 \end{bmatrix}, \begin{bmatrix} 3 & -2 & -4 \\ 3 & -2 & -5 \\ -1 & 1 & 2 \end{bmatrix}$

16. $\begin{bmatrix} 1 & 2 & 4 \\ -3 & 0 & 1 \\ 6 & 2 & 5 \end{bmatrix}, \begin{bmatrix} -1 & 2 & 0 \\ -1 & 2 & -2 \\ 1 & -1 & 1 \end{bmatrix}$

17. $\begin{bmatrix} 4 & -1 & -4 \\ 3 & 0 & -4 \\ 3 & -1 & -3 \end{bmatrix}, \begin{bmatrix} 4 & -1 & -4 \\ 3 & 0 & -4 \\ 3 & -1 & -3 \end{bmatrix}$

18. $\begin{bmatrix} 2 & 1 & 0 \\ 3 & 4 & 2 \\ 5 & 1 & 3 \end{bmatrix}, \begin{bmatrix} 1 & 3 & -5 \\ -2 & -6 & 10 \\ 0 & 4 & 7 \end{bmatrix}$

19. $\begin{bmatrix} 6 & 2 & 3 \\ 5 & 1 & -1 \\ -7 & 8 & 2 \end{bmatrix}, \begin{bmatrix} 2 & -1 & 0 \\ -3 & 1 & -6 \\ -2 & 1 & 4 \end{bmatrix}$

20. $\begin{bmatrix} 1 & 0 & 0 \\ 0 & 2 & 0 \\ 0 & 0 & 3 \end{bmatrix}, \begin{bmatrix} 1 & 0 & 0 \\ 0 & \frac{1}{2} & 0 \\ 0 & 0 & \frac{1}{3} \end{bmatrix}$

In Exercises 21–28, determine the inverses (if they exist) of the matrices.

21. $\begin{bmatrix} 1 & -3 \\ -2 & 5 \end{bmatrix}$ 22. $\begin{bmatrix} 2 & 0 \\ 0 & 4 \end{bmatrix}$ 23. $\begin{bmatrix} \frac{1}{2} & 0 \\ 0 & \frac{1}{6} \end{bmatrix}$

24. $\begin{bmatrix} 1 & 2 \\ 3 & 6 \end{bmatrix}$ 25. $\begin{bmatrix} 2 & 4 \\ 1 & 2 \end{bmatrix}$ 26. $\begin{bmatrix} 2 & -3 \\ 1 & 2 \end{bmatrix}$

27. $\begin{bmatrix} 3 & 4 \\ 2 & 3 \end{bmatrix}$ 28. $\begin{bmatrix} 5 & 2 \\ 3 & 1 \end{bmatrix}$

In Exercises 29–34, determine the inverses (if they exist) of the matrices.

29. $\begin{bmatrix} 1 & 2 & 0 \\ 0 & 1 & 1 \\ -1 & 0 & 1 \end{bmatrix}$ 30. $\begin{bmatrix} 1 & 0 & 2 \\ -3 & 3 & -1 \\ 0 & 1 & 2 \end{bmatrix}$

31. $\begin{bmatrix} 2 & -1 & 0 \\ 3 & 4 & 2 \\ -1 & 6 & 2 \end{bmatrix}$ 32. $\begin{bmatrix} 1 & 2 & 1 \\ 0 & 5 & 3 \\ 2 & -6 & -4 \end{bmatrix}$

33. $\begin{bmatrix} -1 & 4 & 0 \\ -1 & 5 & 1 \\ -1 & 0 & -3 \end{bmatrix}$ 34. $\begin{bmatrix} 1 & 0 & 2 \\ -1 & 2 & 3 \\ 1 & -1 & 0 \end{bmatrix}$

In Exercises 35–42, write each system of linear equations in matrix form. You need not solve the systems.

35. $\begin{aligned} x + y &= 2 \\ 3x - y &= 4 \end{aligned}$ 36. $\begin{aligned} 7x - 2y &= 3 \\ 2x + 5y &= 5 \end{aligned}$

37. $\begin{aligned} 2x + 3y &= 1 \\ 4x \quad &= 2 \end{aligned}$ 38. $\begin{aligned} 5x \quad &= 0 \\ 2x + y &= -2 \end{aligned}$

39. $\begin{aligned} 3x + 2y - 7z &= 1 \\ 4x + 3y - 2z &= 4 \\ x + y - z &= 5 \end{aligned}$ 40. $\begin{aligned} x + 3y - z &= 6 \\ 2x + y + z &= 8 \\ 3x + y + 4z &= 2 \end{aligned}$

41. $\begin{aligned} x + y + z &= 4 \\ 5x - y \quad &= 6 \\ x + y + 2z &= 9 \end{aligned}$ 42. $\begin{aligned} x + 3y - z &= 6 \\ 2x + y + z &= 4 \\ y + 9z &= 3 \end{aligned}$

In Exercises 43–48, write each system of linear equations in matrix form, and solve the system using the matrix inverse method. (The inverse of each matrix of coefficients has already been found in Exercises 21–34.)

43. $\begin{aligned} x - 3y &= 2 \\ -2x + 5y &= 1 \end{aligned}$ 44. $\begin{aligned} 3x + 4y &= 1 \\ 2x + 3y &= 5 \end{aligned}$

45. $\begin{aligned} 5x + 2y &= -2 \\ 3x + y &= 6 \end{aligned}$ 46. $\begin{aligned} -x + 4y \quad &= 4 \\ -x + 5y + z &= 2 \\ -x - \quad 3z &= 1 \end{aligned}$

47. $\begin{aligned} x + \quad 2z &- 1 \\ -3x + 3y - z &= 2 \\ y + 2z &= 3 \end{aligned}$ 48. $\begin{aligned} x + \quad 2z &= -2 \\ -x + 2y + 3z &= 1 \\ x - y \quad &= 3 \end{aligned}$

In Exercises 49–54, compute the inverses of the matrices, if they exist, using a calculator.

49. $\begin{bmatrix} 1 & 0 \\ 2 & 1 \end{bmatrix}$ 50. $\begin{bmatrix} 1 & 2 \\ -1 & 6 \end{bmatrix}$ 51. $\begin{bmatrix} 2 & 1 \\ 4 & 3 \end{bmatrix}$

52. $\begin{bmatrix} 0 & 1 \\ 1 & 3 \end{bmatrix}$ 53. $\begin{bmatrix} 1 & 2 \\ 3 & 6 \end{bmatrix}$ 54. $\begin{bmatrix} 2 & -3 \\ 6 & -7 \end{bmatrix}$

In Exercises 55–60, compute the inverses of the matrices, if they exist, using a calculator. Round answers to four decimal places.

55. $\begin{bmatrix} 1 & 2 & 3 \\ 0 & 1 & 2 \\ 4 & 5 & 3 \end{bmatrix}$ 56. $\begin{bmatrix} 2 & 0 & 4 \\ -1 & 3 & 1 \\ 0 & 1 & 2 \end{bmatrix}$

57. $\begin{bmatrix} 1 & 2 & -3 \\ 1 & -2 & 1 \\ 5 & -2 & -3 \end{bmatrix}$ 58. $\begin{bmatrix} 1 & 2 & -1 \\ 3 & -1 & 0 \\ 2 & -3 & 1 \end{bmatrix}$

59. $\begin{bmatrix} 1 & 2 & 3 \\ 2 & -1 & 4 \\ 0 & -1 & 1 \end{bmatrix}$ 60. $\begin{bmatrix} 1 & 2 & -1 \\ 2 & 4 & -3 \\ 1 & -2 & 0 \end{bmatrix}$

In Exercises 61–66, solve the systems of two equations in two variables by determining the inverse of the matrix of coefficients and then using matrix multiplication, on a calculator.

61. $\begin{aligned} x + 2y &= 2 \\ 3x + 5y &= 4 \end{aligned}$

62. $\begin{aligned} x + 5y &= -1 \\ 2x + 9y &= 3 \end{aligned}$

63. $\begin{aligned} x + 3y &= 5 \\ 2x + y &= 10 \end{aligned}$

64. $\begin{aligned} 2x + y &= 4 \\ 4x + 3y &= 6 \end{aligned}$

65. $\begin{aligned} 2x + 4y &= 6 \\ 3x + 8y &= 1 \end{aligned}$

66. $\begin{aligned} 3x + 9y &= 9 \\ 2x + 7y &= 4 \end{aligned}$

In Exercises 67–70, solve the systems of three equations in three variables by determining the inverse of the matrix of coefficients and then using matrix multiplication, on a calculator. Round answers to four decimal places.

67. $\begin{aligned} x + 2y - z &= 2 \\ x + y + 2z &= 0 \\ x - y - z &= 1 \end{aligned}$

68. $\begin{aligned} x - y \quad &= 1 \\ x + y + 2z &= 2 \\ x + 2y + z &= 0 \end{aligned}$

69. $\begin{aligned} x + 2y + 3z &= 1 \\ 2x + 5y + 3z &= 3 \\ x + 8z &= 15 \end{aligned}$

70. $\begin{aligned} 0.3x + 0.5y + z &= 20 \\ x - 2y + 4z &= 9 \\ 0.6x + y - 2z &= -4 \end{aligned}$

In Exercises 71–75, consider the economies defined by the input–output matrices A and open sectors D. Determine the output levels required of the industries to meet the demands of the industries and of the open sector.

71. $A = \begin{bmatrix} 0.20 & 0.60 \\ 0.40 & 0.10 \end{bmatrix}, \quad D = \begin{bmatrix} 24 \\ 12 \end{bmatrix}$

72. $A = \begin{bmatrix} 0.10 & 0.40 \\ 0.30 & 0.20 \end{bmatrix}, \quad D = \begin{bmatrix} 6 \\ 12 \end{bmatrix}$

73. $A = \begin{bmatrix} 0.30 & 0.60 \\ 0.35 & 0.10 \end{bmatrix}, \quad D = \begin{bmatrix} 42 \\ 84 \end{bmatrix}$

74. $A = \begin{bmatrix} 0.20 & 0.20 & 0.10 \\ 0 & 0.40 & 0.20 \\ 0 & 0.20 & 0.60 \end{bmatrix}, \quad D = \begin{bmatrix} 4 \\ 8 \\ 8 \end{bmatrix}$

75. $A = \begin{bmatrix} 0.20 & 0.20 & 0 \\ 0.40 & 0.40 & 0.60 \\ 0.40 & 0.10 & 0.40 \end{bmatrix}, \quad D = \begin{bmatrix} 36 \\ 72 \\ 36 \end{bmatrix}$

76. Consider the following input–output matrix that defines the interdependency of five industries.

	1	2	3	4	5
1. Auto	0.15	0.10	0.05	0.05	0.10
2. Steel	0.40	0.20	0.10	0.10	0.10
3. Electricity	0.10	0.25	0.20	0.10	0.20
4. Coal	0.10	0.20	0.30	0.15	0.10
5. Chemical	0.05	0.10	0.05	0.02	0.05

Determine **(a)** the amount of electricity consumed in producing $1 worth of steel; **(b)** the amount of steel consumed in producing $1 worth of product in the auto industry; **(c)** the largest consumer of coal; **(d)** the largest consumer of electricity; and **(e)** the industry on which the auto industry is most dependent?

77. Let a_{ij} be an arbitrary element of an input–output matrix. Why would you expect a_{ij} to satisfy the condition $0 \le a_{ij} \le 1$?

78. In an economically feasible situation, the sum of the elements of each column of the input–output matrix is less than or equal to unity. Explain why this should be so.

7.4 Determinants and Cramer's Rule

• Determinant of a 2 × 2 Matrix • Minors • Determinant of a 3 × 3 Matrix • Cramer's Rule

Associated with every square matrix is a number called its determinant. The determinant of a matrix is a tool that is used in many branches of mathematics and science. In this section, we introduce determinants and see how they can be used to solve systems of linear equations. We commence our discussion of determinants by defining the determinant of a 2 × 2 matrix. This leads to determinants of 3 × 3 matrices and then of $n \times n$ matrices.

Determinant of a 2 × 2 Matrix

Let A be a 2 × 2 matrix,

$$A = \begin{bmatrix} a_{11} & a_{12} \\ a_{21} & a_{22} \end{bmatrix}$$

The determinant of A is denoted $|A|$ and is given by

$$|A| = a_{11}a_{22} - a_{21}a_{12}$$

Example 1 **Finding the Determinant of a 2 × 2 Matrix**

Let $A = \begin{bmatrix} 1 & 3 \\ -5 & 7 \end{bmatrix}$. Compute $|A|$.

SOLUTION Using the preceding definition, we get

$$\begin{aligned} |A| &= a_{11}a_{22} - a_{21}a_{12} \\ &= (1 \times 7) - (-5 \times 3) = 7 + 15 \\ &= 22 \end{aligned}$$

Self-Check 1 Find the determinant of $A = \begin{bmatrix} 4 & -1 \\ 2 & 6 \end{bmatrix}$.

Minors

The determinant of a 3 × 3 matrix is defined in terms of determinants of 2 × 2 matrices. The determinant of a 4 × 4 matrix is defined in terms of determinants of 3 × 3 matrices, and so on. For these definitions, we need the following concept of minor. Associated with every element of a square matrix is a smaller matrix (called a submatrix), obtained by omitting the elements that lie in the row and column of the given element. The determinant of this submatrix is called the **minor** of the element.

Consider the 3 × 3 matrix

$$A = \begin{bmatrix} a_{11} & a_{12} & a_{13} \\ a_{21} & a_{22} & a_{23} \\ a_{31} & a_{32} & a_{33} \end{bmatrix}$$

Let us compute the minor of a_{23}, denoted A_{23}. We delete the row and column containing a_{23}, namely, the second row and third column of A, as shown here.

$$\begin{bmatrix} a_{11} & a_{12} & a_{13} \\ a_{21} & a_{22} & a_{23} \\ a_{31} & a_{32} & a_{33} \end{bmatrix}$$

The submatrix corresponding to a_{23} is

$$\begin{bmatrix} a_{11} & a_{12} \\ a_{31} & a_{32} \end{bmatrix}$$

The minor of a_{23}, denoted A_{23}, is the determinant of this matrix. We get

$$A_{23} = a_{11}a_{32} - a_{31}a_{12}$$

Example 2 Finding the Minor of a Given Matrix Element

Consider the matrix $A = \begin{bmatrix} -1 & 2 & 4 \\ 0 & 5 & 1 \\ 3 & -1 & 2 \end{bmatrix}$. Find the minor of a_{12}.

SOLUTION To find the minor of a_{12}, first delete the first row and second column.

$$\begin{bmatrix} -1 & 2 & 4 \\ 0 & 5 & 1 \\ 3 & -1 & 2 \end{bmatrix}$$

We get the submatrix

$$\begin{bmatrix} 0 & 1 \\ 3 & 2 \end{bmatrix}$$

We see that

$$A_{12} = \begin{vmatrix} 0 & 1 \\ 3 & 2 \end{vmatrix} = (0 \times 2) - (3 \times 1) = -3$$

Self-Check 2 Consider the matrix $A = \begin{bmatrix} 5 & 1 & -2 \\ 3 & 6 & 0 \\ 4 & 2 & 1 \end{bmatrix}$. Find A_{22}.

We are now in a position to discuss the determinant of a 3×3 matrix.

Determinant of a 3 × 3 Matrix

Consider the 3×3 matrix

$$A = \begin{bmatrix} a_{11} & a_{12} & a_{13} \\ a_{21} & a_{22} & a_{23} \\ a_{31} & a_{32} & a_{33} \end{bmatrix}$$

$|A|$ can be found by multiplying the elements of the first row by their minors and summed as follows:

$$|A| = a_{11}A_{11} - a_{12}A_{12} + a_{13}A_{13}$$

Note that the signs alternate.

Example 3 **Finding the Determinant of a 3 × 3 Matrix**

Find the determinant of the matrix $A = \begin{bmatrix} 2 & -1 & 3 \\ 0 & 1 & 4 \\ 5 & -2 & -1 \end{bmatrix}$.

SOLUTION We have that $a_{11} = 2$, $a_{12} = -1$, and $a_{13} = 3$. The minors of these elements are as follows:

$$A_{11} = \begin{vmatrix} 1 & 4 \\ -2 & -1 \end{vmatrix} = (1 \times -1) - (-2 \times 4) = 7$$

$$A_{12} = \begin{vmatrix} 0 & 4 \\ 5 & -1 \end{vmatrix} = (0 \times -1) - (5 \times 4) = -20$$

$$A_{13} = \begin{vmatrix} 0 & 1 \\ 5 & -2 \end{vmatrix} = (0 \times -2) - (5 \times 1) = -5$$

Thus,

$$\begin{aligned} |A| &= a_{11}A_{11} - a_{12}A_{12} + a_{13}A_{13} \\ &= (2 \times 7) - (-1 \times -20) + (3 \times -5) = 14 - 20 - 15 \\ &= -21 \end{aligned}$$

Self-Check 3 Compute $\begin{vmatrix} 3 & 4 & -2 \\ 1 & 0 & 5 \\ 2 & -3 & 1 \end{vmatrix}$.

Expansion of a Determinant Using Any Row or Column

We have defined the determinant of a 3 × 3 matrix using the first row. The determinant can be evaluated using any row or column. The rule is to select a row (or column), multiply the elements of that row (or column) by their minors, and add these terms according to the following sign convention.

$$\begin{bmatrix} + & - & + \\ - & + & - \\ + & - & + \end{bmatrix}$$

The following example illustrates how this sign convention works.

Example 4 **Expansion of a Determinant Using Various Rows and Columns**

Consider the following determinant. Expand the determinant using (a) the second column and (b) the third row.

$$\begin{vmatrix} 1 & 2 & -1 \\ 0 & 3 & 4 \\ 1 & -3 & 5 \end{vmatrix}$$

SOLUTION **(a)** The sign convention table gives the signs of the expansion using the second column to be −, +, −. Thus, when we multiply the elements of the second column by their minors, the first and last terms are negated.

$$\begin{aligned}\begin{vmatrix}1 & 2 & -1\\0 & 3 & 4\\1 & -3 & 5\end{vmatrix} &= -2\begin{vmatrix}0 & 4\\1 & 5\end{vmatrix} + 3\begin{vmatrix}1 & -1\\1 & 5\end{vmatrix} - (-3)\begin{vmatrix}1 & -1\\0 & 4\end{vmatrix}\\ &= -2[(0 \times 5) - (1 \times 4)] + 3[(1 \times 5) - (1 \times -1)] + \\ &\qquad 3[(1 \times 4) - (0 \times -1)]\\ &= -2[-4] + 3[6] + 3[4]\\ &= 38\end{aligned}$$

(b) The table gives the signs of the expansion using the third row to be +, −, +.

$$\begin{aligned}\begin{vmatrix}1 & 2 & -1\\0 & 3 & 4\\1 & -3 & 5\end{vmatrix} &= 1\begin{vmatrix}2 & -1\\3 & 4\end{vmatrix} - (-3)\begin{vmatrix}1 & -1\\0 & 4\end{vmatrix} + 5\begin{vmatrix}1 & 2\\0 & 3\end{vmatrix}\\ &= 1[(2 \times 4) - (3 \times -1)] + 3[(1 \times 4) - (0 \times -1)]\\ &\qquad + 5[(1 \times 3) - (0 \times 2)]\\ &= 1[11] + 3[4] + 5[3]\\ &= 38\end{aligned}$$

Observe that the determinant is again found to be 38. The value of the determinant will be the same whichever row or column is used for the expansion. One should expand a determinant using the row or column that has the most zeros. This reduces the computation involved.

Self-Check 4 Compute $\begin{vmatrix}4 & 1 & 0\\2 & 5 & 3\\1 & 2 & 0\end{vmatrix}$ using (a) the second row and (b) the third column.

CRAMER'S RULE

Determinants are useful tools in many areas of mathematics. We shall now see how they can be used to solve systems of linear equations. Consider the following system of two linear equations in two variables having matrix of coefficients A.

$$a_{11}x + a_{12}y = c_1 \qquad (1)$$
$$a_{21}x + a_{22}y = c_2 \qquad (2)$$

Let us eliminate x from these equations by multiplying equation 1 by $-a_{21}$, multiplying equation 2 by a_{11}, and adding. We get

$$\begin{array}{rcr} -a_{21}a_{11}x - a_{21}a_{12}y &=& -a_{21}c_1\\ a_{11}a_{21}x + a_{11}a_{22}y &=& a_{11}c_2\\ \hline -a_{21}a_{12}y + a_{11}a_{22}y &=& -a_{21}c_1 + a_{11}c_2\end{array}$$

Thus,

$$y(-a_{21}a_{12} + a_{11}a_{22}) = -a_{21}c_1 + a_{11}c_2$$

giving

$$y = \frac{a_{11}c_2 - a_{21}c_1}{(a_{11}a_{22} - a_{21}a_{12})}, \quad \text{if} \quad (a_{11}a_{22} - a_{21}a_{12}) \neq 0$$

We can similarly solve for x by multiplying equation 1 by $-a_{22}$, multiplying equation 2 by a_{12}, and adding. The solution to the system of equations is

$$x = \frac{a_{22}c_1 - a_{12}c_2}{(a_{11}a_{22} - a_{21}a_{12})}, \quad y = \frac{a_{11}c_2 - a_{21}c_1}{(a_{11}a_{22} - a_{21}a_{12})}, \quad \text{if} \quad (a_{11}a_{22} - a_{21}a_{12}) \neq 0$$

These solutions can be conveniently written in terms of determinants as follows:

$$x = \frac{\begin{vmatrix} c_1 & a_{12} \\ c_2 & a_{22} \end{vmatrix}}{\begin{vmatrix} a_{11} & a_{12} \\ a_{21} & a_{22} \end{vmatrix}}, \quad y = \frac{\begin{vmatrix} a_{11} & c_1 \\ a_{21} & c_2 \end{vmatrix}}{\begin{vmatrix} a_{11} & a_{12} \\ a_{21} & a_{22} \end{vmatrix}}, \quad \text{if} \quad \begin{vmatrix} a_{11} & a_{12} \\ a_{21} & a_{22} \end{vmatrix} \neq 0$$

We have arrived at Cramer's rule for the solution of a system of two linear equations in two variables.

Cramer's Rule for Two Linear Equations in Two Variables

The solution to the system

$$a_{11}x + a_{12}y = c_1$$
$$a_{21}x + a_{22}y = c_2$$

is

$$x = \frac{\begin{vmatrix} c_1 & a_{12} \\ c_2 & a_{22} \end{vmatrix}}{|A|}, \quad y = \frac{\begin{vmatrix} a_{11} & c_1 \\ a_{21} & c_2 \end{vmatrix}}{|A|}$$

if $|A| = \begin{vmatrix} a_{11} & a_{12} \\ a_{21} & a_{22} \end{vmatrix} \neq 0.$

To remember this rule, observe that the matrix A is the matrix of coefficients of the system. The determinant in the numerator for x is obtained by replacing the first column of A by c_1 and c_2. The determinant in the numerator for y is obtained by replacing the second column of A by c_1 and c_2.

Example 5 Cramer's Rule for Two Equations in Two Variables

Solve the following system of linear equations using Cramer's rule.

$$x + 2y = 8$$
$$-2x + 3y = 5$$

SOLUTION The matrix of coefficients of this system is

$$A = \begin{bmatrix} 1 & 2 \\ -2 & 3 \end{bmatrix}$$

We get

$$|A| = \begin{vmatrix} 1 & 2 \\ -2 & 3 \end{vmatrix} = (1 \times 3) - (-2 \times 2) = 3 + 4 = 7$$

Because $|A| \neq 0$, Cramer's rule can be used. We get

$$x = \frac{\begin{vmatrix} 8 & 2 \\ 5 & 3 \end{vmatrix}}{7} = \frac{(8 \times 3) - (5 \times 2)}{7} = \frac{14}{7} = 2$$

$$y = \frac{\begin{vmatrix} 1 & 8 \\ -2 & 5 \end{vmatrix}}{7} = \frac{(1 \times 5) - (-2 \times 8)}{7} = \frac{21}{7} = 3$$

The solution is $x = 2$ and $y = 3$.

Self-Check 5 Solve the following system using Cramer's rule.

$$\begin{aligned} 2x + 3y &= 4 \\ 3x - y &= -5 \end{aligned}$$

We have derived Cramer's rule for a system of two linear equations in two variables. The rule can be extended to larger systems of linear equations. We now state the rule for a system of three equations in three variables.

CRAMER'S RULE FOR THREE LINEAR EQUATIONS IN THREE VARIABLES

Consider the following system of linear equations with matrix of coefficients A.

$$\begin{aligned} a_{11}x + a_{12}y + a_{13}z &= c_1 \\ a_{21}x + a_{22}y + a_{23}z &= c_2 \\ a_{31}x + a_{32}y + a_{33}z &= c_3 \end{aligned}$$

The solution is

$$x = \frac{\begin{vmatrix} c_1 & a_{12} & a_{13} \\ c_2 & a_{22} & a_{23} \\ c_3 & a_{32} & a_{33} \end{vmatrix}}{|A|}, \quad y = \frac{\begin{vmatrix} a_{11} & c_1 & a_{13} \\ a_{21} & c_2 & a_{23} \\ a_{31} & c_3 & a_{33} \end{vmatrix}}{|A|}, \quad z = \frac{\begin{vmatrix} a_{11} & a_{12} & c_1 \\ a_{21} & a_{22} & c_2 \\ a_{31} & a_{32} & c_3 \end{vmatrix}}{|A|}$$

if $\quad |A| = \begin{vmatrix} a_{11} & a_{12} & a_{13} \\ a_{21} & a_{22} & a_{23} \\ a_{31} & a_{32} & a_{33} \end{vmatrix} \neq 0.$

Observe that the determinant in the numerator for x is obtained by replacing the first column of A by c_1, c_2, and c_3. y and z are obtained by replacing the second and third columns of A.

Example 6 **Cramer's Rule for Three Equations in Three Variables**

Use Cramer's rule to solve the following system of equations.

$$\begin{aligned} x + y + z &= 5 \\ 2x - y + 2z &= 4 \\ 3x + 2y - z &= -3 \end{aligned}$$

SOLUTION We focus on showing how to set up the solutions using Cramer's rule in this example, omitting the actual computation of the determinants because we are familiar with this aspect. The determinant of the matrix of coefficients is found first.

$$|A| = \begin{vmatrix} 1 & 1 & 1 \\ 2 & -1 & 2 \\ 3 & 2 & -1 \end{vmatrix} = 12$$

Because $|A| \neq 0$, Cramer's rule applies. We get

$$x = \frac{\begin{vmatrix} 5 & 1 & 1 \\ 4 & -1 & 2 \\ -3 & 2 & -1 \end{vmatrix}}{12} = \frac{-12}{12} = 1, \quad y = \frac{\begin{vmatrix} 1 & 5 & 1 \\ 2 & 4 & 2 \\ 3 & -3 & -1 \end{vmatrix}}{12} = \frac{24}{12} = 2,$$

$$z = \frac{\begin{vmatrix} 1 & 1 & 5 \\ 2 & -1 & 4 \\ 3 & 2 & -3 \end{vmatrix}}{12} = \frac{48}{12} = 4$$

The solution to the system is $x = -1$, $y = 2$, and $z = 4$.

Self-Check 6 Use Cramer's rule to solve the following system.

$$\begin{aligned} x + 2y - z &= -3 \\ 2x + y + z &= 3 \\ 3x + y - 2z &= -2 \end{aligned}$$

Cramer's rule can be extended to larger systems of four equations in four variables, five equations in five variables, and so on. It is important to realize that Cramer's rule is only applicable if the determinant of the matrix of coefficients is nonzero. One must use a method such as the Gauss–Jordan elimination method if this determinant is zero. Furthermore, the amount of computation involved in using Cramer's rule for larger systems of equations is great. Gauss–Jordan elimination is more efficient and preferable. The real importance of Cramer's rule lies in the fact that it gives a formula for the solution of a system of n linear equations in n variables. However, practice at computing solutions of small systems with it is important for understanding the rule.

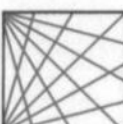

Determinant Using a Calculator

Calculators can be used to compute determinants. We illustrate the steps. You should find out how to implement these steps on your calculator. Let us compute $|A|$ for the following matrix A. We see that $|A| = 78$.

$$A = \begin{bmatrix} 5 & 2 & 4 \\ 1 & 3 & -7 \\ 0 & 1 & 3 \end{bmatrix}$$

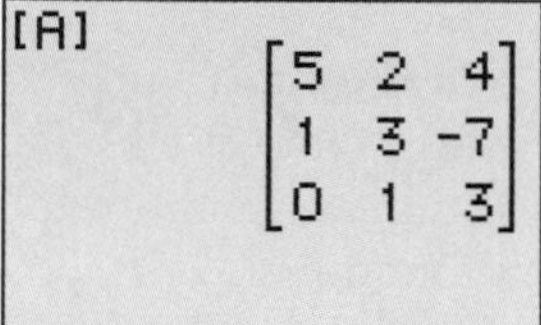

Enter A.

[A]
[5 2 4]
[1 3 -7]
[0 1 3]
det [A]
78

Compute $|A|$.

Self-Check 7 Compute $\begin{vmatrix} 2 & -3 & 4 & 0 \\ -5 & 1 & 6 & 7 \\ -9 & 3 & 2 & -1 \\ 0 & 1 & 3 & 5 \end{vmatrix}$ using a calculator.

Answers to Self-Check Exercises

1. $|A| = 26$.
2. $A_{22} = 13$.
3. $|A| = 87$.
4. $|A| = -21$. Note that there is less computation involved using the third column than in using the second row because of the zeros in the third column.
5. Solution is $x = -1$ and $y = 2$.
6. Solution is $x = 1$, $y = -1$, and $z = 2$.
7. $|A| = -412$.

EXERCISES 7.4

In Exercises 1–8, compute each 2×2 determinant.

1. $\begin{vmatrix} 1 & 2 \\ 3 & 4 \end{vmatrix}$ **2.** $\begin{vmatrix} 2 & 0 \\ 1 & 4 \end{vmatrix}$ **3.** $\begin{vmatrix} 5 & 2 \\ 1 & 3 \end{vmatrix}$ **4.** $\begin{vmatrix} -1 & 4 \\ -2 & 6 \end{vmatrix}$

5. $\begin{vmatrix} 2 & 5 \\ 0 & 1 \end{vmatrix}$ **6.** $\begin{vmatrix} 0 & 4 \\ 0 & 1 \end{vmatrix}$ **7.** $\begin{vmatrix} -2 & -6 \\ -3 & -4 \end{vmatrix}$ **8.** $\begin{vmatrix} 5 & -2 \\ -4 & 2 \end{vmatrix}$

In Exercises 9–12, compute the minors of the elements.

9. $A = \begin{vmatrix} 1 & 2 & -1 \\ 3 & 4 & 2 \\ 5 & 1 & 3 \end{vmatrix}$, minors of a_{11}, a_{12}, a_{21}

10. $A = \begin{vmatrix} 4 & -1 & 3 \\ 2 & 6 & 1 \\ 7 & 4 & 3 \end{vmatrix}$, minors of a_{12}, a_{33}, a_{13}

11. $A = \begin{vmatrix} 5 & 0 & 4 \\ 1 & -2 & 6 \\ 3 & 1 & -5 \end{vmatrix}$, minors of a_{11}, a_{31}, a_{23}

12. $A = \begin{vmatrix} 3 & 4 & 7 \\ 1 & 0 & 2 \\ 6 & -1 & 5 \end{vmatrix}$, minors of a_{21}, a_{33}, a_{32}

In Exercises 13–18, compute the determinants using the first row.

13. $\begin{vmatrix} -1 & 2 & 0 \\ 4 & 1 & 3 \\ 5 & 2 & -1 \end{vmatrix}$

14. $\begin{vmatrix} 7 & -1 & 3 \\ 2 & 4 & 1 \\ 9 & 1 & 3 \end{vmatrix}$

15. $\begin{vmatrix} 0 & 0 & 4 \\ 1 & 2 & 5 \\ 4 & 9 & 1 \end{vmatrix}$

16. $\begin{vmatrix} 5 & 1 & 0 \\ 4 & 2 & 0 \\ 3 & 5 & 0 \end{vmatrix}$

17. $\begin{vmatrix} 1 & -1 & 1 \\ 2 & 4 & 7 \\ 3 & 5 & 1 \end{vmatrix}$

18. $\begin{vmatrix} 1 & 2 & 3 \\ 5 & 1 & 4 \\ -1 & 2 & 6 \end{vmatrix}$

In Exercises 19–24, compute the determinants in two ways, using the suggested rows and columns. Verify that the value of the determinant is the same each way.

19. $\begin{vmatrix} 1 & 0 & 0 \\ 2 & 3 & 4 \\ 1 & 4 & 5 \end{vmatrix}$
Row 1, then column 3

20. $\begin{vmatrix} 4 & -1 & 2 \\ 3 & 1 & 4 \\ 5 & 0 & 1 \end{vmatrix}$
Row 1, then row 3

21. $\begin{vmatrix} 5 & 2 & 1 \\ 0 & 4 & 3 \\ 5 & 1 & 2 \end{vmatrix}$
Column 1, then column 2

22. $\begin{vmatrix} 5 & 0 & 4 \\ 2 & 4 & 1 \\ 3 & 0 & 2 \end{vmatrix}$
Row 1, then column 2

23. $\begin{vmatrix} 0 & 2 & 4 \\ 0 & 4 & 3 \\ 5 & 1 & 2 \end{vmatrix}$
Row 3, then column 1

24. $\begin{vmatrix} 4 & 5 & -1 \\ 2 & 1 & 3 \\ 6 & 1 & 7 \end{vmatrix}$
Row 3, then column 3

In Exercises 25–30, solve each system using Cramer's rule if possible. If Cramer's rule cannot be used, say so. You need not solve the system another way.

25. $\begin{aligned} x + 2y &= 5 \\ -x + y &= 1 \end{aligned}$

26. $\begin{aligned} 2x + 3y &= 9 \\ x - y &= 2 \end{aligned}$

27. $\begin{aligned} x + y &= 0 \\ 2x + 3y &= 1 \end{aligned}$

28. $\begin{aligned} 4x + 2y &= 20 \\ -2x + y &= -10 \end{aligned}$

29. $\begin{aligned} 6x - 2y &= 5 \\ -3x + y &= -4 \end{aligned}$

30. $\begin{aligned} -x + 2y &= 4 \\ 3x + y &= 9 \end{aligned}$

In Exercises 31–36, solve each system using Cramer's rule if possible. If Cramer's rule cannot be used, say so. You need not solve the system another way.

31. $\begin{aligned} 3x - y &= 5 \\ 2x + 2y &= 6 \end{aligned}$

32. $\begin{aligned} 2x + y &= -1 \\ 3x - y &= -9 \end{aligned}$

33. $\begin{aligned} x + 4y &= 9 \\ 3x - y &= 14 \end{aligned}$

34. $\begin{aligned} 5x + 2y &= -1 \\ 15x + 6y &= 7 \end{aligned}$

35. $\begin{aligned} x + y &= 7 \\ x - y &= 1 \end{aligned}$

36. $\begin{aligned} x + 3y &= 0 \\ -x - 4y &= 2 \end{aligned}$

In Exercises 37–42, solve each system using Cramer's rule if possible. If Cramer's rule cannot be used, say so. You need not solve the system another way.

37. $\begin{aligned} x + 2y + 3z &= 6 \\ x - 2y + z &= 0 \\ 3x + y - z &= 3 \end{aligned}$

38. $\begin{aligned} x + y + 3z &= 6 \\ 2x - y + z &= 0 \\ 5x - y + 2z &= 6 \end{aligned}$

39. $\begin{aligned} 2x - y + z &= -1 \\ 3x + y - z &= -4 \\ x + 2y + 3z &= 7 \end{aligned}$

40. $\begin{aligned} 3x - y + 2z &= 5 \\ 2x + 3y + z &= 17 \\ -x + y - z &= 0 \end{aligned}$

41. $\begin{aligned} x - 3y + z &= 6 \\ 2x - 5y + 3z &= 15 \\ -x + 3y + z &= -2 \end{aligned}$

42. $\begin{aligned} 3x - y + z &= 2 \\ 4x + 3y + 2z &= 1 \\ x + 4y + z &= -1 \end{aligned}$

In Exercises 43–48, solve each system using Cramer's rule if possible. If Cramer's rule cannot be used, say so. You need not solve the system another way.

43. $\begin{aligned} 2x - y + z &= 1 \\ 3x + 2y + 3z &= 15 \\ -x - 2y + 4z &= 1 \end{aligned}$

44. $\begin{aligned} 4x - y + z &= 3 \\ 2x + 3y - z &= 1 \\ 2x - 4y + 2z &= 2 \end{aligned}$

45. $\begin{aligned} 6x + 2y + z &= 3 \\ 3x - y - z &= 5 \\ 2x + 2y - 3z &= 3 \end{aligned}$

46. $\begin{aligned} -2x + 3y - z &= 1 \\ 3x - y + 4z &= 3 \\ -x + 5y + 2z &= 5 \end{aligned}$

47. $\begin{aligned} 3x + y - z &= 3 \\ -x - y + 2z &= 5 \\ 5x + 2y - z &= 8 \end{aligned}$

48. $\begin{aligned} x + 3y + z &= 5 \\ -2x + 5y - z &= -9 \\ x + y + z &= 5 \end{aligned}$

In Exercises 49–57, compute the determinants of the matrices using a calculator.

49. $\begin{bmatrix} 1 & 2 \\ 3 & 4 \end{bmatrix}$

50. $\begin{bmatrix} 5 & -3 \\ 0 & 1 \end{bmatrix}$

51. $\begin{bmatrix} 7.2 & 9.3 \\ -6.5 & -8.1 \end{bmatrix}$

52. $\begin{bmatrix} 1.23 & 5.32 \\ -9.78 & 4.32 \end{bmatrix}$

53. $\begin{bmatrix} 8 & 4 & 3 \\ -2 & -7 & 1 \\ 0 & 2 & 9 \end{bmatrix}$

54. $\begin{bmatrix} 2 & 1 & 3 \\ 8 & -7 & -2 \\ 0 & -4 & 3 \end{bmatrix}$

55. $\begin{bmatrix} -8 & 4 & 3 \\ 2 & 4 & 6 \\ -7 & 3 & -9 \end{bmatrix}$

56. $\begin{bmatrix} 6 & 3 & 2 & 6 \\ -2 & 5 & -8 & 5 \\ 1 & 3 & 5 & 6 \\ -6 & -8 & 0 & 1 \end{bmatrix}$

57. $\begin{bmatrix} 1 & 2 & 3 & 4 \\ 4 & 3 & 2 & 1 \\ 1 & 0 & 1 & 0 \\ 2 & 2 & 2 & 2 \end{bmatrix}$

7.5 Symmetric Matrices and Archaeology

• Transpose of a Matrix • Symmetric Matrix • Seriation in Archaeology

In this section, we continue the algebraic development of matrices and we introduce the concept of symmetry of matrices. We have seen how symmetry is a valuable tool for working with functions, and we will see that it is valuable when working with matrices too. We will see how archaeologists use matrix algebra to determine the chronological order of graves and artifacts.

Transpose of a Matrix

The following concept of transpose leads into these discussions.

Exploration 1

Consider the following pairs of matrices labeled A and A^t. The matrix A^t is called the transpose of A. Determine the relationship between A and A^t.

(a) $A = \begin{bmatrix} 1 & 2 \\ 3 & 4 \end{bmatrix}$, $A^t = \begin{bmatrix} 1 & 3 \\ 2 & 4 \end{bmatrix}$

(b) $A = \begin{bmatrix} 0 & -2 & 4 \\ 5 & -3 & 8 \\ 7 & 9 & -2 \end{bmatrix}$, $A^t = \begin{bmatrix} 0 & 5 & 7 \\ -2 & -3 & 9 \\ 4 & 8 & -2 \end{bmatrix}$

(c) $A = \begin{bmatrix} -1 & 2 & 4 \\ 3 & 7 & 5 \end{bmatrix}$, $A^t = \begin{bmatrix} -1 & 3 \\ 2 & 7 \\ 4 & 5 \end{bmatrix}$

Transpose The **transpose** of a matrix A, denoted A^t, is the matrix in which the columns are the rows of the given matrix A.

The first row of A becomes the first column of A^t, the second row of A becomes the second column of A^t, and so on. The (i, j)th element of A becomes the (j, i)th element of A^t. If A is an $m \times n$ matrix, then A^t is an $n \times m$ matrix.

Self-Check 1

(a) If $A = \begin{bmatrix} 3 & 2 & -4 & 5 \\ 7 & 2 & 7 & 0 \end{bmatrix}$ find A^t.

(b) If $A^t = \begin{bmatrix} 1 & 2 & 3 \\ 4 & 5 & 4 \\ 3 & 2 & 9 \end{bmatrix}$ find A.

Symmetric Matrix

We now introduce symmetric matrices. Symmetric matrices are probably the most important class of matrices. They are used in fields such as theoretical physics, chemistry, biology, mechanical and electrical engineering, psychology, and sociology. We shall discuss an application in archaeology.

Exploration 2

Consider the following matrices. Examine the arrangements of the elements. Describe any patterns that they have in common.

(a) $A = \begin{bmatrix} 1 & 3 \\ 3 & 5 \end{bmatrix}$ (b) $A = \begin{bmatrix} 0 & -1 & 2 \\ -1 & 4 & 9 \\ 2 & 9 & 7 \end{bmatrix}$

(c) $A = \begin{bmatrix} 3 & 5 & 2 \\ 5 & 4 & -1 \\ 2 & -1 & 6 \end{bmatrix}$

These observations lead to the definition of a symmetric matrix.

Symmetric A **symmetric** matrix is a matrix that is equal to its transpose.

$$A = A^t$$

Note the symmetry of symmetric matrices about the main diagonal. All nondiagonal elements occur in pairs symmetrically located about the main diagonal. The idea of a symmetric matrix is not as new as it might first appear. The student will have used a mileage chart to look up distances between cities. A mileage chart is in fact a symmetric matrix.

Example 1 | **Mileage Chart**

Consider the following matrix, which represents distances in miles between various U.S. cities.

	Chicago	San Francisco	Houston	New York
Chicago	0	2189	1085	841
San Francisco	2189	0	1955	3025
Houston	1085	1955	0	1636
New York	841	3025	1636	0

Observe that the matrix is symmetric. All elements occur in pairs, symmetrically located about the main diagonal. There is a reason for this; namely, that the distance from city X to city Y is the same as the distance from city Y to city X. For example, the distance from Chicago to Houston, which is 1085 (row 1, column 3), will be the same as the distance from Houston to Chicago (row 3, column 1). All such mileage matrices will be symmetric.

Further Application . . .

The final application in this chapter brings together much of the matrix algebra that we have developed. We introduce the mathematics behind trying to solve the archaeological problem of determining the chronological order of graves.

Example 2 Seriation in Archaeology

A problem confronting archaeologists is that of placing sites and artifacts in proper chronological order. This branch of archaeology, called **seriation**, began with the work of Sir Flinders Petrie in the late 19th century. Petrie studied graves in the cemeteries of Nagada, Ballas, and Hu, all located in what was prehistoric Egypt. (Recent carbon dating shows that all the graves ranged from 6000 BCE to 2500 BCE) Petrie used the data from approximately 900 graves to order them and assign a time period to each type of pottery found.

Let us look at this general problem of seriation in terms of graves and varieties of pottery found in graves. An assumption usually made in archaeology is that two graves that have similar contents are more likely to lie close together in time than are two graves that have contents with little in common. The mathematical model that we now construct leads to information concerning the common contents of graves and, thus, to the chronological order of the graves.

We construct a matrix A, all of whose elements are either 1 or 0, that describes the pottery content of the graves. Label the graves **1**, **2**, . . . , and the types of pottery 1, 2, ... Let the matrix A be defined by

$$a_{ij} = \begin{cases} 1 & \text{if grave } i \text{ contains pottery type } j \\ 0 & \text{if grave } i \text{ does not contain pottery type } j \end{cases}$$

The matrix A contains all the information about the pottery content of the various graves. The following result now tells us how information is extracted from A:

The element g_{ij} of the matrix $G = AA^t$ is equal to the number of types of pottery common to both grave i and grave j.

Thus, the larger g_{ij}, the closer grave i and grave j are in time. By examining the elements of G, the archaeologist can determine the chronological order of the graves. Let us verify this result.

$$\begin{aligned} g_{ij} &= \text{element in row } i, \text{ column } j \text{ of } G \\ &= (\text{row } i \text{ of } A) \times (\text{column } j \text{ of } A^t) \\ &= \begin{bmatrix} a_{i1} & a_{i2} & \cdots & a_{in} \end{bmatrix} \begin{bmatrix} a_{j1} \\ a_{j2} \\ \vdots \\ a_{jn} \end{bmatrix} \\ &= a_{i1}a_{j1} + a_{i2}a_{j2} + \cdots + a_{in}a_{jn} \end{aligned}$$

Each term in this sum will be either 1 or 0. For example, the term $a_{i2}a_{j2}$ will be 1 if a_{i2} and a_{j2} are both 1 (i.e., if pottery type 2 is common to graves i and j). It will be 0 if pottery type 2 is not common to graves i and j. Thus, the number of 1s in this expression for g_{ij}, (the actual value of g_{ij}) is the number of types of pottery common to graves i and j. Let us now look at an example.

Consider the four graves in Figure 7.5. The graves are labeled **1**, **2**, **3**, and **4**, and contain three types of pottery labeled 1, 2, and 3. Let us find the chronological order of these graves.

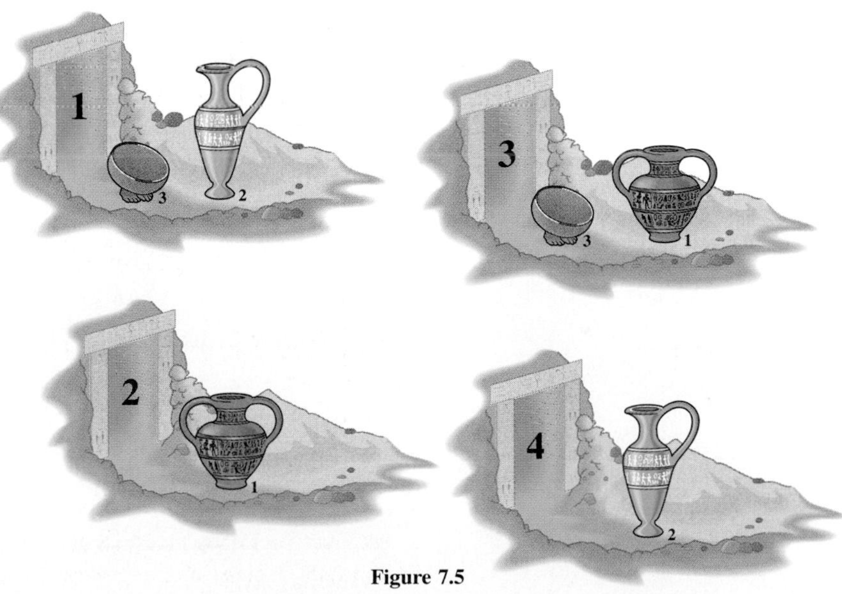

Figure 7.5

Construct the matrix A that describes the pottery contents of the graves. We get

$a_{11} = 0$ because grave 1 does not contain pottery type 1,
$a_{12} = 1$ because grave 1 contains pottery type 2,
$a_{13} = 1$ because grave 1 contains pottery type 3, and so on.

The matrix A is as follows.

$$\begin{array}{cc} & \text{Pottery Types} \\ & \begin{array}{ccc} 1 & 2 & 3 \end{array} \\ \text{Graves} \begin{array}{c} \mathbf{1} \\ \mathbf{2} \\ \mathbf{3} \\ \mathbf{4} \end{array} & \begin{bmatrix} 0 & 1 & 1 \\ 1 & 0 & 0 \\ 1 & 0 & 1 \\ 0 & 1 & 0 \end{bmatrix} \\ & \text{Matrix } A \end{array}$$

We now compute the product of $G = AA^t$.

$$G = AA^t = \overset{A}{\begin{bmatrix} 0 & 1 & 1 \\ 1 & 0 & 0 \\ 1 & 0 & 1 \\ 0 & 1 & 0 \end{bmatrix}} \overset{A^t}{\begin{bmatrix} 0 & 1 & 1 & 0 \\ 1 & 0 & 0 & 1 \\ 1 & 0 & 1 & 0 \end{bmatrix}} = \begin{bmatrix} 2 & 0 & 1 & 1 \\ 0 & 1 & 1 & 0 \\ 1 & 1 & 2 & 0 \\ 1 & 0 & 0 & 1 \end{bmatrix}$$

Observe that G is symmetric. This means that the information contained in the elements above the main diagonal is duplicated in the elements below it. (G will always be symmetric. See Exercise 19.) We systematically look at the elements above the main diagonal, which are 1; these indicate graves that have pottery in common and are therefore close in time.

$g_{13} = 1$ implies that graves 1 and 3 have one type of pottery in common.
$g_{14} = 1$ implies that graves 1 and 4 have one type of pottery in common.
$g_{23} = 1$ implies that graves 2 and 3 have one type of pottery in common.

Graves 1 and 3 have pottery in common; they are close together in time. Let us start with graves 1 and 3 and construct a diagram.

$$1 - 3$$

Next, add grave 4 to this diagram. Because $g_{14} = 1$, grave 4 is close to grave 1. We get

$$4 - 1 - 3$$

Finally add grave 2. Because $g_{23} = 1$, grave 2 is close to grave 3. We get

$$4 - 1 - 3 - 2$$

The mathematics does not tell us which way time flows in this diagram. There are two possibilities:

$$4 \rightarrow 1 \rightarrow 3 \rightarrow 2 \qquad \text{and} \qquad 4 \leftarrow 1 \leftarrow 3 \leftarrow 2$$

The archaeologist usually knows from other sources which of the two extreme graves (4 and 2 in our case) came first. Thus, the chronological order of the graves is known.

The matrices G are in practice large, and the information cannot be sorted out as easily or give results that are as unambiguous as in this illustration. For example, Petrie examined 900 graves; his matrix G would be a 900×900 matrix. Special mathematical techniques have been developed for extracting information from these matrices. These methods are implemented on computers.*

Group Discussion

The Archaeology Model

We presented a model for analyzing the chronological ordering of graves. Mathematical models often represent ideal conditions and do not give results for all situations. It is important to recognize any limitations that can exist for a mathematical model. What complications can you visualize as occurring with some graves in this type of situation. Are there special cases that the model as it stands cannot handle? Discuss and construct examples of these situations. Archaeologists use other information and methods to supplement the information given by this mathematical model.

Answer to Self-Check Exercise

1. (a) $A^t = \begin{bmatrix} 3 & 7 \\ 2 & -2 \\ -4 & 7 \\ 5 & 0 \end{bmatrix}$. **(b)** $A = \begin{bmatrix} 1 & 4 & 3 \\ 2 & 5 & 2 \\ 3 & 4 & 9 \end{bmatrix}$.

EXERCISES 7.5

Determine the transpose of each of the matrices in Exercises 1–8. Indicate whether the matrix is symmetric.

1. $A = \begin{bmatrix} -1 & 2 \\ 2 & -3 \end{bmatrix}$

2. $B = \begin{bmatrix} 1 & 2 \\ 0 & 3 \end{bmatrix}$

3. $C = \begin{bmatrix} 3 & -1 \\ 2 & 4 \end{bmatrix}$

4. $D = \begin{bmatrix} 4 & 5 \\ -2 & 3 \\ 7 & 0 \end{bmatrix}$

5. $E = \begin{bmatrix} 4 & 5 & 6 \\ -1 & 2 & 3 \\ 0 & 1 & 2 \end{bmatrix}$

6. $F = \begin{bmatrix} 1 & -1 & 3 \\ -1 & 2 & 0 \\ 3 & 0 & 4 \end{bmatrix}$

7. $G = \begin{bmatrix} -2 & 4 & 5 & 7 \\ 1 & 0 & 3 & -7 \end{bmatrix}$

8. $H = \begin{bmatrix} 1 & -2 & 3 \\ -2 & 5 & 6 \\ 3 & 6 & 7 \end{bmatrix}$

*Students who are interested in pursuing this topic further should consult "Some Problems and Methods in Statistical Archaeology" by David G. Kendall (*World Archaeology* 1969, **1**:61–76).

In Exercises 9–12, each matrix is to be symmetric. Determine the elements indicated by the * and write down the complete matrix.

9. $A = \begin{bmatrix} -1 & * \\ 2 & -3 \end{bmatrix}$ **10.** $B = \begin{bmatrix} 1 & 2 & 4 \\ * & 6 & * \\ 4 & 5 & 2 \end{bmatrix}$

11. $C = \begin{bmatrix} 3 & 5 & * \\ * & 8 & 4 \\ -3 & * & 3 \end{bmatrix}$ **12.** $D = \begin{bmatrix} -3 & * & 8 & 9 \\ -4 & 7 & * & 7 \\ * & 2 & 6 & 4 \\ * & 7 & * & 9 \end{bmatrix}$

ARCHAEOLOGY

The matrices in Exercises 13–18, describe the pottery contents of various graves. Determine possible chronological orderings of the graves in each case.

13. $\begin{bmatrix} 1 & 0 & 0 & 1 \\ 0 & 0 & 1 & 1 \\ 1 & 1 & 0 & 0 \\ 0 & 1 & 0 & 0 \end{bmatrix}$ **14.** $\begin{bmatrix} 1 & 1 & 0 \\ 0 & 0 & 1 \\ 0 & 1 & 1 \\ 1 & 0 & 0 \end{bmatrix}$

15. $\begin{bmatrix} 0 & 1 & 0 & 0 \\ 1 & 1 & 0 & 0 \\ 1 & 0 & 0 & 1 \\ 1 & 0 & 1 & 1 \end{bmatrix}$ **16.** $\begin{bmatrix} 1 & 0 & 1 \\ 0 & 1 & 0 \\ 0 & 1 & 1 \\ 1 & 0 & 0 \end{bmatrix}$

17. $\begin{bmatrix} 0 & 1 & 1 & 0 \\ 0 & 1 & 0 & 1 \\ 1 & 0 & 0 & 0 \\ 0 & 0 & 1 & 0 \\ 1 & 0 & 0 & 1 \end{bmatrix}$ **18.** $\begin{bmatrix} 0 & 0 & 1 & 1 & 0 \\ 0 & 1 & 0 & 0 & 0 \\ 1 & 0 & 1 & 0 & 1 \\ 1 & 0 & 0 & 0 & 1 \\ 0 & 1 & 0 & 1 & 0 \end{bmatrix}$

19. Let $G = AA^t$, for a matrix A that describes the pottery content of graves. Use this physical interpretation to reason that G is always symmetric. *Hint:* Explain why you would expect $g_{ij} = g_{ji}$.

20. If matrix A describes the pottery contents of graves, the matrix $P = A^tA$ can be used to arrive at the chronological order of the pottery. The element p_{ij} of P is equal to the number of graves that have pottery i and j in common. Thus, the larger p_{ij}, the closer pottery i and pottery j are in time. Compute the matrix P for the situations described by Exercises 13 and 15 and determine the chronological orderings of the pottery.

21. Relationships Within a Group The model introduced here in archaeology is used in sociology to analyze relationships among a group of people. For example, consider the relationship of "friendship" within a group. Assume that all friendships are mutual. Label the people $1, \ldots, n$ and define a square matrix A as follows:

$a_{ii} = 0$ for all i (diagonal elements of A are zero)

$$a_{ij} = \begin{cases} 1 & \text{if } i \text{ and } j \text{ are friends} \\ 0 & \text{if } i \text{ and } j \text{ are not friends} \end{cases}$$

(a) Prove that if $F = AA^t$, then f_{ij} is the number of friends that i and j have in common.

(b) Suppose that all friendships are not mutual. How does this affect the model?

22. Diseases and Symptoms The following matrix A describes the relationship between four diseases and five symptoms.

Diseases \ Symptoms	1	2	3	4	5
1	1	0	0	1	1
2	0	1	0	0	0
3	0	0	1	1	0
4	1	1	1	0	1

Matrix A

For example, $a_{12} = 0$ implies that disease 1 does not have symptom 2, whereas $a_{14} = 1$ means that disease 1 has symptom 4. Compute the matrix $D = AA^t$ and use the elements of D to show which diseases are "close together" according to these symptoms. The model indicates the extra care that must be taken in diagnosing these diseases.

CHAPTER 7 PROJECT | Influence Within a Group

Select some group of people with which you are familiar. For example, the group could be a student committee, a group of professors, or possibly your own current group. Give the list of answers that you think the group would give to the following questionnaire. Construct the digraph that describes the influence structure within the group. Use the distance matrix to rank the members according to their estimated influence within the group.

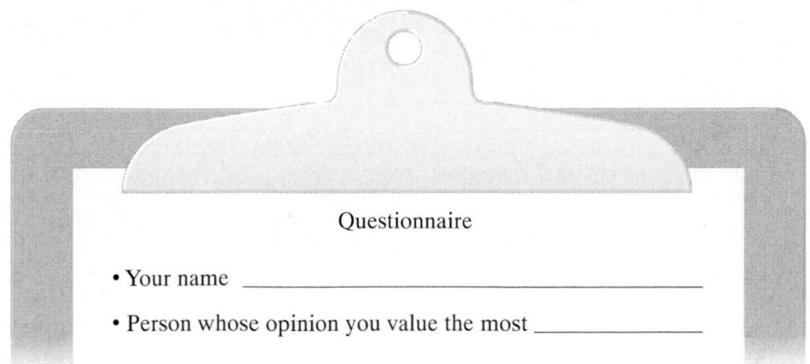

Questionnaire
• Your name
• Person whose opinion you value the most

WEB PROJECT 7

To explore the concepts of this chapter further, try the interesting WebProject. Each activity relates the ideas and skills you have learned in this chapter to a real-world applied situation. Visit the Williams Web site at

www.saunderscollege.com/math/williams

to find the complete WebProject, related questions, and live links to the Internet.

Chapter 7 Highlights

Section 7.1

Matrix: A rectangular array of numbers. The numbers in the array are called the elements of the matrix.

Size of a Matrix: A matrix with m rows and n columns is said to be of size $m \times n$.

Square Matrix: A matrix with the same number of rows as columns.

Row Matrix: A matrix that has one row.

Column Matrix: A matrix that has one column.

Location of an Element of a Matrix: The location is given by the row and column in which the element lies.

Equality of Matrices: Two matrices are said to be equal if they are of the same size and if their corresponding elements are equal.

Addition and Subtraction of Matrices: If matrices are of the same size, add or subtract corresponding elements. If the matrices are not of the same size, sum or difference does not exist.

Section 7.2

Multiplication of Matrices: The element in row i and column j of the product matrix AB is obtained by multiplying the corresponding elements of row i of A with column j of matrix B and adding the products.

Existence of Product: The product AB of two matrices A and B exists if and only if the number of columns in A is equal to the number of rows in B.

Size of Product Matrix: If A is an $m \times n$ matrix and B is an $n \times r$ matrix, then AB is an $m \times r$ matrix.

Properties of Matrix Multiplication: Matrix multiplication is associative. It is not commutative.

Main Diagonal: Elements on the diagonal from top left to bottom right of a square matrix.

Section 7.3

Identity Matrix: A square matrix with 1s on the main diagonal and zeros elsewhere. Identity $n \times n$ matrix is denoted I_n.

Inverse: An $n \times n$ matrix A has an inverse B if and only if $AB = BA = I_n$. Inverse is then written A^{-1}.

Gauss–Jordan Elimination for Finding the Inverse of a Matrix: Let A be an $n \times n$ matrix.

1. Adjoin the identity $n \times n$ matrix I_n to A to form the augmented matrix $[A{:}I_n]$.
2. Compute the reduced form of $[A{:}I_n]$. If the reduced form is of the form $[I_n{:}B]$, then B is the inverse of A. If the reduced echelon form is not of this form, then A has no inverse.

System of Equations: Write system as $AX = B$. If A has an inverse, the solution is unique and is given by $X = A^{-1}B$.

Section 7.4

Determinant of a 2 × 2 Matrix: $|A| = a_{11}a_{22} - a_{21}a_{12}$.

Minor of a Matrix Element: Eliminate the row and column in which the element lies. The determinant of the resulting submatrix is called the minor of the element.

Determinant of a 3 × 3 Matrix: Expand using any row (or column). Multiply the elements of the row (or column) by their minors and add according to the sign convention.

Cramer's Rule: A method that can be used to solve a system of linear equations that has a square matrix of coefficients and a unique solution.

Section 7.5

Transpose of a Matrix A: The matrix whose columns are the rows of the matrix A. Denoted A^t.

Symmetric Matrix A: A matrix which is equal to its transpose. $A = A^t$.

CHAPTER 7 REVIEW EXERCISES

1. Give the sizes of the following matrices. Indicate whether the matrix is square, a row matrix, or a column matrix.

(a) $\begin{bmatrix} 1 & 5 \\ 6 & 9 \end{bmatrix}$ **(b)** $\begin{bmatrix} 4 & 7 & 3 \\ -2 & 8 & -1 \end{bmatrix}$

(c) $\begin{bmatrix} 7 & 4 & -8 & 5 \\ 5 & -2 & 6 & 8 \\ 4 & 6 & -3 & 7 \end{bmatrix}$ **(d)** $\begin{bmatrix} 2 & 4 & -2 & 7 \end{bmatrix}$

2. Give the specified elements of the following matrices:

$$A = \begin{bmatrix} 4 & 7 & 3 \\ -2 & 8 & -1 \end{bmatrix},$$

$$B = \begin{bmatrix} 7 & 4 & -8 & 5 & 9 \\ 5 & -2 & 6 & 8 & -5 \\ 4 & 6 & -3 & 7 & 12 \end{bmatrix}$$

(a) a_{13} **(b)** a_{23} **(c)** a_{22} **(d)** b_{31}
(e) b_{22} **(f)** b_{34} **(g)** b_{25}

3. Solve the following matrix equations for x, y, and z:

(a) $\begin{bmatrix} x & 2y \\ x - y & x + 3z \end{bmatrix} = \begin{bmatrix} 5 & 8 \\ 1 & 11 \end{bmatrix}$

(b) $\begin{bmatrix} x^2 & x - y \\ z & x + y + z \end{bmatrix} = \begin{bmatrix} 9 & 1 \\ 2 & -5 \end{bmatrix}$

4. Perform the specified matrix operations, if possible, using matrices A, B, C, and D.

$$A = \begin{bmatrix} 5 & 3 \\ 3 & 6 \end{bmatrix}, \quad B = \begin{bmatrix} -4 & 5 \\ -2 & 2 \end{bmatrix},$$

$$C = \begin{bmatrix} 4 & 7.1 & 3.5 \\ 3 & 8 & -1.9 \end{bmatrix}, \quad D = \begin{bmatrix} 7 & 4.1 & -8.2 \\ 5 & -2 & 6.8 \end{bmatrix}$$

(a) $A + B$ **(b)** $A + 3D$ **(c)** $5A - 2B$
(d) $C + 3D$

5. Perform the specified matrix operations, if possible, using matrices A, B, C, and D.

$$A = \begin{bmatrix} 1 & 4 \\ -3 & 6 \end{bmatrix}, \quad B = \begin{bmatrix} 0 & 5 \\ -2 & 6 \end{bmatrix},$$

$$C = \begin{bmatrix} 2 & 5 & -1 \\ 3 & 7 & -2 \end{bmatrix},$$

$$D = \begin{bmatrix} 2.1 & 5 & 4.3 \\ 3 & -1.7 & -6 \\ 0 & -4.5 & 1 \end{bmatrix}$$

(a) AB **(b)** AC **(c)** AD **(d)** A^2
(e) $2AB + A^3$ **(f)** $AC - 3BC$

6. A, B, C, D, and E are matrices of the following sizes:

A 2×2, B 2×2, C 2×3, D 3×3, E 3×1

Predict the sizes of the following matrices, if they exist.

(a) AB **(b)** AC **(c)** ABC **(d)** CD
(e) AE **(f)** CDE **(g)** $A^2C + 2BCD$
(h) $D^3E + 4E$

7. Determine the inverse (if it exists) of each of the following matrices:

(a) $\begin{bmatrix} 2 & 0 \\ 0 & 6 \end{bmatrix}$ (b) $\begin{bmatrix} 1 & 2 \\ 1 & 3 \end{bmatrix}$ (c) $\begin{bmatrix} 2 & 4 \\ 1 & 2 \end{bmatrix}$

(d) $\begin{bmatrix} 1 & 1 & 5 \\ 1 & 2 & 8 \\ -1 & 1 & 2 \end{bmatrix}$ (e) $\begin{bmatrix} 1 & 2 & -1 \\ 3 & 1 & 4 \\ 5 & 5 & 2 \end{bmatrix}$

8. Write each of the following systems of equations in matrix form; then solve using the matrix inverse method.

(a) $\begin{aligned} x + 2y &= 8 \\ x + 3y &= 11 \end{aligned}$ **(b)** $\begin{aligned} x + y + z &= 3 \\ y + 2z &= 5 \\ 2x + y - 4z &= 3 \end{aligned}$

9. Compute the following determinants:

(a) $\begin{vmatrix} 2 & 3 \\ 1 & 6 \end{vmatrix}$ (b) $\begin{vmatrix} -3 & 5 \\ -1 & 2 \end{vmatrix}$ (c) $\begin{vmatrix} 1 & 5 & 7 \\ 2 & 3 & -2 \\ 4 & 2 & 1 \end{vmatrix}$

(d) $\begin{vmatrix} -3 & 5 & 7 \\ 1 & 0 & 0 \\ 4 & 2 & 9 \end{vmatrix}$ (e) $\begin{vmatrix} 2 & 7 & 2 \\ 3 & 1 & 0 \\ 0 & 2 & 0 \end{vmatrix}$

10. Solve the following systems of equations using Cramer's rule, if possible.

(a) $\begin{aligned} x + 2y &= 8 \\ 2x + y &= 7 \end{aligned}$ **(b)** $\begin{aligned} x + 2y - z &= 3 \\ 2x - y + 4z &= 7 \\ -x + 3y + 5z &= 6 \end{aligned}$

11. Determine the transpose of each of the following matrices. Indicate whether the matrix is symmetric.

(a) $A = \begin{bmatrix} -1 & 2 \\ 2 & -3 \end{bmatrix}$ **(b)** $B = \begin{bmatrix} -2 & -3 \\ 3 & 4 \end{bmatrix}$

(c) $C = \begin{bmatrix} 4 & 5 \\ -2 & 3 \\ 7 & 0 \end{bmatrix}$ **(d)** $D = \begin{bmatrix} 2 & 3 & -5 \\ 3 & 0 & 4 \\ -5 & 4 & 6 \end{bmatrix}$

12. Population Movement Model Construct a mathematical model of population flows among cities, suburbs, and nonmetropolitan areas of the United States. Their respective populations in 1995 were 60 million, 140 million, and 65 million. The matrix giving the probabilities of the moves is

(from)

City	Suburb	Nonmetro	(to)
0.96	0.01	0.015	City
0.03	0.98	0.005	Suburb
0.01	0.01	0.98	Nonmetro

This model is a refinement of the model of Exercise 87, in Section 7.2, in that the metropolitan population is broken down into city and suburb. Predict the populations of city, suburban, and nonmetropolitan areas for the years 1996 through 2000 (in millions, to two decimal places). Predict also the population for the year 2010 and the long-term prediction.

13. Producers and Consumers The following matrix A decribes the relationship between four producers and six types of consumers.

		Consumers					
		1	2	3	4	5	6
Producers	1	0	1	0	1	0	1
	2	0	0	0	0	1	0
	3	1	0	0	1	1	1
	4	1	1	0	0	0	0

Matrix A

For example, $a_{12} = 1$ implies that the product of producer 1 is used by customer 2, whereas $a_{13} = 0$ means that the product of producer 1 is not used by consumer 3. Compute the matrix $P = AA^t$ and use the elements of P to reveal which producers are in most direct competition.

14. Group Analysis The following table represents information obtained from a questionnaire given to a group of people. Construct the digraph that describes the leadership structure within the group. Use the distance matrix to rank the members according to their influence on the group.

Group Member	Person Whose Opinion Is Valued
M_1	M_4 and M_5
M_2	M_1
M_3	M_2
M_4	M_3
M_5	M_2 and M_4

Chapter 7 Test

1. Give the sizes of the following matrices:

(a) $\begin{bmatrix} 3 & 2 \\ 4 & 5 \\ 6 & 8 \end{bmatrix}$ (b) $\begin{bmatrix} 1 & 0 & -4 \\ 6 & 8 & 7 \end{bmatrix}$

(c) $\begin{bmatrix} 2 & 1 & 3 \\ 4 & 5 & 7 \\ -9 & 0 & 8 \end{bmatrix}$ (d) $\begin{bmatrix} 3 & 2 & 4 & 5 & 6 \end{bmatrix}$

2. Give the specified elements of the following matrix:

$$A = \begin{bmatrix} 2 & 0 & 1 & -7 & 4 \\ 9 & 7 & -2 & 0 & 6 \\ 8 & 3 & 5 & 1 & 2 \end{bmatrix}$$

(a) a_{14} (b) a_{22} (c) a_{31}

3. Solve the following matrix equation:

$$\begin{bmatrix} x & 3y \\ x + y & x + y + z \end{bmatrix} = \begin{bmatrix} 1 & 6 \\ 3 & 8 \end{bmatrix}$$

4. Perform the specified matrix operations, if possible, using matrices A and B.

$$A = \begin{bmatrix} 0 & -1 & 3 \\ 4 & 2 & 8 \end{bmatrix}, \quad B = \begin{bmatrix} 2 & -5 & 2 \\ 5 & 8 & 9 \end{bmatrix}$$

(a) $A + B$ (b) $3A - 2B$

5. Perform the specified matrix operations, if possible, using matrices A and B.

$$A = \begin{bmatrix} 1 & 3 \\ -4 & 2 \end{bmatrix}, \quad B = \begin{bmatrix} 0 & 5 & -1 \\ 3 & 2 & 6 \end{bmatrix}$$

(a) AB (b) BA (c) A^2

6. A, B, and C are matrices of the following sizes:

$$A\ 2 \times 2, \quad B\ 2 \times 3, \quad C\ 3 \times 1$$

Predict the sizes of the following matrices, if they exist.

(a) AB (b) ABC (c) BA
(d) $A^2B + 2AB$ (e) C^tB^tA

7. Use matrix multiplication to show that the matrix B is the inverse of the matrix A.

$$A = \begin{bmatrix} 1 & 2 \\ 2 & 5 \end{bmatrix}, \quad B = \begin{bmatrix} 5 & -2 \\ -2 & 1 \end{bmatrix}$$

8. Find the inverse of the following matrix.

$$\begin{bmatrix} 1 & 2 & 0 \\ 0 & 1 & -1 \\ -1 & 1 & 2 \end{bmatrix}$$

9. Write the following system of equations in matrix form; then solve using the matrix inverse method.

$$x - 4y = -1$$
$$2x - 7y = -1$$

10. Compute the following determinants.

(a) $\begin{vmatrix} 2 & -1 \\ 4 & 3 \end{vmatrix}$ (b) $\begin{vmatrix} 3 & 0 & 5 \\ 3 & 2 & -1 \\ 7 & 2 & 8 \end{vmatrix}$

11. Solve the following system of equations using Cramer's rule.

$$x + 3y = -1$$
$$4x + 2y = 6$$

12. Determine the transpose of each of the following matrices. Indicate whether the matrix is symmetric.

(a) $A = \begin{bmatrix} 1 & 2 \\ 3 & 4 \end{bmatrix}$ (b) $\begin{bmatrix} 2 & 3 & 1 \\ 3 & 4 & -5 \\ 1 & -5 & 0 \end{bmatrix}$

13. **Group Analysis** The table represents information obtained from a questionnaire given to a group of people. Construct the digraph that describes the leadership structure within the group. Use the distance matrix to rank the members according to their influence on the group.

Group Member	Person Whose Opinion Is Valued
M_1	M_2
M_2	M_5
M_3	M_2
M_4	M_3 and M_5
M_5	M_1 and M_3

14. Traffic Analysis The following matrix gives the daily traffic between zones of a city.

		Destination Zone			
		I	II	III	IV
	I	1200	1300	500	800
Origin	II	1300	2000	300	700
Zone	III	500	900	1000	1800
	IV	650	430	3400	1900

Use this O–D matrix to determine the following: (**a**) the daily traffic from zone II to zone IV; (**b**) the daily traffic from zone III to zone I; (**c**) the internal traffic in zone III; and (**d**) the two zones between which the traffic flow is greatest.

8 Sequences and Series

Leonardo Fibonacci (circa 1175–1250) *(Smith Collection, Rare Book and Manuscript Library, Columbia University)*

A sequence is a collection of numbers arranged in a certain order. For example, 2, 4, 6, 8, . . . is a sequence. We encounter sequences every day. A savings account accumulates annual amounts, and the annual amounts form a sequence: A_1, A_2, A_3, ... The velocity of a rocket at regular intervals after launch will be v_1, v_2, v_3, ... The tools that we develop in this chapter can be applied to both of these situations. We discuss special sequences, such as arithmetic and geometric sequences, in some detail. One of the most important sequences, both historically and from the point of view of applications, is the Fibonacci sequence. This is the sequence

$$1, 1, 2, 3, 5, 8, 13, \ldots .$$

Leonardo Fibonacci was perhaps the most talented mathematician of the Middle Ages. He was born about 1175 in the commercial center of Pisa, Italy. His father was a merchant, and Fibonacci traveled in connection with his father's business. These journeys took him through Egypt, Greece, and Syria, which brought him in contact with eastern and Arabic mathematical practices. His writings strongly advocated the Hindu–Arabic number system (base ten) and did much to foster the introduction of this system into Europe. Fibonacci discovered this sequence while searching for a mathematical model that describes the growth of a population of rabbits. Since that time, this sequence has been found to occur in such diverse fields as botany, architecture, archaeology, and sociology.

The following application illustrates a model for e-mail control. Such models are often created to get a feel for a situation and to make some sort of predictions about what can occur. These models are valuable because the numbers can easily be changed to explore different scenarios.

APPLICATION | Accumulation of E-Mail Messages

You've got mail. How good are you at managing it? *(© Superstock/Charlie Hill)*

A business executive receives numerous e-mail messages a day. Because she has been lax in handling her inbox, many messages have accumulated. She decides on a plan to tidy up the inbox. During a day, she will halve the number of messages in her inbox while still adding eight new messages (messages that might need more consideration before answering). You are asked to model this procedure using mathematics. The model will, for example, lead to the following picture of the purging, sorting, and saving of 112 messages.

Day	1	2	3	4	5	6
Start of day (number of messages)	112	64	40	28	22	19
During the day	purge, look at new mail, sort, save					
End of day	64	40	28	22	19	

The contents of the inbox will be reduced to fewer than 20 messages in 5 days according to this plan. The model can be used to look at various scenarios. For example, what would be the effect of reducing the number of messages added daily to the inbox to four instead of eight? What if the initial number of messages had been 1552? How long would it have then taken to get the number of messages in the inbox down below 20 with this plan? The model enables us to answer these and other questions. *(This application is discussed in the Chapter Project on page 576.)*

8.1 Sequences, Series, and the Fibonacci Sequence

• Sequence • Definition Using Recursion • Definition Using nth Term • Series • Sigma Notation • Partial Sums • The Mean of a Sequence of Numbers • Fibonacci Sequence

SEQUENCE

The idea of a sequence is not new to the student. We have come across sequences on numerous occasions in this course. We talked about patterns in sequences of numbers in Section R.1. We discussed a sequence of fish populations in Section 3.6. In Section 4.2, we determined whether a number was in the Mandelbrot set or not by determining whether a certain sequence of numbers became increasingly large. We found the sequence of populations for U.S. suburbs and cities in Section 7.2. We now look closely at the mathematical background of sequences.

A **sequence** is a collection of numbers arranged in a certain order (i.e., there is a first number, second number, third number, and so on). The numbers in the sequence are called **terms.** For example, 2, 4, 6, 8, . . . is a sequence. The first term is 2, the second term is 4, and so on.

Subscripts are used to denote the terms of the sequence. For example,

$$a_1, a_2, a_3, \ldots$$

is a sequence with first term a_1, second term a_2, and so on. A sequence may have a finite number of terms (called a **finite sequence**) or an infinite number of terms (an **infinite sequence**).

There are many ways of defining sequences. We now look at some of these ways.

Definition of a Sequence Using *n*th Term

A sequence may be defined by giving the nth term. Such information enables one to determine any term of interest. For example, consider the sequence with nth term

$$a_n = n^2 + 1$$

Let us determine the first four terms of this sequence. Let $n = 1, 2, 3, 4$ in turn. We get

$$a_1 = (1)^2 + 1 = 2$$
$$a_2 = (2)^2 + 1 = 5$$
$$a_3 = (3)^2 + 1 = 10$$
$$a_4 = (4)^2 + 1 = 17$$

The sequence is 2, 5, 10, 17, ...

Note that a sequence may be interpreted as a function $f(n)$ having as its domain the set of positive integers, 1, 2, 3, ... Then $a_1 = f(1)$, $a_2 = f(2)$, $a_3 = f(3)$, ... In the preceding example, the function would be $f(n) = n^2 + 1$. We see that $a_1 = f(1) = 2$, $a_2 = f(2) = 5$, $a_3 = f(3) = 10, \ldots$.

Self-Check 1 The nth term of an infinite sequence is $a_n = 3n^2 - 2$. Determine the first five terms of this sequence.

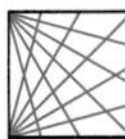

Terms of a Sequence with a Calculator

The terms of a sequence can be computed using a calculator. Calculators have facilities for recalling and editing the last expression used. We illustrate a method that uses this technique. Let us compute the first six terms of the sequence defined by $a_n = 2n^2 - 3n + 4$. The terms will be

$$a_1 = 2(1)^2 - 3(1) + 4$$
$$a_2 = 2(2)^2 - 3(2) + 4$$
$$\vdots$$
$$a_6 = 2(6)^2 - 3(6) + 4$$

We compute these terms as shown to get $a_1 = 3$, $a_2 = 6$, $a_3 = 13$, $a_4 = 24$, $a_5 = 39$, $a_6 = 58$.

Enter and compute the first term.

Recall and edit the first term to get the second term. Compute.

Repeat to get the other terms.

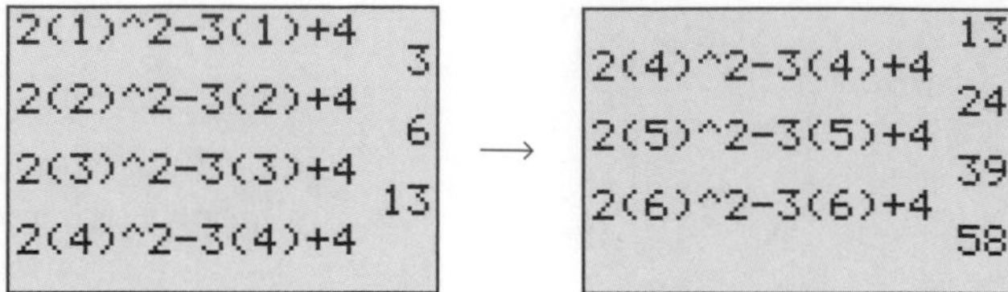

Further Ideas 1 Determine the most efficient way of finding the terms of a sequence on your calculator. Find the first six terms of the sequence $a_n = 2n^2 - 5n + 1$. [If your calculator has a table facility, for example, you might use the above function interpretation of a sequence to get terms. Enter the sequence as $y = 2x^2 - 5x + 1$ and select initial value 1 and increment 1 for the table.]

Recursive Definition of a Sequence

A sequence may be defined by giving the first term together with a formula that gives the nth term as an expression of the preceding term. Such a definition is called a **recursive definition.**

For example, consider the sequence defined by

$$a_1 = 5, \qquad a_n = 2a_{n-1} + 3 \qquad \text{for } n > 1$$

Let us determine the first four terms of the sequence. Let $n = 2, 3, 4$ in turn. We get

$$a_2 = 2a_1 + 3 = 2(5) + 3 = 13$$
$$a_3 = 2a_2 + 3 = 2(13) + 3 = 29$$
$$a_4 = 2a_3 + 3 = 2(29) + 3 = 61$$

The sequence is 5, 13, 29, 61, ...

Self-Check 2 Determine the first six terms of the sequence defined by

$$a_1 = 2, \qquad a_n = 3a_{n-1} - 1, \qquad \text{for } n > 1$$

Series

It is often necessary to find the sum of consecutive terms of a sequence. An expression of the form

$$a_1 + a_2 + a_3 + \cdots + a_n$$

is called a **finite series.** A special notation called the **sigma notation** is used to express such sums. The Greek letter sigma (Σ) is used in the following manner:

$$\sum_{k=1}^{n} a_k = a_1 + a_2 + a_3 + \cdots + a_n$$

The letter k is called the **index of summation.** It takes values ranging from the smallest value written below the Σ sign to the largest value written above the Σ.

Example 1 **Writing a Series Using Sigma Notation**

Write $a_1 + a_2 + \cdots + a_7$ using sigma notation.

SOLUTION Because there are seven terms in the summation, the index ranges from 1 to 7. We write

$$\sum_{k=1}^{7} a_k = a_1 + a_2 + a_3 + \cdots + a_7$$

Example 2 **Evaluating a Sum of Terms**

Evaluate $\sum_{k=1}^{5} k(k^2 - 3)$

SOLUTION Let $k = 1, 2, 3, 4,$ and 5 in turn to get five terms. Add the terms.

$$\sum_{k=1}^{5} k(k^2 - 3) = 1(1^2 - 3) + 2(2^2 - 3) + 3(3^2 - 3) + 4(4^2 - 3) + 5(5^2 - 3)$$
$$= -2 + 2 + 18 + 52 + 110 = 180$$

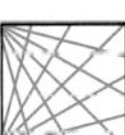

Sums of Terms of a Sequence with a Calculator

Sums can be conveniently computed using a calculator. We illustrate a method that makes use of the earlier technique of recalling and editing the last expression, and then using Ans to add the previous sum. Let us verify the result of the previous example:

$$\sum_{k=1}^{5} k(k^2 - 3) = 1(1^2 - 3) + 2(2^2 - 3) + 3(3^2 - 3) + 4(4^2 - 3) + 5(5^2 - 3)$$
$$= 180$$

We get

Enter and compute the first term.

Recall and edit the first term. Add answer to get the sum of the first two terms.

Repeat.

```
1*(1^2-3)
              -2
2*(2^2-3)+Ans
               0
3*(3^2-3)+Ans
              18
4*(4^2-3)+Ans
```

→

```
               0
3*(3^2-3)+Ans
              18
4*(4^2-3)+Ans
              70
5*(5^2-3)+Ans
             180
```

Further Ideas 2

(a) Evaluate $\sum_{k=1}^{6} k^2(k - 4)$ by hand.

(b) Determine a way of computing sums on your calculator. Verify your answer to (a) using your calculator.

Partial Sums

The sum of the first n terms of a sequence is called the ***n*th partial sum** and is denoted S_n.

$$S_n = a_1 + a_2 + \cdots + a_n$$

We can express a partial sum using the sigma notation:

$$S_n = \sum_{k=1}^{n} a_k$$

Example 3 **Evaluating a Partial Sum**

Find S_5 for the sequence $a_k = 2k^2 - 3$.

SOLUTION

$$\begin{aligned} S_5 = \sum_{k=1}^{5} a_k &= a_1 + a_2 + a_3 + a_4 + a_5 \\ &= (2 \cdot 1^2 - 3) + (2 \cdot 2^2 - 3) + (2 \cdot 3^2 - 3) + (2 \cdot 4^2 - 3) \\ &\quad + (2 \cdot 5^2 - 3) \\ &= -1 + 5 + 15 + 29 + 47 = 95 \end{aligned}$$

Self-Check 3

(a) Find S_6 for the sequence $a_k = k^3 + 1$ by hand.
(b) Verify your answer to (a) using a calculator.

The Mean of a Sequence of Numbers

Sequences and sigma notation are used in many areas of mathematics. One of these areas is statistics, for which the mean of a sequence of numbers is important. The **mean** of the numbers $a_1, \ldots, a_n$, denoted $\bar{a}$, is defined to be

$$\bar{a} = \frac{a_1 + \cdots + a_n}{n}$$

that is,

$$\bar{a} = \frac{\sum_{k=1}^{n} a_k}{n}$$

$\bar{a}$ gives us, in a sense, a middle number associated with the numbers. It is an average.

Example 4 **Computing the Mean of a Number of Terms of a Sequence**

Determine the mean of the first five numbers of the sequence defined by $a_n = 3n - 5$.

SOLUTION We are interested in the first five terms of the sequence. Therefore, we sum from 1 to 5.

$$\begin{aligned}\sum_{k=1}^{5} a_k &= a_1 + a_2 + a_3 + a_4 + a_5 \\ &= (3 \cdot 1 - 5) + (3 \cdot 2 - 5) + (3 \cdot 3 - 5) + (3 \cdot 4 - 5) + (3 \cdot 5 - 5) \\ &= -2 + 1 + 4 + 7 + 10 = 20\end{aligned}$$

Thus,

$$\bar{a} = \frac{20}{n} = \frac{20}{5} = 4$$

Self-Check 4

(a) Determine by hand the mean of the first six numbers of the sequence defined by $a_n = \; - n^2 + 3$.

(b) Verify the mean using a calculator.

We now introduce what is probably the most famous sequence of all time!

Example 5 | The Fibonacci Sequence

The following sequence is called the **Fibonacci sequence:**

$$1, 1, 2, 3, 5, 8, 13, 21, 34, 55, 89, 144, \ldots$$

It was discovered by the Italian mathematician Leonardo Fibonacci in the 13th century. Since that time, mathematicians and scientists have used the sequence repeatedly. There is even a Fibonacci Association.

Each term in the sequence is the sum of the two preceding terms. For example, the next number in the sequence would be 89 + 144; that is, 233. In concise mathematical language, this property can be stated in a recursion formula,

$$a_{n+2} = a_{n+1} + a_n$$

where $a_1 = 1$ and $a_2 = 1$.

We now give the original problem that led Fibonacci to this sequence. Fibonacci was interested in finding a mathematical model that describes the growth in population of rabbits. He made assumptions that a pair of rabbits produces a pair of young rabbits every month and that newborn rabbits become adults in 2 months at which time they produce another pair.

Starting with an adult pair, how many adult pairs will be in the colony after the first month, second month, third month, and so on? Fibonacci showed by counting (left as an exercise), that the first seven terms of the sequence are

$$1, 1, 2, 3, 5, 8, 13$$

He then observed that each term was the sum of the two preceding terms. This enabled him to compute the number of adult pairs in the colony at any time.

Scientists have been puzzled why Fibonacci numbers arise in some fields. Biologists have discovered that arrangements of leaves around stems of plants

often follow a Fibonacci pattern. The leaves of roses, cabbages, and scales of spruce trees, for example, follow a Fibonacci pattern. There are five rose leaves and eight cabbage leaves in a full circle of the stems. There are 21 scales of spruce in a full circle of a tree.

Answers to Self-Check Exercises

1. The first five terms are $a_1 = 1$, $a_2 = 10$, $a_3 = 25$, $a_4 = 46$, $a_5 = 73$.
2. The first six terms are 2, 5, 14, 41, 122, 365.
3. $S_6 = 447$.
4. Mean is $-\frac{73}{6}$.

Answer to Further Ideas Exercise

1. The first six terms are $a_1 = -2$, $a_2 = -1$, $a_3 = 4$, $a_4 = 13$, $a_5 = 26$, $a_6 = 43$.
2. The sum is 77.

EXERCISES 8.1

In Exercises 1–8, determine the first four terms of the sequences.

1. $a_n = 2n + 3$
2. $a_n = 5 - 3n$
3. $a_n = n^3 + 3n + 2$
4. $a_n = 4$
5. $a_n = 2^n$
6. $a_n = 3^{n+1}$
7. $a_n = 2n(n + 1)$
8. $a_n = (-1)^n$

In Exercises 9–16, determine the first five terms of the sequences. Verify your answers using a calculator.

9. $a_n = 2n - 5$
10. $a_n = 3n^2 + 5$
11. $a_n = 2n(4n - 5)$
12. $a_n = (2n + 1)(3n - 2)$
13. $a_n = \dfrac{n - 3}{2 + 3n}$
14. $a_n = \dfrac{n + 3}{2n + 1}$
15. $a_n = 1 + (\frac{1}{2})^n$
16. $a_n = 2^{n-3} - 4n$

In Exercises 17–22, find the stated term of each sequence.

17. $a_n = 5n - 1$, fourth term

18. $a_n = 3n^2 + 4$, fifth term

19. $a_n = \dfrac{n-7}{2n+3}$, seventh term

20. $a_n = (n + 1)(n - 2)$, sixth term

21. $a_n = 3^{n-10}$, 12th term

22. $a_n = (-1)^n + 5n$, 23rd term

In Exercises 23–32, the sequences are defined recursively for $n > 1$. Give the first five terms of each sequence.

23. $a_1 = 3,\ a_n = a_{n-1} + 3$

24. $a_1 = 1,\ a_n = a_{n-1} + 4$

25. $a_1 = -2,\ a_n = 4a_{n-1} - 2$

26. $a_1 = 0,\ a_n = \dfrac{a_{n-1}}{2}$

27. $a_1 = -3,\ a_n = 5a_{n-1} - 2n$

28. $a_1 = 7,\ a_n = -2a_{n-1} + 3n^2$

29. $a_1 = 2,\ a_n = \dfrac{n+2}{a_{n-1} + 3}$

30. $a_1 = 1,\ a_n = \dfrac{2a_{n-1} + 3}{a_{n-1} + 1}$

31. $a_1 = 1,\ a_n = (a_{n-1})^2$

32. $a_1 = -1,\ a_n = (a_{n-1})^{n+1} + 3$

In Exercises 33–44, evaluate the sums.

33. $\sum_{k=1}^{4} (2k + 3)$ **34.** $\sum_{k=1}^{5} (4 - 2k)$

35. $\sum_{k=1}^{6} k(k + 3)$ **36.** $\sum_{k=1}^{7} (k^2 - 4)$

37. $\sum_{k=1}^{4} (2k^2 + 3k - 4)$ **38.** $\sum_{k=2}^{5} (k + 1)(2k - 3)$

39. $\sum_{k=1}^{4} \dfrac{4}{2k+1}$ **40.** $\sum_{k=3}^{6} \dfrac{k-1}{3k+2}$

41. $\sum_{k=1}^{10} (-1)^k$ **42.** $\sum_{k=3}^{5} 2k(k^2 + 4)$

43. $\sum_{k=1}^{8} 27$ **44.** $\sum_{k=1}^{4} 3(2^k + 2k - 4)$

In Exercises 45–52, express the given sums using sigma notation.

45. $1 + 2 + 3 + 4 + 5 + 6$

46. $1^2 + 2^2 + 3^2 + 4^2 + 5^2$

47. $1 - 2 + 3 - 4 + 5 - 6$

48. $\dfrac{1}{2} + \dfrac{1}{3} + \dfrac{1}{4} + \dfrac{1}{5} + \dfrac{1}{6} + \dfrac{1}{7}$

49. $1 + 3 + 5 + 7 + 9 + 11$

50. $\dfrac{1}{9} + \dfrac{1}{16} + \dfrac{1}{25} + \dfrac{1}{36}$

51. $\dfrac{1}{4} + \dfrac{2}{5} + \dfrac{3}{6} + \dfrac{4}{7}$

52. $\dfrac{2}{3} + \dfrac{5}{4} + \dfrac{8}{5} + \dfrac{11}{6} + \dfrac{14}{7} + \dfrac{17}{8}$

In Exercises 53–60, find the partial sums for the given sequences by hand. Verify your answers using a calculator.

53. $S_3,\quad a_k = 2k$ **54.** $S_4,\quad a_k = 3k - 1$

55. $S_5,\quad a_k = k^2 + 3$ **56.** $S_6,\quad a_k = k(k + 1)$

57. $S_6,\quad a_k = k$ **58.** $S_3,\quad a_k = (-1)^k + k$

59. $S_5,\quad a_k = 1/k$ **60.** $S_6,\quad a_k = k^2 + 3k - 4$

In Exercises 61–66, determine the means of the given terms of the sequences by hand. Verify your answers using a calculator.

61. $a_n = 2n + 3$, first five terms

62. $a_n = n^2 - 1$, first four terms

63. $a_n = n(n - 3)$, first five terms

64. $a_n = 3n^2 + n$, first seven terms

65. $a_n = (-1)^n$, first 50 terms

66. $a_n = 3 + 1/n$, first eight terms

Terms of a Sequence with a Calculator

Recall that a sequence may be interpreted as a function $f(n)$ having as its domain the set of positive integers, 1, 2, 3, ... Consequently, $a_1 = f(1)$, $a_2 = f(2)$, $a_3 = f(3)$, ... For example, the terms of the sequence $a_n = n^2 + 1$ would be given by the values of the function $f(n) = n^2 + 1$ at $n = 1$, $n = 2$, $n = 3$, and so on. This interpretation enables us to use the techniques that we developed for functions to analyze sequences. We can, for example, enter the nth term of a sequence as a function into a calculator and use the calculator to compute the terms of the sequence. This is a particularly convenient way to compute the terms of a sequence if your calculator has a table facility.

Use a calculator in the above manner to compute the stated terms of the sequences in Exercises 67–74, rounded to four decimal places.

67. $a_n = 2n - 4$, first six terms

68. $a_n = n^3 - 2n$, first six terms

69. $a_n = 7n^{-1} + n$, first seven terms

70. $a_n = 5n^{1/2} + 3n$, first five terms

71. $a_n = (n + 1)(n - 3)$, a_8 through a_{14}

72. $a_n = n^2 - n + 2$, a_{20} through a_{30}

73. $a_n = \dfrac{2n + 3}{n + 1}$, a_5 through a_{12}

74. $a_n = 2^n - 3n$, a_4 through a_{10}

8.2 Arithmetic Sequences

• Arithmetic Sequence • Common Difference • The *n*th Term of an Arithmetic Sequence • Partial Sums of an Arithmetic Sequence • Simple Interest • Motion of an Object

In this section and the following, we discuss two important types of sequences called arithmetic and geometric sequences.

Arithmetic Sequence

An **arithmetic sequence** is a sequence in which successive terms differ by a constant. The constant is called the **common difference** of the sequence.

Thus the sequence $a_1, a_2, a_3, \ldots, a_n, a_{n+1}, \ldots$ is an arithmetic sequence if

$$a_{n+1} - a_n = d$$

for a constant d. d is the common difference.

Example 1 Showing That a Sequence Is an Arithmetic Sequence

Show that the following numbers are the first four terms in an infinite arithmetic sequence. Find the common difference of the sequence.

$$4, 7, 10, 13, \ldots$$

SOLUTION Let us compute successive differences. We get

$$7 - 4 = 3$$
$$10 - 7 = 3$$
$$13 - 10 = 3$$

Successive terms differ by a constant, namely 3. Thus, the sequence is an arithmetic sequence with common difference 3.

Self-Check 1 Show that the following numbers are the first four terms of an arithmetic sequence. Find the common difference of the sequence.

$$-1, 1, 3, 5, \ldots$$

THE *n*TH TERM OF AN ARITHMETIC SEQUENCE

To get the next term of an arithmetic sequence, we add the common difference. Thus, the relationship between successive terms in an arithmetic sequence can be written

$$a_{n+1} = a_n + d$$

By letting $n = 1, 2, 3, \ldots$ in this formula, we get

$$a_2 = a_1 + d$$
$$a_3 = a_2 + d = a_1 + 2d$$
$$a_4 = a_3 + d = a_1 + 3d, \text{ etc.}$$

This pattern leads to

$$a_n = a_1 + (n - 1)d$$

ARITHMETIC SEQUENCE

The *n*th term of an arithmetic sequence having first term a_1 and common difference d is

$$a_n = a_1 + (n - 1)d.$$

In practice, we are often given information about a sequence and want to determine a complete description of the sequence from that information. We discussed various ways of defining a sequence in Section 8.1. The most useful complete description is usually an expression for the *n*th term. The following examples illustrate various scenarios.

Example 2 Arithmetic Sequence Given the First Term and Common Difference

The first term of an arithmetic sequence is 3, and the common difference is 5. Write down the first four terms of the sequence. Find the *n*th term of the sequence and use it to determine the seventh term.

SOLUTION The information that is given about the sequence is $a_1 = 3$, $d = 5$. Thus,

$$a_2 = a_1 + d = 3 + 5 = 8$$
$$a_3 = a_2 + d = 8 + 5 = 13$$
$$a_4 = a_3 + d = 13 + 5 = 18$$

The first four terms in the sequence are 3, 8, 13, 18.
The formula for a_n gives

$$\begin{aligned} a_n &= a_1 + (n - 1)d \\ &= 3 + (n - 1)5 \\ &= 5n - 2 \end{aligned}$$

The sequence is thus $a_n = 5n - 2$. This expression for a_n can now be used to find any other term of the sequence. Let $n = 7$ to get the seventh term.

$$a_7 = 5(7) - 2 = 33$$

Self-Check 2 The first term of an arithmetic sequence is 23, and the common difference is -2. Find the nth term of the sequence, and then use this general term to determine the sixth term.

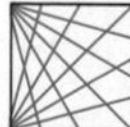

Terms of an Arithmetic Sequence with a Calculator

The terms of an arithmetic sequence can be conveniently computed on a calculator. Let us compute the first 11 terms of the sequence having first term 3.2 and common difference 5.7. Note the use of Ans + 5.7 to add 5.7 to the updated answer. The terms are found to be 3.2, 8.9, 14.6, 20.3, . . . , 54.5, 60.2.

Enter the first term.

Add the common difference to the answer.

Repeated use of the enter key gives the following terms.

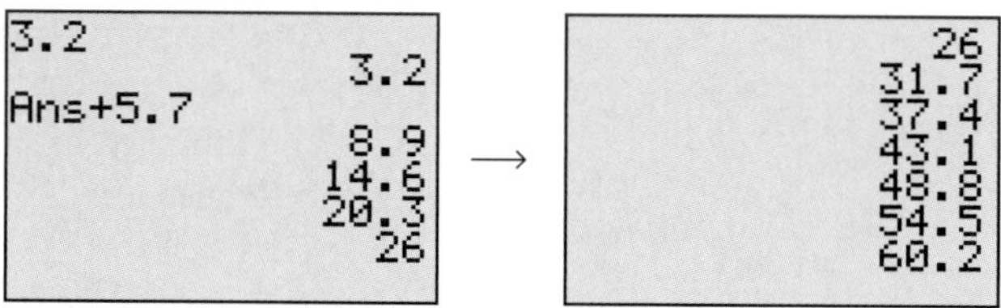

Further Ideas 1 Determine how to compute the terms of an arithmetic sequence on your calculator. Use your calculator to compute the first eight terms of the arithmetic sequence having first term 6.78 and common difference 4.8.

Example 3 **Arithmetic Sequence Given Two Consecutive Terms**

The third and fourth terms of an arithmetic sequence are

$$a_3 = 5, a_4 = 9$$

Determine the nth term, and use it to find the tenth term.

SOLUTION The common difference d is

$$d = a_4 - a_3 = 9 - 5 = 4$$

We can use the formula $a_n = a_1 + (n - 1)d$ with $n = 3$ to find a_1 as follows.

$$\begin{aligned} a_3 &= a_1 + (3 - 1)d \\ 5 &= a_1 + 2(4) \\ a_1 &= -3 \end{aligned}$$

The nth term of the sequence can now be found.

$$\begin{aligned} a_n &= a_1 + (n - 1)d \\ &= -3 + (n - 1)(4) \\ &= 4n - 7 \end{aligned}$$

The sequence is thus $a_n = 4n - 7$. Let $n = 10$ to get the tenth term.

$$a_{10} = 4(10) - 7 = 33$$

The following example illustrates the approach for finding the sequence when two terms are given but the terms are not consecutive. This situation involves solving a system of linear equations.

Example 4 **Arithmetic Sequence Given Two Terms**

If $a_3 = 11$ and $a_7 = 23$ for an arithmetic sequence, determine the nth term and use it to find a_{11}.

SOLUTION Use $n = 3$ and $n = 7$ in the formula $a_n = a_1 + (n - 1)d$.

$$a_3 = a_1 + (3 - 1)d \quad \text{gives} \quad 11 = a_1 + 2d$$
$$a_7 = a_1 + (7 - 1)d \quad \text{gives} \quad 23 = a_1 + 6d$$

We have arrived at a system of two linear equations in the variables a_1 and d. Solve this system.

$$a_1 + 2d = 11$$
$$a_1 + 6d = 23$$

The solution is found to be

$$a_1 = 5, d = 3$$

Thus, $a_n = 5 + (n - 1)(3)$. The sequence is $a_n = 3n + 2$. The 11th term can now be computed.

$$a_{11} = 3(11) + 2 = 35$$

Self-Check 3 If $a_2 = 1$ and $a_8 = 25$ for an arithmetic sequence, find a_1 and the common difference d. Use these values to find a_5.

Partial Sums of an Arithmetic Sequence

We now derive a formula for the nth partial sum (sum of the first n terms) of an arithmetic sequence. Consider an arithmetic sequence $a_1, a_2, a_3, \ldots, a_n, \ldots$. We can express the nth partial sum as follows.

$$\begin{aligned} S_n &= a_1 + a_2 + \cdots + a_{n-1} + a_n \\ &= a_1 + [a_1 + d] + [a_1 + 2d] + \cdots + [a_1 + (n - 2)d] + [a_1 + (n - 1)d] \end{aligned}$$

Now write this same expression in reverse order.

$$S_n = [a_1 + (n - 1)d] + [a_1 + (n - 2)d] + \cdots + [a_1 + 2d] + [a_1 + d] + a_1$$

Add the corresponding terms of these two forms of S_n to get

$$\begin{aligned} 2S_n = {} & [2a_1 + (n - 1)d] + [2a_1 + (n - 1)d] + \cdots + [2a_1 + (n - 1)d] \\ & + [2a_1 + (n - 1)d] \end{aligned}$$

There are n identical terms on the right. Thus,

$$2S_n = n[2a_1 + (n-1)d]$$
$$S_n = \frac{n}{2}[2a_1 + (n-1)d)]$$

We have arrived at a formula for computing S_n for an arithmetic sequence, given a_1 and d. Because $a_n = a_1 + (n-1)d$, S_n can also be expressed in terms of a_1 and a_n as follows:

$$S_n = \frac{n}{2}[a_1 + a_1 + (n-1)d] = \frac{n}{2}[a_1 + a_n]$$

Let us now summarize these results for partial sums.

Partial Sum

The nth **partial sum,** S_n, of an arithmetic sequence $a_1, a_2, a_3, \ldots$ is

$$S_n = \frac{n}{2}[2a_1 + (n-1)d]$$

This sum may also be expressed

$$S_n = \frac{n}{2}[a_1 + a_n]$$

These results are used in various ways. We now give examples to illustrate some of these scenes.

Example 5 Finding a Partial Sum of an Arithmetic Sequence

Consider the arithmetic sequence for which $a_1 = 6$ and $d = 3$. Find the sum of the first six terms.

SOLUTION Because the information given is a_1 and d, we use the formula $S_n = \frac{n}{2}[2a_1 + (n-1)d]$ to find S_6. Letting $a_1 = 6$, $d = 3$, and $n = 6$, we get

$$S_6 = \frac{6}{2}[2(6) + (6-1)3]$$
$$= 3[12 + 15] = 81$$

The sum of the first six terms is 81.

Example 6 Finding a Partial Sum of an Arithmetic Sequence

Consider the arithmetic sequence for which $a_n = 4n - 3$. Find S_5.

SOLUTION Because we are given a_n, we can use it to compute a_1 and a_5 and then use the formula $S_n = \frac{n}{2}[a_1 + a_n]$ with $n = 5$ to compute S_5.

$n = 1$ in $a_n = 4n - 3$ gives $a_1 = 4(1) - 3 = 1$

$n = 5$ gives $a_5 = 4(5) - 3 = 17$

Thus $S_5 = \frac{5}{2}[1 + 17] = 45$

Self-Check 4 Find S_4 for the arithmetic sequence in which $a_2 = 3$ and $d = -2$.

Example 7 Determining the Number of Terms in an Arithmetic Sequence

The first term of an arithmetic sequence is 3 and the last term is 27. If the sum of the terms is 120, how many terms are in the sequence?

SOLUTION This sequence is finite. We are given $a_1 = 3$, $a_n = 27$, and $S_n = 120$. We want to find n. Let us use the formula $S_n = \frac{n}{2}[a_1 + a_n]$. We get

$$120 = \frac{n}{2}[3 + 27]$$

Solve for n.

$$120 = \frac{n}{2}[30]$$

$$n = \frac{120}{15} = 8$$

The number of terms in the sequence is eight.

Applications . . .

Example 8 Simple Interest

Let an amount P be invested in an account that pays simple interest at an annual rate r. Let A_n be the amount present after n years. Show that the sequence of annual amounts $A_1, A_2, A_3, \ldots$ is an arithmetic sequence. Interpret the common difference of this sequence.

SOLUTION The amount is P. In simple interest, the interest is paid on this amount, thus, the interest earned every year is Pr. Therefore

$$A_{n+1} - A_n = Pr$$

The sequence of annual amounts $A_1, A_2, A_3, \ldots$ is an arithmetic sequence with

$$d = Pr$$

The common difference of the arithmetic sequence is the annual interest.

For example, if \$1000 is deposited at 5% simple interest, the annual interest is

$$\$1000(5/100) = \$50$$

The annual amounts will be an arithmetic sequence,

$$\$1000, \$1050, \$1100, \$1150, \$1200, \ldots$$

having common difference \$50.

Example 9 **Motion of an Object**

An object starts with an initial speed u and moves with constant acceleration a. The velocity v_n after n seconds is given by the equation $v_n = u + an$. Show that the velocities $v_1, v_2, v_3, \ldots$ form an arithmetic sequence. Interpret the common difference of the sequence.

SOLUTION We have that $v_n = u + an$. Replacing n with $n + 1$ gives

$$v_{n+1} = u + a(n + 1)$$

Subtracting the expression for v_n from v_{n+1} gives

$$\begin{aligned} v_{n+1} - v_n &= [u + a(n + 1)] - [u + an] \\ &= a \end{aligned}$$

Because a is constant, the velocities form an arithmetic sequence with the acceleration being the common difference.

For example, if an object moved from rest with constant acceleration 20 feet/second^2, the velocities at intervals of 1 second are (in feet/second),

$$20, 40, 60, 80, \ldots$$

Answers to Self-Check Exercises

1. Common difference = 2.
2. $a_n = 25 - 2n$. $a_6 = 13$.
3. $a_1 = -3$, $d = 4$, $a_5 = 13$.
4. $S_4 = 8$.

Answers to Further Ideas Exercise

1. First eight terms are 6.78, 11.58, 16.38, 21.18, 25.98, 30.78, 35.58, 40.38.

EXERCISES 8.2

In Exercises 1–6, show that the numbers are terms of an arithmetic sequence. Find the common difference.

1. $1, 3, 5, 7, \ldots$ **2.** $5, 8, 11, 14, \ldots$
3. $-1, 3, 7, 11, \ldots$ **4.** $12, 10, 8, 6, \ldots$
5. $\frac{1}{2}, \frac{3}{2}, \frac{5}{2}, \ldots$ **6.** $\frac{1}{4}, \frac{7}{12}, \frac{11}{12}, \frac{15}{12}, \ldots$

In Exercises 7–12, determine whether the numbers are terms of an arithmetic sequence.

7. $0, 5, 10, 15, \ldots$ **8.** $0, \frac{3}{4}, \frac{3}{2}, \frac{5}{4}, \ldots$
9. $\frac{1}{2}, -\frac{1}{4}, -1, -\frac{7}{4}, \ldots$ **10.** $-1, 1, 3, 7, \ldots$
11. $-5, 1, 7, 13, \ldots$ **12.** $2, 6, 18, 64, \ldots$

In Exercises 13–18, give the next two terms of each arithmetic sequence.

13. $0, 3, 6, 9, \ldots$ **14.** $-2, 3, 8, 13, \ldots$
15. $-\frac{1}{2}, 0, \frac{1}{2}, 1, \ldots$ **16.** $\frac{1}{3}, \frac{1}{4}, \frac{1}{6}, \frac{1}{12}, \ldots$
17. $41, 48, 55, 62, \ldots$ **18.** $12, 8, 4, 0, \ldots$

In Exercises 19–24, the first terms and the common differences of arithmetic sequences are given. Determine the first four terms of each sequence.

19. $a_1 = 2, d = 3$ **20.** $a_1 = 0, d = 5$
21. $a_1 = -3, d = -1$ **22.** $a_1 = \frac{1}{2}, d = 3$

23. $a_1 = -6, d = \frac{3}{4}$

24. $a_1 = 25, d = -7$

In Exercises 25–30, the first terms and the common differences of arithmetic sequences are given. Use a calculator to determine the first ten terms of each sequence.

25. $a_1 = 2.36, d = 3.2$

26. $a_1 = 5.38, d = 4.6$

27. $a_1 = 2.7689, d = 5.23$

28. $a_1 = -6.982, d = 4.89$

29. $a_1 = 67.92, d = -5.6$

30. $a_1 = -3.546, d = -5.76$

In Exercises 31–36, the first terms and the common differences of arithmetic sequences are given. Find the specified terms.

31. $a_1 = 3, d = 4$; fifth term

32. $a_1 = 5, d = -3$; seventh term

33. $a_1 = -1, d = 8$; fourth term

34. $a_1 = \frac{5}{8}, d = \frac{1}{2}$; third term

35. $a_1 = -5, d = 2$; 20th term

36. $a_1 = -\frac{2}{3}, d = \frac{1}{8}$; ninth term

In Exercises 37–42, consecutive terms of arithmetic sequences are given. Find the specified terms.

37. $a_2 = 3, a_3 = 5$; fifth term

38. $a_5 = 2, a_6 = 7$; ninth term

39. $a_8 = -4, a_9 = 4$; first term

40. $a_9 = \frac{1}{4}, a_{10} = \frac{3}{4}$; seventh term

41. $a_3 = 0, a_4 = -5$; 15th term

42. $a_8 = -5, a_9 = -7$; 100th term

In Exercises 43–48, two terms of arithmetic sequences are given. Find the specified terms.

43. $a_2 = 4, a_4 = 8$; seventh term

44. $a_3 = -3, a_6 = 6$; 12th term

45. $a_4 = \frac{1}{2}, a_{11} = 4$; sixth term

46. $a_7 = 3, a_3 = 2$; first term

47. $a_1 = 27, a_9 = -4$; 20th term

48. $a_5 = 3, a_{11} = 33$; 31st term

In Exercises 49–54, determine the specified partial sums of the given sequences.

49. $a_1 = 3, d = 2; S_5$

50. $a_1 = 2, d = 1; S_7$

51. $a_1 = -5, d = 2; S_4$

52. $a_1 = \frac{1}{2}, d = 4; S_9$

53. $a_1 = 0, d = \frac{3}{2}; S_5$

54. $a_1 = -\frac{1}{3}, d = -\frac{3}{4}; S_{12}$

In Exercises 55–60, determine the specified partial sums of the given sequences.

55. $a_n = 2n + 1; S_8$

56. $a_n = n - \frac{1}{3}; S_7$

57. $a_n = n - 6; S_{15}$

58. $a_n = n; S_{10}$

59. $a_n = -4n + 5; S_4$

60. $a_n = 2n + \frac{1}{3}; S_7$

61. The first term of an arithmetic sequence is 2 and the last term is 10. The sum of the terms is 30. How many terms are in the sequence?

62. The first term of an arithmetic sequence is -3 and the last term is -27. The sum of the terms is -120. How many terms are in the sequence?

63. The first term of an arithmetic sequence is 4 and the last term is 99. The sum of the terms is 1030. How many terms are in the sequence?

64. The sum of the first five terms of an arithmetic sequence is 40 and the common difference is 3. Determine the first term of the sequence.

65. The sum of the first eight terms of an arithmetic sequence is 48 and the common difference is -4. Determine the first term of the sequence. Show that the nth term of the sequence is $a_n = 24 - 4n$.

66. The first term of an arithmetic sequence is 4, and the eighth term is 18. Show that the nth term of the sequence is $a_n = 2n + 2$.

67. The first term of an arithmetic sequence is -3, and the 14th term is 75. Find the nth term of the sequence.

68. Find the sum of the integers from 1 to 100.

69. **Simple Interest** An amount of $2500 is deposited at 6% simple interest. Determine the annual amounts for the first 5 years.

70. **Simple Interest** An amount of $500 is deposited at 4% simple interest. Determine the annual amounts for the first 6 years.

71. **Simple Interest** An amount of $1000 is deposited at 7% simple interest. How long will it take the amount to reach $1490?

72. **Accelerated Motion** An object moves from rest with constant acceleration 30 feet/second^2. Give the velocities of the object for the first 10 seconds.

73. **Accelerated Motion** An object moves from rest with constant acceleration 12 feet/second^2. Give the velocities of the object for the first 8 seconds.

74. **Motion Under Gravity** An object falls from rest under gravity. After the first 3 seconds, the velocities are 32, 64, and 96 feet/second respectively. Determine the acceleration of gravity.

75. Chimes A clock chimes the time on the hour. How many times does it chime in a 12-hour period?

76. Accumulated Savings If you put into savings 1 cent on January 1st, 2 cents on January 2nd, 3 cents on January 3rd, and so on, for a whole year, how much money will you have saved during the year?

8.3 Geometric Sequences, Count of Bacteria, and the Flaw in the Pentium Chip

• Geometric Sequence • Common Ratio • The *n*th Term of a Geometric Sequence • Partial Sums of a Geometric Sequence • Sums of Infinite Geometric Sequences • Repeating Decimals • Functions, Graphs, and Series • Count of Bacteria • Compound Interest • The Flaw in the Pentium Chip

We now continue the discussion of special sequences with an introduction to geometric sequences.

Geometric Sequence

A **geometric sequence** is a sequence in which the ratios of successive terms are constant. The constant is called the **common ratio** of the sequence.

Thus, the sequence $a_1, a_2, a_3, \ldots, a_n, a_{n+1}, \ldots$ is a geometric sequence if

$$\frac{a_{n+1}}{a_n} = r$$

for a constant r. r is the common ratio.

Our development of geometric sequences very much parallels that of arithmetic sequences, with common difference being replaced by common ratio. We shall find, however, that the study of geometric sequences is a little richer than that of arithmetic sequences. There are more interesting possibilities that arise!

Example 1 **Showing That a Sequence Is a Geometric Sequence**

Show that the following numbers are the first four terms of an infinite geometric sequence. Determine the common ratio.

$$2, 6, 18, 54, \ldots$$

SOLUTION Let us compute the ratios of successive terms.

$$\frac{6}{2} = 3, \qquad \frac{18}{6} = 3, \qquad \frac{54}{18} = 3$$

The ratios of successive terms are constant, namely 3. Thus, the sequence is a geometric sequence with common ratio 3.

The *n*th Term of a Geometric Sequence

Observe that the relationship between successive terms in a geometric sequence can be written

$$a_{n+1} = a_n r$$

Thus, by letting $n = 1, 2, 3, \ldots$, we get

$$a_2 = a_1 r$$
$$a_3 = a_2 r = a_1 r^2$$
$$a_4 = a_3 r = a_1 r^3, \text{ etc.}$$

This pattern leads to

$$\vdots$$

$$a_n = a_1 r^{n-1}$$

Geometric Sequence

The nth term of a geometric sequence having first term a_1 and common ratio r is given by

$$a_n = a_1 r^{n-1}.$$

As in the case of an arithmetic sequence, we are often given information about a geometric sequence and want to determine a complete description of the sequence from that information. The most useful complete description is usually an expression for the nth term. The following examples illustrate various scenarios.

Example 2 **Determining a Geometric Sequence Given the First Term and Common Ratio**

The first term in a geometric sequence is 3, and the common ratio is 2. Write down the first four terms of the sequence. Find the nth term, and use it to determine the eighth term.

SOLUTION We are given that $a_1 = 3$ and $r = 2$. Thus,

$$a_2 = a_1 r = 3(2) = 6$$
$$a_3 = a_2 r = 6(2) = 12$$
$$a_4 = a_3 r = 12(2) = 24$$

The first four terms of the sequence are 3, 6, 12, 24.

Using the preceding formula for a_n, we get

$$a_n = a_1 r^{n-1} = 3(2)^{n-1}$$

The sequence is thus $a_n = 3(2)^{n-1}$. Let $n = 8$ to get a_8.

$$a_8 = 3(2)^{8-1} = 3(2)^7 = 3(128) = 384$$

Self-Check 1 The first term in a geometric sequence is -3, and the common ratio is 4. Find the nth and fourth terms.

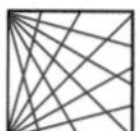

Terms of a Geometric Sequence with a Calculator

The terms of a geometric sequence can be conveniently computed on a calculator. Let us compute the first five terms of the sequence having first term 7.3 and common ratio 3.8. Note the use of Ans * 3.8 to multiply the updated answer by 3.8. The terms are 7.3, 27.74, 105.412, 400.5656, 1522.14928.

Enter the first term.

Multiply the answer by the common ratio.

Repeated use of the enter key gives the following terms.

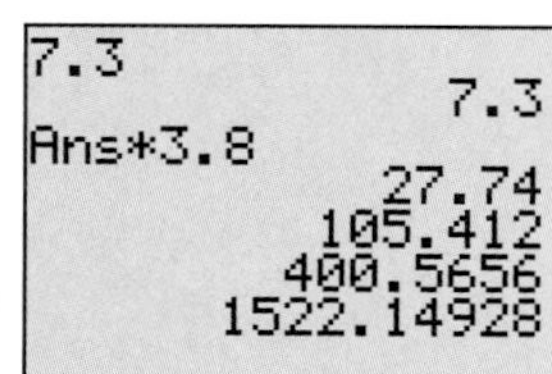

Further Ideas 1 Determine how to compute the terms of a geometric sequence on your calculator. Compute the first six terms of the geometric sequence having first term 6.7 and common ratio 4.8.

Example 3 Determining a Geometric Sequence Given Two Consecutive Terms

The fourth and fifth terms of a geometric sequence are -54 and 162. Find the common ratio, the first term, and the ninth term.

SOLUTION We are given that $a_4 = -54$ and $a_5 = 162$. Thus,

$$r = \frac{162}{-54} = -3$$

The common ratio is -3.

We can use the formula $a_n = a_1 r^{n-1}$ with $n = 4$ to get a_1 as follows:

$$a_4 = a_1 r^{(4-1)}$$
$$-54 = a_1(-3)^3$$
$$-54 = -27a_1$$
$$a_1 = 2$$

The first term in the sequence is 2.

The sequence is thus $a_n = 2(-3)^{n-1}$. The ninth term is now given by

$$a_9 = a_1(-3)^{(9-1)} = 2(-3)^8 = 2(6561) = 13{,}122$$

The ninth term in the sequence is 13,122.

The following example illustrates the approach for finding the sequence when two terms are given, but the terms are not consecutive terms.

Example 4 **Determining a Geometric Sequence Given Two Terms**

If $a_2 = 12$ and $a_5 = 768$ for a geometric sequence, find a_7.

SOLUTION Use $n = 2$ and $n = 5$ in the formula $a_n = a_1 r^{n-1}$.

$$a_2 = a_1 r^{(2-1)} \quad \text{gives} \quad 12 = a_1 r$$
$$a_5 = a_1 r^{(5-1)} \quad \text{gives} \quad 768 = a_1 r^4$$

Divide the second equation by the first to determine r.

$$\frac{768}{12} = \frac{a_1 r^4}{a_1 r}$$
$$64 = r^3$$
$$r = 4$$

Use this value of r in the equation $12 = a_1 r$ to get a_1.

$$12 = 4a_1$$
$$a_1 = 3$$

The sequence is therefore $a_n = 3(4)^{n-1}$. The seventh term can now be computed:

$$a_7 = 3(4)^6 = 3(4096) = 12{,}288$$

Self-Check 2 If $a_2 = 12$ and $a_4 = 108$, with $r > 0$, find a_6.

PARTIAL SUMS OF A GEOMETRIC SEQUENCE

We now derive a formula for the nth partial sum of a geometric sequence. Consider the geometric sequence $a_1, a_2, a_3, \ldots$. The nth partial sum is

$$S_n = a_1 + a_2 + a_3 \cdots + a_n$$

If the common ratio is r, this sum may be written

$$S_n = a_1 + a_1r + a_1r^2 + \cdots + a_1r^{n-1} \qquad (1)$$

If $r = 1$ (a special case), we get

$$S_n = na_1$$

If $r \neq 1$, multiply both sides of expression 1 for S_n by r to get rS_n.

$$rS_n = a_1r + a_1r^2 + a_1r^3 + \cdots + a_1r^n \qquad (2)$$

Subtract expression 2 from expression 1.

$$S_n - rS_n = a_1 - a_1r^n$$
$$S_n(1 - r) = a_1(1 - r^n)$$

Because $r \neq 1$, both sides can be divided by $(1 - r)$ to get

$$S_n = \frac{a_1(1 - r^n)}{(1 - r)}$$

PARTIAL SUM

The nth partial sum of a geometric sequence with first term a_1 and common ratio r $(r \neq 1)$ is

$$S_n = \frac{a_1(1 - r^n)}{(1 - r)}$$

This result can be used in various ways. We now give examples to illustrate some of these scenes.

Example 5 Finding a Partial Sum of a Geometric Sequence

Find S_6 for the geometric sequence defined by $a_n = 2(3)^{n-1}$.

SOLUTION Compare

$$a_n = 2(3)^{n-1} \quad \text{to} \quad a_n = a_1r^{n-1}$$

We see that $a_1 = 2$ and $r = 3$. The previous formula for S_n now gives

$$S_n = \frac{2(1 - 3^n)}{(1 - 3)}$$

$$S_6 = \frac{2(1 - 3^6)}{(1 - 3)} = \frac{2(1 - 729)}{-2} = 728$$

The sum of the first six terms of the sequence is 728.

Example 6 Finding a Partial Sum of a Geometric Sequence

Compute $\sum_{k=1}^{7} 5(2)^{k-1}$.

SOLUTION Observe that this is the sum of the terms of the finite geometric sequence with $a_1 = 5$ and $r = 2$ having seven terms ($n = 7$). Thus,

$$S_n = \frac{a_1(1 - r^n)}{(1 - r)}$$

$$S_7 = \frac{5(1 - 2^7)}{(1 - 2)} = \frac{5(1 - 128)}{-1} = 635$$

Self-Check 3 Compute $\sum_{k=1}^{4} 6(-3)^{k-1}$.

Partial sums are in effect the sums of a finite number of terms. We now take this discussion one stage further. We use the results from partial sums to find expressions for the sums of the terms in an infinite geometric sequence.

Sums of Infinite Geometric Sequences

Consider the infinite geometric sequence $a_1, a_1r, a_1r^2, \ldots, a_1r^{n-1}, \ldots$ having common ratio r. We have seen that the nth partial sum of this sequence is

$$S_n = \frac{a_1(1 - r^n)}{(1 - r)}$$

We now proceed to arrive at a formula for the sum of all the terms in the sequence if the common difference is a fraction; that is, we consider the situation when $|r| < 1$. Write S_n as follows:

$$S_n = \frac{a_1}{(1 - r)} - \frac{a_1r^n}{(1 - r)}$$

If $|r| < 1$, then as n increases r^n gets smaller and smaller. The second term in this expression, namely $a_1r^n/(1 - r)$, gets closer and closer to zero. S_n gets closer and closer to the first term, $a_1/(1 - r)$. We express this

$$S_n \to \frac{a_1}{(1 - r)} \quad \text{as} \quad n \to \infty$$

We call the expression

$$a_1 + a_1r + a_1r^2 + \cdots + a_1r^{n-1} + \cdots$$

an **infinite geometric series.** We have arrived at the following result for the sum of an infinite geometric series.

Sum of a Geometric Series

The sum S of an infinite geometric series

$$a_1 + a_1r + a_1r^2 + \cdots + a_nr^{n-1} + \cdots$$

having common ratio $|r| < 1$ and first term a_1 is

$$S = \frac{a_1}{(1 - r)}$$

Example 7 **Sum of a Geometric Series**

Find the sum of the infinite series having $a_1 = 3$ and $r = \frac{1}{2}$.

SOLUTION The infinite series is

$$3 + 3\left(\frac{1}{2}\right) + 3\left(\frac{1}{2}\right)^2 + 3\left(\frac{1}{2}\right)^3 + \cdots$$

The common ratio is $\frac{1}{2}$. The formula for the sum of a geometric series thus applies.

$$S = \frac{a_1}{(1 - r)} = \frac{3}{\left(1 - \frac{1}{2}\right)} = \frac{3}{\left(\frac{1}{2}\right)} = 6$$

The sum of the series is 6.

Self-Check 4 Find the sum of the following infinite series.

$$1 - \frac{1}{4} + \frac{1}{16} - \frac{1}{64} + \frac{1}{256} - \cdots$$

Repeating Decimals

From our discussion in Section R.1, we realize that a real number in decimal form that repeats a block of digits is a rational number. It can be written as a fraction. For example, the number 0.74291291291 . . . , where the block 291 is repeated, is a rational number. The notation $0.74\overline{291}$ is used to indicate the repeating part. The mathematics of geometrical sequences can be used to write repeating decimals as fractions. The following example illustrates the method.

Example 8 **Writing a Repeating Decimal as a Fraction**

Write the repeating decimal $0.74\overline{291}$ as a fraction.

SOLUTION This decimal can be written

$$\begin{aligned} 0.74\overline{291} &= 0.74 + 0.00291 + 0.00000291 + 0.00000000291 + \cdots \\ &= 0.74 + 0.00291 + 0.00291(10^{-3}) + 0.00291(10^{-6}) + \cdots \end{aligned}$$

We make the interesting observation that the repeating block is the sum of a geometric series having first term $a_1 = 0.00291$ and common ratio $r = 10^{-3}$. Use the formula for the sum of the series to get

$$\begin{aligned} 0.74\overline{291} &= 0.74 + \frac{a_1}{(1 - r)} = 0.74 + \frac{0.00291}{1 - 10^{-3}} = 0.74 + \frac{0.00291}{1 - 0.001} \\ &= \frac{74}{100} + \frac{0.00291}{0.999} = \frac{74}{100} + \frac{291}{99{,}900} \\ &= \frac{(74)(999)}{99{,}900} + \frac{291}{99{,}900} = \frac{74{,}217}{99{,}900} \end{aligned}$$

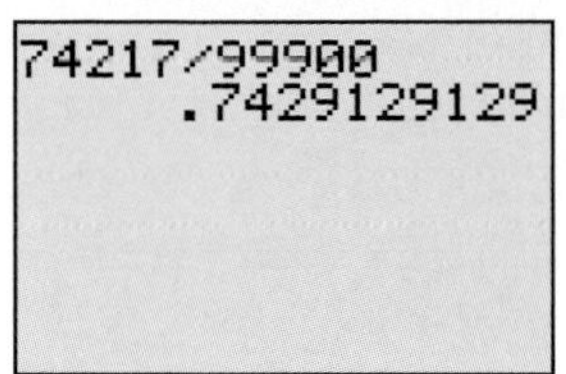

Figure 8.1

Thus, $0.74\overline{291}$ can be written in fraction form as $\frac{74,217}{99,900}$. We can use a calculator to support this result (Figure 8.1).

Self-Check 5 Write the repeating decimal $0.\overline{78}$ as a fraction. Support your answer using a calculator.

FUNCTIONS, GRAPHS, AND SERIES

Functions and their graphs can be used to understand the behavior of series. For example, consider the geometric series having first term and common ratio $a_1 = 3.8$ and $r = 0.76$. The sum of this series is

$$S = \frac{a_1}{(1 - r)} = \frac{3.8}{(1 - 0.76)} = 15.8\overline{33} \qquad \text{(using a calculator)}$$

Let us use the graphing calculator to verify this sum. The nth partial sum of this series is

$$S_n = \frac{a_1(1 - r^n)}{(1 - r)} = \frac{3.8(1 - 0.76^n)}{(1 - 0.76)}$$

We construct the function

$$f(x) = \frac{3.8(1 - 0.76^x)}{(1 - 0.76)}$$

where n has been replaced by x. Observe that $f(n) = S_n$; the values of f at positive integer values of x are the partial sums of the series. We can plot the graph of f and study its behavior as x gets bigger and bigger. This will give us values of S_n as n increases.

If we plot f over the intervals $0 \le x \le 100$, $0 \le y \le 30$, we get the graph shown in Figure 8.2. Tracing along the graph, we see that $f(x)$ gradually increases until it reaches 15.833333; thereafter, it remains constant. This confirms that S_n increases to $15.8\overline{33}$.

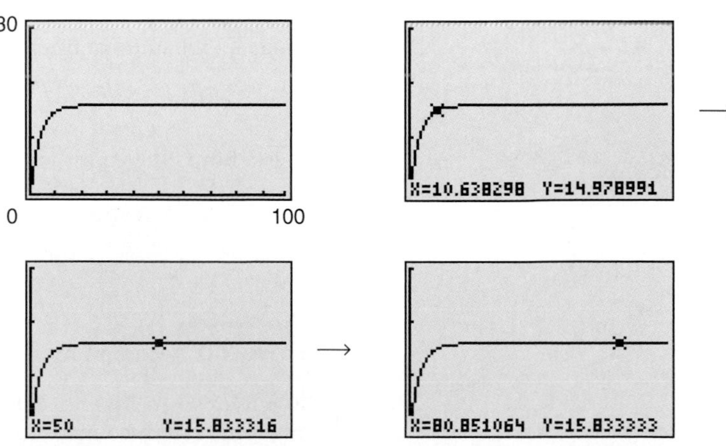

Figure 8.2

Self-Check 6 Use a graph to determine the sum of the terms in the geometric sequence having $a_1 = 7.26$ and $r = 0.35$, rounded to five decimal places.

Applications . . .

We now discuss applications of geometric sequences. These include applications in the worlds of bacteriology and finance.

In Section 5.1, we saw how exponential functions can be used to describe and analyze bacterial growth. We now give an alternative way of analyzing bacterial growth using sequences.

Example 9 Count of Bacteria

A certain bacterium doubles in count every hour. If the initial count is 500, what will the count be in 8 hours?

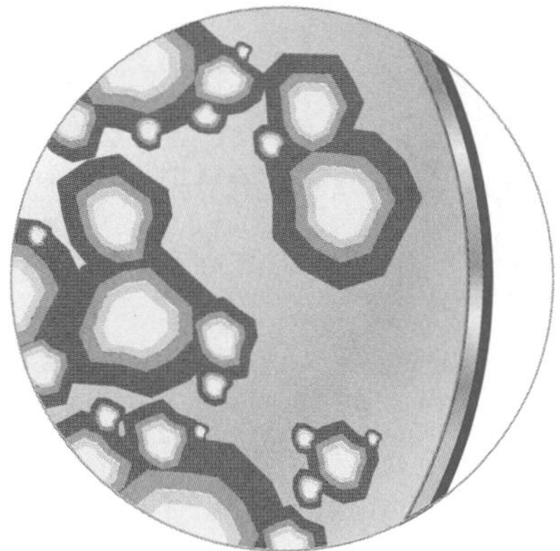

SOLUTION We can describe this bacterial growth by a sequence. The counts of bacteria at hourly intervals will be

$$500, 500(2), 500(2)^2, 500(2)^3, 500(2)^4, \ldots$$

The hourly growth of the bacterium is described by a geometric sequence with first term $a_1 = 500$ and common ratio $r = 2$. The count after n hours will be

$$a_n = 500(2)^{n-1}$$

Thus, after 8 hours the count will be

$$a_8 = 500(2)^{8-1} = 500(2)^7 = 500(128) = 64{,}000$$

Example 10 Compound Interest

In Section 5.2, we saw that if a sum P is deposited in an account that pays interest at an annual rate r, compounded m times per year, the amount A_n in the account at the end of n years will be

$$A_n = P\left(1 + \frac{r}{m}\right)^{mn}$$

This formula holds for all positive integer values of n. It therefore holds if we replace n by $n + 1$.

$$A_{n+1} = P\left(1 + \frac{r}{m}\right)^{m(n+1)}$$

Taking the ratio of these expressions for A_{n+1} and A_n gives

$$\frac{A_{n+1}}{A_n} = \left(1 + \frac{r}{m}\right)^{m}$$

The sequence of annual amounts $A_1, A_2, A_3, \ldots$ is thus a geometrical sequence with common ratio $(1 + \frac{r}{m})^m$.

For example, let a sum of \$1000 be deposited in an account that pays interest at an annual rate of 8%, compounded quarterly. We have

$$r = 8\% = \frac{8}{100} = 0.08 \qquad \text{and} \qquad m = 4$$

Thus

$$\left(1 + \frac{r}{m}\right)^4 = \left(1 + \frac{0.08}{4}\right)^4 = 1.02^4$$

The geometrical sequence is

$$1000(1.02)^4, 1000(1.02)^8, 1000(1.02)^{12}, \ldots$$

The terms are computed to give the following sequence of annual amounts.

$$\$1082.43, \$1171.66, \$1268.24, \ldots$$

Further Ideas 2 *Actuarial Science* Although it is impossible to make exact predictions concerning many future financial situations, investment and insurance companies use the following type of mathematical model to get estimates. A 22-year-old graduate of the University of Wisconsin starts her career earning \$30,000 per annum in the year 2000. If she receives constant raises of 4% per annum, how much money will she have earned in her working lifetime if she retires at the age of 65 in the year 2043? Mathematicians (called actuaries) carry out this type of analysis to construct pension schemes for these companies. Actuarial science is an extremely well-paying profession!

Example 11 Mooring Posts in Venice

There are no cars, buses, or trucks in Venice; all traffic is by boat. There are gondolas, but there are also boat buses and boat taxis. One gets around either by foot or by boat. Groceries and material such as building equipment are all brought in by boat. Wooden posts are used for mooring. Crews work full time putting in new posts and replacing old ones. There are special boats with pile drivers for do-

ing the job. A mooring post is 18 feet long with at least 6 feet below the ground. A pile driver drives the post in about 1 foot with the first drive and then 0.9 as far with each successive drive. (a) How many drives will it take to drive the post in the required 6 feet? (b) If the pile driver were to continue indefinitely how deep could it drive the post?

SOLUTION

(a) The post is driven 1 foot at the first drive. It will go in another 0.9 feet at the second drive, and another (0.9)(0.9) feet at the third drive, and so on. The depths thus form a geometric sequence, with $a_1 = 1$ and $r = 0.9$. The depth after n drives will be

$$S_n = \frac{a_1(1 - r^n)}{(1 - r)}$$

With $a_1 = 1$ and $r = 0.9$, we get the following sequence of depths, rounded to one decimal place.

$$S_1 = 1,\ S_2 = 1.9,\ S_3 = 2.7,\ S_4 = 3.4,\ S_5 = 4.1,\ S_6 = 4.7,\ S_7 = 5.2,\ S_8 = 5.7,\ S_9 = 6.1$$

It will thus take nine pile drives to drive the post in 6 feet.

(b) Let us determine how far into the ground the pile driver can drive the post if it were to continue indefinitely. Using the formula for the sum of an infinite geometric series, we get

$$S = \frac{a_1}{(1 - r)} = \frac{1}{(1 - 0.9)} = 10$$

If the pile driver were to continue, the post would gradually be driven closer and closer to a depth of 10 feet but would not be driven deeper.

We introduced geometric series in this section. There are many other types of series. Series play an important role in implementing mathematics on calculators and computers. The values of exponential and logarithmic functions on calculators, for example, are computed using series in which the sums approximate the values of these functions. Our final example in this section illustrates an infinite series that made the news in 1994. A brief history of these events makes interesting reading. It reveals what research in mathematics can involve, how it can have an unforeseen impact, and the smallness of the electronic news world in which we live.

Example 12 The Flaw in the Pentium Chip

In the fall of 1994 Professor Thomas Nicely of Lynchburg College, Virginia, discovered an error in the result of an arithmetic computation performed by a computer using the Pentium chip manufactured by Intel. His discovery was reported on the Internet, and it made the front pages of the world's newspapers on Thanksgiving Day. The discovery shook the business and scientific worlds.

Professor Nicely was interested in prime numbers. A number is prime if it is divisible only by 1 and itself. The numbers 1, 2, 3, 5, 7, 11, 13, and 17, for example, are prime numbers. The number 6 is not prime because it is divisible by 2 and by 3. Two such consecutive primes are called **twin primes** if they differ by 2. The numbers 3 and 5, for example, are twin primes, as are 5 and 7. Nicely's research involved investigating twin primes and, in particular, the sum of their reciprocals.

$$\left(\frac{1}{3}+\frac{1}{5}\right)+\left(\frac{1}{5}+\frac{1}{7}\right)+\left(\frac{1}{11}+\frac{1}{13}\right)+\cdots + \left(\frac{1}{29}+\frac{1}{31}\right)+\cdots+\left(\frac{1}{p}+\frac{1}{p+2}\right)+\cdots$$

It is known that the sum of this series has a finite value, but nobody knows what this value is. Nicely was performing the computations on a number of different computers—one of his objectives has been to demonstrate that personal computers are just as effective as supercomputers for this kind of research. When he added a machine based on Intel Corporation's Pentium chip to his collection, he noticed inconsistencies in the results he was getting. He traced the problem to the values he was getting for $\frac{1}{p}$ and $\frac{1}{p+2}$, with $p = 824{,}633{,}702{,}441$. He placed the blame at first on his own programs, then on the compiler and operating system, and finally on the bus and motherboard of the machine. By late October, he was convinced that the difficulty was in the Pentium chip itself.

On October 30, in an e-mail message to several other people who had access to Pentium systems, Nicely said that he thought there was a bug in the processor's floating-point arithmetic unit. One of the recipients of Nicely's memo posted it on the CompuServe network. The posting found its way to Terje Mathiesen, a computer scientist at Norsk Hydro in Oslo, Norway. Mathiesen had published articles about programming the Pentium and had posted notes on the Internet about its accuracy. Mathiesen confirmed Nicely's example and initiated a chain of Internet postings about the bug in the newsgroup comp.sys.intel. A day later, Andreas Kaiser of Germany posted a list of two dozen numbers for which reciprocals on the Pentium were not as accurate as expected. The story went from there to *The Boston Globe*, *The New York Times*, CNN, and around the world.

At first, Intel tried to downplay the error, saying that people using the chip would encounter an error only once in every 27,000 years. They offered to replace the chip if an owner could demonstrate that they really needed that much accuracy in their work. Intel badly misgauged the public reaction to the reported error. Customers who had bought or were planning to buy Pentium-based computers were confused and angry. Some experts contended that the flaw in the chip might affect the accuracy of corporate balance sheets or the calculations that banks make to pay interest to depositors. Disagreement reigned over how often an error was likely to occur. IBM carried out its own testing and concluded that an error was likely to occur once every 24 days. IBM announced that it would be halting shipments of all its products containing the Pentium chip. Said G. Richard Thoman, an IBM senior vice president, "We believe no one should have to wonder about the integrity of data calculated on IBM PCs." Easing its earlier hard line, Intel then finally agreed to replace the defective chip on request, after much damage, however, had been done to its image.

The Pentium's flaw involved division problems containing numbers with many digits. For example, if 4,195,835 is divided by 3,145,727 and then multiplied by 3,145,727, the result should be the original number, 4,195,835. Computers using the Pentium chip however gave 4,195,579 as the result! Perform this computation on your calculator to check for a "Pentium-type" error. Explain the way you have keyed these numbers and operations into your calculator, making use of the hierarchy of operations, to get the desired computation.*

Group Discussion

Ancestor Analysis

Suppose you trace your lineage and find that you had an ancestor four generations back who was Greek (the most recent Greek in your line). What part Greek are you? Generalize this result. If you had an ancestor n generations back who was Hawaiian, what part Hawaiian are you? Analyze your lineage. Express the answer in terms of the nth term of a certain geometrical sequence.

Answers to Self-Check Exercises

1. $a_n = -3(4)^{n-1}$. $a_4 = -192$.
2. $a_1 = 4$, $r = 3$, $a_6 = 972$.
3. The sum is -120.
4. The sum is $\frac{4}{5}$.
5. $0.\overline{78} = \frac{26}{33}$.
6. The sum is 11.16923.

Answers to Further Ideas Exercises

1. The first six terms are 6.7, 32.16, 154.368, 740.9664, 3556.63872, 17,071.86586.
2. In her working lifetime, she will have earned \$3,300,371.45.

*Readers who are interested in more detail of this historic event should read an article by Cleve Moler, in *MATLAB News & Notes,* Winter 1995, available from The MathWorks, 24 Prime Park Way, Natick, MA 01760–1415.

EXERCISES 8.3

In Exercises 1–6, show that the numbers are terms of a geometric sequence. Find the common ratio.

1. 3, 9, 27, 81, . . .
2. $-2, 6, -18, 54, \ldots$
3. 1, 5, 25, 125, . . .
4. 8, 4, 2, 1, . . .
5. $2, -2, 2, -2, \ldots$
6. $5, \frac{10}{3}, \frac{20}{9}, \frac{40}{27}, \ldots$

In Exercises 7–12, determine whether the numbers are terms in a geometric sequence.

7. 2, 10, 20, 40, . . .
8. 1, 4, 16, 32, . . .
9. 1, 3, 9, 27, . . .
10. $1, \frac{1}{2}, \frac{1}{4}, \frac{1}{12}, \ldots$
11. $15, -5, \frac{5}{3}, -\frac{5}{9}, \ldots$
12. 3, 7.5, 18.75, 46.875, . . .

In Exercises 13–18, give the next two terms of each geometric sequence.

13. 1, 4, 16, 64, . . .
14. $-3, 6, -12, 24, \ldots$
15. 27, 9, 3, 1, . . .
16. 5, 15, 45, 135, . . .
17. $-12, -6, -3, -\frac{3}{2}, \ldots$
18. $-\frac{1}{36}, -\frac{1}{18}, -\frac{1}{9}, -\frac{2}{9}, \ldots$

In Exercises 19–24, the first term and common ratio of a geometric sequence are given. Give the first five terms of each sequence.

19. $a_1 = 2, r = 4$
20. $a_1 = -3, r = -1$
21. $a_1 = 5, r = \frac{1}{2}$
22. $a_1 = 36, r = -\frac{1}{3}$
23. $a_1 = 1, r = 5$
24. $a_1 = -7, r = -4$

In Exercises 25–30, the first terms and the common ratios of geometric sequences are given. Use a calculator to determine the first six terms of each sequence. Round each term to four decimal places.

25. $a_1 = 4.76, r = 3.8$
26. $a_1 = 4.36, r = 5.2$
27. $a_1 = 2.981, r = 3.25$
28. $a_1 = -7.23, r = 1.35$
29. $a_1 = 93.26, r = -0.6$
30. $a_1 = -2.356, r = -2.862$

In Exercises 31–36, the first term and common ratio of a geometric sequence are given. Determine the specified term.

31. $a_1 = 3, r = 4$; sixth term
32. $a_1 = -1, r = 5$; fourth term
33. $a_1 = 63, r = -\frac{1}{3}$; seventh term
34. $a_1 = -\frac{1}{15}, r = 5$; third term
35. $a_1 = \frac{1}{3}, r = \frac{1}{2}$; fifth term
36. $a_1 = 8, r = 7$; ninth term

In Exercises 37–42, two consecutive terms of a geometric sequence are given. Find the common ratio and first term of each sequence.

37. $a_2 = 6, a_3 = 12$
38. $a_4 = 128, a_5 = -512$
39. $a_7 = 2916, a_8 = 8748$
40. $a_3 = -8, a_4 = -16$
41. $a_2 = -\frac{11}{3}, a_3 = \frac{11}{9}$
42. $a_{28} = -1, a_{29} = 1$

In Exercises 43–48, two terms of a geometric sequence are given. Find the common ratio, the first term, and the specified term of each sequence.

43. $a_3 = 8, a_6 = 64$; fourth term
44. $a_4 = -27, a_3 = 729$; sixth term
45. $a_5 = 32, a_{10} = 1$; 12th term
46. $a_5 = 128, a_3 = 8$; fourth term
47. $a_7 = -1, a_9 = -1$; second term
48. $a_4 = -24, a_9 = 768$; 15th term

In Exercises 49–54, find the specified partial sum for each sequence.

49. $a_n = 3(5)^{n-1}, S_4$
50. $a_n = 5(2)^{n-1}, S_6$
51. $a_n = -2(2)^{n-1}, S_3$
52. $a_n = 128(\frac{1}{2})^{n-1}, S_5$
53. $a_n = (\frac{1}{81})(-3)^{n-1}, S_4$
54. $a_n = 12(-1)^{n-1}, S_{43}$

In Exercises 55–60, find the sums of the infinite geometric series having the specified first terms and common ratios.

55. $a_1 = 2, r = 1/3$
56. $a_1 = 7, r = 1/4$
57. $a_1 = 13, r = 1/8$
58. $a_1 = 1, r = -1/27$
59. $a_1 = 11, r = -1/5$
60. $a_1 = 6, r = 1/15$

In Exercises 61–66, write each repeating decimal as a fraction. Support your answers using a calculator.

61. $0.32\overline{42}$
62. $0.978\overline{31}$
63. $0.58\overline{365}$
64. $0.32\overline{875}$
65. $0.7\overline{345}$
66. $0.365\overline{846}$

In Exercises 67–70, use a graph to determine the sum of the terms in the geometric series having the given first terms and common ratios, rounded to four decimal places. Verify your answer using the formula for the sum of a geometric series.

67. $a_1 = 4.53$ and $r = 0.85$
68. $a_1 = 25.54$ and $r = 0.27$
69. $a_1 = -9.867$ and $r = 0.5$

70. $a_1 = 3.892$ and $r = 0.76$

In Exercises 71–74, let $f(x) = a_1 r^{x-1}$. Use a table of values of f to determine the first term in the given geometric sequences to exceed or fall below the given number.

71. $a_1 = 5$ and $r = 1.25$. Exceed 18.

72. $a_1 = 10$ and $r = 1.5$. Exceed 1000.

73. $a_1 = 1000$ and $r = 0.85$. Fall below 50.

74. $a_1 = 100$ and $r = 0.5$. Fall below 0.1.

75. Count of Bacteria A certain bacterium doubles in count every hour. If the initial count is 100, what will the count be in 5 hours?

76. Population Growth The population of a certain county is estimated to increase at a rate of 10% every 20 years. If the current population is 500,000, what is the prediction for the population in 60 years time?

77. Ancestral Count Commencing with your parents, how many ancestors do you have going back six generations?

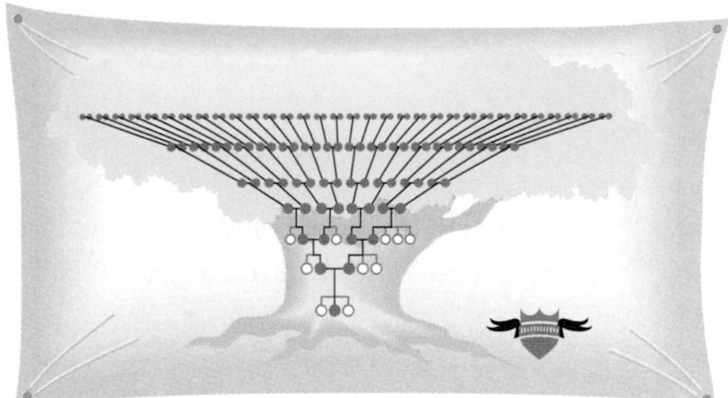

78. Growth in College Enrollment The enrollment at a certain college is estimated to be increasing at a rate of 2% per annum. If the current enrollment is 3500 students, what will the enrollment be in 5 years time?

79. Population Decrease The population of a certain city is decreasing at a rate of 5% per annum. If the current population is 75,000, what will the population be in 5 years time?

80. Compound Interest A sum of $1000 is deposited in an account that pays interest at 5.25%, compounded quarterly. Use a geometric sequence to determine the amount in the account after 5 years.

81. Comparison of Growth Two sums of money, $1000 and $3000, are deposited in accounts that pay interest at 9% and 5%, respectively, compounded annually. Use a graph to determine how long it will take for the money in the first account to exceed that in the second account.

82. Twin Primes Determine the first six terms in the sequence of twin primes $a_1 = (\frac{1}{3} + \frac{1}{5})$, $a_2 = (\frac{1}{5} + \frac{1}{7})$, . . . , investigated by Nicely. Compute the partial sum for these terms as accurately as you can on your calculator. (See the discussion of the flaw in the Pentium chip.)

8.4 Mathematical Induction

• Principle of Mathematical Induction • Extended Principle of Mathematical Induction • Compound Interest Formula Using Induction

In this section, we introduce a very important form of proof called mathematical induction. We first define the method, then give a number of examples to illustrate its application.

Principle of Mathematical Induction Let P_n be a statement involving the positive integer n. If

1. P_1 is true,
2. P_k implies the truth of P_{k+1} for every positive integer k, then P_n is true for all positive integers n.

Our first example illustrates the use of mathematical induction in proving a formula for the sum of the first n positive integers.

Example 1 Using Mathematical Induction

Use mathematical induction to prove that if n is a positive integer, then

$$1 + 2 + 3 + \cdots + n = \frac{n(n+1)}{2}$$

SOLUTION

1. We first show that the formula is valid when $n = 1$. Let $n = 1$. The formula gives

$$1 = \frac{1(1+1)}{2}$$

This result is true. The formula is valid when $n = 1$.

2. For the second step in the induction process, we assume that the formula is valid for some integer k and show that it then holds for the next integer $k + 1$. When $n = k$, the formula gives

$$1 + 2 + 3 + \cdots + k = \frac{k(k+1)}{2} \qquad \text{(assume to be valid)}$$

When $n = k + 1$, the formula gives

$$1 + 2 + 3 + \cdots + k + (k+1) = \frac{(k+1)(k+2)}{2} \qquad \text{(to be proved)}$$

Starting with the result for $n = k$, add $k + 1$ to each side. We get

$$\begin{aligned} 1 + 2 + 3 + \cdots + k + (k+1) &= \frac{k(k+1)}{2} + (k+1) \\ &= \frac{k(k+1) + 2(k+1)}{2} = \frac{k^2 + 3k + 2}{2} \\ &= \frac{(k+1)(k+2)}{2} \end{aligned}$$

The formula is therefore correct when $n = k + 1$.

By induction the formula is therefore valid for all positive integer values of n.

Our next example illustrates the use of mathematical induction in proving a formula for the sum of the first n positive odd integers.

Example 2 Using Mathematical Induction

Use mathematical induction to prove that if n is a positive integer, then

$$1 + 3 + 5 + \cdots + (2n - 1) = n^2$$

SOLUTION We check the formula for the two conditions of mathematical induction.

1. We first show that the formula is valid when $n = 1$. Let $n = 1$. The formula becomes
$$1 = 1^2$$
This statement is true; therefore, the formula is correct when $n = 1$.
2. Assume that the formula is valid when $n = k$, thus,
$$1 + 3 + 5 + \cdots + (2k - 1) = k^2$$
We now have to show that it follows that the formula is correct for $n = k + 1$. The kth term is $2k - 1$; thus, the $(k + 1)$th term is $2(k + 1) - 1$, that is, $2k + 1$. Add the $(k + 1)$th term to each side. We get
$$\begin{aligned} 1 + 3 + 5 + \cdots + (2k - 1) + (2k + 1) &= k^2 + 2k + 1 \\ &= (k + 1)^2 \end{aligned}$$
This is the formula with $n = k + 1$. Thus, the formula is correct when $n = k + 1$.

It follows by induction that the formula is valid for all positive integer values of n.

Example 3 Compound Interest Formula Using Induction

Use mathematical induction to prove that if a sum P is deposited in an account that pays interest at an annual rate i, compounded annually, the amount A_n in the account at the end of n years will be

$$A_n = P(1 + i)^n$$

SOLUTION

1. Let $n = 1$. The formula gives
$$\begin{aligned} A_1 &= P(1 + i) \\ &= P + Pi \end{aligned}$$
This result is correct because P is the original amount and Pi is the interest accumulated during the first year. Therefore, the formula is correct when $n = 1$.
2. Let $n = k$. Assume that the formula is valid when $n = k$. Thus,
$$A_k = P(1 + i)^k$$
We now have to show that it follows that the formula is correct for $n = k + 1$.
$$\begin{aligned} A_{k+1} &= (\text{Amount at beginning of } k + 1 \text{ year}) \\ &\quad + (\text{Interest accumulated in year } k + 1) \\ &= A_k + A_k i \\ &= A_k(1 + i) \end{aligned}$$

$$= P(1 + i)^k(1 + i)$$
$$= P(1 + i)^{k+1}$$

This is the formula with $n = k + 1$. Therefore, the formula is correct when $n = k + 1$. By induction, it follows that the formula is valid for all positive integer values of n.

Extended Principle of Mathematical Induction

The previous type of induction was useful for proving formulas that were valid for all values of $n \geq 1$. We now introduce the so-called extended induction. This type of induction can be used to examine a relationship that might only hold for values of n greater than a certain fixed value. Extended induction might be useful, for example, for showing that a certain formula was true for all values of $n \geq 6$.

Extended Principle of Mathematical Induction

Let P_n be a statement involving the positive integer n. If

1. P_j is true for some positive integer j,
2. P_k implies the truth of P_{k+1} for every positive integer $k \geq j$, then P_n is true for all positive integers $n \geq j$.

Example 4 **Using Extended Induction**

For which natural numbers is $2^n > 2n + 1$?

SOLUTION Let us do some preliminary investigation using a calculator. We construct a table of $Y_1 = 2^x$ and $Y_2 = 2x + 1$ values (Figure 8.3). It appears from the table that $2^x > 2x + 1$ for $x \geq 3$.

X	Y1	Y2
0	1	1
1	2	3
2	4	5
3	8	7
4	16	9
5	32	11
6	64	13

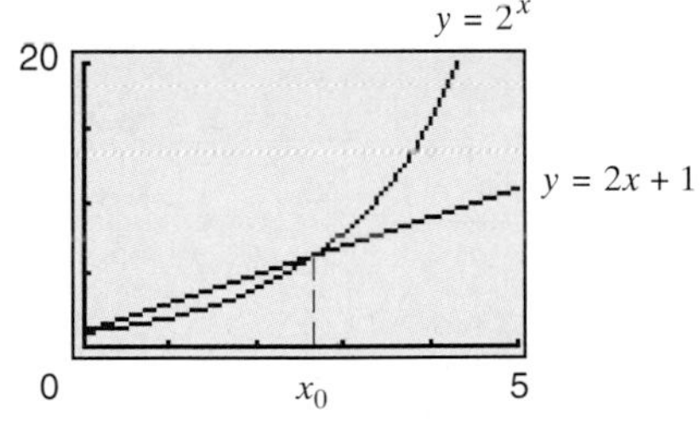

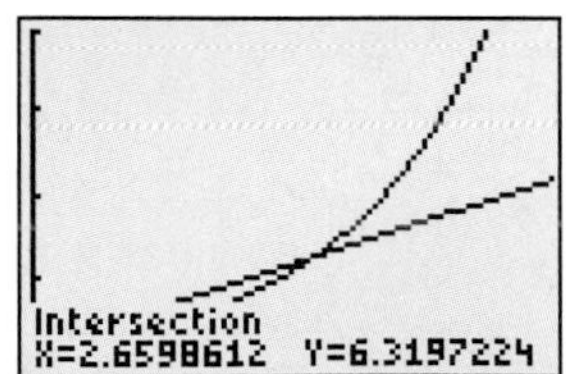

Figure 8.3

Further, we examine the graphs of $y = 2^x$ and $y = 2x + 1$. The graph of $y = 2^x$ seems to lie above the graph of $y = 2x + 1$ for $x > 2.6598612$. It appears that $2^x > 2x + 1$ for $x > 2.6598612$. These results lead us to expect that $2^n > 2n + 1$ for $n \geq 3$. Let us prove this result for n, a natural number, using the extended induction principle.

We check the formula for the two conditions of extended mathematical induction.

1. Let $n = 3$. We get

$$2^n = 2^3 = 8$$

$$2n + 1 = 2(3) + 1 = 7$$

The result $2^n > 2n + 1$ is true when $n = 3$.

2. Assume that the result is valid when $n = k$. Thus,

$$2^k > 2k + 1$$

We now have to show that it follows that the formula is correct for $n = k + 1$; that is, we have to show that $2^{k+1} > 2(k + 1) + 1$. We have

$$\begin{aligned} 2^{k+1} &= 2(2^k) \\ &= 2^k + 2^k \end{aligned}$$

But $2^k > 2k + 1$, by the induction hypothesis; and $2^k > 2$ when $k \geq 3$. Use these results for the first and second 2^ks. We get

$$\begin{aligned} 2^{k+1} &> 2k + 1 + 2 \\ &= 2(k + 1) + 1 \ (k \geq 3) \end{aligned}$$

Thus, when

$$2^k > 2k + 1 \ (k \geq 3) \qquad \text{it follows that} \qquad 2^{k+1} > 2(k + 1) + 1$$

Therefore, $2^n > 2n + 1$ when $n \geq 3$.

EXERCISES 8.4

In Exercises 1–6, use mathematical induction to prove the statements for all positive integer values of n.

1. $2 + 4 + 6 + \cdots + 2n = n(n + 1)$
2. $1 + 4 + 7 + \cdots + (3n - 2) = \dfrac{n(3n - 1)}{2}$
3. $2 + 5 + 8 + \cdots + (3n - 1) = \dfrac{n(3n + 1)}{2}$
4. $4 + 8 + 12 + \cdots + 4n = 2n(n + 1)$
5. $3 + 9 + 15 + \cdots + (6n - 3) = 3n^2$
6. $1 + 2^2 + 3^2 + \cdots + n^2 = \dfrac{n(n + 1)(2n + 1)}{6}$

In Exercises 7–12, use mathematical induction to prove the statements for all positive integer values of n.

7. $1 + 3^2 + 5^2 + \cdots + (2n - 1)^2 = \dfrac{n(2n + 1)(2n - 1)}{3}$
8. $1 + 3^1 + 3^2 + \cdots + 3^{n-1} = \dfrac{3^n - 1}{2}$
9. $1 + 4^1 + 4^2 + \cdots + 4^{n-1} = \dfrac{4^n - 1}{3}$
10. $1 \cdot 2 + 2 \cdot 3 + 3 \cdot 4 + \cdots + n(n + 1) = \dfrac{n(n + 1)(n + 2)}{3}$
11. $1 \cdot 3 + 2 \cdot 5 + 3 \cdot 6 + \cdots + n(2n + 1) = \dfrac{n(n + 1)(4n + 5)}{6}$
12. $\dfrac{1}{1 \cdot 2} + \dfrac{1}{2 \cdot 3} + \dfrac{1}{3 \cdot 4} + \cdots + \dfrac{1}{n(n + 1)} = \dfrac{n}{n + 1}$

In Exercises 13–19, use mathematical induction to prove the statements for all positive integer values of n.

13. $(xy)^n = x^n y^n$
14. $\left(\dfrac{x}{y}\right)^n = \dfrac{x^n}{y^n}, y \neq 0$
15. If $x > 1$, then $x^n > x^{n-1}$
16. If $0 < x < 1$, then $x^n < x^{n-1}$
17. $4^n - 1$ is divisible by 3
18. $5^n - 1$ is divisible by 4
19. $n^3 - n + 3$ is divisible by 3

20. Use mathematical induction to prove that the nth term of an arithmetic sequence having first term a_1 and common difference d is $a_n = a_1 + (n - 1)d$.

21. Use mathematical induction to prove that the nth term of a geometric sequence having first term a_1 and common ratio r is $a_n = a_1 r^{n-1}$.

22. For which natural numbers is $2^n > n^2$? *Hint*: Use the result of Example 4 in the text.

23. For which natural numbers is $n^3 > n^2 + 100$?

The Binomial Theorem

- Binomial Expansion • Pascal's Triangle • The Binomial Theorem
- Factorial • Binomial Coefficients

Binomial Expansion

A **binomial** is an algebraic expression consisting of two terms. For example, $3x + 2y$ is a binomial. In this section, we discuss powers of binomials.

Consider the fourth power of this binomial, $(3x + 2y)^4$. We know that

$$(3x + 2y)^4 = (3x + 2y)(3x + 2y)(3x + 2y)(3x + 2y)$$

By multiplying terms, it can be shown that

$$(3x + 2y)^4 = 81x^4 + 216x^3y + 216x^2y^2 + 96xy^3 + 16y^4$$

This is called the **expansion** of $(3x + 2y)^4$. We discuss the pattern that the terms of such an expansion follow and introduce a formula for writing down such expansions.

Consider the expansion of $(a + b)^n$ to various values of n. By multiplying terms, it can be shown that

$$\begin{aligned}
(a + b)^1 &= a + b \\
(a + b)^2 &= a^2 + 2ab + b^2 \\
(a + b)^3 &= a^3 + 3a^2b + 3ab^2 + b^3 \\
(a + b)^4 &= a^4 + 4a^3b + 6a^2b^2 + 4ab^3 + b^4 \\
(a + b)^5 &= a^5 + 5a^4b + 10a^3b^2 + 10a^2b^3 + 5ab^4 + b^5, \text{ etc.}
\end{aligned}$$

We now investigate the pattern followed by the terms in these expansions.

1. The expansion of $(a + b)^n$ has $n + 1$ terms [e.g., the expansion of $(a + b)^5$ has six terms].
2. The first term is a^n, and the last term is b^n [e.g., the first term in $(a + b)^5$ is a^5 and the last term is b^5].
3. In successive terms, the powers of a decrease by 1, whereas those of b increase by 1 [e.g., in $(a + b)^5$, note how the powers of a decrease, a^5, a^4, a^3, a^2, a whereas the powers of b increase, b, b^2, b^3, b^4, b^5].
4. The sum of the powers of a and b in each term is n [e.g., the third term in $(a + b)^5$ is $10a^3b^2$. Note that the sum of the powers of a and b is 5. The sum of the powers of a and b is 5 in each term].

The coefficients in these expansions follow an interesting pattern called **Pascal's triangle.** This is an array of numbers named after the 17th century mathematician Blaise Pascal. The numbers in the array are arranged as follows:

$$
\begin{array}{ccccccccccccccccc}
 & & & & & & & & 1 & & & & & & & & \\
 & & & & & & & 1 & & 1 & & & & & & & \\
 & & & & & & 1 & & 2 & & 1 & & & & & & \\
 & & & & & 1 & & 3 & & 3 & & 1 & & & & & \\
 & & & & 1 & & 4 & & 6 & & 4 & & 1 & & & & \\
 & & & 1 & & 5 & & 10 & & 10 & & 5 & & 1 & & & \\
 & & 1 & & 6 & & 15 & & 20 & & 15 & & 6 & & 1 & & \\
 & 1 & & 7 & & 21 & & 35 & & 35 & & 21 & & 7 & & 1 & \\
1 & & 8 & & 28 & & 56 & & 70 & & 56 & & 28 & & 8 & & 1 \\
. & . & . & . & . & . & . & . & . & . & \text{etc.}
\end{array}
$$

The first and last numbers of each row are 1. The other numbers in any row after the second are found by adding the two nearest numbers in the row above it. For example, $21 = 6 + 15$ in this table.

The triangle is interpreted in terms of its rows. The coefficients in the expansion of $(a + b)^n$ are given by row $n + 1$. For example, the coefficients in the expansion of $(a + b)^6$ are given by row 7. They will be

$$1 \quad 6 \quad 15 \quad 20 \quad 15 \quad 6 \quad 1$$

Because powers of a decrease and powers of b increase according to the previous rules, we get

$$(a + b)^6 = a^6 + 6a^5b + 15a^4b^2 + 20a^3b^3 + 15a^2b^4 + 6ab^5 + b^6$$

Self-Check 1 Expand $(a + b)^7$ using Pascal's triangle.

THE BINOMIAL THEOREM

Although Pascal's triangle may be used for determining binomial expansions for small powers, it is not too practical for larger powers. A result called the binomial theorem is used.

THE BINOMIAL THEOREM

$$(a + b)^n = a^n + \frac{na^{n-1}b}{1} + \frac{n(n-1)a^{n-2}b^2}{1 \cdot 2} + \cdots + \frac{n(n-1)(n-2)\cdots(n-r+1)a^{n-r}b^r}{1 \cdot 2 \cdot 3 \cdots r} + \cdots + b^n$$

The binomial theorem can be proved using mathematical induction. Observe that the powers of a and b follow the pattern discussed previously. It can be shown that the coefficients follow the pattern of Pascal's triangle.

Example 1 **Expanding Using the Binomial Theorem**

Expand $(3x + 2y)^4$ using the binomial theorem.

SOLUTION Let $a = 3x$, $b = 2y$, and $n = 4$ in the binomial theorem. It gives

$$(3x + 2y)^4 - (3x)^4 + \frac{4(3x)^3(2y)}{1} + \frac{4 \cdot 3(3x)^2(2y)^2}{1 \cdot 2} + \frac{4 \cdot 3 \cdot 2(3x)(2y)^3}{1 \cdot 2 \cdot 3} + (2y)^4$$

$$= 81x^4 + 216x^3y + 216x^2y^2 + 96xy^3 + 16y^4$$

Self-Check 2 Expand $(x - 2y)^5$ using the binomial theorem.

Factorial

The coefficients in a binomial expansion involve certain products of numbers. A convenient formula can be used to express these products. Before we can write this formula, we need to introduce the concept of factorial.

> ***n* Factorial** Let n be a positive integer. Write
>
> $$n! = n(n - 1)(n - 2) \cdots 3 \cdot 2 \cdot 1$$
>
> The symbol n! is called n *factorial.*
> We define $0! = 1$.

For example,

$$5! = 5 \cdot 4 \cdot 3 \cdot 2 \cdot 1 = 120$$

Binomial Coefficients

The coefficient of the general term in the binomial theorem is

$$\frac{n(n - 1)(n - 2) \cdots (n - r + 1)}{1 \cdot 2 \cdot 3 \cdots r}$$

This can be conveniently expressed in terms of factorials. Multiply both the numerator and the denominator by $(n - r) \cdots 3 \cdot 2 \cdot 1$.

$$\frac{(n(n - 1)(n - 2) \cdots (n - r + 1))((n - r) \cdots 3 \cdot 2 \cdot 1)}{(1 \cdot 2 \cdot 3 \cdots r)((n - r) \cdots 3 \cdot 2 \cdot 1)} = \frac{n!}{r!(n - r)!}$$

or

$$= \frac{n!}{(n - r)!r!}$$

(Both forms are used)

The symbol $\binom{n}{r}$ is used for this general binomial coefficient.

$$\binom{n}{r} = \frac{n!}{(n - r)!r!}$$

The binomial theorem can now be written

Binomial Theorem (in Terms of Binomial Coefficients)

$$(a+b)^n = a^n + \binom{n}{1}a^{n-1}b + \binom{n}{2}a^{n-2}b^2 + \cdots + \binom{n}{r}a^{n-r}b^r + \cdots + b^n$$

Example 2 **Computing Factorials and Binomial Coefficients**

Compute each of the following: (a) 7! (b) $\binom{9}{4}$.

SOLUTION

(a) $7! = 7 \cdot 6 \cdot 5 \cdot 4 \cdot 3 \cdot 2 \cdot 1 = 5040.$

(b) $\binom{9}{4} = \dfrac{9!}{(9-4)!4!} = \dfrac{9 \cdot 8 \cdot 7 \cdot 6 \cdot 5 \cdot 4 \cdot 3 \cdot 2 \cdot 1}{(5 \cdot 4 \cdot 3 \cdot 2 \cdot 1)(4 \cdot 3 \cdot 2 \cdot 1)} = 126$

Factorials and Binomial Coefficients with a Calculator

Calculators have built-in functions for computing factorials and binomial coefficients. The symbol ${}_nC_r$ may be used for $\binom{n}{r}$ on your calculator. This notation stems from the fact that $\binom{n}{r}$ is the number of combinations of n elements, taken r at a time. We discuss combinations in Section 9.1. Determine how to compute 7! and $\binom{9}{4}$ (or ${}_9C_4$) on your calculator (Figure 8.4).

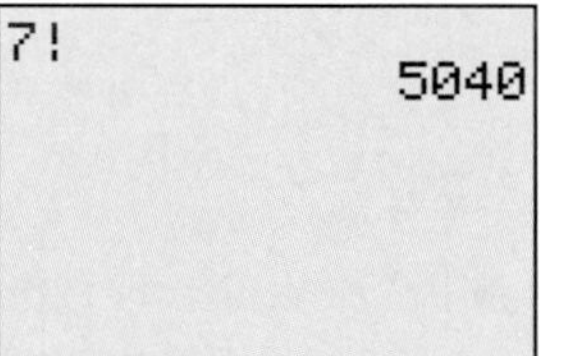

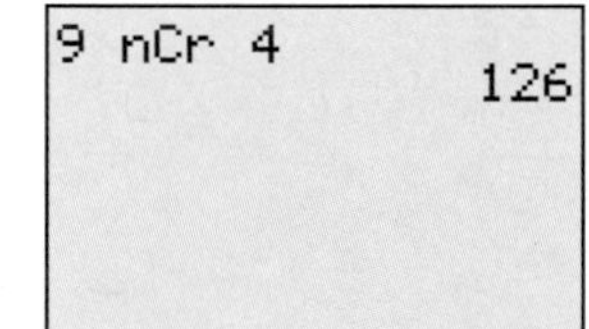

Figure 8.4

Self-Check 3 Compute each of the following: (a) 8! (b) $\binom{7}{3}$.

Example 3 **Expanding Using the Binomial Theorem**

Use the binomial theorem to expand $(x-2y)^6$.

SOLUTION Let $a = x$, $b = -2y$, and $n = 6$ in the binomial formula.

$$(x-2y)^6 = x^6 + \binom{6}{1}x^5(-2y) + \binom{6}{2}x^4(-2y)^2 + \binom{6}{3}x^3(-2y)^3$$

$$+ \binom{6}{4}x^2(-2y)^4 + \binom{6}{5}x(-2y)^5 + (-2y)^6$$

The binomial coefficients are

$$\binom{6}{1} = \frac{6!}{5!1!} = 6, \quad \binom{6}{2} = \frac{6!}{4!2!} = 15, \quad \binom{6}{3} = \frac{6!}{3!3!} = 20,$$

$$\binom{6}{4} = \frac{6!}{2!4!} = 15, \quad \binom{6}{5} = \frac{6!}{1!5!} = 6$$

The expansion is therefore

$$\begin{aligned}(x - 2y)^6 &= x^6 + 6x^5(-2y) + 15x^4(4y^2) + 20x^3(-8y^3) + 15x^2(16y^4) \\ &\quad + 6x(-32y^5) + 64y^6 \\ &= x^6 - 12x^5y + 60x^4y^2 - 160x^3y^3 + 240x^2y^4 - 192xy^5 + 64y^6\end{aligned}$$

Often, only certain terms of a binomial expansion are required. The following result is then useful.

Term in a Binomial Expansion

The rth term in the binomial expansion of $(a + b)^n$ is

$$\binom{n}{r-1}a^{n-(r-1)}b^{r-1}.$$

Example 4 **Finding a Term in a Binomial Expansion**

Determine the fourth term of the binomial expansion of $(2x + 3y)^9$.

SOLUTION Let $a = 2x$, $b = 3y$, and $n = 9$ in the preceding formula. Because we are interested in the fourth term, let $r = 4$. The formula gives

$$\begin{aligned}\text{Fourth term} &= \binom{9}{4-1}(2x)^{9-(4-1)}(3y)^{4-1} = \binom{9}{3}(2x)^6(3y)^3 \\ &= \frac{9!}{6!3!}64x^6 27y^3 = (84 \cdot 64 \cdot 27)x^6y^3 \\ &= 145{,}152x^6y^3\end{aligned}$$

Self-Check 4 Determine the sixth term in the expansion of $(x^2 + 3)^8$.

Answers to Self-Check Exercises

1. $(a + b)^7 = a^7 + 7a^6b + 21a^5b^2 + 35a^4b^3 + 35a^3b^4 + 21a^2b^5 + 7ab^6 + b^7$.
2. $(x - 2y)^5 = x^5 - 10x^4y + 40x^3y^2 - 80x^2y^3 + 80xy^4 - 32y^5$.
3. **(a)** $8! = 40{,}320$. **(b)** $\binom{7}{3} = 35$.
4. Sixth term in the expansion of $(x^2 + 3)^8$ is $13{,}608x^6$.

EXERCISES 8.5

In Exercises 1–8, expand using Pascal's triangle.

1. $(a + b)^5$
2. $(a + b)^8$
3. $(1 + 2x)^7$
4. $(x + 2y)^6$
5. $(3x - 2y)^5$
6. $(4x + 5y)^3$
7. $(x + y^2)^6$
8. $(4x - 2y^3)^4$

9. The first nine rows of Pascal's triangle are given in this section. Write down the tenth row.

10. Give the 11th and 12th rows of Pascal's triangle.

In Exercises 11–22, evaluate the factorials and binomial coefficients.

11. $6!$ **12.** $9!$ **13.** $3!$ **14.** $\dfrac{9!}{6!}$ **15.** $\dfrac{12!}{10!}$

16. $\dfrac{8!}{3!}$ **17.** $\dfrac{100!}{99!}$ **18.** $\dbinom{7}{5}$ **19.** $\dbinom{10}{6}$ **20.** $\dbinom{3}{2}$

21. $\dbinom{5}{5}$ **22.** $\dbinom{9}{4}$

In Exercises 23–33, use the binomial theorem to expand the expressions.

23. $(x + y)^4$
24. $(x - y)^5$
25. $(2 + 2y)^3$
26. $(x - 4y)^6$
27. $(3 - 4y)^5$
28. $(2x + 5y)^3$
29. $(x + 1/y)^4$
30. $(x + y^2)^3$
31. $(x + y^{1/2})^5$
32. $(2x^2 - y^3)^3$
33. $(x^{1/2} + y^{3/2})^4$

In Exercises 34–41, determine the specified terms of each binomial expansion.

34. $(x + y)^7$, fifth term
35. $(2x - y)^9$, sixth term
36. $(x - 3/y)^{12}$, fourth term
37. $(x + 2y^3)^{10}$, eighth term
38. $(3x - 5y)^6$, third term
39. $(x^2 + 5y)^8$, seventh term
40. $(x + y^{1/2})^{15}$, ninth term
41. $(5x - y)^5$, fourth term

CHAPTER 8 PROJECT E-Mail Messages

In this project you are asked to model the accumulation of e-mail messages. Models such as this, albeit very idealistic, are created to get a feel for a situation and to make some sort of prediction as to what can occur. A value of such models is that the numbers can be changed easily to explore different scenarios. We carry this out here.

A business executive receives many e-mail messages a day. Messages have accumulated in her inbox. She decides on a plan to tidy up the inbox. During a day, she will halve the number of messages in the inbox, while still adding eight new messages (messages that might need more consideration before answering).

(a) Let a_n be the number of messages in the inbox at the beginning of day n. Construct a sequence for the e-mail. If the initial number of messages is 208 ($a_1 = 208$), how long will it take the number of messages to drop below 20?

(b) What would be the effect of reducing the number of messages added daily to the inbox to four instead of eight? How long will it now take to drop below 20?

(c) What if the initial number of messages had been 1552 with eight new messages added daily? How long would it have then taken to get the number of messages in the inbox down below 20 with this plan?

(d) Describe your method of handling e-mail! Can it be modeled?

WEB PROJECT 8

To explore the concepts of this chapter further, try the interesting WebProject. Each activity relates the ideas and skills you have learned in this chapter to a real-world applied situation. Visit the Williams Web site at

www.saunderscollege.com/math/williams

to find the complete WebProject, related questions, and live links to the Internet.

CHAPTER 8 HIGHLIGHTS

Section 8.1

Sequence: A collection of numbers arranged in a certain order. The numbers in the sequence are called **terms.**
Recursive Definition: A sequence defined by giving the first term together with a formula for the nth term as an expression of the preceding term.
Sigma Notation: This is used to express the sum of a finite number of terms of a sequence.
nth Partial Sum: The sum of the first n terms of a sequence. Is denoted S_n.
Fibonacci Sequence: 1, 1, 2, 3, 5, 8, 13, 21, 34, 55, 89, 144, ... Each term in the sequence is the sum of the two preceding terms.

Section 8.2

Arithmetic Sequence: A sequence in which successive terms differ by a constant. The constant is called the common difference d. $a_{n+1} = a_n + d$. $a_n = a_1 + (n - 1)d$. nth partial sum, S_n, of an arithmetic sequence is $S_n = \frac{n}{2}[2a_1 + (n - 1)d]$ or $S_n = \frac{n}{2}[a_1 + a_n]$.

Section 8.3

Geometric Sequence: A sequence in which the ratios of successive terms are constant. The constant is called the common ratio r. $a_{n+1} = a_n r$. $a_n = a_1 r^{n-1}$. nth partial sum of a geometric sequence is $S_n = \dfrac{a_1(1 - r^n)}{(1 - r)}$.

Sum S of an Infinite Geometric Series: If common ratio $|r| < 1$, $S = \dfrac{a_1}{(1 - r)}$.
Repeating Decimal: A real number in decimal form that repeats a block of digits. It is a rational number. It can be written as a fraction.

Section 8.4

Principle of Mathematical Induction: Let P_n be a statement involving the positive integer n. If P_1 is true, and P_k implies the truth of P_{k+1} for every positive integer k, then P_n is true for all positive integers n.
Extended Principle of Mathematical Induction: Let P_n be a statement involving the positive integer n. If P_j is true for some positive integer j, and P_k implies the truth of P_{k+1} for every positive integer $k \geq j$, then P_n is true for all positive integers $n \geq j$.

Section 8.5

Pascal's Triangle: Pattern of the terms in the expansion of $(a + b)^n$.
n Factorial: $n! = n(n - 1)(n - 2)\cdots 3 \cdot 2 \cdot 1$.
General Binomial Coefficient:

$$\binom{n}{r} = \frac{n!}{(n - r)!r!}.$$

Binomial Theorem:

$$(a + b)^n = a^n + \frac{na^{n-1}b}{1} + \frac{n(n - 1)a^{n-2}b^2}{1 \cdot 2} + \cdots + \frac{n(n - 1)(n - 2)\cdots(n - r + 1)a^{n-r}b^r}{1 \cdot 2 \cdot 3 \cdots r} + \cdots + b^n$$

or

$$(a + b)^n = a^n + \binom{n}{1}a^{n-1}b + \binom{n}{2}a^{n-2}b^2 + \cdots + \binom{n}{r}a^{n-r}b^r + \cdots + b^n$$

The rth term is $\dbinom{n}{r-1}a^{n-(r-1)}b^{r-1}$.

Chapter 8 Review Exercises

1. Determine the first four terms of each of the following sequences.

(a) $a_n = 4n - 1$ (b) $a_n = 3n^2 + 4$

(c) $a_n = 2n^3 - n^2 + 1$ (d) $a_n = 2^{n-3}$

(e) $a_n = (4n - 1)(2n + 3)$

(f) $a_n = (-2)^n + 4n + 3$

2. The following sequences are defined recursively for $n > 1$. Give the first four terms of each sequence.

(a) $a_1 = 2, a_n = a_{n-1} + 4$

(b) $a_1 = -3, a_n = 2a_{n-1} - 5$

(c) $a_1 = 1, a_n = -3a_{n-1} + 2$

(d) $a_1 = -6, a_n = 3a_{n-1} + 2n$

(e) $a_1 = -1, a_n = (a_{n-1} + 1)^2$

(f) $a_1 = 4, a_n = 2(a_{n-1})^3 + 3$

3. Evaluate the following sums.

(a) $\sum_{k=1}^{4}(3k - 1)$ (b) $\sum_{k=1}^{6}(1 + 2k)$

(c) $\sum_{k=2}^{7}(3k^2 + 2k + 1)$ (d) $\sum_{k=3}^{6}(2^k - 3k + 1)$

4. Determine the specified partial sums for the following sequences.

(a) $S_3, a_k = 3k$ (b) $S_5, a_k = 4k + 2$

(c) $S_2, a_k = k^2 + 3k - 1$

5. Determine whether the following sequences are arithmetic sequences. Find the common differences of the arithmetic sequences.

(a) 1, 4, 7, 11, . . . (b) −10, −2, 6, 14, . . .

(c) 5, 8, 12, 17, . . . (d) 3.5, 4.75, 6, 7.25, . . .

6. The first terms and common differences of arithmetic sequences are given. Find the specified terms of the sequences.

(a) $a_1 = 2, d = 5$; fourth term

(b) $a_1 = -3, d = 6$; fifth term

(c) $a_1 = 8, d = -3$; sixth term

(d) $a_1 = 0, d = 5$; 50th term

7. Two terms of arithmetic sequences are given. Find the specified terms of the sequences.

(a) $a_3 = 7, a_7 = 19$; ninth term

(b) $a_7 = 21, a_2 = 1$; fifth term

(c) $a_5 = 11, a_8 = 23$; second term

(d) $a_4 = 18, a_9 = 43$; 100th term

8. Arithmetic sequences are defined here in various ways. Determine the specified partial sums of the sequences.

(a) $a_1 = 5, d = 4; S_3$ (b) $a_1 = -3, d = 6; S_4$

(c) $a_n = 2n + 1; S_6$ (d) $a_n = 3n - 4; S_5$

9. The first term of an arithmetic sequence is 2, and the last term is 17. The sum of the terms is 57. How many terms are in the sequence?

10. The sum of the first six terms of an arithmetic sequence is −12, and the common difference is −2. Determine the first term of the sequence.

11. The sum of the first 12 terms of an arithmetic sequence is 210. If the first term is 1, determine the common difference.

12. The first terms and common ratios of geometric sequences are given. Determine the specified terms.

(a) $a_1 = 3, r = 2$; fourth term

(b) $a_1 = -1, r = 3$; fifth term

(c) $a_1 = 1, r = \frac{1}{2}$; third term

(d) $a_1 = 64, r = -\frac{1}{2}$; eighth term

13. Two terms of geometric sequences follow. Find the common ratio, the first term, and the specified term of each sequences.

(a) $a_1 = 324, a_7 = 8748$; fifth term

(b) $a_5 = 16, a_7 = 64$; second term

14. Find the specified partial sums of the following geometric sequence.

(a) $a_n = 2(3)^{n-1}, S_5$ (b) $a_n = 2(-3)^{n-1}, S_4$

15. Find the sums of the following infinite geometric series.

(a) $a_1 = 3, r = \frac{1}{2}$ (b) $a_1 = -1, r = -\frac{1}{2}$

16. Write each of the following repeating decimal as a fraction.

(a) $0.4\overline{256}$ (b) $0.93\overline{521}$

17. Use mathematical induction to prove each of the following statements.

(a) $3 + 6 + 9 + \cdots + 3n = \dfrac{3n(n + 1)}{2}$

(b) $1^3 + 2^3 + 3^3 + \cdots + n^3 = \dfrac{n^2(n + 1)^2}{4}$

(c) $1 \cdot 2 + 2 \cdot 4 + 3 \cdot 8 + \cdots + n \cdot 2^n = (n - 1)2^{n+1} + 2$

18. Use Pascal's triangle to expand the following expressions.
(a) $(a + b)^4$ (b) $(2 - 3x)^5$ (c) $(2x - y)^6$

19. Evaluate the following factorials and binomial coefficients.
(a) $5!$ (b) $\dfrac{12!}{10!}$ (c) $\dbinom{7}{4}$ (d) $\dbinom{127}{125}$

20. Use the binomial theorem to expand the following expressions.
(a) $(x + 2y)^3$ (b) $(x - 3y)^5$ (c) $(2x + 4y)^3$

21. Determine the specified term in each of the following expansions.
(a) $(2x + 5y)^7$, fourth term
(b) $(x - 2y)^9$, fifth term
(c) $(5x - 3y)^{17}$, sixth term

22. Consider the arithmetic sequence having first term $a_1 = 4.27$ and common difference $d = 4.63$. Use a calculator to determine the first eight terms of this sequence.

23. Consider the geometric sequence having first term $a_1 = 5.82$ and common ratio $r = 1.93$. Use a calculator to determine the first ten terms of this sequence.

24. Use a graph to determine the sum of the terms in the geometric sequence having $a_1 = 2.97$ and $r = 0.45$. Verify your answer using the formula for the sum of a geometric series.

CHAPTER 8 TEST

1. Determine the first four terms of the following sequence.

$$a_n = 2n^2 - 3n$$

2. The following sequence is defined recursively for $n > 1$. Give the first five terms of the sequence.

$$a_1 = 3,\ a_n = 2(-1)^n a_{n-1} + 5$$

3. Evaluate the sum.

$$\sum_{k=2}^{5} (k^2 - 3k + 2)$$

4. Determine the partial sum S_6 for the arithmetic sequence $a_n = 3n + 4$.

5. If $a_1 = 3$ and $d = 4$ for an arithmetic sequence, find the fifth term.

6. If $a_2 = 4$, $a_8 = 40$ for an arithmetic sequence, find the tenth term.

7. The first term of an arithmetic sequence is 3, and the last term is 42. The sum of the terms is 315. How many terms are in the sequence?

8. The sum of the first eight terms of an arithmetic sequence is 36, and the common difference is 3. Determine the first term of the sequence.

9. The sum of the first ten terms of an arithmetic sequence is 345. If the first term is 3, determine the common difference.

10. If $a_1 = 4$ and $r = 5$ for a geometric sequence, find the sixth term.

11. If $a_5 = 112$ and $a_9 = 1792$ for a geometric sequence, find the 12th term.

12. Find the partial sum S_7 of the geometric sequence $a_n = 5(2)^{n-1}$.

13. Find the sum of the infinite geometric series with $a_1 = 7$ and $r = \frac{1}{3}$.

14. Write the repeating decimal $0.73\overline{48}$ as a fraction.

15. Use Pascal's triangle to expand $(1 + 3x)^5$.

16. Evaluate the following factorial and binomial coefficients.
(a) $7!$ (b) $\dbinom{8}{3}$ (c) $\dbinom{243}{241}$

17. Use the binomial theorem to expand $(2x + y)^5$.
18. Determine the fifth term in the binomial expansion of $(3x - 2y)^8$.
19. Consider the arithmetic sequence having the first term $a_1 = 5.36$ and common difference $d = 3.27$. Use a calculator to determine the first five terms of this sequence.
20. Consider the geometric sequence having the first term $a_1 = 4.91$ and common ratio $r = 1.25$. Use a calculator to determine the first six terms of this sequence.

9

Permutations, Combinations, and Probability

London in the time of John Graunt
(Archive Photos/Popperfoto)

In this chapter, we provide the mathematical tools necessary to discuss the number of ways things can be carried out (permutations and combinations) and the likelihood of the different results occurring (probability). The notion of probability has been around since ancient times, at least as long as games of chance have existed. Dice that resemble contemporary dice have been found in Egyptian and Greek tombs. However, it was not until the 17th century that mathematicians undertook a systematic analysis of the subject. Modern use of probability in mortality tables can be traced back to the work of Captain John Graunt, which was published in London in 1662. Records of deaths were first kept in London in 1592, and Graunt based his deductions on these data. These records were intended to describe the progress of the plague. Graunt was careful to publish the actual figures on which he based his conclusions, comparing himself when doing so to a "silly schoolboy, coming to say his lessons to the world (that peevish and tetchie master), who brings a bundle of rods, wherewith to be whipped for every mistake he has committed."

Since the time of Graunt, the theory of probability has become a major field within mathematics. Probability is used by insurance companies to decide on financial policies, by governments to plan economic policies, and by theoretical physicists to understand the nature of atomic-sized systems in quantum mechanics. Public opinion polls, such as the Gallup Poll, have their theoretical acceptability based on probability theory.

The tools that we introduce enable us to determine, for example, the probability that four men and two women will be hired for a vacancy of six when ten men and five women apply for the positions. We find that in a class of 30 students there is an extremely high probability that at least two students will have their birthdays on the same day. Some of the results that we arrive at are very surprising.

We have all undoubtedly sat at a traffic light and wondered whether the light would ever turn to green. In this chapter, we give you the tools to do a mathematical analysis of the situation while you wait!

APPLICATION Blood Donors

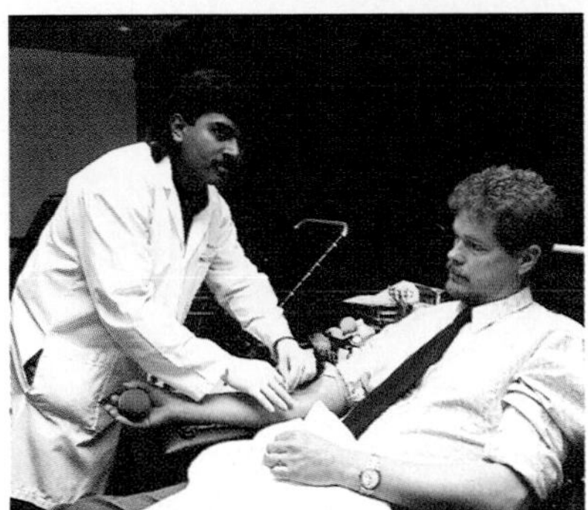

Giving your best. *(© David M. Grossman/The National Audubon Society Collection/Photo Researchers)*

The blood of any one individual contains a number of antigens; blood is classified according to these antigens. There are eight blood groups, labeled AB^+ to O^-, shown in Figure 9.1.

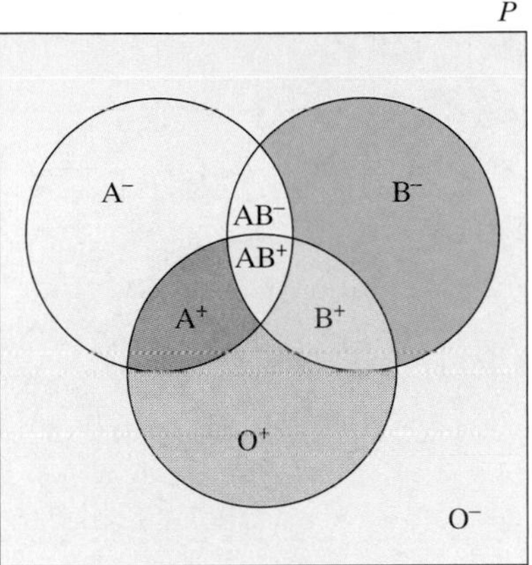

Figure 9.1 Blood groups.

The Red Cross provided the following figures for the percentage of people within each blood group:

AB^+	AB^-	A^+	A^-	B^+	B^-	O^+	O^-
3.3%	0.7%	33.3%	6.7%	9.2%	1.8%	37.4%	7.6%

Thus, O^+ is the most common group, whereas AB^- is the rarest. These figures are used to determine the probabilities of people contacting a donor at random in the street. For example, the author has 0.56 probability of walking up to a person on the street and being able to have blood from that person. What is your blood group? What is the probability that you can receive blood from the first person you walk up to? The model we develop in this chapter enables us to answer these and other questions. *(This application is discussed in Section 9.2, Example 9, pages 600–601.)*

9.1 Permutations and Combinations

- Tree Diagram • Fundamental Counting Principle • Factorials
- Permutations • Combinations

TREE DIAGRAM

It is sometimes necessary to determine the number of different ways things can occur. For example, suppose one wants to drive across central Florida from Day-

tona on the east side to Tampa on the west side, calling at Orlando and Lakeland on route. It is possible to travel directly from Daytona to Orlando along three roads, from Orlando to Lakeland along two roads, and from Lakeland to Tampa along four roads. How many routes from Daytona to Tampa pass through Orlando and Lakeland?

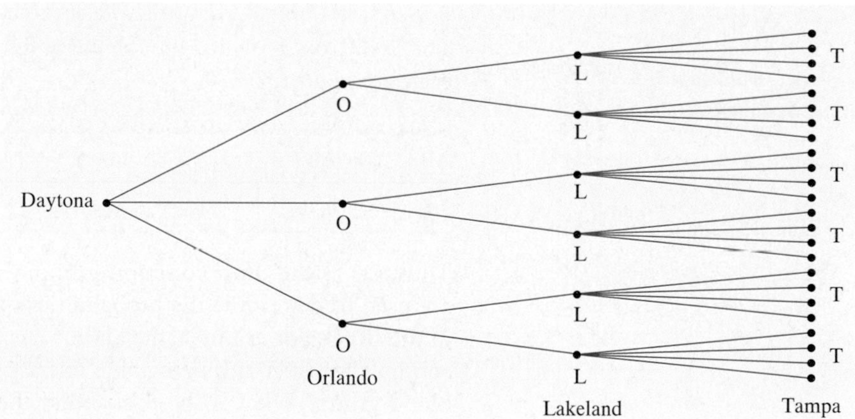

Figure 9.2 Routes from Daytona to Tampa.

One way to solve this problem is to use a **tree diagram,** see Figure 9.2. The diagram is read from left to right. Starting from Daytona, there are three possible ways of getting to Orlando (O), illustrated by the three branches of the tree starting at Daytona. Having arrived at Orlando by any one of these routes, there are then two possible ways of getting to Lakeland. These are illustrated by the branches from Orlando to Lakeland (L). It can be seen that there are six distinct routes for getting to Lakeland, corresponding to the six distinct paths in the diagram, which start from Daytona and end up at Lakeland. Finally, having arrived at Lakeland, there are four ways of getting to Tampa, illustrated by the branches from Lakeland to Tampa (T). We can now see that there are 24 distinct routes from Daytona to Tampa, corresponding to the 24 paths in the diagram that start at Daytona and end at Tampa.

Let us now look at how the number 24 actually arises in this example. (Our aim is to understand the thinking behind this analysis and then to formulate a general mathematical result.)

There are three ways of getting from Daytona to Orlando. There are then two ways that each of these can lead to Lakeland. There are thus $3 \cdot 2$, (that is 6) routes for getting from Daytona to Lakeland.

Finally there are four ways for each of these routes to lead to Tampa. There are therefore $3 \cdot 2 \cdot 4$, that is 24 routes for getting from Daytona to Tampa.

This analysis leads to the following general rule for computing the number of ways different events (such as selecting routes) can happen.

FUNDAMENTAL COUNTING PRINCIPLE

Let $E_1, E_2, \ldots, E_k$ be a sequence of events that can occur in $m_1, m_2, \ldots, m_k$ ways. The total number of ways the events can take place is $m_1 m_2 \cdots m_k$.

Let us now look at an example to get a feel for the rule.

Example 1 **Selecting Clothes**

Timothy Palmer has six shirts, four pairs of trousers, and three pairs of shoes. How many different ensembles can he wear?

SOLUTION Apply the fundamental counting principle. Let E_1 be the choice of a shirt, E_2 the choice of trousers, and E_3 the choice of shoes. The shirt can be selected in six ways, the trousers in four ways, and the shoes in three ways. Thus, the number of different ensembles is

$$6 \cdot 4 \cdot 3 = 72$$

Further Ideas 1 *Eating Options* Stavros Cafeteria offers the following choices: two meats, four vegetables, three salads, two desserts, and three drinks. Assuming a meal contains one of each item, how many meal possibilities are there?

Thus far, it has been readily apparent how many ways each event can occur. In Example 1 a shirt could be chosen in six ways. We now turn our attention to situations in which it has to be determined how many ways each event can occur.

PERMUTATIONS

The chess club at Durham University, consisting of 20 people, is to select a president, a secretary, and a treasurer. No person can hold more than one position. In how many ways can this be done?

The president can be selected in 20 ways. Because one person has been selected as president, there are 19 ways remaining to choose a secretary. Finally, because two people have already been selected for office, there are 18 ways left to choose a treasurer.

Apply the fundamental counting principle with E_1 corresponding to selecting a president, E_2 to selecting a secretary, and E_3 to selecting a treasurer. The number of ways the group can be picked is

$$20 \cdot 19 \cdot 18 = 6840$$

In this example, we have found the number of ways of selecting three people from a group of 20 people. Note that there is a difference between each of the three positions. One is president, one is secretary, and the other is treasurer. We categorize such selections as ones in which *order is of significance.*

Permutation An arrangement of r elements from a total of n possible elements, where order is significant, is called a **permutation.**

Formula for Permutations

We now derive a formula for computing the number of permutations that arise when selecting r elements from a total of n elements. The derivation is based on the reasoning used in the preceding illustration.

There are n elements to select from. The first element can therefore be selected in n ways. There are $(n - 1)$ elements remaining from which to choose the second element. The second element can be selected in $(n - 1)$ ways. There are $(n - 2)$ elements left from which to choose the third element. The third element can be selected in $(n - 2)$ ways. Continuing thus, it can be seen that there are $(n - r + 1)$ ways of selecting the rth and final element.

Applying the fundamental counting principle, we find that there are

$$n(n - 1)(n - 2) \cdots (n - r + 1)$$

ways of selecting r elements from n, with no repetitions. We refer to this as the number of permutations of n elements taken r at a time. The symbol ${}_nP_r$ is used for this number.

Number of Permutations

The number of permutations of n elements taken r at a time, denoted ${}_nP_r$, is given by

$${}_nP_r = n(n - 1)(n - 2) \cdots (n - r + 1)$$

Example 2 Computing a Permutation

Compute ${}_7P_3$.

SOLUTION ${}_7P_3$ is the number of permutations of seven elements taken three at a time.

$${}_7P_3 = 7 \times 6 \times 5 = 210$$

Self-Check 1 Compute (a) ${}_9P_2$ (b) ${}_4P_4$.

FACTORIALS

We introduced factorials to write binomial coefficients in a convenient manner in Chapter 8. Factorials are also useful in working with permutations. We remind the student of the definition.

A number of the form $r(r-1)(r-2)\cdots 3\cdot 2\cdot 1$ is called a **factorial.** We write

$$r! = r(r-1)(r-2)\cdots 3\cdot 2\cdot 1$$

For example, $5! = 5\cdot 4\cdot 3\cdot 2\cdot 1 = 120$. We now find that a permutation ${}_nP_r$ can be conveniently written in terms of factorials.

$${}_nP_r = n(n-1)(n-2)\cdots(n-r+1)$$

Multiply and divide this expression by $(n-r)\cdots 3\cdot 2\cdot 1$.

$$\begin{aligned}{}_nP_r &= n(n-1)(n-2)\cdots(n-r+1)\times\frac{(n-r)\cdots 3\cdot 2\cdot 1}{(n-r)\cdots 3\cdot 2\cdot 1}\\ &= \frac{n(n-1)(n-2)\cdots(n-r+1)(n-r)\cdots 3\cdot 2\cdot 1}{(n-r)\cdots 3\cdot 2\cdot 1}\\ &= \frac{n!}{(n-r)!}\end{aligned}$$

$${}_nP_r = \frac{n!}{(n-r)!}$$

Example 3 | Computing a Permutation

Express ${}_8P_5$ in terms of factorials and compute.

SOLUTION

$${}_8P_5 = \frac{8!}{(8-5)!} = \frac{8\cdot 7\cdot 6\cdot 5\cdot 4\cdot 3\cdot 2\cdot 1}{3\cdot 2\cdot 1} = 8\cdot 7\cdot 6\cdot 5\cdot 4 = 6720$$

We now give an example to illustrate how problems are set up in terms of permutations and then solved.

Example 4 | Horse Racing

Ten horses are entered to run in the Windscale Stakes. How many ways are there of selecting the winner and runner up?

SOLUTION We want to determine the number of ways of selecting a winner and a runner-up from ten horses. This is the number of permutations of ten elements taken two at a time. We get

$$_{10}P_2 = \frac{10!}{(10-2)!} = \frac{10!}{8!} = 10 \cdot 9 = 90$$

There are 90 ways of selecting the winner and runner-up.

Further Ideas 2 *Picture Taking* The five Krishnan children are to have their picture taken. It was decided to have them arranged in a line. In how many ways can this be done? *Hint:* The number of permutations of five elements taken five at a time.

Having discussed selections in which order is important, we now look at selections in which order is not significant.

COMBINATIONS

Three students are to be selected from a college Spanish class of 20 to serve on a committee to organize an open house for high school students. How many selections are possible?

Observe that there is no distinction to be made between the students selected—no first, second, and third in this case. We are interested in knowing how many ways we can select three students from 20 students when the *order is not significant.* Let the number of ways be N.

Consider one such selection of three students. If the order had been important, this one group would have resulted in 3! distinct groups. Similarly, each of the N such groups would have resulted in 3! distinct groups if order had been important. There would thus have been $3!\,N$ selections if order had been significant. But the number of selections of 3 from 20, order being insignificant, is $_{20}P_3$. Thus,

$$3!\,N = {_{20}P_3}$$

This equation gives

$$N = \frac{_{20}P_3}{3!} = \frac{20!}{17!3!} = \frac{20 \cdot 19 \cdot 18 \cdot 17 \cdots 3 \cdot 2 \cdot 1}{(17 \cdot 16 \cdots 3 \cdot 2 \cdot 1)(3 \cdot 2 \cdot 1)} = \frac{20 \cdot 19 \cdot 18}{3 \cdot 2 \cdot 1} = 1140$$

There are 1140 ways of selecting three students from 20, order being unimportant.

In this example, we see that the number of ways of selecting three people from 20 when order is not significant is $_{20}P_3/3!$. This example leads to the following general result.

Combination An arrangement of r elements from a total of n elements, where order is of no significance is called a **combination.** The number of combinations of n elements, taken r at a time, is denoted $_nC_r$. Then

$$_nC_r = \frac{_nP_r}{r!} \quad \text{or} \quad _nC_r = \frac{n!}{(n-r)!r!}$$

Note that the formula for ${}_nC_r$ is the same as for the binomial coefficient $\binom{n}{r}$. Pascal's triangle can thus be used, if desired, to compute combinations.

Example 5 **Computing a Combination**

Compute ${}_7C_3$.

SOLUTION

$${}_7C_3 = \frac{7!}{(7-3)!3!} = \frac{7!}{4!3!} = \frac{7 \cdot 6 \cdot 5 \cdot 4 \cdot 3 \cdot 2 \cdot 1}{(4 \cdot 3 \cdot 2 \cdot 1)(3 \cdot 2 \cdot 1)} = \frac{7 \cdot 6 \cdot 5}{3 \cdot 2 \cdot 1} = 35$$

Self-Check 2 Compute **(a)** ${}_8C_4$ **(b)** ${}_{50}C_3$.

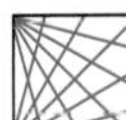

Factorials, Permutations, and Combinations with a Calculator

Calculators have built-in functions for computing factorials, permutations, and combinations. Let us compute 7!, ${}_8P_5$, and ${}_9C_4$ (Figure 9.3).

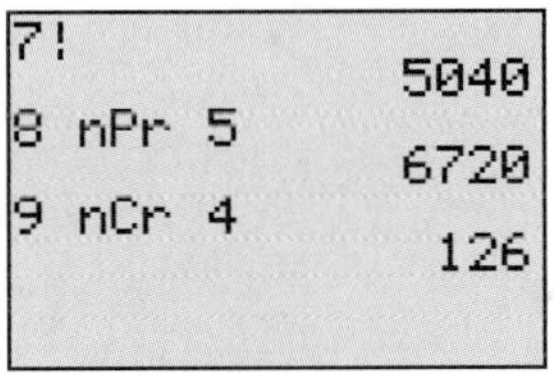

```
7!
                5040
8 nPr 5
                6720
9 nCr 4
                 126
```

Figure 9.3 Factorials, permutations, and combinations on a calculator.

Self-Check 3 Use your calculator to compute **(a)** 9! **(b)** ${}_{10}P_5$ **(c)** ${}_{12}C_5$.

We have seen how problems often are presented initially in words. It becomes a challenge to translate those problems into mathematics. The mathematics then leads to the solution. We now give a number of examples to illustrate the type of problem that lends itself to a solution using combinations.

Example 6 **Scholarship Recipients**

Fifteen students are eligible for four Carter Scholarships. The scholarships are all worth the same amount of money. In how many ways can the scholarships be awarded?

SOLUTION Four students are to be selected from 15. The order of selection is unimportant. The number of awards is thus

$${}_{15}C_4 = \frac{15!}{(15-4)!4!} = \frac{15!}{11!4!} = \frac{15 \cdot 14 \cdot 13 \cdot 12}{4 \cdot 3 \cdot 2 \cdot 1} = 1365$$

Self-Check 4 *Conference Participants* Campus Life funds are available for four students from a group of nine to attend a conference on "The Etiquette of Dating." How many selections are possible?

Example 7 Quiz Team

A quiz team of five children is to be selected from a seventh grade class of 18 children. There are ten girls and eight boys in the class. How many different teams can be made up composed of three girls and two boys?

SOLUTION There are ${}_{10}C_3$ ways of selecting three girls from ten. There are ${}_8C_2$ ways of selecting two boys from eight. Because any one of the ${}_{10}C_3$ selections of girls can be taken with any one of the ${}_8C_2$ selections of boys, there will be ${}_{10}C_3 \times {}_8C_2$ teams made up of three girls and two boys. We get

$$ {}_{10}C_3 \times {}_8C_2 = \frac{10 \cdot 9 \cdot 8}{3 \cdot 2 \cdot 1} \times \frac{8 \cdot 7}{2 \cdot 1} = 120 \times 28 = 3360 $$

There are 3360 possible teams.

Further Ideas 3 *Judiciary Committee* Carlson University Judiciary Committee is to be composed of three graduate students and four undergraduates. The selection is to be made from a pool of six graduate students and 12 undergraduates. How many different committees can be formed?

Answers to Self-Check Exercises

1. (a) ${}_9P_2 = 172.$ (b) ${}_4P_4 = 24.$
2. (a) ${}_8C_4 = 70.$ (b) ${}_{50}C_3 = 19{,}600.$
3. (a) $9! = 362{,}880.$ (b) ${}_{10}P_5 = 30{,}240.$ (c) ${}_{12}C_4 = 495.$
4. Number of selections possible = 126.

Answers to Further Ideas Exercises

1. Number of possibilities = 144.
2. Number of ways = 120.
3. Number of different committees = 9900.

EXERCISES 9.1

In Exercises 1–16, evaluate each expression. Check your answers using the built-in function on a calculator.

1. $4!$ **2.** $7!$ **3.** $2!$ **4.** $13!$ **5.** ${}_4P_2$
6. ${}_5P_3$ **7.** ${}_8P_1$ **8.** ${}_9P_5$ **9.** ${}_6P_5$ **10.** ${}_{11}P_3$
11. ${}_7C_3$ **12.** ${}_5C_4$ **13.** ${}_{12}C_2$ **14.** ${}_3C_2$ **15.** ${}_2C_1$
16. ${}_6C_2$

In Exercises 17–19, use a calculator to find the numbers that satisfy the given conditions.

17. The smallest positive integer n such that $n! > 10{,}000$.

18. The smallest positive integer n such that ${}_{10}P_n > 150{,}000$.

19. The values of n for which ${}_{10}C_n > 200$.

20. Selecting Clothes Fred Palle has seven shirts, five pairs of trousers, and four pairs of shoes. How many different ensembles can he wear?

21. **Class Possibilities** Javier Rodriguez must take one biology course, one chemistry course, and one physics course to satisfy the natural science requirement at Denbigh University. There are three biology classes, four chemistry classes, and two physics classes to choose from. In how many different ways can Javier satisfy the natural science requirement?
22. **Number of Routes** There are four different direct roads connecting Detroit to Lansing, three roads from Lansing to Grand Rapids, and two roads from Grand Rapids to Muskegon. How many routes from Detroit to Muskegon pass through Lansing and Grand Rapids?
23. **Examination Choices** An examination consists of three parts A, B and C. One question must be answered from each part. There are five questions to choose from in A, four in B, and two in C. In how many ways can the examination questions be answered?
24. **Arranging Books** Six different books are to be arranged on a shelf. In how many ways can they be arranged?
25. **Picture Taking** In how many ways can eight children line up to have their picture taken?
26. **Table Arrangements** Six people sit at a round table. How many arrangements are possible?
27. **Batting Orders** In how many ways can a manager arrange the batting order in a nine-player baseball team?

28. **License Plates** How many license plates can be made first using a letter, then three numbers between 0 and 9, then two letters?
29. **Competition Winners** There are seven finalists in the Singer of the World Competition in Cardiff, Wales. In how many ways can a winner and runner-up be selected?
30. **Committee Selections** A president, vice president, secretary, and treasurer are to be selected from a group of 15 people. How many selections are possible?
31. **Olympic Medalists** There are eight runners in the Olympic 100-meter final. In how many ways can the gold, silver, and bronze medals be given?
32. **Number of Routes** There are four different direct routes connecting Tenby to Gainesville. In how many ways can a person travel from one town to the other and return along a different route?
33. **Forming Numbers** How many three-digit numbers can be formed using the digits 1, 2, 3, and 4 if **(a)** no repetition is allowed, **(b)** repetition is allowed?
34. **Examination Choices** Students taking an examination are required to answer three questions out of seven. How many selections are possible?
35. **Class Enrollment** There is room for 25 students in a mathematics class, but 30 students wish to enroll. How many different groups are possible?
36. **Number of Card Hands** How many five-card hands can be dealt from a pack of 52 playing cards?
37. **Distributing Computers** There are ten faculty members and a secretary in the biology department at Lexington College. The department is to be given four new computers (all alike). If the secretary is to be given one of these computers, in how many ways can these computers be distributed?
38. **Filling Job Positions** Brandeis, Inc., has four identical secretarial positions to fill. Twelve people apply for the positions. In how many ways can the positions be filled?
39. **Planning a Meal** A meal consists of one meat, two vegetables, one salad, and one dessert. There are three meats, five vegetables, two salads, and four desserts to choose from. How many different meals can be planned?
40. **Number of Subsets** Consider the set $A = \{a, b, c, d\}$. How many subsets does A have?
41. **Awarding Contracts** In how many ways can three different contracts be awarded to six different firms if no firm is to get more than two contracts?
42. **Seating Possibilities** A car holds six people. **(a)** How many ways can six people be seated in the car? **(b)** How many ways can the people be seated if only two can drive?
43. **A Psychology Study** Six people are to be selected at random from 20 people for a psychology study. In how many ways can this be done?

44. **Bits** The numbers 0 and 1 are used in computing. Each 0 and 1 is called a bit (short for binary digit). A set of eight 0s and 1s, such as 10001101, is called a byte. How many different bytes can be formed?
45. **Telephone Numbers** How many seven-digit telephone numbers can be formed if zero cannot be used for the first digit?
46. **Seating Arrangements** Six adjacent seats have been reserved at a concert. In how many ways can three couples fill those seats if partners are to sit next to one another?
47. **Student Committees** An elected committee is to consist of three seniors, two juniors, and one sophomore. In the election for the posts, there are six seniors, five juniors, and seven sophomore candidates. How many different committees are possible?
48. **Canoeing Arrangements** A group of eight people canoes down the Juniper Springs Run. The distinct canoes each hold two people. How many different seating arrangements are there? (**a**) Distinguish between the front and back seat of each canoe. (**b**) Do not distinguish between the front and back seats.

49. **Department Representatives** The foreign language department at Buckingham College has 200 majors; 100 are Spanish majors, 75 are French majors, and 25 are German majors. Each language discipline is to elect a president and a vice president to represent its students in the department. In how many ways can this be done?
50. **Sleeping Arrangements** Twenty students are to stay in a Llanberis Hostel. The hostel has ten single rooms, three double rooms, and a room that sleeps four people. Each room has a distinct number. In how many ways can the students be placed?
51. **Committee Structures** The Natural Sciences Division at Cranford College has three departments: biology, chemistry, and physics. Biology has 80 majors, chemistry has 20 majors, and physics has ten majors. Each department is to select two representatives to serve on a Natural Sciences Committee. This committee is to then elect a chair and a secretary. In how many ways can this six-person committee with chair and secretary be formed from the 110 students?

9.2 Probability and Blood Groups

• Probability • Sample Spaces and Events • Mutually Exclusive Events • Probabilities Associated with Two Events • Probabilities Using Techniques of Counting • Traffic Lights • Blood Donors • Defective Tires • Student Committees • Card Hands • The Birthday Problem

We know that probability has been around for a long time. In fact, it has been around at least as long as games of chance have existed. Marked cubes that resemble contemporary dice have been found in Egyptian and Greek tombs. Probability came into its own as a field in the 17th century with its use in mortality tables and is now used in an increasing number of applications. Insurance companies use probability to decide financial policies, the government uses probability to determine its financial and economic policies, and theoretical physicists use probability to understand the nature of atomic-sized systems. In this section, we introduce the student to this important area of mathematics.

Consider a situation that has a finite number of **outcomes.** We refer to the situation as an **experiment.** We will develop mathematical techniques for discussing probabilities associated with the outcomes.

To understand some of the concepts involved in probability, let us look at a specific experiment, that of throwing a die. There are six possible outcomes: the die can fall on 1, 2, 3, 4, 5, or 6. Let us assume that the die is balanced, so that each outcome is equally likely. We cannot predict the result of any particular toss of the die. However, if we throw the die many times, because each outcome is equally likely, we can expect each number to appear approximately $\frac{1}{6}$ of the time. This suggests that we assign the probability $\frac{1}{6}$ to each outcome. Let us generalize this discussion.

Probability of a Single Outcome

Consider an experiment that has n equally likely outcomes. The probability of a single outcome is $\frac{1}{n}$.

Now let us return to the experiment of rolling the die to develop additional ideas. Instead of being interested in a single outcome, we could want to know about a class of outcomes. For example, we could be interested in throwing an even number, that is, in throwing 2, 4, or 6. Here there are three outcomes of interest. If the die were thrown many times we would expect 2 to appear approximately $\frac{1}{6}$ of the time, 4 to appear approximately $\frac{1}{6}$ of the time, and 6 to appear approximately $\frac{1}{6}$ of the time. Thus, an even number would be thrown approximately $(\frac{1}{6} + \frac{1}{6} + \frac{1}{6})$ of the time, or $\frac{1}{2}$ of the time. We say that the probability of throwing an even number is $\frac{1}{2}$, and we write $P(\text{even number}) = \frac{1}{2}$. Let us again generalize.

Consider an experiment that has n equally likely outcomes. Let m of these outcomes be of interest. The probability that one of these outcomes of interest will occur is

$$P(\text{outcome of interest}) = \underbrace{\left(\frac{1}{n} + \frac{1}{n} + \cdots + \frac{1}{n}\right)}_{m \text{ times}} = \frac{m}{n}$$

We have arrived at the following result.

Probability of an Outcome of Interest

Consider an experiment that has a number of equally likely outcomes. Then,

$$P(\text{outcome of interest}) = \frac{\text{Number of outcomes of interest}}{\text{Total number of outcomes possible}}$$

We now look at an example to illustrate these basic ideas of probability. The student will observe that the secret is to look at the problem in terms of the concepts that we have developed. We define the outcomes and determine the outcomes of interest. The mathematics then follows.

Example 1 **Selecting Cards**

Consider the experiment of drawing a single card from a deck of 52 playing cards. What is the probability that the card will be (a) a king, (b) a king or a queen?

SOLUTION

(a) The selection of a card is an outcome. There are 52 cards in the deck; therefore, there are 52 possible outcomes. There are four kings in the pack. There are therefore four outcomes of interest. The probability of picking a king is

$$P(\text{king}) = \frac{4}{52} = \frac{1}{13}$$

(b) There are still 52 possible outcomes. We are now interested in a king or a queen. There are four kings and four queens in the pack. There are therefore eight outcomes of interest. The probability of drawing a king or a queen is

$$P(\text{king or queen}) = \frac{8}{52} = \frac{2}{13}$$

Self-Check 1 Tara Panizian rolls a die. What is the probability that the number is greater than four?

Sample Spaces and Events

It is convenient to frame some of the preceding concepts in the language of set theory. This approach enables us to gradually develop new ideas. Let us represent all the outcomes of an experiment as the elements of a set S, called the **sample space** of the experiment. At this time, we assume that each outcome is equally likely to happen. The various collections of outcomes are represented by the subsets of S. Any subset E of S is called an **event.** Let $n(S)$ and $n(E)$ be the number of elements in S and E, respectively. If E is the subset that corresponds to the outcomes of interest, then the probability of E occurring, denoted by $P(E)$, is $n(E)/n(S)$.

Note that if none of the outcomes of an experiment are of interest, then $n(E) = 0$, implying that $P(E) = 0$. On the other hand, if all the outcomes are of interest, then $n(E) = n(S)$, implying that $P(E) = 1$. The smallest probability is thus 0, the largest is 1, and all other probabilities lie between 0 and 1. We have that $0 \leq P(E) \leq 1$.

Probability of an Event

Consider an experiment with sample space S. The probability of an event E occurring is

$$P(E) = \frac{n(E)}{n(S)}, \qquad 0 \leq P(E) \leq 1$$

Tossing a Die

We make these concepts clear by looking at the experiment of tossing a die. There are six possible outcomes. The sample space $S = \{1, 2, 3, 4, 5, 6\}$. Thus, $n(S) = 6$. Suppose that we are interested in tossing an odd number. The event $E = \{1, 3, 5\}$ describes these outcomes of interest. Therefore $n(E) = 3$. The probability of throwing an odd number is

$$P(\text{odd number}) = \frac{n(E)}{n(S)} = \frac{3}{6} = \frac{1}{2}$$

If we are interested in tossing the number 8 on the die, then there are no outcomes of interest: $n(E) = 0$ and $P(8) = 0$. The probability of throwing an 8 is 0. On the other hand, if we are interested in tossing 1, 2, 3, 4, 5, or 6, then $n(E) = 6$ and $P(1, 2, 3, 4, 5, 6) = 1$.

The following example illustrates the art of constructing a sample space. One has to first decide what an arbitrary element looks like and then decide on some convention for listing all the elements in a systematic manner.

Example 2 Tossing a Die Twice

Consider the experiment of tossing a die twice. Use a sample space to determine the probability of getting the numbers 6 and 2, in either order.

SOLUTION Let each element of the sample space be a pair of numbers (X, Y), with X being the number obtained on the first throw of the die, Y the number obtained on the second throw. For example, (1, 4) corresponds to getting 1 and then 4. Let us list all the elements of the form $(1, Y)$ first, then all the elements $(2, Y)$, and so on. We get

$$\begin{aligned} S = \{&(1, 1), (1, 2), (1, 3), (1, 4), (1, 5), (1, 6),\\ &(2, 1), (2, 2), (2, 3), (2, 4), (2, 5), (2, 6),\\ &(3, 1), (3, 2), (3, 3), (3, 4), (3, 5), (3, 6),\\ &(4, 1), (4, 2), (4, 3), (4, 4), (4, 5), (4, 6),\\ &(5, 1), (5, 2), (5, 3), (5, 4), (5, 5), (5, 6).\\ &(6, 1), (6, 2), (6, 3), (6, 4), (6, 5), (6, 6)\} \end{aligned}$$

The sample space consists of 36 elements; there are 36 possible outcomes: $n(S) = 36$. The event of interest is the subset E that corresponds to getting 6 and 2 in either order.

$$E = \{(6, 2), (2, 6)\}$$

There are two elements in the event of interest: $n(E) = 2$. Thus,

$$P(6 \text{ or } 2) = \frac{n(E)}{n(S)} = \frac{2}{36} = \frac{1}{18}$$

Further Ideas 1 *Tossing a Die and a Coin* Consider the experiment of tossing a die and then tossing a coin. Construct a sample space and use it to determine the probability of tossing an even number followed by tossing heads.

The sample space of an experiment is made up of many events. Up to this time, we looked at the probabilities associated with single events. Let us now discuss ways of looking at probabilities associated with two events. The experiment could, for example, be selecting a card from a deck. One event could be selecting an ace. The other event might be selecting a black card. We know how to determine the probability of selecting an ace. We know how to determine the probability of selecting a black card. What is the probability of selecting an ace *and* a black card (a black ace)? What is the probability of selecting an ace *or* a black card? The mathematics that we now develop enables us to answer questions such as these.

Probabilities Associated with Two Events

Consider an experiment with sample space S. Let F and G be two events in this sample space. We could be interested in two types of situations associated with F and G.

1. F *and* G both happening. The outcomes of interest in this case lie in both F and in G. In the language of set theory, we say that the outcomes of interest are described by the elements in the intersection of F and G, written $F \cap G$.
2. F *or* G happening. The outcomes of interest now lie in F or G. They are described by the union of F and G, written $F \cup G$.

These observations lead to the following probabilities associated with the events F and G.

Probabilities of AND and OR

Consider an experiment with sample space S. Let F and G be two events in this sample space. Then

$$P(F \text{ and } G) = P(F \cap G) = \frac{n(F \cap G)}{n(S)},$$

$$P(F \text{ or } G) = P(F \cup G) = \frac{n(F \cup G)}{n(S)}.$$

Example 3 **Probability Associated with Two Events**

Josh Leddin tosses a die. What is the probability that the number will be an even number and that it will be greater than 2?

SOLUTION Let S be the sample space: $S = \{1, 2, 3, 4, 5, 6\}$. Thus, $n(S) = 6$. Let F be the event of tossing an even number: $F = \{2, 4, 6\}$. Let G be the event of tossing a number greater than 2: $G = \{3, 4, 5, 6\}$. The probability of tossing a number that is even and is at the same time greater than 2 is

$$P(F \text{ and } G) = P(F \cap G) = \frac{n(F \cap G)}{n(S)}$$

We see that $F \cap G = \{4, 6\}$. $n(F \cap G) = 2$. Thus,

$$P(F \cap G) = \frac{n(F \cap G)}{n(S)} = \frac{2}{6} = \frac{1}{3}$$

The probability of throwing an even number greater than 2 is $\frac{1}{3}$.

Example 4 **Probability Associated with Two Events**

Deborah Yeager tosses a die. What is the probability that an even number or a number that is greater than 4 will be tossed?

SOLUTION Let S be the sample space: $S = \{1, 2, 3, 4, 5, 6\}$. Thus, $n(S) = 6$. Let F be the event of tossing an even number, therefore, $F = \{2, 4, 6\}$. Let G be the event of tossing a number greater than 4. $G = \{5, 6\}$. The probability of tossing an even number or a number greater than 4 is

$$P(F \text{ or } G) = P(F \cup G) = \frac{n(F \cup G)}{n(S)}$$

We see that $F \cup G = \{2, 4, 5, 6\}$. $n(F \cup G) = 4$; thus,

$$P(F \cup G) = \frac{n(F \cup G)}{n(S)} = \frac{4}{6} = \frac{2}{3}$$

The probability of tossing an even number or a number greater than 4 is $\frac{2}{3}$.

Further Ideas 2 Victoria Peronti tosses a die. Determine the probability of getting (a) an odd number less than 5, (b) an odd number or a number less than 5.

We now return to our introductory experiment involving aces and black cards. Let us see how to structure the problem and arrive at answers using the tools we have developed.

Example 5 **Probabilities Associated with Two Events**

A card is selected from a pack of 52 playing cards. (a) What is the probability of selecting an ace and a black card (a black ace)? (b) What is the probability of selecting an ace or a black card?

SOLUTION

(a) There are 52 cards in the deck; there are therefore 52 possible outcomes. Let F be the event of selecting an ace and G be the event of selecting a black card. There are two cards that are aces and are black; namely, the ace of clubs and the ace of spades. Thus, $n(F \cap G) = 2$. This gives

$$P(\text{ace and black}) = \frac{n(F \cap G)}{n(S)} = \frac{2}{52} = \frac{1}{26}$$

(b) We now need to know $n(F \cup G)$, the number of cards that are aces or black. There are four aces and there are 26 black cards. If we add these numbers, we get 30. However some cards have been counted twice; namely, the two black aces. Thus, $n(F \cup G) = 28$. This gives

$$P(\text{ace or black}) = \frac{n(F \cup G)}{n(S)} = \frac{28}{52} = \frac{7}{13}$$

MUTUALLY EXCLUSIVE EVENTS

Let S be the sample space of an experiment, and let F and G be two events in the sample space. The two events could have nothing in common. F and G are said to be **mutually exclusive** if they have no elements in common. In the language of set theory, F and G are mutually exclusive if they are disjoint, and we write $F \cap G = \varnothing$. We now show that if F and G are mutually exclusive, then the probability of F or G happening can be expressed in a particularly convenient way. We know that

$$P(F \text{ or } G) = P(F \cup G) = \frac{n(F \cup G)}{n(S)}$$

Because F and G are disjoint, the number of elements in $F \cup G$ will be equal to the number of elements in F plus the number of elements in G. Thus, $n(F \cup G) = n(F) + n(G)$, giving

$$\begin{aligned} P(F \text{ or } G) &= \frac{n(F) + n(G)}{n(S)} \\ &= \frac{n(F)}{n(S)} + \frac{n(G)}{n(S)} \\ &= P(F) + P(G) \end{aligned}$$

We have arrived at the following result.

MUTUALLY EXCLUSIVE EVENTS

Consider an experiment with sample space S. Let F and G be two events in this sample space. F and G are mutually exclusive if $F \cap G = \varnothing$. If F and G are mutually exclusive, then

$$P(F \text{ or } G) = P(F) + P(G)$$

Example 6 Mutually Exclusive Events

A die is tossed. What is the probability of throwing an even number or a 3?

SOLUTION The sample space is $S = \{1, 2, 3, 4, 5, 6\}$, thus, $n(S) = 6$. Let F be the event of tossing an even number, therefore, $F = \{2, 4, 6\}$ and $n(F) = 3$. Let G be the event of tossing a 3: $G = \{3\}$ and $n(G) = 1$. Observe that F and G have no elements in common. F and G are mutually exclusive events. Thus, the probability of throwing an even number or a 3 is

$$\begin{aligned} P(F \text{ or } G) &= P(F \cup G) = P(F) + P(G) \\ &= \frac{n(F)}{n(S)} + \frac{n(G)}{n(S)} = \frac{3}{6} + \frac{1}{6} = \frac{4}{6} = \frac{2}{3} \end{aligned}$$

The probability of throwing an even number or a 3 is $\frac{2}{3}$.

Further Ideas 3 Consider the experiment of tossing a die. Show that the event of tossing an odd number and tossing a 6 are mutually exclusive events. Determine the probability of tossing an odd number or a 6.

Example 7 Mutually Exclusive Events

A die is rolled twice. Determine the probability that the sum of the two numbers thrown is 6 or 7.

SOLUTION The sample space S of this experiment is found in Example 2. We see that $n(S) = 36$.

Let F be the event of throwing two numbers that add to 6. The subset of S having numbers that add to 6 is

$$F = \{(1, 5), (2, 4), (3, 3), (4, 2), (5, 1)\}. \qquad n(F) = 5.$$

Let G be the event of throwing two numbers that add to 7.

$$G = \{(1, 6), (6, 1), (2, 5), (5, 2), (3, 4), (4, 3)\}. \qquad n(G) = 6.$$

Observe that F and G are disjoint sets. The events F and G are mutually exclusive. Thus,

$$\begin{aligned} P(F \text{ or } G) &= P(F \cup G) = P(F) + P(G) \\ &= \frac{n(F)}{n(S)} + \frac{n(G)}{n(S)} = \frac{5}{36} + \frac{6}{36} = \frac{11}{36} \end{aligned}$$

The probability that the numbers thrown add to either 6 or 7 is $\frac{11}{36}$.

We have discussed experiments in which all the outcomes were in some intuitive sense equally likely. We now extend our discussion to include experiments that do not have equally likely outcomes.

Example 8 Traffic Lights

Consider the experiment of arriving at a traffic light. There are three possible outcomes to this experiment. The light could be green, yellow, or red. The colors are usually of varying durations; thus, the three outcomes are not equally likely. Each will have a certain probability.

Suppose that the durations of the various lights in the major road direction are green, 85 seconds; yellow, 5 seconds; and red, 35 seconds. Let us determine the probability of getting, upon arrival at the traffic light, a green light, a yellow light, and a red light. The duration of the cycle is $(85 + 5 + 35)$ seconds, or 125 seconds. The green light is on for $\frac{85}{125}$ of this time, the yellow light $\frac{5}{125}$ of the time, and the red $\frac{35}{125}$ of the time. This suggests that we define the probabilities as follows:

$$P(\text{green}) = \frac{85}{125} = 0.68$$

$$P(\text{yellow}) = \frac{5}{125} = 0.04$$

$$P(\text{red}) = \frac{35}{125} = 0.28$$

Thus, the probability of arriving at the light along the major road and finding it green is 0.68. The probability of having a yellow light is 0.04, and that of red is 0.28.

Example 9 Blood Donors

The blood of any one individual contains a number of antigens; blood is classified according to these antigens. The antigens are denoted A, B, and Rh. Every person is classified in two ways. If A or B antigens are present in the blood, these are listed. If the Rh antigen is present, the blood is said to be positive; otherwise, it is negative. For example, if the blood of an individual has all three antigens A, B, and Rh, the classification is AB^+. If the person has both A and B but no Rh antigen, the classification is AB^-; if the person has no A and no B antigen, the classification is either O^+ or O^-, depending on whether Rh is present or not; and so on. It is possible to have all combinations of antigens; thus, there are eight blood groups labeled AB^+, AB^-, A^+, A^-, B^+, B^-, O^+, and O^-.

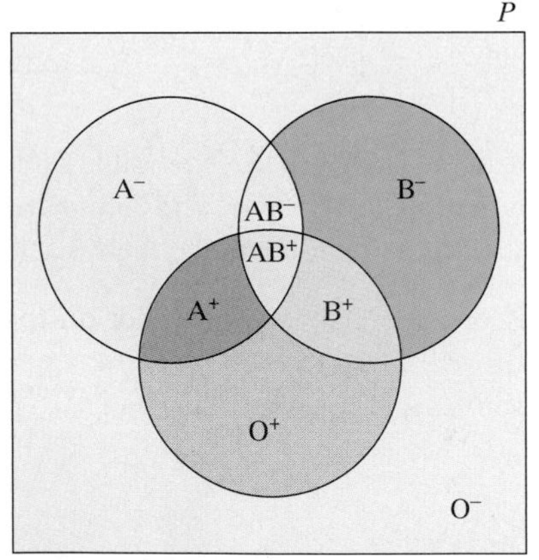

Figure 9.4

Certain restrictions regarding the donation of blood depend on the antigens in the blood of the donor and the recipient. If an antigen is absent from the blood of a person, that person cannot receive blood with that antigen. For example, a person with A^+ blood cannot receive blood from a person with AB^+ blood because the B antigen is in the latter but not in the former.

We now construct a mathematical model of this situation. Our model will enable us to answer the following question: What is the probability that a person you meet at random on the street is able to give you blood? Let us represent the groups of people with various blood groups as sets. Let P be the set of all people. There are eight distinct subsets labeled AB^+ through O^-. These correspond to the eight blood groups (Figure 9.4).

The Red Cross provided the following figures for the percentage of people within each blood group:

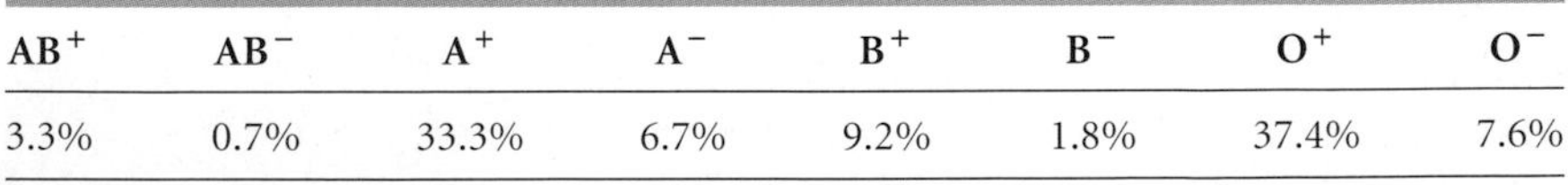

AB^+	AB^-	A^+	A^-	B^+	B^-	O^+	O^-
3.3%	0.7%	33.3%	6.7%	9.2%	1.8%	37.4%	7.6%

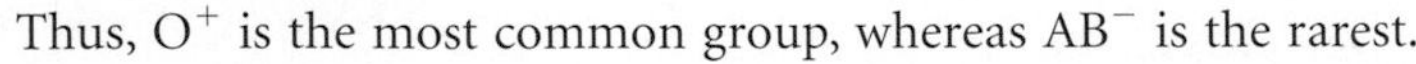

Thus, O^+ is the most common group, whereas AB^- is the rarest.

Consider the experiment of selecting a person at random in the United States with a view to his or her blood group. This experiment has eight possible outcomes, ranging from AB^+ to O^-. The sample space is

$$\{AB^+, AB^-, A^+, A^-, B^+, B^-, O^+, O^-\}$$

On the basis of these statistics, let us define the various probabilities of the simple events as follows:

$$P(AB^+) = 0.033, \quad P(AB^-) = 0.007, \quad P(A^+) = 0.333,$$
$$P(A^-) = 0.067, \quad P(B^+) = 0.092, \quad P(B^-) = 0.018,$$
$$P(O^+) = 0.374, \quad P(O^-) = 0.076.$$

The author's blood group is B^+. What is the probability that a stranger whom he contacts on the street can give him blood? A person with B^+ blood can receive blood from B^+, B^-, O^+, and O^-. These are the groups that contain no additional antigens. Let G be the event $\{B^+, B^-, O^+, O^-\}$, then

$G = \{B^+\} \cup \{B^-\} \cup \{O^+\} \cup \{O^-\}$, giving

$$\begin{aligned} P(G) &= P(B^+) + P(B^-) + P(O^+) + P(O^-) \\ &= 0.092 + 0.018 + 0.374 + 0.076 \\ &= 0.56 \end{aligned}$$

Hence there is a 0.56 probability of the author's contacting a donor at random on the street.

Probabilities Using Techniques of Counting

When the sample space of an experiment becomes large, it is not practical to list all the elements. The counting techniques of the previous section can often be used to determine the numbers of elements in the sample space and in the event of interest, leading to the desired probability. We now illustrate such situations.

Example 10 Tossing a Die Twice

A die is thrown twice. What is the probability of getting an even number on the first throw and a number greater than 4 on the second throw?

SOLUTION Let the sample space S consist of pairs of numbers (X, Y), where X is the result of the first throw and Y is the result of the second throw. There are six possibilities for X and six possibilities for Y. Thus, there are $6 \times 6 = 36$ elements in the sample space.

The outcomes of interest will be the elements (X, Y), having X even and Y greater than 4. There are three possibilities for X: 2, 4, or 6. There are two possibilities for Y: 5 or 6. There will therefore be $3 \times 2 = 6$ outcomes of interest. The probability of getting an even number followed by a number greater than 4 is

$$\frac{6}{36} = \frac{1}{6}$$

Example 11 Defective Tires

Olson's Garage has 12 tires in stock, and one is defective. Out of four tires placed on a car, what is the probability that one will be defective?

SOLUTION The total number of ways of selecting four new tires from the 12 available, order being unimportant, is ${}_{12}C_4$. To determine the number of outcomes of interest, we assume that the defective tire is one of the ones selected. There are ${}_{11}C_3$ ways of selecting the other three tires.

Thus, the probability that the defective tire is among the four put on the car is

$$\frac{{}_{11}C_3}{{}_{12}C_4} = \frac{(11 \cdot 10 \cdot 9)/3!}{(12 \cdot 11 \cdot 10 \cdot 9)/4!} = \frac{11 \cdot 10 \cdot 9}{3 \cdot 2 \cdot 1} \times \frac{4 \cdot 3 \cdot 2 \cdot 1}{12 \cdot 11 \cdot 10 \cdot 9} = \frac{4}{12} = \frac{1}{3}$$

Example 12 Student Committees

A Student Affairs committee consisting of eight students is to be selected from a group of 22 students: seven seniors, six juniors, five sophomores, and four freshmen. What is the probability that the committee will consist of three seniors, two juniors, two sophomores, and one freshman?

SOLUTION The total number of ways of selecting the committee of eight from the 22 students is ${}_{22}C_8$. There are thus ${}_{22}C_8$ outcomes possible.

The committees of interest will be composed of three seniors, two juniors, two sophomores, and one freshman. There are ${}_7C_3$ ways of selecting the three seniors from seven, ${}_6C_2$ ways of selecting the two juniors from six, ${}_5C_2$ ways of selecting the two sophomores from five, and ${}_4C_1$ ways of selecting the single freshman from four. There will thus be a total of ${}_7C_3 \times {}_6C_2 \times {}_5C_2 \times {}_4C_1$ committees of interest. The required probability is

$$\frac{{}_7C_3 \times {}_6C_2 \times {}_5C_2 \times {}_4C_1}{{}_{22}C_8} = \frac{(7 \cdot 6 \cdot 5)/3! \times (6 \cdot 5)/2! \times (5 \cdot 4)/2! \times 4/1}{(22 \cdot 21 \cdots 15)/8!}$$

$$= 0.0657 \text{ (rounded to four decimal places)}$$

Example 13 Card Hands

A hand of 13 cards is dealt from a pack of 52 bridge cards. Determine the probability of getting a single ace in the hand.

SOLUTION There are ${}_{52}C_{13}$ hands of 13 cards that can be dealt from 52 cards. There are thus ${}_{52}C_{13}$ outcomes possible.

The hands of interest contain one ace and 12 other cards. There are four aces and 48 other cards in the pack. There are four ways of selecting the ace and ${}_{48}C_{12}$ ways of selecting the other 12 cards. There are thus $4({}_{48}C_{12})$ hands of interest.

The required probability is

$$\frac{4({}_{48}C_{12})}{{}_{52}C_{13}} = \frac{4(48 \cdot 47 \cdots 37)/12!}{(52 \cdot 51 \cdots 40)/13!} = \frac{4 \cdot 39 \cdot 38 \cdot 37 \cdot 13}{52 \cdot 51 \cdot 50 \cdot 49}$$

$$= 0.4388 \text{ (to four decimal places)}$$

Further Ideas 4 A hand of 13 bridge cards is dealt from a pack of 52 bridge cards. Determine the probability (rounded to four decimal places) of (**a**) getting all four aces in the hand, (**b**) getting no aces in the hand.

Example 14 The Birthday Problem

Consider a room with n people in it. What is the probability that at least two people have the same birthday (same month, same day of the year)? We will find the results of this investigation rather surprising. If the room has 25 people in it, for example, we will find that the probability that at least two people have the same birthday is 0.569, much higher than one might intuitively expect.

We approach this problem by looking at the probability that no two people have the same birthday. There are ${}_{365}P_n$ ways of selecting, in order, n distinct birthdays from 365 possible birthdays. There are thus ${}_{365}P_n$ outcomes of interest. Because each person in the room has 365 possibilities for a birthday, there is a total of 365^n ways that the birthdays can occur for the n people. Thus, the probability that no two people have the same birthday is

$$\frac{{}_{365}P_n}{365^n}$$

We have that

$$\text{Probability}\begin{pmatrix}\text{at least two people have}\\ \text{the same birthday}\end{pmatrix} + \text{Probability}\begin{pmatrix}\text{no two people have}\\ \text{the same birthday}\end{pmatrix} = 1$$

because one of these events has to be true.

The probability that at least two people have the same birthday is thus

$$1 - \frac{{}_{365}P_n}{365^n}$$

The probabilities for the various values of n have been computed in the table of Figure 9.5. In a room containing 35 people, there is a probability of 0.81438 that there are at least two people having the same birthday!

Use your calculator to determine probabilities for 36–40 people in a room. Why does your calculator give an error message for $X = 40$? How can you overcome this and determine the probability for 40?

No. of People Probability

X	Y1
5	.02714
10	.11695
15	.2529
20	.41144
25	.5687
30	.70632
35	.81438

Figure 9.5

Further Ideas 5 Determine the probability of at least two people in your class having the same birthday. Determine the birthdays of the people in your class to see how many have birthdays on the same day—and how soon the first pair arose.

Group Discussion

Birthdays on the Same Day of the Month

Determine the formula for finding the probability that two people in a group of n have their birthdays on the same day of the month (both on the 21st, for example, assuming 31-day months). How many people does it take for the probability to be more than 0.5? Test the formula for your group of students. Draw up a list of the number with their birthdays on the 1st, 2nd, 3rd, and so on.

Answer to Self-Check Exercise

1. Probability is $\frac{1}{3}$.

Answers to Further Ideas Exercises

1. Probability is $\frac{3}{12} = \frac{1}{4}$.
2. **(a)** Probability of an odd number less than 5 $= \frac{1}{3}$.
   (b) Probability of an odd number or a number less than 5 $= \frac{5}{6}$.
3. $S = \{1, 2, 3, 4, 5, 6\}$. $n(S) = 6$. $F = \{1, 3, 5\}$. $n(F) = 3$. $G = \{6\}$. $n(G) = 1$. F and G have no elements in common; they are mutually exclusive.

$$P(F \text{ or } G) = \frac{n(F)}{n(S)} + \frac{n(G)}{n(S)} = \frac{3}{6} + \frac{1}{6} = \frac{4}{6} = \frac{2}{3}.$$

4. **(a)** $\dfrac{{}_{48}C_9}{{}_{52}C_{13}} = 0.0026$ **(b)** $\dfrac{{}_{48}C_{13}}{{}_{52}C_{13}} = 0.3038$

EXERCISES 9.2

1. **Tossing a Coin** A coin is tossed. What is the probability that it will land on **(a)** heads, **(b)** tails?
2. **Drawing a Card** A single card is drawn from a deck of 52 bridge cards. Determine the probability that it is **(a)** an ace, **(b)** a black card, **(c)** a number less than 4.
3. **Selecting a Letter** Consider the word *Mississippi.* A letter is selected at random. Find the probability that the letter is **(a)** *s*, **(b)** *p*, **(c)** *i*, **(d)** *a*, **(e)** *m*, **(f)** *s*, *p*, or *i*.
4. **Tossing a Die** On a single toss of a die, what is the probability of tossing **(a)** a 6, **(b)** a 1 or a 2, **(c)** an even number.
5. **Birthdays** Bruno Rager's birthday is in April. Determine the probability that it is **(a)** April 28, **(b)** in the second week of April, **(c)** in the first half of April.
6. **Selecting Balls** A bag contains ten balls: three black, five white, and two red. Determine the probability of selecting in one draw **(a)** a black ball, **(b)** a black or a white ball, **(c)** a black or a red ball.
7. **Tossing a Die Twice** Consider the experiment of tossing a die twice. Use a sample space to find the probability **(a)** of tossing two 5s, **(b)** of tossing a 3 and a 4 in any order, **(c)** that the sum of the two numbers is greater than 9.
8. **Tossing a Die Twice** Consider the experiment of tossing a die twice. Use a sample space to find the probability of tossing a 2 on the first throw and a 4 on the second throw.
9. **Multiple Coin Toss** A coin is tossed three times. Construct a sample space. What is the probability of tossing three heads?

10. **Tossing a Coin and a Die** A die is thrown and then a coin is tossed. Construct a sample space. What is the probability of getting an even number followed by heads?

11. **Selecting Three Balls** A box contains two black balls and two white balls. Three balls are drawn in succession without replacement. Construct a sample space. Determine the probability of drawing two black balls and then a white ball.

12. **Two-Digit Numbers** A two-digit number is to be constructed at random from the numbers 1, 2, 3, 4, and 5. Construct a sample space. What is the probability of getting (**a**) 22, (**b**) 31, (**c**) a number greater than 40, (**d**) a number whose digits are repeated, (**e**) a number the sum of whose digits is 8?

13. **Tossing a Die** A die is tossed. Determine the probability that the number is (**a**) an odd number less than 5, (**b**) an even number or a number less than 5.

14. **Selecting a Card** A single card is picked from a pack of 52 playing cards. What is the probability that it will be a king or a spade?

15. **Family Probabilities** A family has four children. Find the probability that there are (**a**) either four boys or four girls, (**b**) at least one boy and one girl.

16. **Tossing a Die** Consider the experiment of tossing a die. F is the event of throwing an odd number and G is that of throwing a number greater than 4. Are F and G mutually exclusive events?

17. **Tossing a Die** A die is tossed. F is the event of throwing an even number, and G is the event of throwing a number less than 3. Are F and G mutually exclusive events?

18. **Tossing a Die** A die is tossed. F is the event of tossing an even number. G is the event of tossing a number less than 2. Are F and G mutually exclusive events?

19. **Tossing a Die Twice** A die is tossed twice. F is the event that the sum of the numbers is 9, and G is the event that one of the numbers is 6. Are F and G mutually exclusive events?

20. **Tossing a Die Twice** A die is tossed twice. F is the event that the same number appears on both dice. G is the event that the sum of the numbers is 10. Are F and G mutually exclusive events?

21. **Tossing a Die Twice** A die is tossed twice. Determine the probability of rolling a sum of either 8 or 9.

22. **True–False** Nora Deloso guesses at the answer to five of the questions on a true–false test. Determine the probability that (**a**) all five answers are correct, (**b**) four answers are correct, (**c**) at least three answers are correct, (**d**) four or five answers are correct.

23. **Selecting Cards** A first card and then a second card are selected without replacement from a pack of 52 playing cards. Determine the probability that the two cards are (**a**) the same number or the same face card, (**b**) the same suit.

24. **Selecting a University** In a certain state there are three large state universities, two large private universities, and four small private universities. A student plans to attend one of these schools. Determine the probability that he will attend (**a**) a small private university, (**b**) a private school, (**c**) a large school.

25. **Buying Land in Florida** A resident of New York bought a lot in an undeveloped subdivision in Florida. The subdivision has 20 lots. Of these lots, two are swampy and have no trees, three are swampy with trees, three are dry all year round but have no trees, and the remaining 12 are dry and have trees. Determine the probability that the lot (**a**) is dry, (**b**) is swampy or has no trees, (**c**) has trees.

26. **Tossing a Die Twice** A die is thrown twice. What is the probability of getting an odd number on the first throw and a number less than 5 on the second throw?

27. **Defective Tires** Strazalka' s corner garage has 20 tires in stock, and one is defective. Out of four tires placed on a car, what is the probability that one will be defective?

28. **Student Committees** A committee of nine students is to be selected from a group of 30 students: 12 seniors, eight juniors, four sophomores, and six freshmen. What is the probability that the committee will consist of four seniors, two juniors, two sophomores, and one freshman? Give the answer to four decimal places.

29. **Filling Job Positions** Of 15 people who apply for jobs, ten are men, and five are women. If six people are hired at random, what is the probability that four are men and two are women? Give the answer to four decimal places.

30. **Defective Light Bulbs** A box contains eight light bulbs, two of which are defective. Three bulbs are selected. What is the probability that one of the three bulbs is defective? Give the answer to four decimal places.

31. **Getting off an Elevator** A building with 30 floors is serviced by an elevator. If four people use the elevator at once, determine the probability that they all get off at (**a**) the same floor, (**b**) the 20th floor, (**c**) different floors.
32. **U.S. Senate** Assume that the U.S. Senate has 40 Republicans and 60 Democrats. What is the probability that a committee of 12 members is made up of four Republicans and eight Democrats? Give the answer to four decimal places.

33. **Television Commercials** During the hour from 9:00 PM to 10:00 PM, the following times were devoted to programs and commercials on the three major channels.

	Commercials	Programs
Channel 2	10 minutes	50 minutes
Channel 6	12 minutes	48 minutes
Channel 9	11.5 minutes	48.5 minutes

These times were recorded by three separate investigators on the same evening. On the basis of these results, determine the probability of (**a**) turning the television on to Channel 6 at an arbitrary time between 9 and 10 and getting a commercial. (**b**) turning the television on to one of the major channels between 9:00 and 10:00 and getting a commercial. Give the answer to four decimal places.

34. **Traffic Lights** The durations of the lights in the major road direction for a certain traffic light are green, 90 seconds; yellow, 4 seconds; and red, 40 seconds. Find the probability of getting, upon arrival at the light along the major road, (**a**) green, (**b**) yellow, (**c**) red, (**d**) green or yellow.

Blood Donors

Exercises 35–40 are based on the blood donor example of this section.

35. Determine the groups that can give blood to people in each of the following categories: A^+, A^-, AB^-, and O^-.
36. Determine the probabilities of finding a donor for each of the categories in Exercise 35.
37. What is your blood type? What is the probability of your walking up to a person who can give you blood?
38. Determine the groups that can be *given* blood by each of the following categories: AB^+, AB^-, A^+, O^+, and O^-. This is called the recipient problem. Note that a person with O^- blood can give blood to any group. The O^- group is called the universal donor.
39. What is the probability that a person with B^+ blood will be able to give blood to a person (of unknown group) injured in a street accident?
40. Determine your blood type. What is the probability that you can give blood to a randomly selected person on the street?

Birthdays

41. What is the probability that at least two people in a room of 45 people have the same birthday?
42. Find the probability that the instructor of your class was born in (**a**) April, (**b**) January, February, or March.
43. Find the probability that you and your instructor were born in (**a**) April, (**b**) the same month.

CHAPTER 9 PROJECT | Traffic Flow

Traffic is becoming an increasing problem in our cities, and particularly around university campuses. We have discussed how origin–destination matrices are used by highway departments to study traffic patterns within cities. We have also seen how probabilities associated with traffic lights can be determined. This project is intended to carry the ideas further.

(a) Divide the class into groups and assign each group the task of determining the probabilities of a set of traffic lights near campus.

(b) Contact your local highway department for a street map and the traffic counts for the streets surrounding the campus. (This information should be available. If it is not available, construct a map of the streets near campus and measure the average traffic along the streets in vehicles per hour.)

(c) Traffic flow is greatly determined by the timing of traffic lights. Each group should use all the information from (a) and (b) to make suggestions for improving the flow of traffic, and possibly movement of students (students often have to wait long periods for lights to change) in the area of the campus. Draw up a single comprehensive plan for improvement and submit it to your local highway department.

WEB PROJECT 9

To explore the concepts of this chapter further, try the interesting WebProject. Each activity relates the ideas and skills you have learned in this chapter to a real-world applied situation. Visit the Williams Web site at

www.saunderscollege.com/math/williams

to find the complete WebProject, related questions, and live links to the Internet.

CHAPTER 9 HIGHLIGHTS

Section 9.1

Fundamental Counting Principle: Let $E_1, E_2, \ldots, E_k$ be a sequence of events that can occur in $m_1, m_2, \ldots, m_k$ ways. The total number of ways the events can take place is $m_1 m_2 \cdots m_k$.

Permutation: An arrangement of r elements from a total of n possible elements where order is significant.

$_nP_r$: The number of permutations of n elements taken r at a time.

$$_nP_r = n(n-1)(n-2)\cdots(n-r+1) \qquad \text{or} \qquad _nP_r = \frac{n!}{(n-r)!}$$

Factorial: $r! = r(r-1)(r-2)\cdots 3 \cdot 2 \cdot 1$

Combination: An arrangement of r elements from a total of n elements where order is of no significance.

$_nC_r$: The number of combinations of n elements taken r at a time.

$$_nC_r = \frac{_nP_r}{r!} \qquad \text{or} \qquad _nC_r = \frac{n!}{(n-r)!r!}$$

Section 9.2

Probability: Consider an experiment that has n equally likely outcomes.

$$\text{Probability(single outcome)} = \frac{1}{n}$$

$$P(\text{an outcome of interest}) = \frac{\text{Number of outcomes of interest}}{\text{Total number of outcomes possible}}$$

Sample Space: Set S of all outcomes of an experiment. Any subset E of S is called an event. Probability of E occurring, denoted by $P(E)$, is $n(E)/n(S)$. Also $0 \le P(E) \le 1$.

Two Events in a Sample Space: Let F and G be two events in a sample space S. Then $P(F \text{ and } G) = P(F \cap G) = n(F \cap G)/n(S)$,
$P(F \text{ or } G) = P(F \cup G) = n(F \cup G)/n(S)$.
F and G are said to be mutually exclusive if they have no elements in common. We write $F \cap G = \varnothing$; then $P(F \text{ or } G) = P(F) + P(G)$.

Chapter 9 Review Exercises

1. Evaluate each of the following expressions:
   (a) 6! (b) 3! (c) ${}_5P_2$ (d) ${}_7P_1$ (e) ${}_{12}P_4$
   (f) ${}_7C_3$ (g) ${}_9C_4$ (h) ${}_8C_4$
2. **Number of Routes** There are three different direct roads in Britain from London to Birmingham and four different roads from Birmingham to Chester. There are five different roads from Chester to Colwyn Bay. How many different routes are there from London to Colwyn Bay that pass through both Birmingham and Chester?
3. **Student Committees** A senior, junior, sophomore, and freshman are to be selected from a group of five seniors, three juniors, two sophomores, and six freshmen. In how many ways can the selection be made?
4. **Soccer Teams** A college soccer team is selecting its uniforms. There are four different styles of shirts to select from, three different types of shorts, and five different patterns of socks. How many different uniforms can be selected?
5. **Carpet Selections** A buyer of a new house has narrowed the selections of carpeting to five, wall paper to four, paint to three, and vinyl to six, all of which go together. How many possible selections are there?
6. **Book Arrangements** Eight different books are to be arranged on a shelf. In how many ways can they be arranged?
7. **Picture Lineups** In how many ways can a family of five line up to have their picture taken?
8. **Seating Arrangements** In how many ways can five male students and their girl friends be seated at a round table if men and women are to alternate?
9. **Book Arrangements** Four history books, three mathematics books, and five biology books are to be placed on a shelf. In how many ways can the books be arranged if all the history books are to be together, all the mathematics books together, and all the biology books together.
10. **Contracts** In how many ways can three different contracts be awarded to seven different firms if no firm is to get more than two contracts?
11. **Recitals** An artist plans a recital of six classic pieces and four modern pieces. He plans to play four classics and two moderns before the intermission and the rest after the intermission. How many different arrangements of his program are possible?

12. **License Plates** How many license plates can be made using first two letters, then three numbers between 0 and 9, then a single letter?
13. **Four-Digit Numbers** How many four-digit numbers can be formed using the numbers 1, 2, 3, 4, and 5 if (**a**) no repetition is allowed, (**b**) repetition is allowed?
14. **Subsets** Consider the set $A = \{1, 2, 3, 4, 5\}$. How many subsets does A have?
15. **Test Questions** A test contains 16 questions. A student is instructed to answer ten of the questions. (**a**) In how many ways can this be done? (**b**) In how many ways can this be done if six of the first nine questions must be answered, and four of the last seven questions are required?
16. **Selecting Cards** In how many ways can a set of four cards be selected from a pack of 52 playing cards?
17. **Committees** A president, vice president, secretary, treasurer, and financial secretary are to be selected from a group of 16 people. In how many ways can this be done?
18. **Card Probability** A single card is drawn from a pack of 52 playing cards. Determine the probability that the card is (**a**) a red card, (**b**) a number less than 5, (**c**) a king, (**d**) an ace, a king, a queen, or a jack.

19. **Die Probability** A die is tossed. What is the probability that it falls on **(a)** 6, **(b)** 1, 2, or 6, **(c)** an even number, **(d)** a number less than 4?

20. **Tossing a Die Twice** Consider the experiment of tossing a die twice. Use a sample space to find the probability **(a)** of tossing two 6s, **(b)** of tossing a 1 and a 2, **(c)** that the difference of the numbers is 3, **(d)** that the sum of the two numbers is greater than 7?

21. **Selecting Two Balls** A box contains one red ball and three blue balls. Two balls are drawn in succession without replacement. Construct a sample space. Determine the probability of drawing **(a)** a red and then a blue ball, **(b)** a red and a blue ball in any order.

22. **Selecting a Card** A single card is drawn from a pack of 52 playing cards. What is the probability that it will be an ace or a club?

23. **Filling Job Positions** Out of 16 people who apply for jobs, nine are men and seven are women. If five people are hired at random, what is the probability that two men and three women will be hired. Give the result to four decimal places.

24. **Defective Cellular Phones** In a supply of 50 cellular phones, five were known to be defective. Of the 50 phones, ten have already been sold. What is the probability that two of the ten sold were defective? Give the result to four decimal places.

25. **Tossing a Die** A die is tossed. F is the event of tossing an even number. G is the event of tossing a number less than 2. Are F and G mutually exclusive events?

26. **Tossing a Coin** A coin is tossed twice. F is the event that it lands tails up on the second throw. G is the event that it lands on the same face both times. Are F and G mutually exclusive events?

27. **Telephone Calls** In the hours from 6:00 PM to midnight, the telephone in a girl's dormitory is busy 105 minutes on the average. What is the probability of not getting a busy signal when phoning the dorm?

28. **Foxes and Mice** A log lies across a small stream. A fox and ten mice use the log to cross the stream at night. When the fox arrives at the log, he waits to prey on any mice that arrive after him. His capacity is five mice; if he catches five he leaves. Suppose all orders of arrival (one at a time) are equally likely. Find the probability that **(a)** four mice will be eaten, **(b)** five mice will be eaten, **(c)** no mice will be eaten.

29. **Long Place Name** The longest place name in the world is *Llanfairpwllgwyngyllgogerychwyrndrobwll-llantysiliogogogoch.* It is a small village in North Wales. The name means (in Welsh) the Church of Mary, in a hollow, by a white hazel, near the fierce whirlpool, close to the Church of Tysilio, near to the red cave. There are many long place names in Wales. They describe the area around a village or characteristics of the village. Welsh is the most widely spoken of the ancient Celtic languages. The double "ells" (*ll*) that appear in the name (there are five of them) form a single letter in the Welsh alphabet. Two of these *ll*s appear in *llan,* the Welsh word for church. The first is Llanfair, the Church of Mary, the second Llantysilio, the Church of St. Tysilio, a Celtic saint. In the Welsh alphabet, *ch* is also a single letter.

   A letter (in the English alphabet) is selected at random from the letters that make up this name. Determine the probability that it is **(a)** *g,* **(b)** *l,* **(c)** *a* or *y,* **(d)** *l, g,* or *a,* **(e)** an English vowel, **(f)** a Welsh vowel (they are *a, e, i, o, u, w, y*).

   A letter in the Welsh alphabet is selected at random. What is the probability that it is **(g)** *ll,* **(h)** *ch.* (The shortest place name is *Y,* the name of a French village.)

Chapter 9 Test

1. Evaluate each of the following expressions:
   (**a**) ${}_8P_2$ (**b**) ${}_9C_3$
2. **Different Routes** There are three different direct roads from Gainesville to Ocala and four different roads from Ocala to DeLand. There are two different roads from DeLand to Daytona. How many different routes are there from Gainesville to Daytona that pass through Ocala and DeLand?
3. **Student Committees** A senior, junior, sophomore, and freshman are to be selected from a group of four seniors, five juniors, three sophomores, and two freshmen. In how many ways can the selection be made?
4. **Book Arrangements** Twelve different books are to be arranged on a shelf. In how many ways can they be arranged?
5. **Awarding Contracts** In how many ways can two different contracts be awarded to eight different firms?
6. **License Plates** How many license plates can be made using first three letters, then three numbers between 0 and 9, and finally a single letter?
7. **Test Questions** A test contains ten questions. A student is instructed to answer seven of the questions. (**a**) In how many ways can this be done? (**b**) In how many ways can this be done if four of the first six questions must be answered and three of the last four questions must be answered?
8. **Committee Selections** A president, vice president, secretary, treasurer, and financial secretary are to be selected from a group of 20 people. In how many ways can this be done?
9. **Tossing a Die** A die is tossed. What is the probability that it falls on (**a**) 1, (**b**) an odd number, (**c**) a number less than 5?
10. **Selecting Balls** A box contains two red balls and three blue balls. Two balls are drawn in succession without replacement. Construct a sample space. Determine the probability of drawing (**a**) a red and then a blue ball (**b**) a red and a blue ball in any order.
11. **Job Selection** Out of 20 people who apply for jobs, 12 are men and eight are women. If six people are hired at random, what is the probability that five men and one women will be hired. Give the result to four decimal places.
12. **Defective Phones** In a supply of 120 cellular phone sets, six were known to be defective. Of the 120 sets, ten have already been sold. What is the probability that three of the ten sold were defective? Give the result to four decimal places.
13. **Tossing a Die** A die is tossed. F is the event of tossing an odd number. G is the event of tossing a number greater than 4. Are F and G mutually exclusive events?
14. **Getting off an Elevator** A building with 16 floors is serviced by an elevator. If five people use the elevator at once, determine the probability that they all get off at (**a**) the same floor, (**b**) the tenth floor, (**c**) different floors.

CUMULATIVE TEST CHAPTERS 6, 7, 8, 9

1. Solve the following system of equations using the method of substitution, and verify the answers using a graph.

$$x - 2y = 10$$
$$3x + y = 2$$

2. Solve the following system of equations using the method of Gauss–Jordan elimination:

$$x + 2y + 3z = 3$$
$$x + 3y + 5z = 4$$
$$2x + 7y + 11z = 8$$

3. Solve the following system of nonlinear equations. Verify your answer using a graph.

$$5x^2 + y^2 = 21$$
$$2x + y = 3$$

4. Give the specified elements of the following matrix:

$$A = \begin{bmatrix} 5 & 2 & 0 & -4 & 7 \\ 8 & -6 & 3 & 0 & 1 \\ 3 & 9 & 7 & -5 & 4 \\ 0 & 6 & -9 & -2 & 3 \end{bmatrix}$$

(a) a_{24} **(b)** a_{12} **(c)** a_{45}

5. Perform the specified matrix operations using the matrices A, B, and C.

$$A = \begin{bmatrix} 1 & 0 \\ -2 & 4 \end{bmatrix}, \quad B = \begin{bmatrix} 2 & -1 \\ 3 & 0 \end{bmatrix},$$
$$C = \begin{bmatrix} 1 & 0 & -1 \\ 2 & 1 & 1 \end{bmatrix}$$

(a) $A + 2B$ **(b)** AC

6. Write the following system of equations in matrix form; then solve using the matrix inverse method.

$$x + 3y = 13$$
$$2x + 5y = 21$$

7. Compute the following determinant.

$$\begin{vmatrix} -7 & 2 & 0 \\ 5 & -9 & 0 \\ 3 & 7 & 2 \end{vmatrix}$$

8. Determine the first four terms of the following sequence:

$$a_n = 3n^2 - 2n$$

9. Determine the partial sum S_7 for the arithmetic sequence $a_n = 3n - 1$.

10. If $a_1 = 2$ and $d = -3$ for an arithmetic sequence, find the sixth term.

11. The sum of the first six terms of an arithmetic sequence is 48 and the common difference is 3. Determine the first term of the sequence.

12. If $a_1 = 3$ and $r = 2$ for a geometric sequence, find the seventh term.

13. Find the partial sum S_5 of the geometric sequence $a_n = 3(2)^{n-1}$.

14. Evaluate the following factorials and binomial coefficients:

(a) $6!$ **(b)** $\binom{7}{3}$ **(c)** $\binom{250}{248}$ **(d)** ${}_7P_3$ **(e)** ${}_8C_5$

15. Determine the fourth term in the binomial expansion of $(2x - y)^9$.

16. **Telephone Numbers** How many seven-digit telephone numbers can be formed if 0 or 1 cannot be used for the first digit?

17. **Student Committees** Four seniors, three juniors, two sophomores, and two freshmen are to be selected for a committee from a group of eight seniors, ten juniors, 12 sophomores, and 14 freshmen. In how many ways can the selection be made?

18. **Tossing a Die** A die is tossed. What is the probability that it falls on **(a)** 6, **(b)** an odd number less than 5, **(c)** a number greater than 2?

19. **Getting into a Class** A popular class has a limit of 25 students, but 57 students want to get into the class. Mary and John (good friends) both want to get into the class. Assuming the selection is random, what is the probability that Mary gets into the class but John does not? Give the answer to four decimal places.

20. **Scholarships** Forty-five students apply for eight scholarships. Of the 45 students, 28 are female and 17 are male. What is the probability that four female and four male students are selected? Give the answer to four decimal places.

21. **Production Scheduling** A company manufactures two types of washing machines, the Swirl and Rapid. It takes 6 hours and 4 hours in labor time to manufacture the Swirl and the Rapid, respectively. The cost of manufacturing a Swirl is \$550,

whereas the cost of manufacturing a Rapid is \$700. The company has 2360 hours of labor and \$308,000 in manufacturing costs available each week. The profit is \$55 on each Swirl and is \$50 on each Rapid. What should the weekly production schedule be to ensure maximum profit?

22. **Group Analysis** The following table represents information obtained from a questionnaire given to a group of people. Construct the digraph that describes the leadership structure within the group. Use the distance matrix to rank the members according to their influence on the group.

Group Member	Person Whose Opinion Valued
M_1	M_2
M_2	M_3
M_3	M_5
M_4	M_3
M_5	M_1 and M_4

APPENDIX A

Conic Sections

• Conic Sections • Automobile Headlights • Orbits of Planets • Parabola
• Ellipse • Hyperbola • General Forms

CONIC SECTIONS

In this section, we discuss curves that are obtained by cutting right circular cones at various angles with planes. For this reason, the curves are called **conic sections.** There are three classes of such curves: parabolas, ellipses, and hyperbolas. We have already discussed the parabola. We shall now see how its equation can be derived from a geometrical definition. The curves are illustrated in Figure A.1.

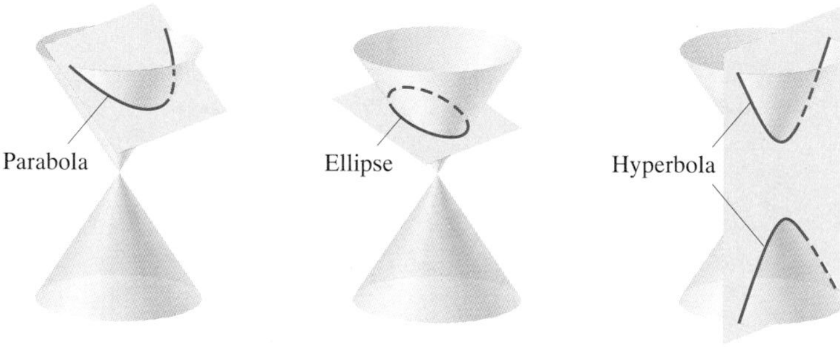

Figure A.1

While the Greeks studied conic sections many centuries ago their importance remains to this date. The paths of projectiles for example, are parabolas. The orbits of planets and artificial satellites are ellipses, and the paths of certain subatomic particles within the nuclei of atoms are hyperbolas.

AUTOMOBILE HEADLIGHTS

Reflecting backgrounds for automobile headlights are made to be parabolic in shape. The bulb is located at the point we call the focus of the parabola so that the light rays are then reflected in parallel beams (Figure A.2).

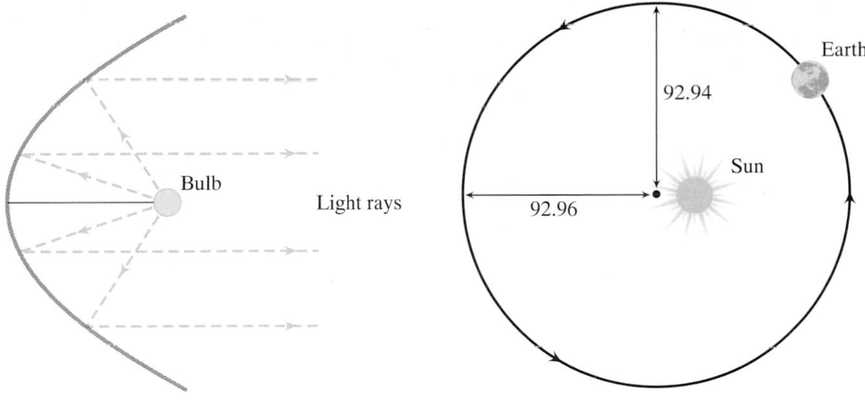

Figure A.2 **Figure A.3**

Orbits of Planets

The orbits of the planets about the Sun are ellipses, and the orbits of comets are parabolas, ellipses, or hyperbolas, with the Sun located at a focus of the ellipse. The equation of the orbit of the Earth has been determined to be the ellipse

$$\frac{x^2}{92.96^2} + \frac{y^2}{92.94^2} = 1$$

where the units are millions of miles. The orbit is almost a circle (Figure A.3).

The orbit of Halley's comet in the same units is the ellipse

$$\frac{x^2}{2{,}778{,}220} + \frac{y^2}{173{,}640} = 1$$

Einstein, in the early part of this century, predicted a very slight rotation of the elliptical orbits of the planets with his theory of relativity. Following that prediction, astronomers discovered that there was in fact such a rotation by studying the orbit of the planet Mercury. This observation is one of the main evidences in support of the Theory of General Relativity.

Each one of the conic sections is defined in terms of fixed distances from points or lines. We start with the geometric definition of the parabola.

Parabola

Parabola A **parabola** is the set of all points in a plane that are equidistant from a fixed point (called the **focus**) and a fixed line (called the **directrix**).

Parabola with Vertex at the Origin and y-Axis as Axis

Let us derive the equation of the parabola having the convenient point $F(0, p)$ on the y-axis as focus and the line $y = -p$ as directrix (Figure A.4). Let $A(x, y)$ be an arbitrary point on the graph. The distance of A from the focus is equal to the distance of A from the directrix. Thus,

$$d(A, F) = d(A, B)$$
$$\sqrt{(x-0)^2 + (y-p)^2} = \sqrt{(x-x)^2 + (y+p)^2}$$

Square both sides, and simplify

$$x^2 + (y-p)^2 = (y+p)^2$$
$$x^2 + y^2 - 2py + p^2 = y^2 + 2py + p^2$$
$$x^2 = 4py$$

The equation of this parabola is

$$x^2 = 4py$$

This parabola has vertex (0, 0) and vertical axis $x = 0$.

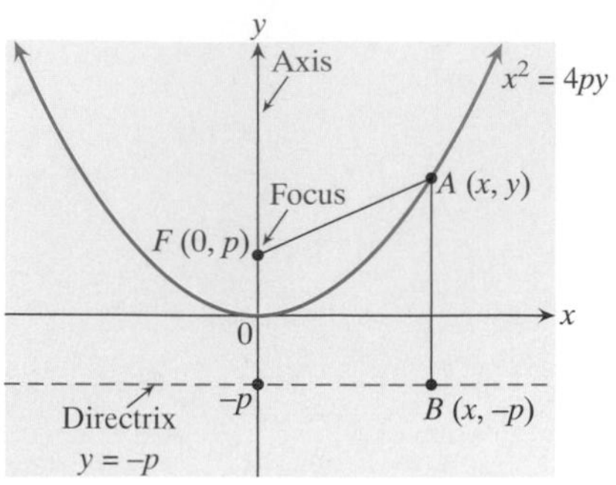

Figure A.4 Parabola with vertex (0, 0) and vertical axis $x = 0$.

Parabola with Vertex at the Origin and x-Axis as Axis

The equation of a parabola that has focus $(p, 0)$ on the x-axis and directrix $x = -p$ can be obtained in a similar manner. Its equation is $y^2 = 4px$. This parabola has vertex (0, 0) and horizontal axis $y = 0$ (Figure A.5).

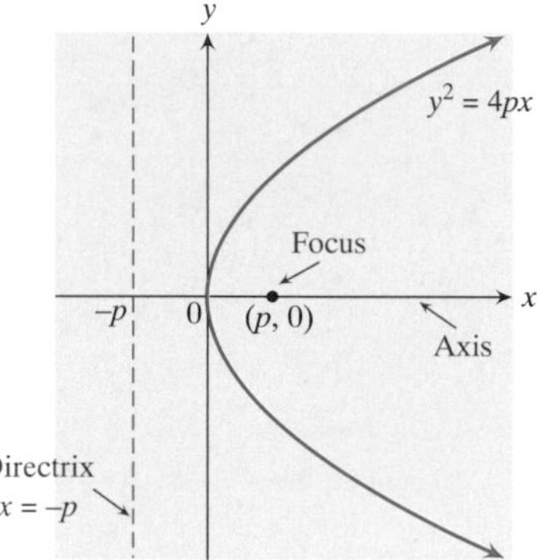

Figure A.5 Parabola with vertex (0, 0) and horizontal axis $y = 0$.

Standard Forms of the Equation of a Parabola

The **standard forms** of the equation of a parabola with vertex at the origin are

$x^2 = 4py$, with vertical axis $x = 0$, focus $(0, p)$, and directrix $y = -p$.
$y^2 = 4px$, with horizontal axis $y = 0$, focus $(p, 0)$, and directrix $x = -p$.

Example 1 Analyzing a Parabola

Find the vertex, axis, focus, and directrix, and sketch the graph of the parabola $x^2 = 8y$.

SOLUTION On comparison with the standard form $x^2 = 4py$, $p = 2$. The vertex is (0, 0), the axis is the line $x = 0$, the focus is the point (0, 2), and the directrix is the line $y = -2$ (Figure A.6).

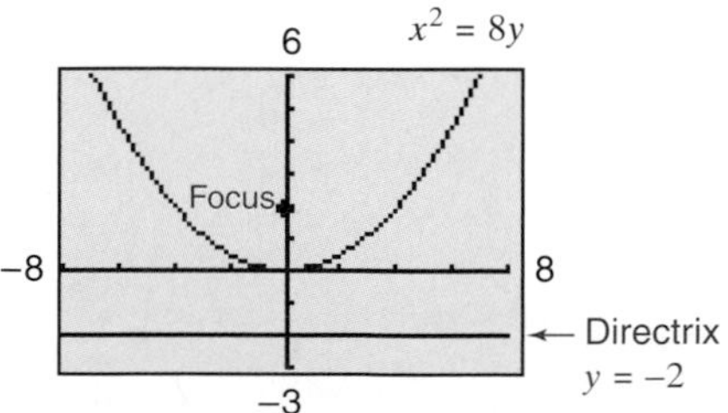

Figure A.6 Enter equation as $Y1 = X\wedge 2/8$.

Self-Check 1 Find the vertex, axis, focus, and directrix, and sketch the graph of the parabola $y^2 = 12x$.

We now discuss the other two conic sections, namely the ellipse and hyperbola.

ELLIPSE

> ***Ellipse*** An **ellipse** is the set of all points in a plane the sum of whose distances from two fixed points, called **foci,** is constant.

We introduce this discussion by looking at ellipses that have both foci, F_1 and F_2, centrally located on either the x-axis or the y-axis (Figure A.7). If P is an arbitrary point on the ellipse, then $d(F_1P) + d(F_2P) =$ constant. The points A, B, C, and D are called the **vertices** of the ellipse. The line segment AB is called the **major axis** of the ellipse, and CD is the **minor axis.** The midpoint of the major axis is called the **center** of the ellipse. The centers of these ellipses are at the origin.

Major axis horizontal

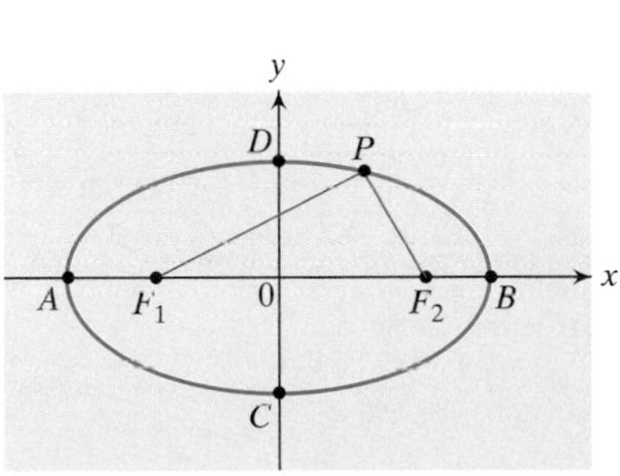

$d(F_1P) + d(F_2P) =$ **constant.**
Major axis AB**, minor axis** CD**.**
Foci F_1, F_2**, vertices** A, B, C, D**.**

Major axis vertical

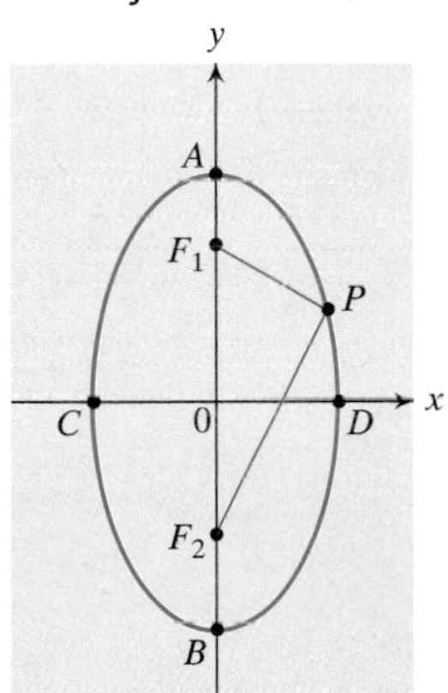

$d(F_1P) + d(F_2P) =$ **constant.**
Major axis AB**, minor axis** CD**.**
Foci F_1, F_2**, vertices** A, B, C, D**.**

Figure A.7 Ellipse.

The equations of these ellipses can be found by applying the distance formula to the definition, as in the case of the parabola. Consider the ellipse with a horizontal major axis. Let the foci be at the points $F_1(-c, 0)$ and $F_2(c, 0)$, $c > 0$. The standard form of the equation of this ellipse is

$$\frac{x^2}{a^2} + \frac{y^2}{b^2} = 1, \qquad a > b > 0$$

where $c^2 = a^2 - b^2$. The vertices A and B are the points on the ellipse where $y = 0$. Letting $y = 0$ gives $x = -a$ or $x = a$. The vertices A and B are thus $(-a, 0)$ and $(a, 0)$. Similarly C and D are found to be $(0, -b)$ and $(0, b)$.

The standard form of the equation of the ellipse with major axis vertical and foci $F_1(0, -c)$ and $F_2(0, c)$ is $\frac{x^2}{b^2} + \frac{y^2}{a^2} = 1, a > b > 0$, with $c^2 = a^2 - b^2$. The vertices are $(-b, 0)$, $(b, 0)$, $(0, -a)$, $(0, a)$. These results are summarized in Figure A.8.

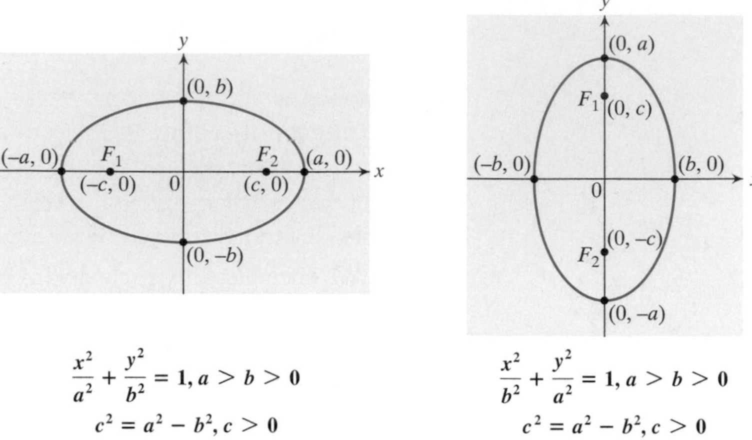

$$\frac{x^2}{a^2} + \frac{y^2}{b^2} = 1, a > b > 0$$
$$c^2 = a^2 - b^2, c > 0$$

$$\frac{x^2}{b^2} + \frac{y^2}{a^2} = 1, a > b > 0$$
$$c^2 = a^2 - b^2, c > 0$$

Figure A.8 Standard form of the equation of an ellipse.

Example 2 Analyzing an Ellipse

Find the center, vertices, and foci and sketch the graph of the ellipse

$$\frac{x^2}{25} + \frac{y^2}{9} = 1$$

SOLUTION This equation is of the standard form

$$\frac{x^2}{a^2} + \frac{y^2}{b^2} = 1$$

with $a^2 = 25$ and $b^2 = 9$. Thus, the major axis is horizontal, and the center is $(0, 0)$, with $a = 5$ and $b = 9$. The vertices are $(-5, 0)$, $(5, 0)$, $(0, -3)$, $(0, 3)$. We compute c to find the foci.

$c^2 = a^2 - b^2 = 25 - 9 = 16$, giving $c = 4$

The foci are $F_1(-4, 0)$, $F_2(4, 0)$ (Figure A.9).

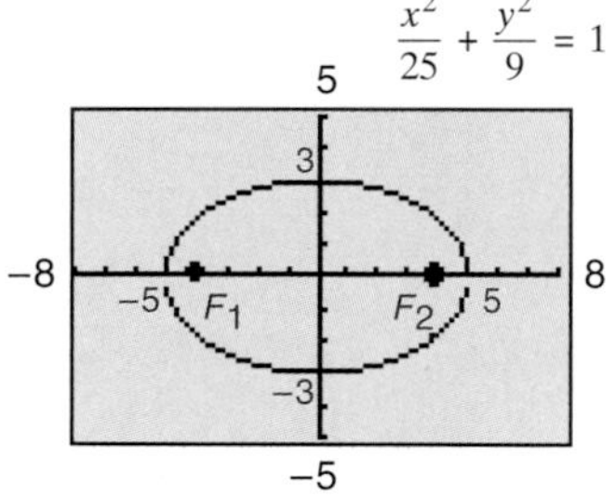

Figure A.9 Enter the equation as $Y1 = 3\sqrt{(1 - X\wedge 2/25)}$, $Y2 = -3\sqrt{(1 - X\wedge 2/25)}$.

Self-Check 2 Find the center, vertices, and foci and sketch the graph of the equation $4x^2 + y^2 = 16$. (Rewrite the equation in the standard form.)

$$\frac{x^2}{b^2} + \frac{y^2}{a^2} = 1$$

HYPERBOLA

> ***Hyperbola*** A **hyperbola** is the set of all points in a plane the difference of whose distances from two fixed points (called foci) is constant.

As in the discussion of ellipses, let us look at hyperbolas that have both foci, F_1 and F_2, centrally located on either the x-axis or the y-axis (Figure A.10). If P is an arbitrary point on the hyperbola, then $d(F_1P) - d(F_2P) =$ constant. The points A and B are called the **vertices** of the hyperbola. The line segment AB is called the **transverse axis** of the hyperbola. The midpoint of this axis is called the **center** of the hyperbola. The centers of these hyperbolas are at the origin.

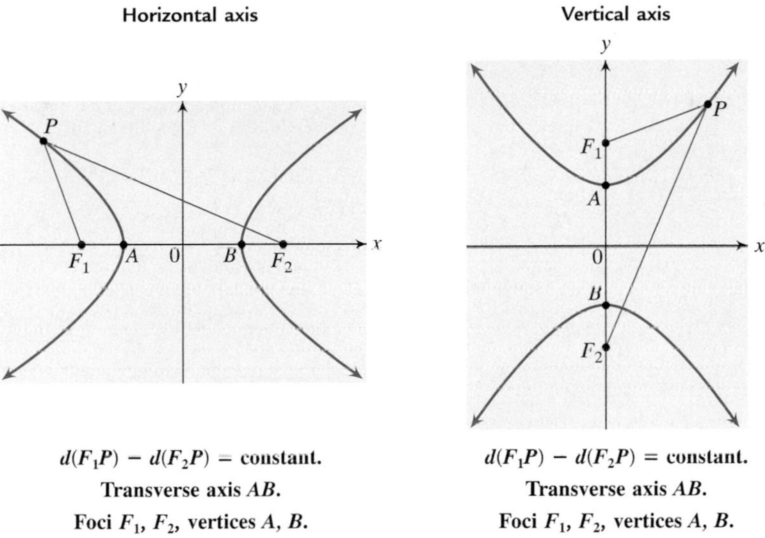

Figure A.10 Hyperbola.

The equations of these hyperbolas can be found by using the distance formula. Consider the hyperbola with a horizontal axis. Let the foci be at the points $F_1(-c, 0)$ and $F_2(c, 0)$, $c > 0$. The standard form of the equation of this hyperbola is

$$\frac{x^2}{a^2} - \frac{y^2}{b^2} = 1, \qquad a > 0, \qquad b > 0$$

where $c^2 = a^2 + b^2$. The vertices A and B are the points on the ellipse where $y = 0$. Letting $y = 0$ gives $x = -a$ or $x = a$. The vertices A and B are thus $(-a, 0)$ and $(a, 0)$. The two distinct curves of a hyperbola are called **branches.** The branches get closer and closer to but never intersect two lines called **asymptotes.** We arrive at the equations of the asymptotes as follows. Solve the equation of the hyperbola for y to get

$$y = \pm\frac{b}{a}\sqrt{x^2 - a^2}$$

Rewrite as

$$y = \pm \frac{bx}{a}\sqrt{1 - \frac{a^2}{x^2}}$$

As the magnitude of x gets larger in comparison to a, the term $\frac{a^2}{x^2}$ gets smaller, and the hyperbola gets closer to the lines $y = \pm\frac{b}{a}x$. The equations of the asymptotes are $y = \pm\frac{b}{a}x$.

The standard form of the equation of the hyperbola with vertical axis and foci $F_1(0, -c)$ and $F_2(0, c)$ is $\frac{y^2}{a^2} - \frac{x^2}{b^2} = 1$, $a > 0$, $b > 0$, with $c^2 = a^2 + b^2$. The vertices are $(0, -a)$, $(0, a)$ and the asymptotes are $y = \pm\frac{a}{b}x$. These results are summarized in Figure A.11.

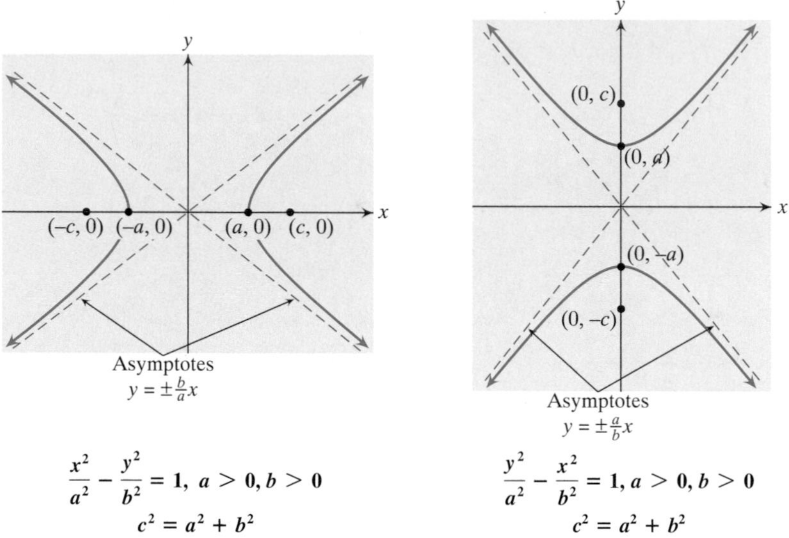

$$\frac{x^2}{a^2} - \frac{y^2}{b^2} = 1,\ a > 0, b > 0 \qquad c^2 = a^2 + b^2$$

$$\frac{y^2}{a^2} - \frac{x^2}{b^2} = 1,\ a > 0, b > 0 \qquad c^2 = a^2 + b^2$$

Figure A.11 Standard form of the equation of a hyperbola.

Example 3 Analyzing a Hyperbola

Find the center, vertices, foci, and asymptotes and sketch the graph of the hyperbola

$$\frac{x^2}{4} - \frac{y^2}{1} = 1$$

SOLUTION This equation is of the standard form $\frac{x^2}{a^2} - \frac{y^2}{b^2} = 1$ with $a^2 = 4$ and $b^2 = 1$. Thus, the axis is horizontal and the center is $(0, 0)$, with $a = 2$ and $b = 1$.

The vertices are $(-2, 0)$ and $(2, 0)$. We compute c to find the foci.

$c^2 = a^2 + b^2 = 4 + 1 = 5$, giving $c = \sqrt{5}$

The foci are $(-\sqrt{5}, 0)$ and $(\sqrt{5}, 0)$. The asymptotes are the lines $y = \pm\frac{b}{a}x$, giving $y = \frac{1}{2}x$ and $y = -\frac{1}{2}x$ (Figure A.12).

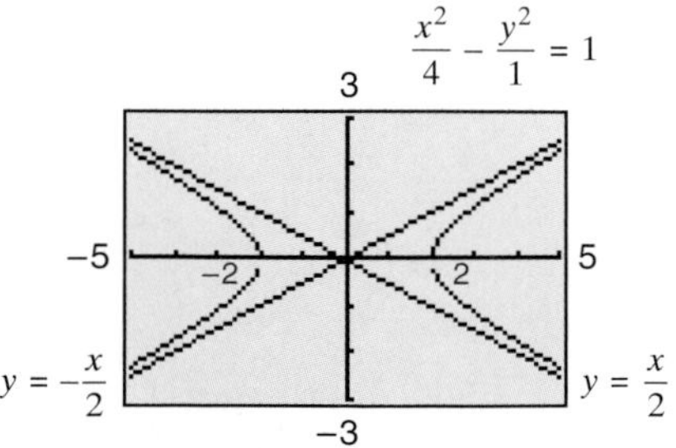

Figure A.12 Enter the equation of the hyperbola as $Y1 = \sqrt{}(X \wedge 2/4 - 1)$, $Y2 = -\sqrt{}(X \wedge 2/4 - 1)$. Enter the asymptotes as $Y3 = X/2$ and $Y4 = -X/2$.

General Forms

All the conic sections that we have discussed have had their centers at the origin. Conic sections having centers at points other than the origin can be interpreted using horizontal and vertical shifts.

Example 4 Analyzing a Parabola

Show that the graph of $(x - 6)^2 = 8(y + 3)$ is a parabola. Find the axis, vertex, focus, and directrix and sketch its graph.

SOLUTION The graph of $(x - 6)^2 = 8(y + 3)$ can be obtained from that of the parabola $x^2 - 8y$ using shifts. The graph is thus a parabola. The equation $x^2 = 8y$ is in standard form. On comparison with $x^2 = 4py$, we see that $p = 2$. This parabola has vertex $(0, 0)$, axis $x = 0$, focus $(0, 2)$, and directrix $y = -2$ (graph of Figure A.13).

To get the graph of

$$(x - 6)^2 = 8(y + 3)$$

shift the parabola $x^2 = 8y$, 6 units to the right, 3 units down. The axis, vertex, focus, and directrix all make corresponding shifts. The axis becomes $x = 6$, and the vertex is $(6, -3)$. The focus is $(6, -1)$, and the directrix is $y = -5$ (Figure A.13).

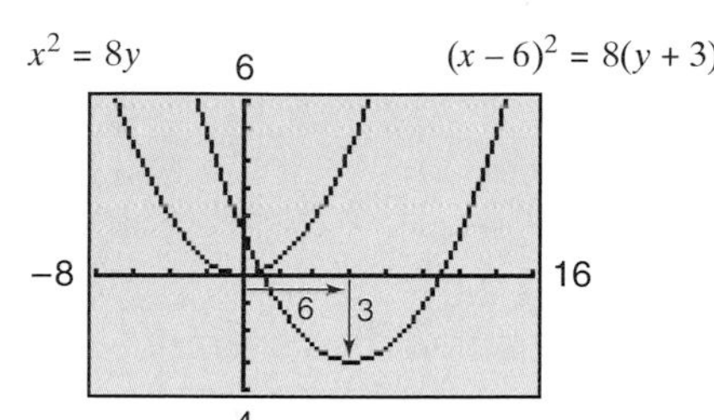

Figure A.13 Enter the equation as $Y1 - X\wedge 2/8$ and $Y2 = (X - 6)\wedge 2/8 - 3$.

Example 5 Analyzing a Hyperbola

Show that the graph of

$$\frac{(x + 2)^2}{16} - \frac{(y - 3)^2}{4} = 1$$

is a hyperbola. Find the center, vertices, and asymptotes, and sketch its graph.

SOLUTION This graph can be obtained from that of the hyperbola

$$\frac{x^2}{16} - \frac{y^2}{4} = 1$$

using shifts. It is thus a hyperbola. The equation

$$\frac{x^2}{16} - \frac{y^2}{4} = 1$$

is in standard form. On comparison with

$$\frac{x^2}{a^2} - \frac{y^2}{b^2} = 1$$

we see that $a^2 = 16$ and $b^2 = 4$, thus, $a = 4$ and $b = 2$. The major axis is horizontal, the center is $(0, 0)$, and the vertices are $(-4, 0)$ and $(4, 0)$. We get the graph shown in Figure A.14a. To get the graph of

$$\frac{(x+2)^2}{16} - \frac{(y-3)^2}{4} = 1$$

shift this graph 2 units to the left and 3 units up (Figure A.14b). The center is $(-2, 3)$ and the vertices are $(-6, 3)$ and $(2, 3)$.

The asymptotes of

$$\frac{x^2}{16} - \frac{y^2}{4} = 1$$

are the lines $y = \pm(b/a)x$ with $a = 4$ and $b = 2$. They are the lines $y = \pm\frac{1}{2}x$. Shift these lines 2 units to the left and 3 units up to get the asymptotes of

$$\frac{(x+2)^2}{16} - \frac{(y-3)^2}{4} = 1$$

These asymptotes are $(y - 3) = \pm\frac{1}{2}(x + 2)$. On simplifying these equations, the asymptotes are $y = \frac{1}{2}x + 4$ and $y = -\frac{1}{2}x + 2$.

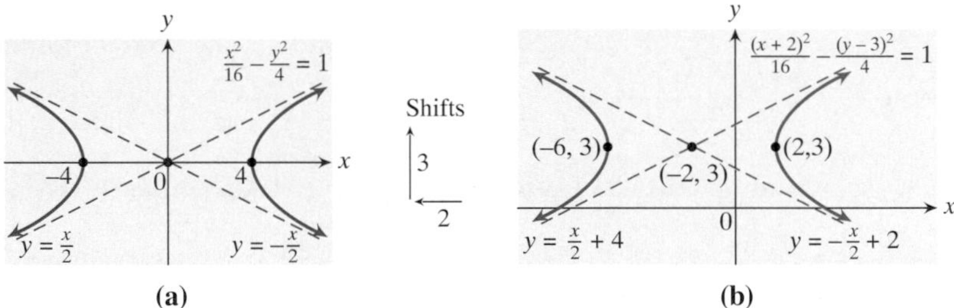

Figure A.14

Self-Check 3 Draw the graphs of

$$\frac{x^2}{16} - \frac{y^2}{4} = 1 \quad \text{and} \quad \frac{(x+2)^2}{16} - \frac{(y-3)^2}{4} = 1$$

on your calculator to verify this result.

SUMMARY OF STANDARD FORMS OF EQUATIONS OF CONIC SECTIONS

The graphs of these more general standard forms can be obtained from the graphs of those centered at the origin by using the following shifts.

Parabola:

$(x - h)^2 = 4p(y - k)$, shift $x^2 = 4py$ until the vertex is at (h, k)
$(y - k)^2 = 4p(x - h)$, shift $y^2 = 4px$ until the vertex is at (h, k)

Ellipse:

$$\frac{(x-h)^2}{a^2} + \frac{(y-k)^2}{b^2} = 1, \text{ shift } \frac{x^2}{a^2} + \frac{y^2}{b^2} = 1 \text{ until the center is at } (h, k)$$

$$\frac{(x-h)^2}{b^2} + \frac{(y-k)^2}{a^2} = 1, \text{ shift } \frac{x^2}{b^2} + \frac{y^2}{a^2} = 1 \text{ until the center is at } (h, k)$$

Hyperbola:

$$\frac{(x-h)^2}{a^2} - \frac{(y-k)^2}{b^2} = 1, \text{ shift } \frac{x^2}{a^2} - \frac{y^2}{b^2} = 1 \text{ until the center is at } (h, k)$$

$$\frac{(y-k)^2}{a^2} - \frac{(x-h)^2}{b^2} = 1, \text{ shift } \frac{y^2}{a^2} - \frac{x^2}{b^2} = 1 \text{ until the center is at } (h, k)$$

If the squared term in any of the standard forms of the conic sections are expanded and simplified, we get an equation of the form

$$Ax^2 + By^2 + Cx + Dy + F = 0$$

where A, B, C, D, and F are real numbers. Conversely, when an equation is given in this form, the method of completing the square can be used to see whether the equation is that of a conic section.

Example 6 Sketching the Graph of a Conic

Sketch the graph of the equation $x^2 + 4y^2 - 2x + 24y + 33 = 0$.

SOLUTION Separate the x and y terms as follows:

$$x^2 - 2x + 4y^2 + 24y + 33 = 0$$
$$(x^2 - 2x) + 4(y^2 + 6y) + 33 = 0$$

Complete the square of both $(x^2 - 2x)$ and $(y^2 + 6y)$. Compensate for the numbers added by adding them also to the right side of the equation. We get

$$(x^2 - 2x + 1^2) + 4(y^2 + 6y + 3^2) + 33 = 1^2 + 4(3^2)$$
$$(x-1)^2 + 4(y+3)^2 + 33 = 37$$

Proceed to arrive at the standard form.

$$(x-1)^2 + 4(y+3)^2 = 4$$
$$\frac{(x-1)^2}{4} + \frac{(y+3)^2}{1} = 1$$

The equation is that of an ellipse with center $(1, -3)$ and horizontal major axis (Figure A.15).

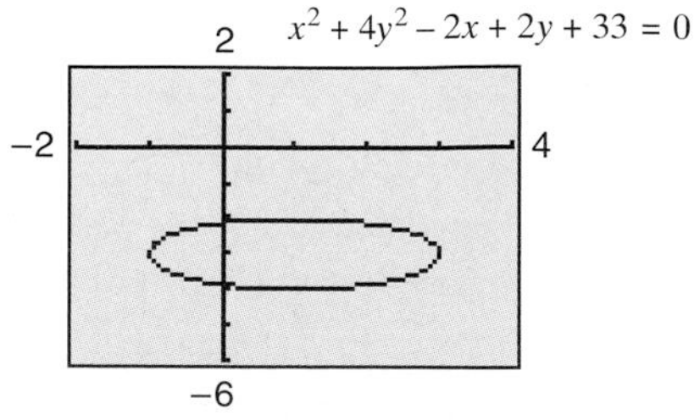

$x^2 + 4y^2 - 2x + 2y + 33 = 0$

Figure A.15 Enter the equation as $Y1 = \sqrt{}(1 - (X - 1)\wedge 2/4) - 3$, $Y2 = -\sqrt{}(1 - (X - 1)\wedge 2/4) - 3$.

Further Ideas 1 Determine whether the following equations are those of conic sections. Identify the conic sections. Graph each equation on your calculator, if possible.

(a) $3x^2 - y^2 + 12x + 2y + 2 = 0$
(b) $x^2 - 8x - 4y - 28 = 0$
(c) $4x^2 + y^2 + 8x - 4y + 12 = 0$

Answers to Self-Check Exercises

1. Vertex (0, 0), axis $y = 0$, focus (3, 0), directrix $x = -3$. See Figure A.16.
2. Major axis vertical. Center (0, 0); vertices $(-2, 0)$, $(2, 0)$, $(0, -4)$, $(0, 4)$; foci $(0, -\sqrt{12})$, $(0, \sqrt{12})$. See Figure A.17.

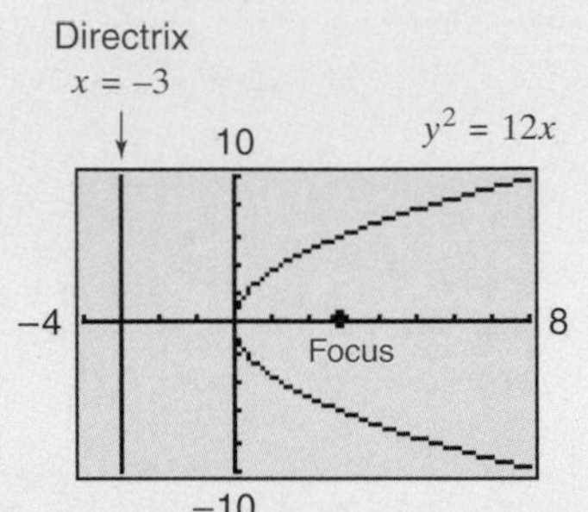

Figure A.16 Enter the equation as $Y1 = \sqrt{12X}$, $Y2 = -\sqrt{12X}$.

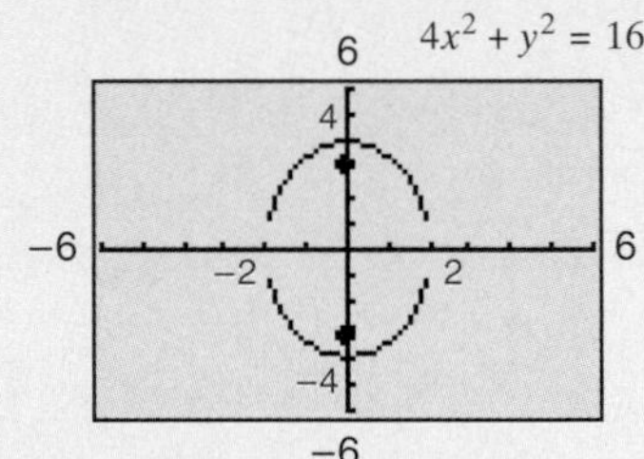

Figure A.17

Answer to Further Ideas Exercise

1. **(a)** Hyperbola

$$\frac{(x + 2)^2}{3} - \frac{(y - 1)^2}{9} = 1$$

(b) Parabola

$$(x - 4)^2 = 4(y + 11)$$

(c) Not a conic section. We get

$$\frac{(x + 1)^2}{1} + \frac{(y - 2)^2}{4} = -1$$

EXERCISES

In Exercises 1–8, find the focus and directrix of each parabola. Sketch a graph of the parabola and its directrix. Support the graph using a graphing calculator.

1. $x^2 = 12y$ **2.** $x^2 = -4y$ **3.** $y^2 = 8x$
4. $y^2 = -10x$ **5.** $x^2 - 8y = 0$ **6.** $y^2 + 3x = 0$
7. $2x^2 = -5y$ **8.** $3y^2 - 25x = 0$

In Exercises 9–18, determine the equation of the parabola that has its vertex at the origin and has the given properties.

9. Focus (0, 3) **10.** Focus $(0, -1)$
11. Directrix $y = -1$ **12.** Directrix $y = 5$
13. Focus (6, 0) **14.** Directrix $x = -3$
15. Focus $(-1, 0)$ **16.** Directrix $x = -7$

17. Directrix $y = -\frac{5}{3}$

18. Focus $(0, -\frac{7}{2})$

In Exercises 19–24, find the vertices and foci, and sketch the graph of each ellipse. Support the graphs using a graphing calculator.

19. $\dfrac{x^2}{4} + \dfrac{y^2}{1} = 1$

20. $\dfrac{x^2}{25} + \dfrac{y^2}{4} = 1$

21. $\dfrac{x^2}{9} + \dfrac{y^2}{16} = 1$

22. $\dfrac{4x^2}{9} + \dfrac{y^2}{1} = 1$

23. $\dfrac{x^2}{1} + \dfrac{y^2}{25} = 1$

24. $\dfrac{9x^2}{49} + \dfrac{25y^2}{36} = 1$

In Exercises 25–30, find the vertices, foci, and asymptotes of each hyperbola. Sketch the graph of the hyperbola and its asymptotes. Support the graphs using a graphing calculator.

25. $\dfrac{x^2}{9} - \dfrac{y^2}{1} = 1$

26. $\dfrac{x^2}{25} - \dfrac{y^2}{49} = 1$

27. $\dfrac{x^2}{4} - \dfrac{y^2}{4} = 1$

28. $\dfrac{y^2}{25} - \dfrac{x^2}{1} = 1$

29. $\dfrac{y^2}{4} - \dfrac{x^2}{9} = 1$

30. $\dfrac{9y^2}{16} - \dfrac{25x^2}{36} = 1$

In Exercises 31–36, determine whether the equations represent ellipses or hyperbolas by writing them in standard form (if possible). Find all vertices, foci, and asymptotes.

31. $x^2 + 4y^2 = 16$

32. $9x^2 - 4y^2 = 36$

33. $16y^2 - 9x^2 = 25$

34. $5x^2 + 7y^2 = 35$

35. $x^2 + 9y^2 = -36$

36. $25x^2 - 4y^2 = -100$

In Exercises 37–44, identify each conic section as being a parabola, an ellipse, or a hyperbola and sketch its graph. Find the axis, vertex, focus, and directrix of each parabola; the center, vertices, and foci of each ellipse; and the center, vertices, foci, and asymptotes of each hyperbola. Support the graphs using a graphing calculator.

37. $(x + 5)^2 = 4(y - 3)$

38. $(x - 4)^2 = -8(y + 7)$

39. $(y + 3)^2 = 12(x - 4)$

40. $(y - 5)^2 = -3(x - 2)$

41. $\dfrac{(x - 1)^2}{9} + \dfrac{(y + 4)^2}{4} = 1$

42. $\dfrac{(x + 6)^2}{25} + \dfrac{(y - 2)^2}{49} = 1$

43. $\dfrac{(x - 3)^2}{1} - \dfrac{(y - 2)^2}{9} = 1$

44. $\dfrac{(y + 2)^2}{25} - \dfrac{(x - 5)^2}{16} = 1$

In Exercises 45–52, determine whether the given equations are those of conic sections. Identify the conics. Sketch the graphs of the conics. Verify the graphs using a graphing calculator.

45. $x^2 - 4x + 16 = 4y$

46. $x^2 + 2x + 33 = 8y$

47. $x^2 + 4y^2 - 2x + 8y + 1 = 0$

48. $4x^2 + 9y^2 - 16x + 54y + 61 = 0$

49. $x^2 - 25y^2 + 2x + 100y - 124 = 0$

50. $-4x^2 + 9y^2 - 16x + 54y + 29 = 0$

51. $y - x^2 - 6x - 13 = 0$

52. $3x^2 + 4y^2 + 18x + 8y + 35 = 0$

In Exercises 53–60, find the equations of the parabolas having the given properties. Confirm your answer by drawing a graph of the equation on your calculator to see that it does indeed represent a parabola with the desired properties.

53. Vertex $(1, 4)$, axis $x = 1$, y-intercept 6

54. Vertex $(-3, 2)$, axis $y = 2$, x-intercept -2

55. Vertex $(1, 2)$, vertical axis, passing through the point $(5, 3)$

56. Vertex $(2, -4)$, horizontal axis, passing through the point $(4, -8)$

57. Focus $(1, 3)$ and directrix $x = -1$

58. Focus $(3, -3)$ and directrix $y = 1$

59. Vertex $(2, -2)$ and focus $(2, -1)$

60. Vertex $(-4, 1)$ and directrix $y = -1$

61. Sketch the graph of the ellipse that has vertices $(3, 0)$, $(-3, 0)$, $(0, 7)$, $(0, -7)$. Use the sketch to arrive at the equation of the ellipse.

62. Sketch the graph of the ellipse that has vertices $(7, 1)$, $(1, 1)$, $(4, 3)$, $(4, -1)$. Determine the equation of the ellipse.

63. Sketch the hyperbola that has vertices $(6, 0)$ and $(-6, 0)$, and asymptotes $y = \pm 2x$. Determine the equation of the hyperbola.

64. Sketch the hyperbola that has vertices $(3, 1)$ and $(1, 1)$ and asymptotes $y = 3x - 5$ and $y = -3x + 7$. Determine the equation of the hyperbola.

APPENDIX B

TI-82/TI-83 Calculator Reference

1. Evaluating a Function Evaluate $f(x) = x^3 - 4x + 5$ at $x = 2.7$.

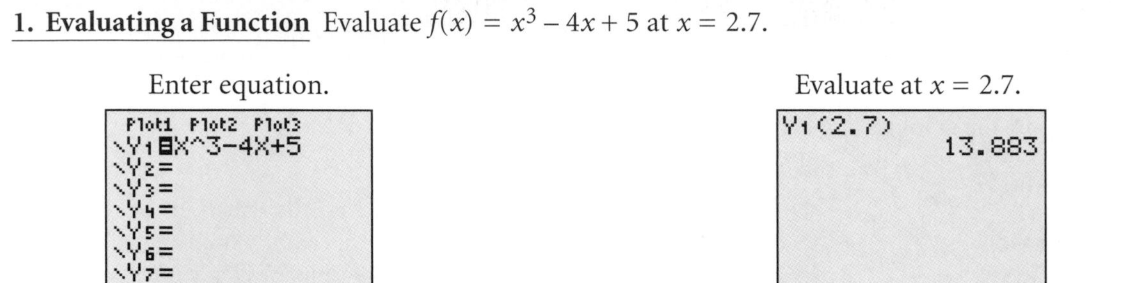

Note: We do not indicate the use of the 2nd key. For example, we use QUIT rather than 2nd MODE. The information is conveyed in this way in a cleaner, more meaningful manner.

2. Table of Values Construct a table of values of $f(x) = x^2 - 7x + 1$ from $x = -5$ to $x = 15$ in increments of 2.

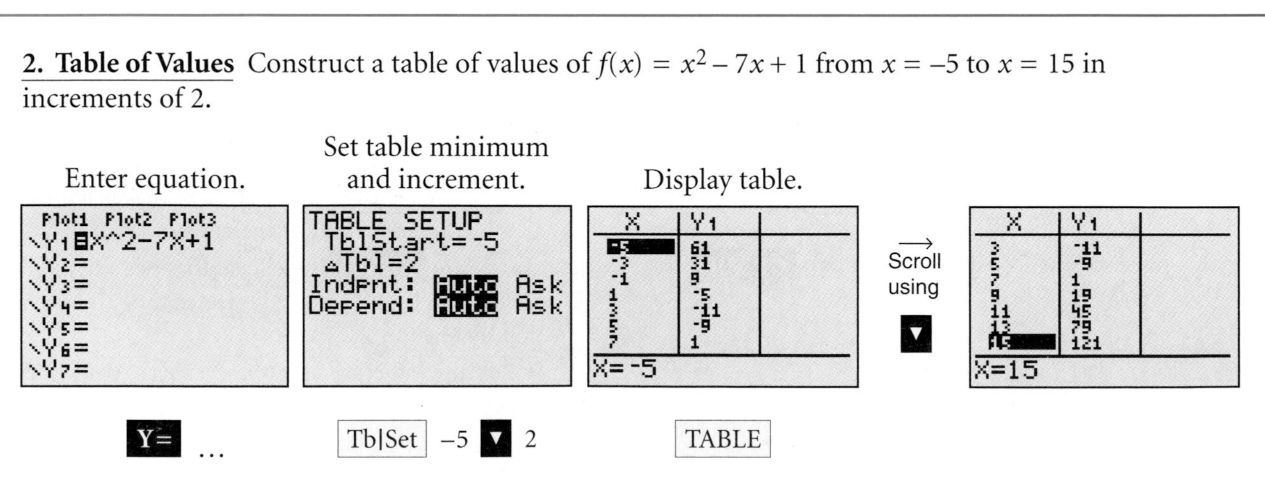

3. Graphing Draw the graph of $y = x^2 - 8x + 20$.

Enter equation.

Y= ...

Select window.

```
WINDOW
 Xmin=-2
 Xmax=8
 Xscl=1
 Ymin=-2
 Ymax=12
 Yscl=2
 Xres=1
```

WINDOW –2 **▼** 8...

Graph.

12

–2 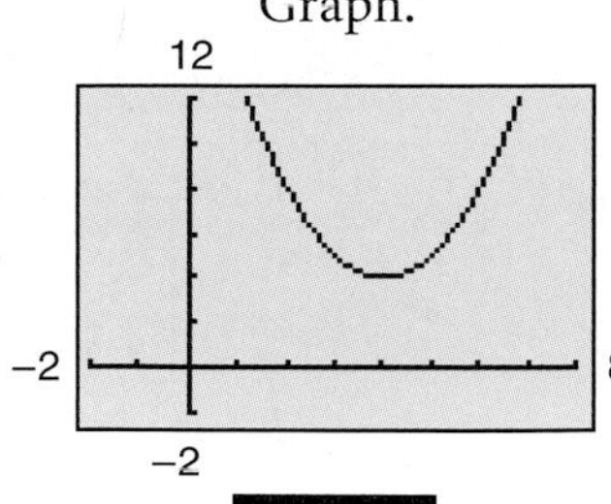 8

–2

GRAPH

4. Special Functions The square root, absolute value, natural exponential, and natural logarithm functions.

Enter equations.

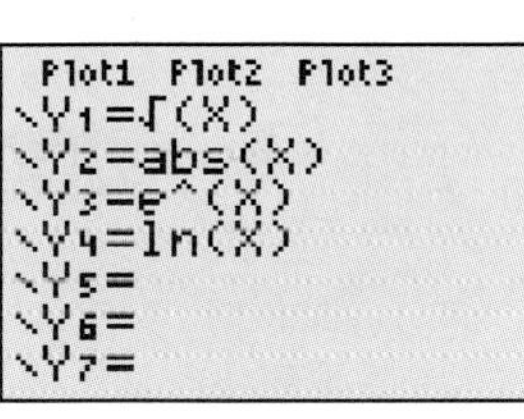

Y= ...

Select a window and display graphs.

5 $f(x) = \sqrt{x}$

–5 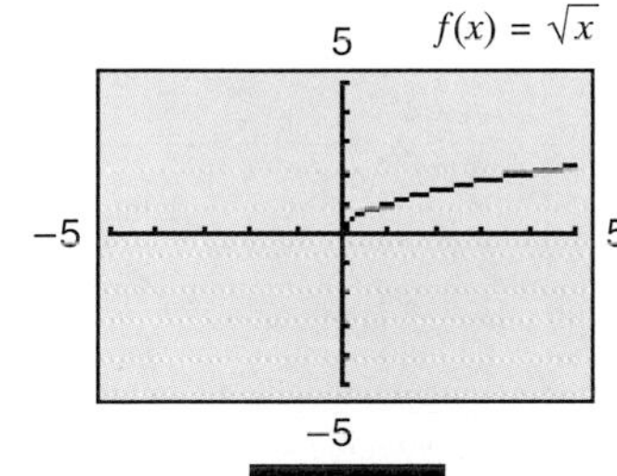 5

–5

GRAPH

5 $f(x) = |x|$

–5 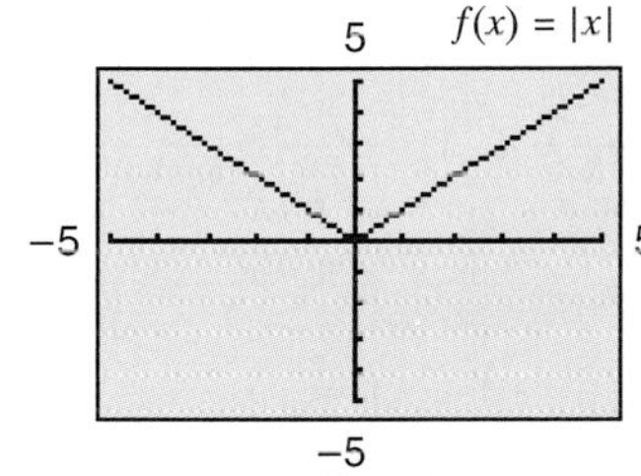 5

–5

GRAPH

5 $f(x) = e^x$

–5 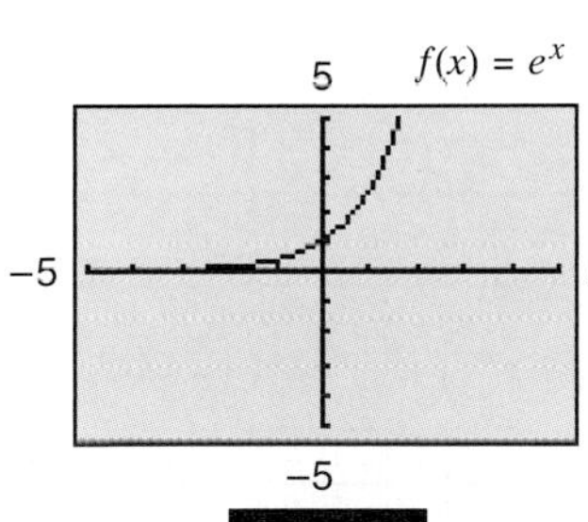 5

–5

GRAPH

5 $f(x) = \ln(x)$

–5 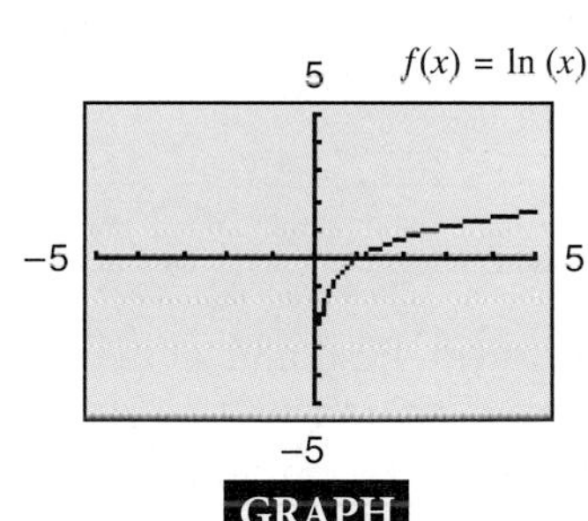 5

–5

GRAPH

Function keys: [√], **MATH** **►** 1 (TI-83), [ABS] (TI-82), [e^x], **LN**

5. Piecewise-Defined Functions Graph $f(x) = \begin{cases} x^2 & \text{if } x \leq 2 \\ x + 4 & \text{if } x > 2 \end{cases}$

Enter two equations.

Plot1 Plot2 Plot3
\Y1=(X^2)/(X≤2)
\Y2=(X+4)/(X>2)
\Y3=
\Y4=
\Y5=
\Y6=
\Y7=

Y= (X^2) **÷** (X TEST 6 2)
(X+4) **÷** (X TEST 3 2)

Select a window and display graph.

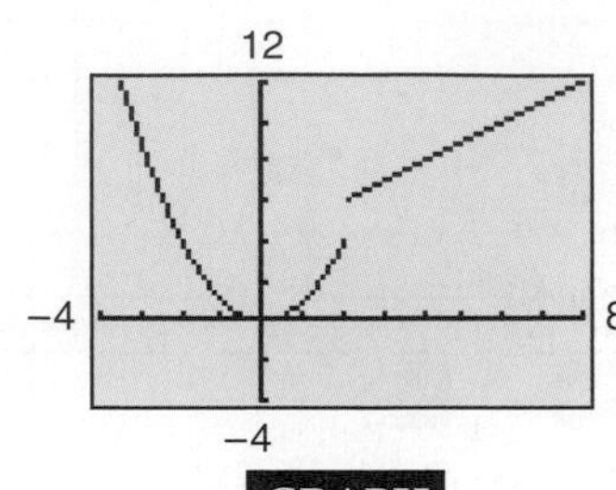

GRAPH

6. Trace/Zoom Zoom in on the vertex of $y = x^2 - 8x + 20$.

Graph.

12
−2
8
−2

GRAPH

Enter trace mode.

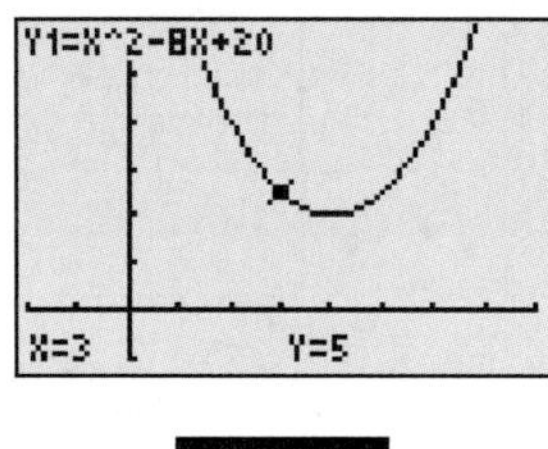

TRACE

Trace to lowest point.

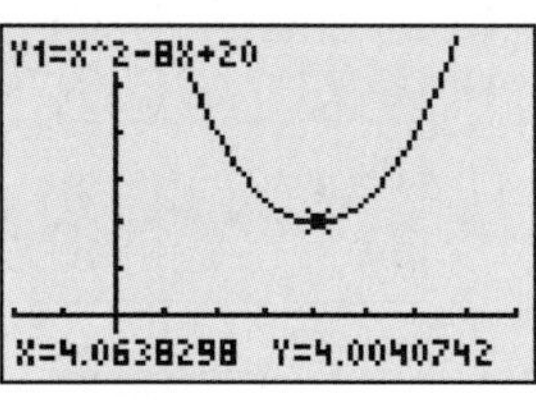

► and ◄

Zoom.

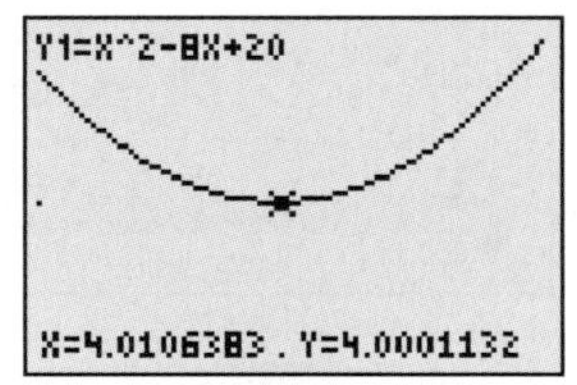

ZOOM 2
ENTER **TRACE**

7. Maximum/Minimum Find a local maximum for $f(x) = x^3 - 7x^2 + 9x + 4$.

Graph.

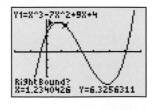

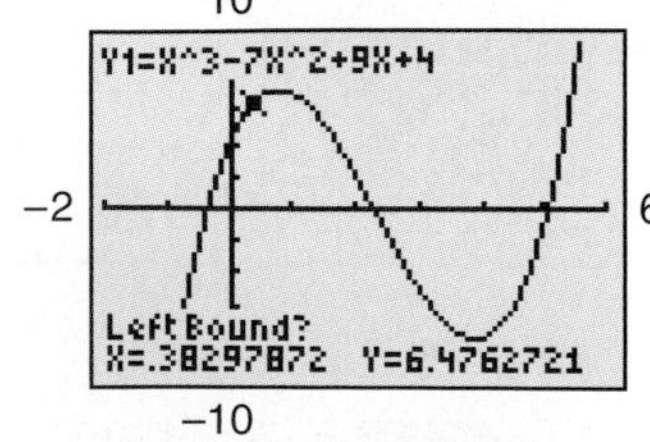

CALC 4 (3 for minimum).
Trace to left of maximum. **ENTER**

Y1=X^3-7X^2+9X+4
Right Bound?
X=1.2340426 Y=6.3256311

Trace to right of maximum. **ENTER**

Y1=X^3-7X^2+9X+4
Guess?
X=.72340426 Y=7.2260097

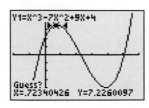

Trace to guess of maximum. **ENTER**

Display maximum.

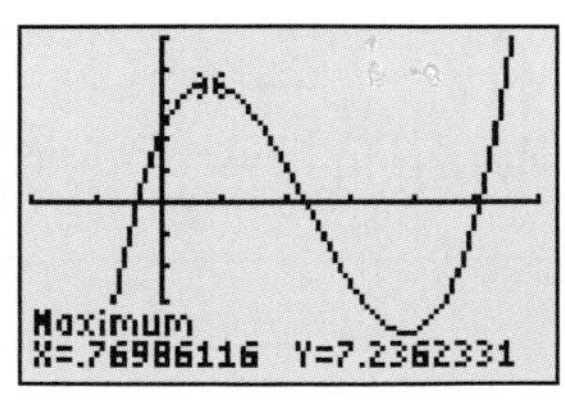

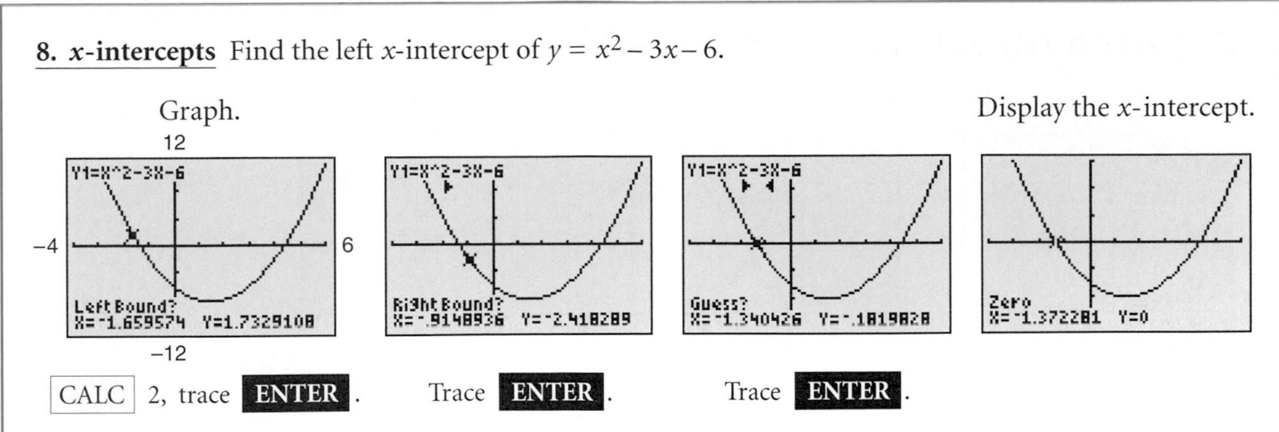
8. x-intercepts Find the left x-intercept of y = x2 – 3x – 6.
Graph.
Display the x-intercept.
12
–4
6
–12
Y1=X^2-3X-6
Left Bound?
X=-1.659574 Y=1.7329108
Right Bound?
X=-.9148936 Y=-2.418289
Guess?
X=-1.340426 Y=-.1819828
Zero
X=-1.372281 Y=0
CALC 2, trace ENTER.
Trace ENTER.
Trace ENTER.

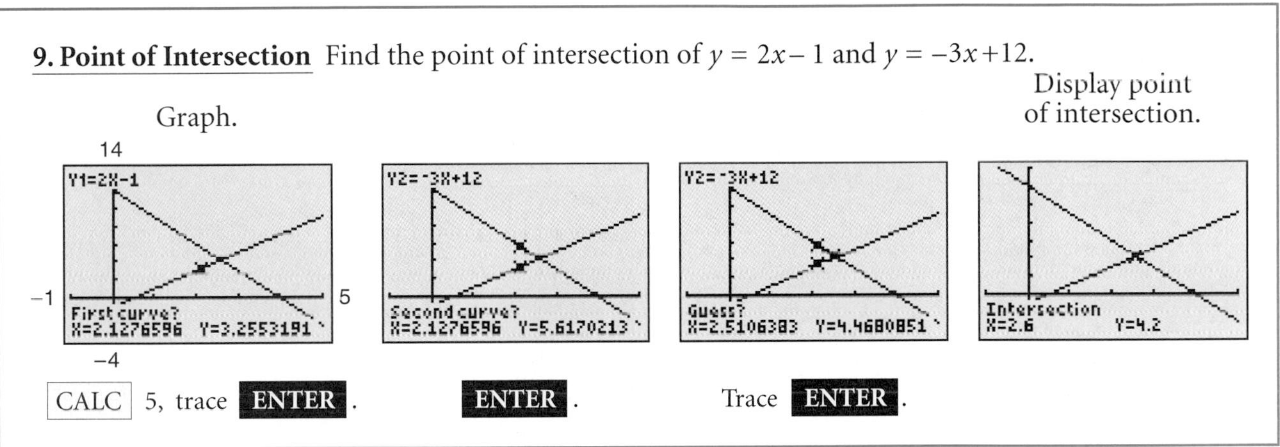
9. Point of Intersection Find the point of intersection of y = 2x – 1 and y = –3x + 12.
Graph.
Display point of intersection.
14
–1
5
–4
Y1=2X-1
First curve?
X=2.1276596 Y=3.2553191
Y2=-3X+12
Second curve?
X=2.1276596 Y=5.6170213
Guess?
X=2.5106383 Y=4.4680851
Intersection
X=2.6 Y=4.2
CALC 5, trace ENTER.
ENTER.
Trace ENTER.

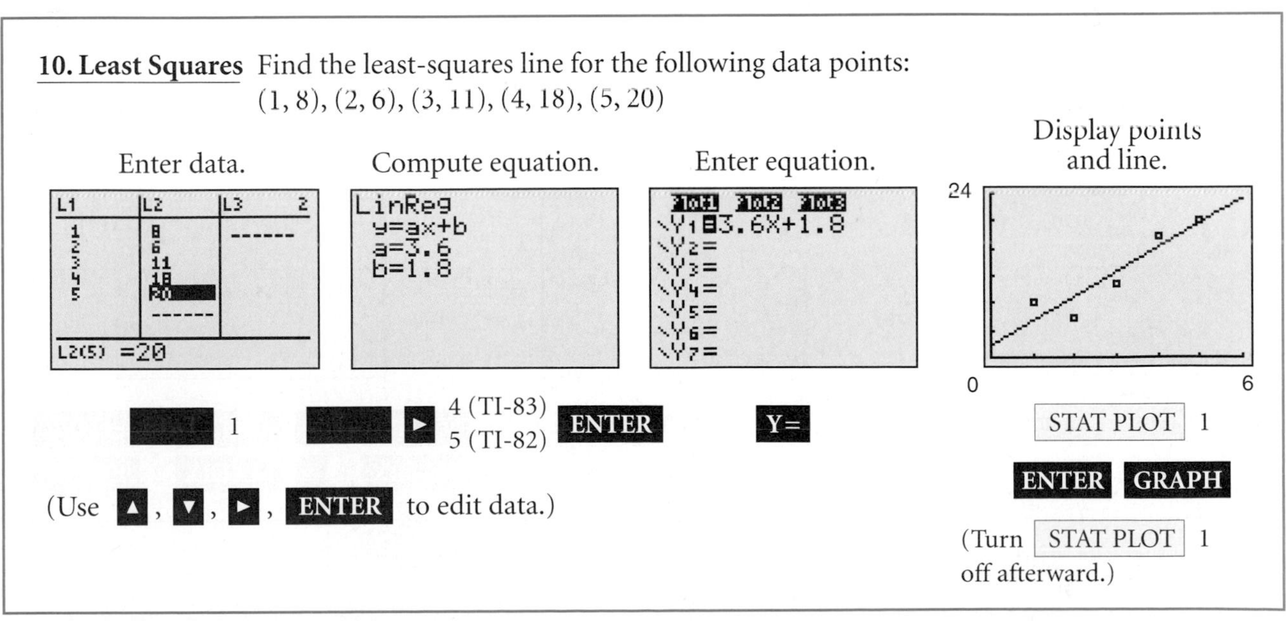
10. Least Squares Find the least-squares line for the following data points:
(1, 8), (2, 6), (3, 11), (4, 18), (5, 20)
Enter data.
Compute equation.
Enter equation.
Display points and line.
L1 L2 L3
L2(5) =20
LinReg
y=ax+b
a=3.6
b=1.8
Plot1 Plot2 Plot3
Y1=3.6X+1.8
24
0
6
1
4 (TI-83)
5 (TI-82)
ENTER
Y=
STAT PLOT 1
ENTER GRAPH
(Use ▲, ▼, ►, ENTER to edit data.)
(Turn STAT PLOT 1 off afterward.)

11. Entering and Displaying a Matrix Enter and display $A = \begin{bmatrix} 2 & 4 \\ 1 & 3 \end{bmatrix}$.

Select screen for editing.

MATRX ◄ ENTER

Enter the matrix.

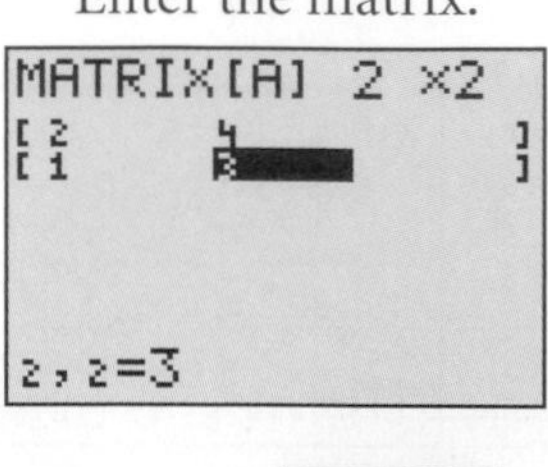

Edit using ENTER.

Display matrix A.

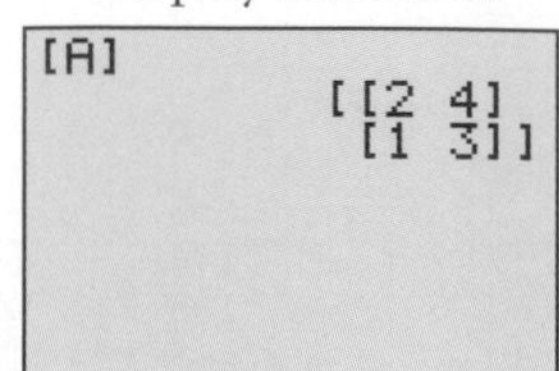

QUIT MATRX 1 ENTER

12. Reduced Form of a Matrix Find the reduced form of $A = \begin{bmatrix} 1 & -2 & 4 & 12 \\ 2 & -1 & 5 & 18 \\ -1 & 3 & -3 & -8 \end{bmatrix}$ (only on TI-83).

Enter the matrix.

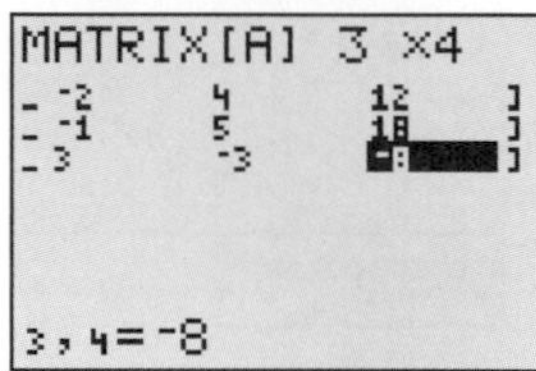

Edit using ENTER.

Display the reduced form.

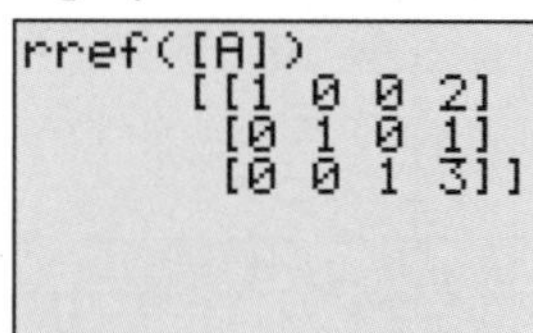

QUIT MATRX ► B ENTER MATRX 1)

13. Matrix Operations Find $A + B$ and AB, where $A = \begin{bmatrix} 2 & 4 \\ 1 & 3 \end{bmatrix}$ and $B = \begin{bmatrix} 5 & -2 \\ 0 & 3 \end{bmatrix}$.

Enter matrix *A*.

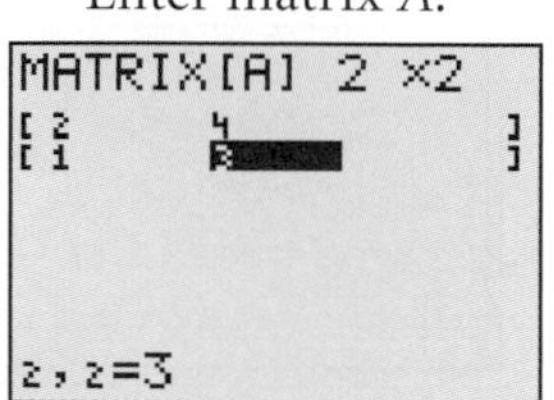

Edit *A* QUIT

Enter matrix *B*.

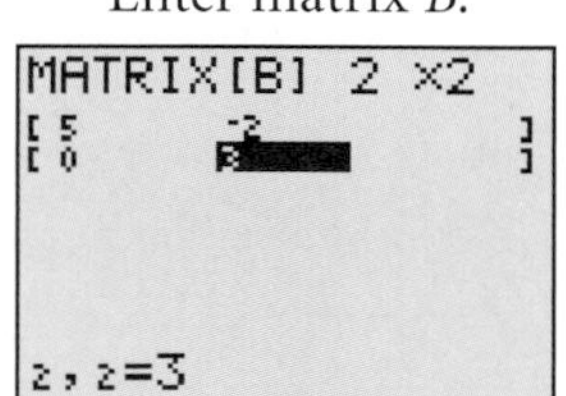

Edit *B* QUIT

Display $A + B$ and AB.

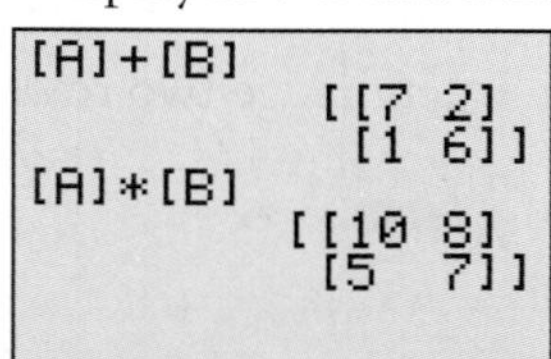

MATRX 1 + MATRX 2 ENTER

MATRX 1 × MATRX 2 ENTER

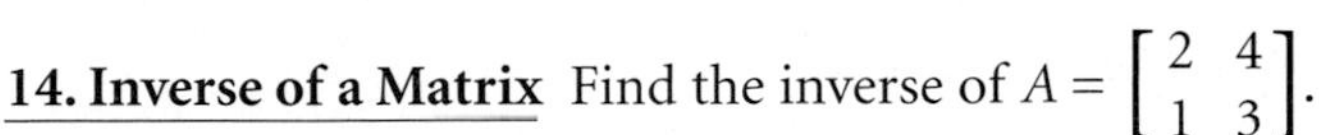

14. Inverse of a Matrix Find the inverse of $A = \begin{bmatrix} 2 & 4 \\ 1 & 3 \end{bmatrix}$.

Enter matrix A.

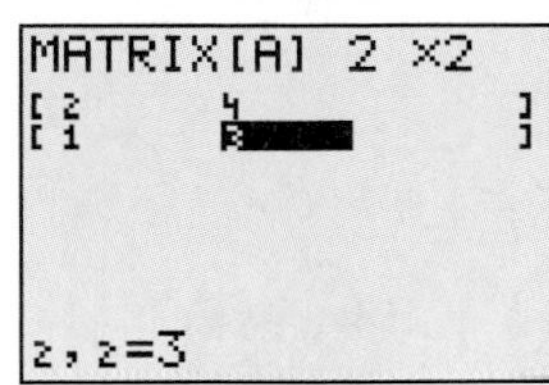

Edit A QUIT

Display A^{-1}.

[A]⁻¹
[[1.5 -2]
[-.5 1]]

MATRX 1 x^{-1} ENTER

15. Determinant of a Matrix Find the determinant of $A = \begin{bmatrix} 5 & 2 & 4 \\ 1 & 3 & -7 \\ 0 & 1 & 3 \end{bmatrix}$.

Enter matrix A.

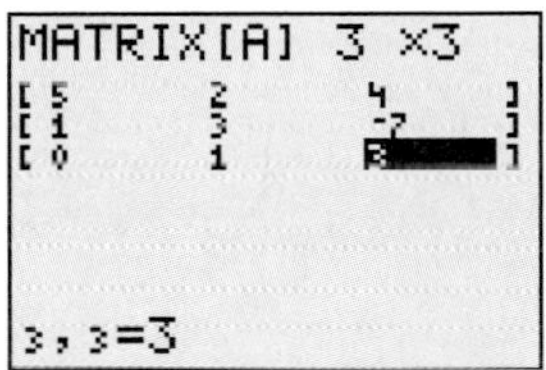

Display $|A|$.

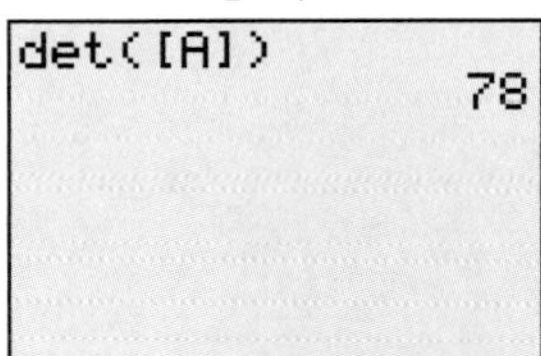

Edit A QUIT  MATRX ► 1  MATRX 1)  ENTER

16. Summary of Formats for Entering Row Operations

Press MATRX ► to get the Matrix Math Menu.

Press C, D, E, F on TI-83 or 8, 9, 0, A on TI-82 to select the operation.

C or 8: Interchange two rows.
rowSwap([A], first row, second row)

D or 9: Add a row to another row.
row+([A], row to be added, row to be added to)

E or 0: Multiply row by a number.
*row(number, [A], row to be multiplied)

F or A: Add a multiple of a row to another row.
*row(number, [A], row to be multiplied, row to be added to)

17. Factorials, Permutations, and Combinations Compute 7!, ${}_8P_5$, and ${}_9C_4$.

```
7!
                5040
8 nPr 5
                6720
9 nCr 4
                 126
```

7 **MATH** ◄ 4 **ENTER** , 8 **MATH** ◄ 2 5 **ENTER** , 9 **MATH** ◄ 3 4 **ENTER** .

ANSWERS TO ODD-NUMBERED EXERCISES

Review Chapter

Exercises R.1

1. (a) $\{2, 9\}$ (b) $\{-9, -2, 0, 2, 9\}$
 (c) $\{-9, -\frac{7}{3}, -2, 0, 2, \frac{8}{3}, 9\}$

3. (a) $\{\frac{4}{2}, 5, 9\}$ (b) $\{-60, -21, -\frac{15}{3}, 0, \frac{4}{2}, 5, 9\}$
 (c) $\{-60, -21, -\frac{15}{3}, 0, \frac{17}{6}\}$

5. 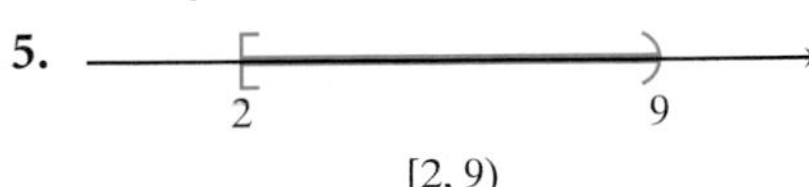

[2, 9)

7. 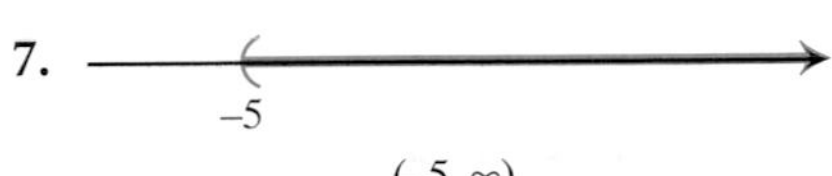

(–5, ∞)

9. 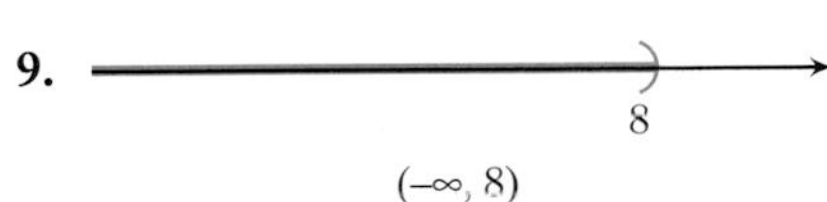

(–∞, 8)

11. 23.4569 13. 29.555556 15. 539.6

17. 1.3694 19. −12.1648

21. commutative property of multiplication

23. distributive property

25. associative property of addition

27. associative property of addition

29. 27.784 31. 1.220652 33. 1.7321 35. 5.56

37. 0.43 39. 7.19 41. 64 43. −64

45. 81 47. $8 + (\frac{9}{3}) - 1 = 10$

49. $5 + [(\frac{9}{3}) \times 4] = 17$ 51. $2 + 7 - (3 \times 5) = -6$

53. $7 + 2^3 = 15$ 55. $1 + (4 \times 5^2) = 101$

57. $(2 \times 7) - 2^5 + 1 = -17$ 59. −7.6220

61. −16.92 63. 5.7689 65. 10.7930

67. 0.5785 69. 4.5723

71. (a) 3.141592654 (b) 3.14159 (c) 3.1415927
 (d) 240.4887

73. (a) 1.7321
 (b) error message, because the square root of a negative number is not a real number
 (c) 1.7321

75. Each term is the square of previous term. The 6th term will be 4294967296.

77. (a) $2^{2^n} + 1$ for $n = 1, 2, 3, 4, 5$ is 5; 17; 257; 65,537; 4,294,967,297.
 (b) 4,294,967,297/641 = 6,700,417. Thus, $2^{2^5} + 1$ is not prime.

79. For example, $\pi = 3.141592654$ on a TI-83 calculator. $\pi - 3 = .1415926536$; one more place revealed. $10(\pi - 3.1) = .4159265359$; another place revealed. $100(\pi - 3.14) = .159265359$. No new additional places revealed. $1000(\pi - 3.141) = .5926535898$. Two new places revealed. $10000(\pi - 3.1415) = .926535898$. No new places revealed. $100000(\pi - 3.14159) = .26535898$. No further places revealed after this stage. TI-83 has $\pi = 3.1415926535898$ in memory. TI-85 does not behave like this. π gives 3.14159265359. $\pi - 3$ gives no new info. $10(\pi - 3.1)$ gives .415926535898. No more info after this. TI-85 has $\pi - 3.1415926535898$ in memory.

Exercises R.2

1. $\frac{1}{8}$ 3. 27 5. 25 7. 5 9. −10

11. 243 13. not a real number 15. 3

17. 4 19. −3 21. 3 23. 64 25. $\frac{1}{9}$

27. −243 29. not a real number 31. $\frac{1}{3}$

33. 36 35. $\frac{128}{7}$ 37. 9 39. 1 41. 9

43. x^2 45. $\dfrac{y^3}{x^5}$ 47. x^5 49. $x^{7/6}$ 51. $x^{7/6}$

53. x^4 55. $2x(\sqrt[3]{x^2})$ 57. $x^2y^3(\sqrt{y})$

59. $|y|\,|z^3|(\sqrt[4]{2x^3yz})$

61. $\dfrac{1}{x^{-n}} = (x^{-n})^{-1} = x^{(-n \cdot -1)} = x^n$

63. $\dfrac{1}{\left(\dfrac{1}{x}\right)} = \dfrac{1}{x^{-1}} = (x^{-1})^{-1} = x$

65. 1158.5620 67. 2276.9310 69. 1.8929

71. 0.4052 73. 2.3707 75. 2.0140

77. −5.9997 79. 2.5×10^7 81. 4.3×10^{-4}

83. 6.34×10^{10} 85. 20,000 87. −4700

89. 0.00093471 91. 1.2×10^6 93. 5.058×10^1

95. (a) $5.873958432 \times 10^{12}$
(b) $2.408322957 \times 10^{15}$ miles

97. 8.7998×10^{-3} grams

99. (a) Amount in account after 1 period = $P(1 + r/n)$. Amount in account after 2 periods = $[P(1 + r/n)](1 + r/n) = P(1 + r/n)^2$. There will be nt such periods in t years. Thus, the amount in the account after t years = $P(1 + r/n)^{nt}$.
(b) \$13,002.04 (c) \$13,036.15

101. 1413.72 cubic inches **103.** 274

Exercises R.3

1. 2, $2x^2$, 1 **3.** 4, $7x^4$, 5 **5.** 8, $4x^8$, no constant term

7. polynomial **9.** not a polynomial

11. polynomial **13.** $7x^2 + 4x + 4$

15. $x^2 + 7x + 3$ **17.** $3y^3 + 5y^2 + y + 5$

19. $4t^2 + 2t + 6$ **21.** $4x^2 + 11x + 6$

23. $7x^2 + 23x + 6$ **25.** $x^4 - 2x^2 - 8$

27. $2x^3 - 6x^2 + 8x$ **29.** $4x^3 + 12x^2 + 2x + 6$

31. $x^3 + 2x^2 - x + 6$

33. $4x^4 - 8x^3 + 15x^2 - 14x + 8$

35. $2y^2 + 10y + 8$

37. $(a + b)(a - b) = a \cdot a - a \cdot b + b \cdot a - b \cdot b = a^2 - b^2$

39.

$$\begin{aligned}(a + b)^3 &= (a + b)(a + b)^2 = (a + b)(a^2 + 2ab + b^2)\\ &= a(a^2 + 2ab + b^2) + b(a^2 + 2ab + b^2)\\ &= a^3 + 2a^2b + ab^2 + ba^2 + 2ab^2 + b^3\\ &= a^3 + 3a^2b + 3ab^2 + b^3\end{aligned}$$

41. $16x^2 + 8x + 1$ **43.** $x^3 + 12x^2 + 48x + 64$

45. $16x^2 + 40xy + 25y^2$ **47.** $x^3 - 3x^2 + 3x - 1$

49. $x^4 + 2x^3y - 2xy^3 - y^4$ **51.** $x + 2 + \dfrac{2}{x + 3}$

53. $x^2 - 2x - 2 - \dfrac{13}{x - 3}$ **55.** $3x^2 - 1 + \dfrac{2}{2x^2 + 1}$

57. $2x^3 + 2x^2 + 2x + 5 + \dfrac{6}{x - 1}$

59. false, $(x^3 + x + 1) + (-x^3 + x + 2) = 2x + 3$, a polynomial of degree 1

61. true, $(ax + b)(cx + d)$ (with $a \neq 0$, $c \neq 0$) = $acx^2 + (bc + ad)x + bd$, and $ac \neq 0$.

63. (a) False, let $P = 2x + 3$, $Q = 5x$; then $P + Q = 7x + 3$.
(b) True, the constant term of PQ is obtained by multiplying constant terms of P and Q.

Exercises R.4

1. $2x^2(2x + 1)$ **3.** $3(x^4 + 2x^2 + 3)$

5. $6x^3(4x^3 + 2x^2 + 3x + 5)$

7. $5y^2(y + 4)(2y^2 + 8y + 1)$

9. $(x + 2)(x - 1)$ **11.** $(x + 3)(x + 2)$

13. $(x + 4)(x + 3)$ **15.** $(3x + 1)(x - 2)$

17. $(5y - 1)(2y + 3)$ **19.** $(x - 2y)(x + y)$

21. $(2x - y)(x + 3y)$ **23.** $(4x - y)(2x + y)$

25. $(x - y)(x + 4)$ **27.** $(3x + y)(x - 4)$

29. $(x + 2y)(5 - y)$ **31.** $(x + 2)(y^2 - 3)$

33. $x(x + 2)(x - 1)$ **35.** $x(x - 3)(x - 1)$

37. $3(2x + 1)(x + 4)$ **39.** $(x + y^3 - 3)(3x + y)$

41. $(u + 3)(u - 2)(v + 2)$

43. $(a + b)(a - b) = a^2 + ba - ab - b^2 = a^2 - b^2$

45.
$$\begin{aligned}(a - b)^2 &= (a - b)(a - b)\\ &= a^2 - ba - ab + b^2\\ &= a^2 - 2ab + b^2\end{aligned}$$

47.
$$\begin{aligned}(a - b)(a^2 + ab + b^2)\\ &= a(a^2 + ab + b^2) - b(a^2 + ab + b^2)\\ &= a^3 + a^2b + ab^2 - ba^2 - ab^2 - b^3\\ &= a^3 - b^3\end{aligned}$$

49. $(3x + y)(3x - y)$

51. $(3x + 2y + 2z)(3x + 2y - 2z)$

53. $(4u^3 + 3v^2)(4u^3 - 3v^2)$ **55.** $(x + y)^2$

57. $(x - 2y)^2$ **59.** $(x + 2y - 3z)^2$

61. $(2x + y)(4x^2 - 2xy + y^2)$

63. $(5a - 2b)(25a^2 + 10ab + 4b^2)$

65. $(x + y - z)(x^2 + 2xy + y^2 + xz + yz + z^2)$

67. $(x + 3)(x - 1)$ **69.** $(x + 3y)^2$

71. $4(t - 2)(t + 1)$ **73.** $(u - 3v)(u^2 + 3uv + 9v^2)$

75. $3(3x + 7)(x - 1)$

77. $[3x + (3y + 4z)^2][3x - (3y + 4z)^2]$

79. $(x + 2y + 3)(x + 2y - 3)$

81. $(u - 2v)(u + 2v + 1)$ **83.** $(x + 2)^2$

85. $(x + 1)^2$ **87.** $(x - 5)^2$

Exercises R.5

1. $\dfrac{3x}{2 + 5x}$ **3.** $\dfrac{1}{2x + 1}$ **5.** $\dfrac{1}{x + 4}$ **7.** $\dfrac{x - 1}{x + 3}$

9. $\dfrac{z - 1}{z - 2}$ **11.** $2(x + 2)$ **13.** $\dfrac{x}{4(4x + 1)}$

15. $\dfrac{(u + 2)(4u - 1)}{(u - 3)(u + 5)}$ **17.** $\dfrac{4x - 1}{x(x - 1)}$

19. $\dfrac{4x^2-3x+1}{(x-1)(x+1)}$ **21.** $\dfrac{x^2-8x+18}{(x+4)(2x-3)}$

23. $\dfrac{4y-9}{3y}$ **25.** $\dfrac{3x^2+4x+2}{(x+1)(x+2)(x-1)}$

27. $\dfrac{7x^2+19x+1}{(x+2)(x-1)(x+3)}$

29. $\dfrac{2z^2+7z-9}{(z+1)(z-3)(z+2)}$ **31.** $\dfrac{x(x-1)}{(x-4)(x-2)}$

33. $\dfrac{(x-2)}{(x-3)}$ **35.** $\dfrac{5v^3-8v^2+11v-4}{v(v-2)}$ **37.** x^3+x^2

39. x **41.** $\dfrac{x-1}{x+1}$ **43.** $\dfrac{x-3}{2x}$ **45.** $\dfrac{s-1}{s}$

47. $\dfrac{6x-1}{(x-1)(2x+3)}$ **49.** $\dfrac{4x^2+3x+2}{x(x-1)(x+2)}$

51. $\dfrac{-5x-2}{(x-1)(x+2)(x-2)}$

53. $\dfrac{9x-7}{(x-3)(x+2)(x+1)}$ **55.** 2

Exercises R.6

1. solution **3.** not a solution

5. solution, solution **7.** solution **9.** 4

11. 10 **13.** $\frac{2}{5}$ **15.** -7

17. linear, in standard form

19. linear, standard form is $2x-11=0$

21. linear, standard form is $x-8=0$

23. nonlinear

25. 8 **27.** $\frac{11}{3}$ **29.** 3 **31.** -6 **33.** -5

35. 2 **37.** -5 **39.** 700 **41.** 65 minutes

43. 68 **45.** 31.2% **47.** 7 inches

49. 7.26 mistakes per page on average

51. Mercury: 87.91 days; Pluto: 90,668.09 days

53. multiplied by 27

55. 10 miles per hour

Exercises R.7

1. **(a)** 0 **(b)** 9 **(c)** -3 **(d)** -9

3. **(a)** 5 **(b)** 32 **(c)** -3 **(d)** 0

5. **(a)** -15.901 **(b)** 7.928 **(c)** 3 **(d)** 16.547

7. **(a)** 2 **(b)** 5 **(c)** not defined

9. **(a)** 6 **(b)** 0 **(c)** not defined

11. **(a)** -0.4 **(b)** -2 **(c)** not defined

13. $4a+1$ **15.** $8a^2+2a+3$

17. $9s+2$ **19.** $(-\infty,\infty)$

21. $(-\infty,\infty)$ **23.** $[3,\infty)$

25. $(-\infty,6)\cup(6,\infty)$ **27.** $(4,\infty)$

29. $(-\infty,\infty),(-\infty,\infty)$ **31.** $(-\infty,\infty),7$

33. $(-\infty,\infty),[8,\infty)$ **35.** $[3,\infty)$

37. $(-\infty,\infty),[0,\infty)$ **39.** $(-\infty,\infty),(-\infty,0]$

41. $(-\infty,\infty),[-2,\infty)$

43. $(-\infty,0)\cup(0,\infty),(-\infty,0)\cup(0,\infty)$

45. **(a)** 24.1000 **(b)** 53.2725 **(c)** 199.2875

47. **(a)** 1.2247 **(b)** 2.5981
(c) $f(0)$ gives an error message. *Check:* When $x=0$, $\sqrt{x-2}=\sqrt{-2}$, not a real number. Thus, 0 is not in the domain of f.

49. **(a)** 2.4495 **(b)** 0
(c) $f(2)$ gives an error message because 2 is not in the domain of f.

51. $2x+h$ **53.** $4x+2h-1$

55. **(a)** 4, 9, 64, 207.36 feet **(b)** mathematical domain $(-\infty,\infty)$ meaningful domain [10, 85]

57. **(a)**

Time	0	1	2	3	4	5	6	7	8	9	10
Height	0	144	256	336	384	400	384	336	256	144	0

(b) 10 seconds **(c)** [0, 10]

59. **(a)**

t	1	2	3	4	5	6	7
$N(t)$	47	184	405	704	1075	1512	2009

(b) 32 months (to nearest month)
(c) 48 months

61. **(a)**

	1989	1990	1991	1992	1993	1994	1995
t	3	4	5	6	7	8	9
$N(t)$	9,270	9,384	9,510	9,648	9,798	9,960	10,134

total 67,704 hundreds **(b)** year 2004

63. **(a)** 65.2% **(b)** 87.31% **(c)** $P(35)=110.0825$. There will be 110% of households with cable in 2015, but it cannot have more than 100%. The function is valid from 1980 to 1995, domain [0, 15]. It will become increasingly unreliable outside interval [0, 15].

65. **(a)** $V(x)=\dfrac{x(64-x^2)}{4}$

(b)

x	0	1	2	3	4	5	6	7	8
$V(x)$	0	15.75	30	41.25	48	48.75	42	26.25	0

(c) [0, 8]
(d) $x=5$. $V(5)=48.75$ cubic inches

Chapter R Review Exercises

1. **(a)** $\{2,12\}$ **(b)** $\{-7,0,2,12\}$ **(c)** $\{-7,0\}$
(d) $\{-7,\frac{3}{4},0,2,\frac{5}{4},12\}$ **(e)** $\{\sqrt{2}\}$

2. **(a)** 

$(-3, 8]$

(b) 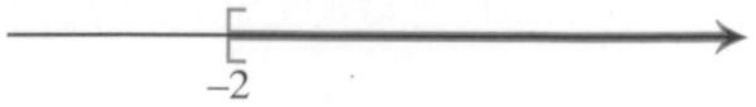

$[-2, \infty)$

(c) 

$(-\infty, 4)$

3. **(a)** commutative property of addition
(b) distributive property
(c) associative property of addition
(d) associative property of multiplication
(e) distributive property

4. **(a)** 34.98972 **(b)** 1.49087 **(c)** 16.65

5. **(a)** $8 + (\frac{9}{3}) = 11$ **(b)** $(7 \times 3) + (\frac{12}{6}) = 23$
(c) $(4 \times 3^2) + (\frac{4}{2}) = 38$

6. **(a)** 11 **(b)** 3 **(c)** $-\frac{5}{2}$ **(d)** $\frac{8}{27}$

7. **(a)** 5 **(b)** -2 **(c)** 3 **(d)** -2

8. **(a)** 9 **(b)** $\frac{1}{125}$ **(c)** 625 **(d)** not a real number

9. **(a)** 1,000,000 **(b)** 2 **(c)** $\frac{1}{80}$

10. **(a)** 8 **(b)** 25 **(c)** 4

11. **(a)** $\dfrac{1}{x}$ **(b)** $\dfrac{1}{x^6}$ **(c)** $x^{7/2}$ **(d)** $\dfrac{x^5}{y^{12}}$

12. **(a)** $2x(\sqrt[3]{4x})$ **(b)** $3|x|y^3\sqrt{y}$ **(c)** $xy^2z(\sqrt[3]{y^2z})$

13. **(a)** 12.0409 **(b)** -91.6464

14. **(a)** 11.714 **(b)** 1.994 **(c)** 213.987

15. **(a)** 5.4662 **(b)** 10.5262 **(c)** -5.7961

16. **(a)** 4.73×10^8 **(b)** 1.25×10^{-5}

17. **(a)** 34,750 **(b)** 0.009473

18. **(a)** 5.67147384×10^3 **(b)** 1.479×10^7

19. **(a)** 21862.9285 **(b)** 20097.1602

20. $2.104004782 \times 10^{29}$ **21.** \$10,794.62

22. **(a)** \$12,655.71 **(b)** \$12,752.15 **(c)** \$12,774.40

23. 1032.5028 cubic inches

24. **(a)** $8x^2 - 3x + 3$ **(b)** $8x^2 + 2x - 15$
(c) $2x^4 - x^3 - x^2 + 11x - 3$

25. **(a)** $4x^2 - 20x + 25$
(b) $125x^6 + 75x^4y + 15x^2y^2 + y^3$

26. **(a)** $4x^6 - 2x^3 + 12x^2 = 2x^2(2x^4 - x + 6)$
(b) $3x^2(2x + 1)^2(1 - 2x)$ **(c)** $4x^2yz^3(2xyz^6 - 1)$

27. **(a)** $(x + 5)(x + 1)$ **(b)** $(x - 3)(x - 1)$
(c) $(2x + 3)(x - 2)$ **(d)** $(5x + 4)(3x - 2)$

28. **(a)** $(2x - 3y)(x + 4y)$ **(b)** $(5x - 2y)(3x + 2y)$
(c) $(2x - y)^2$

29. **(a)** $2(x + 3y)(x - 2)$ **(b)** $(3x + 1)(2x + y)$
(c) $(3x - y)(7x - 4)$

30. **(a)** $(x + 2y)(x - 2y)$ **(b)** $(3x + 4y)(3x - 4y)$
(c) $7(11x + y)(y - x)$

31. **(a)** $(x + y)^2$ **(b)** $(3x - 4y)^2$
(c) $[3 - 2(x + 2y)]^2$

32. **(a)** $(3x + 2y)(9x^2 - 6xy + 4y^2)$
(b) $(4x - y)(16x^2 + 4xy + y^2)$
(c) $(5 - 8x)(124x^2 - 110x + 25)$

33. **(a)** $\dfrac{x - 3}{x - 1}$ **(b)** $\dfrac{x - 1}{x + 2}$ **(c)** $\dfrac{2x - y}{4x + 5y}$

34. **(a)** $\dfrac{6(x - 2)^3}{(x + 5y)}$ **(b)** 1

35. **(a)** $\dfrac{15x + 17}{(x + 3)(2x - 1)}$ **(b)** $\dfrac{9x + 13}{(x + 1)(x + 5)}$
(c) $\dfrac{x(17x - 3)}{(3x - 1)(x + 2)(x + 1)}$

36. **(a)** no **(b)** yes **(c)** yes

37. **(a)** 4 **(b)** 2 **(c)** -7 **(d)** 3

38. **(a)** linear, in standard form **(b)** nonlinear
(c) linear, standard form is $13x - 13 = 0$

39. **(a)** 5 **(b)** $-\frac{14}{5}$

40. 5 units **41.** 6% **42.** 11.25 seconds **43.** 90

44. **(a)** $P = 50d$ **(b)** 6 feet **45.** 2.25 times as far

46. **(a)** 10 **(b)** 19 **(c)** -5

47. **(a)** 13 **(b)** -3 **(c)** 3 **(d)** $18u^2 + 12u - 3$
(e) $8a^2 - 5$

48. **(a)** $(-\infty, \infty), (-\infty, \infty)$ **(b)** $(-\infty, \infty), [-6, \infty)$
(c) $(-\infty, \infty), (-\infty, 1]$ **(d)** $(-\infty, 4], [3, \infty)$

49. **(a)** $R(10) = \$6300$. $R(15) = \$9300$.
$R(20) = \$12{,}200$.
(b) $R(0) = 0$. Expect revenue on 0 sales to be \$0.

50. **(a)** $s(2) = 736$ feet. $s(4) = 1344$ feet.
$s(6) = 1824$ feet.
(b) $s = 0$ when $t = 25$ **(c)** yes

51. **(a)**

	1998	1999	2000	2001	2002	2003	2004	2005
t	10	11	12	13	14	15	16	17
$N(t)$	16,863	20,533	24,565	28,959	33,715	38,833	44,313	50,155

(b) MS-DOS will probably not be around in the year 2005, but viruses will.

52. 2006

Chapter R Test

1. **(a)** $\{3, 15\}$ **(b)** $\{-1, 0\}$

2. **(a)** associative property of multiplication
(b) distributive property

3. 4.0235 **4.** 13 **5.** **(a)** 2.25 **(b)** -243

6. $\dfrac{y^{14}}{x^9}$ 7. $4x^2|y|\sqrt{3x}$ 8. 8.4432

9. $12x^2 + 2x - 2$

10. (a) $(3x - 1)(2x + 3)$ (b) $(4x - y)(2x + 3y)$
(c) $(2x + 9y)(2x - 9y)$

11. $\dfrac{11x - 2}{(x - 1)(2x + 1)}$ 12. -1

13. (a) 21 (b) -1 (c) -3 (d) $12a^2 - 26a + 11$

14. (a) $(-\infty, \infty), (-\infty, 3]$ (b) $[-2, \infty), [-1, \infty)$

15. \$12,517.96 16. 3.35 17. 184.53 pounds

18. (a) $s(1) = 352$ feet. $s(2) = 672$ feet. $s(3) = 960$ feet.
(b) 23 seconds

Chapter 1

Exercises 1.1

1. A, quadrant IV; B, quadrant I; C, quadrant IV; D, x-axis.

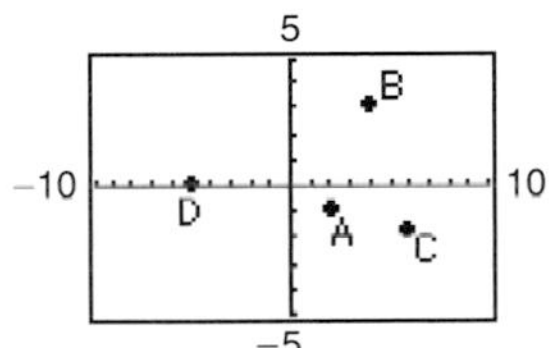

3. A, quadrant I; B, quadrant III; C, quadrant III; D, quadrant II.

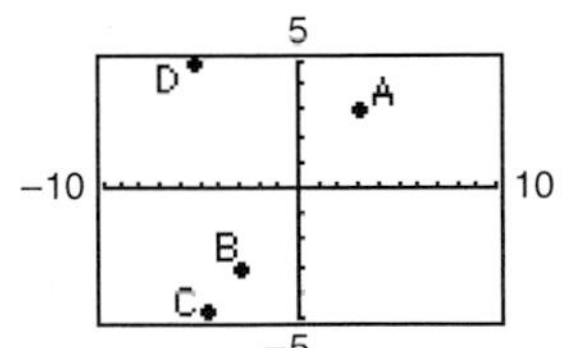

5. (a) yes (b) no (c) yes

7. (a) no (b) yes (c) yes

9. $X_{\min} = -10, X_{\max} = 10, X_{\text{scl}} = 2, Y_{\min} = -4,$
$Y_{\max} = 20, Y_{\text{scl}} = 4$

11. $X_{\min} = -5, X_{\max} = 10, X_{\text{scl}} = 1, Y_{\min} = -20,$
$Y_{\max} = 10, Y_{\text{scl}} = 5$

13. $X_{\min} = -12, X_{\max} = 24, X_{\text{scl}} = 4, Y_{\min} = -300,$
$Y_{\max} = 500, Y_{\text{scl}} = 100$

15. $X_{\min} = -10, X_{\max} = 10, X_{\text{scl}} = 2, Y_{\min} = -12,$
$Y_{\max} = 10, Y_{\text{scl}} = 2$

17. $X_{\min} = -6, X_{\max} = 4, X_{\text{scl}} = 1, Y_{\min} = -6,$
$Y_{\max} = 4, Y_{\text{scl}} = 1$

19. $X_{\min} = -8, X_{\max} = 4, X_{\text{scl}} = 2, Y_{\min} = -10,$
$Y_{\max} = 10, Y_{\text{scl}} = 2$

21. $X_{\min} = -4, X_{\max} = 4, X_{\text{scl}} = 1, Y_{\min} = -4,$
$Y_{\max} = 4, Y_{\text{scl}} = 1$

23. $X_{\min} = -4, X_{\max} = 6, X_{\text{scl}} = 1, Y_{\min} = -10,$
$Y_{\max} = 8, Y_{\text{scl}} = 2$

25. $X_{\min} = -4, X_{\max} = 4, X_{\text{scl}} = 1, Y_{\min} = -6,$
$Y_{\max} = 4, Y_{\text{scl}} = 1$

27. y-int $= -16$, x-int $= 8$

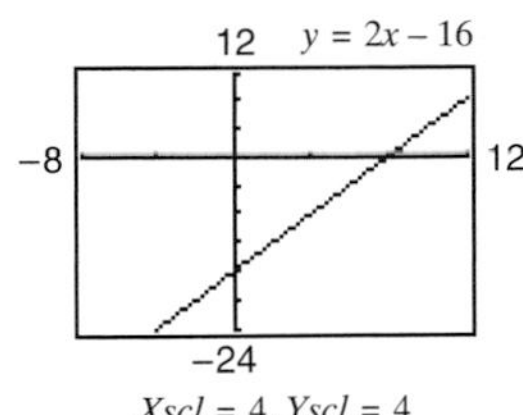

Xscl = 4, *Yscl* = 4

29. y-int $= -15$, x-int $= 3$

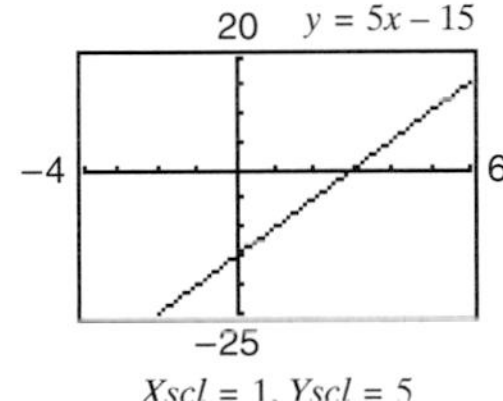

Xscl = 1, *Yscl* = 5

31. y-int $= -27$, x-int $= -9$

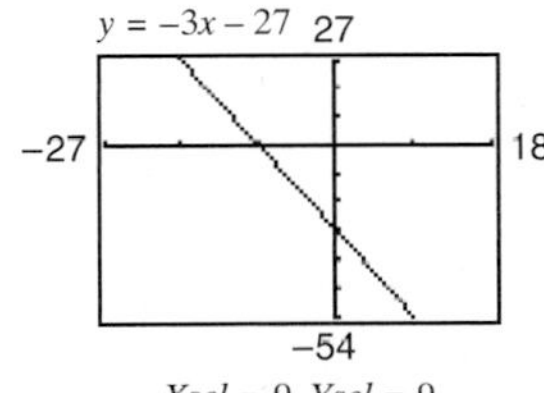

Xscl = 9, *Yscl* = 9

33.

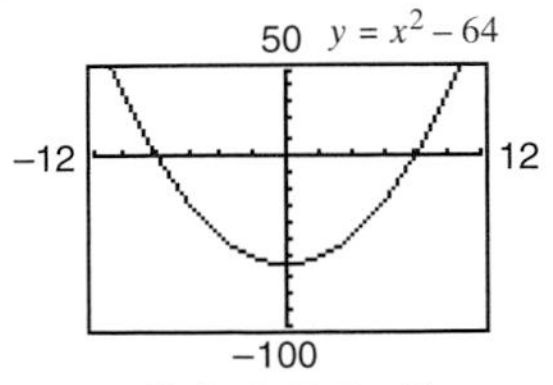

Xscl = 2, *Yscl* = 10

35.

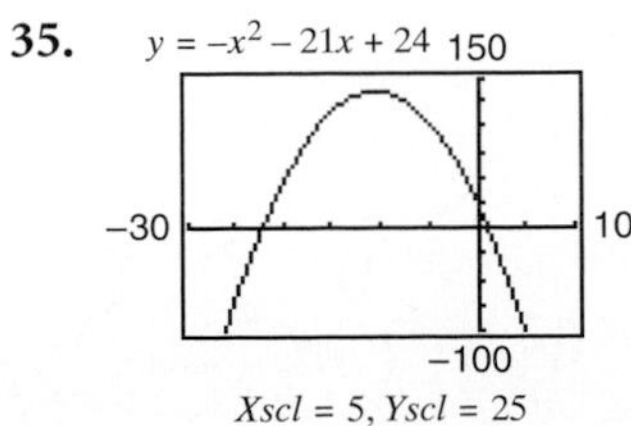

Xscl = 5, *Yscl* = 25

37. 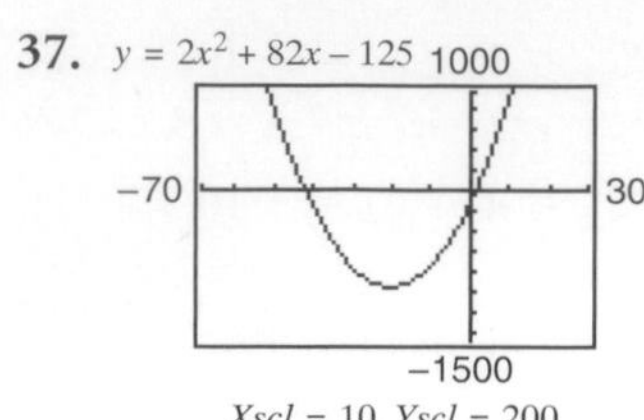

Xscl = 10, Yscl = 200

39. 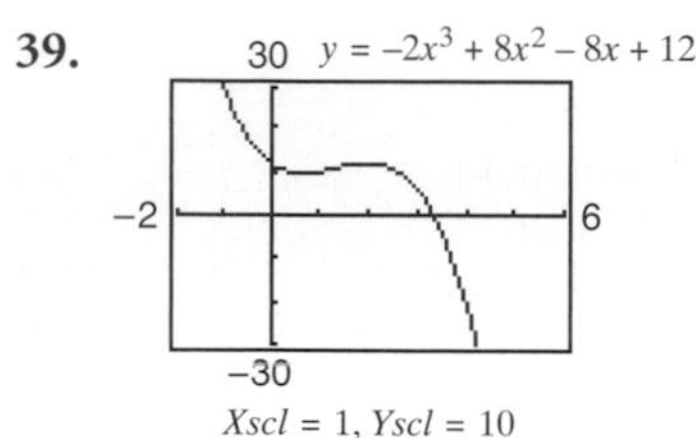

Xscl = 1, Yscl = 10

Section 1.2

1. 2 3. -1 5. undefined 7. 2
9. undefined 11. -3 13. $y = 2x - 4$
15. $y = 4x + 4$ 17. $y = 2x$ 19. $y = 2x + 5$
21. $y = -x - 6$ 23. $y = \frac{3}{2}x + 5$ 25. $y = x + 1$
27. $y = -3x + 10$
29. $y = \frac{3}{2}x - 6$
31. $y = 4$ and $x = 1$
33. $y = -3$ and $x = 6$
35. $y = 0$ and $x = -3$
37. parallel, $y = x$ and $y = x - 2$
39. perpendicular, $y = 3x - 4$ and $y = -\frac{1}{3}x + 5$
41. perpendicular, $y = x$ and $y = -x + 1$
43. $y = 2x - 10$ 45. $y = \frac{1}{2}x$
47. $m = 2$, y-int $= 8$, x-int $= -4$

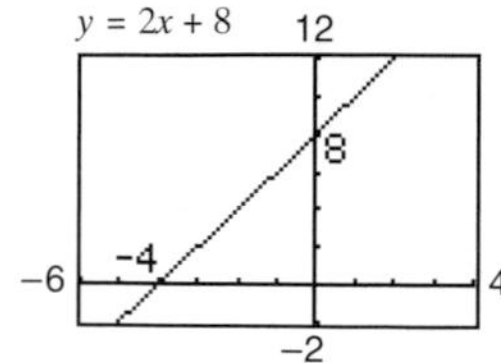

49. $m = 2$, y-int $= 6$, x-int $= -3$

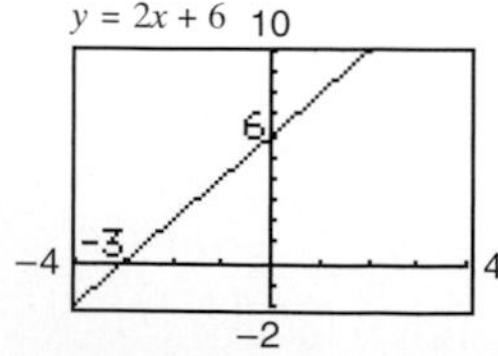

51. $3x + 5y + 9 = 0$, $y = -\frac{3}{5}x - \frac{9}{5}$
$m = -\frac{3}{5}$, y-int $= -\frac{9}{5}$, x-int $= -3$

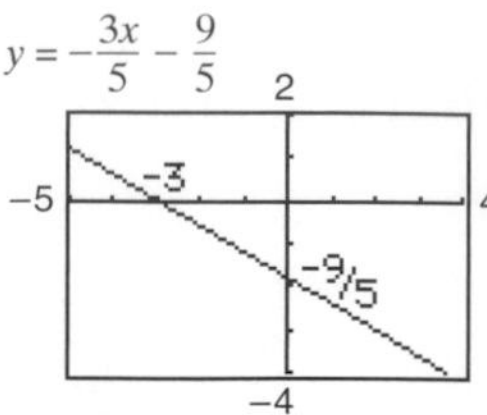

53. $y = 2x + 10$ 55. $y = 15$ 57. $y = -4x - 8$
59. $y = -4x + 7$ 61. $y = -2x + 13$ 63. $m = 1$
65. $A = -4$ 67. $A = -6$ 69. x-int $= -4$
71. Estimate might be $y = 4x + 3$. LSq line $y = 3.9x + 3$. When $x = 5$, $y = 22.5$.
73. Estimate might be $y = 2x - 2$. LSq line is $y = 2.1923x - 1.7692$. When $x = 7$, $y = 13.5769$.
75. (a) $y = -0.204x + 38.74$
(b) 27.52 miles per gallon
77. (a) $y = 0.0062x - 0.2949$ (b) 2.8 (c) 644
79. (a) $y = 13.7001x + 4581.4966$ (b) \$6840
81. (a) $y = -0.6383x + 36.9111$ (b) 18.40%
83. (a) 2000: \$12,033 2005: \$12,438 2010: \$12,842
(b) Prediction for 2000 is below that of 1995. Unsatisfactory. Big change took place in 1992. Let $x = 0$ in 1992. Then get 2000($x = 8$): \$15,364, 2005($x = 13$): \$18,532, 2010($x = 18$): \$21,700.

Exercises 1.3

1. (a) -17.692 (b) 12.44 (c) 25.0508
3. $x = 32$
5. (a) linear, rate of change $= -4$, $f(0) = 3$, $f(x) = -4x + 3$
(b) not linear
(c) linear, rate of change $= 3.5$, $f(0) = 1$, $f(x) = 3.5x + 1$
(d) linear, rate of change $= -2.3$, $f(0) = 1.7$, $f(x) = -2.3x + 1.7$
7. (a) $C(t) = 6t + 300$, when t is the time in months.
(b) Slope is the monthly upkeep; y-int is the initial cost of the clubs.
9. (a) $C(t) = 20t + 470$, where t is the time in weeks.
(b) Slope is the monthly feeding cost; y-int is the initial cost of the dog.
11. (a) $C(t) = 27t + 250$, where t is the time in months.

(b) Slope is the monthly cost of cable, y-int is the price of the television.

13. (a) $A(t) = 2{,}000t + 34{,}000$, where t is time in years from 1997.
(b) In 2033, $t = 36$. $A(36) = 106{,}000$ tons.

15. (a) $s(t) = 65t + 30$ miles from Tampa, where t is the time in hours.
(b) Slope is the speed and s-int is the starting distance from Tampa.

17. (a) 1 meter $= 1.0936$ yards. $f(m) = 1.0936m$ converts m meters to yards. (b) 100 meters is 109.36 yards. 1500 meters is 1640.4 yards. (1 mile is 1760 yards). Both races are shorter in meters.

19. (a) $y = 2.42x + 69.5$, fixed cost $=$ \$69.50.
(b)

No. of items	70	71	72	73	74	75	76
Cost	238.90	241.32	243.74	246.16	248.58	251.00	253.42

(c) 178th item

In Exercises 21–25, numbers are approximations as viewed from the graphs. Answers have been rounded to the nearest integer.

21. $W(t) = 3t/5 + 46$, t years from 1975. 2000: 61%, 2010: 67%.

23. $S(t) = -19t/30 + 42$, t years from 1965. 2000: 20%, 2010: 14%.

25. Female: $E(t) = t/5 + 74$, t years from 1970. 2000: 80, 2010: 82.
Male: $E(t) = t/5 + 68$. 2000: 74, 2010: 76.

Exercises 1.4

1.

x	−3	−2	−1	0	1	2	3
y	15	10	7	0	7	10	15

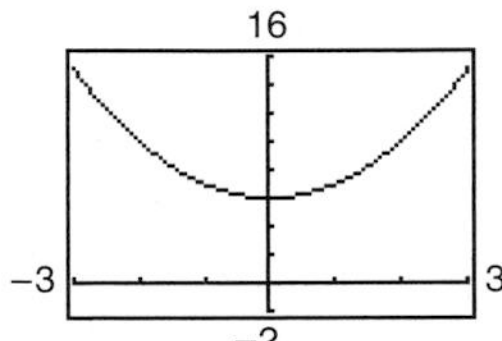

3.

x	−4	−3	−2	−1	0	1	2
y	5	0	−3	−4	−3	0	5

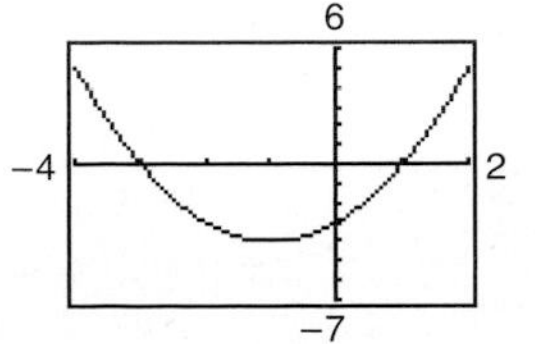

5.

x	−2	−1	0	1	2	3	4	5	6
y	−29	−8	7	16	19	16	7	−8	−29

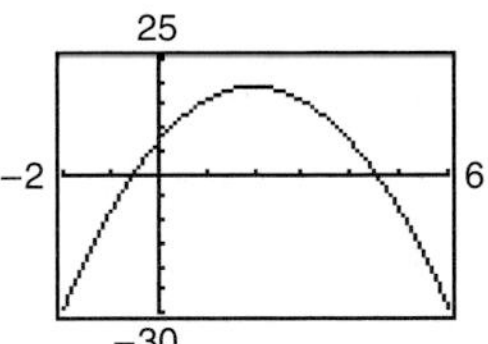

7. $(-4, -1)$, $x = -4$, 15; opens up

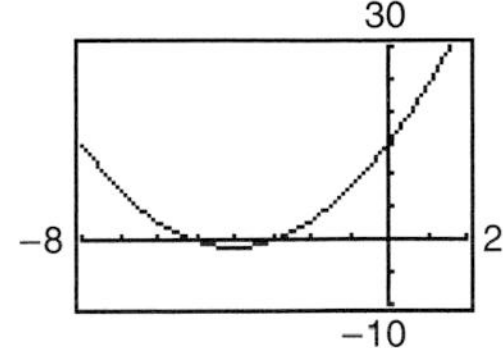

9. $(-3, -59)$, $x = -3$, -5; up

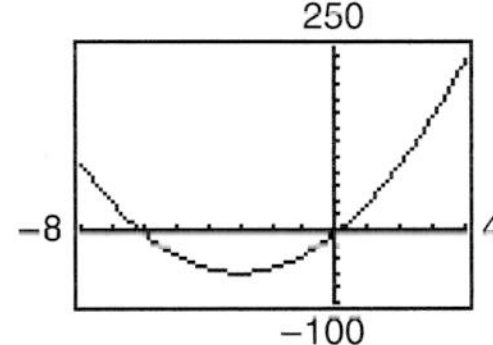

11. $(0, 5)$, $x = 0$, 5; up

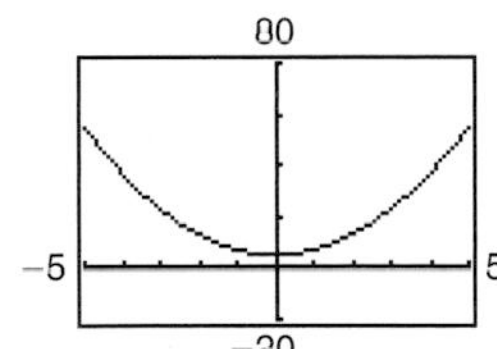

13. $(2.5, 11.5)$, $x = 2.5$, -1; down

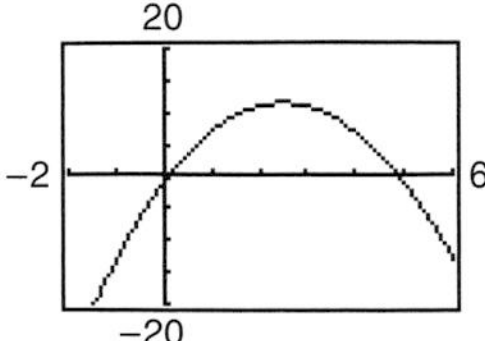

15. $(1.75, -15.25)$, $x = 1.75$, -3; up

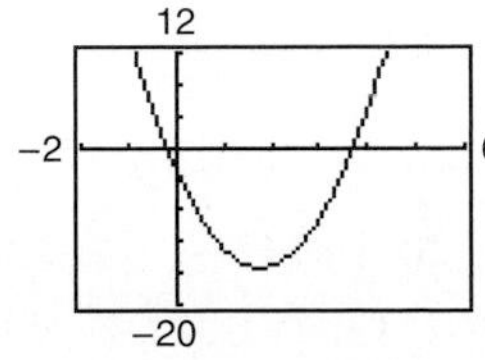

17. $(0, -7)$, $x = 0$, -7; up

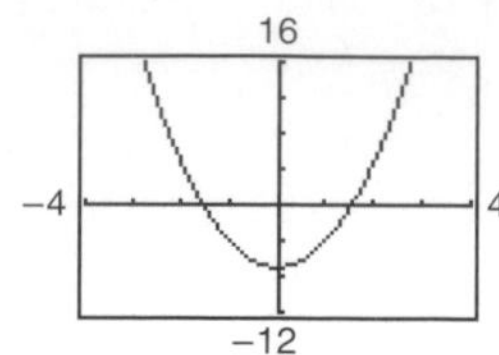

19. $(1.5238, -7.1762)$ **21.** $(-2.9355, -58.7258)$

23. $(26.7000, 20.2890)$ **25.** $(13.6500, -78.3225)$

27. $-5, 3$ **29.** $-1, 4$ **31.** $\frac{5}{2}, 2$ **33.** $1, 4$

35. $0.4105, 6.0895$ **37.** $-0.3333, 1.5000$

39. $-0.7644, 7.0644$ **41.** 4 **43.** 4

45. -7 **47.** 7

49. **(a)** $y = x^2 - 7x + 10$ **(b)** $(3.5, -2.25)$

51. **(a)** $y = 3x^2 + 9x - 30$ **(b)** $(-1.5, -36.75)$

53. **(a)** $y = -2x^2 + 2x + 12$ **(b)** $(0.5, 12.5)$

Exercises 1.5

1. 2304 feet, 12 seconds **3.** 624 feet/second

5. \$276.50

7. **(a)** 1733 units **(c)** $894 \le x \le 2572$

9.

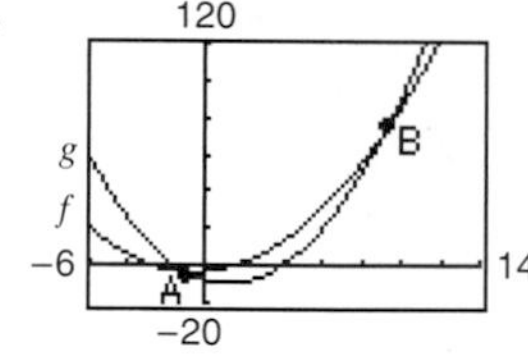

$A(-1.10, -3.13)$, $B(9.10, 72.33)$

11. \$22.11

13. **(a)** $f(x) = 122.1298701x^2 - 171.6056277x + 3109.145455$. Year 2002 high is approximately 18,637.
(b) year 2000

15. **(a)** $R(x) = -0.02x^2 + 145x$ **(b)** \$232,800
(c) \$262,500

17. **(a)** $N(t) = 6t^2 + 72t + 9000$ **(b)** 1,118,400 cars

19. **(a)** $E(x) = -0.6125x^2 + 7.1149x + 30.3542$, to four decimal places;
$I(x) = 0.8548x^2 - 1.8976x + 88.2583$;
$D(x) = -1.4631x^2 + 8.9917x - 57.8917$
(b) $E(13) = 19.3$; $I(13) = 208.1$; $D(13) = -188.3$; in billions of dollars

Exercises 1.6

1. 4.1231 **3.** 7.2801 **5.** 1.4142 **7.** $(2, 4)$

9. $(4.5, 5)$ **11.** $(2.5, -1)$ **13.** $(-1, 2), 2$

15. $(-5, 2), \sqrt{17}$ **17.** $(1, 3), \sqrt{8}$

19. $(x - 3)^2 + (y - 1)^2 = 16$

21. $x^2 + (y - 4)^2 = 25$

23. $(x - 1.5)^2 + (y - 6)^2 = 10.24$ **25.** $(-2, 3), 6$

27. $(3, 1), 1$ **29.** $(1, 3), \frac{1}{2}$ **31.** $(0, 4), 2$

33. circle, $(1, 5)$, 6 **35.** not a circle

37. circle, $(3, 0)$, 1.4142 **39.** not a circle

41. $(x - 20)^2 + (y - 20)^2 = 100$,
$(x - 20)^2 + (y - 20)^2 = 25$

43. $(x - 10)^2 + (y - 10)^2 = 100$

45. $y = \sqrt{100 - (x - a)^2} + 10$ for $a = 10, 30, 50, 70$

47. $y = -\sqrt{100 - (x - a)^2} + 20$ for $a = 10, 70$
$y = \sqrt{100 - (x - a)^2} + 20$ for $a = 30, 50$

49. $D(1.5, 2.5)$, $E(3.5, 2.5)$

51. $B(4.5, 1)$, $C(4.5, 4.5)$, $D(2.75, 2.75)$

53. $(3.7, 2.8)$, 1.9, $(x - 3.7)^2 + (y - 2.8)^2 = (1.9)^2$

55. Face: circle center $(4, 4)$, radius 2.
$(x - 4)^2 + (y - 4)^2 = 4$.
Mouth: circle center $(4, 3)$, radius 0.5.
$(x - 4)^2 + (y - 3)^2 = 0.25$.
Eyes: circles centers $(3, 5)$, $(5, 5)$, radius 0.2.
$(x - 3)^2 + (y - 5)^2 = 0.04$ and
$(x - 5)^2 + (y - 5)^2 = 0.04$.
Nose: line $x = 4$ from $(4, 4)$ to $(4, 5)$.

57. $AB^2 = 16$, $AC^2 = 109$, $BC^2 = 205$. ABC is not a right triangle.

59. $AB^2 = 29$, $AC^2 = 29$, $AD^2 = 58$, $BC^2 = 58$, $BD^2 = 29$, $CD^2 = 29$. Have $AB = AC = BD = CD$. Square $ABDC$ with diagonals $AD = BC$.

61. $(1, 0), (-9, 0)$ **63.** $9, -1$ **65.** $(4, 5)$

67. x-intercepts 7.8990 and -1.8990.
y-intercepts $5, -3$.

69. $(x - 3)^2 + (y - 4)^2 = 9$

71. $(x - 5)^2 + (y - 2)^2 = 4$

73. $(x - 2)^2 + (y - 2)^2 = 4$ **75.** 50.2655, 201.0619

77. $(x - 2)^2 + (y - 1)^2 = 9$ or
$(x - 8)^2 + (y - 1)^2 = 9$

Chapter 1 Review Exercises

1. **(a)**

(b)

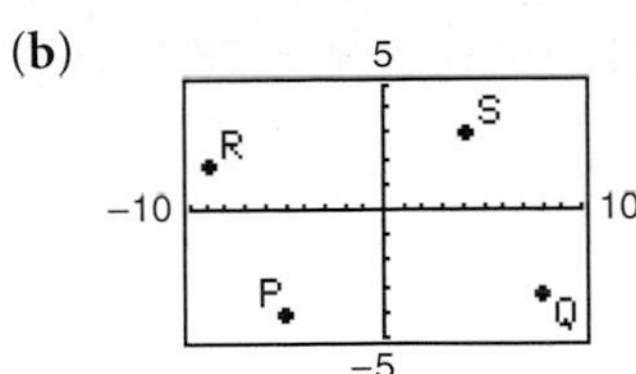

2. (a) yes, no, yes (b) yes, yes, no (c) no, no, yes

3. (a) $X_{\min} = -5, X_{\max} = 5, X_{\text{scl}} = 1, Y_{\min} = 5,$ $Y_{\max} = 25, Y_{\text{scl}} = 5$
   (b) $X_{\min} = -10, X_{\max} = 40, X_{\text{scl}} = 10, Y_{\min} = -8,$ $Y_{\max} = 12, Y_{\text{scl}} = 2$
   (c) $X_{\min} = -25, X_{\max} = 50, X_{\text{scl}} = 5, Y_{\min} = -80,$ $Y_{\max} = 40, Y_{\text{scl}} = 20$

4. (a) x, y-int 7, −7

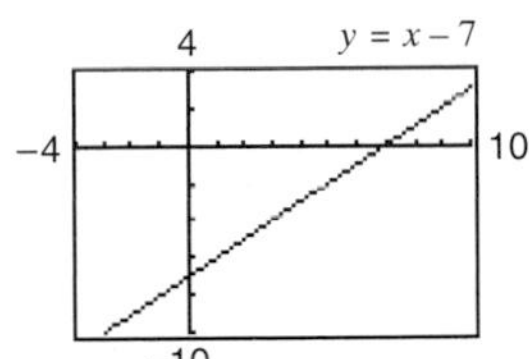

   (b) x, y-int 2.5, 5

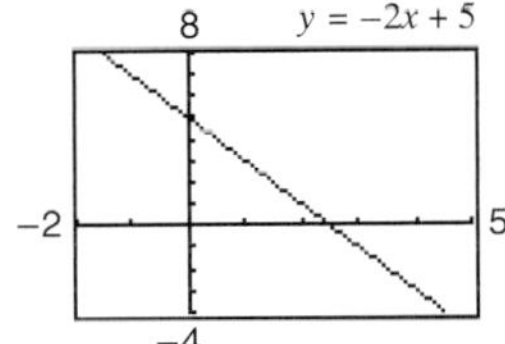

   (c) x, y-int 3, −36

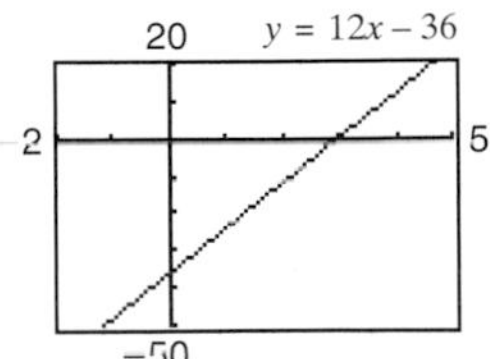

5. (a) $m = -1$ (b) $m = -\frac{9}{10}$ (c) $m = 3$
   (d) Slope is undefined.

6. (a) $y = 4x - 7$ (b) $y = -2x + 3$

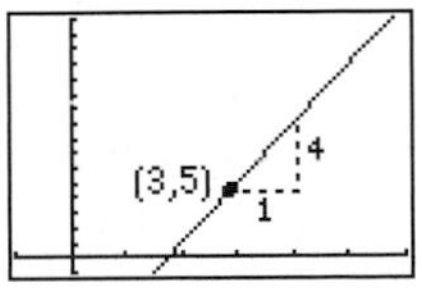

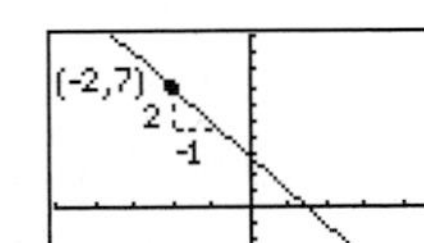

   (c) $y = -2$ (d) $x = 5$

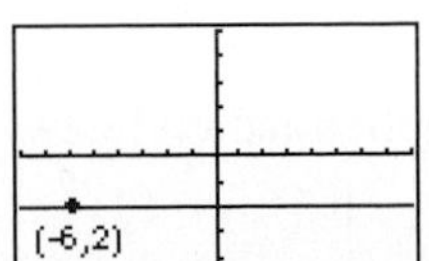

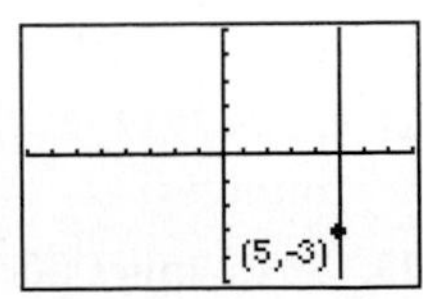

7. (a) $\frac{3}{2}$ (b) $\frac{8}{5}$ (c) $-\frac{7}{2}$

8. (a) $\frac{2}{9}$ (b) -1 (c) $\frac{1}{12}$

9. Points lie on a line.

10. (a) $y = -2x + 6$ (b) $y = 2x + 1$
    (c) $y = -\frac{1}{2}x + \frac{15}{2}$

11. (a) $y = 3x + 4$ (b) $y = -x + 7$
    (c) $y = 6x - 2$

12. $y = 5x + 10$

13. $y = \frac{1}{2}x - 6$

14. $x + 3y - 23 = 0$

15. $y = 3x + 6$

16. −9

17. 11

18. −11.7725, 3.6403, 55.3374

19. (a) linear, rate of change = 4, $f(0) = -2$, $y = 4x - 2$
    (b) not linear
    (c) linear, rate of change = −2.8, $f(0) = 4.1$, $y = -2.8x + 4.1$

20. (a) $f(x) = 15x + 105$
    (b) Slope is the cost of a stud. y-int is the initial cost of the cleats.

21. (a) $y = 3.53x + 454.9$
    (b) $631.40
    (c)

x	70	71	72	73	74	75	76
y	702	705.53	709.06	712.59	716.12	719.65	723.18

    (d) at 98th item

22. Canada, 12,937; Japan, 38,565; United States, 152,874 thousand

23. (a) (2, 5), 9, $x = 2$, opens up

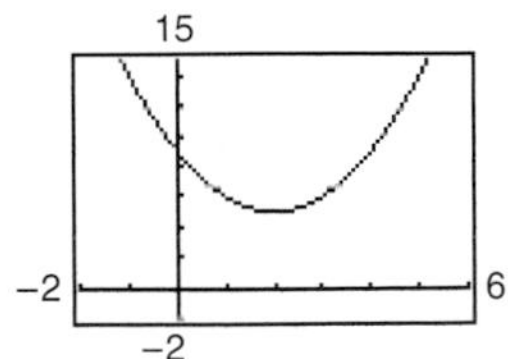

    (b) (−3, −8), 1, $x = -3$, opens up

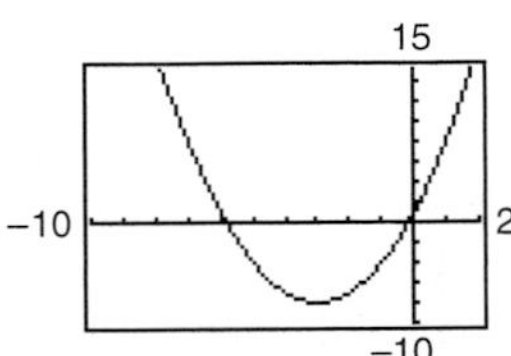

    (c) (2, −11), −3, $x = 2$, opens up

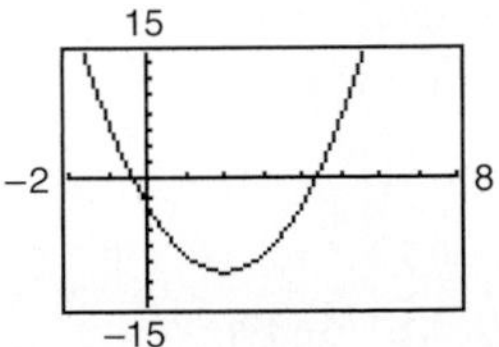

(d) $(3, 20), -7, x = 3$, opens down

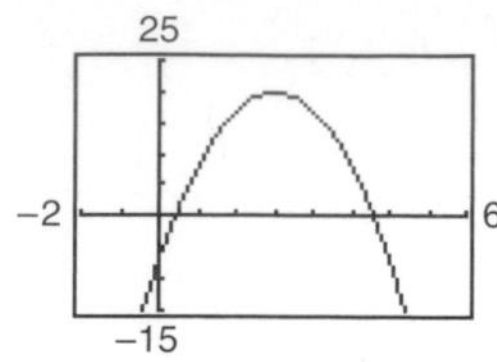

24. $a = 7$ **25.** $y = x^2 - x - 6$

26. $y = x^2 + 3x - 4, y = -x^2 - 3x + 4$

27. 1287.02 feet at 8.97 seconds **28.** \$21.54

29. (a) $v(x) = 0.001x^2 + 0.52x + 316$
(b) 508.5 feet/year
(c) surface 117 years, middle 93 years, base 73 years

30. (a) 4.721 (b) 4.1231 (c) 10.4403

31. (a) $(4, 6)$ (b) $(-6, 2)$ (c) $(-\frac{1}{2}, \frac{3}{2})$

32. $(5, 8)$ **33.** $(0, 3)$ and $(0, -5)$

34. $(\frac{11}{5}, \frac{33}{5})$ and $(1, 3)$

35. (a) $(3, -4), 3$ (b) $(-1, -7), 5$
(c) $(2, 3), 2$ (d) $(-3, -1), 2.4495$

36. (a) $(x - 3)^2 + (y - 6)^2 = 9$
(b) $(x + 1)^2 + (y - 4)^2 = 36$

37. Center $(3, 6)$, radius $\sqrt{5}$, equation $(x - 3)^2 + (y - 6)^2 = 5$

38. $(x + 3)^2 + (y - 1)^2 = 1$

39. 12.5664, 12.5664

40. (a) $y = -2x - 4$ (b) $(x + 8)^2 + (y - 6)^2 = 16$

Chapter 1 Test

1. $(-2, 19)$, yes; $(1, 1)$, yes; $(5, 31)$, no

2. $X_{min} = -4; X_{max} = 16; X_{scl} = 2; Y_{min} = -16;$ $Y_{max} = 12; Y_{scl} = 4$

3. $y = -x + 1$ **4.** -3 **5.** $-\frac{2}{7}$ **6.** $y = -2x - 1$

7. $y = -2x + 3$ **8.** $x + 2y - 5 = 0$ **9.** 4

10. Change of 1 in x causes a constant change of -2 in f. f is linear; rate of change $= -2$; $y = -2x + 3$.

11. vertex $(3, -16)$; axis $x = 3$; y-int $= -7$; $a > 0$; open up

12. $y = -x^2 + 2x + 15$ **13.** 9.4340.

14. $(-1.68, -1.68)$ and $(10.68, 10.68)$

15. (a) $(-5, 1), 6$ (b) $(3, 1), 2$

16. $(x - 2)^2 + (y + 7)^2 = 81$

17. $(x - 2)^2 + (y - 5)^2 = 4$

18. (a) $y = 2.5x + 594$ (b) \$649

19. $P = 0.5432121212t + 18.98545455$; 78.7 per square mile

20. \$18.60

Chapter 2

Exercises 2.1

1. 1, 2 **3.** 0, 4 **5.** 2, −4

7. $\frac{1}{2}, -3$ **9.** $\frac{3}{2}$ **11.** $\frac{1}{7}, -2$

13. 0, 4 **15.** $\frac{1}{4}$ **17.** ± 3

19. ± 3 **21.** $\pm\frac{1}{5}$ **23.** $0, -\frac{4}{5}$

25. $\frac{1}{3}, -3$ **27.** ± 4.1231 **29.** 4.9326, −0.9326

31. −0.5445, −2.1221 **33.** −6, 4 **35.** −2, 6

37. −7, −1 **39.** $3 \pm \sqrt{1/2}$ **41.** $-1 \pm \sqrt{5/2}$

43. $-1 \pm \sqrt{1/3}$ **45.** 1, −7 **47.** $-5/2 \pm \sqrt{37}/2$

49. $2 \pm \sqrt{11}/2$ **51.** −0.2087, −4.7913

53. 0.2247, −2.2247 **55.** −2

57. no real solutions **59.** 0.8309, −0.3309

61. no real solutions **63.** 0.8000

65. $a = 1$: 2.1583, −1.1583; $a = 2.1$: 1.3549, −0.8787; $a = 4.3$: 0.8876, −0.6550

67. $b = 2.3$: no real solutions; $b = 7.3$: −2.1761, −0.5276; $b = -9.2$: 0.3791, 3.0282

69. −0.7803, 3.0156 **71.** −0.8799, 3.4093

73. −0.2670, 1.6255 **75.** 0.3045, 11.5844

77. no real solutions **79.** no real solutions

81. single real solution **83.** single real solution

85. ± 8 **87.** $\frac{1}{2}$ **89.** 0, 8

91. Two real solutions, $a < 1.289285714$; single solution, $a = 1.289285714$; no solutions, $a > 1.289285714$.

Exercises 2.2

1. 5 and 15 seconds

3. 5 seconds (400 feet is the maximum height.)

5. 12 seconds **7.** 21.86 feet **9.** 3 feet

11. width 12 feet, length 22 feet

13. 7, 13 **15.** 11, 13, and −13, −11

17. 6 **19.** −7 or 2

21. 0.3047757271 hours (approximately 18.29 minutes)

23. 234.7 miles per hour

Exercises 2.3

1. ± 1 3. $\pm 1, \pm 2$ 5. $\pm\sqrt{2}, \pm 2$
7. $\pm 1/\sqrt{2}$ 9. $\pm\sqrt{5}$ 11. $2, -2$
13. $4, 7$ 15. $1, 2$ 17. $1, \frac{1}{3}$
19. $\frac{1}{5}, -\frac{1}{2}$ 21. $\frac{3}{2}, \frac{5}{4}$ 23. 9
25. 1 27. $-2, -\frac{3}{2}$ 29. 5
31. 10 33. 3 35. 4
37. 7 39. 6 41. 3
43. 4 45. 243 47. -7
49. $-10, 6$ 51. $-34, 30$ 53. $-\frac{39}{8}$
55. ± 1 57. $\frac{1}{6}, \frac{1}{2}$ 59. $-\frac{3}{4}, -1$
61. ± 8 63. $1, \sqrt[3]{3}$ 65. $-8.24, 0.24$
67. $-1.36, 3.69$ 69. $-4.85, 1.85$
71. $-2.7977, -0.3056, 1.4033$
73. $-1.8487, 0.4124, 1.8363$
75. $-4.1485, -0.9609, 0.9540, 3.1554$
77. $-1.6511, 1.2739, 2.3772$
79. (a) $k < 25$ (b) $k = 25$ (c) $k > 25$
81. (a) $k = 5$ (b) $k = -0.25$ (c) $k = -5$
(d) No, the graph can never have one or no x-intercepts.

Exercises 2.4

1. $(-\infty, 5]$ 3. $(-5, \infty)$ 5. $(-\infty, 3]$
7. $(-\infty, 1]$ 9. $(-\infty, \frac{9}{4})$ 11. $(-\frac{3}{5}, \infty)$
13. $(1, 3)$ 15. $(2, 4]$ 17. $(-1, 5)$
19. $[-4, 1)$ 21. $[-4, -1]$ 23. $[-1, 1]$
25. $(-6, 6)$ 27. $(-4, 2)$ 29. $(-2, 1)$
31. $(-\infty, -3) \cup (2, \infty)$ 33. $(-\infty, -1) \cup (1, \infty)$
35. $(-1, 10)$ 37. $(-\infty, -3) \cup (3, \infty)$ 39. $[-2, 5]$
41. $(1, 4)$ 43. $(-\infty, 2) \cup (6, \infty)$ 45. $(-3, 1)$
47. $(-\infty, 2) \cup (5, \infty)$ 49. $(-\infty, 2) \cup (4, \infty)$
51. $(-2, 1) \cup (3, \infty)$
53. $(-\infty, -3] \cup [-2, -1] \cup [1, \infty)$
55. $(-1.16, 5.16)$ 57. $(-1.90, 7.90)$
59. $(-3.79, 0.79) \cup (3, \infty)$
61. $[-1.71, 1.14] \cup [3.57, \infty)$ 63. $(-2, 1) \cup (3, \infty)$
65. $(-2.67, 2)$
67. $x = 3$. The graphs touch at the point $(3, 9)$; the line is tangent to the curve.
69. $C \geq 15°$ 71. $F > 68°$ 73. $1925 \leq n \leq 2300$
75. (a) over 25 years (b) over \$3750
77. $h > 8$ 79. When $0 \leq t < 3$ or $12 < t \leq 15$
81. between \$129.84 and \$770.16

Exercises 2.5

1. $5, -5$ 3. $10, -2$ 5. $\frac{5}{3}, -\frac{13}{3}$
7. $\frac{5}{3}, -1$ 9. 2 11. $-4, 1$
13. $\frac{1}{2}, 0$ 15. $-4, 4$ 17. $-3, 6$
19. $5, -5$ 21. $\frac{9}{4}, -\frac{5}{4}$ 23. $(-4, 4)$
25. $(-\infty, -5) \cup (5, \infty)$ 27. $(-\infty, -5)$ and $(11, \infty)$
29. $(-\infty, -2] \cup [5, \infty)$ 31. $(-\infty, -3) \cup (4, \infty)$
33. $(-\infty, 1] \cup [5, \infty)$ 35. $(-4, 3)$ 37. $[-1, 6]$
39. $(-6, 4)$ 41. $[0, 7]$ 43. $(-\infty, -1] \cup [\frac{19}{5}, \infty)$
45. $(-6, 6)$ 47. $[-14, 16]$ 49. $(-5, -2) \cup (2, 5)$
51. $(-2, \frac{5}{2}) \cup (\frac{5}{2}, 7)$

Chapter 2 Review Exercises

1. (a) $-2, 1$ (b) $-7, -3$ (c) $\frac{1}{2}, -5$ (d) $-\frac{2}{3}, \frac{7}{2}$
2. (a) ± 4 (b) $\pm\frac{1}{3}$ (c) ± 2 (d) $8, -2$
(e) $0.4114, -0.9114$
3. (a) $3, 7$ (b) $-12, -2$ (c) $4 \pm \sqrt{\frac{1}{2}}$
(d) $1 \pm \sqrt{\frac{11}{3}}$
4. (a) $0.6458, -4.6458$ (b) $3.5414, -2.5414$
(c) $0.7386, -4.7386$ (d) $-0.1156, -2.8844$
(e) no solutions (f) single solution, -3
5. $2.6411, -1.1194$
6. (a) two (b) two (c) none (d) one
7. For example, $x^2 + 3x - 4 = 0$
8. 15 miles per hour 9. 5 and 9 seconds
10. (a) ± 1.7321 (b) ± 0.7071
(c) $-3.8165, -2.1835$ (d) 16
11. (a) 1 (b) 14 (c) $\frac{1}{3}$ (d) no solutions (e) 7
(f) 24 (g) $\pm\frac{1}{32}$
12. (a) $-2.2296, 0.7437, 3.8192$ (b) -2.5270
(c) $-3.8176, -2.8356$
13. (a) $(-\infty, 10]$ (b) $(-\infty, 11)$ (c) $[-\frac{1}{2}, \infty)$
(d) $(-2, \infty)$
14. (a) $[2, 3]$ (b) $(0, 3)$ (c) $[\frac{4}{3}, 7]$ (d) $(-1, \frac{3}{2}]$
15. (a) $[-2, 3]$ (b) $(-\infty, 1) \cup (4, \infty)$ (c) $(0, 5)$
16. (a) $[-4, -\frac{1}{2})$ (b) $(-4, 17]$
(c) $(-\infty, 5) \cup (7, \infty)$
17. between 7 and 11 seconds
18. more than 8 years
19. (a) $6, -8$ (b) $13, -5$ (c) $4, -1$
(d) $-4, \frac{2}{3}$ (e) $-3, \frac{1}{7}$ (f) no solution

20. **(a)** $[-7, 7]$ **(b)** $[-7, 1]$ **(c)** $(-\frac{1}{3}, \frac{11}{3})$
(d) $(-\infty, -4] \cup [9, \infty)$ **(e)** $(-\infty, -\frac{3}{7}) \cup (\frac{9}{7}, \infty)$
(f) $[-7, -1] \cup [1, 7]$ **(g)** $(-9, 1)$

21. **(a)** $(1, 5)$ **(b)** $(-\infty, -6] \cup [4, \infty)$ **(c)** $(-6, 3)$

Chapter 2 Test

1. $5, \frac{7}{2}$ **2.** ± 3 **3.** $2 \pm \sqrt{\frac{5}{2}}$

4. **(a)** $1.1926, -4.1926$ **(b)** no solutions

5. $-2.8457, 1.1672$ **6.** At 4 seconds and 10 seconds

7. $\pm\sqrt{5}$ **8.** 6 **9.** -3.2794 **10.** $(-3, \infty)$

11. $(1, 5)$ **12.** $-8, -\frac{6}{5}$ **13.** $(-10, 4)$

14. width, 10 feet; length, 35 feet

15. between 5 and 12 seconds

Chapters R, 1, and 2 Cumulative Test

1. **(a)** 1.5929 **(b)** 15.7711

2. **(a)** 0.2963 **(b)** 25

3. $\dfrac{y^{22}}{x^{12}}$ **4.** $6x^2 + x - 15$

5. **(a)** $(3x - 2)(4x + 1)$ **(b)** $(3x + 5y)(3x - 5y)$

6. **(a)** domain, $(-\infty, \infty)$; range, $(-\infty, 4]$
(b) domain, $[1, \infty)$; range, $[3, \infty)$

7. $y = 2x - 7$ **8.** $y = -2x + 3$ **9.** -2

10. vertex, $(4, -20)$; axis, $x = 4$; y-int $= -4$; $a > 0$; opens up

11. 4.2426

12. **(a)** center, $(3, -7)$; radius, 5
(b) center, $(-1, 3)$; radius, 3

13. $\frac{3}{2}, -4$

14. **(a)** $1.3166, -5.3166$ **(b)** no solutions

15. ± 2 **16.** -3.3130

17. $(-3, \infty)$ **18.** 86

19. $y = 3x + 490$; \$541 **20.** at 5 and 11 seconds

Chapter 3

Exercises 3.1

1. function **3.** not a function **5.** not a function

7. $x^2 + 2x - 15$; polynomial of degree 2

9. $x^3 + 4x^2 - 7x - 10$; polynomial of degree 3

11. $-1, 3$ **13.** $-1, 3$ **15.** $-1, 5$

17. $0, 1, 2$ **19.** $-5, 0, 2$ **21.** $-3, 0, \frac{7}{3}$

23. x-int $= 1, 3, 6$; y-int $= f(0) = -18$

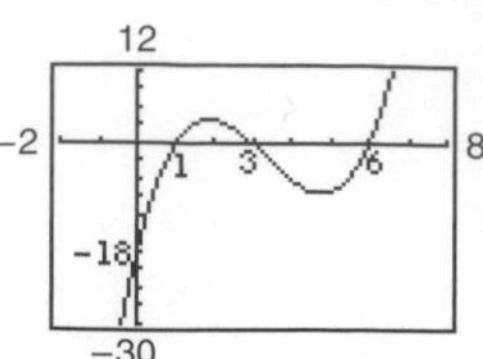

25. x-int $= -4, -1, 2$; y-int $= 8$

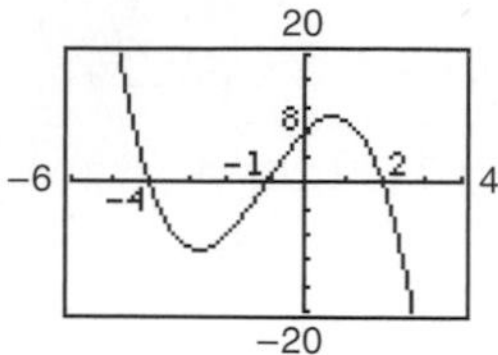

27. x-int $= -3, 0, 2$; y-int $= 0$

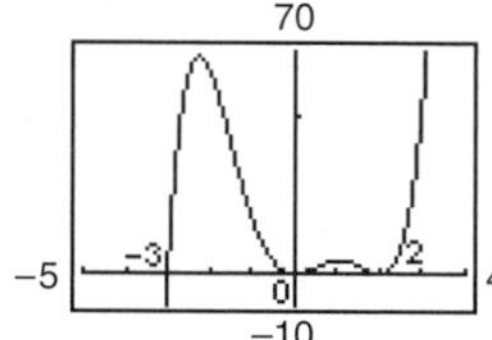

29. x-int $= -2, 0, 2$; y-int $= 0$

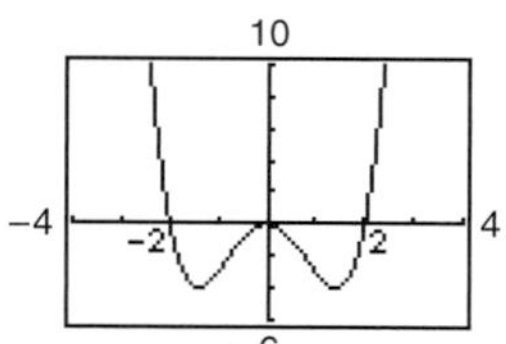

31. x-int $= 0, 2, 5$; y-int $= 0$

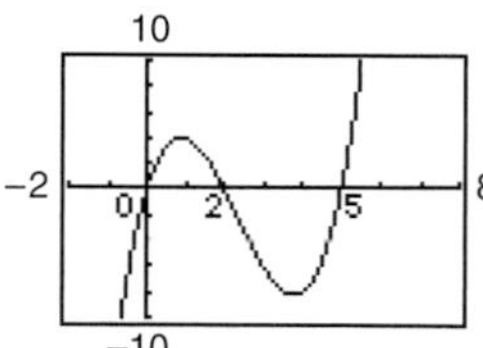

33. x-int $= -1, -3, 0, 1$; y-int $= 0$

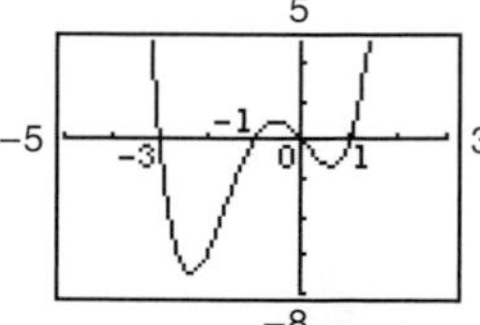

35. $f(x) = 3(x + 3)(x - 1)(x - 5)$

37. $f(x) = -2(x + 3)(x - 2)(x - 6)$

39. $f(x) = k(x + 2)(x - 3), k > 0$

41. $f(x) = k(x + 2)(x - 1)(x - 4), k > 0$

43. $f(x) = (x + 5)(x - 2)(x - 4)$

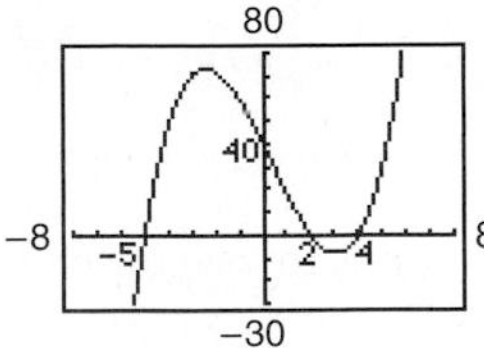

45. $f(x) = 3(x + 2)(x - 1)(x - 5)$

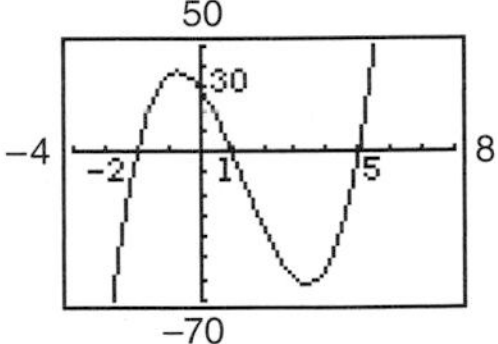

Exercises 3.2

1. increasing: $[b, \infty)$; decreasing: $[a, b]$; constant: $(-\infty, a]$; local max: none; local min: at $x = b$

3. increasing: $(-\infty, a]$; decreasing: $[a, b]$; constant: $[b, \infty)$; local max: at $x = a$; local min: none

5. increasing: $[a, b]$, $[c, \infty)$; decreasing: $(-\infty, a]$, $[b, c]$; local max: at $x = b$; local min: at $x = a$, $x = c$

In the windows of Exercises 7–11 $X_{scl} = 1$, $Y_{scl} = 1$

7. min at $(-4, -6)$; dec $(-\infty, -6]$; inc $[-6, \infty)$

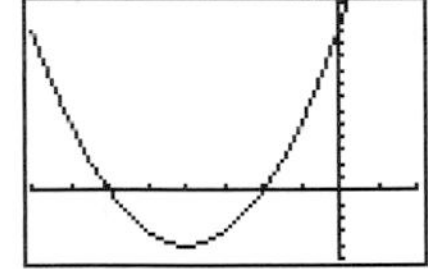

9. max at $(1, 11)$; inc $(-\infty, 1]$; dec $[1, \infty)$

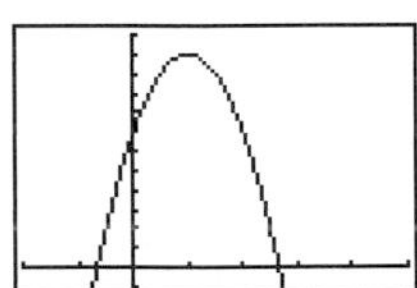

11. max at $(3, -5)$; dec $(-\infty, 3]$; inc $[3, \infty)$

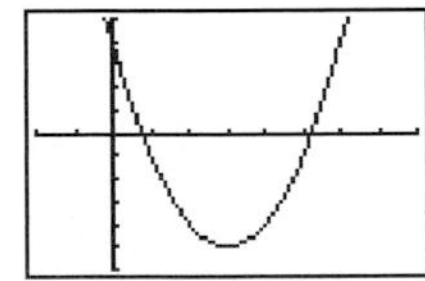

13. max$(-0.6039, 7.8562)$; min$(1.1039, -2.1062)$; inc$(-\infty, -0.6039]$; $[1.1039, \infty)$; dec$[-0.6039, 1.1039]$

15. max$(1.1130, 14.1166)$; min$(-0.1797, 8.7160)$; dec$(-\infty, -0.1797]$; $[1.1130, \infty)$; inc$[-0.1797, 1.1130]$

17. max$(2.8581, 32.2099)$; min$(-0.5248, -6.5062)$; dec$(-\infty, -0.5248]$; $[2.8581, \infty)$; inc$[-0.5248, 2.8581]$

19. max$(-22.0623, 1928.1248)$; min$(-0.6043, -3011.9770)$; inc$(-\infty, -22.0623]$; $[-0.6043, \infty)$; dec$[-22.0623, -0.6043]$

21. Peaked in 1992 with 492 ducks; lowest in 1997 with 367 ducks

23. 10,000 copies

25. Min (13.4803, 64.1086). Interest decreased from 1978 to mid-1991 (corresponding to $x = 13.5$) but has been on the increase ever since.

29. 13.490013 million miles

31. **(a)** 1985 with 69,841 warheads **(b)** 35,375 **(c)** no?

33. 456 yards along the bank and 109 yards under the river

Exercises 3.3

1. domain: $(-\infty, \infty)$; range: $(-\infty, \infty)$

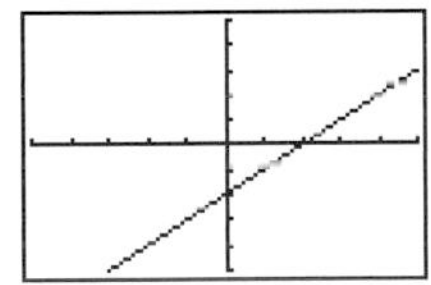

3. domain: $(-\infty, \infty)$; range: $(-\infty, \infty)$

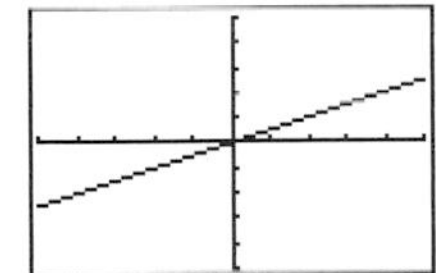

5. domain: $(-\infty, \infty)$; range: $[2, \infty)$

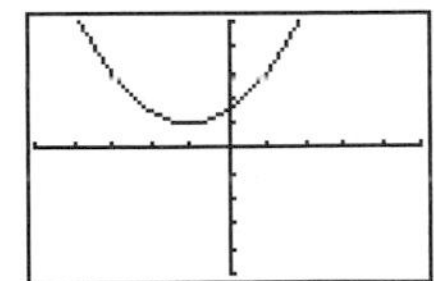

7. domain: $(-\infty, \infty)$; range: $[-3, \infty)$

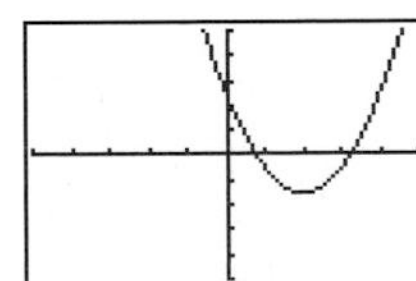

9. domain: $(-\infty, \infty)$; range: $[2, \infty)$

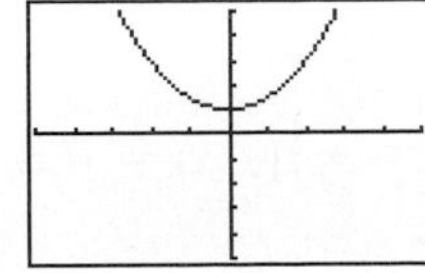

11. domain: $(-\infty, \infty)$; range: $[0, \infty)$

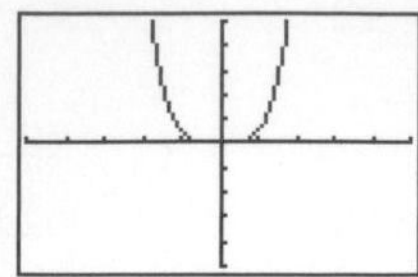

13. domain: $(-\infty, \infty)$; range: $(-\infty, 0]$

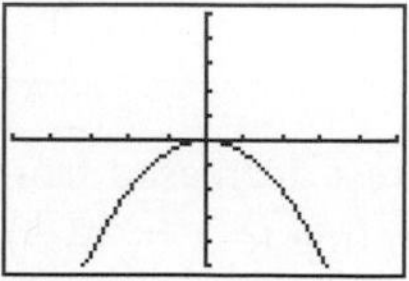

15. domains: $(-\infty, \infty)$; range $(a = 2, 3)$: $[0, \infty)$; range $(a = -5)$: $(-\infty, 0]$

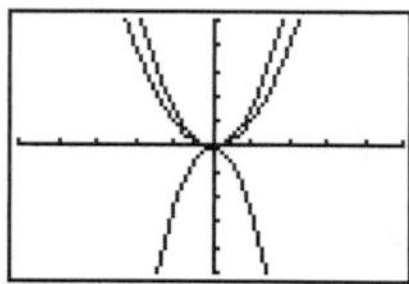

17. range: $[-5, 3]$

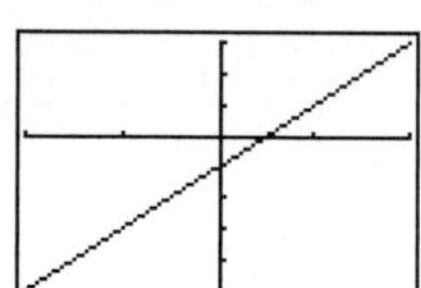

19. $(-2, 8)$; range is an open interval

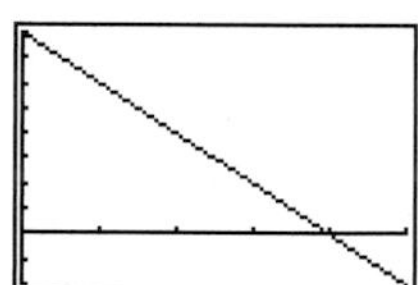

21. range: $[-1, 17]$ **23.** range: $[-8, 8]$

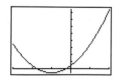

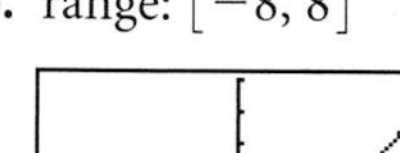

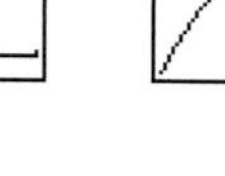

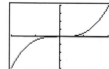

25. range: $[-16, 0]$

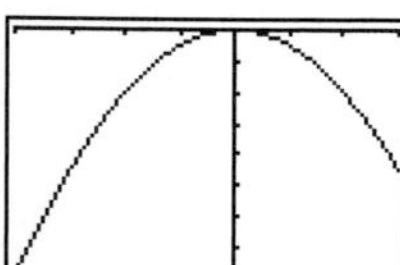

27. **(a)** even **29.** **(a)** neither **31.** **(a)** neither

33. **(a)** even **35.** **(a)** even

37. **(a)** symmetry about origin; odd

(b) $f(-x) = \dfrac{6}{(-x)}$; $-f(x) = -\dfrac{6}{x}$; $f(-x) \neq f(x)$; $f(-x) = -f(x)$; odd

39. **(a)** symmetry about origin; odd

(b) $f(-x) = \dfrac{8}{(-x)^{3}}$; $-f(x) = -\dfrac{8}{x^{3}}$; $f(-x) \neq f(x)$; $f(-x) = -f(x)$; odd

41. **(a)** not symmetric about y-axis or origin; neither even nor odd;

(b) $f(-x) = 3(-x)^3 - (-x) + 5$; $-f(x) = -3x^3 + x - 5$; $f(-x) \neq f(x)$; $f(-x) \neq -f(x)$; neither

43. **(a)** symmetry about y-axis; even

(b) $f(-x) = |(-x)| + 4$; $-f(x) = -|x| - 4$; $f(-x) = f(x)$; $f(-x) \neq -f(x)$; even

45. **(a)** symmetry about y-axis; even

(b) $f(-x) = |(-x)^2 - 4|$; $-f(x) = -|x^2 - 4|$; $f(-x) = f(x)$; $f(-x) \neq -f(-x)$; even

47. even, $f(-x) = f(x)$; for example, $f(-6) = 288 = f(6)$

49. neither; for example, $f(-6) = -45$, but $f(6) = 9$; $f(-6) \neq f(6)$ and $f(-6) \neq -f(6)$

51. even; $f(-x) = f(x)$; for example, $f(-9) = -1320 = f(9)$

53.

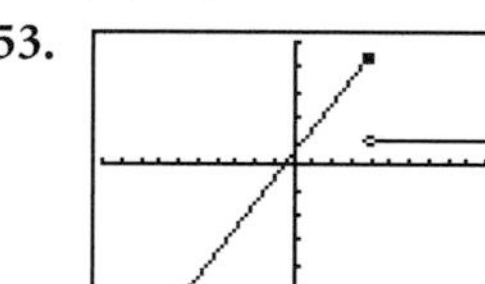

55.

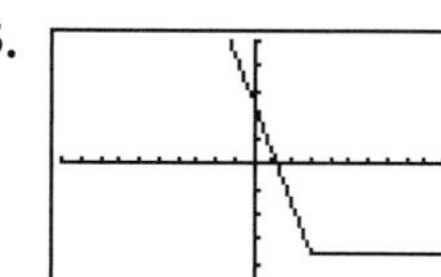

57.

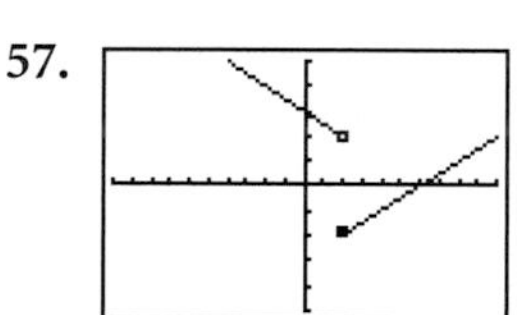

59.

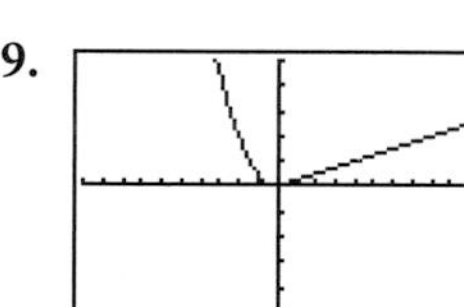

61.

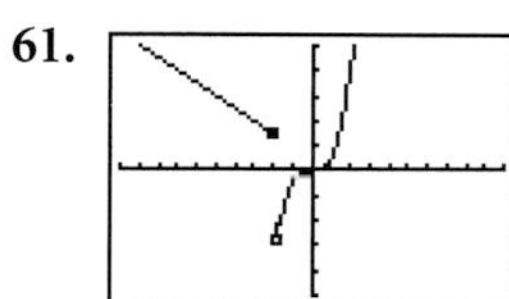

63.

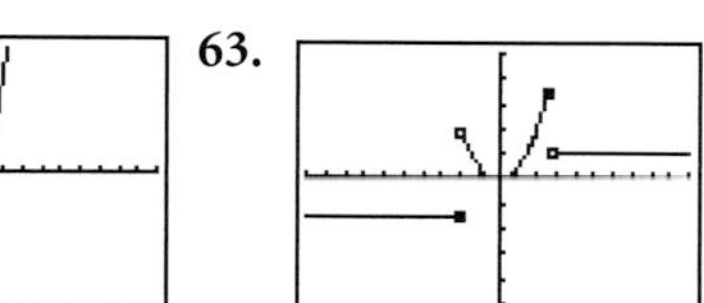

65. $k = -7$

67. Range is $[-25, 56]$.

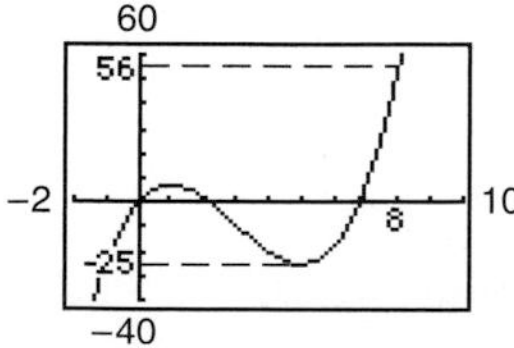

69. **(a)** 12 month: 4 words; 24 month: 316 words

(b) 12/18 months: 120 words; 18/24 months: 192 words

(c) Number of words per year would gradually increase and then start decreasing, giving the curve below. One starts learning slowly, then the rate increases and then the rate decreases.

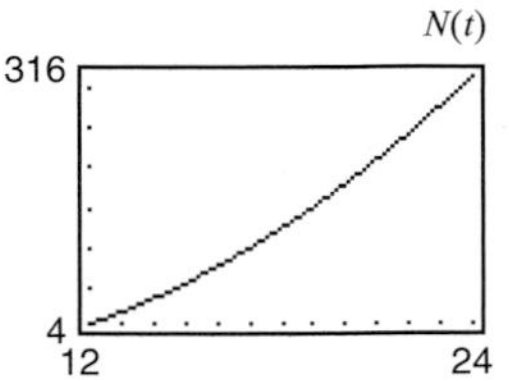

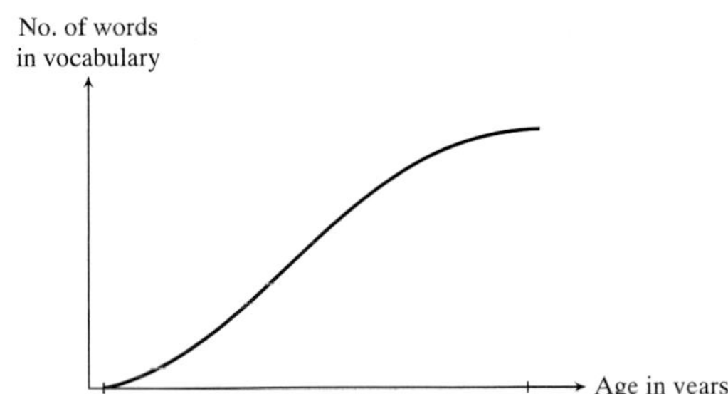

71. **(a)** $C(x) = \begin{cases} 5x & \text{if } 0 \le x \le 50 \\ 250 + 3(x - 50) & \text{if } 50 < x \le 100 \\ 400 + 2(x - 100) & \text{if } x > 100 \end{cases}$

(b) $C(75) = 325 = \$3.25$; $C(120) = 440 = \$4.40$

73. **(a)** See graph below.

(b) show every 24 days

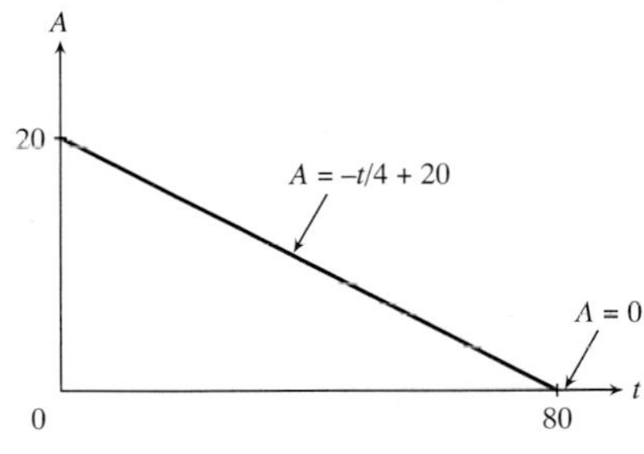

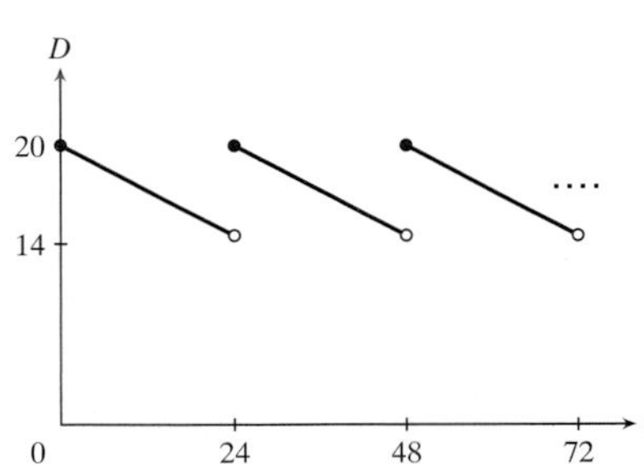

75. **(a)** 9.72 pounds **(b)** 2.90 pounds

77. **(a)** mouse: 405; cow: 4990.84

(b) $S = \dfrac{73.3M^{0.74}}{M}$

As M increases S decreases (see graph). As body weight increases, the restlessness decreases. This indicates that mice are more restless than elephants.

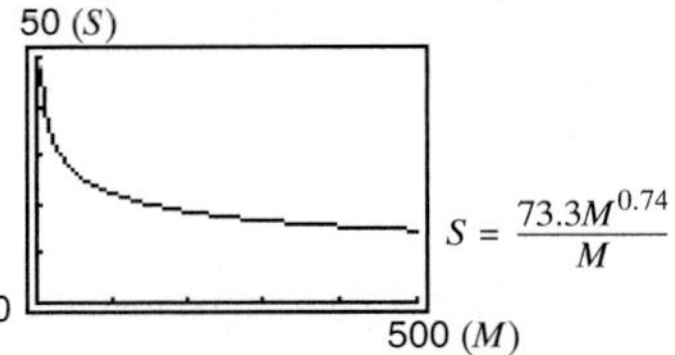

Exercises 3.4

1. Shift $f(x) = x^2$ up 4.

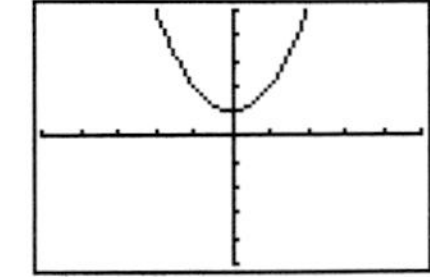

3. Shift $f(x) = x^2$ to the left 3.

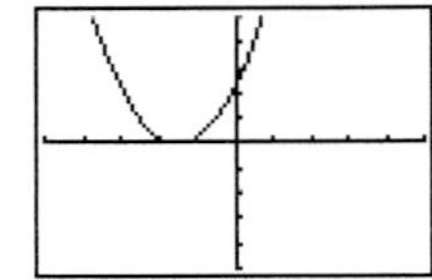

5. Shift $f(x) = x^3$ down 4.

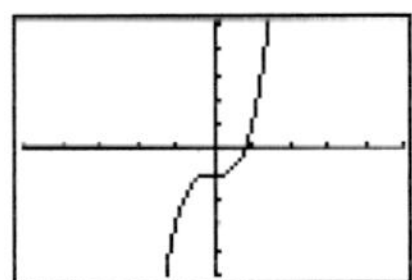

7. Reflect $f(x) = x^3$ in x-axis, up 6.

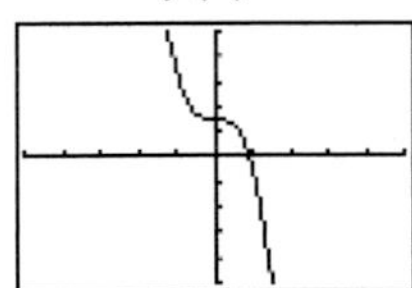

9. Shift $f(x) = x^2$ to the right 4 and reflect in x-axis.

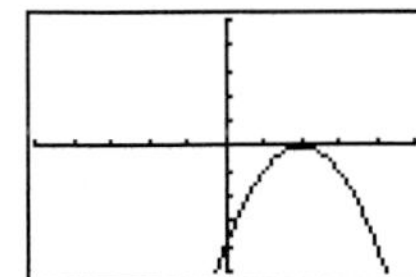

11. Shift $f(x) = x^3$ to the right 3, down 8.

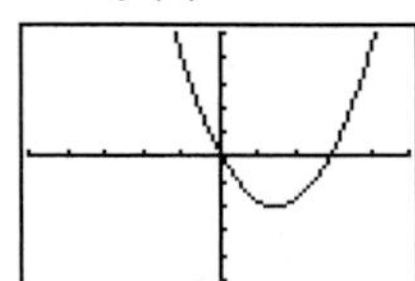

13. Shift $f(x) = |x|$ to the left 3, down 5.

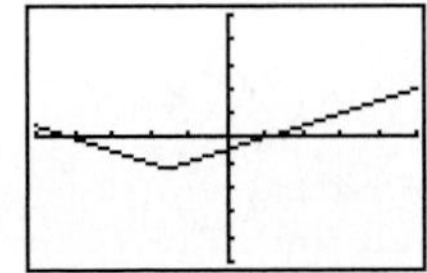

15. Reflect $f(x) = 2x^4$ in x-axis, up 10.

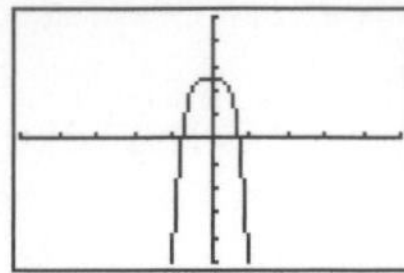

17. Shift $f(x) = \sqrt{x}$ to the left 6, down 8.

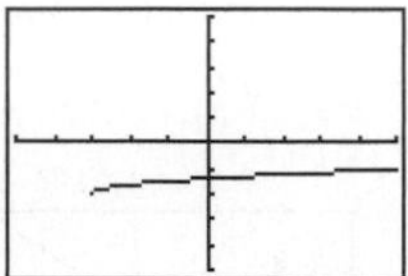

19. $g(x) = (x - 2)^2 - 2$
$f(x) = x^2$
Shift f to the right 2, down 2.

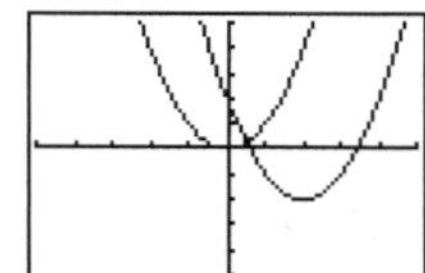

21. $g(x) = (x + 1)^2 + 3$
$f(x) = x^2$
Shift f to the left 1, up 3.

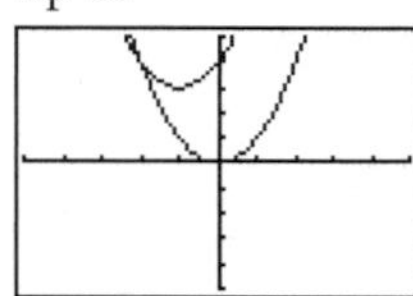

23. $g(x) = -(x - 2)^2 + 3$
$f(x) = x^2$
Shift f to the right 2, reflect in x-axis, shift up 3.

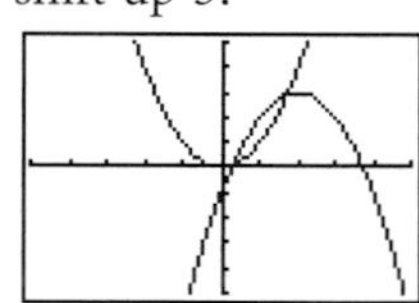

25.

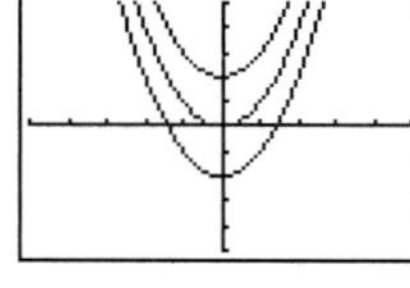

27.

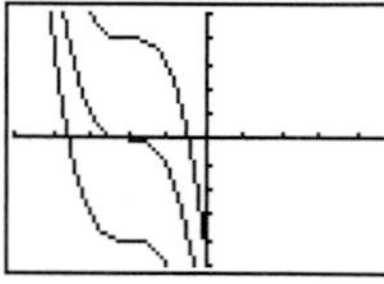

29.

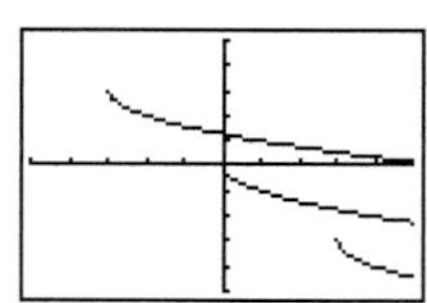

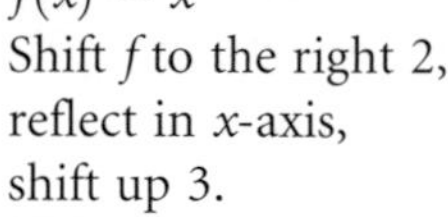

31. $g(x) = x^2 - 5$

33. $g(x) = -(x - 6)^2 + 5$

35. $g(x) = -x^3 + 5$

37. $g(x) = |x| + 3$

39. $g(x) = -|x - 2| + 6$

41. $g(x) = -\sqrt{x} - 1$

43. $f(x) = (x + 5)^2 - 10$
$f(x) = x^2$
$f(x) = (x - 5)^2 - 10$

45. $f(x) = x^2$
$f(x) = -x^2$

47. $f(x) = |x - 4|$
$f(x) = |x|$
$f(x) = |x + 4|$

49. **(a)** The value of g is 2 more than the value of f at each x.
(b) $h(x) = x^2 - 3$ **(c)** $k(x) = x^3 + 5$

51. $f(x) = x^2$; $(a) \leftrightarrow (f)$; $g(x) = -x^2$; $(b) \leftrightarrow (d)$; $h(x) = -x^2$; $(c) \leftrightarrow (e)$

53. **(a)** $f(x) = (x - 2)^2$ **(b)** $f(x) = (x + 5)^2$
(c) $f(x) = -(x - 3)^2$

55. **(a)**

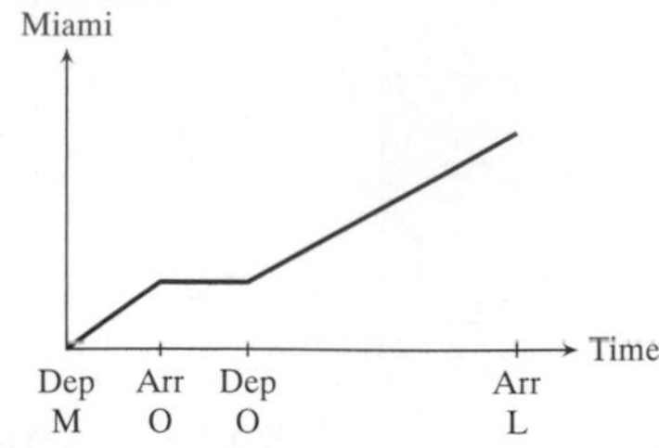

(b)

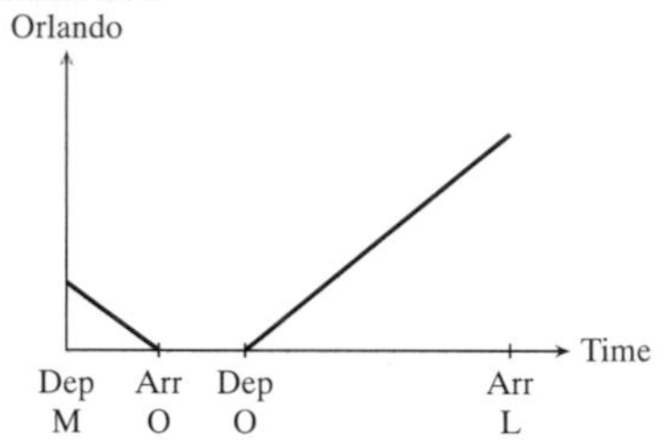

(c)

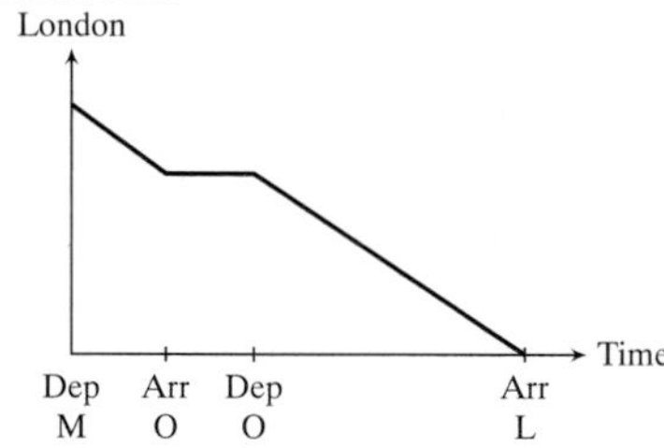

57.

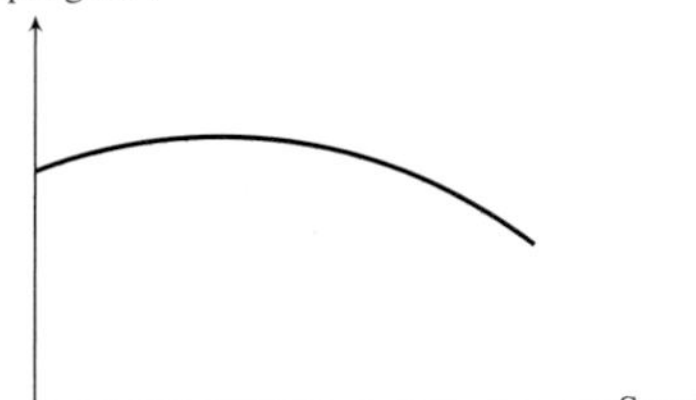

59.

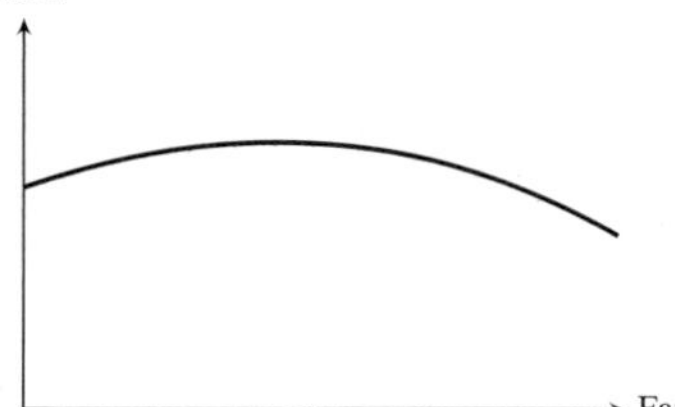

61.

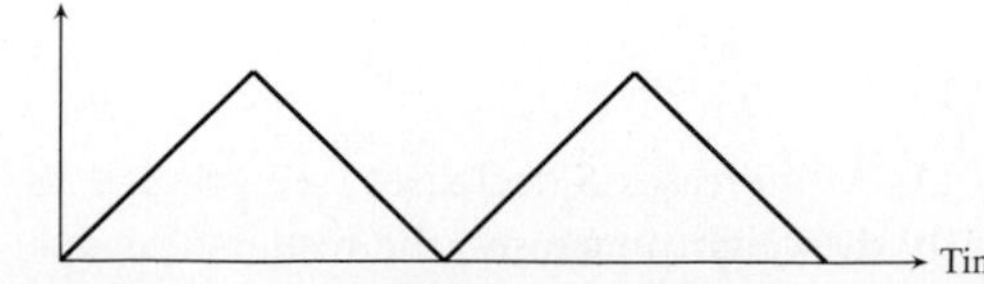

63. **(a)** $s_1(t) = -16t^2 + 288t$ and $s_2(t) = -16t^2 + 288t + 352$
(b) 1296 feet at $t = 9$ **(c)** 1648 feet at $t = 9$

65. **(a)** $s_1(t) = -16t^2 + 288t$ and $s_2(t) = -16(t-5)^2 + 288(t-5) + 352$
(b) 1296 feet at $t = 9$ **(c)** 1648 feet at $t = 14$

67. $d(t) =$

$$\begin{cases} 55t & 0 \le t \le 2 & \text{Cleveland to Toledo} \\ 110 & 2 \le t \le 3 & \text{Meal in Toledo} \\ 70(t-3) + 110 & 3 \le t \le 6.5 & \text{Toledo to Chicago} \end{cases}$$

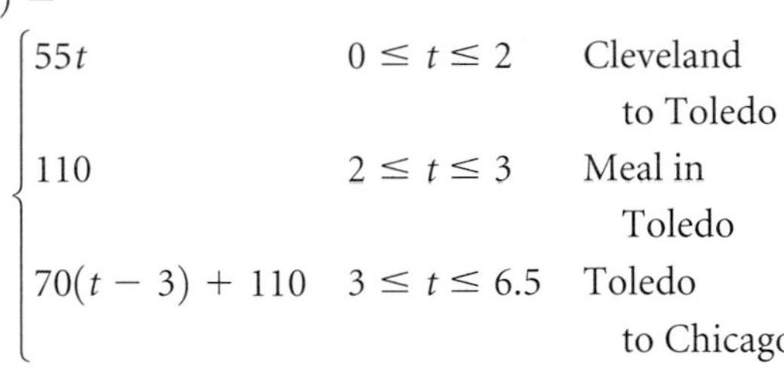

69. **(a)** 15.04 miles per hour
(b) $d(t) = 15.04t/60$ miles, t in minutes
(c)

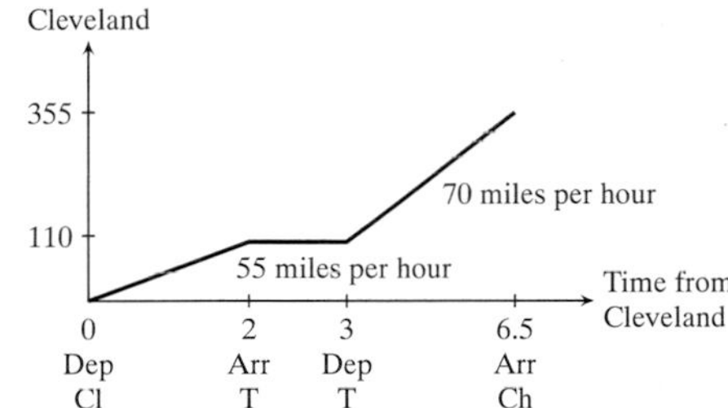

(d) Probable race: Runners jostle for first $\frac{1}{4}$ mile (first lap), run at a constant pace for the middle $\frac{1}{2}$ mile or so, and then sprint the last stretch.

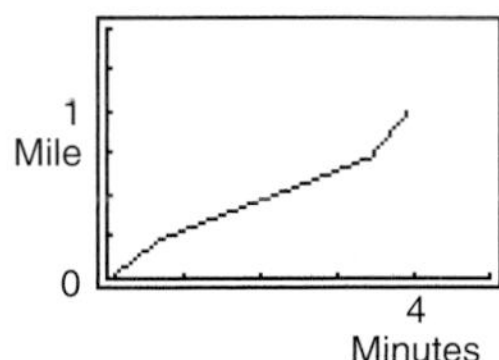

71. ball passes over goal posts at a height of 22 feet

73. ball lands 115.6 feet from pin; does not land on the green

Exercises 3.5

1. all real numbers except 0

3. all real numbers except $-\frac{1}{3}$

5. all real numbers except $-2, 1$

7. all real numbers except $-7, 0, 3$

9. all real numbers except $0, 4$

11. all real numbers except $-4, 1$

13. $f(x) = x, x \neq 2$

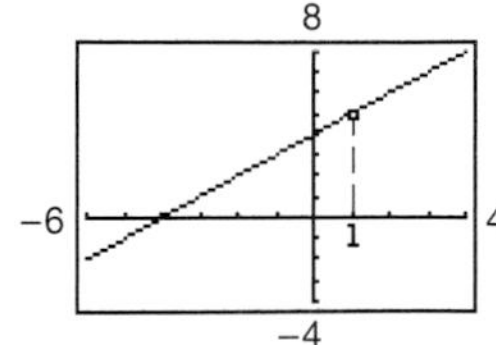

15. $f(x) = 2x + 5; x \neq \frac{2}{3}$

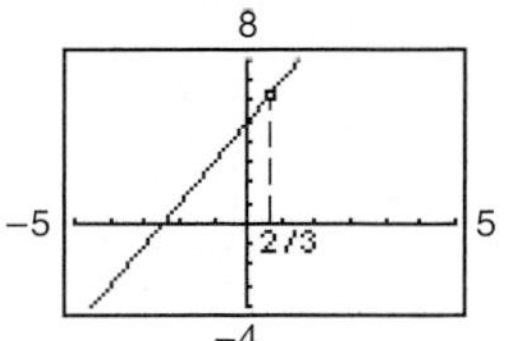

17. $f(x) = x + 4; x \neq 1$

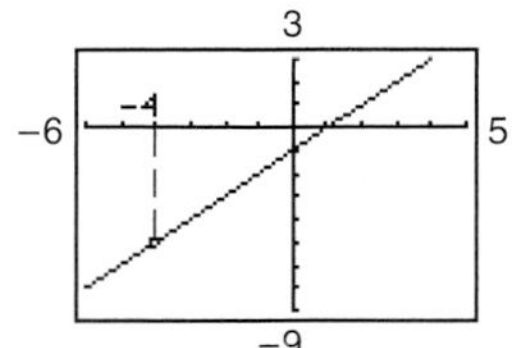

19. $f(x) = x - 1; x \neq -4$

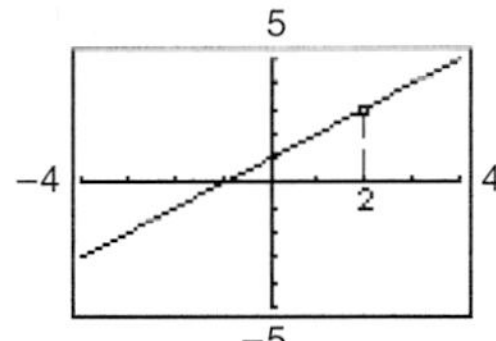

21. $f(x) = x + 1; x \neq 2$

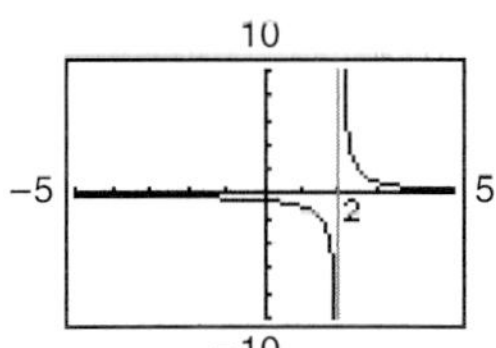

23. vertical asymptote $x = 2$; horizontal $y = 0$

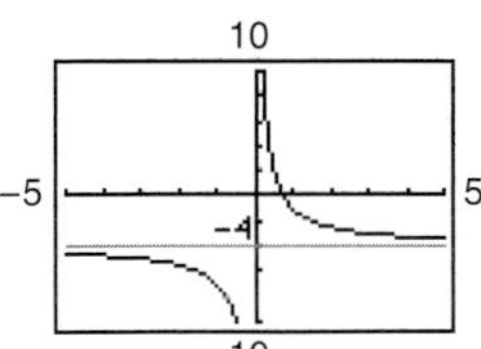

25. vertical $x = 0$; horizontal $y = -4$

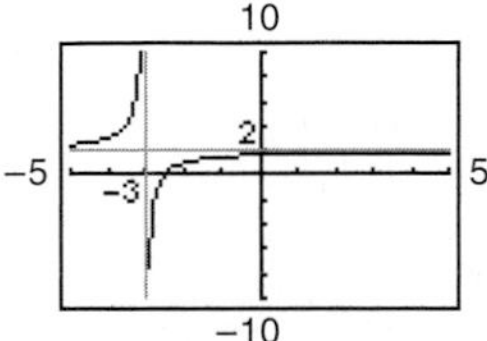

27. vertical $x = -3$; horizontal $y = 2$

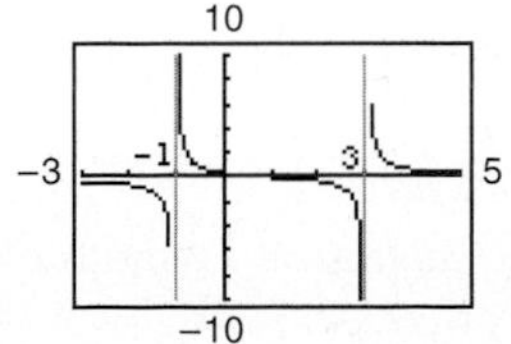

29. vertical $x = -1, x = 3$; horizontal $y = 0$; zero $x = 1$

31. vertical $x = 1, x = 4$; horizontal $y = 0$; no zeros

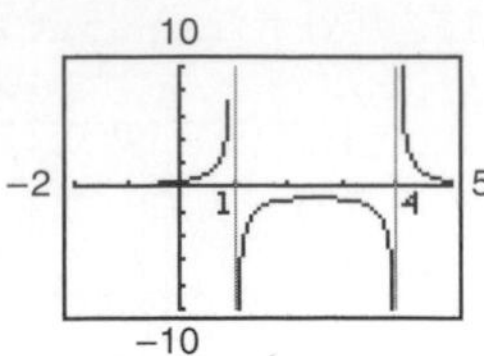

33. vertical $x = -3, x = 4$; horizontal $y = 2$; zeros $x = -\frac{5}{2}, 5$

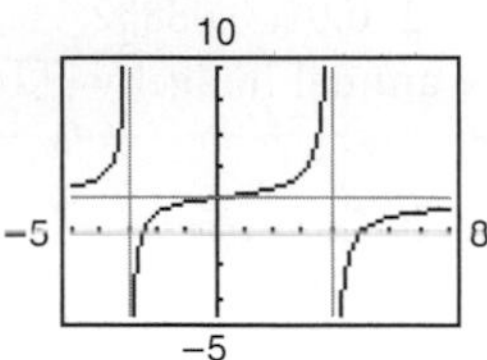

35. vertical $x = -4, x = 2$; horizontal $y = 1$; zeros $x = -1, 3$

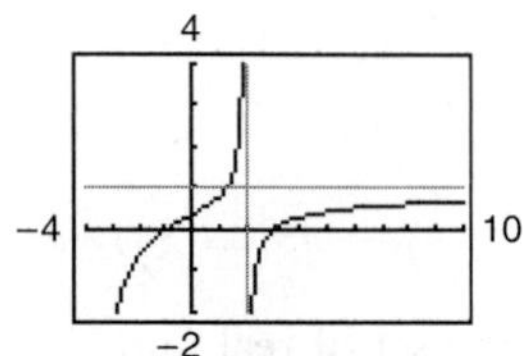

37. vertical $x = 2$; horizontal $y = 3$; zero $x = -\frac{5}{3}$

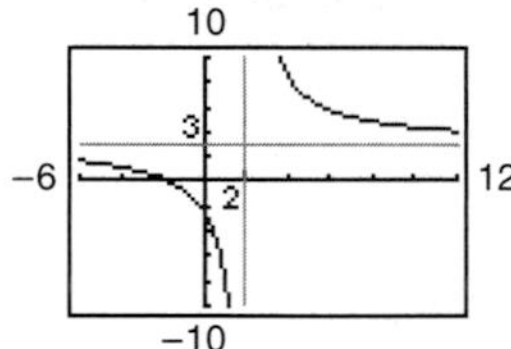

39. **(a)** vertical asymptote $x = 0$; horizontal asymptote $y = 0$; $f(x) = \dfrac{1}{x}$

(b) vertical $x = 0$; horizontal $y = 2$; $g(x) = \dfrac{1}{x} + 2$

(c) vertical $x = 3$; horizontal $y = 2$; $g(x) = \dfrac{1}{x-3} + 2$

41. **(a)** vertical asymptote $x = 0$; horizontal asymptote $y = 0$; $f(x) = \dfrac{1}{x^2}$

(b) vertical $x = 0$; horizontal $y = 3$; $g(x) = \dfrac{1}{x^2} + 3$

(c) vert $x = -1$; horizontal $y = 3$; $g(x) = \dfrac{1}{(x+1)^2} + 3$

43. Slant asymptote $y = x + 3$; vertical asymptote $x = 2$

45. Slant $y = x - 1$; vertical $x = 3$

47. Slant $y = x + 5$; vertical $x = 4$

49. Slant $y = -3x + 1$; vertical $x = 0$

51. Graphs intersect at $(1, 1)$. If $f(x) = x^{-1/n}$ and $g(x) = x^{-1/m}$ with $n > m$, $f > g$ on $(0, 1)$ and $g > f$ on $(1, \infty)$.

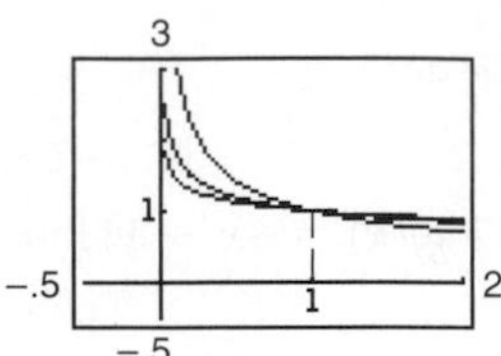

Exercises 3.6

1. $(f + g)(x) = 3x + 1$, real numbers
$(f - g)(x) = x + 7$, real numbers
$(fg)(x) = 2x^2 - 2x - 12$, real numbers
$(f/g)(x) = (2x + 4)/(x - 3)$, all real numbers except $x = 3$

3. $(f + g)(x) = 11x^2$, real numbers
$(f - g)(x) = -5x^2 - 4x + 2$, real numbers
$(fg)(x) = 24x^4 - 10x^3 + x^2 + 4x - 1$, real numbers
$(f/g)(x) = (3x^2 - 2x + 1)/((4x - 1)(2x + 1))$, all real numbers except $x = \frac{1}{4}, -\frac{1}{2}$

5. $(f + g)(x) = 3|x|$, real numbers
$(f - g)(x) = |x| + 6$, real numbers
$(fg)(x) = 2x^2 - 3|x| - 9$, real numbers
$(f/g)(x) = (2|x| + 3)/(|x| - 3)$, all real numbers except $x = -3, 3$

7. $(f + g)(x) = (2x - 7)/(x - 4)$, all real numbers except $x = 4$
$(f - g)(x) = (2x - 9)/(x - 4)$, all real numbers except $x = 4$
$(fg)(x) = 2/(x - 4)$, all real numbers except $x = 4$
$(f/g)(x) = 2(x - 4)$, all real numbers except $x = 4$

9. $(f + g)(x) = (x^2 + 6x + 10)/(x + 3)$, all real numbers except $x = -3$
$(f - g)(x) = (x^2 + 6x + 8)/(x + 3)$, all real numbers except $x = -3$
$(fg)(x) = 1$, domain, all real numbers except $x = -3$
$(f/g)(x) = (x + 3)^2$, all real numbers except $x = -3$

11. $(f + g)(x) = 0$, real numbers
$(f - g)(x) = 14$, real numbers
$(fg)(x) = -49$, real numbers
$(f/g)(x) = -1$, real numbers

13. $(f+g)(x) = 5x^2 + \sqrt{x-4};\ x \geq 4$
$(f-g)(x) = 5x^2 - \sqrt{x-4};\ x \geq 4$
$(fg)(x) = (5x^2)\sqrt{x-4};\ x \geq 4$
$(f/g)(x) = 5x^2/\sqrt{x-4};\ x > 4$

15. $(f+g)(x) = \sqrt{x-3} + \sqrt{x-4} + 4;\ x \geq 4$
$(f-g)(x) = \sqrt{x-3} - \sqrt{x-4} + 4;\ x \geq 4$
$(fg)(x) = \sqrt{(x-3)(x-4)} + 4\sqrt{x-4};\ x \geq 4$
$(f/g)(x) = (\sqrt{x-3}+4)/\sqrt{x-4};\ x > 4$

17. 22 **19.** 40 **21.** $\frac{7}{22}$ **23.** does not exist

25. $(g \circ f)(x) = 6x + 23;\ (f \circ g)(x) = 6x + 11;$ domains: real numbers

27. $(g \circ f)(x) = 3x^2 + 1;\ (f \circ g)(x) = 9x^2 + 6x + 1;$ domains: reals

29. $(g \circ f)(x) = 4;\ (f \circ g)(x) = 49;$ domains: reals

31. $(g \circ f)(x) = 1/x;\ (f \circ g)(x) = 1/x;$ domains: reals except $x = 0$

33. $(g \circ f)(x) = -3;\ (f \circ g)(x) = 3;$ domains: reals

35. $(g \circ f)(x) = -6x^2 + 3x - 10;$
$(f \circ g)(x) = 18x^2 - 21x + 10;$ domains: reals

37. $(g \circ f)(x) = (|x+1|)^2 - 4|x+1| + 3;$
$(f \circ g)(x) = |x^2 - 4x + 4|;$ domains: reals

39. $(g \circ f)(x) = x;\ (f \circ g)(x) = |x|;$ domains: reals

41. $(g \circ f)(x) = \sqrt{\sqrt{x-3}+5};$ domain: $x \geq 3;$
$(f \circ g)(x) = \sqrt{\sqrt{x+5}-3};$ domain: $x \geq 4$

43. 9 **45.** 3 **47.** $4a + 3$ **49.** 19

51. $(f \circ g)(x)$ **53.** $(g \circ f)(x)$ **55.** $(h \circ g)(x)$

57. $(g \circ f)(x)$ **59.** $(f \circ g)(x)$ **61.** $(g \circ h)(x)$

63. Company should raise production to 283 computers per week. $P(283) = \$2345.14$.

65. **(a)** \$3584
(b) Decreases to $x = 240$, increases from $x = 240$ on. Average cost of production originally decreases as production increases, but a stage is reached after which the average cost of producing a single item increases.
(c) 240 units **(d)** \$2720

67. **(a)** $D(t) = 6t^2 - 306t + 2376$
(b) increase or decrease during year t
(c) $D(5) = 996$, $D(20) = -1344$, $D(45) = 756$. There is an increase of 996 visitors during the fifth year (1950), a decrease of 1344 during the 20th year (1965), an increase of 756 during the 45th year (1990).
(d) Min $D(t)$ at $A(25.5, -1525.5)$. The largest decrease in visitors took place in 1970: 1525 fewer visitors than the year before.

69. $f(x) = cx(1-x)$
$c = 2.5$: 0.5, $\to$ 0.6; steady state
$c = 3.2$: 0.5, . . . , 0.513045, 0.799455, . . . Period 2
$c = 3.5$: 0.5, . . . , 0.874997, 0.382820, 0.826941, . . . Period 3
$c = 4$: 0.5, $\to$ 0; steady state

71. $f(x) = 3.8397x(1-x);\ c = 3.8397$: 0.72, 0.77408352, 0.6714799169, . . .
0.149549, 0.488349, 0.959404, . . . Period 3

73. $f(x) = 1.2x(1-x),\ c = 1.2$: 0.08, 0.08832, . . . $\to$ 0.1667. 16.67% of the annual budget will be for the library.

Exercises 3.7

1. $g(f(x)) = \dfrac{(2x-1)+1}{2} = x;$

$f(g(x)) = 2\left(\dfrac{x+1}{2}\right) - 1 = (x+1) - 1 = x;$

domains and ranges of both are set of real numbers

3. $g(f(x)) = 4\left(\dfrac{x}{4}\right) = x;\ f(g(x)) = \dfrac{4x}{4} = x;$ domains and ranges of both are the set of real numbers

5. $g(f(x)) = \sqrt[3]{x^3} = x;\ f(g(x)) = (\sqrt[3]{x})^3 = x;$ domains and ranges of both are the set of real numbers

7. $g(f(x)) = \dfrac{3-(3-2x)}{2} = x;$

$f(g(x)) = 3 - 2\left(\dfrac{3-x}{2}\right) = x;$ domains and ranges of both are the set of real numbers

9. $g(f(x)) = (\sqrt{x-5})^2 + 5 = x;$
$f(g(x)) = \sqrt{x^2+5)} - 5 = x;$ domain of $(g) = [0, \infty) = \text{range}(f)$

11. $f(g(x)) = \dfrac{\dfrac{x}{1-x}}{\dfrac{x}{1-x}+1} = \dfrac{x}{x+(1-x)} = x;$

$g(f(x)) = \dfrac{\dfrac{x}{x+1}}{1-\dfrac{x}{x+1}} = \dfrac{x}{(x+1)-x} = x$

Range of f is $(-\infty, 1) \cup (1, \infty) =$ the domain of g.

13. $f^{-1}(x) = \dfrac{x-2}{5};$ domain $(-\infty, \infty)$

15. $f^{-1}(x) = \dfrac{x}{2};$ domain $(-\infty, \infty)$

17. $f^{-1}(x) = 4x - 3;$ domain $(-\infty, \infty)$

19. $f^{-1}(x) = \sqrt[3]{x-3}$; domain $(-\infty, \infty)$

21. $f^{-1}(x) = \dfrac{1}{x^2}$; domain $(0, \infty)$

23. $f^{-1}(x) = \sqrt{x}$; domain $[0, \infty)$

25. one-to-one

27. not one-to-one

29. not one-to-one

31. one-to-one

33. not one-to-one

35. not one-to-one

37. one-to-one

39. not one-to-one

41. one-to-one

43. not one-to-one

45. $f^{-1}(x) = x^3 - 4$; domain $(-\infty, \infty)$

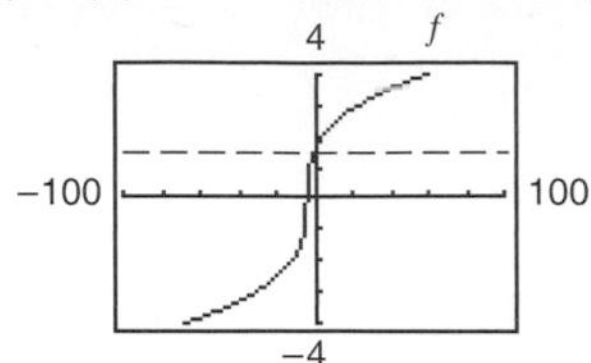

47. $f^{-1}(x) = (\sqrt[3]{x} + 5)^2$; domain $(-\infty, \infty)$

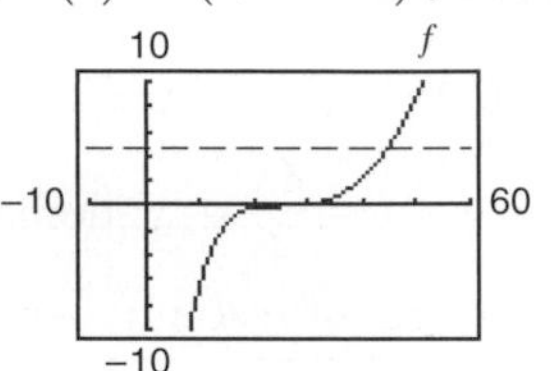

49. $f^{-1}(x) = \dfrac{1}{x} + 4$; domain $(0, \infty)$

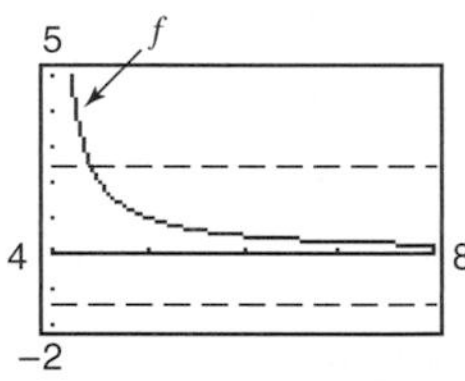

51. $f^{-1}(x) = x - 3$; domain $[0, \infty)$

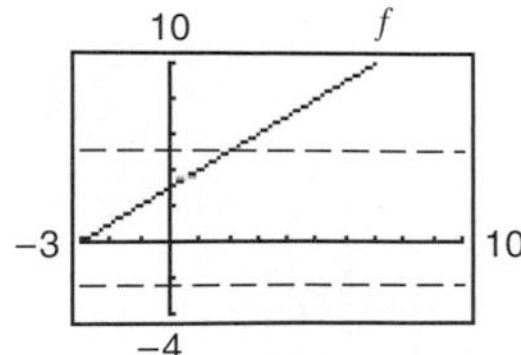

53. g is the inverse of f.

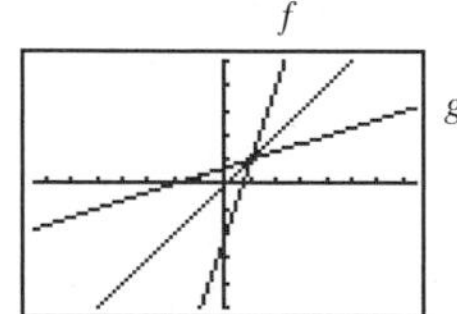

55. g is not the inverse of f. $f^{-1}(x) = (x + 1)/2$.

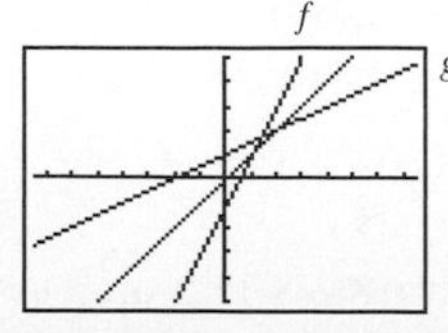

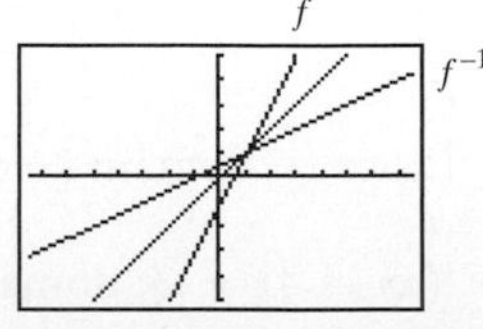

57. g is not the inverse of f. $f^{-1}(x) = \sqrt{x-3}$; domain $[3, \infty)$

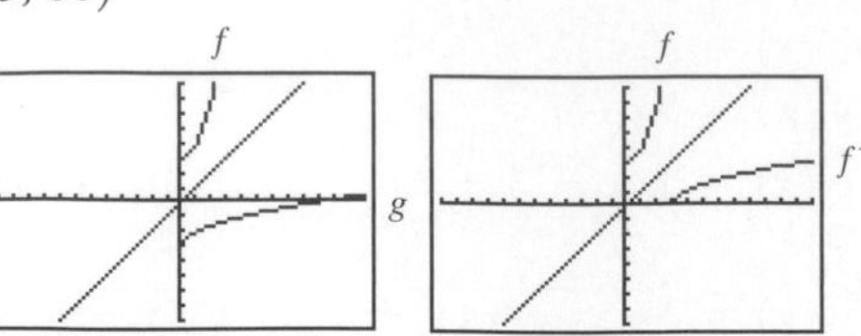

59. $f(x) = \sqrt{x-5}$, $g(x) = x^2 + 5$, $x \geq 0$; symmetry; g is the inverse of f.

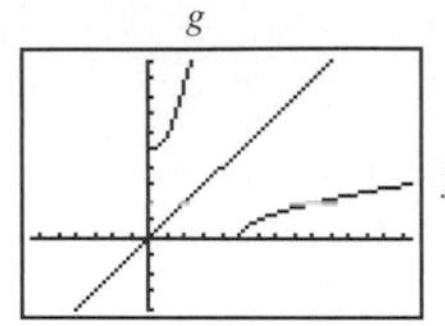

61. f is one-to-one and thus has an inverse for $-3 \leq k \leq 3$.

63. $f^{-1}(x) = \left(\dfrac{1}{a}\right)x - \dfrac{b}{a}$

65. $f(x) = ax^2 + bx + c$, $a \neq 0$. Graph is a parabola. There is a horizontal line that cuts the parabola in two points; thus, f is not one-to-one by the horizontal line test. There is no inverse.

67. $f(x) = ax^3 + b$, $a \neq 0$. If $p \neq q$, $p^3 \neq q^3$, $ap^3 \neq aq^3$, and $ap^3 + b \neq aq^3 + b$. Thus, $f(p) \neq f(q)$. f is one-to-one. It has an inverse: $f^{-1}(x) = \sqrt[3]{\dfrac{x-b}{a}}$.

69. $r = \sqrt{\dfrac{A}{\pi}}$; $r(1) = 0.5642$ inches

71. **(a)** $t = (v - u)/a$ **(b)** 3 seconds

73. **(a)** $x = (C - 700)/2$ **(b)** 265

75. **(a)** $W(d) = 140\left(\dfrac{3956}{3956 + d}\right)^2$. When $d = 0$, $W = 140$. W intercept of graph is 140. As d increases, W gradually decreases. Thus, the graph gradually falls from 140.

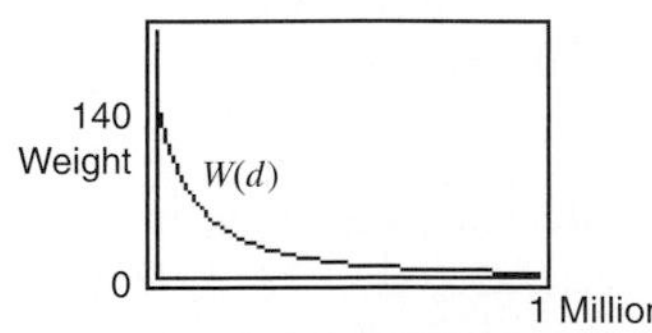

(b) Horizontal lines cut the graph at one or no points. Thus, W is one-to-one and has an inverse. $d = 3956\left(\sqrt{\dfrac{140}{W}} - 1\right)$; 42,852 miles from Earth

77. B U Y - T I - S T O C K

79. S C A T T E R

Chapter 3 Review Exercises

1. **(a)** x-intercepts $= -1, 2, 5$;
y-intercept $= f(0) = 10$

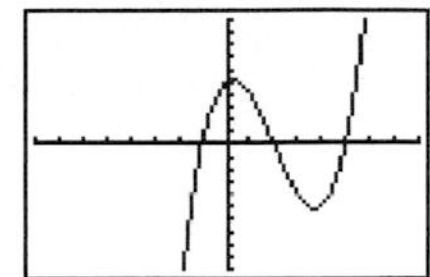

(b) x-intercepts $= -2, 0, 3$;
y-intercept $= f(0) = 0$

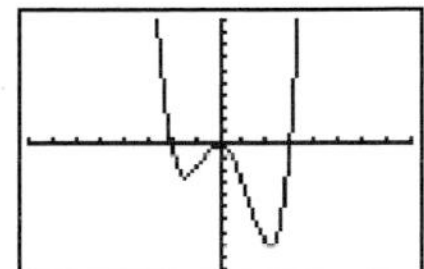

(c) x-intercepts $= -4, 0, 1$;
y-intercept $= f(0) = 0$

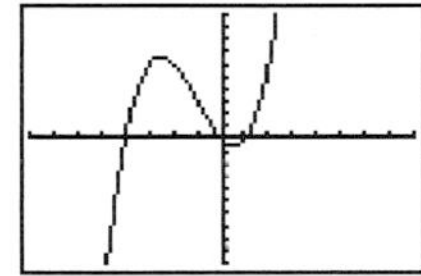

2. $f(x) = 2(x + 3)(x - 2)(x - 4)$

3. **(a)** inc$(-\infty, -1.27]$, $[-0.39, \infty)$;
dec$[-1.27, -0.39]$; max$(-1.27, 1.16)$;
min$(-0.39, 0.47)$; zero $x = -1.83$

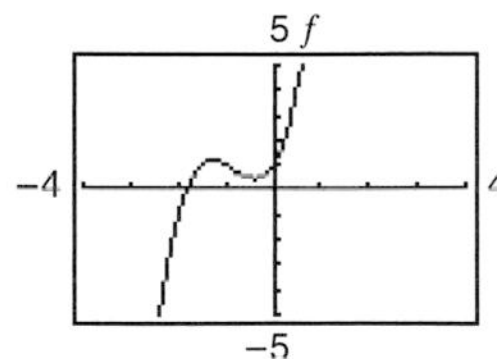

(b) dec$(-\infty, 1.92)$; inc$(1.92, \infty)$;
min$(1.92, -6.04)$; zeros 1.17 and 3.40

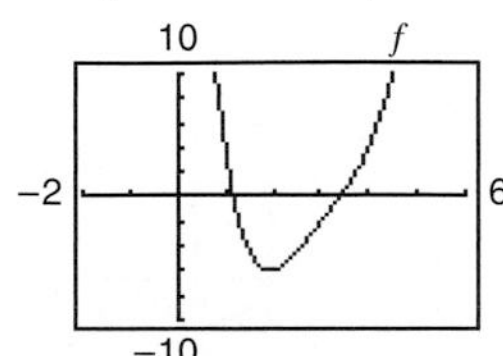

4. Sales after t days of advertising
$s(t) = t^3 - 89t^2 + 1{,}600t + 10{,}000$ units, $t \leq 4$.
Max at (11.0447, 18,162.112). Sales peak after 11 days.

5. **(a)** $s(t) = 0.1t^3 - 16t^2 + 468t$

(b) max(17.4942, 3825.9379); zero when (38.5271, 0). Takes 17.49 seconds to reach max height, 38.53 seconds to come back to ground level.

(c) Height increasing from 0 to 17.49 seconds, decreasing from 17.49 to 38.53 seconds.

6. **(a)** Range $[-11, 13]$.

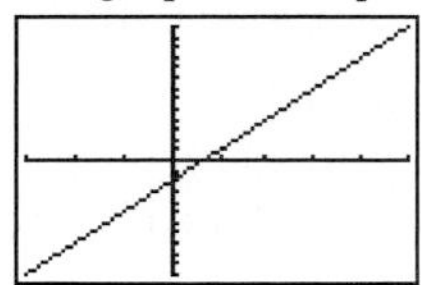

(b) Range $[-6, 14]$

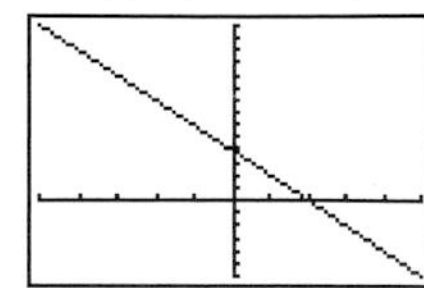

(c) Range $[-5, 13]$.

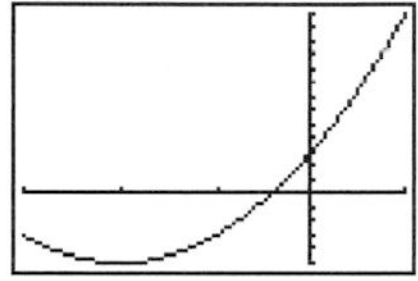

7. **(a)** **(b)**

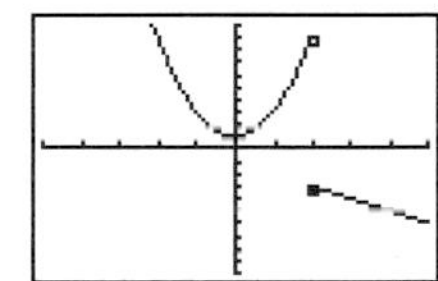

8. **(a)** $P(x) = \begin{cases} 8x & \text{if } 0 \leq x \leq 9 \\ 12(x - 9) + 72 & \text{if } 9 < x \leq 16 \end{cases}$

(b) $P(11) = \$95$

9. 10.6 to 81.5 days

10. **(a)** $f(x) = -\dfrac{2}{x^2}$; $f(-x) = -\dfrac{2}{(-x)^2}$
$f(-x) = f(x)$; $f(-x) \neq -f(x)$; even

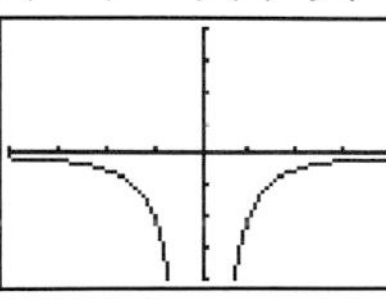

Symmetric about y-axis.

(b) $f(x) = 3$; $f(-x) = 3$
$f(-x) = f(x)$; $f(-x) \neq -f(x)$; even

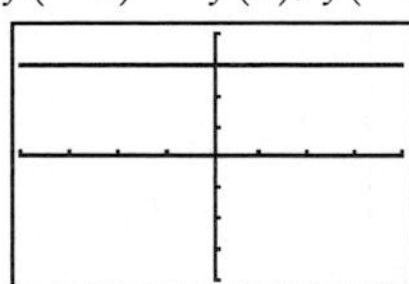

Symmetric about y-axis.

(c) $g(x) = -4x^3$; $g(-x) = -4(-x)^3$
$g(-x) \neq g(x)$; $g(-x) = -g(x)$; odd

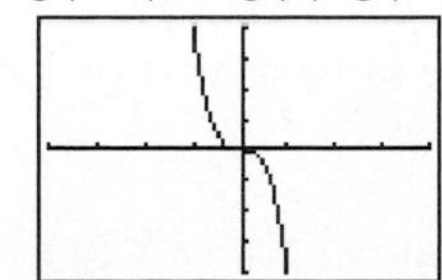

Symmetric about origin.

(d) $h(x) = x^4 + x$; $h(-x) = (-x)^4 + (-x)$
$h(-x) \neq h(x)$, $h(-x) \neq -h(x)$; neither

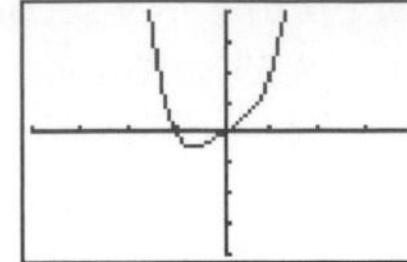

Not symmetric about y-axis or origin.

11. (a) $g(x) = x^2 - 3$. Shift $f(x) = x^2$ down 3.

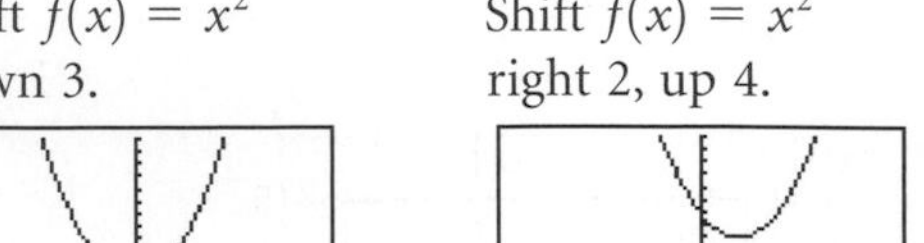

(b) $g(x) = (x - 2)^2 + 4$. Shift $f(x) = x^2$ right 2, up 4.

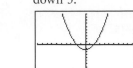

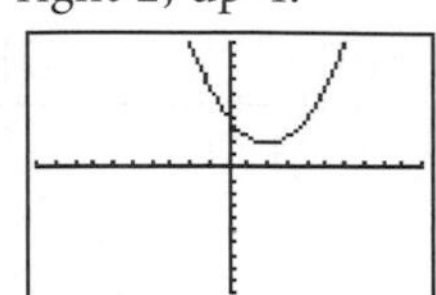

(c) $g(x) = |x - 3| - 5$. Shift $g(x) = |x|$ right 3, down 5.

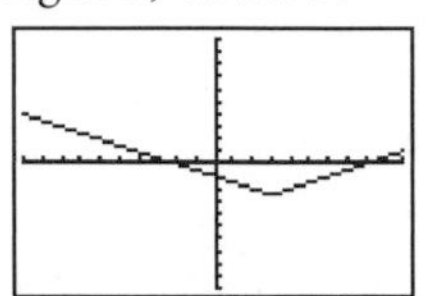

(d) $f(x) = -(x - 3)^2 - 4$. Shift $f(x) = x^2$ right 3, reflect in x-axis, down 4.

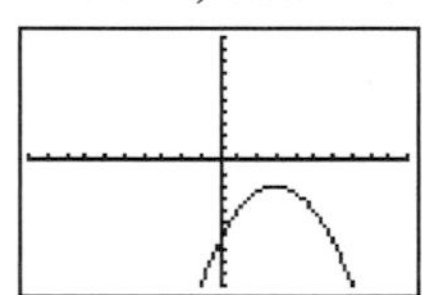

(e) $f(x) = \sqrt{x - 1} + 3$. Shift $f(x) = \sqrt{x}$ right 1, up 3.

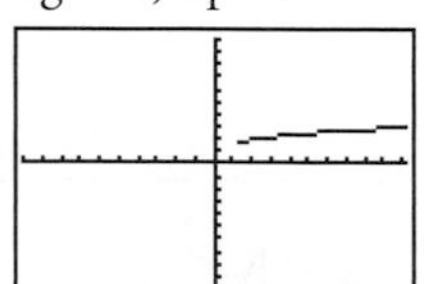

12. g: shift f down 4; h: shift f left 3, up 2

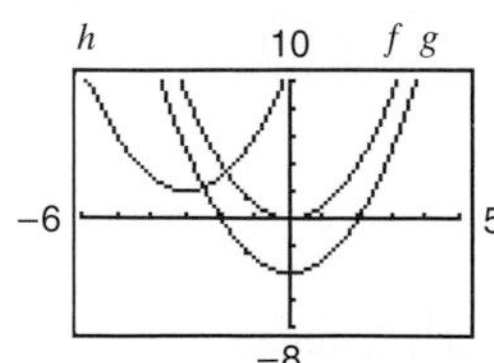

13. (a) $g(x) = x^3 + 5$ **(b)** $g(x) = -x^3 + 5$
(c) $g(x) = (x - 2)^3 - 10$

14. (a) Value of g at x is the same as that of f at $(x - 1)$.
(b) $h(x) = x^2 + 4$
(c) $k(x) = (x - 1)^2 + 4$

15.

Distance
Time
Appears
Cars level
Disappears

16.

No. of items bought
Price

17. $N(t) = -kx^2 + 150$; $k > 0$; domain $x \geq 0$

18. Tom starts from Dallas at 9:00 AM and travels at 60 miles per hour. Let t be time measured from 9:00 AM. $f(t) = 60t$ gives Tom's distance from Dallas at time t. Sally starts from Dallas at 9:15 AM at 75 miles per hour. $g(t) = 75(t - \frac{1}{4})$ gives Sally's distance from Dallas at time t. Sally catches up with Tom at $A(1.25, 75)$ below, 75 miles from Dallas, 1.25 hours after 9:00 AM; that is at 10:15 AM.

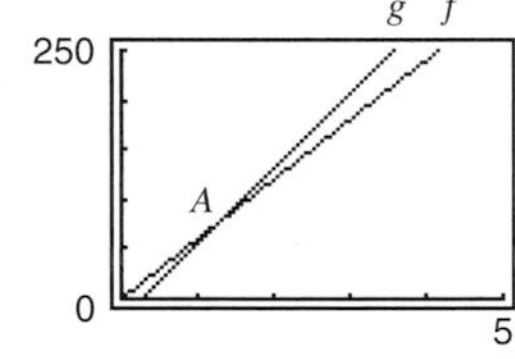

19. (a)

Decreasing $(-\infty, 0]$, increasing $[0, \infty)$

(b)

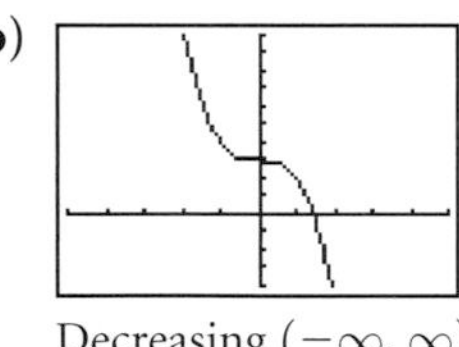

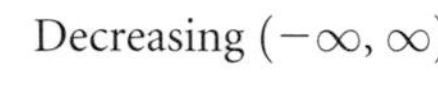

Decreasing $(-\infty, \infty)$

(c)

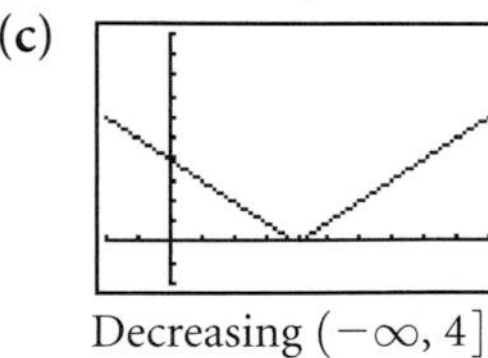

Decreasing $(-\infty, 4]$, increasing $[4, \infty)$

(d)

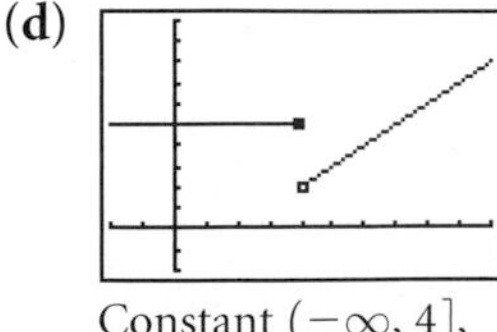

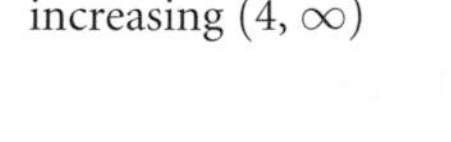

Constant $(-\infty, 4]$, increasing $(4, \infty)$

(e)

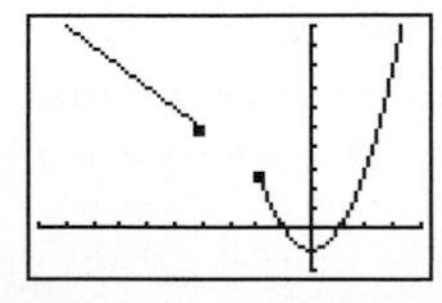

Decreasing $(-\infty, -4]$ and $[-2, 0]$, increasing $[0, \infty)$

20. (a) all real numbers except 0
(b) all real numbers except -4 and 2

21. **(a)** domain: all reals except 1; no asymptotes

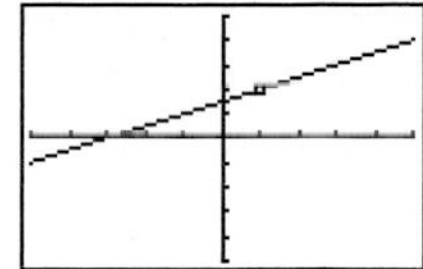

(b) vert asymptote $x = 4$; horiz $y = 3$; domain: all reals except 4

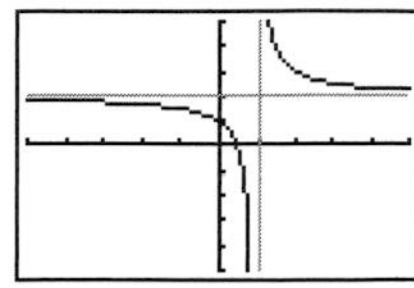

(c) vert $x = -3$, $x = 2$; horiz $y = 0$; domain: all reals except -3, 2

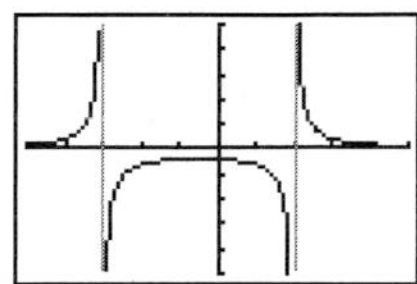

(d) vert $x = -2$, $x = 4$; horiz $y = 4$; domain: all reals except $x = -2$, 4

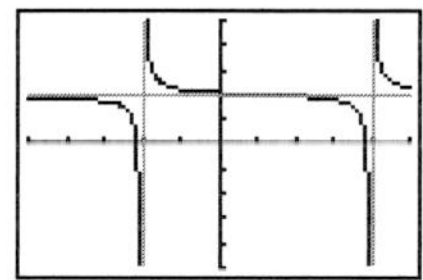

22. **(a)** $(f + g)(x) = -x + 4$; real numbers
$(f - g)(x) = 7x - 6$; real numbers
$(fg)(x) = -12x^2 + 19x - 5$; real numbers
$(f/g)(x) = (3x - 1)/(-4x + 5)$; all real numbers except $x = \frac{5}{4}$

(b) $(f + g)(x) = 4x + 1 + 2\sqrt{3x + 4}$; $x \geq -\frac{4}{3}$
$(f - g)(x) = 4x + 1 - 2\sqrt{3x + 4}$; $x \geq -\frac{4}{3}$
$(fg)(x) = (4x + 1)(2\sqrt{3x + 4})$; $x \geq -\frac{4}{3}$
$(f/g)(x) = (4x + 1)/(2\sqrt{3x + 4})$; $x > -\frac{4}{3}$

(c) $(f + g)(x) = (x^4 + 3x^3 + x^2 + 1)/x^2$; real numbers except $x = 0$
$(f - g)(x) = (x^4 + 3x^3 + x^2 - 1)/x^2$; real numbers except $x = 0$
$(fg)(x) = (x^2 + 3x + 1)/x^2$; real numbers except $x = 0$
$(f/g)(x) = x^2(x^2 + 3x + 1)$; real numbers

23. **(a)** $(g \circ f)(x) = 6x + 3$; $(f \circ g)(x) = 6x - 16$; real numbers

(b) $(g \circ f)(x) = 48x^2 + 40x + 6$;
$(f \circ g)(x) = 12x^2 + 16x - 3$; real numbers

(c) $(g \circ f)(x) = x - 7 - 6\sqrt{x + 4}$; $x \geq -4$ or $[-4, \infty)$
$(f \circ g)(x) = \sqrt{(x - 7)(x + 1)}$; $x \geq 7$ or $x \leq -1$

24. $F(x) = (g \circ f)(x)$ **25.** $H(x) = (f \circ g)(x)$

27. **(a)** one-to-one **(b)** one-to-one

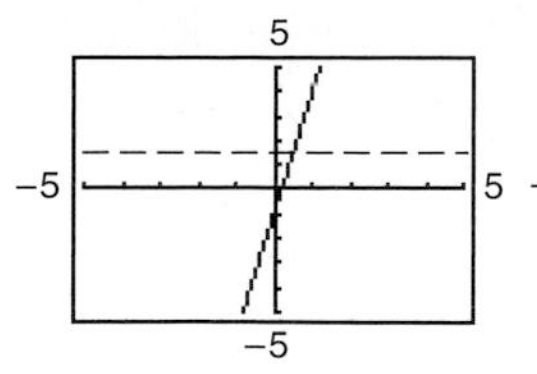

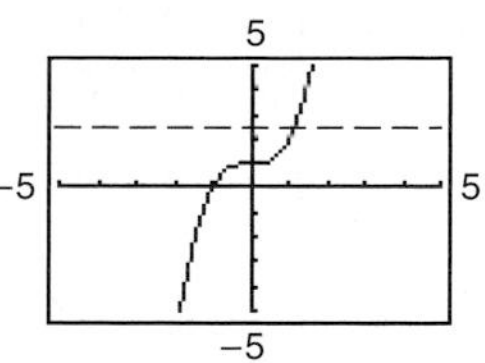

(c) not one-to-one

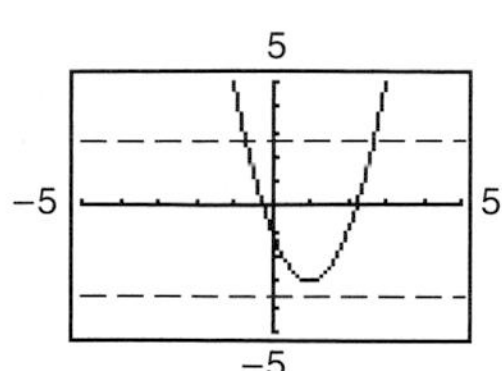

28. **(a)** $f^{-1}(x) = \dfrac{x + 4}{7}$; domain $(-\infty, \infty)$

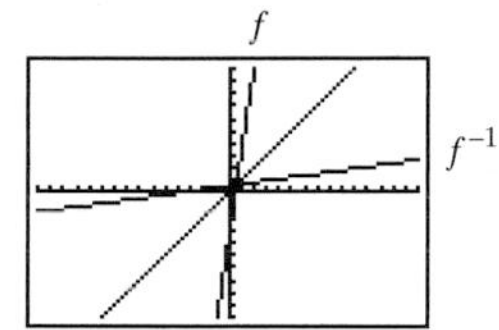

(b) $f^{-1}(x) = \sqrt{4 - x}$; domain $(-\infty, 4]$

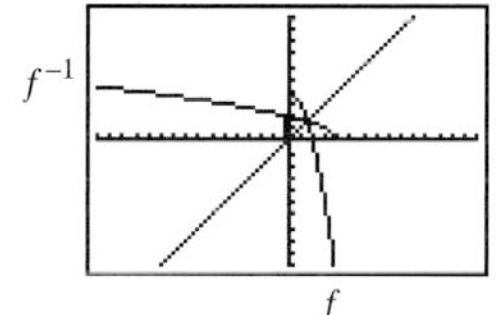

Chapter 3 Test

1. x-intercepts $= -2, 4, 6$; y-intercept $= f(0) = 48$

2. range $[-8, 41]$

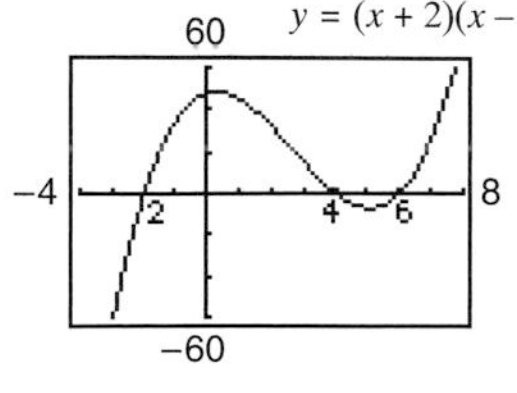

Exercise 1

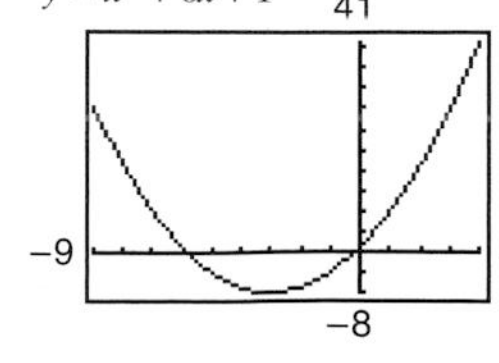

Exercise 2

3. $f(-x) = f(x)$; symmetric about y-axis

4. Shift f right 4, up 3.

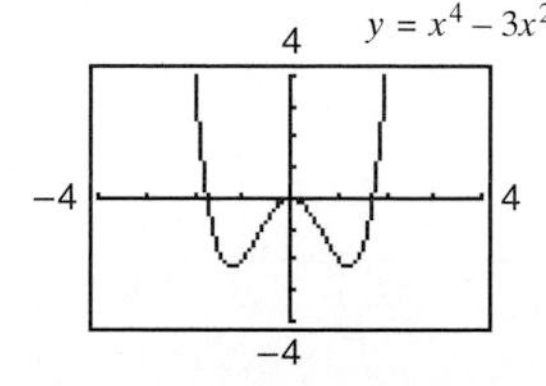

Exercise 3

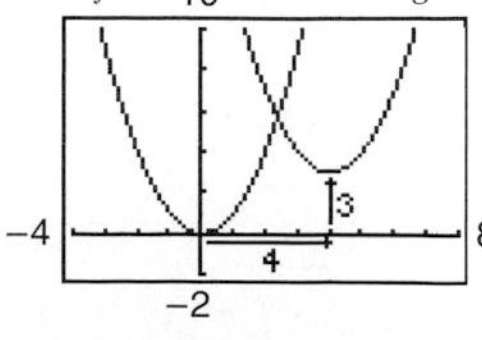

Exercise 4

5. $g(x) = (x + 2)^2$
6. vert $x = -1$, $x = 4$; horiz $y = 0$; domain: all real numbers except -1, 4
7. inc$(-\infty, -1.54]$, $[0.54, \infty)$; dec$[-1.54, 0.54]$; max$(-1.54, 11.51)$; min$(0.54, 2.49)$; zero $x = -2.70$

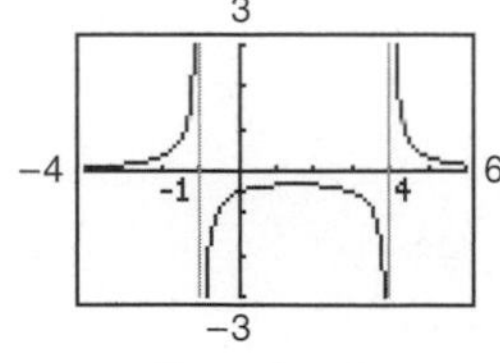

Exercise 6

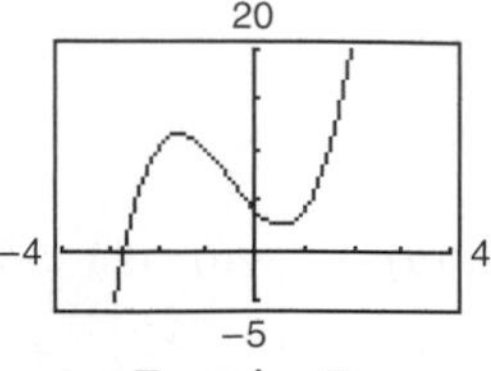

Exercise 7

8. $(f + g)(x) = (3x - 1) + (\sqrt{x - 2}) = 3x - 1 + \sqrt{x - 2};\ x \geq 2$
   $(f - g)(x) = (3x - 1) - (\sqrt{x - 2}) = 3x - 1 - \sqrt{x - 2};\ x \geq 2$
   $(fg)(x) = (3x - 1)(\sqrt{x - 2});\ x \geq 2$
   $(f/g)(x) = (3x - 1)/(\sqrt{x - 2});\ x > 2$
9. $(g \circ f)(x) = x + 4 + 5\sqrt{x - 3};\ x \geq 3$
10. $f^{-1}(x) = \dfrac{x - 2}{3}$
11. (a)

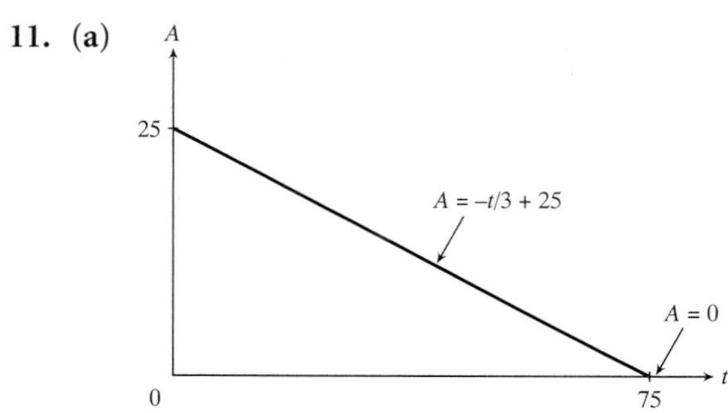

(b) $D(t) = \begin{cases} -t/3 + 25 & \text{if } 0 \leq t < 21 \\ -t/3 + 32 & \text{if } 21 \leq t < 42 \\ -t/3 + 39 & \text{if } 42 \leq t < 63 \\ \ldots \end{cases}$

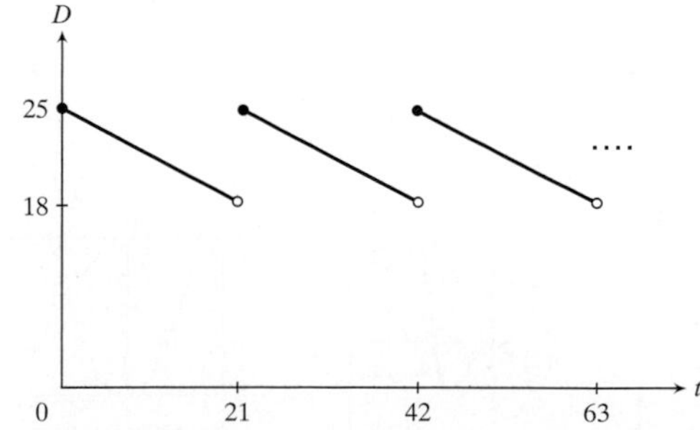

(c) every 21 days

12. max(1.5108849, 1059.2527); min(9.155783, 612.45098); Population peaked with 1059 deer in the middle of 1991. It was at its lowest in 1999 with 612 deer. The population is now increasing again.
13. $x = (C - 520)/6.5$; 2570 items

Chapter 4

Exercises 4.1

1. $q(x) = 2x^2 + 7x + 15$, remainder $= 33$
3. $q(x) = -2x^2 + 3x + 1$, remainder $= -3$
5. $q(x) = x^3 + 7x^2 + 32x + 168$, remainder $= 833$
7. $q(x) = x^3 - 4x^2 + 24x - 137$, remainder $= 825$
9. $q(x) = 3x^2 - x$, remainder $= 1$; $f(1) = 1$
11. $q(x) = 4x^2 - 6x + 15$, remainder $= -22$; $f(-2) = -22$
13. $q(x) = x^3 + x^2 - 3x + 9$, remainder $= -2$; $f(-1) = -2$
15. $q(x) = 3x^3 - 28x^2 + 223x - 1782$, remainder $= 14{,}260$; $f(-8) = 14{,}260$

17. 9 19. -45 21. 22
23. -58 25. zero 27. zero
29. not a zero 31. zero 33. factor
35. factor 37. not a factor 39. factor
41. 9 43. -3 45. $\frac{35}{16}$ 47. -7

Exercises 4.2

1. real and complex 3. real and complex
5. imginary and complex 7. real and complex
9. $2i$ 11. $6i$ 13. 11 15. $7i$
17. $-i$ 19. i 21. -1
23. real part $= 2$, imaginary part $= 3$
25. real part $= 0$, imaginary part $= 7$
27. real part $= 0$, imaginary part $= -8$

29. $3 - 2i$ 31. $3 + 5i$ 33. $-9i$
35. $-1 \pm \sqrt{2}i$ 37. 1 39. -2
41. $\pm 2i$ 43. $1 \pm 2i$ 45. $7 + 5i$
47. $2 - 5i$ 49. $3 - i$ 51. $-6 - 2i$
53. 12 55. $1 + 7i$ 57. $13 - 4i$
59. $-12 - 8i$ 61. $-21 + 20i$ 63. $-7 - 24i$
65. $\frac{1}{2} + \frac{1}{2}i$ 67. $-\frac{1}{2} - \frac{5}{2}i$ 69. $\frac{12}{13} + \frac{5}{13}i$
71. $\frac{1}{5} + \frac{2}{5}i$ 73. $-\frac{3}{2} - \frac{5}{2}i$ 75. 13

77. 5.8310 **79.** 9 **81.** 6.7082

83. Let the complex number be $a + bi$, then $(a + bi) - (a - bi) = 2bi$, an imaginary number.

85. no **87.** yes **89.** no **91.** yes

93. $-2 + 0i$ lies in the Mandelbrot set. $-3 + 0i$ does not lie in the Mandelbrot set. -2.000001 does not lie in the Mandelbrot set. The boundary of the set lies on/between points for 2 and 2.000001. Further exploration shows that 2.000 . . . 001 always gives a point outside the set. This evidence points to the fact that (2, 0) lies in the set and on its boundary.

Exercises 4.3

1. factor **3.** not a factor **5.** factor

7. $29 - 18i$ **9.** $-8 + 7i$ **11.** $-147 + 129i$

13. 2 (multiplicity one), -3 (one), -5 (one)

15. 4 (one), -2 (two) **17.** 5 (three), -7 (four)

19. $2i$ (one), $-3 + 4i$ (two)

21. $f(x) = x^3 - 9x^2 + 23x - 15$

23. $f(x) = x^3 - 4x^2 - 12x$

25. $f(x) = x^3 - x^2 - 8x + 12$

27. $f(x) = x^3 - (6 - i)x^2 + (13 - 7i)x + (-10 + 10i)$

29. $f(x) = x^3 - (2 + 2i)x^2 + (-1 + 4i)x + 2$

31. 1, 2, 3 **33.** $-3, 1, 2$ **35.** 2 (multiplicity three)

37. $-\frac{1}{3}, 2, 3$ **39.** $-2, i, -i$

41. $f(x) = 2x^3 - 6x^2 + 8$

43. $f(x) = -4x^3 + 24x^2 - 48x + 32$

45. $f(x) = -2x^3 - 16x^2 - 40x - 32$

47. $f(x) = (\frac{1}{2})x^4 - 5x^3 + 18x^2 - 27x + \frac{27}{2}$

49. $f(x) = -x^3 + (6 - i)x^2 + (-13 + i)x + 10 + 2i$

51. $f(x) = x^3 - 8x^2 + 21x - 20$

53. $f(x) = x^3 - 4x^2 + 22x + 68$

55. $f(x) = x^4 - 14x^3 + 69x^2 - 134x + 78$

57. $f(x) = x^4 + 2x^3 + 6x^2 + 80x + 200$

Exercises 4.4

1. 1, 2, 3 **3.** $-3, 1, 2$

5. $-4, -3, 2$ **7.** $-2, -1, 1, 4$

9. -1 (multiplicity two), 1, 3 **11.** $-2, \frac{2}{3}, 1$

13. $\frac{1}{2}$, 1 (multiplicity two) **15.** $-\frac{1}{4}, 1, 2$

17. $-2, \frac{1}{3}, \frac{1}{2}$ **19.** $-1, \frac{1}{2}$, 1 (multiplicity two)

21. $-2, -1, \frac{1}{2}$ **23.** $-4, -1, \frac{1}{4}$

25. $-2, \frac{1}{6}, 1, 2$ **27.** $-1, 2, -\sqrt{2}i, \sqrt{2}i$

29. $-1, \frac{2}{3}, -i, i$ **31.** $2, -\sqrt{3}i, \sqrt{3}i,$

Chapter 4 Review Exercises

1. **(a)** $q(x) = x^2 + 4$, remainder $= 11$
(b) $q(x) = 2x^2 + x$, remainder $= 6$
(c) $q(x) = 2x^3 + 5x^2 + 16x + 70$, remainder $= 275$

2. **(a)** 9 **(b)** 393 **(c)** -198

3. **(a)** zero **(b)** not a zero **(c)** zero

4. **(a)** not a factor **(b)** factor **(c)** factor

5. **(a)** $4i$ **(b)** $5i$ **(c)** 6 **(d)** $-13i$

6. **(a)** $-i$ **(b)** i **(c)** $-i$ **(d)** -1

7. **(a)** $1 \pm 2i$ **(b)** $\dfrac{3 \pm \sqrt{23}i}{4}$

8. **(a)** $10 + 2i$ **(b)** $-7 - i$ **(c)** $11 - 11i$
(d) $17 - 9i$ **(e)** $38 + 34i$ **(f)** $-5 - 12i$

9. **(a)** $-\frac{14}{13} - \frac{5}{13}i$ **(b)** $2 - 4i$ **(c)** $\frac{31}{73} - \frac{39}{73}i$

10. 13 **11.** factor **12.** $-21 + 74i$

13. $f(x) = x^3 - 4x^2 - 11x + 30$

14. $-5, 1, 3$ **15.** $f(x) = 3x^3 - 21x + 18$

16. **(a)** $-2, \frac{1}{2}, 3$ **(b)** $-2, -1, \frac{1}{2}$ **(c)** $2, -\sqrt{3}i, \sqrt{3}i$
(d) $-1, -2 - i, -2 + i$

Chapter 4 Test

1. $q(x) = x^2 + x + 8$, remainder $= 22$ **2.** 57

3. not a zero **4.** **(a)** $12i$ **(b)** $-9i$

5. **(a)** i **(b)** $-i$ **6.** $\dfrac{-3 \pm \sqrt{47}i}{4}$

7. **(a)** $5 + 2i$ **(b)** $27 + 14i$ **8.** $-\frac{13}{29} + \frac{11}{29}i$

9. $k = 6$ **10.** factor

11. $-3, 2, 5$ **12.** $-\frac{5}{2}, -1, 2$

Chapter 5

Exercises 5.1

1. **3.**

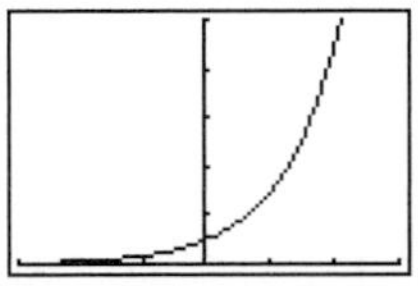

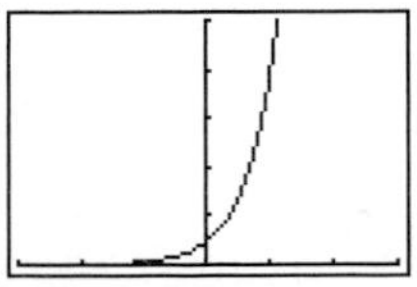

5. **(a)** 3.2490 **(b)** 10.7034 **(c)** 0.2333

7. **(a)** 0.1088 **(b)** 0.0055 **(c)** 1351.1761

In Exercises 9–19 all windows have $X_{scl} = 1$, $Y_{scl} = 1$. D stands for domain, R for range.

9. Shift f up 3.
$D(-\infty, \infty)$, $R(3, \infty)$

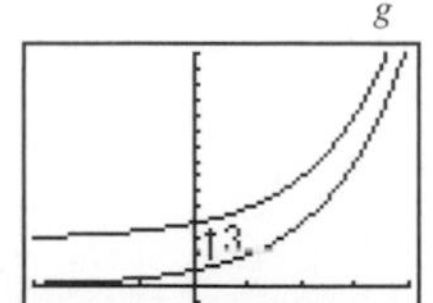

11. Shift f up 5.
$D(-\infty, \infty)$, $R(5, \infty)$

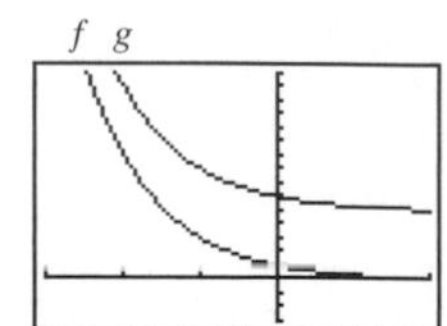

13. Shift f left 1.
$D(-\infty, \infty)$, $R(0, \infty)$

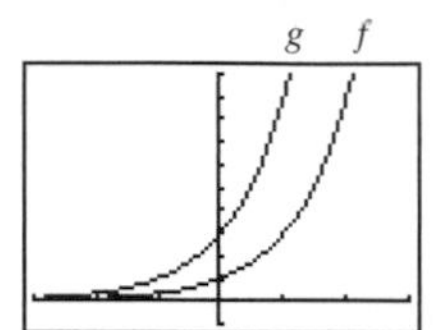

15. Shift f left 3.
$D(-\infty, \infty)$, $R(0, \infty)$

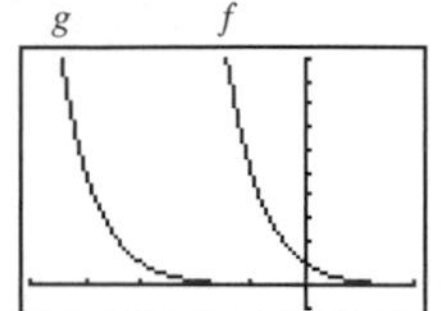

17. Shift f right 5, down 2.
$D(-\infty, \infty)$, $R(-2, \infty)$

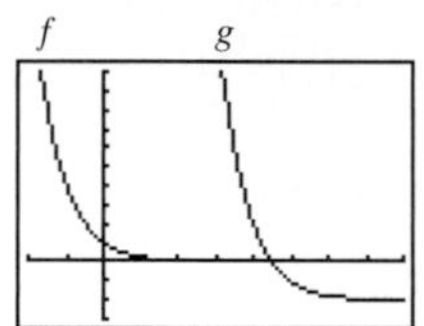

19. Multiply values of f by 3.
$D(-\infty, \infty)$, $R(0, \infty)$

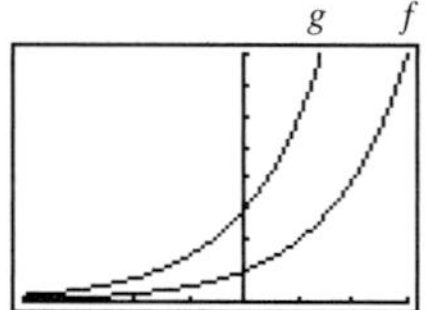

21. $f(x) = 3^x$, $g(x) = 4^x$; $f > g$ when $x < 0$; $f = g$ when $x = 0$; $f < g$ when $x > 0$.

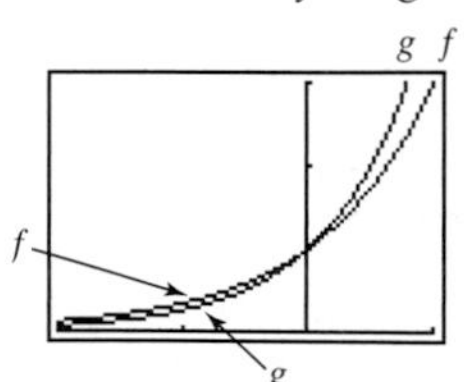

23. $g(x) = 2^x + 1$; shift f up 1.

25. $g(x) = -2^x + 1$; reflect f in x-axis, shift up 1.

27. $g(x) = 2^{-x} - 3$; reflect f in y-axis, shift down 3.

29. $g(x) = 2^{x-1} + 2$ **31.** $g(x) = -2^{x-2} + 5$

33. $f(x) = 8(2^{x/12})$

35. $f(x) = 6(2^{x/9})$; initial value is 6; doubling time 9.

37. **(a)** 8.70 billion **(b)** 3.66 billion

39. 139.7 million

41. **(a)** 25 **(b)** 356 thousand **43.** 2007

45. 25.6

47. $N(t) = 100(2^{t/10})$

t	0	10	20	30	40	50	60
N	100	200	400	800	1600	3200	6400

49. $N(t) = 600(2^{t/13})$

t	0	26	52	78	104	130
N	600	2,400	9,600	38,400	153,600	6,144,000

51. 58 minutes **53.** 26 minutes

Exercises 5.2

1. **(a)** 1.2840 **(b)** 2.3396 **(c)** 0.2019 **(d)** 0.0045

3. **(a)** $e^{-3} = 0.0498$; $e^{-2} = 0.1353$; $e^{-1} = 0.3679$; $e^0 = 1$; $e^1 = 2.7183$; $e^2 = 7.3891$; $e^3 = 20.0855$ **(b)** See graph.

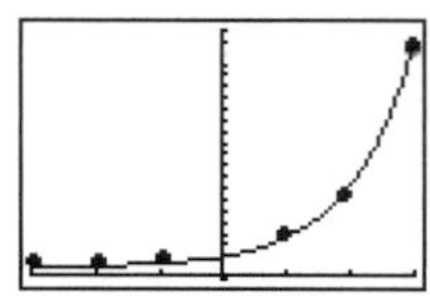

5. Reflect $f(x) = e^x$ in x-axis.

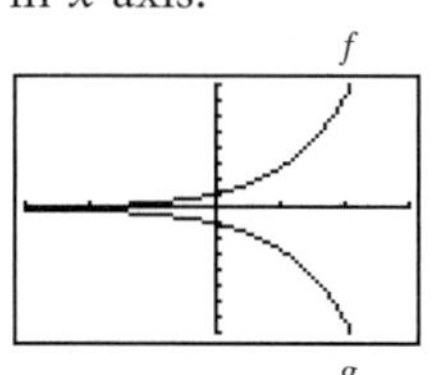

7. Shift $f(x) = e^x$ right 3.

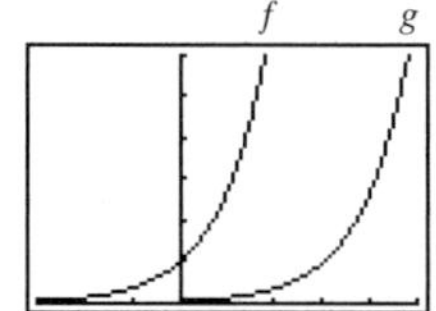

9. Reflect $f(x) = e^x$ in y-axis, shift right 1, up 3.

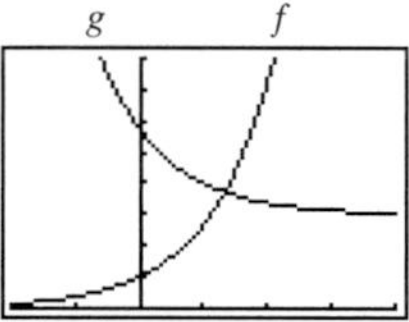

11. 6 **13.** $-\frac{7}{2}$ **15.** $\frac{3}{4}$ **17.** $-\frac{3}{4}$

19. no solution **21.** 3 **23.** 125 **25.** 64

27. 0.2619 (A) **29.** −1.8293 (A)

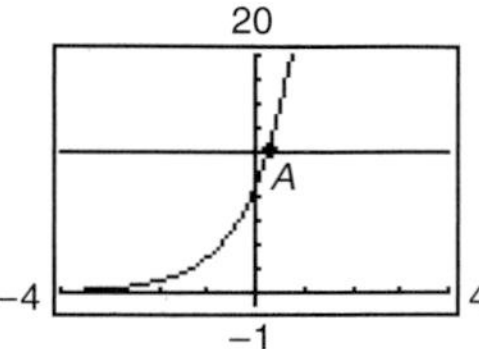

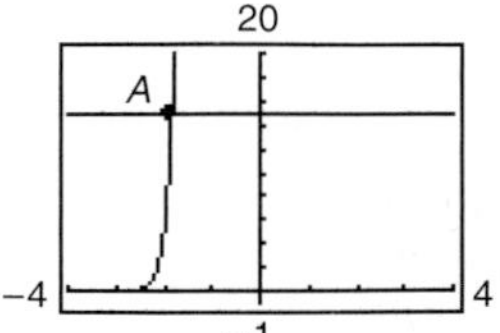

31. −2.0453 (A), 1.3787 (B) **33.** 1.1681 (A)

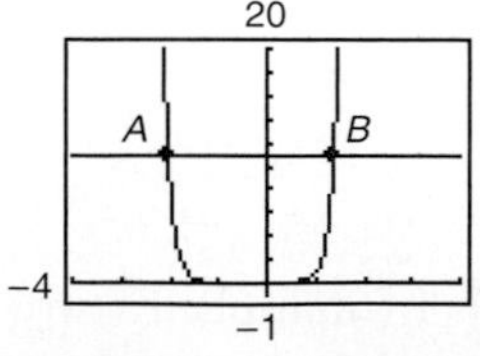

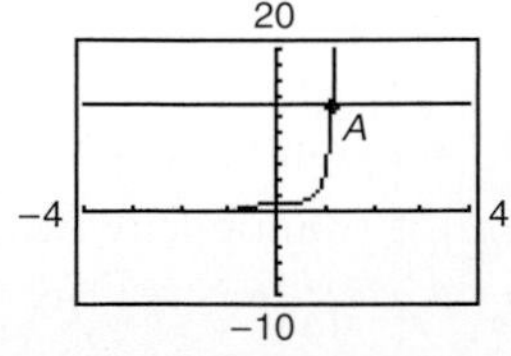

35. -1.8533 (A), 2.1599 (B)

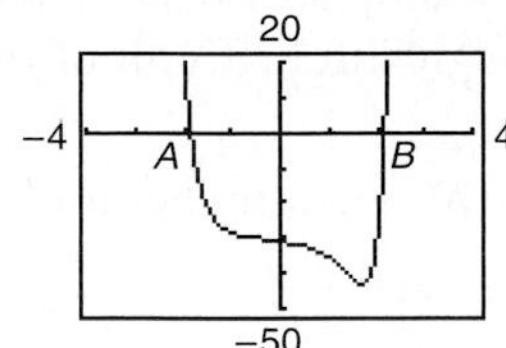

37. -2.4779 (A), 2.4779 (B)

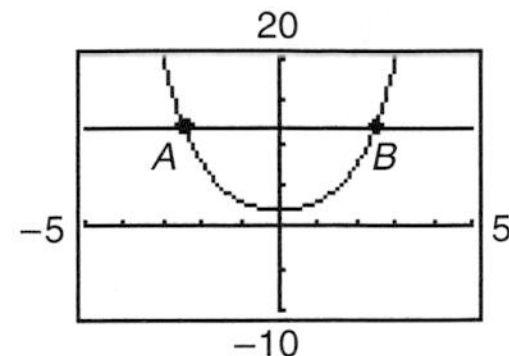

39. **(a)** \$10,099.82 **(b)** \$10,169.99

41. **(a)** \$15,627.12 **(b)** \$15,719.57

43. **(a)** \$10,897.12 **(b)** \$10,994.69

45. **(a)** \$61,879.83 **(b)** \$61,992.32

47. 3 years 8 months **49.** 11 years 7 months

51. 6% **53.** 4.5%

55. 561.26 nanometers

Exercises 5.3

1. $f(x) = \log_4 x$ **3.** $g(x) = \log_{12} x$

5. $g(x) = 8^x$ **7.** $x = \log_4 y$

9. $x = \log_{13} y$ **11.** $5 = \log_2 32$

13. $-3 = \log_2 0.125$ **15.** $x = 5^y$

17. $x = 12^y$ **19.** $8 = 2^3$ **21.** $7^0 = 1$

23. 2 **25.** 0 **27.** 1

29. 3 **31.** -3 **33.** $\frac{1}{4}$

35. **(a)** 0.7561 **(b)** 1.9140 **(c)** -0.8916 **(d)** -2.5481

37. **(a)** -0.6539 **(b)** 1.7783 **(c)** 2.8291 **(d)** 6.1702

39. **(a)** -0.4425 **(b)** 0.6974 **(c)** 1.2050 **(d)** 2.7350

41. Shift $f(x) = \ln x$ up 3.

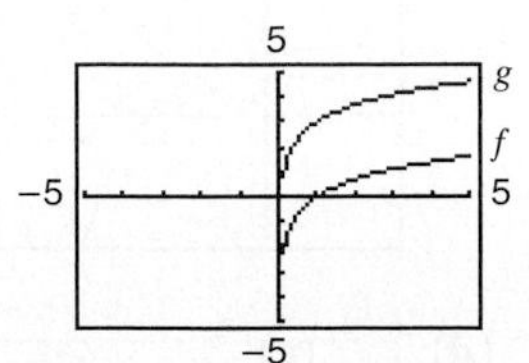

43. Shift $f(x) = \log x$ right 4.

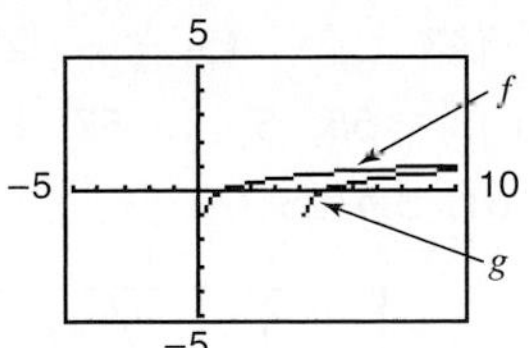

45. Shift $f(x) = \log x$ left 1, down 3.

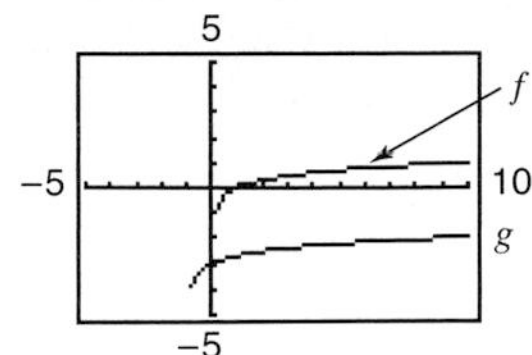

47. Reflect $f(x) = \ln x$ in y-axis.

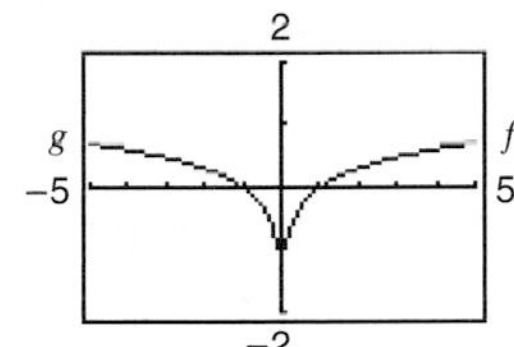

49. $f^{-1}(x) = \log_2(x + 3)$

51. $f^{-1}(x) = (\log_7(x - 3) - 5)/2$

53. $f^{-1}(x) = (6^x - 7)/4$ **55.** $f^{-1}(x) = \ln(x - 1)$

57. $f^{-1}(x) = (\ln(x - 1))/2 - 2$

59. $f^{-1}(x) = \frac{1}{3}e^x - 2$

61. $g(x) = \ln(x + 1)$. Shift left 1.

63. $g(x) = \ln(-x)$. Reflect in y-axis.

65. $g(x) = \ln(-(x + 1))$. Reflect in y-axis, shift left 1.

67. $f(x) = \ln x^2$. Vertical asymptote y-axis; $f(x) = f(-x)$. f is symmetric about y-axis.

69. $f(x) = \ln|x|$. Vertical asymptote y-axis. $f(x) = f(-x)$. f is symmetric about y-axis.

Exercises 5.4

1. 13 **3.** $\frac{19}{2}$ **5.** $\frac{127}{3}$

7. 1 **9.** 3 **11.** 4

13. $x = \log_3(2y) - 1$ **15.** $x = \log_2(4y) + 5$

17. $x = (3^{2y} - 1)/4$ **19.** $x = 6y - 2$

21. 8 **23.** 10 **25.** -16

27. **(a)** 2.4849 **(b)** 1.0987 **(c)** 3.5836 **(d)** -1.0987

29. **(a)** 0.5596 **(b)** 7.7836 **(c)** -0.5596 **(d)** 2.7726

31. $\log_a x + \log_a y - \log_a z$

33. $3\log_b x + 2\log_b y + \log_b z$

35. $-4\log_3 x - \log_3 z + \log_3 5$

37. $\log_a(x^2/2)$ **39.** $\log_c(x^8/(5x+2))$

41. $\log_{10}(x^{-2}y^{-1}(x-2)^4)$ **43.** $\frac{16}{3}$ **45.** $\frac{16}{3}$ **47.** 4

49. 0 **51.** $\frac{14}{9}$ **53.** 1 **55.** 5 **57.** 1, 6

59. 5.5894 **61.** 3.4125

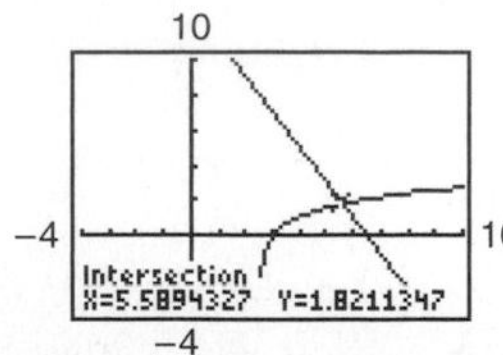

10
−4
10
Intersection
X=3.4125028 Y=3.2375084
−4

63. 6.9226 **65.** 2.7666

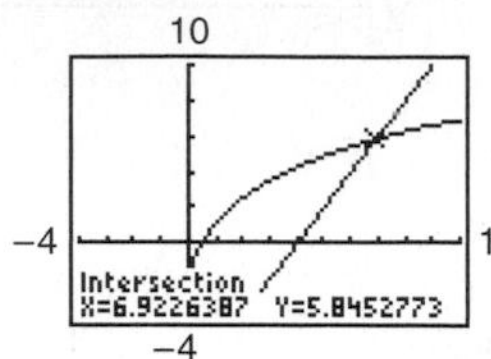

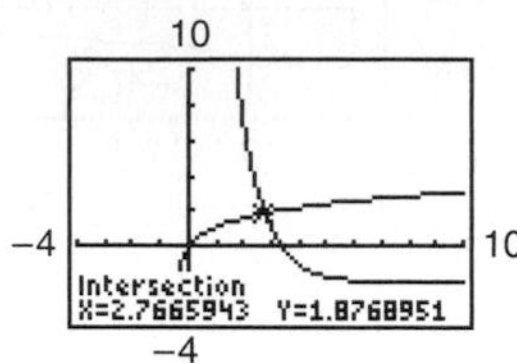

67. 1.4650 **69.** 0.4499 **71.** 2.4128

73. 3.2362 **75.** −1.0969 **77.** 6.1530

79. 0.5365 **81.** 0.3715 **83.** 2.0000

85. 0.9186

87. **(a)** $\log_a(x/y) = \log_a x - \log_a y$: let $p = \log_a x$, $q = \log_a y$. Then $x = a^p$, $y = a^q$. $x/y = a^{p-q}$. $\log_a(x/y) = \log_a a^{p-q} = p - q = \log_a x - \log_a y$.
(b) $\log_a(x^r) = r\log_a x$: Let $p = \log_a x$. Then $x = a^p$. $x^r = a^{rp}$. $\log_a(x^r) = \log_a(a^{rp}) = rp = r\log_a x$.

89. 2530 years **91.** 1690 years **93.** 10,860 years

95. **(a)** $t(N) = 17\ln(N/k)/\ln 2$ **(b)** 169.4 minutes

Exercises 5.5

1. **(a)** $N(t) = 12.23655495(1.034783205)^t$
(b) 95.1967 million

3. **(a)** $E(t) = 892.5321814(1.057320469)^t$
(b) year 2000

5. **(a)** $R(t) = 603.7404884(1.102255883)^t$
(b) \$48,257 million

7. **(a)** Exponential curve does not fit the latest data well.
(b) Use data from 1977 on. The curve fits the data much better.
(c) \$520 billion

9. **(a)** $f(t+p) = k(3^{(t+p)/p}) = k(3^{(t/p+1)}) = k(3^{t/p})(3) = 3k(3^{t/p}) = 3f(t)$
(b) $f(t+q) = k(a^{(t+q)/q}) = k(a^{(t/q+1)}) = k(a^{t/q})(a) = ak(a^{t/q}) = af(t)$
(c) $N(t+1) = 3.256628189(1.047281015)^{t+1} = 1.047281015N(t) = N(t) + 0.047281015N(t)$.
Annual growth is 4.7281015%. If a calculator gives $f(t) = k(a^t)$ as the exponential function of best fit, then an increase of 1 in t multiplies f by a. If we write $a = 1 + b$, then the growth in 1 unit of time is $100b$%.

11. **(a)** $Q(t) = 250(2^{-t/30})$ **(b)** 31.25 milligrams

13. **(a)** $$Q(t) = \begin{cases} 80(2^{-t/3}) & \text{if } 0 \le t < 3 \\ 80(2^{-(t-3)/3}) & \text{if } 3 \le t < 6 \\ 80(2^{-(t-6)/3}) & \text{if } 6 \le t < 9 \\ 80(2^{-(t-9)/3}) & \text{if } 9 \le t < 12 \\ 80(2^{-(t-12)/3}) & \text{if } 12 \le t \end{cases}$$
(b) 12.60 milligrams

Chapter 5 Review Exercises

1. **(a)**

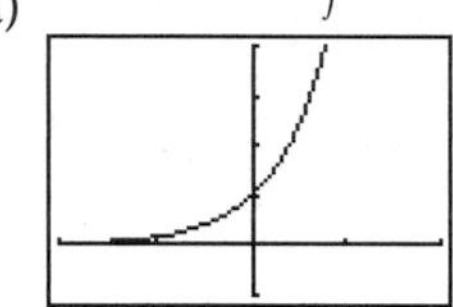

(b) f

2. **(a)** Shift $f(x) = 3^x$ up 2.

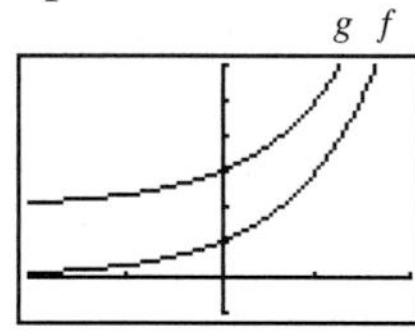

(b) Shift $f(x) = 2^x$ left 1.

g f

(c) Shift $f(x) = 6^x$ right 2, down 3.

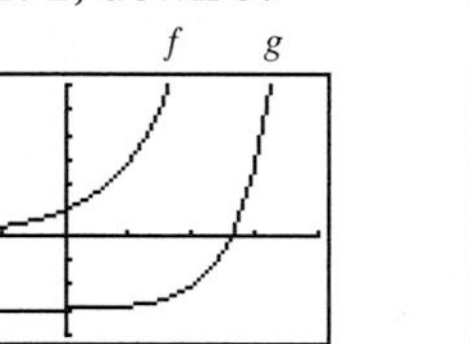

(d) Shift $f(x) = 3^x$ left 2, reflect in x-axis.

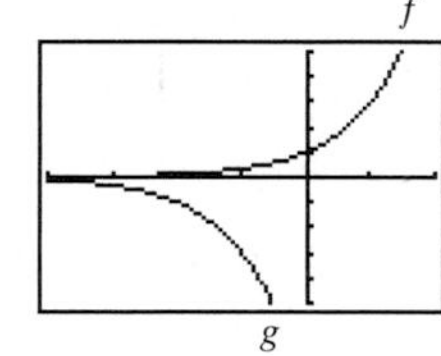

3. **(a)** Shift $f(x) = e^x$ down 3.

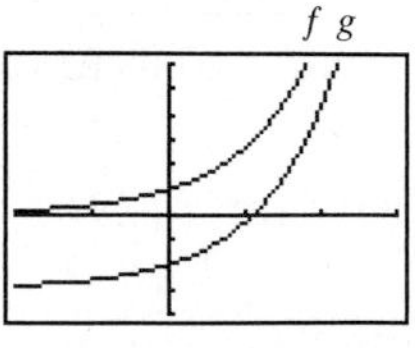

(b) Reflect $f(x) = e^x$ in y-axis, shift up 1.

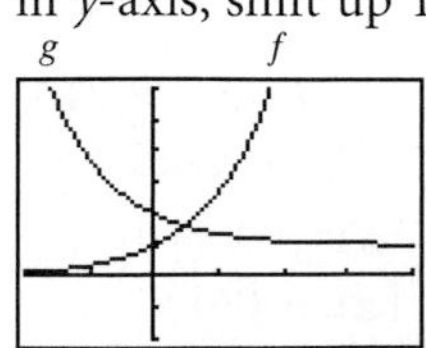

(c) Shift $f(x) = e^x$ left 1, up 2.

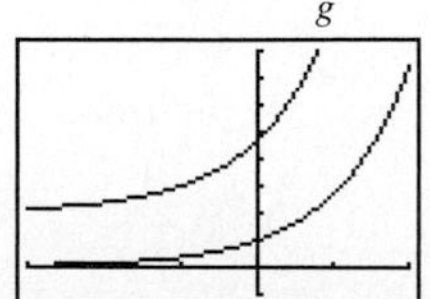

(d) Reflect $f(x) = e^x$ in x-axis, shift up 6.

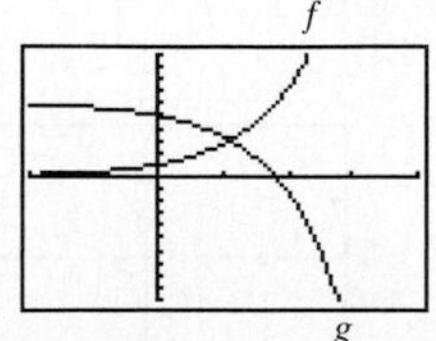

4. **(a)** 8 **(b)** $\frac{4}{3}$ **(c)** $\frac{13}{14}$ **(d)** $\frac{9}{8}$ **(e)** 3

5. **(a)** 5.46 (A) **(b)** −3.86 (A), 1.86 (B)

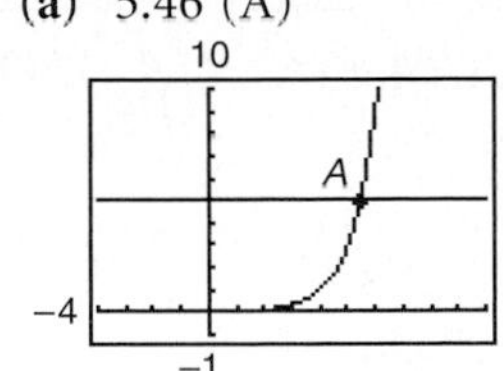

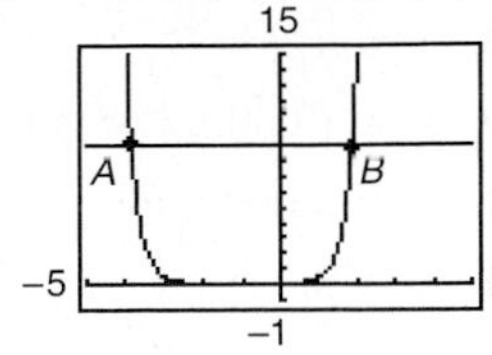

(c) 1.13 (A)

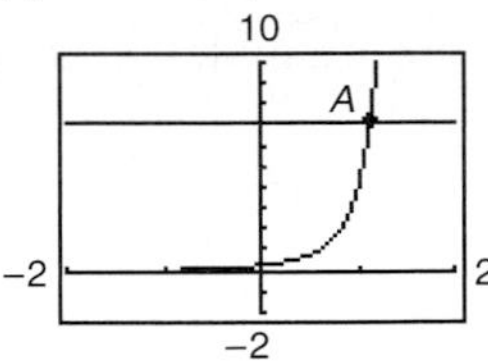

6. **(a)** $x = \log_3 y$ **(b)** $4x = \log_2 y$ **(c)** $2 = \log_3(9)$

7. **(a)** $x = 3^y$ **(b)** $5x = 7^y$ **(c)** $\frac{1}{27} = 3^{-3}$

8. **(a)** 2 **(b)** 3 **(c)** −3 **(d)** $-\frac{1}{3}$

9. **(a)** $f^{-1}(x) = \log_3(x + 5)$
(b) $f^{-1}(x) = (\log_5(x + 7) - 5)/3$
(c) $f^{-1}(x) = (4^{x-8})/5$
(d) $f^{-1}(x) = (3^{x-9} + 5)/2$

10. **(a)** 2 **(b)** 50.1377197 **(c)** 2
(d) $(\log_4 3y - 1)/7$

11. **(a)** $\log_a(x^3(x - 3)^2/5)$ **(b)** $\log_7(5/(x + 1)^2)$

12. **(a)** 40.5 **(b)** $\frac{20}{7}$ **(c)** −5, 1 **(d)** −3.5

13. **(a)** 1.62 (A) **(b)** −1.63 (A), 1.79 (B)

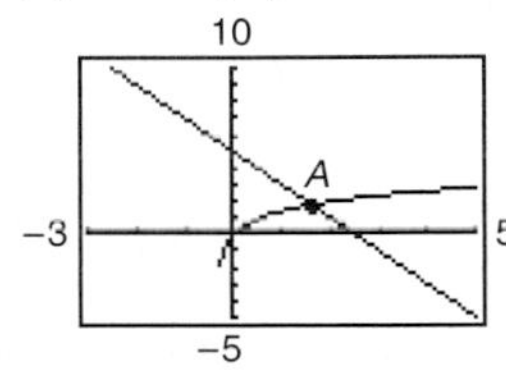

14.

x	−3.2	−1.8	0	4.3	5.7
e^x	0.0408	0.1653	1	73.6998	298.8674
e^{-x}	24.5325	6.0496	1	0.0136	0.0033

15. **(a)**

x	0.00038	0.75	1	3.45	12.78
ln x	−7.8753	−0.2877	0	1.2384	2.5479
log x	−3.4202	−0.1249	0	0.5378	1.1065

(b) Graph of $f(x) = \ln x$ lies below the graph of $g(x) = \log x$ for $0 < x < 1$ and above it for $x > 1$. Graphs cross at $x = 1$.

16. **(a)** 2.5850 **(b)** 0.7784 **(c)** 0.7680
(d) 1.6994

17. **(a)** Shift $f(x) = \ln x$ up 3. **(b)** Shift $f(x) = \ln x$ down 1.

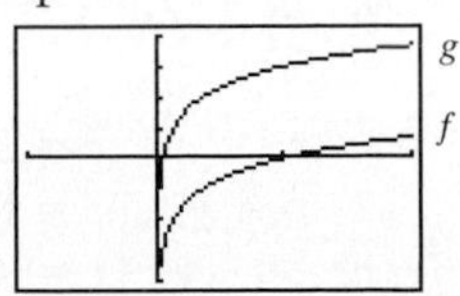

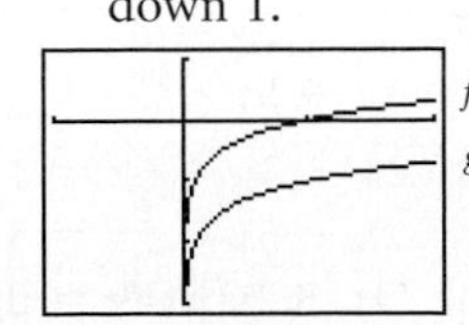

(c) Shift $f(x) = \ln x$ right 1. **(d)** Shift $f(x) = \ln x$ left 1, down 2.

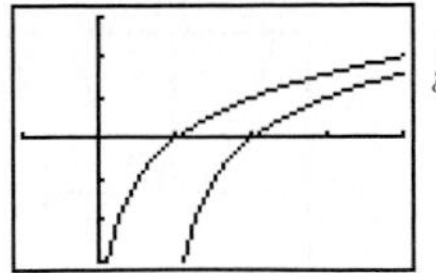

(e) Shift $f(x) = \ln x$ right 2, reflect in x-axis **(f)** Reflect $f(x) = \ln x$ in y-axis, right 3.

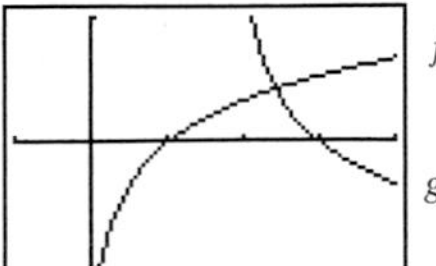

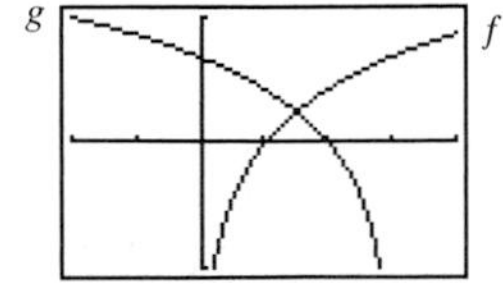

18. **(a)** 2002 **(b)** $t = 25 \ln(P/35)/\ln 2$; $t(50) = 12.86432932$; in 12th year

19. **(a)** $N(t) = 500(2^{t/15})$
(b)

t **(min)**	0	15	30	45	60
N	500	1000	2000	4000	8000

20. **(a)** $N(t) = 200(2^{t/80})$
(b)

t **(hours)**	0	4	8	12	16
N	200	1,600	12,800	102,400	819,200

21. 4 hours 9.14 minutes

22. 16,380 years

23. initial value 8; doubling time 9; $f(x) = 8(2^{x/9})$

Chapter 5 Test

1. $f(x) = 2^x$; $g(x) = 2^{x+4} - 5$; shift f left 4 and down 5.

2. **(a)** 10 **(b)** $\frac{5}{2}$ **(c)** $\frac{9}{4}$

3. **(a)** 3 **(b)** −5 **4.** 2.90

5. **(a)** $x = \log_5 y$ **(b)** $7x = \log_3 y$

6. **(a)** $x = 6^y$ **(b)** $4x = 3^y$

7. −4 **8.** $f^{-1}(x) = 2^{x-7} + 3$

9. **(a)** 745.9894968 **(b)** $x = (\log_3 2y + 1)/5$

10. **(a)** 2 **(b)** $\frac{25}{7}$ **11.** 3.75

12. **(a)** 1.4087 **(b)** 5.5892

13. **(a)** $P(t) = 20(2^{t/53})$ **(b)** 2026
(c) $t = 53\ln(P/20)/\ln 2$; $t(30) = 31.003013$ in 2026

14. $t = -5568\ln(R/15.3)/\ln 2$; approximately 1950 years old

Cumulative Test Chapters 3, 4, 5

1. range $[-2, 34]$

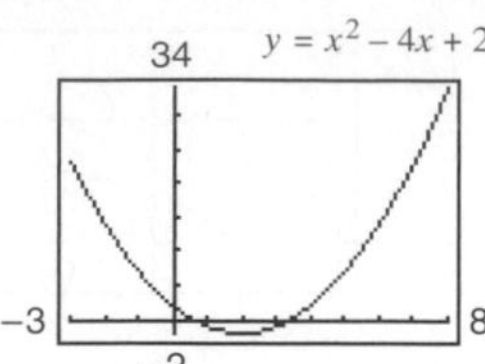

2. Shift f right 3, down 5.

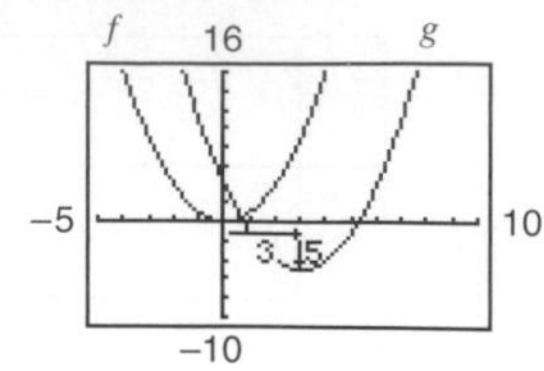

3. vertical $x = -1, x = 2$; horizontal $y = 0$
domain: all reals except $-1, 2$

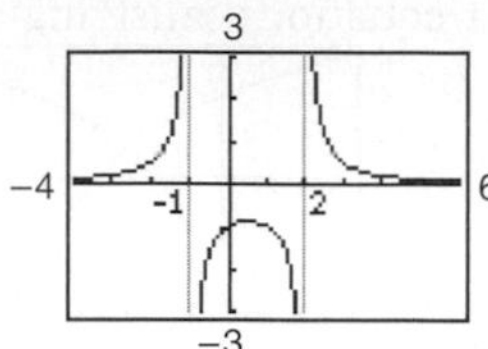

4. increasing $(-\infty, -1.87]$, $[0.54, \infty)$; decreasing $[-1.87, 0.54]$; max $A(-1.87, 15.13)$; min $B(0.54, 1.24)$; zero $C\ x = -3.12$

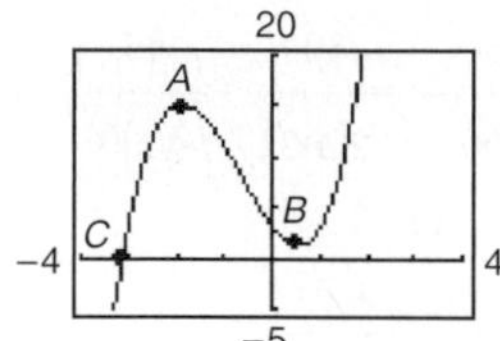

5. $(g \circ f)(x) = x + 1 + 3\sqrt{x - 4}$; domain: $[4, \infty)$

6. f is one-to-one by vertical line test; $f^{-1}(x) = \dfrac{6 - x}{2}$

7. $q(x) = x^2 + 5x + 6$, remainder $= 13$

8. Divide $f(x)$ by $g(x) = x - 4$; $f(4) = 59$

9. **(a)** $9i$ **(b)** $-6i$ **10.** **(a)** -1 **(b)** $-i$

11. **(a)** $7 + 8i$ **(b)** $31 + i$ **12.** $-\dfrac{5}{2} + \dfrac{9i}{2}$

13. Shift f right 3 and up 7.

14. 2.87 **15.** -5

16. **(a)** 218.7266 **(b)** $x = (2 - \log_2 y)/3$

17. **(a)** after about 2 years **(b)** after about 4.5 years

18. **(a)** 12.7 years

Chapter 6

Exercises 6.1

1. $x = 1, y = 2$ **3.** $x = 0, y = 3$

5. $x = \frac{3}{5}, y = -\frac{2}{5}$ **7.** $x = 4, y = 0$

9. $x = 2, y = 2$ **11.** $x = 3, y = 1$

13. $x = 1, y = 1$ **15.** $x = 3, y = -1$

17. $(x, -3x - 4)$ **19.** $x = 4, y = 0$

21. $x = 3, y = 2$ **23.** $(x, (1 - x)/2)$

25. $x = 4, y = 1$ **27.** no solution

29. $x = 1.0000, y = 1.0000$

31. $x = 3.0909, y = -1.2727$ **33.** $(x, -(x + 4)/3)$

35. $x = 3.2210, y = 1.9358$

37. $x = 2.2128, y = -0.0057$

Answers to Exercises 39–47 are examples of systems. There are others.

39. $x + 2y = 3, 2x + y = 0$

41. $x - 2y = 11, 2x + y = 7$

43. $3x - 2y = 18, 2x - y = 12$

45. $x + 2y = -11, x - y = 10$

47. $x - y = 3, 2x - 2y = 6$

49. **(a)** $c \neq 2$ **(b)** $c = 2$ **(c)** They can have many solutions only if they are the same line—lines are distinct because y-intercepts differ.

51. $y = 2x - 1$ **53.** $y = -\dfrac{3x}{2 + 1}$ **55.** $y = 3x - 2$

57. $y = 3x - 4$

59. algebra \$42, trigonometry \$35

61. Roma \$32, Seabreeze \$64

63. 6 quarters, 8 dimes **65.** 16 and 14 years old

67. \$4000 at 8%, \$1000 at 10% **69.** $-38, 46$

71. $L = 60, W = 40$

73. 15 pounds of nuts and 10 pounds of raisins

75. 32 from Kent, 18 from Sanford

77. $a = 8, v_0 = 5$

79. $m = 2.24, b = 580$; \$3,268

Exercises 6.2

1. $\begin{bmatrix} 1 & 3 \\ 2 & -5 \end{bmatrix}, \begin{bmatrix} 1 & 3 & 7 \\ 2 & -5 & -3 \end{bmatrix}$

3. $\begin{bmatrix} -1 & 3 & -5 \\ 2 & -2 & 4 \\ 1 & 3 & 0 \end{bmatrix}, \begin{bmatrix} -1 & 3 & -5 & -3 \\ 2 & -2 & 4 & 8 \\ 1 & 3 & 0 & 6 \end{bmatrix}$

5. $\begin{bmatrix} 1 & 0 & 0 \\ 0 & 1 & 0 \\ 0 & 0 & 1 \end{bmatrix}, \begin{bmatrix} 1 & 0 & 0 & 8 \\ 0 & 1 & 0 & 2 \\ 0 & 0 & 1 & -7 \end{bmatrix}$

7. $\begin{aligned} x + 2y &= 3 \\ 4x + 5y &= 6 \end{aligned}$ **9.** $\begin{aligned} x + 9y &= -3 \\ 5x \quad &= 2 \end{aligned}$

11. $\begin{aligned} 2x - 3y + 6z &= 4 \\ 7x - 5y - 2z &= 3 \\ 2y + 4z &= 0 \end{aligned}$ **13.** $\begin{aligned} x \quad &= 3 \\ y \quad &= 8 \\ z &= 4 \end{aligned}$

15. $\begin{bmatrix} 1 & 3 & -2 & 0 \\ 1 & 2 & -3 & 6 \\ 8 & 3 & 2 & 5 \end{bmatrix}$ **17.** $\begin{bmatrix} 1 & 2 & 3 & -1 \\ 0 & 3 & 10 & 0 \\ 0 & -8 & -1 & -1 \end{bmatrix}$

19. $\begin{bmatrix} 1 & 0 & 0 & -23 \\ 0 & 1 & 0 & 17 \\ 0 & 0 & 1 & 5 \end{bmatrix}$

21. to create zeros below the leading 1 in the first column

23. to get a leading one in the right location for the second row

25. to create zeros above the leading one in the third column

27. to create a leading one in the third row

29. $x = 10, y = -9, z = -7$

31. $x = 1, y = 4, z = 2$ **33.** $x = 3, y = 0, z = 2$

35. $x = 0, y = 4, z = 2$ **37.** $x = 2, y = 3, z = -1$

39. $x = 6, y = -3, z = 2$

41. $y = 2x^2 + x + 1$ **43.** $y = x^2 + 3x + 2$

45. $y = 4x^2 + 2x - 1$ **47.** $y = -x^2 + 5x + 2$

49. 4 ones, 4 fives, 2 tens **51.** \$38

53. 200 nickels, 120 dimes, 160 quarters

55. \$1000 at 7%, \$2000 at 8%, \$5000 at 10%

57. 20 tables, 80 chairs, 10 cupboards

59. (a) $a = 4, v_0 = 20, s_0 = 8$
(b) $a = -1$ (deceleration), $v_0 = 15, s_0 = 21$

Exercises 6.3

1. in reduced form

3. not in reduced form; nonzero element above leading 1 in row 2

5. in reduced form **7.** in reduced form

9. in reduced form **11.** in reduced form

13. not in reduced form; nonzero elements above leading 1 in row 3

15. in reduced form

17. not in reduced form; leading 1 in row 3 not to right of leading 1 in row 2

19. $x = 2, y = 4, z = -3$

21. $x = -3r + 6, y = r, z = -2$

23. $x = -2r + 3, y = 5, z = r$

25. $x = 2, y = -1, z = 1$

27. $x = -2r + 3, y = r + 4, z = r$

29. $x = -3r + 4, y = -2r + 1, z = r$

31. no solution **33.** $x = -2r + 1, y = r, z = -2$

35. $x = 2, y = 3, z = 1$

37. $x = -2r + 4, y = 5r + 6, z = r$

39. (a) Start with appropriate reduced form and work backward. For example,

$$\begin{bmatrix} 1 & 0 & 3 & 1 & 0 \\ 0 & 1 & 2 & -1 & 0 \\ 0 & 0 & 0 & 0 & 1 \end{bmatrix} \text{ leads to}$$

$$\begin{aligned} x + y + 5z \quad &= -1 \\ x + 2y + 7z - w &= 0 \\ -x + y - z - 2w &= 4 \end{aligned}$$

(b) Let the unique solution be $x = 1, y = 2, z = 3$. Construct four equations satisfying these conditions; for example, $x + y + z = 6$, $x - y + z = 2$, $2x + y - z = 1$, $x + 2y - z = 2$.

41. $I_1 = 4, I_2 = 1, I_3 = 5$ **43.** $I_1 = 3, I_2 = 4, I_3 = 7$

45. $I_1 = 2, I_2 = 1, I_3 = 3$

Exercises 6.4

1. $(1, -5), (1, 5)$ **3.** $(4, -3), (4, 3)$

5. $(\sqrt{2}, 3), (-\sqrt{2}, 3)$ **7.** $(-1, 1), (1, -1)$

9. $(-5, -9), (5, 11)$ **11.** $(-2, 5), (2, 1)$

13. $(-1, -2), (-1, 2), (1, -2), (1, 2)$

15. $(0, -2), (0, 2)$

17. $(\pm 4, \pm 1)$ **19.** $(\pm 1/\sqrt{2}, \pm\sqrt{2})$ **21.** $(\pm 4, 0)$

23. $(\pm 1, \pm 5)$ **25.** $(\frac{8}{3}, \frac{1}{3}), (4, 3)$ **27.** $(2, 4), (-2, 0)$

29. $(\pm 1, \pm 3)$ **31.** $(1, 0), (2, \frac{1}{3})$

33. $(1, 16), (-3, 0)$ **35.** $(1, 3), (-1, -3)$

37. $(-2, -1), (1, 2)$

39. $(1, 2), (-1, -2), (2, 1), (-2, -1)$

41. no solutions **43.** $(\frac{1}{4}, \frac{1}{3})$

45. $(-1.6458, -0.2915), (3.6458, 10.2915)$

47. $(-1.1547, 1.8453), (1.1547, 4.1547)$

49. no solution **51.** $(-0.7388, 2.4777)$

53. $(1.5944, -1.6668), (2.6506, 0.8334)$

55. $(-1.8637, -4.4734)$

Exercises 6.5

1. yes **3.** yes **5.** yes **7.** yes

9. no **11.** yes **13.** yes

In the windows of Exercises 15–63, $x\,\text{scl} = 1$, $y\,\text{scl} = 1$.

15. 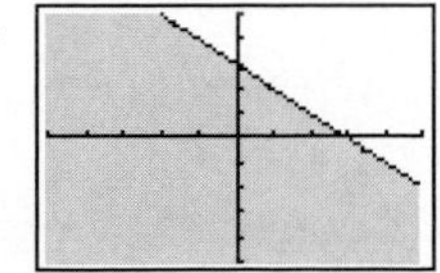

17.

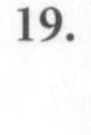

19.

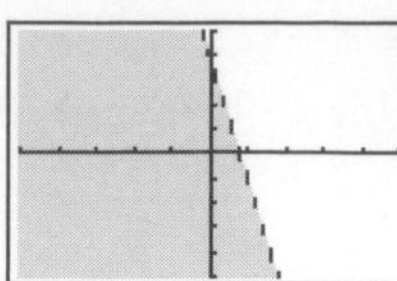

21.

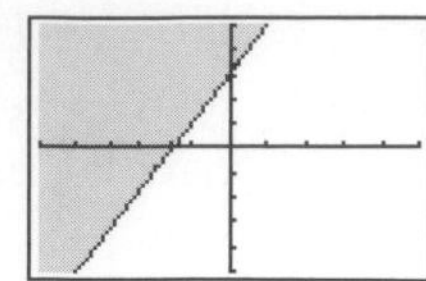

23.

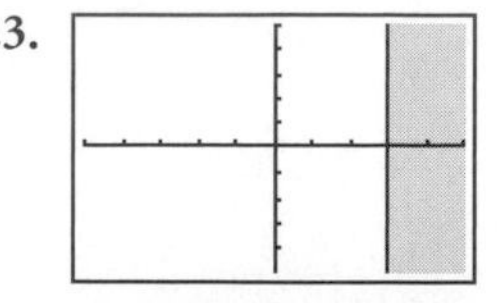

25.

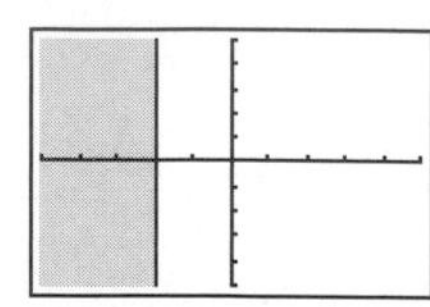

27.

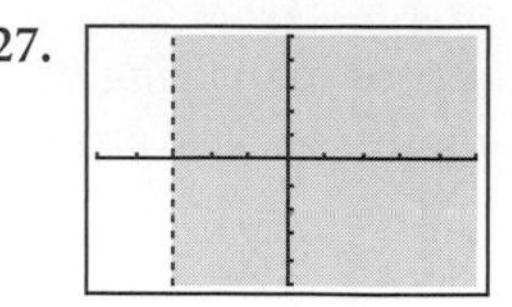

29.

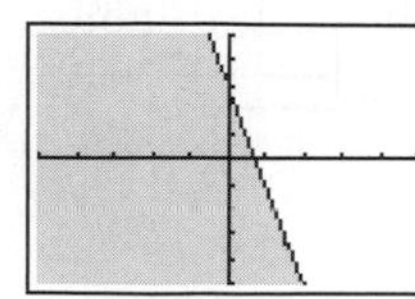

31.

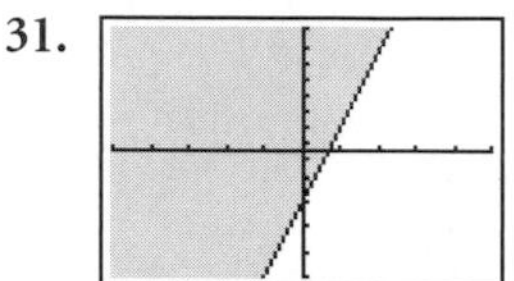

33.

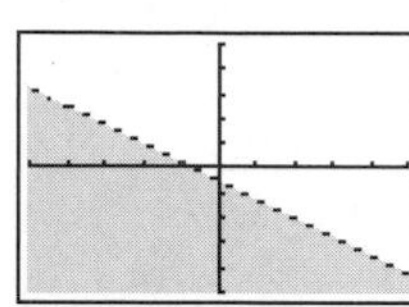

35.

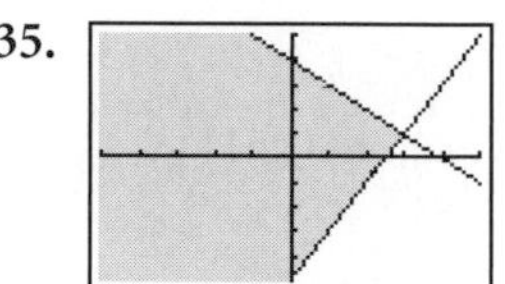

37.

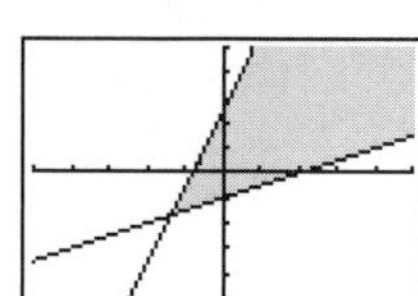

39.

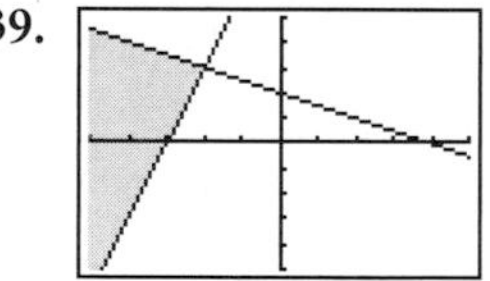

41.

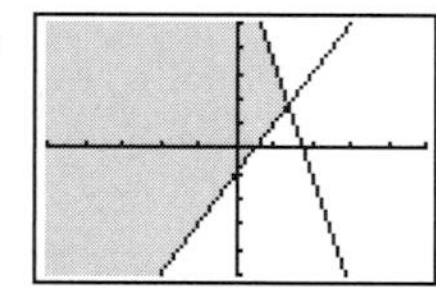

43.

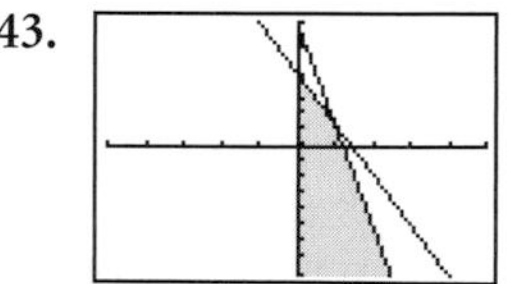

45.

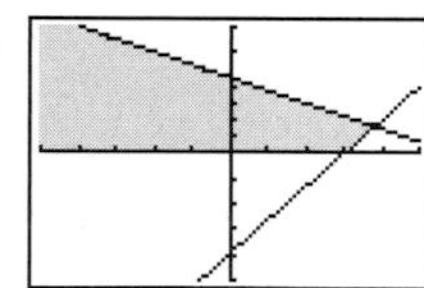

47.

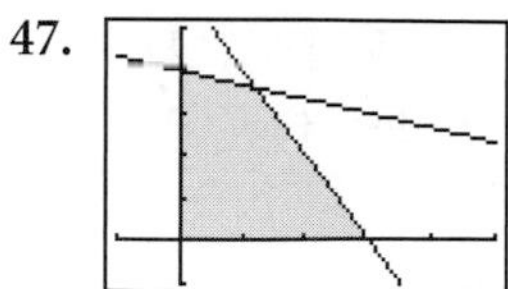

49.

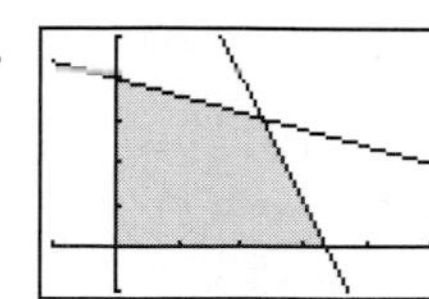

51.

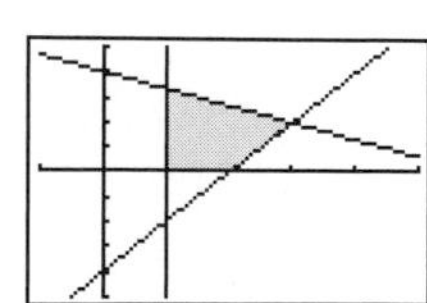

53. vertex $(4, 1)$

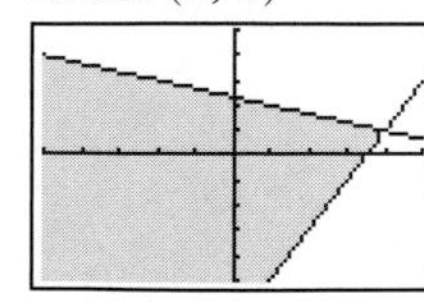

55. vertex $(3, 1)$

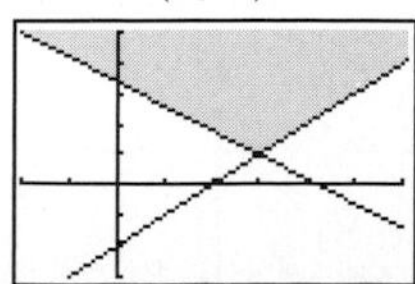

57. vertex $(4, -2)$

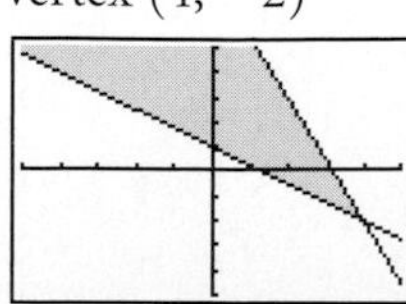

59. $(0, 0), (0, \frac{9}{2}), (5, 2), (6, 0)$

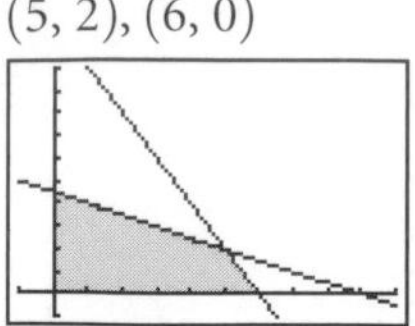

61. $(0, 0), (3, 2), (0, 8), (\frac{11}{3}, 0)$

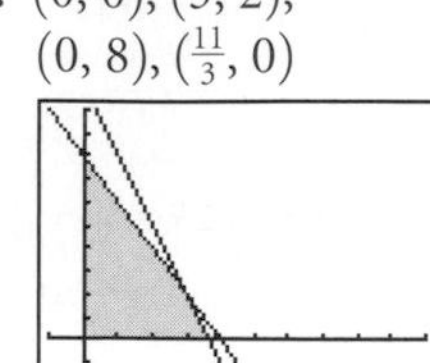

63. $(0, 2), (0, 7), (2, 5), (\frac{7}{2}, 2)$

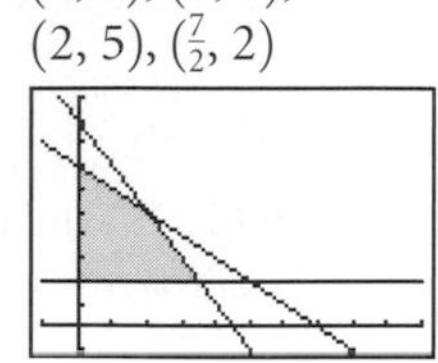

Exercises 6.6

1. 26 at $(4, 3)$ **3.** 6 at $(1, 2)$ **5.** 24 at $(3, 4)$

7. $\frac{64}{3}$ at $(\frac{16}{3}, 0)$ **9.** max $= 12$ along a side

11. 31 at $(2, \frac{9}{2})$

13. Maximum profit is \$6400 when there are 400 of KC-1 and 300 of KC-2.

15. Maximum profit is \$2400. Ship x from Chester and $120 - 2x$ from Crewe, where $0 \leq x \leq 30$. Company has flexibility.

17. Maximum is \$5040 when there are 24 suits and 32 dresses.

19. 11 Torro and 17 Sprite buses; they would carry 785 students.

21. 7 ounces of item M and none of item N

23. **(a)** Maximum profit is \$8000 when there is nothing for Pringle and 2000 for Williams.
(b) \$80

Chapter 6 Review Exercises

1. **(a)** $x = 5, y = 1$ **(b)** $x = 2, y = -3$
(c) $x = 5, y = 2$

2. **(a)** $x = 0, y = 4$ **(b)** $x = -3, y = 1$
(c) $x = 2, y = 4$

3. for example, $x + y = -3, x - y = 7$

4. for example, $-3x + y = -7; 6x - 2y = 14$

5. **(a)** $y = -2x + 5$ **(b)** $y = 2x - 6$
(c) $y = -11x/5 - \frac{3}{5}$

6. 5 quarters, 3 dimes **7.** math \$37; history \$35

8. **(a)** $x = 7, y = 6, z = -3$
(b) $x = 1, y = 2, z = -1$ **(c)** no solution

9. **(a)** $y = 2x^2 + x - 3$ **(b)** $y = 3x^2 + 4x + 1$

10. **(a)** $x = 7, y = 6, z = -3$
(b) $x = 1, y = 2, z = -1$
(c) $x = -3r - 1, y = -2r + 3, z = r$
(d) no solution
(e) $x = -r + 3, y = -4r + 1, z = r$
(f) no solution

11. **(a)** (4, 1) **(b)** (0, −3), (2, 1)
(c) (−2, −3), (−2, 3), (2, −3), (2, 3)
(d) no solution
(e) (4, 3), (−4, −3) **(f)** (2, 5)

12. **(a)** (−0.62, 1.38), (1.62, 3.62) **(b)** (−1.78, 2.60)
(c) (1.16, 2.17), (2.14, 4.13)

13. **(a)** no **(b)** yes **(c)** yes **(d)** yes **(e)** no
(f) no

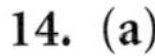
14. **(a)**

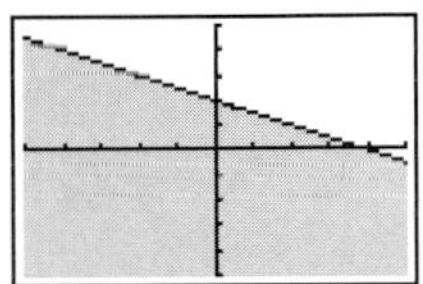

(b)

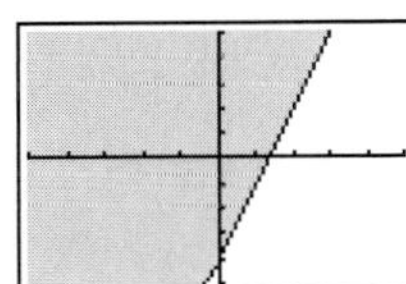

(c)

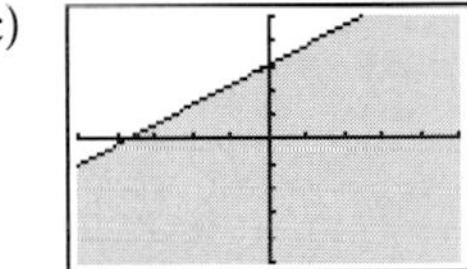

(d)

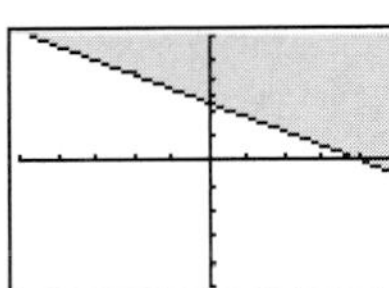

(e)

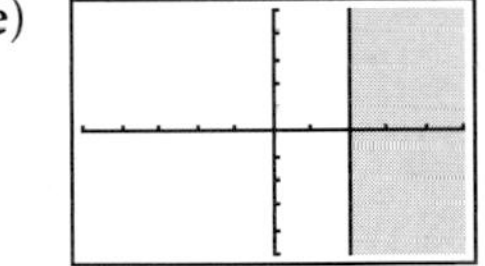

(f)

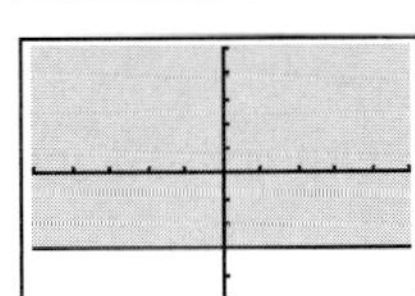

(g)

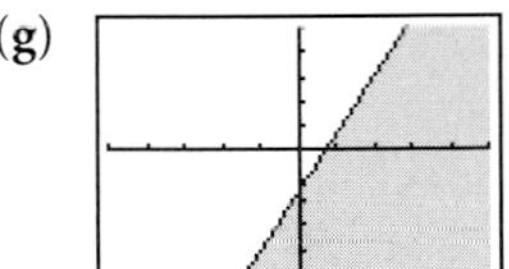

(h)

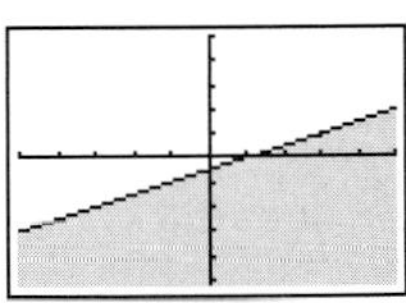

15. **(a)** vertex (4, 1)

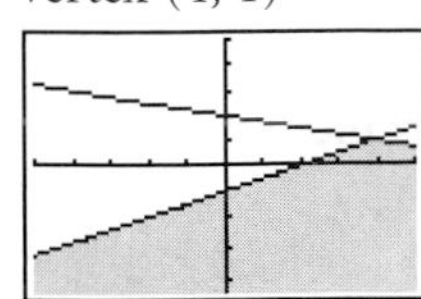

(b) vertex (−2, 5)

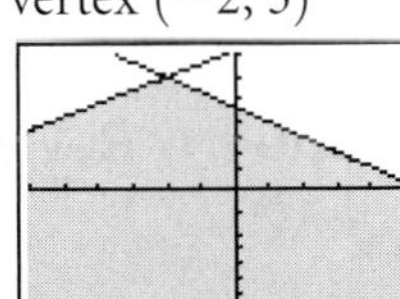

(c) vertex (3, 4)

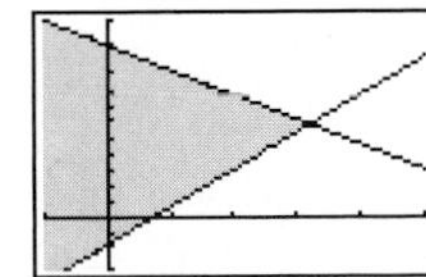

(d) (0, 0), (0, 5), (3, 2), $(\frac{13}{3}, 0)$

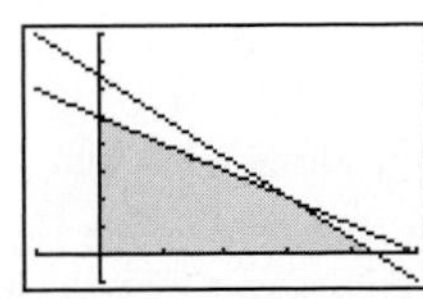

(e) (0, 0), $(0, \frac{14}{3})$, (1, 4), (2, 0)

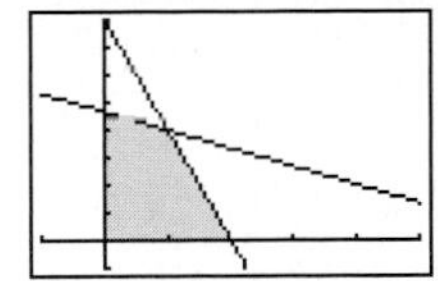

(f) (1, 2), (1, 6), (2, 4), $(\frac{8}{3}, 2)$

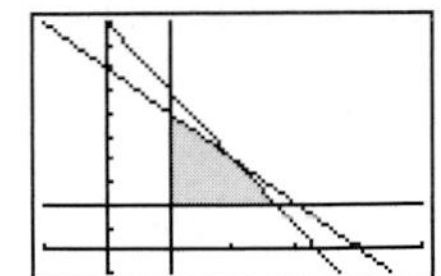

16. **(a)** 24 at (0, 6) **(b)** 18 at (3, 2)
(c) 16 along a side **(d)** 31 at (4, 5)

17. Maximum profit is $26,000 when there are 510 Panoramas and 160 Visions.

18. Maximum profit is $710 when there are 50 cabinets and 30 tables.

Chapter 6 Test

1. $x = 2, y = 4$ **2.** $x = 3, y = -1$ **3.** $y = 3x + 1$
4. $x = 1, y = 2, z = -1$ **5.** $y = 2x^2 - 3x + 5$
6. (−2, 4)
7. **(a)** yes **(b)** no **8.** 12 when $x = 3, y = 2$
9. Maximum profit is $54,100 when 420 Cruisers and 380 Jets.
10. English $38, math $42

Chapter 7

Exercises 7.1

1. 2 × 2, square matrix **3.** 3 × 2
5. 3 × 1, column matrix **7.** 2 × 1, column matrix
9. 2 × 4 **11.** 4 **13.** 3 **15.** 1 **17.** 1
19. 5 **21.** 6 **23.** 3 **25.** 3 **27.** −12
29. −6 **31.** −1 **33.** 2
35. $x = 2, y = 3$ **37.** $x = 2, y = 1, z = 2$
39. no solution **41.** $x = 1, y = 2, z = 3$
43. $x = 3 - r, y = 3, z = r$

45. $\begin{bmatrix} 5 & 4 \\ 4 & 0 \end{bmatrix}$ **47.** does not exist **49.** $\begin{bmatrix} 9 & 6 \\ 9 & -3 \end{bmatrix}$

51. $\begin{bmatrix} 10 & 5 \\ 6 & 0 \end{bmatrix}$ **53.** does not exist **55.** $\begin{bmatrix} 5 & 7 \\ 4 & 6 \\ -1 & 8 \end{bmatrix}$

57. does not exist **59.** $\begin{bmatrix} -3 & -3 \\ 2 & 2 \\ -1 & -4 \end{bmatrix}$

61. $\begin{bmatrix} 8 & -6 & 11 \\ 15 & -3 & 17 \end{bmatrix}$ **63.** does not exist

65. $x = 5, y = 2, z = 1$

67. no solution **69.** no solution

71. **(a)** 29 **(b)** 8 **(c)** 3 **(d)** 45

73. **(a)** $i = 1, j = 2; i = 4, j = 4$ **(b)** $i = 1, 4$
(c) $j = 2, 5$ **(d)** $i = j = 3$
(e) $i = 1, j = 4; i = 2, j = 4$ **(f)** $i = 3, j = 2$

75. **(a)** 200 **(b)** 900 **(c)** 6000 **(d)** 1800
Yes: # rows = # zones = # columns

77.

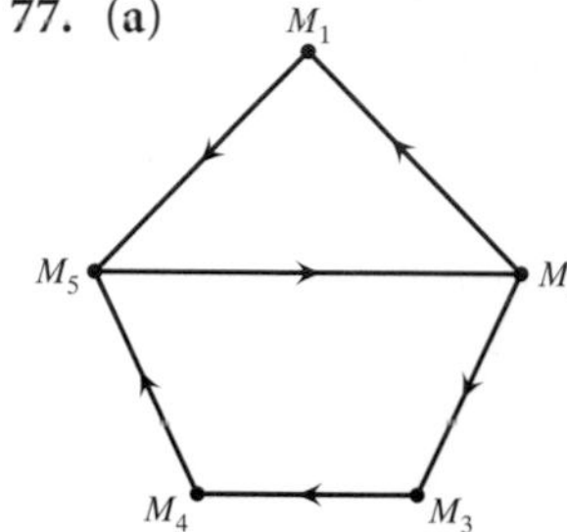

M_2, M_5, M_4, M_1, and M_3.

(b)
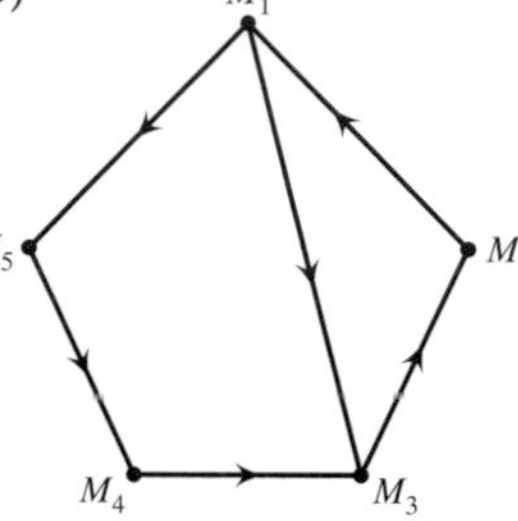

M_1, M_2, M_3, and M_4 and M_5.

(c)

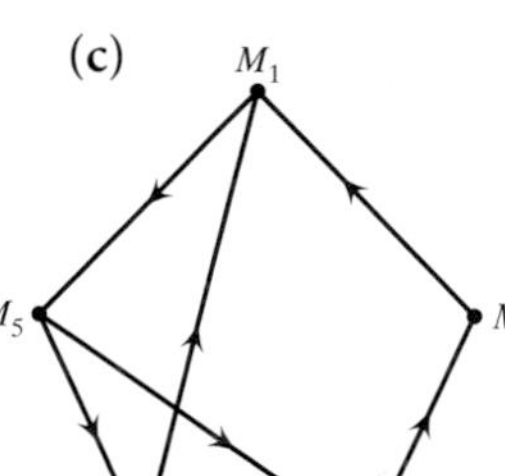

M_5, M_1, M_2, M_3, and M_4.

(d)
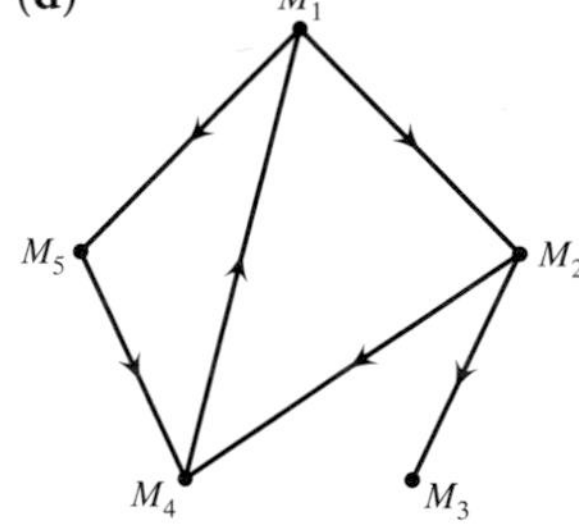

M_1, M_2, M_4, M_5, M_3.

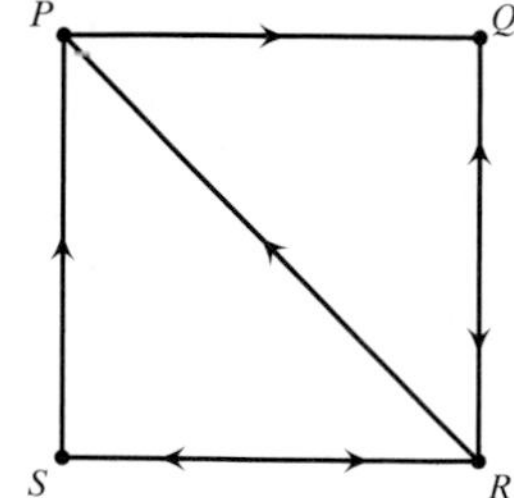

81. **(a)**
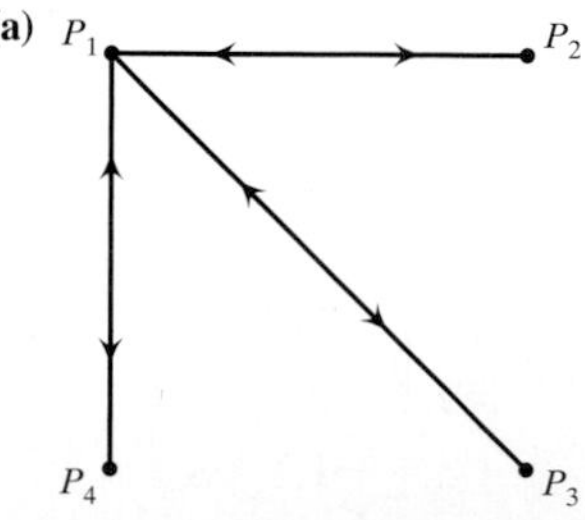

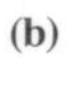

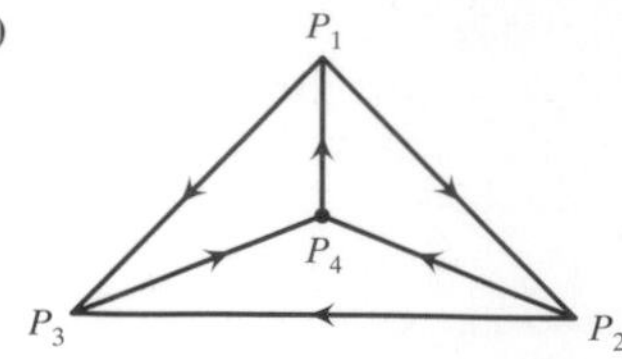

(c)
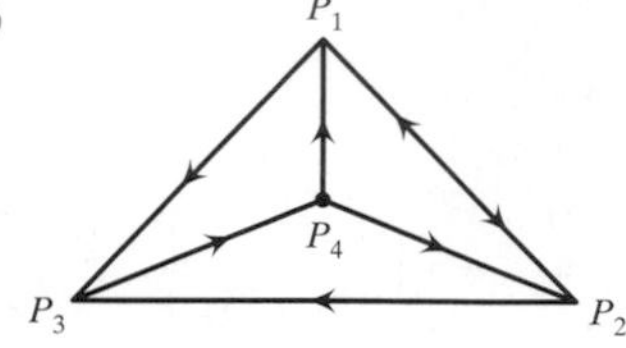

Exercises 7.2

1. $\begin{bmatrix} 2 & 4 \\ 8 & 10 \end{bmatrix}$ **3.** $\begin{bmatrix} 2 & 0 & 4 \\ -4 & 9 & 10 \end{bmatrix}$ **5.** $\begin{bmatrix} 6 & 0 \\ -9 & 12 \end{bmatrix}$

7. $\begin{bmatrix} -1 & 6 & 10 \\ -1 & 12 & 22 \end{bmatrix}$ **9.** $\begin{bmatrix} -2 & 12 & 20 \\ -2 & 30 & 56 \end{bmatrix}$

11. $\begin{bmatrix} 7 & 10 \\ 15 & 22 \end{bmatrix}$ **13.** $\begin{bmatrix} 4 & 4 \\ 7 & 13 \end{bmatrix}$ **15.** $\begin{bmatrix} 3 & 6 \\ 10 & 12 \end{bmatrix}$

17. $\begin{bmatrix} 12 & 6 \\ 18 & -1 \end{bmatrix}$ **19.** $\begin{bmatrix} 5 \\ 5 \end{bmatrix}$ **21.** $[10 \quad 2]$

23. $\begin{bmatrix} 0 & 9 \\ 30 & -14 \end{bmatrix}$ **25.** $\begin{bmatrix} -1 & 12 \\ 2 & 8 \end{bmatrix}$

27. 2×2 **29.** does not exist

31. does not exist **33.** does not exist

35. does not exist **37.** 2×3

39. 2×2 **41.** 2×4 **43.** 2×2

45. does not exist **47.** 4×2 **49.** does not exist

51. $\begin{bmatrix} -5 & -12 & 8 \\ 8 & -16 & 18 \end{bmatrix}$ **53.** does not exist

55. does not exist **57.** $\begin{bmatrix} -15 & -1 \\ -28 & -18 \\ 21 & -47 \end{bmatrix}$

59. does not exist **61.** $\begin{bmatrix} 36 & -13 \\ -46 & 59 \end{bmatrix}$

63. 2×2 **65.** 2×2 **67.** 4×2

69. does not exist **71.** 2×2

73. $\begin{bmatrix} -17.56 & 74.6 \\ 11.02 & -26.1 \end{bmatrix}$ **75.** $\begin{bmatrix} -61.46 & 261.1 \\ 38.57 & -91.35 \end{bmatrix}$

77. $\begin{bmatrix} -22.04 & 86.78 \\ 23.18 & 22.04 \end{bmatrix}$ **79.** $\begin{bmatrix} -33.5148 & -22.3330 \\ 3.9774 & 15.2089 \end{bmatrix}$

81. $\begin{bmatrix} 10.71 & -14.28 \\ -31.4 & 59.64 \end{bmatrix}$ **83.** does not exist

85. does not exist

87.

	1995	1996	1997	1998	1999	2000	...	2010
Metro	200	199.3	198.62	197.96	197.32	196.70		191.44
Nonmetro	65	65.7	66.38	67.04	67.68	68.30		73.56

Long-term prediction: 176.22 for metro, 188.33 for nonmetro.

Exercises 7.3

1. $\begin{bmatrix} 3 & 6 \\ 1 & 0 \end{bmatrix}$ **3.** $\begin{bmatrix} 4 & -5 \\ 2 & 3 \end{bmatrix}$ **5.** $\begin{bmatrix} 0 & 7 & 6 \\ -2 & 4 & 3 \\ 1 & 9 & 8 \end{bmatrix}$

7. $\begin{bmatrix} 1 & 2 & -1 & 3 \\ 4 & 6 & 0 & 5 \\ 7 & 1 & 2 & 9 \\ 12 & 3 & 4 & 1 \end{bmatrix}$

9. yes **11.** yes **13.** yes **15.** yes

17. yes **19.** no

21. $\begin{bmatrix} -5 & -3 \\ -2 & -1 \end{bmatrix}$ **23.** $\begin{bmatrix} 2 & 0 \\ 0 & 6 \end{bmatrix}$ **25.** does not exist

27. $\begin{bmatrix} 3 & -4 \\ -2 & 3 \end{bmatrix}$ **29.** $\begin{bmatrix} -1 & 2 & -2 \\ 1 & -1 & 1 \\ -1 & 2 & -1 \end{bmatrix}$

31. does not exist **33.** $\begin{bmatrix} 15 & -12 & -4 \\ 4 & -3 & -1 \\ -5 & 4 & 1 \end{bmatrix}$

35. $\begin{bmatrix} 1 & 1 \\ 3 & -1 \end{bmatrix}\begin{bmatrix} x \\ y \end{bmatrix} = \begin{bmatrix} 2 \\ 4 \end{bmatrix}$ **37.** $\begin{bmatrix} 2 & 3 \\ 4 & 0 \end{bmatrix}\begin{bmatrix} x \\ y \end{bmatrix} = \begin{bmatrix} 1 \\ 2 \end{bmatrix}$

39. $\begin{bmatrix} 3 & 2 & -7 \\ 4 & 3 & -2 \\ 1 & 1 & -1 \end{bmatrix}\begin{bmatrix} y \\ x \\ z \end{bmatrix} = \begin{bmatrix} 1 \\ 4 \\ 5 \end{bmatrix}$

41. $\begin{bmatrix} 1 & 1 & 1 \\ 5 & -1 & 0 \\ 1 & 1 & 2 \end{bmatrix}\begin{bmatrix} x \\ y \\ z \end{bmatrix} = \begin{bmatrix} 4 \\ 6 \\ 9 \end{bmatrix}$

43. $x = -13, y = -5$ **45.** $x = 14, y = -36$

47. $x = -7, y = -5, z = 4$

49. $\begin{bmatrix} 1 & 0 \\ -2 & 1 \end{bmatrix}$ **51.** $\begin{bmatrix} 1.5 & -0.5 \\ -2 & 1 \end{bmatrix}$ **53.** does not exist

55. $\begin{bmatrix} 2.3333 & -3.0000 & -0.3333 \\ -2.6667 & 3.0000 & 0.6667 \\ 1.3333 & -1.0000 & -0.3333 \end{bmatrix}$

57. does not exist

59. $\begin{bmatrix} -0.4286 & 0.7143 & -1.5714 \\ 0.2857 & -0.1429 & -0.2857 \\ 0.2857 & -0.1429 & 0.7143 \end{bmatrix}$

61. $x = -2, y = 2$ **63.** $x = 5, y = 0$

65. $x = 11, y = -4$

67. $x = 0.7778, y = 0.3333, z = -0.5556$

69. $x = 143, y = -47, z = -16$

71. $\begin{bmatrix} 60 \\ 40 \end{bmatrix}$ **73.** $\begin{bmatrix} 210 \\ 175 \end{bmatrix}$ **75.** $\begin{bmatrix} 165 \\ 480 \\ 250 \end{bmatrix}$

77. a_{ij} is amount from industry i needed to produce one unit from industry j. Thus, no a_{ij} can be negative. If the amount from industry i required for one unit of industry j were greater than 1, the value going into industry j would be greater than the final value, an economically infeasible situation.

Exercises 7.4

1. -2 **3.** 13 **5.** 2 **7.** -10

9. $A_{11} = 10; A_{12} = -1; A_{21} = 7$

11. $A_{11} = 4; A_{31} = 8; A_{23} = 5$

13. 45 **15.** 4 **17.** -52 **19.** -1

21. 35 **23.** -50 **25.** $x = 1, y = 2$

27. $x = -1, y = 1$ **29.** not applicable

31. $x = 2, y = 1$ **33.** $x = 5, y = 1$

35. $x = 4, y = 3$ **37.** $x = 1, y = 1, z = 1$

39. $x = -1, y = 1, z = 2$

41. $x = 7, y = 1, z = 2$ **43.** $x = 1, y = 3, z = 2$

45. $x = 1, y = -1, z = -1$

47. $x = 2, y = 1, z = 4$ **49.** -2 **51.** 2.13

53. -460 **55.** 438 **57.** 0

Exercises 7.5

1. $\begin{bmatrix} -1 & 2 \\ 2 & -3 \end{bmatrix}$; symmetric

3. $\begin{bmatrix} 3 & 2 \\ -1 & 4 \end{bmatrix}$; not symmetric

5. $\begin{bmatrix} 4 & -1 & 0 \\ 5 & 2 & 1 \\ 6 & 3 & 2 \end{bmatrix}$; not symmetric

7. $\begin{bmatrix} -2 & 1 \\ 4 & 0 \\ 5 & 3 \\ 7 & -7 \end{bmatrix}$; not symmetric

9. $\begin{bmatrix} -1 & 2 \\ 2 & -3 \end{bmatrix}$ **11.** $\begin{bmatrix} 3 & 5 & -3 \\ 5 & 8 & 4 \\ -3 & 4 & 3 \end{bmatrix}$

13. 2 − 1 − 3 − 4 (either way)

15. 1 − 2 − 3 − 4 (either way)

17. 4 − 1 − 2 − 5 − 3 (either way)

19. The element g_{ij} of the matrix $G (= AA^t)$ is equal to the number of types of pottery common to graves i and j. g_{ji} is equal to the number of pottery types common to graves j and i. Thus, $g_{ij} = g_{ji}$ implying that the matrix G is symmetric.

21. **(a)** $F = AA^t$

$f_{ij} = (\text{row } i \text{ of } A) \times (\text{column } j \text{ of } A^t)$

$= a_{i1}a_{j1} + a_{i2}a_{j2} + \cdots + a_{in}a_{jn}$

Each term in this sum will be either 1 or 0. Number of 1s is the number of friends common to people i and j.

(b) If friendship is not mutual then person i could consider person j to be a friend, but not vice versa. The matrix F would not then be symmetric.

Chapter 7 Review Exercises

1. **(a)** 2×2, square **(b)** 2×3 **(c)** 3×4 **(d)** 1×4 matrix, row matrix

2. **(a)** 3 **(b)** −1 **(c)** 8 **(d)** 4 **(e)** −2 **(f)** 7 **(g)** −5

3. **(a)** $x = 5, y = 4, z = 2$ **(b)** $x = -3, y = -4, z = 2$

4. **(a)** $\begin{bmatrix} 1 & 8 \\ 1 & 8 \end{bmatrix}$ **(b)** does not exist **(c)** $\begin{bmatrix} 33 & 5 \\ 19 & 26 \end{bmatrix}$

(d) $\begin{bmatrix} 25 & 19.4 & -21.1 \\ 18 & 2 & 18.5 \end{bmatrix}$

5. **(a)** $\begin{bmatrix} -8 & 29 \\ -12 & 21 \end{bmatrix}$ **(b)** $\begin{bmatrix} 14 & 33 & -9 \\ 12 & 27 & -9 \end{bmatrix}$

(c) does not exist **(d)** $\begin{bmatrix} -11 & 28 \\ -21 & 24 \end{bmatrix}$

(e) $\begin{bmatrix} -111 & 182 \\ -117 & 102 \end{bmatrix}$ **(f)** $\begin{bmatrix} -31 & -72 & 21 \\ -30 & -69 & 21 \end{bmatrix}$

6. **(a)** 2×2 **(b)** 2×3 **(c)** 2×3 **(d)** 2×3 **(e)** does not exist **(f)** 2×1 **(g)** 2×3 **(h)** 3×1

7. **(a)** $\begin{bmatrix} \frac{1}{2} & 0 \\ 0 & \frac{1}{6} \end{bmatrix}$ **(b)** $\begin{bmatrix} 3 & -2 \\ -1 & 1 \end{bmatrix}$

(c) does not exist **(d)** $\begin{bmatrix} -4 & 3 & -2 \\ -10 & 7 & -3 \\ 3 & -2 & 1 \end{bmatrix}$

(e) does not exist

8. **(a)** $\begin{bmatrix} 1 & 2 \\ 1 & 3 \end{bmatrix}\begin{bmatrix} x \\ y \end{bmatrix} = \begin{bmatrix} 8 \\ 11 \end{bmatrix}$; $x = 2, y = 3$

(b) $\begin{bmatrix} 1 & 1 & 1 \\ 0 & 1 & 2 \\ 2 & 1 & -4 \end{bmatrix}\begin{bmatrix} x \\ y \\ z \end{bmatrix} = \begin{bmatrix} 3 \\ 5 \\ 3 \end{bmatrix}$; $x = -\frac{5}{2}, y = 6, z = -\frac{1}{2}$

9. **(a)** 9 **(b)** −1 **(c)** −99 **(d)** −31 **(e)** 12

10. **(a)** $x = 2, y = 3$ **(b)** $x = 2, y = 1, z = 1$

11. **(a)** $A^t = \begin{bmatrix} -1 & 2 \\ 2 & -3 \end{bmatrix}$; symmetric

(b) $B^t = \begin{bmatrix} -2 & 3 \\ -3 & 4 \end{bmatrix}$; not symmetric

(c) $C^t = \begin{bmatrix} 4 & -2 & 7 \\ 5 & 3 & 0 \end{bmatrix}$; not symmetric

(d) $D^t = \begin{bmatrix} 2 & 3 & -5 \\ 3 & 0 & 4 \\ -5 & 4 & 6 \end{bmatrix}$; symmetric

12.

	1995	1996	1997	1998	1999	2000
City	60	59.98	59.95	59.94	59.93	59.92
Suburb	140	139.33	138.67	138.02	137.40	136.78
Nonmetro	65	65.7	66.38	67.04	67.68	68.30

In 2010: city: 60; suburb: 131.45; nonmetro: 73.56. Long-term: city: 61.83; suburb: 114.83; nonmetro: 88.33.

13. $P = \begin{bmatrix} 3 & 0 & 2 & 1 \\ 0 & 1 & 1 & 0 \\ 2 & 1 & 4 & 1 \\ 1 & 0 & 1 & 2 \end{bmatrix}$

$p_{13} = 2$, producers 1 and 3 have 2 customers in common.

$p_{14} = 1$, producers 1 and 4 have 1 customer in common.

$p_{23} = 1$, producers 2 and 3 have 1 customer in common.

$p_{34} = 1$, producers 3 and 4 have 1 customer in common.

Thus, producers 1 and 3 are in the most direct competition.

14.

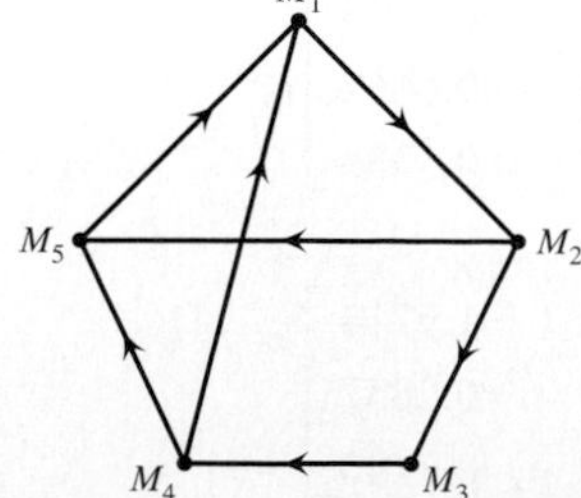

$$D = \begin{bmatrix} 0 & 1 & 2 & 3 & 2 \\ 2 & 0 & 1 & 2 & 1 \\ 2 & 3 & 0 & 1 & 2 \\ 1 & 2 & 3 & 0 & 1 \\ 1 & 2 & 3 & 4 & 0 \end{bmatrix} \begin{matrix} 8 \\ 6 \\ 8 \\ 7 \\ 10 \end{matrix}$$

$M_2, M_4, (M_1 \text{ and } M_3), M_5$

Chapter 7 Test

1. **(a)** 3×2 **(b)** 2×3 **(c)** 3×3 **(d)** 1×5

2. **(a)** -7 **(b)** 7 **(c)** 8

3. $x = 1, y = 2, z = 5$

4. **(a)** $\begin{bmatrix} 2 & -6 & 5 \\ 9 & 10 & 17 \end{bmatrix}$ **(b)** $\begin{bmatrix} -4 & 7 & 5 \\ 2 & -10 & 6 \end{bmatrix}$

5. **(a)** $\begin{bmatrix} 9 & 11 & 17 \\ 6 & -16 & 16 \end{bmatrix}$ **(b)** does not exist

(c) $\begin{bmatrix} -11 & 9 \\ -12 & -8 \end{bmatrix}$

6. **(a)** 2×3 **(b)** 2×1 **(c)** does not exist
(d) 2×3 **(e)** 1×2

7. $AB = BA = I_2$; thus, B is the inverse of A.

8. $\begin{bmatrix} -1 & 4 & -2 \\ 1 & -2 & 1 \\ 1 & -3 & 1 \end{bmatrix}$

9. $\begin{bmatrix} 1 & -4 \\ 2 & -7 \end{bmatrix}\begin{bmatrix} x \\ y \end{bmatrix} = \begin{bmatrix} -1 \\ -1 \end{bmatrix}$; $x = 3, y = 1$

10. **(a)** 10 **(b)** 14 **11.** $x = 2, y = -1$

12. **(a)** $A^t = \begin{bmatrix} 1 & 3 \\ 2 & 4 \end{bmatrix}$; not symmetric

(b) $B^t = \begin{bmatrix} 2 & 3 & 1 \\ 3 & 4 & -5 \\ 1 & -5 & 0 \end{bmatrix} = B$; symmetric

13.

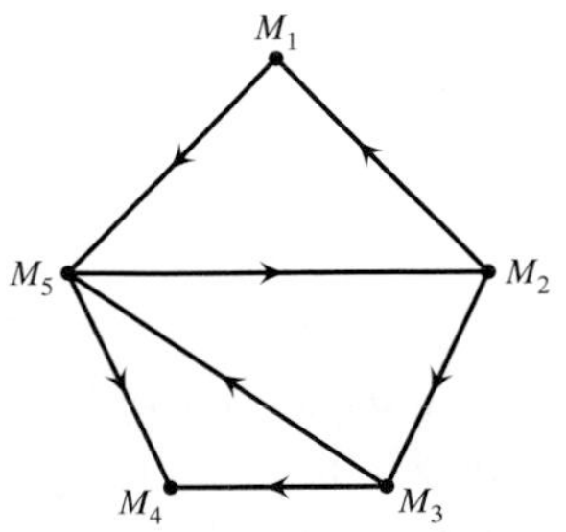

$$D = \begin{bmatrix} 0 & 2 & 3 & 2 & 1 \\ 1 & 0 & 1 & 2 & 2 \\ 3 & 2 & 0 & 1 & 1 \\ \infty & \infty & \infty & 0 & \infty \\ 2 & 1 & 2 & 1 & 0 \end{bmatrix} \begin{matrix} 8 \\ 6 \\ 7 \\ \infty \\ 6 \end{matrix}$$

$(M_2 \text{ and } M_5), M_3, M_1, M_4$

14. **(a)** 700 **(b)** 500 **(c)** 1000
(d) from zone IV to zone III

Chapter 8

Exercises 8.1

1. 5, 7, 9, 11 **3.** 6, 16, 38, 78

5. 2, 4, 8, 16 **7.** 4, 12, 24, 40

9. $-3, -1, 1, 3, 5$ **11.** $-2, 12, 42, 88, 150$

13. $-\frac{2}{5}, -\frac{1}{8}, 0, \frac{1}{14}, \frac{2}{17}$ **15.** $\frac{3}{2}, \frac{5}{4}, \frac{9}{8}, \frac{17}{16}, \frac{33}{32}$

17. 19 **19.** 0 **21.** 9

23. 3, 6, 9, 12, 15 **25.** $-2, -10, -42, -170, -682$

27. $-3, -19, -101, -513, -2575$ **29.** $2, \frac{4}{5}, \frac{25}{19}, \frac{57}{41}, \frac{287}{180}$

31. 1, 1, 1, 1, 1 **33.** 32 **35.** 154 **37.** 74

39. $\frac{992}{315}$ **41.** 0 **43.** 216

45. $\sum_{k=1}^{6} k$ **47.** $\sum_{k=1}^{6} (-1)^{k+1}k$ **49.** $\sum_{k=1}^{6} (2k + 1)$

51. $\sum_{k=1}^{4} \frac{k}{k + 3}$ **53.** 12 **55.** 70 **57.** 21

59. $\frac{137}{60}$ **61.** 9 **63.** 2 **65.** 0

67. $-2, 0, 2, 4, 6, 8$

69. 8, 5.5, 5.3333, 5.75, 6.4, 7.1616, 8

71. 45, 60, 77, 96, 117, 140, 165

73. 2.1667, 2.1429, 2.125, 2.1111, 2.1, 2.0909, 2.0833, 2.0769

Exercises 8.2

1. 2 **3.** 4 **5.** 1

7. yes, common difference 5

9. yes, common difference $-\frac{3}{4}$

11. yes, common difference 6 **13.** 12, 15 **15.** $\frac{3}{2}$, 2

17. 69, 76 **19.** 2, 5, 8, 11

21. $-3, -4, -5, -6$ **23.** $-6, -\frac{21}{4}, -\frac{9}{2}, -\frac{15}{4}$

25. 2.36, 5.56, 8.76, 11.96, 15.16, 18.36, 21.56, 24.76, 27.96, 31.16

27. 2.7869, 7.9989, 13.2289, 18.4589, 23.6889, 28.9189, 34.1489, 39.3789, 44.6089, 49.8389

29. 67.92, 62.32, 56.72, 51.12, 45.52, 39.92, 34.32, 28.72, 23.12, 17.52

31. 19 **33.** 23 **35.** 33

37. 9 **39.** -60 **41.** -60

43. 14 **45.** $\frac{3}{2}$ **47.** $-\frac{373}{8}$

49. 35 **51.** -8 **53.** 15

55. 80 **57.** 30 **59.** -20

61. 5 **63.** 20

65. $S_8 = 48, d = -4$
$S_n = \frac{n}{2}[2a_1 + (n-1)d]$ gives
$48 = \frac{8}{2}[2a_1 + (8-1)(-4)]$
$12 = 2a_1 - 28, a_1 = 20$
$a_n = a_1 + (n-1)d = 20 + (n-1)(-4),$
$a_n = 24 - 4n$

67. $a_n = 6n - 9$

69. \$2500, \$2650, \$2800, \$2950, \$3100

71. 8 years

73. 12, 24, 36, 48, 60, 72, 84, 96 feet per second

75. 78 times

Exercises 8.3

1. $r = 3$ **3.** 5 **5.** -1

7. no **9.** yes, $r = 3$ **11.** yes, $r = -\frac{1}{3}$

13. 256, 1024 **15.** $\frac{1}{3}, \frac{1}{9}$ **17.** $-\frac{3}{4}, -\frac{3}{8}$

19. 2, 8, 32, 128, 512

21. $5, \frac{5}{2}, \frac{5}{4}, \frac{5}{8}, \frac{5}{16}$

23. 1, 5, 25, 125, 625

25. 4.7600, 18.0880, 68.7344, 261.1907, 992.5247, 3771.5940

27. 2.9810, 9.6883, 31.4868, 102.3321, 332.5795, 1080.8832

29. 93.2600, -55.9560, 33.5736, -20.1442, 12.0865, -7.2519

31. 3072 **33.** $\frac{7}{81}$ **35.** $\frac{1}{48}$ **37.** $r = 2, a_1 = 3$

39. $r = 3, a_1 = 4$ **41.** $r = -\frac{1}{3}, a_1 = 11$

43. $r = 2, a_1 = 2, a_4 = 16$

45. $r = \frac{1}{2}, a_1 = 512, a_{12} = \frac{1}{4}$

47. $r = \pm 1, a_1 = -1, a_2 = \pm 1$

49. 468 **51.** -14 **53.** $-\frac{20}{81}$

55. 3 **57.** $\frac{104}{7}$ **59.** $\frac{55}{6}$

61. $\frac{107}{330}$ **63.** $\frac{58,307}{99,900}$ **65.** $\frac{1223}{1665}$

67. 30.2000 **69.** -19.7340 **71.** seventh term

73. 20th term **75.** 1600 **77.** 64 ancestors

79. 61,088 **81.** 30.38 years

Exercises 8.5

1. $a^5 + 5a^4b + 10a^3b^2 + 10a^2b^3 + 5ab^4 + b^5$

3. $1 + 14x + 84x^2 + 280x^3 + 560x^4$
$+ 672x^5 + 448x^6 + 128x^7$

5. $243x^5 - 810x^4y + 1080x^3y^2$
$- 720x^2y^3 + 240xy^4 - 32y^5$

7. $x^6 + 6x^5y^2 + 15x^4y^4 + 20x^3y^6$
$+ 15x^2y^8 + 6xy^{10} + y^{12}$

9. 1 9 36 84 126 126 84 36 9 1

11. 720 **13.** 6 **15.** 132

17. 100 **19.** 210 **21.** 1

23. $x^4 + 4x^3y + 6x^2y^2 + 4xy^3 + y^4$

25. $8 + 24y + 24y^2 + 8y^3$

27. $243 - 1620y + 4320y^2 - 5760y^3$
$+ 3840y^4 - 1024y^5$

29. $x^4 + 4x^3/y + 6x^2/y^2 + 4x/y^3 + 1/y^4$

31. $x^5 + 5x^4y^{1/2} + 10x^3y + 10x^2y^{3/2} + 5xy^2 + y^{5/2}$

33. $x^2 + 4x^{3/2}y^{3/2} + 6xy^3 + 4x^{1/2}y^{9/2} + y^6$

35. $-2016x^4y^5$ **37.** $15{,}360x^3y^{21}$

39. $437{,}500x^4y^6$ **41.** $-250x^2y^3$

Chapter 8 Review Exercises

1. **(a)** 3, 7, 11, 15 **(b)** 7, 16, 31, 52
(c) 2, 13, 46, 113 **(d)** $\frac{1}{4}, \frac{1}{2}$, 1, 2
(e) 15, 49, 99, 165 **(f)** 5, 15, 7, 35

2. **(a)** 2, 6, 10, 14 **(b)** $-3, -11, -27, -59$
(c) $1, -1, 5, -13$ **(d)** $-6, -14, -36, -100$
(e) $-1, 0, 1, 4$
(f) 4; 131; 4,496,185; $1.817868704 \times 10^{20}$

3. **(a)** 26 **(b)** 48 **(c)** 477 **(d)** 70

4. **(a)** 18 **(b)** 70 **(c)** 12

5. **(a)** no **(b)** yes, 8 **(c)** no **(d)** yes, 1.25

6. **(a)** 17 **(b)** 21 **(c)** -7 **(d)** 245

7. **(a)** 25 **(b)** 13 **(c)** -1 **(d)** 498

8. **(a)** 27 **(b)** 24 **(c)** 48 **(d)** 25

9. 6 **10.** 3 **11.** 3

12. **(a)** 24 **(b)** -81 **(c)** $\frac{1}{4}$ **(d)** $-\frac{1}{2}$

13. **(a)** $r = 3, a_1 = 12, a_5 = 972$
(b) $r = \pm 2, a_1 = 1, a_2 = \pm 2$

14. **(a)** 242 **(b)** -40

15. **(a)** 6 **(b)** $-\frac{2}{3}$

16. **(a)** $\frac{2107}{4950}$ **(b)** $\frac{23,357}{24,975}$

18. **(a)** $a^4 + 4a^3b + 6a^2b^2 + 4ab^3 + b^4$
(b) $32 - 240x + 720x^2 - 1040x^3$
$+ 810x^4 - 243x^5$
(c) $64x^6 - 192x^5y + 240x^4y^2 - 160x^3y^3$
$+ 60x^2y^4 - 12xy^5 + y^6$

19. **(a)** 120 **(b)** 132 **(c)** 35 **(d)** 8001

20. **(a)** $x^3 + 6x^2y + 12xy^2 + 8y^3$
(b) $x^5 - 15x^4y + 90x^3y^2 - 270x^2y^3$
$+ 405xy^4 - 243y^5$
(c) $8x^3 + 48x^2y + 96xy^2 + 64y^3$

21. **(a)** $70{,}000x^4y^3$ **(b)** $2016x^5y^4$
(c) $(-3.671103516 \times 10^{14})x^{12}y^5$

22. 4.27, 8.9, 13.53, 18.16, 22.79, 27.42, 32.05, 36.68

23. 5.82, 11.2326, 21.678918, 41.84031174, 80.75180166, 155.8509772, 300.792386, 580.529305, 1120.421559, 2162.413608

24. 5.4

Chapter 8 Test

1. −1, 2, 9, 20 **2.** 3, 11, −17, −29, 63

3. 20 **4.** 87 **5.** 19

6. 52 **7.** 14 **8.** −6

9. 7 **10.** 12,500 **11.** 14,336

12. 635 **13.** 10.5 **14.** $\frac{97}{132}$

15. $1 + 15x + 90x^2 + 270x^3 + 405x^4 + 243x^5$

16. (a) 5040 (b) 56 (c) 29,403

17. $32x^5 + 80x^4y + 80x^3y^2 + 40x^2y^3 + 10xy^4 + y^5$

18. $90{,}720x^4y^4$

19. 5.36, 8.63, 11.9, 15.17, 18.44

20. 4.91, 6.1375, 7.671875, 9.58984375, 11.98730469, 14.98413086

Chapter 9

Exercises 9.1

1. 24 **3.** 2 **5.** 12 **7.** 8

9. 720 **11.** 35 **13.** 66 **15.** 2

17. $n = 8$ **19.** $n = 4, 5, 6$ **21.** 24 **23.** 40

25. 40,320 **27.** 362,880 **29.** 42 **31.** 336

33. (a) 24 (b) 64 **35.** 142,506

37. 120 **39.** 480

41. 210 **43.** 38,760 **45.** 9 million

47. 1400 **49.** 32,967,000,000 **51.** 810,540,000

Exercises 9.2

1. (a) $\frac{1}{2}$ (b) $\frac{1}{2}$

3. (a) $\frac{4}{11}$ (b) $\frac{2}{11}$ (c) $\frac{4}{11}$ (d) 0 (e) $\frac{1}{11}$ (f) $\frac{10}{11}$

5. (a) $\frac{1}{30}$ (b) $\frac{7}{30}$ (c) $\frac{1}{2}$

7. (a) $\frac{1}{36}$ (b) $\frac{1}{18}$ (c) $\frac{1}{6}$

9. $\frac{1}{8}$ **11.** $\frac{1}{6}$

13. (a) $\frac{1}{3}$ (b) $\frac{5}{6}$

15. (a) $\frac{1}{8}$ (b) $\frac{7}{8}$

17. not mutually exclusive

19. not mutually exclusive

21. $\frac{1}{4}$ **23.** (a) $\frac{3}{51}$ (b) $\frac{12}{51}$

25. (a) $\frac{3}{4}$ (b) $\frac{2}{5}$ (c) $\frac{3}{4}$ **27.** 0.2 **29.** 0.4196

31. (a) $\frac{1}{27{,}000}$ (b) $\frac{1}{810{,}000}$ (c) 0.812

33. (a) $\frac{1}{5}$ (b) 0.1861

35. A^+ from A^+, O^+, O^-, A^-; A^- from A^-, O^- AB^- from AB^-, O^-, A^-, B^-; O^- from O^- only

39. 0.125

41. 0.94097758995 [Get an overflow. Split up the arithmetic.]

Chapter 9 Review Exercises

1. (a) 720 (b) 6 (c) 20 (d) 7 (e) 11,880 (f) 35 (g) 126 (h) 70

2. 60 **3.** 180 **4.** 60 **5.** 360

6. 40,320 **7.** 120 **8.** 48 **9.** 103,680

10. 336 **11.** 1,555,200 **12.** 17,576,000

13. (a) 120 (b) 625 **14.** 32

15. (a) 8008 (b) 2940 **16.** 270,725

17. 524,160

18. (a) $\frac{1}{2}$ (b) $\frac{3}{13}$ (c) $\frac{1}{13}$ (d) $\frac{4}{13}$

19. (a) $\frac{1}{6}$ (b) $\frac{1}{2}$ (c) $\frac{1}{2}$ (d) $\frac{1}{2}$

20. (a) $\frac{1}{36}$ (b) $\frac{1}{18}$ (c) $\frac{1}{6}$ (d) $\frac{5}{12}$

21. (a) $\frac{1}{4}$ (b) $\frac{1}{2}$ **22.** $\frac{4}{13}$ **23.** 0.2885

24. 0.2098

25. mutually exclusive **26.** not mutually exclusive

27. $\frac{51}{72}$

28. (a) $1.262626263 \times 10^{-4}$ (b) 0.246969697 (c) 2.505210×10^{-8}

29. (a) $\frac{7}{58}$ (b) $\frac{11}{58}$ (c) $\frac{8}{58}$ (d) $\frac{21}{58}$ (e) $\frac{13}{58}$ (f) $\frac{22}{58}$ (g) $\frac{5}{51}$ (h) $\frac{2}{51}$

Chapter 9 Test

1. (a) 56 (b) 84 **2.** 24 **3.** 180

4. 479,001,600 **5.** 56 **6.** 456,976,000

7. (a) 120 (b) 60 **8.** 1,860,480

9. (a) $\frac{1}{6}$ (b) $\frac{1}{2}$ (c) $\frac{2}{3}$

10. (a) $\frac{3}{10}$ (b) $\frac{3}{5}$ **11.** 0.1635 **12.** 0.0103

13. not mutually exclusive

14. (a) $\frac{1}{65{,}536}$ (b) $\frac{1}{1{,}048{,}576}$ (c) 0.4999

Cumulative Test

1. $x = 2, y = -4$ **2.** $x = 2, y = -1, z = 1$

3. $(-\frac{2}{3}, \frac{13}{3})$ and $(2, -1)$

4. (a) 0 (b) 2 (c) 3

5. (a) $\begin{bmatrix} 5 & -2 \\ 4 & 4 \end{bmatrix}$ (b) $\begin{bmatrix} 1 & 0 & -1 \\ 6 & 4 & 6 \end{bmatrix}$

6. $\begin{bmatrix} 1 & 3 \\ 2 & 5 \end{bmatrix}\begin{bmatrix} x \\ y \end{bmatrix} = \begin{bmatrix} 13 \\ 21 \end{bmatrix}$; $x = -2, y = 5$

7. 106 **8.** 1, 8, 21, 40 **9.** 77 **10.** -13

11. $\frac{1}{2}$ **12.** 192 **13.** 93

14. **(a)** 720 **(b)** 35 **(c)** 31,125 **(d)** 210 **(e)** 56

15. $-5376x^6y^3$ **16.** 8,000,000 **17.** 50,450,400

18. **(a)** $\frac{1}{6}$ **(b)** $\frac{1}{3}$ **(c)** $\frac{2}{3}$ **19.** 0.2506 **20.** 0.2261

21. Maximum profit is \$25,300 with 210 Swirl and 275 Rapid.

22. $D = \begin{bmatrix} 0 & 3 & 2 & 3 & 1 \\ 1 & 0 & 3 & 4 & 2 \\ 2 & 1 & 0 & 1 & 2 \\ 4 & 3 & 2 & 0 & 1 \\ 3 & 2 & 1 & 2 & 0 \end{bmatrix} \begin{matrix} 9 \\ 10 \\ 6 \\ 10 \\ 8 \end{matrix}$

gives M_3, M_5, M_1, (M_2 and M_4)

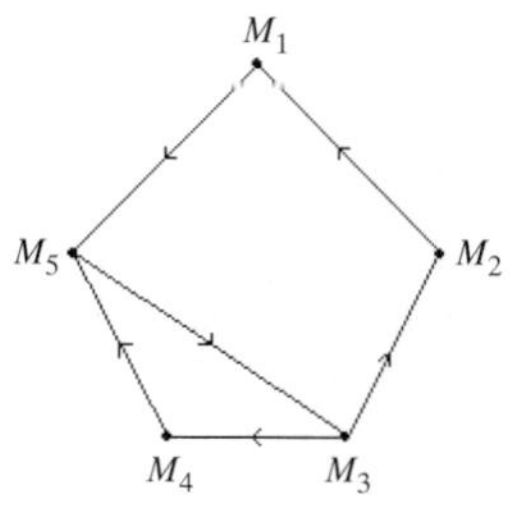

Appendix

1. $(0, 3)$, $y = -3$

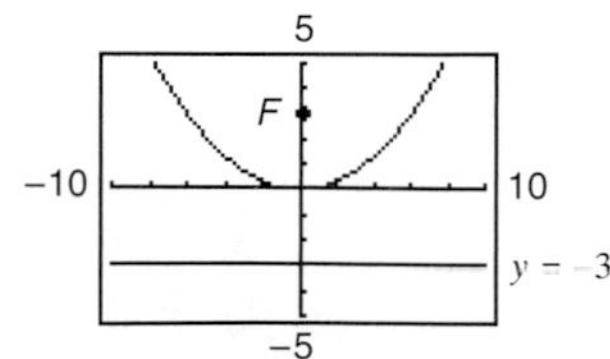

3. $(2, 0)$, $x = -2$

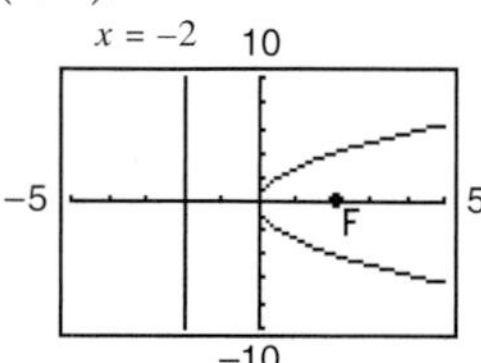

5. $(0, 2)$, $y = -2$

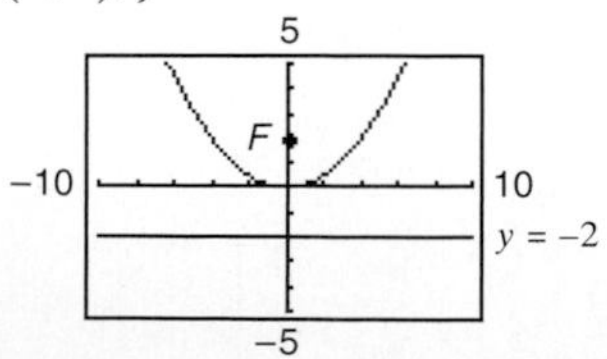

7. $(0, -\frac{5}{8})$, $y = \frac{5}{8}$

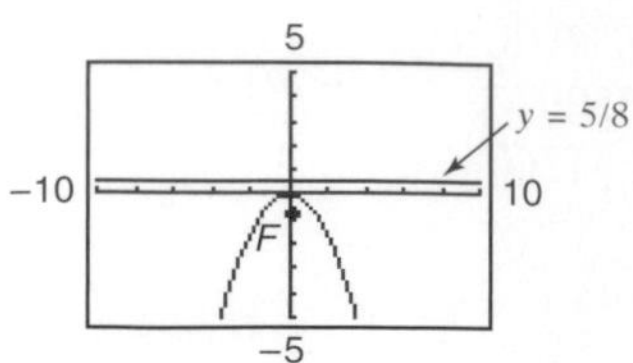

9. $x^2 = 12y$ **11.** $x^2 = 4y$ **13.** $y^2 = 24x$

15. $y^2 = -4x$ **17.** $x^2 = \dfrac{20y}{3}$

19. vertices $(-2, 0)$, $(2, 0)$, $(0, -1)$, $(0, 1)$; foci $(-\sqrt{3}, 0)$, $(\sqrt{3}, 0)$

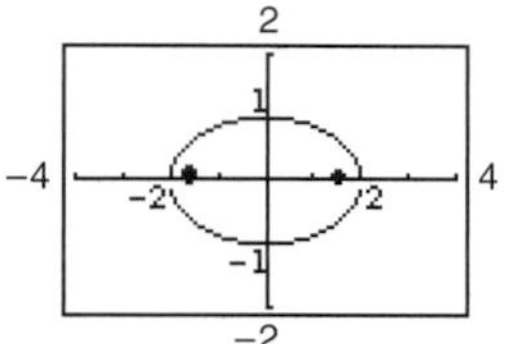

21. vertices $(-3, 0)$, $(3, 0)$, $(0, -4)$, $(0, 4)$; foci $(0, -\sqrt{7})$, $(0, \sqrt{7})$

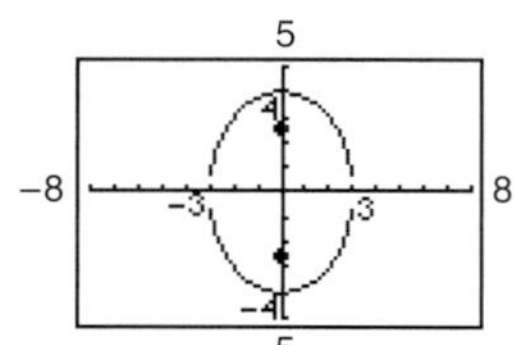

23. vertices $(-1, 0)$, $(1, 0)$, $(0, -5)$, $(0, 5)$; foci $(0, -\sqrt{24})$, $(0, \sqrt{24})$

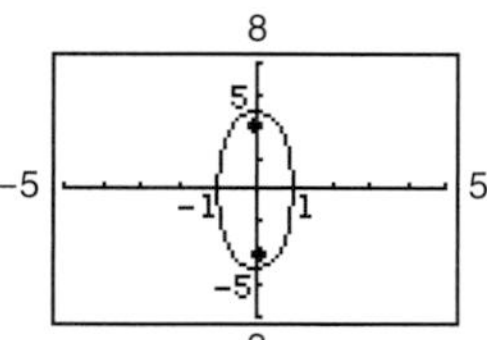

25. vertices $(-3, 0)$, $(0, 3)$; foci $(-\sqrt{10}, 0)$, $(\sqrt{10}, 0)$; asymptotes, $y = \pm\frac{1}{3}x$

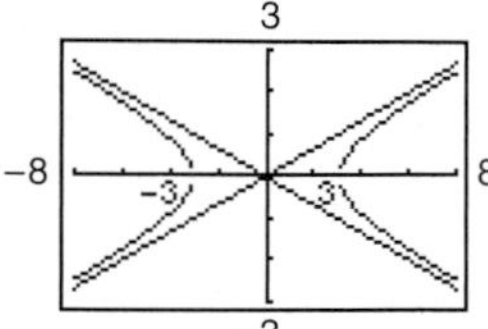

27. vertices $(-2, 0)$, $(2, 0)$; foci $(-\sqrt{8}, 0)$, $(\sqrt{8}, 0)$; asymptotes, $y = \pm x$

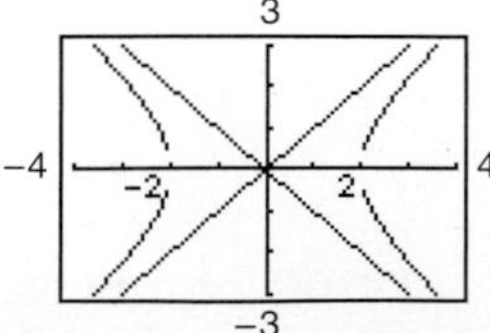

29. vertices $(0, -2)$, $(0, 2)$; foci $(0, -\sqrt{13})$, $(0, \sqrt{13})$; asymptotes, $y = \pm\frac{2}{3}x$

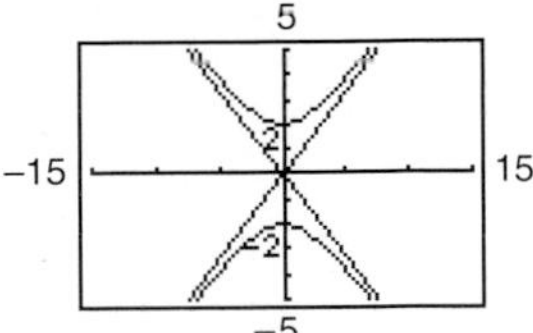

31. Ellipse: vertices $(-4, 0)$, $(4, 0)$, $(0, -2)$, $(0, 2)$; foci $(-\sqrt{12}, 0)$, $(\sqrt{12}, 0)$

33. hyperbola: vertices $(0, -\frac{5}{4})$, $(0, \frac{5}{4})$; foci $(0, -\frac{25}{12})$, $(0, \frac{25}{12})$; asymptotes, $y = \pm\frac{3}{4}x$

35. not an ellipse or hyperbola

37. parabola with vertex $V(-5, 3)$, focus $(-5, 4)$, axis $x = -5$, and directrix $y = 2$

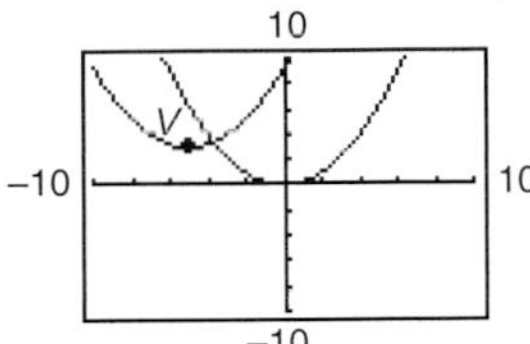

39. parabola with vertex $V(4, -3)$, focus $(7, -3)$, axis $y = -3$, and directrix $x = 1$

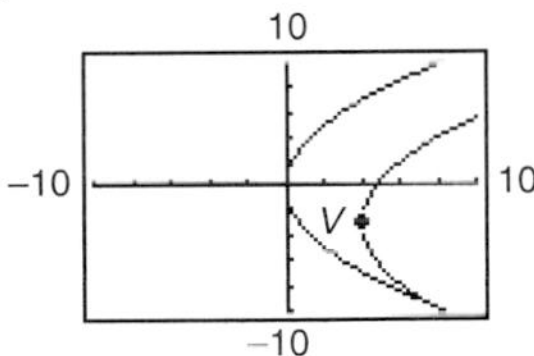

41. ellipse with center $(1, -4)$; vertices $(-2, -4)$, $(4, -4)$, $(1, -6)$, $(1, -2)$; and foci $(1 - \sqrt{5}, -4)$, $(1 + \sqrt{5}, -4)$

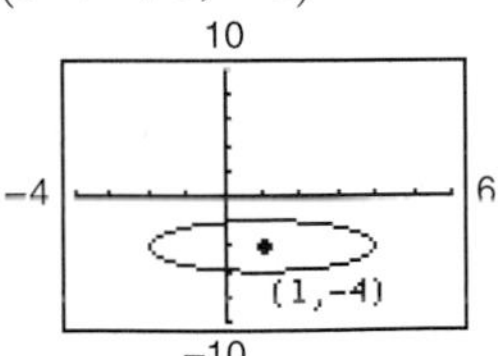

43. hyperbola with center $(3, 2)$; vertices $(2, 2)$, $(4, 2)$; foci $(3 - \sqrt{10}, 2)$, $(3 + \sqrt{10}, 2)$; asympotes $y = 3x - 7$ and $y = -3x + 11$

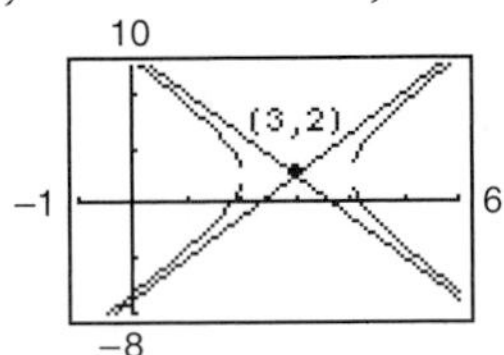

45. parabola, opens up, vertex $(2, 3)$

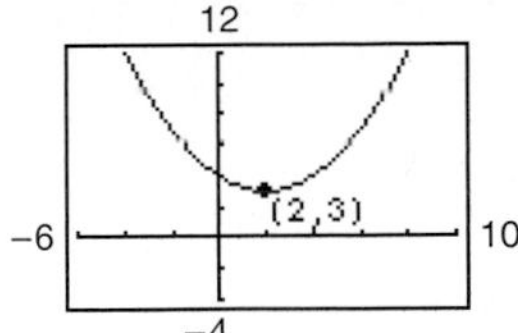

47. ellipse, center $(1, -1)$; vertices $(-1, -1)$, $(3, -1)$, $(1, -2)$, $(1, 0)$

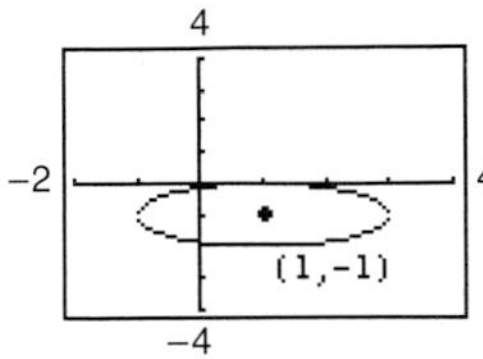

49. hyperbola, center $(-1, 2)$; vertices $(-6, 2)$, $(4, 2)$

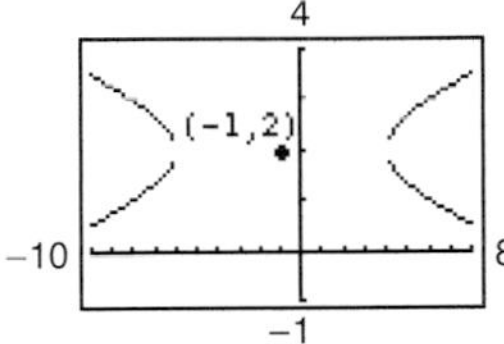

51. parabola with vertex $(-3, 4)$, opening up

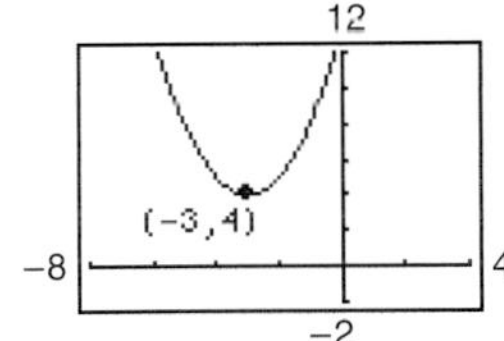

53. $(x - 1)^2 = (\frac{1}{2})(y - 4)$

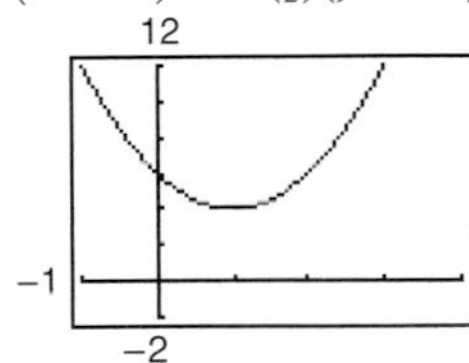

55. $(x - 1)^2 = 16(y - 2)$

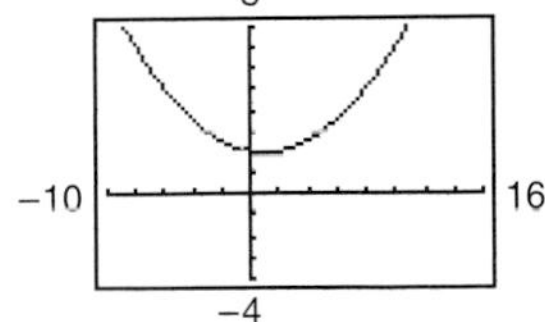

57. $(y - 3)^2 = 4x$

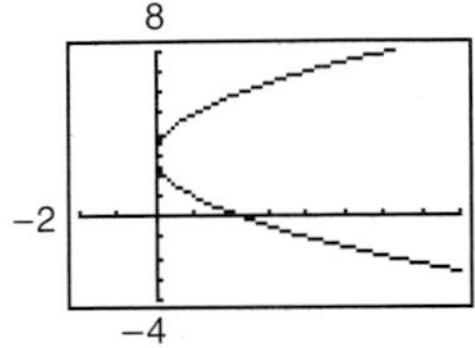

59. $(x - 2)^2 = 4(y + 2)$

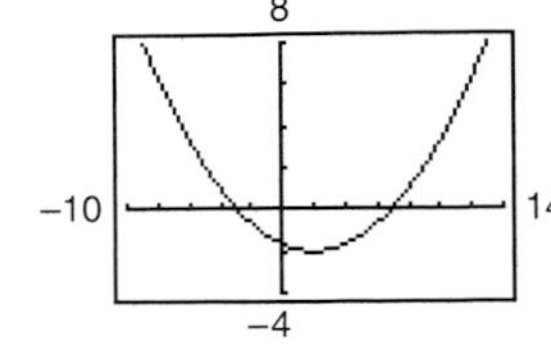

61. $\dfrac{x^2}{9} + \dfrac{y^2}{49} = 1$

63. $\dfrac{x^2}{36} - \dfrac{y^2}{144} = 1$

INDEX

D

E

F